THE PAPERS OF THOMAS A. EDISON

FINANCIAL CONTRIBUTORS

Public Foundations
National Endowment for the Humanities
National Historical Publications and Records Commission
National Science Foundation

Private Foundations
The Alfred P. Sloan Foundation
Charles Edison Fund
The Hyde and Watson Foundation
National Trust for the Humanities
Geraldine R. Dodge Foundation

Private Corporations and Individuals
Alabama Power Company
Anonymous
Association of Edison
    Illuminating Companies
AT&T
Atlantic Electric
Battelle Memorial Institute
The Boston Edison Foundation
Cabot Corporation
    Foundation, Inc.
Carolina Power & Light
    Company
Consolidated Edison Company
    of New York, Inc.
Consumers Power Company
Cooper Industries
Corning Incorporated
Duke Power Company
Entergy Corporation (Middle
    South Electric System)
ExxonMobil Corporation
Florida Power & Light
    Company
General Electric Foundation
Gould Inc. Foundation
Gulf States Utilities Company
Dave and Nina Heitz
Hess Foundation, Inc.
Idaho Power Company
IMO Industries

International Brotherhood of
    Electrical Workers
Mr. and Mrs. Stanley H. Katz
Matsushita Electric Industrial
    Co., Ltd.
Midwest Resources, Inc.
Minnesota Power
New Jersey Bell
New York State Electric & Gas
    Corporation
North American Philips
    Corporation
Philadelphia Electric Company
Philips Lighting B.V.
Public Service Electric and Gas
    Company
RCA Corporation
Robert Bosch GmbH
Rochester Gas and Electric
    Corporation
San Diego Gas and Electric
Savannah Electric and Power
    Company
Schering-Plough Foundation
Texas Utilities Company
Thomas & Betts Corporation
Thomson Grand Public
Westinghouse Foundation
Wisconsin Public Service
    Corporation

THE PAPERS OF THOMAS A. EDISON

Volume 5

*An 1879 photograph of the
Menlo Park laboratory
staff surrounding Thomas
A. Edison, who is seated
in the middle and holding
a hat.*

Volume 5

The Papers of Thomas A. Edison

RESEARCH TO DEVELOPMENT AT MENLO PARK

January 1879–March 1881

VOLUME EDITORS

Paul B. Israel
Louis Carlat
David Hochfelder
Keith A. Nier

EDITORIAL STAFF

Lindsay Frederick Braun
Grace Kurkowski

SPONSORS

Rutgers, The State University of New Jersey
National Park Service, Edison National Historic Site
New Jersey Historical Commission
Smithsonian Institution

THE JOHNS HOPKINS UNIVERSITY PRESS
BALTIMORE AND LONDON

Volume 5 of *The Papers of Thomas A. Edison* appears through the generous publication subsidy of the National Historical Publications and Records Commission, the National Endowment for the Humanities, and Rutgers, The State University.

The Johns Hopkins University Press
2715 North Charles Street
Baltimore, Maryland 21218-4363
www.press.jhu.edu

The paper used in this book meets the minimum requirements of the American National Standard for Information Sciences—Permanence of Paper for Printed Library Materials, ANSI Z 39.48-1984.

Library of Congress Cataloging-in-Publication Data
(Revised for volume 3)

Edison, Thomas A. (Thomas Alva), 1847–1931
   The papers of Thomas A. Edison

   Includes bibliographical references and index.
   Contents: v. 1. The making of an inventor, February 1847–June 1873—v. 2. From workshop to laboratory, June 1873–March 1876—v. 3. Menlo Park. The early years, April 1876–December 1877.
   1. Edison, Thomas A. (Thomas Alva), 1847–1931.   2. Edison, Thomas A. (Thomas Alva), 1847–1931—Archives.   3. Inventors—United States—Biography.   I. Jenkins, Reese.
TK140.E3A2   1989   600   88-9017
ISBN 0-8018-3100-8 (v. 1. : alk. paper)
ISBN 0-8018-3101-6 (v. 2. : alk. paper)
ISBN 0-8018-3102-4 (v. 3. : alk. paper)
ISBN 0-8018-5819-4 (v. 4. : alk. paper)
ISBN 0-8018-3104-0 (v. 5. : alk. paper)

A catalog record for this book is available from the British Library.

TO ROBERT ROSENBERG, who brought us into the digital age;
TO HELEN ENDICK, who kept the lights on; and
TO GRACE KURKOWSKI, who kept us connected.

Contents

Calendar of Documents

List of Editorial Headnotes

Preface

This volume chronicles one of the most important periods in Thomas A. Edison's career. Between January 1879 and March 1881, Edison invented a complete system of electric light and power. Although the carbon-filament incandescent lamp was the most visible element of the system, Edison and his assistants also developed a new generator, a meter, underground conductors, and fuses and lamp sockets. In addition, they developed electric motors and built an experimental electric railway at the laboratory complex in Menlo Park, New Jersey. During the more than two years that Edison experimented on electric light and power, he greatly expanded his laboratory staff and facilities. In the process he created the first full-scale research and development facility in the United States. During 1880, as the work turned from research to development, Edison increasingly directed the research of his assistants who were each assigned a specific project.

The scale of Edison's effort was made possible by the financial largesse of investors in the Edison Electric Light Company and the banking house of Drexel, Morgan & Co. Between October 1878 and March 1881, they provided nearly $130,000 for research and development. By the end of this period the success of the effort was evident as Edison moved his offices into New York and set about commercializing his new system. The electric light investors were willing to fund a central station in New York City to demonstrate the system and established the Edison Electric Illuminating Company of New York for this purpose. Because they were reluctant to expend more funds to manufacture system components, however, Edison used his own funds to establish manufacturing shops in association with his closest assistants. The first of these, the Edison Electric Lamp Company, was in operation by the winter of 1881.

As was typical throughout his career, Edison did not work exclusively on electric light and power, despite the complexity of that project. Throughout 1879, he and Charles Batchelor, his principal experimental assistant, spent considerable time on the electromotograph telephone receiver, which Edison had developed initially for the British market. The sale of foreign rights provided Edison with significant income but intense competition among companies using the Edison, Bell, and other telephones in Britain and France led to consolidations in each of these countries; meanwhile, in the United States the Bell Telephone Company bought out Western Union's telephone interests, including the rights to Edison's telephone patents. Edison was kept abreast of the complex negotiations leading to these consolidations but relied on his representatives to handle them.

Edison also experimented briefly with a variety of technologies besides the telephone, most notably electromagnetic ore separation, which would occupy much of his time during the 1890s. He also conducted experiments on preserving fruit in a vacuum and on heavier-than-air flight. Nonetheless, Edison's primary focus during the twenty-seven months covered in this volume was on electric light and power, and this would continue to dominate his time as he moved to New York and began his effort to introduce his system into the marketplace.

The progress of the Thomas A. Edison Papers depends on the support of many individuals and organizations, including the Sponsors, other financial contributors, academic scholars, Edison specialists, librarians, archivists, curators, and students. Representatives of the four Sponsors have assisted with this volume and the editors thank them for their continuing concern and attention. The strong support of public and private foundations and of their program officers has sustained the project and helped it remain editorially productive.

Preparation of this volume was made possible in part by grants from the Division of Research Programs (Collaborative Research) of the National Endowment for the Humanities, an independent federal agency; the National Historical Publications and Records Commission; the New Jersey Historical Commission; the Charles Edison Fund; the National Trust for the Humanities; as well as through the support of Rutgers,

The State University of New Jersey, and the National Park Service (Edison National Historic Site). The editors appreciate the interest and support of the many program officers and trustees, especially Elizabeth Arndt, Michael Hall, Timothy Connolly, Howard Green, Mary Murrin, John P. Keegan, Ann Orr, and Thomas L. Morrissey. Any opinions, findings, conclusions, or recommendations expressed in this publication are solely those of the editors and do not necessarily reflect the views of any of the above federal foundations or agencies, the United States Government, or any other financial contributor.

The Edison Papers project is indebted to the National Park Service for its multifaceted support. The editors express particular thanks to Marie Rust, John Maounis, and Nancy Waters of the Northeast Region and Maryanne Gerbauckas, Terri Jung, Roger Durham, Leonard DeGraaf, Douglas Tarr, Edward Wirth, Karen Sloat-Olsen, and Linda Deveau of the Edison National Historic Site in West Orange, New Jersey.

Many associates at Rutgers University have contributed significantly to the Edison Papers. The editors are grateful to Holly M. Smith, Dean of the Faculty of Arts and Sciences, and her predecessor Richard F. Foley, along with their dedicated staff, especially Barry V. Qualls, Dean for Humanities, and Robert Wilson, Executive Vice Dean. We would especially like to thank Judith Brooke, Associate Director of Development for the Faculty of Arts and Sciences. The editors value the support of colleagues and staff in the History Department, especially Ann Gordon, Deborah White, Philip J. Pauly, Susan Schrepfer, Michael Adas, James W. Reed, John Gillis, Ziva Galili, Rudy Bell, and John Chambers; Michael Geselowitz and his staff at the IEEE History Center; as well as members of the Rutgers University Libraries, notably Marianne Gaunt, Ruth Simmons, Ron Becker, Thomas Frusciano, and Grace Agnew. The support of the Rutgers University administration has encouraged the editors and facilitated their work. The staff of the project thank Joseph J. Seneca, David A. Rumbo, Ronald Thompson, Karen James, Jacquelyn Halvsa, Robert Thieme, Joseph M. Harrigan, Constance J. Bornheimer, Judith A. Garelick, Irma Lucy Cardinale, and Sameerah Diaab Allen. Roy Marantz and David Cruz of Rutgers Computing Services and John Amodeo, Thomas Vosseler and the staff of the Faculty of Arts and Sciences Computer & Network Operations Group have offered their invaluable computer expertise.

Many scholars have shared their insights and assisted the editors in a variety of ways. For this volume, notable help came from Bernard W. Carlson, Bernard S. Finn, Robert Friedel, Leonard F. Guttridge, Charles Hummel, Peter L. Jakab, Sean F. Johnston, Ronald Kline, René Rondeau, and Marc Rothenberg.

Institutions and their staff have provided documents, photographs, photocopies, and research assistance. The staff of the Henry Ford Museum & Greenfield Village—especially William S. Pretzer, Marc Greuther, Charles Hanson, Judy Endelman, Linda Skolarus, and Jeanine Head—have supplied many invaluable services. The editors appreciate the professional courtesies of John Fleckner and Robert Harding at the National Museum of American History's Archives Center; John Liffen, Robert Bud, and Emily Lewis of the Science Museum in London; Tim Procter and Lenore Symons of the Institute of Electrical Engineers in London, Sheldon Hochheiser, AT&T Corporate Historian; Mary Bowling at New York Public Library; Florence Bartechevski of the Baker Library at Harvard University; the staff at the Historical Society of Pennsylvania; and Irene Kovacs at the Foundation of the Postal and Telecommunication Museum in Budapest. The City of Ft. Myers, Florida, and the trustees and staff of its Edison-Ford Winter Homes—notably Michele Wehrwein Albion, Judy Suprise, Jim Newton, and Les Marietta—have assisted the Edison Papers. Outside computer advice has come from Michael Sperberg-McQueen and the members of the Model Editions Partnership.

Staff members, interns, and students not mentioned on the title page but who have contributed to this volume include Thomas E. Jeffrey, Theresa Collins, Linda Endersby, Janette Pardo, David Ranzan, Brian Shipley, Keith A. Barbera, Elva Kathleen Lyon, and Vicky Chen.

As always, the project has had the benefit of the superb staff of the Johns Hopkins University Press. For this volume, the editors are indebted to Robert J. Brugger, Melody Herr, Julie McCarthy, and Lee Sioles.

Chronology of Thomas A. Edison

January 1879–March 1881

1879

2 January	Has laboratory machine shop in Menlo Park, New Jersey, begin constructing first dynamo of his own design.
19 January	Begins fundamental experiments on properties of heated wires which lead to discovery of occluded gases.
22 January	Unsuccessfully seeks Sprengel mercury vacuum pump in attempt to remove occluded gases from lamp wires.
23 January	Conducts vacuum lamp experiments with hand pump.
30 January	Son Thomas Alva, Jr., leaves for convalescence in Florida.
6 February	Begins several days of vacuum lamp tests using hand pump.
9 February	Drafts caveat on use of vacuum to overcome occluded gases in lamp wires.
13 February	Begins testing first dynamo machine.
14 February	Construction of two demonstration telephones for Great Britain begins at Sigmund Bergmann's shop.
15 February	Devises new arrangement for generators using an odd number of armature coils and commutator blocks.
26 February	Sends nephew Charley Edison to London with two telephones and two new electromotograph receivers.
13 March	Begins designing bipolar dynamo with large field magnets.
14 March	George Gouraud, Edison's agent in London, presents the first public demonstration of Edison's new electromotograph receiver.
c. 18 March	Begins laboratory demonstration, for investors and others, of incandescent platinum lamps wired in parallel.
19 March	Begins planning electric illumination of *New York Herald* Arctic expedition aboard USS *Jeannette*.
25 March	Begins collecting statistics on gas companies in New York City.
26 March	Acquires Geissler mercury vacuum pump.

8 April	Begins experiments on thermo electricity.
12 April	Decides to incorporate new coil and commutator arrangement into the new bipolar dynamo.
29 April	Signs agreement for sale of telephone patents and formation of company on the European continent.
c. 30 April	Conducts tests of first large dynamo.
April	Designs combined Sprengel and Geissler vacuum pump.
c. 1 May	Begins drafting British provisional patent specification describing system of electric lighting.
	Writes circular letter inquiring about platinum deposits.
2 May	Signs agreement for sale of electric light patents and formation of company on the European continent.
4 May	James Adams, Edison's technical representative in Britain, dies in London.
12 May	Agrees to assign to Western Union the U.S. rights to electromotograph receiver.
14 May	Edison Telephone Co. of Europe, Ltd., incorporated in New York.
15 May	Signs agreement to form Edison telephone companies in Great Britain.
16 May	Agrees to assign George Gouraud half of his proceeds from sale of British patent rights for electromotograph receiver.
18 May	Makes first drawings for electric railroad.
20 May	Ships dynamo and other equipment for USS *Jeannette*.
26 May	Dispatches Frank McLaughlin to investigate platinum deposits in Quebec.
28 May	Advertises for an experienced analytic chemist and soon after hires Dr. Otto Moses.
May	Agrees to formation of Edison telephone company in England.
	Begins developing switchboard for telephone exchanges.
12 June	Designs tertiary coil arrangement for electromotograph receiver.
	Begins extensive series of tests to find best compositions for chalk buttons for electromotograph receiver.
1 July	Sends Edward Johnson to London as chief engineer for telephones.
c. 6 July	Agrees to give laboratory assistant Francis Upton 5 percent of profits and 5 percent of Edison Electric Light Co. stock in lieu of salary.
7 July	With Francis Upton drafts a letter published under Upton's name in *Engineering* regarding John Hopkinson's paper on dynamo testing.

9 July	Sends first lot of telephones and a switchboard to London.
23 July	Proposes to Adolph Sutro use of hydroelectric power in Comstock mines.
July	Moves electromotograph receiver from telephone box to adjustable arm and adopts "inertia" arrangement for carbon telephone transmitter in telephones sent to Britain.
	Francis Upton and laboratory assistant Francis Jehl begin extensive series of tests of electro-deposition meter.
	With chemical staff assays ores sent to laboratory and has shop make apparatus for separating and concentrating the ores.
	Makes preliminary arrangements for purchase of platinum ores in California.
6 August	U.S. Patent Office declares interference on inertia telephone transmitter application.
7 August	Assigns Francis Upton to begin analyzing long-distance transmission of electric power for mines near Virginia City, Nevada.
14 August	U.S. Patent Office declares set of six interference cases involving carbon telephone transmitters.
c. 20 August	Hires glassblower Ludwig Böhm.
26 August	Departs for American Association for the Advancement of Science (AAAS) meeting in Saratoga, New York.
c. 28 August	Writes AAAS paper on electric light experiments, to be read by Francis Upton.
30 August	Demonstrates telephone to the AAAS.
	Leaves Saratoga.
6 September	New switchboard exchange demonstrated for newspaper reporters in London.
8 September	Drafts circular letter soliciting loan of sewing machines from U.S. manufacturers for electric motor experiments.
	Francis Upton and Francis Jehl begin month-long series of experiments on and design changes for vacuum pump.
11 September	Sends Frank McLaughlin to California to evaluate platinum-bearing sands.
20 September	Files British provisional patent specification for telephone switchboard.
September	Abandons inertia telephone transmitter.
	Begins hiring inspectors for telephone companies in London and provincial cities.
	Assigns U.S. rights for polyform to Charles Lewis and associates, who organize the Menlo Park Manufacturing Co. to promote and sell Edison's patent medicine.

1 October	Western Union orders 100 electromotograph receivers.
3 October	Authorizes George Gouraud to organize telephone companies in British colonies.
7 October	Begins electric light experiments with molded carbon spirals.
14 October	Executes agreement with José Husbands to establish telephone company in Chile.
	Drafts letter to Edward Johnson analyzing legal status of his British telephone patent.
	Agreement to combine French Edison, Gower, and Blake telephone companies signed in Paris.
15–16 October	Begins series of experiments to solve problems with chalks for electromotograph receiver.
19 October	Charley Edison dies in Paris.
22 October	Conducts first successful carbon filament lamp experiments with carbonized thread in a vacuum.
October	Prepares to amend basic telephone patent specification in Great Britain.
1 November	Executes patent application for high-resistance carbon filament lamp.
8 November	Francis Upton and other laboratory assistants resume experiments with electric meter.
November	Francis Upton begins preparing article on Edison's electric light system, eventually published under Edison's name in February 1880 issue of *Scribner's Monthly*.
1 December	Dissolves all financial connections with former partner Joseph Murray.
9 December	Organizes Edison Ore Milling Co.
11 December	Files patent application for manufacture of horseshoe-shaped cardboard lamp filaments.
21 December	*New York Herald* publishes detailed description of electric lighting system.
c. 29 December	Begins public exhibition of electric lighting system in and around laboratory.
1880	
2 January	Closes laboratory to public after being inundated with visitors during public demonstrations of electric lighting system.
	Begins experimenting with high-speed dynamo and low resistance armature constructed of iron plates.
17 January	Agrees to give Western Electric exclusive rights to his Australian telephone patents.
	Agrees to accept lower royalty on sale of electric pen apparatus.

	Sends telephone receivers operated by electric motors to London *Times*.
19 January	Chemist John Lawson begins extensive series of copper deposition experiments for electric meter.
27 January	Authorizes Edward Johnson to negotiate with Edison Telephone Co. of London regarding his royalties and British interests outside London.
	Consolidation of Edison, Gower, and Blake telephone interests in Paris collapses.
28 January	Executes patent application for his system of electric distribution.
January	Hires engineers Charles Clarke and Julius Hornig.
7 February	Requested to stop hiring telephone inspectors for Edison Telephone Co. of London because of company's weak financial condition.
8 February	Conducts experiments to prevent sparking in commutator brushes of generator.
12 February	Has Francis Upton prepare a preliminary estimate for a central station supplying 10,000 lights.
	With principal laboratory assistant Charles Batchelor begins two weeks of experiments on paper and fiber filaments.
13 February	Plans experiments to prevent "electrical carrying" in lamps.
19 February	Chemist Otto Moses begins working on problem of darkening of lamp bulbs.
10 March	Executes patent application for electric meter.
10–12 March	Visited by professors George Barker and Henry Rowland, who test the thermal efficiency of the lamp.
11 March	Executes patent application for iron plate armature.
14 March	Has Charles Mott begin keeping journal of daily activities at the laboratory.
19 March	Visited by professors Cyrus Brackett and Charles Young, who test the efficiency of the dynamo.
21 March	Discontinues night work because of problem with Charles Batchelor's eyes.
25 March	Begins experiments with magnetic iron ore separator.
27 March	Agreement establishing new combined French telephone company signed in Paris.
Winter–Spring	With laboratory staff tests wide variety of paper, wood, and vegetable fibers as lamp filaments.
1 April	Begins experiments with Francis Jehl and Otto Moses on using single instead of double vacuum pump.
3 April	Executes patent application for magnetic ore separator.
	Sells U.S. patent rights for motograph relay to Western Union.

6 April	*American Machinist* sends engineers to Menlo Park to make tests of the efficiency of Edison's electric light and power system.
23 April	Francis Upton and David Cunningham begin supervising installation of electric lighting system on the steamship *Columbia*.
24 April	Purchases former electric pen factory across the railroad tracks in Menlo Park for lamp factory.
26 April	Granted exclusive eight-year license for telephones by president of Chile.
27 April	Edison, his wife, and several assistants attend reception aboard the *Columbia* in New York.
April	Workmen begin constructing electric railroad. Experiments with frosted lamp bulbs.
1 May	Workmen begin laying nearly five miles of underground electrical cables throughout Menlo Park.
8 May	Charles Clarke and Francis Upton begin designing a large generator to be directly connected to a Porter-Allen steam engine.
13 May	Begins operating electric railroad.
c. 13 May	Agrees to underwrite publication of *Science*.
1 June	Agreement reached to combine British Edison and Bell telephone companies into the United Telephone Co., Ltd.
16 June	With Princeton University professor Charles Young conducts spectroscopic examination of lamps to determine cause of electrical carrying.
21 June	Visited by representative of Corning Glass Works, which subsequently supplies bulbs for lamps.
c. 2 July	Laboratory staff finishes making 400 vacuum pumps for factory.
3 July	Publication of first issue of *Science*.
4 July	Offers to file dynamo patent application for Henry Rowland in order to challenge patent application of Siemens & Halske.
7–19 July	With Charles Batchelor conducts experiments on electric air balloon.
9 July	Begins experimenting with bamboo filaments.
16 July	Francis Upton begins testing underground conductors in Menlo Park and finds insulation defective.
17 July	Workmen begin digging up underground conductors.
23 July	Francis Upton calculates the costs and profits for a central station.
24 July	Begins experiments with depositing volatile hydrocarbons on lamp filaments.

28 July	Executes patent application for lamp filament of carbonized bamboo or similar fiber.
	With Charles Batchelor, executes patent application for device to test lamp filaments after carbonization.
30 July	Reinstates night work at the laboratory.
c. 30 July	Assigns chemist Otto Moses to conduct literature search on carbonaceous materials and processes of carbonization.
31 July	Executes patent application for a symmetrical system of electrical distribution to maintain uniform voltage.
July	Begins survey of proposed central station district in lower Manhattan to collect statistics on the use of gas lights and power.
4 August	Executes patent application for "feeder and main" system of electrical distribution.
c. 9 August	Executes patent application for direct-connected steam dynamo.
16–24 August	Experiments on vacuum preservation of fruit.
17 August	French Edison telephone company merges with Gower and Soulerin companies to form the Société Générale des Téléphones.
18 August	Signs several agreements authorizing sale of patents for telephone, electric light and power, and electric railways in dozens of foreign countries and colonies.
19 August	Drexel, Morgan & Co. agree to arrange a line of credit for Edison's electric railway experiments.
27 August	Sends John Segredor to Georgia, Florida, and Cuba to gather plant samples for possible use as filaments.
c. 30 August	Sends inquiries about bamboo to Brazil, Panama, Cuba, and other Carribean islands.
c. 31 August	Edison gives laboratory assistant Wilson Howell the task of developing a better insulating compound for underground cables.
14 September	Edward Johnson design a screw-in lamp socket as part of a general effort to develop electric lamp fixtures.
23–24 September	With Francis Upton and Francis Jehl conducts tests of gas lighting using lamps at Sigmund Bergmann's New York shop.
24 September	Workmen begin insulating underground conductors with improved compound developed by Wilson Howell.
28 September	With Charles Clarke conducts experiments on heating of copper rods revolved through magnetic lines of force.
30 September	Charles Clarke begins redesigning large dynamo armature using copper bars instead of insulated wire.
	First lot of thirty lamps completed at lamp factory and sent to laboratory for testing.

September	Publishes article drafted by Francis Upton on electric light and power system in *North American Review*.
c. 6 October	Sends William Moore to China and Japan to search for best variety of bamboo to use for lamp filaments.
12 October	Second lot of 100 test lamps completed at factory and sent to laboratory.
21 October	With Francis Jehl begins experiments on treating filaments in hydrocarbon vapor.
22 October	Glassblower Ludwig Böhm leaves his position at the laboratory.
27 October	John Segredor dies of yellow fever in Havana, Cuba.
October	Has Francis Upton make new estimate for the cost of a 10,000-light central station plant.
8 November	Prepares to send John Branner to South America to procure bamboo and cane for use in filaments.
8–12 November	Testifies in telephone interference cases.
16 November	Attorney for Lucy Seyfert demands $7,000 in payment of promissory notes made by Edison in 1874 as part of a plan to raise money for the Automatic Telegraph Co.
c. 18 November	Forms Edison Electric Lamp Co. as a partnership with Charles Batchelor, Edward Johnson, and Francis Upton.
28 November	Assisted by William Hammer begins two weeks of experiments on electrical carrying in lamps.
1 December	Decides that all lamps made by factory will use bamboo filaments.
2 December	Asks Edison Electric Light Co. to authorize payment for services of electrical engineer Hermann Claudius to map out feeder and main system of conductors for first central station district in New York City.
3 December	Glassblowers begin making 200 simplified single vacuum pumps for lamp factory.
11 December	Executes patent application for improved dynamo armature using copper bars in the induction circuit.
15 December	Writes William Crookes about manufacturing Crookes's radiometer at the lamp factory.
16 December	Executes two caveats on voltage regulation.
17 December	Albert Herrick begins experiments on electroplating carbons directly to the lead-in wires.
	Edison Electric Illuminating Co. of New York incorporated.
20 December	Demonstrates electric light and power system at Menlo Park to New York aldermen and other city officials and provides them with a lavish banquet catered by Delmonico.
1881	
1 January	Francis Upton takes over management of the lamp factory.

15 January	Asks Sherbourne Eaton to assume major responsibility for management of the Edison Electric Light Co.
17 January	Has Charles Clarke answer inquiries from Antoine Breguet, director of installation services, regarding his plans to exhibit at the Exposition Internationale d'Électricité in Paris.
27 January	Hinds, Ketcham & Co. printing plant is first manufacturing establishment lighted with Edison system.
7 February	Drexel, Morgan & Co. order Edison to construct a direct-connected steam dynamo for an exhibition installation in London.
c. 10 February	Agrees to supply equipment for exhibition of electric lighting system in South America.
14 February	Appoints Thomas Logan as foreman of the Menlo Park machine shop in place of John Kruesi, who becomes manager of the Edison Tube Co.
26 February	Conducts first test of the Porter-Allen direct-connected steam dynamo.
28 February	Samuel Insull becomes Edison's private secretary.
Feb 1881	Leases office space in New York City, with Edison Electric Light Co.
c. 1 March	Acquires shop at 104 Goerck Street in New York for the Edison Machine Works.
4 March	Incorporates Electric Tube Co. in New York.
8 March	Signs agreement with Edison Electric Light Co. regarding manufacture of lamps.
26 March	Has Yale locks put on doors of the Menlo Park laboratory after staff complain of being overrun with uninvited visitors.
28 March	Asks bookkeeper William Carman to compile an account of the cost of electric light experiments and of money received from the Edison Electric Light Co.
March	Begins spending majority of working days in New York.

Editorial Policy and User's Guide

The editorial policy for the book edition of Thomas Edison's papers remains essentially as stated in Volume 1, with the modifications described in Volumes 2–4. The additions that follow stem from new editorial situations presented by documents in Volume 5. In addition, we describe some of the features incorporated in this volume.

Selection

The fifteen-volume book edition of Thomas Edison's papers will include nearly 7,000 documents selected from an estimated 5 million pages of extant Edison-related materials. For the period covered in Volume 5 (January 1879–March 1881), the editors have selected 422 documents from approximately 9,200 extant Edison-related documents. Most extant documents detail Edison's inventive work and his business relationships; very few are concerned with his family or other personal relationships.

In selecting documents the editors have sought to include key notebook entries and other materials detailing his inventive work on the electric light and power system. More documents of this type are included from 1879, when Edison was personally involved in the research leading to the successful invention of his system of electric light and power, than from 1880, when Edison was directing development work undertaken by a variety of researchers, many of them recently hired. In selecting notebook materials from 1880 we have relied on a set of notebooks kept by Charles P. Mott, a member of the office staff, between March 1880 and March 1881. These "Mott Journals" contain daily summaries of work and events at the laboratory. In several instances the editors have chosen a Mott Journal entry rather than a group of notebook entries to deal with the diversity of laboratory activities. Readers will find references to the notebooks in the notes following these Mott Journal documents.

In selecting correspondence and other documents detailing Edison's business relationships, the editors have typically chosen letters that either provide considerable summary detail or contain information about a wide range of issues and which can be annotated by reference to other related documents. When Edison himself was an active correspondent, as in his exchanges with Edward Johnson and George Gouraud, more of the correspondence is used than in instances, such as the French telephone business, where most of the documents consist of reports to him on actions taken by others. As with notebook entries, there are some documents that the editors have selected because they are important in their own right.

To provide some sense of the public and professional reaction to Edison's efforts, several documents are included that discuss newspaper and journal articles, and extensive reference is made to others in the notes. In addition, the editors have chosen a few articles and letters to the editor as examples of the response to Edison's work.

As in the other volumes, the editors have selected a few key artifacts. For carbon-filament lamps, however, because of constant design changes during the research and development effort—and even after the beginning of commercial manufacture in early 1881—a descriptive list of lamps at the Henry Ford Museum is presented in an appendix in lieu of artifacts (App. 3).

Transcription

The only significant changes in transcription practice from earlier volumes involve numbers. For purposes of clarity the editors have silently supplied decimal points in numbers. Colons have also been added for time designations, most commonly in the dateline of a telegram or in laboratory tests. In a few special instances commas have been supplied in large numbers for clarity.

Annotation

In the endnotes following each document, citations are generally given in the order in which the material is discussed. However, when there are several pieces of correspondence from the same person or a run of notebook entries, these are often listed together rather than in the order they are discussed to simplify the reference.

References to the Microfilm and Digital Editions
The editors have not provided a comprehensive calendar of Edison documents because the vastness of the archive makes preparation of such an aid impractical. Their annotations include, however, references to relevant documents in the microfilm edition. These references are indicated by the acronym *TAEM*. The citations themselves are to reel and frame numbers (e.g., 37:439). Thus, the book edition may serve as an entree into the microfilm edition, just as the latter may serve as an entree into the archives.

For the first time, the editors have included references to documents found on the Edison Papers web site (http://edison .rutgers.edu). The web site contains approximately 175,000 images from the first three parts of the microfilm edition of documents at the Edison National Historic Site. There are also nearly 5,000 additional images not found on the microfilm that come from outside repositories. Citations to images in the digital edition are indicated by the acronym *TAED*. The citations themselves are in two forms. Those to specific documents, such as a letter, are in an alpha-numeric code (e.g., D8120H). Those to documents found in volumes such as notebooks are indicated by both an alpha-numeric code and a set of image numbers (e.g., N117:137, 140). In a few instances one or more specific images in a lengthy document is referred to by an image number or numbers in parentheses (e.g., W100DEC002 [image 7]). All of these images can be seen by going to the Edison Papers homepage and clicking on the link for "Single Document or Folder" under "Search Methods." This will take the user to http://edison.rutgers.edu/ singldoc.htm, where the images can be seen by putting the appropriate alpha-numeric code in one of the two boxes to retrieve either a document or a folder/volume. If retrieving a folder/volume, the user should click on "List Documents" and then "Show Documents" in the introductory "target" for that folder/volume. Then click on any of the links to specific documents in the folder/volume and put the appropriate image number in the box under the "Go to Image" link. Putting image numbers in that box when viewing any document will take you to the specific image number.

The digital edition contains a number of other features not

available in the book or microfilm editions, including lists of all Edison's U.S. patents by execution date, issue date, and subject, and links to their images; a comprehensive chronology and bibliography; maps and images from Edison's early life; a "document sampler" to introduce new users and nonscholars to the collection; and links to related resources on the web. In addition, a collection of essays, images, and patents concerning the African-American inventor Lewis Howard Latimer was mounted on the web site in 2002. Other materials, such as chapter introductions from the book edition, a transcription of the Mott Journals with links to the images, and biographical sketches, will eventually be added. Material from outside repositories, including items cited in this volume, will continue to be added.

Headnotes

Because of the limited scholarly literature and the technical complexities discussed in many of the documents, there are more introductory headnotes in these volumes than is common in historical documentary editions. Each chapter begins with a brief introductory headnote that highlights Edison's personal, technical, and business activities during the period covered by the chapter. Within chapters there are occasional headnotes that appear before documents (see List of Editorial Headnotes). These are used for a variety of purposes. Artifacts and drawings without accompanying text are always preceded by headnotes (e.g., "British Demonstration Telephone"). In addition, the editors sometimes use headnotes to discuss particular technical issues (e.g., "Dynamo Design"); to describe the characteristics of a set of notebooks or other documents (e.g., "Charles Mott Laboratory Journals"); or to provide an overview of activities that are otherwise referenced only in scattered documents and endnote references (e.g., "Preparations for Commercial Lamp Manufacture").

Just as the chapter introductions and headnotes serve as guides for the general reader, discursive endnotes often contain annotation of interest to the general reader. These notes also include business or technical details likely to be of more concern to the specialized reader. In general, the editors have provided more detailed information for technical issues that have received little scholarly attention than for topics that are already well treated in the secondary literature.

Appendixes

As in Volumes 1–4 we include relevant selections from the autobiographical notes that Edison prepared in 1908 and 1909 for Frank Dyer and Thomas Martin's biography of Edison (see App. 1). As in previous volumes, we also include a list of U.S. patents that Edison executed during the period covered by this volume (App. 4).

There are two new appendixes in this volume. The first (App. 2) is a list of Menlo Park employees who were known to have been at the laboratory from 1879 to 1880. Although it is not comprehensive, this is the most complete list we have been able to compile from the available documentation. It is presented primarily to indicate the range of workers at the laboratory. The second (App. 3) lists lamps made at Menlo Park during the period of this volume. These lamps are presently located at the Henry Ford Museum & Greenfield Village in Dearborn, Michigan. As noted above, the list appears in lieu of selecting one or two lamp artifacts to represent design changes.

Errata

Errata for previous volumes can be found on the Edison Papers web site at http://edison.rutgers.edu/berrata.htm.

Editorial Symbols

~~Newark~~ Overstruck letters
  Legible manuscript cancellations; crossed-out or overwritten letters are placed before corrections
[Newark] Text in brackets
  Material supplied by editors
[Newark?] Text with a question mark in brackets
  Conjecture
[Newark?]ᵃ Text with a question mark in brackets followed by a superscript letter to reference a textnote
  Conjecture of illegible text
⟨Newark⟩ Text in angle brackets
  Marginalia; in Edison's hand unless otherwise noted
[   ] Empty brackets
  Text missing from damaged manuscript
[---] One or more hyphens in brackets
  Conjecture of number of characters in illegible material

Superscript numbers in editors' headnotes and in the documents refer to endnotes, which are grouped at the end of each headnote and after the textnote of each document.

Superscript lowercase letters in the documents refer to textnotes, which appear collectively at the end of each document.

List of Abbreviations

ABBREVIATIONS USED TO DESCRIBE DOCUMENTS

The following abbreviations describe the basic nature of the documents included in the fifth volume of *The Papers of Thomas A. Edison:*

AD	Autograph Document
ADf	Autograph Draft
ADfS	Autograph Draft Signed
ADS	Autograph Document Signed
AL	Autograph Letter
ALS	Autograph Letter Signed
D	Document
Df	Draft
DS	Document Signed
L	Letter
LS	Letter Signed
M	Model
PD	Printed Document
PL	Printed Letter
TD	Typed Document
TL	Typed Letter
TLS	Typed Letter Signed
X	Experimental Note

In these descriptions the following meanings are assumed:

Document   Accounts, agreements and contracts, bills and receipts, legal documents, memoranda, patent applications, and published material, but excluding letters, models, and experimental notes

Draft   A preliminary or unfinished version of a document or letter

Experimental Note   Technical notes or drawings not included in letters, legal documents, and the like

Letter   Correspondence, including telegrams

Model   An artifact, whether a patent model, production model, or other

The symbols may be followed in parentheses by one of these descriptive terms:

abstract   A condensation of a document

copy   A version of a document made by the author or other associated party at the time of the creation of the document

fragment   Incomplete document, the missing part of which has not been not found by the editors

historic drawing   A drawing of an artifact no longer extant or no longer in its original form

letterpress copy   A transfer copy made by pressing the original under a sheet of damp tissue paper

photographic transcript   A transcript of a document made photographically

telegram   A telegraph message

transcript   A version of a document made at a substantially later date than that of the original, by someone not directly associated with the creation of the document

STANDARD REFERENCES AND JOURNALS

Standard References

ABD	*Australian Biographical Dictionary, 1851–1890*
ACAB	*Appleton's Cyclopaedia of American Biography*
ANB	*American National Biography*
BJHS	*British Journal for the History of Science*
DAB	*Dictionary of American Biography*
DBF	*Dictionnaire de Biographie Française*
DNB	*Dictionary of National Biography*
DSB	*Dictionary of Scientific Biography*
Ency. Brit.	*Encyclopaedia Britannica, 9th edition*
Ency. NYC	*Encyclopedia of New York City*
Gde. Ency.	*Grande Encyclopédie Inventaire Raisonné des Sciences, des Lettres et des Arts*

NCAB	*National Cyclopedia of American Biography*
NGD	*New Grove Dictionary of Music and Musicians*
OED	*Oxford English Dictionary*
TAEB	*The Papers of Thomas A. Edison* (book edition)
TAED	*The Papers of Thomas A. Edison* (digital edition, http://edison.rutgers.edu)
TAEM	*Thomas A. Edison Papers: A Selective Microfilm Edition*
TAEM-G#	*A Guide to Thomas A. Edison Papers: A Selective Microfilm Edition, Part #*
WGD	*Webster's Geographical Dictionary*
WWW–1	*Who Was Who in America, vol. 1*
WWWS	*World Who's Who in Science*

Journals

AAAS Proceedings	*Proceedings of the American Association for the Advancement of Science*
Am. J. Sci.	*American Journal of Science*
Sci. Am.	*Scientific American*
Sci. Am. Supp.	*Scientific American Supplement*
Teleg. J. and Elec. Rev.	*Telegraphic Journal and Electrical Review;* formerly *Telegraphic Journal*

ARCHIVES AND REPOSITORIES

In general, repositories are identified according to the Library of Congress system of abbreviations. Parenthetical letters added to Library of Congress abbreviations were supplied by the editors. Abbreviations contained entirely within parentheses were created by the editors and appear without parentheses in citations.

CaQMMMcM	McCord Museum of Canadian History, Montreal, Canada
DeGH	Hagley Museum and Library, Greenville, Del.
DLC	Library of Congress, Washington, D.C.
DSI-NMAH	Archives, National Museum of American History, Smithsonian Institution, Washington, D.C.

(HuBPo)	Foundation of the Postal and Telecommunication Museum, Budapest, Hungary
MdBJ	Special Collections, Milton S. Eisenhower Library, The Johns Hopkins University, Baltimore, Md.
MdNACP	National Archives at College Park, Md. (formerly Washington National Records Center, Suitland, Md. [MdSuFR])
MiDbEI	Library and Archives, Henry Ford Museum & Greenfield Village, Dearborn, Mich.
NjWAT	Corporate Research Archives, American Telephone and Telegraph Company, Warren, N.J.
NjWOE	Edison National Historic Site, West Orange, N.J.
NNNCC-Ar	Division of Old Records, New York County Clerk, New York City Archives, N.Y.
NNYNARA	National Archives and Records Administration, Northeast Region, New York City, N.Y.
PHi	Historical Society of Pennsylvania, Philadelphia, Pa.
(UkLS)	Science Museum, London

MANUSCRIPT COLLECTIONS AND COURT CASES

References to documents included in *Thomas A. Edison Papers: A Selective Microfilm Edition* (Frederick, Md.: University Publications of America, 1985–) are followed by parenthetical citations of reel and frame of that work; for example, Cat. 1185:34, Accts. (*TAEM* 22:562). Documents found at NjWOE after the microfilming of contemporaneous material will be filmed as a supplement of the next part of the microfilm edition.

Accts.	Accounts, NjWOE
Argument of T. L. Clingman	*Argument of T. L. Clingman . . . before the Examiner in Chief,* Pat. App. 276,133, RG-241, MdNACP

Batchelor	Charles Batchelor Collection, NjWOE
Böhm v. Edison	*Böhm v. Edison*, Patent Interference File 7943, RG-241, MdNACP
Butler	Benjamin Butler Papers, DLC
CR	Company Records, NjWOE
DCG	Daniel C. Gilman Papers, ms. 1, MdBJ
DF	Document File, NjWOE
EBC	Edison Biographical Collection, NjWOE
Edison Caveat	Edison Caveats, RG-241, MdNACP
Edison Electric Light Co. v. U.S. Electric Lighting Co.	*Edison Electric Light Co. v. U.S. Electric Lighting Co.*, Lit., NjWOE
Edison v. Maxim v. Swan	*Edison v. Maxim v. Swan*, Lit., NjWOE
Edison v. Short	*Edison v. Short*, Patent Interference File 7340, RG-241, MdNACP
Edison v. Siemens v. Field	*Edison v. Siemens v. Field*, Lit., NjWOE
1880 Census	U.S. Bureau of the Census. 1970. *Population Schedules of the Tenth Census of the United States, 1880. National Archives Microfilm Publications Microcopy No. T9.* Washington, D.C.: National Archives. Reel 790, Raritan Township, N.J.
Electric Railway Co. of U.S. v. Jamaica and Brooklyn Road Co.	*Electric Railway Co. of U.S. v. Jamaica and Brooklyn Road Co.*, Lit., NjWOE
EP&RI	Edison Papers & Related Items, MiDbEI
ESP Scraps.	Edison Speaking Phonograph Company Scrapbooks (Incoming Correspondence), UHP
ESP Treasurer's Lbk.	Edison Speaking Phonograph Co. Treasurer's Letterbooks, UHP
Force	Martin Force Papers, NjWOE
HAR	Henry A. Rowland Papers, ms. 6, MdBJ
Harrington v. A&P	*Harrington, Edison, and Reiff v. Atlantic & Pacific, and George Gould et al.,* Equity Case File 3980, RG-276, and Equity Case File 4940, RG-21, NNYNARA

Hummel	Charles Hummel, Wayne, N.J.
Insull Notes	Samuel Insull's Notes for T. C. Martin, Feb. 1909, William H. Meadowcroft Papers, NjWOE
Jehl Diary	Typescript copy of Francis Jehl's diary, EP&RI
Keith v. Edison v. Brush	*Keith v. Edison v. Brush*, Lit., NjWOE
Kellow	Richard W. Kellow File, Legal Series, NjWOE
Lab.	Laboratory notebooks and scrapbooks, NjWOE
Lbk.	Letterbooks, NjWOE
Lit.	Litigation Series, NjWOE
MCR	Miscellaneous Company Records, NjWOE
Meadowcroft	William H. Meadowcroft Papers, NjWOE
Miller	Harry F. Miller File, Legal Series, NjWOE
Newcomb	Simon Newcomb Papers, DLC
Pat. App.	Patent Application Files, RG-241, MdNACP
Pioneers Bio.	Edison Pioneers Biographical File, NjWOE
PPC	Primary Printed Collection, CR, NjWOE
PS	Patent Series, NjWOE
Quad.	Quadruplex Case (Vols. 70–73 and Telegraph Law Cases [TLC]), NjWOE
Reed	Reed Papers, CaQMMMcM
Scraps.	Scrapbooks, NjWOE
Seyfert v. Edison	*Seyfert v. Edison*, Lit., NjWOE
SSW	S. S. White Dental Manufacturing Co. Collection, DeGH
Telephone Interferences	*Edison v. Gray v. Dolbear et al.*, Patent Interference Files 6627, 6628, and 6630, RG-241, MdNACP
TI	Telephone Interferences (Vols. 1–5), NjWOE; a printed, bound subset of the full Telephone Interferences
Time Sheets	Laboratory Time Sheets, Employee Records, NjWOE
TP	Theodore Puskas Collection, HuBPO
UHP	Uriah Hunt Painter Papers, PHi
WJH	William J. Hammer Collection, DSI-NMAH

RESEARCH TO DEVELOPMENT AT MENLO PARK
JANUARY 1879–MARCH 1881

Menlo Park laboratory complex, showing the new brick machine shop behind the main laboratory building, with the new brick office and library building at right.

–1– January–March 1879

At the beginning of the New Year, Edison focused his efforts largely on the project to which he had committed his laboratory in the latter half of 1878—incandescent electric lighting. On the second day of January his assistants began to make a working dynamo for the first time. Edison hoped to obtain the greatest electrical current possible from the least amount of mechanical work, and he and Charles Batchelor incorporated into their design what they had learned from weeks of experiments and study of contemporary generators. During its construction they devised several significant modifications. Tests with this machine began on 13 February but gave poor results indicating problems with the pattern of winding the armature coils and connecting them to the commutator. Edison subsequently considered designs that would eliminate the commutator altogether but decided instead to build and test a large armature with an odd number of commutator connections. This novel arrangement was a major improvement which he later patented. Then in mid-March, for reasons which are not clear, he decided to build an altogether different form of generator in which the armature would rotate between two unusually long and powerful field magnets.

While the first dynamo was being constructed Edison pushed ahead with other components of a complete lighting system. Francis Upton began to elucidate the principles on which a practical electric meter could be designed. Edison himself studied the economic advantages he expected of incandescent lighting compared with arc lighting. In mid-January, having failed to devise a pyroinsulation coating that would permit the wire spirals in his lamps to withstand the heat of incandescence, he embarked on a series of experiments on the behavior of metals at high temperatures. He quickly saw the need for two important additions to his laboratory. One was an oxyhydrogen blowpipe, used to produce the highest temperatures in scientific laboratories, which he improvised from a magic lantern projector. The other was a Sprengel mercury vacuum pump, capable of creating a much stronger vacuum than the mechanical pump he had been using. These experiments led Edison to give brief, albeit public, consideration to substituting nickel for the more costly and refractory platinum wire in his lamp. In the course of several weeks he learned how the absorption of atmospheric gases by a heated wire as it cools affected its structure, and how these gases could be excluded by a cycle of heating and cooling in a strong vacuum. Edison made this process the subject of a draft caveat and included it in U.S. and British patent applications; he later presented a paper on these findings to the annual meeting of the American Association for the Advancement of Science.[1] The mercury vacuum pump proved so valuable that Edison contracted in March with two New York glassblowers to build Geissler pumps, capable of a much higher vacuum, and other apparatus.

Edison continued to experiment with his electromotograph telephone receiver during January. He had promised this instrument to George Gouraud in London, whose efforts to form an Edison telephone company in England depended on this instrument's circumvention of Alexander Graham Bell's patents. In addition John Tyndall, director of the Royal Institution of Great Britain, planned to lecture publicly in mid-February on Edison's acoustical inventions but postponed until March so that he could demonstrate the device. Sigmund Bergmann's shop in New York made two complete telephones with the new receiver under the supervision of Edison's nephew Charley, who had been closely involved with its development. After these were completed Edison's machine shop built two other receivers without transmitters. Edison assigned Charley to carry these four instruments to London on 26 February and remain there because of the chronic ill health of James Adams, his technical representative in England. That same day he began drafting a specification which became the basis of his fundamental U.S. patent on the new receiver but his negotiations with Western Union regarding the instrument dragged on for some time.[2] Gouraud reported that the device was hailed in London as "the greatest invention of the age" and that with its success "people will doubt you less concerning the Electric Light," a reference to a spate of strong criticism in British newspapers and technical journals about Edison's intentions and the feasibility of his lighting system.[3]

In January, Edison's financial backers were negotiating arrangements to provide more funds and to obtain foreign rights to the electric lighting system. A number of them, with the notable exception of John Pierpont Morgan, fretted about reports that he had abandoned the platinum lamp in favor of nickel. In mid-March he set up for the investors a laboratory demonstration consisting of up to twenty platinum lamps, wired in parallel, powered by a Gramme dynamo. The exhibition was viewed by several hundred people over the course of about two weeks. Francis Upton judged it a qualified success but the *New York Herald* was more effusive, reporting that the lamps, using wires treated by the process discovered in February, were nearly ready for practical use and were superior to gas and electric arc lighting. Edison calculated that they would be more economical than gas, a judgment that was hard to justify to skeptics in part because of the lack of photometric standards for commercial gas lighting in the United States.[4] He also began planning an arc lighting system run by his dynamo for the *New York Herald*'s Arctic expedition aboard the USS *Jeannette*.

The phonograph had made Edison a household name in 1878 and reports about the electric light kept him in the news. In January 1879 a Chicago publisher asked for his assistance with a book about him and his inventions. Edison apparently gave no time to this project and the book was printed without his participation. At about the same time he became peripherally involved in efforts to defeat a congressional overhaul of the patent system and to pass another bill restoring lower postage rates for material produced by the electric pen. Edison still received a royalty on the sale of pens, an arrangement that generated both a small income and considerable aggravation from keeping accounts and pursuing infringers. In an unrelated business venture, Edison's long-time associate Josiah Reiff asked him to attend a secret February meeting in Baltimore to help organize a challenge to the Western Union Telegraph Co.'s monopoly. Edison replied that he was "exceedingly busy and have so many men to give instructions that it is out of the question."[5]

During this period Edison was putting finishing touches on his new two-story office and library building and added at least six employees to his staff. These included John Lawson, a self-taught chemist who asked for a job in order to further his studies, and C. E. Mumsell, a graduate of the Columbia School of Mines who worked in the chemical laboratory only during March and April.[6] Grosvenor Lowrey, Edison's attorney, recommended Francis Jehl, who became a laboratory assistant at the beginning of March. A. S. Reinmann was in the laboratory at the end of that month setting up the Geissler pump he built with his partner William Baetz. Edison also began to alter the terms of Francis Upton's employment and his responsibilities in the laboratory, agreeing to give him the duty and public credit for writing about the electric light in Edison's name. Amid the arrival of new men was one absence of personal significance. Edison's oldest son, Thomas Alva, Jr., who had been "quite sick" in January, left at the end of that month with his wife Mary's sister for an extended stay in Florida.[7]

1. Doc. 1796.
2. See Doc. 1697 n. 4.
3. See Doc. 1703.
4. At the end of February the New England Association of Gas Engineers voted to adopt for purposes of comparison the form of standard gas burner certified in London. A trade journal called this decision "a step in the right direction" and expressed hope "that it will be followed throughout the United States. . . . Of course, every man is entitled to use that burner that will give him the most light from a given quantity of gas consumed; but when two or more men come to compare results, they must have a common standard of comparison, in order to understand what basis they are reasoning upon." "The Ninth Annual Meeting of the New England Association of Gas Engineers," *The American Gas Light Journal*, 3 Mar. 1879, Cat. 1012:2, Scraps. (*TAEM* 23:518; *TAED* SM012002a).
5. Reiff to TAE, 9 Feb. 1879, DF (*TAEM* 51:849; *TAED* D7936G); TAE to Reiff, 10 Feb. 1879, Lbk. 4:170 (*TAEM* 80:44; *TAED* LB004170).
6. Lawson to TAE, 6 Jan. 1879; Charles Chandler to TAE, 27 Feb. 1879; both DF (*TAEM* 49:865, 883; *TAED* D7913B, D7913P); N-79-03-10.2, Lab. (*TAEM* 31:1129–51; *TAED* N032:3–25).
7. TAE to Leslie Ward, 22 Jan. 1879, DF (*TAEM* 49:939; *TAED* D7914B).

–1652–

Charles Batchelor to James Adams

[Menlo Park,] Jan 2nd [187]9

Dear Jim,[1]

I hope by this time Ellen is with you all right.[2] I find I have two of your letters to answer.[3] So will go at it. I think there are patents on Private line printer in France but I think Lefferts[4] sold them   have the company[5] write to Edison direct about it. Puskas made you a present of it[6] and I should think he ought to too the only thing I am sorry for is that Bliss[7] gets something out of it too—

We cannot send you a tasimeter[8] we have not one, Edison has given them all away. Browning[9] is ~~giving~~ making them in London I believe. The Pope telephone[10] is new to me I nev[er][a] heard of it before. I always send you a paper when we send to Puskas and frequently when we dont send to them I send one to you. New receivers progressing slow Charley Edison is working it all time.[11] Edison and I are constantly on the 'Light.' In regard to your job with the Company Edison hopes you will not let it interfere with you going to different countries for Puskas. I told him I thought you had fixed it so that it would be no detriment, is that so? There that answers your letters now for something else:— You will have got from Ellen a statement[12] in which you see $500 as your share of money coming from Puskas. I have got a thousand but you will also see that Edison says on the bottom that that is not all but as soon as he finds out how much there is to divide he will pay us the balan[ce][13a]   It has never been calculated up yet but Will Carman[14] is going to get at it as soon as he can   You see Bliss comes in on this and I wanted Edison to keep back his share and pay it on a/c to Edison, Batchelor, & Adams on account of Foreign Pen Royalty,[15] but he would not as he said Bliss was hard up.[16] Has Bliss sent you statement? Rosa[17] is anxiously waiting a letter from Ellen.

Jim will you send me all the circulars and information of the Gramme, Siemens, and Meritens Magneto Electric Generators.[18] Shop is all finished and the Engine running and place heated by steam.[19] For the last 8 weeks we have worked all night right straight along and sleep in the day time we intend to do this until this is perfected. We are making right along 400 carbons per week for Gold and Stock.[20] You remember Badger of Montreal,[21] well he had a claim on telephone in Canada, owing to Edison's arrangement with him some time ago, I daresay you remember it; well, the Gold & Stock have paid him a thousand dolls. for his claim, and are making a contract with Edison for $1 per year royalty on each telephone there,[22] and they are going to start central systems; so we can expect something from that source also.

Very cold here   10 below zero tonight outside— Yours as ever

Chas Batchelor[23]

Rosa desires love to Ellen Martha & Kids.[24] C.B.

ALS (letterpress copy), NjWOE, Batchelor, Cat. 1330:41 (*TAEM* 93: 255; *TAED* MBLB2041). ᵃEdge of original not copied.

1. James Adams had been one of Edison's principal experimental assistants since 1874. In March 1878, Edison sent him to London to help introduce the carbon telephone in Britain; later in the spring he undertook a similar task in France. He thereafter worked alternately in London and Paris. See *TAEB* 2:250 n. 4; Docs. 1258 and 1338.

2. Adams's wife, Ellen, had sailed for England about two weeks earlier. TAE to Theodore Puskas, 16 Dec. 1878, DF (*TAEM* 19:993; *TAED* D7840ZEH); "Sailed," *New York Herald*, 19 Dec. 1878, 10.

3. Not found.

4. Marshall Lefferts, Edison's friend and promoter from 1869 until his death in 1876, had been president of the Gold and Stock Telegraph Co. (see *TAEB* 1:170 n. 6). He and other company officials had owned foreign rights to Edison's Universal private-line printer which they used to set up a stock-reporting system in London in 1871 (see Doc. 168). Nothing is known of their activities in France.

5. The Société du Téléphone Edison had been organized in December 1878 for the commercial development of telephone patents held by Edison and by Elisha Gray, whose interests were represented by Joshua Bailey. See *TAEB* 4:515 n. 1.

6. Hungarian promoter and inventor Theodore Puskas was the agent for Edison's telephone, phonograph, and electric light patent rights in continental Europe. He had probably given Adams stock in the recently formed French telephone company. See *TAEB* 3:678 nn. 4–5 and Doc. 1498.

7. George Bliss (see *TAEB* 4:255 n. 8), general manager for Edison's Electric Pen and Duplicating Press, had a one-ninth interest in the sale price of Edison's French telephone patent (see *TAEB* 3:557 n. 1, 678 n. 5). In a letter of 13 January, Bliss asked Edison for a statement of the cash and stock due him from the sale of rights to the Société du Téléphone Edison to which Edison replied "$425 $200 been sent   be bet[ween] 3 & $4000 in stock   think stock will be worth par in short time" (DF [*TAEM* 52:249; *TAED* D7940A1]; see also note 13).

8. The tasimeter was an instrument devised by Edison in 1878 for measuring minute changes in temperature. It operated on the same principle as his carbon transmitter—the variable resistance of carbon. Edison applied the instrument to astronomy and thought it might be of use in other scientific investigations as well. See *TAEB* 4:231 n. 3 and Docs. 1316, 1329, 1401, and 1420.

9. John Browning was a London instrument maker to whom Edison had given the right to manufacture the tasimeter free of royalties; Philadelphia telegraph manufacturers Partrick & Carter had rights to manufacture them in the United States. Only a few were made. See Docs. 1444 and 1533.

10. Nothing is known of this telephone. It may have been an instrument designed or represented by Franklin Pope, Edison's former business partner and a prominent electrical engineer and inventor. At this time he had charge of Western Union's patent department, in which capacity he also acted as attorney for Amos Dolbear in the Telephone Interferences. In 1878, Pope had briefly represented Elisha Gray's tele-

phone interests in England. See *TAEB* 1:115 n. 1, 4:332 n. 2; Docs. 1270 and 1365.

11. Charles P. Edison (1860–1879), son of Edison's brother, William Pitt Edison. After Charley had rejoined the laboratory staff in June 1878, Edison placed him in charge of the new receiver experiments (see Doc. 1661 n. 1).

12. Not found.

13. Edison apparently made arrangements for sharing French telephone royalties with Adams and Batchelor similar to those they had for American telephone royalties and for other inventions on which they had assisted. Batchelor's signed receipt for $1,000, dated 15 December 1878, is in DF (*TAEM* 19:992; *TAED* D7840ZEF). On 31 January, Batchelor received credit for $1134.74 from the French telephone sale, Adams for $567.37, and Bliss for $810.55. Ledger #3:111, 205, 351, Accts. (*TAEM* 87:58, 102, 153; *TAED* AB001:52, 96, 147); see also Cat. 1318: 56, Batchelor (*TAEM* 93:880; *TAED* MBA:24); for Edison's other arrangements with Batchelor and Adams, see Docs. 637, 1190, and 1345.

14. In *TAEB* 3:6 n. 5 the editors stated that "William Carman's name does not appear in any records until he signed two letters for Edison on 5 February 1878." However, according to his Edison Pioneers file he began working at the Ward Street shop in early 1876 as a sales agent for the electric pen and then in the spring became general office manager at the Menlo Park laboratory. "Carman, William," Pioneers Bio.

15. Edison had received no foreign electric pen royalties for some time. On 2 January 1879, Bliss sent a statement showing gross foreign royalties of $544.25 from September 1877 to 12 August 1878, when he stopped receiving reports from George Beetle. He told Edison that Beetle "professes to have made certain sales since then on which it is proposed to pay a royalty too small for me to consider & the sales, as reported, are insignificant." DF (*TAEM* 162:1058; *TAED* D7923A1).

16. Edison had paid Bliss two hundred dollars on this account in December, which Bliss gratefully acknowledged as "the first ray of genuine sunshine which has fallen on me in a long time." Bliss to TAE, 16 Dec. 1878, DF (*TAEM* 19:992; *TAED* D7840ZEG).

17. Batchelor's wife, Rosanna.

18. For descriptions of the Siemens, Gramme and de Meritens generator designs, see *TAEB* 4:585 nn. 5–6, and 802 n. 1, respectively.

19. Regarding the steam engine, see *TAEB* 4:600 n. 1. Batchelor later estimated the cost and coal consumption of steam heating for the weeks of 3 and 10 January (N-79-01-01:105, Lab. [*TAEM* 30:310; *TAED* N013:51]).

20. The Menlo Park laboratory had been making carbon buttons for Edison's telephone transmitter in large quantities since summer 1878 (see *TAEB* 4:635 n. 1). Button orders can be found in folder 79-38, DF, NjWOE (*TAED* D7938) and Cat. 1308, Batchelor (*TAEM* 90:693; *TAED* MBN003); a summary of Gold and Stock's payments for buttons from November 1878 to March 1879 is recorded in Ledger #3:15, Accts. (*TAEM* 87:18; *TAED* AB001:12).

The fragile buttons were easily broken in transit, prompting complaints from Gold and Stock and Western Electric, which manufactured the telephones (Enos Barton to TAE, 6 and 16 Jan. 1879; George Scott to TAE, 22 Jan., 27 Mar., 2 and 10 Apr. 1879; George Walker to TAE,

5 Mar. 1879; Gold & Stock to TAE, 5 Mar. and 14 Apr. 1879; all DF [*TAEM* 52:4, 7–8, 30, 33–34, 23–24, 35; *TAED* D7937B, D7937E, D7937F, D7937W, D7937Y, D7937Z, D7937P, D7937Q, D7937ZAA]). This prompted efforts to develop a better method of packing the buttons for shipment and to make the buttons themselves less friable (N-78-12-20.3:33, N-79-03-10.2:9–23, both Lab. [*TAEM* 30:534, 31:1129–36; *TAED* N015:16, N032:3–10]; Cat. 1308:117 [Order No. 50], Batchelor [*TAEM* 90:725; *TAED* MBN003:32]). The Gold and Stock Telegraph Co. of California sought to overcome this problem by making its own buttons from carbon supplied by Edison. When George Ladd, president of the California company, complained in June about the cost of the carbon, Edison explained that "the reason why the charges have been increased is that heretofore we have been using imported Vienna chimneys but the American competition has driven the former out of the market while the American chimneys are very cheap, they are also exceedingly poor, and we now break 10 chimneys where we formerly broke one. For instance last week we broke 35 chimneys and only made 900 buttons" (Walker to TAE, 25 and 28 Jan. 1879; Ladd to TAE, 27 June 1879 [misdated as 1880 in *TAEMG2*]; all DF [*TAEM* 52:9–10, 55:635; *TAED* D7937G, D7937H, D7937ZBN1]; TAE to Ladd, 7 July 1879, Lbk. 4:451 [*TAEM* 80:92; *TAED* LB004451]). In 1880, Edison noted that "the Carbon we make for telephones is excessively costly it would take us three months to make ten lbs & take 25 bbls of oil= only ¹⁄₁₀ of 1 per cent made is suitable for telephones" (TAE to W. H. Van Ornum, 24 Sept. 1880, DF [*TAEM* 55:657; *TAED* D8043ZAT]).

21. Franklin Badger, superintendent of the Montreal Fire Alarm System, had held the exclusive Canadian rights to Edison's telephone inventions. At Edison's request he agreed in the latter part of 1878 to sell these rights to the Gold and Stock Telegraph Co. (see *TAEB* 4:455 n. 1). On 18 January 1879, Badger wrote that he had signed the papers and received his $1,000 payment the previous November (DF [*TAEM* 52:172; *TAED* D7939C]).

22. This agreement was executed on 21 January 1879. DF (*TAEM* 52:174; *TAED* D7939D).

23. Charles Batchelor was Edison's chief experimental assistant. See *TAEB* 1:495 n. 9, 2:72.

24. "Martha & Kids" are unidentified. Adams reportedly had two daughters, but their names and ages at this time are unknown. Obituary, *New York Herald*, 6 May 1879, Cat. 1241, item 1186, Batchelor (*TAEM* 94:483; *TAED* MBSB21186).

DYNAMO DESIGN   Doc. 1653

In December 1878, Edison and his assistants had begun sustained research into the operating principles of contemporary dynamo-electric generators and started to design new machines incorporating what they had learned. Drawing on this research, which continued during the first week of January, Edison designed his first experimental generator intended specifically to meet the requirements of his projected electric light and power system. He sought to produce an economical machine that would provide as much energy as possible by completely containing the magnetic field so that all of it could be used to generate electricity.[1]

On 30 December (Doc. 1646), Batchelor had described a modified Siemens generator in which the stationary inner iron core was wired to become an electromagnet that supplemented the power of the external field magnets; the hollow armature rotated between that core and the field magnets. In notes written on 1 January, Batchelor described a design for commutator springs that seems to specify the same sort of moving armature and commutator. (Edison's drawing of 4 January shows the attachment of the coils to the commutator.) In the drawings of the present document, the external field magnets are gone, replaced by an iron shell, and it is not clear which components move. Judging by commutator drawings from later in the first week of the year, the dynamo design had shifted to one having a stationary armature and commutator with a moving internal field magnet and commutator springs. When the field magnet rotated inside the armature drum, the springs rotated around the commutator, allowing current to flow from the armature coils to a circuit outside the machine. By having only the small–

Edison's 4 January drawing shows details of his first dynamo design, including the shell case (top left), the armature (top center) with an exploded view of its connections (bottom), and the rotating field magnet (top right).

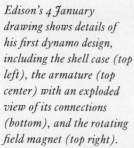

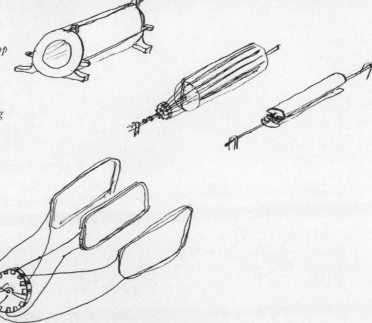

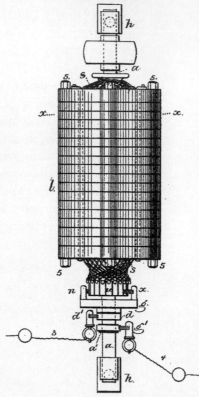

est part of the generator—in this case, the field magnet—move, minimum work was required to provide the highest rotational speed.[2]

During the first week of January, Charles Batchelor specified the dynamo's dimensions and windings and prepared some measured drawings for it. Designed to fit on a base eighteen and one-quarter inches wide by thirty-six and one-eighth inches long, the machine was to be a little over a foot and a half high with a diameter of twelve inches and an armature five inches across. There are several measured drawings of commutator designs from the week following this document, and the machine was completed in mid-February (see Doc. 1682).[3]

1. Docs. 1621, 1627, and 1641; Vol. 16:320–25, 328–46, 348–50; N-78-12-11:71–126; N-78-12-16:232–47; all Lab. (*TAEM* 4:752–57, 760–79, 781–83; 29:714–39, 1180–87; *TAED* NV16:274–79, 282–301, 303–5; N007:36–61; N010:106–13).

2. N-79-01-01:25, 45, 50–51, 55; Vol. 16:349; all Lab. (*TAEM* 30:270, 280, 283, 285; 4:782; *TAED* N013:11, 21, 24, 26; NV16:304).

3. N-79-01-01:25–59; Unbound Notes and Drawings (1879); Oversize Notes and Drawings: Machine Shop Drawings (1879–1880); Cat. 1146; all Lab. (*TAEM* 30:270–87, 44:974–5, 45:44, 6:638; *TAED* N013:11–28, NS79:2–3, NS7986C:1–2, NM014:16).

–1653–

Notebook Entry:
Electric Lighting

[Menlo Park,] Jan 2nd 1879

Edisons Magneto Electric Mach.

Have begun to make a practical working machine after a few weeks hard study on magneto electric principles.[1]

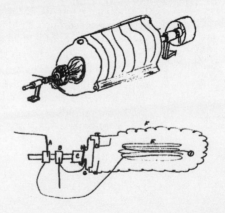

Chas Batchelor

ADDENDUM[a]

[Menlo Park,] Feb 15 1879

We have not been able to take off these currents as yet as the pull on the shell has ruptured the wire and we were obliged to cut it up

X, NjWOE, Batchelor, Cat. 1304:25 (*TAEM* 91:27; *TAED* MBN004:25). Written by Charles Batchelor. [a]Addendum is an X, written by Batchelor.

1. See headnote above.

Menlo Park NJ Jan 3, [1879]

Dear Sir

Referring to yours of the 30th ult relative to Griest matters,[1] I do not agree with you. I do not believe in any cheap substitute but feel confident the Electric pen can stand on its own merits. Until I see something more alarming than what has yet appeared I prefer to let my royalty remain as it is. Very Truly

T. A. Edison   G[riffin][2]

L (letterpress copy), NjWOE, Lbk. 4:81 (*TAEM* 80:34; *TAED* LB004081A). Written by Stockton Griffin.

1. In a letter of 30 December 1878, Bliss had complained about the low prices offered by John Griest, the Chicago manufacturer of a duplicating pen whom he was suing for infringement of Edison's electric pen patent (see Doc.1430). At that time he asked Edison to take a smaller royalty on a cheap pen outfit that he planned to put on the market in competition with Griest. In his 7 January response to Edison's letter, Bliss warned that Griest would earn enough to pay his legal expenses but "when you get a judgement it will be good for nothing, because he has nothing to pay with. The only way to fight that class of competition is to make it unprofitable for them to stay in the business." On 15 January, Bliss again proposed reducing the royalty, this time suggesting three dollars per outfit rather than the five dollars provided by contract (Doc. 817). In marginal notes on this letter, Edison declared that he would "accept for a few months my proportion of $3—on a treadle pen but not on the electric pen." As the Griest suit dragged on, Edison wrote Bliss on 18 March that he did not want to settle with Griest and "would much prefer fighting it out." Finally, in April, Bliss settled the suit and purchased Griest's patent, which Edison apparently did not realize had been issued (see Doc. 1718). Bliss to TAE, 30 Dec. 1878; 7 and 15 Jan. and 4 Apr. 1879; Bliss to Lemuel Serrell, 4 Apr. 1879; all DF (*TAEM* 18:376; 50:372, 392, 442, 449; *TAED* D7822ZCV, D7923F, D7923R, D7923ZBC, D7923ZBH); TAE to Bliss, 18 Mar. 1879, Lbk. 4:222 (*TAEM* 80:64; *TAED* LB004222A).

2. Stockton Griffin, a former telegraph operator, had been Edison's personal secretary since May 1878. See *TAEB* 1:297 n. 1, 4:276 n. 1.

Menlo Park N.J. Jany 3d 1879

My Dear Puskas

I have been so very busy since our last exchange of cables that I have not had a moments time to write you.[1] I am working entirely at night with six of my assistants.[2] I found that nothing could be accomplished during the day on account of the numerous visitors.[3]

I am sorry we cannot carry out the negotiations with Drexel Morgan & Co[4] as foreshadowed in my cable telegram of the

3d ult.[5] If you still feel disinclined to enter into such negotiations you must without delay make preparations to have parties there furnish some money for experimental purposes. The fund I have here is being rapidly exhausted, as it is very expensive experimenting. I bought last week $3000 worth of copper rods alone, and it will require $18 000 worth of copper to light the whole of Menlo Park ½ mile radius.[6] It is also essential that Serrell[7] should have at least $3000 in hand. A second set of patents on the Continent ought to have been taken out over a month ago but which has been neglected owing to a lack of funds.[8]

Please write me immediately if you can arrange so that your parties shall contribute some aid in experimenting and how much.

We are making rapid progress and the whole thing is going to be a success, but it will require more time than I at first anticipated. Dont fail to supply Serrell with at least $1 500 <u>immediately</u> for a second set of patents. Yours Truly

T. A. Edison   G[riffin]

L (letterpress copy), NjWOE, Lbk. 4:79 (*TAEM* 80:33; *TAED* LB004079). Written by Stockton Griffin; the original is in TP.

1. Edison is apparently referring to an exchange of cables in early December 1878 regarding the negotiations with Drexel, Morgan & Co. over foreign rights to his electric light patents. See Doc. 1596.

2. The six experimental assistants were probably Charles Batchelor, Francis Upton, Charles Edison, Martin Force, John Knight, and chemist Henry McIntire; another twenty men were also working at Menlo Park at the beginning of the year. See *TAEB* 4:749 n. 8; no complete set of laboratory time sheets exists for the first month and a half of 1879.

3. In a newspaper interview published on 28 December 1878, Edison explained,

> I work nights in order to escape from visitors. It is very nice and still here nights! I fled out to this uninhabited spot so as to be alone, and I would get well at work about eight in the morning, resolved to put in a day, when I would see a lot of heads coming over the hill from the depot—amiable and delightful people, ministers, teachers, scholars, farmers, doctors, who wanted to know, you know, and those excellent people would just devour two straight hours of my time and pay for it with expressions of admiration. I generally got a chance to work half an hour before dinner and an hour after dinner, and I found it wouldn't do. Now I work eleven straight hours a day—a night, you know. ["Two Hours at Menlo Park," *New York Daily Graphic*, Cat. 1241, item 1091, Batchelor (*TAEM* 94:444; *TAED* MBSB21091)]

4. The New York banking house of Drexel, Morgan & Co., formed in 1871 by the combination of the London-based Morgan and Philadelphia-

based Drexel firms, had just agreed to commercialize Edison's light in Great Britain. Carosso 1987, 136–74; Doc. 1649.

5. Doc. 1596.

6. There is no evidence of the purchase of $3,000 worth of copper rods. In December, Edison had based his calculations for copper conductors on a price of twenty cents per pound (Doc. 1597).

7. Lemuel Serrell was Edison's primary patent attorney from 1870 until the early 1880s. See *TAEB* 1:196 n. 2.

8. Edison had executed his first European electric light patent applications in October 1878 (see Doc. 1498). He repeated this request for money at the end of January (see Doc. 1673).

–1656–

Telegrams: To/From Wallack's Theatre

January 7, 1879[a]
[Menlo Park,] 10:04 AM

Ticket office   Wallacks Theatre[1]   NY

Will call at box office tonight for three[b] orchestra seats well forward next to aisle[2]   Can I have them. Pls reply

Thomas A Edison

NY 11 [A.M.][c]

Thos A Edison
Yes

Wallacks

L (telegrams), NjWOE, DF (*TAEM* 49:43–44; *TAED* D7903E, D7903F). [a]Date taken from document, form altered. [b]Obscured overwritten text. [c]Received on letterhead of T. A. Edison.

1. Wallack's Theatre was located at 844 Broadway in New York. Wilson 1879, 1512.

2. On 6 January laboratory employee John Knight had telegraphed Stockton Griffin: "Theatre Tickets OK." The theater was staging *At Last*, a new romantic melodrama by English playwright Paul Merritt. Some information regarding Merritt's play may be found in a substantial synopsis and unflattering review in the *New York Herald*. Edison evidently made other theater outings the same month. On 13 January he wired Wallack's, "Please reserve three orchestra seats well forward next to aisle tomorrow night" for a performance of *Ours*, an 1866 romantic comedy by noted English playwright Thomas Robertson. Wallack's billed the play as a "popular military drama" and advertised that "a splendid military band, in addition to the usual orchestra of the theatre, has been engaged, in order to give increased effect and beauty to the general picture" of the final scenes, which take place behind the British lines in the Crimean War. DF (*TAEM* 49:43, 55; *TAED* D7903D1, D7903O); "Amusements," *New York Herald*, 31 Dec. 1878, p. 10, 7 Jan. 1879, p. 1, and 13 Jan. 1879, p. 1; *DNB*, s.v. "Robertson, Thomas William."

WASHINGTON, D.C. JAN 9 1879[a]

My Dr E

I enclose you a copy of P[ost]O[ffice] Bill with provision to cover the Pen—[1] I will follow it up & have no doubt of my being able to get it through OK—

I am printing in a little book form The Phono' Contract, Incorporation By Laws, &c,[2] & want a letter from you waiving your right to require the Co to have full paid up capital of $40,000 (we paid up 10,000) (or paid you)[3] You might waive it for 1 year if you dont want to do it entirely: When will you give "Standard Phono"[4] a new wrestle? Yrs

U H Painter[5]

⟨All right— Of course if ~~the~~ I waive right on the Capital Phono will on the $10,000. am going to tackle new phono in three weeks Batch has drawings ready & its going to be a big success for dictating.[6]⟩

ALS, NjWOE, DF (*TAEM* 49:54; *TAED* D7903N). [a]Place and date from *Philadelphia Inquirer* handstamp.

1. Painter's enclosure has not been found. The bill was an attempt to reverse a Post Office decision that material produced by electric pen be accepted only with first-class postage, rather than at the lower rates for printed matter. George Bliss asked Edison to support the legislation because the higher postage was "a gross outrage, in view of their former decisions & the fact that no change has been made in the law. It will result in the ruin of the Pen business unless modified. . . . We must get a law through Congress this winter making Electric Pen matter third class, if the right action can not be obtained from the Dept." Painter reported on 14 January that he had "got the Dep't to suspend exclusion of E Pen matter pending the action of Congress." Congress settled the issue in March by passing a bill that defined third class printed matter as including "the reproduction upon paper by any process except that of handwriting, of any words, letters, characters, figures, or images, or of any combination thereof, not having the character of an actual and personal correspondence." Bliss to TAE, 14 Jan., 8 and 10 Mar. 1879; Painter to TAE, 14 Jan. 1879; all DF (*TAEM* 50:390, 417–18; 49:64; *TAED* D7923P, D7923ZAI, D7923ZAJ, D7903S).

2. Painter and Edward Johnson had engineered a reorganization of the Edison Speaking Phonograph Co. in November 1878 (see Docs. 1509 and 1583). In February, Johnson also promised to send promptly the "pamphlet containing acc't of the organization, agreements, By-Laws, resolutions &c" (Johnson to TAE, 8 Feb. 1879, DF [*TAEM* 51:695; *TAED* D7932H]). The pamphlet has not been found. A later printed collection of Edison's phonograph contracts (Cat. 1402, NjWOE) contains his 1878 contracts, the Certificate of Incorporation of the Edison Speaking Phonograph Co., and his 21 January 1879 release (DF [*TAEM* 51:762, 19:69, 62, 51:771, 766; *TAED* D7932ZBF, D7830ZAE, D7830Y, D7932ZBM, D7932ZBG]).

3. These sums were stipulated in the 18 January 1878 phonograph contract (Doc. 1190). Edison had conditionally accepted similar terms in October 1878, before the reorganization of the Edison Speaking Phonograph Co. (see *TAEB* 4:591 n. 2 and Doc. 1499). On 18 January 1879, Painter made a formal proposal of these terms and asked that the necessary papers be prepared; Edison executed the release on 21 January (Painter to TAE, 18 Jan. 1879; TAE agreement with Edison Speaking Phonograph Co., 21 Jan. 1879; both DF [*TAEM* 51:688–89; *TAED* D7932D, D7932E]). A signed copy of this agreement in Painter's hand is in Unbound Documents (1879–1882), UHP.

4. That is, a phonograph for general use by the public rather than one for exhibition purposes. See *TAEB* 4:4, 134, 211 n. 1, 537; Docs. 1199, 1227 figs. 16–21, 1341, and 1481.

5. Uriah Painter, Washington correspondent for the *Philadelphia Inquirer,* had connections to the Republican leadership. He first became acquainted with Edison in the early 1870s in connection with the Automatic Telegraph Co. He had helped form the Edison Speaking Phonograph Co. and was a member of its board of directors. See *TAEB* 2:661 n. 2; Docs. 1190 and 1583.

6. Nothing is known of this new phonograph and no drawings have been found. However, Edison did execute what became U.S. Patent 227,679 on 19 March for a disc phonograph design that dated back to October (see Doc. 1481) and that marked Edison's last effort to develop a "Standard Phono." It is possible that he worked further on the design at this time.

–1658–

Notebook Entry:
Electric Lighting

[Menlo Park,] Jany 9 1879

Question:[a] How can voltaic cells be arranged in multiple arc and what effect will they produce.[1b]

Laws. The votaic cell indicates only the current, or as it is called the quantity of Elec. passing through it. It is only another form of galvanometer. A number may be put in ~~line~~ series[c] and each one will give an equal amount of gas no matter what the surface of the platinum or the resistance in any one is. This will be only true when the current has been passing ~~to~~ for a time sufficient to saturate all the various platinum electrodes with gas.[b]

The "hindrances" ~~may~~ to the passage of the current are two fold. First the resistance of the water as a conductor. Secondly the opposing electromotive forces set up ~~in~~on the surfaces of the immersed plates.[b]

The resistance may be measured in two ways, making it first say one inch and measuring the current, then two inches and measuring. The difference will ~~only~~ be due to the added resistance for the greater part. It can also be measured ~~in~~ by the method used by   That is to send rapidly reversed currents through the water and measure the by an electro–dynamometre. This method takes for granted that the polarization is zero for such currents.[b]

The resistance of the decomposing cells will directly be a dead loss for the current will be used in heating it. Indirectly it may be utilized ~~in~~ for the[d] heating of[c] the water ~~thus making~~ makes the gas come off more readily and also warms the gas and makes its combustion better.[2]

The electromotive force of a water cell when the poles are covered with H&O is about 1.2 Daniells   Sprague gives it at 1.464 Volts a Daniell being 1.079[3]   1.8 Encyl Brit[4]

1.464	.165541
1.079	.031004
1.363	.134537

1.363 Daniells

The H is carried with the current the same as a metal and thus the O will be found at the pole connected with the + of battery and H with that of the −

TAE

X, NjWOE, Lab., N-78-12-11:160 (*TAEM* 29:752; *TAED* Noo7:74). Written by Francis Upton. [a]Obscured overwritten text. [b]Followed by dividing mark. [c]Interlined above. [d]"for the" interlined above.

1. In this entry Francis Upton returned to the problem of a practical electric meter for Edison's lighting system. He followed a line of inquiry suggested by Edison in Doc. 1622, which was to use the amount of gas evolved in a decomposing cell to calculate the quantity of current passing through the cell over a given interval. With the cells in multiple arc (that is, parallel) rather than in series the meter would have a larger capacity. Upton does not answer the question he initially asks regarding the effect of such an arrangement on the cells but instead discusses other issues regarding the use of electrolytic cells to measure the amount of current.

2. A portion of the current's energy would be dissipated in overcoming increased resistance in the cell. This increase was due to polarization, or the accumulation of hydrogen on the negative electrode. Upton suggests (in the previous paragraph) that polarization could be eliminated, at least for purposes of experimental measurement, by rapidly reversing the polarity of the electrodes with an alternating current. For Edison's ideas regarding reversing the polarity of meter electrodes, see *TAEB* 4:791–92 nn. 1–3; for a much earlier Edison design for overcoming polarization by current reversal, see Doc. 659.

3. As Upton noted above, the process of electrolysis established an electromotive force counter to the flow of current through the cell due to polarization and acted as a "hindrance" to the main current. Here

Upton seeks a practical measure of this force using information provided by Sprague 1875. Sprague gives a value of 1.464 volts to the force needed to free hydrogen. With a Daniell battery cell of 1.079 volts (according to Sprague), this force is slightly greater than that provided by 1.2 cells, as Upton had estimated (Sprague 1875, 224, 227–29). In the following calculation Upton employed logarithms to derive a more exact figure of 1.363 Daniell cells. On 14 January, he tested a Callaud cell relative to a Daniell cell and found it to be 1.87 when fresh and 1.9 when "played out." (N-78-12-20.3:31, 35, Lab. [*TAEM* 30:533, 535; *TAED* No15:15, 17]).

Later in the year, probably in the late winter or spring, Upton calculated the volume and latent energy of gas that could be evolved from water. In another extensive series of notes and calculations that probably date from the same period, he concluded that the "method of measuring resistances is very faulty" due to polarization in the battery (N-79-01-19:135, Lab. [*TAEM* 31:240; *TAED* No23:119]). These pages include notes on the equivalence of heat and electrical energies, particularly the energy of burning zinc and hydrogen, and also the resistance of Condit & Hanson battery cells. Upton was evidently prompted by the general discussion in Sprague 1875 (pp. 207–19) of energy and equivalents in the chapter on electromotive force, though he refers in addition to the *Encyclopedia Britannica* and other sources. N-78-12-20.1:241–47; N-79-01-19:92–137, all Lab. (*TAEM* 30:235–38, 31:197–242; *TAED* No12:121–24, No23:76–121).

4. Upton wrote this as a marginal comment to the right of the calculation. The meaning of the number 1.8 and the specific reference in the *Encyclopedia Britannica* have not been determined.

–1659–

Notebook Entry:
Electric Lighting

[Menlo Park,] Jan 13 1879

2½ Jablochkoff[1] costs 2½ H[orse].P[ower].[2]

say   25 cts. for power[3]
      50 cts   "   carbon[4]
      75 cts.

will pay for 7½ H.P.[5]

    7½
    6 lights per horse power[6]
    45
    2 Glass globes round the Jablochkoff[7]
    90 Lamps[a]

    90[b]
    15[8]
    1350 candles   2½ Jab = 2500[9]

But 1350 candles can be distributed   add ⅓ or more[c] 1750[10]

Labor not counted[11][d]

Jabholkoof[12]

2⅕5 h.p.   ~~2550~~

Carbons   ~~5~~0̶100

                                               ~~75~~150

Light unobscured   ~~18~~3500 candles

  " modified   in—35 places.   ~~900~~1750 candls

E[dison]

15 h.p.   cost 150.—

Total light 1350.

  Distribution 90 as to 5.[13]

    1350[e]
     450[14]
E   1800.
Jabk. 1750.

X, NjWOE, Lab., N-78-12-04.1:57, 56 (*TAEM* 29:591; *TAED* N006:
29). Document multiply dated. [a]Multiply underlined. [b]Calculation of
15 × 6 = 90 in left margin. [c]"or more" interlined below. [d]Entry to here
by Upton, remainder of document written by Edison on facing page.
[e]Calculation of 15 × 90 = 1350 in left margin.

    1. In this set of calculations Edison and Upton compare the cost of
Edison's incandescent lamp to a Jablochkoff candle, a modified type of
arc light invented by the emigré Russian engineer Paul Jablochkoff that
had been installed in Paris in 1878 to much public acclaim. Although
Jablochkoff candles provided larger and brighter lights for outdoor ex-
panses at a cost apparently lower than gas systems, like other arc lights
they were unsuitable for most interior uses. Edison had been making
comparisons to the Jablochkoff since November 1878, and a *New York
Herald* article of 17 January 1879 reported that he had been making tests
with the "most improved Jablochkoff candles of Paris." In that article
the *Herald* reporter copied some figures from a set of Edison calculations
that are somewhat different than those used in this notebook entry
("Electric Light," *New York Herald*, 17 Jan. 1879, Cat. 1241, item 1105,
Batchelor [*TAEM* 94:450; *TAED* MBSB21105a]). For Edison's earlier
comparisons with the Jablochkoff, see *TAEB* 3:442 n. 2; Docs. 1577 and
1651; N-78-12-11:19–21, N-78-12-16:65–66, both Lab. (*TAEM* 29:
688–89, 1109–10; *TAED* N007:10–11, N010:35–36); "Content at
Menlo Park," *New York World*, 5 Dec. 1878, Cat. 1241, item 1057 Batchelor (*TAEM* 94:434; *TAED* MBSB21057). For comparisons to gas lighting from around this time see N-78-12-20.1:111, 263, Lab. (*TAEM* 30:
170, 245; *TAED* N012:56, 131).

Up to this time it was generally thought that the Jablochkoff candle required one horsepower to operate. However, at the end of January reports on the installation of Jablochkoff lights at the Place de l'Opera in Paris concluded that each light required about one and a quarter horsepower. Stayton 1878; "The Electric Light," *Engineering,* 31 Jan. 1879, and "Electric v. Gas Illumination," ibid., 7 Feb. 1879, Cat. 1012:12, 4, Scraps. (*TAEM* 23:535, 524; *TAED* SM012034a, SM012014a).

2. This was the horsepower typically required for a type "A" Gramme generator, a moderate-sized model designed for use in arc lighting. Dredge 1882–85, 1:155–69; King 1962c, 382–91.

3. The 17 January issue of the *New York Herald* (see note 1) reported a figure of 20 cents. On the following page of this notebook, in comparing the costs of gas and electric lighting, Upton wrote that horsepower "cost ⅕ ct per hour   say 1 ct per hour." At that rate, 25 cents would be the cost for running two and a half Jablochkoff candles for ten hours. At about this time the cost of horsepower was typically between $0.02 and $0.03 per hour but much higher for small engines and much lower for the most efficient ones; 500 horsepower engines were reported to operate at $0.0083 per horsepower per hour. NS-78-005, Unbound Notes and Drawings, 1873–1878; N-79-01-01:105; both Lab. (*TAEM* 7:842, 30:310; *TAED* NS7805:37, N013:51); Emery 1883, 426–28 and Schedule A, esp. col. 28; Thurston 1884–85, 453; cf. Doc. 1707.

4. In a 17 December notebook entry Batchelor had written, "Taking into consideration the consumption of carbon as twice as much as H.P. in Jab. candle it would be 4⅓ cheaper than gas"; the 17 January issue of the *New York Herald* (see note 1) reported a figure of 40 cents. In fact, contemporary reports stated that the carbons cost between 12 and 15 cents each and lasted only ninety minutes, making their cost more than ten times that of horsepower at the rate calculated by Upton. N-78-12-11:19, Lab. (*TAEM* 29:688; *TAED* N007:10); "Electric and Gas Light," *New York Sun,* 18 Oct. 1878, Cat. 1241, item 956, Batchelor (*TAEM* 94:378; *TAED* MBSB20956X); Stayton 1878, 166.

5. That is, the 50 cents required to replace the carbons in the Jablochkoff arc light would pay for five additional horsepower to run Edison's incandescent lamps.

6. In a note of 2 November 1878, Edison had indicated that one horsepower could operate six incandescent lamps equal to six gas jets or one Jablochkoff equal to sixty-six gas jets (NS-78-005, Lab. [*TAEM* 7:842; *TAED* NS7805:37]; in *TAEB* 4:719 n. 2 it was incorrectly stated that Edison had treated a single incandescent lamp as "equivalent to six gas burners"). In his 17 December notebook entry Batchelor had remarked, "We get 6 lights per HP" and "90 can[dles]. per HP"; that is, six lamps each equivalent to a fifteen-candlepower gas lamp (N-78-12-11: 21, Lab. [*TAEM* 29:689; *TAED* N007:11]).

7. Although Upton calculated that the cost of operating 45 incandescent lamps was equivalent to 2 Jablochkoff candles, he factored in the need, as Edison put it, to "use a shade to tone down the glare" of the arc light. Translucent glass globes were often used (except in very high exterior fixtures) to diffuse and absorb the light and, as Edison noted, "a great deal of the illumination is thus lost." According to a newspaper account, Edison considered this waste of light to be 50% of the total produced. The French electrical expert Theodose du Moncel reported a

figure of 43% and British engineer G. H. Stayton indicated that "the globes take off about one-third of the light." "Invention by Accident," *New York Sunday World*, 17 Nov. 1878, Cat. 1241, item 1012, Batchelor (*TAEM* 94:411; *TAED* MBSB21012); du Moncel 1879a, 2; Stayton 1878.

8. Edison sought to make his lamps equivalent in candlepower to that of a typical gaslight. A report of August 1878 on the comparative costs of Jablochkoff and gas lighting reported that ordinary gaslights on London streets were equal to twelve to fifteen candles, while the *Encyclopaedia Britannica* indicated that the standard for a gas lamp in much of Europe and America was just over fourteen candlepower; Edison commonly used fifteen candlepower as his standard measurement for a gas lamp. Stayton 1878; *Ency. Brit.*, s.v. "Gas and Gas-Lighting."

9. To this point Edison considered the Jablochkoff candle to be equivalent to sixty-six gas jets, or 990 candlepower. The figure "2500" probably represents a rounding off of the 2,475 candles that would be produced by two and a half Jablockhoff candles at this rate. Doc. 1577; "The Genie of Menlo Park," *New York Sun*, 19 Dec. 1878, Cat. 1241, item 1066, Batchelor (*TAEM* 94:436; *TAED* MBSB21066).

10. Because the intensity of light decreases as the square of its distance from its source, distributing many small lights around an area provides a given minimum level of effective illumination to all parts with less total candlepower than is needed for a single large central light, which puts excess illumination close to its source. Here, Upton calculates that this distribution effect for incandescent lights produces about 30% greater illumination than that from a single Jablochkoff candle.

11. Since Jablochkoff candles lasted only about ninety minutes, even a fixture burning several of them automatically one after another needed daily refilling, creating a significant labor cost. Labor involved in producing the power was probably included in the estimated cost of horsepower.

12. The above calculations by Upton appear to be based on the assumptions contained in an interview with Edison that appeared in the 19 December 1878 issue of the *New York Sun*:

> With a one-horse power the Jablochkoff candle gives a light equal to sixty-six gas jets. With the same power I get only six lights, each equal to one gas jet. People look at each other and ask how I can hope to compete with the candle. The answer is easy. The Jablochkoff candle consumes carbon that costs three times more than the power that supplies the electricity. Hence, at the same expense, I could use eighteen of my lights. The candle cannot be subdivided. Its light is so intense that a ground glass globe is used to modify its power. This involves a loss of fifty per cent. of light, reducing the value of the sixty-six gas jets to thirty-three. Thus at the same expense we have eighteen of my lamps of one gas jet each to one carbon candle equal to thirty-three gas jets. These eighteen lights, judiciously distributed over the area to be lighted, double their value when compared to a single lamp of eighteen gas jets. What I mean is this: Here is a room with one gas jet. The gas jet is equal to fifteen candles; but fifteen candles distributed around the room would give more light than the gas jet. On the same principle I say

that one carbon candle, equal to thirty-three gas jets, would give only about one-half the light that eighteen of my lamps would give. They would actually surpass the Jablochkoff candle in economy when used in lighting up a given area. And all this despite the fact that I can obtain only a total light of six gas jets per horse power to sixty-six by the use of a carbon candle. ["The Genie of Menlo Park," *New York Sun,* 19 Dec. 1878, Cat. 1241, item 1066, Batchelor (*TAEM* 94:436; *TAED* MBSB21066)]

In the set of calculations that follow, however, Edison changes the figure for the light produced by a Jablochkoff candle from 990 candlepower to 700, which is the figure given in the 17 January *New York Herald* (see note 1) as well as in Stayton 1878.

13. This is the ratio of Edison incandescent lamps to Jablochkoff arc lights without taking into account the distribution effect for incandescent lamps (see note 10).

14. This calculation does take into account the distribution effect (see note 10) and treats it as equivalent to 33% more candlepower; Upton calculated it at about 30%.

–1660–

From Rhodes &
McClure

Chicago, Ill., Jan. 14th 1879[a]

Dear Bro.

We are completing a nice book concerning yourself & your inventions & are O.K. on most of your well known discoveries, except the Electric Light—[1]

Is it requesting too much to ask some light from you on this subject?[2]

We shall be exceedingly thankful for a few words—more or less—adapted to the popular mind, from you on this important topic—

Dont fail, please, to send as soon as possible—though it may be asking too much—a few words, or a short chapter—

Would also like to know real names of little "Dot"[3] & "Dash,"[4] & names of other members of your family—

Any incidents would be thankfully received; also account of any new inventions—

We desire & intend to make the volume interesting, instructive & entertaining— All this may seem too bold; & asking too much, but Col. Wilson[5] of the W[estern].U[nion].T[elegraph]. Co. here, encouraged me to write you, & he said he thought you would respond.

We sincerely hope you may, & will reciprocate, if possible— Your Friends & Bros.

Rhodes & McClure.[6]

Friend E— This is OK= Give me a Good lift on this Yours   Firman[7]

L, NjWOE, DF (*TAEM* 49:656; *TAED* D7906C). Letterhead of Rhodes & McClure. ª"Chicago, Ill.," and "187" preprinted.

1. McClure 1879 was a pastiche of adaptations and republications from a variety of sources, including newspapers, scientific journals, Prescott 1878a, and Bliss 1878 (*TAEB* 4, App. 3). It contained a lengthy biographical section, based largely on Bliss, numerous anecdotes about Edison, and descriptions and histories of his inventions.

2. There is no record of a reply from Edison. McClure devoted the last eleven pages of the book to Edison's electric light. He apparently wrote several paragraphs himself, defending Edison's claim to have subdivided the electric light and explaining the Jablochkoff arc light. However, McClure's descriptions of the Wallace dynamo and Edison's platinum lights are closely adapted from newspaper articles published in the latter part of 1878. The final section concerning lamp regulators and the tuning-fork dynamo consists of extensive excerpts and two illustrations from translations of Edison's application for a French patent that appeared in the 14 February 1879 issue of the *Engineer* and the 22 March issue of *Scientific American*; they were reportedly printed on 10 February without any drawings by the *Standard* (London). Edison filed substantially the same specification in April for his British Patent 4,226 (1878), Cat. 1321, Batchelor (*TAEM* 92:107; *TAED* MBP013A). The parts of these patents published in the British press and later in McClure included the lamp regulator found in Edison's first electric light patent, U.S. Patent 214,636, executed on 5 October 1878, and his tuning-fork generator patent, U.S. Patent 218,166, executed on 3 December 1878. "Edison's Electric Light," *New York Sun*, 20 Oct. 1878, "Edison's Electric Light," *New York Herald*, 11 Dec. 1878, "Edison's Electric Light," *New York Herald*, 26 Feb. 1879, Cat. 1241, items 963, 1048, 1139, Batchelor (*TAEM* 94:382, 431, 461; *TAED* MBSB20963, MBSB21048X, MBSB21139X); "The Edison Electric Light," *Engineer* 47 (1879):113; "The Edison Electric Light," *Sci. Am.* 40 (1879): 185.

3. Edison's daughter Marion, born 18 February 1873.

4. Edison's son Thomas Alva, Jr., born 10 January 1876.

5. Col. J. J. S. Wilson had been Western Union's district superintendent in Chicago since 1866. Reid 1879, 248, 275, 561.

6. Richard Rhodes and James McClure were partners in the Chicago publishing firm (Hutchinson 1879, 907). Rhodes subsequently devised the audiphone, a hearing-aid that transmitted auditory vibrations through the user's jaw. An advertisement for the audiphone was included in McClure 1879. Edison recommended the audiphone to persons inquiring about his own projected hearing aid, leading Rhodes to send Edison an instrument in December 1880 and ask him to write a letter of recommendation for an advertising pamphlet ("The Audiphone," *Chicago Tribune*, enclosed with A. Ayers to TAE, 24 Sept. 1879; Rhodes to TAE, 9 Dec. 1880, Rhodes & McClure to TAE, 24 Dec. 1880; all DF [*TAEM* 49:11; 53:21–23; *TAED* D7901C, D8002B, D8002C]).

7. Leroy Firman was general manager of the Telephone Exchange and the American District Telegraph Co. in Chicago (Hutchinson 1879, 399). During a 5 January 1879 test of Edison's telephone on a Western Union line between Chicago and Indianapolis, a connection was made through the Telephone Exchange to Firman's house (McClure 1879, 112–14).

Notebook Entry:
Telephony

New Receiver   Design No 1[1]
Fig 1[2] Design No 1

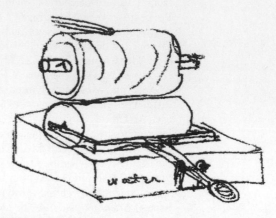

Fig 2[3] Des 1

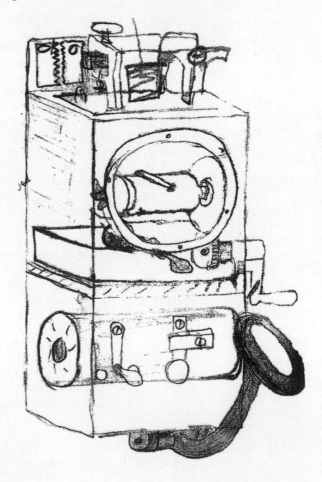

Fig 3 Des 1

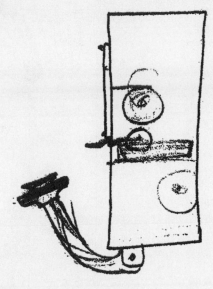

Fig 4 Des 1

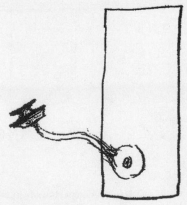

Chas P Edison

X, NjWOE, Lab., N-79-01-14:1 (*TAEM* 31:260; *TAED* No24:001). Document multiply signed and dated.

1. Edison's new receiver was based on his 1874 discovery of the electromotograph principle—a change in friction caused by electrochemical decomposition (Doc. 419). In the spring of 1877 Edison had devised a telephone receiver employing this principle in which the varying electric signal made the sounding diaphragm vibrate by changing the friction between a metallic contact arm and a chemically treated moving surface (see Doc. 873). He revived the idea in June 1878 in an effort to get around Alexander Graham Bell's basic patent on an electromagnetic telephone receiver (see Docs. 1362–1363). Charley Edison had been conducting most of the experiments on the electromotograph receiver since the end of summer 1878. Most of his work up to this time, includ-

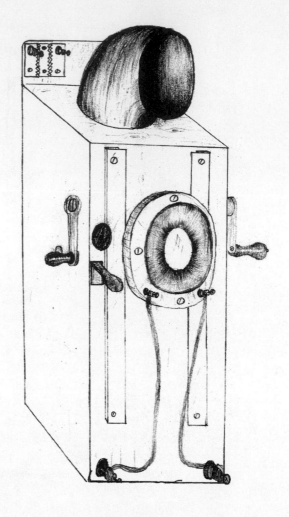

ing some experiments in early January 1879, concerned the composition of the chalk button that provided the moving surface and methods for keeping it wet. In October 1878, Charley had also begun to consider ways of arranging the components of a combined telephone set for easy use in both receiving and transmitting (see Docs. 1440, 1473, 1476, 1495–1496, 1529, 1553, 1560–1561, 1563, 1573, 1575, 1578, 1581, 1591, 1616, and 1650).

These 14 January drawings and another series labeled "Design 2," as well as several undated sketches that follow them in the notebook, are part of an effort by Charley to work out the arrangement for the demonstration telephone sets to be sent to Gouraud (see Doc. 1681). N-79-01-14:4–29, N-78-12-02:235–49, both Lab. (*TAEM* 31:263–75, 29:550–58; *TAED* N024:4–16, N005:109–16).

2. In this drawing the rotating chalk is shown at the top with a wetting roller underneath that would transfer moisture to the chalk cylinder when the user lifted it up by pressing the lever; the figure label is "water." Resting on top of the chalk cylinder is a contact arm that would

be attached to the receiving diaphragm. A similar design is shown in Edison's U.S. Patent 221,957, executed on 24 March 1879.

3. The chalk cylinder, wetting roller, and contact arm are shown here in a box containing a full telephone set. The receiving diaphragm that reproduced the sound of the incoming message would fill the circular opening in the middle of the box, with the contact arm attached to its center. Other parts of the set include a telegraph sounder (in place of a call bell) at the top, a hand crank at the right to rotate the chalk cylinder while listening, an induction coil inside the lower section to transfer the signal from the local to the line circuit, control switches, and Edison's carbon telephone transmitter mounted on an arm projecting from below. What appears to be a lightning arrester is at the top left; not shown are the battery and the electrical connections.

–1662–

From George Gouraud

[London,] 18 ~~Feb~~ Jan 79

Important!!!

My dear Edison—

"Receiver"— Pray do not lose a moment unnecessarily in forwarding this as until it is recd I can make no substantial progress with my negotiations as such people as we want will not go into this thing with the certainty of litigation as must be the case if we use the so-called Bell telephone as a receiver[1]— We might upset their patent which would [~~wi~~?][a] be worse for us than consolidating with them   Both of these are unnecessary once we have the new receiver— And so much has been said in the papers about this that people are getting incredulous— Meanwhile besides we are exposed to the danger of other equally good telephones appearing in the field— That fellow Bailey[2] is now here talking everywhere about a telephone he has which he calls a "Carbon receiver & microphone transmitter"[3] no doubt all rubbish as I have no doubt that he is an imposter—but still [~~ho~~?][a] he can make trouble by circulating false reports—

The moment I have this new receiver I shall move with vigour & have no doubt of most satisfactory results— Tyndalls lecture as advised by cable[4] will be the best possible opening for the financial campaign & I am sure you will not fail to support me at this critical moment with your usual promptness & efficiency—

Please cable me name of steamer bringing the receiver—[5] Also the word "Tyndall" if you will have all he wants in time for his lecture on your "accoustical discoveries" for 17th Feb. He ~~prom~~ says "I promise to do ~~you~~ Edison full justice & will give one entire evening to the Telephone"—

I will if possible have a wire from the Hall of the Royal Institution[6] to Liverpool—a Midland Railway[7] wire—<u>not</u> one one of Preeces[8]   Yours ever

G E Gouraud[9]

ALS, NjWOE, DF (*TAEM* 52:472; *TAED* D7941D). [a]Canceled.

1. That is, a magneto telephone, which could be used as either a receiver or a transmitter. Edison had been promising to send his new receiver to George Gouraud in London for several months. Gouraud and others seeking to market a telephone system in Great Britain using Edison's carbon transmitter became particularly concerned about the need for a non-infringing receiver after the British Bell telephone company obtained an injunction based on Bell's British patent in October 1878 (see Doc. 1497).

2. Nothing is known of Joshua Bailey prior to his involvement, beginning in early 1878, with marketing Elisha Gray's telephone in Europe. He and Theodore Puskas subsequently agreed to combine the Gray and Edison telephone interests in France and Bailey was attempting to combine their interests throughout Europe. See Docs. 1213, 1449, and 1541.

3. A number of variants of the carbon telephone were being promoted around this time, including at least one on which Bailey had made the application for the British patent; it is not clear to which one he referred (Brit. Pat. No. 4,367 [1879]; "4367," *Teleg. J. and Elec. Rev.* 7 [1879]: 254). Gouraud wrote again in February that Bailey and Puskas were in London "making a great deal of noise" about "no end of 'other telephones' they claim to have." He reported that they were testing for the benefit of British Post Office officials "a <u>carbon</u> transmitter which they <u>claim</u> to be <u>independent of yours</u>. They are trying to effect a combination with me, and with the Bell Company. I do plenty of listening and little talking but am anxious more than I can tell you for the receipt of the receiver" (Gouraud to TAE, 21 Feb. 1879, DF [*TAEM* 52:484; *TAED* D7941I]).

4. John Tyndall was professor of natural philosophy and director of the Royal Institution of Great Britain. He researched a wide variety of diamagnetic and acoustic phenomena. Tyndall was known as a vigorous and articulate popularizer of scientific ideas, and delivered many of the Institution's Friday evening public lectures. Two days before writing this letter, Gouraud had cabled Edison, "Tyndall lectures seventeenth February your accoustical discoveries   Send complete instruments especially receiver." *DNB*, s.v. "Tyndall, John"; Gouraud to TAE, 16 Jan. 1879, DF (*TAEM* 52:468; *TAED* D7941B).

5. Edison cabled on 31 January, "Manufacturers delay delivery Receiver   impossible reach Tyndall in time." Gouraud responded the next day that he would ask Tyndall to postpone the lecture but reminded Edison of "the importance of the introduction to the scientific world which meeting his wishes will insure"; Tyndall rescheduled the lecture for late March (see Docs. 1703 and 1715). In a second letter that day, Gouraud also promised to mail a copy of the *Correspondence Scientifique* "from which you will see that somebody appears to have made the receiver which you are promising us. This is the great danger of the situation. A good lump of money can be got for you immediately we can move

without the certainty of legal conflict with the Bell Company." He urged Edison to give the receiver his full attention and "really push it through with your usual vigour even if it does delay a little bit something connected with the light." Gouraud indicated that the article to which he referred was written by Count du Moncel, but nothing further is known of it. TAE to Gouraud, 31 Jan. 1879; Gouraud to TAE, both 1 Feb. 1879; all DF (*TAEM* 52:476–79; *TAED* D7941E, D7941F, D7941G).

6. The Royal Institution's Lecture Theatre was noted for its superior acoustics and audience sight lines. The Institution's building was completed in 1801 at 21 Albemarle St. in London's Picadilly section. Berman 1978, 25–26; Cunningham [1857?], 192.

7. The Midland Railway was formed in 1844 by the amalgamation of several small lines into what at the time was Britain's largest railway. The Midland remained one of the major rail systems in central and northern England although it lacked access to London over its own rails until 1867, when it completed a freight line into the city. The following year that line was extended for passenger service into the still-unfinished St. Pancras station. Simmons 1978, 25–26; Barnes 1966, 229–31, 248–67.

8. A longtime engineer with the British Post Office telegraph system, William Preece had been appointed Electrician and Assistant Engineer-in-Chief in February 1878. Edison had enjoyed a collegial relationship with him until shortly after that time, when they had a bitter falling out over Preece's role in advancing David Hughes's claims to the invention of the microphone. Baker 1976, 176–77; Docs. 1346 and 1348.

9. When Edison first met George Gouraud in 1873 he was the London representative of the Automatic Telegraph Co. in London. In 1878, Gouraud, who was then resident director in Europe for the Mercantile Trust Co., became Edison's agent in regard to the telephone in Britain. He also had an interest in Edison's electric light inventions in Britain. See *TAEB* 1:280 n. 7; Docs. 1344, 1365, 1532, and 1612.

–1663–

From Grosvenor Lowrey

New York, Jan 18th 1879[a]

My dear Edison

I am back again, & feeling pretty well— The news came to me while absent that you were also out of sorts but now Mr Goddard[1] tells me that you are at work again— I am extremely curious to see what you have been doing & some day next week I shall try to go to Menlo.

There is a directors meeting on Tuesday next and I think you should come in. It will encourage them all to see you and talk with you; & it is due to them to be kept informed— I advise you decidedly to come—[2]

Goddard gives me a good account of the state of progress— I hope either you or Upton[3] will from time to time write me what is being done—as the question is one which all parties ask me, with the impression apparently that I can answer—

You will remember that I have not seen you since the visit by Mr Drexel & others & my news is therefore rather old[4]

Please give my compliments to the ladies, & Mr Griffin & believe me   Truly Yours

G. P. Lowrey[5]

P.S. I have some letters from New Zealand Australia & India which have suggested to me that we may perhaps use those countries if necessary to raise additional funds, by selling them to this Company.

ALS, NjWOE, DF (*TAEM* 50:224; *TAED* D7920J). Letterhead of Porter, Lowrey, Soren & Stone. [a]"New York," and "187" preprinted.

1. Nothing is known of Calvin Goddard apart from his role as secretary of the Edison Electric Light Co.

2. Tracy Edson, another Edison Electric Light Co. trustee, asked Edison to attend this meeting, scheduled for the afternoon of 21 January in New York. Edison telegraphed in reply, "Will try and be there." It is not known whether Edison attended, but that evening he telegraphed Stockton Griffin from a depot in Jersey City, the terminus from which numerous railroad lines served Manhattan by ferry across the Hudson River. Edson to TAE and TAE to Edson, both 20 Jan. 1879; TAE to Griffin, 21 Jan. 1879; all DF (*TAEM* 50:226, 49:80; *TAED* D7920K, D7920L, D7903ZAB).

3. Francis Upton was hired by Edison in November 1878 to review all scientific literature related to electric lighting, and he soon became part of the Menlo Park laboratory staff. Trained as a mathematician at Bowdoin College, Upton received in 1875 the first master of science degree conferred by Princeton; he then studied in Berlin for a year with Hermann von Helmholtz before returning to the U.S. in the fall of 1878. Grosvenor Lowrey was a friend of Upton's family. See *TAEB* 4:702 n. 1 and Doc. 1610.

4. Anthony Drexel of Philadelpia, dominant partner in the investment firm of Drexel, Morgan & Co., had visited Menlo Park on 9 December 1878. See *ANB*, s.v. "Drexel, Anthony Joseph," and *TAEB* 4: 751 nn. 4–5.

5. Grosvenor Lowrey was general counsel for Western Union and since 1875 had acted as Edison's attorney in patent matters pertaining to that company. In the fall of 1878, Lowrey became Edison's principal advisor regarding financial support for electric light experiments and was instrumental in establishing the Edison Electric Light Co. See *TAEB* 2: 696 n. 2; Docs. 1459, 1465, 1471, and 1494.

–1664–

From George Barker

Philadelphia, Jan. 19th 1879.[a]

My dear Edison:—

I was very sorry to hear yesterday from Mr. Robinson[1] that you were sick again and confined to your house. You must remember, my dear fellow that the laws of physiology are as inflexible as the laws of electricity. No man can work as much as you do and eat as little and as irregularly, without paying for it sooner or later. You cannot do as much in a week as you do now, to be sure, if you eat and sleep regularly; but you will have in the long run more weeks to work in.

Your long telegram[2] I found on my table the other night when I came home. I am glad you have imported a Gramme and a Siemens machine,[3] and I now hope you will fit up a suitable dynamometer[4] and make an accurate test of the electromotive force and current strength which each machine you have gives when worked under the best conditions. We are all on the lookout for your machine; and are all ready to "stand from under"[5] when it shall appear. The problem to be solved would seem to[b] be a very simple one. Given the resistance of the external circuit (made up of the resistance of a given number of lamps) which of course will be equal to the internal resistance of the machine, you have simply to get an electromotive force which divided by the total resistance, will give you the current strength necessary to maintain your lamps at the desired temperature.[6] I fear however, if the papers report you correctly, that you cannot make incandescent lighting a profitable thing if you can get only six or eight lamps of 15 candles each, or 90 to 120 candles to the horse power. The theoretical efficiency is 3000 candles to the horse power, nearly; one candle light being about 13.1 foot pounds per minute. Of course this can never be reached in practice; but 1000 to 1500 candles, or one half the theoretical, is easily attainable with the carbon points.[7] By the way, Mr. Wallace tells me that his machine has suffered from the disparaging things the Sun reporter said about it in your name.[8] Excuse me, but really I think your reputation is being lessened by the foolishness of the men whom you allow to write you up.

I received the phonograph all right.[9] Many thanks. It is not as elegant as the other, but it works quite as well. How long may I keep it? I have not yet received the chalk cylinders you said you would ask your nephew to make and send me.[10]

I write to ask you to come over and spend next Sunday with me. You need a play day and you promised me you would come this fall. I want you to see if I get the best out of the phonograph it can do. And I also want to show you some things of mine. Twill do you good. So come and we will have a time. Cordially yours

George F. Barker.[11]

ALS, NjWOE, DF (*TAEM* 49:71; *TAED* D7903X). Letterhead of University of Pennsylvania. ᵃ"Philadelphia," and "187" preprinted. ᵇ"seem to" interlined above.

1. Probably Heber Robinson, manager of the Western Union office in Philadelphia. "The Philadelphia New W.U. Office," *Operator,* 1 July 1876, 7; Reid 1879, 178.

2. Not found.

3. Edison had obtained Gramme and Siemens generators in December 1878. See *TAEB* 4:774 n. 2, 802 n. 1.

4. Edison had a Prony brake dynamometer at the laboratory. Like other forms of absorptive dynamometers, which measured torque by means of straps or jaws clamped tightly around a rotating shaft, the Prony destroyed in friction the power it measured and was an impractical instrument for dynamos. Around this time Edison was evidently investigating designs of transmission dynamometers, which transmit power through belts or springs so that it may be applied to useful work, such as generating electricity. Measured drawings by John Kruesi labeled "Edison's Dynamometer" and dated 14 and 29 January show the use of springs to convey power from the driving wheel to the receiving wheel in a manner similar to that of other contemporary transmission dynamometers. Francis Upton later explained why Edison abandoned this form of dynamometer:

> When a spring is used to transmit power, there is always some play in the bearings of the apparatus used to translate the motion of the spring to an index. The constant of the spring is also liable to change from strain. As this constant also is determined when the machine is at rest, it is extremely difficult to feel sure that when in rotation centrifugal force will not disturb some of the parts, thus altering it. The small amount of deflection that can be readily obtained from a spring tends to render all readings uncertain. [Edison then designed an electric dynamometer in which, according to Upton,] magnets were used attached to the driving wheel to drag after them armatures attached to the receiving wheel. The number of pounds it took to detach the armatures was found with the various strengths of current while the machine was at rest and the results tabulated. It was hoped that when in motion the table thus proposed could be utilized, but it was found that the jar of the belt detached the armatures long before a correct measurement could be obtained. [Upton 1880b, 179]

This device was developed in late February and built in March. Although there are several drawings by Edison and Batchelor of this design, its principle of operation is unclear. Knight 1876–77, s.v. "Dynamometer"; Thompson 1902, 754–56; Oversize Notes and Drawings: Machine Shop Drawings (1879–1880), Vol. 16:381–83, N-79-02-24.1:7, N-79-02-10:119–21, N-79-03-20:165–67, all Lab. (*TAEM* 45: 45–46, 4:817–20, 30:973, 32:1198–99, 33:77–78; *TAED* NS7986C: 3–4; NV16:339–41, No20:4, No46:58–59, No47:76–77).

5. A British and American slang expression meaning to escape something or get out of the way. It apparently came from the Royal Navy and the London Fire Brigade, where it was used routinely as a warning when

a heavy object was dropped from above. Franklyn 1975, s.v. "Stand from under"; Mathews 1951, 1636.

6. Many contemporary electrical experts thought a generator would work best when its internal electrical resistance was equal to the external resistance of the circuit. This view was based on the understanding that the maximum power output for any given battery occurred when its internal resistance matched that of the rest of the circuit (Prescott 1877, 88; Sabine 1867, 262–63). Thus, most electrical experts agreed with the president of the Society of Telegraph Engineers, German-born inventor and engineer C. William Siemens, who argued that "in order to get the best effect out of a dynamo-electric machine there should be an external resistance not exceeding the resistance of the wire in the machine" (Higgs and Brittle 1878, 80). Barker had sent a copy of Siemens's statement to Edison and had reiterated this point in a November 1878 letter (see Doc. 1592). Although this view of best practice for batteries, and by analogy generators, was widely held, maximum power output was not the only concern in the practical use of batteries in telegraphy, the main electrical industry of the day (see Pope 1869, 71–72; Prescott 1877, 88; Sprague 1875, 203). Furthermore, as a result of their recent comparative tests of some commercial dynamos at the Franklin Institute in Philadelphia, electrical scientists Edwin Houston and Elihu Thomson had concluded that "the greatest efficiency will, of course, exist where the external work is much greater than the internal work, and this will be proportionately greater as the external resistance is greater" (Houston and Thomson 1878, 61; these tests are discussed in Carlson 1991, 82–87). In his own study of dynamos, Edison and his assistants concentrated on the effects of making the external resistance higher than the internal resistance (Docs. 1572 and 1634; N-78-12-04.1:12, Lab. [*TAEM* 29:569; *TAED* N006:7]; see also N-78-12-20.1:144, 213, 221–29, 232, Lab. [*TAEM* 30:187, 225–29, 231; *TAED* N012:73, 107, 111–15, 117]). Francis Upton explicates the mathematical relationship between internal and external resistance in Doc. 1832.

7. In the *New York Sun* of 25 November 1878, Edison had publicly predicted that he could build a generator that would operate six fifteen-candlepower lamps per horsepower. This assumption appears as early as a note of 2 November. In his cost calculations, Edison often made comparisons to the Jablochkoff candle, assuming its output to be between 600 and 1,000 candlepower. Published tests results of other arc lights reported production of about 1,000 to 2,000 candlepower per horsepower. In the report of their Franklin Institute tests, Houston and Thomson gave similar results, although they argued that the effective light produced was lower. Nonetheless, they believed that arc lights would be two to three times less expensive than gas lighting. However, they argued that a system of incandescent electric lighting, heating, and power of the sort Edison envisioned would be uneconomical. Although the transmission of power looked "quite promising," they argued that incandescent elements such as platinum would be much less economical than arc or gas lighting and that since the maximum theoretical efficiency of a dynamo in converting steam power to electricity was only 50% electric heating would be "totally impracticable." "The New Electric Lights," *New York Sun*, 25 Nov. 1878, Cat. 1241, item 1021, Batchelor (*TAEM* 94:416; *TAED* MBSB21021a); NS-78-005, Lab. (*TAEM* 7:842; *TAED*

NS7805:37); Doc. 1659; Higgs and Brittle 1878, 45, 47, 57; King 1962c, 365; Prescott 1879, 441; Houston and Thomson 1878, 60–61.

8. See Doc. 1636. William Wallace (1825–1904), British-born wire manufacturer and inventor, established Wallace & Sons at Ansonia, Conn., with his father and brothers in 1848 (*DAB*, s.v. "Wallace, William"). With electrical inventor Moses Farmer he constructed dynamos based on Farmer's design. At Barker's suggestion, Edison visited the Ansonia shop in September 1878 and subsequently bought two Wallace dynamos (see Docs. 1423, 1436, and 1474).

9. See *TAEB* 4:814 n. 2.

10. See Doc. 1635.

11. George Barker was professor of physics at the University of Pennsylvania. See *TAEB* 2:329 n. 8.

–1665–

*Notebook Entry:
Electric Lighting[1]*

Menlo Park NJ Jany 1~~8~~9=79

Experiments with Platina & Platinum-Iridium alloys 20 per cent Ir—at the incandescent point with galvanic battery to determine any changes that may take place.[2]

Platina wire 004 does not permanantly elongate to the extent that the pt-Ir-alloy does. the latter is very soft at the incandescent point so much so that with 004 Pt-Ir wire 20 pc is unable to support its own weight when placed horizontal between supports 80 mm. apart,[3] it slowly bows downward from a slight curve thus

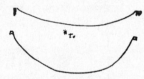

It [--][a] has[b] not anything like the strength of platinum when[c] at the same temperature=

When the Pt-Ir wire is allowed to remain incandescent for 30 minutes a remarkable change is effected   It str~~e~~atches, grows smaller in diameter ie[d] from 004 to 003. & breaks   upon examination under the microscope it has this appearance

Platina wire thus

If the platina wire is only brought to incandescence for say 2 minutes it sh~~e~~ws these cracks   they first are not open but just as if a "mole" had been burrowing around the wire. after a longer time these break & sharp edged cracks make their appearance.[e]

when the platinum wire is fused after being cracked the piece upon the end when it does not break off short has this appearance;[4]

Just before Pt or Pt-Ir wire fuses between the supports it contracts violently and the cause is apparant when we place it under the microscope. it has drawn up into globules   if it breaks quickly the globular tendency is just perceptible but if it contract considerable before breaking these globules a[re] plain

To ascertain loss of weight we placed 005 Pt wire in supports weight 0.0296 m̶. g.ramme in 5 minutes we raised heat so it fused   upon weighing the whole it was 0.0293. g showing loss .0003. .gramme.

Another piece weighed 0.0266. after being incandescent for 20 minutes it weighed 0.0265, shewing loss of .0001 .g. at this rate it would all be dissipated in 122 hours, but organic matter may account for some of this loss.

The same piece was then made incandescent for 20 more minutes and weighed 0.0262 gramme shewing a further loss of weight of 0.0003. or a total loss of .0004. gramme   after 10 minutes more we again weigh it and find weight to be 0.0260 shewing a further loss of weight of 0.0002^c or a total of 0.0006. gramme.

After 20 minutes more we again weigh it and find its weight to be 0.0258 shewing a further loss of 0.0002^c or a total of 0.0008. gramme.

Yesterday I noticed a peculiar phenomenon when the platinum of 004 wire was placed between the supports 80 mm apart and brought to incandescence for 1 hour a portion only was incandescent after the lapse of that time thus

with large Eaton spectroscope[5] no lines were visible, but perhaps if larger wire were used lines could be seen— It is barely possible that the platinum is contaminated with Osmium, Ruthinum, Rhodium or other more oxidizable metal of the platinum group= or that there is an immense amount of oxygen condensed in the platinum which is driven out by the heat.

I think that I smell ozone coming from the incandescent wire, at least the smell is peculiar.

Another hypothesis is that chloride of sodium in the air combines with the Pt to form a volitile Pt chloride.

I find that the 001 wire made (Pt) by Johnson Mathay & Co London[6] melts in the ordinary gas jet—under the microscope it shews all broken up in short lengths and clearly fused at these breaks   a piece of wire held in the flame bends with a[c] jerk thus

with wire 002 made from Raynors[7] pt. it fuses readily but these short breaks are not so perceptible as in JM & Co wire. all the breaks are smooth   thus

whereas JM wire

with Raynors of 010 Pt. after being incandescent 5 minutes on battery, shewed fine cracks just commencing but the sections bounded by the cracks were very much smaller than J.M. & Co small 004 wire.

I now put in 010. 10 p.c. Pt Ir—JM & Co wire.—taken out and experiments resumed with the peice which we have repeatedly weighed—

I notice that these peice does not drop or bow downwards nearly as much as it did when first put in— There is just 70 mm of this wire, after 3 minutes I notice it commences to drop considerable—[e]

With platina wire I notice that the heating always commences exactly in the middle of the wire and reaches towards[c] the two supports simultaneously but the action is not so rapid but that the eye can easily trace it.

JM & Cos 005 platina if held in gas flame for 30 seconds shews cracks under microscope; also little round raised drops as if gas were underneath but tension was insufficient to burst the bubble.[e]

heated air can be seen rising from 70 mm of 004 wire 2 feet over it—

The 0.0266 wire after be subjected again to incandescence for 20 minutes weighs—0.0252. shewing a loss since last weighing of 0.0006[c] or a total of ~~0.0014~~ .0014.[c]

It is again made incandescent for 20 minutes & weight found to be 0.0249, shewing loss of 0.0003[c] or total of 0.0017— The wire fused.

we now take ½ of this fused wire which weighs 0.0132 gramme and subject it to incandescence; again for ~~+~~60 minutes continuously—

note— I notice the wire although growing smaller keeps perfectly bright= [f]

the loss in 60 minutes 0.0130 or .0002.[c] This loss is not very great but may be due to the fact that I put a book up on one side of the wire within ¼ inch of it to shield it from air drafts which cooling it in spots raises the temperature of the rests of the wire by reducing resistance & this tends or sometimes causes fusion

we again put the wire to the incandescent point for. 20 minutes. note I did not suceed in fusing the 001 wire in flame from burning paper though it became incandescent.[g] It weighed afterwards .0129, shewing loss of .0001 after 20 more minutes of incandescence it weighs. 0.0129, shewing no loss The scale (Beckers best)[8] does not work as sensitive as it should be. We again put it to the incandescent point, for 20 minutes and I take the book down to allow draughts of air to strike the platinum as it is not highly incandescent, and if the action is due to oxygen this will hasten it. we weigh & find no loss—[h]

we again bring it to the incandescent point, slightly higher than before for 20 minutes— we weigh and find no loss I now at 7 oclock pm bring it to incandescence to be left all night.

X, NjWOE, Lab., N-79-01-19:27 (*TAEM* 31:134; *TAED* No23:13). [a]Canceled. [b]Interlined above. [c]Obscured overwritten text. [d]Circled. [e]Followed by dividing mark. [f]"note . . . bright" inserted between dividing marks. [g]"note . . . incandescent" inserted between dividing marks and enclosed by left bracket. [h]Multiply underlined.

1. This entry is continued in Doc. 1666.
2. On 3 January, Edison had conducted a short series of experiments and recorded the visible changes made by heat in 20 % platinum-iridium wire. He noted that it softened enough when incandescent to sag under its own weight and become permanently elongated. He also saw that if the wire were drawn tight again while incandescent it would, "after a moment of quiet commence to vibrate like a tuned string reaching a vibration of ⅛ of an inch apparent, & then it sags by this effort owing to elongation The effect of this vibration is also to cool the wire producing contraction, a blue light appears in the centre of the neutral point of vibration" (N-78-12-31:45, Lab. [*TAEM* 30:398; *TAED* No14:22]). Immediately preceding Edison's 3 January entry are several pages of notes by Francis Upton, dated the same day, on the number of battery cells required to bring platinum-iridium wire ("stretched between two small

vices") to increasing intensities of light, and the resistance of the wires at each stage. Upton continued these tests for the next two days, and on 14 January Charles Batchelor prepared tables summarizing the changes in resistance and appearance of platinum and 10% and 20% platinum-iridium alloy wires as increasing numbers of Condit & Hanson battery cells were applied. N-78-12-31:30–44, 48–78; N-78-12-20.1:86–106; Unbound Notes and Drawings (1879); all Lab. (*TAEM* 30:391–98, 400–415, 158–68; 44:977–80; *TAED* N014:15–22, 24–39; N012:44–54, NS79:5–8).

3. On 9 January, Batchelor made several drawings of an "Experimental Inst for heating effects in wires," a device for holding and stretching long wire segments; Edison's undated drawing of a similar apparatus is in N-78-12-15.1:111, Lab. (*TAEM* 29:998; *TAED* N009:58). Batchelor also roughly sketched what appears to be a concave reflector intended for "concentration of heat." N-78-12-15.1:121–23, Lab. (*TAEM* 29:1003–4; *TAED* N009:63–64).

4. Text is "smooth."

5. Asahel Eaton was a Brooklyn chemist and manufacturer of spectroscopes and other optical instruments. Edison had acquired a spectroscope in 1875, but it is not known whether this is the Eaton instrument he refers to here. *TAEMG2*, s.v., "Eaton, Asahel K"; *TAEB* 2:427 n. 2.

6. Johnson Matthey & Co. was a London company that refined precious metals and manufactured various platinum devices. Edison had ordered a considerable quantity of platinum and platinum alloy wires and sheets from the firm in November 1878, and some of the material had reached Menlo Park by this time. See *TAEB* 4:762 n. 5.

7. Hiram Raynor was a New York platinum dealer and supplier of wire and other laboratory apparatus. Letterhead, Raynor to TAE, 21 Mar. 1878, DF (*TAEM* 17:394; *TAED* D7811H); Wilson 1879, 1196.

8. Christian Becker, a noted Dutch manufacturer of precision balances at mid-century, established Becker & Sons in New York. Turner 1983, 65, 251; Wilson 1879, 89.

–1666–

Notebook Entry: Electric Lighting[1]

Menlo Park, Jany 20[–22]=79

Found that battery had run down on platina cold. It probably kept incandescent for 2 or 3 hours then by midnight became invisible   we find at 8.30 am this morning on weighing that it had lost nothing=[a] We again bring it to incandescence for 1 hour

note— I notice a flash of smoke go off when the Pt was heated this morning shewing that dust had been deposited over night. the weight must have been very small. after one hour it weighs no less[b]

note— The whole loss of this peice with the one broken from it amounts to 8.4 percent. 4 hours & 20 minutes then stopped losing— one hour more no change.

I placed a peice of Pt-Ir 10 per cent foil in stand and brought to incandescence for 2 minutes. under microscope it showed slight cracks   thus.

and some of the surface of the foil bounded by these cracks showed a tint of iridescence   I again brought the foil to incandescence for 5 minutes. under the microscope the cracks were large and spotted all over with sections that were iridecent thus

This shews oxidation and I think it is <u>Iridium</u>   after causing it to become incandescent for 5 more minutes these iridescent spaces lose their line appearance and approach that of the iridescence of lamp black   The iridicence is blue & ultra violet.[c]

Palladium becomes incandescent and I think melts at about the same temperature as platina. It become greatly annealed. do not see any iridescence   notice that hair points on strip have melted.[c]

with the 0.0132 which weighed 0.0129 by 2 hours more heat it loses 0.0004 ie[d] goes to 0.0125. we put on a fresh battery but this did not make it more incandescent as we used 1 cell less unfortunately on putting it in stand again it fused. under the microscope the sharp edges of the cracks are rounded they have not materially increased   the wire is as bright as ever. The end is a beautiful ½ globule very smooth. thus   I notice what appears to be an air bubble   thus   I noticed on taking away from the stand that under the microscope a large peice of some salt, or organic matter was attached to it= This might have caused the fusion by reducing the temperature at that point.[c]

I again put the paladium to incandescence for 15 minutes. under the microscope the whole surface is crystaline shewing the crystals perfect these are <u>line</u> crystals & look like Antimony. they do not Iridesce like the Iridium in the Pt–Ir alloy= between the crystals something has apparantly been eaten away as the top & bottom of crystals require a different focus of the microscope. These crystals can be seen with the naked eye.= crystals may be thrown upwards= prisms—

I now having broken our small 0.0132 piece I substitute for the battery the Gramme machine, and use 410 mm of Raynors Pt wire, which I have flattened out so as to be nearly ³⁄₆₄[e] wide.

I brought it to incandescence for 3 hours. Its first weight after cleaning by a momentary incandescence was 0.3430. after 3 hours it weighs 0.3210

Gold cannot be brought (.002 wire) to red heat at least can not see it in daylight (fuses)—

Ruthenium is probably the impurity in platina because it quickly tarnishes and becomes iridicent while ~~platinum~~ iridium does not appear to. Iridium (pure) keeps bright. Rhodium becomes iridicent but not near as much as Ruthenium.

Upon putting Iridsomine to incandesce it becomes covered with crystals which ~~s~~shoot out from it & are[f] like crystals of permanganate Potash to the naked eye. They iridesce beautifully

Everything about Ruthenium looks[e] rounded[e] & resemble fused carbon. I also notice these long needle like crystals & I think they are either osmium or Ruthenium[e]

oxygen has no other effect than to cool the wire like wind.

It looks as if platinum was volatilized at a high temperature= The 410 mm wire after 1 more hour weighs 0.3103.[c]

When Pt-Ir wire has been in the Hydrogen flame for ½ an hour the green coloration disappears but by increasing the battery current passing through it, It again appears. of course the extra battery raised the temperature the first cell having polarized

we put the 410 mm wire 1 hour more under incandescence, it weighs 0.3075. we incandesce it again for 1 hour and find that it weighs 0.3050. We again put it to 1 hours incandescence after which [it?][g] weighs 0.3018.[c]

note— I notice that Bichloride Platinum colors flame ultra violet or greyish violet, also green at the edge. This may be due to vapor of platinum—

We broke the 410 wire ½ weighs 0.1477. after 1 hours incandescence it weighs 0.1474.

I seal a piece of iron wire in glass tube and I find that I can bring it to a bright yellow color. when the tube broke, I find that the oxidation skin is very thin, owing the expansion of the iron in the glass it touches and fuses the latter.

we put the ½ of the old 410 mm in and bring to incandescence. it weighs cold 0.1541. It busted=[c]

we now roll out a piece of 20 pc Pt-Ir alloy .020 and after bringing it to incandesce we weigh and find weight to be 1.0147   we the[n] put it to incandescence with gramme machine. there is about 400 mm & it is about ¹⁄₁₆ wide— 1 oclock & 20 minutes pm January 22 1879—

X, NjWOE, Lab., N-79-01-19:43 (*TAEM* 31:147; *TAED* N023:26).
[a]Multiply underlined. [b]Multiply underlined; followed by dividing mark.
[c]Followed by dividing mark. [d]Circled. [e]Obscured overwritten text. [f]"&
are" interlined above. [g]Obscured by inkblot.

1. This entry is a continuation of Doc. 1665 and is continued in Doc.
1669.

–1667–

Telegrams: To/From
Henry Morton

January 22, 1879[a]
[Menlo Park,] 3:46 PM

Prof Henry Morton[1]
Have you a Sprengel pump[2] I could borrow a few days   Pls
reply my expense

T A Edison

Hoboken N.J. 4:55 PM

T A Edison
We have only the original form no better than a good valve
pump.[3] Prof A W Wright[4] of New Haven has a good one

H Morton

L (telegrams), NjWOE, DF (*TAEM* 50:11–12; *TAED* D7919I,
D7919J). [a]Date taken from document, form altered.

1. Henry Morton, a professor of chemistry, had been appointed in
1870 as the first president of the Stevens Institute of Technology in
Hoboken, N.J., a position he held until his death in 1902. Morton first
met Edison in January 1878 in connection with the phonograph. *ANB*,
s.v. "Morton, Henry"; *TAEB* 4:8 n. 7, 70 n. 4.
2. The air pump devised in 1865 by Hermann Johann Philipp Sprengel produced a vacuum of nearly one-millionth of an atmosphere, several orders of magnitude better than other contemporary devices. The
basic Sprengel apparatus employed a long, thin tube (the fall tube)
placed vertically between reservoirs of mercury at the top and bottom; a
branch near the top connected it to the vessel to be exhausted. As the
mercury flowed from the upper reservoir into the fall tube it broke into
droplets, and air trapped between these droplets was carried to the lower
reservoir and expelled into the atmosphere. Although extraordinarily
effective, a Sprengel pump with a single fall tube worked slowly because
of the tube's small volume. The most important modifications to the
basic Sprengel design by this time involved arrangements for removing
entrapped air bubbles from the mercury before it was reintroduced into
the fall tube. Thompson 1887, 612–13, 633; Webb 1965, 569–71; Turner
1983, 98–105.
Immediately before telegraphing Morton, Edison had wired George
Barker, "Where can I borrow a Sprengel pump" (DF [*TAEM* 50:11;
TAED D7919H]). Edison had apparently begun conducting experiments using the mechanical pump that he had acquired in January 1875
and first used in electric light experiments in 1877 and again in October

1878 (see note 3; *TAEB* 3:547 n. 2; Doc. 1491). The only extant records of vacuum experiments at this time are from 23 January, but the pump failed and was repaired the following day (see Doc. 1669 n. 4; Reinmann & Baetz receipt to TAE, 24 Jan. 1879, DF [*TAEM* 50:498; *TAED* D7925F1]). Edison made do with it until late March, when he bought a Geissler pump (see Doc. 1714). The Geissler instrument operated on the opposite principle of the Sprengel design, by raising a column of mercury to force air from the top of a glass tube (Thompson 1887, 587–89). Francis Jehl states in his reminiscences that Francis Upton acquired a Sprengel pump from Princeton in winter 1879 (Jehl 1937–41, 252–53); however, Jehl was not employed at the laboratory until early March (see Doc. 1685 n. 2) and Upton remembered only that he borrowed a Geissler pump from Princeton University, which Edison recalled as being after the purchase of a Geissler pump from Reinmann & Baetz (Upton's deposition, Complainants's Rebuttal, pp. 3240–41, 3251–52, *Edison Electric Light Co. v. U.S. Electric Lighting Co.*, Lit. [*TAEM* 48:127, 132; *TAED* QD012F:125, 130–31]; Edison's testimony, p. 43, *Böhm v. Edison* [*TAED* W100DED032, image 12]).

3. This is a form of mechanical air pump. In the basic form of this instrument, a piston tightly fitted in its cylinder draws air through a valve from the vessel to be exhausted. On the return stroke this valve is closed and another opened, allowing the piston to expel air to the atmosphere. A two-cylinder instrument was commonly in laboratory use at this time and was the form of mechanical pump owned by Edison. Turner 1983, 98–103; Jehl 1937–41, 797.

4. Arthur Williams Wright was a professor of physics and chemistry at Yale University, where he later built and directed the Sloane Physical Laboratory. There is no record of a request to Wright. *ANB*, s.v. "Wright, Arthur Williams," and *NCAB* 13:348.

–1668–

From Calvin Goddard

New York, Jany 22d 1879[a]

Dear Sir

I enclose stock ctfs as follows:

#1	T R Edson[1]—	30	shrs.
2	J H Banker[2]—	30	"
3	C Batchelor	130	"
4	S. L Griffin	20	"
5	E. Fox[3]	8	"
6	W. C Croffut[4]	5	"
7	E H Johnson[5]	8	"
8	U H Painter	20	"
9	Jas Adams	30	"
10	T A Edison	219	"
		500	"

in place of temporary ctf No. 2. cancelled.[6] Yours truly

C. Goddard

ALS, NjWOE, DF (*TAEM* 50:227; *TAED* D7920M). Letterhead of Edison Electric Light Co. ª"New York," and "187" preprinted.

1. Tracy Edson was one of the original directors of the Edison Electric Light Co. Edson had amassed a fortune on royalties from patents used in printing United States currency. As a major shareholder of the Gold and Stock Telegraph Co. he had helped guide the company into its affiliation with Western Union. At this time he also sat on the boards of directors of Gold and Stock and the American Speaking Telephone Co. See *TAEB* 1:235 n. 1, 4:592 n. 4.

2. James Banker, another original director of the Edison Electric Light Co., was a long-time Western Union director and member of that company's executive committee. Edison Electric Light Co., Articles of Incorporation, 16 Oct. 1878, DF (*TAEM* 18:38; *TAED* D7820ZAM); Western Union 1873–78.

3. Edwin M. Fox, a former telegrapher friend of Edison's, was a writer for the *New York Herald* and contributed most of the paper's articles on Edison. He contributed articles on Edison's inventions to *Scribner's* as well. Fox was also an attorney with a private practice in New York City and later in the year he offered Edison counsel regarding *Harrington v. A&P* and Edison's desire to reach a compromise with Jay Gould on the matter. *TAEB* 4:195; Fox 1879a, 1879b, 1879c; Doc. 1764 n. 2.

4. William A. Croffut was an established newspaper writer who at this time worked for the *New York Daily Graphic.* He and Edison began a friendly relationship in 1878. He wrote or arranged the publication of favorable stories about Edison in the *Graphic* and other newspapers and dubbed Edison the "Wizard of Menlo Park" (see *TAEB* 4:133, 195, 197 n. 1). In thanking Edison for the stock Croffut promised, "If I can do anything in the world for you at any time in every way, order me up & I'll go it alone" (Croffut to TAE, 3 Feb. 1879, DF [*TAEM* 50:242; *TAED* D7920W]).

5. Edward Johnson was a former telegraph operator who met Edison while working for the Automatic Telegraph Co. The two became good friends with Edison providing personal loans to Johnson in 1875. Johnson, who would remain associated with Edison for many years, had given public exhibitions of Edison's telephones and phonographs in 1877–78 and was at this time treasurer of the Edison Speaking Phonograph Co. See *TAEB* 1:505 n. 13; 2–4, passim.

6. Under the terms of his 15 November 1878 agreement with the Edison Electric Light Co. (Doc. 1576), Edison had received 2,500 of its 3,000 stock shares, each having a face value of $100 (Edison Electric Light Co., Certificate of Incorporation, 16 Oct. 1878, DF [*TAEM* 18:38; D7820ZAM]). Calvin Goddard had advised Edison on 18 January that the company was prepared to issue permanent stock certificates and that if he brought his temporary certificates to the quarterly board meeting on 21 January, "you can make such transfers as you wish" (Goddard to TAE, 18 Jan. 1879, DF [*TAEM* 50:223; *TAED* D7920I]). The transfers to Edson and Banker were presumably sales; those to Fox, Croffut, Johnson and Painter were outright gifts and were acknowledged as such. There is no information about the transfers to Griffin or Batchelor but these likely were intended as compensation for work on electric light matters. When Batchelor wrote James Adams in Paris to expect his 30 shares he did not indicate if they were a gift or fulfillment of an agree-

ment, but he promised that "we expect to make them worth \$1000 each before another year is over." Edson to Stockton Griffin, 24 Jan. 1879; Banker to Griffin, 30 Jan. 1879; Fox to TAE, 26 Jan. 1879; Croffut to TAE, 3 Feb. 1879; Johnson to TAE, 23 and 25 Jan. 1879; Painter to TAE, 25 Jan. 1879; all DF (*TAEM* 50:229, 241, 237, 242, 228, 230; 49:87; *TAED* D7920N, D7920V, D7920R, D7920W, D7920M1, D7920P, D7903ZAG); Batchelor to Adams, 1 Feb. 1879, Cat. 1330:44, Batchelor (*TAEM* 93:258; *TAED* MBLB2044).

–1669–

Notebook Entry: Electric Lighting[1]

[Menlo Park,] Jany 23=79

I find this wire has been rolled thinner in one place abou[t] ⅓ of it is brilliantly incandescent while the remainder is yellow white— This piece busted last night at 7.20 after having been incandescent for 6 hours. It weighs ie[a] both pieces 0.9188. Under the microscope I find that the cracks are still visible but within the cracks are innumerable hole with a crater like appearance   The whole wire is honeycombed and to the naked eye has the appearance of frosted silver. one of the parts I roll out 3 times its length. After rolling it has its old appearance. it is bright. We weigh and find that its weight 0.3527. time 10:30 am. after being partially incandescent at 3 oclock PM it weighs 0.3325 gramme. The end which was incandescent The most is greatly honeycombed. The proves that platinum volitilizes at at yellow and probably at a red heat and that the green flame seen when held in hydrogen flame is volatized Pt.[2]

We now commence some experiments with the oxyhydrogen flame,[3] to determine fusing points of non conductors and metals. We make a rough guess as to fusibility by noticing the time and trouble it requires to fuse the substance

Platinum fuses quite readily— Rhodmium fuses easier than Platinum. Ruthinium does not fuse but something volitile comes off which colors the flame and the piece becomes light and porous.

Iridium[b] does not fuse or color the flame or appear to get smaller. I will mention that the gases are not mixed before consumption. The gases are burned from the orifices of two common blow pipes.

Iron instantly melts and oxidizes but I think the extra heat due to its oxidation is the cause of its melting. Cobalt is very difficult to meelt, it oxidizes slightly.

Nickel is exceedingly difficult to melt and I think it has[b] as high a melting point as platinum and possibly higher if fused in a vacuum and prevented from oxidizing—[4] a globule of nickel in molten state was brilliantly incandescent and did not appear to oxidize or grow smaller. Silicum does not fuse but a a certain temperature ~~wit~~ it oxidizes with explosion

Boron does not fuse or apparantly oxidize

Aluminium becomes incandescent without oxidizing or melting but if the oxygen pipe be brought near it it instantly oxidizes—

Rutile[5] is fused with exceeding difficulty   it is a conductor both before and after fusion. after fusion a small piece thus .x.

has with powerful pressure a resistance of 62 ohms   when powdered and a platinized block pressed upon the powder it had a resistance of 17 ohms

Metallic Zircon oxidizes at very high temperature, becomes coated on outside with white oxide—

Metallic Chromium from Johnson Mathay and Co does not fuse that we can see. we get a current through it after fusing and by great pressure bring it down to 18 ohms

The chromium from Brooklyn given me by Beardsley[6] is slightly attracted by the magnet showing iron— It fuses tolerably easy. The fused mass probably an oxides. its resistance by great pressure is 20 ohms—

We now mould up a block of Titanic Acid for fusion in the O.H. flame. we find that it fuses at very high temperature and the moulded piece contracts and cracks, owing to the volatiztn of the (something— = It does not conduct. It has in[b] places a vitrous appearance color azure=

Alloy Titanic Acid, Nickel= do not appear to mix—

Alloy Iron & Nickel. Iron burns up[b] but think some has alloyed with the nickel. resistance of a button ½ ohm its 1/16 thick= its very brittle

Molybdenium all goes off in fumes

Chromium & Nickel do not alloy as we cannot perfectly fuse the nickel or touch the Chromium. I notice chromic colors on lime shewing oxidation.

We take a piece of nickel and roll it out   cut a narrow strip and pass a current through it and very strange to say it becomes brilliantly incandescent <u>without</u>[c] fusing. I think it nearly if not equal to platinum— It slowly oxidizes but we shall prevent this by sealing the burner= when it does fuse it acts like the Pt-Ir 20 pc alloy—it remain hard when incandescent. It is very probable that absolute chemically pure nickel will have a very much higer fusing point than the sample we have which is probably only commercial. This is a great discovery for electric Light[7]—in the way of economy[d]

X, NjWOE, Lab., N-79-01-19:54 (*TAEM* 31:158; *TAED* N023:37). [a]Circled. [b]Obscured overwritten text. [c]"out" added later. [d]"in . . . economy" interlined above; probably added later.

1. This entry is a continuation of Doc. 1666 and is continued in Doc. 1670.

2. Edison subsequently reported this conclusion to Grosvenor Lowrey. When Lowrey erroneously told Edward Dickerson, Sr., an eminent New York patent attorney and technical expert retained by the Edison Electric Light Co., that Edison had discovered that an electric current would volatilize platinum in a vacuum, Dickerson replied that this fact was already well established. Edison then wrote directly to Dickerson to clarify his claim. He acknowledged that "all the books state that platina is easily volatilized in vacuum tubes by the electric disruption charge," but asserted that "no record can be found in any work . . . that platina volatilizes at a continuous yellow heat—the volatilization of which increases as the temperature rises and that platina loses weight in the flame of a candle or of Hydrogen and colors the flame green." Dickerson replied that he had misunderstood Edison's discovery and now believed that it had not been anticipated in any printed work. TAE to Dickerson, 8 Feb. 1879, Lbk. 4:166 (*TAEM* 80:43; *TAED* LB004166); Dickerson to TAE, 13 Feb. 1879; Lowrey to TAE, 10 and 17 Feb. 1879; all DF (*TAEM* 50:246, 244, 247; *TAED* D7920Y, D7920X, D7920Z).

3. An oxyhydrogen flame is generally one produced by the combustion of hydrogen with pure oxygen, which produces a higher temperature than hydrogen burned in the atmosphere. Because of the inherent danger a ready mixture of the two gases was not widely used; instead the gases generally were conducted separately to or immediately prior to the flame. Apparatus for producing the oxyhydrogen flame varied and it is not clear what Edison had but he may have used the one from the lamp of his magic lantern which was modified the next day (*Ency. Brit.*, s.v. "Oxyhydrogen flame"; see Doc. 1670 and John Ott Notebook, p. 12, Hummel). On this day, Edison and Francis Upton tested the fusibility of approximately seventy-five additional minerals, ores, and metals. Most of the substances fused, though Edison noted that molybdenite (number 77), did so only with difficulty and "gives a splendid light does not apparently fuses after having been fused" (N-79-01-21:11–33, Lab. [*TAEM* 30:592–603; *TAED* N016:7–18]). A set of undated notes by Upton is probably related to these experiments or those made during following days (N-78-11-28:61–67, Lab. [*TAEM* 29:50–53; *TAED* N001:32–35]).

4. Another notebook entry from this date indicates that they were conducting some experiments in a vacuum (N-79-01-21:35–37, Lab. [*TAEM* 30:604–5; *TAED* N016:19–20]). The normal melting point of platinum (1,773.5°C) is higher than that of nickel (1,455°C).

5. Titanium dioxide.

6. George Beardslee was a Brooklyn electroplater. See *TAEB* 2:85 n. 14.

John Ott's 30 January 1879 drawing of a modified oxyhydrogen light.

7. In his own notebook, Charles Batchelor described how he obtained a sample of .003 inch thickness by repeated annealing and hot rolling, and noted that he could reduce it to this thickness before "the characteristic cracking of its edges" took place. He measured the fusing point at 1,500 degrees Centigrade, compared with only 1,050 for iron, and noted that he had put in "Hessian crucible a small lot of nickel but could not melt it in our furnace whilst with cast iron turnings I could melt comparatively easy." He reported that "nickel stands better or as good as platinum for the Light" (Cat. 1304:27–28, Batchelor [*TAEM* 91:29–30; *TAED* MBN004:27–28]; N-79-01-21:9, Lab. [*TAEM* 30:591; *TAED* N016:6]).

–1670–

Notebook Entry:
Electric Lighting[1]

[Menlo Park,] Jany 24 1879—

Batchelor has succeeded in rolling nickel in sheets .002 thick. the piece was somewhat crack.

With another piece he has rolled it without a flaw or crack .003 and is now going to roll it thinner between steel plates. his success is due to annealing after each rolling—

a piece of 002 nickel is brought to yellow heat for 10 minutes on Gramme machine   when it melts in a weak spot—a perfect globule is upon the end of the broken point= upon bending ist gives a crackling sound shewing considerable oxidation= it is black & shining just like Russia Iron.[2]

This oxide is extraordinarily tenacious   the nickel may be bent several times nearly at right angle without shewing cracks in the oxide, but if bent 20 or 30 times exceedingly fine cracks may be detected under the microscope.

we have now riged up a regular Oxy-Hy blowpipe use the lantern blowpipe.[3] We melt in two minutes a pin head of Iridium (J[ohnson]&M[atthey])—on flat lime— Ruthenium boils at once something goes off leaving a porous something which is infusible.[a] Osmium Iridium—(Iridsomine) is infusible but something comes off but not a great deal. Nickel is not very easily fused it only fuses apparantly when it oxidizes   Chromium is a little harder to fuse than nickel—after fusing nickel it is quite malleable.

Osmium-Iridium after 1 hour with the O.H. flame does not fuse

X, NjWOE, Lab., N-79-01-19:62 (*TAEM* 31:166; *TAED* N023:45).
[a]Obscured overwritten text.

1. This entry is a continuation of Doc. 1669 and continues in Doc. 1672.

2. Sheet iron made in Russia that had a "smooth, glossy surface of a purplish color." Knight 1876–77, s.v. "Russia Sheet-Iron."

3. The magic lantern for the projecting stereopticon slides was a common application of the oxyhydrogen flame (see Doc. 1669), in conjunction with a nonvolatile and infusible solid, such as lime, to radiate its intense heat in the form of dazzling light (*Ency. Brit.*, s.v. "Oxyhydrogen flame"; Knight 1876–77, s.v. "Oxy-hydrogen Lamp."). Edison had acquired such slides in 1874 (see Doc. 560 n. 2). A common arrangement in this application was to mix the hydrogen and oxygen in a chamber just prior to the nozzle outlet whereas in the blowpipe the two gases remained separate until the hydrogen emerged from the nozzle, where a stream of oxygen was directed into it. The purpose of the blowpipe being to maximize heat rather than light, it lacked the radiating solid used in the lamp (Gage and Gage 1914, 100–107). The blowpipe was introduced earlier in the century and was the apparatus most used at this time to attain very high temperatures in the laboratory. Earlier in the decade French chemist Henri Étienne Sainte-Claire Deville had concentrated the oxyhydrogen blast in a small lime furnace, thus maximizing its effectiveness, while conducting with Henri Jules Debray his prominent studies of platinum and platinum-iridium alloys. George Matthey, the British metallurgical scientist who was a partner in Johnson Matthey & Co., had adapted the Deville furnace to commercial production of platinum. Deville had found that the highest temperature attainable from his furnace was that at which combustion products dissociated and stopped reacting (estimated at around 2,500–2,800°C); the practical limit was considerably lower. The effects of very high temperatures on various substances was an active area of scientific and technical research in this era; however, no reliable, precise methods had yet been established for measuring these very high temperatures, and Edison did not attempt to do so in these trials (Child 1940, 149–50; F. W. Gibbs 1958, 118; Moissan 1904, 1; Partington 1964, 4:498–99; Siemens 1889, 2:221–22, 3:258). Edison later discussed his use of the oxyhydrogen blowpipe in Doc. 1796.

–1671–

From Grosvenor Lowrey

New York, Jan 25th 1879[a]

My Dear Edison:

I went into Mr Fabbri's[1] to day and found that Griffin had been there this morning. They all gathered round me and, in a joking way, asked if I knew anybody who wanted to buy their stock. I enquired what was meant, and they said "Haven't you heard the news?"— They then told me that you had been obliged to give up platinum. Mr. Fabbri looked serious and, of course, so did we all, although they said something about nickel being found to be a substitute. I told them what you had said to me about past experience of the use sometimes to be made of obstacles, and said these were the things that they, and everybody, must look for, of course; that no great end was attained, so far as my experience goes, without considerable doubt and tribulation, and this was just the time when we must all stand by the inventor and the enterprise &c. &c. &c.— Mr Morgan[2] stood by listening without saying anything. After a while somebody said "Nothing has ever yet been done with Mr Puskas," and I said "Mr Edison informs me that he has called, or will soon call, upon Mr Puskas for $1800 more for patent fees &c and perhaps you will not care to take Mr Puskas' place in these expenditures, if you are losing confidence"— Mr Morgan spoke for the first time and said that he had been looking with some anxiety for just such a time as this, because this was the time to settle with Puskas on fair terms, and he (M), Wright,[3] & Fabbri all said they were perfectly ready to go on[b] as had always been talked about. I tell you this because it produced a very pleasing impression on my own mind for I saw that there was a true ring in it and that these gentlemen were likely, in a stress, to turn out—as I always supposed they would—not to be very easily frightened away from a thing they once made up their mind to. All they, or I, shall ask of om you is to give confidence for confidence— Express yourself, especially when you come to a difficulty, freely. You naturally, having an experience of difficulties and of the overcoming of them, in your line, (which none of the rest of us can have,) may feel that it would be prejudicial, sometimes, to let us see how great you difficulties are, lest we, being without your experience in succeeding, might lose courage at the wrong time. This will be true sometimes of all people; but every active mind greatly interested in a particular subject works in its own way when a difficulty is presented in finding out the causes, and reasoning against the probability of their being insuperable; and with our friends I think that would be the result in almost every instance where you yourself should show that you still believe in a possible success.

With this in view I would like to have a talk with you right down to the bottom of everything and would like to have Mr Morgan join in it if I could; but I could not get him out to Menlo now, and I do not want to take your time a moment. I may perhaps go out to Menlo this afternoon at 3o'c, but more likely not. In that case I may telegraph you this afternoon to see if either you or Mr Upton could come in tonight; but again I fear taking Upton from you at a time when you may need him, as well as of interrupting your studies and of fatigueing you by the journey, for I can well understand that you are at the highest tension just now.

I have an engagement on Monday which will prevent my getting out,[4] and very soon I hope to get away to the Hot

Springs.[5] If nothing more convenient presents itself I will send Mr Griffiths[6] to you on Monday and ask you to state to him carefully, (and he will take it down in short-hand) what the present situation is; whether the experiments which you substitute for those in platinum require so large a force of men &c. &c. Very truly Yours

G. P. Lowrey

P.S.[c] Since the above was written your letter of the 24th[7] is received— One embarrassment which you no doubt feel is that the public has been taken into confidence in this matter too early and the opportunity is afforded for those who do not believe to make capital because of the certainty with which your friends have always spoken as to what you had done— Be sure you are right about Nickel and everything else before having anybody know much about it— Naturally, our people while not at all blaming you or me, will say that we have been rash in setting up expensive buildings ~~wh~~ while it was uncertain whether the minor details were yet secured— In other words that so large an engine was not necessary, for the ascertainment of what you now know about platina— But on the whole I think they will be reasonable; ~~an~~ but they will withhold their opinion on nickel until it is completely proved[d]   L.

LS, NjWOE, DF (*TAEM* 50:232; *TAED* D7920Q). Letterhead of Porter, Lowrey, Soren & Stone. [a]"New York," and "187" preprinted. [b]Interlined above. [c]Postscript written by Lowrey. [d]"me . . . proved" enclosed by brace at right.

1. Egisto Fabbri, a partner in Drexel, Morgan & Co., was among the original directors of the Edison Electric Light Co. Drexel, Morgan & Co.'s offices, like those of Lowrey's firm, were located in the Drexel Building on Broad Street in lower Manhattan. *TAEB* 4:592 n. 4, 621 n. 5.

2. John Pierpont Morgan (1837–1913) was a co-founder and principal partner of Drexel, Morgan & Co. *ANB*, s.v. "Morgan, John Pierpont."

3. James Hood Wright was a partner in Drexel, Morgan & Co. *NCAB* 33:443.

4. Lowrey instead arranged to visit Menlo Park on Tuesday, 28 January. Lowrey to TAE and TAE to Lowrey, both 27 Jan. 1879, DF (*TAEM* 50:239–40; *TAED* D7920T, D7920U).

5. Hot Springs Reservation, established by Congress in 1832, was a region in the Ouachita mountains of central Arkansas noted for dozens of hot mineral springs. It became known as "The American Spa" and attracted both wealthy and indigent health seekers; at this time more than 50,000 people visited each year, most of them enduring a day-long stagecoach ride from Little Rock. Though each of the bathhouses constructed there was devoted to the cure of a specific disease, these springs were known generally for the relief of muscular and skeletal afflictions (Lowrey suffered from gout; see Doc. 1711). In 1921, Congress desig-

nated the area Hot Springs National Park. *WGD*, s.v. "Hot Springs"; Altman 2000, 45–47, 124–25; http://www.hot.springs.national-park.com/info.htm#esta (a site maintained by U.S. National Parks Net, a private enterprise).

6. E. W. Griffiths was probably a clerk in Lowrey's firm. Porter, Lowrey, Soren & Stone to TAE, 19 Feb. 1879, DF (*TAEM* 49:115; *TAED* D7903ZAW).

7. Not found.

–1672–

Notebook Entry:
Electric Lighting[1]

[Menlo Park,] January 287th [1879]

Last 3 days been arranging our apparatus for conducting experiments—

with O[xygen]. H[dryogen]. Batchelor is now experimenting[2]

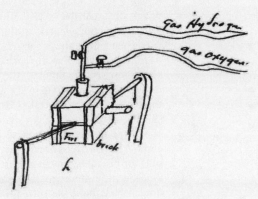

Lime crucible inside—[3]

I have rigged up two carbon points with the gramme machine and concentrate the arc in a lime crucible—[4]

Melted nickel and preserved it in bottle for future experiments—[5] also alloy Ca & Ni   also Ni & Fe, also Large quantity of Fe with small quantity Ni

Owing to the enormous power of the light my eys commenced to pain after 7 hours work and I had to quit[6]

X, NjWOE, Lab., N-79-01-19:65 (*TAEM* 31:169; *TAED* N023:48).

1. This entry is a continuation of Doc. 1670.

2. See Doc. 1670 n. 3.

3. The lower labels on the figure are "Fire" and "brick." The rods behind and in front of the apparatus suspending it were able to rotate in their supports while attached to the frame that held the blocks of firebrick together. An undated drawing of this arrangement, labeled "Oxy. H. blowpipe to melt nickel &c and pour in moulds," follows a 22 January note of alterations to be made at the laboratory. N-78-12-04.2:51, Lab. (*TAEM* 29:354; *TAED* N004:27); for other undated drawings see

N-79-03-20:145–51, Lab. (*TAEM* 33:67–70; *TAED* N047:66–69); for a discussion of the blowpipe, see Doc. 1670 n. 3.

4. Within a day or two of this, Edison told a *New York Herald* reporter that he had that day

> produced the highest temperature that has ever been made by artificial means. I concentrated the electricity from a thirteen horse power machine into the space of half an inch by enclosing carbon points in a block of lime. Pieces of iridium, one of the hardest metals to melt, dropped into the flame volatilized immediately with an explosion. A small screw driver passed across the flame would be cut in two, the part touched by the heat melting instantly. Even parts of the lime crucible fused under the intense heat, and the light from it was so glaring that it painfully affected my eyes. ["Edison's Light," *New York Herald,* 30 Jan. 1879, Cat. 1241, item 1119, Batchelor (*TAEM* 94:455; *TAED* MBSB21119)]

Edison's report indicates the temperatures produced were above those practically attainable with the oxyhydrogen flame (see Doc. 1669 n. 3) since the melting point of iridium is 2454°C and volatilization only occurs when temperatures surpass 4800°C.

Edison's electric-arc furnace closely resembled his oxyhydrogen furnace, shown above, although these brief mentions of it provide little detail; it is shown in an undated rough sketch possibly made sometime after these experiments (Oversize Notes and Drawings [1879–1886], Lab. [*TAEM* 45:37; *TAED* NS7986:44]). Edison's design was clearly the same type of electric furnace that was independently reinvented by Henri Moissan over a decade later and was cited as one of the major reasons for his 1906 Nobel Prize in chemistry. Electric arcs were the only available means for surpassing the temperatures created with an oxyhydrogen blast until the turn-of-the-century development of acetylene technology. Investigators occasionally used the heat of an electric arc, together with its electrical effects, for experimental purposes, and some had enclosed an arc, but development of electric arc furnaces for regular laboratory and industrial uses essentially began with the 1882 work of William Siemens, who used larger enclosed spaces than Edison and only about a third as much power. That and later studies were apparently independent of Edison's work. He made no known further effort to develop or claim this furnace design, apparently announcing it only to the popular press. The only person actively involved in the later development of such apparatus who might have obtained any detailed knowledge of Edison's design and its operation was Edward Acheson, but he only started work at Menlo Park in 1880 and his designs followed much more those of Siemens. Moissan 1904, 1–11; Finlay 1978, 437–40, 446; Leddy 1989, 489–92; Pring 1921, 1–9; Westbrook 1978, 122–24; Siemens 1889, 2:221–27, 3:258–62; Szymanowitz 1971, 54–61.

5. In the notebook entry that followed Edison wrote: "In fusing nickel the button is never apparantly covered with an oxide still there is a thick shell of something that is infusible covers the molten globule yet when I take it out, it appears to be metallic nickel. There is something very curious about this matter." N-79-01-19:66–67, Lab. (*TAEM* 31:170–71; *TAED* N023:49–50).

6. In his notebook entry of the next day Edison wrote: "Suffered the pains of hell with my ey[e]s last night from 10 PM till 4 AM when got to sleep with a big dose of morphine. Eyes getting better & do not pain much at 4 PM—but I lose today." Edison conducted only a few additional fusibility tests over the next day or two. N–79–01–19:66–68, Lab. (*TAEM* 31:170–72; *TAED* N023:49–51).

–1673–

To Theodore Puskas

MENLO PARK, N. J.[a] Jany 28. 1878[1879]

My Dear Puskas,

Have you any objection to my giving Gouraud a power of atty to collect royalties from the London Stereoscopic Co? They do not perform their agreement as they will not sell any phonographs and I have evidence of the fact.[1] My income from royalties in this country has averaged $275 per week and still continues.[2] I learn there is a great demand for them in England. I propose if you have no objection to give Gouraud a power of attorney and 10 per cent of all collections. He will look after the Stereoscopic Co sharp and put the law on them if they do not sell the instruments.

I enclose you power of atty as asked for.[3] I cannot do anything in regard to Telephones in Great Britain because I have already committed myself to Gouraud.[4]

Please send me statement up to January 1st of how we stand. Also my portion of the French Telephone Cos stock certificates if they have been issued.

Electric Light progressing and Serrell should have money to take out another set of patents. About $1500 will answer. We have had 480 lamps on one circuit each could have been made equal to one gas jet. They were made of iron wire and were only brought to a[b] red heat. 16 Horse power was required. We used iron wire to make this test as platina wire would have been too expensive= Of course owing to oxidation and the melting point of iron we did not dare to bring these 480 lamps above a full red heat, but fairly demonstrated the fact that the light could be subdivided and all that now remains is an unexceptionable and practical lamp. We have about reached this point and have now a lamp that would last several months but there is a defect which requires time to remedy.[5] I am now supplied with every appliance to carry out experiments. The total expense thus far has been about $35,000 and the weekly expenses are about $800.[6] If I have good luck, perhaps I shall not call on your people for any money but if I am unfortunate I must call on you to contribute something to carry out the work. This would have been done by Drexel Morgan & Co but as you seem to have other parties there willing to take hold I earnestly request you to have matters so arranged that there will be no delay should I require financial assistance to carry on experiments because there is no question it will be a success. The creation of a new industry with all its ramifications is a very tedious, _expensive_ and long process. Soon after the thing has been thoroughly proven at Menlo Park it will require to be put into practical operation before the public will be perfectly satisfied.

Please write me at once your views on these matters. Very Truly

T. A. Edison   G[riffin]

P.S. I have just recd your favor of the 15th   The check for £200 has not yet come to hand.[7]

The check which you sent Dec 12th was rec'd and should have been acknowledged.[8]

Please continue my subscription to L'Électricité, I am taking all the foreign & domestic scientific papers now.[9]

I have opportunities frequently to sell phonographs for Russia, Belgium Italy Spain &c   Cannot I not sell them & account to you for the same?[10] Please answer _all_ of the points in this letter when you write.   E. G

L, NjWOE, Lbk. 4:135 (_TAEM_ 80:35; _TAED_ LB004135). Written by Stockton Griffin; the original is in TP. [a]Place from Edison's laboratory handstamp. [b]"made . . . brought to a" illegible in letterbook and missing from TP photocopy; text taken from TP transcription.

1. The London Stereoscopic and Photographic Co. held rights to Edison's phonograph in the United Kingdom and was to pay royalties to Edison through Puskas. Edison had recently received reports that the company was not fulfilling its contract with him. See Docs. 1237 (esp. nn. 1 and 7), 1340, and 1645.

2. Edison received an average of about $370 per month in combined phonograph sales and exhibition royalties from April to December 1878. Royalties for the months of November and December were, respectively, $213 and $273. Edison Speaking Phonograph Co. account for 1878, DF (_TAEM_ 19:205; _TAED_ D7831AA [image 32]).

3. Puskas had cabled on 16 January, "Send French power to prosecute phonograph infringers   leave name blank." The next day he sent Edison details of the situation and the legal advice he had received, and enclosed a draft power of attorney. Edison acknowledged this letter on 31 January and indicated that he had mailed a power of attorney three days prior; the executed power has not been found. DF (_TAEM_ 51:780, 781, 783; _TAED_ D7933C, D7933C1, D7933C2); TAE to Puskas, 31 Jan. 1879, TP (_TAED_ Z400BL).

4. Puskas had cabled Edison on 24 January, "Persons negotiating for

colonial telephone patents desire purchase English dont know your arrangements with Gouraud are you free receive offers during few days I claiming nothing English proceeds you settling with Gouraud." Edison replied the same day, "Have given Gouraud full powers England only." DF (*TAEM* 52:182; *TAED* D7939G, D7939H).

5. Edison's description of these iron wire lamp tests appeared in the *New York Herald* two days later (this account referred to 448 lamps). Edison reportedly declared that "The electric light is an accomplished fact, and it is more economical than gas," despite public skepticism which he attributed to the gas companies. All that remained, he promised, was to choose the final form of generator and lamps; within a year there would be a full-scale demonstration at Menlo Park. He added that when placed in commercial operation, as many as 20,000 lamps could be put in each circuit ("Edison's Light," *New York Herald*, 30 Jan. 1879, Cat. 1241, item 1119, Batchelor [*TAEM* 94:455; *TAED* MBSB21119]). During the second week of January, Francis Upton had tested the resistance of several iron spiral lamps and compared them to platinum spirals (N-78-12-20.1:86–106; N-78-12-31:78–91; both Lab. [*TAEM* 30: 158–68, 415–21; *TAED* N012:44–54, N014:39–45]).

6. Weekly summaries of electric light expenditures since November 1878, including costs of the new machine shop and engine house, total approximately $15,000 to this date but they do not include some major expenses such as Edison's recent copper purchases, platinum bought from Johnson Matthey & Co., or patent fees paid to Lemuel Serrell, some of which are itemized in Edison's personal and laboratory accounts. With the new buildings completed and largely equipped, Edison's weekly expenses dropped considerably; in recent weeks they had varied between approximately $650 and $707. The weekly summaries are found in Electric Light Co. Statement Book, Accts. (*TAEM* 88:412; *TAED* AB031); for an example of weekly summaries see Doc. 1562. Other related accounts are in Ledger #3, Accts. (*TAEM* 87:5; *TAED* AB001); see also two scrapbooks of bills, Cat. 1161 (Mar. 1878–Nov. 1879) and Cat. 1162 (Nov. 1879–Dec. 1881), NjWOE.

7. Puskas wrote on 15 January that he had enclosed a £200 check with a letter two days earlier but had neglected to mail it (DF [*TAEM* 49:65; *TAED* D7903T]). In his 31 January letter (see note 3) Edison told Puskas that he had received the check, "which materially alters the question regarding a power of atty for Gouraud to look after the Stereoscopic Co. It does not now appear to be necessary."

8. Puskas had sent a check for $142.25 to pay for tin foil. Since he had not received an acknowledgment he mentioned the matter again in his 15 January letter (see note 7). Puskas to TAE, 11 Dec. 1878, DF (*TAEM* 19:332; *TAED* D7832ZBX).

9. In his 15 January letter (see note 7), Puskas indicated that in the absence of instructions to the contrary, he had extended Edison's subscription to *L'Électricité* "although I by no means look upon it as a remarkably good paper but if you don't throw a bone occasionally to a hungry dog he is apt to, if not bite, at least annoy you by barking at you." Edison received numerous scientific and technical journals from the U.S., England, and the Continent (see *TAEB* 4:659, 747; George Van Ness bill to TAE, 30 Nov. 1878, DF [*TAEM* 17:372; *TAED* D7809ZAD1]). Around this time he was also inquiring about obtaining back issues of the

Philosophical Magazine, Popular Science Monthly, and the *Comptes Rendus.* A. E. Foote to TAE, 3 Mar. 1879, with attached list of requisitions (undated); both DF (*TAEM* 49:767, 775, 771; *TAED* D7910C, D7910H).

10. On 18 January, Uriah Painter had asked Edison to have legal papers drawn up specifying the territories in which the Edison Speaking Phonograph Co. could not sell phonographs. The same day, Edison signed an agreement allowing the company to manufacture and sell phonographs in "any country except—India, Australian Colonies, Russia, England, Norway, Sweden, Austria, Germany, France, Italy, Spain and Belgium." These were territories Edison had reserved under other agreements (see note 1 and *TAEB* 3:678 n. 5, 4:284 n. 6). Painter complained about this arrangement a week later, telling Edison: "If you will make Puskas 'put up or shut up,' on Phono' or let us in on some of European Territory we can sell good many there <u>right away</u>— People tell us they cant get any over there, & if we dont sell them they will get of infringers." Edward Johnson, now treasurer of the Edison Speaking Phonograph Co., protested that the "enlargement of the European territory wherein I cannot sell" had "cheated" him out of a lucrative sale. "In my judgment," Johnson wrote, "its only a matter of time when you will throw open to us at least half of Europe   When does Puskas' time run out? When he resigns—or you discharge him we want a hack at it." Agreement with Edison Speaking Phonograph Co., 18 Jan. 1879; Painter to TAE, 18 and 25 Jan. 1879; Johnson to TAE, 25 Jan. 1879; all DF (*TAEM* 51:766, 688; 49:87; 51:691; *TAED* D7932ZBG, D7932D, D7903ZAG, D7932F).

-1674-

To Hattie Van Cleve[1]

[Menlo Park,] Jany 28 79 3:50 PM

Mrs Van cleve

Alice[2] & Tommy start to Florida[3] Thursday P.M. Dont you want to come down Thursday morning[4]

T A Edison

L (telegram), NjWOE, DF (*TAEM* 49:940; *TAED* D7914B1). Written by Stockton Griffin.

1. Hattie Van Cleve was Mary Edison's half sister. She lived in Newark and was married to Cornelius Van Cleve, who worked in the laboratory during 1880. See the Edison Family Genealogy prepared by John Deissler for the Charles Edison Fund, EBC; Time Sheets; Jehl 1937–41, 545.

2. Alice Stilwell, Mary Edison's older sister, had apparently lived with the Edisons in Newark. It is not clear whether she continued to live with them at this time or resided with other family members in Brooklyn, where she had been at least temporarily the previous October. See *TAEB* 3:203; TAE to Alice Stilwell, 25 Oct. 1878, DF (*TAEM* 17:580; *TAED* D7813V).

3. On 22 January Edison had summoned his physician, Dr. Leslie Ward, from Newark to attend his son, Thomas Alva, Jr., who was "quite sick." The Florida trip was evidently intended to help Tommy recover

from this unidentified illness. Josiah Reiff wrote Edison on 11 February that he was "real sorry to learn that you had to send Tommy away & only trust he will rapidly improve." Neither Alice and Tommy's destination nor the duration of their stay is known, but Reiff's inquiry on 8 March about "What news from Florida" suggests that their absence may have been a lengthy one. Probably around this time, Edison absently doodled Alice's name repeatedly on a piece of his letterhead, along with "Master Tommy," "Florida," Mary's name, and his own. DF (*TAEM* 49:939, 99, 149, 597; *TAED* D7914B, D7903ZAP, D7903ZBS, D7903ZMT1).

4. Van Cleve replied on Wednesday, 29 January, that she had "been very sick since Wednesday   Will come if possible." DF (*TAEM* 49:940; *TAED* D7914B2).

–1675–

Notebook Entry:
Electric Lighting[1]

[Menlo Park,] Feby 76 7 8 & 9th 1879[2]
Experiments with wires in a Vacuum— Platinum Iridium alloy 20 pc Ir=

On mechanical pump with $\frac{1}{10}$ of an inch vacuum=[3] Pt-Ir wire loses about 3 $\frac{2}{10}$ of a milligramme per hour which is deposited upon the glass— The surface on which this loss occurs is about $\frac{1}{4}$ of an inch. Resistance .75 of an Ohm cold size wire .020 of an inch   If the platinum wire is brought to a dull red and the vacuum made $\frac{1}{10}$ of an inch & allowed to remain red for 1 minute & then made yellow and the vacuum made perfect again then wire brought gradually up making vacuum perfect after each increment of heat the wire may be brought to 21 candle power ie[a] $\frac{1}{4}$ of inch radiating surface will give 21 c.p.   the wire grows exceedingly bright, rivilling polished silver and under the microscope shews no cracks, whereas if brought to incandescence suddenly it shews great cracks. I think from our experiments that the melting point is determined greatly by the amount of gas within the pores of the metal which by its expansion disrupts the metals and makes it fuse easier.[4] by gradually increasing the heat the gas gradually comes out of the metal without disrupting or cracking it= Roughly speaking I think that if the melting point of platina in the air by suddenly bringing to incandescence is 2000 .FC then its melting point is raised to at least 5000 C by subjecting it to the process of occluding its gas by heat in a vacuum   3 to 4 candle power is all that we can on an average obtain from $\frac{1}{4}$ inch in open air= I think all metals which hold gas in their pores or even other metals or metalloids have a lower melting point because the unequal expansion disrupts the metals & crack it thus allowing it to fall an[b] easy prey to the heat=

The earliest dated vacuum lamp drawing, 3 February 1879.

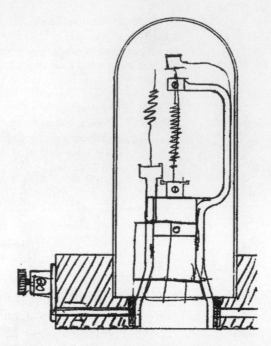

X, NjWOE, Lab., N-78-12-31:97 (*TAEM* 30:424; *TAED* N014:48).
[a]Circled. [b]Obscured overwritten text.

1. This entry is continued in Doc. 1678.

2. Edison dated this document, probably retrospectively, using a different pencil from that in the body of the text. The notebook entry summarizes Edison's interpretation of research on platinum-iridium wires in vacuum conducted in early February, probably using apparatus like that drawn by Batchelor on 3 February. The only other extant notes on these experiments were made by Francis Upton and Charles Batchelor on 8 February. In trials conducted that day, Upton and Batchelor recorded the appearance of a thin coating on the glass and the change in weight of wires about which Edison hypothesizes here. They calculated a slower rate of loss of a wire's mass than Edison does in this document. Edison probably wrote these notes before drafting his 9 February caveat (Doc. 1676). Vol. 16:368, N-78-11-22:75–89, both Lab. (*TAEM* 4:803, 29:190–97; *TAED* NV16:325, N002:39–46).

3. It is not known what kind of mechanical pump Edison had, but several pumps were capable of producing such a vacuum, equal to about 2.5 millimeters of mercury (air pressure is equal to 760 millimeters). Turner 1983, 98–103.

4. Two days earlier, in an entry witnessed by Edison and Batchelor, Upton had written: "An explanation of the changes wh[ich] occur in Pt. may be the following. The Pt absorbs an enormous amount of H[ydrogen] gas which is given off at high temperatures." N-79-01-21:41, Lab. (*TAEM* 30:607; *TAED* N016:22).

Draft Caveat:
Electric Lighting

[Menlo Park,] Feby 9 1879.

Caveat[1]

The object of this invention is to economically produce & subdivide the electric light=

The invention consists in the use of an ~~oxidizable~~ incandescent conductor of very[a] pure[b] iron, nickel or cobalt in a vacuum

The invention further consists in the use of an incandescent conductor of Platinum-Iridium Rhodium Osmium Palladium or alloys of either in a vacuum within a second vacuum or ~~closed~~ seal globe[a] at the atmospheric pressure. the heat of the first vacuum tube due to the heating of the conductor communicating with the air of the second globe[a] to expand it for the purpose of giving motion to a current regulator as already shewn in my[a] patent ~~of~~ applied for _____ 1879=[2]

The invention further consists in the method of manipulating the incandescent conductor before final sealing within the first vacuum tube=

I have discovered ~~by observing the anomalus behavior of platinum at high temper~~ that chemically[c] pure iron when drawn in wires and formed into a spiral, & then placed in a vacuum may be brought to ~~a brilliant~~ incandescence. I have also discovered that many metals which have gas within their pores have a lower melting point than when free from such gas—for instance Iron wire has several volumes of[d] Hydrogen & Carbon Monoxide gas within its ~~poo~~res. If such wire be placed in a vacuum and the ~~eu~~full current be put upon it it raises to a ~~brilliant~~ incandescence   If[a] the[a] current is increased it reaches a brilliant incandescence & then melts but if after the vacuum has been made the wire is brought to a faint dull red for several minutes gas is occluded,[a] then vacuum is then made again perfect & the metal ~~made bright~~ brought to a yellow heat by the current & allowed to remain so for several minutes & the vacuum again made perfect, then by adding more & more battery manipulating the vacuum at each increase of temperature the iron may be brought to the most dazzling incandescence which is perhaps 1000 degrees higher than its fusing point with the gases within its pores. Thus iron may be brought to a much higher incandescence than platinum or platinum-Iridium alloy in the open air=

The same effect takes place with platinum & other metals— With platinum in the open air if the current is placed on suddenly the gas within its pores expands explosively & disrupts the platinum so that it shews myrids of cracks where the exsudden expansion of the gas has caused the pmetal to expand beyond a point where it naturally would were there no gas=

But in a vacuum where reabsorption of the gases are prevented & by putting on the current[a] gradually & manipulating the vacuum[e] as before mentioned a Platinum Iridium spiral having but ¼ of an inch radiating surface may be made to give a light equal to 21 candles steadily for hours whereas in the open air it requires nearly 1 inch of radiating surface to produce the same effect because the platina melting point of the platina alloy & gas within it is enormously lower than as described= Once the gas has been ocel excluded from the metal the current may be put on or off suddenly= of course porous substances conductors such as carbon would be enormously benefited by this process as the great sudden expansion of the gases when the current is put on suddenly disrupts the conductors and devices have to be employed to put it on gradually to prevent the destruction of the conductor which is unnecessary by my process= I prefer to use iron for the reason that its resistance increases at a greater ratio with heat than platinum hence I can obtain a greater resistance with a smaller radiating surface than by the use of platinum or platinum iridium=

C is the burner th D the first vacuum glass sealed with two platinum conducting wires n. & .m weilded[a] in the tube & passing to the conductor. This tube is secured to the pillar X. The heat from the tube expands the air & this causes the aneroid diaphragm to work the circuit manipulating device substantially as shewn in my application filed[3] _____

Claims—

The combination with a thermic regulator sub as described of a sealed vacuum tube containing an incandescent conductor.

2nd   The use of iron nick nickel or cobalt[f] as a material for giving light by incandescence when placed in a sealed vacuum & having its gas freed from gas substantially as described.

3rd   The use of the metals of the[g] platinum group as a material for of any conducting material for lighting by incandescence when the same is enclosed[a] in a sealed vacuum and freed from gas as set forth=

TAE

ADfS, NjWOE, Lab., Unbound Notes and Drawing (1879) (*TAEM* 44:988; *TAED* NS79:16). [a]Obscured overwritten text. [b]"very pure" interlined above. [c]Interlined above. [d]"several volumes of" interlined above. ᵉ"& manipulating the vacuum" interlined above. ᶠ"nickel or cobalt" interlined above. ᵍ"~~the metals of the~~" interlined above.

1. Edison evidently never filed this caveat. His caveats 86 through 89 are not extant; according to Lemuel Serrell, number 86 was filed prior to 5 February, number 87 was an electric light caveat with at least six drawings, and numbers 88 and 89 were sent for Edison's signature on 15 April, three days after the execution of a patent application which incorporated the substance of this draft (Doc. 1695). Serrell to TAE, 5 Feb., 8 Mar., 10 and 15 April 1879, all DF (*TAEM* 51:406, 424, 463, 469; *TAED* D7929L, D7929ZAC, D7929ZBG, D7929ZBK).

2. Edison executed a patent application on 6 February, filed the day after he drafted this caveat, in which a glass globe is sealed around the burner. This application issued in May 1880 as U.S. Patent 227,227.

3. In this application (U.S. Patent 227,227) Edison stated that the atmosphere within the globe was permitted to communicate through an orifice with two "flexible chambers, similar to those used in an aneroid barometer. The expansion of the air by the heat of the burner causes these chambers to bulge outwardly and gives a downward motion" to a lever which would disconnect the lamp from the circuit. This is a variation of the lamp regulator he had included in an 1878 patent application that issued in April 1879 (U.S. Patent 214,637; see Doc. 1462).

Lamp with air regulator from Edison's U.S. Patent 227,227.

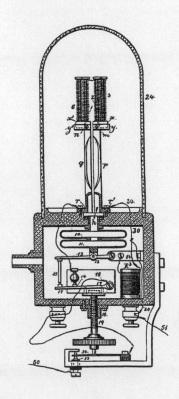

Caveat Electric Light[1]

The object of this invention is to economically produce light by Electricity—

The invention consists in various methods and devices upon which I am now engaged in investigating with a view of adopting that which proves itself the most practical and economical

When using the incandescence of a moulded material[a] to produce the light and where the burner is to have a large resistance to admit of practical subdivision I use a Coil of the material as in fig 1[2]

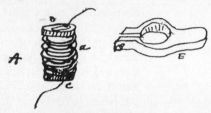

A is a cylinder of the best carbon, cut in a lathe like a spiral spring with two large ends B. C which are used to secure a perfect connection with the leading wires. E is the clamp for making such connections. By this means a resistance of thirty to fifty ohms may be obtained and with scarcely one inch of radiating surface. Of course an moulded conductor may be made in the form of a cylinder and made in this form either by cutting or grinding in a Lathe or by moulding directly—

metallic silicon or Boron may be moulded together like Carbon by powdering the silicon or Boron moistening it with a easily decomposed compound of these substances—moulding and submitting them to high temperature just as carbon is moulded= Various finely divided metals may be combined with infusible oxides such as the oxides of Ziricon Erbia Magnesia Calcium Cerium & analagous Earthy metals can be moulded or worked in this form

In fig 2 is shown another method of forming a burner of high resistance with small[a] radiating surface.

Fig 2

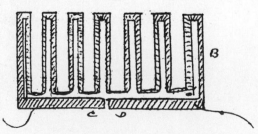

B shows the continuous conductor flat. C & D are the conducting plates of large surface   A shows the flat plate bent around to form the burner

Fig 3   instead of bending to form a round cylinder it may be bent as in fig

fig 3

It may be formed in a volute spiral[a] each turn being insulated from the other by an infusible non conductor as in fig 4[3]

Fig 4

In fig 5 is shown a method of producing light by means of vacuum tubes having capilliary bores it being known that the luminosity of the light from a vacuum tube through which a jump spark is passing is at the square of the diameter of the bore of the tube=[4] Fig 5 shews a length of tubeing arranged like the conductor in fig 2

Fig 5

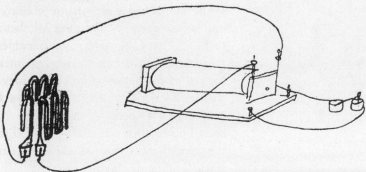

owing to extreme smallness of the bore and the concentration of a long length of tube in a very small space a very moderate coil will give a light of several candlepower= of course these capilliary tubes may be wound in a spiral form

Another method is shown in fig 6[5]

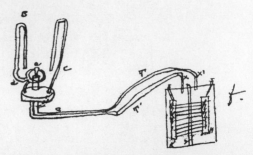

f is a decomposing cell with two electrodes H & K   the liquids are separated from each other by a porous diaphragm which serves to allow the passing of the current but does not allow the gasses to mix to any hurtful extent. the gas from Electrolyes which consists of Oxogen and Hydrogen are collected through the tube X X X' and conveyed by tubes T T' to the lamp= a is a ball of lime, Zircon Magnesia or other infusible substance. d[a] the Jet which is formed of a tube within a tube one of which conveys the Oxogen and the other the Hydrogen   a double chimney B is used through which either the Oxigen or Hydrogen passes from the cell f before it passes into the Jet—d   this is for the purpose of utilizing the heat from the Jet to warm the gas or Gasses before passing to the Jet=

I propose to construct a large Double cell with Electrodes composed of several bushels of coke and divided by a porous diaphragm—with two gasometers the oxigen and Hydrogen are collected for use. A Dynamo electric machine is to be used to effect the Electrolysis= of course any number of lamps may be used according to the power of the machine   this system is applicable to Country Houses where a water power or wind mill power is attainable= a concentration of light nearly equal to the Electric Light is obtainable and the economy is nearly as great the only loss being due to the heating in the decomposing cell by the polorization of the Electrodes and the decomposition of the Ozone into inactive oxygen thus releasing a portion of the ~~energys~~ Energy[b] into the form of unavoidable heat. This effect is reduced to the minimum when a single cell is used with a Dynamo machine of great quantity—[6] Instead of the jet of flame playing upon the cylinder of Zircon the later may be fastened as a cup and placed over the tube from which the Jet precedes as no atmospheric oxogen is required to support combustion as required= a small hole in the Zircon cup serves to allow the steam to escape=

Fig 7 Shews a new form of Thermal regulator for regulating the passage of the current through an incandescence material.

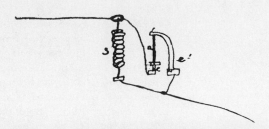

S is the incandescent conductor— A is a cylinder or plate of protooxide of Iron or rather magnetic or black[a] oxide of Iron, which is a non conductor when cold but has a conductivity as great as the metal Iron when red hot. this plate is placed between the uprights c c′ and receives the radiant heat from the spiral= It is shunted around it. If the temperature increases to much in the spiral the plate a at a certain point becomes a comparatively good conductor and thus shunts the current from the spiral. its heat drops and with it the conductivity of the oxides to be again renewed= The whole may be placed over the spiral to obtain the heat of convection as well as radiation= The breadth and mass of the oxide determines the point when a certain heat will render it suddenly a conductor

Small illuminated points may be obtained by employing Electrodes of Pure lampblack—[7]

Df (copy), NjWOE, Lab., Cat. 997:89 (*TAEM* 3:396; *TAED* NE1695:45). Written by William Carman. [a]Obscured overwritten text. [b]Interlined above.

1. It is not known whether Edison filed this caveat (see Doc. 1676 n. 1). It is possible that this draft became caveat 87, 88, or 89; however, these were not completed until at least mid-April by which time Edison had suspended research on the subjects discussed here.

2. Edison first experimented in November 1878 with moulded burners of finely divided metals (see Doc. 1555). Drawings related to this caveat, which include numerous variations of the burners discussed below and shown in figures 1–4, appear on three unbound pages, one of which is dated 28 January 1879 (Unbound Notes and Drawings [1879], Lab. [*TAEM* 44:981, 985–86; *TAED* NS79:9, 13–14]). On 10 March 1879, Edison applied for a patent on incandescent burners of powdered oxide and other finely divided conductors (Case 172). The application was subsequently abandoned and only the claims are extant (E-2536:28, PS [*TAEM* 45:700; *TAED* PT020028]).

3. In November 1878, Edison had begun research on what he called pyroinsulator coatings, substances intended to protect the wire burner from fusing and short-circuiting (see *TAEB* 4:730 n. 3, 755 nn. 1–2).

4. Cf. Doc. 1556.

5. Figure labels on the lamp at left are (counterclockwise) "B," "d," "a," "C," and "a"; for the tubes they are "T," "T′," "X," and "X′"; for the electrolytic cell they are "K," "H," and "f." Edison had first sketched out designs for lighting by electrolysis in early November 1878

(see Doc. 1548). He elaborated on this design in a caveat drafted on 17 March in which he specified lime produced by burning lime acetate, which "puffs the lime out greatly so that a cube 1 inch square will weigh less than ⅒ of a grain." He noted that "a surprising small flame will bring the lime when in this form to a vivid incandescence whereas if allowed to play upon ordinary lime as used in the Oxyhydrogen flame there would be scarcely any light." Cat. 997:94, Lab. (*TAEM* 3:399; *TAED* NE1695:48).

6. That is, a large current.

7. This may be related to the design shown in Doc. 1567.

–1678–

Notebook Entry:
Electric Lighting[1]

[Menlo Park,] Feby 10 1879

Experiments with Steel Piano forte wire in vacuum about .020 wire ¼ inch radiating surface=[2]

We coated one with chalk in water & dried placed in vacuum Some of the strands were brought to ~~white~~ impure white heat— there was large amount of molten oxide that run from one turn to other & cross connected. The decomposition of the chalk making free carbonic acid probably assisted the oxidation= We have noticed previously that piano wire may be brought gradually to pure white incandescence but the oxide formed has a much lower melting point than the steel itself— ~~ha~~Could we[a] occlude all the gases & have a 1,000 000 of an atmosphere vacuum[3] & pure iron wire doubtless we could bring it up to the most dazzling incandescence which would last for months & I do not think it would vo~~l~~latilize=[4b]

We now put in a spiral with lime in H_2O & dried   very little oxide formed but it melted straight connecting wire before the spiral= the lime does not stick well but cakes off= I notice where it melted & flowed where lime was on it. Beautiful scales of violet & gold color probably a compound of iron oxide & lime.[b]

We now put in a plain steel spiral considerably streatched out so turns[a] will not come in contact= & we are going to raise the heat gradually— in 2 minutes we raised it to a full white in attempting to go higher it melted in all parts simultaneously

We now put in another steel spiral with its convolutions further apart to prevents great variation in heat at parts of the spiral= brought it to clear white when it melted in one spot shewing a flaw there— it should have softened in all parts before melting had there been no flaw—

Discovery

Tried another peice brought it up very gradually it was coated with black or blue oxide ~~a~~with 2 cells when it melted it went in all parts at once & then strange to say the oxide volatilzed on the glass & left the whole of the spiral which was incandescent as <u>bright as silver</u>.

I found that this brightened wire was <u>entirely</u> decarbonized & couldnt be hardened whereas a part which was only dull red was hardened by water until it was as brittle as glass[b]

We put a steel wire in atmosphere of Hydrogen   it required a very much greater current to raise it to whiteness and it melted at a impure white heat shewing apparantly that the Hydrogen had more influence on the light emitted than upon the metal which melted precisely at the same point as the iron that was in a vacuum yet the latter gave <u>twice the light</u>—

Perhaps there is an infinitesimal quantity of free Hydrogen in the atmosphere which ~~acts in place~~ fills the place of the hypothetical ether[5]—& thus accounting for the diminished intensity of the platinum in air as against the vacuum; of course taking into consideration the gain due to lessening of connection currents—

Put in new spiral brought to yellow white at 4:35 PM on Gramme machine—added current 4:50 brought to incandescence fused 4:52

Magnesium wire was brought to a dull red before melting ~~o~~in a vacuum hence the melting point as given in the books ie[c] 447. Fahr or 230 C must be wrong or Draper is wrong in attributing 900 to a dull red=[6b]

We now put a wrought[a] iron spiral in and brought it to yellow white on battery   it gradually falls to dull red in 25 minutes owing to polarization of the battery— ~~we~~ Martin[7d] go to supper & takes battery off at 6:55, we keep~~ing u~~ up vacuum— We brought it at 7:20 to a dazzling white by Gramme but in 1 minute it melted because we tried to bring it too high—

X, NjWOE, Lab., N-78-12-31:105 (*TAEM* 30:428; *TAED* No14:52). Expressions of time have been standardized for clarity. [a]Obscured overwritten text. [b]Followed by dividing mark. [c]Circled. [d]Interlined above.

1. This entry is a continuation of Doc. 1675.

2. The following pages in this notebook contain Francis Upton's notes regarding additional experiments that he, Edison, and Batchelor conducted with iron wire on 10 February and with nickel, iron, steel, and platinum wire and nickel and platinum foil on 11 and 12 February. During these experiments they also tried using hydrogen gas rather than a vacuum and tested materials in the oxyhydrogen blowpipe. They produced current for the experiments using both batteries and a Gramme generator. These experiments were then followed by a series of coating experiments (see Doc. 1679). Upton also noted cross references to other notebooks containing drawings of the lamps they were using for the vac-

Edison's drawing of an experimental vacuum lamp.

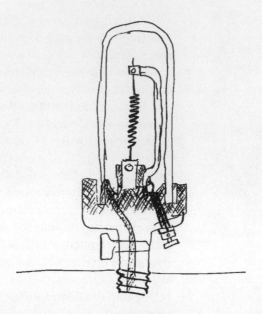

uum experiments. On 12 February Edison drew two vacuum lamp designs, one of which he ordered to be constructed, "1 first= then 3 more." A 12 February measured drawing by Batchelor, on which John Kruesi noted that the screw at the base "must fit Air pump," appears to be a modified version of this design. N-78-12-31:125–63, Vol. 16:372; N-79-01-01:117–21; N-78-12-04.2:61–67; N-78-12-15.1:163–64; all Lab. (*TAEM* 30:438–57; 4:808; 30:316–18; 29:359–62, 1024–25; *TAED* N014:62–81, NV16:330, N013:57–59, N004:32–35, N009:84–85).

3. This would have been possible with a Sprengel pump, which could reach .0006 mm. of mercury. Turner 1983, 105.

4. In his 10 February notes (see note 1), Upton observed that after one trial an iron spiral was "taken out and polished to see if the oxydation could have arisen from the O. in the iron   again badly oxydised." N-78-12-31:125, Lab. (*TAEM* 30:438; *TAED* N014:62).

5. Regarding nineteenth-century concepts of ether space, see Cantor and Hodge 1981 and Whittaker 1989 [1951–53].

6. John William Draper, a professor of chemistry at University of the City of New York (later New York University), had demonstrated in an 1847 *Philosophical Magazine* article that all solid substances become luminous at the same temperature, which he approximated at 977° F (Draper 1847). Francis Upton had summarized Draper's paper as part of his literature review for Edison but made no remark about this observation. *ANB* and *DSB*, both s.v. "Draper, John William"; Literature Search Notebook #2:60, Upton (*TAEM* 95:439; *TAED* MUN002:48).

7. Martin Force had worked as a carpenter on the construction of the Menlo Park laboratory, and had been employed since late 1877 or early 1878 as a general handyman in the lab. See *TAEB* 3:534 n. 3.

[Menlo Park,] Feb. 12, 1878[1879].

Coating Pt with oxides by decomposition of the Nitrates by means of heat[1]

Ex No. 17   Nitrate of Zirconium does not coat the wire at bright red or at dull red heat. Perhaps little Zr in it

Ex. No. 18   Zircon apparently combines. Merck's Zr[2] very impure. melts instantly.

Ex No. 19   Nitrate Calcium coats the wire without apparently dulling it. [~~The purer the lime?~~][a] Under the microscope coated all over but in globules and very thin between the globules. Lime probably very impure. Spots could be found which[b] the coating did not cover. Does not seem to dim the wires much. Wire rubbed with fingers. Coated make contact through the coating at places. Tried at a lower heat. At very dull heat an elegant coating could be obtained. Contact could be obtained ~~by~~ but not easily through the coating.

The idea is to obtain a coating of an oxide to see if it will prevent the volatilization[b] of Pt. Tried the solution.

Used the wire warmed with current and then heat with alcohol lamp. Rubbed and then heated   A[a] perfect coating obtained and the ~~same~~ wire heated by the side of an uncoated one and no difference seen in the light. Does not seem to lower the melting point and coating does not come off at high heats The wire appeared glassy and very irregularly on. Dipped the wire in Nitric[b] acid

Ex. No. 19   Acetate of Magnesia solid form run along the wire slowly, when the wire is at a heat just visible, coats the wire permanently[b] with a smooth coating. Gives more light than the Pt. direct because fine points require less heat to bring them to incandescence in the same way that the Pt wire becomes polished by the melting of the fine pts. on its surface on its surface wh. are unable to withstand[b] the temp of the main wire, while mag is able. Hence gain of light by coating the wire with an extremely thin coating of an infusible oxide   An organic salt of a[b] substance decomposed by the heat.

Ex No 20   Acetate of Alumina does not coat the wire at any heat

Ex. No. 21   Acetate of Lime gives a very fine coating and seemingly continuous under the microscope. Tried to burnish under micros. ~~showed that~~ part was seen to be knocked off. Seems to combine with the Pt. Magnesia ~~does not~~ appears to.

Ex. No. 22. N seems to have no effect other than the air when Pt. Spiral is heated in it and also in vacuum

~~Ex No. 23.~~

Ex. No. 23. Tried Alcohol vapor could see no change

Ex No 24 filled globe with tobacco smoke to watch the air convection.

Ex. No. 25 Put a piece of paper in contact with the spiral so that it would burn no very good smoke

Ex. No. 26 Ammonia and HCl tried moistened paper and waste[3b] and put in the base no ~~goo~~ result.

Ex. No. 27 Spiral coated with Acetate of magnesia in vacuum Current put on at 3-30 P.M. Very brilliant[b] incandescence. Vacuum loses very easily after about an hour fussing over it sent to shop. Spiral left in open air.

The top of the glass blackened and the sides became streaked with a light precipitate after being [---][a] in the air for some time. No ~~double~~ image from reflection to be f seen

In about 4 hours 3 hours white hot.

Ex. No. 28 Spiral coated with Lime[c] 8-30 P.M. Same spiral white 10-30 vacuum ceased was melted.[d] Noticed a white deposit on the glass which reached to within half an inch of the bottom. The brass support had a white metallic lustre towards the spiral.

In the air for ½ hour a thin black deposit noticed which the pillar screened away so that none was deposited [----][a] directly behind the support.

The wire was bare in places inside of the ~~coil~~ spiral and as the deposit was very much less, it is hoped that by completely covering the wire it may be entirely stopped[4]

TAE

X, NjWOE, Lab., N-78-12-31:143 (*TAEM* 30:447; *TAED* N014:71). Written by Francis Upton; document multiply signed and dated. [a]Canceled. [b]Obscured overwritten text. [c]"Spiral . . . Lime" written on opposite page. [d]"was melted" interlined above.

1. In Doc. 1796 Edison stated that he formed an oxide coating by dusting the wire with an acetate, then heating it to oxidize the substance on the wire.

2. Unidentified, but apparently commercial zircon supplied by the chemical firm Merck & Co.

3. Soiled cotton waste was commonly used for cleaning machinery. Knight 1876–77, "Waste."

4. The following day Upton noted regarding this experiment that "the coating under the microscope showed large cracks." He followed this comment with notes on additional coating experiments Nos. 29–35. When the numbered experiments resumed on 14 February they were concerned with the new Edison generator rather than with lamps. N-78-12-31:159–65, Lab. (*TAEM* 30:455–58; *TAED* N014:79–82).

New York [February] 14th 79

Dear Al

Got the other movement[1] in the telephone and it worked "Bang up," which shows that the worm is N[o].G[ood]. Bergman[2] has started to make patterns for the iron box and movements. [-][a] he is not rushing things as much as I would like. I spoke to him today about putting another man on the Job, as there ~~was~~ is[b] only one now but he thought he could not work to advantage. but we have clear work now and there is no reason why we should not have ~~a coup~~ 2 Telephones done by latter part next week. will not be out[c] saturday unless we get patterns done[c] and I hardly think we will. I am hurrying it up as much as possible

Charlie

ALS, NjWOE, DF (*TAEM* 52:16; *TAED* D7937J). [a]Canceled. [b]Interlined above. [c]Obscured overwritten text.

1. The movement was probably a toothed gear drive like that shown in Doc. 1681 and specified in a draft patent specification (Doc. 1693). Edison noted in the draft specification that "a worm and pinion may be substituted" for the mechanism indicated; worm gears had been used on Edison's receivers since the preceding fall (see Doc. 1527).

2. Sigmund Bergmann had worked as a machinist in Edison's Ward St. shop in the early 1870s. In 1876 he opened his own shop on Wooster St. in New York City (see *TAEB* 1:579 n. 1). The shop had been working on the receivers at least since 6 February, when Charley reported to Edison that Bergmann was "getting along fairly" with them and he expected to test one the next day (DF [*TAEM* 49:98; *TAED* D7903ZAO]).

BRITISH DEMONSTRATION TELEPHONE
Doc. 1681

The instrument shown in Doc. 1681 was one of two complete telephone sets taken to England by Charles Edison to demonstrate Edison's new electromotograph receiver; each set was housed in an iron box. Construction of these telephones had begun by 14 February at Sigmund Bergmann's New York shop under Charley's supervision.[1] They were delivered on 23 February to Menlo Park where they were "to be finished immediately." They were ready the next day when Charles Batchelor ordered the laboratory shop to "Make two 2 New Receivers in walnut boxes to be used only as receivers" (without transmitters) which Charley also took with him.[2] These additional receivers were probably made for John Tyndall's lecture at the Royal Institution.[3]

For unknown reasons Edison chose to mount his carbon-telephone transmitter (figure 1 A) directly in front of the receiver diaphragm (figure 1 C, figure 2 D) rather than below it as Charley had indicated in preliminary sketches.[4] The receiver as built incorporated a gear train (figure 2 W, W) rather than the worm drive used in previous experimental models.[5] The chalk button (figure 2 K) was described in the *Electrician* as a mixture of "chalk and potassium hydrate with a small quantity of acetate of mercury, moulded round a flanged reel of brass, lined with platinum on the parts in contact with the mixture."[6] A thumbscrew (figure 2 B) adjusted the tension spring (figure 2 S) which maintained the pressure of the metallic contact spring on the chalk. A roller suspended in a water reservoir (figure 2 M and figure 4) could be raised by a lever (shown on the left side of figure 1) into contact with the chalk button in order to moisten it.[7]

1. See Doc. 1680.
2. Cat. 1308:101, 105 (Order Nos. 6 and 11), Batchelor (*TAEM* 90:717, 719; *TAED* MBN003:24, 26); Doc. 1691. Also on the 24th Batchelor made drawings of circuit arrangements for the new receiver and Charles Edison wrote out the recipe for making the chalk buttons (N-79-02-24.2:3–5, N-79-01-21:44–45, both Lab. [*TAEM* 31:995–96; 30:609; *TAED* N031:3–4, N016:24]). The Science Museum in London has one of the receivers (Inv. 1880-68).
3. See Doc. 1715.
4. See Doc. 1661. The generally favorable article about Edison's telephone in the *Engineer* (from which the Doc. 1681 drawings are taken)

Alternative placement of transmitting and receiving apparatus in the telephone, sketched by Charles Batchelor on the day the demonstration instrument went to England.

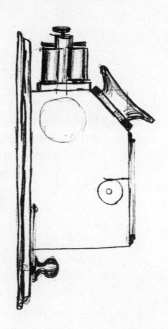

commented that "Unfortunately, the transmitter in these experimental instruments is placed directly in front of the receiving diaphragm, whereas it will ultimately be placed elsewhere" ("Edison's Telephone Receiver," *Engineer* 47 [1879]: 213–14). On 26 February, Batchelor sketched several ideas relating to the telephone, including placing the transmitter at an angle above the receiver apparatus (N-79-02-24.2:11–23, Lab. [*TAEM* 31:999–1005; *TAED* N031:7–13]).

5. See Doc. 1680.

6. "Edison's New Telephone Receiver," *Electrician* 2 (1879): 225–26.

7. Two drawings accompanying a 21 March article in *Engineering* show a slightly different mechanism for lifting the roller. This article included drawings of the telephone's exterior and a schematic diagram of the circuit arrangement, as well as two other drawings illustrating the motograph principle. "Edison's Electro-Chemical Telephone," *Engineering* 27 (1879): 238–40; see also "Edison's New Telephone Receiver," *Teleg. J. and Elec. Rev.* 7 (1879): 112–14 and "Edison's Electro-Chemical Telephone," *Sci. Am.* 40 (1879): 260.

Complete telephone, as viewed from right front.

Rear of receiver apparatus showing alternative arrangement for lifting the roller.

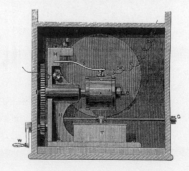

Demonstration Model:
Telephony¹

FIG. 1

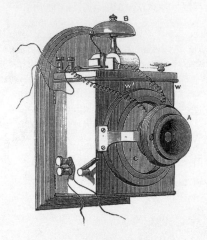

FIG. 2

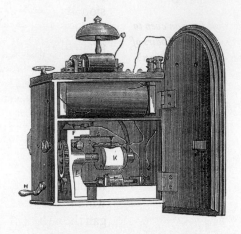

FIG. 3

FIG. 4

M (historic drawing) (est. 15 × 25 × 10 cm), *Engineer* 47 (1879): 213.

   1. See headnote above.

ARMATURE WINDINGS AND COMMUTATOR CONNECTIONS   Doc. 1682

The power drawn from a dynamo depended on the current induced in the armature windings, the armature's connections to the commutator, and the placement of the contacts that drew that current from the commutator. Different dynamo manufacturers—Siemens, Gramme, Weston, Wallace-Farmer—solved these related problems with different designs. In the machine Edison was making (Doc. 1653), he connected the loops of armature wire in a continuous circuit, connected each side of each loop to a separate commutator bar, and designed the commutator contacts so they could touch the commutator

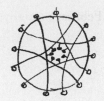

The shift from an even to an odd number of pairs of coils is shown in Edison's 10 January drawings.

bars in various configurations. This document describes some of the first trials of the new dynamo and illustrates the changes Edison made after those tests proved unsatisfactory.

Although Edison had been largely occupied with lamp experiments since early January, he had thought at least occasionally about the dynamo. In mid-January, perhaps as a result of the completion of the commutator springs, he drew a diagram of an armature winding and its connections to the commutator. On 10 February he drew several pages of armature connection diagrams, some of which were significantly different from the dynamo which was nearing completion. The new pattern used a commutator connection like the Siemens machine—rather than connecting each loop to two bars, one connection was made from the wires linking each pair of loops. Most of the armatures in the sketches are also drawn with odd numbers of loops, an important change that allowed Edison to wind them symmetrically. The same day he and Batchelor began the design of a small generator (see Doc. 1696), apparently to test these new arrangements.[1]

On 13 February, Edison conducted the first tests with the large dynamo. Although the staff found some configurations that produced enough power to heat lengths of iron wire, they were clearly dissatisfied.[2] They worked extensively on the 15th and the following two days, testing various arrangements of the commutator springs and drawing alternative armature connections. They also began drawing what might be called "current tracings" with arrows to show the path of the current

Charles Batchelor's 15 February drawing shows the opposing currents set up when an even number of commutator connections was used.

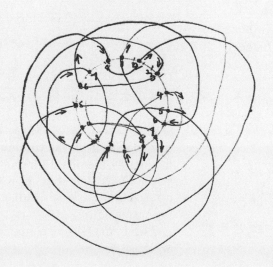

through the armature, something they had done sparingly at the beginning of the year. By the 16th they had returned to the ideas sketched on the 10th, namely, an odd number of coils with a commutator connected to the linking wires. That conclusion is reflected in the "good" drawing appended to this document.[3]

Having arrived at that understanding, Edison designed a new armature with 49 coils for the large dynamo (Doc. 1687). At the same time, he was not willing to abandon the present armature without testing it thoroughly, and the notebooks show repeated attempts to devise wiring schemes for eight coils. The seven-coil design was clearly superior, though, and became part of the patent model specified on 22 February.[4] The draft patent application (Doc. 1694) describes this arrangement as a "new & novel method" and includes it in the claims, although it was not claimed in the filed application.

1. Vol. 16:364, 370–71; N-79-02-10:1–31; both Lab. (*TAEM* 4:797, 804–7; 32:1141–57; *TAED* NV16:319, 326–329, N046:1–17).
2. N-78-12-31:165, Lab. (*TAEM* 30:458; *TAED* N014:83).
3. N-78-12-04.2:77–111; N-78-12-11:169–79; N-78-12-15.1:181–211; N-78-12-31:196–219, N-79-02-15.1:1–97; all Lab. (*TAEM* 29:367–85, 756–61, 1031–44; 30:474–85; 31:659–705; *TAED* N004:40–58, N007:78–83, N009:91–104, N014:99–110, N028:2–48).
4. Cat. 1308:99 (Order No. 1), Batchelor (*TAEM* 90:716; *TAED* MBN003:23); Machine Shop Drawings (1879–1880), Lab. (*TAEM* 45:47–48; *TAED* NS7986C:5–7).

–1682–

Notebook Entry:
Electric Lighting[1]

[Menlo Park,] Feb. 15th 1879.

Magneto Electric Machine

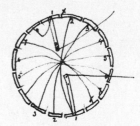

On page 25 is described a magneto Electric machine[2] which is now finished, and of[a] which the coils 1 2 etc are connected up as shown in sketch with 2 commutator springs to take off the currents[3]

This is all short circuited in the machine. Very large spark at X[4]

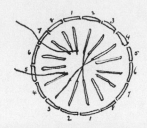

By connecting commutator springs as in this sketch we get lots of current although it is high tension owing to having in 4 coils or 80 ohms.[5] This readily heats iron wire in 2 feet lengths even when turned by hand

ADDENDUM[b]

[Menlo Park, c. February 16, 1879][6]

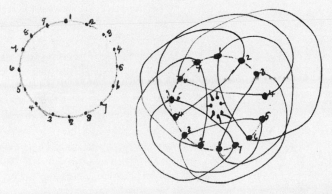

good[7c]
Chas Batchelor

X, NjWOE, Batchelor, Cat. 1304:35 (*TAEM* 91:34; *TAED* MBN004: 32). Written by Charles Batchelor. [a]Interlined above. [b]Addendum is an X, written by Batchelor. [c]Multiply underlined.

1. See headnote above.
2. Doc. 1653. Edison patented this design in U.S. Patent 264,643; see Doc.1694, esp. n. 2. Both of these documents and the illustrations accompanying them show the machine and its parts.
3. The first two drawings in this document show the commutator-end view of the stationary armature's wiring. The rotating internal field magnet, external case, and connections are not shown. The slightly curved numbered rectangles around the exterior represent cross-sections of the commutator bars, to which are attached the coils wound around the drum. Each coil is connected at its ends to the commutator bars and then to the adjacent coils to produce one continuous circuit. That is, the sections numbered 2 represent several loops of wire, and the lines across the circle represent connections at one end to coil 1 and at the other to coil 3. These coils could apparently be connected in many ways at the commutator end of the drum (see note 5). The manner in which Edison

This 2 January drawing shows how Edison connected the armature coils between the commutator bars (the numbered rectangles).

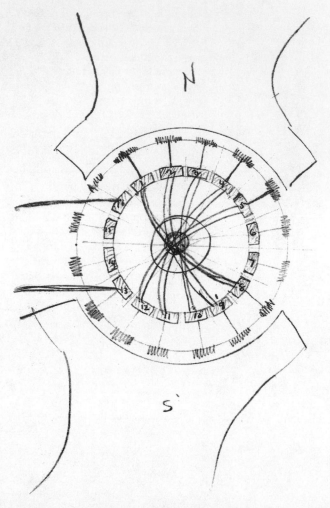

connected the armature coils between the commutator bars is shown in N-78-12-11:109, Lab. (*TAEM* 29:732; *TAED* N007:54).

In this machine the commutator springs rotated on the same shaft that spun the field magnet inside the armature. These springs are arranged radially, as also shown in N-78-12-31:167, Lab. (*TAEM* 30:459; *TAED* N014:83). In U.S. Patent 264,643, figures 1 and 8, they are shown touching the circumference of the commutator.

4. In a note regarding the first tests of this machine Upton noted a "fearful spark, resembling an arc." Sparking at the commutator was a serious problem in dynamo design. It was exacerbated by windings in which current flow was not smooth, a problem Edison attacked with his new windings (see headnote above). Batchelor also noted a "cross" of two coils caused by a loose wire. N-78-12-31:165, N-78-12-04.2:77, both Lab. (*TAEM* 30:458, 29:368; *TAED* N014:82, N004:40).

5. There were 16 commutator springs—one for each bar, or two for each coil. Only half of the springs are connected here. This technique—varying the number and position of springs connected—was the basis of experiments recorded this day by Batchelor and Upton (see headnote

above, n. 3). In these experiments, battery power usually activated the field magnet, but sometimes the Gramme machine was used. Output was gauged primarily by the spark or arc that could be produced in the external circuit between carbon points, and sometimes various wires were put in the circuit to see how readily they could be heated.

"Tension" and "intensity" referred to voltage; it is not clear what "high" means in this context. (Voltage is determined by the strength of the field magnet, the speed of rotation, and the length of the actively affected wire in the armature where the induction occurs.) Different connections produced different voltage and current levels. Since the four coils were connected end to end, or in series, their individual resistances were evidently 20 ohms.

6. See headnote above.

7. In the completed (right-hand) drawing, Batchelor sketched seven coils whose ends are shown as numbered dots. The outside loops represent the coils themselves wrapped around the armature. The inner lines connecting the numbered dots show the connections between the coils. The seven central dots shown linked to those connecting wires represent the seven sections of the commutator.

–1683–

Notebook Entry:
Electric Lighting

[Menlo Park,] Feby 16 1879

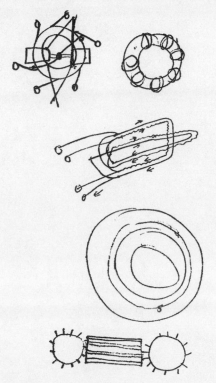

Machine without Commutator[1]

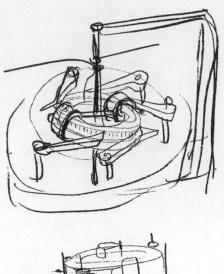

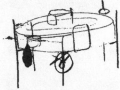

Magneto   Machine without Commutator

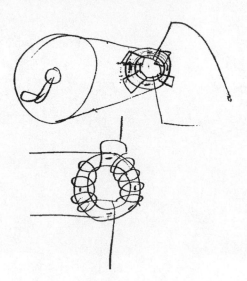

ditto

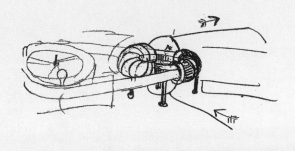

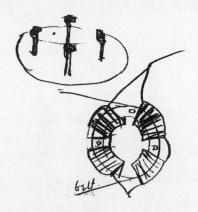

Non-commutator Dynamo Mac

TAE

X, NjWOE, Lab. N-79-02-15.1:53 (*TAEM* 31:684; *TAED* N028:27). Document multiply signed and dated.

1. During the second half of February 1879, Edison was attempting to design a generator that could produce direct current without commutators and their attendant problems, something he had considered during his first investigations into generator design the previous December (see e.g., N-78-12-04.2:117–29, N-78-12-31:234–35, all Lab. [*TAEM* 29:387–93, 30:493–94; *TAED* N004:60–66, N014:117–17]; *TAEB* 4:804 n. 4). Edison sometimes referred to these designs, in which the conductor is significantly affected by only one pole of the field magnet, as "Faraday" machines because they drew on arrangements investigated by Michael Faraday and reported in his *Experimental Researches* (Faraday 1965 [1855]: 28th Series, sections 3084–3122). Other electrical investigators (principally in Germany) had studied what were often called unipolar generators, and Werner Siemens built several for commercial use about 1878. In the 1880s and 1890s unipolar machines attracted considerable interest from physicists and electricians studying induction phenomena; however, they never found wide use in practice because they produced strong currents at low voltages and required high rotational speeds, which made collecting current from them on a commercial scale problematic (Miller 1981; for the general design features of unipolar dynamos, see Hawkins 1922, 1:23–28, 127–30, 2:57–59, 298–302).

MENLO PARK, N.J.[a] February 17. 1879—

Dear Sir.

My ordinary avocations are such that the current events of politics and other matters not immediately connected with my pursuits sometimes escape my attention   This will account for the fact that although an amendment to the patent laws, has, as I am informed, been for sometime under the consideration of Congress.[2] Certain special features of the proposed amendment have but recently been brought to my attention.[3] If I am rightly informed one provision of the law is intended to limit the time within which an inventor may have protection for inventions not fully ~~pr~~oerfected or put in use, and this I am told is accomplished by requiring the payment of additional fees from time to time upon patents for such inventions   I am sure that this provision will not only act oppressively upon many inventors, but will strongly tend to discourage and prevent the perfection of useful inventions by those most fitted for that purpose, and most likely to accomplish it. My own practice for many years—a practice not adopted as the result of any plan or purpose, but arising from the natural habit of my mind, has been to study a subject for a time, and then taking out patents for such parts of a general system I may succeed in making, to leave it again for a considerable time. When returning to it, I have often been able almost at once to eliminate difficulties or overcome obstacles which before seemed impossible. I have many inventions now in practical use, the most important[b] of which have been perfected in the way I have described— As completed they are a system[c] based on different[c] inventions or discoveries, some of which have been made years before the others, and as I went along, finding my way from day to day and year to year.

I am compelled to protect myself by patents upon each patentable part of the system as I find out or devise it, otherwise I might lose the full benefit of the perfected system by losing the control of some fundamental feature of it.[4]

It would be very burthensome to me if I were required ~~by~~ to make[d] additional payments on a great number of patents which I have not yet been able to put into such practical use as to yield me a pecuniary return. I think I may say that the operation of such a rule in the past might by embarrassing and discouraging the inventor who had got hold of the clue to some secret of nature, have delayed for a hundred years the beneficial use by mankind of inventions which are now indispensable to modern life and commerce.

I have spoken of myself and my inventions only in order to protest in the interest of all other inventors against any legislation calculated to make our traditional struggle against the capitalist any more difficult, and I take the liberty of asking you to consider these suggestions and give them such weight in your official action as you think they deserve. Yours Truly

Thomas A Edison

ALS, DLC, Butler, (*TAED* Xo42A). ^aPlace from Edison's laboratory handstamp. ^bObscured overwritten text. ^cMultiply underlined. ^d"to make" interlined above.

1. A controversial former Union general, Benjamin Butler was now a prominent member of the House of Representatives from Massachusetts. He had previously acted as an attorney for Edison in the Quadruplex Case. *ANB*, s.v. "Butler, Benjamin Franklin"; *TAEB* 2:493 n. 11.

2. The House of Representatives took up Senate Bill 300 in early February, after having debated similar measures in the previous session. The bill provided for a number of modifications to the patent system, among them the imposition of new fees. In addition to the one-time $35 fee upon issue, inventors would be required to make payments of $50 and $100 (later amended to $20 and $50) at the end of four and nine years, respectively, to keep a patent alive. Similar periodic fees were required in many European countries; this provision was explicitly modeled on the British patent system and was designed to annul what the House committee on patents described as an increasing number of "annoying, frivolous, and unimportant" patents. "The New Patent Bill as Passed by the Senate" and "How the Patent Bill Died," *Sci. Am.* 40 (1879): 96, 177; U.S. House of Representatives 1879, 1, 5.

3. Uriah Painter had sent Edison a copy of the Senate bill in December and asked him to "write a letter I can show" but Edison did not reply (Painter to TAE, 12 Dec. 1878, DF [*TAEM* 18:843; *TAED* D7828ZFA]). Several parties brought the matter to Edison's attention again in early February when it reached the House. The Newark Board of Trade and a group of that city's manufacturers apprised Edison of their efforts to defeat it and requested his help, including the possibility of testifying before the House (Anson Searls to TAE, 4, 11, 14 and 19 Feb. 1879; Newark Board of Trade to TAE, 11 Feb. 1879; Andrew Albright to TAE, 17 Feb. 1879; all DF [*TAEM* 51:646, 656, 657, 663, 654, 660; *TAED* D7930C, D7930G, D7930G1, D7930J, D7930F, D7930I]). On 7 February, Painter sent Edison another copy of the bill, telling him "I dont like it— It is not half as bad as it was, & next Congress [may] make a worse one." He also related a conversation with Senator Roscoe Conkling of New York, who he said had opposed it "because it made your fees excessive, & onerous, that many of your patents might be wiped out before they were ready for market." He reported that Conkling "did not think the next Congress would do much worse & did not think it would run much risk to beat it." Painter noted, however, that "opinions differ widely." Gardiner Hubbard, for example, reportedly feared passage of a more damaging measure in the future if this one were defeated, an argument that Hubbard made in his own letter to Edison on

10 February. Painter asked Edison to read the bill "over carefully & if you say bust it—I can do it I think— Letters to Ben Butler & Conger are enough—Conger used to be chairman of Pat' Com'." Omar Conger was a Michigan Republican then completing his fifth term in the House (*NCAB* 12:394); there is no record of Edison having written him on this subject (Painter to TAE, 7 Feb. 1879; Hubbard to TAE, 10 Feb. 1879; both DF [*TAEM* 51:647, 650; *TAED* D7930D, D7930E]).

Painter wrote again on 16 February that he thought he had secured enough votes to defeat the measure but "in talking to [Grosvenor] Lowrey we decided that you should write a letter something like the enclosed & send to me to be used if required in the House— It must be written <u>entirely</u> by you, in your <u>own hand</u>. Do at once & send to me, so I can be ready for any emergency." The draft by Painter and Lowrey has not been found. Butler replied to Edison on 27 February, "I do not think the patent law will come up at all, and if it does it won't pass without much amendment." The House voted down the bill on 1 March. Painter to TAE, 16 Feb. 1879; Butler to TAE, 27 Feb. 1879; both DF (*TAEM* 51:659, 664; *TAED* D7930H, D7930L); "The Patent Bill Defeated," *Sci. Am.* 40 (1879): 176.

4. According to contemporary U.S. patent law, "two or more separate and independent inventions cannot be claimed in one application" except when they are "*necessarily* connected each with the other." In general, however, it was unusual to receive a single patent "to cover the whole" and inventors usually confined their specifications to a particular invention. In contrast, Great Britain, Canada and Continental European countries did generally issue patents that included several separate inventions put together in a system. U.S. Patent Office 1875, 4–5.

–1685–

From Grosvenor Lowrey

[New York,] Feby 18th 1879—[a]

My Dear Edison:

Can you make use of a sturdy strong boy about 18 years old who has been for several years in our office, and, upon my recommendation, for nearly a year now in the Western Union shops, under Mr Phelps.[1]

This young fellow is a German, named Francis Jehl,[2] and although he has a rather awkward appearance, and manners, and is rather slow and might seem to some stupid, he is quite an intelligent, industrious faithful, honest, high minded young fellow.

He has always been greatly interested in electricity, and while an office boy used to make magnets & little electrical machines which he brought to the office. They were, of course, only imitations of others, but showed a mechanical turn of mind, and a strong love for the subject of electricity. We should have kept him as long as he wanted a place, but his enthusiasm overcame him and he begged me to get him a place with Mr

Phelps. He has been greatly disappointed there & is very unhappy. He has been kept at the most monotonous and uninteresting work (I think boring holes & washing bottles, & that sort of thing), and although he would be perfectly willing to do that if he was surrounded by men or things which interested him, he cannot do it there, for, he says, the men and boys are all flatterers of the foreman and do not work honestly and right, and as he will not join in that he is sent to clean out cess-pools & that sort of thing and at last he is unable to bear it any longer.[3] He has very fine and gentle feelings although he is a big coarse looking young fellow. I think he would develop all the faithfulness and staying power of your man Martin,[4]—in fact he, in some respects, reminds me of Martin. He at present receives $4 a week, and lives with his mother, in town. I have promised him to write you.

I do not think Francis' dissatisfaction of Mr Phelps is because of the meanness of the work which he is set to do, but because, having entered the shop with a mind full of interest in the subject of electricity, his hopes are all disappointed and sickened by discovering that he might as well be in a coal yard for any chance there is there for him ever to hear of the subject of electricity or to come any nearer to high mechanical work than drilling holes & washing bottles—

You will doubtless sympathize with & understand this—Yours Very truly

G.P.L.

LS, DF (*TAEM* 49:875; *TAED* D7913L). Letterhead of Porter, Lowrey, Soren & Stone. [a]Place taken from letterhead; "18" preprinted.

1. Electrical inventor George Phelps was superintendent of Western Union's manufacturing shop in New York City. See *TAEB* 1:135 n. 2.

2. Francis Jehl started work at Menlo Park on 3 March 1879 (Jehl testimony, p. 67, and Francis Upton testimony, p. 97, both *Keith v. Edison v. Brush* [*TAEM* 46:147, 162; *TAED* QD002:34, 49]). Nevertheless, in his reminiscences of the Menlo Park laboratory, he claimed that he began working for Edison in November 1878, shortly before the arrival of Francis Upton, and offered a first-person account of lamp experiments that took place before this letter (Jehl 1937–41, 16, 19–20, 221–22, 238, 241, 250–53; see also Doc. 1667 n. 2). As a laboratory assistant, Jehl took care of the batteries, operated the vacuum pumps, and helped with experiments on lamps, pumps, meters, and batteries.

3. Jehl subsequently recalled his apprenticeship in New York somewhat differently. At Western Union, he was first put to work on the Phelps pony telephone receiver "drilling holes in a piece of curved steel." According to his reminiscences, however, his assigned tasks changed periodically and he worked on a wide variety of instruments and also observed experiments at the shop which "were most interesting

to a young student." Jehl also stated that while working at Western Union he studied chemistry, natural philosophy, and algebra at the Cooper Union night school, where he volunteered for such tasks as "the chore of cleaning the chemical apparatus and the glass jars holding the wet batteries, in order to be on the stage during the lectures and assist in the work." Jehl 1937–41, 16, 124–25.

4. Martin Force.

-1686-

From Pitt Edison

Port Huron, Mich., Feb 19 1879[a]

Dear Bro

No paper come yet   I sent them last week Wednesday[1]   I need to close up the trade with Symington[2] as soon as posable for I must pay Taxes and some interest this month or they will come down on me but I suppose you are busey on the Electc light and have[b] no time to bother with this and I would not bother you only I need the money so bad   Daddy[3] was down to day and I told him that Charley was going to Europe soon and he wants go with him   he thinks he could Help Charley over there and he wanted me to ask you if he could go with him and I told him I would write you about it   if he could go I would like you to write to me or him and let him know   he wants to look over Holland about the fortune which he says is a shure thing if it was looked after[4] so answer as soon as you get this if he can go or not[5]   love to all

WPE[6]

ALS, NjWOE, DF (*TAEM* 49:941; *TAED* D7914C). Letterhead of Huron House. [a]"Port Huron, Mich.," and "187" preprinted. [b]Obscured overwritten text.

1. These probably included the papers mentioned in an undated 1879 letter from Pitt to Thomas Edison dealing with Pitt's family finances and street railway company stock. DF (*TAEM* 51:803; *TAED* D7934G); see also note 2.

2. Thomas and J. S. Symington were in the dry goods trade in Sarnia, Ontario, and were the controlling stockholders in the Sarnia Street Railway Company; at the railway's start in the mid-1870s Pitt had been involved in its operation and Edison had invested in the company (see Docs. 651, 794, 956, 1024). Pitt had obtained a loan from the Symingtons secured by some of Edison's stock (see Docs. 871 and 894). Discussions about selling Edison's stock to the Symingtons recurred intermittently over much of 1877 and 1878 (see e.g., Pitt to Thomas Edison, n.d. [1878], DF [*TAEM* 19:412; *TAED* D7834I1]). In an arrangement reached sometime late in 1878, Edison sold 140 of his 180 shares for $2500, part of the payment for which was cancellation of the outstanding balance on the Symingtons' loan to Pitt (T. and J. S. Symington receipt to T. A. and W. P. Edison, 1 Jan. 1879, DF [*TAEM* 51:794; *TAED*

D7934A]). Sale of Edison's remaining shares was being prepared but required extra documentation before apparently being completed in March 1879 (Pardee & Garvey to Pitt Edison, 5 March 1879, DF [*TAEM* 51:796; *TAED* D7934C]). How these transactions affected any other claims on Pitt by the Symingtons is not clear.

3. Edison's father, Samuel.

4. Samuel Edison had expressed an interest in traveling to the Netherlands the previous year, but nothing is known about the supposed family fortune. James Symington to TAE, 14 May 1878, DF (*TAEM* 15:658; *TAED* D7802ZLL).

5. No answer is extant; no such trip took place until 1886. James Symington to TAE, 24 Nov. and 21 Dec. 1885; 20 Jan. and 20 Nov. 1886; Samuel Edison to TAE, 30 Oct. 1885; all DF (*TAEM* 77:468, 470, 478; 79:249, 272; *TAED* D8514U, D8514W, D8514ZAF, D8614A, D8614O).

6. Edison's brother, William Pitt Edison.

<table>
<tr><td>

-1687-

*Equipment
Specification:
Electric Lighting*

</td><td>

[Menlo Park,] Feb 22 1879

</td></tr>
</table>

7.[1] Alter the Large Magneto machine[2] in the following particulars:—

1. Make iron packings underneath the bearings and dowell both the bearing and the packing—[3a]

2. Make new shell as shown[b] in drawing marked order No 7a[c] of Feb 16, 1879. Shell to be 5/16 thick and have 3 german silver rings inside to stiffen it each 1/8 thick by 15/16 in length.[4] This shell must be wound with (49) forty nine complete coils of wire as in drawing marked No 7B[5]   The shell will be cut in ~~for~~ with 106 divisions but 8 of these divisions will be occupied by 4 feathers which[6d] run all the length of the shell the other 98 will be occupied by top and bottom of the 49 coils

3   The feathers must be wrought iron and turned up at the end so as to screw to the large casting[7] so

4   Large cast iron shell must be bored larger to suit wire shell and have the feathers grooves cut in it

5   The commutator must be made with 49 blocks as shown in drawing marked 7B[8e]

6   Commutator brushes must be made of wires and both insulated from the shaft instead of as before one fast to shaft

7   Connections from shell to commutator[9f]

8   Resistances of shell must be[10f]

9   Resistance of armature[11] must be [~~16?~~][g] .6 ohms   This will be got by winding 76 complete turns of 16 wire round it[12h]

10. A New Armature must be made in the following manner:—

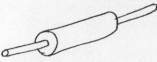

Length 14 inch   Diameter 4.75   Rubber end[i]   A solid cylinder of iron with wire wrapped round it in the manner of that on model No. 1[13]

Batchelor

ADS, NjWOE, Batchelor, Cat. 1308:101 (Order No. 7) (*TAEM* 90:717; *TAED* MBN003:24). Written by Charles Batchelor. [a]Followed by check mark, added later. [b]Check mark, added later, appears above. [c]"a" interlined above. [d]Obscured overwritten text. [e]"as shown . . . 7B" probably added later. [f]Followed by several blank lines. [g]Canceled. [h]"This . . . round it" probably added later and followed by several blank lines. [i]"Length . . . end," at left of drawing, probably added later.

1. This is the seventh order of nine recorded on this day in the shop order book. These mark the beginning of a series of numbered orders for apparatus and models made in the laboratory machine shop. Some of the first entries may have included orders placed earlier than 22 February; for example, part of this order goes back to 16 February (see note 4) and the specifications and drawings for order No. 8 date from the period 10–18 February (see Doc. 1696 n. 2).

2. See Docs. 1653, 1682, and 1694. The machine was not a magneto in the sense of using permanent rather than electromagnets for its field magnet, but it was also not a dynamo in the sense of a self-excited generator and often had its coils energized by batteries rather than power from a separate generator. The applicability of names for various types of generators was in flux in this era; see, e.g., Thompson 1886, 1–4.

3. That is, the packing and bearings were to be held with interpenetrating pegs.

4. The drawing, dated 16 February and headed "New Shell for Edisons Dynamo Electric Machine," and calculations made on it (recto and verso) indicate that the shell was to be about 14 inches long with a diameter of 5 9/16 inches; the German silver rings (made of an alloy of copper, zinc, and nickel) were to be 15/16 inches wide with a circumference of just over 15 1/4 inches. In the patent stemming from this design, Edison stated that he preferred having the shell made of vulcanized fiber. Machine Shop Drawings (1879–1880), Lab. (*TAEM* 45:47; *TAED* NS7986C: 5–6); U.S. Patent 264,643.

5. The drawing, dated 21 February, shows that the shell coil connections make a continuous circuit with each coil connected to the second coil away from it on either side. It also specifies that each groove be filled "with 2 layers of 4 turns each" of "#20 wire silk cov[ered]." The wire specification is probably to No. 20 Birmingham (U.K.) gauge wire, of .035 inch diameter, rather than to American gauge wire of .032 inch diameter (.89 mm. and .81 mm. respectively), since Birmingham gauge was apparently used in specifying the magnet wiring below (see note 12). However, a calculation related to these specifications treated the wire in

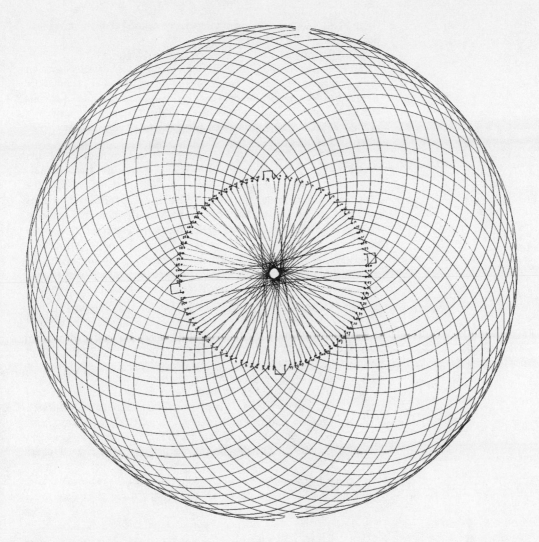

The new shell winding for Edison's large magneto generator.

use as having a diameter of .034 inch. Machine Shop Drawings (1879–1880), Lab. (*TAEM* 45:48; *TAED* NS7986C:7); on wire gauges, see Prescott 1877, 965.

6. That is, the feather would cover the equivalent of two divisions. These are shown between wires 1 and 49 and 23 and 24 on both sides of the shell in drawing 7B.

7. These iron reinforcing rods, like the German silver rings, were intended to make the new shell to hold the wires in which the current was induced much more rigid than the previous shell. A need for such improvements may have become evident as early as 15 February, when a "fearful pulling" was noted during one test of the first version of the machine (N-78-12-04.2:105, Lab. [*TAEM* 29:382; *TAED* N004:55]). The large casting referred to was the end of the cylindrical iron case that enclosed both the shell and its central field magnet.

8. The drawing (see note 5) only showed the wiring sequence, not the

commutator blocks and connections that embodied those connections, but see Docs. 1682 and 1694.

9. No drawing of these connections has been identified.

10. Although no known document gives these, calculations based on the dimensions and wiring noted above and a value of 8.44 ohms per 1000 feet for No. 20 Birmingham gauge wire indicate each of the 49 coils was have to about 0.22 ohm resistance, and that the electrical resistance of the whole shell when connected as a combined coil would have been about 10.75 ohms.

11. "Armature" is used here to refer to the part of the machine that moved, i.e., the field magnet, which rotated in the middle. In more common usage, an armature of a generator was (and is) the opposite of the field magnet; namely, the part holding the wire where the current was induced, which in this machine was the inner of the two cylinders that Edison at various times called shells, the copper-wire-carrying shell rather than the iron case.

12. This would have been Birmingham gauge No. 16 wire with a diameter of .065 inch, with a resistance of 2.45 ohms per 1000 feet. The dimensions given below require a turn to be 37.5 inches long and 76 such turns of this wire would have a resistance just over .58 ohms if perfectly made or somewhat higher if less than perfect, whereas the same length of No. 16 American gauge wire, with a diameter of .05 inch and resistance of 4.3 ohms per 1000 feet, would provide a bit over 1 full ohm total resistance. Prescott 1877, 967.

13. A solid cylinder without large grooves to contain the wrapped wires, in contrast to the rotating field magnet design used in the original machine which resembled a Siemens shuttle armature (see Doc. 1694), matches what was specified in the patent design related to this machine. "No. 1" was the order for a patent office model for this design, entered the same day in the same notebook, two pages earlier (Cat. 1308:99 [*TAEM* 90:716; *TAED* MBN003:23]). The wiring of the field magnet is not shown in the order but presumably matches that shown in U.S. Patent 264,643, figure 3; see also Doc. 1694, figs. 2 and 3.

–1688–

Notebook Entry:
Electric Lighting

[Menlo Park,] February 22 1879

Conceived the idea of using a bobbin of iron wire wound on lime, with platinum for the last layer. The iron and platinum to be pyro-insulated by acetate of magnesia,[1] and the size and rest'nce of the iron made so that when it was at a bright yellow approaching white that the platinum should be incandescent. I effect this by using an iron wire slightly larger than the platinum. made some tests with iron an platina wire joined together & placed between supported and heated with the battery   I find not the slightest difficulty in so selecting the size of the wires that the Pt shall be brilliantly incandescent while the iron is yellow— Hence I can make a lamp with 1 inch of incandescent surface (pt) with a large resistance, the inner wire of iron losing all of its heat through the platina.[2] this is a gain in economy of plant.

I noticed the jerk in the contraction and expansion of the iron wire when it reached certain temperatures.[3] It is very noticable and strong. thereey expansion is different from the contraction if but a slight strain is put on the wire when red it elpermantly elongates with ease. I think electricity passing through the wire in conjunction with this case of elongation has something to do with this phenomenon= I also notice two sudden[a] expansions while the wire was cooling ɪ very striking when it reached a dull red the other after heat was invisible. These jerks are sometimes so sudden that the wire is set vibrating

Covering iron with with acetate of magnesia raises its fusing point in air by probably by preventing oxidation

TAE

X, NjWOE, Lab., N-79-01-19:73 (*TAEM* 31:177; *TAED* N023:57). [a]Obscured overwritten text.

1. The same day, Charles Batchelor ordered construction of "one Patent Office model of New Electric lamp with spool of wire in vacuum, spool being filled with platina wire or iron and platina insulated by some such substance as acetate of manganese." The model was completed on 1 March. Doc. 1695 is a draft of the patent application. Cat. 1308:99, Batchelor (*TAEM* 90:716; *TAED* MBN003:23).

Batchelor's notes regarding further "Experiments for covering Wires" with acetate of magnesia follow in this notebook. He noted that after repeatedly dipping platinum and iron wires in the solution the laboratory staff found that the wires did not become visibly heated on battery current even though "there is no signs of any deposit of Magnesia on these wires after dipping." He also observed that after the iron wire slightly oxidized and the resulting oxide "melted and run in globules along the iron, it becomes more infusible, probably because this prevents further oxidisation." N-79-01-19:77–79, Lab. (*TAEM* 31:182–84; *TAED* N023:61–63).

2. Edison sketched this idea two days later, showing platinum joined to heavier iron and nickel wires, and platinum foil with iron wire; he also showed the use of nickel joined to platinum. On 22 February, when Batchelor put .014 inch iron and .010 inch platinum wires of equal lengths on battery current, he found that the platinum became bright yellow while the iron was barely red. After several incremental increases of current the platinum was "full white with tinge of yellow" while the iron was yellow. Batchelor reported similar results after testing other gauges of iron and platinum-iridium wire. N-79-02-24.1:1, N-79-01-19:78–84, both Lab. (*TAEM* 30:970, 31:183–89; *TAED* N020:1, N023:62–68).

3. Batchelor noted the following day that when soft iron wire was heated on battery current and allowed to cool, "in cooling and contracting it will give a kind of jump. Sometimes when heating it I can plainly

hear crackling noises both when heating up and when cooling due no doubt to expansion and contraction." Cat. 1304:39, Batchelor (*TAEM* 91:38; *TAED* MBN004:36).

–1689–

Notebook Entry:
Electric Lighting

[Menlo Park,] Feb 23rd 1879

Resistance of Metals at different heats[1]

Our experiments show that Platinum increases its resistance between the ordinary temperature and brilliant incandescence only 3[a] three times

Platinum Iridium alloy (Ir 20%) increases nearly 4 four[b] times

Iron in vacuo and covered with Acetate of Magnesia to make it bear greater heat increases its resistance 15[a] fifteen times or more.

Chas Batchelor

X, NjWOE, Batchelor, Cat. 1304:38 (*TAEM* 91:37; *TAED* MBN004: 35). Written by Charles Batchelor. [a]Circled. [b]Obscured overwritten text.

1. The laboratory staff had conducted extensive tests on the resistance of platinum and platinum-iridium wire at different temperatures during the first two weeks of January. Batchelor produced three tables summarizing the results of some of those tests on 14 January. On 22 and 23 February that staff had conducted tests to determine "the increase of resistance under heat of iron wires"; an earlier set of tests on iron wires is dated 12 January. N-78-12-20.2:70–106; N-78-12-31:27–32, 65–87; Unbound Notes and Drawings (1879); N-79-01-19:85–91; N-78-12-31:88–95; all Lab. (*TAEM* 30:150–68, 389–419; 44:977–80; 31:190–96; 30:420–23; *TAED* N008:36–53; N014:13–16, 32–43; NS79:6–9; N023:69–75; N014:44–47).

–1690–

Notebook Entry:
Electric Lighting

[Menlo Park,] Feby 24 1879

I find that Acetate Lime, Magnesia Chloride of lime, Magnesia or Alumina, will not coat iron wire by imersing in liquid & then heating & repeating the operation many times   But platina is coated easily— A Beautiful hard Coating that very smooth is put on platina with imersing it in a syrupy solution of Magnesia Acetate & putting in flame after each time[1]

X, NjWOE, Lab., N-78-12-11:215 (*TAEM* 29:779; *TAED* N007:101).

1. On this day Charles Batchelor had the shop make "4 small troughs for covering wire in solution." Also on this day, Edison sketched and Batchelor made an apparatus for applying successive coats of "a saturated solution of a salt of Zircon, Cerium, Magnesia, Titanium, Lime,

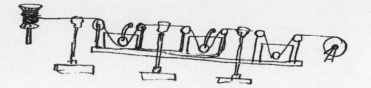

Edison's drawing of apparatus for coating wires. The finished instrument had four wooden troughs, each with a tin dish to hold the solution.

Alumina, or other infusible oxide." Edison noted that "instead of heating by the direct flame a thin tube may be heated & the wire pass through that or through a coil of platinum wire kept Incandescent by a battery." Cat. 1308:105 (Order No. 10), Batchelor (*TAEM* 90:719; *TAED* MBN003:26); N-79-02-24.1:9, Lab. (*TAEM* 30:974; *TAED* N020:5).

–1691–

To George Gouraud

[Menlo Park,] February 25th 79—

Friend Gouraud.

Owing to your numerous Telegrams[1] I thought best to make two rough receivers and two Telephones with transmitters, and send by my nephew C P Edison=[2] Please understand & have others understand that these instruments were made in five days[3] and are only experimental and do not set the invention off to any advantage by any means;[4] The whole arrangement will however show you what the invention is and I think that it will replace every other kind of telephone—in time= We are having some first class ones made and will ship you some shortly. Yours

T. A. Edison.

P.S. Excuse the ugly shape of The telephone[s][a] sent you— they were thrown together= E.

ALS (facsimile copy), Kingsbury 1915, 194. [a]Illegible.

1. Gouraud had cabled Edison about the new receiver on 16 January and 3 and 22 February. Edison would by this time also have received Gouraud's letter of 18 January (Doc. 1662) and two letters dated 1 February, each urging him to rush the instrument. DF (*TAEM* 52:468, 482, 488, 472, 477, 479; *TAED* D7941B, D7941H, D7941J, D7941D, D7941F, D7941G).

2. Edison had advised Gouraud by cable on 23 February, "Messenger probably take receivers per Algeria Will notify." The following day Edison asked Charlie Edison to go to London. He sailed on the *Algeria* on 26 February with the two receivers, two complete telephones, and this document (see headnote p. 73 and Doc. 1703). That same day Gouraud telegraphed Edison, "Messenger capable relieving Adams seriously ill indispensable" to which Edison replied "Messenger sailed Algeria." Gouraud acknowledged this cable exchange in a 27 February letter to Edison in which he elaborated on the state of James Adams's

chronic ill health (see *TAEB* 4:295 n. 2), noting that it had become "very precarious. He is consequently unreliable for the purpose for which I understand him to be here." TAE to Gouraud, 23 Feb. 1879, DF (*TAEM* 52:492; *TAED* D7941L); Charlie Edison to Samuel Edison, 2 Apr. 1879, EP&RI; "The Electromotograph," *New York Herald,* 27 Feb. 1879, Cat. 1241, item 1138, Batchelor (*TAEM* 94:461; *TAED* MBSB21138X); TAE to Gouraud, 26 Feb. 1879; Gouraud to TAE, 26 and 27 Feb. 1879, all DF (*TAEM* 52:494, 496; *TAED* D7941N, D7941O, D7941P).

3. They were actually made in about ten days (see headnote p. 73).

4. Gouraud evidently circulated this letter, parts of which were quoted favorably by the *Electrician,* the *Engineer,* and two London papers. "Edison's New Telephone Receiver," *Electrician* 2 (1879): 206; "Edison's Telephone Receiver," *Engineer* 47 (1879): 213–14; "The Edison Telephone," *Times* of London, 17 Mar. 1879, and "A Loud-Speaking Telephone," *Metropolitan,* 15 Mar. 1879, both Cat. 1241, items 1151 and 1152, Batchelor (*TAEM* 94:466–67; *TAED* MBSB21151X, MBSB21152X).

–1692–

To George Prescott

[Menlo Park,] Feby 25. 79

Friend Prescott[1]

Yours regarding the Quad[2] was duly rec'd.[3] The price named is low. Do they mean circuit created or the wire itself?[4] I think it would be a shame to take less than £20 per annum on at least 30 circuits, and it ought to be 40 circuits (guaranteed) Not less than that.[5] It is either worth that or nothing and I think they will have to take it, if not now, within a year because it must in time replace the Wheatstone Automatic.[6] These figures I give are low but I would be satisfied with the 30 circuits guaranteed at £20 each providing you cannot get more   In fact I will do anything you say.[7] Very Truly

T. A. Edison   G[riffin]

L (letterpress copy), NjWOE, Lbk. 4:191 (*TAEM* 80:48; *TAED* LB004191). Written by Stockton Griffin; circled "C" appears in top and bottom left corners.

1. George Prescott became chief electrician of the Western Union Telegraph Co. in 1870; he joined Edison in 1874 in what became a highly contentious partnership to develop and market Edison's designs for multiple telegraphy, particularly quadruplex telegraphy. The controversy over the quadruplex strained their relationship and affected Edison's relationship with other electricians at Western Union. *TAEB* 1:258 n. 4; Docs. 466, 727, 1173, 1185, 1204; *TAEB* 2, App. 3.

2. Edison's 1874 quadruplex telegraph designs provided the first practical way for four independent messages (two each way) to be transmitted simultaneously on a single line wire. See *TAEB* 2:20 n. 9, 21 n. 16, 270, 361 n. 2.

3. Prescott wrote Edison on 22 February that Henry Weaver (president of the Anglo-American Cable Co., the main Atlantic telegraph cable firm), acting as their agent, had obtained a proposal from the British Post Office, which operated the telegraph system of the United Kingdom, that in lieu of buying the patent rights would pay a royalty of £20 annually "for each circuit worked by means of" quadruplex telegraphy (DF [*TAEM* 51:853; *TAED* D7936I]; on Weaver, see Prescott to TAE, 23 Dec. 1878, DF [*TAEM* 19:577; *TAED* D7836ZCW] and Reid 1879, 798). Edison, Prescott, and Gerritt Smith (another Western Union electrician and inventor who had worked with them in developing the technology) owned the key British patents (Brit. Pats. 384 [1875], Batchelor [*TAEM* 92:35; *TAED* MBP005] and 197 [1877]) and had coordinated their interests through an agreement of 31 May 1877, giving Prescott 45% of all resulting income, Edison 35%, and Smith 20% (DF [*TAEM* 14:683; *TAED* D7717ZAD]). On 12 January 1878 they had assigned their rights to Western Union president William Orton (DF [*TAEM* 19:452; *TAED* D7836C]). Edison's British rights to the quadruplex were claimed by the London firm Smith, Fleming & Co. under an 1873 agreement (Doc. 350), and this had led to extensive negotiations during 1878 (*TAEB* 4:10 n. 5; Docs. 1209–1210 and 1336).

4. Applying quadruplex devices and wiring resulted in four effective circuits for a wire line, thus creating three new circuits (or two if the line had already been duplexed). The question then was whether the proposed royalty would be paid for each extra working circuit, or just for each wire on which the technology was applied.

5. In his 22 February letter (see note 3) Prescott had quoted Weaver's comments that the offer was "absurdly inadequate" and "the price must be raised and a minimum number of circuits fixed so as to ensure a fixed income."

6. The system devised by Sir Charles Wheatstone was extensively employed on those lines that were most heavily used in Britain. As quadruplex telegraphy was introduced in the United Kingdom, it was often used in combination with (rather than instead of) the Wheatstone automatic. On automatic telegraphy, see *TAEB* 1:149; on the Wheatstone system, see Prescott 1877, 702–11 and Preece and Sivewright 1891, 147–64, 200.

7. An agreement was completed only in late 1880, after resolution in court of the issue of the assignment to William Orton, whose estate had to be dealt with as a consequence of his death in April 1878. Grosvenor Lowrey and Lemuel Serrell were made trustees in place of Orton and this was made retroactive to 31 December 1879. The royalty agreed to was £25 per circuit (apparently meaning line wire) per year until the expiration of the later patent (in January 1891), due at the end of each calendar year, with a guaranteed minimum of £300 per year (equivalent to use on twelve lines). After deducting taxes and Weaver's commission, Edison's annual share at the prevailing rates of exchange was at least $400 and grew to well over $600 by the end of the 1880s. Prescott to TAE, 17 June, 15 Sept., and 15 Oct. 1879, 6 and 10 Jan. 1880; TAE to Prescott, 28 Nov. 1879; all DF (*TAEM* 51:894, 549, 922; 55:494, 497; 51:933; *TAED* D7936ZAD, D7929ZDY, D7936ZAX, D8042A, D8042B, D7936ZBG); Agreement between Grosvenor Lowrey and Lemuel Serrell, trustees; TAE, George Prescott, and Gerritt Smith; and Henry

Fawcett, Her Majesty's Postmaster General, 20 Nov. 1880, Miller (*TAEM* 86:314; *TAED* HM800133); Prescott to TAE, 4 Jan. 1883, with enclosed royalty statement; Weaver to Prescott, 2 Jan. 1884; Prescott to TAE, 7 Feb. 1890, with enclosed royalty statement; all DF (*TAEM* 70:1120, 1119 ; 130:606; *TAED* D8373B, D8373A, D9062AAA).

–1693–

Draft Patent Specification: Telephony

[Menlo Park,] Feby 26 1879 to Mch 5 1879

L. W. Serrell.[1]

A U.S. patent is to be taken out on this for W.U. ~~andfor~~ which I furnish you a model—[2] I also furnish you a second model to go with the Canadian patent—[3]

This is the U.S. specification[4]

The object of this invention is to transmit and reproduce sounds telegraphically at a distance with great power and without loss of volume—[a]

~~The invention conists in principle of the~~[b] The peculiar action upon which ~~advanta~~ this invention is based was patented by me (original electromotograph patent) & numbered ____.[5] An application of this action to telephony was also patented or applied for [-----][c]____ (LWS[d] This is the patent where a band of paper was used)—[6] The present application consists more[e] particularly in devices which make the invention perfectly practicable for u[se][f] in commerce and infinitely more reliable & effective

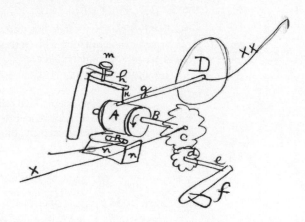

A is a cylinder of ~~chalk~~ compressed chalk soaked in an electrolytic solution such as ~~ana~~ caustic alkali that it may become a conductor of electricity

I will mention that any finely divided non[e] conducting material or porous body having capilliary pores & which has no chemical[g] action upon the absorbed solution may be used—

This cylinder is secured to a shaft .B ~~wh~~ & the whole is rotated ~~by~~ by the operator by means of^h the toothed wheels c. & .d. and shaft & handle e. & f   ~~ana~~ worm & pinion may be substituted and, ~~the hand of the operator made any prime mover~~ motor power of a^i clock work or analagous motor made to replace the hand of the operator. Resting upon the cylinder A is a flat spring g connected to the diaphragm D. this spring is pressed upon the chalk with a pressure of several pounds by means of the wire K & spring .h. the screw m serving to increase or decrease the amount of pressure— The line wire XX^g is connected to the spring g while the earth wire is connected to the shaft on which the cylinder A is secured. If now the cylinder be rotated in the direction shown by the arrow & no current passes the normal friction of the spring g upon the surface of the cylinder A will be very great and the spring will be carried forward in the direction of the rotation of the cylinder thus pulling inwardly the diaphragm D. if now a current passes whose direction is such that Hydrogen will be produced upon the surface of the spring g in contact with the chalk the friction will be reduced to an extent proportionate to the strength of the current ~~and the~~ hence the diaphragm will regain its natural position and continue^e there as long as the current passes as the friction between the spring g & A is too small to produce sufficient traction ~~to~~ [----]^c   if now the current ceases the normal friction at once obtains. the traction increases & the diaphragm is instantly pulled^e outwardly   if now a current in the opposite direction is transmitted the effect is ~~when~~ scarcely noticeable as the evolution of Oxygen upon the surface of the spring g does not except with a [~~po?~~]^c few ~~the~~ solutions decrease the friction like Hydrogen^e but on the contrary generally tends to increase the normal friction, hence I allow a constant current to circulate on the line and am thus enabled to ~~used~~ utilize the opposite wave by causing them to weaken the constant Current which is always in a direction to evolve Hydrogen on the surface of the spring .g.   for instance if no constant current was upon the line the only ~~one of the~~ waves in one direction would produce an effect whereas if a constant current is circulating one of the waves is^g add~~s~~ed to the~~at~~ ~~strength~~ of the current and the other wave neutralizes or nearly neutralizes the constant current hence the tendency of 1 wave is to produce a lessening of friction of say 5 units which added to that produced by the constant current say 5 together give 10 units   if now an opposite current is sent it neutralizes the constant current & we have the normal friction of the cylinder whereas if no constant current was used one[e] current would lessen the friction toby 5 units & the opposite current would perform no work.

n is a receptible of containing water   resting in the water is a roller R composed of any flexible absorbant substance such as sponge or felt.

TA lever to which the roller is attached & by which it the roller may be brought in contact with the cylinder A extends through the box to the outside. The object of this device is to supply the cylinder whichwith water lost by evaporation & make it a conductor to the electric current

In practice when anall the apparatus is enclosed in a cast iron case the loss of eby evaporation is very small and if the roller be held against the cylinder & if the cylinder be rotated three or four times the chalk will take up sufficient moisture so that it will perform its functions for a week or more without again wetting it=

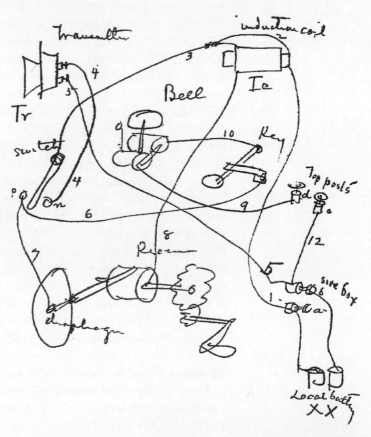

The connections are as follows[7]   Tr is the Regular Carbon telephonic transmitter placed in a local circuit with an induction coil Ic with local battery   If the position of the switch lever is on the button n. then the current proceeds from the local battery XX via wire 1 thence through the primary coil thence via wire 2 to 3 thence to the switch lever thence via wire 4 through the carbon transmitter thence by wire 5 to the post b.[8] thence to the battery   The ~~w~~sonorous waves of the voice are translated into electric waves of a positive & negative character in the secondary coil of the Inductorium Ir[9] in the well known manner.

The connection of the main line is as follows   the line from the distant station enters at the post .d. thence via wire 9 to the bell thence via wire 10 to the key lever thence via wire .6. to the point p ~~then~~ of the switch thence via wire 7 to the spring g of the receiver thence through the moist chalk to wire 8 to & through the secondary coil thence via wire 3 to the switch lever which it will be remembered is on n thence via wire 4 through the transmitter to wire 5 thence to post b & thence via wire 12 to the post c which is connected to the earth. I will mention that either of the posts d c may be connected to the ground or line. It will be noticed that the main line passes through the transmitter which is connected to the ~~b~~local battery hence a portion of the current leaks into the main line & it is this portion which thus leaks into the line that serves as a constant current for short lines but if the line be to long or has too great a resistance this small^e leakage current is so weakened as^e to be insufficient to produce a lessening of friction between the spring g & the chalk hence I ~~add~~ insert^g 1 or more cells ~~to the~~ in the main line=

~~I will now~~ while the switch is in this position ie^d the lever in contact with n transmitting & receiving can go on simultaneously   ~~If~~ by turning the switch lever to the point .p. the telephone apparatus is disconnected & the call bell apparatus become operative. the line enters at the post d passes through the bell via wire 9 thence via 10 to key thence via .6. to .p. thence via lever to wire 3 thence via 2 through the primary coil & via 1 to Local battery through that to post b thence via 12 to post c to earth. by depressing the key the circuit is opened & closed & the two local batteries 1 at each end of the line become operative to ring the bell.

I will mention, that the two receivers herein described & the two transmitters may all be connected in one line & operated without the aid of induction coils but the results are not^e equal to [t]hat from the use of the coil,^{e} also that the Receivers will act as transmitters by ~~reasing~~ reason of the fact that when the spring g & chalk is[e] at rest the resistance of the whole is generally about 2500 ohms & this resistance is reduced instantaneously to 2 or 300 ohms by the slightest movement of either the chalk or the spring ~~the effect~~ hence the movement of the diaphragm D by the voice produces the[e] same[e] result or nearly so as the carbon transmitter. ~~and advantage and~~ if advantage[e] is to be taken of this fact the surface of ~~th~~ the spring g should be reduced to increase the effect and an induction coil having a primary coil of high resistance used in connection with the apparatus although it is not absolutely necessary as the direct results are nearly as good

In preparing[e] the chalk I prefer to use a salt of mercury mixed with caustic soda. The action takes place no[e] matter what the proportions. The mercury[g] salt I prefer to use is the acetate of mercury=[10]

I will mention that the moisture of the chalk may be regulated automatically by taking advantage of the fact that the normal friction of the chalk increases at it becomes[e] dryer. a spring resting on the chalk is connected to the wetting roller & has such a pressure & counteracting spring that when the chalk has its proper moisture the friction during rotation is insufficient to provide sufficient traction to lift the roller but when it becomes dryer the traction becomes sufficient to lift the wetting roller & it supplies moisture until the traction is reduces below a certain point & it falls—

~~I claim the use in telephony of a cylinder disc or other form composed of an absorbant substance moistened with an electrolytic solution which cylinder is set in rotation by any suitable[g] motor~~

I claim the method herein described[e] of producing motion at a distance for reproducing sound waves ~~by~~

2nd   The use of a ~~rotating~~ moving[j] cylinder disc or its equivalent composed of an inert porous substance moistened with an electrolytic solution with an ~~spr~~ electrode connected to a diaphragm plate or other body capable of rendering the motion produced by electrolysis[e] ~~ele~~ audible—

3rd   In combination with such cylinder disc or equivalent in a closed receptable of a ~~roll~~ moistening roller or equivalent & supply of water[e] sub as set forth

4. ~~Translating sound waves into electric waves by depolarizing~~

& any other claims that may strike you as pertinent[11]

T A Edison

ADfS, NjWOE, Lab., N–79–01–21:57 (*TAEM* 30:615; *TAED* N016:30). Unrelated calculations and sketches omitted. Document multiply signed and dated. [a]Followed by "over" to indicate page turn. [b]"~~principle of the~~" interlined above. [c]Canceled. [d]Circled. [e]Obscured overwritten text. [f]Document damaged. [g]Interlined above. [h]"by . . . of" interlined above. [i]"~~any prime mover~~" interlined above; "motive power of a" interlined above that. [j]"~~rotating~~ moving" interlined above.

1. No correspondence with Lemuel Serrell regarding the preparation of this application has been found.

2. On 24 February, Charles Batchelor had ordered construction of "one model for new Receiver for Serrell to be sent over directly it is done"; John Kruesi indicated that it was finished on 4 March. On 28 February Batchelor ordered a wooden model of the complete receiver "just as we shall decide to make them," and on 6 March he asked Kruesi to make a second patent office model, identical to the first, that was completed the next day. Cat. 1308:105, 111, 117 (Order Nos. 12, 29, 46), Batchelor (*TAEM* 90:719, 722, 725; *TAED* MBN003:26, 29, 32).

3. Edison's 21 January 1879 agreement with Gold & Stock regarding Canadian telephone patents covered those patents "for which he has applied or is about to apply" (DF [*TAEM* 52:174; *TAED* D7939D]). Batchelor ordered "one model of new Receiver for Canada" on 28 February; there is no indication when this model was completed (Cat. 1308:111 [Order No. 28], Batchelor [*TAEM* 90:722; *TAED* MBN003:29]).

4. The body of this draft is substantially the same as the application Edison executed on 24 March 1879, which issued in November 1879 as U.S. Pat. 221,957. Edison did significantly elaborate the claims (see note 9).

5. U.S. Pat. 158,787. See *TAEB* 2:252 n. 2.

6. In the executed version of this document, Edison referred to this as the application "in which there is a band of paper moving beneath a point connected to the diaphragm; this feature therefore is not broadly claimed herein." Edison filed that application (Case 141) on 20 July 1877; it was involved in the Telephone Interferences and subsequently divided and issued in 1892 as U.S. Patents 474,231 (covering the electromotograph) and 474,232. Pat. Apps. 474,231 and 474,232.

7. Circuit diagrams drawn by Edison on 26 February are in N–78–11–21:102–104, Lab. (*TAEM* 29:278–79; *TAED* N003:52–53); see also drawings of 10 March in N–79–03–10.1:1, 5, Lab. (*TAEM* 31:3, 5; *TAED* N022:1, 3).

8. In the drawing Edison indicated that the post was on the "side [of] box."

9. Edison meant the induction coil "Ic"; inductorium was another name for an induction coil.

10. A 24 February notebook entry by Charles Edison gives the following recipe for the chalk buttons: "take 6 oz chalk in mortar and add to it 18 grains of Acetate Mercury grind this up finely then add 1 fluid oz Saturated Solution Caustic Soda." N–79–01–21:45, Lab. (*TAEM* 30:609; *TAED* N016:24).

11. In the final specification Edison claimed the combination of a diaphragm, a moving surface soaked with electrolytic solution, a point pressing on that surface, a means of adjusting the tension, and the mois-

tening roller. He also claimed the electrical connections and circuits, the placement of a local battery upon the line, and the case enclosing the apparatus; this last item was objected to and eventually omitted. Lemuel Serrell to TAE, 14 June 1879, DF (*TAEM* 51:510; *TAED* D7929ZCV); TAE to Serrell, 18 June 1879, Lbk. 4:417 (*TAEM* 80:86; *TAED* LB004417); Pat. App. 221,957.

–1694–

*Draft Patent
Application:
Electric Lighting*

Menlo Park February 27–March 1 1879[1a]

The object of this invention is to generate electric currents by the application of power[2]

The invention consists

1st   In the method of winding and connecting the wire to the commutator.

2nd   in the method ~~of making~~ of winding the wire upon[b] the[c] field magnet—

3rd   The method of generating the current by ~~caus~~ rotating the lines of force and causing the them to cut the wire while standing still

4   The use as a armature[c] to the field magnet of a sheet[3] composed of many ~~was~~ washers of iron seperated from each other by an insulator=

Fig 1[4]

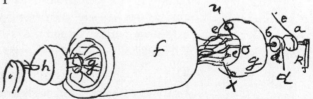

F is the cylinder within which the ~~wire is ordered~~ wire bobbins are placed. This cylinder is made up[c] of a number of discs clamped together by[c] three rods. These discs are insulated from each other by pasting paper upon their sides. The object of this insulation being to prevent Foucault[d] current[5] from circulating within the iron   ~~The field~~ I will mention that this cylinder may be made by coiling iron wire in the form of a bobbin which will prevent to a greater extent the formation of these currents, but is more expensive=

Fig 2[6]

in fig 2 is shewn the field magnet ~~the wire~~ which consists of either a solid or hollow cylinder closed at both ends. The wire is wound longitudinally over its entire circumference starting from any one points it is wound until the ending wire come paralell with the starting wire. these two ends are then taken by insulating wire[c] ~~laid in g~~ to two insulated rings on the shaft ~~upon~~   The wires being laid in grooves cut in the shaft= ~~a modification~~   The form of field magnet shewn in fig 3[e] may also be used

Fig. 3[7]

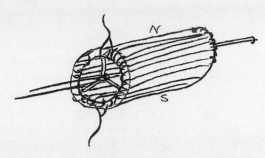

in figure 3 the field magnet consists of a cylinder of iron or iron wire wound in the form of a cylinder. ~~two seperate coils of wire are wound on this~~ brass arms at each end connect it to the shaft which rotates it=

Two seperate[f] coils of wire are wound around the shell[c] passing over the top & coming back underneath, these ~~wires~~ coils are wound longitudinally and the two extremities of the coils ~~one~~on one side are connected together as well as the two extremities on the other side— ~~if a battery current be pa~~

If these ends be connected to a source of Electric Energy the ~~g~~ iron shell at the point where the wires are connected become respectively the N. &. S. poles of the magnet. Whe~~ther~~n either[f] the form of magnet shewn in fig 2 or 3 is inserted and revolved within the ~~shell~~ cylinder of discs it carries its field & lines of force[g] with it Reversing the polarity of the cylinder of discs at each revolution.

between the cylinder and the revolving field magnet of figs ~~1~~2 or .3. is the shell ~~fig~~ A shewn in fig. 4.[8]

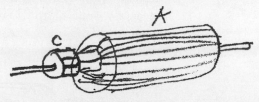

This shell may be of Vulcanized fibre ebony Hard rubber glass or other non metallic substance or it may be of metallic wire if attention is paid to the prevention of the circulation of any currents which might be induced into it.

~~The field magn~~ the wires are wound over the shell longitudinally and are connected to the blocks of the commutator C in a manner to be hereafter described=

Within this shell the field magnet revolves   The whole making a compact machine in which ~~the w~~ concentration of the field is very great, and the magnetic resistance of the iron of the magnetic circuit very low and where every line of force must cut the wire upon the shell .K.[9] thus utilizing the whole field, which[c] leads to great economy as it requires expenditure of energy to form & hold[h] the field and if ~~the a portion of the field doe~~ any of the lines of force do not cut the wire, then to this extent is energy lost.

I will now describe the new & novel method of winding the wire and arranging the commutator so that the [~~many?~~]i whole of the wire is utilized ~~and no part of the energy~~ the loss of energy by shortcircuiting within the machine reduced to a minimum—

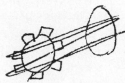

(fig 5 is not shewn in model & marked underneath I cant draw it)[10]

~~Supposin~~ by referring to fig ~~4~~5 and considering the commutator springs to be resting upon the blocks 1 & ~~4~~5 and the ~~top~~ field magnet with has its no[r]t[h] pole on top & its south pole on the bottom then the[c] current from the brush resting on block 1 will[c] go via wires ~~1 & 6 over the north pole returning under south pole via wires 8 & 13~~ 1 & 6, over[f] north pole returning by wires 8 & 13 over south pole thence over the north pole by wires 3 & 4 returning over the south pole by wires 10 & 11 thence over the north pole by wires 5. & .2 then returning under the south pole by 9 & 12, the 9 being connected to the block 5 upon which other commutator brush rests   hence[j] we have[f] a double wire circuit   the wire 12 proceeds to over the north pole by no 7 wire, thence by number 7 south ~~joining~~ joining the other wire at commutator ~~K~~ block .5. This coil is at the neutral point between the north & south poles of the field magnet hence is of no use & is shortcircuited   If[c] the brushes are rotated[j] to blocks 2 & 6 the same connection or circuit is formed but by another set of wires, so that no matter what the position of the brushes a complete circuit is formed cutting all the lines of force. The combination can only be obtained by employing an odd number of coils & an odd number of commutation blocks. The method of winding & the use of an odd number of blocks & coils ~~constitute~~ so that the wires may be used in tension & not in multiple arc consistitutes the novelty of this part of the invention.[11] In fig 5 is shewn another method of winding with an even number of coils but only^c one half can be utilized without loss of energy by shortcircuiting[12]   with this form if the brushes rest upon ~~No~~ block No 1. & 4—& the north pole is ~~upon~~ upwards then the wire—then the wire .4. passes around the cylinder ~~around twice~~ twice &^{k} ends at in the air at^l ~~120~~ hence is not used in this position; but the wire 14 passes over the north pole & return via wire 6 under the south pole thence over the north pole by wire 16, returning under south pole by wire .8. then over north pole by wire 3 returning under^c S pole by wire 11, which is connected to block 4   all the rest of the wire is not used   if the brushes are on blocks 2. & .5. then the wire 6 over the north pole passes around the cylinder 3 times & ends in the air at 10 & is not used— but the wire 16 passes over north pole returning under S pole by wire .8. thence over N pole by wire .3. returning under S pole by wire .11. thence over N pole by 5 returning under S pole by 13, which is connected to ~~block~~ commutator block .5. The wire .7. which connects to this block also passes twice around the cylinder & ends in the air at .1. so no matter what position the brushes may be when they are a^m proper distance^n apart ½ of the wire is always connected properly for receiving the current while the other is thrown out of circuit   I will mention than an extra block may be inserted^c in the commutator & divided in two parts one part of which the wire No 1 may be connected to the other part wire no 10 is connected[13]   this will to some extent prevent sparks.

Referring to fig 1 x .n are the commutator springs secured to a disc g which rotates with the shaft=

b. & .a are insulated collars. The collar 6 is connected to the brush n. while the outer one is connected to one end of the field magnet within the shell   The other end of the field magnet is connected to the brush .X by a wire passing through a grove in the shaft, hence the wire upon the [alt?]^o outer shell is placed in the same circuit [--]^o with the field magnet & the ~~current~~ resultant current is taken from the machine by the brushes d. c resting on the collars b. & .a.

I will mention^c that the field magnet might stand still & the the outer shell be made to rotate.

~~or the inner sh wire upon the fa~~   Also that the ~~fi~~ Iron cylinder could be stripped of its wire & held rigid & the ~~field mag-net~~ wire wound around the outer[g] iron shell ~~eying~~[g] & the wire connected to the commutator be made to rotate

In practice ~~is~~I shall put at fan upon one end of the shaft to [-][o] create a draught of air through the cylinder to carry away any heat which may be generated when the machine is made to exceed its nominal capacity, ~~allthough~~ but this will not be necessary under ordinary conditions as the field magnet unlike similar magnets used in this class of machines where several layers of wire is wound one over the other & thus prevent dissapation of heat, has only one layer of wire hence all of it is in contact with the air & in rapid motion so that ~~an~~ any great[p] accumulation of heat is impossible=[14]

Claims

~~The combination~~ The method herein described of making a cylindrical magnet by winding wire over its entire surface or nearly so[q] sub[stantially] as described[15]

2nd   The shell of wire ~~wound with~~ connisting of[r] an odd number of coils and connected to an odd number of commutator points or blocks all arranged connected & operated substantially as set forth.[16]

3rd   The method of winding the shell[c] as shewn & described in ~~number~~ fig. 45

4—In combination with ~~the field ma~~ revolving field magnet sub as described of the shell of fig 4 or 5 and the outer iron cylinder for the purpose set forth.[17]

5th= The use of Vulcanized fibre to form the shell upon which the wire of the commutator is wound—[18]

T. A. Edison   Chas Batchelor

ADfS, NjWOE, Lab., N-79-01-01:133–65, 173–87 (*TAEM* 30:322–38, 342–49; *TAED* No13:64–80, 84–91). [a]Date taken from document, form altered. [b]"of . . . upon" interlined above. [c]Obscured overwritten text. [d]Interlined above with an "X" written in an unknown hand. [e]"3" written in underlined gap. [f]Interlined above. [g]"& lines of force" interlined above. [h]"& hold" interlined above. [i]Canceled and interlined above. [j]"X" interlined above in an unknown hand. [k]"twice &" interlined above. [l]"in the air at" interlined above. [m]"are a" interlined above. [n]"see page 173" written as page turn. [o]Canceled. [p]"any great" interlined above. [q]"or nearly so" interlined above. [r]"connisting of" interlined above.

1. Edison composed this draft over a few days, beginning and dating it on 27 February, completing and dating it again on 1 March. At several points he found it necessary to leave space for the numbers of wires or commutator blocks, which he probably filled in after examining the actual windings on the machine. William Carman copied this onto pages 71–76 of the missing Experimental Researches Volume 3 on 1 March.

2. Edison executed the application resulting from this draft (Case 177) on 21 April 1879 but Lemuel Serrell continued to modify the specification and drawings (see note 10 and Doc. 1735) and it was not filed until 12 May. It issued as U.S. Patent No. 264,643 on 19 September 1882.

3. Should be "shell."

4. This remained figure 1 in the patent but an additional cross-section view showing the outer shell, armature, and the field magnet was included as figure 2.

5. That is, local currents, more commonly known as eddy currents, that cause heat to develop in a dynamo and interfere with its efficient operation.

6. This became figure 3 in the patent and figure 4 was a cross-section view.

7. This alternative winding was not included in the filed application or the patent.

8. This became figure 5 in the patent and figure 6 was a cross-section view.

9. Should be "A."

10. Line drawn to "fig 45" on following line. Charles Batchelor ordered the model to be modified to show this principle by 22 February (Cat. 1308:99 [Order No. 1], Batchelor [*TAEM* 90:716; *TAED* MBN003: 23]). Lemuel Serrell, concerned that the text and drawing for the shell winding and commutator connections were unclear, sent new ones to Edison 24 April. The filed application and the issued patent contained

The shell winding and commutator connections from Edison's U.S. Patent 264,643.

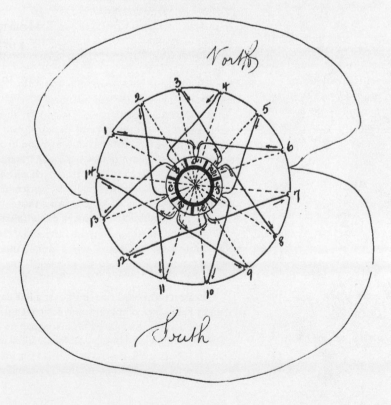

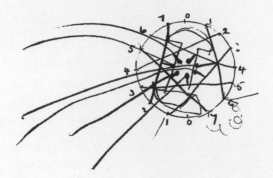

Serrell's modifications; the drawing became figure 7. An additional drawing (figure 8) was also added to show the circuit connections (Serrell to TAE, 24 Apr. 1879, DF [*TAEM* 51:473; *TAED* D7929ZBN]).

11. That is, they are wound in series so that their voltages are added together.

12. This winding was not included in the filed application nor was a modification that Batchelor ordered for the patent model that showed "8 coils with 7 commutators." This latter arrangement first appeared in connection with the tests of the large dynamo on 17 February. Cat. 1308:99 (Order No. 1), Batchelor (*TAEM* 90:716; *TAED* MBN003:23); N-78-12-15.1:211, Lab. (*TAEM* 29:1044; *TAED* N009:104).

13. This arrangement first appeared in a notebook entry of 20 February in connection with tests of the large dynamo. N-78-12-31:225, Lab. (*TAEM* 30:488; *TAED* N014:112).

14. Edison probably referred here to machines with drum armatures; in contrast to most previous designs, some machines with this feature generated fields and currents of sufficient strength for overheating to become a problem. King 1962, 371.

15. This claim, slightly modified, was the first claim in both the filed application and in the issued patent.

16. In the filed application this and the following claim were combined into the fourth claim and were dropped from the issued patent. The claim in the application made no mention of the use of an odd number of coils and commutator blocks and merely referred to figure 7, which showed such an arrangement.

17. This claim, slightly modified, became the second claim in both the filed application and in the issued patent.

18. This claim did not appear in either the final application or the issued patent.

Patent[1]

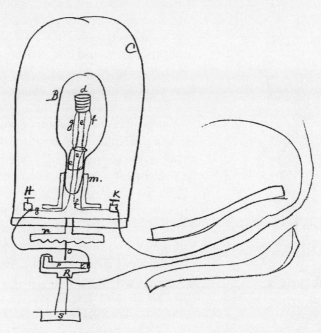

The object of this invention is devise an ~~practical~~ electric lamp which shall admit of being worked in great numbers on one electric circuit

The invention consists in ~~sea~~ placeding in a recepitable ~~of gl~~ made entirely of glass a bobbin of pyro insulated wire which is to be brought to incandescence by the passage of the current through it

The invention further consists in ~~caus~~ making a vacuum in such receptiable while the conductor is ~~h~~gradually heated.

The invention further consists in the use of salt[a] of lime, magnesia zircon, cerium erbia ~~or other~~ for coating the wire which salt is decomposible by heat. The salts which are easily[a] decomposed are the acetates of these metals[2]

The invention further consists of a vacuum receptable made entirely of glass and sealed by melting the same in combination with an incandescent continuous[b] conductor, pyro insulated

The invention further consists in winding ~~bobb~~ pyroinsulated wire upon a bobbin of ~~ana~~ compressed[b] infusible substance such as lime.

The invention further consists in placing the vacuum bulb within another glass receptable also closed from the air and employing the expansion of the air between the two receptables due to the heat of the incandescent bobbin, to produce[a] a movement which shall disconnect the lamp from the electric circuit when its temperature is too great.

B is the vacuum bulb   This bulb is opened at the smaller end and the burner d inserted and placed in connection with a mercury vacuum pump. The platina wires g & f passing through The opening are connected with a battery and variable resistance coil, while the vacuum is being made the heat of the bobbin d is ~~gradually~~ in the course of 1 hour brought ~~from~~ gradually from the temperature of the air upon to ~~brill~~ vivid incandescence   when the vacuum is considered practically perfect the ~~tube~~ open end of the tube is melted & sealed the platina wires ~~being at these~~ passing through the glass are also sealed— Thus I am enabled to obtain a nearly perfect vacuum[a] which is permanent and at the same time give the platinum wire a new and unknown property of great value in electric lighting which is that a platina wire which melts in the open air at a point where it emits a light equal to[c] four candles will when operated upon as described emit ~~25~~ a light equal to 25 candles without fusion. The reason why the melting point of the metal is thus raised is that in the act of making the vacuum with the metal under heat all the gases which are contained in its pores is withdrawn, and when the receptable is sealed cannot reenter when cold hence unequal ~~exp~~ & sudden expansions cannot take place and the wire is never cracked but if left uncovered becomes as bright as the most polished silver, an appearance which cannot be given it in any other way— on the other hand ~~pt~~ it is known that the metals of the platinum group have in a surprising degree[d] the peculiar power of absorbing within their pores many volumes of gas, and it is th~~is~~e sudden expansion of this gas upon a sudden accession of heat that disrupts the wire & produces cracks which extend nearly to its centre when the wire is brought to a moderate incandescence in the open air ~~where the gas~~   These cracks ~~pcause~~ set up[e] a great resistance to the passage of the current and at[a] these points become abnormally heated hence the platina wire easily melts, whereas no such cracks are noticed when the wire has been operated upon in the vacuum & all its gases pumped out— e is a cylinder of ~~l~~Lime with a small ~~bob~~ spool ~~on~~ its extremity on which the wire is coiled   about 30 feet of platinum iridium wire coated with ~~li~~Magnesia oxide is coiled upon the spool[3] The ~~size of~~[a] wire may be of any size but I prefer to use wire[b] .005 of an inch diameter which will give a resistance when incandescence of about 750 ohms— by the use of such high resistant lamps I am enabled to place a great number in multiple arc without bringing the total resistance of all the lamps ~~too lo~~ to such a low point as to require a large main conductor but on the contrary am enabled to use a main conductor of very moderate dimensions— ~~No loss in economy occurs in having~~ another important point is gained by the use of lamps of high resistance as the resistance of the wires leading from the main conductors may be of very moderate dimensions hence can be placed in the pipes already used for gas & at the same time effect a great saving in the cost of wire—

Still another point gained is that the high resistance of the lamps allows all to be placed in multiple arc which is the only method where the maximum economy is attainable as the lamp when connected to the circuit draws from the central station just sufficient current to maintain it at the proper temperature and if by accident of want of regularity in the main current the ~~temperature~~ strength of the current should increase abnormally the access of heat sets the thermal regulator in motion & disconnects the lamp entirely from the circuit, thus stopping all further consumption of energy until the temperature of the lamp is reduced to its normal conditions. I will state that these changes are not perceptable to the eye hence the lamp cannot consume ~~only su~~ any more energy than that required to ~~keep it~~ cause it to emit a certain light

No loss in economy occurs by using[a] so large a resistance because the loss of ~~h~~energy is proportionate to the radiating[a] surface exposed to the air & its temperature and is independent of the resistance of the wire forming[f] such surface— m is a lime cup into which the small end of the vacuum burner is held the platina wires pass under it to the binding posts H. K. n is the thermal regulator operated by the expansion of the air ~~and has already be~~   when the temperature of the air between the bulbs becomes too great the diaphragm bulges outwarded & the point .o. seperates the spring p from R & disconnects the lamp from the circuit where it remains until the temperature is reduced to the normal condition. The spark upon the point is very small as I employ ~~contain~~ constant field magnets at the central station hence the powerful sparks due to the secondary circuit set up by the weaking of the powerful field magnet is avoided

I will mention that the second globe C might be made entirely of glass and the aneroid diaphragm provided with a platina tube be sealed in the glass or the aneroid itself[a] be made of glass ~~of a th~~   I also state that the globe C may be be opened an a light bulb placed between the two other ~~which this bulb~~ the tube from this bulb passing down through the [-]ᵍ base of the instrument has a sealed expanding chamber or coil attached to it—

Claim—[4]

1st   In combination with a vacuum chamber made entirely of glass of ~~an~~ continuous incandescent metallic[b] conductor ~~se~~and sealed as set forth.

2nd   The method herein described of freeing[a] metallic conductors of their gases in a vacuum & afterwards sealing the same

3rd   ~~A coil~~ In combination with a ~~sele~~ sealed vacuum of a bobbin of pyroinsulated wire wound upon an infusible substance.

4   The combination of a vacuum bulb containing a continuous conductor within a second closed chamber for the purpose set forth

5th— The use of a ~~decomposible~~ salt of an~~y~~ metal whose oxide is not readily ~~infus~~ fusible & which salt is decomposed by heat.

6th— The combination of the sealed conductor .d bulb .B. chamber .c. regulator n. o p. r all arranged & operated substantially as set forth

T A Edison

ADfS, NjWOE, Lab., N-79-02-24.2:51 (*TAEM* 31:1018; *TAED* No31:26). Document multiply dated. ᵃObscured overwritten text. ᵇInterlined above. ᶜ"equal to" interlined above. ᵈ"in a surprising degree" interlined above. ᵉ"set up" interlined above. ᶠ"the wire forming" interlined above. ᵍCanceled.

1. Edison wrote this draft the same day that the Patent Office model of Edison's "New Electric lamp" was completed (see Doc. 1688 n. 1) and William Carman then copied it onto pages 65–69 of the missing Experimental Researches Volume 3. The beginning of the application was altered before Edison executed it on 12 April 1879 (see note 2) and additional changes were required by the Patent Office before it issued on 4 May 1880 as U.S. Pat. 227,229 (see note 4). The provisional specification for Edison's British Patent 2402 (1879), which was filed on 17 June, also included the basic elements of this patent, with the final specification having similar claims (Cat. 1321, Batchelor, *TAEM* 92:118; *TAED* MBP017).

2. In the filed U.S. application Edison replaced all of the paragraphs to this point with the following two introductory paragraphs:

> When Platina and other metals that fuse at a high temperature
> are exposed to high heat and then cooled in the atmosphere they are

injured so that they are not well adapted to use in electric lights for a long period of time.

I enclose the conductor that forms the electric candle in a transparent case, and heat the same gradually to expell any gases from the material of the candle. I form a vacuum in the transparent case, and then seal the same hermetically so that all injurious atmospheric influences are avoided. [Pat. App. 227,229]

3. Edison later instructed Lemuel Serrell regarding the "patent & model sent you in lamp with globe within a globe" to add the sentence, "Chemically pure iron cobalt or nickel drawn in wires may take the place of platinum where the light is not required to be intense, as these metals where perfectly pure can be freed from gas as described & be brought to vivid incandescence." However, this sentence was not included in Edison's 12 April application. TAE to Serrell, 10 Mar. 1879, Lbk. 4:207 (*TAEM* 80:57; *TAED* LB004207).

4. The Patent Office examiner objected to the first three claims, which Edison then amended. After further objections to the revised second claim, Edison provided another more acceptable claim. At some point the fifth claim was erased and the sixth renumbered accordingly. He also added three paragraphs at the end of the specification to further distinguish between the invention embodied in this patent and the prior art, including a statement that he did not claim a lamp formed of pyroinsulated metal, which was the subject of a previous application (Case 166) that he subsequently abandoned (see *TAEB* 4:755 n. 1). See application and correspondence in Pat. App. 227,229.

–1696–

Equipment Specification: Electric Lighting

[Menlo Park,] March 1st 1879

32[1] Alter Small Magneto machine[2] as shewn in drawing marked No 8[3] by closing spaces all round to make an outside shell   Book 24 page 31.[4a]

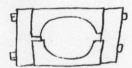

<u>33</u>.[b] Alter Small Magneto machine so as to rotate the core and hold the field mag. still (drawing no.   )   Book 24 page 33.[5a]

<u>34</u>[b] Alter Small Magneto Machine by making an outside shell of iron, winding wire round it, keeping it and the armature still and rotating the shell.[6a]

35.[b] After exp. 34 alter to rotate inner shell and armature cylinder together. Book 24 p 33.[7a]

36   Make an outside shell of rings of permanents magnets ¼ inch thick[8]   This wants to work so as to rotate the shell Also rotate the armature cylinder along with shell[9]

37. Make an inside shell of permanent magnets forming a ring and make so that you can either rotate the shell or the armature[10]

Batchelor[11]

ADS, NjWOE, Batchelor, Cat. 1308:111 (Order Nos. 32–37) (*TAEM* 90:722; *TAED* MBN003:29). Written by Charles Batchelor. Document multiply initialed and dated. [a]Order later overwritten with a double-lined "X." [b]Multiply underlined.

1. Numbers 32–37 are order numbers. See Doc. 1687 n. 1.

2. Edison began planning in December 1878 to build small generators of various designs (*TAEB* 815 n. 1 and Doc. 1646). It is unclear how many such machines had been constructed before the shop began making this machine. Francis Jehl later testified that when he entered the laboratory in early March 1879 "there were many small experimental machines of almost an endless variety." Edison also testified that "there were a great many small machines." However, there is no other evidence of such machines. Testimony of Edison and Jehl, pp. 40, 72, *Keith v. Edison v. Brush*, Lit. (*TAEM* 46:134, 150; *TAED* QD002:21, QD002:37).

The machine detailed in this document was designed by Edison and Batchelor between 10 and 12 February. Batchelor prepared a measured drawing by 18 February and entered an order for the machine into the shop order book on 22 February. The original design specified a shell wound with wire in which an "armature"—an electromagnet that served as a field magnet for the generator—rotated; this was similar to

Charles Batchelor's measured drawing of the modified small magneto.

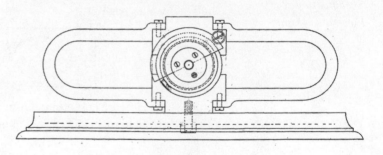

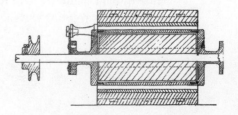

the design of the larger generator with which Edison was then experimenting (see Docs. 1653, 1682, and 1694). Edison appears to have designed this small generator to test his new armature winding and commutator arrangement (see headnote p. 76), but after making tests of these new designs on the larger generator he instead decided to use the small machine to experiment with different arrangements of the armature (field magnet) in the shell. N-79-02-10:7–31, Vol. 16:437, both Lab. (*TAEM* 32:1145–57, 4:873; *TAED* N046:5–17, NV16:394); Cat. 1308:103 (Order No. 8), Batchelor (*TAEM* 90:718; *TAED* MBN003:25).

3. The "drawing marked No 8" is the measured drawing made by Batchelor, which was dated 18 February by John Kruesi and which Batchelor dated 22 February, the same day he entered the order for it in the shop order book. Vol. 16:437, Lab. (*TAEM* 4:873; *TAED* 394); Cat. 1308:103 (Order No. 8), Batchelor (*TAEM* 90:718; *TAED* MBN003:25).

4. The reference is to the start of four undated pages of rough drawings and notes by Edison that sketch the program of experiments and tests entered in these orders. The first page says only "No 1" and shows an outside view of the case for the large generator which had been built in January and February, followed by a version of the drawing included here in order 32. N-79-01-14:31–37, Lab. (*TAEM* 31:276–9; *TAED* N024:17–20).

5. The reference is to a note without any drawing that reads "No. 2. hold the field magnet still & rotate the wire." N-79-01-14:33, Lab. (*TAEM* 31:277; *TAED* N024:18).

6. The original version of the following drawing is a faint sketch headed "No. 3." with the center circles labeled "shell" and a note saying "shell rotate." N-79-01-14:33, Lab. (*TAEM* 31:277; *TAED* N024:18).

7. The referenced note reads: "No. 4. rotate inner cylinder & wire together." N-79-01-14:33, Lab. (*TAEM* 31:277; *TAED* N024:18).

8. The absence of "X" marks and of John Kruesi's initials from items 36 and 37 suggest that these permanent magnet shells were not made (see note 11).

9. The original version of these instructions, along with a rough, faint sketch, are at N-79-01-14:35, Lab. (*TAEM* 31:278; *TAED* N024:19).

10. An addendum to the original experimental plans reads: "See if shell can be made of paper maché" (N-79-01-14:37, Lab. [*TAEM* 31:279; *TAED* N024:20]). Batchelor's next entry, order 38 on 2 March, called for "a hand power with a dynamometer on it for these small machines" and referred to Book 24 page 38 (N-79-01-14:35, Lab. (*TAEM* 31:278; *TAED* N024:19); however, this page shows only what appears to be a wooden frame structure. Kruesi noted that order 38 was finished on 15 April (Cat. 1308:113 [Order No. 38], Batchelor [*TAEM* 90:723; *TAED* MBN003:30]).

11. Batchelor dated and either initialed or signed each item except 34. On 6 March John Kruesi initialed and dated orders 32–33; he initialed but did not date orders 34–35.

To George Walker[1]

[Menlo Park,] Mch 3d [187]9

Dear Sir.

Your favor of 28th ult was duly rec'd   I have no objection to allowing the parties mentioned the freedom of the laboratory. I cant say however that I will be able to see them.[2]

The new receiver is getting on finely. I had the whole force of the shop, 15 men,[3] working on it two days last week and I think I have now the best form of instrument and the best telephone extant. How about the Contract? Mr Edson said he would have it ready for me but I have not seen it as yet.[4] Very Truly

T. A. Edison   G[riffin]

L (letterpress copy), NjWOE, Lbk. 4:197 (*TAEM* 80:52; *TAED* LBoo4197). Written by Stockton Griffin; circled "C" written at top of page.

1. George D. Walker was vice president of the Gold and Stock Telegraph Co. and president of the American Speaking Telephone Co. He had a long association with Western Union, including terms as a director and vice president; he had previously been a lawyer and politician in Massachusetts. Reid 1879; 533, 539, 626, 632.

2. Walker hoped to arrange a laboratory visit and meeting with Edison for "Mrs. Quincy Shaw a da[ughter]. of Prof. Agassiz & Mr. & Mrs. John Bancroft—(a son of Mr. Geo. Bancroft the historian) all of Boston" on their return from Philadelphia on 4 March. Walker described them as "people of scientific tastes & not merely curious visitors" (Walker to TAE, 28 Feb. 1879, DF [*TAEM* 49:139; *TAED* D7903ZBK]). Pauline Shaw, youngest child of naturalist Jean Louis Agassiz and the wife of copper magnate Quincy Adams Shaw, was a philanthropist, educational reformer, and champion of public kindergartens (*DAB*, s.v. "Shaw, Pauline Agassiz").

3. There are no extant time sheets for this week or for most of 1879. A set of time sheets from the second week of February (Time Sheets), which may or may not be complete shows twenty-five employees (not including Charles Batchelor). Most of these are for the same men mentioned in *TAEB* 4:749 n. 8 and Doc. 1655 n. 2. Ten of these men are known to have worked in the machine shop. Among the newer employees were Matthew Hankins, who was probably a day laborer, and O. Weisman and Arthur Williams, about whom nothing else is known; they may have worked in the shop as may other employees whose time sheets are missing. In addition, other members of the staff may have been helping out in the shop as well.

4. Negotiations with Western Union for the assignment to Gold and Stock of Edison's U.S. rights to the electromotograph receiver had begun in the latter part of 1878 (see Docs. 1585, 1637, 1640, and 1643). Several weeks after writing this letter Edison made notes for a reply to Samuel White's 27 March inquiry concerning the contract. He complained that the proposed contract offered no incentive to push ahead with telephone research and "tie the whole system by pat[ent]s without any delay . . . as it stands whether I del[ive]r one or 20 imp[rovemen]ts

in any year me & my ass[istan]ts" would receive the same compensation. In its final form, the contract provided for a one dollar annual royalty on each receiver, not to exceed $3,000 per year. It is dated 12 May 1879 but as Tracy Edson wrote on 13 May that he had just reviewed a draft, it was evidently completed sometime thereafter. TAE marginalia on White to TAE, 27 Mar. 1879; Edson to TAE, 13 May 1879; both DF (*TAEM* 52: 263, 49; *TAED* D7940K, D7937ZAM); Agreement with Western Union, Miller (*TAEM* 86:11; *TAED* HM790061).

–1698–

To George Gouraud

[Menlo Park,] March 5, 79

Dear Sir

Your favor of the 22d ult was duly rec'd.[1] Regarding "the Telephone in countries where there are no patents" I do not feel at liberty to authorize you to go ahead and occupy the territory but I do not see why you cannot do so without my sanction— After you are established in such countries I will give you my support. Very Truly

T. A. Edison   G[riffin]

My object you see is to avoid complications with Puskas.

L (letterpress copy), Lbk:4:203 (*TAEM* 80:54; *TAED* LB004203). Written by Stockton Griffin; circled "C" written at top of page.

1. Gouraud had inquired about the status of Edison's telephone in countries where no patents could be obtained, particularly Holland: "Please advise me immediately if in such cases I am at liberty to do anything I please. I think I can get money for Holland by simply giving all your rights whatever they are, which in the case of Holland per se are nothing as I understand it—hence the transaction would be based upon your authority to use your name and your telephone." He added that he was communicating with "first class people, banker friends of mine who want to make a telephone exchange." Gouraud to TAE, 22 Feb. 1879, DF (*TAEM* 52:490; *TAED* D7941K).

–1699–

From Theodore Puskas

London. W. 5th of March 1879[a]

My dear Edison,

There are many things that concern our mutual interests, which have passed here in Europe, that I have not liked to put on paper for reasons which will be explained to you by Mr. Bailey to whom I have given a memorandum covering the points referred to, with the request to read it over and explain it to you.—[1] I desire to say to you that Mr. Bailey and I have worked together in entire harmony and that he has my entire confidence.—

Will you do me the favor to talk with Mr. Bailey fully about the probable amount of money that you will want me to put up for the E. Light and the time or times at which you will want it—,[2] and generally as to our matters, as we can come to a clearer understanding than by any amount of correspondence. Yours very truly

Theo. Puskas

ALS, NjWOE, DF (*TAEM* 49:143; *TAED* D7903ZBN). Letterhead of Langham Hotel. [a]"London. W." preprinted.

1. Joshua Bailey began an extended stay in New York sometime before 25 March, when Edison told George Bliss of his arrival. Puskas's memorandum has not been found but probably pertains to telephone matters about which Edison had been concerned for some time. On 9 February, Charles Batchelor had written to James Adams in London asking him to "Tell us all you know about the French Co, as Edison wants to know who the company is, what they propose to do, what they are doing, whether they have any telephones going or not, we have to depend on you for all information as he very seldom gets any information or letters from either London or Paris. I think it would be a good idea to answer the questions just as I have put them." Two days before Puskas wrote this letter, Edison had written to him asking about the arrangements made with the French telephone company and the status of the company stock owed him. He also indicated that money should be sent to Lemuel Serrell to pay for French patents on two new telephone improvements. Puskas replied on 17 March that although he and Bailey were bound by contract to pay for the patents, the company was "willing to bear the expense for this on any future improvement, but would greatly prefer having the Patents taken out here [in Paris] by a french agent" to save costs. He wrote that if Edison would "forward the necessary drawings and specifications in English the translation will be made by own officials and draughtsmen here and the Patent taken out at once." Marginalia on this letter indicates that Edison replied, "Mr. Bailey is arranging so that all the future Tel pats will be taken out as you desire." TAE to Bliss, 25 Mar. 1879, Lbk. 4:234 (*TAEM* 80:68; *TAED* LB004234A); Batchelor to Adams, 9 Feb. 1879, Cat. 1330:46, Batchelor (*TAEM* 90:260; *TAED* MBLB2046); TAE to Puskas, 3 Mar. 1879, Lbk. 4:200 (*TAEM* 80:53; *TAED* LB004200); Puskas, to TAE, 17 Mar. 1879, DF (*TAEM* 52:255; *TAED* D7940E).

2. Two days before Puskas wrote this letter, Edison complained to Grosvenor Lowrey that he had "not heard anything from Puskas and have concluded to wait ten days longer, if by that time he fails to let me hear from him I want to see Drexel Morgan & Co and have a new set of patents secured." Serrell informed Edison on 1 April that Bailey had met with him to discuss expenses for the Continent, and he enclosed an itemized estimate of fees for a second set of European patents totaling $2,185. He stated that Bailey planned to stop at Menlo Park to ascertain Edison's wishes regarding individual countries but had instructed him to "go ahead with France Belgium and Germany which will be done.— so as to be ready for end of next week." TAE to Lowrey, 3 Mar. 1879,

Lbk. 4:196 (*TAEM* 80:51; *TAED* LB004196A); Serrell to TAE, 1 Apr. 1879, DF (*TAEM* 51:449; *TAED* D7929ZAX).

<table>
<tr><td>–1700–

Notebook Entry:
Telephony</td><td>[Menlo Park,] Mch 9 1879
Try stiffer springs from Diaphragm to chalk.[1]
Dry chalks out.
Try Motograph receivers in an ~~w~~secondary coil.
shunt Receivers with plain resistance
shunt with a magnet.
wet with caustic soda water

                           TAE</td></tr>
</table>

X, NjWOE, Lab., N-79-03-10.1:3 (*TAEM* 31:4; *TAED* N022:2).

1. To facilitate telephone experiments, Charles Batchelor had requested the machine shop on 3 March to build a new receiver instrument like the ones taken to England by Charles Edison, this one to "be made to change diaphragms   change for all kinds spring. Change chalks easy." It was completed on 13 March (Cat. 1308:112–13 [Order No. 40], Batchelor [*TAEM* 90:723; *TAED* MBN003:30]). Also on 13 March, Batchelor placed an order to make springs tipped with 39 different materials (Cat. 1308:121 [Order No. 61], Batchelor [*TAEM* 90: 727; *TAED* MBN003:34]). Having settled on a standard chalk surface (see Doc. 1693 n. 10), Edison probably wanted to see if changing the material in contact with it made any difference to the electromotograph action. His previous experiments suggested that the arrangement of the spring and the action between its contact and the surface material were important variables affecting the operation of the electromotograph (see Docs. 424–425, 461–462, 525, 607, 613, 775, 777, 780–781, 783, 787, 792, 1362).

<table>
<tr><td>–1701–

Agreement with
Henry Davies</td><td>[Menlo Park,?] March 13, 1879[a]
    This agreement entered into this thirteenth day of March 1879, by and between Thomas A Edison of Menlo Park Middlesex Co state of New Jersey party of the first part & ~~J~~ [J-][b] Henry J[c] Davis[1] of the City County & state of New York party of the second part   witnesseth=
    Whereas the said Edison has discovered a compound which resembles gold & which may be moulded in ingots, hammered rolled in thin plates & drawn in wires, and which is very light, and whereas the said ~~metal~~ compound[d] if it can be made cheap and worked conveniently will be of great service in the arts and whereas the said Davis being connected with the Ansonia Clock[e] Co[2] who use an large quantity of brass & other metal</td></tr>
</table>

which in many cases this compound would or could replace, and further that the said Davis has influence with the firm of Phelps Dodge & Co.[3] and the Ansonia Brass Co who could[e] manufacture the said compound & whereas the said Davis is anxious to become interested with the said Edison in the introduction of the said compound in the market for his own use & for others, and whereas the said Edison owing to a defect in the patent laws relating to chemical compounds does not desire to protect the same by Letters patent but prefers that it be kept a trade secret[4] that be it agreed that the said Edison & Davis is are[f] to be the sole proprietors of the said trade secret of manufacturing & working the said compound for a period of five years from the date hereof. The said Edison to continue to experiment at his own expense to perfect processes for working & cheapening the said compound and the said Davis to render him any requisite assistance; that when the process its complete & satisfactory the said Davis is to have it[d] manufactured in various firms for sale on the market, the profits from the same to be divided equally   the said profits shall not be less than 2 cents per lb pound=

The said Davis agrees not to divulge the processes of manufacture, to any other person without the consent in writing of the said Edison;

This whole contract is to be construed as a joint partnership in a trade secret of making the compound herein mentioned.

2 parts of Sulphur
1 part of Sal ammoniac
1 part of tin.

Thos A Edison   Henry J Davies
Witness   S. L. Griffin

ADS, NjWOE, Miller (*TAEM* 86:4; *TAED* HM790057). [a]Date taken from text, form altered. [b]Canceled. [c]"Henry J." interlined above by Stockton Griffin. [d]Interlined above. [e]Obscured overwritten text. [f]Interlined above by Stockton Griffin.

1. Edison meant Henry J. Davies, general superintendent of the Ansonia Clock Co., to whom he had previously licensed the manufacture of phonographic clocks. See *TAEB* 4:56 n. 10; Wilson 1879, 336.

2. The Ansonia Clock Co. had been a branch of the Ansonia Brass Co. until it was incorporated as a separate company in 1877. The firm had offices in Manhattan and factories in Brooklyn and Ansonia, Conn. See *TAEB* 4:58 n. 1.

3. Phelps, Dodge & Co. began in 1833 as a metal business partnership between William Earl Dodge, his father-in-law Anson Greene Phelps, and Daniel James. The Ansonia Brass & Copper Co. and the Ansonia

Clock Co., both of which Dodge served as president, were subsidiaries of this firm (see *TAEB* 4:58 n. 3).

4. The substance described in this document is similar in composition to a bisulphide of tin known as mosaic gold. Formed in delicate scales, mosaic gold was used as a gold substitute and was identified in two chemical reference works used at Menlo Park (Pepper [1869?], 114; Bloxam 1869, 29). Its properties may have come to Edison's attention as laboratory assistant John Lawson, an aspiring chemist whom Edison had hired in January, began alloying scores of metals and metallic sulfides in early March (Lawson to TAE, 6 and 13 Jan. 1879, DF [*TAEM* 49:865, 869; *TAED* D7913B, D7913E]). There is no record of tin sulphide until 13 March, when Batchelor included it in a list of materials to be tried in telephone receiver experiments (Cat. 1308:121 [Order No. 61], Batchelor [*TAEM* 90:727; *TAED* MBN003:34]). Edison varied the proportions of tin, sulphur and sal ammoniac (ammonium chloride) from standard formulations of mosaic gold but it is highly unlikely that this would constitute a patentable process. Edison and Davies apparently did not pursue the commercial development of this compound. Gmelin 1848–66, 5:78–80; N-79-03-31:3–105, Lab. (*TAEM* 32:608–37; *TAED* N040: 2–31).

–1702–

Equipment Specification: Electric Lighting

[Menlo Park,] March 13th 1879

58[1] Alter small magneto machine[2] by taking the shell[3] and putting it in between two powerful poles. So:—

Cores 24 inch long   Cores 4 in thick   Put the machine on the wooden base used by big induction magnets before & fasten down by Z[4]   see book 46. 105 page[5] ⟨Completed April 12th 79   J Kruesi⟩[6a]

Batchelor

ADS, NjWOE, Batchelor, Cat. 1308:119 (Order No. 58) (*TAEM* 90:726; *TAED* MBN003:33). Written by Charles Batchelor; double X later written across entry. [a]Marginalia written by John Kruesi.

1. This is the order number; see Doc. 1687 n. 1.

2. See Doc. 1696.

3. The original shell had 25 coils but used No. 26 silk-covered wire (N-79-02-10:7, Lab. [*TAEM* 32:1145; *TAED* N046:5]). It was, however,

apparently altered before being placed between the bipolar field magnets. In the same notebook entry where Charles Batchelor specifies the dimensions of the field magnets he indicates that the shell was to be 6 inches in diameter with 25 coils of No. 18 silk-covered wire (N-79-01-21:119, Lab. [*TAEM* 30:645; *TAED* No16:61]).

4. The following order (No. 59), also dated 13 March, was to unwind and cut the cores of the two large induction magnets to the same size as that of their magnetizing magnet. There are also other orders from around this time for alternative magnet designs. Batchelor entered another order (No. 60) on 13 March that indicated a bar of iron was scheduled to arrive the following day and should be cut in half to be made into a magnet for experimental purposes while Order No. 55, dated 10 March, was to make a permanent steel magnet with two poles. Cat. 1308: 119–21 (Order Nos. 55, 59–60), Batchelor (*TAEM* 90:726–27; *TAED* MBN003:33–34).

5. A similar drawing, later dated 25 March, appears on that page and shows the dimensions of the field magnets as 2½ feet high and 4 inches thick. In another notebook Batchelor indicated that the two field magnets were each to be 2 feet high and 4 inches in diameter with a total resistance of 1 ohm. After calculating the resistance of various wire sizes that might be used for winding them he decided to order No. 13 cotton covered wire. The measured drawing of this machine is dated 13 March. N-79-02-10:105, N-79-01-21:121, Oversize Machine Shop Drawings (1879–1880), all Lab. (*TAEM* 32:1193, 30:646, 45:53; *TAED* No46:53, No16:62, NS7986CAI).

6. John Kreusi, Edison's principal machinist. *TAEB* 2:633 n. 6.

-1703-

From George Gouraud

London E.C 14th March 1879

Dear Edison,

I need not tell you how delighted I was to receive your letter of the 25th of February[1] by the hand of your nephew and at the same time what I think the "Herald" of the ensuing day justly described "the greatest invention of the age.[2] It was put up yesterday and today my office has been thronged with Representatives of the Press. What they thought about it you will be able to see by the Papers which will appear tomorrow and which I shall forward with this.[3] Tyndall himself was the first to see it coming immediately on the receipt of my telegram and he was really delighted. I lunch with him on Monday when we arrange the details of the lecture which takes place on Thursday next.[4] I congratulate you with all my heart on this beautiful achievement and realization of your promises. After this people will doubt you less concerning the Electric Light.[5] Do not loose a moment in forwarding what you call "the first class machine." Let them be models for the commercial instruments if possible. I shall now proceed with my negotiations and shall hope to give you early reports of satisfactory results. Yours very truly

Geo. E Gouraud

P.S. There was no time for many of the papers to get the article in todays issue notably the "Times" whose reporter has just been here with proof, for correction of a good article twice the length of any enclosed.[6]

LS, NjWOE, DF (*TAEM* 52:499; *TAED* D7941Q). Written by Samuel Insull.

1. Doc. 1691.

2. The *New York Herald* used this phrase to describe the instruments taken to England by Charles Edison. "The Electro-Motograph," *New York Herald,* 27 Feb. 1879, Cat. 1241, item 1138, Batchelor (*TAEM* 94: 461; *TAED* MBSB21138X).

3. In a letter dated 22 March, Gouraud enclosed four clippings from 15 March London newspapers. Three of these were subsequently reprinted, presumably by Gouraud for publicity purposes, as were the 27 February *New York Herald* article (see note 1), the 17 March London *Times* article (see note 7), and articles published in several other British newspapers and journals between 15 March and 5 April. Gouraud to TAE, 22 March 1879, DF (*TAEM* 52:501; *TAED* D7941R); Cat. 1241, items 1151–61, Batchelor (*TAEM* 94:466–73; *TAED* MBSB21151X, MBSB21152X, MBSB21153X, MBSB21155X, MBSB21156X, MBSB21157b, MBSB21158X, MBSB21160X, MBSB21161X).

4. Tyndall's lecture, originally scheduled for 17 February, was set for 27 March (see Docs. 1662 n. 4 and 1715).

5. British technical journals had published blunt editorial criticisms of Edison's electric light research in February and early March. Contrary to Gouraud's expectation, another critical article appeared soon after. On 22 March the London *Times* published a dispatch from New York, dated 6 March, which proclaimed that "Mr. Edison has failed in his experiments." Asserting that the U.S. Patent Office had rejected fourteen of his sixteen applications, leading to the "discouragement which reigns at Menlo-park," the author claimed not only that Edison had been unable to sustain his burners at incandescent heat without melting, but that "the public have never seen so much as one of his lights yet. A favoured few who have been admitted to his laboratory at Menlo-park have beheld it—a single lamp, enclosed in a glass globe, beautiful as the light of the morning star. But he has refused to let any one inspect it closely, and has never allowed the exhibition of it privately to last long. He has never been able to depend upon its durability. His apparatus is as far from perfection as it ever was, and, in fact, well-informed electricians in New York do not now believe that Mr. Edison is even on the right line of experiment." "The Electric Light in the United States," *Times* (London), 22 Mar. 1879 and "The Edison Electric Light," *English Mechanic,* 21 Feb. 1879, Cat. 1241, items 1148 and 1137, Batchelor (*TAEM* 94:464, 459; *TAED* MBSB21148X, MBSB21137X); "The Edison Light," *Journal of Gas Lighting, Water Supply, & Sanitary Improvement,* 18 Feb. 1879; "Edison's Electric Light," ibid., 25 Feb. 1879, Cat.1012:5–6, Scraps.

(*TAEM* 23:521; *TAED* SM012006a, SM012005a); "The Edison Electric Light," *Engineer* 47 (1879): 113; "The Edison Electric Light and Edison's Patent for Electric Lighting," *Electrician* 2 (1879): 164, 193.

After the *Times* article reached the U.S., Edison responded in at least two published interviews. On 11 April he told the *New York Daily Graphic* that he had "never before read a statement containing so many lies" and declared that such unfavorable reports had the "effect to thin out the crowd of visitors who usually come to see me, which to me is a blessing in disguise. I have prayed for an earthquake or something of the sort to keep some of them away, and if I had put in a standing advertisement in THE GRAPHIC for this purpose it could not have answered the purpose better than these flying rumors. I rather like it, and it wouldn't bother me a particle if they kept up the cry—at least until I am ready to show what I have accomplished." In response to the claims of failure he noted that "my electric lamps have been on exhibition, the electric current is all right, and it has burned continuously for two weeks. There are twenty of them burning now in my laboratory" (see Doc. 1705 n. 1). Far from being discouraged, he said, his employees were "as happy as clams" and he dismissed reports of other electricians whose opinions "do not bother me in the least. I reiterate that I have accomplished what I set out to accomplish and believe that it is theoretically possible to advance a great deal farther." In an interview published the next day by the *New York Sun*, Edison offered written rebuttals to sixteen statements made in the *Times*. "Edison Still Hard at Work," *New York Daily Graphic*, 11 Apr. 1879; "Edison's Electric Light," *New York Sun*, 12 Apr. 1879, Cat. 1241, items 1163–64, Batchelor (*TAEM* 94:474–75; *TAED* MBSB21163, MBSB21164).

6. After providing a detailed description of the instruments, the *Times* reporter described Gouraud's demonstration:

the instruments were placed half a mile apart by wire, one being at 6, Lombard Street [Gouraud's office], and the other at 1, Princess Street, the wire passing by way of the Exchange Telegraph Company's offices in Cornhill, which gave a circuit of about half a mile. Conversations were maintained between the two extreme points, songs were sung and tunes were whistled, and in all cases everything was most distinctly heard some 15 feet from the instrument, and could have been heard at a greater distance had the size of the room permitted it. The trial was a thorough success, notwithstanding that the instruments were the first which had been made, and were by no means highly finished, but, on the contrary, rather roughly turned out. In fact, they were 'thrown together' by Mr. Edison (as he puts it) at the urgent request of Colonel Gouraud, in order that they might be used in Dr. Tyndall's forthcoming lecture on the subject of modern acoustics. They were made in five days, and are purely experimental. They, however, satisfactorily demonstrate the simple nature of the invention, and give fair promise that telephonic communication will shortly become as general in the commercial and manufacturing centres of England, as it already is in those of the United States of America. ["The Edison Telephone," *Times* (London) 17 Mar. 1879, Cat. 1241, item 1151, Batchelor (*TAEM* 1241:466; *TAED* MBSB21151X)]

W[ashington] DC 3/14 1879.[a]

My Dr E

Yours of 3-13 to hand—[1] Bliss was here the last few days of session & expressed himself as highly gratified with the work I had done in getting the Electric Pen work in the law as "printed matter" & said it had doubled the value of the Pen & I should have a $1000. & he would see to it at once—[2]

I offered him stock in the Phono' Co on the ground floor when we got up the Co & he declined it. This I thought ought to release you—[3] But I am now willing to let him have the $1000. he is to send me & call it square if you think that is about right— It is certainly acting very liberal towards him—[4]

You have omitted to charge up to this account the expense incurred on "megaphone,"[5] & "aerophone,"[6] which I agreed to pay out of this fund—

I do not want to draw on it at present— Will see you about it before long—

What do you think of my proposition on Bliss? Yours

U H Painter[b]

⟨GH.B. You should not be so enthusiastic   TAE⟩
⟨Please return this letter. E   G[riffin]⟩[c]

ALS, NjWOE, DF (*TAEM* 49:162; *TAED* D7903ZCA). Letterhead of Edison Speaking Phonograph Co. [a]"1879." preprinted. [b]"Over" follows as page turn, written by Edison. [c]Second postscript written by Stockton Griffin.

1. Not found.

2. See Doc. 1657. When George Bliss wrote to Edison on 8 March about passage of this legislation, he praised Painter for having "displayed rare judgement in his manipulation of the matter, and beyond all question the sections affecting us would have been a dozen times killed had it not been for his acquaintance and personal work. I think him entitled to an immense amount of credit. And when one finds such a friend as he is it will do to tie to him for life. I hope you wont forget the services he has rendered." DF (*TAEM* 50:417; *TAED* D7923ZAI); see also note 4.

3. Edison wrote Painter on 1 March on the back of an Edison Speaking Phonograph Co. royalty summary, "You remember Bliss was to have a small share of this for freezing him out= Do you need some money" (Lbk. 4:215 [*TAEM* 80:63; *TAED* LB004216]). Although excluded from Edison's U.S. phonograph contract (Doc. 1190), Bliss expected to receive some interest in the company, to which Painter had agreed in principle (see Docs. 1180–1181, 1183, 1192). Painter later protested that Bliss had been given an opportunity to buy stock in the company but did not have "any money to put in it— I did not believe he knew where to get a $. & that he merely wanted to trade off his influence with you— I put up my own, & made others put up theirs while he was trifling & beat-

ing around trying to get something for nothing!" (Painter to TAE, 1 Apr. 1879, DF [*TAEM* 49:198; *TAED* D7903ZCX]); see also note 4.

4. After receiving this letter, which Edison sent to him, Bliss gave his own account of his conversation with Painter regarding the long delayed settlement of his promised interest in the phonograph, which he suggested that they resolve by "consolidation of the Pen and Phonograph business." In this regard he had agreed "to try and squeeze $1,000.00" for Painter, who promised to "share any expense," from the various interests in mechanical duplicating processes that had benefitted from the passage of the Post Office bill. Bliss also told Edison that he was "perfectly willing you should make any reasonable reduction from my Phonograph interest on this account. I wont quarrel with a man who voluntarily befriends me." Painter responded on 1 April with his account of the matter, including a letter from Edward Johnson concerning the original offer of stock. Three days later, Bliss apologized to Edison for any embarrassment he had caused him in regard to the phonograph and stated that "if Painter will consider what he did on the classification bill an offset to the interest promised me, I shall be perfectly satisfied; only let me know what its equivalent value in money would be and it is all I ask." The issue dragged on to mid-April, when an exasperated Painter complained that he did not know "how to fix this to get it to get rid of Bliss! & at the same time not let him take anything out of you— How shall I?" Stockton Griffin's notation on this letter indicates that Edison suggested "250 or $300 be abt right for Bliss." Painter agreed to credit Bliss $250 and "give him a receipt for the $1000 on his paying the $750. & call it all square! Will that do?" Edison approved these terms on 25 April. Bliss to TAE, 28 Mar. and 4 Apr. 1879; Painter to TAE, 27 Mar., 1, 10, 14, and 18 Apr. 1879; all DF (*TAEM* 49:191, 203, 189, 198, 206, 208, 210; *TAED* D7903ZCS, D7903ZDA, D7903ZCR, D7903ZCX, D7903ZDD, D7903ZDF, D7903ZDH); TAE to Painter, 25 Apr. 1879, Unbound Documents (1879–1882), UHP.

5. Regarding the megaphone, an acoustic device for receiving sound from great distances, see Doc. 281 and *TAEB* 4:407 n. 10.

6. Regarding the aerophone, a device for greatly amplifying sound by steam or compressed air, see *TAEB* 4:89 n. 1 and 171 n. 2.

–1705–

*Notebook Entry:
Electric Lighting*

[Menlo Park,] March 18th 1879

Electric Light[1]

This evening we lit our shop up with 12 spirals[2] giving each about 22 candles[3] from a gramme machine running 960 revolutions [per minute] and taking about 2 horse power[4]   The spirals were Plat. Iridium 20 p[er].c[ent]. and were ⅜ long and .080 [inch] thick having a surface of ³⁄₁₆ square inch.[5] Resistance of machine 1 ohm;[6] resistance of each spiral .33 ohms.

They were put in Multiple arc 4ᵃ four in each arc thus:—[7]

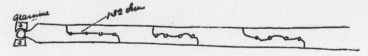

Total outside resistance .44 ohm
"   "   & inside mach. 1.44 ohms
We find also that we can put resistance in equal to 12 elamps more.[8]

Chas Batchelor

X, NjWOE, Batchelor, Cat. 1304:43 (*TAEM* 91:42; *TAED* MBN004: 40). Written by Charles Batchelor. [a]Circled.

1. This marks the beginning of Edison's first significant demonstrations of his basic incandescent lighting system. Over the course of the next few weeks he showed between 12 and 20 lamps, lit by a Gramme generator, to some two hundred people. The initial exhibitions were primarily for stockholders in the Edison Electric Light Co. (see Doc. 1716). According to Francis Jehl's later confused accounts at least one of the demonstrations to the stockholders was a failure, but what he describes is an arrangement of lamps entirely in series rather than in parallel as indicated by this document. It may be that during the course of his demonstrations to the stockholders Edison demonstrated such a system in order to compare it with his own and in one account Jehl quotes a statement by Edison that supports such an inference: "You see, the system is not practical. When one burns out, it is as bad as if the whole circuit had broken. The lamps are not practical, either. Low resistance lamps will never work." At the time Edison considered the demonstrations to be very successful in demonstrating that "the problem of subdivision is solved. The question of economy answered." Although he took down the lamps about 29 March, when he had to put off a visit by Edward Dickerson, he planned to "have a new exhibition pretty soon" and on 12 April the *New York Sun* reported that Edison had been giving "a semi-public exhibition for two weeks" that was seen by "over two hundred persons." It is unclear how long they continued thereafter. "Edison Altogether Serene," *New York World*, 23 March 1879; "Edison Still Hard at Work," *New York Daily Graphic*, 11 April 1879; "Edison's Electric Light," *New York Sun*, 12 April 1879; Charles Lever letter to the editor, "Electricity and Electric Lighting," *Manchester Courier and Lancaster General Advertiser*, 29 May 1879; Cat. 1241, items 1146, 1163, 1164, 1197, Batchelor (*TAEM* 94:463, 474–75, 486; *TAED* MBSB21146, MBSB21163, MBSB21164, MBSB21197X); Jehl 1937–41, 246–47; Dyer and Martin 1929, 289; TAE marginalia on Calvin Goddard to TAE, 29 March 1879, DF (*TAEM* 50:251; *TAED* D7920ZAD); see also Francis Upton to Elijah Upton, 23 March 1879, Upton (*TAEM* 95:514; *TAED* MU009).

2. Batchelor ordered the construction of twelve lamps, each containing a single spiral element, on 10 March. They were completed a week later after Edison reported "We was all night bringing up 12 lamp in vacuum worked all day Sunday all night Sunday night all day Monday." These lamps appear not to have contained electro-thermal regulators;

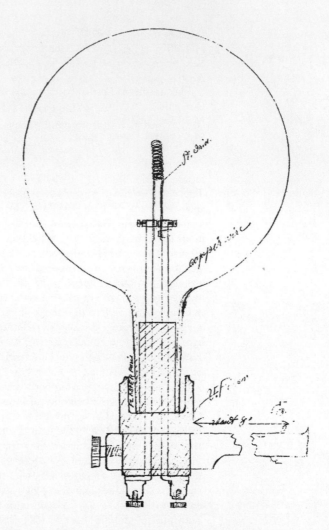

by this time Edison thought that the vacuum treatment of the wire increased its melting point sufficiently to prevent it from melting when brought to incandescence. Cat. 1308:119 (Order No. 54), Batchelor (*TAEM* 90:726; *TAED* MBN003:33); N-78-12-04.2:195, 199; N-78-11-21:119; both Lab. (*TAEM* 29:424, 426; 29:286; *TAED* N004:97, 99; N003:59).

3. In addition to Batchelor's estimate here of 22 candlepower the lights were variously said to range from that of an ordinary gas jet (i.e., 15 candlepower) to 20 candlepower. In fact, the lights were probably adjustable as indicated by a report from the end of April which described the "device for turning the light off and on" as "a circular piece of metal in the shape of a diminutive wheel attached to the under part of the lamp. . . . By turning the little wheel at the bottom of the electric lamp all the way around the full power is obtained; by turning only half way around only half the full light is given. In this manner the current can be regulated to any degree of light, from a dull red, giving only the faintest glimmer, to a beautiful, brilliant white, shedding light in every corner of

the largest room." Francis Upton to Elijah Upton, 23 March 1879, DF (*TAEM* 95:514; *TAED* MU009); see Doc. 1716; "Edison Altogether Serene," *New York World*, 23 March 1879; "Edison Still Hard at Work," *New York Daily Graphic*, 11 April 1879; "Edison's Electric Light," *New York Herald*, [27] April 1879; Cat. 1241, items 1146, 1163, 1173, Batchelor (*TAEM* 94:463, 474, 479; *TAED* MBSB21146, MBSB21163, MBSB21173); "Notes," *Teleg. J. and Elec. Rev.*, 15 May 1879, Cat. 1012: 52, Scraps. (*TAEM* 23:546; *TAED* SM012052a).

4. In various tests, Gramme generators of the type Edison used (the "A" or *type d'atelier* pattern) were typically operated at 800 or 900 revolutions per minute, requiring around two to two-and-a-half horsepower. This could produce a potential of around 60 volts. There were many slightly variant machines of this type over the years; they weighed from 350 to over 400 lbs. and stood about half a meter high and somewhat more than that wide. King 1962, 384–85, 390–91; Prescott 1884, 45, 64, 68; an illustration of this type of generator is in *TAEB* 4:585.

5. Presuming the wire used was .020 as in Doc. 1675 (see also N-79-03-10.1:37, Lab. [*TAEM* 31:21; *TAED* N022:19]), Batchelor probably meant a radius rather than a diameter of .08, thus giving a surface of ³⁄₁₆ square inches. The 23 March *New York World* quoted Edison as saying that the burner of his lamp was "only ³⁄₁₆ of an inch in height and the light is about the same dimensions" ("Edison Altogether Serene," *New York World*, 23 March 1879, Cat. 1241, item 1146, Batchelor [*TAEM* 94:463; *TAED* MBSB21146]).

6. The Gramme generator of this type tested at the Franklin Institute by Edwin Houston and Elihu Thomson measured about 1.7 ohms internal resistance. It is not known whether various Gramme type A dynamos differed that much or whether this reflects uncertainties in the measurements. Prescott 1884, 64.

7. "Multiple arc" refers to connecting the lamps in parallel circuitry, not to arc lighting. In this arrangement, each lamp was placed in series with three others in a branch circuit and the branches were wired in parallel.

8. Such a test directly refuted the analyses of various electrical experts claiming that connecting more than a very small number of lights at once would dim them all to invisibility. This probably involved connecting more parallel branch wires, but it is not known whether the loads on each were 1.32 ohms or whether the distribution of lamps and equivalent resistances was altered. Manufacture of ten more lamps was ordered on this day (Order 65) and completed the next and Edison soon reported success with as many as twenty lights at a time; yet another ten may have been ordered on 4 April when Kruesi wrote that date on the original Order 65. Cat. 1308:123 (Order No. 65); "Edison Still Hard at Work," *New York Daily Graphic*, 11 April 1879, Cat. 1241, item 1163; both Batchelor (*TAEM* 90:728, 94:474; *TAED* MBN003:35, MBSB21163); see also Doc. 1722.

Memorandum: New
York Herald *Arctic*
Expedition Lighting
Plant

Herald[a] Arctic Expedition box[1b]

1	~~M~~Edison Faradic machine[2]
1	Wallace Lamp[3]
1	Candle holder for carbons
1	Belt[c]
1	Pulley[c]
6	Laces
1	Edison Lamp[4]
2	Oil cups
20	lb connecting wire

TAE

D, NjWOE, Lab., N-79-02-10:169 (*TAEM* 32:1223; *TAED* N046:83).
Written by Charles Batchelor. [a]Interlined above. [b]Followed by dividing
mark. [c]Obscured overwritten text.

1. Batchelor listed these items for the lighting plant Edison had
agreed to supply for the voyage of the USS *Jeannette* to the North Pole,
an expedition sponsored by *New York Herald* publisher James Gordon
Bennett. It is not clear how Edison became involved with the venture. In
April 1879, while the *Jeannette* was being outfitted at a naval yard near
San Francisco, her commander, Lieutenant George De Long, wrote him
of his wish "to illuminate the ship from time to time during the long
Arctic Winter, and subject our crew to the benefits morally and physi-
cally arising from that light." De Long to TAE, 21 Apr. 1879, DF
(*TAEM* 50:44; *TAED* D7919Z); Guttridge 1986, 31–33, 42–43, 59.

According to a laboratory order book, the following items were sent
by express on 20 May: "Edison first Faradic machine marked Leggett &
Co. N.Y. for Arctic expedition, in same box 1 Pulley for Baxter engine
18. × 4 [inches] 1,400 bare & 400 ft of No 11 B[are]. copper wire also
33 ft of 4 in Belt & 50 ft of Lacing." The wire was intended for use with
telephones (see below). "300 E.L. carbon points" were sent to the *Jeannette* two days later. On 24 May, Francis Upton wrote his father that "the
machine for lighting the North pole had been sent away and a large number of carbons had been prepared for the purpose." Cat. 1308:231,
Batchelor (*TAEM* 90:773; *TAED* MBN003:80); Upton to Elijah Upton,
24 May 1879, Upton (*TAEM* 95:529; *TAED* MU015).

Edison also helped secure other equipment for the expedition. On 16
May, Jerome Collins, the *Herald* meteorologist whom Bennett attached
to the expedition, inquired if he would order the steam engine for the
dynamo and also requested help securing 100 miles of telephone wire
and transmitter carbons. Edison replied that he would "send my foreman to NY tomorrow to attend to whole business and inform you of
result." Collins later reported difficulty getting telephones from Gold
& Stock and the Bell company, and Edison asked George Walker if he
would "let Collins of NY Herald have six pony crown telephones for
Heralds Artic expedition. If there is any charge how much will it be."
Four days later James Bennett remitted $24.07 to Edison, presumably

for the telephones or related expenses. Guttridge 1986 (pp. 59–60) says the *Jeanette* had Bell telephones. Collins to TAE, 16 and 22 May 1879; TAE to Collins, 16 and 20 May 1879; TAE to Walker, 22 May 1879; Bennett to TAE, 26 May 1879; all DF (*TAEM* 50:58, 69; 52:54; 50: 70–71; *TAED* D7919ZAM, D7919ZAU, D7919ZAN, D7937ZAT, D7919ZAU1, D7919ZAW).

The *Jeannette* sailed from San Francisco on 8 July 1879 and was crushed by ice in June 1881. Although the crew escaped before it sank, most did not survive the consequent three-month trek to Siberian settlements. Guttridge 1986, chaps. 13–18.

2. At the beginning of May Collins asked for estimates of the time and cost to build a "two man power electric generator." On 15 May, Edison wired that it was "doubtful if apparatus will be satisfactory with hand power" and invited Collins to see the one in operation at Menlo Park that afternoon. The next day Collins telegraphed his consent to use a Baxter steam engine for power instead. John Kruesi's measured drawing of a generator belted to a $4 \times 3\frac{1}{2}$ inch drive pulley for "E L. Arctic Exp." is in Vol. 16:430, Lab. (*TAEM* 4:864; *TAED* NV16:385). Collins to TAE, 2 and 16 May 1879; TAE to Collins, 15 May 1879; all DF (*TAEM* 50:47, 58; *TAED* D7919ZAC, D7919ZAM, D7919ZAK1).

A notation appended to the laboratory order for items shipped to the *Jeannette* (see note 1) indicates that a "letter with explanation" was sent to Collins in San Francisco on 26 May. This letter has not been found but presumably it incorporated Francis Upton's copious undated instructions for the operation, maintenance, and repair of the ship's lighting system, particularly the dynamo. N-79-04-09, Lab. (*TAEM* 32: 1121–31; *TAED* N045:75–85).

Collins attempted to use the lights after the *Jeannette* was firmly trapped in ice that autumn, but the dynamo failed to produce any measurable current. Fearing that sea water had saturated the insulation, the ship's engineer completely re-insulated and re-wound the armature coils but found no improvement. De Long and his crew gave up on the machine and used its parts in makeshift pumps in early 1880. Guttridge 1986, 104–7, 131, 145.

3. William Wallace's arc lamp employed large carbon plates, which lasted far longer than the carbons in other arc lights, and its simple mechanism for regulating the carbon minimized the attention they required (see *TAEB* 4:774 n. 1). On 6 March John Kruesi instructed the shop to make four Wallace lamps which were completed on 29 March, though it is not clear that these were intended for the expedition (Cat. 1308:117 [Order No. 51], Batchelor [*TAEM* 90:725; *TAED* MBN003117B]). Edison subsequently recommended Brush lights; this series-wired system provided an automatic shunt around each lamp so that the failure of one would not affect the others (Bright 1972, 30–1). However, Jerome Collins reported that the Brush company insisted that Edison's generator be sent to them in Cleveland so the lamp regulators could be adjusted to it. It is not known what kind of arc lights ultimately were used on the expedition but they reportedly were to provide three thousand candlepower (Collins to TAE, 7 May 1879, DF [*TAEM* 50:50–51; *TAED* D7919ZAE, D7919ZAF]; Guttridge 1986, 60–61).

4. These were presumably platinum-spiral incandescent lamps,

Edison's 19 March sketch of a lamp with reflector.

possibly arranged with a shade such as that shown in Edison's drawing of 19 March that follows Batchelor's list in the notebook. N–79–02–10:171, Lab. (*TAEM* 32:1224; *TAED* N046:84).

–1707–

Notebook Entry:
Electric Lighting

[Menlo Park, March 19, 1879?[1]]

H.P. 3000.   18 000 lamps[2]   Daily 10 hours[3]

Engineers	9.00
Stokers	67.00
Assistants.	8.00.
Inspectors.	8.00.
Coal 27¼ tons.[4]—	81.50
Extra coal[5] 3 tons—	9.00.
Oil	3.00
Waste	1.00
Repairs.	10.00.
One day 10 Hours[a]	136.50.[6]

¾ of 1 cent for 10 hours[7]

Gas. 18 000 burners consume 900,000 feet in 10 hours[8] which taking the actual cost of producing the ~~light~~ gas at 90 c[ents] per 1000 feet,[9]

Gas	$810.00
Electric.	136.00

or as compared with ~~gas the Electric Light would produce~~ Electric Light[b] in economy gas must be made for 15⅛[a] cents per 1000 feet=[10]

in England the E Light would be

Engineers	6.00.[11]
Stokers.	4.00
Assistants.	6.00
Inspectors	6.00
Coal 27 tons	54.00[12a]
Extra Coal	6.00
Oil.	2.00
Waste	1.00
Repairs	8.00
	93.00

This brings cost to about 10½ cents per 1000 feet.[13]

X, NjWOE, Lab., N–79–03–20:213 (*TAEM* 33:100; *TAED* N047:100). Some decimal marks added for clarity. [a]Obscured overwritten text. [b]"Electric Light" interlined above.

1. This document appears between notebook entries by Charles Batchelor dated 19 March. It follows several undated pages of estimates and ancillary calculations of capital and operating expenses for an electrical generating station, and is one of Edison's recurring attempts to compute the costs of generating electricity for a given number of electric lights. N-79-03-20:206–11, 216, Lab. (*TAEM* 33:97–99, 102; *TAED* N047:97–99, 102); see Doc. 1716.

2. Edison had been using a ratio of six lamps of fifteen candlepower apiece lit by one horsepower in his calculations regarding the cost of electric lighting since the fall of 1878 (see e.g., Docs. 1651 and 1659).

3. This was the standard length of a workday used in various published estimates of the cost of power, not necessarily an estimate of lighting use. Emery 1883, Thurston 1884–85.

4. This amount of coal works out to a rate of 2.03 lbs. per horsepower per hour, the tons in question here (and in other cost calculations) being long tons, 2,240 lbs., the normal measure then for coal supply. A specific basis for Edison's calculation here has not been identified; however, it was not merely a projection of his earlier assumptions since the rate of 2 lbs. per horsepower used in previous estimates would correspond to 26.75 tons rather than 27.25. The only steam engines able to attain such rates of coal use per horsepower per hour were those of the most efficient and largest designs such as the Corliss steam engine. In a January interview Edison had reported that the figure for one recently installed Corliss engine was 1.74 lbs. per horsepower per hour; in contrast, typical 200 to 500 horsepower engines required over 2.5 lbs. of coal per horsepower while smaller ones used anywhere from 3 to 8 lbs. N-79-01-21:257, Lab. (*TAEM* 30:712; *TAED* N016:130); Siemens 1889, 3:191, 201, 245; Emery 1883, 427 and schedule A: col. 11; "Edison's Light," *New York Herald*, 30 Jan. 1879, Cat. 1241, item 1119, Batchelor (*TAEM* 94:455; *TAED* MBSB21119); see also Menlo Park Notebook No. 36 (box 13, EP&RI), which gives figures on the amount of coal used weekly in tests conducted at Menlo Park between 3 January and 7 March 1879.

The preceding labor figures also assume that very large engines would be used. According to an 1874 study of American steam power costs, engines of 150 to 200 horsepower output required engineers and firemen costing $2.50 and $1.50 respectively. An engine of 500 horsepower—the largest covered in the study—only increased the expense for engineers and firemen each to $3.00 per day, but producing 3,000 horsepower with six such engines would have involved a payroll much larger than that estimated by Edison. Although not in common use, engines producing more horsepower were available; the Corliss engine that powered exhibits at the 1876 U.S. Centennial Exhibition in Philadelphia produced 1,400 horsepower. On 25 March, Edison had written the Corliss Engine Company and requested that a representative call on him regarding motive power for his electric lighting system. Emery 1883, Schedule A, cols. 15, 17; Briggs 1982, 152; TAE to Corliss Engine Co., 25 Mar. 1879, Lbk. 4:234 (*TAEM* 80:68; *TAED* LB004234).

5. Some fuel was needed to bring the steam boilers to operating temperature prior to each working period. Just over 11% extra is assumed here; in a similar estimate William Siemens assumed 33%. Siemens 1889, 3:378.

6. This amounts to a total cost of less than half a cent per horsepower,

i.e., $0.0045, composed of $0.0027 for fuel and $0.0018 for other costs combined. The most economical rates reported in the 1874 review, those for a 500 horsepower engine adjusted for the difference in coal prices ($2.99 per ton instead of $4.17), come to $0.0042 for fuel and $0.0024 for other costs, totaling $0.0066 (Emery 1883, 428, Schedule A, cols. 24, 26). Cf. calculations made by Charles Batchelor in January regarding the cost of running a 1200 H.P. condensing steam engine (N-79-01-21:257, Lab. [*TAEM* 30:712; *TAED* No16:130]).

7. This is the cost per lamp.

8. This is based on the standard consumption rate of 5 cubic feet of gas per hour for each 15 candlepower burner; see *Ency. Brit.*, s.v. "Gas" and Docs. 1589 and 1651; cf. Edison's calculations of 9 March comparing the amount of horsepower and coal gas produced by a given amount of coal (N-79-02-10:34–41, Lab. [*TAEM* 32:1159–62; *TAED* No46:19–22]).

9. Commercial supply in this era was gas manufactured from coal rather than natural gas. Edison's basis for taking $0.90 as the production cost is not known. In New York City around this time gas sold at retail for about $2.50 per thousand cubic feet; beginning in July a rate war drove the price to $2.00 for small users and substantially lower for large commercial customers. Doc. 1439; "Reduction in Price of Gas in New York," *American Gas Light Journal*, 16 July 1879, Cat. 1022, item 40, Scraps. (*TAEM* 24:445; *TAED* SM022040a).

10. According to a newspaper interview published on 23 March, Edison compared the cost of his light to gas at 15½ cents per thousand cubic feet. Other estimates made by Edison and his staff used different rates of gas consumption than five cubic feet per hour per light. Francis Upton had noted a case in January in which a 16 candlepower lamp burned 4 cubic feet in 72 minutes. In April Edison estimated the annual operating and amortized interest costs of an electric light system consisting of 24 central stations lighting 400,000 lamps with 4,000 generators using 68,000 horsepower. Comparing this system with gas burners using 4 feet per hour, he concluded that his light would cost about ⅔ as much. "Edison Altogether Serene," *New York World*, 23 Mar. 1879, Cat. 1241, item 1146, Batchelor (*TAEM* 94:463; *TAED* MBSB21146); N-78-12-20.1:111, N-79-03-25:72, both Lab. (*TAEM* 30:170, 32:176; *TAED* No12:56, No34:37).

11. Edison's basis for British wage and price estimates has not been identified. In a similar exercise, British industrialist and scientist William Siemens estimated expenses for "wages, repairs, and sundries" at a rate equivalent to $34 per day to produce 3,000 horsepower per hour. Although comparable to the $33 for non-fuel expenses estimated by Edison, Siemens allowed for operating the engines only eight hours each day. Siemens 1889, 3:378.

12. The cost of coal in London was almost $5 per ton around this time, however, Edison apparently based his figure on the cost of coal for gas companies, which were able to resell some of the coke produced in the course of manufacturing the gas and thus cut their net coal cost close to $2 per ton. Siemens 1889, 3:378; *Ency. Brit.*, "Gas."

13. In comparison, retail gas prices in Britain ranged from the equivalent of $0.44 in some towns to $0.65 in London. Thompson 1910, 363; Siemens 1889, 3:378.

Notebook Entry:
Electric Lighting

Vacuum Experiments.[1]
    4 iron— wire common[a]

4 of pure iron—
4 " Piano steel
4 " Copper
4 " Aluminum
4 " Magnesium
4 " Nickel
~~34~~ " $^{10}/_{1000}$—platinum
~~34~~ $^{210}/_{100}$—Pt-Ir 10 pc
4 $^{10}/_{1000}$—Pt Ir—20 p.c.
4 $^{10}/_{1000}$— ~~wit~~Pt Ir—20 p.c. with Magnesia
4 $^{5}/_{1000}$—Pt Ir—20—
4 $^{5}/_{1000}$ pt Ir 20—with Nitrate Magnesia

    ~~one~~2 lime cylinders with 2 layers $^{5}/_{1000}$—Pt Ir 20 pc pyro-insulation $^{12}/_{1000}$—arranged thus

previously brought up in hand pump—[2b]

Two of these—[3]

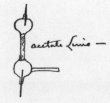

Acetate Lime—[c]

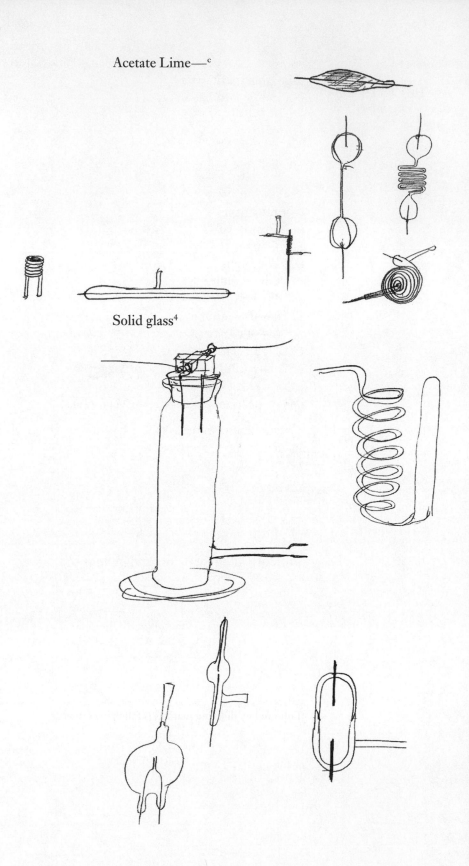

Solid glass[4]

ground joint [5]

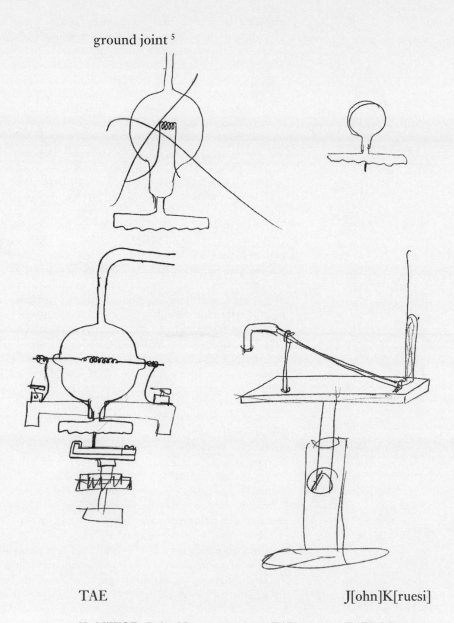

TAE   J[ohn]K[ruesi]

X, NjWOE, Lab., N-79-03-10.1:19 (*TAEM* 31:12; *TAED* N022:10).
Document multiply signed and dated; two small, unclear figures omitted. [a]Drawing followed by dividing mark. [b]Followed by dividing mark.
[c]Followed by dividing mark as is following drawing.

1. The metals on hand in the laboratory (piano wire, pure iron, copper, aluminum and nickel) were drawn into the indicated wires on or
about 31 March. Experiments with these samples began on 1 April, the
same day that Charles Batchelor recapitulated Edison's instructions and
produced a nearly identical list including all the metals designated below, with the addition of platinum. Batchelor indicated that spirals of

each wire were to be sealed in glass bulbs. In separate notes on these experiments, he remarked that a spiral of .02 inch platinum "same as shop lamps but uncovered and in vacuo" produced a "splendid light, better than anything we have seen on platinum without coating, I should say about 25 candlepower." The test of chemically pure iron was inconclusive because although the spiral remained intact, the wire melted at the joint to the platinum lead wires "where the air had not been got out of it." Copper proved similarly inconclusive because the glass globe cracked. Aluminum, the only other spiral for which Batchelor recorded results, "would not come up at all above a whitish yellow when it seemed to get weak in the knees and each spiral showed a tendency to rest on its neighbor." Cat. 1308:131 (Order No. 95), Batchelor (*TAEM* 90:732; *TAED* MBN003:39); N-79-03-10.1:37–39, N-79-03-25:15–17, Lab. (*TAEM* 31:21–22, 32:147–48; *TAED* N022:19–20, N034:8–9).

2. Charles Batchelor requested that ten bobbins like those sketched by Edison be made on 20 March and eleven days later requested another two dozen "of the little lime bobbins for experiment." Cat. 1308:129, 131 (Order Nos. 89, 93), Batchelor (*TAEM* 90:731–32; *TAED* MBN003: 38–39).

3. The purpose of this instrument is unclear; probably it was a test apparatus from which the glass globe could be removed, and the spool or spiral quickly replaced.

4. The coiled devices below are probably variations on Edison's idea of producing light by passing a spark through a glass capillary tube (see Doc. 1677). The large apparatus is likely one of the "2 Instruments for testing spools, glass to be made out of battery bottles" which Batchelor included in his 1 April list of things to make, which followed his observation that "in our experiments with the small spools we find when we fuse them they always give out at the bottom   May be due to point of greatest potential." N-79-03-25:13, 19, Lab. (*TAEM* 32:146–49; *TAED* N034:7, 10).

5. Label on drawing identified as "ground joint" is "3$^{20}/_{1000}$ twisted together." Among the items that Batchelor had indicated to be made in his 1 April list for these experiments was "1 Aneroid instrument for experiment." The following three drawings appear to be for this design, which is shown in the laboratory order book as a "Regulator for lamp with aneroid plate" that was completed on 8 April. Another lamp design using a "platinum sleeve on corrugated diaphragm to hold bulb" that Batchelor included in both his 1 April list and in the laboratory order book, does not appear amongst Edison's drawings. N-79-03-25:19, Lab. (*TAEM* 32:149; *TAED* N034:10); Cat. 1308:131–32 (Order Nos. 96, 99), Batchelor (*TAEM* 90:733–34; *TAED* MBN003:39–40).

Francis Upton to Elijah Upton[1]

Dear Father:

I felt sorry to hear of George's trouble,[2] but was not astonished as I could not well see how he could get on always as his affairs were. I think it much better that the anxiety should be over with and a new start made. You have always said that George could only learn by hard knocks, and I think he has had a lesson.

I had a talk with Mr. Edison about wages and the future.[3] He does not offer me any higher wages, but makes me to a certain extent a sharer in any success he may have. He has promised me to give[a] me the copyright of any work he may publish on the Electric Light. In doing this I am thus to have the editorship of the work, and to pay for the expense of getting up the drawings. I shall also have the proceeds of any article he may write. He does not like the trouble of writing and yet is willing that another should dress his thoughts for the press. You may smile at the prospect and think that promises will never buy bread and lodging, yet there is a certain degree of hope in such[a] conditions. The electric light has been on exhibition during the past few days. It shows a fair promise of being ultimately a success, and of course then giving me all the money I may need. He has lit up his shop with 18 small lights made from platinum. Each light was about equal to a gas burner. I am fully satisfied with my prospects even if the light does not succeded for I shall have no trouble getting a job somewhere else on Edison's recommendation.

Lizzie[4] writes me last from Gibraltar, she says that she has enjoyed her[a] journey through Spain very much. When she wrote she had just returned from a donkey ride to the fortress. I was astonished when I saw the stamp, a two and half penny English, for I did not know but that something was wrong and they were on the way home.

I expect to go home in four or six weeks and spend a few days.[5] I want to ask the united wisdom of the family to advise me how I had better get married, and what sort of place I had better have after marriage.[6]

I am with very much love   Your Son

Francis R. Upton.

ALS, NjWOE, Upton (*TAEM* 95:514; *TAED* MU009). [a]Obscured overwritten text.

1. Elijah Wood Upton, father of Francis Upton, operated a large glue manufactory in Peabody, Mass. and a nearby bleachery. "Upton, Francis," Pioneers Bio.; Vinton 1874, 348–50.

2. George Upton was Francis's eldest brother and, like his father, a glue manufacturer. The nature of his "trouble" is not known. Vinton 1874, 433; Francis Upton to TAE, 16 Sept. 1879, DF (*TAEM* 50:128; *TAED* D7919ZCL).

3. On 2 March, Upton had written his father:

I feel that I am able to take care of myself, trusting to the future. I find my work very pleasant here and not much different from the times when I was a student. The strangest thing to me is the $12 that I get each Saturday, for my labor does not seem like work but like study and I enjoy it. The electric light I think will come in time and then be a success, how long that time will be is quite a question. I think that it will be successful eventually, and then my place will be secure. I learn much every week and soon, that is in a year or two think I will be an expert in electrical questions. My pay I know is very small in dollars but the chance to get knowledge is beyond measure. [Upton to Elijah Upton, 2 Mar. 1879, Upton (*TAEM* 95:512; *TAED* MU008)]

On 30 March he reiterated this point when he told his father that "I am perfectly contented as matters are now, I keep making progress in my profession and acquainting myself more and more with the science of electricity. There is no need of taking any thought or worrying as to the future since I cannot alter it. I have made up my mind to stay here as long as I can and take my chances at whatever may turn up. I know I can make a living in various ways, so I feel that my future is in good enough hands if I leave it to itself without much thought" (Upton to Elijah Upton, 30 Mar. 1879, Upton [*TAEM* 95:517; *TAED* MU010]).

4. Upton was engaged to Elizabeth Perry, who was in the midst of an extended European tour. Upton autobiographical sketch, Pioneers Bio.

5. See Doc. 1750.

6. In his 30 March letter to his father Upton indicated that he preferred to purchase the large house that was available near the laboratory for about $450. Although he apparently had some difficulty obtaining this property, he and Lizzie did settle in Menlo Park after their marriage. Upton to Elijah Upton, 30 Mar., 15 June, and 19 Oct. 1879, Upton (*TAEM* 95:517, 534, 563; *TAED* MU010, MU017, MU031).

–1710–

Notebook Entry: Electric Lighting

[Menlo Park,] March 23 1879

Plaintum and Plat. Iridium wire can be coated by an Insulator which will stand heat   I have coated wires thirty inches long with a coating .001 thick of Acetate of Magnesia and wound two layers on a lime bobbin; bringing the wire up to incandescence afterwards without fear of burning[1]

I have used for this purpose:—

> Acetate of Magnesia
> Acetate of Lime
> Oxalate of Cerium

Just dampening the fingers with the solution and afterwards rubbing it on the wire and passing it through a candle flame to make it cake on hard. I find that the coating breaks off a little when brought up to incandescence after winding on the spool owing I suppose to the short turns on the spool but I can fix that very easily (and I do so) by covering it to .010 thick and then coiling it up not less than 2 inches in diameter and bringing it to incandescence in a vacuum.

the result of which is to melt the Magnesia down to a hard glassy mass on the wire which does not crack when wound on the spool only ⅛ inch diameter but seems perfectly homogeneious under the microscope

In all the solutions so far tried the Acetate of magnesia is by far the best but I believe the magnesia attacks the platinum slightly and if would get an[a] oxide of Cerium or Zirconium on it would be much better. Cerium however is very difficult to put on[2]

Chas Batchelor.

X, NjWOE, Cat. 1304:44, Batchelor (*TAEM* 91:43; *TAED* MBN004:41). Written by Charles Batchelor. [a]Obscured overwritten text.

1. See Docs. 1688 and 1690 regarding earlier coating experiments.

2. On 25 and 26 March, Batchelor conducted some additional experiments with acetate of magnesia. On the 25th he noted that acetate of magnesia "makes very good coating, but we want it thicker than we get this and easier to put on," while nitrate of magnesia "goes on very easily but it is very rough and when wound on spool it seems to break off on top side showing the platinum underneath entirely free from coating." This led him to combine the two on the following day, finding that "this makes good and thick coating the acetate making it stick to the Platina and the nitrate thickening it." In addition he experimented with rubbing on coatings of nitrate of barium, tungstate of soda, acetate of lime, nitrate of lime, chloride of calcium, and aluminic nitrate. On 31 March he also tested acetate of zinc and nitrate of aluminum. The coating experiments continued in April; see Doc. 1720. N-79-03-25:1–9, Lab. (*TAEM* 32:140–44; *TAED* No34:1–5).

–1711–

From Grosvenor Lowrey

New York Mar 24th 1879[a]

My Dear Edison:

We have succeeded to day, in the case of the New England Iron Co agst the Metropolitan Elevated Ry Co,[1] in having the Court dismiss the plaintiffs complaint. This releaves me; and I am off for the Hot Springs of Arkansas in the hope there to find permanent cure for the irritating and embarrassing tendency to Gout which lays me up so often.

Mr. Morgan goes to Europe in about three weeks, and I am very desirous to close matters between you and his firm before I go in respect to England,[2] and, if there is anything to be done, in respect to the Continent, also.

I have sent you once a copy of the Agreement contemplated by me,[3] and I now return it to you that you may refresh your recollection and your ideas generally, and then I should like, if possible to have you come in on, say, Wednesday, it being next to impossible to get any one of that firm to Menlo.

Won't you on Wednesday bring Mrs Edison in, and we will go & see Her Majestys Ship "Pinafore,"[4] if that is agreeable. Dine with Mrs Lowrey & I and we will all go together.

I have never heard anything from Col Gouraud since his letter which I sent to you in which he said that before stating his views as to the interest he ought to have he would confer with you.[5]

Let me know by telegraph tomorrow if you can come.[6] Truly Yours

G. P. Lowrey

Please send the agreement back by mail—[b]

LS, NjWOE, DF (*TAEM* 49:182; *TAED* D7903ZCM). Letterhead of Porter, Lowrey, Soren & Stone. [a]"New York" and "187" preprinted. [b]Postscript enclosed by right brace.

1. This was a breach of contract suit concerning construction of the Metropolitan Elevated Railway Co. in New York, for which Lowrey served as counsel. For Edison's prior but unrelated involvement with this company, see Doc. 1371 esp. n. 1. "A Suit for Four Millions" and "The Elevated Railway Suit," *New York Times*, 5 and 25 Mar. 1879, both p. 3.

2. Although Gouraud's interest in Edison's light in England with respect to Drexel, Morgan & Co. had been the subject of considerable correspondence among Edison, Lowrey, and himself through the latter part of 1878, no specific proposals seem to have been made by this time (see Docs. 1461, 1502, 1504, 1522, 1532, 1551, 1612, and 1648). In a 31 January reply to Lowrey after receiving the enclosure to Doc. 1648, Gouraud deferred suggesting terms for himself pending further consultation with Edison. Gouraud closed by asking Lowrey to "keep me advised of the progress of the electric light invention, as Edison is a bad correspondent you know." Lowrey's office forwarded this letter to Edison on 19 February; Stockton Griffin's docket note indicates that a copy had already been received from Gouraud. DF (*TAEM* 49:117, 115; *TAED* D7903ZAX, D7903ZAW).

3. Lowrey is referring to Doc. 1649 (see esp. n. 1); the agreement evidently was not signed on 31 December 1878 although its terms may have been agreed upon and its duration calculated from that date.

4. William Gilbert and Arthur Sullivan's 1878 operetta, *HMS Pinafore*, firmly established the collaborators' popularity in England and in the United States, where numerous unauthorized productions were mounted. One of these, billed as the show's first and "only perfect repre-

sentation" in New York, had been playing at the Standard Theatre since January. "Amusements," *New York Tribune*, 15 Jan. 1879, 3; "Amusements," ibid., 24 Mar. 1879, 3; *NGD*, s.v. "HMS Pinafore"; Wyndham 1926, 145.

5. See note 3.

6. There is no record of a reply by Edison.

–1712–

To Calvin Goddard

[Menlo Park,] March 25th [187]9

Dear Sir

I wish to obtain the last annual report to the stockholders of the Gas Companies in N.Y. and if possible for 2 or 3 years back— Please ask Mr Galloway[1] if he can give you information on the following points.

The total capacity of the gasometers[2] of the NY companies each by themselves

The amount of actual capacity employe[d][a] in every day work to the total capacity

The number of retorts necessary to manufacture the gas of any given Co and used in daily practice and the extra number of retorts necessary bu[t][a] not used.

The number of employees in the [retort?][b] house.[3] Their average wages and average amount of gas made.

Amount of coke used for firing per 1000 or 10 000 feet and also amount of salable coke.

Amount of oil used for enriching per 1000 ft. of gas.

Amount of horse power used in the works per 100 000 feet of gas made— Very Truly

T. A. Edison   G[riffin]

L (letterpress copy), NjWOE, Lbk. 4:237 (*TAEM* 80:70; *TAED* LB004237). Written by Stockton Griffin. [a]Edge of original not copied. [b]Faint letterpress copy.

1. Robert Gallaway, an associate of J. P. Morgan, was a trustee of the Edison Electric Light Co. He was connected with the Atlantic Dock Iron Works, a firm that built many large gas-making plants, and during his career was also an officer of the New York Gaslight Co. There is no record of a reply from either Gallaway or Goddard to Edison's request. Obituary, *New York Times*, 14 Nov. 1917, 15; Articles of Incorporation of Edison Electric Light Co., 16 Oct. 1878, DF (*TAEM* 18:38; *TAED* D7820ZAM).

2. This was a British term for a holding tank for gas that consisted of a large cylinder with its bottom edge immersed in a water tank. The term gasometer arose because the known capacity of the tank allowed a ready calculation of the quantity of gas. Knight 1876–77, s.v. "Gasometer"; see also Doc. 1548.

3. This was the part of the gas plant in which coal or other carbona-

ceous matter was placed in chambers made of iron or clay known as gas-retorts and heated in a brick furnace to high temperatures in order to distill the gas. A retort required three men to charge it and afterwards the coke byproduct had to be raked out. Knight 1876–77, s.vv. "Gas-retort," "Gas-retort Charger."

–1713–

Draft to Josiah Reiff

[Menlo Park, March 25, 1879[1]]

JCR[2]

My new office has put me 1100 in debt.[3] I have no income but what is anticipated   Eckert[4] is now in a position to appreciate your efforts in trying to obtain your <u>own</u>.[5] I hope to god he will never get a cent, in fact he seems to have a sure thing this time

E

ADfS, NjWOE, DF (*TAEM* 51:860; *TAED* D7936M).

1. Edison wrote this draft on the reverse of Josiah Reiff's letter of 21 March (see note 2), which Stockton Griffin indicated was "ansd Mch 25 79."

2. Josiah Reiff was a railroad financier who had provided most of the funds for Edison's work on automatic telegraphy and had loaned him money for personal purposes (see *TAEB* 1:243 n. 7, 497 n. 3, 620). On 21 March he had written Edison from Washington asking to borrow $1000: "It may seem strange for you to assume the role of <u>capitalist</u> instead of receiver, but try it, it will do you good" (DF [*TAEM* 51:859; *TAED* D7936L]).

3. In recent weeks Edison had accepted bids for finishing work on the new two-story brick office and library building, one of the structures erected when he expanded the Menlo Park compound in the latter part of 1878. Weekly summaries of laboratory expenses from 15 November 1878 to 5 April 1879 show the progress of construction and indicate that the office cost approximately $2,400. Isaac Fenn to TAE, 22 Jan. and 11 Feb. 1879; E. P. Edwards to TAE, 10 Feb. 1879; all DF (*TAEM* 50:493, 509, 508; *TAED* D7925E, D7925S, D7925R); Electric Light Co. Statement Book, Accts. (*TAEM* 88:433–65; *TAED* AB031:22–54).

4. Thomas Eckert was president of the Atlantic and Pacific Telegraph Co., and Edison blamed him for his difficult dealings with that company (see, for example, Doc. 853). Eckert had expected to become general manager of Western Union when the two companies merged in 1878 but was disappointed. On 11 March Eckert abruptly resigned amid speculation that he was defecting to a rumored telegraphic consortium of eastern railroads in opposition to Western Union. He rescinded his resignation but quit again in December to become president of the American Union Telegraph Co., a company organized by Jay Gould with cooperation from the Baltimore & Ohio and railroads controlled by Gould. See *ANB*, s.v. "Eckert, Thomas Thompson"; *TAEB* 2:120 n. 6, 3:472 n. 1; Klein 1986, 234–5, 277; "Resigning the Presidency," *New York Tribune*, 13 Mar. 1879, 8; "Resignation of General Eckert," ibid., 12 Dec. 1879, 8.

5. At this time Reiff was in Washington trying to secure passage of legislation favorable to the proposed new telegraph syndicate (see note 4). He had earlier promised Edison that his success would "be worth $100,000 each to you & me," evidently because he intended to broker Edison's services and the assignment of telegraph patents, including those for the automatic system, to the company. Reiff to TAE, 9, 12 and 23 Feb. and 4 May 1879, DF (*TAEM* 51:849, 851, 855, 866; *TAED* D7936G, D7936H, D7936J, D7936R).

-1714-

From Reinmann &
Baetz

New York, March 26th 79

Dear Sir:

I like to call your attention once more ~~of~~ to the perfectness of the mercury, what you want to use with the air pump.[1] The best way, to clean and dry it, is, to put the mercury in a vessel, and mix it with good Plaster of Paris. Than pour it through a funnel from paper, with an opening of about ½ to 1 m/m, continually[a] and keep the last drops in the funnel. The mercury must be kept perfectly free from[a] all metallic substansces, otherwise the pump will be coated in very short time with moisty[a] mercurial (metallic)[b] parts. As it is a very difficult and dangerous job ~~t~~ho clean[a] and dry such a pump again, great care has to be taken especially with[b] the first mercury used, so the rubber tube does not get wet.[2] One of us would rather like to come over to you once more, to get the instrument in right work. The mercury, what you had there, should have been washed with the Plaster of Paris anyway, as it showed some remainders in the pump allready.

Enclosed please find our bill for $48.00 for which your check is wellcome—[3]

Awaiting Your further [---][c] commands, we remain   Yours Very Rspfly

Reinmann & Baetz[4]

L, NjWOE, DF (*TAEM* 50:511; *TAED* D7925U). [a]Obscured overwritten text. [b]Interlined above. [c]Illegible.

1. This was a Geissler vacuum pump, which Reinmann & Baetz had promised to deliver on this day. They also advised Edison to have about 25 pounds of mercury ready if he wished to try the instrument immediately. Machinist John Kruesi wrote an order on 28 March to have an "Iron pan made for Geissler Vacuum machine," which was probably to be placed under the pump to collect spilled mercury. Reinmann & Baetz to TAE, 25 Mar. 1879, DF (*TAEM* 50:510; *TAED* D7925T); Cat. 1308:127 (Order No. 82), Batchelor (*TAEM* 90:730; *TAED* MBN003:37).

2. A thin film of air would cling to the glass tubing under the best cir-

cumstances and the additional vapor pressure of contaminating moisture would seriously impair the pump's operation; see Doc. 1816. Thompson 1887, 588; Tilden 1919, 93–95.

The problem of water and mercury vapors in the pump was one addressed by the British chemist William Crookes, on whose techniques Edison made extensive notes about this time, and a number of whose publications were placed in a laboratory scrapbook devoted to the radiometer, vacuum pumps, and phenomena of high vacua (N-79-04-21:1–25, Lab. [*TAEM* 30:726–39; *TAED* N017:1–13]; Cat. 1050, passim, Scraps. [*TAEM* 26:394; *TAED* SM050]). Edison noted Crookes's observation that sulphuric acid used as a drying agent could itself evolve water vapor; however, as he had not yet started using a desiccant in his own pumps, he did not adopt Crookes's preference for phosphorous anhydride until later in the year (see Doc. 1816 n. 5). Edison drew on several of Crookes's publications (including Crookes 1873, 1876, 1876–77, and 1878a), summarizing a number of his techniques and observations relating to other chemical methods of reducing vapor pressure, Geissler spark tubes, and the preparation and operation of pumps. He later referred in passing to Crookes in the patent application executed on 28 January 1880 for his vacuum pump arrangement (U.S. Pat. 248,425). DeKosky 1983 provides a useful discussion of the apparatus and procedures developed by Crookes.

3. A. S. Reinmann offered to come to Menlo Park again on 28 March "to bring everything about the pump in right shape. There are few more

Edison's drawing of a combined Geissler and Sprengel vacuum pump.

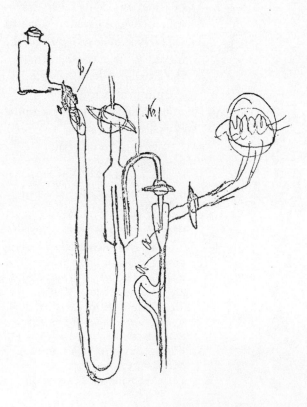

little points, which I have to explain by experimenting with the pump."
Edison subsequently hired William Baetz to build additional pumps of
the Geissler, Sprengel and other designs, which he did at Menlo Park
and his New York shop. Reinmann & Baetz to TAE, 27 Mar., 7 and 19
Aug. 1879, all DF (*TAEM* 50:514, 532; *TAED* D7925W, D7925ZAK,
D7925ZAL); Edison's testimony, pp. 43–44, *Böhm v. Edison* (*TAED*
W100DED032, images 12–13).

Edison later testified that shortly after receiving the Geissler pump
he conceived the idea of combining its action with that of a Sprengel
pump, to achieve the latter's efficacy with the former's rapidity as shown
in a drawing he presented as evidence in the patent interference. He had
Reinmann & Baetz build four or five of these instruments (*Böhm v. Edison* (pp. 43–44; Edison Exhibit No. 5 [*TAED* W100DED032, images
12–13; W100DED059E]). It is not clear if Edison's payments to the firm
of $89.30 in April for "Mercury Pump & work" and $23.00 in May for a
"vacuum pump and attach[me]nt" were for Geissler or combination
pumps (Electric Light Co. Statement Book, Accts. [*TAEM* 88:467, 474;
TAED AB031:56, 61]).

4. This firm's letterhead identified it as "Manufacturers of Chemical
and Physical Glass Instruments" on Fulton Street in New York City.
Reinmann & Baetz had repaired Edison's mechanical vacuum pump in
January (see Doc. 1667 n. 2).

–1715–

From George Gouraud

[London,] Thursday—27th Mar [1879]

My Dear Edison

I have only a moment after the close of Tyndalls lecture
& before the close of the mail to write you a line of felicitation
at the happy manner in which the[a] Professor brought the Telephone before the first British audience at the Royal Institution
today & the enthusiastic reception it met—

Nothing could have been more satisfactory in every respect.
You will see by the papers I will send you[1] what effect will be
produced by this Scientific Endorsement of the invention—
I[b] had a wire run to the Piccadilly Circus which you will remember & Charley performed in various ways   I sat in the
Gallery of the the lecture room & heard distinctly everything
that was said—[2] The audience was <u>enthusiastic</u> & Tyndall
himself quite <u>warmed</u> over the beauty of the results & the simplicity of its means.

I am sure you will appreciate now if you have not before the
importance of the opportune arrival of the instrument—& for
the roughness of which Professor T. said you had no need to
make apology—

I beg you will not lose a moment in forwarding the models
for the mftr of commercial instruments. The demand for in-

struments is showing itself in every mail rec.d & from all parts of the country & if we are too long in meeting it dissatisfaction[b] will result.

Do support me promptly in this & I will soon show you some results.

Hastily with my best congratulations   Truly Yours

Geo. E Gouraud

ALS, NjWOE, DF (*TAEM* 52:506; *TAED* D7941S). On paper with unidentified emblem. Stockton Griffin's transcription is in DF (*TAEM* 52:510). [a]Interlined above. [b]Obscured overwritten text.

1. The original clippings have not been found but see note 2.

2. The Royal Institution is located about one-half mile by road from the famous London crossroads of Piccadilly Circus. According to the *Illustrated London News*, "conversation between the Professor and Mr. [Charles] Edison at Piccadilly Circus was carried on with much vivacity. Readings from Emerson, Shakespeare, and other writers were repeated with echo-like exactness, and distinctly heard all over the theatre. In singing the peculiar qualities of the voice were remarkably observable, and the loudness exceeded that of any telephone hitherto exhibited at the Royal Institution." James Adams also assisted at this lecture. After Tyndall demonstrated the receiver on 2 April over a line from the Royal Institution to a meeting of the Royal Society, Gouraud reported that "the interest and commendations were excessive." "The Royal Society," *Times* of London (reprint) 4 Apr. 1879 and "Edison's Carbon Telephone," (reprint) *Illustrated London News*, 5 Apr. 1879, Cat. 1241, items 1158 and 1161, Batchelor (*TAEM* 94:471, 473; *TAED* MBSB21158X, MBSB21161X); Gouraud to TAE, 3 Apr. 1879, DF (*TAEM* 52:512; *TAED* D7941T).

–1716–

Anonymous Article in the New York Herald

[New York, March 27, 1879]

EDISON'S ELECTRIC LIGHT.[a]

SEVERAL RECENT IMPROVEMENTS IN THE SYSTEM—A DISCOVERY OF IMPORTANCE—LESS THAN HALF THE COST OF GAS.

The first practical illustration of Edison's electric light as a system has just been given. For the past two nights his entire laboratory and machine shop have been lighted up with the new light, and the result has been eminently satisfactory. In the chemical portion of the laboratory two of the incandescent lights supplied the place of the fine[1] gas jets ordinarily in use there, and in the machine shop, a building 125 feet long by 25 feet wide, twelve of the new lights did the work of the eighteen gas burners previously used. The light given was clear, white and steady, pleasant to the eye, and of such character that colors could be readily distinguished. None of the unpleasant glare so noticeable in the carbon light was visible. The electric lamps were regulated so that each gave a light of from eighteen to twenty candle power in intensity, but the purity of the light made the intensity appear even higher. The generator used was the ordinary Gramme machine, said to consume two and one half horse power, but which in reality has been found to consume three horse power. The exhibition was witnessed by some of the leading directors of the company controlling the patents, together with a few scientific friends of the inventor. All were much pleased at the result.[2]

THE LATEST IMPROVEMENTS.

The progress made during the past two months by Edison has been quite marked. The light, instead of being at a standstill, as was generally believed, has in reality been greatly developed. The most prominent of the improvements is one relating to the platinum and iridium coil. By a discovery, pronounced by experts who have seen it as of great scientific importance, Edison has succeeded in practically creating a new alloy—one that possesses properties hitherto unknown. Its melting point is much higher than that of any known metal. This discovery has proved most valuable in connection with the electric light. Previously the best result which the inventor could obtain per horse power was four lights, but by the use of this new alloy he now obtains with ease six lights per horse power, and makes it possible, under certain conditions, to obtain as high as eleven.

A BETTER GENERATOR NEEDED.

One of the chief difficulties Edison has had to contend against has been the want of a proper generator for his peculiar method. In the course of his experiments with the light he has used several, and constructed two or three of his own, but none of them have come up to his requirement. The one giving the best satisfaction thus far is the Gramme machine; but even that does not develop electricity with sufficient economy to satisfy him, although from that machine he has obtained, as above stated, at the rate of six lights per horse power. He continues confident that in a short time he will have succeeded in constructing a generator that will give much better results. But even with the Gramme, imperfect in many respects for incandescent lighting as it is, he manufactures the light at a cost of a trifle over one cent per hour for six electric lights. His mechanics are now constructing a new generator, which promises to yield much better results, so far as economy is concerned.

THE STYLE OF LAMPS USED.

Since beginning work on the electric light Edison has devised more than a score of lamps of different styles, but one by one they have been laid aside as not coming up to his conception of what he ought to have. There now remain only two styles, which closely compete with each other in point of excellence. In both there is a small glass chamber through which passes the little coil which is made incandescent by the electricity. One of the lamps resembles the ordinary student's lamp.[3] The other is of globular form and hardly as handsome. The essential construction of each is practically the same, although there are some few minor points of difference. The regulator or contrivance by which the electricity is turned off and on has been improved and is now much more simple than it was originally, the expansion of the air in the globe serving to work it.

LESS THAN HALF THE COST OF GAS.

The recent improvements which have been made by the great inventor have necessitated the changing quite materially of the figures upon which the cost of electric lighting has previously been based by him. Lately he has obtained estimates from different builders of the cost of engines, boilers, dynamo machines and the other necessary appliances.[4] With their guarantees and with the experiments already made by the lamps now in use (the horse power consumed being measured by a dynamometer and the resultant electricity by an electro-dynamometer, the whole reduced to so many foot pounds per candle power), he has been able to make the necessary estimates. The results are certainly startling, as compared with gas, as they show that gas must be made and stored in the gasometer at fifteen and a half cents for each 1,000 feet to compete with his system, and with a possibility of a further reduction by improvement. In one of Edison's electric burners equal to sixteen candles there is consumed 2,560 foot pounds per minute, whereas a gas jet of equal power consumes 48,000 foot pounds, or eighteen times more energy to produce the same effect as is produced by the electric light. To put it in more popular terms, the present estimates as to the expense of electric lighting under Edison's system as made by him figure the cost of electricity as compared with gas at less than one-half, or, supposing gas to be furnished at the rate of $2.25 per 1,000 feet,[5] an equal supply of electric light can be furnished for $1, leaving at that price a fair profit for the company supplying the electricity. The figures put forward by the gas companies Edison claims are based on error. Those who make them do not, he says, understand his system, and he further claims that in many instances they estimate twenty-four hours per day as the daily consumption, when, in point of fact, the consumption of gas or electricity will not average for all the burners more than three or four hours at the most.

INCANDESCENCE VS. CARBON.

Among the experiments frequently tried of late in the laboratory is one showing the essential difference between light produced by the carbon process and that by incandescence. In the centre of the laboratory, suspended from the ceiling by wires, is a carbon apparatus, said to emit a light of 4,000 candle power.[6] Last night the twelve incandescent lights were frequently compared with it to show the relative quantities of light. The light from the carbon gave apparently not more than three or four times as much light as that given by the incandescent lights. What it did give, however, was not nearly as effective for practical purposes as were the twelve incandescent lights.

By the latter the mechanics were enabled to do their work with as much facility as if it was day, but during the time the carbon light was shining it was noticed that many of the workmen were unable to work, the flickering, unsteadiness and sharp shadows from the carbon process being painful to the eye. Immediately in the vicinity of the carbon the light was brilliant and intense, but at a distance of thirty or forty feet it was very much decreased. The scientists who witnessed the comparisons were unanimous in pronouncing in favor of incandescence as much better adapted for practical uses.

GETTING READY FOR THE PUBLIC.

The work in the laboratory during the past month has been assiduous, the machinery and mechanics being kept going until midnight every night. Only two things, Edison says, are now necessary before the light can be given to the public. The first is the standard lamp to be used and the second a better generator than the one now in operation. Neither of these requirements is regarded by him as difficult of attainment. Indeed, both might be dispensed with, he states, and the system be now put in practical operation everywhere, and electricity supplied at less than half the cost of gas, but feeling satisfied that a delay of a few months will enable him to get the system down to even greater economy he prefers to send it from the laboratory as near complete as possible. The undertaking to light up Menlo park will be commenced in a few weeks. Mr.

Batchelor, Edison's chief assistant, who has full charge of the details of the work, has already made the estimates and given the preliminary orders for the accomplishment of the same.[7]

PD, *New York Herald*, 27 Mar. 1879, 5. In Cat. 1241, item 1147, Batchelor (*TAEM* 94:464; *TAED* MBSB21147X). [a]Followed by dividing mark.

1. Probably intended to be "five."

2. It is not known who saw the lamps in use.

3. This was a common form of Argand lamp having an elevated reservoir from which oil flowed by gravity through a check-valve to the burner. Knight 1876–77, s.v. "Student's Lamp."

4. Two days earlier, Edison had requested the Corliss Engine Co. to send a representative to Menlo Park "in relation to motive power for my Electric Light system." No agent visited by the time this document was published. In January the Buckeye Engine Co. had promised to submit estimates for a 1250 horsepower central station plant but these have not been found (TAE to Corliss Engine Co., 25 Mar. 1879, Lbk. [*TAEM* 80:68; *TAED* LB004234]; Corliss Engine Co. to TAE, 26 Mar. 1879; Buckeye Engine Co. to TAE, 7 Jan. 1879; both DF [*TAEM* 50:513, 8; *TAED* D7925V, D7919E]). Edison evidently was considering condensing steam engines, for which Charles Batchelor itemized anticipated savings over other types. Edison stated in a preliminary British patent specification filed in June that he planned to use compound condensing steam-engines to drive the generators at large central stations (N-79-01-21:251–61, Lab. [*TAEM* 30:709–15; *TAED* No16:127–132]; Brit. Pat. 2,402 [1879], Batchelor [*TAEM* 92:118; *TAED* MBP017]).

5. See Doc. 1707 n. 9.

6. This may have been the Wallace arc light that Edison purchased at the end of 1878 (see Doc. 1615), or one of the four Wallace lights that the machine shop completed on 29 March (see Doc. 1706 n. 3).

7. A few days earlier the *New York World* reported that Edison had given up Menlo Park as a test site because

> there are so few houses here the experiment would not be conclusive. Therefore some time this summer I intend to light up Metuchen, two miles off, which is a large place and which will offer tests that could not be found here. I will run the current over poles. It is my object to present the lamp in such a way that all kinds and classes of people may have an opportunity to judge of it and to compare it with all other modes of illumination. It is more important to me that the servant in the kitchen should express her mind on the lamp, with regard simply to its lighting power, than that her mistress should wonder over it as a novelty without comparing it properly with gas. When I have brought it to that degree that it is popular my work will be done and the company will take charge of the whole matter. ["Edison Altogether Serene," *New York World*, 23 Mar. 1879, Cat. 1241, item 1146, Batchelor (*TAEM* 94:463; *TAED* MBSB21146)]

[Menlo Park,] March 30th [187]9

Dear Jim,

I have now a little more time to take up your long letter in detail.[1] First of all about the <u>Phono royalty</u> you say you got 10% on the money advanced by Nottage, I don't think that is so, because I know I only got 10%   I think you are mistaken and also in the 10 per cent in the states, you know that is 5 per cent in three different companies which is only 5 per cent of the whole, and 5 per cent on England is the same, as if there were 6 companies selling different applications of the phonograph and you had 5 per cent in each; that is only 5 percent of the whole.[2]

The Phono Co have by the way made a small phono. which they sell for $10— and which they have a guaranteed order for 10 000 and they think the company who are going to handle them will take 25,000 for the fall trade. It is a very good little phono, has a zinc cylinder, and altogether talks just about as good as the large ones, they are going to be made by the Remingtons or Colts or some such firm.[3]

In regard to Gouraud's assertion that he kept him and Jack Wright at a great cost to himself[a] and friend it is not so as Edison says he advanced about £200— and when the money came he took out £2000 for it so I think he was amply repaid[4]   his letters to Edison never show any such spirit.

As you have probably seen in paper sent you we have lighted our shop up every night this week.[5]

You dont need to be afraid of such men as G[ouraud]. P[uskas] & B[ailey] in future as we have got quite a different class of men in with us now and any man that gets in now will have to plunk down money first. Bailey is here and wants to get a hand in in European Electric light, well there is a chance for him but not such a chance as Puskas got for telephone, oh no we dont do things that way now, if he wants to come in on electric light he will have to guarantee $25,000—for experimental purposes whenever called on by Edison to be used by Edison as he sees fit without any questions being asked. Bailey has been here three times with different gentlemen but I have not had much talk with him.[6] The light is the biggest success we've had, and although it will be some time before it is out, at present I have refused $200—per share for my stock; I heard a man offer Edison W.U. Tel Stock for Electric Light stock and you know W.U. is 106 now. Dont let on of this to anybody.

I have seen it stated in London papers that Edison's backers had gone back on him, well if you had seen them here the other night you would not think so;[7] besides all the money for this thing was put up before we started. But you can bet your life that B. will not enjoy such priviledges without paying something to get into the game. Puskas of course is already in[a] as he paid for patents before we made this arrangement, but I'll tell you one thing, that our negotiations in the future will be with men of standing and solidity and I think G. & B. & P and I may add G.H.B[liss]. will have a decided hard time of it if they have no money   Yours

Chas Batchelor

ALS (letterpress copy), NjWOE, Cat. 1330:47 (*TAEM* 93:261; *TAED* MBLB2047).

1. Not found.

2. Early in 1878, Edison had assigned Batchelor and Adams shares of the net American royalties for the phonograph, Batchelor getting 10% and Adams 5% of the money Edison received through three separate business arrangements (see *TAEB* 4:18 n. 7 and Doc. 1190). Similar agreements covered Edison's American telephone interests (see Doc. 1345). There was only one firm (the London Stereoscopic and Photographic Co.) involved in the phonograph in Britain (*TAEB* 4:148 n. 1). Evidently Edison made similar arrangements with Adams and Batchelor regarding royalties due him from that company, but formal records of them have not been found.

3. This small instrument (later called the "parlor phonograph") was finished in the same manner as the large phonograph made by Sigmund Bergmann, and an advertising circular promised that both models were in fact made by the same firm. It was priced at fifteen dollars and a limited number were sold in early 1879; later in the year the price was reduced to ten dollars (Rondeau 2001, 62–65; J. R. Holcomb & Co. undated circular in collection of René Rondeau; Edison Speaking Phonograph Co. royalty statements enclosed with Johnson to TAE, 5 Mar., 10 Apr., and 26 Aug. 1879, DF [*TAEM* 51:697, 702, 746; *TAED* D7932K, D7932O, D7932ZAT1]). In late February Johnson began receiving bids from several firms, including C. Remington & Sons and the Colts Patent Fire Arms Manufacturing Co., for the manufacture of as many as ten thousand, and in mid-April the phonograph company awarded the contract to the Philadelphia firm Brehmer Bros. (C. Remington & Sons to Johnson, 14 and 19 Feb. 1879; Colts Patent Fire Arms Manufacturing Co. to Johnson, 28 Mar. and 2 Apr. 1879; Brown & Sharpe Mfg. Co. to Edison Speaking Phonograph Co., 29 Mar., 4 and 7 Apr. 1879; Brehmer Bros. to Johnson, 9 Apr. and 19 June 1879; all ESP Scraps.; Painter to Brehmer Bros., 15 Apr. 1879, ESP Treasurer's Lbk. [Nov. 1878–July 1879], 570). Surviving models indicate that those manufactured by Bergmann had top-mounted speakers while the Brehmer Bros. machines had them side-mounted (Rondeau 2001, 62–65).

4. Jack Wright was a telegrapher with the Automatic Telegraph Co. who had gone with Edison in 1873 to assist in demonstrating and testing Edison's automatic system in the United Kingdom and who continued to work on that project for some time after Edison returned to the

U.S. (see Docs. 318 and 405). Gouraud had acted as the British business agent for the Automatic Co. and paid Wright his salary, the amount of which was disputed at the time (see Doc. 420).

5. The particular newspaper story or stories sent to Adams have not been identified, but see Doc. 1716 and the articles referred to in Doc. 1705 n. 1. The day before Upton had written his father, "Mr. Edison has arranged during the past week or two to show his light as he has it now. It has made a very good impression on those who are chiefly interested in its success. There is no doubt but that he can have money enough to carry on his experiments if he needs more, from many sources. I know so many troubles ahead since I see how the lamp works experimentally that I do not expect as much for a long time to come." Upton to Elijah Upton, 30 Mar. 1879, Upton (*TAEM* 95:517; *TAED* MU010).

6. Joshua Bailey had recently come to the U.S., on behalf of his own and Theodore Puskas's interests, to take up issues regarding the business arrangements involving Edison's telephone, electric light, and other inventions in European countries. His stay in the U.S. lasted almost two months. Among those who probably accompanied Bailey to Menlo Park were the men who subsequently incorporated the Edison Telephone Co. of Europe and the Edison Electric Light Co. of Europe—James Banker, Samuel White, Robert Cutting, and Robert Cutting, Jr. Puskas to TAE, 5 March 1879; Bailey to TAE, 13 May 1879; both DF (*TAEM* 49:143, 52:285; *TAED* D7903ZBN, D7940X); Docs. 1731 and 1736.

7. It is not known to what specific items Batchelor referred, but in an interview appearing in the 23 March *New York World* Edison had responded to the rumor that "my backers went back on me" by stating that "if such is the case I don't know it, and I don't believe any one can buy any of the company's stock now at any price." "Edison Altogether Serene," *New York World,* 23 March 1879, Cat. 1241, item 1146, Batchelor (*TAEM* 94:463; *TAED* MBSB21146).

–2– April–June 1879

By the beginning of April Edison believed that he had succeeded at what was called subdividing the light, or operating multiple electric lamps simultaneously and independently. His laboratory demonstration in late March convinced him that one mechanical horsepower could sustain six lamps, each equal to a conventional gas jet. From this he calculated the operating expense of his lighting system as one-third that of gas. He wrote his associate Theodore Puskas in Paris that he had only to devise a generator adapted to his system and to refine the lamp to make it suitable for manufacture and everyday use. He began making plans for a large demonstration in Menlo Park which would last for several months, after which he would permit the Edison Electric Light Co. to place his system in commercial use in New York City.

In fact, far more work remained in these and other details than Edison recognized or would acknowledge. The dynamo based on the March design with two-foot field magnets was not completed until 12 April, immediately after which its armature was reconstructed to provide an odd number of commutator connections. On its first trial, the armature windings were torn apart by the strong magnetic forces inside the machine. Edison had to wait for a new armature and it was the end of April or early May before he could try again. The results were gratifying, however, showing it produced more electric current per mechanical horsepower than any other form of generator. Francis Upton designed numerous tests to evaluate the machine more fully but found it difficult to obtain reliable results. At the end of June the machine shop began building a new form of dynamometer for measuring the mechanical power transmitted to the dynamo pulley. Even without complete data, however, Edison adopted this form of bipolar generator as the standard and planned to build thirty machines for the anticipated Menlo Park demonstration. Although the long field magnets gave the new generator its distinctive appearance and superior performance, Edison never tried to patent this design in the United States. He did include it in the British patent specification he drafted at the beginning of May.[1] In that draft he began to articulate for the first time the advantages of a low-resistance armature circuit, which he obtained in part by removing the field magnet coils from the circuit and connecting them instead to an outside power source. In the meantime, Edison ordered an extensive series of experiments on bimetallic thermo-electric generators by which he hoped to recover energy from the waste heat of steam engines. Edison evidently provided his first bipolar generator, along with arc lights, telephones, and other equipment, to the *New York Herald* Arctic expedition in May.

During this period Edison was also trying to perfect his lamp, which consisted of a fine platinum wire "burner" in a high vacuum. One of the remaining difficulties was to control the amount of current to each lamp so the wire would not melt. He tried a number of different regulator designs before settling in early May on a mechanical governor, driven by an electric motor, to momentarily interrupt the circuit to a lamp or household. He also continued to work on formulating a suitable pyroinsulation and thoroughly coating the burner with it. Confident of his lamp, he described its form and method of construction in his draft British patent specification. The greatest obstacles to its commercial use, however, were the relative scarcity of platinum and the difficulty of drawing this refractory metal and its alloys into fine wire. At the end of April Edison received a candid assessment of the situation from Johnson Matthey, & Co., one of the world's foremost metalworking firms, which had been supplying the laboratory with wire. The company promised that they could meet his need for large quantities of wire but that the costs of doing so would be high. Edison immediately took into his own hands the problem of increasing the supply of ore. He requested published geologic surveys and maps of the United States and its territories and inquired about deposits elsewhere in North America. About the first of May he began sending out 2,000 electric pen copies of a letter to "Postmasters and other public men in mining regions" requesting information about ore deposits in their locales;[2] later he dispatched Frank McLaughlin to investigate deposits in Quebec. These activities were widely reported in the newspapers, prompting one former acquaintance in Chicago to call him "about the 'busiest' man alive."[3] Edison's inquiries brought scores, if not hundreds, of replies and ore samples throughout the late spring and summer. At the end of June he began making plans to purchase the platinum-bearing waste sands from hydraulic gold mines in California, which he thought "would undoubtedly give us all the crude metal we desire."[4]

In anticipation of a large-scale demonstration of his system, Edison began to work on details like water-tight junction boxes and insulating compounds for the underground conductors. He also included the basic design for his meter in the draft British patent specification and the shop began to build one for testing. The device operated by the electro-deposition of copper onto a metal rod which could be removed and weighed to give an indirect measure of the amount of current consumed. It also had a fusible link to protect the household circuit from becoming overloaded. While occupied with these seemingly mundane arrangements, Edison also began to develop for the first time his ideas for electrified freight railroads. During his trip West in 1878 he considered the feasibility of pulling light freight trains with electric locomotives, but it was not until May that he began to examine the requirements of such a system and the form it might take. After making a series of drawings, however, he dropped this project until 1880.

By the middle of April George Gouraud had concluded negotiations to create an Edison telephone company in London and similar ones in provincial cities. However, unexpected contingencies impeded efforts to place the telephone in commercial service. James Adams, Edison's technical representative, died in early May after a long period of poor health. Edison selected his longtime associate Edward Johnson to replace him but Johnson withdrew under pressure from the Edison phonograph company. Edison had difficulty finding someone else and eventually prevailed on Johnson to go. His nephew Charley, in London since March, did not get along with Gouraud or the manager of the new company. Edison called him home in May but Charley abruptly left for Paris and apparently engaged with one of Edison's rivals. Despite all this, the nascent company gave successful demonstrations to leaders of the British government and prominent members of the business, press, and scientific establishments. In the United States, the instrument was exhibited successfully at the Franklin Institute and Edison finally reached agreement with Western Union regarding rights to the receiver in May.[5]

Gouraud pressed Edison to set up a working telephone exchange before the Bell interests could do so. Edison devised a manual telephone exchange switchboard, and on 23 May Charles Batchelor instructed the machine shop to build a switchboard and sixty-five complete telephones in order to erect a trial exchange in the laboratory before shipping the equipment to London. Although the electromotograph receiver worked well in the laboratory Edison and Batchelor continued experimenting to improve its durability and reduce the effects of induction in the outside line. In June Edison decided to address the latter problem by placing the chalk cylinder in the tertiary circuit of an induction coil, effectively isolating it from transient disturbances.

Edison arranged at the end of April and beginning of May to sell telephone and electric lighting patents to prospective companies on the European Continent, although no money changed hands at that time. In mid-May, however, he received a large advance royalty payment from the London company's investors. Edison promptly ordered about five hundred journal volumes and books for his library and gave some money to his wife. The payment must have alleviated considerable anxiety, for he reportedly had been "dead broke" and having trouble meeting payroll.[6]

Edison probably added nearly a dozen names to his payroll between April and June, although a gap in the records makes this difficult to ascertain. Among those for whom records first appear in this time are David Cunningham,[7] a machinist; James Seymour,[8] a laboratory assistant; and Charles Wurth,[9] a machinist who had previously worked for Edison. Another significant addition was Alfred Haid, a Ph.D. analytical chemist (probably hired to replace a departed chemist) who conducted numerous ore assays. During this period Edison contracted for the services of New York glassblower William Baetz. He also agreed to give Francis Upton a 5% share in the profits from the electric light.

1. Doc. 1735.

2. Doc. 1734; "Wanted, A Platinum Mine," *New York Sun*, 7 July 1879, Cat. 1241, item 1224, Batchelor (*TAEM* 94:494; *TAED* MBSB21224X).

3. Charles Leyenberger to TAE, 6 May 1879, DF (*TAEM* 49:225; *TAED* D7903ZDQ).

4. See Doc. 1767.

5. See Docs. 1697 n. 4 and 1761.
6. See Doc. 1750.
7. See Doc. 1950 n. 5.
8. See Doc. 1855 n. 9.
9. See *TAEB* 2:519 n. 2.

–1718–

From George Bliss

Chicago, Apl 4th 1879[a]

Dr Sr.

I have made the Western Electric Mfg Co several propositions all of which declined. They have made me none based on
what appears to me justice. I am entitled to large credit from
them for the delays with which they have filled my orders; also
because of the poor quality of goods delivered.[1] Now I dont
want to ruin the business & have worked ceaselessly to build it
up. In this they have not cordially co-operated. I am considering whether it will not be better to turn the whole thing over to
them, if they are disposed to make any reasonable arrangement, & in some way to entirely separate my interests from
them.[2] If you have any wish in the premises you had better
write me fully at an early date[3]   Respy

Geo. H. Bliss   Gen. Mgr

LS, NjWOE, DF (*TAEM* 50:444; *TAED* D7923ZBD). Letterhead of
Edison's Electric Pen and Press. [a]"Chicago," and "1879" preprinted.

1. Bliss's dispute with the Western Electric Manufacturing Co.,
which manufactured electric pens under contract with Edison (Doc.
817), dated back to the fall of 1878 (see Docs. 1425 and 1486). By January 1879 Bliss was running out of pen supplies, and on 26 March he reported that "Pen matters are at a stand still just now owing to the fact
that the Western Electric refuse to deliver any more goods unless their
account is paid in full. I claim damages for the quality of goods furnished
and the repeated hindrances in my getting goods which have caused me
great loss." Bliss to TAE, 13 Jan. and 26 Mar. 1879, DF (*TAEM* 50:383,
432; *TAED* D7923M, D7923ZAV).
2. In his 26 March letter (see note 1), Bliss stated that he wanted to
"go ahead and make Electric Pens and any other kinds which will sell regardless of them [Western Electric] if they do not withdraw from their
present position." He told Edison that "If you have that Reed pen in
shape I wish you would send me a sample. Or if the Gramme pen, which
I saw when there, is better than the old form send one of those." There
are two notebook drawings by Edison from 15 February and another
from the following day that show a perforating device operated by a
small Gramme dynamo run as an electric motor; Charles Batchelor
made two measured drawings of this device on 22 February. The "Reed"
pen may be related to a faint sketch that Edison made on 14 February of
an "Electric Pen vibrator" designed "To utilize a current by breaking

the connection and having a jump spark" pass between two electrodes, presumably burning a small hole in the paper. N-79-02-15.2:1, 3, N-78-12-15.1:175, Machine Shop Drawings (1879–1880), N-78-12-31:183, all Lab. (*TAEM* 31:776–77, 29:1030, 45:49; 30:467, *TAED* N029:1–2, N009:90, NS7986C:8, N014:91).

Bliss reported in another 4 April letter to Edison that he had concluded the purchase on Edison's behalf of the Griest mechanical duplicating pen patent (see Doc. 1654). Bliss received a license to manufacture the device on which Edison was to receive a five dollar royalty; Bliss asked that Edison waive this royalty on the first 400 outfits. He additionally wanted an unrestricted license to manufacture mechanical pens under the Huffman patent he had acquired in January (see Doc. 1559 esp. n. 3). Edison refused both requests, the latter because he feared it might be used against Western Electric and draw him into Bliss's troubles with the company. Edison and Bliss had to agree on terms of the licenses before the patents could be assigned to Edison and they negotiated through Lemuel Serrell, who informed Edison on 11 May that agreement had been reached. However, the matter had not been closed up by 17 May, when Bliss wrote that he planned to visit Menlo Park to complete the arrangements, and it is not known when the transfers were executed. Bliss to TAE, 4 Apr. and 26 Mar. 1879; Bliss to Lemuel Serrell, 4 Apr. 1879; Serrell to TAE, 29 Mar., 8, 9, and 25 Apr., 8 and 10 May 1879; Bliss to TAE, 17 May 1879; all DF (*TAEM* 50:442, 436, 449, 440, 448, 450, 461, 465, 467–68; *TAED* D7923ZBC, D7923ZAX, D7923ZBH, D7923ZBB, D7923ZBG, D7923ZBJ, D7923ZBT, D7923ZBU, D7923ZBV, D7923ZBW).

3. On the reverse of another letter from Bliss on this date, Edison wrote: "I am thoroughly disgusted with the pen biz and all at sea about it— You had better fix up with the WE Mfg Co somehow" (Bliss to TAE, 4 Apr. 1879, DF [*TAEM* 50:442; *TAED* D7923ZBC]). After having been forced by a lack of supplies to close his pen agencies in mid-April, Bliss and his partner, Charles Holland, agreed in May with Western Electric on terms for paying $14,200 they owed the company. Bliss secured his portion of the debt with his entire interest in the 1877 contracts with Edison regarding European telephone and phonograph rights (see *TAEB* 3:678 n. 5) and reported the settlement to Edison on 17 May. Edison formally assented to this transfer on 13 June 1879, with the stipulations that the first $600 due Bliss from the sale of French telephone patents be applied to the payment of foreign electric pen royalties, and that Bliss accept twenty-five shares in the Edison Telephone Co. of Europe in lieu of cash as payment for his telephone interests on the Continent outside of France (Bliss agreement with Western Electric, 14 May 1879, Box 92, NjWAT; Bliss to TAE, 17 Apr. and 17 May 1879, both DF [*TAEM* 50:457, 468; *TAED* D7923ZBQ, D7923ZBW]; Bliss agreement with Western Electric, 22 May 1879 with TAE addendum of 13 June 1879, Miller [*TAEM* 86:20–21; *TAED* HM790065]).

[Menlo Park,] April 4th 79[a]

Model[1]

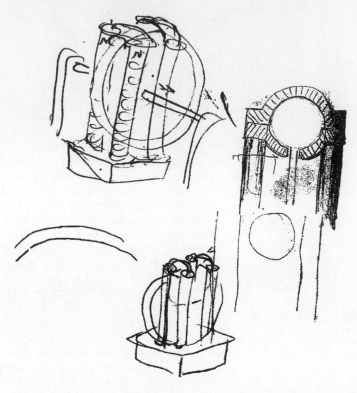

J[ohn] K[ruesi]

X, NjWOE, Lab., N-79-01-14:83 (*TAEM* 31:298; *TAED* N024:39).
Drawn by Edison; date written by John Kruesi.

1. Edison's dominant line of work on generator design during 1879 concentrated on machines with drum armatures, but for several months he continued to design machines with ring armatures similar to those found in the Gramme machine used at Menlo Park at the time. These sketches for a patent model show a generator with a ring armature that differed significantly from prior generators of this type in both its spatial and electrical configurations. Edison arranged the armature so each loop of the coil passed through the center of the field magnet pole pieces in contrast to most ring dynamos in which the coils rotated within the space between the poles. Although probably envisioned as a way to increase the effect of the field on the armature, this resulted in the coils moving parallel to rather than cutting across the lines of force and thus induced very little current. The electrical connections specified in the patent application (U.S. Patent 219,393), which was not executed until 7 July, indicate two significant variations from other ring generators. In an effort to reduce the circulation of extraneous currents Edison constructed his ring so that it was not a continuous circuit. This required a special arrangement for the commutator to compensate for the breaks.

Drawings from Edison's patent application for a ring dynamo (U.S. Patent 219,393).

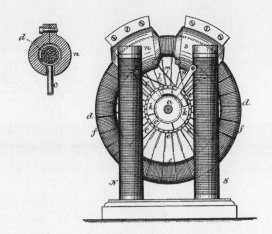

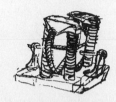

Edison's drawing of his ring dynamo.

The other difference was the use of an intermittently connected shunt around part of the field magnet coils that repeatedly cut and restored the strength of that magnet. It was this latter feature that made the design at all operative. In a standard generator the weakening of the field would make the cutting of the lines of force by the armature less effective in inducing current, but by changing the field in this case Edison was providing an alternative means than motion to induce the current, similar to that employed by Faraday in his seminal experiment on induction.

The first sketches showing a generator with a ring armature passing through the pole pieces date from 16 February. On 7 March measured drawings were prepared and a machine ordered that embodied a version of this arrangement. In this design each magnet pole was composed of a single electromagnet coil with the armature at right angles to the mass of the magnet. It is unclear if this machine was actually built. Although a completion date of 12 April was recorded, the sketch accompanying it shows two coils for each pole of the magnet similar to the design shown in Edison's 4 April sketches for the patent model. Since Edison had continued to explore alternative designs, it is possible that the 12 April date refers instead to the completion of the patent model, the order and measured drawing for which are both dated 5 April. Cat. 1308:117, 133 (Order Nos. 53, 105), Batchelor (*TAEM* 90:725, 733; *TAED* MBN003:32, 40); Machine Shop Drawings (1879–1880); N-79-02-15.1:53–59; N-78-12-04.2:178–81; N-78-12-20.3:47–53; N-79-01-14:49–57, 87; N-79-02-10:57–63, 77, 94; N-79-04-03:3–9; all Lab. (*TAEM* 45:51–52, 58; 31:684–87; 29:416–17; 30:542–45; 31:285–89, 300; 32:1170–73, 1180, 1189; 31:343–46; *TAED* NS7986C:10–14; NS7986C:22; N028: 27–30; N004:89–90; N015:24–27; N024:26–30, 41; N046:30–33, 40, 49; N025:3–6).

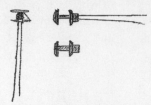

Spiral made of 40 inches of Plat Ir .005. Covering and all .013   when wound on the spool this breaks all up so that you can see all the platina—[b]

Think it would be good to mix Nitrate of Magnesia and Acetate of Lime as Acet Lime sticks so fast to the wire and coats so thin that it would hold the nitrate covering together[c]

12 noon   Put two platina wires in bottle containing Chloride of Calcium solution and connected them to 2 cell of Daniels battery to see whether it would coat on the metal[d]

12 noon   Put piece platinum in boiling Sulphate Magnesia to see if it will coat on

12 noon   Put a2 pieces of platina wire in Chloride Magnesium with battery 2 cells[e] on to see if it would deposit on it

Coating Platinum Wires[c]

Try.[2]

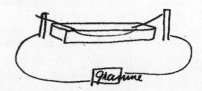

Acetate Mag solution   Keep wire hot with Gramme mach[c]

Try Coating the $8/1000$ covering of Acetate of Magnesia with rubber dissolved in Benzine to make pliable in winding

Platinum Pyroinsulation[c]

Deposit from Chl Calcium or Chl Magnesia by battery

Chloride Mag. solution[f] on Gramme machine with about 60 ohms resist. deposit very thick on one wire, looks crystalline in form   full of cracks— When cold dry[g] it cracks by contratction so that it leaves large spaces in between[h]

Chl Calcium works better apparently finer coating dont crack but when bent & rubbed comes off completely   perhaps a coating of rubber would do this good

Platinum Pyroinsulation[c]

Try amalgam of Mg. and Hg volatilizing Hg with Heat[c]

Try rubbing metallic elCa or Mg on wire

Suspend wire in boiling Carbonate of Magnesia or any salt of Mg. and see if it becomes coated in time.[c]

Make a piece $^{\text{139}}/_{1000}$ thick straight and bring to incandescence and see if it cracks when winding on a $\frac{1}{16}$ mandril.

OK. This is the boss yet ~~goes on~~ bringing to incandescence condenses it so that it looks solid and hard and crystalline   it bends easily without cracking on $\frac{1}{32}$ mandril but it seems to [~~he?~~][i] eat platina bad[c]

I must get something that wont[c]

Make a split mould with $^{12}/_{1000}$ hole   put wire put wire through and stretch   filter water through keeping mould hot[3c]

Cover wire to $^8/_{1000}$ and serve[4] with silk[f] fibre & wind   the heat will eat off the fibre and leave wire and covering on loose

Mix Magnesia with Rubber dissolved in Benzine in variable proportions

Mix Magnesia with Gutta Percha dissolved in chloroform & coat

Try plaster of paris on a wire and note shrinkage

Coat silk fibre with ~~H~~Mg.O. by dipping in alcoholic solution of same and cover wire regular way.

Mix Carbonate of Baryta with rubber in Benzine and coat wire   by great heat it fuses to lead colored slag. nonconductor[c]

Coat with very fine ~~Woodbridge~~ pipe clay[j] and ~~Fire clays very finely worked and~~ mixed with Mg.O.

Coat Silk fibre with pipe clay and Mg.O. and wind in usual manner[c]

Deposit from Sulphate of MgO.

No good[k]   too fluffy & rubs off

Try Alcoholic sol. of MgO. for coat, bring up in vacuum before bending.[c]

This is good, it is a little difficult to put on and it breaks up by contraction a little however these are not bad faults and can be got over but it still eats into the platina[c]

Try Alcoholic Solution of ~~Cerium~~ Rutile[e] for coat   bring up in vacuum before bending (Titanium)

Try Cerium[f] Oxide & Oxalate[c]

Try Zirconium Oxide

TAE   Chas Batchelor
                                                       J[ohn] K[ruesi]

X, NjWOE, Lab., N-79-03-25:21 (*TAEM* 32:150; *TAED* N034:11). Written by Charles Batchelor; document multiply signed and dated. [a]Date taken from document, form altered. [b]Paragraph preceded and followed by dividing marks. [c]Followed by dividing mark. [d]After this and

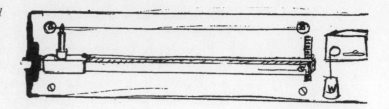

each subsequent entry Batchelor left space for recording results; followed by dividing mark. e"2 cells" interlined above. fObscured overwritten text. gInterlined above. h"Chloride Mag. . . . in between" enclosed by wavy line on left. iCanceled. j"Pipe clay" interlined above. k"No good" multiply underlined.

1. Charles Batchelor dated the first page of this entry, comprising the first two paragraphs, on 6 April. At noon the next day he began experimenting and drawing up the subsequent list of pyroinsulator coatings and processes to be tried, dating the remaining pages 7 April. Probably at a later time, Edison signed and dated several of these pages 6 April. Batchelor left space after each item coating and process to indicate the results of his experiments; it is not clear when he recorded these results.

2. On 1 April, Batchelor had sketched another device for "covering wire" by heating it with a candle flame pulled by clockwork along a track below the wire. N-79-03-25:11, Lab. (*TAEM* 32:145; *TAED* N034:6).

3. A drawing dated 5 April and labeled "No 102 Lime mould E. light" may be the mould in question (Vol. 16:384, Lab. [*TAEM* 4:821; *TAED* NV16:343]). Batchelor's order for No. 102 gives dimensions and describes it as a punch and die for limes "to be made on Edison's plan"; Kreusi dated this 10 April, which may be the date of completion (Cat. 1308:132 [Order No. 102], Batchelor [*TAEM* 90:733; *TAED* MBN003:40]).

4. That is, to wrap the silk fiber tightly to keep it from unraveling. *OED*, s.v. "Serve," 1(54).

–1721–

From Theodore Puskas

Paris, le April 7th 1879[a]

My dear Edison,

I have sent you to-day the following cable "want motographs very badly."[1]

The french Compy. here is very anxious to get the instrument which just now would be particularly useful as a great deal of noise is being made about the new loud speaking Gower Telephone.[2] I had a few days ago an excellent opportunity of making a comparative test of the two instruments at an exhibition given by a society of Engineers[3] here and wrote to Col. Gouraud about it asking him to let me have his two electromotographs for the occasion, but could unfortunately not get them. I have also promised to show it to the Academy of Science here.[4]

Please let me know which receiver you think the best for ła long line with much inductions and also if on such lines you use stronger Batteries and bigger induction coils than under ordinary circumstances.

I am very sorry to hear from you nephew that Adams iłs so ill now again;[5] when I was in London the last time I persuaded him according to a Doctors advice to go to Bournemouth[6] and told Col. Gouraud to let him have some rest, but in spite of my admonitions Adams would not stay in Bournemouth but came back to London where he fell ill again. His illness may interfere with us here as we have only a press but no carbon, and in his present state I feel delicate about writing him to send the carbon as I know that if he could only creep he would come over to help us without giving a thought to the conqsequences to his health. I asked Mr Bailey to inform you of all this begging him to cable me when you would be able to send me carbon directly here to[b] Paris, but he has neither cabled nor written and I do not know if he has spoken to you at all on the subject.

A gentleman of the name of Fleury of Innsbruck in Tyrol[7] has written to me and offered his services in Austria, Germany or Italy referring to you; do you know him, or can you give me any information about this person? Very truly yours

Theo. Puskas

Another reason why I am anxious to get the new receiver is because it might influence the decision about your german Patent which is still pending in the courts.[8]

ALS, NjWOE, DF (*TAEM* 52:266; *TAED* D7940M). Letterhead of Theodore Puskas. [a]"Paris, le" and "18" preprinted. [b]Obscured overwritten text.

1. Puskas cabled again four days later, "When shipping motographs urgent   Answer" to which Edison replied the same day, "Soon as can manufacture." On 21 April, after receiving this and another letter from Puskas inquiring about the receiver, Edison wrote that he had "hoped to have it out before this time but find by experimenting that it is susceptible of much improvement. My men are now working at 4 or 5 new styles which will be far superior to those exhibited in England. . . . Now while these refinements are being added it would be poor policy to put out the inferior insts such as I sent Gouraud." Puskas to TAE, 7, 9, and 11 Apr.; TAE to Puskas, 11 Apr. 1879; all DF (*TAEM* 52:265, 268–69; *TAED* D7940L, D7940N, D7940O, D7940P); TAE to Puskas, 21 Apr. 1879, TP (*TAED* Z400BP).

2. Frederick Allen Gower, New England agent of the Bell Telephone

Co., had adapted the Bell instrument to produce louder speech by employing a large and heavy diaphragm and a stronger electromagnet, which in some demonstrations was reportedly energized by a Gramme dynamo. Its other novel feature was a tube connected to a metallic reed below the diaphragm; blowing air into the tube produced a call signal in the receiving instrument. The telephone was powerful enough to be used at a distance of several feet at both the transmitting and receiving ends and had been successfully exhibited to the Académie des Sciences in Paris in February. John Tyndall lectured on it in London a few days after introducing the electromotograph, prompting George Gouraud to urge Edison to hurry his commercial model. Gouraud cautioned that Gower's was "apparently an important improvement," at least temporarily overshadowed by enthusiasm for the Edison receiver. Bruce 1973, 227, 235; "Gower's Telephone," *Teleg. J. and Elec. Rev.*, 7 (1879): 75; "A New Telephone," *Times* (London), 25 Feb. 1879, 10; "Gower's Telephone," *English Mechanic* 29 (1879): 206; "Recent Improvements in Bell's Telephone," ibid, p. 209, Cat. 1241, item 1192, Batchelor (*TAEM* 94:485; *TAED* MBSB21192X); Gouraud to TAE, 22 Mar. 1879, DF (*TAEM* 52:501; *TAED* D7941R).

3. Puskas probably meant the Société des Ingénieurs Civils de France, the first and foremost French professional engineering organization. Shinn 1980, 194–96.

4. The Académie des Sciences was officially created in 1816 and remains the most prestigious general scientific society in France and a unit of the national government. Its lecture and meeting rooms are in Paris on the left bank of the Seine. Crosland 1992, 14–19, 50–56, 83.

5. Charley Edison had cabled Puskas four days earlier, "Adams very sick for week past and no improvement in his condition   Dont expect telephone for exhibition   Have written." His letter has not been found. TP (*TAED* Z400BN).

6. A resort town on the English Channel. *WGD*, s.v. "Bournemouth."

7. A. L. Fleury was connected with an Innsbruck mining firm. During the previous year he had kept up such a voluminous correspondence of suggestions and queries to Edison that Stockton Griffin "designated him the 'Terror of the Tyrol' on account of his being sadly afflicted with the Caeoethes Scribendi," and Griffin docketed his letters with that nickname. On 18 March, Fleury had written Edison that he was planning to inquire of Puskas "at once" about the electric light. In reply to Puskas, Edison stated that Fleury "is a perfectly competent person to act as agent in the countries named and I strongly advise you to procure his services. He has been recommended to me by strictly first class parties, and I have corresponded with him frequently during the past year." Griffin to James Redpath, 5 Sept. 1878, ESP Scraps. 7:25; Fleury to TAE, 18 Mar. 1879, DF (*TAEM* 49:172; *TAED* D7903ZCH); TAE to Puskas, 21 Apr. 1879, TP (*TAED* Z400BP); for Fleury's correspondence with Edison see *TAEM-G1*, s.v., "Fleury, A. L."

8. See Doc. 1625. Puskas advised Edison on 19 March that patent attorneys in Germany had informed him "that nothing further has been decided about your Patent for Phonograph and Telephone in that country and the matter is still pending." Puskas to TAE, DF (*TAEM* 52:256; *TAED* D7940F).

MENLO PARK, N.J. April 8th 1879[1a]

Dear Sir

The fact to be accomplished by me is the invention, perfection, and introduction into practice of a complete system of illuminating by electricity which shall effect every object and take the place of the present method of lighting by gas. That the cost of any light, or any number of lights equal in candle power to gas is to be but one third that of the latter.[2] That the cost of the plant shall be no greater than that of gas. That after introduction it shall work as satisfactorily as regards disarrangement as the gas systems of New York. To the present date (April 8, 78) I have sub-divided the light into 20 burners each burner being equal to 16 candles, or one gas jet. There is no obstacle to further sub-division the requirements being, more lamps, generators, and an additional supply of power which is analogous to more retorts furnaces &c in gas making. That any electric lamp may be turned up or down the same as gas. I have demonstrated that for every horse power taken from the engine that six electrical jets are obtained each equal to a gas jet which brings the cost to ⅓ that of gas. That there is not as much waste of energy by leakage or otherwise as in gas. that the major portion of plant is salable at a small sacrifice after the lapse of any length of time. That I can measure by means of extreme simplicity the amount of electricity consumed by each customer.

That remaining to be accomplished is,

1st The proper generator both in regard to economy in construction and amount of electricity generated per horse-power applied, its reliability and adaptability to my system.

2d A standard lamp free from all mechanical objections, and of the most extreme simplicity that is attainable.[3]

A test at Menlo Park with 500 lights placed in the surrounding houses and upon posts over the whole of the Park about ½ mile square of from 3 to 5 months duration 24 hours daily, to ascertain the permanency of the lamps and to bring out under my supervision the inevitable disarrangements which occur in all new systems of industry.

All these defects being eradicated the system will be turned over to the Company in New York. Very Truly

Thomas A Edison

LS, HuBPO, TP (*TAED* Z400BO). Written by Stockton Griffin. [a]Place from Edison's laboratory handstamp.

1. An incomplete autograph draft of this letter dated 9 April that was torn from an unidentified laboratory notebook is in WJH (box 25, folder 8).

2. The basis of this estimate is uncertain; see Doc. 1707.

3. Edison presumably meant to differentiate the essential principle of his incandescent burner from the details of its construction, and also from ancillary apparatus such as the regulator. In a newspaper interview published on 23 March he reportedly claimed that the only thing remaining was "to do away with mechanical defects. . . . The burner is no longer affected by chemical action and it acts perfectly well in the sealed globe." "Edison Altogether Serene," *New York World*, 23 March 1879, Cat. 1241, item 1146, Batchelor (*TAEM* 94:463; *TAED* MBSB21146).

–1723–

Notebook Entry:
Telephony

[Menlo Park,] April 8 1879

1 cell constant on line[1]

We find this morning that it is lower probably due to loss of water.[2] I Edison can hear Martin reading 3 feet away but do no understand what is said yet can hear "Mary" etc plain. with induction coil in circuit it reduces it ½ in volume

Zinc pen C[hemically].P[ure]. I hear nothing   Batch hears something ie[a] reads it.

Silver   nearly as good as platinum with coil out, but ½ as good with[b]

Ruthinium   not so good as platinum   calling Platinum 100 then Ru 65.

Bismuth C.P.— When 1st put on it is not so good as platina but after turning ½ a minute it gets louder & louder until it equals or nearly equals platinum   we now put in Pt and find Bismuth is louder. the latter does not squeak while Pt does Bismuth is worn a little.

Cadmium. Batch does not hear a mortal sound

Cobalt. about same as platinum perhaps better than Pt with the coil in. Squeaks less than Pt. not wormn—[c]

Arsenic about 80 to Pt 100—squeaks a little[c]

Antimony. about same as Pt—perhaps slightly less volume. Squeaks about like Pt.[c]

Tellurium   about 50 to Pt 100.[c]

Aluminum   about 20 to Pt 100   Squeaks fearful.[c]

Thallium— Batch hears nothing[c]

Iridium— about same as platina[c]

Nickel. 65 to Pt 100. Squeaks badly[c]

Iron—Chem Pure. 55 to Pt 100[c]

Chromium. Nothing at all[c]

Platinum Iridium alloy   not so good as platinum   we now put in platinum to ascertain and find that it is as good as platinum

The chalk is getting dry and uneven surface   Platinum

comes quite low so I can just hear him read but not to understand when the coil is out—[c]

Sulphide of Lead. about 40 pc to Pt 100—[c]

Brass. about same as platina[c]

Babbitt Metal   exceedingly low   Batch just hears it. it appears to be as loud with coil in as out.

It got so dry that we had to wet it then Pt came out loud.[c]

Magnesium— Batch hears absolutely nothing. current passes through ok—[c]

Gold. about 55 to Pt 100[c]

German Silver. works easier even ~~Dutch~~ with stronger pressure   about 75 to pt 100

DiSulphide[d] Tin (mosaicum)[3]   Very low. Very very very low low![c]

We find German Silver to be slightly louder than Platinum on dryer chalk and a little louder on wetter chalk[c]

Tin. low—very low   I dont hear it 3 feet away[c]

Note   we find it exceedingly difficult to make comparative tests owing to drying of chalk between the tests. We now have put in Tin again but find it exceedingly low

Lead— Batch can scarcely hear it.[c]

Copper— not so good as Platina[c]

Sulphide Copper   Phenomenal   Light pressure loud as Plat. Try again[c]

Carbon— Nothing[4c]

We now try without any battery on. no constant—

Platinum not so loud[e] as with battery—[c]

Zinc. about 60 to Pt. 100[c]

Tin— Low—about 20 to Pt. 100[c]

German Silver— about same as platina—[c]

Lead— Nothing[c]

No need going further on this—

We now put[e] 2 cells constant in the line—[c]

Lead— nothing—[c]

We put platina back and try 1 2 3 & 4 cells constant= We find 2 cells weakens but 1 cell is just right[f]

We now put in Chalk 2[g]

3 oz Chalk
5 grs Acet. Hg.
½ fluid oz Caustic Soda[h]

Platina spring— With 2 cells Callaud on primary and 1 cell in line permanently   Talking ~~not~~ [-------][i] just as[j] loud as on No 1 chalk [~~but very good?~~][i] I see no difference[c]

The ordinary width $^3/_{16}$[k] of spring is best   The $^3/_8$ wide being the worst and there is apparently no difference between $^1/_8$ and $^3/_{16}$ ~~This~~

I turned off the roller perfectly smooth and started on the $^3/_8$ but the result was the same   Think this button is equally as good as No 1 difference of 4 grs Mercury making no difference[c]

Made No 3 button—

3 oz chalk
$^1/_2$ fluid oz Na.O[g]

With no Acet Hg but it is far inferior to 1 & 2. low talking which no wetting will make as good as 1 & 2

Chas Batchelor

X, NjWOE, Lab., N-79-04-08.2:3 (*TAEM* 31:895; *TAED* N030:1). Document multiply signed and dated. [a]Circled. [b]Page torn. [c]Followed by dividing mark. [d]"Di" added in left margin. [e]Obscured overwritten text. [f]Remainder of text written by Charles Batchelor. [g]Multiply underlined. [h]Formula enclosed by braces. [i]Canceled. [j]"just as" interlined above. [k]Interlined above.

1. This statement may have been added after Edison and Batchelor began to experiment with different numbers of cells.

2. These notes evidently represent the resumption of tests of the electromotograph receiver. No record of experiments from the previous day(s) have been found but Charles Batchelor had ordered all of these materials except one (sulphide copper) to be prepared as electromotograph contact springs on 13 March (see Doc. 1700 n. 1).

3. That is, stannic sulfide, also known as artificial gold, mosaic gold, and tin bronze.

4. The next day Charles Batchelor recorded tests with five additional chalks (Nos. 4–8) and also ordered fabrication of "Platina faced springs for new receiver" in five widths. N-79-04-08.2:20–33, Lab. (*TAEM* 31: 905–10; *TAED* N030:10–16); Cat. 1308:135 (Order No. 119), Batchelor (*TAEM* 90:734; *TAED* MBN003:41).

—1724—

Notebook Entry:
Thermo Electricity

[Menlo Park,] April 8 1879[1]

Experiments on Thermo electricity[2a]

We have cast about 100 alloys and are casting several hundred more[3]   these alloys vary by 10 per cent and besides we are to make alloys of the metals with the arsenides, selenides, phosphides sulphides, Tellurides, and also bars of powdered metals oxides etc moulded under hydraulic pressure. These bars are 1203 mm in length and 12 mm thick with holes in the ends   The bars to be tested are connected together by a copper strap, Thickness 2¾ mm length 28½ mm breadth 14mm copper screws are used to secure the bars together on the strap. the distance between the bars after they are coupled is 3 mm at the othe extremities holes are drilled in the sides 10mm from the end into which the wires are inserted holes drilled in the end have screws which clamp the conducting wire. The sulphides and brittle metallic alloys have clamps instead of screws both to secure them together and to secure the wires this device will be explaned further on.

As a source of heat I used a cube of copper with boiling water and the couple was screwed to the side of the cube by a projecting screw which entered a threaded hole in the copper strap connecting the Thermocouple together. A Thompson Mirror Galvanometer[4] with coarse wire was used the resistance of which is ___ ohms—distance between the face of the galvanometer case and the scale 5500 mm. directing magnet

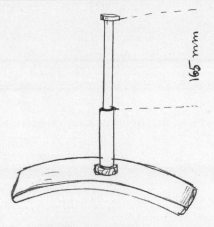

as shewn above. under these conditions one Daniells cell gave a deflection of 100 degress through 1820 ohms 150 degrees through 1245 ohms deflection of 200 through 930 ohms deflection of 250 degrees through 757 ohms deflection of 300 degrees through 642 ohms deflection of 350 degrees through 559 deg. deflection to the left=

After taking these readings we find we had a thermo battery in circuit we take the readings again

Deflection	100	1870	ohms
"	150	—1270	"
"	200	—965	"
"	250	—780	"
"	300	670	"
	350	570	"

We test every metal against copper and allow it to gain a constant temperature before taking the final reading— we have concluded to use the copper clamp both for securing the wires and the bars together.[5]

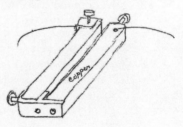

Boiling water

Copper and Brittania	Reading 1st	202 deg
This had been on over 1 hour—	2nd—	25 "
	3rd=	25 "[a]

We now put them on for 10 minutes & then take the reading, copper on constantly

No 77[6a]   Copper & Copper 20 pc Iron 80 per cent put on 9:35 pm. Copper to the right—

Deflection to right	1st reading	90
	2nd "	88 92[a]

No 68.— 70 per cent of copper 30 pc of Antimony— 9:45 pm—

Deflection—To the left=[a]	1st Reading	60
	2nd "	65[a]

No 78. 10 per cent copper 90 pc of iron 10 pm 10:15

Deflection to the right.	1st reading	115
		115

I noticed that after 5 minutes it gave deflection of 132 This shews that the bars gradually become heated at the binding post ends.[a]

No 23— Sulphide ~~Lead~~ Copper 10:23 pm
    Deflection—   1st Reading

            35° through 5100[b] ohms
            65   "   50
            128   "   25   "
            300   "   10   "

Keeps same deflection after 10 minutes. The Sulphide Lead is a very poor heat conductor and is scarcely warm even close to the juncture while the copper is hot at the extreme end. The galvanometer is thrown violently off the scale with no resistance in—[7]

X, NjWOE, Lab., N-79-04-08.1:5 (*TAEM* 30:1105; *TAED* N021:1). [a]Followed by dividing mark.

1. This notebook entry describes procedures and some results of Edison's experiments on thermoelectric batteries which began on or before 8 April. Edison re-numbered four pages of these notes, which he apparently wrote after the "resumption" of these experiments at 7:15 that evening, whose results are on the pages immediately following. No earlier dated records have been found. N-79-04-08.1:29–63, Lab. (*TAEM* 30:1115–32; *TAED* N021:11–28).

2. Francis Upton suggested Edison's rationale for these experiments in undated notes probably made about this time: "In thermo piles, current does not cost directly it is the E.M.F. that costs. The heat that is required for a thermo must be so much in excess of that required that if a fraction only can be used the thermo must be very profitable on low resistance" (N-78-12-20.3:58, Lab. [*TAEM* 30:548; *TAED* N015:30]). According to the preliminary specification of the British electric light patent he filed in June 1879 (Brit. Pat. 2,402 [1879], Batchelor [*TAEM* 92:118; *TAED* MBP017]), Edison planned to use the heat radiated by the surface condensers of the compound condensing steam engines that would drive the generators at a large central station to produce sufficient current in the thermoelectric piles to energize the field magnets. He described the arrangement of surface condensers and thermoelectric piles as consisting of "a great number of iron pipes whose surfaces are painted with a thick coating of a non-conducting substance. The pipes after painting are about five inches in diameter. Over these pipes are slipped rings of the double sulphide of lead and copper. These rings are cast in a mould, and both the inner and outer edges are either covered or plated with copper. The inner copper ring of one disk connects with the outer copper ring of the adjoining disk throughout the entire series of disks." By using these thermoelectric piles to control the strength of the field magnets he could combine them with variable resistances to regulate the strength of the outgoing current. Although he claimed that "owing to the exceedingly low resistance of these disks and the considerable electro motive force which they give between the low range of temperature very powerful currents are obtained, and nearly as much energy is thus obtained in the form of an electric current from the waste heat of the en-

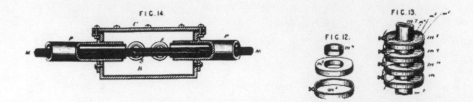

gine as can be obtained from the engine itself through the medium of the Faradic machines," Edison very likely found it impossible to scale the effect sufficiently to power his generators and never employed thermoelectric piles in his system. An undated measured drawing of Edison's "Electric Thermo Battery" is in Oversize Notes and Drawings (Undated), Lab. (*TAEM* 45:36; *TAED* NS7986B:43).

3. Francis Upton made preliminary calculations related to thermopiles and notes on a cadmium-antimony-copper thermopile on 9 March; preparations for systematic experiments evidently began about two weeks later. On 25 March Edison drew a thermopile arrangement and prepared a list (labeled "No 75") of several dozen metals and alloys to be tried in thermopiles, including many already designated for telephone experiments (see Doc. 1700 n. 1). John Kruesi then entered an order for these in the laboratory order book. In a series of undated notes from about the same time Edison identified approximately a dozen metallic sulfides to be alloyed with metals and other sulfides in proportions varying by 10%, then specified dozens of permutations of triple alloys. He also listed other "special compositions" to be cast in large and small bars, the latter about the size indicated in this document. Laboratory assistant C. E. Munsell spent the second half of March researching metallic sulfide compounds and eventually summarized the opinions of chemical authorities and his own findings in an indexed notebook entry dated 15 March to 12 April. The casting of metal bars began on or before 31 March, when John Lawson prepared alloys of copper and zinc. Lawson listed other metals including lead, iron, zinc, cadmium and antimony to be compounded in 10% gradations. His notes describe his method of fusing the metals and also refer to various authorities regarding their composition and characteristics. Lawson continued this work for an unspecified period, at some time shifting his attention to metallic sulfide compounds. N-78-12-20.3:56–59, N-79-02-10:47, N-78-11-21: 129–33, all Lab. (*TAEM* 30:547–48, 32:1165, 29:291–93; *TAED* No15: 29–30, No46:25, No03:64–66); N-79-01-12:7–55, N-79-03-10.2:25–55, N-79-03-31:1–67, all Lab. (*TAEM* 30:858–80, 31:1137–51, 32: 607–31; *TAED* No18:4–28, No32:11–25, No40:2–25).

4. William Thomson developed this type of galvanometer for submarine telegraphy. It consisted of a small mirror, with a magnetized needle on its back, suspended vertically in a thin circular coil of insulated wires. Passage of a current through the wires deflected the needle, causing a beam of light reflected by the mirror to move along a horizontal graduated scale. *DSB*, s.v. "Thomson, William"; Prescott 1877, 148–54.

5. This arrangement differed from the apparatus employed for the

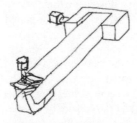

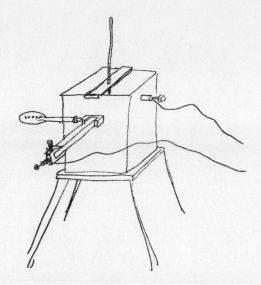

The new test arrangement for the thermoelectric experiments in which the metal bar was clamped to the copper cell.

"resumption" of tests that evening, in which the clamp held a single bar and Edison "use[d] the copper cell itself as the other bar." N-79-04-08.1:29–31, Lab. (*TAEM* 30:1115–16; *TAED* N021:11–12).

6. It is not clear how tests of this and other numbered bars in this document are related to those made with bars of the same composition and identifying number that Edison recorded in his other notes this day. The latter reflect substantially smaller galvanometer deflections, perhaps due to a different experimental arrangement. In most instances Edison tested two bars of each compound. In one case he attributed substantially different results from two ostensibly identical samples to variations in the homogeneity of the bars, a conclusion consistent with John Lawson's prior remarks on the difficulty of obtaining some of these alloys. N-79-04-08.1:33, Lab. (*TAEM* 30:1117; *TAED* N021:13).

7. On 9 April Edison made "Miscellaneous Experiments with rough Masses   not very reliable" on dozens of other materials, including electromotograph telephone points and various ores and minerals; he continued this work the next day (N-79-04-8.1:65, Lab. [*TAEM* 30:1133 *TAED* N021:29]). There are no further dated records of thermoelectric experiments until John Lawson returned to the subject in October 1879. All of these notes by Edison and Lawson, as well as some undated notes by an unknown experimenter on sulphide and sulphate of copper alloys, take up the remainder of this notebook (N-79-04-8.1:65–285, Lab. [*TAEM* 30:1133–1208; *TAED* N021:29–105]).

–1725–

To George Prescott

Friend Prescott

Your favor of this date is just at hand.[1] I agree with you to let Herz[2] sell the Continental patents to the highest bidder.

New Receiver— Your folks do not seem to take much interest in this instrument so I have been working up the English field. Rec'd a cable order yesterday for 2000 complete outfits.[3] They find it works on their worst wires and where no other telephone can be made to work. Just as soon as your people take an interest in the matter I will see what I can do.[4] Very Truly

T. A. Edison   G[riffin]

L (letterpress copy), NjWOE, Lbk. 4:273 (*TAEM* 80:72; *TAED* LB004273). Written by Stockton Griffin. [a]Place from Edison's laboratory handstamp.

1. Earlier in the day Prescott had written Edison that because they had already "paid so much for our Continental patents I feel unwilling to give the matter up without another effort. I have therefore requested Mr. Serrell to pay the French & Italian annuities, and if you agree will write to Herz to dispose of them upon the best terms he can get and we will give him half. What do you say?" In March Lemuel Serrell had advised Edison that the fifth annuity on his French duplex and quadruplex patents would come due on 28 April; Edison indicated on this letter that "Prescott should be notified." Prescott to TAE, 10 Apr. 1879; Serrell to TAE, 19 Mar. 1879; both DF (*TAEM* 49:207, 51:442; *TAED* D7903ZDE, D7929ZAR).

2. A year earlier Cornelius Herz had tried unsuccessfully to sell the quadruplex in Italy. A promoter and confidence man who was reputedly also a physician and member of the San Francisco Board of Health, Herz and inventor Stephen Field had bought rights to the quadruplex in Austria, Spain, France and Belgium from Edison and Prescott in September 1877. See *TAEB* 3:526 n. 7, 4:23 n. 16; Docs. 1043 and 1236.

3. See Doc. 1726 n. 3.

4. In his letter of this date (see note 1) Prescott had asked, "What is the reason that we dont get your new telephone receiver which is spoken so well of in London?" Edison's response here is substantially the same as his draft reply on Prescott's letter.

–1726–

From George Gouraud

London 10 Apl. 1879[a]

My Dear Edison

I dont know what rubbish Adams may have written you but I merely wish to say that he is the unqualifiedly the most difficult individual to get on with that it has ever been my lot to be [--][b] thrown with. But for the diseased condition of the mans body which naturally affects his mind I could not [--][b] [-][b] &

would not have taken the trouble I have to keep him in a good humor. He has been worse ever since yr nephew arrived. He is really dangerously ill and utterly unfit in consequence for the purpose for wh. he is here. You will fully appreciate the of my having here a man whose health will allow of his being able to attend regularly to the business of the Telephone. Hence I have found it indispensable to ~~n~~ detain yr nephew. This decision will probably explain Adams sudden departure for Paris yesterday. He did not seem to think it necessary to consult me as to my wishes or convenience in the matter & I only learned of his intention through Charley, whom he had told that he was going. Charley has had much to bear from Adams, & will no doubt do quite as well without him. Adams is really ~~u~~ unfit for any kind of business & the Dr says any excitement may easily kill him, his heart is so diseased. If he can get quiet work in connexion with the Paris Co.—to [~~with?~~]ᵇ wh. I presume heᶜ has gone he may do well. I have for a long time felt that the expense he has been to me has been justified only on the grounds of charity—& this expense has not been trifling either as you will see. Until last week I gave him money in amts & at times as he requested—& finding him spending more than I do myself I suggested that he shd rec. a regular allowance for his expenses of $5 a day the same I give Charley— I now remember that he seemed not to like this & that may have something to do with his leaving me. I have ~~that~~ thought it well to trouble you with these remarks lest in his diseased state of mind & body he shd distort matters any wise. Now that he has chosen to go I hope he will remain away & not do as he has several before done only come here when he was not wanted in Paris or when he was out of money as he has more than once done— getting money from me just before going to Paris & immediately after his return   If he repeats that performance he will not find that I will—

You must expect yr nephew to remain till the definite model for yr telephone arrives & long enough thereafter to put into practical operation the Telephone Exchange System—which I presume he is competent to do? Is he? Please answer this question. If he is not then send some one who is as it is important that the first Tel. Ex. in the first city of the world shd be a pronounced success.

I have made a first class combination for a London Compy —upon terms that will greatly gratify you— The contract will be engrossed next week—& a round sum of money will be forthcoming immediately the definitive ~~intr~~ struments are rcd

& yr signature is affixed to the document which I may possibly take over to you next week myself & I will at the same time be able to see "the wonderful lamp"[1]

The Presdt. of the Royal Society has asked me to the Grand Soireé of the year 30th inst. with yr telephone.[2]

I hope you approve the name I gave it "The Edison loud-speaking Telephone"—as it is now every where called—very appropriately I think—as at once indicating its essential feature of success wherever the Bell telephone was equally[d] conspicuous as a failure.

The Cable for quotation 2000 telephones[3] was at request of the proposed Company to fix cost & save time, as immediately you decide upon model they want a large number put in hand so as to enable them to occupy the field before the Bell Co do with the Gower improvement   So hurry up!!! Yrs

GEG

ALS, NjWOE, DF (*TAEM* 52:517; *TAED* D7941V). Letterhead of George Gouraud. [a]"London," and "187" preprinted. [b]Canceled. [c]Repeated at end of one page and beginning of next. [d]Interlined above.

1. Gouraud cabled on 18 April that he was sailing for New York. He also urged Edison to hurry a reply to his 9 April message (see note 3). The telephone contract was not completed until mid-May, pending continuing negotiations with the English investors (see Doc. 1743). Gouraud to TAE, 18 Apr. 1879, DF (*TAEM* 52:529; *TAED* D7941X).

2. Gouraud and Charley demonstrated Edison's receiver at this annual Royal Society function on 30 April before Society Fellows and British and foreign dignitaries. The previous morning, Charley and Arnold White, in Gouraud's stead, operated it for the president of the Royal Society, the Speaker of the House of Commons, and other notables over a line from the Royal Institution lecture room to the Royal Society. "The Soirée of the Royal Society," *Teleg. J. and Elec. Rev.,* 7 (1879): 164; "The House of Commons and the Telephone," London *Globe* and London *Standard,* both 30 Apr. 1879, Cat. 1241, items 1177 and 1183, Batchelor (*TAEM* 94:481–82; *TAED* MBSB21177X, MBSB21183X).

3. Gouraud had cabled the previous day, "Quote price 2000 Telephones complete and when deliverable." Gouraud to TAE, 9 Apr. 1879 DF (*TAEM* 52:516; *TAED* D7941U).

–1727–

Equipment Specification: Electric Lighting

[Menlo Park,] Apl 12 [1879]

127[1]   Alter small Edison Faradic[2] so that it will have 49 coils of #   wire in   layers of   turns each.

Make new commutator with 49 points and wind coil so as to be perfectly symmetrical   Bevil off the edges of the poles so as the wire will not catch in turning

The new commutator must be independent from the cylinder.

New cylinder must have 1½shaft and the wood core must be pinned through besides being driven on tight the iron must go clean down to shaft so:—

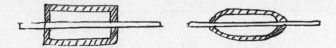

or down so as I shall direct further on

C[harles] B[atchelor].

ADS, NjWOE, Batchelor, Cat. 1308:137 (Order No. 127) (*TAEM* 90:735; *TAED* MBN003:42). Written by Charles Batchelor; double X later written across entry.

1. This is the order number; see Doc. 1687 n. 1.

2. This was order No. 58 (see Doc. 1702), which was completed this day. They apparently decided to incorporate the arrangement of an odd number of coils and commutator blocks that Edison had arrived at in mid-February (see headnote, p. 76) before experimenting with it. A measured drawing of the new generator, dated 15 April, is in Machine Shop Drawings (1879–1880), Lab. (*TAEM* 45:59; *TAED* NS7986C:23–24); it was finished at the end of April. This may be the machine designated as "Edison first Faradic machine" that was installed on the *Jeanette* (see Doc. 1706 n. 2; Vol. 16:430, Lab. [*TAEM* 4:864; *TAED* NV16:385–386]).

The *New York World* reported that during the first test this machine generated so much force that it tore apart the coil of the armature and a new one had to be made ("What Edison Has Done," *New York World*, 30 Apr. 1879, Cat. 1241, item 1174, Batchelor [*TAEM* 94:480; *TAED* MBSB21174]). There are no dated tests of this machine prior to the second week of May (N-78-12-20.2:154–91, Lab. [*TAEM* 29:875–93; *TAED* N008:77–95]), but according to newspaper accounts of a 2 May test reported in the *English Mechanic*,

> Wires were connected with five of Wallace's voltaic lamps, and the moment steam power was applied the lamps were lit. The machine did not become hot, but it gave off a few sparks. The lamps gave a steady light, although currents of air were not excluded, and the five lamps were lit with one machine with one-half of 1-horse power for each lamp. The experiment, although imperfectly made, was entirely satisfactory, and showed that Edison's generator would allow of nine lamps in a circuit, or would develop more dynamo-electric power per horse-power than any machine yet constructed. . . .
> These results are obtained by getting a larger magnetic field, and employing a superior method of winding the bobbin with the insu-

lated wires. Edison will now have 30 of these machines constructed for the formation of his model station, which will probably be at Menlo-park. It is estimated that he can get at least 400 lights by means of his 80-horse power steam-engine and 30 of his dynamo-electric machines. He will now devote his time to producing a lamp which will be without flaw. ["The Electric Light," *English Mechanic*, 16 May 1879, Cat. 1241, item 1203, Batchelor (*TAEM* 94:488; *TAED* MBSB21203X)]

The only order at this time was for one new machine "same stile as improved 58 but larger" that had longer field magnets (3 feet long and 6 inches in diameter) and a bigger armature (9 inches long and 9 inches in diameter); the brushes were also altered (Cat. 1308:143 [Order Nos. 149–50], Batchelor [*TAEM* 90:738; *TAED* MBN003:45]). Measured drawings of the new machine are in Machine Shop Drawings (1879–1880), Lab. (*TAEM* 45:62–65; *TAED* NS7986C:27–35).

-1728-

From James Green

New York, 17 April 1879[a]

Dear Sir

At an adjourned meeting of the Board of Directors of the Edison Electric Light Company held today a Committee was appointed, consisting of myself you, Mr Banker and Mr Cutting[1] to consider and report upon the expediency of taking steps to occupy some of the public buildings in New York with the Edison Electric Light.

Will you kindly indicate a day when you will be in New York and have time to meet and confer with the other members of the Committee upon the subject. If convenient I should prefer next Monday at 2 or 2.30 p.m.[2]

An early reply so that I comay consult the convenience of the other members, will greatly oblige. Yours Very Truly

J. O. Green[3]

ALS, NjWOE, DF (*TAEM* 50:252; *TAED* D7920ZAF). Letterhead of Western Union Telegraph Co. [a]"New York," and "187" preprinted.

1. Robert L. Cutting, Jr., a wealthy New York broker, was among the incorporators of the Edison Electric Light Co. Cutting was also a municipal reformer who had participated in the removal of the Tweed Ring from Tammany Hall. Articles of Incorporation, DF (*TAEM* 18:38; *TAED* D7820ZAM); Obituary, *New York Times*, 14 Jan. 1879, 7.

2. Edison's response has not been found but Cutting wrote again on 19 April that he would "communicate the substance of your letter to the other members of the Committee who will doubtless be content to await your convenience in the matter." Green further explained that "it was not proposed to in any way interfere with the proposed practical exhibition of the light at Menlo Park, but to consider the expediency of ob-

taining a foothold in Public Buildings before they are occupied by rivals; which is said to be the case with the Post Office." Beginning on 26 April five Maxim arc lamps were used to light the main room of the New York Post Office in place of gas jets. Green to TAE, 19 Apr. 1879, DF (*TAEM* 50:253; *TAED* D7920ZAG); "The Electric Light in the Post Office," *New York Herald*, [27 April 1879, p. 8], Cat. 1241, item 1173, Batchelor (*TAEM* 94:479; *TAED* MBSB21173b); "Electric Lights in the New York Post Office," *Scientific American* 24 (1879): 330.

3. Like his father, Western Union president Norvin Green, James Green was a physician who left that profession to enter the telegraph business. Around this time he superintended the Western Union office responsible for railroad contracts. His stake in the Edison Electric Light Co. is unknown. Obituary, *New York Times,* 11 Mar. 1924, 19; Reid 1879, 560–61; Green to TAE, 7 Feb. 1880, DF (*TAEM* 54:460; *TAED* D8033O).

-1729-

From Johnson Matthey & Co.

London. 19 April 1879[a]

Dear Sir

We are favoured with your letter under date the 31st ulto,[1] and are much interested in your remarks.— The faults you point out as incidental to the platinum and iridio platinum wire hitherto prepared are to be readily overcome.—

It is simply because the necessity of overcoming them has never risen before that it has not been done.— These faults would exist to an exaggerated degree we imagine in the wire prepared on the Continent, especially in the iridio-platinum which as there made is only a mechanical mixture of the prepared iridium and platinum sponge compressed and forged, and yet it has been greatly adopted as available for ordinary commercial employment— Our mode of preparation is by fusion—and by giving a second fusion and great care in the subsequent drawing, we can doubtless arrive at a perfection which we should before have done had the want of such perfect homogeneity existed or occured to us.—

We shall be very happy to undertake any experiments with this view.—

You will readily understand that in obtaining absolute uniformity in size of long lengths of iridio-platinum we have great difficulties to contend with, as no jewels will stand a long draft—the excessive hardness of the wire so quickly cuts the holes that we are obliged to pass each length of [---][b] wire backwards through a fresh hole and sometimes backwards and forwards several times to secure a good result, and by this means though with a great sacrifice of jewels and labour we can we think guarantee to you no greater variation in any part of each ½ lb length (the size of reel we should propose to supply) in a wire of .005 than from .0048 to .0052.—[2]

Our price for these alloys in ordinary sizes is 45/ per <u>per troy oz</u> for 10% alloy, and 62/6 <u>per oz</u> for 20%.— We would be willing to contract for a quantity of 100 lbs of each of the extra fineness required .005 at the rate respectively of 42/6 and 60/ per oz Troy[c] on ½ lb reels—and at the rate of 25 lbs per week after one months notice.—[3]

As platinum must advance considerably in price in consequence of the diminished supplies from the mines and the increasing demand, we cannot too strongly recommend for the consideration of any association that may be formed for this business that it will be to their interest to contract with us for a supply upon present market rates, and we are willing to book for their acceptance a quantity of 1000 lbs to be taken up at the rate of not less than 100 lbs per month, the purchase of which will give them virtual control of this material[4]—for which our price to others will be very considerably advanced.—

If this offer meets with consideration we should ask for a reply within one month and for a deposit of 5%.— We are Dear Sir   Yours faithfully

Johnson and Matthey[5d]   P.T.O.[6]

P.S. Since writing the above we have received your telegram "Send following 20% alloys 6 ozs each—six seven and ten thousandths wire"— This we will put in hand instantly and dispatch with the least possible delay.—[7] JM

L, NjWOE, DF (*TAEM* 50:41; *TAED* D7919Y). Letterhead of Johnson Matthey & Co. [a]Place taken from letterhead; "18" preprinted. [b]Canceled. [c]"per oz Troy" interlined above. [d]P[lease].T[urn].O[ver]. written as page turn.

1. Not found.
2. Johnson Matthey had been experimenting for Edison with high concentrations of iridium in platinum alloys since late 1878 (see *TAEB* 4:762 n. 5) and in early February had written Edison that they were sending "a specimen containing <u>30%</u> Iridium and we shall hope to work up to 40% if not to 50% but the cost of labour and destruction of tools will be enormous." Later in the year he asked them to draw wire from an alloy of platinum and osmium, which they had great difficulty in doing, and also asked them to examine a piece of platinum-iridium wire so that they could overcome the defects in its manufacture. Johnson Matthey to TAE, 7 Feb., 7 July, and 13 Sept. 1879, DF (*TAEM* 50:21, 82, 125; *TAED* D7919O, D7919ZBE, D7919ZCH).
3. At the contemporary exchange rate of somewhat less than five U.S. dollars for one British pound sterling, the offered prices equal approxi-

mately $28 and $15, respectively (*Ency. Brit.*, s.v. "Money"). One troy ounce of 9⁄10 platinum-iridium wire is equivalent to approximately 378 feet of this diameter wire, or about 86 miles per hundred pounds troy (105 miles per hundred pounds avoirdupois).

4. This would constitute a substantial portion of the world's production of the metal which, as Edison noted in his 1874 paper on platinum for the *Operator* (Doc. 489), "does not exceed two tons per year," principally from the Ural Mountains.

5. The London firm of Johnson Matthey & Co., refiners of precious metals and manufacturers of platinum apparatus, advertised that its fusion process produced alloys "of the most perfect compactness, strength, and durability, and of increased resistance to the action of acid." The company continued its experiments for Edison during the summer on platinum alloys with iridium and osmium. Johnson Matthey & Co., Paris Universal Exposition circular, Cat. 30,102, Scraps. (*TAEM* 27: 951–52; *TAED* SB796B:2–3); Johnson Matthey & Co. to TAE, 7 July 1879, DF (*TAEM* 50:82; *TAED* D7919ZBE).

6. Unidentified; presumably a clerk.

7. Edison had sent this telegram at 5:50 P.M. the previous day. Johnson and Matthey did not ship the alloys until 5 May. TAE to Johnson Matthey, 18 Apr. 1879; Johnson Matthey to TAE, 5 May, 1879; both DF (*TAEM* 50:40, 49; *TAED* D7919X, D7919ZAD).

–1730–

Francis Upton to Elijah Upton

Menlo Park. April 27 1879.

Dear Father:

This month is nearly through and I hardly know where the time has gone, though I have learned and done quite an amount.[1] Mr. was sick during the past week for three days and during that time I had a fine chance to experiment to my satisfaction.

One thing is quite noticeable here that the work is only a few days behind Mr. Edison, for when he was sick the shop was shut evenings as the work was wanting to keep the men busy.[2] I had a note from Louisa[3] saying that she ~~expected~~ would like to make me a visit. I expected her here yesterday but she did not come.

There still is hope that this summer will see a public exhibition of the electric light. There are thousands of difficulties to be overcome yet before it can be given to the public and Mr. Edison will overcome them if any does. I have not in the least lost my faith in him for I see how wonderful the powers he has, are for invention. He holds himself ready to make anything that he may be asked to make if it is not against any law of nature   He says he will either have what he wants or prove it impossible.

If he does not have a lamp to use electricity he will show that with present knowledge it cannot be had

I hope to go home for a visit in about ~~th~~ three weeks, and hope to find the family health good.

I am with much love   Your Son

Francis R. Upton.

ALS, NjWOE, Upton (*TAEM* 95:527; *TAED* MU014).

1. Five days previously, Upton had reported to his father that he was "busy now with measuring currents, using two instruments that have just been made after my drawings. You see I have a chance to learn how to make instruments; and see them made. I shall be an expert on dynamo-electrical machines in the course of a short time for I see so many various trials of different devices   Nearly each week there is a new experiment to be tried." Upton's experiments were probably related to his effort to determine the voltage of a standard cell for use in dynamometer tests of generators. Upton to Elijah Upton, 22 April 1879, Upton (*TAEM* 95:525; *TAED* MU013); Upton 1880b, 180; Cat. 1308:139 (Order Nos. 134–35), Batchelor (*TAEM* 90:736; *TAED* MBN003:43); N-79-04-21:50–153, N-78-12-20.2:16–35, both Lab. (*TAEM* 30:748–96, 29:807–16; *TAED* N017:23–75, N008:9–18).

2. The *New York Herald* reported that Edison had been confined to bed by a severe cold. His principal assistants continued to work under Batchelor's direction but the *Herald* likened his absence to that "of the star performer from the cast of the play—the performance may go on, but the life of the acting is not there." "Edison's Electric Light," *New York Herald*, 27 April 1879, p. 8, Cat. 1241, item 1173, Batchelor (*TAEM* 94:479; *TAED* MBSB21173a).

3. Probably Louisa Farley, the twenty-year-old daughter of Upton's sister, Maria. Vinton 1874, 432–33.

–1731–

*Agreement with
Samuel White,
Theodore Puskas,
Joshua Bailey, and
James Banker*

[Menlo Park,][1] April 29, 1879[a]

This Agreement made this Twenty ninth[b] day of April, 1879, by and between Thomas A. Edison of Menlo Park, in the State of New Jersey; Samuel S. White,[2] of Philadelphia, in the State of Pennsylvania; Theodore Puskas, of 45 Avenue de l' Opera, Paris, France, and Joshua F. Bailey,[3] of 45 Avenue de l' Opera, Paris, France, parties of the first part, and James H. Banker of the City of New York[c] of the second part, witnesseth:

Whereas the parties of the first part, collectively, own or control the use in the following countries of Europe, to wit: Belgium, Austria, Denmark, Germany, Russia, Italy and Spain, of certain inventions of Thomas A. Edison, George M. Phelps,[4] Elisha Gray,[5] and Frederick K. Fitch,[6] in or relating to Speaking Telephones, and of letters patent which have been or may be granted in said several countries, or any of them, for such inventions, or for any other further improvements or additions thereto, or further inventions of the like character, which may be made within five years from the date hereof.

And whereas the parties[d] of the second part desire to acquire an interest in all the rights of said several parties of the first part collectively in such inventions and letters patent aforesaid, and the parties of the first part are willing to sell and convey to them[d] such interest in consideration of the agreements on their[d] part as herein set forth.[7]

Now in consideration of the premises and of the mutual agreements of the parties hereto as herein set forth, they agree with each other as follows:

First: The parties of the first part agree to sell, convey and secure to the parties[d] of the second part, by instruments good and sufficient in the law of the several countries above named, for the sum of twelve thousand dollars, one undivided sixth part of all the right, title and interest of all and each of them in each of the countries above named, in all the inventions of said Edison, Phelps, Gray, and Fitch, which they now own or control and in all additions to or improvements of said inventions and in all further inventions of said Edison and others above named of the like character which they may acquire within five years from the date hereof, and in all letters patent for any of such inventions, improvements or additions.

And the parties of the first part agree that all or any part of said sum of twelve thousand dollars which may be paid by the parties[d] of the second part, shall be applied to the procuring of, and bringing into use, as soon as possible, by sale or other most advantageous mode, the inventions and patents aforesaid in the said countries of Europe.

Second: The parties[d] of the second part agree to pay, for application in Belgium and Austria to the uses set forth in the first clause hereof, the sum of six thousand (6000) dollars[e] of the twelve thousand dollars above named; and to pay Two thousand dollars of said Six thousand on the execution of this agreement, and the remaining Four thousand dollars in installments of $1000 each on the expiration of each thirty days from the date of this instrument.[f]

Third: It is mutually agreed that if the said parties[d] of the second part, after paying said six thousand dollars[g] shall elect not to pay the residue of said twelve thousand dollars, mentioned in the first clause hereof, on notice after payment of said six thousand dollars[g] to so elect (and option to this effect is hereby expressly reserved to the parties[d] of the second part)—then the parties[d] of the second part shall be entitled to an undivided sixth[d] interest in the inventions and patents aforesaid in and for the countries of Belgium and Austria only—and upon payment of said six thousand dollars[g] as above provided, the parties of the first part agree to sell, assign and secure to the parties[d] of the second part, by good and sufficient instruments as aforesaid, one undivided sixth part[h] of all the right, title and interest of all and each of them in the said inventions and patents in and for Belgium and Austria.[8]

Fourth: The parties to this agreement agree to form together a corporation[9] under the act of the state of New York, entitled "An Act to provide for the organization of certain business corporations," being Chapter 611, of the Laws of 1875, and that such corporation shall be such "limited liability company" as is provided by the thirty third section of that Act, with a capital of one hundred thousand dollars, divided into one thousand shares of one hundred dollars each; that shares to the amount of seventy two thousand dollars of the capital shall be issued as full paid stock in payment of the patents of the parties of the first part hereto in the European Countries named in the preamble hereof, and the rest of the capital shall remain in the Treasury of the company, and subject to its control.

Fifth: The seven hundred and twenty shares issued in payment for patents shall be distributed as follows:

To Thomas A. Edison, and to Theodore Puskas, jointly, three hundred and sixty shares; to Samuel S. White, one hundred and twenty shares; to Joshua F. Bailey one hundred and twenty shares;[10] and to the parties of the second part, one hundred and twenty shares[i] but if they shall elect to pay only $6000. under the option in the 3rd clause then they shall receive only 60 of said 720 shares and the rest shall be distributed as follows, viz to said Edison & Puskas jointly[j] 390 shares; to said White 135 shares & to said Bailey 135 shares.[k]

Sixth: The agreements of the parties hereto shall bind, and enure to the benefit respectively, of their respective executors, administrators, and assigns.

The words parties of the second part being changed to the words party of the second part and the corresponding pronouns being changed accordingly throughout [entire?][j] foregoing instrument and the clause beginning but if they shall elect being also added to the fifth article all before execution[l]

In witness whereof the parties hereto have hereunto set their hands and seals the day and year first above written.

Thomas A Edison[m]   Samuel S. White[m]

J. F. Bailey[m]   James H Banker[m]

In presence of Wm Carman

ADDENDUM[11n]

Menlo Park NJ Apl 30. 79[o]

It is understood by all the parties to the within instrument and they hereby severally agree that it is part thereof that on the formation of the corporation within provided for they will convey and assign to it their respective interests in all the patents and inventions within referred to including the interest which may be a~~q~~cquired by the party of the second part under the within agreement. The shares allotted by the fifth clause being designed to be given to the parties there named respect~~fu~~lively in exchange for the interests so to be transferred by them to the company

Thomas A Edison   Samuel S. White

J. F. Bailey   James H Banker

Witness   G E Carman[12]

TDS, NjWOE, DF (*TAEM* 52:270; *TAED* D7940P1). Notarization omitted. An executed copy of this document is in SSW. [a]Date taken from text, form altered. [b]"Twenty ninth" written in unknown hand. [c]"James H. Banker . . . New York" written in unknown hand. [d]Written in unknown hand. [e]"six . . . dollars" written in unknown hand. [f]"and to pay . . . date of this instrument" written in unknown hand. [g]"six thousand dollars" written in unknown hand. [h]"sixth part" written in unknown hand. [i]"the parties of the second part, one hundred and twenty shares" written in unknown hand. [j]Obscured overwritten text. [k]"but if they shall . . . 135 shares." written by William Carman. [l]Paragraph written in left margin by William Carman. [m]Followed by wax seal. [n]Addendum is a DS written by William Carman. [o]Place and date written by Stockton Griffin.

1. This was notarized the same day by Stockton Griffin; agreements notarized in New York used notaries located in the city.

2. Samuel White headed a successful and highly respected Philadelphia firm of the same name that manufactured artificial teeth and other dental appliances. In recent years he had invested heavily in the telegraph and telephone inventions of Elisha Gray and was also a large stockholder in the American Speaking Telephone Co. *DAB*, s.v. "White, Samuel Stockton"; Davis 1995, v–vi, xxii–xxiii.

3. Joshua Bailey owned rights to the telephone inventions of George Phelps; he also had authority to represent both Puskas and Elisha Gray. Bailey agreement with Edison Telephone Co. of Europe, Ltd., 15 May 1879, SSW (*TAED* X078AG4).

4. George Phelps devised several variations on the Gray magneto telephone employing a single mouthpiece with two diaphragms and various arrangements of permanent or electro-magnets. See *TAEB* 4:26 n. 3; Prescott 1878b.

5. Elisha Gray was a noted electrical inventor connected with Western Electric. His claim to the invention of the telephone provided one basis for Western Union's entry into the telephone business. See *TAEB* 1:402 n. 5 and Hounshell 1975.

6. The electrician Frederick Fitch owned a New York electrical supply shop, successor to the firm of Fitch and Meserole, which Edison had patronized. Fitch devised a telephone transmitter employing carbon plates which he claimed worked with the speaker standing several feet away. At this time he was also at work on a carbon receiver. *TAEM-G1,* s.v. "Fitch, Frederick K."; Fitch to Samuel White, 21 May 1879, SSW (*TAED* X078AI).

7. This agreement superceded the December 1877 contracts with Puskas regarding telephone rights in Continental Europe (see *TAEB* 3:678 n. 5). Sometime in April Edison had prepared an agreement with Puskas amending those contracts, which stipulated that Puskas receive one half of the proceeds remaining after the payment to Edison of a one-time royalty on the sale of each patent; it was never executed. In May, after completing arrangements for his European electric light patents (Doc. 1736), Edison agreed with Puskas (through Bailey) to annul the 1877 contracts. Draft agreement between TAE and Puskas, April 1879, DF (*TAEM* 52:274; *TAED* D7940Q); TAE agreement with Puskas, 14 May 1879, Legal (*TAEM* 144:1197; *TAED* HK203AAA).

8. Under terms of an unsigned contract dated May 1879, this paragraph also would have required White and Bailey to give Banker a certain share of their British rights to the Gray, Phelps and Fitch telephones. SSW (*TAED* X078AG5).

9. The Edison Telephone Co. of Europe Ltd. was incorporated in New York on 14 May 1879 by Edison, White, Banker, Robert L. Cutting, Jr. and his father, Robert L. Cutting. Certificate of Incorporation, DF (*TAEM* 52:278; *TAED* D7940R1).

10. Bailey subsequently sold to White five thousand dollars of his stock in the European telephone company. Bailey agreement with White, 14 May 1879, SSW (*TAED* X078AG3).

11. An executed copy of this addendum written by Edison is with the contract in SSW.

12. Handyman George Carman had worked for Edison since about April 1877, doing odd jobs and some experimental work in the laboratory. See *TAEB* 4:30 n. 6.

Equipment
Specification:
Electric Lighting

[Menlo Park,] Apl 29th 1879

144[1]   Take small Dynamo[a] (141)[2] and make ~~an~~ a solid iron armature on a shaft ~~a~~ ¹⁄₆₄ less than the hole and perfectly in the middle. The armature must be exactly the same length as the heads   I dont any gear wheel on it or commutator

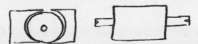

145   Take the armature that Flammer[3] made for (141) and take off the wire and fill up with fine iron that George[4] will bring today. Wind the wire as full as the magnet head will allow

146   After 145 has been made and tried take the armature and shaft and put on shaft a wooden armature[5] the right size for winding on the ~~wire~~ copper so as to make an armature with only copper wire and no iron.[6]

C[harles]. B[atchelor]

ADS, NjWOE, Batchelor, Cat. 1308:141 (Order Nos. 144–46) (*TAEM* 90:737; *TAED* MBN003:44). Written by Charles Batchelor; document multiply signed and dated. [a]Obscured overwritten text.

1. The numbers 144, 145, and 146 are the order numbers; see Doc. 1687 n. 1.

2. This was a machine that had been altered from a small magneto, identified as "Bentleys"; possibly a device designed by Henry Bentley or given by him to Edison. They had begun to work with it on 20 April and two days later it was modified to "be just like our Dynamo," with 5 inch long field magnets of 1½ inch diameter. Cat. 1308:141 (Order Nos. 140–41), Batchelor (*TAEM* 90:737; *TAED* MBN003:44).

3. Charles Flammer was one of the laboratory machinists. See *TAEB* 4:135.

4. George Carman did odd jobs around the laboratory. See *TAEB* 4:30 n. 6.

5. Earlier in the month the laboratory had begun using wooden cylinders to experiment on the best method for winding the armature. Francis Jehl later recalled that the difficulty of drawing the armature windings on paper had led Edison to suggest that "Kruesi make up a few small wooden models of the drum. . . . Then you can take string and actually wind it round the block, instead of drawing imaginary lines." Jehl remembered that the laboratory staff then had winding competitions. Cat. 1308:136 (Order No. 126), Batchelor (*TAEM* 90:735; *TAED* MBN003:42); Jehl 1937–41, 292–94.

6. In September Edison applied for a patent specifying a wooden cylinder with a helix of iron wire over which the copper wire was wound longitudinally. He described this as a means to "increase the effectiveness and cheapen the construction of the revolving armature." U.S. Pat. 222,881.

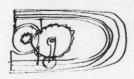

Charles Batchelor's drawings of the "Bentley" magneto and the alterations made to it on 22 April 1879 incorporating the long field magnets and armature from Edison's dynamo.

Notebook Entry:
Electric Lighting

[Menlo Park,] May 1st 1879

Electric Light.
House Meter.
Measuring current by weight of copper deposited[1]

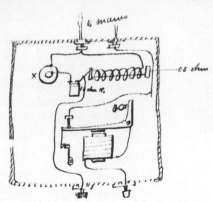

X Depositing Cell

Chas Batchelor

X, NjWOE, Batchelor, Cat. 1304:48 (*TAEM* 91:47; *TAED* MBN004:45). Written by Charles Batchelor.

1. Charles Batchelor's drawing is the first extant 1879 design of a complete meter, though there are some drawings from 2 February of an electrolytic cell for a meter. The design drawn by Batchelor is similar in principle to one Edison had conceived in December 1878 (see Doc. 1622). Edison himself made similar drawings on 4 May and John Kruesi entered an order to make the meter on 8 May; the measured drawings date from the next day. There are no dated records of tests of this meter until the end of July. These tests continued until mid-September and then resumed briefly in mid-November. N-79-03-25:117, 120–21; Cat. 1146; N-79-04-03:101–16, 167–70, 197–205; all Lab. (*TAEM* 32:197, 199; 6:639–40, 643, 645; 31:392–400, 422–24, 437–42; *TAED* N034: 58, 60; NM014:17–18, 21, 23; N025:53–61, 84–86, 98–104); Cat. 1308: 143 (Order No. 151), Batchelor (*TAEM* 90:738; *TAED* MBN003:45).

Edison had included this basic design in a February 1879 U.S. patent application but the description and claims related to the meter as well as part of a drawing showing the meter were deleted before the patent issued (Pat. App. 227,227). He did include the same description and the full drawing of the meter in a final British specification filed in April 1879 (Brit. Pat. 4,226 [1878], Batchelor [*TAEM* 92:107; *TAED* MBP013A]). According to that specification, the apparatus consisted of a box containing a "coil of very large wire, whose resistance is proportioned to the number of burners used in the house. This resistance is but the fractional part of the resistance of a single lamp." The actual measuring device was an electrolytic cell

which contains a solution of copper, [and] has two electrodes of copper, one of which is very thick, while the other is very thin.

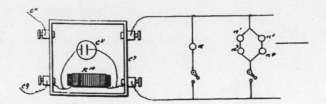

The meter box shown in Edison's British Patent 4,226 (1878) appears at left in the drawing; the resistance coil is R10 and the electrolytic cell is C8.

The small portion of the current which passes through the cell carries over copper and deposits it upon the thin plate. If one lamp is placed in connection, it draws current from the main, and a proportionate quantity passing through the cell effects a deposit upon the thin plate; if another lamp is connected, double the quantity is deposited, and so on. At the end of any period, say one month, the plate is taken by the inspector to the central office and accurately weighed. As the deposit of copper upon the thin plate will be proportioned to the total amount of current passing into the house, the same becomes a correct measure or standard for the charge for the electricity supplied. [Brit. Pat. 4,226 (1878)]

In a variation Edison described in the U.S. application, one plate could be suspended from a spring balance and its weight read from an external dial "but this requires delicate mechanism in every house, which is rendered unnecessary by weighing the plates" (see Doc. 1622 for this design). The May design includes a feature that was not included in either the U.S. or British applications. This was a circuit breaker (the magnet and lever shown at bottom in Batchelor's drawing) that would interrupt the circuit in case of an unusually heavy current, such as that caused by a short circuit.

–1734–

Platinum Search Circular

MENLO PARK, N.J.[a] [c. May 1, 1879][1]

Dear Sir:

Would you be so kind as to inform me if the metal platinum occurs in your neighborhood? This metal as a rule, is found in scales associated with free gold, generally in placers.

If there is any in your vicinity, or if you can gain information from experienced miners as to localities where it can be found and will forward such information to my address I will consider it a special favor, as I shall require large quantities in my new system of Electric lighting.[2]

An early reply to this circular will be greatly appreciated. Very truly

Thomas. A. Edison.

ALS (electric pen copy), NjWOE, DF (*TAEM* 50:589; *TAED* D7928G). [a]Place from laboratory handstamp.

1. Although Edison had received by this date letters from two individuals responding to newspaper stories about his need for platinum, this circular letter marks the beginning of his intensive and highly public search for the mineral. He may have been prompted to write it by Doc. 1729, which he would have received at the end of April. The *New York Sun* subsequently reported that he sent approximately 2,000 copies to "Postmasters and other public men in mining regions" around 1 May, enclosing with them specimen cards sprinkled with platinum and iridosmine. Edison wrote the circular in his fine telegraphic hand and had it reproduced with the electric pen; the first extant reply to it, dated 7 May from the Dakota Territory, was addressed to Edison on the copy from which the text of this document is taken. William Ross to TAE, 3 Apr. 1879; Fred Gerhard to TAE, 8 Apr. 1879; A. J. Cogan to TAE, 7 May 1879; all DF (*TAEM* 50:581, 583, 589; *TAED* D7928A, D7928B, D7928H); "Wanted, A Platinum Mine," *New York Sun*, 7 July 1879, Cat. 1241, item 1224, Batchelor (*TAEM* 94:494; *TAED* MBSB21224X).

2. Edison received scores of letters about platinum throughout the late spring and summer, some replying to this circular and others to specific inquiries. On 14 May he instructed Stockton Griffin to "get a box marked Platina & keep all the correspondence etc in it." Many of the incoming letters include Edison's marginal comments, some of them extensive and illustrative of the extent of his search. Marginalia on A. R. C. Selwyn to Edwin Pope, 14 May 1879; DF (*TAEM* 50:593; *TAED* D7928L); for a sample of the replies see the folder "Mining–Platinum Search," DF (*TAEM* 50:580; *TAED* D7928).

Edison, who was familiar with existing sources of platinum (see Doc. 489), made numerous personal inquiries regarding as-yet-undeveloped supplies. He solicited help on 30 April from Wexel & De Gress, a munitions firm operating in Central and South America, and the next day from Edison Electric Light Co. secretary Calvin Goddard. He contacted other acquaintances, including A. E. Foote, a Philadelphia mineral dealer, James Dwight Dana, and George Barker. Barker replied that the amount of platinum produced from known sources "is very limited at least for general use," and cautioned that "An indiscriminate examination for it in all possible rocks, would therefore be a great waste of time." Wexel & De Gress to TAE, 3 May 1879; Goddard to TAE, 2 May 1879; Dana to TAE, 28 May 1879; Barker to TAE, 30 May 1879; all DF (*TAEM* 50:587, 586, 605; 49:252; *TAED* D7928E, D7928D, D7928U, D7903ZEM); TAE to Foote, 26 May 1879, Lbk. 4:358 (*TAEM* 80:80; *TAED* LB004358A).

Somewhat later, Edison contacted state officials concerning geological surveys. In July, after having turned his attention to ferrous black sands found in California and coastal Oregon and Washington (see Doc. 1767), he ordered detailed maps of those states as well as Nevada, Mexico, Canada, and other countries. He also wrote to the U.S. Minister to Russia in St. Petersburg that he was "endeavoring to collect statistics regarding the product of platinum from all parts of the world" and desired answers to four specific questions about the production, price, and export of it in Russia, "which at present is the only market in the world." TAE to G.W. & C.B. Colton & Co., 9 and 15 July 1879, Lbk. 4:456 and 482; TAE to U.S. Minister, Russia, 18 July 1879, Lbk. 5:2 (*TAEM* 80:93, 104,126; *TAED* LB004456, LB004482, LB005002); replies from state officials and publishers concerning survey maps include Howard Chauncey to TAE, 4 June 1879; Joseph Miller to TAE, 5 June; E. H. Gove to TAE, 5 June; William Strong to TAE, 6 June; Frank Putney to TAE, 6 June; J. A. T. Hull to TAE, 6 June; George Martin to TAE, 6 June; W. E. Woodruff to TAE, 6 June; Bowen & Kingsbury to TAE, 7 June; John Collett to TAE, 7 June; James Galloway to TAE, 9 June; George Drew to TAE, 9 June; James Gaines to TAE, 9 June; A. M. Nelson to TAE, 10 June; John Proctor to TAE, 11 June; and F. P. Thompson to TAE, 11 June 1879; all DF (*TAEM* 50:628, 633–45, 651–52, 656–57, 661, 663–66; *TAED* D7928ZAJ, D7928ZAM, D7928ZAN, D7928ZAN1, D7928ZAO, D7928ZAP, D7928ZAQ, D7928ZAR, D7928ZAS, D7928ZAT, D7928ZAX, D7928ZAY, D7928ZAZ1, D7928ZBD, D7928ZBF, D7928ZBG).

–1735–

Draft British Patent Specification: Electric Lighting

[Menlo Park, c. May 1, 1879[1]]

~~Preliminary Sp~~ ~~Provisional~~ Specification[2] for ~~ana~~ new English patent—

I have ascertained that[a] When wires or sheets of[b] platina Iridium or other metallic conductors of Electricity ~~which~~ that fuse at a high temperature are exposed ~~for[c]~~ to a high temperature near their melting point in air ~~either~~ for several hours by passing a current of electricity through them and then are allowed to cool the metal is found to be ruptured and under the microscope there is revealed myrids of cracks in every direction, many of which are seen to reach nearly to the centre of the wire. hence ~~if~~ ~~I have also discovered that platinum or platinum alloyed with Iridium or metals of the platinum group~~

~~If the heat is long continued these cracks~~ I have also discovered that contrary to the received notion platinum or platinum & Iridium alloy looses weight when exposed to the heat of a candle that ~~the~~Even heated air causes it to lose weight, that the loss is so great that it ~~colors~~ tinges a hydrogen flame Green, and under the influence of an Electric current ~~that~~ and at a yellow white heat ~~an~~the loss is very[c] great, after a time the metal falls to pieces. hence: wire[d] ~~and~~or sheets of[e] platinum or platinum & Iridium alloy as now known in commerce[f] ~~is~~are useless for giving light by incandescence ~~as our~~ 1st because Its[d] loss of weight makes it expensive, &[c] unreliable and causes the burner to be rapidly destroyed 2[g] because its electrical resistance changes by loss in weight and its ~~melting point as a whole~~ light giving power for the total surface is greatly reduced by the cracks or ruptures the melting point being determined by the weakest spot where the greatest difference of potential of the electric current is present which ~~abnormally heats t at~~ causes this point to be brought to a higher heat than the rest of the surface of the wire. again as it is essential to obtaine a[c] steady light the platinum burner must be screened from the air, and when thus screened by being placed in a glass vessell the glass soon becomes coated with a black deposit of platinum.

~~I have discovered a re~~   If A platinum ~~Iridium~~ spiral brought to incandescence under these conditions may be made to give a light of three ~~sperm~~ standard candles when near its melting point & when the radiating surface is $^{53}/_{16}$ of an inch, but this amount of light will be gradually reduced as before described.

I now come to that part which constitutes my discovery and invention.

I have found that the cause of the rupturing of the metal when brought to incandescence is due to the expansion of the gases contained both[c] in the physical pores & also[c] in the mass ~~bei~~the later[h] air being probably placed there in the act of drawing the wire or rolling the sheet   now this air when subjected to these high heats greatly expands and ruptures the metal,[i] and it cannot be driven out by slowly heating as it is[c] soon absorbed. I have also discovered that the loss of weight and apparent volatilzation of the metal is due entirely to the attrition of the air molecules against the highly heated surface   having thus ascertained the cause of fracture & loss of weight[d] I have conducted experiments to obviate these defects and have succeeded by the following method[3]

~~If~~ a spiral of platinum wire ~~be~~ is[c] placed in a glass bulb ~~and connected to the source of electricity~~ with its ~~either~~ends passing through & sealed in the glass[j] and the air exhausted from the bulb by a Sprengel pump[k] until the ~~spark~~ discharge from a three inch Induction coil will not pass between[l] subsidinary[c] wires in the bulb 4 millimetres apart.[4] the wires of the spiral ~~is~~are then connected to magneto[d] electric machine or battery whose current can be controlled by the addition of resistance. ~~A~~ sufficient ~~quantity~~ current is allowed to pass through the wire to bring it to about[c] ~~250~~150 ~~or 300~~ deg Fahr ~~when~~   it is allowed to remain at this temperature for 10 or 15 minutes. while thus heated both the air mechanically mixed in the metal by drawing or rolling as well as that in the physical pores[d] ~~e~~is expelled on the one hand by heat & on the other hand tends to pass outwards to equalize the pressures. while this air or gases[m] is passing out of the metal the mercury pump is kept continuously working. after the expiration of the 15 minutes the current passing through the metal is augumented so that its temperature will[d] be about 300 degrees Fahr, and it is allowed to remain at this temperature for another ten or 15 minutes. ~~Thus~~

If the mercury pump be worked continuously and the temperature of the spiral ~~gradu~~ raised at intervals of 10 or 15 minutes until it attains ~~mo~~ to vivid[n] incandescence and the bulb be then sealed we then have ~~a~~the metallic wire in a state heretofore unknown for it may have its temperature raised to the most dazzling incandescence emiting a light of 25 standard candles whereas before treatment the average melting point of a number of the spirals ~~having~~ of the same length & size of wire and having the same radiating[d] surface was but 3 standard candles.— The wires ~~in the Vacuum~~ subjected to the process of ~~excluding~~ freeing them from air, are found after the process to[d] have a polish exceeding that of silver and obtainable by no other means.[h] no cracks can be seen even after the spiral has been raised suddenly[c] to incandescence many times by the current and no volatilization takes place as there[d] is no deposit upon the glass bulb, nor does ~~the~~ a delicate[d] balance shew any loss of weight after burning for many hours continuously because if the spiral is in high vacuo the attrition of the gasous molecules is[d] reduced to the minimum and were it possible to obtain an absolute vacuum there[d] could be no loss whatsoever Even with the residual gas the total loss ~~w~~after a years use[o] would scarcely be noticed.

I have further discovered that if an alloy of platinum & Iridium or even platinum[p] be coated with ~~eith~~ the[c] oxide of magnesium ~~by~~ in the manner here~~a~~inafter stated, ~~t~~and ~~brought~~ subjected to the[c] vacuum process described that ~~a~~combination takes place between the metal and the oxide~~s lik~~ giving ~~the all~~ the former remarkable properties= ~~w~~With a spiral ~~of the same size as that of platinum already described~~ having a radiating surface of 3/16 of an inch, light equal to that given by 40 standard candles may be obtained whereas the same ~~w~~spiral not[d] passed through ~~the~~my process would melt before giving a light of 4 candles. The[d] [-----][q] effect of the oxide of magnesia is to harden the wire to a suprising extent, to render it more refractory ~~and~~   A spiral spring made ~~of~~[r] of the wire ~~loses~~ is just as elastic & springy[s] when at[c] dazzling incandescence as when cold=

~~It is ob~~ I have found that chemically[g] pure iron & nickel[t] drawn in wires and subjected to the vacuum process may be made to give a light equalling that of platinum in the open air & not subjected to the vacuum process=

Carbon sticks may be also freed from air in this manner and be brought to a temperature that the carbon becomes pasty and if then allowed to cool is very homogeneous & hard= rods or plates made of[u] mixtures of finely divided conducting & non~~n~~conducting[v] materials may thus be freed from air.

It is also obvious that the metal might be heated ~~have its temp~~[w] ~~by~~ by subjecting the containing bulb to a considerable temperature but this does not produce but partially free the wire.[h]

I will now describe the form of burner or lamp[x] which I propose to use   ~~It~~ ~~To obtain practically unlimited subdivision of the~~   ~~be able to bring to~~   To obtain practically several hundred electric lights each equal to the ordinary gas jet upon one circuit ~~It~~it is essential for many reason both on the score of economy, facility, and[g] reliability, to place them all in multiple arc and to prevent the combined resistance of several hundred lamps from falling to such a low point as to require a main conductor[d] of immense dimensions with low resistance[y] and generating[z] machines of corresponding character~~s~~, ~~I~~it is essential to reverse the present ~~pra~~ universal practice of having lamps which have but one or two ohms resistance[d] and construct lamps which shall have when ~~burning~~ giving ~~l~~their proper light [resistance] of several hundred ohms ~~and~~   I have ascertained by experiment, ~~that no loss~~ that the loss of energy is in proportion ~~of~~ to[c] the extent of the radiating surface & independent of the resistance. ~~he~~of the conductor[5]   ~~hence two~~ hence whether we have 1000 lamps each of ¼ of an inch raidiating surface & ~~r~~each of 1 ohm resistance or 1000 lamps having the same radiating surface & ~~500~~1000 ohms resistance each the loss of energy ~~on~~from each lamp when giving each[c] a light of 15 candles will be the same, but the combined resistance of the 1 ohm lamps will be ¹⁄₁₀₀₀ of an ohms requiring an enormous main[c] conductor—whereas the combined resistance of the 1000 ohm lamp will be 1 ohm requiring a conductor of very moderate dimension.[6] In practice ~~the~~ a[c] resistance of 200 to 300 ohms in the burner will be sufficient. again with lamps of low resistance the lamp connections & leading wires must be large[d] to prevent ~~l~~great loss of energy by resistance. the leading wires from the main conductors are large ~~and~~ expensive & bulky to handle. the low resistance of the burner ~~requires the larger~~ or incandescent conductor requires large terminals to convey the current and these ~~after~~ offer by their conduction a medium for the rapid dissipation of energy without producing any effect whereas with a lamp of high resistance ~~the leading~~

~~wires from the main condr does may be small and the~~ all these objections are obviated. =

My burner consists of a bobbin ~~of co~~ composed of an infusible oxide ~~either~~ such as oxide of Calcium Cerium Zirconium, Magnesium, ~~m~~freed from Silica & turned in a lathe from sticks[aa] & moulded by hydraulic pressure in a manner hereinafter set forth   This bobbin is shewn in fig 1

It is secured to two platinum wires as[d] in fig 2
Fig 2

These wires serve to hold the bobbin in the centre of the sealed glass vacuum bulb and at the same time serve as conductor of the electric current to the wire coiled upon the bobbin

Fig. 3 shews the manner of connecting the wire upon the bobbin to the platina supports
Fig 3.

The bobbin is sealed in the vacuum bulb as shewn in fig 4
Fig 4[7]

The two platinum wires 1 & 2 are ~~firess~~t passed through a small tube of glass X which is softened around the wires. the ~~bobbin~~ wires 1 & 2 pass out through the sides of the narrow tube leading from the bulb and are sealed to prevent leakage by having smaller glass tube softened upon the wires bent at right angles which tubes are sealed to the glass of the bulb=[d]

Before the burner is permanently sealed at .g. before being taken from the air pump (the wire upon the bobbin which previous to winding thereon has passed through the vacuum process heretofore described) is brought to incandescence for about one hour to heat the bobbin and thereby expell its air, whereupon the bulb is permanently sealed is &[c] is ready to be placed in the lamp= This is shewn in fig 5[bb]
fig 5[cc]

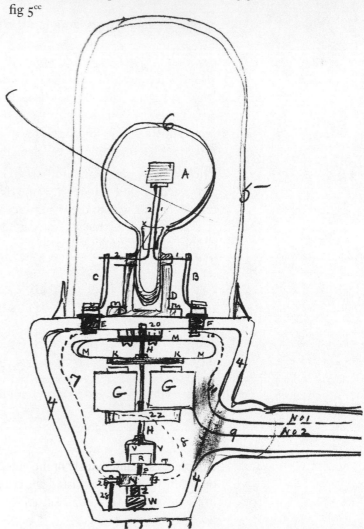

6 is the bulb containing the burner A supported by the platinum conducting wires 1 & 2   The bulb lays in the socket D the extremeties of the conducting wires 1 & 2 are connected by copper wires to the insulated clamping screws E & F   5 is an

outer glass vessel to prevent injury to the vacuum bulb. This may take any form such as an opal shade, large^c bulb—

G is an electro magnet having a resistance of but a fraction of that of the lamp, and serves to regulate the ~~amount~~ total energy consumed or given out by the lamp   this it does by being adjusted in such a manner that it will serve to disconnect the lamp from the circuit should the current increase materially^c from any cause beyond what is necessary to produce the normal amount of light.[8] The magnet ~~when is provided with an~~ is ~~rr~~ in the same circuit as the incandescent conductor of the bobbin .A. The lamp being in multiple arc where alone this form of regulation is admissible the current passes from wire No 1 to post F thence via B^d to the wire 1 to the bobbin A through its wire to wire 2 thence via C to the post E via wire 7^{dd} to the ~~s~~contact spring S through R to the spring T thence to the left to the magnet .G. & from the right of magnet .G. via wire 9 to No 2 wire. The circuit admits of being broken by the effect of the attraction of the armature .K. which is secured upon the wide spring^{ee} M, the upper part of which is secured to the base of the lamp fixture. connected to the armature is a rod H which slides in bearings at 20 & at 22 so that the motion of the armature to & from its magnet gives this rod an up and down movement. upon the lower end is secured a cup ~~w~~ .V. whose periphery serves to come in contact with the springs S .&. T and seperate them from the metal disc .R. which is secured to the disc of insulating or non conducting material N (to which the springs S & T are secured) by a pin .P. a pin 28 passing through a hole in a projection 29 from N serves to prevent the latter from turning when it is adjusted upward ~~in~~ or downward by the double threaded screws Z .&. Z.[9] by means of the latter the springs S .&. T may be brought in so close a proximity to the edge of the cup .V. that a very slight current passing through the magnet .G will attract its armature sufficiently to cause a downward motion of the ~~R~~rod H & seperate the springs S. T. simultaneously thus opening the circuit in two places simultaneously ~~of~~ the object of which will hereafter be set forth. If thus adjusted the magnet will continue to vibrate automatically with great rapidity but ~~owing to both~~ quite noiselessly when the case is closed. in this condition ~~no increase of the current~~   ~~will allow~~ only a certain amount of energy will pass to the lamp and if the current is increased while thus adjusted the effect will be only to increase the amplitude of the vibrations rendering the closing of the ~~sh~~ circuit shorter, so that in a given time the total energy will be about the same whether the current is strong or weak   If^d now the disc N is adjusted downward more and more current will pass to the lamp as the amplitude of the armature of the magnet^{ff} is thus^d decreased ~~w~~ until it reaches a point where the power of the current is insufficient to ~~anna produce~~ energize the magnet to such an extent as to overcome the constantly increasing power of the circular spring M ~~hence~~ and seperate the springs S. T. from .R. If the Tensions of the current [is?]^q are^c properly regulated, The lamp will give its normal amount of light and the springs S .&. T will be in such proximity to the edge of the cup .V. that any abnormal accession of current from accident of otherwise will cause the latter to seperate the springs S .&. T & prevent further rise of temperature in the burner .A ~~band~~ it will continue to vibrate as long as the abnormal flow of current continues & falling back to a quiescent position when the current is at its normal strength. As there is no difficulty in keeping the temperature of a 1~~5~~6 candle power burner within the limits of 12 to 18 candles the regulator ~~will~~ may be set ~~at 18 candle~~ so as to vibrate & reduce the total energy when^d the current reaches a point sufficient to produce a light equal to^{gg} 18 candles.

The main object is to produce even illumination and ~~not particularly~~ afford a method of increasing & decreasing this amount ~~between~~ from a certain limit to any lesser amount & not particularly to prevent the fusion of the incandescent conductor because^g ~~The~~ in practice ~~but h~~ I shall not have the burner give more than a light equals to 16 Standard Candles whereas it will give forty without fusion.

~~Heretofore~~   ~~In workin~~ when it is desired to disconnect the light altogether the upward movement of N. in the act of turning out the light ~~as with gas~~ reaches a point which is the normal position of the ~~m~~armature magnet hence the springs S & T are forced against the edge of the cup V and the circuit is permanently opened   ~~heretofore it has been difficult to obviate the spark due to the break in the circuit when the lamp is disconnected without which is the la~~

With lamps worked in series I have regulated the temperature by thermal regulators which served to operate devices which shunted the burner and gave a new route for the passage of the energy.— Regulators worked in this manner ~~h~~are very little troubled^d with sparks at the contact points, as lamps of low resistance may be used and the difference of potential is very small. ~~and the~~ But when lamps are to be worked in multiple arc it is essential that the lamp should be disconnected entirely from the circuit at[d] intervals following each other with some degree of rapidity and with dynamo electric machines where the field magnet is secured by the system of reciprocative accumulation,[10] the break in the circuit causes the latter to discharge ~~and~~ or when many lamps are used be reduced ~~fr~~ in strength for an instant & this gives a very powerful spark sufficient to fuse the platinum contact springs & point by producing a momentary arc. by dispensing with the varible field magnet ~~& making it~~ & ~~con~~ excluding it from the circuit, and energizing it by extraneous energy[11] the spark is so reduced as to be inconsiderable   ~~For but~~ being proportionable to the discharging time of the induction cylinder of the magneto electric generator not withstanding the fact that the spark is small I have found that time brings out a defect not at first perceptible which[d] is that ~~a small point~~ the action of the spark as in the electric arc causes a small point not larger than a needle to shoot out from one of[hh] the platina contact points and although the contacts are still as perfect as at first the effect is to change the relative position of the point ~~& thus~~ of contact & thus alter the adjustment   I have f devised a method of ~~obviating the spark~~ reducing the spark between contact points[ii] from a strong current of great electromotive ~~fr~~ force such as that given by a dynamo machine on short circuit between metallic points, to a point almost inapreciable and without the consumption of energy by the use of resistances & other devices heretofore used for such purposes   This device is applicable to all kinds of electrical apparatus ~~and I shall use it in my lamp~~ [----][q]

~~The~~ It is known that the resistance of the spark in the electric arc as well as between metallic points decreases with the increase of energy. hence the spark between a pair of single points will have 4 times the fusing power as when the same circuit is broken simultaneously at two different parts, and[d] in a corresponding ratio for every additional point where the circuit is broken the difference of potential ~~is~~ reduced in the resistance[h] increased. ~~so that with~~ Thus when[ji] a spark between 2[c] platina terminals is[c] 2 inches in length, The same may be reduced so as to be invisible in daylight if the circuit is broken simultaneously at twelve different places in the circuit, and to obtain ~~ana~~ simple & cheap device for this purpose I have devised the mechanism shewn in fig

fig. 6.[12]

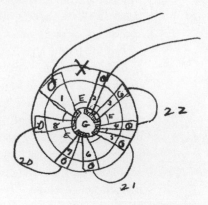

X is a ring of insulating material such as Vulcanized rubber 1 2 3 4 5 6 7 & 8 are contact springs of steel or other elastic metal tipped upon their ends with Iridium platinum or other unoxidizable metal— The ends of 2 & 3 rest upon a metallic sector inlaid or resting upon the ~~insulating~~ disc G which is also of non conducting material. These sectors are 4 in number 4 & 5 rests[d] rests[kk] upon the 2nd sector 6 &[c] 7 on[d] the 3rd & 7 & 8 on the last. These sectors serve to connect the springs together ~~hence~~ The circuit for which these[d] ~~p~~ devices are to asct ~~afor~~ opening & closing enters say spring 1 passes across the sector to spring 8 via wire 20 to spring .7 thence via sector to spring .6. via wire 21 to spring 5 thence via sector to spring 4 thence via wire 22 to spring 3 thence via sector to the spring .2. to the other end of the circuit wire. When the springs are all in contact with their sectors the circuit is complete. ~~by means of~~

If the ~~edg~~ ~~of the~~ edge of the ring E ~~the b~~ is brought against the springs[ll] all of the springs ~~may be seperated~~ are seperated from their sectors simultaneously and the circuit is broken in 8 different places simultaneously   fig .7. shews the arrangement with 4 springs

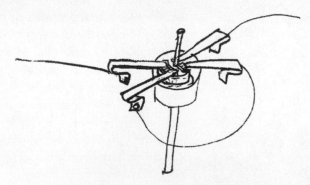

Having now described my Lamp I will describe the devices which I propose to employ upon the premises of the consumer for measuring the energy consumed in a given period and sto provide against accidental crossing of the conductors leading from the mains.[13]

for measinguring[mm] the curre energy consumed I place in a box along with other devices a cell containg two electrodes one of of copper one of which is the cell itself and the other a hollow[d] rod imersed in the placed in the centre of the cell and immerse which I fill with a neutral solution of Cupric Sulphate. And The amount of energy consumed being ascertained by weighing[d] the depos central rod at ati any given interval of time. as these rods are of very little weight, and the amount of deCopper deposit may be due to any fractional part of the total energy passing & still be very acurate it is obvious that for this purpose the device is extremely practical as the rods may be numbered and many hundred[nn] carried to the central office by a single call inspector to be weighed and returned or others substituted when taken out.

The arrangement of the curcuit for passing but a fraction of the energy consumed is shewn in fig in fig 8

fig 8

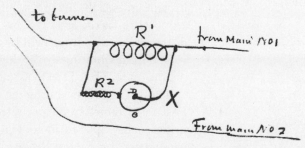

X is the copper cell   D the central electrode. .R. a Resistance[d] but the fractional part of an ohm R^2 a resistance of several ohms R^2 & the cell X obtains energy by forming a derivation[h] from R^1— The Resistance of R^1 will depend upon the Number[d] of lamps lawhose energy consumption is to be measured. ROnly an infinitesimal loss of energy occurs by the interpolation of the resistances R^1 R^2 & cell as the resistance of R^1 may be but $1/100$ of the total resistance[d] of all[c] the lamps combined & still afford reliable measurements with one lamp—

R^2 is for the purpose tof causing any change in the resistance of the cell to become a small factor in the calculation= In the same box I place a device, which I call a[oo] safety magnet. This device is shewn in[pp] fig. 89.

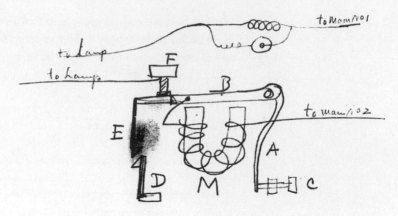

The object of this magnet is to ~~open the circuit~~ disconnect the ~~lam~~premises from the mains should the leading wires ~~from~~ to[c] the lamps be accidentally crossed or get in electrical connection other than through the bobbins of the lamps. the crossing of the wires will tend to draw a powerful current from the mains ~~and if~~ tThe lever .B. is so adjusted by the ~~sprin~~ retractile spring A by means of the screw[qq] .C. that the current will be insufficient to cause the magnet to attract the lever when ~~the~~ all the lamps are on but any great accession of current such as would follow if the lLeading wires to the lamps become crossed the magnet would then be powerful enough to attract .B. ~~as~~ as the lever B & back contact point ~~E~~F form part of the circuit the circuit will be opened. ~~and~~ the lever is prevented from vibrating by the downward movement of B causing the catches E & D to lock together[pp]

The method of generating the current will now be explained. I have ascertained by means of the Dynamometer ~~an~~for measuring energy applied in foot pounds and by the electrodynamometer[g] for energy obtained in the electric form reduced to foot pounds that the practice of making the leading wires & lamps composing the extraneous resistance equal to the internal resistance of the machine although the most effective for obtaining the maximum current[rr] but is by no means so economical as when the extraneous resistances are many times the ~~r~~internal resistance of the machine especially with machines which have a constant field magnet

second that great economy is attained by increasing the electromotive force of the current by the use of[ss] powerful field magnets in lieu of high speed of the induction cylinders or rings, such ~~that~~magnets being kept up by extraneous energy

3rd   The use of ~~abnormal~~ field ~~am~~magnets ~~haveing an~~

~~abnormal amount of iron for in thei~~ of great mass. The machine which I prefer to use is shewn in fig 10[14]

~~4th~~ [tt] ~~10~~ [uu]

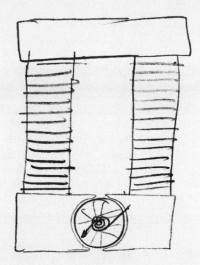

Make good drawing ~~here~~

Serrell is preparing The Dynamo patent now   Wait for tracing=[15]

~~Shew~~   To follow in this patent figs from Caveat 89——[16] 1 4. 7. 8 9 10 11 12 13. 14. <u>16</u>. 15. 17. 19. 20 & close the patent.

ADf, NjWOE, Lab., N-79-02-15.2:31 (*TAEM* 31:787; *TAED* N029:12).
[a]"I . . . that" interlined above. [b]"wires or sheets of" interlined above; "or" separately interlined. [c]Interlined above. [d]Obscured overwritten text. [e]"wire . . . of" interlined above. [f]"as . . . commerce" interlined above. [g]Added in margin. [h]"#" appears in margin. [i]Starting here, four pages facing the text have vague sketches, possibly for a table lamp, probably unconnected with this draft; the first page also has an unrelated calculation. [j]"with . . . glass" interlined above. [k]"by a Sprengel pump" added later, "Sprengel pump" interlined above. [l]"#" appears in each margin. [m]"or gases" interlined above. [n]"~~mo~~ to vivid" interlined above. [o]"~~w~~ after a years use" interlined above. [p]"or even platinum" interlined above. [q]Canceled. [r]"made ~~of~~" interlined above. [s]"& springy" added in margin. [t]"& nickel" interlined above. [u]"rods . . . of" interlined above. [v]"& non~~n~~conducting" interlined above. [w]"~~have its temp~~" interlined above. [x]"or lamp" interlined above. [y]"with low resistance" interlined above. [z]Interlined below. [aa]"& . . . sticks" interlined above. [bb]Followed by diagonal line down to the right, filling remainder of page; figure on next page. [cc]Drawing canceled by large "X". [dd]"via wire 7" interlined above. [ee]Illegible canceled text interlined above. [ff]"of . . . magnet" interlined above. [gg]"light equal to" interlined above. [hh]"one of" interlined above. [ii]"between contact points" interlined above. [jj]"Thus when" interlined above. [kk]Interlined above in a different hand. [ll]"is . . . springs" interlined above. [mm]"uring" interlined above. [nn]"many hundred" inter-

lined above. [oo]"device, . . . a" interlined above. [pp]Followed by dividing mark. [qq]"means . . . screw" interlined above. [rr]"for . . . current" interlined above. [ss]"the use of" interlined above. [tt]Final sentence of preceding paragraph added after and written around this ordinal. [uu]Canceled false start of Fig. 10 not reproduced.

1. Edison probably authored this draft during the first week of May. It contains a meter design that dates from 1 May (see Doc. 1733) but does not include the 9 May version of the regulator (see note 8), which was included in the final specification. It is unclear what Lemuel Serrell, Edison's patent attorney, had in hand when he wrote on 8 May, "Shall I enter provisional specification on the new electric light patent in England and then finish up specification, as quick as possible so as to get the seal immediately without waiting for the completion of the specification?" DF (*TAEM* 51:483; *TAED* D7929ZBW).

2. Under British patent law, an applicant could obtain a patent by filing either a very general "provisional specification" or a more detailed "complete specification" of an invention. A provisional specification provided patent protection until the submission of the complete specification, which had to be done within six months after the filing of the provisional (Davenport 1979, 30). Edison's draft is the text for the more detailed complete specification of British Patent 2,402 (1879), Batchelor (*TAEM* 92:118; *TAED* MBP017). The provisional specification was filed on 17 June and the final specification on 17 December.

3. The preceding discussion is based on lamp experiments that took place between mid-January and mid-February (see Docs. 1665, 1666, 1669, 1670, 1672, and 1675) and the following discussion of improvements in lamp technology are based on Edison's 9 February draft caveat (Doc. 1676) and his 1 March draft patent (Doc. 1695).

4. An approximate way to gauge the degree of evacuation, extensively used and reported by William Crookes and his assistant Charles Gimingham, is by the strength of spark discharge that a vacuum will inhibit. A three inch induction coil was one that produced a spark powerful enough to extend that far through air at normal pressure and temperature. As Edison wrote in his notes on Crookes's vacuum technique and designs: "The nearer he gets to a vacuum the more power it requires to drive the spark through the tube." Spark gauges usually involved not only extra wires as noted here but also a subsidiary chamber on a branch off the main space being evacuated. N-79-04-21:3, Lab. (*TAEM* 30:727; *TAED* N017:2).

5. This is Edison's "electric light law" (see Doc. 1577). Edison's "law" appears to have been rediscovered by Irving Langmuir during his early twentieth century investigation into the problem of lamp blackening in connection with tungsten lamps. According to Reich 1985 (p. 121), Langmuir showed "that the light radiated from a filament was a direct function of its surface area, but that filament shape significantly affected heat loss. Knowing this, he designed a closely spiraled filament that emitted light from its total surface but lost heat only from the surface of the spiral."

6. See Doc. 1608.

7. Figure labels, clockwise from left, are "2," "2," "1," "x," "1," and "g."

8. Prior to developing vacuum lamps Edison had focused his efforts on thermal regulators designed to prevent the metal filaments (primarily platinum and platinum-iridium) from reaching their melting point. Beginning in late February he began to develop regulators of the type shown in this patent designed to prevent a sudden increase in current from damaging the lamp. Edison made two sketches on 20 February of an incomplete device, similar in principle to the one described in this document, consisting of a spring-loaded rod moved by an armature responding to two electromagnets. In March he drew several regulators employing a magnetic motor to turn a governor which would interrupt the lamp circuit at a particular speed. In early April Edison briefly considered interposing variable resistances in the circuit; one design of a "Sunflower regulator for Electric Light" consisted of loops of wire that could be switched into the circuit although this drawing does not show any mechanism for doing so automatically. Edison returned to the idea of a solenoid regulator on 4 April and in the next few days he also sketched several designs apparently intended to operate by thermal expansion; he had four of these and one solenoid built. He drew two regulators of the design shown in this document on 7 April and John Kruesi made a measured drawing ten days later. In the completed provisional and final specifications of this patent, however, Edison replaced this regulator with one actuated by a governor driven by a magnetic motor. That design was essentially complete by 9 May and a regulator of that design built two days later. N-79-03-10.1:41–51; N-79-01-14:73–81; N-79-04-03:31–35; N-79-01-21:129–33, 143–49; N-78-12-04.2:201–205; Vol. 16:386; N-79-01-21:159–67; N-79-04-03:75; all Lab. (*TAEM* 31:

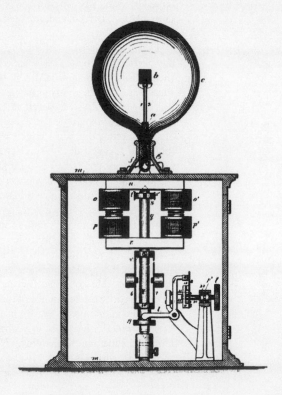

A lamp with governor regulator based on Edison's 9 May design, from his British Patent 2,402 (1879).

23–28, 293–97, 356–58; 30:650–52, 656–59; 29:427–29; 4:822; 30: 663–67; 31:379; *TAED* No22:21–26; No24:34–38; No25:16–18; No16: 66–68, 73–76; Noo4:100–2; NV16:344; No16:81–85; No25:40); Cat. 1308:132–33 (Order Nos. 99, 156), 145; Brit. Pat. 2,402 (1879); both Batchelor (*TAEM* 90:733–34, 739; 92:118; *TAED* MBNoo3:40, 46; MBPo17); see also note 12 and Doc. 1770 n. 9.

9. Should be "W."

10. Edison apparently meant by "reciprocal accumulation" that the field magnet was in the same circuit as the lamps.

11. That is, a separately excited field.

12. In the regulator design of 9 May that was included in the final specification, Edison replaced this device for minimizing sparks with a circuit-opening lever. When the collar of the regulator governor moved up, a block on the end of this lever would push against a set of springs causing them to break contact with a metal block and open the circuit. Brit. Pat. 2,402 (1879), Batchelor (*TAEM* 92:118; *TAED* MBPo17).

13. For the meter, see Doc. 1733.

14. The specifications resulting from this draft are Edison's first detailed description of his distinctive dynamo design. The armature consisted of a wooden cylinder with an iron head at each end, between which fine iron wire was wound until flush with the edge of the heads. Over this the insulated induction wire was wound longitudinally, "the ends of which are connected to the commutator. The wire of the parallel induction helix is substantially endless, and it is wound with reference to obtaining a continuous current. The number of parallel coils may be more or less in number, but I find the desired object can be obtained by using an even number . . . and an odd number of the commutator plates." Brit. Pat. 2,402 (1879) (p. 6), Batchelor (*TAEM* 92:118; *TAED* MBPo17).

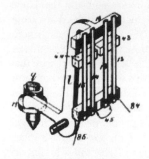

Arrangement for breaking the dynamo circuit.

Edison's explanation of the large field magnets depended on an analogy between magnetic and electrical circuits. At this time there was little theoretical understanding of the behavior of the magnetic circuit but there was a growing body of design practices on which he could draw, as did later theoreticians. Based intellectually on Faraday's analogy between lines of magnetic force and battery current flowing through an electric circuit, this knowledge was developed by what has been called "dynamo theory," or the semi-empirical tradition of machine design (Jordan 1990, 131–38, 161–69, 172–73). This analogy guided some of Francis Upton's early dynamo work (see *TAEB* 4:768 n. 3) and was elaborated upon by Edison in an undated table in which he developed an extended comparison between magnetic and electrical circuits (Oversize Notes and Drawings [1879–1886], Lab. [*TAEM* 45:32–34; *TAED* NS7986B:36–41]). It seems to have led Edison to conceptualize the iron parts of the dynamo as the magnetic equivalent of electrical conductors, suggesting that greater mass and good surface contacts would reduce the "resistance." He explained that by the large magnet cores he was "enabled to reduce the resistance of the magnetic circuit to a very low point, and at the same time by reason of this large mass of iron of great magnetic conductivity the use of a single layer of wire is possible, thus obtaining the maximum economy." He also noted the iron bar connecting the magnet cores "is greater in mass than in the same length of core. The surface of both the back and the end of each core are ground together, and are permanently secured by bolts and nuts. By thus employing a large mass of iron for the back, and grinding the faces of contact to a point where air suction becomes powerful, I reduce the resistance of this point to a minimum and prevent the appearance of free magnetic poles" (Brit. Pat. 2,402 [1879] [p. 6], Batchelor [*TAEM* 92:118; *TAED* MBP017].

Edison did not attempt to secure a U.S. patent on his field magnet design until 1880 (see Doc. 1890). His September 1879 application for an improved armature (Case 184, which issued as U.S. Pat. 222,881) illustrated but did not describe the complete dynamo with its distinctive magnets, and he was later obliged to make an explicit statement limiting that application to the armature alone (see Doc. 1851). Edison's reluctance was evidently due to concern that the Patent Office would hold the design to be merely an alteration in the proportions of well-known machines. In October Zenas Wilber reported that he thought first assistant examiner Henry "Townsend and his ass't Buckingham are inclined to my view about the magneto machine, viz that when change of proportion between the parts is attended by a new result or marked increase of old result, then is involved invention and patentability" (Wilber to TAE, 24 Oct. 1879, DF [*TAEM* 49:454; *TAED* D7903ZJR]).

15. It is unclear to which patent application Edison refers. He executed a U.S. application (Case 177) on 21 April but it was not filed until 12 May and covered his earlier, entirely different dynamo design in which the field magnet rotated within the stationary induction windings, the whole surrounded by an iron shell (see Doc. 1694). There are no other known generator patents originating at this time but there is an undated drawing with figure labels, possibly for a patent, showing the three major Edison generators designed up to 1 May—the one shown in Case 177, the first bipolar design with 2 foot field magnets (see Doc. 1702) and the modification of this with 3 foot magnets (see Doc. 1727 n. 2). Oversize Notes and Drawings (1879–1886), Lab. (*TAEM* 45:29; *TAED* NS7986:33).

16. This caveat has not been found; it was completed by 15 April, when Serrell sent it (and Caveat 88 and extra copies of the drawings) to Edison for his final signature. Serrell to Edison, 15 April 1879, DF (*TAEM* 51:469; *TAED* D7929ZBK).

–1736–

Agreement with Theodore Puskas, Joshua Bailey, and James Banker

[New York,[1]] May 2, 1879[a]

THIS AGREEMENT made this second day of May in the year 1879, by and between THOMAS A. EDISON, of Menlo Park in the State of New Jersey: THEODORE PUSKAS, of 45 Avenue de l'Opera, Paris, France; Joshua F. Bailey, of 45 Avenue de l'Opera, Paris, France, of the first part, and James H. Banker, of the City of New York, of the second part,

WITNESSETH:[b]

WHEREAS the parties of the first part, collectively, own or are entitled to acquire and control the exclusive use in the following counties of Europe, to wit: Belgium, Austria, France,

Denmark, Germany, Russia, Italy and Spain,[2] of certain existing and future inventions of Edison in or relating to means for
the development of electric currents for[c] and the application of
electricity to the uses of lighting, power and heating—and already own certain and may hereafter acquire other letters
patent of said countries or some of them for said inventions;
the interests of said parties in such inventions and patents being respectively the following, namely—Edison having 30 undivided 60ths, Puskas 27 undivided 60ths, and Bailey 3 undivided 60ths, thereof.—[3b]

AND WHEREAS the party of the second part desires to acquire an interest in the said inventions and patents, which the
parties of the first part have agreed to sell to him, on the conditions hereinafter set forth:[b]

NOW, in consideration of the premises and of the mutual
agreements of the parties hereto, they agree with each other as
follows:[b]

FIRST:— The parties of the first part hereby agree to sell,
convey and secure to the party of the second part, by instruments good and sufficient in the law of the several countries
above named, for the sum of Twenty five thousand dollars, ten
undivided sixtieth parts of all the right, title and interest in all
the patents of Edison aforesaid,[4] in all or any of the countries
above named, which they or either of them now own or may
acquire within five years from the date hereof, and the use,
benefit and enjoyment of his inventions aforesaid in the said
countries which they now have and in all additions to or improvements thereof which they may hereafter acquire: provided however that of said sum of twenty five thousand dollars,
five thousand dollars be paid as provided in the second article
hereof and the remaining twenty thousand dollars on or before
the expiration of seven months from the date of this instrument.[5b]

And the parties of the first part agree to apply so much as
may be necessary of said sum of twenty five thousand dollars,
or of such part thereof as they shall receive, to the expense of
Edison in developing said inventions or in making improvements or additions thereto of a like character, and of procuring
letters patent in said European countries for the inventions
above named and to the bringing into use as soon as possible,
by sale or by other most advantageous mode, of the said inventions and patents.[b]

SECOND. The party of the second part agrees to pay to the
parties of the first part on the execution of this agreement

three thousand dollars of the twenty-five thousand dollars mentioned in the First clause, and the further sum of two thousand dollars as soon as specifications now being prepared by L. W. Serrell, Edison's solicitor of patents, for further European patents are ready for filing in such European countries.[b]

THIRD. The parties mutually agree that if the party of the second part, after paying said five thousand dollars, shall, at any time within the seven months provided in the First Clause, elect not to pay the rest of said twenty five thousand dollars—and option to this effect is expressly reserved to the party of the second part—then the party of the second part shall be entitled to receive, instead of the ten undivided sixtieth parts,—only two undivided sixtieth parts of and in the right, title and interest of the parties of the first part in the said inventions and patents for said Countries of Continental Europe,—and after payment of said five thousand dollars and on the declaration of such election by the party of the second part—the parties of the first part will sell, assign and secure to him by instruments good and sufficient in the law of said several countries such two undivided sixtieth parts in such inventions and patents in and for said several countries.[b]

FOURTH: The parties of the first and second parts agree with each other respectively—that if the party of the second part shall pay the full twenty five thousand dollars as herein provided, they will, immediately upon such payment, form together a corporation under the act of the Legislature of the State of New York entitled "An Act to provide for the organization of certain business corporations," passed June 21, 1875;[6] such corporation to be such limited liability company as is provided by the 33rd Section of said Act, with a capital of three hundred thousand dollars, divided into three thousand shares of one hundred dollars each, and that upon formation of such corporation they will severally sell, assign and convey to it their respective rights, titles and interests in all the inventions and patents referred to in this instrument—as they shall stand after such assignment to the party of the second part and as provided by the first clause hereof—to wit, the following undivided interests, namely: Edison ^{30}/₆₀ths Puskas ^{17}/₆₀ths, Banker ^{10}/₆₀ths and Bailey ^{3}/₆₀ths thereof:—and the said parties respectively covenant that they will hold their said several interests meantime for the purpose in this instrument declared, to be assigned to said corporation immediately on the formation thereof as herein agreed—provided however that shares of the stock of said Company shall be issued to them respectively in proportion to their respective interests in the patents and inventions so assigned to it.[b]

In Witness Whereof the parties hereto have hereunto set their hands and seals the day and year first above written.

Thomas A. Edison   (seal)   J. F. Bailey.   (seal)
James H. Banker.   (seal)   (seal)[7]

In the presence of   as to J. F. Bailey and James H. Banker Randolph Hurry.[8] As to Thomas A. Edison   Stockton L. Griffin.

ADDENDUM[d]

[New York?] May 10, 1879

The parties to the foregoing agreement hereby severally agree that the words Sweden and Norway were omitted by mistake from the list of countries named in the preamble thereof, and therefore and for the consideration recited in said agreement shall be henceforth read and construed and have the same effect in all respects as if the words "Sweden and Norway" had been written therein next after the word Spain where it first occurs, before execution thereof, and said agreement is hereby amended to the foregoing effect.

In Witness Whereof the parties hereto have hereunto set their hands and seals the   day of   one thousand eight hundred and seventy nine. In presence of

(seal)[9]   (seal)

(seal)   (seal)

TD (copy), NjWOE, DF (*TAEM* 50:308; *TAED* D7921111). Notarization omitted. [a]Place taken from oath; date taken from text, form altered. [b]Followed by horizontal line to right margin. [c]Written in left margin, probably by Griffin. [d]Addendum is a TD (copy).

1. Although Edison's signature to this agreement was later notarized by Stockton Griffin in Menlo Park, on 2 May Edison was in New York where he signed a separate agreement regarding the formation of the Edison Telephone Co. of Europe. Agreement with Robert Cutting, Samuel White, Robert Cutting, Jr., and James Banker, 2 May 1879, CR (*TAEM* 97:527; *TAED* CH001AAA).

2. An unsigned copy of an undated addendum to this agreement (p. 6; not transcribed) stipulated that Sweden and Norway had been inadvertently omitted from this list, and that they were to be considered in the same respects as the other countries named.

3. Shortly afterwards, Bailey sold Samuel White one-half of his interest in this agreement. Bailey agreement with White, 14 May 1879, SSW (*TAED* X078AG1).

4. An ancillary agreement executed eight days later gave Banker the option to purchase in addition, at par value, 1/10 of Edison's interest in the company formed under this contract. This option was to expire in De-

cember but Edison later extended it to 1 January 1880. Banker subsequently paid Edison $15,000 for this interest. Agreement between TAE, Banker, Puskas, and Bailey, 10 May 1879; agreement between TAE and Banker, 29 Nov. 1879; TAE receipt to Banker, 10 Jan. 1880; all DF (*TAEM* 50:314, 316; *TAED* D7921222, D7921333, D7921444).

5. This contract superseded Edison's prior commitments to Puskas regarding the electric light, from which Bailey, acting as Puskas's attorney, had released Edison on 28 April. Edison subsequently agreed to transfer Banker's interest to Puskas if Banker failed to fulfill the terms of the contract. TAE agreement with Puskas, 14 May 1879, Miller (*TAEM* 86:15; *TAED* HM790062). For the original manuscript copy of this agreement, see Kellow (*TAEM* 144:1197; *TAED* HK203AAA).

6. Banker paid the balance of the twenty-five thousand dollars on 29 November 1879; the Edison Electric Light Co. of Europe, Ltd. was incorporated in January 1880. Banker to Bailey and Banker to Puskas, both 29 Nov. 1879; Certificate and Articles of Incorporation; all DF (*TAEM* 52:430, 54:139; *TAED* D7940ZDJ, D7940ZDK, D8024A).

7. The 10 May agreement with Banker (see note 4) referred to the fact that Puskas had not yet signed. It is not clear why Bailey did not sign this contract on behalf of Puskas, who instead had to await its arrival in Paris.

8. Randolph Hurry was an attorney in New York City. Wilson 1879, 702.

9. The addendum was apparently unsigned at the time this copy was made.

–1737–

From William Kirk & Co.

Newark May 6th 1879

Sir

Mr Murry[1] has moved out of No 12 Ward st. and we would like to have a settlement of your acct[2]   Yours Respectfully
W H Kirk & Co[3]

L, NjWOE, DF (*TAEM* 49:859; *TAED* D7912E).

1. Joseph Murray had been Edison's manufacturing partner between 1872 and 1875, after which he established his own firm in the shop at 10–12 Ward St. in Newark where he continued to make instruments for Edison. See *TAEB* 1:282 n. 1, 3:642 n. 1, 4:34 n. 2.

2. Edison retained an undetermined financial interest in Murray's business and in 1878 had advanced about $1,100 on his behalf to forestall eviction and seizure of his machinery but at this time Murray was selling off his machinery (see Docs. 1294 and 1306). After the landlord again requested a settlement of the account in August 1879, Edison instructed Stockton Griffin to verify that Murray had taken down the shop's overhead equipment because "Kirk wants me to pay his bill—& I will not until that is removed" (Murray to TAE, 6 Sept. 1879; William Kirk & Co. to TAE, 7 Aug. 1879; both DF [*TAEM* 49:364, 860; *TAED* D7903ZHN, D7912F]).

3. This firm was owned by William Kirk, a Newark builder and landlord (see Doc. 157).

*Charles Batchelor to
Charles Edison*

Dear Charley

Yours of April 21st just to hand.[1] We are exceedingly sorry
to hear of Adam's death through your telegram[2] and hope you
will do all you can to fix matters for his wife. As Gouraud reported he was so sick, Edison had decided to recall him and
had already sent the telegram, thinking he would have a better
chance to get well here.[3]

Gouraud is here at present and we are making 50 telephones
for Central System there, which he will probably take with
him.[4] You are right in supposing the width of the spring has
considerable to do with the proper working of the motograph
the longer the line the narrower the spring and even in Laboratory out of 10 different widths I find ⅛ to 3/16 wide by far the
best. One of our first ideas on the motogh was that it formed
H. gas under the stylus but later we have come to the conclusion that it must be due to capillary action; the passage of the
current, altering the capillary force, acting in the pores of the
chalk so as to drive the moisture away from the surface.[5]

The new telephone is simply immense. Next Scribner will
have a picture of it and I shall send you one.[6] We have no difficulty in working both ends equally loud and in cases like central system we put the line battery on at central station so as to
have it in same direction for both receivers. We have also devised for the call a 12 cell battery that goes inside a Leclanche
battery jar and is just the best thi[ng][a] you ever saw, it is an open
circuit battery and will las[t][a] I should judge 12 months.[7] This
is the connection of the instrument working with central station as far as I know at present[8]

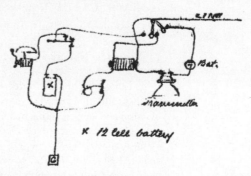

Transmitter will be ma[de][a] with short funnel so that if you
talk anywhe[re][a] within 6 inches it is all OK. We shall put on a
bell instead of a sounder and as you see we use a double point
key in order to prevent any man from having his battery closed
when not working

The permanent line battery is put on at Central Station in this manner; so that [it is?][b] polarity i[s][c] always ~~working~~ right for both receivers, the magne[t][a] you see is a 500 ohm Germ[an] silver magnet which gives resistance to the leg and also tells when they have done talking.[9]

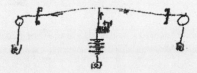

I sent you a book entitled 'Edison [and][a] his inventions'[10] did you get it? Send me by mail [if?][a] you can Du Moncels new book on Tel. Phon etc[11]   Sent you Herald with Obituary of Jim[12] what do you thin[k][a] of it   I should like to know the address of his relations in Scotland so I could send some. Will send you [---- ---- --------][d] in next   Yours

Chas Batchelor

ALS (letterpress copy), NjWOE, Batchelor, Cat. 1330:49 (*TAEM* 93:263; *TAED* MBLB2049). [a]Copied off edge of page. [b]Canceled. [c]"polarity i" interlined above and copied off edge of page. [d]Illegible.

1. Not found.

2. Charley cabled from London on 5 May "Adams dead   Cable transfer me Brown Shipley 100 pounds immediately   Unprocurable here." Edison did so the same day, noting that the sum should be charged to Adams's account. DF (*TAEM* 49:220, 52:531; *TAED* D7903ZDN, D7941ZAA).

3. On the morning of 2 May Edison had wired Theodore Puskas in Paris, "Tell Adams return immediately to America Gouraud and Bailey having engaged experts here." DF (*TAEM* 52:277, 530; *TAED* D7940Q1, D7941Z).

4. See Doc. 1742 n. 4.

5. Cf. Doc. 1693. Batchelor recorded a different understanding in a notebook entry on 19 May: "The best theory of the motograph that we can give to account for the action is this:—   That the passage of the current has such an effect on its capillary power as to draw the solution to surface and lubricate it." He remarked that a chalk cylinder, after having dried out, could be restored only by turning it on a lathe, which he attributed "to the fact that the NaO crystallizes on the surface and these crystals must be turned off or dissolved out, in order to get good talking." This led Edison to consider at the end of May how capillary action might be applied to what he termed a "Motograph Engine ie Capilliary Motor." It is not clear how this device would operate. On 1 June, however, Batchelor reported that "We have accidentally found out that one spring we used (which was much better than any other) was made of Paladium instead of Platina and taking into consideration the fact that Palladium occludes 700 times its own bulk of Hydrogen and the motograph shows itself at the pole where Hydrogen is, we think it is nothing more than a battery action." Cat. 1304:55, 57, Batchelor (*TAEM* 91:54–55;

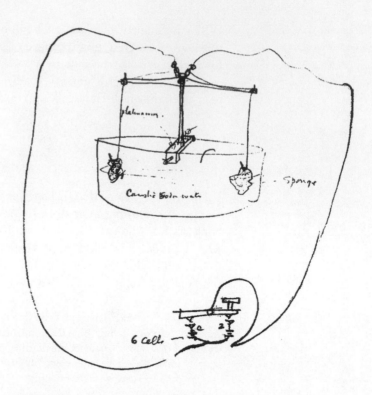

Edison's sketch of an engine working on the capillary principle of the motograph receiver.

TAED MBN004:51–52); N-79-02-24.1:41, Lab. (*TAEM* 30:990; *TAED* N020:21).

In a portion of Batchelor's 19 May notebook entry that he probably added several weeks after that date, he observed that rubbing a platina point across the surface of the chalk caused current to flow through a galvanometer. Presumably about this time Francis Upton made several pages of notes on "Chalks   Test of E.MF on" (N-79-02-24.2:152–63, Lab. [*TAEM* 31:1068–73; *TAED* N031:76–81]). The *Scientific American* reported in its 26 July issue that

> Mr. Edison discovered some peculiar freaks in the receiving instrument which at first puzzled him; but on connecting the binding posts of the telephone with a galvanometer, he found to his surprise that the chalk and platinum rubber of the telephone formed a generator of electricity of no mean order, as it equaled in electromotive force a half of a Daniell cell. He therefore arranged four of the chalk cylinders upon a non-conducting shaft, and connected the platinum rubber of one chalk cylinder with the metallic boss of the next, the terminals being a rubber on one end, and a spring touching the metallic boss of the chalk cylinder at the other end. A series of four chalk cylinders thus mounted and connected . . . is equivalent to two Daniell cells, but the power varies somewhat with the speed at which it is rotated. . . . Whether the current is due to the decomposition of the solution with which the chalk is moistened, or whether it is due to capillarity or some other cause, has not been definitely determined. ["Progress at Menlo Park," *Sci. Am.* 41 (1879): 52]

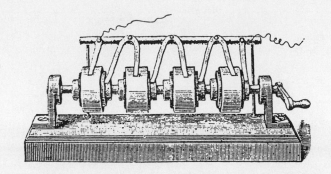

6. In June, in the first of a series about Edison's inventions that appeared in *Scribner's Monthly* (Fox 1879c), Edwin Fox described Edison's receiver and prospective applications of the electromotograph principle to machine bearings, the Atlantic cable, and stethoscopes. The accompanying engraving of the receiver had been published by *Scientific American* in its 26 April issue ("Edison's Electro-Chemical Telephone," *Sci. Am.* 40 [1879]: 260).

7. Figure labels are "LINE," "Bat.," "Transmitter," and "X 12 cell battery." There is no evidence about the development of this particular battery. On 9 May Batchelor made repeated galvanometer measurements with a "10 couple Leclanche" battery; between then and 5 June, he and George Carman tested almost two dozen different batteries, having some of them ring call bells hundreds of times over many hours. Batchelor instructed the laboratory shop to make "65 10 cell batteries for new receiver order" on 23 May. N-78-11-28:154–214, Lab. (*TAEM* 29: 97–127; *TAED* N001:79–109); Cat. 1304:51–52, 56, Cat. 1308:147 (Order No. 168), Batchelor (*TAEM* 91:50–51, 55; 90:740; *TAED* MBN004: 48–49, 53; MBN003:47).

8. This arrangement was completed on or before 1 May, when Batchelor made several related drawings, including one of "connections for each station when worked with Central Station." The same day he also sketched the locations of the connections in each telephone set. Edison's notation on this latter page indicates that the call bell, induction coil, key, switch and binding post were to be made by Western Electric, and the transmitter itself by Sigmund Bergmann's shop. N-79-03-25:85, 90–95, 99, Lab. (*TAEM* 32:182, 185–87, 189; *TAED* N034:43, 46–48, 50); Cat. 1304:47, Batchelor (*TAEM* 91:46; *TAED* MBN004:44).

9. On Edison's rationale for a constant battery current over the line to each receiver, see Doc. 1693. Batchelor made more distinct drawings on 1 May of these connections from "Central Station battery." N-79-03-25: 95, Lab. (*TAEM* 32:187; *TAED* N034:48); Cat. 1304:47, Batchelor (*TAEM* 91:46; *TAED* MBN004:44).

10. See Doc. 1660.

11. English editions of Count Theodose du Moncel's 1878 book, *Le Telephone, le Microphone, et la Phonographe*, were separately published in 1879 in London and New York; the two versions were not identical (du Moncel 1879b, 1879c).

12. The *New York Herald* reported that Adams had died in Paris on 4 May, at age thirty-three. It stated that he had left his native Scotland

as a youth and principally made his living at sea for fourteen years until beginning his association with Edison. "James Adams," *New York Herald*, 6 May 1879, Cat.1241, item 1186, Batchelor (*TAEM* 94:483; *TAED* MBSB21186).

-1739-

To Theodore Puskas

MENLO PARK, N.J.[a] May 10, 1877[1879]

Dear Sir

Since my letter of April 8th 1879[1] I have completed the standard generator which exceeds in simplicity and economy all generators previously devised by any one;[2] The size being less and from 88 to 90 per cent of the horse power applied is thrown into the form of current.

Nothing more remains but the construction of the standard lamp, and the working up of the mechanical details for a large central station. Thirty generators are now being made to start the stations at Menlo Park[3] for 500 burners with a radius of ½ mile, 300 lamps being placed in the streets and 200 distributed in the eight houses on the Park,[4] all provided with metres and the whole system worked as if in actual practice. About 4 to 6 months continuous running will be sufficient time after the system is started to eradicate all defects and prove beyond doubt the economy and practicability of the system for replacing gas as an illuminant   Very Truly

Thomas A Edison

LS, HuBPO, TP (*TAED* Z400BR). Written by Stockton Griffin. [a]Place from Edison's laboratory handstamp.

1. Doc. 1722.
2. See Doc. 1727 n. 2.
3. See Docs. 1727 n. 2 and 1749.
4. Press accounts from the previous year reported that Menlo Park at that time was "composed wholly of Edison's laboratory and half a dozen houses where his employes live" (Doc. 1277). Other sources refer to an inn, a tavern, and a store in the village. "The Tally-Ho's Journey," *New York World*, 5 May 1878, Cat. 1240, item 582; "A Night with Edison," *Scribner's Monthly*, Nov. 1878, p. 95, Cat. 1241, item 973; both Batchelor (*TAEM* 94:194, 390; *TAED* MBSB10582X, MBSB20973X); Jehl 1937–41, 1:25–28, 285, 3:1121–22.

From Tracy Edson

New=York, May 13th 1879[a]

Dear Sir,

I duly recd. & ansd. Mr. W. C. Noyes' letter as requested—[1]

I also recd. today the Draft of Contract about the New Receiver,[2] and the same is in Mr. Cary's[3] hands to be written out for signature

I have today seen the Patent issued in April last for the "Fitch Carbon Transmitter,"[4] and have tried the Instrument—it seems to work well—but I suppose it infringes you does it not? Yours Truly

Tracy R. Edson

⟨Yes. Today I send to patent office as a model the so called Blake transmitter[5]   This will [of?][b] be a fine joke when I get a patent on the instrument exactly as they use it, for I had it that way over a year ago—[6]⟩

ALS, NjWOE, DF (*TAEM* 52:49; *TAED* D7937ZAM). Letterhead of Celluloid Manufacturing Co. [a]"New=York," and "18" preprinted. [b]Canceled.

1. Not found.

2. Edson had written to Edison on 1 May informing him that he and Hamilton Twombly, possibly accompanied by James Banker, were planning to come to Menlo Park on 3 May and that he would bring the new receiver contract at that time. As Edson requested Edison replied the following day by telegram that they could come out. The agreement between Edison and Western Union assigned the electromotograph receiver to the company in exchange for a royalty of one dollar per year for each receiver put into use up to the amount of $3,000. The contract specified that the receiver would be "at least six times louder than any electro-magnetic receiver now in use. . . . be as simple and durable and not more expensive to construct than the best magnetic receivers now in use," and would work with Edison's carbon transmitter. On 21 May, George Walker wrote Edison to ask "when are we to have your wonderful receiver? . . . There are a great many inquiries about it & we want it here as soon as possible." He also informed Edison that the contract executed by Western Union had been sent him several days previously. The contract was dated 12 May and signed by company president Norvin Green; a copy is in Miller (*TAEM* 86:11; *TAED* HM790061). Edson to TAE, 1 May 1879; TAE to Edson, 2 May 1879; George Walker to TAE, 21 May 1879; all DF (*TAEM* 52:37–38, 56; *TAED* D7937ZAC, D7937ZAD, D7937ZAW).

3. Clarence Cary, a Western Union attorney, headed the company's Claim and Legal Bureau. Reid 1879, 560.

4. See Doc. 1731 n. 6.

5. Francis Blake had been for many years an officer in the United States Coast Survey where he had been involved in the electrical determination of longitudes. In the summer of 1878 he devised what became known as the Blake transmitter; he filed a patent application for this design on 3 January 1879. This was a form of inertia transmitter with two

spring-mounted electrodes—a platinum bead and a carbon disk—that remained in contact with each other while being vibrated by the diaphragm. This arrangement eliminated the need for frequent adjustment of the pressure on the carbon button and also reduced the mechanical vibration from the transmitter mounting. After Emile Berliner made several improvements to this design, most notably by substituting a harder carbon disk, it was used extensively by the American Bell Telephone Co. Edward Johnson had tested the Blake instrument in April and reported to Edison that it did not produce "the natural tone of the Edison proper or the Bell magneto" but "only a pure vibration when its low— It's horrible when loud." Brief for Francis Blake, 338; Blake's Patent Specification; both TI (*TAEM* 11:850, 935; *TAED* TI4:76, TI6540); *ANB*, s.v. "Blake, Francis"; "The Blake Transmitter," *Sci. Am.* 41 (1879): 274; Wile 1974, appendix 2; Fagen 1975, 70–71; King 1962b, 330; Prescott 1972 [1884], 368–70; Johnson to TAE, 19 Apr. 1879, DF (*TAEM* 49:211; *TAED* D7903ZDI).

The patent model was probably for Edison's patent application Case 178, which he filed on 2 June 1879. This application was subsequently placed in interference with Blake's 3 January 1879 application (see Doc. 1792). When Edson wrote again on 27 May that he thought it "of great importance that you press your application for your Patent on the 'Blake' and the 'Fitch' Transmitting Models," Edison noted on the letter that "Blake & Fitch cant come out till theres an interference then I will have pat Blake have none & Fitch annulled." Two days later Edison, apparently in reference to this application, wired Lemuel Serrell to "Hurry up last telephone case should be in office immediately to stop another application." During his testimony in the interference proceeding Edison testified that he had made an instrument like that shown in Case 178 in September 1877 and he produced both an instrument and a sketch from October 1877 of a modification of this design. Edson to TAE, 27 May 1879; TAE to Serrell, 29 May 1879; both DF (*TAEM* 49:246, 51:496; *TAED* D7903ZEI, D7929ZCI); Edison's Patent Specification Case 178; Edison's testimony, 12–13, 16–17; both TI (*TAEM* 11:985, 1002, 1004; *TAED* TI7:8, 25, 27).

6. Edison is probably referring to the subject of Telephone Interference No. 1 between his Case 141 (filed 20 July 1877, which was later divided and issued as U.S. Patents 474,231 and 474,232) and Francis Blake's 3 January 1879 application. Edison claimed to have reduced this design to practice by 5 March 1877. Brief in Behalf of Thomas A. Edison, 19, 26–35, Miscellaneous Interferences, TI (*TAEM* 11:870, 874–78; *TAED* TI5:18, 22–26).

–1741–

From Arnold White

[London,] 13th May 1879

Dear Sir

Telephone   I have to report to you that on Saturday the 10th inst, at the House of the President of the Royal Society, the Prince & Princess of Wales, the Crown Prince of Denmark, and many members of both Houses of Parliament, were present at a demonstration of your telephone.

As your nephew, in the face of my protest absented himself in Paris, I placed an instrument in the hands of Mr Conrad Cooke, author of the article in "Engineering."[1] To his attention and energy we are indebted for the occasion proving an entire success.

The Prince of Wales desires to convey to you his congratulations on your invention. He was greatly pleased and amused at the experiments. I am glad to say that on this occasion I secured the support of Sir John Lubbock[2] who will put £2000 into the Company, subject to his approval of the other names.

<u>House of Commons</u> I have to-day received from the Speaker copy of my letter to him <u>printed as a Parliamentary paper</u>.[3] This coupled with the success of the letter to the Times and the leading article in that journal,[4] will I am informed, secure the success of the Company.

<u>C. P. Edison</u>. I st took yesterday the extremely strong step of requesting Colonel Gouraud to secure your nephews recall.[5] Any one bearing your name must, in this country, give us great assistance or the reverse. Caprice and stubbornness might easily be surmounted and overcome in one whose name is not Edison. But the Butcher affair[6]—a matter requiring very delicate handling—your nephew's departure to Paris; his refusal on his return to assist with the instruments, and, I may add, his language, decided me to request that he may be recalled. We can ill-spare one who could, if he liked do so well, but as he will neither do anything himself or consent to other people aiding in the matter, I have come to the conclusion to request his recall, and to lay a plain statement of the matter before you. Copies of telegrams & correspondence passed between us will in due course be sent you.[7] I am, Dear Sir, Yours faithfully

Arnold White[8]

ALS, NjWOE, DF (*TAEM* 52:538; *TAED* D7941ZAM).

1. Conrad Cooke was an engineer known at this time chiefly as the author of an *Engineering* article describing the first demonstration of the Bell telephone in Britain in 1876. Cooke had also designed an electric light system for the clock tower of the Parliament buildings and later helped introduce incandescent gas lighting in London. A frequent contributor to *Engineering*, he co-authored the first volume of Dredge 1882–85. Obituary, *Teleg. J. and Elec. Rev.* 98 (1926): 109; Obituary, *Journal of the Institution of Electrical Engineers* 66 (1926): 1203.

2. Sir John Lubbock was a member of Parliament and a prominent banker. He was also a noted entomologist and member of the Royal Society. *DNB*, s.v. "Lubbock, Sir John."

3. Not found.

4. White's letter to the editor of the *Times* described the widespread

adoption of the telephone in American cities and extolled the benefits of the exchange system for British life and commerce. In a lengthy editorial comment the same day, the *Times* cautiously endorsed White's optimistic outlook and expressed its hope that Parliament would not readily grant telephone monopolies to private companies. "The Economics of the Telephone" and untitled editorial, *Times* (London), 10 May 1879, 6 and 11.

5. News of difficulties involving Charley reached Edison on 6 May, when Gouraud telegraphed from New York that White had cabled him "implying that Charley is giving trouble   Please cable him immediately to implicitly obey White." Edison may have done so; Charley wired from London on 9 May that he had returned from two days in Paris and would "run things smoothly" in the wake of Adams's death but that White was "exasperated." Edison did cable his nephew on 12 May to "Return here immediately." Charley protested that his recall was "premature   some misrepresentation   await explanatory letters enroute   [Prof. William] Barrett upholds me" (Gouraud to TAE, 6 May 1879; Charles Edison to TAE, 9 and 13 May 1879; TAE to Charles Edison, 12 May; all DF [*TAEM* 52:533, 535, 537, 536; *TAED* D7941ZAD, D7941ZAF, D7941ZAK, D7941ZAI]). Those letters have not been found, but a few days later Charley sent a detailed defense of his recent actions:

> The only reason that I can attribute for your cable is that Mr White, as I have heard from outside parties also, has conveyed to you the idea that when I left London for Paris I deserted the telephone in England. I wish it distinctly understood I did not. Had no such intentions, notified Mr White twice by cable that I would be in London the day before the Prince of Wales exhibition he knowing I was in London & waiting for him at Lombard St nevertheless he took the instruments and put them in other people's hands, thereby giving me no chance whatsoever. now if he persists ~~that~~ in saying that I deserted the telephone, I will leave it to your good reason to judge for yourself   It is plain for me to understand the course taken by Mr White as I have seen enough of this man since Gouraud left to be convinced that he is a perfect failure in his position, and if I had went on and done everything he wanted me to do there would have been ~~reports of~~ some discouraging reports of the telephone. [Charles Edison to TAE, 15 May 1879, DF (*TAEM* 52:546; *TAED* D7941ZAR)]

6. This refers to a purported September 1878 agreement between George Gouraud and William Butcher concerning the use of Edison's telephone on railroads and in factories. Gouraud and Edison repudiated the contract over the summer and took steps to indemnify the company from claims based on it; Butcher was jailed on forgery charges. Gouraud to Edward Bouverie, 14 July 1879; TAE agreement with Bouverie, 1 Aug. 1879; Renshaw & Renshaw to Gouraud, 20 Sept. 1879; all DF (*TAEM* 52:616, 648, 800; *TAED* D7941ZCU, D7941ZDF, D7941ZFV).

7. Not found.

8. Arnold Henry White was a London journalist who contributed to a number of newspapers and magazines. He was apparently recommended to George Gouraud by William Barrett and became manager of

the Edison Telephone Co. of London. White later stood for Parliamentary elections several times and achieved considerable recognition as an outspoken proponent of eugenic principles. Charles Edison to TAE, 15 May 1879, DF (*TAEM* 52:546; *TAED* D7941ZAR); Searle 1976, 118.

–1742–

George Gouraud to Arnold White and Dillwyn Parrish?[1]

Menlo Park, May 15, 79 9:05 PM

Brunswick   London[2]

Cable Drexels authority pay Edisons sixty days draft my order 5,000   signed agreement attached altered as cabled[3]   If company want exceeding fifty telephones ordered must order specially[4]   the 500 referred to are for provincial exchanges England[5]

Gouraud

L (telegram), NjWOE, DF (*TAEM* 52:545; *TAED* D7941ZAQ). Written by Stockton Griffin.

1. Dillwyn Parrish, a London merchant, was the largest investor in the Edison Telephone Co. of London, Ltd. Edison Telephone Co. of London list of stockholders, 28 June 1879, DF (*TAEM* 52:603; *TAED* D7941ZCK).

2. Much of the cable correspondence regarding Edison's telephone in England was addressed or signed in code. This is the only known use of the "Brunswick" cipher; the Edison Telephone Co. of London was designated as "Quephone." Gouraud's secretary transcribed the signatures on the reply to this cable as "Parrish   White," which suggests that "Brunswick" was registered in both their names on the company's behalf. Copy of Dillwyn Parrish and Arnold White to Gouraud, 17 May 1879, appended to Gouraud to TAE, 17 May 1879, DF (*TAEM* 52:553; *TAED* D7941ZAU).

3. Earlier in the day Gouraud had telegraphed Edison from New York, "London negotiations closed   I am on five fifteen train with contract." The provisional agreement signed this day provided for the formation of the Edison Telephone Co. of London, Ltd. Neither this version of the agreement nor Gouraud's cable has been found. Two days later Gouraud wrote Edison that he had received a reply from London advising him to "Consider the matter finally settled." The agreement was modified after Gouraud's return to England and finally executed in London on 14 July (see Doc. 1765). The company was capitalized at £200,000 in 2,000 shares. The contract assigned Edison a 20% royalty on each telephone sold or leased by the company, with an advance of £5,000. On 24 May, William Carman recorded the payment from Gouraud of $24,500 (equivalent to £5,000) for Edison's advance, which was to have been contingent upon assignment of patents to the company and completion of the first 50 of 500 telephones. A copy of the receipt for this advance was written on the cover of an undated printed copy of the contract. Gouraud to TAE, 15 and 17 May 1879; Agreement with Edward Bouverie, 14 July 1879; Agreement with Edison Telephone Co. of London, Ltd., n.d.; all DF (*TAEM* 52:544, 553, 617, 1046; *TAED*

D7941ZAP, D7941ZAU, D7941ZCV, D7941ZKV); Ledger #3:397, Accts. (*TAEM* 87:165; *TAED* AB001:159).

4. Gouraud had reported to Edison the day before that he was pressing Arnold White to complete the negotiations and had "concluded that by cabling them [the London investors] that I had ordered 50 instruments which are soon to be finished & fwd quite independently of any negotiations they are having that it wd give confidence in the belief that the definite instruments are not a myth." Edison drafted a reply stating that "Its all nonsense to talk about 50 instruments by 21st= it will be about the 28 or 30th= Again these instruments should be all put up or at least a part in a system here & to do that we have to make switchboards indicators batteries etc." The laboratory shop was instructed on 15 May to make "65 Complete Telephones" but this was probably superceded by Charles Batchelor's orders on 23 May for the construction of "central station switch complete for 50 fifty subscribers with 12 exchange strips as per sketch" and "65 new Edison Telephones complete for Central Station system." Edison received fifty telephone boxes from Sigmund Bergmann by 31 May and on 2 June he wired Bergmann to "make up the complete supply for sixty five Telephones." Several of the instruments were damaged in transit to London. Gouraud to TAE, 14 May 1879; TAE to Gouraud, 15 May 1879; both DF (*TAEM* 52:541, 543; *TAED* D7941ZAN, D7941ZAO); Cat. 1308:145–49 (Order Nos. 160, 168–172), Batchelor (*TAEM* 90:739–41; *TAED* MBN003:46–48); Bergmann to TAE, 31 May 1879; TAE to Bergmann, 2 June; both DF (*TAEM* 52:62, 63; *TAED* D7937ZBB, D7937ZBC).

5. The contract permitted the formation of independent Edison companies to operate in Parliamentary boroughs or postal districts outside London's East Central Postal District. These firms were to lease or buy telephones from the London company, to which Edison would give 10% of the royalties he received. Agreement with Edward Bouverie, 14 July 1879, DF (*TAEM* 52:617; *TAED* D7941ZCV); see Doc. 1765.

–1743–

Agreement with George Gouraud

[Menlo Park,] May 15, 1879[a]

This agreement entered into this 15th day of May 1879 by and between Thomas Alva Edison of Menlo Park Middlesex County New Jersey and George E Gouraud of No 6 Lombard St London England,

Witnesseth. Whereas the said Edison has signed this day a certain Provisional agreement looking to the formation of a Telephone Company in London a copy of which is hereunto annexed and entitled "Heads of Provisional agreement, Edisons Patents Loud Speaking Telephones," transferring his patents for telephones in the United Kingdom of Great Britain and Ireland and the Channel Islands and the Isle of Mann and Whereas the said Gouraud, in consideration of one dollar in hand the receipt of which is hereby acknowledged and of service already performed and [promises][b] already made in the [exploitation?]][b] of the said inventions in the said countries and in confirmation of previous understandings is entitled to one half of all moneys or other benefits derived by said Edison from said patents in said countries and improvements thereof. Now I do by these presents assign unto the said Gouraud all my right title and interest in one half of all the benefits derived by me now and hereafter from the said contract and from all subsidiary contracts resulting thereto both directly & indirectly and I agree to execute any further papers which may be necessary by way of authorizing said company or any companies to pay to said Gouraud directly said[c] one half of all monies [received?]][b] or other benefits as may be due to me as above described[1].

It is further agreed and understood that the said Gouraud is to continue to use due diligence and his best endeavors to form the subsidiary companies according to the provisions of the contract here to appended these services being a part of the consideration from the said[2] Gouraud to the said Edison

Witness   S. L. Griffin[3]   Thomas A Edison

ADDENDUM[d]

[Menlo Park, May 15, 1879?]

And I do hereby further agree said Gouraud shall during the term of the said patents continue to be my sole agent in the countries aforesaid in regard to said patents.[4]

DS, NjWOE, Miller (*TAEM* 86:17; *TAED* HM790063). Written by Stockton Griffin. [a]Date taken from text, form altered. [b]Document damaged. [c]Written in left margin. [d]Addendum is a D, written in an unknown hand.

1. On 23 May, Edison paid for half of the £1,000 worth of shares in the Edison Telephone Co. of London to which Gouraud was required to subscribe. The same day he signed a power of attorney that nominated Gouraud to the company's board of directors and gave him voting power for Edison's shares. Edison also agreed to give Gouraud first option to buy any of his shares. Gouraud's receipt, 23 May 1879, DF (*TAEM* 52:560; *TAED* D7941ZAX); Power of Attorney, 24 May 1879, Miller (*TAEM* 86:23; *TAED* HM790067).

2. For the agreement regarding subsidiary companies see Docs. 1742 n. 5 and 1765.

3. Griffin notarized this agreement.

4. On 24 May a similar clause was also added to the 23 May power of attorney (see note 1). Power of Attorney, 24 May 1879, Miller (*TAEM* 86:23; *TAED* HM790067).

May 16, 1879[a]
London 8:25 AM

[Thomas] Edison
    Important Exhibition Society Arts[1]   White refuses me instruments   Cable me order for them immediately[2]

[Charles] Edison

Menlo Park NJ 9:30 AM

[Charles] Edison
    Stop cabling and come home[3]

[Thomas] Edison

L (telegrams), NjWOE, DF (*TAEM* 52:552; *TAED* D7941ZAS, D7941ZAT). Written by Stockton Griffin. [a]Date taken from document, form altered.

1. Established in 1754 as the Society for the Encouragement of Arts, Manufactures and Commerce (later the Royal Society of Arts), this was a leading institutional supporter of popular scientific, and at this time, technical education. Cardwell 1972, 75–76, 80–86, 126–29.

2. The previous day, Charley protested Edison's recall order (see Doc. 1741 n. 5) and wrote that Conrad Cooke, with whom White had made arrangements for the upcoming demonstration, had approached him for assistance but "I told him that as it was I was powerless to do anything but if I could get the instruments I would do all I could to make it a success   I just wrote a note to Mr White asking him for the use of the instruments for that occasion, but up to closing of Post I have nothing from him and am very anxious." Charles Edison to TAE, 15 May 1879, DF (*TAEM* 52:546; *TAED* D7941ZAR).

3. In Charley's stead, Conrad Cooke surveyed the history of various telephonic instruments at the Society on 21 May, giving considerable attention to the explication and demonstration of Edison's receiver. Cooke sent Edison a copy of his lecture from the Society's journal; he intended to republish the lecture "as a separate illustrated book" but this has not been located. Arnold White also sent Edison two press accounts of Cooke's demonstration and excerpts of John Tyndall's remarks and reported that "considering the worn state of the instrument its performance was remarkably good." Cooke 1879; "The Telephone," *Teleg. J. and Elec. Rev.* 7 (1879): 219; White to TAE, 22 May 1879; Cooke to TAE, 26 May 1879; both DF (*TAEM* 52:556, 565; *TAED* D7941ZAW, D7941ZBD).

Technical Note:
Electric Railway

E[lectric]. Tramway
Reversible ~~Electr~~ Commutator[1]

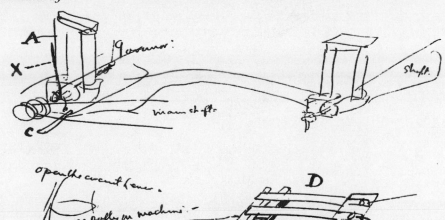

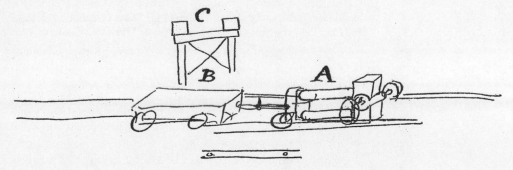

18 in of gr[a]d[e].[2]

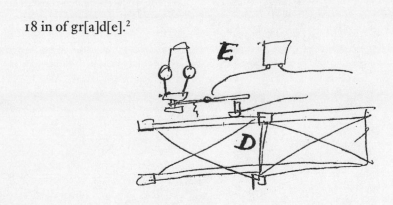

$$1.10 \times .80 = 1.90^{a} \quad \$1.90^{3}$$

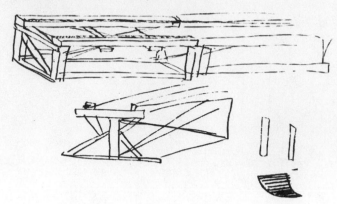

1250   10 Miles pr hour.[4]

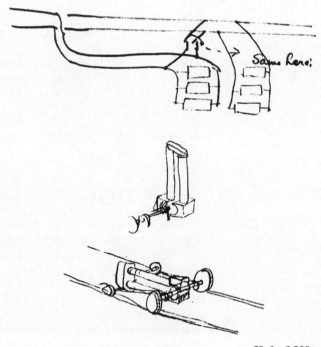

J[ohn] K[ruesi]

X, NjWOE, Lit., *Electric Railway Co. of the U.S. v. Jamaica and Brooklyn Road Co.*, Edison's Exhibits No. 1–5 (*TAEM* 115:824; *TAED* QE001A001). Drawings by Edison; text by John Kruesi. Multiply initialed and dated. [a]Form of equation altered; decimal points added for clarity.

1. These drawings were presented as exhibits during an 1881 patent interference hearing and then again in an 1892 court case in connection with testimony by Edison and his associates regarding work at Menlo Park on electric railways. Edison testified that while passing through

Edison's sketch of an electric railroad using wind power to generate the current.

Iowa on his 1878 Western trip (see *TAEB* 4:375, map) "I conceived the idea that if an electric railroad could be made to operate economically it would be of great value for drawing grain to the main lines of railroad and thus extend the radius of economical grain production." He proposed to develop a remotely controlled system suitable for "the flat of the Northwest where the winds continually sweep over them, and I thought that I could utilize those winds to run windmills which, in their turn, would give motion to dynamo machines and furnish electricity to operate the small motors and cars which I proposed to deploy." A sketch dated 21 May shows "a station for an electric railway with a turn-out track . . . the track on trestles and a station with windmill for obtaining power with wires running from the station out to the track." Edison recalled that he did "a great deal of studying on the subject" after his return to Menlo Park and noted that "the idea that large railroads could be operated by electricity cheaper than by steam was many times speculated upon, and figured upon, and talked over in the laboratory, but at that date the knowledge in the art was not sufficiently advanced to permit us to see our way clear to displace steam on a large railroad." Edison's testimony, 74, 99, *Edison v. Siemens v. Field;* Edison's Exhibit No. 6, Edison's testimony, 497–99, *Electric Railway Co. of U.S. v. Jamaica and Brooklyn Road Co.;* both Lit. (*TAEM* 46:42, 54; 115:829, 752, 783–84; *TAED* QD001:38, 50; QE001:93, 47–8).

Edison testified that the first of these drawings represented "a dynamo machine driven by belting, and a motor with a governor connected to it for governing the speed" and that the "words 'reversible commutator' refer to reversing the position of the brushes on the commutator, so as to permit the motor running in either direction." The governor, designed to be placed on top of the motor, functioned by opening the circuit when its speed exceeded a certain limit, necessary in the absence of a train crew; the circuit could be entirely opened with the "open the circuit lever," C. John Kruesi identified D as "a device for reducing the sparking at the breaking of the circuit." It was designed to interrupt the circuit simultaneously at a number of points, each of which would have lower voltage and produce a smaller spark than the circuit as a whole. Edison recalled that sometime prior to making these drawings they had tested a device based on this principle but in which "the levers of several sounders were used in place of the spring contacts"; this worked "satisfactorily." Edison articulated the principle behind multiple-contact circuit breakers and described another such device for reducing sparking in the British provisional specification he had drafted around the beginning of May (Doc. 1735). In at least some designs of this system, copper conductors parallel to the rails carried current to the locomotive. Edison's and Kruesi's testimony, 497, 45, 51, *Electric Railway Co. of U.S. v. Jamaica and Brooklyn Road Co.;* Edison's testimony, 78, *Edison v. Siemens v. Field;* both Lit. (*TAEM* 115:783, 758, 761; 46:44; *TAED* QE001:47, 22, 25; QD001:40).

2. Edison identified these drawings as "a form of governor; a part of a trestle which I proposed to put the track on; also a locomotive drawing a car . . . [and] a section of trestle which I proposed to mount the rails on." He testified that "we made several models of the trestle work which we proposed to employ on these little wheat railroads. I also remember trying to get money for the erection of an experimental railroad at Menlo Park, but did not succeed." The trestles were to elevate the tramway so cattle would not interfere with its automatic operation. Edison's testimony, 497, 499, *Electric Railway Co. of U.S. v. Jamaica and Brooklyn Road Co.,* Lit. (*TAEM* 115:783–84; *TAED* QE001:47–48).

On 16 May John Kruesi drew a cross section of a trestle with two braces, carrying an Edison electric locomotive with governor and estimated the cost of materials for an unknown length of this type of trestle, as well as one of a Y-construction (N-79-04-09:70–5, Lab. [*TAEM* 32:1078–80; *TAED* N045:32–34]). The former design was rejected as "Too costly," although a nearly identical structure was drawn on 25 May (Vol. 16:435, Lab. [*TAEM* 4:870; *TAED* NV16:391]). Kruesi testified that around 25 May he had made several models of trestles as well as a full-scale trestle at Menlo Park (*Edison v. Siemens v. Field,* 42–43, Lit. [*TAEM* 46:26; *TAED* QD001:22]). Related drawings of the trestle and locomotive were submitted as Exhibits 8, 10–11 in *Electric Railway Co. of U.S. v. Jamaica and Brooklyn Road Co.,* Lit. (*TAEM* 115:831–33; *TAED* QE001:96–98); an estimate of capital expenses was included in Exhibit No. 9 (Lit. [*TAEM* 115:831; *TAED* QE001:96]).

3. This figure may be related to the cost for a section of trestle.

John Kruesi's measured drawings of 18 May show the construction of the trestle at a scale of 1 inch per foot. At right the motor car is pulling a freight car; at left the motor car (labeled "E.E.T. No. 1") is shown with its governor visible at the upper right.

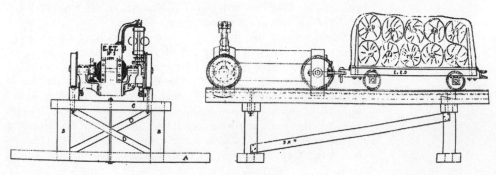

4. The meaning of "1250" is unclear. The following drawing shows "dynamos connected to the track in sections with a reversing switch for reversing the direction of the current on any particular section." Edison explained separately that "the wires connected to the dynamo system on the left is a circuit reversing switch, whereby the direction of the flow of the current in a certain section of the track near the station may be changed. The system of dynamos on the right was connected to another section of the track, and provided with a circuit reversing lever as implied by the words 'same here.'" Edison testified that he proposed two methods for reversing the direction of the locomotive: "one reversing the direction of the flow of the current through the bobbin, the other the reversing of the commutator brushes." Dynamo stations were to be located at seven to ten mile intervals along the line. In some designs of this system copper conductors parallel to the rails carried current to the locomotive while in others the rails themselves were electrified by means of copper strips placed beneath the fish plate connecting the rails; the wheels would have their rims insulated from the hub. The system was designed to operate with a minimal workforce and Edison proposed "the use of a telephone between two stations, a part of the circuit being formed by the electric railway track" to enable "the attendants of the stations to communicate with each other, to run trains, or to govern trains from the stations without having anybody on the trains." Edison's and Kruesi's testimony, 78–81, 85, 37, *Edison v. Siemens v. Field;* Edison's and Kruesi's testimony, 497–98, 49–50; Edison's Exhibit No. 7, *Electric Railway Co. of U.S. v. Jamaica and Brooklyn Road Co.;* both Lit. (*TAEM* 46:44–45, 47, 23; 115:783–84, 760–61, 830; *TAED* QD001:40–1, 43, 19; QE001:47–48, 24–5, 94).

–1746–

Notebook Entry:
Telephony

[Menlo Park,] May 18 1879

Edison's Telephone   New receiver.[a]

The remarkable effects produced by the chalk cylinder in the receiver and the extreme loudness and clearness of articulation have led me to try a great number of experiments on it. At a distance from the transmitting[b] instrument of 100 feet, everything can be heard perfectly on the receiver in an ordinary lecture room. If different people are talking and some playing musical instruments or whistling at the transmitting end each one comes out clear and distinct. With the carbon transmitter adjusted lightly on the button the sounds of a small musical box can be reproduced with 3 times their loudness by simply placing it on the table where the transmitter stands.[1] In ordinary talking such is the amplitude of the diaphragm of the receiver that its record on the phonograph is a simple matter and the retransmission over the telephonic circuit by the phonograph receiving it again on another phonograph have frequently been done by me; the reproduction of the records on the 'phono' being almost as loud as the original talking and its articulation exceedingly clear

Chas Batchelor.

X, NjWOE, Batchelor, Cat. 1304:53 (*TAEM* 91:52; *TAED* MBN004: 50). Written by Charles Batchelor. [a]Multiply underlined. [b]Interlined above.

1. *Scientific American* noted in a 26 April article about Edison's receiver that "no one who has heard this new telephone can fail to have been astonished at its clear, articulate, and loud tones; it might appropriately be called 'The Shouting Telephone,' for its 'voice' is louder than that of any ordinary speaker. . . . Where it is in use it is of course unnecessary to go at all near the instrument, for it may be fixed against the wall of an office, and its messages heard at any part of the room spoken in a loud clear tone." "Edison's Electro-Chemical Telephone," *Sci. Am.* 40 (1879): 260.

–1747–

Patent Model Specification: Electric Lighting

[Menlo Park,] May 20 1879

Patent ofs model—[1]

Kruzi make the following model to illustrate the system of carrying wires through the Hard rubber lined pipe—[2]

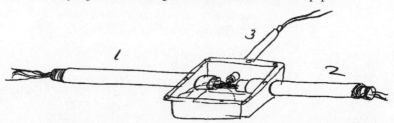

Have a top put on with soft rubber sheet to prevent water coming in ~~& use~~ use bolts. Make the whole of iron   1 & 2 to be with box & all 1 foot long, 3 to be 6 inch long & smaller than 1 & 2.

⟨½ in pipe Iron box lined with h[ard]. rubber all over inside packing rubber cascett⟩[3a]

⟨Made by Geo Jackson⟩[4b]

T.A.E   J Kruesi

ADS, NjWOE, Lab., Vol. 16:390 (*TAEM* 4:823; *TAED* NV16:345). [a]Written by John Kruesi. [b]Written by George Jackson.

1. Edison filed an application covering underground conductors and junction boxes on 7 July (Case 179) that was rejected and subsequently abandoned; only two figures and the claims survive. Edison claimed metallic tubes and junction boxes insulated with a non-conducting substance such as hard rubber, and "the pulleys and cords for passing the conductors through such tubes." Patent Application Casebook E-2536: 34, PS (*TAEM* 45:701; *TAED* PT020034).

2. Figure labels (written by John Kruesi) are "1 ¼ pipe," "3 ⅛ pipe," and "2." Kruesi made the measured drawing for the model the same day. He recorded the order in the laboratory order book on 31 May; it is unclear when George Jackson completed the model. Vol. 16:390 and Machine Shop Drawings (1879–1880), Lab. (*TAEM* 4:823, 45:67; *TAED* NV16:345, NS7986CAS); Cat. 1308:149 (Order No. 178), Batchelor (*TAEM* 90:741; *TAED* MBN003:48).

3. Kruesi probably meant gasket.

4. Machinist George Jackson, who may have worked briefly for Edison and Murray in 1875, had been working at Menlo Park since June 1878. See *TAEB* 4:414 n. 10.

–1748–

From Edward Johnson

New York, May 22 1879.[a]

My Dear Edison

See enclosed—another Bulldozer—[1]

I have carefully considered the proposition made me—and am inclined to accept it in spite of all opposition.[2] I think however that as I am compelled to practically abdicate the certain salary of the Phonograph Co—I should substitute it by a full equivalent as to amount & certainty—$300. is not that when I am abroad & my family here. I am going therefor to ask you to make the terms as follows

$500. per month for 6 months.

5% of advance Royalties for the 6 months with a provision which shall secure me in cases where all the work has been done and my part has been fulfilled but the negotiations ~~have not~~ shall not have[b] been closed—for instance, for 360 or ~~690~~ Days after the expiration of my 6 months—

I presume of course such[c] extraordinary expenses as are entailed by travelling about from Pillar to Post will be borne by the Co.—

If these provisions are satisfactory embody them in one of your characteristic Legal Doc's Send them to Col Gouraud for his sig. & I will sign[3] Yours

E H Johnson

PS. If you will put yourself in my place—& remember the fact that since I went with Palmer in 1868, I have been always seeing the bushy tail of the Fox just around the corner[4]—while another Fox has drawn me off in another direction, only to repeat the story—you will see that I have just reason this time to make myself "Solid[d]" Yours E.H.

I'll work for Gouraud & Edison there as here—& never for EHJ I'm doing that now Jo

ALS, NjWOE, DF (*TAEM* 51:711; *TAED* D7932V). Letterhead of Edison Speaking Phonograph Co. a"New York," and "1879." preprinted. b"shall not have" interlined above. cInterlined above. dObscured overwritten text.

1. Not found. To bulldoze was to intimidate or coerce; a bulldozer was a bully or thug. Lighter 1994–97, s.v. "Bulldoze," "Bulldozer."

2. Johnson was to be Edison's chief engineer for the commercialization of telephones in England. He telegraphed Edison on 13 May, presumably referring to his wife, "She consents ergo I go unless the company veto." Edison wired George Gouraud two days later that he had "got you the best telephone man in country who will go for 2 months for $1000, taking a man with him who he will leave in charge on his return" (Johnson to TAE, 13 May 1879; TAE to Gouraud, 15 May 1879; both DF [*TAEM* 52:537, 543; *TAED* D7941ZAL, D7941ZAO]). The "opposition" apparently involved the Edison Speaking Phonograph Co. On 13 May Painter wrote concerning a meeting that he expected Johnson to call of the company's shareholders, cautioning him that "There is some trap Cheever is setting for you— I oppose either Cheever or [Hilborne] Roosevelt getting any figures or information of any kind from the Co while they are in opposition to us & backing [Oliver] Russell." Charles Cheever had been the phonograph company's first treasurer; at this time Painter and Gardiner Hubbard were trying to recover thousands of dollars they alleged he and fellow shareholder Hilborne Roosevelt had misappropriated (Painter to Johnson, 13 May 1879, ESP Scraps. 11:80; Hubbard to Cheever, 28 May 1879; Hubbard to Painter, 28 May 1879; Hubbard to H. Roosevelt, 28 May 1879; Painter to Cheever, 7 and 13 June 1879; all ESP Treasurer's Lbk. [Nov. 1878–July 1879], 632–35, 667, 671). Hubbard also argued that Johnson had capably filled in when Edison broke his pledge to give his personal attention to perfecting the phonograph, and was now indispensable to the company. He wrote Johnson that "If you now leave us part of the money we have paid you will have been worse than wasted, for we must employ other parties to learn what you have been taught at our expense." He urged him to "state to Mr Edison my view of the case that you cannot for his benefit honorably leave us at the present time, & that he has no right to ask anything of this kind of you" (Hubbard to Johnson, 28 May 1879, ESP Treasurer's Lbk. [Nov. 1878–July 1879], 628).

Presumably about this time Edison wired Johnson in New York, "As there is such a row I advise that you stay   Have sent for Louderback and will send man from here as assistant." Edison had asked Henry Bentley in Philadelphia his opinion of De Lancy Louderback, a former operator now the eastern sales agent for Western Electric (see *TAEB* 4:107 n. 2), who had inquired earlier about the job, and Bentley gave an enthusiastic reply. However, Louderback declined the job on 30 May and Edison wired Gouraud in New York two hours later: "Will not be able to get expert   Johnson withdraws" (TAE to Johnson, n.d. May 1879; TAE to Bentley, n.d. May 1879; Bentley to TAE, n.d. May 1879; Louderback to TAE, 17 and 30 May 1879; TAE to Gouraud, 30 May 1879; all DF [*TAEM* 52:571, 570, 555, 569; *TAED* D7941ZBM, D7941ZBK, D7941ZBL, D7941ZAV, D7941ZBH, D7941ZBI]). Edison then made an offer to Charles Walton, a Gold and Stock Telegraph Co. employee, but quickly rescinded it after receiving unflattering information, apparently regarding Walton's sobriety and the nature of his duties (Walton to TAE, 4 and 9 June 1879, DF [*TAEM* 52:573, 585; *TAED* D7941ZBN, D7941ZBX]; TAE to Walton, 9 June 1879, Lbk. 4:384 [*TAEM* 80:84; *TAED* LB004384]). Francis Upton wrote his father on 1 June that Edison was seeking "a good business man who understands a little about electricity, and who has a record of having managed something well" but was finding "very few who will leave this country for England to take charge of an enterprise which must be of necessity somewhat risky being entirely new and almost untried" (Upton to Elijah Upton, 1 June 1879, Upton [*TAEM* 95:532; *TAED* MU016]).

3. Edison's original offer to Johnson evidently was for the English telephone company to pay him $300 per month with a 5% royalty. Gouraud termed this "a liberal proposition" and feared that the company "will be subject to criticism" for increasing it. He promised that Johnson would "have a fair opening for his talents but shd not be over persuaded to go." Gouraud nevertheless accepted Johnson's terms and Edison drafted and signed a memorandum of agreement embodying them on 11 June, the day Gouraud departed New York for England. This agreement was for one year and otherwise incorporated the provisions above. It granted Johnson royalties created by negotiations completed within sixty days of the end of his employment and also provided for a $500 advance to provide passage for his family, which Gouraud paid immediately. Johnson subsequently signed a contract with the telephone company which set his salary at £100 per month (about $500) but made no mention of royalties. Gouraud to TAE, 24 May and 11 June 1879; Johnson agreement with Telephone Co. of London, Ltd., 24 Sept. 1879; all DF (*TAEM* 52:562, 591, 742; *TAED* D7941ZAZ, D7941ZCB, D7941ZEM); TAE agreement with Gouraud and Johnson, 11 June 1879, Miller (*TAEM* 86:28; *TAED* HM790069).

4. Johnson had been chief assistant to William Jackson Palmer in the survey of a rail line to the Pacific through New Mexico and Arizona in 1867–1868. As an indirect result of this survey Palmer also made plans for a line from Colorado to Mexico and subsequently organized the Denver & Rio Grande to develop this route. Johnson assisted in the early stages of its construction before Palmer dispatched him to the Automatic Telegraph Co. in New York. At this time Palmer was president of the Denver & Rio Grande, a successful developer of coal and iron mines in Colorado and Utah, and a founder of Colorado Springs. *ANB*, s.v. "Palmer, William Jackson"; *NCAB* 33:475, 23:399.

–1749–

From Hamilton Twombly

New York May 23, 1879[a]

Friend Edison,

What progress on the system for exchange switches and signals? ⟨will have system set up here by ~~May~~June 8th⟩[1b] also on the patent to cover the Blake transmitter, ⟨Used Blake Trans as model   its in patent ofs⟩[2b] also the Electric machine? ⟨we are making our 30 machines,[3] did not make your six[4] as recd no orders⟩[b] I write these few lines to remind you that I am much interested in all these things, and perhaps a little impatient for them

Hoping you will have success in all your undertakings   I am   Yours truly

H M K Twombly[5]

⟨You asked if I could get around the Brooks Cable. I answer that I have got a system which I am going to use for my E Light that is infinity better than Brooks⟩[6]

ALS, NjWOE, DF (*TAEM* 49:236; *TAED* D7903ZEA). Letterhead of Western Union Telegraph Company Executive Office. [a]"New York" and "187" preprinted. [b]Marginlia interlined above by Edison.

1. See Docs. 1738 n. 8 and 1742 n. 4. Edison's switchboard system is described in a British provisional patent specification filed on 20 September 1879 (Brit. Pat. 3,794 [1879], Batchelor [*TAEM* 92:129; *TAED* MBP018]). The system consisted of a grid of horizontal and vertical metal bars, with the upright members wired to subscribers' telephones. When alerted by an annunciator to an incoming call, the attendant could insert pins at appropriate junctions in the grid to make electrical connections from any upright to the operator's telephone and, after learning the caller's wish, to any other subscriber's bar.

There is scant evidence for the development of the switchboard prior to this. John Kruesi's 25 March measured drawing of vertical and horizontal bars for one to be built for Edison's new office appears to embody the essential design. On 13 May Edison drew schematically the switchboard frame and wiring and also sketched in more detail the wiring in a representative segment of the switchboard and a detail of the frame's construction. Charles Batchelor drew additional details on 19 May, including signaling apparatus for identifying incoming calls to the switchboard attendant. Machine Shop Drawings (1879–1880); N-79-04-09:25–27, 33; N-79-02-10:165; all Lab. (*TAEM* 45:57; 32:1058–59, 1062, 1221; *TAED* NS7986C:21, N045:12–13, 16; N046:81).

2. See Doc. 1740 n. 5.

3. This apparently refers to the thirty generators that Edison planned to make for the demonstration central station at Menlo Park. See Docs. 1727 n. 1 and 1739.

4. "Your six" apparently refers to a standing request for telephones with the electromotograph receiver for Gold and Stock. On 21 May, George Walker wrote Edison about the "wonderful receiver," stating that "there are a great many inquiries about it & we want it here as soon as possible." He asked if one could be ready by 1 June for use by a professor at Cornell. Edison replied the following day that he would "in all human probability deliver you six new Telephones complete before the 3d of June." Walker to TAE, 21 May 1879, DF (*TAEM* 52:56; *TAED* D7937ZAW); TAE to Walker, 22 May 1879, Lbk. 4:342 (*TAEM* 80:76; *TAED* LB004342B).

5. Hamilton McKown Twombly was an investor and manager in many companies, primarily in the railroad industry. The son-in-law of William H. Vanderbilt, Twombly served on Western Union's Executive Committee and at this time was in charge of its telephone business.

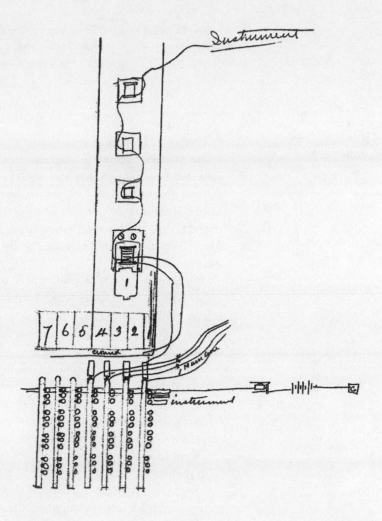

Twombly's middle name was often abbreviated as "McK." and he was often referred to as "Hamilton McKay Twombly." See *TAEB* 4:253 n. 8.

6. Twombly's prior inquiry has not been found. On 6 May, Edison wired Twombly that the "Brooks patent may not hold" and, at the least, a "broad claim cannot be taken on liquid insulator" by David Brooks. Brooks was a distinguished inventor and engineer who held several patents (the most recent issued in December 1878) for insulating underground lines with oil (see *TAEB* 2:466 n. 8). Tracy Edson wrote Edison on 27 May that although Western Union now indirectly controlled the Brooks patents he thought it "important that you should perfect your new plan as soon as possible." Edison filed a patent application (Case 179) on 7 July for insulated underground conductors and junction boxes (see Doc. 1747); it was rejected partly on the basis of an 1875 patent held by Brooks (U.S. Pat. 165,535). TAE to Twombly; 6 May 1879; Edson to TAE, 27 May 1879; Lemuel Serrell to TAE, 19 July 1879; all DF (*TAEM* 49:224, 246; 51:534; *TAED* D7903ZDP, D7903ZEI, D7929ZDK); Patent Application Casebook E-2536:34, PS

(*TAEM* 45:701; *TAED* PT020034); on Western Union and the Brooks underground cable system see Israel 1992, 144–45.

–1750–

Francis Upton to
Elijah Upton

Menlo Park, May 24, 1879.

Dear Father:

I went to Lawrence as I expected and found Mr. Fallon[1] ready to give me any information that I might wish.[2] He also said that he would answer any questions that I should ~~answer~~ ask by letter,[a] which I thought would be the best way of getting information. I could not get to Lowell as I expected on account of the trains, and have sufficient time. Yet as I could find from Mr. Fallon all I wanted it made but little difference whether I went or not.

Mr. Edison seemed pleased to see me back and glad of the information I brought with me. In the afternoon he and I had a long talk about matters pertaining to the light. The telephone system has been progressing since I left, and a large amount of work had been done.

The machine for lighting the North pole had been sent away and a large number of carbons had been prepared for the purpose.[3] Mr. Fallon seemed to think that the electric light might be made in time to take the place of gas in mills. He said that the gas they used cost nearly as much as the water.

I went to lunch with him, and saw his place   he is very deep in gardening.

I judge from certain indications that Edison hauled in some money last week from his telephone contract in England, as he has just ordered about 500 books,[4] and my landlady said given $1000 to his wife.[5] He seemed to be in good spirits which looked as if he had some money in the bank, for he was dead broke when I left, for I heard him say so, and knew he had trouble to pay his pay-roll as he had over drawn his account. There is still a mint of money in telephones yet. I should like very much to have a share of it. Yet if the Electric light succeeds there is far more money in it and I feel sure that I shall have a share in that.

There is the same artist here now painting a picture of Mr. Edison that there was in Brunswick to fill the panels in the chapel. He comes[a] from Scribner's Monthly to give them a head from which to engrave a portrait.[6]

With much love   I am   Your Son

Francis R. Upton

ALS, NjWOE, Upton (*TAEM* 95:529; *TAED* MU015). [a]Obscured overwritten text.

1. Probably John Fallon, employed in an unknown capacity by the Everett Mills in Lawrence. Everett advertised a few years later that it employed more than 1,000 looms and a similar number of workers in producing 200,000 yards of cloth each week. *The Lawrence Directory* 1883, 111, 422.

2. The mills of Lawrence, Mass., had expanded greatly during the Civil War, making the city one of the world's leading textile centers by this time (Cameron 1993, 1–29). On 19 May, Upton had written Edison from his hometown of Peabody, about 15 miles distant, that "the agent of the largest mill in Lawrence says that his gas bill for one month last winter was $30,000 (thirty thousand dollars) and that he would be glad to have your light tried in his mill." Upton suggested that on Friday, 23 May, he should "visit Lawrence and Lowell and make such inquiries as may be of value regarding the probable cost of a plant. As we can now at this date beat gas for lighting a mill I think it worth while to glance at the chance." Upton wrote again the same day that although he had been mistaken and the figure of $30,000 referred to an entire year, "still that is a sum worth asking about." DF (*TAEM* 50:63, 65; *TAED* D7919ZAP, D7919ZAQ).

3. See Doc. 1706.

4. On 26 May, Edison ordered from Bernard Quaritch, a London book dealer, sets of seventeen periodicals and reports encompassing nearly 500 individual volumes. Most were journals of (chiefly British) scientific and technical societies. This order was evidently prepared from Quaritch's inventory and price sheet; the legible prices on Edison's copy of the order total approximately £140, the equivalent of about $700. The same day, Edison requested seven books about mineralogy and mining in the Americas from the Scientific Publishing Co. in New York. Lbk. 4:350, 358 (*TAEM* 80:79–80; *TAED* LB004350, LB004358).

5. Like several other laboratory employees, Upton lived at the boarding house in Menlo Park operated by Sarah Jordan, a stepdaughter of Mary Edison's father (Jehl 1937–41, 512; *TAEB* 4:538). There is no record of a large sum given to Mary around this time.

6. *Scribner's Monthly Magazine* tried at the end of March to have Edison sit for a Mr. Chase, a New York artist, but nothing seems to have come of this. On 16 May, the magazine wrote a letter of introduction for Francis Lathrop to Edison, which Lathrop enclosed with his own letter the next day. Lathrop was a noted portraitist and muralist who had painted several panels in the chapel of Bowdoin College, Upton's alma mater, in Brunswick, Maine. *Scribner's* published an engraving made from his portrait with Edwin Fox's article on "Edison's System of Fast Telegraphy" in its October 1879 issue (18:840). On 25 May Alexander Graham Bell replied to a request from Edison (not found) "for my photograph for the Editor of Scribner's Monthly." Bell declined, citing his "very great personal objection to having my photograph published." Scribner's to TAE, 27 Mar. and 16 May 1879; Lathrop to TAE, 17 May 1879; Bell to TAE, 25 May 1879; all DF (*TAEM* 49:691, 701, 700, 238; *TAED* D7906ZAB, D7906ZAH, D7906ZAG, D7903ZEC); *ANB,* s.v. "Lathrop, Francis Augustus."

London. 27 May 1879.

My dear Sir.

I cabled you to-day:—"You continue Puskas Telephone contract countries not included in assignment to Company"; and a few minutes since received your reply,—"Yes."[1]

On sitting down with Mr Puskas to go over the contracts made with you and with the Company, we found that we had overlooked Australia, India, & some other minor ~~contracts~~ countries[a] in which Mr Puskas had paid for patents. The release which I signed in your favor gave up to you all Mr Puskas's interest in the contract of 17th December 1877, whereas this release should have applied only to the countries embraced in the assignment made by you to the "Edison Telephone Company of Europe." In order to make matters straight between you & Mr Puskas, will you kindly send him[b] a memorandum contract continuing to him his rights to the countries in question, as defined in the contract of December 17th.[2]

In regard to the preliminary payments to be made to you in the countries embraced in the assignment to the Company, we would suggest to leave this point open a few days, during which we will have opportunity to consider the manner in which we will carry out the exploitation in Europe, and will make you a proposition which you will accept as equivalent to the original agreement with Mr Puskas. We would also propose that you hold 4,000 dollars of the stock of the Company as collateral security for these advances. I have agreed with Mr Puskas that this deposit shall come in part from our interest, and in part from his; to wit 2,000 dollars of his stock to be held by you, and 2,000 dollars of ours. I have written to Dr White in this mail in reference to this, and have no doubt that he will agree to the proposition. In all events the 2,000 dollars either of his and my stock, or of mine, shall be placed with you as collateral, if this is satisfactory to you. We propose this, because it is agreed between Mr Puskas & myself for considerations lying between us, that the amount of these preliminary payments to you shall be borne equally between his interest and Dr White's and mine.[3]

Mr Puskas desires me to explain to you that on account of the suits in which he is engaged in Colorado, he wishes to have the stock coming to him, issued in the name of his brother Francis Puskas, residing at Buda-Pesth in Hungary. He desires also that you will explain to the other gentlemen concerned, the reason why his stock is issued not in his name. You know more or less of his affairs in Colorado, & will understand the reason of this action. Will you please to send to him at once the amount of stock coming to him, withholding from it the one half of either 4,000 or 5,000 dollars, as you may see fit; and I write to Dr White to make up the other half from our interests.[4]

I shall go to Belgium to commence operations early next week, and Mr Puskas's Brother, who is a very intelligent Gentleman, will commence operations in Vienna; and probably in the course of a fortnight Mr Puskas will go to Germany.

I have skirmished about considerably yesterday and to–day, picking up opinions about Electric Light, and I find that coming events have decidedly cast their shadow here, and that the Experts are "hedging," and those who six months ago found all propositions as to the substitution of Electric Light for Gas ridiculous, begin to consider it quite probably that the change will take place. You have doubtless seen the report of the testimony of Sir William Thomson before the Parliamentary Committee; and this represents the general drift of opinion.[5]

My cable address is always "Bailey, London," and my post office address care of "Munro & Co, 7 rue Scribe, Paris."[6]

I desire to present my kind remembrances to Mr Griffin, Mr Batchelor and Mr Upton, and to remain— Very sincerely yours,

J. F. Bailey   by G.H.[7]

L, NjWOE, DF (*TAEM* 52:289; *TAED* D7940ZAF). [a]Interlined above. [b]Obscured overwritten text.

1. Edison wrote his reply on Stockton Griffin's transcription of Bailey's cable, which differs slightly from the above text. DF (*TAEM* 52:288; *TAED* D7940ZAE).

2. The assignment of patents is Doc. 1731 (see esp. n. 7). The 1877 contract granted Puskas rights in specified European countries and also any foreign country (except Canada and Great Britain) in which he paid patent fees on Edison's behalf (see *TAEB* 4:678 n. 5). Edison agreed to correct the error in the release and wrote to Bailey to ask that he or Puskas prepare an appropriate contract "for my signature and for execution" (TAE to Bailey, 11 June 1879, Lbk. 4:390 [*TAEM* 80:85; *TAED* LB004390]).

3. It is not clear what "preliminary payments" Edison was to receive. In his 11 June reply to Bailey (see note 2), Edison consented "to all the other propositions as contained in your letter."

4. Edison promised in his 11 June letter (see note 2) that he would "have the stock issued to Mr Puskas Brother as requested and will explain the reason to the other gentlemen concerned. The stock has not been issued yet, I have just written Mr Banker in regard to it." Puskas's business in Colorado was likely related to his earlier involvement in gold mining there and the reported appropriation of his tract by claim-jumpers. Gábor 1993, 43–5.

5. Sir William Thomson (later Lord Kelvin), a professor of natural philosophy at Glasgow University and one of the world's foremost contemporary physicists and electrical engineers, was among the experts who appeared before a House of Commons Select Committee investigating the practicality of electric lighting (*DSB*, s.v. "Thomson, William"). On 23 May, Thomson testified enthusiastically about the advantages of electric light and power and predicted that they would soon be used widely in homes and other confined spaces. *Nature* published a lengthy excerpt of a report on what it termed Thomson's "fanatical view of the electric light" in its 29 May issue and the *Journal of Gas Lighting, Water Supply, & Sanitary Improvement* also printed a critical account on 24 June. Portions of the transcript of his testimony were later entered into evidence in *Edison Electric Light Co. v. U.S. Electric Lighting Co.* "Notes," *Nature*, 20:110–11; "The Select Committee on the Electric Light," *Journal of Gas Lighting, Water Supply, & Sanitary Improvement*, Cat. 1012:86, Scraps. (*TAEM* 23:564; *TAED* SM012086c); *Edison Electric Light Co. v. U.S. Electric Lighting Co.*, p. 4118, Lit. (*TAEM* 48:569; *TAED* QD012G:75).

6. Unidentified.

7. Unidentified.

–1752–

Telegrams: From/To Grosvenor Lowrey

May 27, 1879[a]
New York 10:07 AM[b]

Thomas A Edison

Can you come in town tomorrow Wednesday to talk over foreign matters and perfect agreement.[1] Mr Morgan has just arrived from London   answer

G P Lowrey

[Menlo Park]

G. P. Lowrey

Yes will be in tomorrow

Edison

L (telegrams), NjWOE, DF (*TAEM* 50:326; *TAED* D7922F, D7922G). First message written by Stockton Griffin; second message written by William Carman. [a]Date taken from document, form altered. [b]Written on Western Union message form.

1. This concerns Edison's arrangements with both George Gouraud and Drexel, Morgan & Co. for the electric light in Britain (see Docs. 1649, 1711). On the afternoon of 9 May, Gouraud telegraphed Edison from New York to ask if he might visit Menlo Park the next day. A few minutes later, George Soren, a partner in Lowrey's firm, wired: "Will it not be prudent to delay settling or talking interests with London visitor till Lowrey returns   He will be here Tuesday" (13 May). Edison promised in reply the same day, "will do nothing with him on Light until Lowrey returns." Gouraud wrote on 10 May that he could not keep his

appointment with Edison that day but asked him to "Please wire Lowrey to show me Drexel Morgan agreement Light Pats for England if you see no objection." Two days later, Edison advised him "Do not come today will be away. Cannot do anything with Contract until Lowrey returns." DF (*TAEM* 52:535–36, 50:326, 49:226; *TAED* D7941ZAG, D7941ZAJ, D7922D, D7903ZDR).

On 2 June Soren inquired if Edison wished him to "draw the agreement between yourself & Col Gouraud which was spoken of at Drexel M & Co. the other day." He indicated that he had reviewed Edison's contract with Drexel, Morgan "to see what modification of it is necessary in respect to your arrangement with G.; & also for the purpose of drawing a <u>mem.</u> of the agreement of the D.M. & Co. to repay you half of what you are to give G." but wanted "more light" before proceeding (DF [*TAEM* 50:327; *TAED* D7922H]). In an agreement dated 9 June 1879, Edison granted Gouraud one-tenth of his net income from electric light and power patents held in Britain or obtained within five years, as compensation for prior services on behalf of the light. An unsigned copy of this agreement is in DF (*TAEM* 50:328; *TAED* D7922I).

–1753–

From Edwin Fox

New York May 29 1879[a]

My dear Edison:

I had hoped long before this to be able to return you the $125 you so kindly let me have to meet some pressing engagements some months ago,[1] but I have found it next to impossible to make collections beyond sufficient to meet current expenses. Clients are profuse in promises but rather backwards in furnishing the cash. My books show some $1500. due all of which will come in good time but, Gracious! how slowly!

Thinking that perhaps in your multitudinous expenses a little money might be serviceable at the present time[b] I send you herein $25 on a/c.

I trust that within a few weeks I can remit the balance.

Accept the assurance of my sincere gratitude for your kindness in the premises   Truly Yours

Edw. Fox

PS. As soon as my "wicked partner"[2] returns from Kentucky with his big fee for defending Col. Buford[3] I'll be rich and treat. E.M.F.

ALS, NjWOE, DF (*TAEM* 49:249; *TAED* D7903ZEK). Letterhead of Curtis & Fox, Attorneys & Counsellors at Law. [a]"New York" and 187" preprinted. [b]"at . . . time" interlined above.

1. William Carman recorded a payment of this amount to Curtis & Fox on 6 December 1878, during a period when Fox spent considerable time at Menlo Park writing for the *New York Herald* (see Docs. 1545 n. 1 and 1613). It is not clear if this is the same transaction to which Fox re-

ferred in late January when he explained to Edison "why our firm note which is over due was not taken up in season." Fox stated that he had been ill for some time and in his absence "at the office everything seems to go by default." Ledger #3:71, Accts. (*TAEM* 87:38; *TAED* AB001: 32); Fox to TAE, 26 Jan. 1879, DF (*TAEM* 50:237; *TAED* D7920R).

2. George M. Curtis, a New York City judge until 1874, was tried by the state Senate in 1873 on several counts of judicial corruption, including "grossly improper conduct and scandalous and indecent language upon the bench." He was acquitted and later served in the state Assembly. Edison had consulted Curtis about an unknown matter in 1878. "The State Legislature," *New York Times*, 22 Jan. 1873, 2); Obituary, ibid., 15 May 1915, 13; TAE to Curtis, 6 Dec. 1878, DF (*TAEM* 16:501; *TAED* D7802ZZMM).

3. Colonel Thomas Buford, brother of a prominent former Confederate general, was charged with the assassination of a Kentucky appellate judge in March 1879. Buford was convicted but later re-tried and acquitted on grounds of insanity. "The Record of Murder," *New York Times*, 27 Mar. 1879, 1; "Gen. Abe Buford of Kentucky," ibid., 10 Feb. 1882, 2.

–1754–

From Otto Moses

Charleston, S.C. June 1, 79

My dear Sir:

Your advertisement in the Herald for "an experienced analytical chemist—very expert in qualitative analysis and assaying,"[1] has been brought to my notice by a thoughtful friend in New York. This person, having frequently heard me express a longing to be surrounded by such an atmosphere of energy and genius as I had enjoyed in your laboratory[a] (if you remember) about the time of your departure on the Draper expedition, resolved the opportunity should not escape my attention; and indeed I am very glad she did so.[2] For there is, I think, nothing in the world could yield me greater pleasure and satisfaction than to know you desired a chemist among your assistants, and that I might fill the place. But to the point: my experiences during seven years in the conduct of a chemical laboratory in this city (~~during~~ a considerable portion of the time serving as State Inspector of Phosphates) based upon a four years course at the Mining Academy in Freiberg, Saxony, studies in Paris and extensive travel abroad, will justify me in addressing a reply to your advertisement. From the rapid and exhaustive character of your investigations I imagine blowpipe analysis would enter most largely into your method; for my ability in its use I beg to refer you (as most easy of access) to Prof. H. B. Cornwall of Princeton—Editor of Richter on the Blowpipe,[3] and to Dr Chandler of the School of Mines,[4] to whom, by the way, I think I showed the first silver assay ever made in Columbia College.

My laboratory is equipped with delicate balances of Becker and of Lingke's make,[5] apparatus, &c.

As it is of more importance to secure a contact with a fertile suggestor of work, than for me to be paid for it, I will not now propose that question. My income is sufficient for the present, and my philosophy is <u>to know</u>, and to conquer Nature rather than Fortune.

Hoping to hear from you, dear Sir, at an early opportunity I am with kindest regards to Mr Batchelor and your other assistants, Very truly yours,

Otto A Moses[6]

ALS, NjWOE, DF (*TAEM* 49:914; *TAED* D7913ZAH). [a]"in your laboratory" interlined above.

1. Edison's advertisement for this position appeared in the *New York Herald* on 28 May 1879 (p. 16). He also received responses to an advertisement in the *New Yorker Staats-Zeitung,* a German-language daily, and corresponded with several applicants about wages. In a draft reply to a 16 July letter concerning an assay of placer mine waste sand, Edison proposed chemical treatments to remove platinum and additional gold from waste sand, noting that he had "some very smart German chemists." Besides Moses the only German chemist that Edison is known to have had on his staff at this time was Alfred Haid, a Ph.D. analytical chemist, who began working in the laboratory by the first week of June 1879, probably replacing Henry McIntire. Two other chemists, John Lawson and C. E. Mumsell, were both American-born. Lawson had no advanced chemical training before joining Edison's staff while Mumsell had a Ph.D. from the School of Mines at Columbia College. Mumsell may have already left the staff by this time. Otto Kretzchmer to TAE, 28 and 29 May 1879; Mr. Ertmine to TAE, 28 and 29 May 1879; George Smith to TAE, 29 May 1879; W. C. Hendricks & Co. to TAE, 16 July 1879; all DF (*TAEM* 49:902, 911, 906–7, 909; 50:938; *TAED* D7913Z, D7913ZAF, D7913ZAC, D7913ZAD, D7913ZAE, D7928ZHB); for Haid, see letterhead of Haid to Insull, 12 Sept. 1884, DF (*TAEM* 71: 262; *TAED* D8403ZGK); for Mumsell, see Charles Chandler to TAE, 27 Feb. 1879, DF (*TAEM* 49:883; *TAED* D7913P) and N-79-03-10.2: 25, Lab. (*TAEM* 31:1137; *TAED* N032:11); for McIntire, see *TAEB* 4: 749; for Lawson, see "Lawson, John W.," Pioneers Bio.

2. After seeing the *Herald* advertisement, Otto Moses's mother-in-law wrote Edison to recommend him for the job, noting that he "last summer had the pleasure of spending a few days with you" and was now preparing to move from Charleston for New York. Nothing is known of Moses's presence at Menlo Park in 1878. Mrs. Da Vega to TAE, 28 May 1879, DF (*TAEM* 49:898; *TAED* D7913X).

3. Henry Cornwall studied at the Royal Mining Academy in Freiberg and in 1873 became professor of applied chemistry and mineralogy at Princeton (*WWW–1*, s.v. "Cornwall, Henry Bedinger"). He translated

Plattner 1875, which had been revised by Theodore Richter, from the German.

4. Charles Chandler, professor of chemistry at Columbia University, cofounded the School of Mines there in 1864. *ANB*, s.v. "Chandler, Charles Frederick."

5. August Lingke & Co. manufactured a variety of instruments in Freiberg, beginning in 1791. de Clercq 1985, 132–33, 136.

6. Otto Moses was a native of Charleston. In addition to the training described above, he was reportedly also a graduate of the University of Leipzig. It is not known when he began to work at Menlo Park; the earliest evidence of his work at the laboratory dates from January 1880. He later served as one of Edison's representative at the Paris Exhibition of 1881. Obituary, *New York Times*, 6 Jan. 1906, 9; unidentified Obituary, Cat. 1339:10, Batchelor, NjWOE; Time Sheets, NjWOE.

–1755–

To Theodore Puskas

MENLO PARK, N.J.[a] June 2d 1879

Confidential
My Dear Puskas

I was very much surprised today by the receipt of a letter from my nephew dated from Paris,[1] as I cabled him some two weeks ago to return to Menlo Park[2] as his mission was accomplished and I wish to say that if he remains in[b] Europe he does so on his own responsibility and that he is not acting in any capacity for me.

I will also caution you that he has an ungovernable temper which when aroused requires about 2 weeks to become calmed down. This is for your information and I would prefer you would not mention what I have written to him   Very Truly
T. A. Edison   G[riffin]

L, NjWOE, Lbk. 4:374 (*TAEM* 80:83; *TAED* LB004374). Written by Stockton Griffin; the original is in TP. [a]Place from Edison's laboratory handstamp. [b]Obscured overwritten text.

1. Not found.
2. See Doc. 1744.

–1756–

Notebook Entry: Mining

[Menlo Park,] June 6 1879—

McLaughlin has just returned from the Chaudiere River & the Leloup[1] Canada where there are placer-mines[2]   This alluvial is said to contain a minute trace of Platinum & Iridiosmine   McL brings some of the black sand from the sluces before the fine gold is panned from it. By the aid of the microscope I ascertain that the platina is combined with the gold also in seperate pieces. The Pt is not alloyed with Au but is in contact with it. The amount of Platinum is about ¼ that of the gold. Iridosmine is very plentiful. It forms a large blulk of the sand there is probably 6 oz of Iridosmine to every oz of gold. I have found that aSodium amalgman will take up the Iridismine[a] & platinum, the Au being previously taken out by pure mercury— There is undoubtedly enough Pt & Iridium in the Canada district to supply the world   instead of the black sand containing a trace of the platinoid The mine should be called a platanoid mine with traces of gold—

T.A.E.

X, NjWOE, Lab., N-78-11-22:149 (*TAEM* 29:218; *TAED* N002:67). [a]Obscured overwritten text.

1. The Chaudière and the Rivière du Loup are tributaries of the St. Lawrence in southeastern Quebec.

2. Frank McLaughlin was a telegrapher acquaintance of Edison's working in New York as an electric pen agent; he had been Edison's representative for the phonograph in Australasia in 1878 (see Doc. 1328). Edison dispatched him to Quebec on 26 May after corresponding with Edwin Pope of the Montreal Telegraph Co. about gold deposits there (Edwin Pope to TAE, 6, 13, and 15 May 1879; McLaughlin to TAE, 1 June 1879; all DF [*TAEM* 50:588, 591–92, 614; *TAED* D7928F, D7928J, D7928K, D7928ZAC]). A brief description of McLaughlin's trip and findings is in "Wanted, A Platinum Mine," *New York Sun*, 7 July 1879, Cat. 1241, item 1224, Batchelor (*TAEM* 94:494; *TAED* MBSB21224X).

–1757–

To Hamilton Twombly

[Menlo Park,] June 11, 79 8:[2930?][a] AM[b]

Will have system set up soon as get my switch board done[1] Will notify— Kinney[2] is getting on finely— will have my underground system in[c] few days—[3d]

Standard Dynamo machine completed.

Edison

ALS (telegram), NjWOE, DF (*TAEM* 49:262; *TAED* D7903ZES). [a]Illegible. [b]Time and date written by Stockton Griffin. [c]Obscured overwritten text. [d]"(over)" follows as page turn.

1. Hamilton Twombly had wired the previous day, "Have you the exchange system ready." Edison wrote this reply on the same paper on which Twombly's message was transcribed. Twombly to TAE, DF (*TAEM* 49:262; *TAED* D7903ZES).

2. Edison meant Patrick Kenny, a former superintendent of the Gold and Stock Telegraph Co.'s manufacturing shops, who had begun collaborating with Edison on facsimile telegraphy in the spring of 1878 and started working at the laboratory that December. See Docs. 1328 and 1638; *TAEB* 4:406 n. 6.

3. Apart from a patent application covering underground conductors and junction boxes (see Doc. 1747), there is no evidence of work on an underground system for electric lighting at this time.

–1758–

From William Croffut

New York, June 11—1879[a]

My Dear Edison:—

Now haven't I exercised a good deal of patience in this matter of staying away from you? Like the fellow who went to sleep in sermon time, it is only because I "have perfect confidence in the minister."

Johnson tells me you are to have an exhibition of either the[b] electric light or telephone soon. I want to "be thar," & don't you fail to notify me. I shall do it up for the Indianapolis paper,[1] besides which W. R. wants it for the Tribune,[2] & Mr. Ford (one of the editors here,)[3] wants it for his letter to the London News. I am to notify him. You will find the Tribune quite inclined to do full justice to all your inventions. If Griffin reads this before you do, or you never see him at all ask him to write me about the experiments & when I can come down.[4] Can't Ford & I come two[b] or three days before the others, as our papers are so distant? Very Truly Yrs

W. A. Croffut

Do you read "Bourbon Ballads"?[5]

ALS, NjWOE, DF (*TAEM* 49:707; *TAED* D7906ZAO). Letterhead of the *New York Tribune.* [a]"New York," and "187" preprinted. [b]Obscured overwritten text.

1. Croffut's unsigned dispatch appeared as "The Chemical Telephone" in the *Indianapolis Journal* on 5 July 1879 (p. 7). The article described Johnson's demonstration of the telephones he took to England and, more briefly, Edison's dynamo, new dynamometer, and electric light. Croffut had contributed an article on the light to this newspaper in October 1878 (see Doc. 1530).

2. Whitelaw Reid was editor of the *New York Tribune,* having succeeded Horace Greeley (*ANB,* s.v. "Reid, Whitelaw"). Croffut apparently had moved to the *Tribune* from the *New York Daily Graphic* in recent months.

3. Probably Isaac Ford, identified as an editor at the *Tribune's* address. Wilson 1879, 484.

4. On 19 June Croffut wrote Griffin that he had "asked half a dozen correspondents of prominent papers to go down to M.P. to see the new telephone" in the next day or two but had not yet received a definite invitation. Croffut promised that a demonstration for the press would give Edison "a beautiful setting up. W.R. wants me to go down for Tribune." According to Croffut's article in the 1 July *Tribune,* Edward Johnson exhibited fifty receivers, described as "the first we have finished," which he planned to carry to England on 1 July. Edison also described his gener-

ator and reportedly claimed of his electric light that "it has come. I have demonstrated, to myself and to my friends, that I have accomplished all I ever expected to in this matter. . . . I claim that this solves the question." Croffut also sent a dispatch to the London *Times* aboard Johnson's ship; it was published on 15 July and reprinted elsewhere in slightly edited form three days later (see Doc. 1780 n. 13). The *Daily News*, probably the London paper to which Isaac Ford contributed, published a cabled account of the demonstration on 2 July; the event was also covered by the *New York Herald*. Croffut to Griffin, 19 June 1880; Croffut to TAE, 5 Aug. 1879; both DF (*TAEM* 49:713, 719; *TAED* D7906ZAR, D7906ZAV); "Local Miscellany. Edison's New Telephone," *New York Tribune*, 1 July 1879; "Edison's Work," *New York Herald*, 1 July 1879; "Mr. Edison and the Electric Light," *Daily News* (London), 2 July 1879; Cat. 1241, items 1212, 1214–15, Batchelor (*TAEM* 94:492; *TAED* MBSB21212X, MBSB21214X, MBSB21215X); "The Edison Telephone and Electric Light," *Times* (London), 15 July 1879, 8.

5. Croffut wrote "Bourbon Ballads," a series of mock ballads satirizing the policies of the southern Democrats known derisively as "Bourbons." They appeared on an irregular but frequent basis in the *Tribune* and other papers throughout the spring and summer and reportedly were published as a collection in 1880. Obituary, *New York Times*, 2 Aug. 1915, 9.

–1759–   [Menlo Park,] June 12 1879.

Notebook Entry: Experiments on Chemical Telephone
Telephony   Connections   Transmitting end[1]

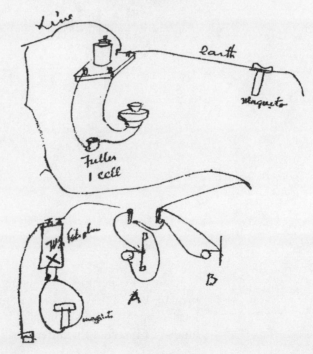

When A is in only & X is out it is exceedingly loud. when X is in it reduces it to ¼ of the Volume—

When both A & B are in the cutting out of X does not materially alter the volume of sound—[a]

I have just found a bug of large dimensions which is that sometimes when B is cut in scarcely a sound can be heard on A at other times the cutting in makes no difference; This phenomenon must be due to polarization.[2a]

I find that with 2 chalks the current that will polarize ~~wone~~ will not always sufficiently polarize the other hence one chalk will be loud & the other low. ~~o~~If the battery is increased so as to over polarize then the chalks may have the pressure increased on one & decreased on[b] other & this will bring them ok so that ~~inserting~~ when one is taken[b] out of ckt the other will go low owing to increase in strength of polarizing current. ~~hen~~ when thus very low the insertion of the other chalks makes it loud again as it weakens the polarization current. On the other hand when one is out & the other is over polarized & low increasing the ~~tens~~ pressure on the spring ~~w~~so that it will drag as before A loudness will be nearly restored

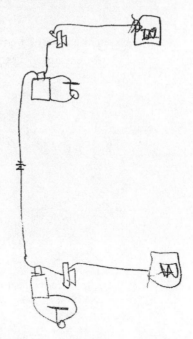

Just tried experiment of putting induction coil in line ie[c] its primary coil of $^{40}/_{100}$of an ohm in line & putting the chalk in a local formed of the secondary of 160 ohms it worked quite loud—[3] I inserted the other chalk right in the line & it didnt

make any difference, but putting in the Induction coil of the regular telephone X lowered it greatly— I now try putting carbon transmitter in line with 4 cells carbon battery, placing primary of induction in ~~Coi~~ line & putting chalk in secondary, I find that you can scarcely hear it with 100 ohms in circuit with 4 ~~e~~small carbon cells in line   with 12 ohms it is fair

We not use 1 cell carbon in line & shunt the transmitter around it to see effect. it works loud on short ckt but 12 ohms inserted makes it low—[a]

We now wind 4 layers of 23 wire on the outside of the Induction[b] Coil forming a tertiary coil[4]   in the circuit of this we put the chalk & a polarizing battery 1 Daniel cell & find that it comes strong & loud in fact as loud as the old way when both induction coils in also that 12 500 ohms may be inserted in the line & it is still readable several feet away= I then wind 6 layers of No 3 wire[5] on a ~~W E Mfg Co~~ Bergman coil for a tertiary & find it as loud. This extra Resistance over that which would be given by a 23 wire prevents the chalk ckt shortcktg the induction from the coil when talking= The greater the resistance that can be put in this tertiary & have it work loud the better it will be when transmitting for the other man—

By this means we make each chalk receiver entirely independent of everything[b] else & have a perfectly constant polarization of the chalk   we can use a Chloride of Silver battery to polarize the chalk & this will probably last 2 years

We propose to do away with the switch & ring the bell through the coil by our little multiple battery=[6a]

TAE

X, NjWOE, Lab., N-79-06-12:1 (*TAEM* 35:485; *TAED* No80:2). Multiply signed and dated. [a]Followed by dividing mark. [b]Obscured overwritten text. [c]Circled.

1. This drawing of an experimental telephone arrangement shows the transmitting circuit at top with an induction coil on the left and the carbon transmitter and a local battery ("Fuller 1 cell") directly underneath. Edison apparently intended to draw the receiving circuit to the right of the transmitting circuit but, finding a lack of space after drawing in the "magneto" receiver, instead designated that line as the "earth" or ground wire and drew the "line" wire to the receiving circuit at bottom. The receiving circuit includes two electromotograph receivers ("A" and "B") at right and a "magneto" receiver as part of a standard "W[estern].E[lectric]. Telephone" set marked "X" at left.

2. Edison appears to be analogizing from the polarization of batteries, a phenomenon in which the accumulation of ions on one electrode in an electrolyte increasingly impedes the flow of current. See also Doc. 1738 n. 5.

3. Each receiver is represented by a horizontal line representing the metal stylus of the electromotograph touching a small circle representing the chalk. These are in circuit with the secondary (outer coil) of the induction coil while the transmitters are connected to the small inner primary coil and to the earth. A battery is shown in the line between the induction coils.

4. Charles Batchelor had sketched a receiver in the tertiary circuit of an induction coil on 1 June and Edison made a similar sketch on 15 June. In a portion of a notebook entry dated 19 May but evidently written later (probably about this time), Batchelor commented that "We find it much better to work the button for receiving in a tertiary circuit in telephone as its polarization is always constant and independent of the other end." Cat. 1304:57, 55, Batchelor (*TAEM* 91:56, 54; *TAED* MBN004:54, 52); N-79-01-01:204, Lab. (*TAEM* 30:356; *TAED* No13:98).

The tertiary circuit was included in the commercial telephone designed in July (Doc. 1784). In a patent application executed on 17 July, Edison described the rationale for placing the receiver in a tertiary circuit instead of directly upon the line: "the object of placing it in a tertiary circuit is to keep the apparatus free from earth currents, which cause alterations in the volume of sound, owing to the marvellous delicacy of the apparatus." According to the patent application the primary coil was connected in a local circuit with the transmitter and battery, the secondary with the line wire, and the tertiary with the receiver local circuit and battery. Edison also specified the need for a battery "to keep a constant current in tertiary circuit" but noted that the battery could be omitted when "certain chemicals are used in the porous or chalk cylinder" and the receiver spring was tipped with platina, presumably because of the generative effect described in Doc. 1738 n. 5. This paragraph was later amended to state that the battery was unnecessary when the spring was tipped with palladium but was still needed when platinum was used. Edison included in a later British patent the palladium-tipped spring, "whereby the degree of frictional variation is increased under the action of a given strength of electric current, and a corresponding augmentation in the volume of sound is obtained." Pat. App. 231,704; Brit. Pat. 5,335 (1879), Batchelor (*TAEM* 92:141).

5. Edison meant the thinner No. 30 gauge wire, whose resistance is many times greater and which he specified for a tertiary winding of 6 layers in an undated note made around this time. N-79-01-01:201, Lab. (*TAEM* 30:354; *TAED* No13:96).

6. The multiple battery referred to is probably a 10 cell battery indicated in a 15 June drawing of a telephone board (N-79-01-01:205, Lab. [*TAEM* 30:356; *TAED* No13:98]); see also Doc. 1738 n. 7. Following this document are four pages of notes concerning tests of a variety of chalk button solutions, which include Edison's notion that "It is probable that every solution will require a different polarizing current, chalk seems to require ¾ or ½ Daniel, salt 1/20 of a Daniel roughly speaking" (N-79-06-12:19–25, Lab. [*TAEM* 35:494–97; *TAED* No80:11–14]).

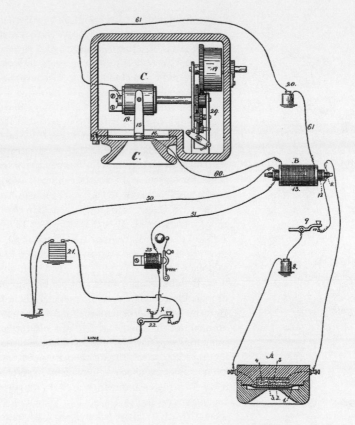

Edison's patent drawing (U.S. Patent 231,704) shows the electromotograph receiver, C, at top, connected to the tertiary coil, 13, which is the outer layer of the induction coil, B; the transmitter, A, is at bottom.

–1760–

To James Banker

[Menlo Park,] June 19, 79 9:15 AM

James H Banker[a]

Cant come in today.[1] Have to send Gramme machine away[2] must conduct some experiments before it goes[3]   Been up all night.

T. A. Edison

L (telegram), NjWOE, DF (*TAEM* 49:266; *TAED* D7903ZEX). [a]Sent "Care R L Cutting & Co."

1. Banker had requested that Edison have Stockton Griffin stop at Robert Cutting, Jr.'s office on 14 June, where he planned to "explain European light Company to him." On the day of that proposed meeting, Cutting invited Edison to go with him and Banker "some day next week to Coney Island and dine there quietly and have a little spree which we are sure you need." Shortly before noon on 18 June, Banker wired: "Cannot go today to Menlo   If you can come tomorrow to Cuttings at noon can have meeting before going to Coney Island. Better bring Griffin." Banker to TAE, 12 and 18 June 1879; Cutting to TAE, 14 June 1879; all DF (*TAEM* 50:317, 49:266, 264; *TAED* D7921A, D7903ZEW, D7903ZET).

2. Edison had arranged in December 1878 to borrow a Gramme dy-

namo through Dr. C. C. Soulages, a partner of Cornelius Herz (see Docs. 1595 and 1647). Soulages wrote in early May that the Gramme was needed in San Francisco and asked for its return, and also remarked on press reports that Edison planned to begin using his own dynamo. Edison wired Calvin Goddard to "ask him the price of it and Telegraph me" but Soulages declined to sell. Goddard reported on 10 May that "Soulages will let us know when he must have Gramme." In mid to late May, John Kruesi recorded an order to "Take measurements of Gram machine"; the measured drawings of the machine and its parts are in N-79-02-10:158–59 (Lab. [*TAEM* 32:1218; *TAED* N046:78]). About the date of this telegram Kruesi noted the order to "have Gramme mach. ready for shipment Friday afternoon," probably 20 June. It was addressed to Soulages at 30 Lafayette Place in New York. Soulages to TAE, 4 May 1879, Soulages to Goddard, 4 May 1879, TAE to Goddard, 5 May 1879, Goddard to TAE, 10 May 1879; all DF (*TAEM* 50:255, 257–58; *TAED* D7920ZAI, D7920ZAL, D7920ZAJ, D7920ZAO); Cat. 1308: 149, 155 (Order Nos. 174, 193), Batchelor (*TAEM* 90:741, 744; *TAED* MBN003:48, 51).

3. Beginning on 9 June and continuing until 17 June, Edison and Francis Upton made several tests of Edison's new generator using the Gramme to excite the field magnet. On 17–18 June they conducted additional tests in which the field of the Gramme was excited by the Edison. One set of the Gramme tests, which Upton reported in his paper to the August 1879 meeting of the American Association for the Advancement of Science, showed that a drop in voltage occurred as the current taken from the Gramme increased. In his paper Upton called this effect "analogous to the case of a battery which becomes polarized as the strength of the currents taken from it are increased. This fall in electromotive force may be called the polarization of the armature and is due to the saturation of the ring of iron wire which forms it." N-78-12-20.2: 192–281, N-79-06-16.1:1–101, Lab. (*TAEM* 29:894–938, 35:161–211; *TAED* N008:96–140, N077:1–51); Upton 1880b, Table 1, 182.

–1761–

From Henry Bentley

Phila' 6/19 187[9][a]

Friend Edison

Franklin Institute packed last night[1] ~~aisles~~passageways and all—hanging on all around the stairs &c   Prof Rogers[2] said in a brief address after I got through that it was the[b] success of the year just closed. at the Institute. Many big guns there and various curious remarks were made—publicly—and all of the most gratifying character. Everybody was thunderstruck at the perfection with which it went off, and at its loudness. [~~The?~~][c] A vote of thanks was presented to Prof Edison "and his able co-adjutor" Henry Bentley for the opportunity afforded of seeing and hearing the the wonderful ins't.[3] It was at its very best and it filled the auditorium with its tones most completely. The cornet on it was simply wonderful to hear.

I am glad I stuck to it and brought it over the other night. I send you <u>some</u> papers—no use to send all, you will get the general run of it from these.[4] You have added another wreath to your brow now sure.

I find when I hold the door of the box open the sound is clearer and louder. Please see if in the arranging of this box it cant be left quite open or full of holes. Yours

Bentley[5]

ALS, NjWOE, DF (*TAEM* 49:268; *TAED* D7903ZEZ). Message form of Philadelphia Local Telegraph Co. [a]"187" preprinted. [b]Multiply underlined. [c]Canceled.

1. Founded as a Philadelphia mechanics' institute, the Franklin Institute was by this time a nationally prominent center for science and technology (Sinclair 1974). On 2 May its secretary had asked Edison, for the second time, to demonstrate his electromotograph telephone and electric light at one of the Institute's monthly meetings. Later that month, George Barker wrote that he had heard Edison planned to show the telephone there but "I could not credit the statement, because in the first place, I could not believe that a society which is controlled scientifically by [Edwin] Houston & [Elihu] Thomson, who have themselves said so many shameful things about you, and a society in whose meetings you have been held up to ridicule & against whom I have had to defend you often, could have the cheek to ask such a thing of you; and in the second, because I did not believe you would accede if they did ask it." Edison wrote on Barker's letter, "Lord no I have not promised them any thing nor would not= I am going to send Bently two pair & if possible you a pair." Franklin Institute to TAE, 2 May 1879; Barker to TAE, 16 May 1879; both DF (*TAEM* 49:219, 229; *TAED* D7903ZDL, D7903ZDU).

2. Probably Robert Rogers, a noted chemist and physician and former dean of the medical college of the University of Pennsylvania. Rogers was now professor of medical chemistry and toxicology at Jefferson Medical College in Philadelphia. *ANB*, s.v. "Rogers, Robert Empie."

3. Before exhibiting the telephone, Bentley asked Edison to have Stockton Griffin write out "in your own words—any new points on it that you may desire or the general tenor of what you wish printed." Because Bentley expected his remarks to be published in the *Journal of the Franklin Institute* he wanted "if possible to give it your own language as nearly as possible," and arranged to visit Menlo Park on 17 June to "get the bearings" of the instrument so he could "give it the biggest kind of a send-off at the Institute." The *Journal* printed only a brief account of Bentley's demonstration, but a full account was published in the *Philadelphia Inquirer.* The Institute's secretary wrote Edison a formal letter of thanks on 19 June. Bentley to TAE, both 14 June 1879; Franklin Institute to TAE, 19 June 1879; all DF (*TAEM* 52:71, 73; 49:270; *TAED* D7937ZBL, D7937ZBM, D7903ZFA); "Proceedings, etc.," *Journal of the Franklin Institute* 108 (1879): 71; "The Telephone," *Philadelphia Inquirer,* 19 June 1879, 3.

4. Not found.

5. Henry Bentley was founder and president of the Philadelphia Local Telegraph Co. See *TAEB* 2:433 n. 4.

-1762-

*Francis Upton to
Elijah Upton*

Dear Father:

I have broken my usual rule today, for I have been working nearly all this[a] Sunday.[1] Mr. Edison has been very hard at work on his new telephone receiver during the past week, and has succeeded in getting it into very good order. He has been working night and day and I have been up two nights and parts of four. He sees about $100,000 in cash to come to him from England if he gets it to work satisfactorily. He has ⅔ of this clear after he has paid such shares as he had given away.[2]

I had a talk with him this afternoon and he spoke of five per cent on the electric light, he did not make me a direct offer but hinted that I might have so much.[3] He said that he would give it me the same as he gave to Mr. Batchelor only half the amount he gets. It will be five[b] per cent of the profits, yet I shall have no papers except that my name will be placed on the agreement he has entered into with the electric light Co. I take no risk beyond losing my time for a year or two for it is not a partnership, Mr. Edison runs all risk puts in all the money and gives me five per cent on all his sales clear. There is about $100,000 worth of property here and some twenty to thirty men employed so I think five[b] per cent on the total product is a fair show. I think with some pertinacity I could get him to make is seven and a half, for he is very easy in such matters, yet as it is pure generosity on his part I think it is not becoming in me to try and jew him. Besides if I grumble perhaps he may say no.

This is what I come in for, fifty shares of Electric light stock, worth today over $100, free from assessment. This comes to me immediately but of course not for me to sell. Then when the Electric light is accepted $5,000 in cash and $1500 a year for 17 years, so why should I find any fault with my chances. I am going to say yes, I will take five per cent, would not you? The only trouble in the arrangement is that the light is not yet out and far from perfection, yet as I am only getting $600 a year I do not lose much if I go two years without pay.[4] I do not say I can get five per cent, yet things look that way. I am with much love   Your Son

Francis R. Upton

ALS, NjWOE, Upton (*TAEM* 95:536; *TAED* MU018). [a]Interlined above. [b]Obscured overwritten text.

1. Upton had spent the day working out "the elements of uncertainty" in testing generators. All of these had to do with the test apparatus or the generator and belting except the first, which Upton described as "Change in the brain of the operator especially in his memory." This

was the only element upon which Edison commented, writing "You bet old boy." N-78-12-28:182, Lab. (*TAEM* 30:73; *TAED* No11:74).

2. For Edison's obligation to George Gouraud, see Doc. 1743. Nothing certain is known about Edison's agreements with Batchelor, James Adams, and Charley in regard to the English telephone but see Docs. 1345 and 1652 and *TAEB* 4:816 n. 1 regarding his arrangements with them for U.S. and French telephone royalties. Between 20 and 23 May Edison credited \$1631.78 to Charles Batchelor, \$815.89 to Mrs. James Adams, and \$500 to Charley Edison, presumably to apportion his half of the \$24,500 advance received from London about this time (Ledger #3: 112, 205, 178, 335, Accts. [*TAEM* 87:59, 102, 89, 146; *TAED* AB001:53, 96, 83, 140]; Doc. 1742).

3. The previous week Upton reported to his father that he

had a talk with Mr. Edison today regarding my future. He seems to think favorably of giving me a share in the business instead of wages. The only trouble is that I may have to wait a long time before the profits come. . . .

I think it would be quite a fine thing if I could come into partnership, if I only had a little more capital to live on until returns appear. Do you not think I have done well to be able to talk of ten per cent of the total profits at the end of six months [at Menlo Park], and to have any proposition considered as it is now? [Upton to Elijah Upton, 15 June 1879, Upton (*TAEM* 95:534; *TAED* MU017)]

4. Upton wrote his father again the following Sunday, "I do not understand why you should object to my taking the share I thought of taking in the Electric Light. I have not had a good chance to talk alone with Mr. Edison during the past week so have not said anything more about the matter. I only risk my pay if I accept his terms to take a share, for he will only put my name on the agreements he has with the Electric Light Co. and give me some shares of the stock. The market value of the stock will be more than my salary for eight years and the light will be a success or failure long before that time. If I take the share, it is only in the profits which will be very large in case of success, \$30,000 a year guaranteed royalty." Upton to Elijah Upton, 29 June 1879, Upton (*TAEM* 95:541; *TAED* MU020).

–1763–

From Joshua Bailey

Paris 25 June 1879.

Dear Sir,

We cabled you yesterday to telegraph when Motographs will be sent.[1] Mr. Griffin wrote some time ago that they would be shipped from New York June 3rd— We are in great need of the motograph here & cabled in order to know how to shape our operations. Enclosed herewith you will find Memorandum of Contract in accordance with your suggestion—[2] We have engaged your Nephew with the French Company in order to prevent his engaging with Dr Herz.[3] He has an excellent opportunity which we hope he will improve— We advised him strongly to return to you, & offered to advance him money for that purpose, but he did not see fit to accept the advice or the offer. Very truly Yours,

J. F. Bailey

ALS, NjWOE, DF (*TAEM* 52:297; *TAED* D7940ZAJ).

1. The previous day Theodore Puskas had wired Edison, "Cable date Motographs sent." Bailey telegraphed "Date motographs important" on 26 June, the same day Edison cabled Puskas, "Ten days." After making another inquiry, Puskas instructed Edison on 16 July to "Ship motographs French line Wednesday" (the same day). On 23 July, in reply to another inquiry from Bailey that day, Edison cabled that the instruments had been "Shipped care Johnson Six Lombard" (George Gouraud's office). Puskas to TAE, 24 June, 15 and 16 July 1879; Bailey to TAE, 26 June and 23 July 1879; TAE to Puskas, 26 June and 15 July 1879; TAE to Bailey, 23 July 1879; all DF (*TAEM* 52:296, 308–9, 302, 310; *TAED* D7940ZAI, D7940ZAQ, D7940ZAS, D7940ZAL, D7940ZAT, D7940ZAR, D7940ZAM, D7940ZAR, D7940ZAU).

2. Enclosure not found but see Doc. 1751 n. 2.

3. Charley wrote Edison from Paris on 26 June that "Not having received an answer from my letters after waiting a considerable length of time I anticipated that you were displeased with opperations in England and proceeded as I had intimated in my last letter. I accordingly made arrangements to stay in Paris for Puskas and do the telephone work for the French company." He acknowledged Edison's request that he return home but said that "before acting I will give you the situation here and await your reply." In closing, he asked, "Who is Dr Herz—? when I was hanging around waiting for Bailey he came to me and offered me $300 month to work for him." Charley Edison to TAE, DF (*TAEM* 52:298; *TAED* D7940ZAK).

On 5 June, George Prescott had sent a letter from Cornelius Herz (not found) which he asked Edison to "return to me with your views on the requests about your nephew &c." Edison wrote on Prescott's letter, "Undoubtedly Herz is a perfect fraud. . . . I would be a little careful about being int[erested?] with him. CPE does not und[erstand] anything abt quad or Printer If Herz wld pay in adv for a man for a certn length of time you might send an op[erato]r from up st[ai]rs who unds Quad— thats only proper way do it." Prescott to TAE with marginalia, DF (*TAEM* 51:873; *TAED* D7936W).

–1764–

To Hamilton Twombly

[Menlo Park,] June 26th [187]9

Dear Sir.

I learn that Gould is trying to effect an arrangement towards using the Automatic system on the new lines.[1] Cannot something be done so the Western Union would take all the Automatic patents?[2] I have a large interest in them and of course would like to get paid something for my four years labor in perfecting it. While the value of these patents for use upon the W.U. lines are at present small, owing to unlimited wire capacity the system is of great value to a new company without wire capacity because it gives one wire the capacity [of]ᵃ ten without an expenditure of capital, hence it can always and will probably always be used as a club to hurt the W.U. The patents contain the base of all that is valuable in automatic telegraphy and cover it for years to come. Very Truly

Thos A Edison

LS (letterpress copy), NjWOE, Lbk. 4:430 (*TAEM* 80:89; *TAED* LB004430). Written by Stockton Griffin; circled "C" written at top of page. ᵃFaint letterpress copy.

1. The lines were those of the American Union Telegraph Co., launched by financier Jay Gould (see *ANB*, s.v. "Gould, Jay" and *TAEB* 2:369 n. 12) on 15 May to oppose Western Union (see Doc. 1713 n. 4). Gould, whose dealings made him a public symbol of business corruption, was the principal figure behind an earlier challenge to Western Union that ended in 1877 with the company's purchase of the Atlantic and Pacific Telegraph Co. (see Doc. 985 n. 1). American Union ultimately did not rely on automatic telegraphy although another new enterprise, the American Rapid Telegraph Co., did so using another system (see Doc. 1790 n. 9). Reid 1886, 577–80, 778–81; Harlow 1936, 409–11; Israel 1992, 146–48.

Edison's system of automatic telegraphy employed punched-tape automatic transmitters and electrochemical receiving instruments to transmit messages at high speed so that each occupied the line for only a small fraction of the time needed by ordinary manual transmission (see *TAEB* 1–3 passim). Gould took control of Edison's automatic patents in 1875 without paying him or most other investors in the Automatic Telegraph Co.; this became the subject of a lawsuit by Edison and investors in the Automatic Telegraph Company. Gould also never assigned the patents to the Atlantic and Pacific Telegraph Co. although the company used the system under a separate agreement, which gave Western Union some claim to the automatic when it subsequently acquired Atlantic and Pacific. See *TAEB* 2:464 n. 4, 3:10 n. 5; Docs. 522, 561, 676 and 750.

2. Edison was apprised of the prospect of the new company using his automatic system on 14 June by Edwin Fox who had begun to provide legal advice to Edison in his suit with George Harrington against Atlantic and Pacific regarding the automatic. Fox also simultaneously explored a compromise with Gould but these overtures went nowhere and the suit dragged on for decades. Fox to TAE, 10, 12, 14, 16, 24 June, and 16 and 19 July 1879, DF (*TAEM* 51:874, 884, 890, 892, 896, 907; 49:296; *TAED* D7936X, D7936ZAA, D7936ZAB, D7936ZAC, D7936ZAE, D7936ZAN, D7903ZFQ); *TAEB* 2:469 n. 2.

On 24 June, Josiah Reiff also alerted Edison to this situation and sent him a draft letter to Twombly on which Edison loosely based this document. Reiff had hoped that the passage of a bill allowing railroads to compete for commercial messages would provide a new market for Edison's automatic system, but after the formation of American Union he encouraged Edison to settle with Western Union. Twombly wired on 27 June, "Letter received, how much do you want for your interest." Edison replied, "For my interest will take sixteen thousand five hundred cash." Reiff continued his efforts to reach a settlement and in mid-July Edison prompted him to "Hurry up your negotiation   I want $16,500. sixteen thousand five hundred dollars For my share= delay is dangerous." Grosvenor Lowrey and Western Union president Norvin Green also became involved but talks reached an impasse in late July, at least in part over attempts to separate Edison's interest from Reiff's. In a draft letter to Twombly from around this time Edison noted that his "offer was not for an indefinite period, and if it was thought to use my interest as a leverage to cut him out all together it was a mistake   I can not consent, but will do anything to help you make him come to reasonable terms." Negotiations continued through September but no settlement was reached. Reiff to TAE, 4, 9, 12, and 23 Feb., 8 Apr., 4 May, 24 June, with Reiff draft to Twombly, and 26 June 1879; Twombly to TAE, 27 June 1879; TAE to Twombly, 27 June and 29 July 1879; Reiff to TAE, 18 July, with TAE marginalia, 22 and 28 July, 5 and 29 Aug., 17 and 19 Sept. 1879; all DF (*TAEM* 51:847, 849, 851, 855, 865–66, 897, 903–5, 916, 909–10, 912, 917–19, 921; *TAED* D7936F, D7936G, D7936H, D7936J, D7936Q1, D7936R, D7936ZAF, D7936ZAJ, D7936ZAK, D7936ZAL, D7936ZAS, D7936ZAO, D7936ZAP, D7936ZAQ, D7936ZAT, D7936ZAU, D7936ZAV, D7936ZAW).

–1765–

From George Gouraud

London 27th June 1879

Dear Edison

When I got back, I found that steps had already been taken by the London Company to the end of forming Local Boards in both Liverpool, and Manchester, it having been their understanding that they could do this without consulting you or me,[a] and surely there was nothing in the Document as originally drawn which implied anything to the contrary. You will see that it was very important that there should be no ambiguity on this point, as once the London Company had formed Local Boards in the large provincial towns, and established a Central System it would have been very difficult, if not indeed impossible, for us to have successfully formed a separate Company. I have accordingly withheld the original signed Agreement, and substituted for it another one with alterations defining this point as you will observe in Article 8.[1] There were also several other clauses which required further modification in our interests. Opening these questions naturally opened one, or two other points on the other side, but so obviously just and equitable, and ~~on the whole~~ entirely[b] unobjectionable as regards ourselves, that they were therefore assented to in order to secure the main points of the clause above referred to. You should therefore receive by this mail the Definitive Agreement made in the name of the Right Honorable Edward Pleydell Bouverie[2] for himself, and the others, who have signed the Preliminary Agreement[3]

I have received the assignment ~~and~~ of[c] the Patents[4] but Counsel here prefer that they should be made in a somewhat different manner to comply[d] to with[d] the absolute requirements of the English law, so that we are obliged to trouble you to sign your name once or twice more. This will finish the whole thing up, and make everything to the liking of everybody. See that they are duly certified by the Consul as before, which will give our friend Griffin another chance of airing himself on Broadway   Yours very truly

Geo. E Gouraud

LS, NjWOE, DF (*TAEM* 52:598; *TAED* D7941ZCF). Written by Samuel Insull. [a]"or me" interlined above by Gouraud. [b]Interlined above by Gouraud. [c]Interlined above. [d]Obscured overwritten text.

1. This article stipulated that the director appointed by Edison should approve the formation of any local companies. Under an ancillary document Gouraud signed on 2 July, Edison was to receive a one-time royalty of £5,000 each for Liverpool and Manchester in lieu of forming independent companies there. The principal contract specified this sum as Edison's royalty for each Metropolitan Postal District which the London company might reserve to itself instead of an independent company. Edison had already designated Gouraud as his representative to the board and he subsequently sent this nomination to the company and drew up a power of attorney specifically authorizing Gouraud to establish district companies. Details of Edison's relationship to the district companies were enumerated in a supplemental agreement dated 1 August 1879. Gouraud to TAE, 4 and 29 July 1879; Agreements with Edward Pleydell Bouverie and the Edison Telephone Co. of London, Ltd., 2 and 14 July and 1 Aug. 1879; TAE to Gouraud, 29 July 1879; all DF (*TAEM* 52:609, 646, 607, 617, 648, 647; *TAED* D7941ZCO, D7941ZDC, D7941ZCN, D7941ZCV, D7941ZDF, D7941ZDD); Powers of Attorney to Gouraud, 24 May and 21 July 1879; TAE to Telephone Co. of London, 21 July 1879; all Miller (*TAEM* 86:23, 35, 32; *TAED* HM790067, HM790073, HM790071).

2. Edward Pleydell Bouverie was a prominent Liberal member of the House of Commons for thirty years until his defeat in 1874 after breaking with the government of William Gladstone. At this time he was involved in financing the debt of Turkey and other countries and was a director of several British companies. *DNB*, s.v. "Bouverie, Edward Pleydell."

3. Gouraud mailed one copy of the contract on 28 June. He later asked Edison to cable the name of the steamer returning the signed agreement, "as upon the receipt of such a cable certain measures may be

proceeded with here without further delay." Having received no answer by 14 July he cabled again, "Are documents signed   rival exchanges forming." Edison promptly replied, "Telephone contracts signed and go first steamer"; Edward Bouverie executed the contract in London the same day. Gouraud to TAE, 28 June, 1 and 14 July 1879; TAE to Gouraud,14 July 1879; all DF (*TAEM* 52:601, 606, 614–15; *TAED* D7941ZCI, D7941ZCL, D7941ZCS, D7941ZCT).

4. The company's official registration was contingent on Edison's assignment of his British Patents 2909 (1877) and 2396 (1878). Just before Gouraud left New York, Edison instructed him to "go to Serrell & have assignments made to Company of two telephone patents subject to Nottage contract" (see *TAEB* 4:148 n. 1). Lemuel Serrell sent the documents for Edison's signature on 14 June. Gouraud to TAE, both 10 June 1879; Serrell to TAE, 14 June 1879; all DF (*TAEM* 52:588–89, 594; *TAED* D7941ZBZ, D7941ZCA, D7941ZCD).

–1766–

From Pitt Edison

Port Huron June 29/79

Dear Bro

I suppose that you got Report for Month of May   I would like to have you let me know if they come every month Reguallor[1] the amt that the road owes[a] is as follows[2]

Bonded debt to Sanborn estate[3]   Payable 3 years from July 1st   5000.00

Savings Bank note which we will reduce the first of this month at least one half or more   posable pay $750.00[4]   1050.00

We owe W. Wastell[5] for two hors[es] bot of him $225.00[b]

The floting dept is not more than $100.00   I think less[c]

but Al this road was in a awfful[a] condition when I took holt of in April last[6]   I have had men working on the track ever since I took of[f] the double cars and now running one hors cars which dos all the business for travel is dull out here you know   the new[d] compay had a lot and barn near the City Hall I have sold it to a Mr Smith[7] for $750.00 and will make the papers as soon as we can get a meeting of the Directors   he makes a cash payment the bal all comes due in time to pay on the bonded dept to Sanborn   I will send you Statement for month of June which will look better than the May Report and if we have no bad luck we[a] will get out of Dept before the Sanborn Bond becomes due and have the road and Stock in splendid condition   So[a] have patiance and we will pull through and come out in good shape yet

now Al abot Charley   since I got your letter his mother is wild to know what the trouble is[8]   is he not acting for you or is he on his own Hook and what trouble did he have in Europe and who with   now do let us know as soon as you get this for we will look for answer every day.

WPE

ALS, NjWOE, DF (*TAEM* 51:799; *TAED* D7934E). [a]Obscured overwritten text. [b]"over" follows as page turn. [c]Followed by dividing mark. [d]Interlined above.

1. There are itemized monthly statements of the receipts and expenses of the Port Huron Railway Co. for 1879 for all months except February, March, and October (DF [*TAEM* 51:807–24; *TAED* D7934ZZA]). Edison had become a major stockholder in this horsedrawn street railroad as a result of financing his brother Pitt's involvement with one of its predecessors; see *TAEB* 1:306 n. 4, 1:497 n. 3, 4:447 n. 2; Docs. 530, 841, and 1148.

2. Stockton Griffin had written Pitt on 26 May that Edison wanted "a statement of entire debt of the R.R. and the money in the treasury and moneys due on the 1st of June 79." In his undated response, Pitt supplied approximate figures and promised to send more precise information later. Griffin to Pitt Edison, Lbk. 4:349 (*TAEM* 80:78; *TAED* LB004349); Pitt Edison to TAE, n.d., DF (*TAEM* 51:801; *TAED* D7934E1).

3. The railroad had borrowed this amount in July 1877 from the estate of J. W. Sanborn, which was administered by company president John Sanborn, to pay construction expenses related to the merger of its predecessors. In return, Sanborn held a mortgage on all of the railroad's property. Jenks n.d., 14–15; Docs. 841 and 960.

4. Pitt soon reported that this entire note was paid on 3 July, and that he hoped to begin paying down the Sanborn debt. Pitt Edison to TAE, 4 July 1879, DF (*TAEM* 51:802; *TAED* D7934F).

5. William Wastell, a Port Huron druggist, was one of the original organizers of the Port Huron and Gratiot Street Railway, a predecessor of the Port Huron Railway. Jenks n.d.; 2, 4, 7; *History of St. Clair County, Michigan* 1883, 585–86.

6. After having operated a competing livery service for some time Pitt became the Port Huron Railway's superintendent; the May 1879 statement shows his salary for April as $46.18. *TAEB* 4:447, n. 2; Jenks n.d., 14; Port Huron Railway Co. statement, May 1879, DF (*TAEM* 51:810; *TAED* D7928ZAU).

7. Unidentified.

8. Edison's letter has not been found. In his letter of 4 July, Pitt asked Edison to "let me know something abot Charley." Edison made a notation on that letter instructing Stockton Griffin to "Write and say Charley is not acting right he is in Paris. If possible I would advise that you get him to come home but do not mention that I said so." DF (*TAEM* 51:802; *TAED* D7934F).

[Menlo Park, June 30, 1879?][1]

Write say letter referred to me for whom information is being obtained[2]   Say have reced black sand from several sources in California all contain platinum & Iridosmine ~~some~~ some more & others less.[3] ~~Sh~~Ask him if ~~we found method if~~ he thinks the Hydraulic miners would sell their black sand for nominal sum after gold has been extracted. if they would this would undoubtedly give us all the crude metal we desire. Would he give us an idea of how much gold is extracted from a barrell of black sand in an average hydraulic claim   what we desire is to get some idea of the amount of black sand accumulated per 100 dollars in gold

We find that a large part of the platinoid metals is in the grains of sand themselves & only obtainable by crushing. We send 3[a] cards[b]   the platinum is from Cherokee flat[4] the Iridosmine from Rogue River Oregon

ADf, NjWOE, DF (*TAEM* 50:647; *TAED* D7928ZAU). [a]Interlined above. [b]"3" interlined above later.

1. Stockton Griffin's docket notation indicates that the completed letter was sent on this date to the "Hon. E. T. Hogan" in Quincy, located in the Sierra Nevada mountains. Nothing more is known of Hogan or what position he may have held.

2. Edison prepared this draft in response to Hogan's 7 June 1879 letter to James Crawford, the superintendent of the U.S. Mint at Carson, Nevada, to whom Edison had written in May for information about platinum in that state. Crawford reported that he knew of none there but promised to make other inquiries; Hogan's response was apparently among several he forwarded to Edison on 21 June. Hogan wrote that a number of miners told him platinum could be found in the area "in more or less quantities in various mining sections in Plumas [County] but, as they have always regarded it as valueless, they have never saved any of it." He promised to send Crawford any platinum samples received from local miners. Hogan to Crawford, 7 June 1879; Crawford to TAE, 24 May and 21 June 1879; all DF (*TAEM* 50:646, 600, 720; *TAED* D7928ZAU, D7928R, D7928ZCQ).

Edison wrote extensive marginalia and draft replies on a number of other letters concerning platinum around this time, typically inquiring about the quantity and composition of sand in various locales (see Mining—Platinum Search [D-79-28], passim, DF [*TAEM* 50:580; *TAED* D7928]). He instructed Stockton Griffin to answer one Oregon correspondent and "ask for sample & say that it is no object to purchase small quantities but if we could be sure of obtaining at least 50 000 oz yearly we would errect works for purifying" (TAE marginalia on A. Pershbaker to TAE, 15 June 1879, DF [*TAEM* 50:683; *TAED* D7928ZBS]).

Edison's platinum circular (Doc. 1734) was printed in the *New York Herald* on 5 July and again by the *New York Sun* two days later. The *Herald* article stated that Edison had received "hundreds" of responses and "samples in large quantities" showing that platinum could be found in the U.S. "in over abundance" for as little as one dollar per ounce. Despite this apparent plenitude, Edison reportedly personally prospected for platinum sand near Menlo Park in late June and declared himself willing to spend up to $20,000 to develop sufficient supplies. "Edison's Electric Light," *New York Herald,* 5 July 1879; "Wanted, A Platinum Mine," *New York Sun,* 7 July 1879, Cat. 1241, items 1219 and 1224, Batchelor (*TAEM* 94:493–94; *TAED* MBSB21219X, MBSB21224X).

3. Iridosmine, a native iridium–osmium alloy, usually also contains platinum. The week before Edison drafted a reply to a man who had sent a sample of it from California, in which he stated that he "could not use the ore unless I could get an unlimited supply say 1000 lbs annually as the purification of platinum from its kindred metals Ruthenium—Palladium is the most difficult process in metallurgy & has only been successfully accomplished on a commercial scale by two firms in Europe." He added that Russian platinum ore sold for $1.84 per ounce, or for $6.85 per ounce in ingots. TAE marginalia on Monroe Thomson to TAE, 14 June 1879, DF (*TAEM* 50:678; *TAED* D7928ZBO).

4. Edison routinely sent samples of black platinum-bearing sand for his correspondents to distribute to miners and prospectors. This one may have been from sand sent by Louis Glass on 12 June from the Spring Valley Mining & Irrigation Co. in Cherokee, Calif. (see Doc. 1776 n. 2).

3

July–September 1879

Encouraged by the progress of his electric light experiments during the spring, Edison told several investors in early July that "Everything looks bright" and that he was making plans for a large demonstration of his system at Menlo Park.[1] Using a dynamometer designed and built at the laboratory, he and his assistants measured the efficiency of his new dynamo. They found that it converted about 80% of mechanical power into useful electric current (far higher than the best contemporary generators) and two similar machines were built in July and August. By early July Edison also professed full confidence in his lamp but promised to continue trying to increase its efficiency. His platinum circular letter[2] having prompted an "overwhelming number of replies," he anticipated no difficulty "obtaining all the platinum requisite for the light."[3] He began to make arrangements to purchase platinum-bearing waste sands from hydraulic gold mines in California and, in September, dispatched Frank McLaughlin there. He was satisfied for the first time with the current regulator for the lamp circuit but did continue to experiment throughout the summer on pyroinsulation coatings for the wire spiral. Francis Upton continued experiments on the household meter to improve its accuracy and reliability under various conditions. Edison also worked out the arrangement of conductors to provide for a nearly uniform voltage drop throughout a distribution system. Edison made this arrangement, which was the basis of what later became known as the "feeder and main" system, the subject of a lengthy caveat in August in which he described the entire central station circuit from the dynamos to individual lamps. At the end of August Edison traveled with

Upton and Charles Batchelor to Saratoga, New York, for the annual meeting of the American Association for the Advancement of Science. He prepared a paper[4] (read by Upton) on his method of treating wires in a vacuum and also demonstrated his telephone; Upton gave his own paper on the recent Menlo Park dynamo tests.[5]

Edison also gave serious consideration for the first time to electric power and transmission. In July he initiated correspondence with Adolph Sutro about generating hydroelectricity on the Virginia River in Nevada to run motors for ventilating and draining shafts in Comstock Lode mines several miles away. Francis Upton made detailed calculations for various configurations of such a system. During August and September Edison directed investigation into small electric motors designed on the principle of his dynamo for use in sewing machines, elevators, and small shops. In early September he solicited sewing machine manufacturers for equipment to test small motors.[6] Upton began testing motors in early September and continued to do so at intervals into November.

With the electric light progressing, Edison and Charles Batchelor put considerable time and effort into the telephone. They made several significant changes to the design adopted in February. During early July they successfully tested a new composition for the chalk button of the electromotograph receiver. A dozen of the new chalks were included with the first lot of telephones sent to England on 9 July (this shipment also included Edison's first switchboard). Because the new chalk seemingly did not dry out, Edison eliminated the water reservoir in the telephone box and moved the receiver from the box to a pivoted arm so that it could be placed conveniently against the user's ear. This became the standard design for his commercial telephone in England, with subsequent minor modifications to the mechanism for rotating the chalk. In mid-July he also decided to use a different form of carbon transmitter based on experimental designs made in 1877. This device, called the inertia transmitter, was more sensitive and seemed never to need adjustment. Edison promised to incorporate it in half the telephones sent to London so that Edward Johnson could reach his own conclusions but confided to Johnson on 21 July that for the first time he felt "perfectly satisfied" with the telephone.[7]

Johnson reported in August that the new chalks did not hold up well. Edison and Batchelor devised a new theory of the chemical action inside the chalks, and accordingly suggested a new regimen for moistening them before they were placed in service. Johnson also complained that the inertia transmitter was unreliable and had to be adjusted frequently. After further experiments, Edison concluded in September to abandon it and use his standard carbon transmitter instead.

By early September Edison had contracted for the manufacture of thirty telephones per week (later fifty), which were to be assembled at the laboratory and shipped to England. The telephone company asked him to hurry still more, because the public demand for instruments since a demonstration to the press amounted to "hundreds a day."[8] He was also asked to manufacture fifty switchboards for London and companies in Liverpool and Manchester. In late September he began the process of hiring a half dozen inspectors to supervise the installation and maintenance of all this equipment.

Edison had reason to be less sanguine about the telephone situation in France. At Joshua Bailey's insistence, he sent two complete instruments to Paris in July to capitalize on public interest in the nascent Edison company there. Soon afterward, however, he began receiving allegations that Theodore Puskas, a trusted associate and partner in the prospective company, intended to sell out his interest to a rival telephone concern. A spate of accusations, denials, and counter allegations stalled efforts to have the company legally organized and clouded its prospects for months. Some of the charges involved Charley Edison, who had ignored Edison's order to return home and seems to have become connected with a competing group.

During this time Edison was also being urged to construct receiving instruments for the Gold and Stock Telegraph Co. He made a small number in July and promised several more for exhibition at the American Institute in September, but did not undertake large-scale manufacture for the U.S. market. He also promised telephones for demonstration in Japan in preparation for selling his rights in the Far East.

In business affairs, Edison received notice from his attorney in late August that the Patent Office had declared six new patent interference cases involving carbon telephone transmitters. Similar individual interference cases were previously declared in January, February, and on 6 August, the latter for an inertia transmitter. Edison arranged to postpone until October the deadline for a formal response to the set of six cases. In an unrelated matter, Grosvenor Lowrey proposed that Edison raise funds for electric light experiments by selling his rights to additional foreign countries but nothing came of it at this time. Edison did approve in August a plan by the Edison Speaking Phonograph Co. to manufacture at least five hundred small phonographs to be sold as toys.

Edison seems to have hired fewer men during this period than in the preceding months, although incomplete records make this difficult to ascertain. He is known to have added Albert Herrick, a laboratory assistant who worked primarily on the electric light.[9] He also advertised for a full-time glassblower and hired Ludwig Böhm, a highly skilled craftsman who had worked with Heinrich Geissler in Bonn, Germany. Böhm arrived about 20 August and within a few weeks was fabricating experimental vacuum pump designs.[10] Samuel Mott, a draftsman, may have started in early summer (or as late as October) making patent drawings.[11] During the summer Francis Upton began to assume the authorial responsibilities which Edison had delegated. He revised the technical portions of three articles on Edison's inventions written for *Scribner's Monthly* by a New York reporter. He also helped Edison write a letter to the editor of the British journal *Engineering* about dynamo tests conducted by John Hopkinson; the letter was published in Upton's name.[12]

1. Doc. 1770.
2. Doc. 1734.
3. Doc. 1770.
4. Doc. 1796.
5. Upton 1880b.
6. See Doc. 1800.
7. Doc. 1781.
8. Arnold White to TAE, 11 Sept. 1879, DF (*TAEM* 52:710; *TAED* D7941ZEA).
9. See Doc. 1972 n. 5.
10. See Doc. 1786 n. 7.
11. See Doc. 1985 n. 7; also Time Sheets, NjWOE.
12. See Doc. 1772.

–1768–

To George Gouraud

[Menlo Park,] July 1, 79

Menlo park London[1]
  Johnson sailed Wisconsin   System worked here perfectly had 15 lines up[2]

L (telegram), NjWOE, DF (*TAEM* 52:606; *TAED* D7941ZCM). Written by Stockton Griffin.

1. "Menlo Park London" was George Gouraud's cable cipher.

2. The *Wisconsin* left New York for Liverpool on this day. Gouraud acknowledged the text of Edison's message on 4 July, adding that he was "very glad to receive your cable announcing successful working of the Central System and Johnsons departure." Nothing further is known about the test of the telephone system. "Shipping News," *New York Herald*, 1 July 1879, 10; Gouraud to TAE, DF (*TAEM* 52:610; *TAED* D7941ZCP).

–1769–

Notebook Entry: Telephony

[Menlo Park,] July 3rd[a] 1879

Edison's Telephone—

We find that 2 oz chalk and ¼ fluid oz of Phosphate Soda makes a far superior chalk in every way; it needs very little wetting, it glazes beautifully, it wears very little, and talks louder.[1]

We find our old inertia telephone works excellently on this receiver so think it possible we might bring it into use.[2] We make it so:—[3]

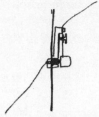

It works elegantly even when you hold[a] it in your fingers. One great thing about this is that it never gets out of adjustment You can talk right into the diaphragm, and then 6 inches away and it is always all right at that distance although to talk right into it vibrates the diaphragm so powerfully that you cannot tell what is said.

Chas Batchelor

X, NjWOE, Batchelor, Cat. 1304:59 (*TAEM* 91:58; *TAED* MBN004059). Written by Charles Batchelor. [a]Obscured overwritten text.

1. On 15 June, amid a long series of receiver chalk experiments, Edison noted that a button composed of 1½ ounces of chalk and ½ ounce of neutral phosphate of soda was "as good as regular   solution saved" (N-79-06-12:35, Lab. [*TAEM* 35:503; *TAED* N080:19]). He and Charles Batchelor had tested scores of chalks and wetting solutions, beginning on 28 May then resuming from 12 June to near the end of that month. They tried different pressure in making the chalks, noting that those pressed hardest were the loudest, and observed the effect of the button's surface texture on the volume and character of sounds produced. Edison and later Francis Upton conducted several electrical tests on many of these specimens, measuring resistance and also apparently looking for

electrochemical battery action which might affect their performance (N-79-06-12:19–43; N-79-02-24.1:47–135, 159–67, 179–203; N-79-03-10.2:61–77; all Lab. [*TAEM* 35:494–506; 30:993–1036, 1046–49, 1055–67; 31:1153–61; *TAED* N080:11–23; N020:24–67, 80–84, 90–102; N032:27–35]). On 4 July, Upton compared the new chalk to a "regular" one, observing that a "very wet" new button was "not quite so loud as regular" (N-79-06-12:41, Lab. [*TAEM* 35:505; *TAED* N080:22]) but additional tests the next week gave more encouraging results (see Doc. 1773 n. 1). To facilitate these experiments, about the second week of June Batchelor had made "a testing machine for chalks" on which the buttons could be easily mounted and tried (Cat. 1308:153 [Order No. 189], Batchelor [*TAEM* 90:743; *TAED* MBN003:50]).

Another form of "inertia telephone," attributed to September 1877.

2. This design is similar in principle to those Edison had devised in September 1877 for "Working by momentum no adjusting screw" (Doc. 1062, figures Nos. 1–4). These were among several methods he had experimented with in 1877 for regulating by inertia the force of a vibrating diaphragm on the carbon (see Edison's testimony, TI 1:51–56, [*TAEM* 11:46–49; *TAED* TI1011 (images 43–46)]). He included one of them in his final specification for British Patent 2,909 (1877), fig. 17 (Batchelor [*TAEM* 92:56; *TAED* MBP008]) and another was described in Prescott 1879 (pp. 528–29, fig. 257), which noted that "it is hardly certain that its action is to be attributed solely to inertia."

3. On 13 July, John Kruesi drew a similar instrument and recorded an order for it to be made. Batchelor then instructed the laboratory shop to "alter one of our bracket telephones to hinge top of box, and put inertia diaphragm on with 3 springs with rubber under them" (Cat. 1308:159 [Order No. 211], Batchelor [*TAEM* 90:746; *TAED* MBN003:53]). In the early morning of 16 July he recorded that the transmitter "works excellent and does not require adjustment even when you shout in the diaphragm" (N-79-06-12:57, 245, Lab. [*TAEM* 35:513, 580; *TAED* N080:30, 97]). On 1 August, Edison executed a patent application (Case 182) for a telephone consisting of this form of transmitter and a magnetic receiver. The application was subsequently divided, with that portion covering the transmitter issuing in October 1882 as U.S. Patent 266,022 (Pat. App. 266,022; Lemuel Serrell to TAE, 16 Sept., 10 Oct., 17 Dec. 1879, DF [*TAEM* 51:552, 574, 619; *TAED* D7929ZEB, D7929ZEQ, D7929ZGB]; TAE to Serrell, 22 Oct. 1879, Lbk. 5:294 [*TAEM* 80:165; *TAED* LB005294B]).

An article on "Edison's Latest Telephone" in the 27 September 1879 *Scientific American* described and illustrated the inertia transmitter:

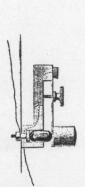

Illustration of inertia telephone in 27 September 1879 Scientific American.

A vulcanite arm is secured to the center of the mica diaphragm by means of a small bolt, which is connected with one pole of the battery by a piece of metallic foil or very thin copper wire. The head of this bolt is platinum-faced, and sunk deeply in the vulcanite arm, the same cavity containing also a piece of carbon pencil, such as is used for electric candles. The carbon fits the cavity loosely and is rounded at both ends. Its outer end is pressed by a platinum-faced spring secured to the outer end of the vulcanite arm. The spring carries at its free end, exactly opposite the piece of carbon, a brass weight, and the pressure of the spring upon the carbon is regulated by the small set screw. A wire or piece of copper foil, connecting with the spring, completes an electrical circuit. [*Sci. Am.* 41 (1879): 198]

–1770–

Draft to Tracy Edson[1]

Dear Sir.

Your favor of the 30th ult was duly rec'd.[2] Regarding the demonstration of the electric light at Menlo Park, I think I will start it on a small scale without making any great outlay. I propose to put up a dynamo machine and 30 lights when I get the standard lamp, and when a 2d dynamo is done I will add that and 30 more lights and so on till the money is exhausted.[3]

I am happy to inform you that the standard dynamo machine goes beyond ~~the~~ our most sanguine[b] expectations. I have completed a dynamometer for measuring the horse power which is the most simple and most[c] effective which has yet been[d] devised, and in fact the only apparatus which will give correct readings. This has been one of our great hindrances. The Scientific American of next week or the week following.[4] During the past we have been busily engaged making tests with the new dynam~~ometer~~ machine[e] and we ~~and~~ find contrary to all expectations that deducting friction the machine as a machine delivers in form of current about 96% of horse power applied to it.[5] Upton thinks that if there is any error in the above estimate it will be in favor of the machine. He thinks that we get the whole of the horsepower in form of current. This will give us in practice about 80% of the total horse power applied that will be useful outside in the lamps 20% being lost in friction and in the machine owing to the fact that certain portions of the current must necessarily be lost in wire around the machine

The best machine of this nature heretofore constructed is that of Siemens which is said to translate 90% of power applied deducting friction but the great ~~difficulty~~erence between our machine and this or of any other heretofore constructed is, that of the 90% no more than 55% could be used outside of the machine 35% being lost in wire around it whereas in our machine 75 to 80% of the 90% will be obtained outside of the machine. These results are greatly in favor of greater economy than calculated.

I have now 80 feet of iron piping coated inside with rubber very beautiful and as soon as I get my boxes cast and coated we will put them down and pass wires through them and show the system in operation. We[c] have received a list of prices from the Rubber Co[6] for coating these pipes and the whole system will be one of remarkable cheapness and efficiency beyond what we had even calculated upon. We have also[c] recd[c] a private price list from Phelps Dodge & Co for pure copper wire very much below that which I had been basing estimates on.[7]

We have sent out 1500 circulars[8] to parties in the mining regions of the Pacific Coast regarding platinum and have rec'd an overwhelming number of replies which go to show that there will not be the slightest difficulty in obtaining all the platinum requisite for the light.

I have also a satisfactory regulator for the standard lamp for the first time.[9] After several months continuous efforts we are now getting out pure oxide of zirconium for the lamps and in 2 weeks will have a sufficient supply to go ahead with the experiments on the burner.[10]

Everything looks bright and as long as there is any hope of obtaining more burners per horsepower than six I shall continue to experiment and produce a lamp equal in efficiency to the standard dynamo machine

We have the forgings for 3 more dynamos which we are now making like the ~~samp~~ model.

The expenses are not over $400 per week and the money on hand will last several weeks.

This is all the information I have at the present time which I think will be of interest. Very Truly

T.A.E.

DfS, NjWOE, DF (*TAEM* 50:264; *TAED* D7920ZAU). Written by Stockton Griffin. [a]Place from Edison's laboratory handstamp. [b]"our most sanguine" interlined above. [c]Obscured overwritten text. [d]"which has yet been" interlined above. [e]Interlined above.

1. Stockton Griffin's docket note indicates that the completed letter was sent to Edson, Calvin Goddard, and Grosvenor Lowrey.

2. Edson had asked for "an Estimate of the cost of the proposed demonstration at Menlo Park, giving an idea also of the extent of the demonstration, as to the number of Lights &c. I should also like to know at what time you expect to make the exhibition— I learn from Mr Banker that you consider the Light and the Generator finished." Edson later explained that he made this request "not because I am getting impatient, but so that I might be able to answer numerous enquiries intelligently, and for the purpose of taking early measures to provide the necessary funds to put the matter through so that there should be no delay after you were ready." Edson to TAE, 30 June and 9 July 1879, DF (*TAEM* 50:263, 268; *TAED* D7920ZAT, D7920ZAV).

3. In his 9 July reply (see note 2), Edson endorsed this proposal because "I think it quite likely that after you have set up one Dynamo Machine and 30 Lights, you will wish to make some changes before putting up the second 30 Lights, and that could be done more economically than if they had all been set up at once."

4. The 26 July issue of *Scientific American,* which would have been published about the middle of the month, included a diagram and explanation of the new dynamometer, which it reported Edison had devised only after he had "tried every dynamometer within reach, and condemned them all." The magazine published an engraving of the instrument in use with Edison's dynamo in its 18 October issue. "Progress at Menlo Park," *Sci. Am.* (41:52), "Edison's Electric Generator," *Sci. Am.* (41:242), Cat. 1241, items 1232, 1332, Batchelor (*TAEM* 94:503, 513; *TAED* MBSB21232, MBSB21332X).

There is no record of the development of Edison's belt dynamometer before John Kruesi's measured drawing on 1 June and his order for its construction. The machine was completed by 30 June, when Edison demonstrated it for reporters, and may have been the reason for Kruesi's order on 29 June to "Get a 8 inch double belt & cement joint alter counter shaft accordingly." Machine Shop Drawings (1879–1880), Lab. (*TAEM* 45:69; *TAED* NS7986CAU); Cat. 1308:155 (Order No. 199); "Local Miscellany. Edison's New Telephone," *New York Tribune,* 1 July 1879 and "Edison's Work," *New York Herald,* 1 July 1879, Cat. 1241, items 1212, 1214; all Batchelor (*TAEM* 90:744, 94:492; *TAED* MBN003: 51, MBSB21212X, MBSB21214X).

Francis Upton described the operation of the new instrument in his paper on "Methods for Testing Faradic Machines" read at the annual meeting of the American Association for the Advancement in August:

> The driving side of the belt is carried under the pulley attached to a weight, from which is placed so far below the line of the driving and driven pulley, that nearly double the strain on the belt is lifted, as in the case of an ordinary block. To weigh this amount, the weight is placed on a platform scale and the diminution due to the pull from the belt measured. A large load of 900 pounds was used so that the jar due to the movements of the belt is so distributed as to produce very little effect on the scale.
>
> The belt ran about 1000 feet in a minute so that 100 pounds difference represented about a horse and a half power. Since the scale showed half a pound, the results were easily measured within one per cent., making the maximum which could be transmitted 13.5 H.P., though the same weight could be made to answer for transmitting larger amounts of power by increasing the speed. [Upton 1880b, 180]

In this Scientific American *illustration of Edison's dynamometer, A is the driving pulley,* **B** *the driven pulley, and* **C** *the tension wheel; the weight to be raised and the scale are at bottom.*

5. This figure is the last and best of several calculations of the efficiency of Edison's "standard dynamo machine" made by Francis Upton during the first week of July as he and Edison tested and calibrated the new dynamometer. The actual figures given in Upton's 7 July notebook entry were "96.3% excluding friction" and "90% including friction" (N-79-06-16.1:219, Lab. [*TAEM* 35:269; *TAED* N077:109]). During the week of tests Upton compared the mechanical force applied to the

EDISON'S ELECTRIC GENERATOR.—[See page 242.]

This Scientific American *engraving of the dynamometer setup in the Menlo Park machine shop shows George Barker observing a test.*

dynamo, as measured by the dynamometer, with the amount of current generated, as indicated by an electrodynamometer and converted to mechanical units. N–79-02-15.1:163–71, N–79-04-08.2:35–49, N–79-06-16.1:120–223, all Lab. (*TAEM* 31:738–42, 911–18; 35:221–71; *TAED* N028:81–5, N030:17–24, N077:61–111).

The electrodynamometer was essentially a high-resistance galvanometer able to handle heavy current, thereby avoiding the measurement errors caused by shunting only a portion of the current through a conventional galvanometer; its copper construction also made it impervious to magnetic interference. Edison's electrodynamometer, which had been completed on 15 April (Cat. 1308:129 [Order No. 90], Batchelor (*TAEM* 90:731; *TAED* MBN003:38), was a modification of the instrument designed by Harvard professor John Trowbridge; in early July, Upton abstracted an article by Trowbridge (Trowbridge 1879) describing the electrodynamometer and the results of tests with it on various generators (N–79-07-07.1:85–103, Lab. [32:290–99; *TAED* N037:42–51]). Beginning on 21 April, Upton had conducted a series of comparative measurements with a standard galvanometer, presumably to evaluate and calibrate the new instrument. On 22 April he conceded, "I do not understand why these results are so far apart. I think the method with

The Scientific American *illustration of Edison's electrodynamometer.*

galvanometer of high resistance is the best" (N-79-4-21:103, Lab. [*TAEM* 30:773; *TAED* No17:49]). His "Testing E.M.F." and "measurement of the dynamometer" continued until he determined, probably in late April, an algorithm for calculating the amount of current from the instrument's deflection (N-79-4-21:131, 138–51, Lab. [*TAEM* 30:787, 791–96; *TAED* No17:64, 68–74]). At intervals from late April to mid-June, Upton used both the electrodynamometer and galvanometer in extensive tests of the Gramme and Edison dynamos, conducted in part to measure the effects of varying field magnet strength (N-79-04-21:75–153, Lab. [*TAEM* 30:760–97; *TAED* No17:35–75]; Doc. 1760 n. 3).

6. Edison sent two pieces of wrought iron pipe to the Novelty Rubber Co. in New Brunswick on 6 June. The firm replied on 30 June with a price estimate and an offer "to take an order to line fifteen or twenty lengths at these prices, and if we find it can be done for less than our present estimate, will give you whatever advantage there may be. It is impossible to give exact figures without lining several of them, as we can tell in no other way, what proportion of them will turn out perfect." Cat. 1308:232, Batchelor (*TAEM* 90:774; *TAED* MBNoo3:81); Novelty Rubber Co. to TAE, DF (*TAEM* 50:525; *TAED* D7925ZAG).

7. Not found.

8. Doc. 1734.

9. This device was presumably the same type as the governor regulator designed in the second week of May and included in the provisional British specification that Edison filed in June (see Doc. 1735 n. 8). John Kruesi built one on 11 May and on 30 June he finished "another lamp regulator same as first but only ½ the resistence" (Cat. 1308:145, 155 [Order Nos. 156, 200], Batchelor [*TAEM* 90:739, 744; *TAED* MBNoo3: 46, 51]). What may be tests of the May regulator, conducted on 23 June, can be found in N-79-06-16.2:5–19, Lab. (*TAEM* 35:370–77; *TAED* No79:3–10).

10. See Docs. 1787 and 1802 n. 3.

From Joshua Bailey

Paris 7 July 1879.

Dear Sir,

We duly received your telegram saying that the Motographs would be sent in ten days & we sent you another today making enquiry whether they have been sent.[1] It is of great importance to us at this moment to have the motograph in hand as it would enable us to attract a great deal of public attention and aid us in the moves we are about making. The day before yesterday the Telephones were put on at the Ministry of Telegraphs in the office corresponding in the office corresponding in France to the opperating room at the top of the Western Union Building & we worked them with great success in the presence of all the Chiefs of Bureaux of the Administration between Versailles & the office— Each one of these persons seated himself in turn at the Telephone & conversed with the Officers at Versailles & on rising made the comment "perfect, perfect." This is the only case in which any Telephone has been able to work under such circumstances. The Gower & the Bell as well as Siemens[2] have utterly failed in all such tests. The wire given to us was a single wire in a Telegraph Cable. We are now pressing our application for a Telegraph Cable & Concession & are promised immediate & favorable action. We are also pushing actively for the organisation of the Telephone Exchange & expect by the close of next week to have our first Instrument up. Our experience with your Nephew is not satisfactory. We find that all the time he has been in our employ he has been in communication with the Herz party and as we understand has accepted an engagement with them.[3] In speaking to him a couple of days ago in regard to the matter & referring to the Contract which we had made with him for 6 months he showed himself quite a good lawyer in saying that it was true that he had made an agreement with us, but that the agreement was binding on us and not on him. We think this does more credit to his legal accuteness than to his sense of honour. I am, Very truly Yours,

J. F. Bailey.

⟨~~Write Bailey say I want it understood that I am not responsible for anything that my nephew may do   I shall have nothing to do with him~~⟩

ALS, NjWOE, DF (*TAEM* 52:306; *TAED* D7940ZAP).

1. Edison's cable was sent on 26 June (see Doc. 1763 n. 1); Bailey's 7 July message has not been found.

2. The Berlin firm of Siemens and Halske manufactured a form of magneto transmitter in which both poles of the horseshoe magnet acted

at a short distance on one side of the metallic diaphragm, reportedly producing great sensitivity; this instrument was supposed to transmit speech from several feet away. "Siemens and Halske's Telephone, with Horseshoe Magnet and Call-Apparatus," *Electrician* 3 (1879): 260–62.

3. In an 11 August 1879 letter to Edison, Enos Barton excerpted a letter sent him by George Beetle from Paris on 28 July stating that "Charlie Edison left the Edison Telephone Co. some two weeks since, and is now putting up the quadruplex between Paris and Brussels for Dr. Herz." DF (*TAEM* 52:316; *TAED* D7940ZAY).

–1772–

Draft to the Editor of Engineering

[Menlo Park, c. July 7, 1879[1]]

~~In~~ Your journal of May 9 1879, contains a paper on Electric Lighting Read before the ~~society~~ institution of Mechanical Engineers[a] by Dr John Hopkinson F.R.S.[2] in which some very ~~e~~interesting tests on the efficiency of Dynamo Electric Machines is described   In these tests the Siemens machine is set down as ~~returning to~~ giving to the circuit 90 per cent of the power applied to[b] the strap.[3] I desire to call ~~the~~ attention to the fact that the writer has assumed the EMF of a Daniel Cell to be ⅞ of a volt, from which he calculates the energy in circuit,[4] while the value of the horse power ~~or foot~~[c] is ~~should be has been~~ should be[d] calculated from a Daniel Cell having an EMF of 1.08[e]   This is as it is given by ~~in B~~ Clerk Maxwell,[5f] and by Kolrousch[6f] at 1.09——[g] taking the Daniel element at the correct EMF this would make the efficiency of the Siemens machine _____ instead of _____.[7] again 0.28 of a horse power seems to be a large amount to run the bobbin ~~without~~ on open circuit.

~~again he uses a calorometer which he states to have a resistance of about,~~

It seems impossible ~~from the number of webers pr~~ that the calorometer described & used could have given correct results. ~~such an ins~~ The[h] apparatus ~~as described would be totally inadequate to g~~ owing to the small amount of wire used & powerful[b] circuit passing through it,[i] thus introducing ~~errors~~ ,the error of unequal temperature between the conductor & fluid & hence a variable resistance all in favor of the higher efficiency of the machine   If this is over estimated the efficiency of the machine is still further reduced.[8] ~~I h~~[j]

Mr Hs paper is of great value as the only test yet made in[b] this direction in a truly scientific manner and, I have no other object in calling attention to these, ~~matters~~ minor defects[k] than that the true efficiency of the machines shall be ascertained in order than any one who may hereafter produce better results shall not[h] have ~~the proper credit for the same~~, a[h] fictitious efficiency quoted against him= ~~and~~

TAE

ADfS, NjWOE, Lab., N-79-07-07.1:45 (*TAEM* 32:270; *TAED* N037:22). [a]"Read . . . Engineers" interlined above; "society" interlined seperately. [b]Obscured overwritten text. [c]"~~or foot~~" interlined above. [d]"~~should be has been~~ should be" interlined above. [e]Number written in underlined gap. [f]Name written in underlined gap. [g]"at 1.09—" interlined above. [h]Interlined above. [i]"owing to . . . through it," interlined above. [j]Followed by dividing mark. [k]"minor defects" interlined above.

1. Francis Upton elaborated on Edison's draft in the same notebook, dating his version 7 July. Upton's calculations and notes on Hopkinson's experiments and other dynamo tests appear throughout his draft and elsewhere in the notebook. The letter was published under Upton's name in the 25 July issue of *Engineering* (Upton 1879). In his reply, dated 26 July and which appeared in the 8 August issue, Hopkinson stated that "my paper does not pretend to minute accuracy, but rather to give rough easily obtained results *of the right sort*" and went on to suggest some possible sources of error in the experiments. Hopkinson also thanked Upton for the "courteous tone of his criticism" and looked "forward with great interest to the publication of the mass of valuable results which have been accumulating in Mr. Edison's laboratory." N-79-07-07.1:35–41, 51–79, 105–11, Lab. (*TAEM* 32:266–69, 273–87, 300–303; *TAED* N037:18–21, 25–39, 52–5); "The Siemens Dynamo-Electric Machine," *Engineering,* Cat. 1025:80, 96, Scraps. (*TAEM* 24:662, 670; *TAED* SM025080b, SM025096a).

2. John Hopkinson, the noted British electrical engineer, gave his first paper "On Electric Lighting" before the Institution of Mechanical Engineers on 24 April in which he demonstrated for the first time that the relationship between a dynamo's current and voltage follows a predictable curve. This paper appeared under its original title in the 9 May 1879 issue of *Engineering* (27:403–404). Hopkinson presented his second paper "On Electric Lighting" a year later. Hopkinson 1879; *DNB,* s.v. "Hopkinson, John."

3. Hopkinson actually made only the more modest claim that "the efficiency of the machine is about 90 per cent., exclusive of friction," implying a 10% loss due to internal electrical effects. By deducting from his dynamometer readings the 0.28 horsepower needed to turn the dynamo under no load, he based his calculations of efficiency solely on the amount of mechanical power available for conversion to electricity (Hopkinson 1879, 245–49).

4. The voltage produced by the Daniell cell was so reliably constant that it was widely used as a standard (see *TAEB* 1:615 n. 3). Here Hopkinson used it to provide a counter electromotive force in a shunt circuit. When the dynamo current was applied the resistance in the circuit was adjusted until the galvanometer showed no deflection, providing a proportion of electromotive force from which the current could be calculated. In his reply to Upton (see note 1), Hopkinson noted that "the value of a Daniell's element was taken as nine-eighths of a volt as a rough means of published results."

5. Renowned British physicist James Clerk Maxwell (see *TAEB* 2:84 n. 3), a member of the British Association's committee (disbanded by this time) for the establishment of standard electrical measurements, assumed in Maxwell 1873 (art. 272) an electromotive force for the Daniell equivalent to 1.079 volts. This reference to Maxwell was subsequently dropped (see note 6).

6. Edison was probably referring to Kohlrausch 1870, a laboratory reference available by this time in three German editions and an English translation by Thomas Waller and Henry Proctor of the 1872 second edition. This translation gives Kohlrausch's value for the Daniell EMF as 1.08 (p. 162) and the editors' own value, based on Maxwell, as 1.079 (table 24c). Upton replaced Edison's references to Maxwell and Kohlrausch with a discussion of several other values ascribed to the Daniell's electromotive force. In print, Upton declared that "The value taken by Messrs. Houston and Thompson, in their tests for the Franklin Institute, is 1.079 volts for the electro-motive force of a Daniells. This value is the one given by Mr. Latimer Clark and is generally received as the normal one." He was probably referring to Clark and Sabine 1871, 95. Upton had written "Houston Thompson 1.079" on the page opposite Edison's discussion (N-79-07-07.1:46 [*TAEM* 32:271; *TAED* N037: 23]); on the Franklin Institute tests by Edwin Houston and Elihu Thomson see *TAEB* 4:558 n. 1.

7. Upton stated in his printed letter that:

> As the energy given off in any circuit is as the square of the electromotive force, to compare the figures given by Dr. Hopkinson with those in the Franklin Institute report, one must multiply his results by $(^{1.079}/_{1.125})^2 = 0.92$. If this correction is made the Siemens machine would give back 83 per cent. instead of 90 per cent. of the power applied in the shape of electricity. . . .
>
> If nine-eighths of a volt be taken as the value of the electromotive force of a Daniells, a curious anomaly presents itself, that a new machine made by Mr. Edison gives back more power in the current than it receives through the belt, unless some very carefully made tests are at fault.

8. In addition to the galvanometer, Hopkinson used a calorimeter in several trials to make indirect measures of the dynamo current. It contained a coil "of uncovered German-silver wire nearly 2 m. long, 1½ mm. in diameter, and having a resistance of about 0.2 ohm" and was placed in the circuit with resistance coils (Hopkinson 1879, 243). Hopkinson made allowances for two sources of experimental error but in two experiments reported values showing more energy produced by the dynamo than was applied to it. In the published letter, Upton elaborated on Edison's criticism of this arrangement:

> the resistance of the wire used was given as 0.2 of an ohm. As a very slight difference in this would be largely multiplied in the complete circuit, it seems strange to read that the resistance was about one-fifth of an ohm, and that the conducting wire was estimated as having 0.01 ohm resistance. No allowance is made for the heating of the wire in the calorimeter during the passage of the current through it.

. . . A wire, stretched between two supports and immersed in water, may be seen to sag slightly when a powerful current is passed through it, thus showing expansion from heating, and consequently increase of resistance.

In his reply to Upton (see note 1), Hopkinson noted that "The calorimeter lost heat too rapidly, and the consequent correction cannot be supposed to eliminate all error from this cause. The heating of its wire, which was German silver, introduces a much smaller error."

–1773–

Charles Batchelor to Edward Johnson

[Menlo Park,] July 9th—[187]9

Dear Johnson—
We send you today the following:—

12[a] twelve complete telephones
12[a] twelve extra chalks
1[a] one switch board etc
12[a] twelve call batteries[b]

The 12 extra chalks are made of chalk and phosphate of Soda, and must not be [~~wett?~~][c] wet by dipping; but by dipping a piece of paper in the solution and letting it soak into the chalk from that.

We find this chalk never actually needs a second wetting, as we have left one out to dry seven days and it was just as loud at the end of that time. They take a magnificent polish and of course do not appear to wear away at all. The solution we wet this chalk at first with is a saturated solution of Phosphate of Soda[1]

This letter is not official information for the company but private information for yourself; Carman will always notify the company of any shipments but alterations in chalks such as described he knows nothing about[d] of course   You will not find the phosphate of Soda quite so loud as the other at first but far superior in the fact that it needs no wetting and is always good   Yours

Chas. Batchelor

ALS (letterpress copy), NjWOE, Lbk. 4:457 (*TAEM* 80:94; *TAED* LB004457). Circled "C" written at top of page. [a]Circled. [b]List enclosed by brace at right and followed by dividing mark. [c]Canceled. [d]Interlined above.

1. Edison and Batchelor made several receiver buttons on 3 July that they tested during the week preceding this letter. Two composed of chalk and ¼ ounce of the phosphate of soda solution were allowed to dry

in the open air until 7 July, when Edison and Batchelor found that they "worked well" at about half their original loudness. One was then soaked in the phosphate of soda "but crumbled down in the solution"; the other was allowed to remain in the air and two days later it "was Bully   talking good without any time being wet." In the early morning of 10 July Batchelor reported that it "still talks good but pretty low." He also noted that another button of phosphate of soda made on 3 July produced good talking four days later and on 9 July was "same as on 7th. This is simply wonderful." N-79-06-12:45–53, Lab. (*TAEM* 35:507–11; *TAED* N080:24–8).

-1774-

Equipment Specification: Electric Lighting

[Menlo Park,] July 10 1879

204[1]   Make 1 Standard Faradic Machine[2]   for sizes of wire see 46 page 185 for armature[3]   on the cores put 3 layers of No 10[4]

C[harles] B[atchelor].

ADS, NjWOE, Batchelor, Cat. 1308:159 (Order No. 204) (*TAEM* 90:746; *TAED* MBN003:53). Written by Charles Batchelor.

1. This is the order number; see Doc. 1687 n. 1.
2. Edison had arrived at his standard bipolar generator design by the beginning of May at which time he ordered a larger version to be built (see Doc. 1869 n. 2). After tests with that machine during June and the first week of July (see Doc. 1770 n. 5), Edison decided to build the machine described here. This was the first of four generators he had constructed over the course of the next six months in an effort to refine the design (see Docs. 1843, 1862, and 1867). It is uncertain when this machine was finished but Upton apparently conducted tests with it on 25 August in preparation for his paper on "Methods of Testing Faradic Machines," which he presented at the annual meeting of the American Association for the Advancement of Science at the beginning of September (see Doc. 1793 n. 6). A second machine was also built in August but its armature was damaged during tests prior to the AAAS meeting. Both machines were subsequently repaired in mid-September (N-79-10-18:33, N-79-06-16.1:259–65, Lab. [*TAEM* 32:800, 35:289–92; *TAED* N301:18; N077:129–32]; Cat. 1308:171, 173 [Order Nos. 225, 230, 232], Batchelor [*TAEM* 90:752–53; *TAED* MBN003:59, 60]).
3. These pages are missing from the book (N-79-02-10, Lab. [*TAEM* 32:1140; *TAED* N046]). The wire size and windings for the armature are given in a notebook entry dated 27 October. The wire was designated as ".032 in average" with a covering of .013 resulting in a wire of .045 inches. This was to be wound in 39 coils of six layers around an armature 8¾ inches in diameter, which had a wooden core of 6¼ inches and an iron ring on either end that was 1¼ inches thick. N-79-10-18:30–31, Lab. (*TAEM* 32:799; *TAED* N301:17).
4. The magnet cores were 6 inches in diameter and three feet in length. N-79-10-18:31, 33, Lab. (*TAEM* 32:799–800; *TAED* N301:17–

8); "Edison's Electrical Generator," *Sci. Am.* 41 (1879): 242, Cat. 1241, item 1332, Batchelor (*TAEM* 94:513; *TAED* MBSB21332X).

–1775–

*Francis Upton to
Charles Farley*

Menlo Park July 11, 1879.

Dear Charles.[1]

I have entered into a new arrangement with Mr. Edison as you may know.[2] I thought it as well to take a share in the light, although it may never bring me anything. I My[a] reasons were largely these. I knew that if I could get an interest my place here would be far pleasanter, for I should be free to do as I like. Since I changed my agreement I notice Mr. Edison takes it for granted that I can direct my own and other's work, [far?][b] much more than when I was working for fixed wages.

Another[a] reason was that if the light does not succeed, I shall have the run of the place here for experimenting if ever I may want to it[c], and stand a much better[d] chance of getting any positions that my be offered in which Mr. Edison is interested. This is prospective, actually we have a [---][b] now a machine for generating currents that is bound to come into great use for transmitting power and for plating.[3] I admit that my reasons were largely personal, that I would rather lose a years wages, in order to run the risk of gaining a large amount in company with a man like Edison. Just think a moment of the amount that a year will decide one way or the other $5,000 in cash and $1,500 a year for 17 years and nothing to do except to draw it.[4] Then I know that the electric light can be in all probability be made to work if sufficient time and money are expended, for there is nothing in the nature of things against it, and all theory is in its favor. We have not as yet what we want, but we have as good if not better[d] than any one else in the world. The natural result of my change of base is that I must draw on my money at home. I wish you would send me a check for $50 drawing[a] the money from the bank.

I do not feel that I have made a mistake for I have great hopes of ultimate success, and success that is far beyond the dreams of experimenters of the past, for light in itself does not cost anything scarcely, it is the enormous amount of heat that by know[n] processes goes with it that costs.

With love to your family   I am   Your Brother

Francis R. Upton

ALS, NjWOE, Upton (*TAEM* 95:545; *TAED* MU022). [a]Obscured overwritten text. [b]Canceled. [c]Interlined above. [d]Hyphenated; first syllable repeated and canceled at start of new line.

1. Charles Farley was married to Upton's eldest sister, Maria Louisa. This letter was addressed to him at Peabody, Mass., Upton's hometown. Vinton 1874, 350, 432–33.

2. Upton informed his father on 6 July that he had "entered into the agreement I spoke of with Mr. Edison to have 5% of the profits from the light, and 5% of the stock he now holds in the electric light" (see Doc. 1762 n. 4). Upton expected to receive 37 shares of Edison's stock in the Edison Electric Light Co. and received 36 at the end of July. This number is far lower than 5% of Edison's original bloc of 2,500 shares; apart from the allocation discussed in Doc. 1668 it is not known how Edison disposed of the balance of his stock. An agreement between Edison and Upton which may be that referred to here is missing from WJH (series 2, box 33). Upton to Elijah Upton, 6 and 27 July 1879, Upton (*TAEM* 95:543, 551; *TAED* MU021, MU025).

3. Upton had written his father on 6 July, one day before revising Doc. 1772, that "We have now the best generator of electricity ever made and this in itself will make a business." Upton to Elijah Upton, 6 July 1879, Upton (*TAEM* 95:543; *TAED* MU021).

4. Upton was presumably referring to Edison's contract with the Edison Electric Light Co. (Doc. 1577). The fifth article provided for the cash payment; the sixth article set forth conditions for a guaranteed minimum royalty of $15,000 per year during the life of the patents.

–1776–

To Louis Glass[1]

[Menlo Park,] July 14, 79

Dear Sir.

The box of sand was rec'd last week.[2] I found but a few scales of platinum in the chunk of blue gravel. I ran it through my stamp mill.[3] There is plenty of platinum in the sand as it comes from the flume (the middle compartment of the box) and a considerable amount, about $\frac{1}{5}$ as much platinum in the sand after it has passed through the amalgamating pans. I am now having an accurate analysis made. Perhaps there is some mistake about the sand which has passed through the amalgamion pans as I find when treated with acid there is a great deal of gold in it, but it takes a long time to make its appearance being in the centre of the fine magnetic particles. How many barrels or tons[a] of sand do you work per month or year?[4] That is I want to get an idea of the quantity of sand taken from the flumes, the total product which I can figure on after I have the analysis. I have 4 of my young men working up processes to get the gold and platinum out of the sand after it has passed your pans.[5] If we succeed and you will let us have the sand we will give you the gold as I dont care anything about that and do not want to make any money except scientifically. Am going to start a man for your place soon who will carry good letters and who is a thoroughly first class fellow.[6]

I rec'd a sample of black sand Saturday from Calpella Mendocino Co— This sample had not been worked and I found 50 times more platinum and iridsomine than gold in it.[7] The writer says there is plenty of the sand. My agent will probably want to hire a prospector to travel with him through the state do you think he will have any difficulty in hiring one?

In regard to the platinum near Oroville to which you refer what would be the probable cost of prospecting for[b] that vein which you speak of as being discovered several years ago?[8]

I send you a "Daily Graphic" with a cartoon which may interest you.[9]

I will also send you a pony crown Telephone in a day or two—tried to get two but[c] failed—[10]

In the course of 2 or 3 weeks I expect to have some of the new style chemical receivers. Not having the right to sell or give them away I will loan you 2 (permanently) and anything else I can do for you do not hesitate to call on me. Thanking you for your kindness in giving information and for the sample ores   I remain   Very Truly Yours

Thomas A Edison

LS (letterpress copy), NjWOE, Lbk. 4:477 (*TAEM* 80:102; *TAED* LB004477). Written by Stockton Griffin. [a]"or tons" interlined above. [b]Added in left margin. [c]Interlined above.

1. Louis Glass was secretary of the Spring Valley Mining & Irrigating Co. in Cherokee, a rich hydraulic mining site north of Oroville in north-central California. He subsequently became general manager of the Pacific Phonograph Co. in San Francisco (Gudde 1975, s.v. "Cherokee"; *TAEMG*3, s.v. "Glass, Louis"). He first wrote on 12 June in reply to Edison's inquiry addressed to Wells Fargo & Co. because "it is from our mine the Platinum etc has been obtained in this locality" (DF [*TAEM* 50:667; *TAED* D7928ZBH]).

2. In his draft of a letter sent on 21 June, Edison had asked for 25 pounds of black sand from the Spring Valley mine. Glass forwarded one sample of gold-bearing gravel and two of black sand, one from which the gold had been removed and the other still untreated. Edison later reported that the untreated sand contained "181 oz per ton of the platinoid metals of which 110 oz is pure platinum." TAE marginalia on Glass to TAE, 12 June 1879; Glass to TAE, 2 July 1879; TAE to Glass, 1 Sept. 1879; all DF (*TAEM* 50:667, 779; 51:147; *TAED* D7928ZBH, D7928ZEA, D7928ZOE).

3. The *Scientific American* reported in its 26 July issue that "Mr. Edison has a stamp mill and all the apparatus required for reducing ores of

various kinds. His facilities for reducing refractory ores and metals are particularly good." Nothing is known of the mill except for John Kruesi's 4 June measured drawing of a "Platinum Stamp Mill." Its operation is not apparent but evidently a five-toothed cog wheel would crush or grind ore against a flat surface. At least two other devices for handling platinum ores were to have been built later. On 15 July, Batchelor wrote an order to "make a platinum separator to get fine scales out of the black sand." His accompanying sketch is unclear but may represent some sort of pan sifter. Ten days later Kruesi ordered an "ore concentrator," evidently a form of centrifuge. "Progress at Menlo Park," *Sci. Am.* 41 (1879): 52; Machine Shop Drawings (1879–1880), Lab. (*TAEM* 45:70; *TAED* NS7986CAV); Cat. 1308:161, 165 (Order Nos. 212, 213), Batchelor (*TAEM* 90:747, 749; *TAEB* MBN003:54, 56).

4. Glass had previously given a detailed description of the hydraulic mining process used by his firm, which produced "an immense torrent absolutely saturated with water and sand" from the crumbling hillsides, the great bulk of which was carried away in flumes and deposited in the Sacramento River valley. He explained that approximately every three months the flumes were cleared of "a large quantity of black sand but I do not suppose we save an ounce in a ton, we get only what may be left in the flumes after the water is turned off," and he offered "no method of separating the black sand, or of even ascertaining how much there may be of it." At this time hydraulic mine operators faced strong challenges (and even the prospect of a complete shutdown) by downstream farmers over the release of large volumes of gravel and silt. Glass to TAE, 30 June 1879, DF (*TAEM* 50:770; *TAED* D7928ZDX); Kelley 1959, 85–123.

5. This work was probably done primarily by Edison's chemical staff, which at this time included Alfred Haid, John Lawson, Otto Moses, and possibly C. E. Mumsell, who is known to have been on the staff at least through mid-April. In his draft reply on W. C. Hendricks & Co.'s letter of 16 July, Edison indicated that

Sodium Amalgam will do better than any process reducing the heaviest coating of peroxide Iron from the Gold. would you be will[ing] to treat your clean up matter by the addition of sodium to the mercury if we will furnish it free   this will give you more gold and of the platinum saved we will pay you 50¢ per oz. treat your gold and free it absolutely of silver, and do it at your mill. I have some very smart German chemists. If you will send 25 lbs of sand that has passed through the Silver pan & thrown away we will give you the amount of gold in it that would be saved to you. [DF (*TAEM* 50:939; *TAED* D7928ZHB)]

About this time Edison also dictated notes for a reply to Colorado ore dealer Frank Ballou that he had "just finished process works elegantly for taking out gold from black sand a sub for chlorination   nothing used but horsepower." On a subsequent letter from Ballou, Edison indicated that the process for separating the gold from the black sand "is an electrical one using the Dynamo machines of the electric light for furnishing current. Am perfecting it." TAE marginalia on Ballou to TAE, 1 and 24 July 1879, DF (*TAEM* 50:778, 1016; *TAED* D7928ZDZ, D7928ZIW); for the subsequent development of this electromagnetic ore separator see Doc. 1921.

6. This was Frank McLaughlin, who departed on 11 September. Glass subsequently promised to provide "a trustworthy and experienced prospector." McLaughlin to TAE, 13 Sept. 1879; Glass to TAE, 26 July 1879; both DF (*TAEM* 51:197, 50:1036; *TAED* D7928ZPH, D7928ZJH1).

7. P. R. Klein had sent Edison a sample of black sand from the vicinity of Calpella, in northwestern California, on 3 July. Edison replied that although it contained much platinum and iridosomine, "I hardly think it would pay to mine for platinum alone the gold is in such great excess." Edison requested additional specimens and sent $20 to pay for preparing them because "My present impression is that we will get all the platinum we want of the Spring Valley Mining & Water Co Cherokee Cala but I desire to investigate all parts." Klein to TAE, 3 July 1879, DF (*TAEM* 50:784; *TAED* D7928ZED); TAE to Klein, 14 July 1879, Lbk. 4:479 (*TAEM* 80:103; *TAED* LB004479).

8. Glass had described in his 30 June letter (see note 4) minerals taken from a lode in the nearby mountains "fifteen or more years ago" which a local chemist had determined to be platinum. He reported that he had recently sent "a man up there prospecting, but he found the gulch filled in about six or eight feet with earth and debris washed down from above, and was unable to get down to the lead, provided there is one there, which I believe to be the case."

9. "The Wizard's Search" appeared in the *New York Daily Graphic* on 9 July 1879, two days after a *Graphic* column skeptical of Edison's electric light noted that "Maybe there is a great future before the platinum burner, but its inventor prospecting for a mine of the metal in order to give it a firm start in life is not a very hopeful sign of immediate usefulness." In its explanation of the cartoon, however, the *Graphic* offered a more optimistic outlook. Contrasting Edison with alchemists and other practitioners of "effete superstition," it described "the hero of our cartoon" as "simply a man of our time masquerading in medieval robe" and noted in conclusions that "Diamonds are so cunningly imitated as to deceive the very elect among jewellers, and in case the sage of Menlo Park should soon be found by some inquisitive reporter to have filled his backyard with manufactured platinum, so neatly done that old Mother Nature would readily endorse it as genuine, the operation in his hands would be deemed so much a matter of course that most of the newspapers would hardly think the fact worth mentioning." "The Wizard of Menlo Park," *New York Daily Graphic*, 9 July 1879, Cat. 1241, item 1225, Batchelor (*TAEM* 94:499; *TAED* MBSB21225X); "What Edison Still Wants," *New York Daily Graphic*, 7 July 1879, p. 28; "Pictures of the Day," ibid., 9 July 1879, p. 44.

10. The crown telephone was a form of magneto devised by George Phelps which used curved bar magnets arranged to resemble a crown. The "pony" may have been a small version, perhaps with only one bar (Prescott 1878b, 601–602). In his 12 June letter (see note 1) Glass had asked Edison for a hand-held carbon transmitter to use with his Phelps receivers because the instruments available from Gold and Stock in San

Francisco "are a cumbersome apparatus for fastening to the walls of a room." Edison promised that he was "going to 'hook' a pair of telephones and send to you so when you get them you need not indicate the source from which they came." Shortly thereafter Charles Batchelor directed that "one Pony Crown Telephone" be sent to Glass (TAE to Glass, 11 July 1879, Lbk. 4:461 [*TAEM* 80:95; *TAED* LB004461]; Cat. 1308:161 [Order No. 214], Batchelor [*TAEM* 90:747; *TAED* MBN003:54]).

To F. G. Lockhart[1]

Dear Sir

Your favor of June 20th was duly rec'd.[2] Many thanks for the sample sent. It contained a small quantity of platinum. That in the bottle was first class ore. I have found a source of platinum which is practically inexhaustible. It is in the hydraulic placer mines of California. The matter taken from the flumes in cleaning up of the Cherokee Flat Mining Co or rather the Spring Valley Water & Mining Co of the above place carries 80 oz of platinum to the ton. I have just finished a process for extracting it[3]   They get the gold out by the silver pan process and the residue of sand they throw away. By my process I extract a large quantity of gold and platinum from the tailings. I think if all your sand is as clean as sample that my process would take out every particle of gold and platinum, as I get every scale of platinum; you can judge that anything that will do this will catch the gold. I am now about testing it on a larger scale. Of course it is only a laboratory process and you probably know that laboratory processes do not always work in practice in fact not one process in 100 work in practice. I wish I could get a larger quantity of your sand but first I will use the Cherokee sand which is much more difficult.[4]

Platinum like your sample is worth from $1.80 to $2.20 per oz. This is the range of prices at the Russian mines at present. $32 per pound is the price at Baranquella U.S. of Colombo S.A.[5] I will pay the same but I hardly think it would pay you to save it as it apparently occurs there in such small quantities. I want 100 000 lbs or more. If you will send another box I would appreciate it.[6] If I get my process working you can have it in exchange for the platinum you save. Very Truly

T. A. Edison   G[riffin]

L (letterpress copy), NjWOE, Lbk:4:472 (*TAEM* 80:99; *TAED* LB004472). Written by Stockton Griffin. ªPlace from Edison's laboratory handstamp.

1. Nothing is known of F. G. Lockhart, whom Edison addressed in Randolph, Ore.

2. In reply to Edison's letter of 30 May 1879 (not found), Lockhart had supplied information about his gold-extraction process and the availability of platinum-bearing sand in the area; he reported that "platinum and iridium are not found in quantities sufficient to justify the working of the sands for them." The draft reply Edison wrote on Lockhart's letter is substantially the same as this document. Lockhart to TAE, 20 June 1879, DF (*TAEM* 50:711; *TAED* D7928ZCN).

3. See Doc. 1776 n. 5.

4. Edison dictated a draft reply (sent 9 July) to a California postmaster, stating that he wanted 100,000 pounds of platinum "after all the metals such as osmium pal[ladium] ruth[enium] rhod[ium] irid[ium] & iron" had been removed. Unlike the Colombian ores containing more than 90% pure platinum, Edison reported that the California sands were only 55% pure and "cont[ai]n all platinoid metals & they are so nearly alike its one of most dif problems in chem to separate them." TAE marginalia on J. S. Williams to TAE, 30 June 1879, DF (*TAEM* 50:766; *TAED* D7928ZDT).

5. Barranquilla is situated on the Magdalena River near Colombia's Caribbean coast.

6. Lockhart did so on 28 July. Lockhart to TAE, DF (*TAEM* 50:1073; *TAED* D7928ZJY).

–1778–

To Tracy Edson

[Menlo Park,] July 16, 1879

My Dear Mr Edson

Your favor of yesterday is just at hand.[1] I have made a lot of new receivers, several styles of them all of which proved in some way defective. I made 50 of different kinds at an expense of from 15 to $1800 and threw them all away. I want to be satisfied with the instrument before it goes out of my hands, it saves money afterward. Of course as I am to receive $1 for each instrument in use by the Company its to my interest to place them in their hands as soon as possible. As long as I saw a chance to improve them I kept at work. I expect now in two weeks to deliver[a] model instruments which will work right in on your telephone exchange[b] system and be perfectly satisfactory.[2] Very Truly

T. A. Edison   per G[riffin]

L (letterpress copy), NjWOE, Lbk. 4:484 (*TAEM* 80:105; *TAED* LBoo4484). Written by Stockton Griffin; circled "C" written at top of page. [a]Obscured overwritten text. [b]Interlined above.

1. Edson had inquired "Where is the hitch that prevents our getting some of your new Receiving Insts? Is it our fault? Can you not make some for us, or put us in the way of having them made?" Stockton Griffin's notes for this reply are on the back of Edson's letter. George Walker had made a similar request for the receiver on 14 July, when he wrote Edison that he heard "flaming accounts of them in the newspapers and from persons who have visited your factory; one gentleman reported that you had your entire force engaged in the construction of 500 for Europe. It seems to me that we should have the benefit of at least a sample, before the outside market is supplied." Edison's 16 July reply to Walker's letter is similar to this document except that he expressly denied making 500 instruments and stated additionally that he had "only sent 4 unsatisfactory Telephones to Europe." Edson to TAE, 15 July 1879; Walker

to TAE, 14 July 1879; both DF (*TAEM* 52:82, 80; *TAED* D7937ZBT, D7937ZBS); TAE to Walker, 16 July 1880, Lbk. 4:483 (*TAEM* 80:104; *TAED* LB004483).

2. Edison sent six receivers to Gold and Stock on 16 August and reported to Arnold White two days later that ten instruments were working satisfactorily there. TAE marginalia on Thomas David to TAE, 16 Aug. 1879; TAE to White, 18 Aug. 1879; both DF (*TAEM* 52:91, 671; *TAED* D7937ZBZ, D7941ZDL).

–1779–

Charles Batchelor to Edward Johnson

[Menlo Park,] July 16th 1879

E. H. Johnson   E[s]q[uire]

We ship you today four 4[a] telephones of a new pattern; the receiver on an independent arm which swings down to your ear and the transmitter stationary, in a box screwed to the wall.[1] Edison has decided on this as the telephone best adapted to do the work, for this reason: that we stand in a disadvantageous position in regard to the other telephones, inasmuch as they put the instrument to the ear; (thus shutting out all extraneous sounds), when they want to catch weak sounds, whilst our diaphragm is so far away, that weak sounds are difficult to get in a noisy room. Now by putting the receiver on an arm, we get it just as loud, and also have it close to the ear. In this shape as you will see it is a much cheaper instrument to make. These telephones are supplied with chalks made of chalk and phosphate of Soda, we having adopted that as the standard chalk for the reason that it is always the same, nothing comes from it,[2] needs very little wetting, and glazes so nicely.

The receivers we send, are on arms that are unscrewed from base, and fastened against the side of box. The back of receiver unscrews out, thus giving you a chance to wet it the chalk[b] with a camels hair brush if necessary. If you wish to take a chalk out you will find on the worm wheel shaft, a long sleeve, which pull towards the worm, and the shaft divides in [two?][c] lengths, now move the piece that has the wheel on out of the [way?][c] unscrew the setscrew in chalk mandrel, and pull short shaft out, thus getting your chalk out.[3]

On the 21st we shall send you eight 8[a] more 4[a] of which will have a new transmitter on which I think you will like because it can never get out of adjustment.[4] we call it the inertia Telephone because it works by inertia, the whole working part being fast on the centre of the diaphragm. Page 4 of this contains a diagram of the connections of the new bracket[5]

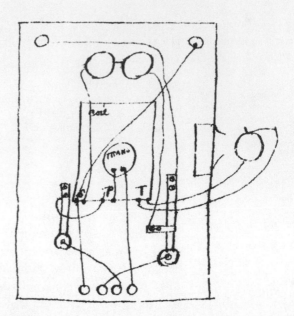

Yours faithfully,

Chas Batchelor   for T.A.E.[d]

ALS (letterpress copy), NjWOE, Lbk. 4:489 (*TAEM* 80:107; *TAED* LB004489). Circled "C" written at top of page. [a]Circled. [b]"the chalk" interlined above. [c]Faint copy. [d]Followed by dividing mark.

1. See Doc. 1784. These instruments were made according to an order (apparently misdated 9 July) for "4 more new Recvrs. of rubber & everything to complete 4 new Telephones"; this followed a 10 July request for the first two swinging arm receivers. They may have been modified and not actually have been shipped until 19 July. Cat. 1308:157, 161 (Order Nos. 202–3, 216), Batchelor (*TAEM* 90:745, 747; *TAED* MBN003:52, 54); see Doc. 1781 esp. n. 1.

2. That is, no scale or crystallization deposited on the chalk surface. Batchelor noted in the early morning of 15 July that he saw "no efflorescence" on the surface of a phosphate soda button made eight days previously, whereas when Edison tested a caustic soda button made the same day he first "wiped the efflorescence" from it. N-79-06-12:45, 49, 55, Lab. (*TAEM* 35:507, 509, 512; *TAED* N080:24, 26, 29).

3. Batchelor made three sketches on 14 July illustrating a similar procedure for a "Hand Motograph receiver." The details are not clear but the worm shaft appears to interlock with the chalk shaft, being held in position by the moveable sleeve. Batchelor's accompanying instruction is to "Push back the sleeve and lift out" the chalk shaft. N-79-02-10:189, Lab. (*TAEM* 32:1225; *TAED* N046:85).

4. They were shipped at the end of July (see Doc. 1782).

5. Figure labels are "coil" and "TRANS."

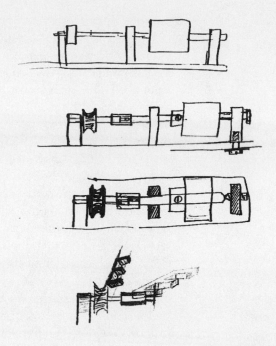

–1780–

From Edward Johnson

London England July 20/79.

My Dear Edison

Sunday morning= After one weeks work= A very meagre report to make— When I arrived I found that the hurry had greatly subsided— The London Co. during Col G[ouraud]'s absence had been down to Liverpool & Manchester trying to organize a company[1]—under the impression as they assert— that they had an equal right with you to go into the provinces and form local Boards— Here was a point of serious difference between you & them & the Col. was compelled to compromise the matter by giving them the Lancashire District— including Liverpool—Manchester & surrounding towns—in addition to the London Districts—they to pay you 10,000£ advance royalty for the same   This had all to be submitted to you for approval—and the papers formally signed must be in hand here before this London Co. can legaly organize—and until organized they do not exist—to rent rooms—build lines—employ labour &c &c—[2] I am consequently as yet without other avocation than simply to keep in good exhibition order the 4 Instruments I brought with me— I met the "Board"—and Col G. says made a splendid impression   He remarked "Thank the Lord we have now some one from America who is a credit to us"—& much more to the same purport. In short—to be brief in this branch of the subject, the entire concern seems to be well satisfied with their bargain. I believe however my coming has given mortal offense to one Prof. Conrad Cook= who aided White in the Exhibition to the Prince of Wales—[3] It seems that he lent his aid to that enterprise—at Mr Whites solicitation and, though he actually did nothing—an understrapper who had helped Charley being the real genius of the occasion—White presented him with a silver cup suitably inscribed in commemoration of the occasion— He thought the inscription was not sufficiently flattering and sent it back with a request that it be altered after the manner which he suggested—this was done—though they secretly laughed at the mans vanity. Now it appears this man sets up the claim that White promised him £1000. per year as Consulting Electrician of the Co—and learning of my arrival he came in to see White, & ascertain whether or no my^a coming was going to interfere— Of course White repudiated any such promise—as he unequivocally asserts he made no such promise— Cook then puts in a claim of 600£ for his services on the occasion—and upon White asking him to be good enough to state precisely what he had done to the Telephones—to this he replied "I cannot do so in such manner as would enable a layman like yourself to understand—the work was of a character entirely too much in the realm of the scientific" from which I conclude that Mr Conrad Cooke is a charlatan= White asked me what I would advise, I told him to simply ignore the mans existence   He said he would do so, but that he Cook—had the entreé to places & people with whom he could do us a deal of harm— I told him that my experience was that such people generally exposed their motives pretty freely and were counted at their proper value by such men as we wanted to interest—& that our work would do more than our influence in building us up— It seems that Cook went straightway to circulating stories about White—saying among other things that White had said that Col Gouraud did not represent you in the Telephone matter &c &c—and circulating some Cock & Bull story in regard to me—which is not worth repeating—except to show that a line of enemies is already opened up for me outside of any action of mine—solely from the jealously of a Local would be scientist= who probably thought he had a soft thing on this Co. of Green electrical dabblers— I gave the Col & White to understand that the only electrician I would consult—because the only one I could to advantage—would be Mr Thomas A Edison—but that if they wanted to employ a lot of people by way of propitiating local jealousies they were as much[b] at liberty to make donations in such directions as any other—but it must be clearly understood that no service would be called for by me— At this summary sort of treatment they all seemed to be well pleased because as White expressed it—"we seemed to have all the talent we required within ~~the~~ our own boundaries" so I guess that matter is effectually[a] disposed of—though you may probably have another anti-Electric Light writer

Now as[a] to this man White— I have of course watched him with some interest & I think have taken his measure— Charley disliked him because of his fahdy-dawdy-ways—as he termed it—and he has the English of McLaughlin[4] to a pretty considerable extent— Outside of that he is evidently[a] a man of brains & of business— He is the proposed secretary of the Co— & here Secretary means business manager— He has authority and asserts it—he probably did so with Charley, who had made up his mind to take a trip to Paris— when he told him he must not go as an Exhibit to the Prince of Wales was arranged for—Charles felt insulted at not being consulted—and was quick to resent Whites tone—by flatly refusing to defer his trip— White having a full consciousness of the fearful cost to Gouraud of allowing his man to scare him—& keep him scared—as was Gourauds case with Adams—determined then & there to assert his independence, and told Charles to go ahead—the instruments could be made to work by others—& they were— The clerks and the boys about the office who liked Charley endeavored to dissuade him from persevering in his purpose of leaving White in the lurch—but No he was insulted & go he would— That is all I can gather of the story from the clerks who hob-nobbed[a] with him. I have not asked White for his views, as I treat the matter with him as of no interest to me. I think Charlie failed to subordinate his personal feelings to the business in hand— Therein was his failure— Had he simply told White to Go to Hell—& then went on doing the work & using his utmost to have made the thing a success He would have been entitled to your sympathy & the insult would probably have been taken by the other fellow— It's a case of failure to send the right man to the Devil— I haven't seen Charles—as yet—nor Puskas— Brown[5] was in the London office just before I arrived—setting up a claim to having rendered you valuable assistance in[a] having given you the "Design" for the present Telephones   Its wonderful what a number of valuable assistants you have had. They are so common that for the sake of being original I've a great mind to disclaim any knowledge of you until the moment of sailing.

Adams it seems has his cards printed as full partner with you— Even White & his associates seem still to be under the impression that such was the case—until I explained the peculiar[a] form of partnership—and his real status with you= He by-the-way—seems to have launched out heavily here—ordering Telephones—Magnets—Experiments—Batteries &c &c with a prodigality simply amazing   Gouraud & White each have an office full of stuff which never came from your establishment   Magnetos of designs & shapes I never beheld— hundreds of Batteries Battery boxes &c &c—for which he could possibly have had no earthly use— I asked for an explanation of his apparent carte-blanche in this direction and White gives it, by saying that Adams had Col Gouraud completely scared he was so afraid of him that he addressed him always in terms he would not use towards high functionaries here— These Adams used to still further enhance his credit with the shops—& near drove the Col to distraction, the[a] net[a] result being to pile up a lot of Lumber here, which is probably only equalled by Phelps cellar[6]—& which gives the impression that your Electro Chemical Jordan has been an extremely hard & expensive road to travel.

Now if you please we will consider that day as having passed, and all these nightmares as having performed their part in promulgating the present day. Col Gouraud and I will agree   White I shall have no possible difficulty in managing= He is weak in so many directions that I am totally indiferent to—that I can throw him a sop every now & then & thus keep him in my trail, & not myself have to fall into his.

Of the Board of Directors, theyre is only this to be said— They are too Rich—too well known—and too entirely men of Brains—for me to Buck against if I were so inclined or fancied, I had occasion—but since they are such I like them and shall make strong personal friends of each of them without doubt. They impress me as men in their line of the Sir Wm Thomson class—in his= too entirely conscious of their standing to be solicitous of concerning any trespassing upon it—

Now for Business.

I had for two days some difficulty in restoring my Buttons to a working state—and still find it impossible to make good ones out of some of them though I have managed to get a few excellent ones, by experimenting with a varying degree of soaking   some seem to have gone entirely others only slightly and others again requiring a vast deal of wetting—and very careful nursing to prevent crystalization—or abraision of the surface—for when they lose the smooth surface, by the smallest possible amount of wear—they cannot be restored—[7]

Another rather singular feature is that when a thoroughly good chalk—one which has been working for 3 Days perfectly without moistening—is suddenly started it fails to alter a word until the handle has been turned rapidly a few times each way— Then it may be turned constantly in one direction for hours & it will work O.K. = This is the case with all of them = so I have instructed the users to always give it a few turns before using—which obviates all difficulty—but how about clockwork = The chalks which I brought inside the Telephone panned out much the best—those wrapped in tin foil were all bad— I theorize that as they had never been wet, they became absolutely dry—while those in the Telephones having been thoroughly moistened retained a sufficient quantity of moisture to give a foundation for re-moistening = Can a chalk which has been once thoroughly dried—be remoistened to the core = Do not we only wet from the surface a short distance in & depend upon the prime moisture therein for giving conductivity to the inner body—? If this inner body once becomes dry then to reach it from the surface requires such long soaking that the outer scale is destroyed? Is not the remedy for this either to thoroughly wet the chalks before leaving America— or make the chalk body so much thinner that it can be readily penetrated by water from the surface when once it has become dry[a]—or make the chalks here? In further corroboration of my notion = take this   I find by wetting a Dry chalk just all it will stand without destroying ie—softening—the scale—& putting it on the spindle in the usual way, it is absolutely silent— but if you turn[a] it around so the Pen runs on the extreme edge & consequently close to the metal rim, it works splendidly— Now is not the current carried through but a thin scale of the surface in this case side wise to the rim. Can't these chalks be given a prime moistening of something that will not dry out— nor interfere with the surface action—

In re. to the necessity for turning the handle a few times before speaking—is this not to clean the surface of the decomposition produced by the last operation of the Instrument— and which has undergone some change meanwhile which has converted it into a sort of non conducting paste? This notion has just occurred to me— I'll put a sort of soft scraper on tomorrow & see if I can't keep it clean—

Now in regard to the call Battery   Its a failure I'm afraid—
I'll tell you why— They Dry out in about 2 weeks—some earlier— Now wetting them again is impracticable for the reason
that the Parafine you have poured in them prevents the moisture from getting in where it was wanted—except in a very
slight degree—sufficient in some cases to operate them for 2
or 3 days & in others not enough to restore them to action at
all— I shall try and recusitate them but am afraid they are
gone for good   I have been looking around for small
Leclanche[8] and find some very small ones used in series of 30
or 40 by Physicians. for made   I got a dozen of them and am
going to try them tomorrow   If they will ring[a] the Bell
O.K.—they will answer our purpose as 1 dozen of them can be
put up in a small box about 6 inches square for about $2.50 or
3.00[9]   If they fail to work satisfactory I think I'll write on to
Roosevelt[10] at Paris and see if he can't find[a] me just what I want
there— I went all through Elliott Bros[11] place they have some
magnificent scientific apparatus—some measuring devices
especially adapted to your Electric Light work— I got a catalogue from them which I send you. No doubt you will find
some very interesting things in it.

See the present no of the Engineer it has a slur at you—[12] I
told the people here that you simply laughed at those things—
The London Times published Crofuts[a] Letter about the
Menlo Telephone exhibit—it was very good—only he made
a fearful blunder—getting the Phonograph & Telephone in-
extricably mixed up— You will find it in full copied into the
Engineer[13]   The Daily News had a cablegram from you on
the reduction of the price of your burners from several dollars
to 56 cents by an alloy— The Engineer slurs you on it—[14] I
hope you will keep me posted on your E.L. work as well as
other things, I have already had some tall controversies on the
subject, but I got away with my man each time— I see one
St. George Lane Fox—has a long series of claims in ahead of
you here some of which are very broad— They look to me very
much as if he had been thinking out a multitude of things for
the purpose of anticipating you in the Patent Office— this is
given color too—by the rumor which I hear that he is anxious
to compromise with you[15]—as far as I can see however I don't
see that he has a single device which in actual practice would
work. I wonder what your cablegram means when it says "we
have now removed all defects"   Have you sent me Telephones different from those I brought? Have you sent me new
chalks? or what is it= Tuesday or Wednesday will tell.[16]

Give my regards to Mrs Edison   Tell her that my wife has written her a good long letter telling her what a horrid country this is. Certainly it cannot compare with our Native Land—

Whats become of the jester[17]—is he still with you?

My wife sends her regards   My next letter—not having to cover so much ground will not tax your time so much to read. Yours Truly

E. H. Johnson

ALS, NjWOE, DF (*TAEM* 52:628; *TAED* D7941ZCZ). [a]Obscured overwritten text. [b]"as much" interlined above.

1. See Doc. 1765. On 3 July, after Gouraud's return from New York, Arnold White demonstrated Edison's receiver at the Manchester city hall for the mayor and about 70 others. He evidently used the new receivers carried by Gouraud, reportedly over a 20 mile line. "Edison's Electro-Chemical or Loud-Speaking Telephone," *Manchester Courier and Lancashire General Advertiser,* 4 July 1879, Cat. 1241, item 1217, Batchelor (*TAEM* 94:493; *TAED* MBSB21217X).

2. The company was incorporated on 2 August 1879. Memorandum and Articles of Association, DF (*TAEM* 52:652; *TAED* D7941ZDF1).

3. See Doc. 1741.

4. Perhaps a reference to the speaking style of Frank McLaughlin.

5. Robert G. Brown was a former Automatic Telegraph Co. operator and an independent builder of telegraph lines. He was also chief operator on Gold and Stock's Merchants Exchange telephone system in New York until about this time, when he became chief electrician for the Société Générale des Téléphones in Paris. On his way there he stopped in London between 15 and 21 June and demonstrated Fitch telephone transmitters to "private individuals representing large houses, To the Stock exchange Telegraph Co representatives, and the 'Edison' people of London." Taltavall 1893, 227–28; Brown to Samuel White, 2 July 1879, box 284, SSW (*TAED* X078AN).

6. Johnson presumably meant the Western Union manufacturing inventories maintained by George Phelps.

7. Johnson reported the next day, in a letter to Edison mistakenly dated 20 July, that "the Telephones talk splendidly this Monday am—one wet last on Friday morning—the other on Thursday morning—so that after once fairly going they hold out well." DF (*TAEM* 52:625; *TAED* D7941ZCY).

8. The Leclanche battery had a zinc anode in a sal ammoniac solution separated by an earthen partition from a carbon cathode resting in a mixture of manganese peroxide and other materials. It produced about 1.5 volts with a moderate current. The French telegraph service adopted this battery, and its suitability for intermittent use led to its adoption for police and fire alarms in the United States. Knight 1876–77, s.v. "Leclanche Battery"; Schallenberg 1978, 348–51; "Constant Galvanic Batteries," *Telegr.* 5 (1868-69): 53; King 1962a, 247; Moise and Daumas 1978, 4:315-16.

9. Johnson explained in a letter to Edison three days later that he believed the problem was caused by having increased the resistance of the

batteries and the line but not that of the magnets in the telephone call bells and switchboard drops—the apparatus for signaling the switchboard attendant. He wrote that the next day he would "have a Bell & two Drops on the switch board made to 100 ohms each—then if your Battery or my little series of Leclanche operate it all right—Gouraud will authorize me to have the entire lot changed." Johnson remarked in closing: "The new Telephones will reach here tonight— I hate like the Devil to make this change—but after full experimenting, and full discussion with Gouraud, he thinks it wise to do so." Johnson to TAE, 23 July 1879, DF (*TAEM* 52:641; *TAED* D7941ZDA).

10. Cornelius Roosevelt, a first cousin of Theodore Roosevelt, was a promoter of Bell's telephone in France. He also had a license from Theodore Puskas for the sale of Edison's phonograph in France and its territories. See *TAEB* 4:56 n. 3, 103 n. 2 and Docs. 1218 and 1333.

11. Elliott Bros. was an instrument shop on the Strand in London. The catalog sent by Johnson has been not found but see *TAEB* 2:178 n. 5.

12. The *Engineer* published in its 11 July issue a detailed report on Edison's final specification for British Patent 5,306 (1878), Batchelor (*TAEM* 92:102; *TAED* MBP015) filed on 28 June. It dismissed the patent's significance, contending that Edison's claims were either unoriginal or of no practical use, and commented that

> Altogether the specification fails to convince us that any great step has been made towards perfecting the electric light by anything that it makes known. We fail indeed to see any indications of progress beyond what has long ago been accomplished. That the electric light has a great future before it cannot be doubted; that the light by incandescence is beautifully soft, pure, and steady, must be admitted; but it brings us face to face with the great enemy, cost. It is the most expensive form of the electric light, and in order that it may be used, Mr. Edison must devise some better way than resorting to incandescent metal coated with insulating material—he must find something cheaper than platinum, iridium, cerium, or zirconium. ["The Edison Electric Light," *Engineer* 43 (1879): 32]

13. See Doc. 1758 n. 4. After describing Edison's new receiver, William Croffut's dispatch erroneously spoke of the phonograph instead of the electromotograph receiver: "The visitors experimented with the new phonograph in every conceivable way, and the conclusion shared by all was that it is so much more powerful than any other phonograph as to lift it almost above rivalry. Experiment also showed that the voice is really magnified in transmission, and delivered in a louder tone than that employed by the speaker." This paragraph was omitted when the *English Mechanic* and the *Electrician* republished the article, and Croffut later apologized for the error. The article has not been found in the *Engineer*. "The Edison Telephone and Electric Light," *Times* (London), 15 July 1879, 8; "The Edison Telephone and Electric Light," *English Mechanic*, 18 July 1879, Cat. 1241, item 1231, Batchelor (*TAEM* 94:502; *TAED* MBSB21231X); "The Daily Papers and Mr. Edison," *Electrician* 3 (1879): 104; Croffut to TAE, 5 Aug. 1879, DF (*TAEM* 49:719; *TAED* D7906ZAV).

14. The *Daily News* (London) published a dispatch from its New York correspondent on 16 July stating that:

Mr. Edison has partially overcome the obstacle to his electric light offered by the high price of platinum. His lamps, instead of costing several dollars a piece, as at first, can now be made of an alloy of platinum with inferior metals, so as to cost only fifty-six cents. He announces that he can now produce the spiral coil for incandescence at a price which all who use gas can easily afford, and that his efforts to find platinum are only induced by the desire to reduce the cost of burners still further. ["Mr. Edison's Electric Light," *Daily News* (London), 16 July 1879, 5]

No reference to this announcement has been found in the *Engineer*. Johnson may have been thinking of an editorial notice in the 5 July issue of the *Electrician* stating that "We have a firm conviction that Mr. Edison is not answerable for one-thousandth part of the newspaper paragraphs containing statements which are frequently as absurd as they are incorrect. The editor of the *Daily News* ought to be better informed and take more care than to go to the expense of a special telegram, and flaunt before the faces of his readers in big type such a paragraph as he inserted on July 2" reporting Edison's claims for the efficiency of his dynamo. "The 'Daily News' and Mr. Edison," *Electrician* 3 (1879): 73–74; "Mr. Edison and the Electric Light," *Daily News* (London), 2 July 1879, Cat. 1241, item 1215, Batchelor (*TAEM* 94:492; *TAED* MBSB21215X).

15. St. George Lane Fox (who later adopted the surname Pitt) was the son of noted archaeologist and anthropologist General Augustus Henry L. F. Pitt-Rivers. He filed three British patent applications related to electric lighting in October and November 1878 (British Patents 3,988, 4,043, and 4,626) and one application in March 1879 (British Patent 1,122). The first three covered a method of coating wire burners, a parallel system of distribution including conductors, meters, and regulators, and wire-burners sealed in nitrogen-filled globes. The most recent application, filed 20 March, was for applying a hard carbon coating to a filament made from an amalgam of conducting and non-conducting substances such as plumbago and magnesia, which was to be operated in a sealed globe free from oxygen. Lane Fox continued his lighting experiments and subsequently organized the Lane Fox Electrical Co., Ltd., which eventually licensed patents to the Edison interests in Britain. Dredge 1882–85, 2:l, li, lix, lxxviii; Swinton 1929; Bright 1972, 49–50; Obituary, *Times* (London), 7 Apr. 1932, 14.

George Gouraud had already forwarded to Edison a letter from David Chadwick, M.P., outlining similarities between the Lane Fox and Edison patents and offering to broker a compromise in order to avoid future litigation. Gouraud soon afterward also sent copies of the first three Lane Fox patents. In marginal notes on a letter from London patent solicitors Brewer & Jensen to Gouraud about this matter, Edison instructed Stockton Griffin to "write Gouraud to say that in case we should ever use anything in Mr Lane Fox patent which is doubtful that we shall of course be ready to talk business." Gouraud to TAE, 14 and 24 July 1879; David Chadwick to Gouraud, 14 July, 1879; TAE marginalia on Brewer & Jensen to Gouraud, 12 July 1879; all DF (*TAEM* 50:337, 342, 339, 336; *TAED* D7922M, D7922P, D7922N, D7922L).

16. In his letter to Edison the next day (see note 7), Johnson stated that he had received "by this mornings mail Batch's letter on Phosphate

Soda—which largely answers my Questions of yesterdays long one to you— Its good news." Batchelor's letter is Doc. 1773.

17. Unidentified.

–1781–

To Edward Johnson

[Menlo Park,] July 21. 1879

Dear Sir

I shipped you Saturday 4 instrument which I am absolutely certain will prove perfectly satisfactory in every respect.[1] They have the merit of being simple, convenient and cheap and not at all liable to get out of order. You will perceive that the chalk is now in a perfectly closed receiver.[2] The chalk is a phosphate of Soda one and of all substances that is the only perfect one. It seems to be absolutely perfect lasting days when exposed to the air   The salt is perfectly neutral, does not change. You may find a crystallization on the surface of the chalk when they arrive as I fear they were soaked too thoroughly. They work just the same whether ¼ wet or fully wet. We make a concentrated solution of Phosphate of Soda then dilute with ½ bulk water, wet with letter copying brush.[3] The surface is the smoothest of all chalks and never changes without you saturate the chalk too much then it crystalizes out. With the receiver in the form sent we gain all the <u>margin</u> possible, as when on the old machine it was hard to hear owing to distance away we can with this get the same loudness.[4] The handle goes easier and the whole thing is better. We send 8 more next week 4 of which will have a different kind of handle and polished Mahogeny boxes and 4 others will contain the Inertia Telephone transmitter that <u>will not</u> get out of adjustment, is adjusted in a second, costs 20 cts and can whisper 25 feet away and get talk 3 inches from it.

Hereafter we will send ½ Inertia and ½ regular until we use up the balance of Bergmanns transmitters then by that time we can hear your decision as to the one you want.

For the first time I feel perfectly at ease on the Telephone. We are going to set up a pair in NY next week as standard— You know I would not do this if I was not <u>perfectly</u> satisfied.

If there are crystalizations just sponge them off with water and it will thereafter be all right. I stake my existence that these chalks will not need wetting more than twice each year.

Batchelor writes you further details.[5] Very Truly

T. A. Edison   G[riffin]

L (letterpress copy), NjWOE, Lbk. 5:9 (*TAEM* 80:127; *TAED* LB005009). Written by Stockton Griffin.

1. Charles Batchelor instructed the laboratory shop on 18 July to "Alter 4 telephones for England to have a coil in them of primary 40 ohms secondary 150 ohms tertiary 400 ohms and make a new box and base to suit them. hinge the top of box so that we can get to transmitter to adjust it." The shop completed the order on Saturday, 19 July. Cat. 1308:161 (Order No. 216), Batchelor (*TAEM* 90:747; *TAED* MBN003:54).

2. In its article on Edison's new telephone, *Scientific American* noted that the reservoir and roller previously used to moisten the chalks had been eliminated and instead "the chalk cylinder is inclosed in a vulcanite box . . . at the end of the movable arm. The cylinder, when once moistened, remains in that condition for an indefinite period, as the box is practically airtight." "Edison's Latest Telephone," *Sci. Am.* 41 (1879): 198.

3. The brush was used to dampen a leaf of tissue paper in a letterpress copy book so that the ink from a letter or other document could be transferred to it to make a copy. The damp leaf and the original were placed between oiled sheets (in order to prevent the transfer of ink to other pages in the book) and then the closed book was placed under pressure in a letterpress in order to effect the transfer of some of the ink from the original to the tissue paper.

4. That is, the receiver produced an equal volume regardless of the signal strength from the distant transmitter.

5. Doc. 1782.

–1782–

Charles Batchelor to Edward Johnson

[Menlo Park,] July 31th 1879[1a]

Dear Johnson,

We got your letter and glad to hear you had pleasant voyage.[2]

We send you today the eight instruments promised in my last:[3] they all have inertia transmitters in them but not plumbago ones, we put in[b] a small piece of carbon cut from one of the Jablochkoff candles,[4] they are simply immense and the beauty of it is you can talk either into the mouthpiece or six inches away and it is equally clear and does not lose its adjustment.— In the last four that were packed of this lot which have rubber mouthpieces,[5] you will find that the rubber ~~do~~ pressure spring has a pin from screw pressing on it; this we find, really gives better result than cap so we have sent you a number of the pins to put in the others.—[6] We have also sent you a number of transmitter carbons[7] although I don't see how you will ever need them.— We also send you 8[c] eight extra chalk buttons which have not been wet at all.— Edison says to please deliver two complete instruments to Puskas as only six out of this eight are billed to London.—[8] Will soon now give you a lot of these. Very respectfully yours

Chas Batchelor for T.A.E

ALS (letterpress copy), NjWOE, Lbk. 5:37 (*TAEM* 80:134; *TAED* LB005037). [a]Circled "C" written at top of page. [b]Interlined above. [c]Circled.

1. Charles Batchelor's original date was not transferred to the letterpress copy, which was dated in an unknown hand.

2. Not found.

3. Doc. 1779. Edison cabled Gouraud on this day, "Shipped today lot Telephones thats simply perfect   telegraph through Drexel seven thousand five hundred for 500   delay money delays Telephones, manufacturers require advance and instant payment delivery." Gouraud replied that he had immediately requested the Drexel house in London to credit Edison $8,000. This arrangement was in accordance with Gouraud's suggestion in a 15 July letter in which he also advised Edison to "make your contracts for a thousand instead of 450 as several hundreds will be required in connection with the formation of the provincial companies," which now included a prospective Glasgow firm. When he acknowledged this letter on 29 July, Edison stated that "I must ascertain what style will be adopted as I am sending different styles and do not yet know what kind will suit your people." TAE to Gouraud, 30 July 1879; Gouraud to TAE, 2 Aug. and 15 July 1879; Drexel, Morgan & Co. to TAE, 1 Aug. 1879; all DF (*TAEM* 52:647, 666, 621; 49:22; *TAED* D7941ZDE, D7941ZDG, D7941ZCW, D7902C); TAE to Gouraud, 29 July 1879, Lbk. 5:27 (*TAEM* 80:133; *TAED* LB005027A).

4. Batchelor observed on 25 July that "We find that very hard carbon such as is used in Jablochkoff candles is best for transmitting, giving clear talking whether you talk into the mouthpiece or far away." On 29 July he wrote a lengthy set of directions for preparing the eight instruments for shipment, instructing John Kruesi that "The four inertia telephones with small Jablochkoff candle carbons on your desk are all right and may be packed immediately— The four Bergmann transmitter telephones on [Charles] Flammer's bench must have new boxes on, with inertia telephones of thick mica and connection strips pasted to diaphragms . . . In the inertia telephones make Jablockoff candle buttons and make them . . . rounded at both ends." Cat. 1304:59; Cat. 1308:165, 167 (Order No. 214), Batchelor (*TAEM* 91:58; 90:749–50; *TAED* MBN004:56, MBN003:56–57).

5. In his written instructions (see note 4), Batchelor told Kruesi that the telephones "must have a rubber mouthpiece from the Bergmann telephones screwed into the wood . . . instead of turning the wood into the shape of a mouthpiece." See also headnote, p. 312 n. 7.

6. An undated sketch of what appears to be a threaded pin on a screw is in Cat. 1308:168, Batchelor (*TAEM* 90:751; *TAED* MBN003:58). Its use is unclear.

7. Kruesi wrote a work order about 15 July for 100 buttons to be made and sent "to Johnson with the next shipment of Telephones," which also included 50 Bunnell sounders. Cat. 1308:159, 161 (Order Nos. 210, 213), Batchelor (*TAEM* 90:746–47; *TAED* MBN003:53–54).

8. Batchelor instructed Kruesi: "Do not address any to Puskas" (see note 4).

–1783–

Notebook Entry:
Electric Lighting

[Menlo Park,] July 31st 1879

~~Coat~~ Pyro insulation[1a]

Acetates.

Acetate of Calcium is decomposed by heat into Acetone and Carbonate Calciu[m].[b]

All ~~carbonates~~ acetates which decompose easily and form stable carbonate do same thing—

If so, then our coating is a Carbonate of Magnesia & the carbonate has to go off under high heat as

Carbonate	Calcium	strong red heat
"	Barium	white heat
"	Magnesium	intense heat

the decomposition of Carb. Calcium commences at a low red heat so that in[a] ~~bring up a~~ heating a wire of this covering it should be left at low red for a long time
Pyroinsulation

Nitrates.

Nitrate Calcium is decomposed at a comparatively low temperature.[c]

Nitrate of Lanthanum at red heat is completely decomposed leaving light brown oxide[c]

Acetate Zinc— will not coat a wire after 30 dippings—but the surface of Pt is slightly roughened as if zinc reduced and combined   perhaps the lampblack of candle reduced it. there is a slight change in the metallic color but not the slightest trace of an oxide—[c]

Acetate Cerium— No oxide—attacks platinum, like zinc, but there are iridescent colors—[d]

Acetate Calcium   coats easily at low heat but is very easily rubbed off and when bent it peels off in shreds

Nitrate of Calcium   Will not coat at all and after 20 dippings does not seem to affect the wire

Nitrate of Cadmium   Difficult to get anything on at all and after you get the brown oxide on a red heat takes it all off again—

Nitrate Magnesium   Considerable trouble to get it on even, though a coating can be got on   affects platina as much as the acetate[2c]

TAE   Chas Batchelor

X, NjWOE, Lab., N-79-07-31:1 (*TAEM* 33:544; *TAED* N052:1). Written by Charles Batchelor. Document multiply signed and dated. [a]Obscured overwritten text. [b]Written off edge of page; followed by dividing mark. [c]Followed by dividing mark. [d]Followed by dividing mark; "acetate zinc . . . iridescent colors—" written by Edison.

1. This entry marks the resumption of Edison's research into pyroinsulation coatings for electric lamp burners. Charles Batchelor recorded these results (except for lanthanum nitrate) into his own notebook the same day. Cat. 1304:60, Batchelor (*TAEM* 91:59; *TAED* MBN004:57).

2. Batchelor continued experimenting the next day with zirconium acetate, which he reported "coats very well indeed." He covered a piece of .004 inch wire "and made all the spirals touch and then put 9 cells C[ondit]&H[anson] battery through it. It gave good light (white) and on examining under the microscope after it seemed to have lost nothing but the covering was slightly browned   The acetate must be evaporated down to a syrupy solution." He also recorded these results in his own notebook. N-79-07-31:9, Lab. (*TAEM* 33:548; *TAED* N052:5); Cat. 1304:60, Batchelor (*TAEM* 91:59; *TAED* MBN004:57).

BRITISH COMMERCIAL TELEPHONE
Doc. 1784

By 10 July Edison had made several significant changes to the design of the telephone sets he had sent to Britain at the end of February (Doc. 1681). The earlier model had been built for demonstrations but the new one was intended for commercial service.

The most visible alteration was the location of the receiver. One of the laboratory criteria for judging the effectiveness of the receiver had been the distance over which speech remained audible, but the loud-speaking qualities of the electromotograph receiver apparently prevented weaker signals from being heard in a noisy room.[1] The receiver was therefore moved from the telephone box to the end of a hinged arm that could be brought level with the listener's ear. This design emerged during the first week of July and was made possible by a new chalk composition that remained moist indefinitely after being wet. Edison found that he could keep the chalks moist by placing them in an airtight vulcanite container and do away with the wetting roller and water reservoir.[2] The earliest designs for the new swing-arm receiver placed the crank for turning the chalk on the receiver housing, connected to the chalk shaft by a worm gear; later the shaft was repositioned and connected directly to the crank.[3] One of these designs may have been used in an order of 10 July for "two new receivers with swinging armes." The same day, however, Batchelor ordered the shop to "make new receiver with worm motion to drive on the arm joint," suggesting that the crank was to be placed at the base of the arm and connected to the chalk by a worm gear and shaft as shown in a measured drawing of 12 July for a Patent Office model. Also on 10 July another order was entered for a receiver with a "clockwork motion and flywheel" similar to that described in the patent application executed by Edison on 17 July (U.S. Patent 231,704).[4] The commercial telephones sent to England in mid-July had the crank on the base of the arm connected to the chalk by a worm gear and shaft.[5]

Two additional modifications were also incorporated into the new receiver. A rubber block at the end of a thumbscrew replaced the earlier tension spring arrangement to maintain pressure between the metallic contact spring and the chalk. The receiver itself was placed in the tertiary circuit of an induction coil.[6] During July the new inertia transmitter was substituted for the standard Edison transmitter, and the wooden mouthpiece was replaced by a rubber one.[7] And at the end of September the inertia transmitter was dropped in favor of a return to the standard carbon transmitter commonly used in the United States with the carbon button held in a cup fixed on the transmitter body.[8]

1. See Doc. 1779.
2. See Docs. 1769 and 1773; "Edison's Exhibit Motophone," TI 2:545 (*TAEM* 11:673; *TAED* TI2:494).
3. N-79-03-10.1:175–83, N-79-07-07.2:157–73, both Lab. (*TAEM* 31:75–79, 32:453–61; *TAED* N022:73–7, N038:79–87).

A vulcanite case contained the receiver, which consisted of a mica diaphragm, from which projected the contact spring. The spring rode on the surface of the rotating chalk cylinder under pressure from the rubber block, adjusted by a thumbscrew.

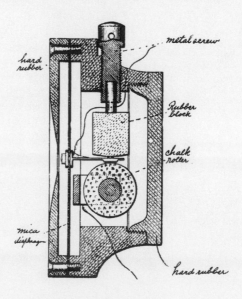

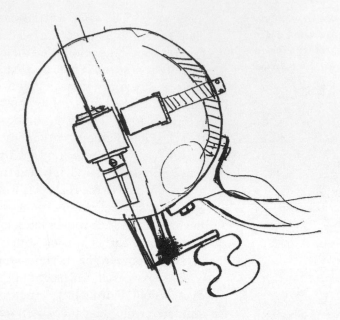

4. The patent model found at the Edison Institute, Dearborn, Mich., (Acc. 29.1980.1389) for Edison's U.S. Patent 231,704 is based on the drawing for that patent, which shows the crank on the box. A slightly different version of the 12 July patent model drawing (Vol. 18:104, Lab. [*TAEM* 4:1138; *TAED* NV18:93]) appears in Edison's British Patent 5,335 (1879), fig. 2, Batchelor (*TAEM* 92:141; *TAED* MBP021). The shop orders are Cat. 1308:157, 159 (Order Nos. 202–3, 205–6), Batchelor (*TAEM* 90:745–46; *TAED* MBN003:52–53). There are also two measured drawings made by John Kruesi in August (Vol. 18:107–108, Lab. [*TAEM* 4:1141–42; *TAED* NV18:96–7]).

5. See Doc. 1779.

6. See note 3 and Doc. 1759 n. 3.

7. See Docs. 1769, 1779, 1781, and 1782. One of the early commercial models, identified by the wooden mouthpiece (see note 8), is in the Science Museum, London (UkLS Inv. 1915-229).

8. See Doc. 1813. It is not known if the instrument shown in Doc. 1784 has the inertia or the standard carbon transmitter. *Scientific American* described and illustrated an instrument with the inertia transmitter in its 27 September 1879 issue but by the time laboratory draftsman Samuel Mott made measured color drawings in October 1879 the inertia transmitter had been supplanted. "Edison's Latest Telephone," *Sci.*

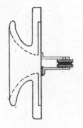

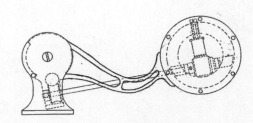

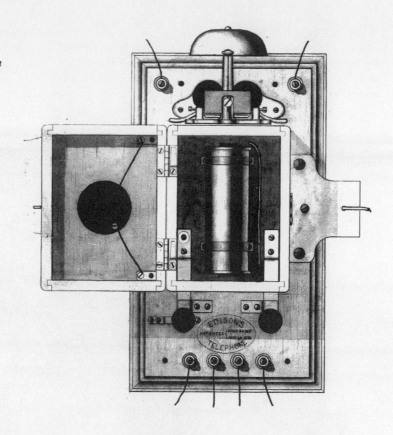

Samuel Mott's October 1879 drawing with the front of the telephone open to show the induction coil.

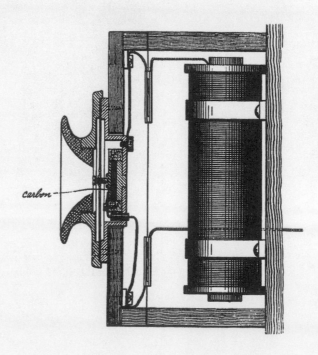

Samuel Mott's October 1879 drawing showing the standard carbon transmitter that replaced the inertia transmitter; it is connected to the induction coil.

carbon

Am. 41 (1879): 198; Oversize Notes and Drawings (1879–1886), Lab. (*TAEM* 45:2; *TAED* NS7986BAB); see also "Edison's Exhibit Motophone," TI 2:545 (*TAEM* 11:673; *TAED* TI2:494).

–1784–

Production Model:
Telephony[1]

[Menlo Park, July 1879[2]]

M (box 27.3 cm × 15.9 × 10; swingarm 27.3 long), UkLS, Inv. 1953-117.

1. See headnote above.

2. This model shows the rubber mouthpiece adopted sometime during the latter half of July (see headnote above).

Watch Hill RI[1] 8-2 1879[a]

My Dr E

Letter from EHJ leaves him fighting chalk bugs but keeping Johnnie Bulls eye shut on it so far! Says everything there is higher than in U.S. but clothes— Where is McLoughlin?— I want some lettering done— Wonder if he can make a map?[2]

If you approve of it, I will get 500 little Phono's made like best model & then advertise & sell direct to public; these Toy houses are hogs![3] They want it all! If you say go ahead I'll wire me here— 500 can be got up first class ready to ship for 3.50 to 4.00 & I think will sell for $10.00 express paid & I think I can make contract with express companies to send anywhere for .50 cts— We can certainly sell some & I am in favor of trying it if you say "go ahead"—

Cant you do something for us on the T̶o̶y̶ commercial machine—?[4]

How about Toy & Clock contracts—now is the time to get them up, while they dont set any store by them—[5]

We have milked the Exhibition cow pretty dry—only sold 4 in July one of those to party named Edison, had to charge him $100 for it, to keep show of kindness to inventor—[6] If we let the cow go dry he will think we are not industrious— Yours

U H Painter

⟨Have[b] sent EHJ some telephones that are perfect= he will have no more trouble— I think 500 Little fellows would sell as you say= Just let me give my exhibition of 100 E lights[7] then I am going on that phono   McL working here going away to California few days⟩

ALS, NjWOE, DF (*TAEM* 49:316; *TAED* D7903ZGH). Letterhead of Edison Speaking Phonograph Co. [a]"18" preprinted. [b]Obscured overwritten text.

1. Watch Hill is a summer resort on Block Island Sound in southern Rhode Island, where Painter sometimes vacationed. *WGD*, s.v. "Watch Hill."

2. As a former telegraph operator, Frank McLaughlin had developed a fine calligraphic hand. Painter had been trying to acquire government maps for Edison's platinum search. When Painter reported on 18 July that the federal government did not publish state maps west of the Mississippi, Edison noted on that letter that he wanted "the book containing description of <u>postal routes</u> that are furnished contractors in bidding." Painter promised to try to find maps of those routes and on 30 July he forwarded a reply to "your request for data about PO Routes west of Miss River" stating that the Post Office had forwarded "returned copies of all the advertisements," presumably meaning those soliciting bids for carrying mail in the West. Painter to TAE, 21 May and 18, 25,

and 30 July 1879, DF (*TAEM* 50:595, 953; 49:302; 50:1094; *TAED* D7928O, D7928ZHI, D7903ZFW, D7928ZKL).

3. In Edward Johnson's absence, Painter made arrangements in early July for the Edison Speaking Phonograph Co. to distribute small phonographs (see Doc. 1717) through Strasburger, Pfeiffer & Co., a New York City toy importer. He agreed to supply instruments with tinfoil and instructions, beginning with a lot of 1,000, for $5.50 each to be sold to wholesalers for no more than $7.50 or to the public for no more than $10.00 apiece. The firm withdrew for unknown reasons, prompting Painter to report on 25 July that the "toy folks have gone back on the little Phono'. & I dont see what we can do except to make up 500 or 1000 on our own ac't & take the chances of selling them ourselves— They say people cant make it go without elaborate tuition & they wont touch anything that the millions cant use at once." Stockton Griffin noted on Painter's letter that Edison thought "Plenty Toy Houses in NY wld take hold of that thing   think Co cld sell them if advtsd." Manufacturing had begun by 6 August, when Painter told Edison that he had "set a shop to work on the 500 little Phonos—& if I cant get any regular House to handle for half the profits will sell 'em ourselves, same as we did the Exhibition Machines." The following week Painter proposed having Frank McLaughlin sell them in New York's Union Square in October. Painter to Strasburger, Pfeiffer & Co., 5 and 13 July 1879, ESP Treasurer's Lbk. (Nov. 1878–July 1879), 689; *Wilson's Business Directory of New York City* 1879, 722; Painter to TAE, 25 July, 6 and 15 Aug. 1879, DF (*TAEM* 49:302, 319; 51:743; *TAED* D7903ZFW, D7903ZGK, D7932ZAR); see Doc. 1848.

4. This inquiry followed the recent failure of Edward Johnson's design of a large phonograph. When Edison received Sigmund Bergmann's prototype, he dictated a note to Griffin on the back of a letter from Painter that it "dont seem to have been made right   as it now stands don't think it wld be a success." Painter responded on 6 August that "Johnson wants to see the wreck of his hopes in the commercial machine— So if you dont think it is worth your while to keep it please box it up, so it will stand an ocean voyage & send over to him." Painter to TAE, 13 July and 6 Aug. 1879; TAE marginalia on Painter to TAE, 25 July 1879; all DF (*TAEM* 49:281, 51:742, 49:302; *TAED* D7903ZFG, D7932ZAQ, D7903ZFW).

5. Edison had signed separate contracts in January 1878 for the commercial development of the phonograph in clocks and toys. Despite considerable work by Edison and the Ansonia Clock Co., no satisfactory clocks had been made to date (see *TAEB* 4:18 n. 7, 49 n. 1, 66 n. 4, 101 n. 1; Docs. 1246 and 1265). Oliver Russell, a principal in the toy agreement, dutifully submitted monthly reports showing that he had produced no instruments (*TAEMG1–2*, s.v. "Russell, Oliver D."). Russell had become partners with Hilborne Roosevelt, a New York organ manufacturer and electrical inventor with several patents to his credit (see *TAEB* 4:56 n. 3), and recently transferred to him a dominant interest in the contract with Edison (Docs. 1499 and 1618; Roosevelt to TAE, 21 June 1879; Russell agreement with Roosevelt, June 1879, both DF [*TAEM* 51:730, 732; *TAED* D7932ZAJ1, D7932ZAL]). Painter believed that Roosevelt had fraudulently negotiated a royalty concession from Edison and urged Edison to cancel the toy contract if Russell had

not paid the full annual royalty as originally scheduled. He asked Edison to "then make the Toy license so it will come to the Phono' Co when we are ready to push it— The same request will apply to the clock contract, both of which are forfeited & both of which we are prepared to develop according to our agreement when we made the contract in Jany 1878" which stipulated that if those contracts were abrogated the Phonograph Co. "should have a license broad enough to cover the whole Phonograph." Edison did not void the contract and later allowed Roosevelt to purchase Russell's entire interest in it (Painter to Edward Johnson, 26 May 1879; Painter to TAE, 1 June and 5 July 1879; Roosevelt to TAE, 15 and 18 Nov. 1880; all DF [*TAEM* 51:715, 720, 735; 55:298–99; *TAED* D7932X, D7932ZAE, D7932ZAM, D8038M, D8038N]).

6. More than a year previously, Edison had promised a phonograph to Henry Edmunds, Jr., a British inventor and entrepreneur who helped to publicize the invention in England (see Doc. 1249 and *TAEB* 4:95 n. 2). He apparently gave it to Edmunds when he visited Menlo Park while traveling in the U.S. during the summer of 1879. Painter charged Edison "for the mach' same as others [$100], & pay you royalty on it [$18]— This is less than though I had charged you what it has cost us to make & ship it." Tritton 1993, 10–33, 53, 192–93; Edward Johnson to TAE, 18 July 1879; Painter to TAE, 26 August 1879; both DF (*TAEM* 51:741, 745; *TAED* D7932ZAP1, D7932ZAT).

7. Painter had advised Edison a few weeks earlier that if "you will not let too much get into the papers about your Electric Light, there can be a large sum made selling gas stocks short here & in London when it is ready to bring out— I can [get] parties to go in & sell 'em blind & give us half the spoils— You have already made em fluctuate in the millions & not made a cent of it." Edison wrote on this letter: "never mind the money. will let you know some time before I make a demonstration." Painter to TAE, 13 July 1879, DF (*TAEM* 49:281; *TAED* D7903ZFG).

–1786–

From Ludwig Böhm

New York, Aug. 10th 1879.

Sir,

Reading yesterday the advertisement in the "Staats Zeitung" in which you want a first class Glassblower who is able to blow instruments for scientifical purposes I permit myself to write to you.[1] I was employed at Dr. Geissler's own working rooms for almost two years in Bonn on the Rhine where I blew apparatus for experiments which Dr. Geissler made.[2] I blew Geisslers vacuum tubes to show the electric light. I think you mean this tubes in your advertisement with "Giesler Tubes." I also blew Potash apparatus spectral tubes, ~~cholir~~ chloride of calcium tubes,[3] Radiometers,[4] alcalimetric apparatus[5] pippets and other things for the "Maass Analyse" for[a] measuring[b] purposes[b] etc. I blew vacuum tubes in all seizes and shapes. I know well how to melt in platinum in tubes etc etc. Now at present I am employed in a New-York bussiness[6] where I chiefly make[b] clinical thermometers, sometimes hydr[o]meters,[b] combined or plain. My present boss engaged me while travelling in Germany last year. I stay here since the 13th of Sept 1878. Now it is one year that I left Dr. Geissler. I you wish to see me you will send a letter to L. K. Böhm c/o Mr. Englehardt Brooklyn President st. 45. where I will get it. It would be disagreeable for me if a letter came to my bussiness place.

I am, sir   most respectfully   yours   obedient

L. K. Böhm[7]

ALS, NjWOE, DF (*TAEM* 49:926; *TAED* D7913ZAO). [a]Written in left margin. [b]Obscured overwritten text.

1. The advertisement appeared in the *New Yorker Staats-Zeitung*, a German language daily newspaper. Edison had been employing William Baetz on a part-time basis. He subsequently recalled that "Desiring the services at all times, night and day, of a glass blower, I advertised for one." Edison's testimony, p. 34, *Böhm v. Edison* (*TAED* W100DED032 [image 3]).

2. Heinrich Geissler was an experimentalist and renowned maker of delicate scientific instruments at the University of Bonn. He was known particularly for the eponymous evacuated tubes used in research on electrical discharge phenomena and for the mercury vacuum pump he devised in 1855 that also bears his name. *DSB*, s.v. "Geissler, Johann Heinrich Wilhelm."

3. The simple "potash bulb" apparatus, which included a calcium chloride tube, had been invented by Justus Liebig as a faster method of organic combustion analysis. Partington 1964, 4:238.

4. The radiometer, a demonstration apparatus devised by William Crookes in 1871, is an evacuated glass bulb containing four vanes coated black on one side and white on the other. Since more radiant energy is absorbed by the black side, the vanes will rotate when placed near a light or heat source. Atkinson 1910, 666–69.

5. An alkalimeter is a graduated tube used to determine the strength of an alkali based on the amount of acid required to neutralize it. Knight 1876–77, s.v. "Alkalimeter."

6. The firm of Bahmann Brothers, identified in a New York directory as thermometer manufacturers. Böhm's testimony, p. 8, *Böhm v. Edison* (*TAED* W100DEC002 [image 7]); Wilson 1879, 58.

7. Ludwig Böhm later testified in a patent interference case that while at Geissler's workshop he "tried to obtain knowledge, theoretical, as well as practical, and became regarded a skillful glass-blower of philosophical instruments" by the time he left at about age nineteen. Stockton Griffin noted on this letter, "ansd & sent ticket   Aug 12, 1879." Böhm recalled that he visited Edison on 13 August and promised that he could blow a particular vacuum pump of which Edison showed him a drawing; this design likely was among the several pumps and a "Device for Bringing up wires for lamps in Vacuo" that Batchelor sketched on 12 and 13 August. One week later he came to work at Menlo Park with his own tools, bellows, and a glassblowing table. Böhm initially worked in the first floor of the laboratory and sometime later moved to a small outbuilding that

became known as the glasshouse. During his first month he seems principally to have blown duplicates of apparatus made by A. S. Reinmann and William Baetz, including combination pumps, after which he constructed different experimental designs (see Doc. 1803). In addition to mercury pumps, during his tenure he made other experimental equipment and bulbs for Edison's lamps. Böhm's testimony, pp. 2, 8–10; Edison's testimony, pp. 34–35, 47; *Böhm v. Edison* (*TAED* W100DEC002 [images 1, 7–9], W100DED032 [images 3–4]); N-79-07-31:264–80, Lab. (*TAEM* 33:673–81; *TAED* N052:131–9); Jehl 1937–41, 325.

–1787–

Notebook Entry:
Electric Lighting

[Menlo Park,] Aug 10th 1879

Electric Light   Coating Wires for Lamps.[1]

1   Took 32 in. of .004 $^{20}/_{[1]00}$ Platinum Iridium wire and brought it up in vacuum taking 20 minutes, then covered it with .001 Zirconia. After commencing to coat, this wire broke in my hands and the two ends of wire shewed structure like this 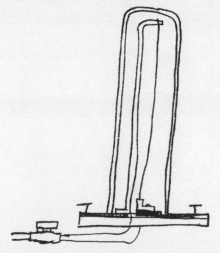 as if one half of section was crystals of a blackish grey color.[a]

Instrument for bringing up the wire in Vacuo

We This happen'd a number of times and Edison thinks it not so much due to the wire having flaws, as to some change produced in the wire during the process I put it through. We now began to investigate it & found that .004 20/ platinum Iridium would stand a breaking strain of 2½ lb, and every time it broke, it would either show a well drawn out (center punch) point or a chisel point showing that the wire stretched down till it finally broke. There so:—

These breaks were not at all like the breaks made when they broke in my hands without pressure which showed a crystalline structure, and broke off sharp—[2]

If we brought up a spiral of one layer on a lime cylinder it gave a splendid light, and would never cross in the spiral; but when heated too much, the leading wires would go first; but if we made the spiral of two layers, there seemed to be something that <u>crossed</u> the bottom and top in between before it had got very hot, always cutting out half the spool almost immediately, and gradually cutting out till there was nothing left but the few bottom turns in circuit. Whatever it is that crosses, seems to move around between the layers as if it was a liquid conductor: It seems as if the Zirconia itself at such a heat was a conductor, which shows itself plainer on the 2 layers, because, there are points between the bottom and top layer of considerable tension especially at bottom of spool[3b]   It may be however that it is due to impurities such as silica in the Zirconia and also from the lime[4]

We tried to coat the wire with light coating of Zirconia Acet.[c] and then put on coat of Acetate Magnesia   after bringing up the spiral and looking at under the Micro. we found that the Magnesia was almost all gone but the Zirconia was on in a dense vitreous mass.—

The best way I find to coat the wire is to evaporate the Acetate of Zirconia down till it shows a slight milkiness with some white sediment and then to rub between your fingers slightly ~~u~~till it feels slightly sticky and apply to the wire by rubbing between your fingers till they become almost dry then pass the wire through the candle flame not in the very[c] hottest part but [~~lower down?~~][d] just above the wick so that it never gets above a low red.

I am always careful that the wire is never so wet that the heat of candle makes it spurt in the least[e]

We find that equal mixtures of Acetate of Zirconia and Acetate of Magnesia make a good coating which looks very fine after being under the action of great heat[5]

TAE   Chas Batchelor

X, NjWOE, Lab., N-79-07-31:13–21, 25–29 (*TAEM* 33:550–53.1, 555–57; *TAED* N052:7–11, 13–15). Written by Charles Batchelor. [a]"See page 16" written to indicate page turn. [b]"See Page 25" written to indicate page turn. [c]Interlined above. [d]Canceled. [e]Followed by dividing mark.

1. This entry marks the resumption of experiments described in Doc. 1783 n. 2.

2. Charles Batchelor also entered results of this day's experiments in his own notebook, where he recorded that "it looks to me as if there were places where the iridium was not properly alloyed but Edison thinks it is due to some chemical change produced during the process I put it through." After describing the trials made to determine the tensile strength of the wire he commented that "These breaks were not at all like the ones made in my hands which makes me still think that if we tried long enough we should have found bad places." Two weeks later Edison sent a sample of broken wire to his London supplier, Johnson Matthey & Co. The firm answered that upon careful examination the "black spots or specks distinctly visible upon the surface of the fracture are oxide of Iridium—the non amalgamation of which with the alloy would account for its weakness [at?] that point.— The fault being known we can readily overcome it and will give the matter our attention." Cat. 1304:61–62, Batchelor (*TAEM* 91:60–61; *TAED* MBN004:58–59); Johnson Matthey & Co. to TAE, 13 Sept. 1879, DF (*TAEM* 50:125; *TAED* D7919ZCH).

3. At this point in the entry is Batchelor's suggestion, on a separate page dated the same day but evidently written at a different time, to "Try baking on the coats at about 200 Fahr. and when on thick enough bring up very slow indeed   Take 10 short pieces and coat thin and put in oven" N-79-07-31:23, Lab. (*TAEM* 33:554; *TAED* N052:12).

4. On 12 August Batchelor took two wires covered with zirconia acetate, twisted them together, and connected one end of each to opposite poles of a battery. The circuit remained open "but if a spirit lamp was applied so as to heat them up, they would make circuit as soon as it got a little above a red. The point where it would make connection always showed as if some impurity in the covering had been present and attacked the platina changing it in some way." Edison repeated this experiment using wires coated with alumina and found that "these did not cross under the same conditions nor yet when heated by the blowpipe." The next day Batchelor again tried zirconia acetate and found that the six pairs of wires he used "all acted alike and crossed at a little above a red heat. Now as our wire is pretty pure and our Zirconia we have taken particular pains to have pure we must come to the conclusion that the Zirconia must be a conductor at that temperature." N-79-07-31:33, 39–45, Lab. (*TAEM* 33:559, 562–65; *TAED* N052:17, 20–3); Cat. 1304:64–66, Batchelor (*TAEM* 91:63–65; *TAED* MBN004:61–3).

Arrangement for testing pyroinsulation properties of zirconia coating.

5. Batchelor and Edison tested more than a dozen other wire coatings between 11 and 14 August. They noted a number of compounds which "all form gelatinous oxide and therefore will fuse down to a vitreous mass." They also observed that zirconia hydrate produced only a thin coating but a "spiral made of it stood very intense heat without crossing." Chloride of aluminum and a solution of powdered quartz also gave promising results, the latter producing a "perfectly clear glass-like coating" that seemed to be "perfect insulation." On 19 August Batchelor found that powdered corundum in solution made a good coating but required high heat to dry it on the wire. He then tried mixing in various quantities in the hope that "if we could coat with very low heat (such as holding over a kerosene lamp) it would save the trouble of first bringing up the wire and also prevent the cracking of the wire due to the heat of gas jet." These mixtures proved unsatisfactory, as did those made with tragacanth and isinglass. N-79-07-31:31, 35–37, 47–51, 59, Lab. (*TAEM* 33:558, 360–61, 566–68, 571; *TAED* No52:16, 18–9, 24–6, 29); Cat. 1304:64–65, 67–68, Batchelor (*TAEM* 91:63–64, 66–67; *TAED* MBN004:61–2, 63–4).

–1788–

Francis Upton to
Elijah Upton

Menlo Park, Aug. 11, 1879.

Dear Father:

I should have gone home Saturday but just at the present I am very busy here with experimenting. Mr. Edison is giving all his attention to the electric light and finding out a great deal new about it. I feel anxious to be on the spot so that I can see why the various contrivances are changed.

Besides this we intend to make a series of tests regarding the transmission of power, and these will be entirely under my charge, as it is nearly all calculation which is needed.[1] I think I have found a law, but I have yet to test it before I can [wit?][a] with certainty say whether it holds in all cases.[2] Mr. Sutro of the Sutro Tunnel has written to Mr. Edison regarding transmitting power to the mines in Virginia City.[3] He say there is a river power of 3000 or 4000 horse power only four or five miles away and that he thinks this could be utilized.[4] They now burn wood at the mines and it costs $10 or $12 a cord. Mr. Edison thinks that the power from Lake Tahoe could be utilized, it is only 20 or 25 miles away and power can be taken that far by means of electricity. I feel very confident that it can be done. This will be a chance for me in case the E—— L—— plays out. I expect to get married in about five weeks, and I am going to come home for arrangements in a week or two. You may expect some hints for a wedding present in the sh shape of the wedding garments. Much love   Your Son

Francis R. Upton.

ALS, NjWOE, Upton (*TAEM* 95:553; *TAED* MU026). ªCanceled.

1. Edison had considered the generation and transmission of hydro-electric power during his return from California the previous summer, when he thought about using the Platte River to make electricity for mines in the region, and soon thereafter concluded that Niagara Falls could be used for a similar purpose (*TAEB* 4:501 n. 5, App. 2; "A Great Triumph," *New York Mail,* 10 Sept. 1878, Cat. 1241, item 878, Batchelor [*TAEM* 94:349; *TAED* MBSB20878X]). In what appears to be an effort to simulate the conditions of power transmission, Upton made several experiments on 8 August in which he placed Edison's generator on the dynamometer and connected it through a high resistance to another dynamo operated as a motor (N-79-06-16.1:224–33, Lab. [*TAEM* 35:272–76; *TAED* N077:112–6]).

2. It is not known to what "law" Upton refers. However, a notebook entry that probably dates from about 7 August contains a statement by Edison that "The strength of a Dynamo machine is as the square of its size   If you double ~~your~~ the size of your machine it is four times as powerful." Upton wrote on this page the title "Law" and on another page, dated 12 August, he noted, "For example   A machine with the same resistance giving twice the E.M.F. will be four times as effective. In order to do this four times the amount of wire must be used on the armature." N-79-06-16.2:75, 120, Lab. (*TAEM* 35:404, 427; *TAED* N079:37, 60).

3. Adolph Sutro, a merchant and stamp mill operator, had constructed a tunnel nearly four miles into Nevada's Mt. Davidson to ventilate and drain the mines of the Comstock Lode, which filled rapidly with hot water. The main bore was completed in July 1878, about a month before Edison visited nearby Virginia City and one of the mines, but disputes with the mine owners delayed its use until the following spring, well after the richest ore had been depleted. Edison had written Sutro on 23 July 1879: "I understand you have considerable water power from your tunnel or from the Carson River. if this is a fact I can deliver, by means of my new Dynamo Machine, 65 per cent of the power in Virginia City with a very inexpensive plant. Should this prove of interest to you please reply and I will give you further particulars." Sutro evidently answered at the end of July and on 8 August Edison responded that he would send additional information "at the earliest practical moment." *ANB,* s.v. "Sutro, Adolph Heinrich Joseph"; *NCAB* 21:126–27; Elliott 1973, 129–132; Docs. 1393, esp. n. 3, and 1475; TAE to Sutro, 23 July and 8 Aug. 1880, Lbk. 5:12, 66 (*TAEM* 80:128, 138; *TAED* LB005012, LB005066B).

4. Upton began analyzing numerous configurations of the proposed power installation on or after 7 August. He considered different transmission voltages and various numbers of dynamos and motors ("receivers"). He prepared a chart for tabulating the weight and cost of one-, two-, and three-ohm copper conductors ranging in length from one to sixty miles, and he also made a number of other capital and operating cost estimates. On 12 August he made lengthy and detailed notes for converting approximately 1,125 horsepower at the river into electric power and transmitting it five miles. He specified sixty 150-volt dynamos and an equal number of "receivers" (plus four others to excite the field magnets). He presumed that each would convert 92 per cent of applied power into useful work for a net efficiency of 84 per cent, although

he estimated the system's overall efficiency at just 50 per cent. He considered requirements for insulating the conductors to carry 9,000 volts, different arrangements for placing the motor field magnets in circuit, and the need for motor governors. Upton estimated the capital cost of this system at about $50,000; this figure excluded machinery for water power and assumed five miles of conductors (no return grounding wire). He also figured the expense of a similar system designed to use only 732 horsepower at the river. There are no further dated notes from this period on the proposed Sutro Tunnel project. N-79-06-16.2:75–145, Lab. (*TAEM* 35:404–39; *TAED* N079:37–72).

-1789-

Caveat: Electric Lighting

[Menlo Park,] August 14, 1879[a]

To all whom it may concern.

Be it known that I, Thomas A. Edison, of Menlo Park, in the State of New Jersey, have invented an Improvement in Conductors, and apparatus for Electric Lights, of which the following is a specification.

The object of this invention is to economically subdivide the electric light.—

The invention consists in devices for producing an even or equal electromotive force in the circuit, to moderate the power of the light without the interposition of devices wasteful of the electric energy.

The method of arranging the conductors is peculiar; supposing that a section of a quarter of a mile is to be lighted, and 1000 electric lamps are to be used all arranged in multiple arc; I first run two conductors ¾ the distance from the central station of a certain size and electrical resistance, to this point I put no lamps across the two wires; the remaining distance I enlarge the conductor commencing very large and tapering down to the size of those from the station or below that size; on these conductors of increased size I place say 250 lamps; again from the central station I run two other conductors to ½the distance or ⅛ of a mile and there connect enlarged conductors until a point is reached where the large conductors commence on the first circuit, on this conductor I put 250 more lamps, then another circuit ⅛ of a mile is run from the central station of full size; thus up to the points where the lamps are put on I am enabled to keep the pressure or electromotive force constant, hence the difference of fall in electromotive between the lamps will not take place gradually from the first lamp to the last lamp of the 1000 lamps as it would if they were all on one circuit, but will only fall between 250 lamps in two cases of 500 lamps, in one case but if the circuit containing 500 lamps is split up into two circuits then the fall will be only between the extreme of 250 lamps.[1]

Fig. 1.

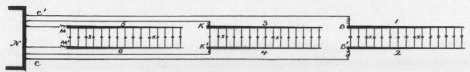

Figure 1. represents three circuits.— 1. and 2. Fig. 1.[b] represent the lamp wires of the longest circuit. N. is the central station. c. c.′ are the smaller leading wires, connecting at B. B′ with the enlarged tapering conductors. x. x. x. are lamps, 3. and 4. the second circuit and 5. and 6. the shortest circuit.

The electromotive forces, will be the same at B.B.′ K. K.′ M.M.′ hence the difference in the amount of light given off by the lamp nearest the station and the one furtherest from the station will only be 12 per cent, under the most unfavorable condition. The drop in the electromotive force, between the circuits with no lamps upon them and when they are all on, is independent of the unavoidable drop between the lamps and may be controlled by hand, or by automatic devices. When controlled by hand, a small wire should be connected to the end of every circuit and returned to the central station and there connected permanently with an electrometer, by which any drop in the electromotive force can be seen at once,[2] and the rise in the electromotive force may be obtained either by increasing the speed of the engine, or of the Dynamo Machine or "Excitor" which serves to supply a permanent current to the field magnets of the inductor bobbins of the machines of the main conductors, the rise may also to a certain extent be effected automatically. Thus

The field magnets being supplied with current from ~~one~~ an excitor, and the induction bobbin connected to the line in multiple arc, the latter has one end disconnected therefrom and the field magnet wound with a separate coil, over the regular coil, and the bobbin wire connected to one end ~~of~~ the other end being connected to the main conductor. This extra coil has but a fraction of the resistance of the induction bobbin, thus without this bobbin the current from the excitor, would, acting in the field magnet coils, give sufficient magnetism to cause the induction bobbin to show an electromotive force of 95 volts which is raised to 100 when the extra coil is in circuit.

If now lamps are placed on circuit they tend to cause a drop in the electromotive force of the induction bobbin, but, the amount of current taken from the machine being increased this reacts on the extra coil to create a stronger field magnet and the drop is prevented, this device is efficacious when the field magnets do not approach any where near their point of saturation.—

Another method consists in winding an extra coil on the field magnet of the excitor itself, of very low resistance, and interpolate it in one of the main conductors, so that all the current which comes from the combined bobbins shall pass through this extra coil.—

Another plan consists in connecting all the extra coils of all[c] the main machines in multiple arc and interpolating the combined conductor in the main circuit so that the whole current shall pass through the extra coils.—[3]

The method of increasing and decreasing the amount of light from the lamp is as follows, and the device employed is shown in fig. 2.[4]

Fig. 2.

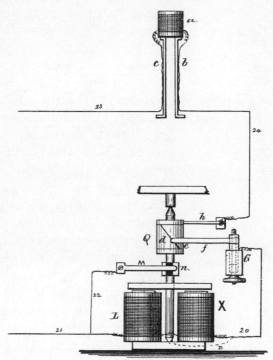

a. is the burner. b. c. the platina wire supports. X. is an electric engine having an resistance 200 times less than the lamp when the latter is giving maximum light. K. is the armature, which is kept in rotation by having the magnet short circuited by break

n. spring m. and shaft every time the armature approaches the poles, the short circuit is by wire 22. and p. (dotted line to represent the base)

Q. is a break wheel on the revolving shaft, at the bottom of this wheel there is but little metal the most part of its circumference being insulating material   the amount of metal on the perhiphery gradually grows greater until near its top when the whole circumference is metal   on this rests permanently a contact spring or roller h. connected to the lamps by wire 24. on the lower part rests a spring f. which by means of an adjustant can be slid up or down Q. by the screw G. the current entering at 23. passes through the lamp thence to h. through the metal of Q. and if the engine is revolving passes to f. each time the metallic portion of Q. comes in contact with f. thence by 20. through the ~~wrought~~ magnet and to 23. while the engine is revolving the amount of metal on Q. being 5 times less than the insulating substance the lamp only receives ⅕ of the energy which it would receive ~~where~~ were[c] the surface of Q. wholly metal, and this amount of energy is wholly independent of the speed with which Q. revolves, this will cause the lamp to be brought only to say a dull red.

If now the spring f. be raised the amount of total contact will be increased and more energy will pass to the lamp, and by this means the spring f. may be raised until it is at a point where the surface of Q. is wholly of metal.

In practice, I arrange Q[d]. so that f. when down to its lowest limit rests wholly in insulation and an extra appliance or lever is arranged with the screw G. whereby an initial motion or start is given the shaft of the engine when the lamp is to be turned on. It is obvious that any form of magnetic motor may be used to give a constant rotation to Q. Even a clock work would serve to give sufficient motion it being wound up in the act of turning on the lamps, or a hot air motor deriving its heat from the lamp might be used, or a vibrating fork-lever, reed may be employed.—

My claims will probably hereafter be

First. The system of electric conductors substantially as herein described.

Second. The methods herein described for regulating the electromotive force, in an electric lighting system.

Third. In combination with an electric burner of a moving break wheel or analogous device of variable conducting surface, so that the circuit may be closed for a greater or lesser period at each revolution or movement by moving a contact spring over its perhiphery whereby the total energy passing in a circuit may be regulated substantially as specified.—

Fourth. The electric engine X. in combination with the revolving break wheel Q. for the purpose set forth.—

Signed by me this 14th day of August A.D. 1879.

Thos A Edison

by Witnesses; Frank McLaughlin   S. L. Griffin[5]

DS, MdNACP, RG-241, Edison Caveat 91. Written in an unknown hand. [a]Date from text; form altered. [b]"Fig. 1." interlined above. [c]Interlined above. [d]Obscured overwritten text.

1. This caveat is the first coherent statement of what later became known as the "feeder and main" system of electric distribution. Although Edison filed a patent application in February 1880 (Doc. 1890, U.S. Pat. 369,280) describing generally his planned central station and distribution system, he did not apply for patents on the feeder and main distribution system for nearly a year. The principle of tapered feeder lines from the central station supplying tapered local main conductors to equalize the electromotive force in a circuit was embodied in a patent for which Edison applied on 9 August 1880 for "equalizing the tension or 'pressure' of the current through an entire system of electric lighting . . . preventing what is ordinarily known as a 'drop' in those portions of the system the more remote from the central station" (U.S. Pat. 264,642).

Francis Upton seems to have begun sketching recognizable elements of the system and producing calculations and numerical tables related to it on 7 August. Some of the calculations that Upton made about this time regarding power transmission to the Comstock mines may also pertain to this system (see Doc. 1788 n. 3). Probably during the summer months, Upton also made a series of undated calculations, headed "Fall E.M.F." (that is, the fall of electromotive force or drop in pressure) in a circuit with a 100 ohm lamp operated by a generator having an internal resistance of ½ ohm connected via a conductor having 10 ohms and by a generator having an internal resistance of 1 ohm with a conductor also of 1 ohm. N-79-06-16.2:60–73, N-79-01-21:185–200, Lab. (*TAEM* 35: 397–403, 30:676–84; *TAED* N079:30–6, N016:94–102).

Edison specified that the conductors should be "composed of several single wires, of different lengths, one or two of which extend the whole length of the conductor, others ending at various points. . . . These wires are not insulated, but merely grouped in a bunch, which have transverse fastenings at intervals, or which may be fastened together by branch conductors passing around where connections are formed. . . . It is preferable to form all conductors which vary in size, decreasing from some point in this manner." These main circuits would then supply branch circuits for each house or business. He also suggested the use of grounding wires at intervals throughout the main circuits in order to reduce the size of the return conductor but this provision was deleted before the patent issued. U.S. Pat. 264,642.

In another patent application executed a few days earlier but also filed on 9 August 1880 (U.S. Pat. 239,147), Edison described a service district

*A drawing from Edison's
U.S. Patent 239,147,
showing the feeder and
main lines arranged to
provide uniform voltage
throughout a square service
district.*

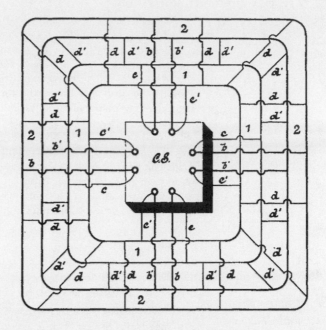

consisting of main wires laid concentrically around a generating station and supplied through feeder conductors from the station. Because the electrical "pressure" in the feeder lines "is apt to be greatest nearest the central station, and to diminish gradually toward the end of the conductors," he proposed to equalize the voltage throughout the system by linking the main wires together with subsidiary conductors so that each main was not wholly dependent on its associated feeder. Also, because the main lines grew successively longer, increasing the overall circuit resistance, he specified that their size should increase in proportion to their distance from the central station. This would maintain throughout the system a constant ratio between the resistance of the main conductors and that of the aggregate of all the lamps or other devices in each main circuit.

Edison refined this idea in a patent application filed in October 1880 (U.S. Pat. 264,645), in which he specified that the feeder conductors should "lead to and connect with the service-conductors, but upon opposite sides—that is, one conductor (say the one from the + pole) connects at some point to one service-conductor of a block, while the one from the − pole connects to the other service-conductor at a point exactly opposite to the other, so that the terminals of all house-circuits of a block, or the points where the house-circuits connect to the service conductors, all have the same mass of conductors between them and the central station." These three patents were substantially embodied in a British specification sealed in March 1881 (Brit. Pat. 3,880 [1880], Batchelor [*TAEM* 92:187; *TAED* MBP028]). See also Doc. 1943.

2. In Doc. 1890 and the February 1880 British provisional specification Edison called for the use of an electrodynamometer, galvanometer, or electrometer for this purpose. Brit. Pat. 602 (1880), Batchelor (*TAEM* 92:158; *TAED* MBP025).

3. Cf. Doc. 1890.

4. Cf. Doc. 1735 nn. 8 and 12. On 29 July, John Kruesi labeled sketches that appear to be by Edison of a regulator with a break wheel, essentially like that described above, for interrupting the circuit at variable frequencies and the next day he drew related components and what appear to be alternative break wheel mechanisms. On 31 July Charles Batchelor drew numerous other variations of this type of regulator. Vol. 16:392–93; N-79-07-31:282; N-79-07-25:5–15; Lab. (*TAEM* 4:827–28; 33:682; 862–67; *TAED* NV16:349–50, N052:140, N056:3–8).

5. The signatures on this document, including Edison's, were overwritten on initials which appear to have been written by Stockton Griffin and then erased so as to be barely visible. Edison signed over "TAE," McLaughlin over initials which may have been his or Griffin's, and Griffin over "C.D," probably machinist Charles Dean.

–1790–

Edward Johnson to Edison and Charles Batchelor

London Aug 18/79.

My Dear Edison—and Batch—

Allowing one week to pass—with[a] a Saturday & Sunday of non-use to intervene, I now write to tell you the result=as to the 6 you have last sent me—

They hold out well—not loud—nor brilliant like the old style when new and at their highest point of efficiency—but in[b] good solid Dog Trot. style. Always ready to go—and always capable of being understood by any chance passerby.

There is the 1st report I have been able to send you of satisfactory work—for the simple reason that these are the 1st Instruments I have had which would stand over from Saturday night till Monday morning and then "Speak when they were spoken to"=[1] I have managed to keep a pair of the old style on Deck through thick & thin for the purpose of showing to friends & Directors of the Company— Saturday night I left a pair of them in the best condition any of us ever heard them—clear loud and articulate EMG action perfect   This—Monday morning—they were as Dumb as the proverbial oyster[2] while their mediocre companions were there all right, mediocre still, but "ready for business"—

Up to this time I have simply sought to delay matters—or rather have <u>not</u> sought to Push them in order to give you time— Now I am ready to <u>Push</u>—and have so notified Gouraud and White. You were therefore Telegraphed today to Rush the 500—and the words "Model your own discretion" were added—because of the discussion between me & White as to the advantages of this over the old style.[3] He slowly takes to the new model for the simple reason that he is not aware of the obstacles in the way of the practical application of the old one—and I do not explain because we do not wish to cast a doubt upon the practicability of this—by confessing that its immediate predecessor was <u>not</u>— I told them I had written you giving the impressions produced upon the various parties in interest by the new model. He therefore feared you might hesitate to go in for 500 without further <u>Parley</u> I told them that these things could be safely left to you—as you had the amplest knowledge of what was requisite from your experience in applying the Telephone to the wants of the thousands of customers of the Western Union— Hence the words in the cablegram[c]

Now Batch writes me you are testing a new lot—Aug 4—& that they are "Bang up"—[4] of course every new lot will improve—as did the Carbon—so I am now only exercised as to how fast they can be had—

The Transmitters are decidedly better than the "Blake"— but they occasionally require readjustment[5] I think you can —if you have not already—surmount that— I think a longer spring would be an improvement as the merest touch of the screw in this—one way or the other will make it either too low—or Microphony[6]

Please advise me as to what you intend doing in re—to the Lathe & Tools If you are not going to send them I'll get the best I can here—[7] I must start a little shop of my own somewhere for repairs—a good many of the earlier ones were broken—but none of these new kind My Idea is to hire a good Instrument maker & keep him on repairs making me odds & ends tools &c

What have you done about the Five Hundred for which Gouraud cabled you the money?— Did you give the contract to Bergmann Dont forget that I have an interest in that little Dutchman which you promised to keep alive by throwing this into his lap—& keeping up his W.U. orders—[8]

I hear that Garrett Smith has gone over to the Craig outfit—isn't that a mistake? Don't they mean Jay Goulds Co—or have the two consolidated—[9]

I read an <u>interesting</u> article in the N.Y. Sun on your Electric Light I presume it will be copied over here— If it is I shall take the Liberty of going for it— I am informed its author is "G" B. Prescott—probably his Tool—Ashley?—[10]

You recall the Little squib about our Co in the journal of the Telegraph?— Col G has asked me to write a note to the Editor denying the soft impeachment wherein I am made to appear as your representative   of course I told him I should be happy to do so.[11]

Batch—

I'll present your Letter in a few days— am going down to Manchester with Col. G—& will take no little pleasure in going to see those famous "Spindles"—& shall also try and see the Old Folks—[12]

The Boys tempted me to go down to Brighton[13] on Saturday & stay till this am which I did taking wife & Edna[14] along— Had a splendid day. Like the town immense—but its no bathing place— The Beach is no where by the side of our famous places— In fact we miss the American "immensity" everywhere   As for the Girls—I haven't seen a pretty one yet.

Oh by the way Old Man (Edison)—I was down to the Silvertown works—went all through the Cable Dept—[15] They were running out a 600 mile cable— Its a Big Thing—that establishment— I saw just one thing there I should liked to have bought for you—if I get rich out of this Telephone I will— It was a 200 mile artificial cable gotten up thus

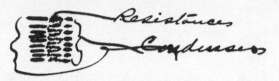

The whole thing in a neat mahogany[a] case about 2 feet wide & 6[a] feet high[d]   The condensers are not quite as large as this sheet of paper and about ½inch thick case & all— They slide in on little shelves—like tiny little drawers— It is so compact & ornamental that it would make a handsome addition to your Laboratory—

Speaking of cables—J E Brown[16] give me a call today— I was very busy and didn't chin him much—but I mean to see more of him and learn something of Automatic affairs— What about my Photographs— I learn that Gouraud has also written to you for some—but he don't expect to pay   I do— You send me a number & the bill & I'll pay it—then if he wants any I'll sell 'em to him.[17] I should like to have a thoroughly good one of Mrs Edison too—as frequent inquiry is made concerning her—& since you have such a presentable wife I should like to be able to point to her. I'll hang it with yours in the very nobby office they have provided for me— Which of's by the way I'm afraid wont see much of me—as it is not what I wanted that is one on[e] the Top Floor—and was the result of a special requisition   No mahogany furniture adorns it—

It's mostly furnished with White Pine Packing Boxes from Menlo Park and assorted Telephones—Tools etc— I made them buy me a Wheatstone Bridge & Galvanometer   I'm going to test all my wires &c—& keep a record of them as they do everywhere here now—frequent Bad joints troublesome— Charley Edison took the ones Adams left here—

See the scourging poor old Reid gets in the journal—on his Book= note also the connection of the name of Edison—[18] Regards to all   Truly Yours

E. H. Johnson   Engr

ALS, NjWOE, DF (*TAEM* 52:673; *TAED* D7941ZDN). [a]Obscured overwritten text. [b]Added later. [c]Followed by dividing mark [d]Sentence enclosed by brace to right of sketch. [e]Interlined above.

1. Johnson is referring to the phosphate buttons sent at the end of July (see Doc. 1782). He had reported that the old caustic soda chalks worked "splendidly" after several days of inactivity but the first phosphate buttons proved less satisfactory. On 5 August he wrote Edison, "This morning on my arrival at the office I find every Phosphate utterly dumb, & all alike decline to be resurrected no matter what the treatment. I take up an Old Caustic soda—and it goes off tip top at the very first touch." Johnson to TAE, 20 July and 5 Aug. 1879, DF (*TAEM* 52:625, 668; *TAED* D7941ZCY, D7941ZDI).

2. A slang expression for a person who speaks little. Wentworth and Flexner 1975, s.v. "Oyster."

3. The message as taken by Stockton Griffin reads "Rush 500 Telephones. Model your discretion." Edison replied the same day, "Are rushing   tell Johnson ten new model working at Gold and Stock." White also repeated this message in a letter that day, in which he expressed his desire to have "at least 300 or four hundred instruments leased out before the end of the year, as with heavy outgoings the Shareholders are naturally desirous to see a good result for their expenditure." Griffin's notes on White's letter for Edison's reply, sent on 1 September (see Doc. 1797 n. 3), indicate that Edison was having "the 500 telephones made in parts at dif shops—just getting them together   will make shipment next week & probably continue." John Kruesi had instructed the shop on 15 August to "Make 500 Telephones." White to TAE, both 18 Aug. 1879; TAE to White, 18 Aug. 1879; all DF (*TAEM* 52:671–72; *TAED* D7941ZDK, D7941ZDM, D7941ZDL); Cat. 1308: 171 (Order No. 229), Batchelor (*TAEM* 90:752; *TAED* MBN003:59).

4. This letter has not been found.

5. Johnson reported a few weeks earlier that one hundred Blake transmitters had arrived at the Bell company's office and that he had told Edison's backers to begin legal proceedings for patent infringement because the Blake "is Carbon pure & simple—and that without it we have nothing to fear from the Bell Tel Coy—nor with it for that matter." Arnold White subsequently made a formal request of the Bell company to stop using the imported Blake instruments. Johnson to TAE, 20 July 1879; White to Telephone Co., Ltd., 23 Aug. 1879; both DF (*TAEM* 52:625, 683; *TAED* D7941ZCY, D7941ZDQ).

6. Johnson probably meant distortions sometimes created by a microphone's amplification.

7. See Doc. 1797.

8. Johnson had apparently entered into a silent partnership with Sigmund Bergmann. On 3 June, he wrote Edison that he had "been making an effort to help Bergmann through with his big contracts but find it somewhat difficult to raise money. The Phono business too dull to advance any." When he submitted an order to Edison on 14 September for 2,500 telephones for Britain, Johnson remarked that he had "written to Bergmann today to go see you at once—& to take up this work independent of anything else if necessary—I will see that the money is forthcoming always in time   Under my agreement with him this work goes to him or my agreement lapses.— I have arranged with Painter to loan him any money he may need on my a/c—to enlarge his plant." At the end of September Johnson asked permission to contract for an additional three thousand telephones but Edison instructed him to "Send orders through me   Bergmann shall have all he can do." Johnson to TAE, 3 June, 14 and 30 Sept. 1879; TAE to Johnson, 30 Sept. 1879; all DF (*TAEM* 49:256; 52:712, 749–50; *TAED* D7903ZEN1, D7941ZEB1, D7941ZES, D7941ZET).

9. Daniel Craig (see *TAEB* 1:182 n. 1), former president of the Associated Press and a backer of Edison's automatic telegraph research, formed the American Rapid Telegraph Co. in February 1879 to compete with Western Union using automatic telegraph technology developed by Theodore Foote and Charles Randall. Gerritt Smith (see *TAEB* 2:295 n. 4), until this time a Western Union assistant electrician, became chief engineer of the new company. American Rapid was not associated with Jay Gould's telegraphic ventures. Harlow 1936, 415–18; Reid 1886, 778–81; Israel 1992, 146, 214 n. 66.

10. In a letter to the editor of the *Sun* signed "G." from New York, the writer stated that not only had Edison failed to solve the problem of subdividing the electric light, "he has not even approximated the solution, and he will not solve it, at the rate of his progress thus far, in a century." The letter specifically criticized Edison for making untenable claims about his generator:

> An inspection of the machine itself shows it to be merely a slight modification of the Siemens machine, the first one in the field, and in no respect its superior. When will the public cease to believe a thing just because Edison says it is so? Those who know him most intimately can't understand it.
>
> Other inventors besides Edison have been pursuing the phantom of indefinite subdivision, but with little better success than the "Professor of Duplicity and Quadruplicity." Claims have been advanced without number, but performance has not followed. ["The Electric Light. Has the Chevalier Edison Really Accomplished Anything?" *New York Sun*, 3 Aug. 1879, Cat. 1241, item 1270, Batchelor (*TAEM* 94:504; *TAED* MBSB21270X)]

Edison's former business partner James Ashley had coined the derisive "professor" nickname in the pages of the *Telegrapher*, which he edited, after his relationship with Edison soured. At this time Ashley was editor of Western Union's *Journal of the Telegraph* (see *TAEB* 2:305–6).

11. Johnson apparently used "soft impeachment," slang for an accusation, ironically. A brief notice about the incorporation of the London telephone company and Edison's receiver in the 1 August issue of the *Telegraphic Journal* declared that "Mr. E. H. Johnston, the able assistant of Mr. Edison, has come over from America as Edison's European representative." A correction in the 1 September issue stated that he was Edison's engineer and that "Colonel Gouraud is Edison's only representative in England." *OED*, s.v. "Impeachment," 4; "Edison's Telephone," *Teleg. J. and Elec. Rev.* 7 (1879): 251; "Notes," ibid., 7 (1879): 287.

12. Charles Batchelor (see *TAEB* 1:495 n. 9) grew up in Manchester and worked in the textile industry there, including a number of years with a thread-manufacturing firm, which is probably the subject of Johnson's reference to "Spindles." Johnson had asked Edison in a 20 July letter to "Tell Batch send me some Letters to his Manchester people— I'm going down there in a few weeks" (DF [*TAEM* 52:625; *TAED* D7941ZCY]).

13. A resort town on the English Channel, about fifty miles south of London. *WGD*, s.v. "Brighton."

14. Johnson's first child, at this time a young girl. *NCAB* 33:475.

15. Silvertown was the popular name for the India-rubber, Gutta-percha, and Telegraph Works Co., a major manufacturer of submarine cables located in Silvertown, on the Thames near London. Bright 1974, 157.

16. James Brown assisted Edison in his shop until 1873, when he went to Great Britain to help with automatic telegraph experiments. He then left Edison's employ to work at the submarine telegraph cable station at Aden. See *TAEB* 1:509 n. 2.

17. Edison had his photograph taken at Mora's Photographic Gallery in New York sometime in the spring. When that negative proved unsatisfactory, the studio asked him to come for another sitting, which he evidently did sometime after 8 May. Presumably in reference to this portrait, Johnson complained in his 20 July letter that Gouraud had taken his print and was trying to have it copied "by the 100.— It is certainly the best Photo. I ever saw—and one that I want to see circulated here— The Board inspected it—and passed eulogies both upon you, & it—as a Photo. They expressed such an eager desire to have one each—that I have determined to make a little cheap capital for myself by presenting them with one." He asked Edison to have fifteen made for this purpose and also "ascertain what Mora will produce them by the 100 for—I want to do a little speculating in them   also send me the bill." Gouraud soon reported that "everybody is falling in love with it. I have consequently had to promise no end of them. . . . I think I could manage a hundred of them with great facility." On 19 August Mora's studio wrote to Stockton Griffin confirming his order for 100 large portraits and promising to deliver 20 that day. The Mora portrait has not been identified. Johnson to TAE, 20 July 1879; Mora's Photographic Gallery to TAE, 6 May 1879; Mora's to Griffin, 19 Aug. 1879; Gouraud to TAE, 25 July 1879; all DF (*TAEM* 52:625; 49:222, 340, 304; *TAED* D7941ZCY, D7903ZDO, D7903ZGZ, D7903ZFX).

18. The *Telegraphic Journal*'s review of Reid 1879 in its 15 August issue severely criticized the book's literary deficiencies while conceding the usefulness of its "vast amount of historical and other information"

about the telegraph in America. The *Journal* noted that "Very finely-executed engravings of Morse, Orton, Elisha Gray, and other telegraphic celebrities, profusely illustrate the work, besides many more portraits of less famous electricians, including Edison, Mr. Pope, and Faraday, who is honoured with a vignette woodcut." "Review," *Teleg. J. and Elec. Rev.* 7 (1879): 268–69.

–1791–

Notebook Entry:
Electric Lighting

[Menlo Park,] Aug 18th 1879

Electric Light   Coating Wires for Lamp insulation[a]

In bringing up the small lime spools with 2 layers of .004 platina wire, we always find it hottest in the middle, owing to the conduction of heat away by the lime bobbin; I took one and made thin top and bottom so:—

and now, when brought up,[b] the heat seems to be much better distributed; at the low red even little difference can be seen[1a]
TAE   Chas Batchelor

X, NjWOE, Lab., N-79-07-31:57 (*TAEM* 33:570; *TAED* N052:28). Written by Charles Batchelor. [a]Followed by dividing mark. [b]Interlined above.

1. Charles Batchelor made a similar entry in his own notebook on this date. A few days later he tried making spools out of powdered quartz but found that the shims needed in the mould to press the silica into a ½ spool shaft, or "button," interfered with the operation of the press. He therefore proposed to compact the spools initially in the mould and then reduce them another ¼ inch with a die in which "the enormous pressure comes direct on the material and is not transmitted through a number of steel buttons, which have possibly untrue surfaces." Cat. 1304:68–69, Batchelor (*TAEM* 91:67–68; *TAED* MBN004:65–66); N-79-07-31:61–63, Lab. (*TAEM* 33:572–73; *TAED* N052:30–31).

New York, Aug. 20th 1879.[a]

From Lemuel Serrell

Dear Sir:

I have just received notice from the Patent Office, declaring several new interferences on your Telephone applications, Cases 141-158-178.[1]

I send enclosed a list of the parties to these cases, and the subject matter involved.[2]

The preliminary statements in each case, must be filed before September 30th 1879.[3] Yours truly

Lemuel W. Serrell pr Pinckney[4]

L, NjWOE, DF (*TAEM* 51:546; *TAED* D7929ZDV). Letterhead of Lemuel Serrell. [a]"New York," and "187" preprinted.

1. On 14 August the Patent Office declared a set of six separate patent interferences (quasi-judicial proceedings in the Patent Office) to sort out the competing applications of Edison and others for patents on various aspects of telephone technology. This was the second set of multiple interferences involving Edison's telephone applications, the first having been declared in March 1878 (see Doc. 1270, esp. nn. 1 and 3). The August 1879 proceedings were designated Interferences 1 through 6. Interference 1, concerning an application that Edison had filed substantially earlier (Case 141), was subsequently combined administratively with the 1878 interferences. Records pertaining to Interference 1 are in TI 1–2 (*TAEM* 11:3, 181; *TAED* TI1, TI2); portions of the records of 2 through 6, including the contested specifications, are in TI 6–7 (*TAEM* 11:919, 977; *TAED* TI6, TI7). An additional interference, designated B[3], was declared on 18 October 1879 and later combined with the others in this set. About the middle of August, the Patent Office also declared a separate interference (designated A[3]) involving an application that Edison filed on 6 August for another form of carbon transmitter (Case 183); this remained a separate proceeding (see Doc. 1804). Other interferences, one involving an application for a carbon transmitter (Case 161) and another for a receiver (Case 159), had been declared in January and February 1879, respectively. The former issued in December as U.S. Pat. 222,390; the latter interference was dissolved several months later (Lemuel Serrell to TAE, 20 Jan., 14 Feb. and 30 June 1879, all DF [*TAEM* 51:395, 411, 525; *TAED* D7929B, D7929R, D7929ZDE]).

2. Serrell's enclosure has not been found. Interference 1 was declared in respect to Edison's Case 141, filed 20 July 1877, and applications by Francis Blake, W. L. Voelker, and J. H. Irwin. Interference 2 involved Edison's Case 158, filed 11 November 1878, and applications of Blake and Charles Chinnock; George Phelps was originally a party but he conceded to Edison and was dropped in October 1879. The remaining proceedings concerned Edison's Case 178, filed 2 June 1879, and applications by one or more of Blake, Chinnock, and Edward Wilson. Each interference concerned arrangements for modulating the vibration of the diaphragm or the pressure of the carbon on the diaphragm. Case 141 was subsequently divided and issued in 1892 as U.S. Patents 474,231 and 474,232; Case 158 was also divided in October 1881 and issued as U.S. Patents 257,677 and 266,021. The disposition of Case 178 has not been

determined but it did not issue as a patent; a typescript of the application is in TI 6:ix (*TAEM* 11:985; *TAED* TI6:8).

3. A preliminary statement was a sworn affirmation stating when the inventor claimed to have accomplished the point in question and what kinds of evidence would be presented to support that assertion. This deadline was extended to 20 October 1879. Serrell to TAE, 29 Sept. 1879, DF (*TAEM* 51:559; *TAED* D7929ZEG).

4. George Pinckney was a clerk in Serrell's office.

–1793–

From George Barker

PHILADELPHIA, Aug. 22. 1879.[a]

My dear Edison:—

I was sorry to have to leave you so abruptly this morning but unless I came away at 8, I would not get here till 2 & that would not give me as much time as I needed to have.

I told you last night that you could have either Thursday or Friday evening for the Electro chemical Telephone.[1] But a letter from Prof. Marsh[2] which I found here, informs me that it has been decided to give his address as retiring President on that evening. As Saturday evening the members will be tired from their excursion to Albany, I have written to Chandler[3] to ask him if he can stay over so as to give his capital lecture on the "Chemistry & Geology of Saratoga Springs" on Monday evening. In that case we will have the conversazione on Friday evening. But if Dr. Chandler should have to give it up unless he could have Friday evening, how would it be about you on Monday evening? Of course I mean the following Monday Sept. 1st. Can you stay in Saratoga over Sunday?[4] It will do you good I know to have a little rest, especially as you are not very well now. I really hope you will bring Mrs. Edison with you. We will all do all that we can to make it pleasant for her. Please present my compliments to her and say to her that as President of the Association for the meeting, I give her a special invitation to be present.

I have written Putnam[5] and given him the titles of your three papers: the Tuning fork, the heating of the wire, and the Elec. chem. Telephone.[6] You will bring to the meeting I suppose:[b] 1st. One or two Resonant Tuning Forks (the hollow ones I mean) 2d. Two of your Electro-chemical Phones:[c] one the loud speaker with large diaphragm for illustration of its loudness and the other the precise arrangement we used last night as it is to go into practical use. This shows the inertia transmitter too. 3d. Those cylinders of the pure earths after heating[d] on cards in a frame, labeled, 4. Some of the cylinders before heating. 5th. Some of the Zirconia covered wire to look at under the microscope. 6th. Some platinum sands.[7] What day will you go up? Do not forget to telegraph to the United States Hotel for a room. I go up on Monday. Cordially yours

George F. Barker.

Hunt up some other good things & bring them too.[e]

ALS, NjWOE, DF (*TAEM* 49:849; *TAED* D7911D). Letterhead of the University of Pennsylvania. [a]"PHILADELPHIA." preprinted. [b]"I suppose" interlined above. [c]Obscured overwritten text. [d]"after heating" interlined above. [e]Sentence written in right margin of final page.

1. At the previous year's meeting of the American Association for the Advancement of Science in St. Louis, Edison had proposed to Barker that they hold a "conversazione" during the meeting in Saratoga, New York, to which he would contribute some "nice things." On 27 July, Barker reported that he had asked "various people who have devised or discovered new things in Science to bring them to the meeting. Prof. Marsh cordially approves the plan and will contribute as also will Profs. Draper, Chandler, Hastings & others." He inquired what Edison could bring besides the loud speaking telephone and whether he would "show a model of your new magneto or one of your electric lamps? I think I will take up your tasimeter, voltameter, and motograph model from here. . . . Will you arrange to have talk on the chemical phone from a distance? Will you read a paper on it? I mean give a description of it to the section as you did of the tasimeter at St Louis: You will see that I am exerting myself to have a good meeting, as I am President this year. I am depending on you to give me a good support and I will help you all I can." He also asked Edison to "enlist [Edwin] Fox in the work, have him write up the coming meeting and your connexion with it, and then come to Saratoga himself and make one of those excellent reports for the Herald which he knows so well how to do." Barker wrote again on 30 July that he would "be quite disappointed if you do not come up there and honor my Presidency with your presence, even if you can stay only a few days." Barker to TAE, 27 and 30 July 1879, DF (*TAEM* 49:306, 844; *TAED* D7903ZFZ, D7911B).

2. Othniel Marsh was the first professor of paleontology in the United States, at Yale University from 1866 to 1899, and the organizer of numerous expeditions that uncovered large fossil beds in the western United States. *ANB*, s.v. "Marsh, Othniel Charles."

3. Charles Chandler (see Doc. 1754, n. 4).

4. Edison traveled to Saratoga with Charles Batchelor and Francis Upton during the night of Tuesday, 26 August, arriving on Wednesday, and leaving on Saturday night, 30 August. On Saturday evening he shared the stage with Barker and Alexander Graham Bell at a public telephone lecture and demonstration. Barker described the development and principles of magneto transmitters and receivers and the carbon transmitter and then Edison, with Batchelor assisting, demonstrated and explained the electromotograph receiver (see Doc. 1797). Francis Upton to Elijah Upton, 24 and 31 Aug. 1879, Upton (*TAEM* 95:557, 560; *TAED* MU028, MU029); "Mr. Edison's Latest Invention," *New*

York Tribune, 1 Sept. 1879; "A New Telephone," *New York Herald,* 31 Aug. 1879; Cat. 1241, items 1321 and 1328, Batchelor (*TAEM* 94:510, 512; *TAED* MBSB21321X, MBSB21328X).

5. Frederic Putnam, a noted naturalist and archaeologist, was curator of the Peabody Museum at Harvard and permanent secretary of the AAAS. Putnam later became professor at Harvard and a curator at the American Museum of Natural History in New York; he is credited with helping to establish anthropology as an academic discipline. *ANB,* s.v. "Putnam, Frederic Ward"; AAAS annual meeting program, 6 June 1879, DF (*TAEM* 49:845; *TAED* D7911C).

6. Edison did not present a formal telephone paper. He did write Doc. 1796 on his high-temperature wire research, which Francis Upton read for him. He also wrote a paper on a type of tuning fork he devised in 1878 that did not require a resonator box to amplify its sound (see Doc. 1413), which Barker read and which was published as Edison 1880b (see Doc. 1810). This paper was also published in the *American Journal of Science* and solicited for the *American Journal of Otology,* although the latter publication has not been found ("On a Resonant Tuning Fork," *Am. J. Sci.* ser. 3, 18 [1879]: 395–96; Clarence Blake to TAE, 20 Sept. and 22 Oct. 1879, DF [*TAEM* 49:729, 737; *TAED* D7906ZBE, D7906ZBI]; a manuscript draft is in Miller [*TAEM* 86:71; *TAED* HM790088]).

At Barker's invitation Upton also presented his own paper on "Methods of Testing Faradic Machines," which included a discussion of Menlo Park generator tests and the development of Edison's dynamometer. Upton told his father that he planned to conduct experiments with Edison's dynamo on 25 August in connection with his paper but no records of these trials have been found. The *New York Tribune* reported that during his presentation Upton attributed an efficiency of 92% to Edison's machine and suggested using the known resistance of a carbon arc to study the laws applicable to long-distance power transmission. In the printed version of the paper, however, Upton did not broach this latter subject and confined himself largely to a discussion of dynamometers and his tests with the Gramme machine, which Edison had been obliged to return before completing the new dynamometer (see Doc. 1760 n. 3). He stated that "No tests are given of Mr. Edison's machines since he does not wish to put on record what he knows are experimental results, for he feels sure that he can better nearly all that he now has. . . . One result which a machine of Mr. Edison's has shown may be mentioned. At 700 revolutions in a minute with one-sixth of a horse power consumed on the magnet he has obtained 130 volts from an armature having a resistance of .65 ohm." Upton also used this occasion to present Edison's suggestion that the term "Faradic" machine be used to refer generally to generators and dynamos, in honor of Faraday. Francis Upton to Elijah Upton, 24 Aug. 1879, Upton (*TAEM* 95:557; *TAED* MU028); "Faradic Machines," *New York Tribune,* 3 Sept. 1879; Cat. 1241, item 1324, Batchelor (*TAEM* 94:511; *TAED* MBSB21324X); Upton 1880b, 183, 178–79.

7. The *New York Times* reported that after Edison gave his paper George Barker

gave a brief description of the various specimens which Dr. Edison had sent for examination and to illustrate his paper. Many of the metallic earths were shown, melted and utterly changed under the combined influences of high heat and extreme pressure. There were also very numerous specimens of platinum from mines on this continent. . . . Of various metals and substances that had been exposed to the high temperatures of the electric arc, several which were shown have been so greatly condensed that they would not now melt at temperatures enormously greater than those hitherto assigned as the melting-point. Dr. Barker believes that the incandescent electric light of Dr. Edison is successful, and that it will meet even the expectations of its sanguine inventor. This is the conclusion of Dr. Barker, after a recent visit to Menlo Park, and a careful examination of the apparatus there. ["Giving Metals New Properties. Some of the Wonderful Results of Prof. Edison's Experiments," *New York Times*, 3 Sept. 1879, Cat. 1241, item 1314, Batchelor (*TAEM* 94:506; *TAED* MBSB21314X)]

-1794-

From William Le Gendre

New York, 22 August 1879

Dr Sir,

I should be pleased to know whether you have the telephone for Japan completed.[1a]

I am anxious now to get it over as soon as possible & avoid unnecessary delay—[2] Very truly yours,

Wm. C. Le Gendre[3]

P.S. I wish you would have drawn up such preliminary form of agreement as you think will be necessary—in the premises— Le Gendre

ALS, NjWOE, DF (*TAEM* 52:189; *TAED* D7939L1). [a]Obscured overwritten text.

1. Edison had taken steps the previous summer toward making arrangements for another party to introduce telephones in Japan and China but no agreement was reached (see Doc. 1372). This letter is the first extant correspondence with William Le Gendre. When he did not receive a reply, Le Gendre wrote again on 30 August. Edison made notes on that letter for a response sent on 4 September (not found) that "one cause delay   made a large no found they were objectionable had to throw them away   mkg a new lot 500 expect to have them in 2 weeks. If you will put me in mind of it wld like have you come down see them in operatn   be able to furnish some." DF (*TAEM* 52:190; *TAED* D7939M).

2. Le Gendre wrote again later in the year about not having received a sample telephone which Edison had evidently promised. Edison's notes on this letter instructed Stockton Griffin to "write say we are not satisfied to let the new class of telephones go into the hands of inexperts, although we have 1000 working in England. Advise that the other form be sent to Japan & arrangements made with the W E Mfg Co." In May 1880, Edison assigned Le Gendre the right to sell telephones manufac-

tured by Western Electric in Japan, Korea, China, and Hong Kong for five years from January 1880. The following March, after the Oriental Telephone Co. had been formed to operate in some of those countries (see Doc. 2056), Gouraud learned of this contract and wrote Edison in alarm. It was probably in reply that Edison sent an undated cable stating "Le Gendre done absolutely nothing gives me no information although repeatedly requested. I consider contract void." In any case, the agreement providing for the establishment of the Oriental Telephone Co. specifically excluded Edison's patent rights in Japan and the Australian colonies. Le Gendre to TAE, n.d. 1879; TAE agreement with Le Gendre, 14 May 1880; Gouraud to TAE, 29 Mar. 1881; TAE to Gouraud, n.d. 1881; agreement with Oriental Telephone Co., Anglo-Indian Telephone Co., Alexander Graham Bell, and Samuel Insull, 25 Jan. 1881; all DF (*TAEM* 52:244; 55:709; 59:1009, 1088, 991; *TAED* D7939ZAX, D8046ZAH, D8150J, D8150ZBH, D8150B).

3. William Le Gendre indicated below his signature his association with Brown Brothers & Co., a Wall St. banking firm advertising "commercial and travelers' credit, available in all parts of the world." Nothing more is known of him. *Wilson's Business Directory of New York City* 1879, 37.

–1795–

Notebook Entry:
Electric Lighting

[Menlo Park,] Aug 22 1879

A spiral of .005 in. in diametre was was wound loosely [——][a] and sealed into a globe of glass. The wire which by which the current was conducted through the glass was made larger. The spiral was the placed in the bridge made as drawn on page 1[1]

The glass around the Pt. wire cracked ~~and~~ enough to spoil the vacuum[2]

3.14	1.5
1.5	1.5

$$3.14 \qquad 0.4969$$
$$1.5 \qquad 0.1761$$
$$1.5 \qquad 0.1761$$
$$44.3 \qquad \underline{1.6464}$$
$$313 \qquad 2.4955$$

Reddish yellow   A spiral in a low vaccuum having a resistance at reddish yellow consumes with current of 1.5 Webers 313 foot. lbs. required per minute

The experimant was continued ~~after~~   With the same no. of Webers the spiral rose to a bright yellow[b]

Around a resistance of 3.65 Ohm a deflection[3]

$^{198}\!/_{200}\times 5$

3.65 Ohm   .99 $\times$ 5 = 4.95[c]

$$\cancel{.99}$$

4.95	0.6946	33,000[5]	4.5185
comp[lement][4] 3.65	9.4377	297	$\underline{2.4733}$
$\underline{4.95}$	0.6946		2.0452
44.3	$\underline{1.6464}$		$\underline{.0334}$
	32.47373		2.0118 ~~111 per H.P.~~
			102 per H.P.[6]

$$\underline{.0334}$$
$$\cancel{297}322 \qquad 2.5067$$

4.95	0.6946
$\underline{3.65}$	$\underline{9.4377}$
	.1323
	$\underline{.0334}$
	.1657 ~~1.35 Webers~~ 1.46 Webers

322	2.5067
$\underline{44.3}$	$\underline{1.6464}$
	$|.8603$
	4301 2.7 cells of carbon[7]

~~3.65~~   4.95

Around 3.05 ~~Cold~~ black

Spiral ~~cold~~ black[d]   3.05 ohms

cold   4.95[e] ohms

Spiral 6 inches .007 in. 20% Pt.Ir

$$3.14$$
$$\underline{.007}$$
$$0.02198$$
$$\underline{6}$$
$$.13188$$

$^{13}/_{100}$ in surface

$100\,)\,13 = 87.6^{c}$ 8 lamps to an inch

Conducting wires .11 ohms[8b]

Resumé Experiment No. 1

A spiral of wire made of 20% Pt. Ir. alloy .007 in. [~~dia?~~]ᵃ diametre and six inches long was sealed into a glass bulb.

The resistance of the spiral

cold was	1.95 ohms
just red or black	$3.05 - .11 = 2.94$
reddish yellow	$3.14 - .11 = 3.01$
bright yellow	$3.65 - .11 = 3.54$
whitish yellow	$3.78 - .11 = 3.67^{b}$

~~It took~~ The current was passed through the tan[gent].
galv[anometer]. but it was found that the measurement were
not to be relied uponᵉ as the induction coil was too near the
galv

$198 \div 2 = .99^{c}$ $92 \times .13 = 11.96^{c}$

	5	
	4.95	0.6946
	44.3	1.6464
	4.95	0.6946
comp	3.54	9.4510
	1.08	.0334
	1.08	.0334
		2.5534 356

4.5185
2.5534
.9651

Around theᵉ arm of the bridge containing the large wire the
high resistance galv. was placed and the deflection from compared with that from five cells of standard Daniells.[9]

In the case measured $(^{198}/_{200} \times 5)^2$ $\times\!\!\!\!/\!\!\!\!44.3)^2\!\!\times\div$ $3.54 \times 44.3 =$
the no. of ft. lbs. of energy consumed 356 per minute or $^{1}/_{92}$ of
a horse power   that is 92 spirals having each .13 in. radiating
surface could be kept at a bright yellow by one horse power. or
11.96 inches of radiating surface

~~Where the~~ This result was obtained in a vacuum good
enough so that the phosphorescent light made its way into the
bulb from the [~~con?~~]ᵃ terminals of the guageingᶠ tube

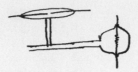

When the air was let into the spiral was barely red and after the glass had cooled it was black.[10]

X, NjWOE, Lab., N-79-08-22:3, 1, 5, 7, 6, 8–9, 11–13, 15 (*TAEM* 35:785, 784, 786–91; *TAED* No85:1–8). Written by Francis Upton; some decimal points added for clarity. [a]Canceled. [b]Followed by dividing mark. [c]Form of equation altered. [d]Interlined above. [e]Obscured overwritten text. [f]"e" interlined above.

1. These experiments and those following in the notebook from 24 August (see note 10) are efforts to determine the power required to raise a spiral to different intensities of light. The circuit drawing is unclear but appears to represent a test arrangement derived from that shown in Hopkinson 1879, fig. 3. The galvanometer is at the top left and below it is a Daniell cell. The component at the bottom center is probably a set of

Diagram of test circuits for measuring current used by John Hopkinson.

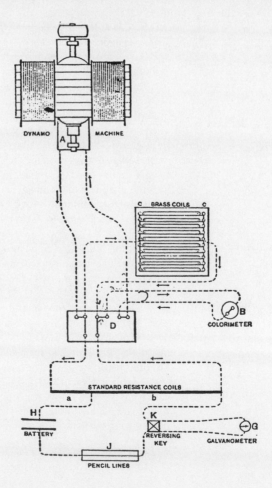

resistance coils; to its right is a three-terminal resistance box like the one shown in an undated drawing by Edison. The circles at the far right may represent batteries. The lamp spiral is not shown, nor is the Wheatstone bridge which was probably the one made by Elliott Brothers that Charles Clarke later recalled was often used "with a Thomson mirror galvanometer and standard Daniell cells for the measurement of voltage—the only means we had for this purpose until good direct-reading voltmeters were developed some years later." See also note 9 and Doc. 1815. N-78-12-28:154, Lab. (*TAEM* 30:60; *TAED* No11:60); Jehl 1937–41, 860–61.

2. In the following logarithmic calculation, 3.14 is the lamp resistance (R) in ohms; as indicated below, 1.5 is the current (C) in webers. The figure 44.3 represents the number of foot pounds exerted "from an ohm when a veber is circulating through it" (roughly equivalent to one watt, given by C²R). Upton 1880b, 181.

3. Rather than measuring the current directly as in the previous experiment, here Upton uses a galvanometer in a shunt circuit with Daniell cells to determine the difference in potential between the two ends of the main circuit. This is similar to the arrangement described in Hopkinson 1879 (pp. 242–45) and referred to in Upton 1880b (p. 181), in which the galvanometer circuit is balanced with the outside test circuit (see also Doc. 1772 n. 4). In a "Plan of tests" from about this time Upton explained the method for determining the current. The fraction is the ratio of the galvanometer deflection obtained to that obtained from the standard Daniell cells. Upton explained that he would multiply this "by 1.08 to reduce to Volts, divide by Ohms in shunt to find no. of Webers   multiply by the resistance in total circuit to obtain Volts. Then multiply the Volts by the Webers and by 44.3 to obtain the no of foot-lbs in circuit." In this particular logarithmic computation the method was somewhat different. He calculates the energy in the circuit, which varies as the square of current, times resistance (E = I² × R). He silently uses Ohm's law to substitute V_R for I; skipping the elementary manipulations he simply squares the voltage and divides by the resistance. He then multiplies to obtain foot-pounds, then corrects this by belatedly multiplying by 1.08. N-79-06-16.1:236–43, Lab. (*TAEM* 35:278–81; *TAED* No77:118–21).

4. Upton is using a conventional logarithmic shortcut designed to aggregate subtraction operations. He recasts the log of 3.65 (.5623) as the complement of 10; then after completing the addition he subtracts the 10.

5. The number of foot-pounds per horsepower.

6. The number of lamps per horsepower.

7. This appears to be a check on his calculation of 322 foot-pounds instead of 297. This also suggests that Upton was using carbon cells to heat the spiral.

8. In this statement and in the following table, Upton wrote "11" but clearly meant ".11"ohms.

9. In an undated note, titled "The use of a high resistance for measuring strong currents," Upton wrote that

a high resistance differential galvanometer of Elliot Bros has been at my disposal during the past year. With this were resistance boxes

containing 60,000 ohms and a box which contained resistances varying from one hundredth of an ohm to five thousand which were
used to make a shunt around the galvanometer. The size of the resistances 60,000 ohms made it possible to measure electromotive
force by placing various batteries of various strengths in the current
and taking the deflection to represent the E.M.F. of the cell by ne-
glecting its resistance. It also made the assumption true that if the
ends of the circuit containing the high resistance galvanometer were
placed at the extremities of a resistance where a strong current was
flowing not enough would escape to make any practical difference in
the flow. Thus the galvanometer readings would denote the difference of potential between the ends of the circuit containing it and
for all practical uses in measuring currents may be considered as an
electrometer. [N-79-02-15.1:199–201, Lab. (*TAEM* 31:756–57;
TAED N028:99–100)]

10. These experiments continued on 24 August with Spiral No. 2, "A
spiral wound very closely and kept in a vacuum over night without heating. The wire .007 in. in diam six inches long covered with Corundum."
Vacuum experiments with this spiral showed "the air escaping from the
Pt." After a full vacuum was obtained the laboratory staff tested the spiral and measured its resistance and the number of foot pounds required
to heat it to different degrees of brightness. They also tested the spiral
in the air. They also brought "a spiral size of grain wheat" to white heat
in five hours and it gave "what was estimated as eight candles by means
of a battery." This spiral was taken to the AAAS meeting in Saratoga,
presumably for use in connection with Edison's paper on his process for
treating platinum by heating it in a vacuum (see Doc. 1796). N-79-08-
22:17–27, Lab. (*TAEM* 35:792–97; *TAED* N085:9–14).

–1796–

*Draft Paper for the
American Association
for the Advancement
of Science*

[Saratoga, N.Y., c. August 28, 1879][1]

~~Some time ago~~ in the course of my experiments upon the
subject of E. Lighting I have developed some striking phenomena arising from the heating of metals especially of wires
of pure[a] Platinum and ~~on~~ alloys of platinum and Iridium. These[b]
experiments are still in progress.[2]

~~The Lamp placed in~~ My system of Electric Lighting is
~~based upon the heating of spirals of by of~~ consists of ~~several
feet of~~ The first phenomena observed was that ~~loss of weight
which~~ platina wires lost weight in a flame of pure Hydrogen,
that the metal colored the flame green and that these[b] [----][c]
two results continued until the whole of the platina [~~was?~~][c] disappeared. That 306 miligrams of ~~4 thousandths~~ platinum
wires 4 thousands of an inch in diameter[d] bunched together &
suspended in [~~an?~~][c] a hydrogen flame lost a fraction less than 1
m.g. per hour for several hours, ~~but no loss of weight occurred~~

when an equal weight amount[a] of wire was suspended over the flame from a[e] Bunson burner flame. If a straight peice of platinum wire is stretched between two supports and made to pass through a hydrogen flame, it will color it tinge the flame green, but if the temperature of the wire is raised above that of the flame[f] by passing a current through it the flame is colored deep green   to ascertain the loss of diminution in the[g] weight of a platinum wire when heated by the electric current, I stretched between two clamping posts a wire five thousandth of an inch diameter weighing 29666[h] m.g. This wire[i] after being brought to[j] incandescent for [-][c] 20[a] minutes by the electric current[k] its weight was 293 265[a] mg showing a loss of [-][c] one[a] m.g.— The same wire was then raised to incandescence for 20 more[a] minutes its[b] weight was[a] 262 m.g. shewing a further loss of weight of 3 mg or a total after 10 minutes additional its weight was 260 mg—in 20 minutes more 258 mg or a total loss of 8 mgrams in one hour and ten minutes. after being kept[b] incandescent for 40 minutes longer it loss [+?][c] 2[a] mg additional or a total loss of 7 8[a] m.g. in 1 hour & 50 minutes[3]   Afterwards a wire strip of platinum offering large surface to the air &[l] 343 mg[a] was kept [-----][c] moderately[m] incandescent by the current for 39 hours after which time[a] it weighed 310 m.g. 301 mg[n] shewing a loss of 33 42[a] m.g. A platinum[a] wire 20 thousandths in diameter was wound in[b] [----][c] the form of a spiral ⅛ of an inch in diameter & about ½ an inch in diameter length.[a] This was secured in clamping posts and covered with a glass shade 2½ inches in diameter & 3 inches high. upon passing the current through the spiral upon bringing the spiral to incandescence by the current for 20 minutes that part of the glass globe in line with the sides of the spiral become slightly darkened   in 45 hours the deposit of platinum was so thick that the incandescent spiral could not be seen through it. This deposit was most perfect and I have no doubt but large plates of glass might be coated economically by being placed on each side of a sheet of platinum kept incandescent by a Dynamo Electric machine.

The loss of weight in[a] together with theis deposit upon the glass became a very serious obstacle in the way of its use for producing the Electric light by the incandescence, but it was easily surmounted when the cause was ascertained. I coated a spiral wire with the wire[o] forming a the[a] spiral p by passing with the oxide of Magnesium by[p] dusting powdered acetate of magnesia on the wire while incandescent. The salt was decomposed by the heat and left a strongly adherent coating of the oxide of magnesia upon the wire. This spiral placed in the glass shade was brought to incandescence [---][c] ~~& allowed to remain in that st 20 minutes an opalescent~~ for some time but ~~no dep~~ instead of a deposit of platinum there was a deposit on the glass of the oxide of magnesia, proving that the [high?][c] effect was due entirely to the molecular[a] attrition ~~or the attrition of the rapidly circulating air in the glass receiver globe cover and produced by the rapid movement of gaseouses[q] molecule when in proximity to the highly incandescent was washing o~~ the[r] that the loss of weight & coloration of the hydrogen flame was also due to the wearing away of the surface of the platina by the attrition produced by the impact of the stream of gases upon its highly incandescent surface, and not to volatilization as I understand it, and I venture to say although I have not tried the experiment that sodium cannot be volatilized in high vacua[b] by the heat derived from incandescent platinum, and if[a] any effects are obtained they[s] will be produced by the action of the residual gases.[4] After being satisfied[b] that the deposit & loss of weight was due to the [air?][c] attrition of the air, I placed a spiral of platinum in a glass receiver of a common air pump & arranged ~~int~~ so that the current could pass through it while[b] the receiver was exhausted, at[b] an exhaustion of 2 millimeters. ~~it look~~ the spiral was kept incandescent for over 2 hours before the deposit was sufficiently thick to become noticable.[5]

by another experiment & at a higher exhaustion it[b] required 5 hours before a tint was observed. in a sealed bulb exhausted by a sprengel mercury pump to a point where a ¼ inch ~~induction coil~~ spark from an induction coil[t] will not pass between wires 2 mm apart I have not detected any deposit whatever although the spiral contained therein had been kept incandescent for many hours, ~~and I feel certain that I shall be able to produce~~. I will now describe the most important phenomena which I have observed. If a piece of platinum wire 1 thousandth of an inch in diameter[u] be held in the flame of a bunsen burner at[b] some point it turns a sharp angle, and under the microscope shews a globule formed at the point where it is bent. in some cases the effect of the heating is to throw the wire into a zig zag shape & melting in many places. with a wire 4[b] thousandths of an inch in diameter this effect does not take place as it temperature cannot be raised to equal that of the small wire, but if th~~is~~e wire be examined under a powerful[b] microscope [its?][c] that part of its surface which has been incandescent will be found ~~to be~~ covered with innumerable cracks. If a peice of wire be streatched between clamping posts and is brought to incandescence for 20 minutes by the passage of the electric current, ~~This effect will be greatly magnified and the wire will present a~~ these cracks may be seen with the naked eye. The wire under the microscope will present the same appearance as a stick of moistened pipe clay after drying. If the current be continued for several hours[v]—the whole wire[b] will become so cracked & shrunken that it will fall to peices. This[b] disintregation[b] has been noticed by Prof.[a] John W Draper.[6] ~~and~~ It[a] was also[a] the cause of the failure of the process of lighting[w] devised by the eminent French chemist Tessie Du Motay[7] who ~~caused~~ raised platinum to incandescence by the combustion of[x] hydrogen on its surface. ~~Now I have discovered that~~ I have discovered the cause of this phenomena ~~an in eliminating this and have succeeded in bring platinum wire~~ and have suceeded in eliminating that which produces it, and in doing so have produced a metal in[b] a state hitherto unknown, a metal which ~~stands~~ is absolutely stable at ~~which~~ temperature where nearly[y] all substances melt or are disintegrated, a metal which is a homogenous as glass ~~and nearly as brittle~~, as hard as steel wire, in[b] the form of a spiral is as springy and elastic when[b] dazzling incandescent as when cold, and which cannot be annealed, by any process generally[a] known ~~to me~~[z] The ~~method~~ cause of the cracking & shrinking of the wire is ~~entirely~~ due entirely[a] to the sudden expansion of the air in the mechanical & physical pores of the platinum wire and the subsequent contraction[b] of the metal when the air has escaped. Platinum ~~wire~~ as delivered from the manufac~~tori~~ures may be compared to a piece of[aa] sandstone, it being made up of innumerable particles seperated at places by air spaces. The sandstone on being melted becomes homogenous & no air spaces exists while with the platinum the air spaces may be eliminated and the wires made homogenous by a simple process. ~~If a~~ [--][c] This process I will now describe[bb] I made a large number of platinum spirals[8] all of the same size a presenting to the air a radiating surface of $\frac{3}{16}$ of an inch. 5 of these spirals ~~were brought to incandescence by the electric current, and~~ had[a] their temperatures slowly increased by the current[cc] until thes melted, the ~~lights~~ which they gave off being measured by a photometer[dd] up to the melting point— This varied from 3 to 4 candles, the melting[b] point of the metal being determined by the ~~depth of the large crack in the wire wire~~[a] ~~due to the~~ [-][c] flaws produced ~~by~~ by sudden heating[ee] one of the same kind of spirals was then placed in the[b] receiver of an air pump and the air exhausted to two mm. a very weak current was then passed through the wire to slightly warm it. For the purpose of assisting the passage of the air from in from[a] the pores of the metal into the vacuum, the temperature of the[ff] wire was then gradually augmented at intervals of 10 minutes[9] until it became red. upon [coat?][c] The object of slowly increasing the temperature so as to [reduce?][c] was to allow the air to pass out of the metal gradually without disrupting it—[gg] afterwards the current was increased at intervals of fifteen minutes   each time before increasing the current the wire was allowed to cool   this expansion & contraction of the wire caused it [to melt together at the points previously containing air. In one hour and forty minutes this spiral had reached such a temperature without melting that it was giving a light of twenty-five standard candles, whereas it would undoubtedly have melted before it gave a light of five candles had it not been put through the above process. Several more spirals were afterwards tried with the same result. One spiral, which had been brought to these high temperatures more slowly, gave a light equal to thirty standard candles. In the open air this spiral gave nearly the same light, although it required more current to keep it at the same temperature.][hh]

[Upon examination of these spirals which had passed through the vacuum process, by the aid of a microscope, no cracks were visible, the wire had become as white as silver and had a polish which could not be given it by any other means. The wire had a less diameter than before treatment, and it was exceedingly difficult to melt][hh] in the oxyhydrogen blow pipe[10] as compared to untreated platinum, that it was as hard as pianoforte the steel wire used by pianoforte makers and could not be annealed at any temperature,[11] and my experiments with many metals treated by the process has proved to my satisfaction and I have no hesitation in stating that what is known as annealing of a wire by heating to make it soft and pliable is nothing but cracking of the metal as in every case when a hard wire has been annealed a powerful microscope shows myrids of cracks. Since the experiments above, Since these experiments I have by the use of sprengel pumps produced higher[b] exhaustions, and have succeeded by consuming 5 hours in excluding the air from the wire[b] & in intermitting the current[ii] by obtaining a light of eight candles from radiating surface of 3 hundreds of an inch, which is equal to 264 candles per inch of radiating surface,[12] divided upon 33 parts   The utmost average [amount?][c] quantity of[ji] light obtainable from a surface of $3/100$ of an inch without passing through this process is 1 standard candle.

Thus I am enabled ~~to~~ by the increased capacity of platinum to withstand high temperatures to employ small radiating surfaces, and consequently require less energy ~~to give th~~ per candle light than if I were compelled to use a radiating[b] surface several times greater in consequence of lower melting points, and I ~~can~~ am thus enabled to[kk] obtain eight seperate[a] lights each giving a light of sixteen candles by the expenditure of 33 000 ft lbs of energy[ll] notwithstanding[mm] ~~with~~ all the losses which occur in the passage from the engine to the ~~belt~~ lamp, ~~but all that is required and and as to the impossibility of subdividing the Electric light with economy as has been asserted by some scientific men~~ a result which is quite sufficient to make the Electric Light a commercial success.[13nn] As a matter of curiosity I have made spirals of other metals and ~~passed them~~ and excluded the air from them in the manner described. Common iron wire may be made to give a light greater than that obtainable from platinum not treated ~~its~~ becomes as hard as steel[14oo] Nickel is far more refractory than Iron. Steel wires used in piano becomes decarbonized remains hard[pp] and ~~very~~ as[a] white as silver.[qq] Aluminum melts only at a white heat. Magnesium melts only at a bright red   The color of the arc at the moment of melting an iron wire in high vacua is deep violet.[15] ~~Palladium requires a longer t~~ In conclusion it may be interesting to state that the melting points of many oxides is dependent upon the manner of applying the heat. For instance, pure ~~Z~~oxide of Zirconium does not fuse in the flame of the[rr] oxyhydrogen blow pipe, while it melts like wax on an incandescent platinum spiral, ~~at a~~ which is at a[ss] far lower temperature, on the other hand ~~alumina~~ the oxide of aluminum [--][c] easily melts in the flame of the oxyhydrogen blow pipe, while it only becomes vitrous at the melting point of platinum.[16]

ADf, NjWOE, Miller (*TAEM* 86:72; *TAED* HM790088a). [a]Interlined above. [b]Obscured overwritten text. [c]Canceled. [d]"4 . . . diameter" interlined above. [e]"~~the flame from a~~" interlined above. [f]"above that of the flame" interlined above. [g]"diminution in the" interlined above. [h]"66" interlined above. [i]"This wire" interlined above. [j]"brought to" interlined above. [k]"by the electric current" interlined above. [l]"offering . . . air &" interlined above. [m]"[-----] moderately" interlined above. [n]"301 mg" interlined above. [o]"the wire" interlined above. [p]"with the oxide of Magnesium by" interlined above. [q]"gases" overwrites "gaseous." [r]"~~produced by . . . washing o~~ the" interlined above; "~~washing o~~ the" interlined separately. [s]"are obtained they" interlined above. [t]"from an induction coil" interlined above. [u]"in diameter" interlined above. [v]"for several hours—" interlined above. [w]"of lighting" interlined above. [x]"the combustion of" interlined above. [y]"temperature where nearly" interlined above. [z]"by any . . . me" interlined above.

any . . . ~~me~~" interlined above. [aa]"piece of" interlined above. [bb]"This process . . . describe" interlined above. [cc]"by the current" interlined above. [dd]"by a photometer" interlined above. [ee]"flaws . . . heating" interlined above. [ff]"The temperature of the" interlined above. [gg]"The object . . . disrupting it—" interlined above; "was to allow" interlined above that. [hh]Page missing, text taken from published paper. [ii]"& in intermitting the current" interlined above. [jj]"average [~~amount?~~] quantity of" interlined above. [kk]"am thus enabled to" interlined above. [ll]"by the . . . energy" interlined above. [mm]Interlined below. [nn]"a result . . . success." interlined above. [oo]"its . . . steel" interlined above. [pp]"remains hard" interlined above. [qq]"as silver"interlined above. [rr]"flame of the" interlined above. [ss]"which is at a" interlined above.

1. In Doc. 1810 George Barker indicated that Edison wrote this paper while in Saratoga. Edison arrived there on 27 August and left on the evening of 30 August, leaving Francis Upton to "read the paper from the writer's exquisite manuscript" on 2 September ("Mr. Edison's Discoveries," *New York Tribune*, 3 Sept. 1879, Cat. 1241, item 1322, Batchelor [*TAEM* 94:511; *TAED* MBSB21322X]). This draft is essentially the same as Edison 1880a with some changes in wording and different paragraphing; text from the published version has been used to replace the text from the missing page nine of the draft. The paper was reprinted within days by three New York newspapers. One of these, the *New York Times*, also reported George Barker's enthusiastic remarks on Edison's electric light and the paper itself, which he praised "as one of the highest importance" ("Giving Metals New Properties. Some of the Wonderful Results of Prof. Edison's Experiments," *New York Times*, 3 Sept. 1879; "Mr. Edison's Experiments," *New York Sun*, 3 Sept. 1879; "Mr. Edison's Discoveries," *New York Tribune*, 3 Sept. 1879; Cat. 1241, items 1314, 1318, and 1322, Batchelor [*TAEM* 94:506, 509, 511; *TAED* MBSB21314X, MBSB21318, MBSB21322X]). The paper was also reprinted in the technical press ("On the Phenomena of heating Metals in Vacuo by means of an Electric Current," *Sci. Am. Supp.* 194:3098; "The Action of Heat in Vacuo on Metals," *Chemical News* 40:152–54; "The Action of Heat in Vacuo on Metals," *Nature* 20:545–46; "Heating Metals in Vacuo by the Electric Current," *Teleg. J. and Elec. Rev.* 7:320–21).

2. On the wire experiments described in the first part of this document see Docs. 1665, 1666, 1669, 1670, 1672, 1675, and 1678; see also Docs. 1676 and 1735.

3. Edison apparently had with him the notebook containing the notes of experiments he had conducted during the last two weeks of January as the precise weights and times given here are found in that book (Doc. 1665). The final paper has some differences. The first twenty minute loss of 1 mg. is incorrectly omitted but the total loss is correctly summed as 8 milligrams after one hour and ten minutes.

4. This statement regarding sodium is not contained in the final paper.

5. See Docs. 1679 and 1735.

6. The work in which John Draper's observation appears has not been identified.

7. Nothing is known of this particular apparatus. Cyprien Tessié du Motay was a French industrial chemist noted for inventing a system of producing pure oxygen with which he planned to modify carbureted gas

systems in Paris and New York. More recently he had experimented with incandescent lighting and, with a partner, filed for a U.S. patent in January 1879 for an electric lamp employing a "pencil formed of earthy material in combination with a metal or metallic oxide." The Patent Office declared an interference with Edison's Case 164 and rejected a related application from Edison (Case 172). The interference was decided in Edison's favor in June; Case 164 issued as U.S. Patent 218,866 and Edison subsequently abandoned the other application. "Death of a Noted Chemist," *New York Times*, 7 June 1880, 2; *Gde. Ency.*, s.v. "Oxygène"; *Ency. Brit.*, s.v. "Gas and Gas-Lighting"; Lemuel Serrell to TAE, 7 Feb., 14 Mar., 9 and 15 Apr., and 14 June 1879, DF (*TAEM* 51:407, 435, 462, 470, 507; *TAED* D7929M, D7929ZAL, D7929ZBF, D7929ZBL, D7929ZCT); TAE to Serrell, 16 Apr. 1879, Lbk. 4:283 (*TAEM* 80:74; *TAED* LB004283A); see also Doc. 1802 n. 1.

8. In the final paper Edison states that these were platinum-iridium spirals of the same quality.

9. In the final paper Edison states that the intervals were fifteen minutes.

10. See Doc. 1670 n. 3.

11. In the final version Edison noted that "it was also scarcely attacked by boiling aqua regia."

12. In the final paper Edison states that the spiral of wire had a "total radiating surface of one-thirty second of an inch, or a surface about equal to a grain of buckwheat." To the side of this statement he wrote the calculation $33 \times 8 = 264$.

13. In the final paper the discussion of economy ends after Edison states the figure of 30,000 foot pounds, which he notes is "less than one horse power"; the rest of the sentence is omitted.

14. In the final paper Edison also noted that common wire became as elastic as steel.

15. The statements about magnesium and the color of the arc when melting iron wire in vacua are omitted from the final paper.

16. See Doc. 1787.

–1797–

Charles Batchelor to
Edward Johnson

Menlo Park NJ Sept 1st 1879

My dear Johnson,

We have just got your gratifying information that you have "tumbled" to the beauty of the 'new design.'[1] Also received yours of Aug 9th and short one, day before.[2]

We are now making thirty telephones, which we shall send you next week direct to London, and we shall endeavour to give you thirty per week thereafter; the parts are being made by different parties, some by Bergman and some down East; but the whole thing[a] is put together by us, so as to make sure of your getting them all alike.[3]

We shall send you the chalk cylinders all pure, without any phosphate of Soda in at all, and you must wet them with a solution of one thimblefull of a saturated solution of phosphate of Soda to four fluid oz of water. When applying to the chalks, about four camels-hair-brushfulls ought to be sufficient for each; this may not appear loud enough at first, but they will come up loud in two or three days.—

The action of the chalk as we understand it is as follows:—[4] the water evaporates from the surface, and that from the inside works outwards to replace it; thus always making a more concentrated solution at the surface. In time all the phosphate will be on the surface in the shape of dry crystals when it will no longer work; as a cylinder of pure dry phosphate of Soda will not talk at all. Now this action must be prolonged as much as possible, and therefore it wants, "the smallest possible amount of phos. of Soda with the greatest amount of water," & wet the chalk as much as it will stand, and talk good.

As there is no place for the water to go except by evaporation through the small hole where the shaft runs out, I cannot conceive but that it must take a very long time (if properly wetted) before these cylinders play out.

You mention that you have to adjust the transmitter occasionally, this should not be, and I presume it is due to your using a carbon battery which gives a spark and heats the carbon when adjusted too far out. Would it not be far preferable to use Leclanche? We always use Callaud and consequently dont have that trouble.[5] We have had a pair of telephones between Edison's house and the office for last 3 weeks, and Edison himself acknowledges they are just as loud now, as when put up although we have never touched them at all.

Edison thinks it much better for you to get what tools you want there instead of sending them from here and I endorse that, for there are many makers of good amateurs apparatus there, and they can get just what you need.

Edison and I have just returned from Saratoga where the meeting of the A.A.A.S. has been this year. He spoke to 2000 people on Saturday night & exhibited the telephone. This is the first time I believe he has taken the stage and he did it excellently. I send you Tribune with account of it.[6] The telephones in Gold & Stock are all in excellent trim except the one in little Phelps shop,[7] which Mart[8] always has to adjust, because "it never worked after he left before," this of course is what we expected there; but reports of them are very good With kind remembrances from all   Believe me   Yours truly
                                                                            Chas Batchelor

ALS, NjWOE, Lbk. 5:123 (*TAEM* 80:140; *TAED* LB005123). Circled "C" written at top of page. ªObscured overwritten text.

1. Doc. 1790.

2. Neither letter has been found. Johnson wrote Edison on 14 September, probably just before this letter reached him, complaining that he had received "Not a word from you or Batch for 2 weeks. Which implies that my letters are read and thrown aside—one to await the other, in the expectation doubtless that I will extricate myself from the difficulties I report without any assistance from you. . . . I have to report in this letter that which might have been done 2 or 3 weeks earlier If you had written me the full details of your experiments with both the chalk and the New Transmitter so that I might have saved the time necessary to go over the same ground you have— If you do not pay this letter any more respect than former ones—I shall recommend the Co to employ a shorthand man for you." DF (*TAEM* 52:712; *TAED* D7941ZEB1).

3. In a reply sent this day to Arnold White's 18 August inquiry about the English telephones, Edison stated that "Owing to the fact that all Telephone manufacturers are driven to fill orders for the W.U. Telegraph Co. I am having the pieces made in different shops, and will put them together here. I will make the first shipment of the 500 lot this week and other shipments as soon thereafter as possible." The colloquial term "down east" refers to coastal New England (*OED*, s.v. "down" [adv. 30]) but there is no evidence that Edison contracted this work to any firm in that region. William Carman's account entries for the new receiver orders show small payments throughout the summer to several firms, most in New York or Newark. Separate entries indicate much larger payments in the fall for the receiver to Bergmann's shop and Claudius Decreux, a New York screw manufacturer. In November Edison also paid approximately $300 for an unspecified purpose to Partrick & Carter, a Philadelphia firm which had previously done manufacturing for him. White to TAE, 18 Aug. 1879, DF (*TAEM* 52:671; *TAED* D7941ZDK); TAE to White, 1 Sept. 1879, Lbk. 5:127 (*TAEM* 80:142; *TAED* LB005127); Ledger #3:411–14, 141, 526, Accts. (*TAEM* 87:170–72, 70, 204; *TAED* AB001:164–6, 64, 198); *Wilson's Business Directory of New York City* 1879, 637.

4. Cf. Doc. 1738.

5. The carbon or Bunsen cell was a modified Grove battery which produced about 1.9 volts, compared with the 1.5 and 1.1 volt output of, respectively, the Leclanche cell and the Callaud, a form of Daniell battery. Because of its low internal resistance the Grove also produced a heavier current than most batteries (*TAEB* 1:599 n. 3, 98 n. 5; 3:94 n. 6). In a letter to Edison on 19 September Johnson blamed the transmitter's "Small Carbon points," which he said "will not stand either a sufficient tension of Batty power—or sound impact to raise the electrical tension to a point of sufficient effectiveness to overcome the obstacles of Induction & long wires" (DF [*TAEM* 52:728; *TAED* D7941ZEC]).

6. The *New York Daily Tribune* published a lengthy account of Edison's telephone demonstration with George Barker and Alexander Graham Bell at the Saratoga town hall which was "crowded with people, who were all greatly interested and amused." The article concluded with the observations that

Mr. Edison's explanation pleased the people greatly. His quaint and homely manner, his unpolished but clear language, his odd but pithy expressions stirred and attracted them. Mr. Edison is certainly not graceful or eloquent. He shuffled about the platform in an ungainly way, and his stooping, swinging figure was lacking in dignity. But his eyes were wonderfully expressive, his face frank and cordial, and his frequent smile hearty and irresistible. If his sentences were not rounded they went to the point, and the assembly dispersed with great satisfaction at having seen and heard the renowned inventor, and having seen and heard his most recent invention. ["Mr. Edison's Latest Invention," *New York Daily Tribune,* 1 Sept. 1879, Cat. 1241, item 1321, Batchelor (*TAEM* 94:510; *TAED* MBSB21321X)]

7. George Phelps, Jr., was the superintendent of Western Electric's New York factory. Phelps, Jr. to TAE, 27 Sept. 1879, DF (*TAEM* 52:116; *TAED* D7937ZCW).

8. Probably Martin Force.

<table>
<tr><td>–1798–</td><td>Paris 2 Sept. 1879</td></tr>
</table>

From Constant Rousseau

Dear Sir,

As legal adviser of the firm Alexis Godillot[1] I have had the honor of being consulted from the commencement of the business with the Telephone which bears your name, & have been instructed to draw up the Articles of Association of which you are aware after the necessary preliminary negociations had taken place & in which I naturally had constantly to take part.

More particularly I was and still am the Adviser of Messieurs Brancy,[2] Chatard,[3] & Berton,[4] and although I cannot say that I am the private Adviser of Dr Evans[5] I have attended at his place numerous Conferences with the above gentlemen & Mr Bailey as to the course to be pursued by the Society in view of the future position [on?][a] which the Society is justly entitled to count in France.

I am still the Adviser of the present Society & take part at almost all the Conferences having for object the transformation of the Society into one with a Capital of 3,500,000 francs.

On account of my so taking part in the Conferences the Members of the Society, especially Messrs. Bailey, Berthon, & Chatard have asked me to inform you of certain incidents which if they are not insurmountable ought nevertheless to be brought to your knowledge for the purpose of obtaining your aid in smoothing them.

I have already had the honor of conferring as to these facts here with Mr Walker during his stay in Paris and he was also of opinion that you ought to be informed of what is going on.[6]

The following are the incidents in question;

It appears that Messrs Alexis Godillot & Brancy have sold out their shares in the Society to a Mr Pellorce;[7] down to the present time however no regular document has been notified to the Society and therefore we must still consider Messrs Alexis Godillot & Brancy as being still parties in it.

Though legal adviser of Messrs Alexis Godillot & Brancy I have not been acting for them in this negotiation between them and Mr Pellorce; but for Mr Berthon and incidentally for Mr Chatard, I have called upon this Mr Pellorce.

This Mr Pellorce who is a financial agent wished to buy at par the shares of these gentlemen but was unwilling to state what end he was aiming at—

I therefore broke off all negociations, not being willing to allow my clients to treat blindfolded & seeing that they might afterwards find that they had been sold.

It has since been ascertained that this Mr Pellorce is nothing more or less than the financial Agent of the Gower Telephone Co of which Mr Rosevelt is one of the partners— Mr Rosevelt is the friend of Mr Puskas & Mr Puskas is in continual relation with these parties; these relations have induced Mr Puskas to pursue a course which is to say the least open to suspicion; for instance Mr Puskas urged Mr Berthon to sell his part to Mr Pellorce and offered even a premium to Mr Berthon if he would sell his part. The Gower Telephone, according to Mr Puskas, is very superior, the Coy formed to work it will eclipse the Edison Telephone Coy &c &c! From the short facts above, you will understand how the dealings of the other partners, Messrs Berthon, Bailey, Chatard, & Evans are necessarily limited to precautions and envelloped in doubts.[8]

Now is Mr Puskas a partner in the sense in which he ought to be as with the other partners, or has[b] he on the contrary become the tool of Messrs Rosevelt & Pellorce, & is he not more interested in the Gower Coy than the Edison?

However this may be it is certain that at the present time when attempts are being made to effect the transformation of the Company, it is necessary to provide against any event which may be hatched up by Mr Puskas. In the state of uncertainty in which he places us, is it not well to have in hand another Power of Attorney from you to Mr Bailey in order that if at the time any obstacle were placed in the way by Mr Puskas, the same might be removed and the transformation of the Society proceeded with without regard to the previous power given by you to Mr Puskas. I need not add that the pecuniary interests of Mr Puskas would not be affected. I enclose the form of Power necessary to be prepared against any obstacle as above,[9] & Mr Bailey will only make use of this Power in case of necessity, & he would not use it if Mr Puskas acted in accord with all the other Members of the Society— I am, dear Sir, Your very obt Servant

Rousseau[10]

LS, NjWOE, DF (*TAEM* 52:322; *TAED* D7940ZBB). [a]Paper damaged. [b]Obscured overwritten text.

1. Georges Alexis Godillot was a Paris merchant and an original partner with Theodore Puskas and Joshua Bailey in the commercial development of telephones in France. See *TAEB* 4:515 n. 1, where he was incompletely identified as Georges Godillot.

2. Alphonse Brancy was a partner in the Société du Téléphone Edison; he was identified in the partnership agreement as a "propriétaire" and a member of the Legion of Honor. See *TAEB* 4:515 n. 1.

3. Alfred Chatard was a civil engineer and Société partner. See *TAEB* 4:515 n. 1.

4. Alfred Berthon, a civil engineer, was a partner in the Société and its chief engineer. See *TAEB* 4:515 n. 1; letterhead, Robert Brown to Samuel White, 23 Oct. 1879, SSW (*TAED* X078AY).

5. Thomas Evans was a Paris dentist and partner in the Société. See *ANB*, s.v. "Evans, Thomas Wiltberger" and *TAEB* 4:515 n. 1.

6. George Walker wrote to Samuel White from Paris on 30 August after having spent a day with Joshua Bailey and conversing also with Rousseau, whom he described as "a hardheaded shrewd little lawyer." Walker's description of events is more concise but essentially the same as that given by Rousseau. He also told White that he agreed with Rousseau's recommendation that a power of attorney to Bailey from Edison was "absolutely necessary to protect your & Edison's interests." A copy of this letter which White sent to Edison on 12 September is in DF (*TAEM* 52:319; *TAED* D7940ZBA); the original is in SSW.

Edison already had some knowledge of the troubles in Paris. The first intimation was a cable from Bailey on 1 August advising: "Dont cable nor write anybody about French or European telephone before receiving letters mailed today." No letters from him on that date have been found but on 4 August he wrote that "three or four days more will unravel an intrigue that has been going on, and of wh. I dont like to write till I am sure of all the facts." Then on 11 August Enos Barton excerpted for Edison a letter from George Beetle reporting that "the Herz combination, who control a half interest in the Gower telephone for Europe, have lately bought a two-thirds interest in the Edison Telephone Co. for France" and were "now pushing telephone district matters in Paris, and have an authorization from the government to erect overhead as well as underground lines for this purpose." Bailey to TAE, 1 and 4 Aug. 1879; Barton to TAE, 11 Aug. 1879; all DF (*TAEM* 52:312, 314, 316; *TAED* D7940ZAV, D7940ZAX, D7940ZAY).

7. Nothing more is known of Pellorce than the information in this document.

8. Bailey sent Edison his first full report of events concerning the French company in a long and detailed letter of 5 September. According to his account, matters came to a head after the Société received authorization to begin building telephone lines in late August. Puskas and Pellorce refused to contribute toward construction costs or to the surety bond required by the government, and Bailey offered no hope that they would underwrite any of the working capital needed to establish the Société as a public share company. Bailey criticized at length Puskas's "disregard . . . of all obligations of honour & contract" but promised that "Against his plotting & obstructions & violence we are now sure of a brilliant result." He outlined several plans for raising the necessary funds to form the new company, asserting that "In fighting the Battle here we are in fact fighting it for the whole Continent & the prestige of success here assures us success elsewhere." DF (*TAEM* 52:326; *TAED* D7940ZBC).

9. The enclosure has not been found. Bailey subsequently indicated that he would enclose with his 5 September letter to Edison (see note 8) a draft power of attorney which would supercede that held by Puskas. Edison evidently received the power but refused to execute it (see TAE to Samuel White, 16 Oct. 1879, SSW [*TAED* X078AV1]).

10. Nothing more is known of Constant Rousseau.

-1799-

From Edward Johnson

London Sept 6th 1879.[a]

My Dear Edison

I write hastily this morning (7 am) to tell you that my inertia Transmitters are too unreliable to be practical   Occasionally one is found which works very well—but an equal number simply wont work at all without that horrible local Vibration caused by the friction of the receiver acting on the Transmitter which in turn reacts on the Receiver to produce a squeak—if a $\frac{1}{100,000,000}$ part of a hairs breadth adjustment of the spring to tighten it is had, the voice at one sinks away from a comparatively good volume to that[b] of a Damn bad Old style Transmitter—[1]

You remember the first four you sent me of the arm pattern—[2] I have taken them, & put Box wood mouthpieces on them not quite so deep as the ones you had on the Big Iron machines[3] but with about the same flare and only ½ inch hole at the bottom— They are pronounced by everybody to be a great success   I do not talk close enough to the mouthpiece to sweat the inside of it—& thus make it offensive.[4] I do not speak loud—scarcely loud enough to be heard more than 10 or 15 feet away— Yet I am able to carry on a whispered conversation with Mr Waterhouse One of our Directors[5]—whose wire—

the longest I have—is 2 miles[b] long— I used this arrangement yesterday in a trial at the Times office where I was compelled to read an article through the Telephone to an Operator at a Type setting machine—which makes a noise like an Automatic Perforator— He playing on his Keyboard & thus setting up Type from[b] the dictation of his Telephone, at the rate of 60 words per minute—[6] This was a severe test—but was accomplished to the entire satisfaction of Mr MacDonald the manager of the Times[7] who has given me the[b] use of a machine— & a Carte Blanche to go ahead and adapt the Telephone to it—& it to the Telephone[8] so as to have the thing in working shape by the time Parliament meets—when Gladstone[9] will speechify—his remarks be taken down in short hand and read in a Telephone to this Typo in the Times office thus avoiding transcription in long hand—and the further delay of sending it from House of Com to the Times[b]— Times says will be of immense value to them & they will give us their hearty support. Same thing has been tried by Bell—but proved utter failure— Now at yesterdays test—everything would have been a dead failure but for the Old reliable Edison Lamp Black Transmitter— The Only True Carbon Telephone[c]   There if my judgment is not sufficiently emphazsized on that subject—Ill print it in letters 5 feet high & send it to you to paste on your Laboratory Wall—

I am sorry you have yielded[b] to the clamor for whispering Telephones—by making a Transmitter less perfect than your Old Original   Far better to have spent a ¼₄ section of your time in adding the mote of sensitiveness to the Old one— which has all the requisites of a perfect Transmitter except extreme sensitiveness—

If you will make me a dozen 1st class Carbon Transmitters Old pattern & send them on[b] I'll give you a different report on your new Telephone than you are getting from the Western Union—

Beside I am supplying the thunder here for the big fight over[b] the carbon— I have interviews every day with Lawyers—Directors &c—& have written an elaborate Pamphlet on the Telephone for their enlightenment   The ground I have taken is that you had the microphone early—but found it impracticable as a Telephone until you had ridden it of its uncertain character &c &c— I'll send you a copy of the paper tonights mail—[10] I want you to send me every particle of data you have on this subject—with samples of early apparatus if you have any— Collect all you can to show you understood, & operated upon, the principle of "imperfect contact"

Dumoncel[11] makes the point On you that you & Hughes[12] differ in this that while you exerted more or less pressure upon things in contact under some degree of initial pressure simply varying that degree of pressure—Hughes has discovered that if two bodies are in extreme superficial contact—sonorous vibrations cause a more intimate relationship between the mMolecules of the two substances &c &c—a distinction without a difference as I will show by explaining that precisely that is the action of your Lamp black Telephone, only with this important addition that the greater number of molecules of the finer Carbon permits a more intimate initial[d] contact without initial pressure—and consequently a greater variation of the intimacy— upon under[d] the action of the sonorous waves— &c &c   The exhibition of the 2—side by side in their 1st stage (Blake) & last stage (your Lampblack) will demonstrate that a more perfect graduation of this intimacy under the varying graduation of the sonorous waves is obtained in yours—or in other words the electrical waves are in more perfect Harmony with the sound waves—

This confirms the ground I take that you were not satisfied with Hughes microphone

We are to have a demonstration of[b] the exchange today to the London Press—big swell— I am to make them an address, a copy of my pamphlet it to be given each—and I'll send you the net result. Yours Hastily

E H Johnson

P.S.—I kept this Letter to send with it, in[b] order to give you result of todays exhibition—a report of the action of the Telephones—

The exhibition was an entire success in every particular—

The working of the exchange was tested from Whites office—where an ample Luncheon was amply partaken of—

Present—representatives of the following Papers— Times, Daily News, The Globe Central News, Morning Post, City Press Morning Advertizer, Standard Observer Spectator[13]

Also—

Telephone Directors

T. Waterhouse   a splendid man
Mr Fowler[14]—The great objector
Sir Phillip Woodhouse[15]—An honest old chap
Lord Anson[16]—A young man anxious to get into business ways[e]

Geo. E. Gouraud—a heavy swell

Parish[17]—a millionare (Gourauds backer)

Arnold White—A nervous, ambitious man (a la Cheever except as to personal deformity[18]) & the man with whom I have the most to do—

I was introduced as Mr. Edisons scientific representative in Europe—and the Chief Engineer of the Company—

The affair was very formal—

I made an address—and supplemented it by a copy to each of the article on the Telephone— (a copy of which I send you todays mail)—

Much interest was manifested by the majority of those present—though there was the usual stragler[b] or two who preferred to discuss the viands and the veal—

After thoroughly testing the Telephone and the exchange—by calling at random for communication with ~~the~~ any of the following[19]

1. D Parish   Copthall Bldng's   ½ mile
2. Car syndicate[20]   56 Old Broad   ½ "
3   Renshaw & Renshaw[21]   Gresham St   ¾ "
4   Munkitrick[22]   1 Prince St   ¼ "
5   Gouraud   6 Lombard St   ¼ "
6   Waterhouse   1 New Court Carey St.   2 "
7— Ofs in which we were—   ¼ mile from
                                              Exchange
8   Kingsbury & Co.[23]   Georges Yard—   ½ "
9   Arthur Anderson[24]—Throgmorten St.   ¼ "
10   London Times   Printing House Sqr   1½ "

I led the party to the Exchange to witness the operation of the switch   After it had been fully tested & explained—the general hub bub of question & cross question began—and I was drawn into quite an extended dissertation on Telephones—Telegraphs Electricity & Edisonianism's in general— This lasted for half an hour— The whole exhibition consuming upwards of 2½ hours—[25]

The Times man then called his Brothers to order and proposed a vote of thanks—which was given— I responded—happily of course— The thing then adjourned. Our own people could not express their[b] pleasure at the manner in which I managed the entire thing—sufficiently—and all went off happy—

Now as to the working of the Insts. The regular carbon

came to [true?]^f <u>every time</u>—& did their work splendidly—
The Inertia had to be apologized for in 2 instances   The^b difference was extremely noticeable— Dont imagine that all this was because of the fact that it being an exhibition we could talk loud at the distant station—for my assistant who was at the Times office had one of your Lampblack Transmitters—and his whispered conversation was distinctly audable throughout the room   His breathing was remarkably so— This too as you will see by my distance schedule—on the next longest line I had— That Settles the Inertia for me   I want only 1st Class Lamp Black Transmitters—

I used the Box wood mouthpiece about this angle

I'll send you the papers as they come out & you^b can "see ourselves as others see us"[26]   Yours Tired (12 mid) E H Johnson

ALS, NjWOE, DF (*TAEM* 52:695; *TAED* D7941ZDV). Letterhead of Edison Telephone Company of London, Ltd. ^{a}"London" and "18" preprinted. ^{b}Obscured overwritten text. ^{c}Multiply Underlined. ^{d}Interlined above. ^{e}"A young . . . ways" enclosed by brace. ^{e}Illegible.

1. That is, a poorly adjusted or malfunctioning standard carbon transmitter.

2. Those instruments were sent in mid-July and used Edison's standard carbon transmitter. It is not certain what type of mouthpiece was used but they may have been made of rubber. See Docs. 1779 esp. n. 1 and 1782 n. 5.

3. The iron demonstration telephones sent in February; see Doc. 1681.

4. Johnson later sent diagrams and measurements (not found) of the mouthpiece so that Edison could make them from "any hard wood that wont split—& that dont require to be painted." He also suggested that an improvement "would be to cut a slit in the under side near the outer edge for escape of spittal—thus protecting the chin of a speaker." Johnson to TAE, 21 Oct. 1879, DF (*TAEM* 52:828; *TAED* D7941ZGN).

5. Theodore Waterhouse was a London solicitor who also provided legal advice to the telephone company. Alfred and Edwin Waterhouse also held stock but Theodore is identified below as a director. Edison Telephone Co. of London list of stockholders, 28 June 1879, DF (*TAEM* 52:603; *TAED* D7941ZCK); see also Doc. 1829.

6. Johnson sent part of the resulting proof slip to Uriah Painter with a description of this test, and the instruction to "Send this to Edison." Johnson to Painter, 5 Sept. 1879, DF (*TAEM* 49:379; *TAED* D7903ZHU).

7. John Cameron MacDonald was a *Times* reporter of considerable mechanical knowledge who, after working as Chief Engineer for the mechanical and printing shops, became the paper's manager in 1873. *The History of the Times*, 2:347–48, 492–93, 502–504.

8. It is unclear if Johnson used a clockwork telephone for this trial; Edison had one sent to Gouraud in July (TAE to Gouraud, 14 July 1879,

DF [*TAEM* 52:615; *TAED* D7941ZCT]; see headnote, p. 310). On 21 October Johnson instructed Edison to

> send me 2 clock work instruments the first moment you an manage it. We have agreed with the Times to do a job for them in which the clockwork is absolutely indispensible Viz: Dictate to a compositor= We must do this ere Parliament meets again. The times has been good to us—& it is a power we cannot—in fact dare not ignore— You must let me have them at once—no matter what sort= whether weight or spring= The Times has put in your pocket thousands of pounds— To do this thing for them will give us another big Leader—& be a big help in the District Organizations   See to it [Johnson to TAE, 21 Oct. 1879, DF (*TAEM* 52:828; *TAED* D7941ZGN)]

After a request on 8 November to "Report progress Times clockwork" Edison cabled that they were "almost finished." With Parliament scheduled to sit in February, Edison received two more anxious inquiries before he shipped the instruments in mid-January (TAE to Johnson, 19 Nov. 1879; White to TAE, 29 Dec. 1879; Edison Telephone Co. of London to TAE, 8 Nov. 1879 and 6 Jan. 1880; all DF [*TAEM* 52:929, 1034, 901; 56:342; *TAED* D7941ZIM, D7941ZKO, D7941ZHV, D8049B]); see Doc. 1880.

9. William Gladstone had been Prime Minister from 1868 until the defeat of his Liberal Party in the 1874 elections. At this time he was Member of Parliament for Greenwich and an outspoken critic of the Conservative government of Benjamin Disraeli, whom he succeeded as Prime Minister in 1880. *DNB,* s.v. "Gladstone, William Ewart."

10. Johnson's pamphlet was a 4 September memorandum to the directors of the London company reviewing the history and operation of the magneto telephone, carbon transmitter, and electromotograph receiver. Johnson stated that Edison's earliest form of carbon transmitter "was simply what Professor Hughes has since elaborated and to which he has given the name of the 'Microphone.' In fact Mr. Edison's chief difficulty for some months was in ridding it of its Microphone character. It was too delicate, too sensitive to extraneous sounds, and too liable to be thrown out of adjustment." Johnson to directors of Edison Telephone Co. of London, 4 Sept. 1879, Unbound Documents (1879–1882), UHP.

11. See du Moncel 1879c, 143–44, 165–70; Johnson used the somewhat different London edition (du Moncel 1879b).

12. David Hughes was a London-born but American-educated electrician and inventor now living in London whose claim to have discovered the principle of the microphone set off a bitter controversy among Edison and partisans on both sides of the Atlantic. On Hughes see *TAEB* 4:287 n. 3; on the ensuing dispute see *TAEB* 4, chaps. 3–5 passim.

13. The *Times, Daily News, Globe, Central News, Morning Post, City Press, Morning Advertiser,* and *Standard* were all London dailies; the *Observer* and *Spectator* were weekly journals.

14. Robert Fowler, a former Member of Parliament, was at this time a London alderman; he was elected lord mayor in 1883. This nickname may derive from his reputation as a determined Conservative organizer

and vociferous supporter of the party's policies. *DNB*, s.v. "Fowler, Sir Robert Nicholas"; Edison Telephone Co. of London list of stockholders, 28 June 1879, DF (*TAEM* 52:603; *TAED* D7941ZCK).

15. Philip Wodehouse, a retired colonial governor. *DNB*, s.v. "Wodehouse, Sir Philip Redmond"; Edison Telephone Co. of London list of stockholders, 28 June 1879, DF (*TAEM* 52:603; *TAED* D7941ZCK).

16. Thomas Francis Anson (subsequently Earl of Lichfield) was twenty-three years old at this time. He later invested in Edison's electric light in England and was connected with the Hudson's Bay Co. Cokayne 1929, s.v. "Lichfield. VIII. 1892"; "Death of Lord Lichfield," *Times* (London), 31 July 1918, 3.

17. Dillwyn Parrish.

18. Charles Cheever had been crippled from birth. See *TAEB* 4:27 n. 11 and Doc. 1190.

19. According to the *Times* article on the test, which described the operation of the central station system, "the stations, or more properly speaking the private offices, which are connected with the exchange are situated—No. 1 in Copthal-buildings, No. 2 in Old Broad-street, No. 3 in Suffolk-lane, No. 4 in Lombard-street, No. 5 in Princes-street, No. 6 in Carey-street, Lincoln's-inn, No. 7 in Queen Victoria-street (the offices of the company), No. 8 in George-yard, Lombard-street, No. 9 in Throgmorton-street, No. 10 being our own establishment." "The Edison Telephone," *Times* (London), 8 Sept. 1879, 12.

20. Probably the Pullman European Car Association, identified in the telephone company's 1880 subscriber list at 57 Old Broad St., E.C. The Edison Telephone Company of London, Limited [pamphlet], 20 Feb. 1880, DF (*TAEM* 56:429; *TAED* D8049ZAQ1).

21. Patent solicitors Alfred and George Renshaw. Their address was on Suffolk Lane; the only subscriber listed (see note 19) on Gresham Street was the accounting firm of Price, Waterhouse & Co. TAE Power of Attorney to Alfred Renshaw and Arthur Renshaw, 14 July 1879, Miller (*TAEM* 86:30; *TAED* HM790070).

22. Probably an individual associated with the Equitable Life Association of the United States, the only subscriber at this address (see note 19).

23. An advertising agency. Letterhead of Kingsbury & Co. to TAE, 1 July 1882, DF (*TAEM* 60:214; *TAED* D8204ZEA).

24. Unidentified.

25. The *Times* referred to this exhibition as the inauguration of practical telephone service in London. It reported that "A great many tests were applied by those present in order to prove the system in various ways, but in no case was there any failure. . . . Communications were opened, maintained, and closed with the various stations in rapid succession, including our own, and with every success. And here we may mention that a paragraph was recently set in type in our office which was dictated through the telephone, the result being a perfectly correct reproduction of the transmitted subject." "The Edison Telephone," *Times* (London), 8 Sept. 1879, 12.

26. No clippings have been found.

Menlo Park, N.J. Sept 8th 1879[a]

Dear Sir

Regarding the introduction of my new system of electric lighting I contemplate in addition the use of electricity as a motive power   I propose to sell it for the purpose of running elevators, small shops, sewing machines etc.[1] In lighting the city of New York there will be about 24 stations each station will supply about 1200 horse power distributed over a certain area, each horse power will supply a certain light according to size. The same power that will supply a light will run two sewing machines   A small dynamo machine has been completed for running Sewing Machines and have applied it to the Wilcox & Gibbs.[2] Tests have shown that this machine requires about $\frac{1}{20}$ of a horse power to run it. When running 500 stitches a minute through leather sixteenth of an inch in thickness and when run by the motor $\frac{1}{16}$ of a horse power.[3] I should like if not contrary to your business principles to obtain one of your standard sewing machines that I may obtain measurements and make the application of the motor.[4] After testing power required and obtained measurements will return it to you. As the Electric Light Co can furnish electricity to run a sewing machine 10 hours daily for less than 3 cts by mechanism requiring absolutely no attention it will undoubtedly lead to a more extensive use of the Sewing Machine   Very Truly

T A Edison

Df, NjWOE, DF (*TAEM* 50:115; *TAED* D7919ZBZ). Written by Stockton Griffin; letterhead of Edison's laboratory. [a]"Menlo Park, N.J." and "18" preprinted.

1. Edison had previously considered designs of battery-powered motors for sewing machines in 1871 and again in 1876. See Docs. 135D, 138A, 139A, 139B, 800, 802, and 829.

2. Inventor James Gibbs and machinist Charles Willcox formed the Willcox & Gibbs Sewing Machine Co. in 1857. The New York firm was known for its light, low-cost machine, one of the first models manufactured for home use. Cooper 1976, 45–48, 76, 123.

John Kruesi sketched parts for a sewing machine motor on 4 August and made measured drawings the next day. On 17 August he wrote a shop order specifying the construction of its armature and field magnet windings. Samuel Mott's undated drawing from September or early October of "Edison's Electric Motor" (designated "No. 1" by Francis Upton) was the basis for an engraving published in the 18 October *Scientific American*. This article briefly described "Mr. Edison's new electric motor intended for running sewing machines, small elevators, lathes, and other light machinery. . . . Its construction differs but slightly from the electric generator. The armature is arranged parallel with the magnet in-

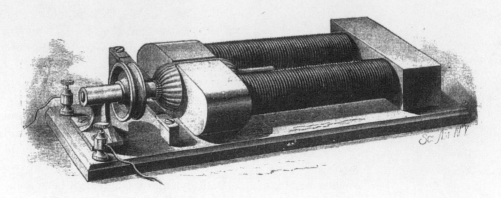

"Edison's Electric Motor" illustrated in the 18 October 1879 Scientific American.

stead of transversely, and the magnet is formed of a single casting. In other respects it is like the generator, having the same form of armature, also commutator cylinder and brushes." Machine Shop Drawings (1879–1880), Lab. (*TAEM* 45:74–75; *TAED* NS7986CAZ, NS7986CBA); Cat. 1308:169 (Order No. 223), Batchelor (*TAEM* 90:751; *TAED* MBN003:58); N-79-10-18:15, Lab. (*TAEM* 32:791; *TAED* N301:9); "Edison's Electrical Generator," *Sci. Am.* (41:242), Cat. 1241, item 1332, Batchelor (*TAEM* 94:513; *TAED* MBSB21332X).

3. On 5 September Upton began testing the electrical characteristics and power output of the motor operated without a load (N-79-06-16.1:268–72, Lab. [*TAEM* 35:294–96; *TAED* No77:134–36]). In undated notes made about this time, Upton also calculated the electrical energy consumed and mechanical power produced by the motor when working the machine through leather and when the pulley was held stationary. In his summary of results he noted that it operated at about 90% efficiency when running free (using ⅕ horsepower), 75% while sewing leather (using ¹⁄₁₃ horsepower), and 60% when stalled. He noted that seven battery cells "ran the motor slowly." Upton concluded the entry with the remark: "Wanted— A measurement of the power that various sewing machines take to drive" (N-79-10-18:2–11, Lab. [*TAEM* 32:785–89; *TAED* N301:3–7]; incorrectly dated in *TAEM-G2* as October 1879).

4. Stockton Griffin indicated on this draft that the letter was sent to the presidents of seven sewing machine manufacturing companies— Singer Mfg. Co., Wheeler & Wilson Mfg. Co., Willcox & Gibbs, E. Remington & Sons, Wilson Sewing Machine Co., Howe Sewing Machine Co., and Domestic Sewing Machine Co. (copies are in Lbk. 5:152–61, 163–66, NjWOE). Edison received five extant affirmative replies. One of these, from Willcox & Gibbs, cautioned him about the difficulty of determining the power needed since "No two machines of any make run alike. A well fitted machine requires more power than a loose jointed one, or the same well fitted machine will run very much harder with bad oil. There are so many contingencies surrounding the use of the sewing machine in families, that the best you can hope to do, is to supply power enough to run an ordinary machine in the condition it is usually found. This done successfully with any machine, would be a great success with ours." Willcox & Gibbs to TAE, 9 Sept. 1879; Wheeler & Wilson Mfg.

Co. to TAE, 10 Sept. 1879; E. Remington & Sons to TAE, 12 Sept. 1879; Wilson Sewing Machine Co. to TAE, 12 Sept. 1879; Singer Mfg. Co. to TAE, 15 Sept. 1879; all DF (*TAEM* 50:117, 119, 122, 123, 127; *TAED* D7919ZCA, D7919ZCB, D7919ZCE, D7919ZCF, D7919ZCJ).

Edison and Kruesi wrote a machine shop order on 12 September for a second, slightly larger motor. In the latter part of October Upton measured the resistance in this "Faradic Motor No. 2" and made comparative tests, including some on Edison's dynamometer, with the first motor; he also attempted to remove a "dead point" on the commutator. Around the end of October he resumed tests on "Sewing Machine Motor No. 1," recording results similar to those obtained in September. On 30 October Kruesi instructed the shop to "Wind the Sewing m. motor with double covered wire put on as much of it as possible   fix comutator so that the motor will start at any point." Upton dated two additional pages of related notes on 5 November. Around 17 November he tried to derive a general "motor equation" relating the amount of work performed by a motor to current and voltage in the circuit (Cat. 1308:173, 181 [Order Nos. 231, 263], Batchelor [*TAEM* 90:753, 757; *TAED* MBN003:60, 64]; N-79-10-18:19–29; N-79-06-16.1:272–85, 266–67; N-79-04-21:218–25; Lab. [*TAEM* 32:793–98; 35:296–302, 293; 30: 830–33; *TAED* N301:11–16; N077:136–42, 133; N017:108–11]).

In December Edison drafted a U.S. caveat and British provisional specification that included applications of electric motors. He specified that in motors for sewing machines there should be "a switch for disconnecting the machine from the circuit, a belt to connect the motor with the sewing machine, using a very small pulley on the motor & a large pully on the Sewing Machine so as to allow of high speeds in the motor. I regulate the speed of sewing by a friction lever pressing on the driving pulley of the sewing machine . . . the pulley of the motor being a friction pulley allows the motor shaft to revolve at a slower speed even when the belt is stopped" (Cat. 1146, Lab. [*TAEM* 6:650; *TAED* NM014U]; Brit. Pat. 33 [1880], Batchelor [*TAEM* 92:146; *TAED* MBP022]).

–1801–

Notebook Entry:
Electric Lighting

[Menlo Park,] Sept. 8, 1879.

The ~~induction~~ large induction coil was brought in order. The[a] condenser was crossed [~~inside?~~][b] so that it had to be rewound. On putting on the whole of a W.U. condenser 1.6 M[icro]F[arads]. ~~it~~ [---][b] a spark ¾ in long only could be obtained.

Thinking the trouble might be in the coil we ~~unwound it and brought it~~ cut off the covering and found all good. On putting back the wires a good spark was obtained with less condenser. 3 inches spark with six plating cells.

The three fall pump arranged thus

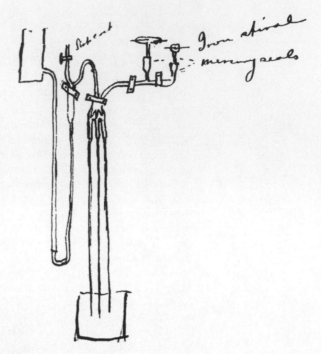

Pump No. 1[1]

A .015 in. iron wire spiral[c] was ~~sealed onto~~ fastened to[d] .020 in. Pt. wire and ~~pl~~ sealed into a glass bulb

Difficulties met with.

Vacuum spark tube cracked

Piece of dirt in one of the small jets preventing the Hg from flowing out from it so that as the other two exhausted the air could enter through the tube which had little Hg. in it. The dirt washed partly away and heating the tube very hot ~~made en~~drove enough away so the sufficient Hg. fill to seal the tube.

About 11-30 a good vacuum obtained. The pouring of Hg into the mercury seals seemed to help very much. A piece of dirt got[a] into the middle tube and stopped its working.

Phenomenon noticed. When at a blue stage the color red could be made to appear on heating. The lower part containing Hg. was strongly heated so that the pump was well filled with Hg. vapor. The aluminum electrode became bright yellow in the blue light with some green phosphorescence. At one time the yellow electrode could be made to change place by heating. It would jump over on heating[e] and then dance as if uncertain where to stay after the flame was taken away.

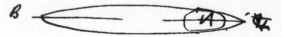

A a blue[a] light. Heat and B grows smaller, ~~end~~ end A larger.

Large coil started without any connection with [~~induction?~~][b] terminals   blue phosphorescence seen faintly. One of the terminals connected and the whole tube filled. Both terminals connected phosphorescence small, both electrodes yellow.

X, NjWOE, Lab., N-79-08-22:29 (*TAEM* 35:798; *TAED* No85:15). Written by Francis Upton. [a]Obscured overwritten text. [b]Canceled. [c]Interlined above. [d]"fastened to" interlined above. [e]"on heating" interlined above.

1. This is the first apparatus built by Ludwig Böhm for which any experimental records exist. Unlike the Sprengel pumps previously used at Menlo Park it employed multiple fall tubes, a modification developed several years earlier by Charles Gimingham, an assistant to William Crookes (Böhm's testimony, pp. 2–3; Francis Jehl's testimony, pp. 4–6, both *Edison v. Böhm* [*TAED* W100DEC002 (images 1–2), W100DED002 (images 3–5)]; Thompson 1887, 633). Upton broke this pump the next day while trying to remove a sealed spiral but "Böhm began to put the new pump together   He worked all night and in the morning the parts were ready." This new pump was designated "Pump No. 2" and was similar to the first but with the additions of a stopcock to

Francis Upton's drawing of the modified three-tube Sprengel pump ("Pump No. 2") assembled on 10 September, with detail of mercury jets.

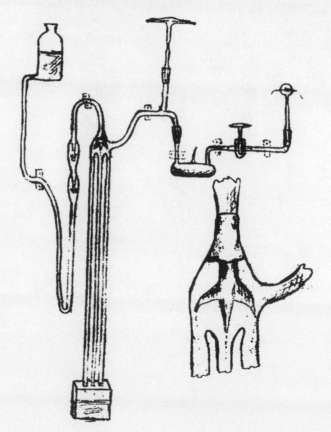

isolate the wire spiral and a "drying tube" containing sulphuric acid. A double "air catcher" was also included to allow trapped air to escape from the mercury before reaching the fall tube (N-79-08-22:29–51, Lab. [*TAEM* 35:798–809; *TAED* No85:15–26]).

When Upton began working with the new pump on 10 September he noted as a "fault" the fact that two of the mercury jets did not extend as far as the middle one into their respective fall tubes but it nevertheless produced sufficient vacuum to stop the spark from a small induction coil and permitted him to test wire spirals heated by dynamo current. (Edison reportedly claimed later that even a "coarse" vacuum would prevent a nine inch spark from jumping the ⅛ inch spark tube gap.) Although again temporarily blocked by dirt the pump worked well until midnight, when the rubber supply tube (at left in sketch of "Pump No. 2") became so swollen with mercury that the weight "broke the pump at the top of the upper air catcher." The pump was repaired and Francis Jehl began using it on the morning of 11 September but could not get a good vacuum. He cleaned the mercury several times but noticed throughout the afternoon that "the Hg. would collect & fill up the fall tubes. The Hg. could not get through that portion of the pump where cloth covering was . . . which seem to be covered with the oxide and other impurities." The cloth may have been attached as a filter to one or both "air catcher" tubes although Jehl's sketch does not clearly indicate its location. The pump did not work properly the rest of the day. N-79-08-22:47–53, Lab. (*TAEM* 35:807–810; *TAED* No85:24–7); "Edison's Vacuum Apparatus," *Sci. Am.* 42 (1880): 34–35.

–1802–

To Thomas Clingman[1]

MENLO PARK, N.J., *Sept.* 10, 1879.

DEAR SIR:

Your favor of the 8th instant is just at hand. You are mistaken about zirconia being a substitute for platinum.[2] It has no value in that line. I used it merely to make bobbins to wind the platinum wires on, but as I find it attacks the wires at a very high temperature, it is useless for electric light purposes.[3] But, as I have expended $1,500 in experimenting to obtain a process for economically extracting the pure oxide, I thought if I could get the crystals cheap enough, I would go into the manufacture of the basic oxide and make all the application in the arts possible; but the high price of the ore, and expense of working it, make it impracticable.[4] Very truly,

T. A. EDISON.

PD, MdNACP, Pat. App. 276,133, *Argument of T. L. Clingman* (*TAED* W100B007).

1. Thomas Clingman of North Carolina had been an influential member of the U.S. House of Representatives and the Senate, where he built a reputation as a strident proponent of Southern interests in the ante-

bellum years, and he served as a general in the Confederate Army. Finding his political ambitions frustrated after the war, he turned to promoting the putative medicinal uses of tobacco and attempting to develop an incandescent electric lamp using zirconia. He met with Edison in December 1879 to discuss his ideas, which Edison thought impractical, but after several years, during which time he found himself in interference with one of Edison's own patents, Clingman finally received two U.S. patents on an incandescent zirconia pencil. On Clingman see Jeffrey 1998 (his relationship with Edison is detailed on pp. 273–85); *ANB*, s.v. "Clingman, Thomas Lanier."

2. Clingman had for many years owned the rights to a large zircon deposit in the mountains of Henderson County, N.C., for which there was no known commercial use until Cyprien Tessie du Motay and Edward Stern substituted zirconia, a metallic oxide, in the calcium lights known as "limelights" widely used in theaters. Clingman agreed to sell 1,000 pounds of the mineral to du Motay and Stern but they ran out of money in 1869 after taking only a small portion of that, leaving him with the balance stored in New York. He contacted Edison on 26 March 1879 to suggest that the oxide was incombustible and might be more durable than platinum for electric lighting. Edison replied that he had "no difficulty with my platinum and iridium electric lamp," but asked Clingman to send a sample of the zircon. After receiving the sample in mid-May, Edison wrote Clingman that he was "now using lime and cerium in my lamp, but if I could get zircons in quantity and cheap, I might use the oxide, as it is better for my purpose." He then asked Clingman to send ten pounds "to see what I can do with it." By the end of May Edison had made tests with the ten pounds and notified Clingman that "it tests very well." While he was not ready to "enter into any arrangements at present about the 600 pounds you have in New York," Edison did want "to know if I conclude to use it, if you can furnish me large quantities without delay." Clingman replied that he could supply twenty thousand pounds over the course of the summer. By mid-June Edison had "a chemist at work cannot say what will do until our experiments are completed Am afraid the high price of purifying as well as 1st cost of ore will prevent its adoption by me." Clingman reiterated his offer in July, prompting Edison to note that he had finally extracted pure oxide from the sample and found it to be the "worst thing ever had to deal with here Cant tell abt getting more oxide out until I have tested it thoroughly for purpose for wh we intend to use it." Clingman inquired yet again about Edison's intentions on 21 August. Encouraged by the prospect of much less expensive sources (see note 4), Edison drafted a reply on this letter stating that the "prohibitory" cost had forced him to abandon zirconia, which was "a great pity because I have now a process for extracting oxide & could create a great demand for it outside of my uses." Clingman responded on 8 September, giving as one reason for his insistence on the stated price his belief that "If Zirconia can be used instead of platinum I think there will be a demand for it." Edison's notes on that letter are substantially the same as this document except for his remark of having spent a thousand dollars to develop an extraction process. Jeffrey 1998, 274; Clingman to TAE, 26 Mar., 17, 16, and 29 May, 12 June, and 22 July 1879; marginalia on Clingman to TAE, 21 Aug. 1879; Clingman to TAE, 8 Sept. 1879; all DF (*TAEM* 50:34, 61, 59, 73, 78, 88, 104, 111;

TAED D7919T, D7919ZAO, D7919ZAN1, D7919ZAY, D7919ZBB, D7919ZBL, D7919ZBU, D7919ZBY); TAE to Clingman, 27 Mar., 15 and 28 May, 31 July 1879, *Argument of T. L. Clingman*, pp. 5–6 (*TAED* W100B005, W100B005A, W100B006, W100B006A).

3. See also Doc. 1787. The previous week, Charles Batchelor had conducted high-temperature experiments on wires coated variously with zirconia and alumina. He noted that zirconia vitrified when heated by an electric current but not the oxyhydrogen flame (an observation made in Doc. 1796), leading him to conclude that "As to the infusibility therefore of Zirconia we can say little as with much less heat where no chemical action is present it has fused." On 10 September Batchelor listed eleven different substances, including "Pure Zirconia," from which to make lamp spools. There are two pages of undated drawings related to these spools but no extant records of experiments with them. Cat. 1304:70–71, Batchelor (*TAEM* 91:69–70; *TAED* MBN004:67–8); N-79-07-31:67, N-79-07-07.2:9–11, Lab. (*TAEM* 33:575, 32:379–80; *TAED* N052:33, N038:5–6).

4. Edison had also been investigating cheaper supplies of the mineral. He arranged for Thomas Arthur Jr., a South Carolina attorney, to deliver five hundred pounds from North Carolina. This arrangement apparently collapsed when Edison lowered his offering price after William Hidden, a mineralogist whom he dispatched to search for platinum in the southeastern United States, reported on 31 July that high-quality zircon might be available for as little as fifty cents per pound. Edison's marginal note on this letter instructed Hidden to purchase five hundred pounds and "arrange so we can get what we want thereafter. . . . Have Zircons sent at once, we are using ours up daily." Hidden visited Henderson County at Edison's direction, reporting that the zircons there were "practically inexhaustible" and might be had at a price "so low as to frighten you at its cheapness," and soon after he began to make preliminary arrangements with a third-party supplier. At the beginning of September Edison instructed Hidden that he wanted "a good big supply of Zircons & Thoria for mfg oxide for different purposes which I have in mind outside of E. Light." However, this letter evidently crossed in the mail with Hidden's report that his prospective deal had fallen through and Edison would have to depend on Clingman, who controlled the entire supply and refused to lower his price except on small quantities for experimental purposes. TAE to Arthur, 14 July 1879, Lbk 4:470 (*TAEM* 80:98; *TAED* LB004470); TAE to Hidden, 13 Aug. 1879; Hidden to TAE, 31 July, 11, 17, and 28 Aug. 1879; all DF (*TAEM* 51:67; 50:1119; 51:58, 83, 120; *TAED* D7928ZML, D7928ZKY, D7928ZMI, D7928ZMW, D7928ZNO); TAE to Hidden, 3 Sept. 1879, Lbk. 5:132 (*TAEM* 80:144; *TAED* LB005132).

[Menlo Park,] Sept 12 1879

The pump was fixed in the morning,[1] ~~an~~ new chamber was put in instead of the old one[2]

new one   old one

which gave very good results. we commenced at half past ten oclock, a lamp made by Mr Bachlor (⅛)[2] of inch, ~~made of~~ covered with[a] Aluminum[b] and Dextrine, ~~and~~ wound in two layers .006 wire. got a ~~p~~ vacuum so that the ⅛ inch spark coil, would not give any light at ~~half~~ one oclock we then put on the large coil and got a kind of red illumination, a half[b] an hour afterwards we got a white kind of light[c]

we put one cell of the chromic ~~cell~~ battery[3] on the bobbin at about half past one oclock, with two ohm resistance in the circuit. it turn black (bobbin) owing to the organic matter (Dextrine)) which was mixed the the aluminum.[c]

at about two oclock we got a bright white light, one of the electrodes had a yellow light on it. at two oclock I put a half an ohm out of the circuit.

At ten minutes past two Mr. Bachlor cut out another half an ohm. At twenty minutes of three Mr Bachlor ~~p~~cut out an half an ohm. This make one cell carbon on the lamp and half an ohm at ten minutes to three we put two cell on with two ohms resistance. It just heated it up to a very dark red. there was a little change in the light, the yellow was fainter, and a very light green spot was observed near the yellow. At 3.20 pm we made resistance ½ ohm less. Thus it is:—

			Remarks
1½ pm.	1 cell	2 ohm res.	Blackened spool.
2 pm	1	1½ "	
2.10	1	1 "	
2.40	1	½ "	
2.50	2.	2 "	Very dark red
3.20	2	1½	Dark red but little brighter

Vacuum has not altered during last hour owing to gas coming out of spool[d]

3.35[c]	2.[c]	1 ohm.	
4.[c]	3[c]	2 ohm	—red—
5.50[c]	4[c]	2 ohm	yellow
5.55	4[c]	1 ohm	" white[e]

Some visitors came[b] in the afternoon, (Mr. Banker & others)[4] and the spool was brought up[b] with seven cell then it bust, giving at the time about 4 candles.

At eight oclock .P.M. we put the pump again in operation, a new spool was put in $(\frac{1}{28})^2$ of an inch covered with Alumina & Dextrine.

At quarter past nine we put one cell on the spool

~~9.15 .P.M.   1 cell.   Resistance.~~

			Remarks
9.15 P.M.	1 cell	3 ohms Res	Blackened it
9.45	1 "	2 "	
10.10	1 "	1	
10.15.	2 "	1	
10.30	2 "	0	cross. showed red only.
11 "	3	2	the spool began to get red. Dark
11.20	3.	1	Red. goes down & up owning to the cross
12			cross prevented from going any further The[f] spool away

Put in a new spool at about eleven oclock & commenced a vacuum. Stop the pump at dinner, sealed it at the bottom with Hg. commenced ¾ of an hour afterward again, vacuum no good. Stoped pump.

~~Commenced again at four oclock & got a pretty good vacuum at 6 oclock~~

			~~Remarks~~
~~7.45.~~	~~2 cell~~	~~3 ohms~~	~~Light Red~~
~~8.~~	~~" "~~	~~2 "~~	~~Red in the middle cool on[b] the ends[c]~~

~~Mr Edison brought[g]~~ put on cells after this untill it busted (7 cells.)[5]

X, NjWOE, Lab., N-79-08-22:55 (*TAEM* 35:811; *TAED* No85:28). Written by Francis Jehl. [a]"covered with" interlined above. [b]Obscured overwritten text. [c]Followed by dividing mark. [d]Sentence enclosed by horizontal lines above and below, and wavy lines at left and right. [e]From "Thus it is" to end of table written by Charles Batchelor; followed by dividing mark. [f]Paper damage before "The" may contain text. [g]Previously canceled.

1. This was "Pump No. 2" (see Doc. 1801 n. 1). Jehl's sketch of the pump as it stood on 12 September shows that the double "air catcher"

Francis Jehl's drawing
of the pump used for the
experiments described
in this document.

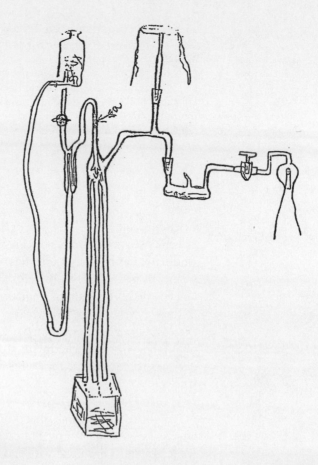

had been replaced by a trap and valve like that used in the first pump. Another departure was the addition of a slender contraction (labeled **a** with an arrow in the drawing) in the supply tube immediately above the jets. This innovation became the subject of a patent interference between Böhm and Edison, in which Böhm testified that this method of regulating the mercury flow enabled him to use somewhat wider jets that were less prone to clogging. A further alteration is discussed in note 2. Böhm's testimony, pp. 3–4; Jehl's testimony, pp. 5–6; Edison's testimony, p. 32; *Böhm v. Edison* (*TAED* W100DEC002 [images 2–3], W100DED002 [images 4–5], W100DED032 [image 1]).

2. Jehl evidently was confused about the "new chamber." His sketch of the "old one" depicts the mercury jets which characterized all of the triple Sprengel pumps. This could not have been replaced by the "new one" although another, unspecified, modification may have been made to the jets. The "new one" appears to be a system of baffle tubes like that shown in Jehl's 12 September drawing in place of the earlier "air catcher," presumably for the same purpose.

3. A form of Bunsen cell, used widely on Gold and Stock Telegraph Co. lines in New York City. See *TAEB* 2:565 n. 3.

4. James Banker planned to visit with Tracy Edson. According to the later testimony of Ludwig Böhm, physicist Albert Michelson was also

among those present. Lowrey to TAE, 10 Sept. 1879, DF (*TAEM* 50:271; *TAED* D7920ZAY); Böhm's testimony, p. 11, *Böhm v. Edison* (*TAED* W100DEC002 [image 10]).

5. Jehl resumed these experiments with Edison on 14 September but late that evening "the pump busted being heated to much on[e] of the fall tubes broke." He returned to work after it was repaired the next morning and continued these experiments intermittently through September. N-79-08-22:67–91, Lab. (*TAEM* 35:817–18; *TAED* N085:34–45); see Doc. 1815.

–1804–

From George Prescott

New York Sep 15 1879[a]

Friend Edison,

I enclose a telegram just received from Mr. Short.[1] I think it would not pay to go into interference[2] in this case now that the Company has bought the patents and I have advised Short that he need do nothing about it unless further advised.

I have a letter from Mr. Weaver[3] saying he will try to hurry the Post Office Authorities in our Quadruplex matter,[4] and also saying he will be satisfied with 5 per cent for his services and declines 10![5] Score one for a conscientious Englishman. Yours truly

Geo. B. Prescott

⟨good for Weaver. I shall certainly go on with the interference as I cannot withdraw. I of course have nothing to do with what Mr Short has done with the Co as I knew nothing about it. I shall only follow out my contract and if Shorts instrument is worth 3000 dolls & I get priority of invention I shall want pay for what I have previously invented— TAE[6]⟩

⟨Hold this until Banker comes wednesday[7]⟩

ALS, NjWOE, DF (*TAEM* 51:549; *TAED* D7929ZDY). Letterhead of Western Union Telegraph Co., George Prescott, electrician. [a]"New York" and "18" preprinted.

1. At this time Sidney Short was a student at Ohio State University in Columbus, his hometown. After graduating in 1880 he became vice president of the University of Denver, where he also taught physics and chemistry. He subsequently did important work on electric motors and street railways, receiving over 500 U.S. and foreign patents on electrical machinery (*NCAB* 13:247). In his 15 September telegram Short asked Prescott for instructions in regard to the interference notice he had received, which he understood to be "only the result of the rules of the office That Edison is not pushing it" (DF [*TAEM* 51:550; *TAED* D7929ZDZ]).

2. This interference proceeding, designated A[3], was declared between Edison's Case 183 for a telephone transmitter, filed on 6 August 1879,

and a patent issued six days later to Sidney Short (U.S. Patent 218,582). Prescott had brought the application to Edison's attention on 7 July, after the patent had been allowed but not issued, while Short awaited foreign patents. Prescott asked if Gold & Stock should purchase the transmitter and arranged for Edison to examine it; he and Short came to Menlo Park for this purpose on 12 July. In a 24 July letter, Prescott expressed his fear that "Short will get his patent issued before your application is acted upon. If you have not filed it I hope you will do so without out delay." Edison stated in reply that the Patent Office model had just been finished and "I am afraid Short will be ahead of me." He asked Prescott to contact Lemuel Serrell and "have a communication sent to Patent Office calling for interference between Short case and my case 158. That will delay the matter and keep it in the Patent Office until I get the other patent in. Have him send the communication in at once." The interference proceeding took place at the beginning of April 1880 and at the end of June the Patent Office ruled in Short's favor. This was affirmed by another ruling in September. Edison appealed to the Commissioner of Patents who once again affirmed the original decision in November 1881. Edison subsequently amended the claims in his application so that it did not conflict with Short's patent and it issued as U.S. Patent 252,442 on 17 January 1882. Prescott to TAE, 7, 10, and 24 July 1879, DF (*TAEM* 51:530, 52:77, 51:537; *TAED* D7929ZDH, D7937ZBP, D7929ZDM); Short's testimony, 21–22, *Edison v. Short;* TAE to Prescott, 25 July 1879, Lbk. 5:22 (*TAEM* 80:132; *TAED* LB005022); Pat. App. 252,442.

The contested invention was a form of inertia transmitter in which a carbon button or disk was mounted between two arms extending from the diaphragm. Edison claimed to have made devices based on the same underlying principle as early as 1877 and as recently as May or June 1878 at which time Charles Edison made a transmitter essentially like that of Case 183 in connection with the microphone controversy (Edison's testimony, pp. 3–9, *Edison v. Short* [*TAED* W100DBA003 (images 1–4)]). In the specification, Edison explained that "the pressure on the carbon button is varied by the outward movement of the diaphragm causing the ends of the arms to move slightly outward and lessen the pressure on the carbon, and this pressure is increased when the ends of the arms are brought toward each other by the return movement of the diaphragm." Although Edison made no mention of inertia in the application, each of these related designs displays a distinguishing characteristic of the inertia transmitter in having the carbon wholly supported on the diaphragm, independent of the instrument case. Edison executed a caveat on 7 August which covered numerous alternative arrangements for varying pressure on the carbon button through the momentum of a mass attached to the diaphragm (U.S. Pat. 252,442; Edison Caveat 90). A sketch containing the patented design is dated 12 July, the same day that a patent office model was drawn by Batchelor. There is also a series of alternative arrangements that follow in the same notebook. On 25 July, John Kruesi recorded the order for the patent office model along with a drawing (N-79-02-10:230–35; Vol. 18:104; both Lab. [*TAEM* 32:1241–43, 4:1138; *TAED* N046:101–3, NV18:93]; Cat. 1308:163 [Order No. 212], Batchelor [*TAEM* 90:748; *TAED* MBN003:55]).

3. Henry Weaver was a pioneer of the British telegraph industry. He had been secretary of the Electric Telegraph Co. and then managed the Anglo-American Cable Co. (Kieve 1973, 67; Bright 1974 [1898], 144). His letter to Prescott has not been found.

4. Weaver had been negotiating with the British Post Office on behalf of Edison, Prescott, and Gerritt Smith since the beginning of the year. The only formal agreement was a licence, dated 20 November 1880, which Edison signed on 1 December 1880. According to its terms the Post Office agreed to a royalty of £25 per circuit, with a minimum payment of £300 pounds per year. Prescott to TAE, 23 Dec. 1878, 17 June and 15 Oct. 1879, DF (*TAEM* 19:577; 51:894, 922; *TAED* D7836ZCW, D7936ZAD, D7936ZAX); Agreement between Grosvenor Lowrey and Lemuel Serrell (as trustees for TAE, Prescott, and Smith) and Henry Fawcett, British Postmaster General, 20 Nov. 1880, Miller (*TAEM* 86: 314; *TAED* HM800133).

5. Prescott reported in June that Weaver had been negotiating with the British authorities but "has had a hard time of it. My recommendation is to allow him a percentage upon all he gets. This will make him always alive to our interests. If you agree to this what shall we put the percentage at? Suppose we ask him to name it?" Prescott to TAE, 17 June 1879, DF (*TAEM* 51:894; *TAED* D7936ZAD).

6. The letterbook copy of Edison's response has only minor modifications of this draft reply to Prescott (Lbk. 5:231 [*TAEM* 80:150; *TAED* LB005231]). Prescott responded on 29 September that the company believed the Short patent had no "intrinsic value whatever. We bought it so as to prevent its falling into other hands and forming the pretext for the manufacture of carbon telephones—which it would take us several years to stop through the ordinary operation of the courts" (DF [*TAEM* 51:560; *TAED* D7929ZEH]). In March 1880 Charles Clarke made a series of experiments on the Short telephone and prepared an undated report for Edison. He concluded that "Wherein this telephone is like the Edison it is a success, wherein it differs I can find no improvement; but the contrary" (N-80-03-31:215–38, Lab. [*TAEM* 34:991–1003; *TAED* N072:123–35]; DF [*TAEM* 55:267; *TAED* D8036ZGQ]).

7. It is not known if James Banker came out to Menlo Park on Wednesday, 17 September.

-1805-

From Tracy Edson

New York Sept 17, 1879

My Dear Sir:

I am requested to inform you that Mr. Lowrey and the members of the Executive Committee of the Edison Electric Light Company, propose making[a] you a visit by the Eleven O'clk AM Train on Monday next, for the purpose of seeing the result of the experiments thus far made, and of conferring with you in regard to the best method of providing the additional means that may be required for completing the experiments, and for making a demonstration in New York, after you have made one in the way you now propose, to your and our satisfaction at Menlo Park.[1]

This course has been suggested by Mr. Lowrey at the meeting held today, as being desirable for the purpose of having it where it will be easily accessible to all who may take an interest in the subject, and where they can witness[a] the light in successful operation on a practical scale night after night, and study the manner of introducing and working it to their entire satisfaction, which they could not do without inconvenience at Menlo Park.—

I would suggest that if possible you have one or more of the new Lamps ready to exhibit at that time, and in the meantime, will you kindly think the matter over, and be prepared to give us an approximate estimate of the cost of the necessary Dynamo Machines, Copper conductors, tubes, Lamps, Engines and other apparatus, for on of your proposed "Half mile square[b] Central Stations," with say 500 Lamps, & oblige Yours Very Truly,

Tracy R. Edson

ALS, NjWOE, DF (*TAEM* 50:274; *TAED* D7920ZAZ). [a]Obscured overwritten text. [b]Interlined above.

1. In anticipation of an imminent visit by Edson and James Banker to Menlo Park, Lowrey had written Edison a week earlier about possible means for raising additional funds. He suggested that "the parties in interest may be very willing to advance money to the company but this would not suit you nor me for if there should be considerable delay the lender might foreclose." Lowrey suggested that the electric light had become "immensely more valuable than when we sent in Fifty thousand Dollars. The countries which you have not disposed of are therefore more valuable than they were then. Whatever amount may be required (and let us suppose that it may be One hundred thousand Dollars) I should propose that it should be raised by your selling those countries to the Company for precisely the same considerations as before, that is to say, you receiving one half of whatever stock is issued and the subscribers receiving say three shares for one." By so doing, "the company will not be owing any money to anybody." Stockton Griffin noted that no reply was made to this letter and there is no evidence that Lowrey or others visited on Monday, 22 September; Edison intended to be in Tarrytown, where Lowrey maintained a residence, the following day. Lowrey to TAE, 10 Sept. 1879; TAE to Porter, Lowrey, Soren, and Stone, 22 Sept. 1879; both DF (*TAEM* 50:271, 49:388; *TAED* D7920ZAY, D7903ZID).

[Menlo Park, c. September 17, 1879[1]]

44 is Edisons Experimental Inst [2a]

We find that the reason we cannot adjust the hammaer of the transmitter out far so[b] as to get it loud is that when adjusted to get very loud talking that there is a reciprocal action between the[b] receiver and the transmitter locally & also between the two stations, and this action causes squeaks, & microphonic effects leaving the impression that the fault is all in the transmitter being to delicately adjusted whereas its not the fact because when the transmitters at both ends are adjusted so that the talking is all scrape squeak & rattle & scarcely intelligible the stoppage of the Receiver handle bwhen transmitting Its comes clear & loud and the transmitt[er][c] may still be further adjusted out. The action is this when turning Receiver, on closing push Button the receiver diaphragm gets a knock   this is transmitted through the air and through the bracket to box & thence to transmitter when gives a wave   this wave reacts on the receiver which retransmits to the transmitter on so on giving a perfect squeak   in the act of talking the receiver diaphragm of course works and these waves retransmitted through air & arm to transmitter produce a confusion of rattle & squeak rendering the talking inarticulate at the distant station & also creating the impression that the transmitter is adjusted out too far. Again[b] in addition to this reciprocal action locally there is reciprical action between the terminals   thus on[b] closing push button the receiver sends a knock locally & the distant receiver sends a knock to its transmitter the two transmitters retransmit this knock hence squeaking   If Batchelor is turning & I am also my push button acts on his own instrument the same as his own though not so strong.

We took arm off of instrument held it in hand 2 feet away the[b] talking was loud & perfect upon an adjustment which we couldnt understand scarcely a word with the arm screwed to base. We then worked push button and brought receiver within one[b] inch of mouth of transmitter still holding in hand on[b] closing push button and giving an initial knock we obtained a musical note which only stopped when receiver was 2½ inch away   the closer the receiver to transmitter the higher the note. on putting the arm on wall 3 feet from transmitter base we got a reciprical note. it interfered badly with talking when put on wall 10 feet away. All these effects are done away with when a back point is put on push button to open receiver when talking   no reciprocal action can take place the margin is enormously increased   we find that the idea of wanting facilities for interruption is nonsense as[b] in NY your own voice drowns the interruption[b] & they always wait until you are through with sentence. In England where ~~theye~~ ~~have~~ public has to be taught[b] they can be taught one way as well as another= Another increase in margin by this device is due to the opening of the tertiary thus preventing to some extent the shortcircuiting of the transmitter waves in its own coil.[3][d]

We find that the Wallace Carbons have more margin for adjustment & work better than Jobolkoff[4]   this is explained by looking at them under the microscope   the Wallace Carbons are composed of particles more finely divided than Jobolkoff— on a fire test we find the latter contains impurities there oozes out a brown stuff totally obscuring the carbon particles, in Wallace only a very little white stuff comes out. the resist of Jobolkoff is $^{640}/_{100}$ of an ohm while wallace is but $^{23}/_{100}$ another point in its favor[d]

We find that when the springs on the chalk form a groove sideways ie[e] more on one side than on the other that the rubber presser may have a slight cant given it so as to increase the pressure on the side where it shows deficiency   in this way we make perfect grooves[d]

We find that it is preferable to wet the chalk in the receiver as when damp the[b] chalk before the water fairly gets in is in a softer state & the palladium makes a nice groove in the act of wetting   when if wet first they palladium does not generally as least for a long period make a nice groove.[d]

~~W~~

Chas Batchelor   Page 1–21[f]

X, NjWOE, Lab., N-79-09-18:1 (*TAEM* 35:919; *TAED* No86:1). [a]Sentence preceded by "No 86" written by Francis Upton to identify notebook; followed by dividing mark. [b]Obscured overwritten text. [c]Written off edge of page. [d]Followed by dividing mark. [e]Circled. [f]Signature and "Page 1–21" written by Charles Batchelor.

1. This document precedes the first dated entry in this notebook, which Charles Batchelor started at 2 A.M. on 18 September (N-79-09-18:27, Lab. [*TAEM* 35:932; *TAED* No86:14]). The discussion of the Wallace and Jablochkoff carbons at the end is related to other Batchelor notes of 18 September (see Doc. 1807).

2. Charles Batchelor tested the operation of telephones manufactured for England before they were shipped, identifying each by number (extant records begin on 30 September). Presumably Edison took "44" from the current lot for these experiments. N-79-09-18:39, Lab. (*TAEM* 35:938; *TAED* No86:20).

3. Edison cabled George Gouraud on 19 September, "We open tertiary on back of push button   stops squeak completely increases margin."

Batchelor subsequently informed Edward Johnson that he would send tracings of the coil connections before and after this alteration "in order to show you that if you do not like the arrangement, then you will find inside the coil box two wires [stripped?] up which if connected together leave it in the state as it was at first." TAE to Gouraud, DF (*TAEM* 52: 734; *TAED* D7941ZEF); Batchelor to Johnson, 3 Oct. 1879, Lbk. 5:248 (*TAEM* 80:151; *TAED* LB005248).

This arrangement was subsequently described in a provisional patent specification filed for Edison in London at the end of December (and also in other countries; see Doc. 1835). With the receiver solely in the circuit of the tertiary induction coil,

> the full effect of the current from the distant station is not obtained upon the receiving apparatus, a portion of it passing from the secondary to the primary circuit. In order to obviate this defect a key is placed in the primary circuit, so arranged as to keep that circuit open while the current from the distant station is acting upon the receiving apparatus, the said key being depressed for the purpose of closing the primary circuit when it is desired to actuate the transmitting aparatus. It has been found, however, that upon opening or closing this key a powerful inductive discharge is directed upon the receiving apparatus at the same station. . . .
>
> [This defect] is obviated by opening the tertiary circuit prior to the closing of the primary circuit, and closing the tertiary circuit to leave it in condition to be acted upon by a current from the distant station immediately after the closing of the primary circuit, and also by opening the tertiary circuit prior to the opening of the primary circuit, and closing the tertiary circuit immediately thereafter. [Brit. Pat. 5,335 (1879), Batchelor (*TAEM* 92:141; *TAED* MBP021)]

For sketches of this arrangement from early November see N-79-09-18: 1–15, 107–109, 113–15, Lab. (*TAEM* 35:919–26, 972–73, 975–76; *TAED* N086:1–8, 54–5, 57–8); another undated drawing is in N-79-04-09, Lab. (*TAEM* 32:1115; *TAED* N045:69) and an undated 1879 measured drawing by Samuel Mott of the connections for a transmitter and receiver pair with tertiary coil is in Machine Shop Drawings (1879–1880), Lab. (*TAEM* 45:77; *TAED* NS7986:55).

4. Nothing is known of the chemical makeup of the carbons used in the Wallace arc light but Dredge (1882–85, 1:412) notes that for some time Wallace & Sons had difficulty obtaining suitable carbons as crude retort carbon proved unsuitable in common use; see *TAEB* 4:774 n. 1 for a description of their form. A description of the process for manufacturing Jablochkoff candles is found in Dredge 1882–85, 1:358–61; see *TAEB* 4:719 for a drawing of the candles.

*Edison's British Patent
5,335 (1879) showing the
push-buttons (**G** and **H**)
in the circuit of the tertiary
induction coil (**D**).*

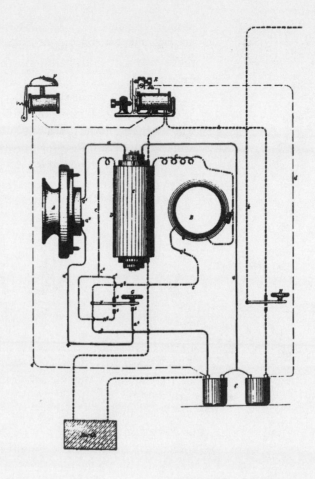

–1807–

*Notebook Entry:
Telephony*

[Menlo Park,] Sept 18th 1879

Edisons Telephone

We noticed that with our new telephones a fault in that the chalk never had so clear a track as noticed in former ones, after following this up persistently for four nights we find that there is considerable difference inbetween our chalks now and previous lots, but on getting some "Cretae precipitae" found we were all right.[1a]

In our inertia telephones with Jablochkoff candle carbons in we noticed a deterioration and on investigation under microscope found ends coated with brown stuff.

I now took a piece of Jablochkoff carbon and heated it in bunsen burner and on examination under microscope it showed itself covered with brown fluffy stuff; scraped this off and heated again and same stuff came out in less quantity.

A piece of Wallace carbon under same treatment gave a white stuff and also showed some pieces of Silica scattered over its surface   also in telephone we noticed that some of the Wallace carbons were perfect insulators, probably due to one of these pieces of silica being in contact with platina instead of carbon

A piece of carbon stick that came from Dr Cleland[2] and looking just like Thompson carbon[3] gave out[b] no other stuff when heated but the grain was much coarser than either of the others   Used[b] all these for transmitters also[a]

Used Plumbago which is very good but there is little margin of adjustment

We now altered the inertia cup to $\frac{5}{16}$ and I cut a carbon button of our old Style to size and put it in so:—

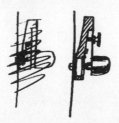

This worked well but not so loud as old transmitter   we now altered the weight making it four times heavier but without any great difference.

We now took the ordinary inertia again & put in a Platinaum black button   We made a brass sleeve to hold the carbon and forced into the inertia cup in order to make one end of brass sleeve and also one end of carbon touch one[b] platina permanently and get our vibration all from the other face, these were very sensitive at a distance but were no improvement close too We tried this with Jablochkoff candle carbon, Wallaces, Dr Clelands, plumbago, platinum black.

We now made some more tests on the old carbon transmitter and found that it kept its resistance more uniform, was louder, could stand a larger battery power, and in many ways was preferable to the inertia and decided to put them on our English telephones   This telephone transmitter which has never been beaten in loudness or distinctness by anybody else's make owes its superiority to the fact that its adjustment is perfectly rigid and its ~~suf~~ surfaces perfectly flat— As the movement of the diaphragm at any time during talking cannot be more than the .001 of an inch, it stands to reason that if there is any[b] shake in adjustment or any rounding of surfaces, it uses this movement all up in taking up such slack etc   We propose to use it so:—[4]

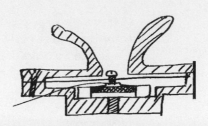

One of the great difficulties with our inertia telephones is that after using some time a number were taken down and tested for resistance with same pressure on each. They varied from 40 ohms to 6 ohms and some were insulators. This variation would kill them entirely.

Chas Batchelor

X, NjWOE, Lab., N-79-04-09 (*TAEM* 32:1106; *TAED* N045:60). Written by Charles Batchelor. Document multiply signed and dated. [a]Followed by dividing mark. [b]Obscured overwritten text.

1. In another 18 September notebook entry, Charles Batchelor described experiments with wetting one-ounce buttons that had been pressed to different degrees. He moistened the first batch with 90 minims of a 75% solution of phosphate soda in water, noting that "as we turn away a great part of the chalk we must put in the same percentage of the 90 minims as there is percentage of the chalk left after turning." He then obtained somewhat better results with "5 pair of chalks of old formula" wet with 15 minims of a 50% phosphate soda solution. In the latter case, a button made using seven spacer washers to increase the effective force of the press performed "red hot" (N-79-09-18:27–33, Lab. [*TAEM* 35:932–35; *TAED* N086:14–7]). None of these notes clarifies the meaning of "Cretae precipitae" but it was most likely precipitated chalk with Batchelor playing on the Latin word *creta* ("chalk"), the root of the English "cretaceous" (*OED*, s.v. "Cretaceous").

2. T. Cleland, a medical doctor, was the Eastern agent for Edison's electric pen and for Western Electric Manufacturing Co.'s electromedical department with offices at 267 Broadway in New York. See *TAEM-G1–2*, s.v. "Cleland, T."

3. Elihu Thomson and Edwin Houston devised an arc light in which two opposing solenoids regulated the distance between the carbon rods in order to maintain a steady arc; nothing is known of the carbons used in this device. Carlson 1991, 124–27; Prescott 1879, 498–501.

4. Figure label is "straight glass plate [in?] here for screw to strike against." On 20 September John Kruesi wrote an order for the shop to "Make a new carbon Transmitter for english Telephone order   See drawing"; the measured drawing of the transmitter was made the same day by either Kruesi or Samuel Mott. Cat. 1308:175 (Order No. 242), Batchelor (*TAEM* 90:754; *TAED* MBN003:61); Vol. 14:171, Lab. (*TAEM* 4:327; *TAED* NV14:179).

London Sept 19, 79 9:37 A.M.

From George Gouraud[1]

Edison[a]

Indispensable send immediately six superior inspectors pay to commence 900 increasing if satisfactory[2] Cable steamer

L (telegram), NjWOE, DF (*TAEM* 52:734; *TAED* D7941ZEE). Written by Stockton Griffin. [a]"Cable" appears above.

1. George Gouraud wrote to Edison a few days later confirming the text of this document, as he often did for his cables. Although Stockton Griffin noted on the back of this telegram that it was from Arnold White he apparently was confused by the ciphers used for cable correspondence about the English telephone; on this day he also erroneously identified the Telephone Co. of London as the recipient of a cable addressed to Gouraud's code name. Gouraud to TAE, 22 Sept. 1879; TAE to Gouraud, 19 Sept. 1879; both DF (*TAEM* 52:737, 734; *TAED* D7941ZEI, D7941ZEF).

2. In his letter confirming this cable (see note 1) Gouraud stated the salary offer in dollars. He explained in another 22 September letter that he desired "the six inspectors for the District Companies. We are very much in want of such men" (DF [*TAEM* 52:739; *TAED* D7941ZEJ]). The first inspectors known to have gone to England were H. C. Rector, who sailed on or about 3 October, and John Crawford and John Wohltman, both recommended by Sigmund Bergmann, who left a few days later. Others known to have followed in 1879 were James Lighthipe, O. A. Enholm, and J. H. Gibson (Rector to TAE, 3 Oct. 1879; Bergmann to TAE, both 30 Sept. 1879; Wohltman to TAE, 7 Oct. 1879; Crawford to TAE, 7 Oct. 1879; TAE to Edison Telephone Co. of London, 7 Oct. 1879; Enholm to TAE, 31 Oct. 1879; Lighthipe to TAE, 4 Nov. 1879; and Gibson to TAE, 13 Dec. 1879; all DF [*TAEM* 52:771, 756, 757, 776, 777, 882, 893, 971; *TAED* D7941ZFC1, D7941ZEV, D7941ZEW, D7941ZFF1, D7941ZFF2, D7941ZFG, D7941ZHH2, D7941ZHN1, D7941ZJM2]; "Lighthipe, James A.," Pioneers Bio.). Edison received a number of other inquiries and recommendations, including two more referrals from Sigmund Bergmann (Osgood Wiley to TAE, 20 Sept. 1879; John Lenhart to TAE, 24 Sept. 1879; William Yackly to TAE, 27 Sept. 1879; John Little to TAE, 2 Oct. 1879; John Wright to TAE, 5 Oct. 1879; Bergmann to TAE, 16 Oct. and 8 Dec. 1879; and G. L. Wiley to TAE, 3 Nov. 1879; all DF [*TAEM* 52:736, 740, 748, 767, 773, 822, 948, 892; *TAED* D7941ZEG, D7941ZEK, D7941ZER, D7941ZFB, D7941ZFE, D7941ZGF, D7941ZJA, D7941ZHL]).

After acquiring rights to the British colonies and some other countries in October (see Doc. 1742), Gouraud anticipated that he would need to hire a number of inspectors and other employees familiar with telephone and electrical technology. He therefore had an application form printed that consisted of twenty-six questions, including the applicant's facility with foreign languages and preferred country of assignment. The form was apparently for New York City applicants as it asked for local references and was to be returned to Gouraud through an office at 120 Broadway (undated application form, DF, Supp. III [*TAEM* 162:1062; *TAED* D7941ZKX]).

Shortly before Gouraud sent this document, Edward Johnson cabled "Want Rose one year subject three months notice pay fifteen hundred must sail immediately See Bergmann." A. W. Rose left on 4 October, evidently to become Johnson's assistant in London rather than a district company engineer. Nothing is known of Rose or his associations with Johnson and Bergmann. Johnson to TAE, 19 Sept. 1879; Rose to TAE, 25 and 29 Sept. 1879; Gouraud to TAE, 26 Sept. and 14 Oct. 1879; TAE to Edison Telephone Co. of London, 2 Oct. 1879; all DF (*TAEM* 52:734, 745, 749, 746, 204, 766; *TAED* D7941ZED, D7941ZEN, D7941ZER1, D7941ZEP, D7939X, D7941ZFA); N-79-09-18:45, Lab. (*TAEM* 35:941; *TAED* N086:23).

–1809–

To Arnold White

[Menlo Park,] Sept 22d 1879

Dear Sir.

Referring to you favor of the 8th inst in reference to patenting the Switch board[1] I think all that is necessary is to have Mr Johnson ta[ke][a] the board, or exhibit it to Brewer & Jensen [patent][a] agents Chancery Lane and have them take o[ut a][a] patent as a communication.[2] Mr Johnson can [undoubtedly][a] write out the claims for the same. There should [be][a] a combination claim made [for?][b] between the board [and][a] the annunciator blocks—also the method of construct[ing][a] the board. Two boards should be shown in the patent with transfer strips, and claims made there[fore.][3a]

In the same patent the new instrument should be included All that will be necessary will be to have one of the new instruments taken to Brewer & Jensen.

I send herewith a copy of my U.S. patent, the difference between this patent and the instrument you have should be patented.[4] Very truly

L (letterpress copy), NjWOE, Lbk. 5:221 (*TAEM* 80:149; *TAED* LB005221). Written by Stockton Griffin. [a]Faint letterpress copy; text taken from draft. [b]Canceled.

1. In a letter of 8 September, White had informed Edison that the directors of the London company wanted "the necessary information for patenting the Switch Board recently sent by you to this Company. For this purpose it will be necessary to have a full technical description of the instrument together with a clear definition of the novelties or improvements upon previous devices." Edison's draft reply on that letter is essentially the same as this document. White to TAE, 8 Sept. 1879, DF (*TAEM* 52:707; *TAED* D7941ZDW).

2. Edward Brewer and Peter Jensen were London patent solicitors used by Lemuel Serrell in connection with Edison's British applications

(see *TAEB* 3:527 n. 1). In his reply to this document White stated that the directors "are not particularly impressed, with the manner in which Messrs Brewer & Jensen have protected your interests under the English Patent Law" and would not engage them for this purpose (White to TAE, 11 Oct. 1879, DF [*TAEM* 52:794; *TAED* D7941ZFR]).

3. Edward Johnson reported that on 18 September he had written "a long spec. for Switch Board   We are going to essay a Patent—to secure ourselves from the possible danger of other parties getting the thing patented—it is so new in this country it is not safe to allow it to run loose." Arnold White filed it in his name as a communication from Edison on 20 September. The provisional contained no formal claims or drawings but did describe in detail the electrical connections between the switchboard proper and the annunciator (a device to automatically transmit a subscriber's number so that the attendant can connect with him or her); it also described the physical layout of the switchboard grid. Each of these elements was embodied in the claims of the final specification. The two patent drawings show an elevation of the entire apparatus and a cross-section of a portion of the switchboard. Johnson to TAE, 19 Sept. 1879, DF (*TAEM* 52:728; *TAED* D7941ZEC); Brit. Pat. 3,794 (1879), Batchelor (*TAEM* 92:129; *TAED* MBP018).

4. Edison's enclosure has not been found. It is not clear to what U.S. patent he is referring but he probably meant his recent U.S. application Case 175, which covered the design of the first electromotograph receiver and combined transmitter sent to Britain in February 1879 (see Doc. 1681). When the Patent Office allowed this application in October Edison instructed Lemuel Serrell not to pay the issue fee immediately "as I am going to take out foreign patents and dont want it published"; it issued as U.S. Patent 221,957 on 24 November (Serrell to TAE, 20 Oct. 1879, DF [*TAEM* 51:585; *TAED* D7929ZEZ]; TAE to Serrell, 22 Oct. 1879, Lbk. 5:294 [*TAEM* 80:165; *TAED* LB005294B]). Arnold White filed a provisional specification covering recent modifications to the receiver on 31 December 1879. It described the substitution of palladium for platinum at the tip of the contact spring, the use of the tertiary coil, and the key for disconnecting the primary circuit and interrupting the tertiary circuit (Brit. Pat. 5,335 [1879], Batchelor [*TAEM* 92:141; *TAED* MBP021]).

–1810–

From George Barker

Phila. Sept. 23, 1879.

My dear Edison:—

Having reached home all right and having got my classes into working order I write to thank you personally for your kindness in contributing so much to the success of the meeting at Saratoga. I hope you enjoyed your trip and that you felt better for your little vacation. Your success in presenting the telephone should render you less reluctant to present your valuable papers in person.

With regard to the papers you presented there, I want to make a few suggestions. The one on the Resonant Fork you handed to me and I presented it.[1] It may go to the Sec. for pub-lication just as it is. The paper on the "Electro-chemical tele-phone" I wish you would be kind enough to write out at some length, giving some account of the motograph principle and the various ways it has been utilized, discussing somewhat its principle of action, and describing its use in the telephone.[2] The paper on "Heating the wires in a vacuum" by far the most important, I am very anxious to have more full and complete. The account you wrote in Saratoga was published in full everywhere.[3] The value of our Proceedings lies in its having the papers read at the meetings not too widely published out-side. Because of the great importance of the subject, and be-cause the paper read was so widely published, I want very much to have you re-write the manuscript you hastily drew up at S. making it at least twice as long and going more into detail. Do not trouble yourself to write it. Let Griffin take it down as you dictate it and then write it out in full. If you will send both these papers to me I will look them over before they go to the printer. Or if you say, I will come over & help you get them up. I am so anxious to have the Proc. enriched by the paper that I will do anything to help it on. I had intended to have it pub-lished in the Am. J. Sci. and credited to the A.A.A.S. But I have just received an urgent letter from Prof. Remsen of Balti-more, asking me to get it for his "American Chemical Journal"[4] and saying that Prof. Gibbs of Harvard wants to add a note to it, as he thinks the results very important.[5] I am inclined to fa-vor Prof. Remsen & to ask for him rather than for myself. If you will rewrite it more fully, giving more of the details, and can do it within ten days, he can get it in the next number of his Journal. Either or both of these publications if they credit the paper to the Assoc. would not interfere with publishing in the Proceedings. If you want me, I will come over & spend Satur-day or Sunday with you (or both) and we will fix the thing up right away.[6] Any wood cuts that you have or can get to illustrate the papers, I hope you will secure.

I brought home my motograph receivers that you gave me all right. But I have no transmitters to work them with nor coils (tertiary). Do send me a pair of the "bang up" carbon transmitters that you promised[7] and a pair of the tertiary coils to use them with on the motograph phones. I want to show them to my class & I can do nothing without the tertiary coil.[8] When you get a pair of the new phones for talking that you can spare, I wish you would send them along too.

Have you any pieces of wire which have been treated by

your process and which are large enough to take the density and to measure the tenacity of? I want to see what physical constants are altered and by how much.

Our friend Draper reached home in first class condition having had great success on his hunt. Struve was very sorry not to see you but he waited till the last moment before telegraphing you that he was coming[9] & when you answered that you would not be at Menlo that afternoon, he had no time to make another appointment.

Mrs B. desires regards to Mrs E. & yourself. Wishing you all success I remain   Cordially yours

George F. Barker.

ALS, NjWOE, DF (*TAEM* 49:389; *TAED* D7903ZIE).

1. On 9 October Barker returned a copy of this paper which Edison had given him in Saratoga and stated that he had arranged for its publication in the *American Journal of Science.* He also asked if Edison had any relevant woodcuts but the article appeared with a single line drawing. It was subsequently published in the *AAAS Proceedings.* Barker to TAE, 9 Oct. 1879, DF (*TAEM* 49:406; *TAED* D7903ZIR); "On a Resonant Tuning Fork," *Am. J. Sci.* 3rd ser. 18 (1879): 395–96; Edison 1880b.

2. Edison failed to revise the paper and it was not published in the *AAAS Proceedings.*

3. The draft paper is Doc. 1796. Barker repeated his request for Edison to expand it in his 9 October letter (see note 1) and it was subsequently published without significant revisions in the *AAAS Proceedings* (Edison 1880a); for its reprinting by newspapers, see Doc. 1796 n. 1.

4. Ira Remsen, noted professor of chemistry at the Johns Hopkins University, founded the *American Journal of Chemistry,* the first continuing journal of American chemical research, in 1879. There is no extant Edison correspondence with Remsen and the article was not printed in the *Journal* or the *American Journal of Science. DSB,* s.v. "Remsen, Ira."

5. Wolcott Gibbs, the Rumford professor of chemistry at Harvard since 1863, established his reputation in part by his research on platinum metals. *DSB,* s.v. "Gibbs, (Oliver) Wolcott."

6. Stockton Griffin noted on this letter that Edison telegraphed a reply on 26 September (Friday) but this has not been found.

7. Presumably the inertia transmitter.

8. Barker reiterated this request in his 9 October letter and again on 13 October. DF (*TAEM* 49:406, 418; *TAED* D7903ZIR, D7903ZIU).

9. Otto Struve, the eminent astronomer and director of the Pulkovo Observatory in Russia, arrived in the United States in mid-August to arrange for the construction of a new telescope. He visited New York briefly before sailing home on or about 10 September. His telegram to Edison has not been found. *DSB,* s.v. "Struve, Otto Wilhelm"; Batten 1988, 196–99.

New York, Sept. 23d 1879[a]

My Dear Sir,

I understand Mr. Wiley[1] has applied to you, on behalf of the G.&S. T. Co. for two of your extra loud Motophones for Exhibition at the Fair of the American Institute in this City.[2]

If possible, I hope you will comply with this request, as the Public is much interested in this wonderful novelty, and it would greatly extend your reputation to gratify the national desire to see it.

All expenses will of course be borne by the Company. Very truly Yours

Tracy R. Edson

⟨We will ~~ma~~ do as you request & furnish chem telephones for[b] the fair & send our boys to fix it OK   TAE⟩

ALS, NjWOE, DF (*TAEM* 52:112; *TAED* D7937ZCQ). Letterhead of Gold & Stock Telegraph Co. [a]"New York," and "187" preprinted. [b]Obscured overwritten text.

1. George L. Wiley was superintendent of the Gold and Stock Telegraph Co.'s New York Private Line and Telephone Department. Wiley to TAE, 4 June 1879, DF (*TAEM* 52:574; *TAED* D7941ZBO); Reid 1879, 626.

2. Wiley asked Edison on 17 September to provide "very promptly, say within 8 or 10 days, one pair of extra loud 'Electro Motographs'" for the American Institute Fair, an annual exhibition in New York of manufacturing, invention, and agriculture. Edison indicated in a draft reply that he had "none of ex loud on hand only mkg regular kind now" and offered to supply those if desired. Wiley wrote again on 22 September, reminding Edison of the Fair's prestige and urging him "either to make a pair of the loud speakers or to alter a pair of the regular kind and make loud speakers of them." Edison agreed to have a pair of each made and promised to "send my men over to get them going." The instruments were to be demonstrated over wires from two downtown offices to the exhibition building on 63rd St. After testing them, Wiley wrote on 24 September that "the cheml. receiver receives better from the regular carbon transmr than from the transmr that accompanies it." Edison noted on this letter that "we have also thrown away the trans & gone back to the old carb trans." Wiley to TAE, 17, 22, and 24 Sept. 1879, DF (*TAEM* 50:486, 52:110, 113; *TAED* D7924B, D7937ZCO, D7937ZCR) on the American Institute see *TAEB* 4:516 n. 4.

Notebook Entry:
Electric Lighting

Electric Light   Lamp spool.[1a]

New Method of winding spools: I[b] wind one layer on the spool and fill up between the convolutions (which are slightly apart) with a thick solution of Acetate of Magnesia and Alumina[2] put on by a brush, this is then baked at about 250° fahr, and then another coating is put on & baked again; these are continued until the wire is entirely covered and about .01 inch above the wire; another layer is then put on and this is set by a few brush-fulls[c] of the solution—

T.A.E   Chas Batchelor

X, NjWOE, Lab., N-79-07-31:73 (*TAEM* 33:578; *TAED* N052:36). [a]Multiply underlined. [b]Interlined above. [c]Obscured overwritten text.

1. The same day Batchelor drew up a list of lamps to be made with different coatings. In another notebook he also wrote instructions for making a brass spool and indicated that six lime and six alumina spools should be made, apparently in addition to the other spools he had listed this day. Experiments with different spool coatings continued into early October. N-79-07-31:77–83; N-79-07-25:17; N-79-08-22:85–91, 111, 117–23; all Lab. (*TAEM* 33:580–83, 868; 35:825–28, 838, 841–44; *TAED* N052:38–41; N056:9; N085:42–5, 55, 58–61); see Doc. 1819.

2. On 8 September, Batchelor had listed eleven experimental spool coverings including one that was 50% alumina and 50% magnesia and another that was 80% alumina and 20% magnesia. It is not known what percentages he used for the new method. N-79-07-31:67, Lab. (*TAEM* 33:575; *TAED* N052:33).

Menlo Park, N.J., Sept 26 1879[a]

To Edison Telephone
Co. of London, Ltd.

~~Menlo park~~ Quephone London[1b]

Johnson exhaustive experiments proved old transmitter only reliable one[2]   hundred instruments ready waiting transmitters nearly done   balance order follow rapidly   making arrangements give hundred weekly[3]   go for Blake ~~transmitter~~ same thing except shape of parts[4]   Bergmann on Switchboards[5]   Impure chalk explains your differences[6]   I guarantee absolutely reliable hereafter

L (telegram), NjWOE, DF (*TAEM* 52:747; *TAED* D7941ZEQ). Written by Stockton Griffin; letterhead of T. A. Edison. [a]"Menlo Park, N.J.," and "18" preprinted. [b]"Cable" written above.

1. Edison would have just received Arnold White's 13 September reminder that "the Register of Cables addressed to this Company is Quephone London." George Gouraud's registered address was "Menlo Park London." DF (*TAEM* 52:711; *TAED* D7941ZEB).

2. Having received no reply to Docs. 1790 and 1799, Edward Johnson wrote a letter on 14 September (which probably would have just reached Menlo Park) chastising Edison for failing to respond to earlier reports of difficulties with the inertia transmitter. Johnson stated that he was

in utter ignorance as to whether you count my experience as of any value as against your conclusions— It is of the utmost importance to the Company that they at least should know of these things— I'll now give you the result of a thorough and complete test of your two forms of Transmitters and you will then understand the anxiety of everybody to know what you have decided upon—much depends upon it. No one here feels disposed to telegraph you to do so & so in ignorance of your grounds for a contrary course— You can readily understand that.

He then described at considerable length his recent experiences with the carbon and inertia transmitters. Of the four telephones with the swing-arm receiver and regular carbon transmitter, Johnson had placed one at the telephone exchange office, used another in the *Times* demonstration, and taken the remaining two for exhibition in Liverpool and Manchester. He reported "Better talking invariably received from the Exchange than from any station no matter who speaks from there—and all our subscribers frequently do—in visits they pay us." He also claimed that the *Times* demonstration succeeded "only by virtue of this Inst. the others failing to give a voice of sufficient volume & uniformity." Finally, he reported that the other instruments performed admirably numerous times for prospective investors in Manchester and Liverpool. With the company needing 2,500 instruments for London and three other cities, "The question now is what Transmitters shall be put on them. If my judgment is to decide the question I now decide it—put on the Lamp-black Button & it only—never mind the increased cost. It is the only True Transmitter   A pin point will work—and is very sensitive—but if it fails as it frequently does—no amount of lung power will resurrect it." DF (*TAEM* 52:712; *TAED* D7941ZEB1).

3. This represented an acceleration of the delivery schedule. Arnold White cabled Edison on Wednesday, 10 September to "Expedite dispatch Telephones," to which Edison replied the same day, "We ship Tuesday fifty London direct   50 weekly thereafter." At the beginning of October Edison arranged for Sigmund Bergmann to manufacture two thousand telephones with lampblack transmitters at the rate of one hundred per week, each one to cost $13.82. Bergmann was to have the first telephones ready within four weeks; he sent "the balance of first hundred telephones" on 12 November with one hundred per week to follow thereafter. At that time Bergmann asked Edison for reimbursement of the $7,000 that he had spent fulfilling the manufacturing contract. White to TAE, 10 Sept. 1879; TAE to White, 10 Sept. 1879; Edward Johnson to TAE, 30 Sept. 1879; TAE to Johnson, 30 Sept. 1879; White to TAE, 1 Oct. 1879; Edison Telephone Co. of London to TAE, 2 Oct. 1879; TAE to Edison Telephone Co. of London, 2 Oct. 1879; Bergmann to TAE, 12 Nov. 1879; all DF (*TAEM* 52:708, 709, 749, 750, 764, 766, 133; *TAED* D7941ZDY, D7941ZDZ, D7941ZES, D7941ZET, D7941ZEY,

D7941ZEZ, D7941ZFA, D7937ZDO); TAE to Bergmann, 4 Oct. 1879, Lbk. 5:250 (*TAEM* 80:152; *TAED* LB005250).

4. On his arrival in Liverpool Johnson found that a Bell company had been organized there which intended to use the Blake transmitter. However, the local investors agreed to witness a head-to-head exhibition of the Edison telephone with the Bell and Blake instruments. Johnson reported in his 14 September letter (see note 2) that with the motograph receiver and "Lamp Black Transmitter I held my own & in the judgment of everybody won the day."

5. Fifty switchboards were on order for the provincial companies. Edison Telephone Co. of London to TAE, 10 Sept. 1879, DF (*TAEM* 52:708; *TAED* D7941ZDX).

6. It is unclear if Edison was responding to the difficulties that Edward Johnson had described in Doc. 1790 or to Johnson's 14 September report (see note 2) that the receiver had more difficulty with induction on short London lines than with induction of apparently equal intensity on a longer line from Manchester. Johnson speculated that the induction in London was "more continuous thus keeping the chalk constantly lubricated—and offering no margin of difference for the talking currents."

October–December 1879

October marked the start of a period of extraordinary activity at the Menlo Park lab, even by Edison's unconventional standards. He confronted serious legal, administrative, and technical obstacles to the commercial use of his telephone; undertook a wholly different line of research for his electric lamp; and, in late December, mounted the first large-scale public demonstration of incandescent electric lighting.

Legal experts engaged by the telephone company in London concluded that Edison's basic patent on the carbon transmitter in that country was unsound and could not be defended. He delved into English patent case law, then spent considerable time preparing instructions to Edward Johnson to amend the specification. During this process he invited the help of Zenas Wilber, an interference examiner in the U.S. Patent Office. Convinced that the original specification had been improperly drawn, he arranged for Wilber to become his patent attorney after the first of the year. George Gouraud engaged in complex negotiations on Edison's behalf to resolve a misunderstanding with the company's investors over formation of new companies in provincial cities. Edison also granted him authority to dispose of his patents in British colonies and a number of other countries. Edward Johnson, however, worried that Gouraud was not acting in good faith and began reporting his activities to Edison. Affairs were less congenial in Paris. Efforts to consolidate the Edison interests with rivals into a unified French telephone company failed at the last moment amid ill–will and suspicion, although Joshua Bailey promised to try again. Almost simultaneously, Edison's nephew Charley died in Paris on 19 October after a brief but agonizing illness. Edison arranged transportation of Charley's body but did not attend the funeral in Michigan on account of having to prepare testimony for the telephone interference cases.

In mid-October Edison learned from Edward Johnson that electromotograph chalks from several different shipments had all stopped working. Edison replied that this "would puzzle the Deity himself" and acknowledged that the news had "sent a little leaden pain right in the pit of my stomach."[1] He promised to send the equipment Johnson would need to make his own chalks and also prepared to tackle the problem himself. Eleven days later he reported that after sleeping only fifteen hours in a week he (with seven assistants) had devised a solution. This proved only temporary, however. In mid-November he advised Johnson to throw away the chalks just sent because only those made and moistened with distilled water would perform satisfactorily. This seems to have solved the problem. During this time Edison also had to address unrelated difficulties with the telephone batteries and operation of the call bells.

In other matters, Edison received an order from the Gold and Stock Telegraph Co. at the beginning of October for one hundred receivers, which he turned over to the Western Electric Manufacturing Co. On 14 October, he executed an agreement with José Husbands to establish a telephone company in Chile.[2] He also continued to recruit and train inspectors for Great Britain. He was again forced to deal with a long-running dispute with E. Baker Welch, who had given Edison financial support for multiple telegraph experiments a decade ago and subsequently laid claim to the quadruplex telegraph. Welch rejected a settlement offer and Edison's attorney, Grosvenor Lowrey, advised him to transfer assets out of his name in case of a lawsuit but Edison replied that Western Union would be obliged to provide him legal protection. In December the Edison Ore Milling Co. was incorporated and Edison entered into contracts to extract gold and platinum from the waste sands of several California hydraulic claims.

On the electric light, Edison continued in early October to push experiments in the same direction as he had previously. Ludwig Böhm built several new vacuum pump designs with multiple mercury jets and various attachments, all intended to increase the pumps' efficiency and extend their capability. Francis Upton continued to work on electric meters. Upton also carried on a dispute in the pages of the *Scientific American* with several electrical experts who contested Edison's claims for the efficiency of his dynamo.

At the beginning of October Edison had not decided on a

definite design for his electric lamp. In fact, although he had been making arrangements to obtain enough platinum ore to manufacture it on a large scale, he had yet to surmount two fundamental problems. One was the lamp's relatively low resistance, about 10 ohms, which required proportionally more current and larger (and more expensive) conducting wires. One possible solution was a form of circuit regulator in which a motor-driven commutator successively interrupted the current to each lamp in a household. Reducing the amount of energy consumed by each lamp effectively raised its resistance, from the standpoint of supplying the household circuit. Edison included this promising approach in a caveat that he apparently drafted during the fall. The other problem was the lack of an insulating compound which would prevent the platinum burner from short circuiting or volatilizing. By virtue of his novel process of removing occluded gas from the metal Edison could raise platinum wire to unprecedented temperatures but he had not, after months of work, discovered a suitable insulator which did not volatilize at the same heat.

There matters stood on 7 October when Charles Batchelor recorded experiments with spirals formed of soft pressed or extruded carbon. Material readily at hand was used: arc light carbons and the lampblack (mixed with tar) used in making telephone transmitter buttons. Batchelor noted that the spirals were wound on lime bobbins just as were the thin wires in the platinum lamps. It was quickly apparent that they gave much higher resistance than platinum. One five-inch length of carbon was measured that day at 100 ohms. This was not Edison's first effort to make a lamp burner of some form of carbon, and he had investigated using other substances in a similarly finely divided state to increase their resistance.[3] The laboratory notebooks give no clue to what may have prompted him to turn again to carbon at this time.

Edison directed related experiments for several weeks using carbon in different shapes and forms. Still, work continued on platinum spirals until 13 October and Francis Upton worked on insulating compounds as late as 17 October. Edison was preoccupied with the telephone and then Charley's death but on 21 October his lamp research took another direction. On that night Charles Batchelor recorded new ways of baking and attaching what he called the "carbon wire."[4] He reported "some very interesting experiments" that night on carbonized fibers of various types.[5] A length of carbonized thread showed high resistance; other materials, including fishing line, paper, cardboard, and various threads were promptly carbonized and placed in evacuated lamp globes. Batchelor matter-of-factly noted that a number of these slender carbons produced a bright light, gave high resistance, and significantly, did not appear to oxidize. During the succeeding weeks dozens of substances were carbonized and tried in lamps. Batchelor's notes indicate a gradual shift from thread to slender strips of paper or cardboard bent into a horseshoe shape. They also show attention to practical design problems such as attaching the carbon to the lead-in wires. On 1 November Edison executed a patent application for a high resistance carbon lamp (with a spiral burner). By the middle of the month he had settled on a design employing a narrow horseshoe of carbonized cardboard and spring clamps to hold it in place; a few weeks later these were replaced by tiny platinum screw clamps. Less than a month after testing his first carbon filament he began preparing the laboratory complex for a public exhibition at the end of the year. News of Edison's discovery began appearing in the newspapers but the first full account, published prematurely on 21 December by the *New York Herald,* created a sensation. By the next week, without any official inauguration, the exhibition was underway as hundreds of people poured nightly into Menlo Park to see scores of electric lamps burning there. An exhausted Edison reportedly looked "as if he had really worked half his life out in searching for this electric light, and was ready to sink into a premature grave" but he availed himself to reporters and guests while employees kept a watchful eye over the crowded laboratory.[6]

Edison added about a dozen and a half men to his payroll, a number of them apparently in late December. They included William Hammer, who had worked with electrician Edward Weston;[7] machinist James Bradley, who had worked with Edison in Newark;[8] and Theodore Carman, a teamster who had worked on construction of the new office building in 1878.[9] William Andrews, a machinist or mechanic, also joined the staff.[10] Several of the new assistants were put to the time-consuming task of operating vacuum pumps; among these were George Crosby[11] and Wilson Howell.[12] Arthur Poinier helped with carbonization experiments.[13] Charles Hughes worked principally as a purchasing agent, acquiring materials for the laboratory.[14] Samuel Mott, a draftsman, was at the laboratory by October.[15] In addition, George Carman, who had been Edison's purchasing agent since 1877 and sometimes helped in the laboratory, began working only as a laboratory assistant in December.[16]

1. Doc. 1825.
2. Doc. 1823.
3. See, e.g., Docs. 1455, 1479, and 1555.
4. Doc. 1830.
5. Doc. 1831.
6. "Edison's Electric Light," *Times* (London), 14 Jan. 1880, 8.
7. See Doc. 1972 n. 7.
8. See Doc. 1939 n. 8.
9. See Time Sheets, NjWOE and *TAEB* 4:694 n. 5.
10. See Time Sheets, NjWOE.
11. See Doc. 1926 n. 3.
12. See Doc. 1985 n. 1.
13. See Doc. 1850 n. 6.
14. See Doc. 1965 n. 5.
15. See Doc. 1985 n. 7.
16. Carman's testimony, *Edison v. Maxim v. Swan* (*TAED* W100DID [image 4]).

–1814–

From George Prescott

New York, October 1st 1879[a]

Dear Sir

Please manufacture, for this Company, One hundred Motophones, of your latest and most approved style.[1] Yours respy

George B. Prescott   Vice President.

Please inform me when they are ready for inspection, and I will send an inspector to Menlo Park to inspect them if this arrangement will be satisfactory to you. G.B.P.

⟨I will inform you when they will be ready & when to send Inspector⟩[2b]

ALS, NjWOE, DF (*TAEM* 52:120; *TAED* D7937ZDB). Letterhead of Gold and Stock Telegraph Co., George Prescott, Vice President. [a]"New York," and "187" preprinted. [b]Marginalia written by Edison.

1. On 1 October, James Green of Gold and Stock telegraphed Edison: "What will be the price of the hundred Chemical Telephones." Edison replied the same day: "Will give you estimates in a day or two." On 8 October, Green wrote that the company's executive committee had agreed to Edison's price of $13.82 per telephone for this order; this was the same price at which Bergmann was manufacturing telephones for Britain (see Doc. 1813 n. 3). Green to TAE, 1 and 8 Oct. 1879; TAE to Green, 1 Oct. 1879; all DF (*TAEM* 52:119, 122; *TAED* D7937ZCZ, D7937ZDF, D7937ZDA).

2. Gold and Stock had not received the telephones by 20 February 1880 when Prescott again wrote Edison about the order. Edison responded that he "had no facilities to perform the work and turned the order over to the WE Mfg Co N.Y. who are making them and doubtlessly will finish ere the next centennial." On 4 March, George Phelps Jr., superintendent of Western Electric's New York factory, where the telephones were being manufactured, notified Edison that the parts were all

finished and they were in the process of putting the telephones together. However, those they had finished were not working as well as desired and he asked Edison to "send one of your men skilled in motophone over here to give our boys some points in setting up & final adjustments." Dissatisfied with the man Edison sent, Phelps wrote again on 18 March to ask him to send another. By 25 March the telephones were apparently ready for delivery. Prescott to TAE, 20 Feb. 1880, with TAE marginal note; Phelps to TAE, 4, 18, and 25 Mar. 1880; all DF (*TAEM* 55:603, 609, 623, 627; *TAED* D8043I, D8043O, D8043X, D8043ZAC).

<table>
<tr><td>

–1815–

Notebook Entry:
Electric Lighting

</td><td>

[Menlo Park,] Oct. 1, 1879

Mr. Edison claims eight candles from this spiral. George[1] read at the end of the room. A rough estimate, assuming 3 ohms as resistance and 2 Webers of current gave 65 per Horse Power. This was from 2½ inches (Batchelor) of .006 Pt. Ir. 20% wire

There is millions in it!!!

Shall put the spirals in bridge

</td></tr>
</table>

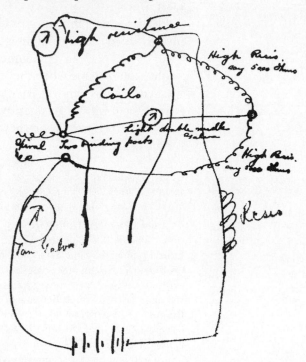

Should have another Galva[nometer] with light needles.[a]

X, NjWOE, Lab. N-79-08-22:92 (*TAEM* 35:829; *TAED* No85:46). Written by Francis Upton. [a]"with light needles" apparently added later.

1. Probably George Carman, who did odd jobs around the laboratory and office. Jehl 1937–41, 1:318.

[Menlo Park,] Oct 2d 1879

Changes in [---]ᵃ pump in two weeks.[1]

Combination with Geissler.[2]
Macleod Guage.[3]
Cu. tube for absorbing Hg[4]
Mercury bottles at bottom of fall tubes thus

so that the escaping Hg. could be collected in one vessel.
Putting pure rubber in white rubber tubes.
Using one fall tube
Doing away with cloth strainer.

No. 3

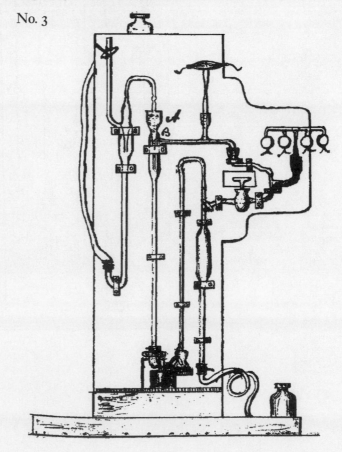

This pump produced a good vacuum in 20 minutes (F[rancis]. J[ehl].)

A. The fall tube sealed at this point with Hg so that in case it became clogged it could be taken out and cleaned.

One tube used as it was found easier to manage[b] than two or more.

B. The pump broke here. No reason given.

No. 4.[5]

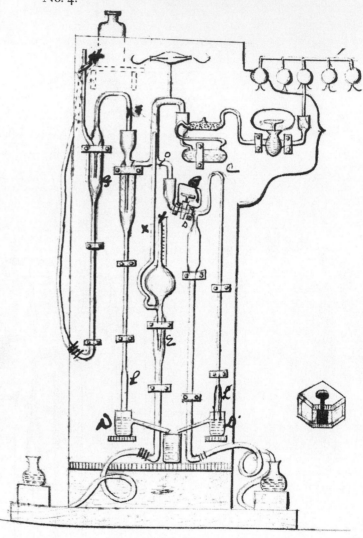

S. D. Mott

X, NjWOE, Lab. N-79-08-22:95, 97, 96, 99 (*TAEM* 35:830–32; *TAED* No85:47–49). Written by Francis Upton; drawings of vacuum pumps No. 3 and No. 4 by Samuel Mott, who signed and dated them. [a]Canceled. [b]Second "a" interlined above.

1. See Doc. 1803 for earlier pump designs.

2. In his 1887 paper on "The Development of the Mercurial Air Pump," Silvanus Thompson noted that a combination of the Geissler and Sprengel pumps such as that suggested by Edison "consists merely in sealing the exhaust tubes of each pump together and to the lamp, [and] cannot be recommended. If the Geissler exhausts more perfectly than the Sprengel, or *vice versa*, then the other pump is useless." Thompson 1887, 663.

3. This gauge, designed by H. McLeod in 1874 to measure extremely low pressures in a vacuum pump, is still in use. The McLeod gauge works by condensing a known volume of the residual gas in a vacuum pump into a much smaller volume and then using Boyle's law to measure the pressure of the gas in the pump by determining the ratio between the level of mercury in the closed tube containing that smaller volume and the level in an open tube. In Edison's McLeod gauge he used a paper scale rather than etching dividing marks on the glass tubes to represent these ratios. In the drawing of pump No. 4 the mercury is raised in tube E until it traps a sampling volume of gas in the small tube f. The volume of mercury continues to be raised in the adjoining open tube x so that it reaches a known mark. Nothing is known about how Edison graded his scale but see N-79-08-22 Lab. (TAEM 35:876–93; *TAED* No85:93–96) for an example of the readings. McLeod 1874.

4. In notes accompanying a 3 October drawing like that of pump number 4 (see below), Samuel Mott referred to the addition of a chamber of "granulated Copper (to take up mercurial fumes)" (N-79-09-20: 8, 11, Lab. [*TAEM* 36:32; *TAED* No96:5]). By January 1880, when *Scientific American* illustrated Edison's standard pump at that time (see note 5), gold leaf had replaced the copper; the use of gold was one of the practices adopted by William Crookes (see Doc. 1714 n. 2).

5. On 8 October Upton noted that "Pump No. 4 was broken and is slightly changed and repaired. The stop cock B is now close to the main tube so that no air is held inside tube. The vessels D, D' were made longer so that the Hg. would not splatter over. In starting the pump the

The new method of sealing the bulbs on the vacuum pump involved using platinum leading wires from the spiral to point A and then copper wires, sealed in small tubes, from A to B.

*Experimental vacuum
pump No. 5.*

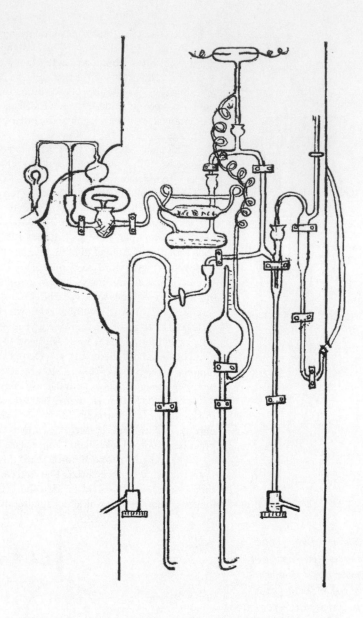

Geissler worked well, but the Sprengel pump required a long time."
The laboratory staff found its operation impaired by dampness in the
mercury and also by air in the copper dust, which was driven off by heating. Two days later they also altered the manner in which the spirals were
sealed with the "leading" (lead-in) wires sealed in small tubes. Ludwig
Böhm also made a new mercury seal. They continued to experiment
with the pump design over the next few days. On 13 October they experimented with using the Geissler pump first until they had achieved
good light from both poles of the lamp and then turned on the Sprengel
pump to run at very fast rate. This seems to have produced a "Good vacuum to stop small spark in an hour." Over the next day or two they made

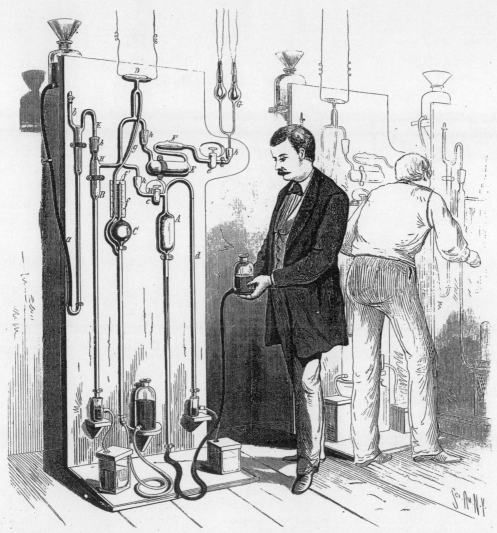

A.—Geissler's Pump. B.—Sprengel's Pump. C.—McLeod's Gauge. D.—Geissler Tube. E.—Bulb containing Phosphorous Anhydride. F.—Bulb containing Gold Leaf. G.—Electric Lamp Bulbs. a.—Mercury Supply Tube. b.—Air Trap. c. and e.—Mercury Sealed Stop Cocks. d.—Discharge Tube. f.—Scale. f'.—Gauge Tube. g.—Connecting Tube. h.—Mercury Sealed Joints.

Edison's vacuum pump as illustrated in the 17 January 1880 Scientific American. *The Sprengel pump is B and the Geissler tube is d. The McLeod gauge is C, with the scale f, the closed "gauge tube" f', and the connecting tube g.*

additional modifications that were incorporated into a fifth pump design. However, the standard Edison vacuum pump illustrated and described in the 17 January 1880 issue of *Scientific American* is identical in construction to No. 4. In operation the pumps differed in Edison's adoption of gold leaf (see note 4) and phosphorous anhydride for absorbing water vapor (see Doc. 1714 n. 2). N–79-08-22:131–35, 139, 159–61, 165–69, Lab. (*TAEM* 35:848–50, 852, 862–63, 865–67; *TAED* No85:65–7, 69, 79–80, 82–4); "Edison's Vacuum Apparatus," *Sci. Am.* 42 (1880): 34–35.

October 3, 1879[a]
London 5:30 PM

Telegrams: From/To
George Gouraud

Edison

Cable authority organize British Colonies same basis London   Glasgow Lancashire closed[1]

[George Gouraud]

Menlo Park 5:45 PM

[George Gouraud]

You may negotiate India all of South America except Argentine[2] Good Hope Turkey Norway Sweden Egypt Switzerland China Holland   Have 30 day option out on Australasia[3]

[Thomas Edison]

L(telegrams), NjWOE, DF (*TAEM* 52:192–93; *TAED* D7939O, D7939P). Written by Stockton Griffin. [a]Date taken from document, form altered.

1. The London telephone company agreed to pay Edison an advance royalty of £10,000 for Lancashire. Investors in the Edison Telephone Co. of Glasgow agreed to pay Edison half the company's net profits. Shortly after this Gouraud drafted a standard contract for other prospective district companies "so as to put an end to these interminable negotiations." Agreement with Edison Telephone Co. of London, 8 Oct. 1879; W. B. Huggins & Co. Memorandum, n.d. 1879; Gouraud to TAE 13 Oct. 1879; all DF (*TAEM* 52:787, 1050, 795; *TAED* D7941ZFK, D7941ZKW, D7941ZFS).

2. On 8 October, after negotiating rights for Chile with José Husbands (see Doc. 1823) Edison wired Gouraud "Peru Chile sold." DF (*TAEM* 52:198; *TAED* D7939U).

3. On 7 October, Gouraud acknowledged this exchange of cables and indicated that he was sending separate powers of attorney for Edison to sign for each of these countries, including Australasia in case the thirty-day option was not taken up; he sent the powers two days later. On 23 July Edison had written Theodore Puskas to ask if he could "make arrangements with a Company operating telephonic exchanges in Australia on an advance of 20 per cent royalty with an advance of $10 000. You had better accept this as you will not do anything with it in Paris and the time will go by with the opportunity." Under the terms of his 17 December 1877 agreement with Edison, Puskas had rights to the telephone in any country other than Great Britain and Canada for which he paid patent costs. No reply from Puskas has been found, and Joshua Bailey subsequently reminded Edison that this provision of the contract had expired. It is not known with what party Edison was negotiating but on 6 October the New York firm of Gifford & Beach (a hardware firm of that name is identified as being located at Murray and Park in Wilson 1879, 533) wrote Edison that they had received his letter of 4 October giving them a thirty day option. They indicated that they expected to hear from their Australian correspondent when the next mail arrived on the 16th; Edison signed Gouraud's Australasian power on 20 October. In addition

to the powers, Gouraud's solicitors were preparing deeds of patent assignment, which he sent on 23 October along with a letter that detailed his plans for establishing companies in each of these countries (Doc. 1833). Gouraud also wrote Edison on 14 October to ask him to order "say at least 500 telephones for use in Norway Sweden Holland Switzerland China Egypt &c &c as it will be absolutely indispensable in order to occupy these fields to place immediately in operation in those countries a certain number of telephones." Gouraud to TAE, 7, 9, and 14 Oct. 1879; Bailey to TAE, 29 Nov. 1879; Gifford & Beach to TAE, 6 Oct. 1879; copy of signed power of attorney for Australasia, 20 Oct. 1879; all DF (*TAEM* 52:197, 203–4, 431, 195, 208; *TAED* D7939T, D7939W, D7939X, D7940ZDL, D7939R, D7939Z1); TAE to Puskas, 23 July 1879, Lbk. 5:14 (*TAEM* 80:129; *TAED* LB005014); TAE agreement with Puskas, 17 Dec. 1877, Kellow (*TAEM* 28:1195; *TAED* HK001H).

–1818–

Notebook Entry:
Electric Lighting

[Menlo Park,] Oct 7 1879

Electric Light   Carbon Spirals[1]

Spiral of Carbon   Mould for Carbon Spiral
  Spiral must be   .18 long
  Inside diameter   .1875 or $\frac{3}{16}$
  outside   "   .207

Made[a] a mould for squeezing   put in some of Wallace soft carbon[2] and squeezed it out of a hole .02 diameter getting it out a yard long if required—

Could make more even sticks by rolling on glass plate with piece of very smooth wood. These sticks could be rolled down to .01 and then wound in spirals. We made some and baked them at a red heat for 15 minutes in a closed tube— When taken out they were hard and solid much more so than we expected and not at all altered in ~~sp~~hape— A spiral made of 'burnt lampblack' mixed with a little tar was even better than the Wallace mixture—[3b] With[c] a spiral having 5 inches of wire of .01 we can get 100 ohms[b]   We now made a double spiral on brass so as to wind the carbon so similar to ~~the~~ some of the[d] first platina spirals we made—

Chas Batchelor

X, NjWOE, Lab., N-79-07-31:85 (*TAEM* 33:584; *TAED* No52:42). Written by Charles Batchelor. [a]Small drawing of spiral overwritten by text. [b]Followed by dividing mark. [c]Obscured overwritten text. [d]"some of the" interlined above.

1. The *New York Herald* gave the following account of Edison's decision to experiment with carbon at this time:

> Sitting one night in his laboratory reflecting on some of the unfinished details, Edison began abstractedly rolling between his fingers a piece of compressed lampblack mixed with tar for use in his telephone. . . . until it had become a slender filament. Happening to glance at it the idea occurred to him that it might give good result as a burner if made incandescent. A few minutes later the experiment was tried, and, to the inventor's gratification, satisfactory, although not surprising results were obtained. Further experiments were made, with altered forms and composition of the substance, each experiment demonstrating that at last the inventor was upon the right track. ["Edison's Light. The Great Inventor's Triumph in Electric Illumination," *New York Herald*, 21 Dec. 1879, Cat. 1241, item 1379, Batchelor (*TAEM* 94:537; *TAED* MBSB21379X)]

In a draft caveat that he probably wrote later in October, Edison described methods of moulding spirals of carbon and finely divided metals, such as silicon or boron. Stockton Griffin docketed this caveat with a date of 14 January 1880, but that was probably the date on which he wrote "TAE says N.G.   File away as evidence." The caveat also includes the lamp regulator described in Doc. 1820. Vol. 16:354, Lab. (*TAEM* 4:786–95; *TAED* NV16362).

In this caveat Edison described in great detail one method of moulding the carbons:

> to form a spiral of carbon wire $^{10}/_{1000}$ of an inch in diameter such spiral being $^{3}/_{16}$ of an inch diameter a mould is made as in fig 1 consisting of a die, C with a bushing B screwed into the die, a taper hole is

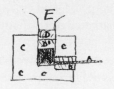

A mould for making carbon spirals from Edison's draft caveat.

bored through this bushing having an orifice of about 7/1000 of an inch. Through this the Carbon which is mixed with an ag[g]lutinate substance such as tar is forced by great pressure   A is the Carbon mixture. D D'; are steel washers   E the end of the press screw. The object of the washers is to allow of a greater pressure than would otherwise be possible, were a punch used the latter bending under great pressure— The Carbon as it issues may if there be sufficient tar mixed with it be wound in spiral form on a mandril, with thread cut in it into which the Carbon presses. This is baked and the mandril unscrewed to allow of the further baking of the spiral at a much high temperature, but if there be but little tar mixed with the Carbon an attatchment is made to the orifice of .B. fig 1—which gives the soft carbon a spiral twist as it issues from the attatchment the latter being only an extension of B curved to represent nearly one turn of the spiral to be formed. Spirals of various sizes may be formed so that one spiral may fit inside the other & thus allow of a great resistance with a small radiating surface when employed as an electric burner.

Other methods included forcing soft carbon through a thread cut in "a rod of nickel or other refractory metal," forming the spiral around a piece of wood and then carbonizing it in a mould, or working the soft carbon or other material into "worsted or other fibrous thread" and then winding it into a spiral and carbonizing it.

On 11 October, Francis Upton noted that they were still "Trying to mold sticks of .010 in diameter and to make them into spirals" and Charles Batchelor was still attempting to make spirals from carbon on 19 October. Between 11 and 21 October they also experimented with heating straight carbon sticks to incandescence. N-79-08-22:135, 169, Lab. (*TAEM* 35:850, 867; *TAED* No85:67, 84); Doc. 1830.

2. See Doc. 1806 n. 4.

3. In his draft caveat (see note 1) Edison wrote that "The best material for making the plastic Carbon consists in exposing in a sealed crucible to a high temperature for several hours ordinary lampblack, afterwards this is mixed with ag[g]lutinat substances such as honey, Tar, turpentine camphor sugar & other viscous substances containing a large amount of Carbon."

–1819–

Notebook Entry: Electric Lighting

[Menlo Park,] Oct. 8, 1879.

At this date the trouble is to get an insulation for the platinum wire. All the ~~known~~ common oxides melt and volatilize at a far lower temperature than the Pt. after the process.

Again[a] many of the oxides become conductors at the temperature of incandescent Pt. All the known oxides are to be tried in various vacuums for it is thought that some one may answer the purpose or a hint be given as to a method of preparing a [~~spi?~~][b] the oxides, as [~~in?~~][b] Pt. has been prepared.

X, NjWOE, Lab. N-79-08-22:129 (*TAEM* 35:847; *TAED* No85:64). Written by Francis Upton. [a]Obscured overwritten text. [b]Canceled.

[Menlo Park,] Oct. 11, 1879.

Regulation[1]

To make a ten Ohm lamp equal to an[a] 100 Ohm[b]

The current may be made to divide by a commutator in the cellar, revolved by an additional circuit so as to turn all the time Edison[b]

With[c] a motor put[c] in the circuit

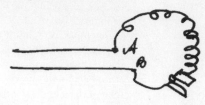

~~The same current~~ If[c] this is so arranged that nine lamps are put in circuit with the motor in each house a regulation may be effected, but the economy would be the same as if a resistance were ~~insert~~ inserted when the motor was taken out. ~~The~~ Reason the same amount of energy must be given of[f] from each lamp by the same current so that the drop of the E[lectro].-M[otive].F[orce]. must be the same ~~from the two~~ between the two extremities of the lamp, so that ~~whe as~~ the total[c] fall must be the same for the whole circuit ~~in order th~~ 100[c] Volts from A to B. If the law holds true that in any circuit the distribution of work is proportional to the fall of the E.M.F., ~~as~~ since ~~with~~ the ~~same~~ current is ~~the same resistance~~ the same when ~~a~~the motor is running as when ~~the the to~~ there is only resistance in the circuit[c], ~~and~~ the same fall of E.M.F. will occur around a lamp to denote the same amount of energy given off, that is 10 Volts, then the 90 Volts fall in the remainder of the circuit will represent the same work given off in either case.[2b]

Mr. Edison proposes to have in the celler of the house an arrangement ~~to~~ with say 10 commutators each one connected with a lamp   If each lamp is on $\frac{1}{10}$ of the time it will receive $\frac{1}{10}$ of the current that it would if it were on all the time   that is a ten ohm lamp would take the place of a hundred Ohm lamp with such an arrangement.[3]

This commutator must have the same number of breaks as there are lamps in the house. If all the lamps are on there will be no spark.

In this way the general average of the houses is taken. That taking a large number of lamps only $\frac{1}{10}$ of them will be on at once.

If there are 1 Ohm conductors for every 10 lamps $\frac{1}{10}$ of the energy will be lost in the conductors when all the lamps are on and a ~~proportionate amount for~~ less proportion when less are on, for then each lamp will have the Cu[rrent?].[4] ~~from~~ for the neighboring houses to use.

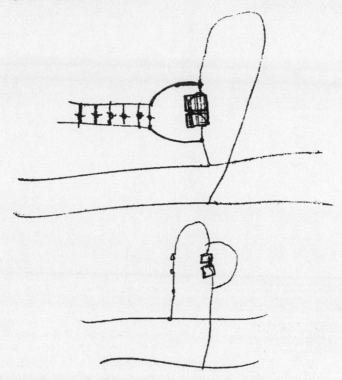

X, NjWOE, Lab. N-79-04-21:178 (*TAEM* 30:810; *TAED* N017:88). Written by Francis Upton. [a]Interlined above. [b]Followed by dividing mark. [c]Obscured overwritten text.

1. About the time Upton wrote this notebook entry, Edison drafted a caveat more clearly describing this method of regulation, which was a result of their inability to create a satisfactorily high-resistance lamp. In this entry Upton is working out on paper some answers to the problem, posed by Edison in the first two paragraphs. However, without accurate diagrams the descriptions in both this note and the caveat are often unclear.

Stockton Griffin docketed the caveat with a date of 14 January 1880, but that was probably the date on which he wrote "TAE says N.G.   File away as evidence." The caveat encompasses the work from mid-October on this regulator and on methods of making carbon spirals (see Docs. 1818 esp. n. 1 and 1830). Vol. 16:354–62, Lab. (*TAEM* 4:786–95; *TAED* NV16362).

2. In this paragraph Upton considers nine lamps in a series circuit with a motor-driven distributor. When a lamp is off the current is shunted through an equivalent resistance to keep the voltage drop constant.

3. Although Upton begins this sentence with "That is," the thought does not follow directly from the preceding sentences but rather sum-

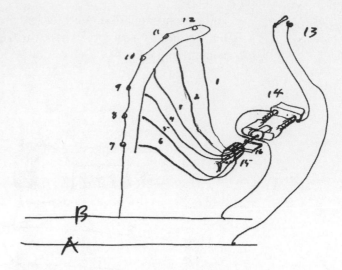

Edison's regulator for a series of lamps as shown in his caveat, 14 the motor and 15 the commutator.

marizes Edison's idea, which is laid out in the caveat. The motor in the cellar has

> one or more subsidiary commutators with 10 blocks in each. . . . one of the main wires rotates with the shaft and sweeps over the face of the 10 commutator blocks . . . to each commutator block a wire proceeds to a lamp of 10 ohms resistance the other ends of all the lamps coming together and are connected to the other main wire; . . . one lamp at a time is placed in multiple arc thus keeping the resistance of the House or branch constantly 10 ohms even if the whole of the lamps are burning, and this is just equivalent to using 10 lamps of 100 ohms each constantly in multiple arc   The Electromotive force required in both cases is the same, and all that the engine accomplishes is to reduce the resistance of the lamp from 100 to 10 ohms the latter being an easier problem. [Vol. 16:358–59, Lab. (*TAEM* 4:790–91; *TAED* NV16:312–3)]

4. It is unlikely that Upton meant "copper" here.

–1821–

Notebook Entry: Electric Lighting

[Menlo Park, October 13, 1879]

Results of Oct. 13 ~~Sp~~

Spiral covered with silica so that there was a thick coating completely hiding the form of the wire. Dead white fluffy. Commenced at 5 P.M very weak currents[1a]

No. per H.P.		Color	Resistance
		Cold	1.22
6 P.M.	687	Red dark on edges	1.56
8-15	404	All red	1.71
8-40	307	Yellow red supporting wires red	1.78

	225	Yellow	1.86
	139		1.99
	114		2.18
	86		2.21
9–20	69	1 Candle about	2.28
	58	Fair light 2 candles about	2.38

The spiral broke off at the leading[b] wires. The spiral was bright[b] and seemed under the microscope to be covered with a bright glaze.

The spiral ~~broke~~ burnt with the same battery as was on for the last when after a few minutes of rest it was put on suddenly.

X, NjWOE, Lab., N-79-08-22:155 (*TAEM* 35:860; *TAED* No85:77). Written by Francis Upton. [a]Elements of following table divided by vertical lines. [b]Obscured overwritten text.

1. Experiments with a silica covered spiral had begun earlier in the afternoon. The results from those earlier experiments as well as the raw data from which Upton derived the following table are found in the pages preceding this notebook entry (N-79-08-22:139–53, Lab. [*TAEM* 35:852–59; *TAED* No85:69–76]). This is the last notebook entry of experimental work with platinum wire lamps.

–1822–

*Draft to
Edward Johnson*

Menlo Park, N. J., [October 14, 1879][1a]

I have been reading English patent decisions & find that they construe patents very liberally to inventors.[2b]

Can any reference be found previous to my patent where ~~sounds~~ waves were translated[c] into electric waves of corresponding character as to pitch & quality.= I say no—[b]

Can Clearac claim it. I say no because it might have been known since the flood that carbon or other finely divided matter varied its resistance by pressure. In the same way as moving a plate in front of a permanent magnet was known previous to Bell but I applied this property of finely divided conductors to translate sound waves into electric waves, and produced a practical invention, and it has nothing to do with the case who was the discoverer of the fact that carbon varied its resistance by pressure. ~~F again~~ ~~I speak a~~

Induction coil. I speak of using the coil ~~and~~ in connection with this carbon transmitter, ~~now~~ in the provisional. now there is only one practical way to use this coil and when I have completed my invention I show how I apply it. I do not understand that the provisional patent is to be full, its only a provisional statement of what the patent is to be a brief so to speak.[3]

No person has used this coil in connection with a carbon transmitter or a water telephone. In fact the water telephone cannot be practically used in connection with it, or in fact can it be used practically at all, its is an inoperative[c] invention. ~~To~~ With the carbon transmitter it is essential that the coil should be used and I say in the provisional that it was to be used in well known manner. I effected a Combination, one of a well known thing & the other a new thing in the art.[b]

In Bells patent you say he mentions "that electric undulations may be caused by alternately increasing & diminishing the resistance of the circuit or by alternately increasing & diminishing the power of the battery= He shows no means. The bare statement is well known in the art that you can produce electric undulations by increasing & decreasing the strength of the battery   a method of doing it is shewn in my U.S.[d] patent No 141,777 of 1873[e]

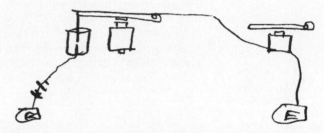

~~Clerac   The Bell people cannot set up the claim that to vary the resistance of the circuit by sound waves acting on finely divided conductors is not new~~.[b]

You say that Clerac, Hughes Sir Wm. T. will testify that ~~tohe~~ ~~vary~~ principle of varying,, resistance by varible pressure on carbon is not new.[4] In the first place, That[c] has nothing to do with it, in the second place, they cannot produce <u>any printed</u> publication of the fact before my application and If I understand the English patent law right, One[c] cannot go back and take testimony of who was the first to conceive. I understand that only a printed publication can be cited against the patent.

The whole effect of varying the resistance with carbon is due to ~~surf~~ a var~~y~~iation in strength of surface pressure. One can hardly design a machine without this taking place= See Richards letters sent in pkg by express.[5]

You say that if the broad principle of varying the resistance of the circuit by the voice has to be given up etc,.[6] now varying the R by the voice by means of water is of no earthly value. It could be given up without the slightest harm.[7] 2nd   As a patent is a contract between the inventor & the public pure &

simple, the law says that you must ~~put your ideas~~ embody your ideas in mechanism, describe the[d] means by which to carry out your ideas, (the idea we cannot allow you a patent for) but if you will embody it in material form we will grant you a monopoly for that form, and to enable us[c] the public to make your invention after the expiration of your monopoly you must give us[c] sufficient information as to design of mechanism in your contract to allow our skilled workmen in that line of business to carry it out, ie[f] make it.[b]

I read in one English decision, that an experiment on iron was tried in a laboratory at a certain time. Some time afterwards an inventor worked out a process ~~and~~ in all its[g] details, and after many years put it in practice   it proved a great success, but it got into court by infringers[h] and the courts decided that an ~~long~~ unknown long dormant[i] mere laboratory experiment could not be cited against a [~~success?~~][j] practical reduction to practice.

Your first and 2nd question put to the lawyers, I believe, should be answered no, and I am backed up in this opinion by the liberal construction of patents in the English Courts.[8] Such a mere allusion cannot act ~~against~~ prejudically against one who clearly works out & practically a method & clearly set its forth in his contract with the public.

You say that the temporary reply is that such device can only be patented as a device.[9] Even then, There is the claim of varying the resistance by an an electric tension regulation ie[f] carbon, peroxide Lead Cyanide Copper or semi conducting material all clearly set forth in the provisional   Suppose the worst, that we were to be kept narrowed down to ~~varying the resistan~~ the use of semi conductors manipulated by diaphragms. There is no other known way of effecting the variation in the resistance so as to make a practical telephone, hence even this claim of varying resistance by "agitating" (thats a clear word) a semi conductor forming part of a circuit by a diaphragm would cover all that is practical and of value

Anything[c] that any expert may testify as to how or the principle upon which this or that thing works does not & cannot alter the fact, that a semi conducting material forming part of a circuit when agitated by a diaphragm varies the resistance & this is clearly shewn both in my provisional & full specification in fact 9/10 of the provisional is a reiteration of this fact.

A piece of semi conductor (I mention several) may be arranged in a million ways so that the sounds waves can act upon it to agitate it, but no matter how it is agitated there will be a disturbance at the surface contacts and this will translate the sound waves. no instrument ~~that~~ can be designed that works but what this effect will take place.

I cannot for the life of me see how any patent lawyer can for an instant suppose & assert after a careful analysis of this patent that the instrument called a microphone is not the carbon telephone. Take the Blake microphone, make an analysis of this Reduce[c] it down to the simplest possible form & you will find that it is the carbon telephone exactly   the only difference you have a bed piece on which is the carbon this is connected to one pole of[k] the circuit   on the carbon is a piece of platina on a spring & this forms the other pole of the circuit. there is a prolongation of the platina towards the diaphragm and the diaphragm presses the two poles of the circuit together. Now: The platina towards the diaphragm could be ivory for that matter or that need not be any prolongation. The platina point on the spring could ~~be reduced~~ have a $\frac{1}{32}$ fact or a $\frac{1}{4}$ inch face that is a thing of mere proportion   if it was a $\frac{1}{4}$ of an inch face you would have our transmitter, hence: a[c] microphone consists in our carbon transmitter with narrowed electrode or electrodes on the surface of which there are fewer points to be [fr?][j] brought into play in translating. ~~If we should use a qu $\frac{1}{4}$ in~~   Look through the old volumes of the English mechanic see the hosts of things called microphones. some of the designs are[c] our carbon transmitter precisely and are made probably by men who are or were entirely ignorant of the existance of our patent or invention. They call it a microphone You ought to have good lawyers in this case & I suppose you have, ~~but their replies~~ but—

If Any[c] combination claim in any patent is good then our claim of the combination of the semi condr transmitter with the induction coil is good— I say in my provisional "These sound transmitting & Receiving instruments are applied[c] to single or multiplex telegraphs, the connections through the wheatstone bridge or the induction coils being used in the well known way.

~~Now If I had not made the combination~~   Of course th~~eye~~ new[l] instruments were connected to the induction coil in the well ~~n~~known way but here was a combination nevertheless, as the thing that I connected to the induction coil was new, & I effected a new combination but I effected the combination in the well known way   The public could not have made any connection in well known way as they did not know of the existence or have one of the elements of the combination—

These are rough notes only as they fly through my head, and are for your own information. The more I read that patent ~~209~~ 2909, the better I think it is & the better I think it will stand in the English Courts.

One thing I would like to know in disclaiming can you alter the phrasiology of your claims add & take from.[10]

ADDENDUM[m]

Menlo Park N.J. Oct 14 79

E. H. Johnson

Idea just struck me   Up to the time of filing of the Prov' Spec' the microphonic biz had not attained to the dignity of a diaphragm

2d   Blake and all those nice people in our pat ofs have been ref'd to my older 77 Pats as prior to their brilliant scintillations.

3d   Our folks over there ought to throw 3 or 4 retainers out to some of best pat men there. I have a letter somewhere from Sir Wm T asking me to do smtg for some relation of his who was greatly interested in financial prosperity of the Bell Tel Co[11]   You ot tk time by the forelock & get acqtd with some of those Sc fellows then get them interested in C Tel by loaning insts &c so that when time comes they will be good experts to call in a court.[12]

Pls notify us the days you receive our shipments of Tele want to see how much time it takes from MP to Victoria St[ation]

I suppose you can get those mouth pcs done in London cheaper than we can here as all box wood here is imptd fm London   I do not think that box wood is essential almost any porous wood wld do cedar dogwood etc.[b]

Gold & S & Bell Tel Co have come together. G&S has priv lines Bell Tel Co exchgs   G&S gets 20% rects fm exchgs—[13] This is the biggest mstk WU ever made but trouble was some of those fellows over there got calls on Bell Tel Stock & wanted to bring it about & you know my sources of inf are genlly more reliable than Reiffs—[14]

mem   the last 20 insts sent you every trans as it came from bench was absolutely the same loudness without any tinkering of any character which were as loud as our old trans— we are testing 25 more tonite wh will be shipd tomw

ADf, NjWOE, DF (*TAEM* 52:804, 802; *TAED* D7941ZFX, D7941ZFW). Letterhead of T. A. Edison. ª"Menlo Park, N. J.," preprinted; date from docket. ᵇFollowed by dividing mark. ᶜObscured over-written text. ᵈInterlined above. ᵉ"No 141,777" and "1873" apparently added later. ᶠCircled. ᵍ"in all its" interlined above. ʰ"by infringers" interlined above. ⁱ"~~long~~ unknown long dormant" interlined above. ʲCanceled. ᵏ"one pole of" interlined above. ˡMultiply underlined. ᵐAddendum is an ADf.

1. Stockton Griffin dated this draft on the reverse of the last page; see also Doc. 1825.

2. These notes were apparently prompted by Edward Johnson's 30 September letter discussing the company's legal strategy (see Doc. 1825). Johnson worried that Bell's patent reference to an undulating current produced by variable resistance, even though "he does not claim it—or show any device by which it may be applied his Publication prevents you from Broadly claiming the principle   You know it is delicate sort of work any way—holding a principle   And you are going to have the opposition of a strong clique—Experts to testify that the Principle is not new will be abundant." Edison was probably reading Johnson and Johnson 1879, the "book of precedents" co-authored by John H. Johnson, a patent attorney engaged by the English telephone company (see Doc. 1828). Edward Johnson had sent this in September with the suggestion to "read particularly the chapters on Claims, Disclaimers, Provisional specification & final spec" in anticipation that, as John Johnson recommended, Edison would have to amend his patent (British Patent 2,909 [1877], Batchelor [*TAEM* 92:56; *TAED* MBP008]) before the company could take legal action against the Bell interests and others for using carbon transmitters. Edward Johnson to TAE, 30 and 19 Sept. 1879; John H. Johnson to Waterhouse and Winterbotham, 4 June 1879; all DF (*TAEM* 52:752, 728, 578; *TAED* D7941ZEU, D7941ZEC, D7941ZBR).

On 1 October Edison had drafted for Johnson "some notes on the Brewer & Jansen opinion made to Valence & Valence" about the soundness of this patent (see *TAEB* 4:416 n. 4). He reiterated the uniqueness of his transmitter in "controlling by s[ou]nd the transmission of elec currents, whereas in Bell its the energy of the snd itself that produces the currents." The majority of Edison's comments concern the motograph receiver and phonograph. TAE to Johnson, 1 Oct. 1879, DF (*TAEM* 52:758; *TAED* D7941ZEX).

3. Johnson complained in his 30 September letter (see note 2) that he feared it would be impossible to "hold the Induction Coil—because you allude to it so vaguely in the Provisional—and only as if it were a well known method of using it."

4. Johnson actually wrote in his 30 September letter (see note 2) that "Dumoncel—Hughes Preece & probably Sir Wm. Thomson will testify" to this effect.

5. Edison presumably was referring to Charles Richards, a prominent consulting engineer, who was assistant superintendent of the Colt Armory in 1877 when experiments were conducted there on the conductivity of graphite under varying pressure (see Docs. 1045 and 1366). The letter has not been found. Johnson requested Edison to "send me Everything bearing on that Hughes Preece controversy— It will come up in court—& I propose to be your mouthpiece." Johnson to TAE, 19 Sept. 1879, DF (*TAEM* 52:728; *TAED* D7941ZEC).

6. Johnson wrote on 30 September (see note 2) that Edison's patent was insufficiently specific and that counsel had advised "that if the Broad principle of varying the resistance of the circuit by the voice—has to be given up—the fact will tend to contract our claims on the devices rather than enlarge them— Unless you can come to my assistance in some formidable way soon I will be overmatched—and will be shown that our Patent is only a Patent for a particular form of Carbon Transmitter."

7. Edison's sixth claim (of thirty) was for "The combination with a diaphragm or tympan, of electrolytic fluid and electrodes, the latter being vibrated by the diaphragm and varying the resistance in the electric circuit"; this form of transmitter was not mentioned in the provisional specification. Brit. Pat. 2909 (1877), Batchelor (*TAEM* 92:56; *TAED* MBP008).

8. In his 30 September letter (see note 2), Johnson stated two questions he had submitted to the company's lawyers. The first was whether "the mere meagre allusion to a principle without clear elucidation; or without attempt to show a method for its application constitute a Publication." He also asked "If such Publication is so ineffective in its power to convey an idea of its importance, to the Public—will it hold as a Publication—as against the rights of a Bona fide inventor who has clearly worked it out and practically applied it by effective devices."

9. Johnson reported in the 30 September letter (see note 2) that it appeared that "such practical device can only be patented as a device   You are therefore in all probability to be thrown back upon your Carbon Button pure & simple—and your mechanism."

10. Johnson replied that "No you cannot amend—by adding to— you can only emend by taking from= you will observe that you can add words—whose sole function is to narrow your boundaries= anything that tends to broaden your boundary lines is Encroachment and will not be allowed." Johnson to TAE, 30 Oct. 1879, DF (*TAEM* 52:867; *TAED* D7941ZHH).

11. No such letter from William Thomson has been found.

12. Johnson arranged two interviews with William Thomson through Josiah Reiff, who "seems to have access to any & everybody over here as elsewhere." He reported that he had also recently invited Sir James Anderson for a private telephone demonstration. Johnson to TAE, 17 Nov. 1879, DF (*TAEM* 52:927; *TAED* D7941ZIL).

13. In September 1878 the National Bell Co. mounted a legal challenge to Western Union's telephone business by suing a Gold and Stock agent in Massachusetts for patent infringement. The two companies negotiated intermittently during the course of the proceedings and by the time trial arguments concluded in September 1879 they were close to a broad settlement. Western Union formally agreed in November to leave the telephone business altogether and assign its relevant patents to the Bell company, in return for 20% of the Bell rental receipts for seventeen years. Bruce 1973, 260–71; Tosiello 1979, 452–87.

14. This may be a reference to the intelligence sent Edison by Josiah Reiff that "I gather that Bell people have scared the W.U. into a settlement." Reiff to TAE, 20 Sept. 1879, DF (*TAEM* 52:108; *TAED* D7937ZCM).

[Menlo Park,] October 14, 1879[a]

Memorandum of Agreement made this 14th day of October 1879[1] by and between Thomas A. Edison of Menlo Park, County of Middlesex, in the state of New Jersey, United States of America, hereinafter called Edison, and José D. Husbands[2] of Valparaiso, Chile, South America hereinafter called Husbands,

Witnesseth: That for and in consideration of services and experience of said Husbands to be rendered & used in the organisation of a joint stock company[3] for the prosecution of the telephone business in chile, South America, and the payment of the sums by the said company to be formed, to the said Edison, as provided in a memorandum of agreement this day executed between the parties hereto, a copy of which is hereto attached and forms a part of this agreement, The said Edison hereby agrees to pay the said Husbands, the sum of three (3000—) Thousand dollars in gold from the first fifteen thousand dollars paid thereon and five thousand dollars ($5000—) in stock from the first fifteen thousand dollars of stock thereon provided to be paid the said Edison, and a pro rata amount on all subsequent payments. Provided that if the said Edison shall accept a gross sum of forty thousand dollars in gold[b] for all his interest in and to the said agreement and the company thereunder organized, the said Husbands shall be entitled to and shall receive in gold[b] the gross sum of ten thousand ($10,000.00) dollars in lieu of all claims and demands

Provision shall be made by the said Edison for the payment of said sums of money and[c] transfers of stock at Valparaiso by the Hon L. H. Foote[4] American Consul or the Bank of Valparaiso as case may be, And the said Edison further agrees to execute to the said Husbands an exclusive conveyance of all his rights in telephones now or hereafter owned by him[c] or controlled by him for the territory of Peru, Ecuador and Bolivia in South America,[5] such conveyance to be deposited in trust with the Hon L. H. Foote or the Bank of Valparaiso aforesaid, to be delivered to the said Husbands on the payment by the company hereinbefore mentioned of fifteen thousand dollars gold and fifteen thousand dollars in stock.

Witness signature Edison   H H Eldred[6] Thomas A Edison[d]
Witness signature Husbands   S. L. Griffin

José D. Husbands[d]

ADDENDUM[e]

[Menlo Park,] October 14, 1879[a]

This memorandum of agreement, entered into this 14th day of October in the year 1879 by and between Thomas A Edison of Menlo Park, Middlesex County in the state of New Jersey, United States of America, hereinafter called Edison and José D Husbands of Valparaiso Chili South America, hereinafter called Husbands;

Witnesseth: Whereas the said Husbands being engaged in merchandizing American products in Chili and other countries on the west coast of South America is about to proceed to Chili as the agent of several well known American manufacturing houses for the introduction of their goods;[7] and

whereas; the said Husbands having seen in various cities in this Country the operation and marked success of the telephone as applied to what is known as the "Central Exchange system," and believing that the said system of telephony is applicable and of great value in the cities of Valparaiso & Santiago Chili and other parts of the West Coast of South America, and that if introduced in a proper manner, that they would prove a success financially, and

whereas; the said Husbands is desirous of effecting an arrangement with the said Edison, securing his cooperation and assistance, his inventions in telephony now made and to be made, and all rights which the said Edison might obtain, and the exclusive right and authorization from the said Edison to purchase and[f] use the telephones of his invention in Valparaiso, Santiago and other parts of Chili with the object of forming a joint Stock Company in the city's of Valparaiso ~~and~~ or[g] Santiago for the exploitation of the said telephonic inventions within such territory and:

whereas; the said Edison is desirous of securing the services and experience of the said Husbands to exploit such inventions, promote such company, and develope the business of telephony in such territory to the exclusion of competitors: Therefore be it agreed by and between the said Edison and Husbands, that the said Edison will furnish the said Husbands with the necessary appliances and telephones to exhibit experimentally in the cities of Valparaiso or Santiago the workings of a telephone system,[8] eight in number, and to furnish and pay the expenses of a telephone expert to put such system in practical operation under direction of the said Husbands, and continue it for a period of six months, and to furnish thereafter, all the necessary telephones and appliances, at the same price as charged by the manufacturer.

The said Husbands agrees to form a stock Company in the said city of Valparaiso or Santiago, with a capitol of not less than one hundred thousand ($100,000) dollars gold to work the telephonic system in Chili, to which Company the said Edison agrees to make over and convey, any and all rights which he has or may have in telephony for the said territory of Chili, and to sign all necessary papers that the Company may desire to secure to them the exclusive use by them in Chili of all his telephone inventions now made or to be made by him.

For the performance of the above stipulations, and agreements on the part of the said Edison: the said Husbands agrees to pay or cause to be paid the said Edison by special agreement with the said Company, so to be formed the sum of fifteen thousand ($15,000) dollars in gold and the further sum of fifteen thousand ($15,000) dollars, in the stock of said Company, such stock to be fully paid up and not subject to assessment, and in equal proportion should the capital stock of the Company be greater than the one hundred thousand ($100,000) dollars contemplated; these sums so paid the said Edison are to be in full satisfaction of all claims, and to be for the exclusive right, and the risks taken in exploitation experimentally in said territory, except as to the material supplied in the exploitation, and the salary, which shall not exceed one hundred dollars per month, and the expenses of the expert furnished by said Edison,[9] which are to be refunded by the said Company to the said Edison.

And it is further agreed by the parties hereto that the contracts and agreements herein provided to be made by the said Edison shall be, within sixty days from the date hereof deposited with L. H Foote American Consul at Valparaiso or with the Bank of Valparaiso to be delivered to the legally authorized representative of the said Company on payment of the aforesaid sum of fifteen thousand dollars ($15,000) gold, and the delivery of three certificates of the stock of said Company of five thousand ($5000) dollars each, duly issued by the said Company in the name of Thomas A Edison

And the said Edison further agrees that in case the said Company contemplated to be formed in this agreement, desires to form auxilliary so companies in the within named territory, or to dispose to others of the rights and franchises herein provided to be conveyed to it exclusively, it shall cause to be paid for such privilege to the said Edison a gross sum of

($25,000) Twenty five thousand dollars[h] in gold in addition to the cash payment of ($15,000) fifteen thousand dollars hereinbefore provided for which said gross sum of twenty five thousand dollars ($25,000) shall be in lieu of the within named stock interest and all other claims of the said Edison under this Contract except reimbursement for material and expenses as herein before stated

Witness to Husbands signature   H. L. Storke[10]

José D. Husbands[e]

DS (copy), NjWOE, Miller (*TAEM* 86:46, 48; *TAED* HM790081, HM790082). [a]Date from document, form altered. [b]"in gold" interlined above. [c]Obscured overwritten text. [d]Followed by representation of seal. [e]Addendum is a DS. [f]"purchase and" interlined above. [g]Interlined above. [h]"Twenty five thousand dollars" interlined above.

1. Edison and Husbands had drafted a single memorandum of agreement on 8 October that encompassed the terms found in this agreement and the addendum. DF (*TAEM* 52:199; *TAED* D7939V).

2. Nothing more is known about Husbands beyond the information provided in this agreement.

3. See Doc. 1886.

4. Lucius Foote, a California lawyer, had been appointed American consul at Valparaiso in March 1879, marking the start of his diplomatic career. *ANB*, s.v. "Foote, Lucius Harwood"; *NCAB* 7:267.

5. On 8 January 1880 Edison assigned to Husbands the rights for Chile, Bolivia, Peru, and Ecuador for a period of ten years upon payment of $40,000 in gold. Miller (*TAEM* 86:102; *TAED* HM800096).

6. Horace Eldred was the general manager of Western Union's telephone department. He accompanied Husbands on subsequent trips to Menlo Park and appears to have been involved in the enterprise from the beginning. On 18 November Edison agreed to give Eldred one-third share of his profits under his contract with Husbands in exchange for providing $750 of the expenses of a telephone expert. The following day Edison signed another agreement with Eldred that indicated this amount was half of the expenses of the expert, Walter Morgan, who was to be paid a salary of $75 per month plus reasonable expenses, including his passage from New York to Valparaiso; see also an undated memorandum regarding the 18 November agreement. Letterhead of Eldred to TAE, 20 and 22 Oct. and 5 Nov. 1879; receipt, 29 Dec. 1879; TAE memorandum, n.d. 1879; all DF (*TAEM* 52:206, 211, 390, 240, 242; *TAED* D7939Y, D7939ZAB, D7939ZAH1, D7939ZAT1, D7939ZAV); agreements with Eldred, 18 and 19 Nov. 1879, Miller (*TAEM* 86:57, 59; *TAED* HM790083, HM790084).

7. Nothing is known of these firms but one of them may have been what Horace Eldred described in a letter to Edison as "an extensive firm of metal dealers" in New York City. He indicated that W. H. Brown of that company was the capitalist backing Husbands. Eldred to TAE, 20 and 22 Oct. 1879, DF (*TAEM* 52:206, 211; *TAED* D7939Y, D7939ZAB).

8. On 20 October Eldred wrote Edison that he and Husbands would

"be pleased to call on you on Sunday next 26th inst. to see the instruments in working order that you have ready for Chili." It is not certain if they came out at that time as Edison indicated he did not think he could get the switchboard from Sigmund Bergmann in time and thought it better that they wait until it arrived. Husbands apparently did come out on 5 November but Eldred was unable to accompany him and sent his assistant H. L. Storke instead. Eldred to TAE, 20 Oct. and 5 Nov. 1879, DF (*TAEM* 52:206, 390; *TAED* D7939Y, D7939ZAH1).

9. See note 6.

10. H. L. Storke was Horace Eldred's assistant in the Western Union telephone department. Eldred to TAE, 5 Nov. 1879, DF (*TAEM* 52:390; *TAED* D7939ZAH1).

–1824–

Notebook Entry:
Telephony

[Menlo Park,] Oct 15th 1879

Telephone—

We pressed some chalks from new mould (that is the one made for the small chalk) and in order to get the right pressure we took ~~an old~~ a chalk moulded in large mould and turned the diameter to fit the small mould and weighed the chalk[1a]

We now put in same weight and pressed to same thickness This however seemed to make a softer chalk and it did not work loud   Weight exactly 1 oz ~~T~~troy

We now put in 1 oz 3 dwt and pressed to same thickness These talked right loud before wetting and fairly belched afterwards.[a]

We now took 2 more chalks of 1 oz weight and the first talked very low before wetting (much lower before wetting) after wetting it came up very loud in fact a belcher compared to the 36 tested yesterday[2]   the second I put more pressure in the polishing operation and it talked moderately before wetting but was exceedingly loud after[a]

The difference between these two[b] chalks and the two on page 49[3] of same weight is only that those on page 49 were used immediately after moulding and the latter two were allowed to stand 1½ hours before turning up[a]

We now put in 1 oz 1 dwt and it came loud before wetting and after wetting was perfectly damned horribly loud

We now put a 1 oz 2 dwt and it ~~came loud b~~ talked[c] moderate but uneven before wetting after wetting it did not increase it much[a]

We now put in 1 oz 3 dwt   these dont 'belch' like the ones of same weight on page 49— We cannot understand the reason why— These were turned by steam instead of foot power & had marks on one side owing to the lathe being out of balance— These gave a sound similar to a phonograph when[b] after ~~sett~~ making a good record was not set in deep enough in reproducing[4a]

X, NjWOE, Lab., N-79-09-18:49 (*TAEM* 35:943; *TAED* No86:25). Written by Charles Batchelor. [a]Followed by dividing mark. [b]Obscured overwritten text. [c]Interlined above.

1. Nothing is known of the earlier larger mould. Samuel Mott's measured drawing of the new small mould is in Oversize Notes and Drawings (1879–1886), Lab. (*TAEM* 45:1; *TAED* NS7986BAA).

2. Thirty-five of these telephones were sent to London on 15 October. Batchelor had noted after testing "we find them all a little lower than they ought to be but very good. . . . We cannot at present account for the fact that these 35 are a little lower than standard." N-79-09-18:47, Lab. (*TAEM* 35:942; *TAED* No86:24).

3. That is, the chalks discussed in the first two paragraphs of this document.

4. At 10:00 P.M. on 16 October Batchelor recorded that "Wc now began a series of experiments to extend over a length of time, making 2 chalks of a kind and keeping them under glass jars." Four sets of chalks were tested; after the third one, Batchelor noted that "turning the chalks by hand and getting no tool marks on surface, gives us the increased friction for turning which is so necessary to good working chalks." They also considered the issue of glazing the chalks. N-79-09-18:59–65, Lab. (*TAEM* 35:948–51; *TAED* No86:30–3).

−1825−

To Edward Johnson

MENLO PARK N.J.[a] Oct 16, '79

My dr Johnson

The telephone notes mailed you a couple of days ago[1] were called forth by yours of Sept 30[2]   Tonight I have rec'd yours of the 3d with the letter of Waterhouse & report of Johnson.[3] I did not receive the blue book—when it comes I will go over it more carefully.[4]

Dont be troubled about me altering things in the patent not practical. I would just as soon throw overboard everything which is not essential to make the telephone a financial success.[5] I want our company there to control telephoning as effected by finely divided conductors and and intimacy of contact and motograph action. With that I feel perfectly safe that we shall always have the best telephone and only practical apparatus, in other words control Telephony because I know if there is any other thing that will be better the chances of our getting it are 1000 to 1 in our favor.

Your letter of the 4th to Batch[6] rec'd tonight stating that chalks unanimously[b] played out after working so long is so utterly incomprehensible that for the first time I acknowledge myself stuck. How a number of telephones made at different times under different conditions could unanimously play out at the same time would puzzle the Deity himself. Tonight we have started complete apparatus whereby you can make mould & turn your own chalks[7] so that you [----- -----]^c [mould &?]^c turn your own chalks so that even [----- ----- --------]^c difficulty you know they will last at least 2 months and 50¢ will cover cost of a new one. 3 weeks before I went to Saratoga I put up a pair of new Telephones between my house and the office with instructions above that under no consideration should they be touched. We find that those instruments tonight are just as loud as they were when first put up. This is a 2½ months record—of course these time bugs could not be [foreseen?]^c and it will be a difficult thing even now to [prevent?]^c the cause of this sudden depreciation.[8]

There is one thing I will mention— That if you put a chalk in a telephone put regular pressure on it and wet it that it will not be only ½ as loud as it would be if you put the chalk in the telephone & put ⅓ more than normal pressure on it and turned it back & forward rapidly for at least 3 minutes until a perfect polished groove was made then decreasing the pressure to ½ of normal am't used in practice and wetting afterwards putting on regular pressure and grinding for 2 more minutes under these conditions the^d results are very loud— This manipulation was first brought out when we tested 23 telephones in one lot. At first we put all the chalks in the 23 [telephones]^c and put normal pressure on them, wet them and tried them— The loudness did not come up enough to satisfy us. I then remembered how I had [--------]^c several previously that had been loud   I then [went to?]^c work took all the chalks out threw them away [---- ---]^c chalks and experimented by longer grinding [-----]^c press [-----]^c the result was the required loudness in every one.[9] The reason why the chalks sent in tin foil do not give satisfaction has been discovered. The galvanometer shows that there is a strong current between the tin foil resting on the chalk and the brass holder of the chalk and this tends to disturb the position of the moisture in the chalk by drawing it all inward   we do not now put tin foil around them. Another thing, if a chalk has been t[urned?]^c an[d]^c allowed to remain in the open air ½ a day or so and then put in the telephone it does not give good results, for the reason that when any porous matter has been wet and water evaporated the mass holds together and is quite hard, although this is not so when alcohol is used   apparently the affect of the water is to interlock the particles the consequence is the chalk has a harder surface. The face of the chalk should not be touched by the fingers—drawing a finger across the chalk will cause it to be low at that place altho' nothing may be visible.

John Ott would like to go to London and take charge of your repairing shop.[10] I think he alone with a lathe and a few tools would be able to do all the repairing [required?]ᶜ for a long time. He would be invaluable. I think he will go for the same pay he gets here $3.75 per day.

Mem— Perhaps you have unscrewed the rubber case too often previous to [----- that the?]ᶜ air within [the?]ᶜ case [----]ᶜ become saturated with moisture; if you open it there will be an equalization   it will be replaced with air with normal moisture which must be again moistened by evaporation from the chalk.

Do you hear much about the negotiations. You know Gouraud has a power of atty from me and can sign contracts and do things without my knowledge hence I do not get the information that I would were everything to be sent here for signature   So please keep me posted as far as lies in your power.

I will mention that next week we shall send you 50 instruments and perhaps Bergmannn will have some switch boards ready to go with this lot of telephones.

We have been to work continuously on telephones and E. Light for 40 hours with the exception of 3 hours for sleep and are now on the track again for another big hunk of continuous work. We are striking it big in E. L. better than my vivid imagination at first conceived. We got 48 4½ candles lights last night by the expenditure of 1 horse power[11]—where this thing is going to stop Lord only knows— See Oct No Philos Magazine paper by Schwendler[12] wherein he shows the absurdity of Preeces paper & mathematical statements showing the impossibility of sub division of the E.L.[13]

How do you and Preece get along together do you have anything to do with him?[14] Is Webster Q.C. the [   ]ᶜ The English courts from all I can read lately in their decisions do not hold the poor inventor exactly to the cast iron rules of patent law apparently where any doubt or mistiness, decisions have been invariably in favor of the inventor.

That chalk catastrophe has sent a little leaden pain right in the pit of my stomach. I'll make those chalks last for 10 years before I get through or I'll eat a ton of them. More anon. Very truly

T. A. Edison   G[riffin]

L (letterpress copy), NjWOE, Lbk. 5:280 (*TAEM* 80:156; *TAED* LB005280). Written by Stockton Griffin; circled "C" in top left corner. [a]Place from Edison's laboratory handstamp. [b]Obscured overwritten text. [c]Faint letterpress copy. [d]Interlined above. [e]Remainder of sentence illegible due to faint letterpress copy.

1. Doc. 1822.

2. DF (*TAEM* 52:752; *TAED* D7941ZEU).

3. Neither Johnson's letter nor the one from Theodore Waterhouse has been found. John Johnson had written a brief opinion of two Edison telephone patents, including 2,909 (1877); Batchelor (*TAEM* 92:56; *TAED* MBP008), for Waterhouse in June; the report to which Edison refers has not been found but was evidently much longer (see Doc. 1828). Edison cabled George Gouraud that he thought the report was "sound except perhaps combination claims induction coil. In provisional I speak of combination of transmitter with coil in final specification I more fully describe combination. the words well known ways means circuit contacts in primary   This point will bear my analysis" (John Johnson to Waterhouse and Winterbotham, 4 June 1879; TAE to Edison Telephone Co. of London, 20 Oct. 1879; DF [*TAEM* 52:578, 827; *TAED* D7941ZBR, D7941ZGM]).

4. The term "blue book" refers to official reports of Parliament or the Privy Council and, more generally, to British government publications (*OED*, s.v., "blue-book"; MacFie 1879–83, I:ixxx). Johnson reported on 30 October that after conferences with John Johnson and Theodore Waterhouse (in which, he promised, Edison's opinions expressed in Doc. 1822 were fully considered), "we are all ready to apply for this amended Patent—& can get it—if at all—in a fortnight— But if we file the Blue Book as now sent you—& subsequently hear from you that we have overlooked some vital point—we cannot withdraw from file—and alter— Hence Mr Waterhouse wants your comments before we file." Johnson explained each proposed change at some length and asked for comments on each. The same day, John Johnson sent the revised specification to Waterhouse; Edison cabled "Blue book all right" on 9 November. The amended specification, filed in February 1880, excised all but four of the original thirty claims. The remaining claims incorporated (with minor modifications) the original claims 1 and 2 relating to the transmitter, which covered the use of a mica diaphragm and "electric tension regulators" (such as carbon buttons), in combination with the diaphragm, for varying the resistance of a closed circuit; claim 18 covering the electromotograph receiver; and claim 20 covering the phonograph. Johnson to TAE, 30 Oct. 1879; copy of John Johnson to Waterhouse, 30 Oct. 1879; TAE to Johnson, 9 Nov. 1879; all DF (*TAEM* 52:867; 56:796; 52:902; *TAED* D7941ZHH, D7941ZHH1, D7941ZHW); Brit. Pat. 2,909 (1877) and Memorandum of Alteration, 10 Feb. 1880, Batchelor (*TAEM* 92:56; MBP008).

5. The inclusion of the phonograph in this patent had long been a source of concern (see Doc. 1259 esp. n. 1). In a set of notes to Johnson regarding the 1878 legal opinion by Brewer and Jensen Edison maintained that "the phonog is a legit part of the Telep as much a part of tel as the paper is in the Morse register"; he conceded that "if it is found that it wld seriously jeopard the pat in lit[igation] then I shld be willing

to have phono part cut out & disclaimed but I shld want to have Gouraud with me that in case old [George] Not[ta]ge shld kick at this." The claim covering the phonograph was retained. TAE to Johnson, 1 Oct. 1879, DF (*TAEM* 52:758; *TAED* D7941ZEX).

6. Not found.

7. Charles Batchelor instructed the machine shop on 15 October to "make a chalk mould with hole in   diameter for Johnson London" and another one for Paris; he also ordered "15 more large chalks for Johnson." The next day Samuel Mott made measured drawings of "The Chalk Cylinder Press," which Edison planned to send on 4 November. Cat. 1308:177 (Order Nos. 246 and 247), Batchelor (*TAEM* 90:755; *TAED* MBN003:62); Oversize Notes and Drawings (1879–1886), Lab. (*TAEM* 45:1; *TAED* NS7986BAA); Johnson to TAE and TAE to Johnson, both 30 Oct., DF (*TAEM* 52:866; *TAED* D7941ZHE, D7941ZHF).

Arnold White asked Edison's permission on 17 October for the telephone company to have an English firm make a small number of receivers "with a view of testing the capacity of English manufacturers, for the construction of your telephone." Edison's notes on this letter are essentially the same as his formal reply that he did not object but suggested the order be for "a small lot because there are certain things about the instrument which the workmen have to be educated up to. I send a man next week with a lathe and everything necessary for you to make your own chalks." White to TAE, 17 Oct. 1879, DF (*TAEM* 52:823; *TAED* D7941ZGG); TAE to White, 31 Oct. 1879, Lbk. 5:316 (*TAEM* 80:174; *TAED* LB005316); see also Doc. 1854.

8. Edison had shipped telephones to London on 1 and 2 October which would have just arrived and the previous day he had cabled Johnson, "Hows chalks. ten weeks here show not slightest diminution. shall send outfit for moulding turning." Johnson replied the next day, "Played new lot OK   send extra chalks   fix push button old way." On 17 October, Edison cabled Johnson that he could "hereafter make instruments loud as old wooden box if you will use new push [button] and interrupt with bell, otherwise powerful knock of push tears chalk. answer." Later that day he cabled again, "Dont answer Telegram   got device obviate tearing chalk." The following morning Johnson telegraphed, "All inoperative   vital suggest temporary resurrection today   send new device first steamer." By the early afternoon, however, Edison was no longer so sure about his new device and cabled: "owing resistance inertia transmitter knock harmless   regular transmitter, powerful knock eventually destroys chalk, louder chalk quicker destruction. we always used new push hence bug unsuspected. Temporary expedient, new push, reverse chalk holder, polish new groove gradually. new device allows interruption merely weakens knock." TAE to Edison Telephone Co. of London, 2 Oct. 1879; TAE to Johnson, 15, 17 and 18 Oct. 1879; Johnson to TAE, 16 Oct. 1879; all DF (*TAEM* 52:766, 817, 824–25, 820; *TAED* D7941ZFA, D7941ZGA, D7941ZGI, D7941ZGK, D7941ZGD); for the push button arrangement, see Doc. 1806.

9. On 15 October Charles Batchelor noted that "We sent 23 [to London] on Oct 9th which were all low on testing but when we put in the ribbed spring they all belched right out." N-79-09-18:47, Lab. (*TAEM* 35:942; *TAED* N086:24).

10. John Ott was a machinist who first worked for Edison in Newark and who had been employed at Menlo Park since at least 1878 (see *TAEB* 4:276 n. 1). Johnson cabled on 28 October for Edison to "Send Ott with mould" and wrote Batchelor a few days later that he had rented a room for him, but Ott did not go (Johnson to TAE, 28 Oct. 1879; Johnson to Batchelor, 2 Nov. 1879; both DF [*TAEM* 52:865, 885; *TAED* D7941ZHD, D7941ZHK]).

11. There are no extant records of this trial, which was probably with a platinum spiral (see Doc. 1821).

12. Carl Louis Schwendler, a Prussian native, began his electrical career working for Siemens and Halske in Berlin, where he became a research assistant to Werner Siemens. He was involved in manufacturing, laying, and testing several undersea telegraph cables and in 1868 he went to India as assistant to the Director-General of Telegraphs there. He was subsequently employed by the British government in a similar capacity, and remained in India until shortly before his death in 1882. "The Late C. L. Schwendler," *Electrician* 8 (1882): 149.

13. The January issue of the *Philosophical Magazine* contained William Preece's mathematical analysis of the production of heat by electrical resistance. Preece argued that "*joined up either in series or in multiple arc, the heat generated in each of a number of resistances varies inversely as the square of their number.*" Since he stipulated that "what is true for the production of heat is equally true for the production of light beyond certain limits," he concluded that operating more than a few lights in a single circuit was "a possibility which this demonstration shows to be hopeless, and which experiment has proved to be fallacious." Preece 1879a, 32, 29, 34.

On the basis of tests for the electric illumination of Indian railroad stations, Louis Schwendler examined the causes of the apparent loss of energy in a circuit with multiple lights. He identified a number of difficulties particular to the operation of arc lights and argued generally that "*The economical solution of the division of the electric light is theoretically quite possible, but practically difficult to obtain.*" Schwendler specifically addressed Preece's contention

that the division of the electric light is impossible. This it certainly is under the conditions introduced by Mr. Preece, viz. that the resistance of each voltaic arc, or each incandescent wire, is maintained *constant*. But it is unfair to the electric light to introduce this condition, especially as it does not at all represent the question at issue.

When a number of lights are connected in series, the resistance of each must be diminished, and when a number of lights are joined parallel, the resistance of each must be increased in proportion to their number, so as to maintain the total external resistance constant. If Mr. Preece will introduce this condition into his equation, he will find that theoretically the division of the electric light is quite possible, *i.e.* that, theoretically, however the lights be arranged, the unit of light will always be produced by the same expenditure of energy. Inventors should not, therefore, be downhearted. On the other hand, investors in gas need not hasten to get rid of their shares; for there are many questions involving practical difficulties which still remain to be solved. [Schwendler 1879, 339]

14. Johnson wrote at the end of October that he had

been to see Preece for the 1st time—had a long interview. . . . I am
in with him now & have shown him my guage—am invited to make
myself at home there &c &c— Guide me— I shall have something
to write you in re to your controversy with him—after a little   It
is not unlikely you have gone somewhat astray on him— We shall
presently see—as I purpose taking up the whole matter with Mr
Waterhouse with a view of finding out the truth of the matter—&
the proper course for us to adopt in the treatment of it—when it
comes up in court—as also for your own government   This is all
independent of Preece himself—as until yesterday I never saw
him—& yesterday the subject was only hinted at. [Johnson to TAE,
30 Oct. 1879, DF (*TAEM* 49:478; *TAED* D7903ZKC]

–1826–

From Joshua Bailey

Paris, 17 October 1879

My dear Sir,

On the 10th instant I wrote you briefly saying that an
arrangement had been made for the fusion of the 3 Telephone
Cos which have commenced operations in France.[1] A proposition was made nearly 4 weeks ago by the Credit Mobilier[2] &
Erlanger[3] who are interested in the Gower Telephone for a fusion of the Comps. and this proposition was under consideration pending 10 days. The proposition made by them was that
the 3 Companies should be united on a given Capital to be divided equally between the three Companies. Two weeks ago
we went into the final Meeting for the consideration of this
proposition and declined it on two grounds   1st that as
brought forward it was intended as a great stock operation and
not as a serious exploitation of the Telephone and 2nd because
the division of interest proposed did not seem so fair as far as
we were concerned. The following Monday I called on Mr.
Rosevelt and told him the reasons that we had refused the fusion, urged the adoption of a different policy and proposed the
basis of division of interest at 45 per cent for the Edison, 30%
for the Gower and 25% for the Blake. This proposition was accepted on the 10th with the trifling modification of 44¼ for the
Edison 31½ for the Gower & 24¼ for the Blake—[4] A point
which I regard as more important than the ratio of division of
the several interests in the plan now adopted is that the control
of the Company remains entirely in the hands of the holders of
the Patents and a Board of Directors has been agreed upon
which secures the non speculative administration. The Erlanger & Credit Mobilier interests are in the background, hav-

ing only three representatives in a Board of 12 Members, the 9 others being solid for an economical administration and the pushing of the Telephone without regard to operations on the Bourse. Mr. Rosevelt has worked with me in securing the adoption of this plan and it is probable that we shall work together throughout Europe as a consequence of this operation if the persons interested in the States in the Edison Telephone Compy of Europe Limited accept the basis which has been talked of between us   we propose to form a combination Company in Paris for operating in all the Countries on the Continent with a Capital of 2 Millions of francs and to call over from the States 2 or 3 additional Experts and to bring the necessary material over for starting the Telephone Exchanges at once in the principal Cities in Spain, Italy, Austria, Germany & Belgium. The prorata of division that is proposed between us is that of 60% for the Edison and 40% for the Gower. I do not think that this proposition represents the value of the two Telephones correctly, but the Gower people have made a good deal of stir, are active and are bound to have a hand in the Telephone Exploitation, in these Countries. There is no question whatever that it will be more profitable for us to join hands and be relieved from the opposition of the only knot of persons that so far has undertaken actively the pushing of the Telephone in Europe. For the Capital of 3,000,000 of francs I think we should have to give nearly a half interest in our Patents, but we should be able to start at once, and on a considerable scale the exploitation in a sufficient number of Cities in each Country to give us the prominence that would enable us to hold the field for the future. I have been unable up to the present time to make any move myself outside of France on account of the circumstances connected with Mr. Puskas of which I have heretofore written you. I will not go into any detail as to the circumstances that have taken place since the report made in my letters of the latter part of August and first September.[5] I will only say that Mr. Puskas ignorance of business is only equalled by his deceitfulness and treachery and that he is a man entirely unfitted for connection with any such business as that which we have in hand and his presence in any Capacity whatever is only an obstruction. He has entirely lost the confidence of every person connected with our own Company and is no better liked by those to whom he tried to sell us out. Everything that has been done during the last two Months has been done against his opposition or entirely without his knowledge. I cabled yesterday giving the points of the proposed fusion and stating that Mr. Puskas makes trouble and requesting that the power of Attorney which Dr. White advised me in his letter of September would be forwarded to me on application, should be sent forward. Mr. Puskas stated to me the day before yesterday that he would oppose all attempts to carry out the fusion unless he were named on the Board of Directors and also brought forward the proposition that the new company should be called the Edison Telephone Coy. As to the first proposition there is no person connected with either of the 3 Companies that would consent to sit at a Board where Mr. Puskas should be found as a Member. In regard to the second it would be very pleasant to us as we have no doubt it would be to Mr. Edison that the rival Companies should consent to take the name of our Company and this was proposed be me in the first interviews. A suggestion was made in reply that the name of Mr. Edison should be coupled with the names of the other persons and this as a business proposition was entirely fair, but I do not think that this would be a proposition so agreeable to Mr. Edison as that which was made by the Gower Coy to drop the names of all the interested Companies and to take another, to wit, "La Société Générale de Téléphones" and this last I have agreed to. The whole subscription of the Capital stock of the new Coy will be made privately without calling on the public and no one who is in the operation doubts that the shares will command a large premium from the date that they are first issued. No more than about ⅓ of the 8 Millions will be for sale— The greater part of it being held by the subscribers as an investment. It is believed by those who are inside that the shares now issued will be worth 200 or 300 francs per cent within 12 months from the present date. The reason of this is that the 3 Cos together hold the only Concessions that have been granted by the Government for the exploitation of the Telephone in France and almost absolutely everything that is known today that could be used in the Central system it is almost certain that the Government would grant no additional Concessions even if another good Telephone should be brought forward. The Wires in Paris are obliged to be laid in the sewers and the Administration of the City and the national administration of the Telegraphs are very jealous of allowing interlopers to come into the Sewers with them. The situation amounts to a virtual monopoly of the Telephone in France. Within a month from this time we shall commence the formation of Telephone Exchanges at Lyons, Bordeaux, Marseilles & Lille. Beside the central system the demand for an isolated application of Telephones is already enormous and beyond our power to meet with the actual means of manufacture. We shall at once build a Telephone Factory and commence turning out Telephones for isolated application as well as for central systems. Mr. Rosevelt & I have worked together during the last fortnight in entire accord and it is due entirely to the manner in which he met the propositions made by me that the fusion has been effected. Both he and I have had a great deal of difficulty with some persons in our respective Companies. Mr. Berthon in our own Coy has stood very much opposed to the fusion because we declined to give him the position of director. We took however [---]ᵃ immediate steps to put the Coy into liquidation & yesterday he expressed his acceptance of the fusion— Dr. Evans has more than doubled his investment in the new Company as has also Mr. Chatard the other silent partner remaining with us. We have brought back the part of Mr. Godillot which he had sold out. Mr. Rosevelt subscribing for 200,000 francs out of the 427 000 and the remainder being taken by Dr. Evans, Mr. Chatard and several persons of influence that for one reason or another we desire to have directly connected with us. I regret that so much time has passed in concluding the French business, but the result warrants all the time that has been spent in gaining it. In regard to the shares that are coming to our interests I desire your instructions as to whether I shall allow them to be sold or shall hold on to them. By the statutes of the new Company all the shares of the Company will be syndicated for a period of 6 months, each person having shares will declare whether he desires to sell or to hold. If he elects to hold his shares he cannot during the period of 6 months make any sales. If he elects to sell the sale is arranged by a Committee of one person from each of the 3 Societies— Our representative in this Committee of the Syndicate will be Mr. Chatard who is a good financier and each of the other Companies will be represented by a Banker so that the composition of the Committee guarantees the proper handling of the shares. The shares that are coming to me I shall hold on to. The only circumstance that creates any doubt in my mind as to the future of our Enterprise in France is the possibility that the Government may buy in our Enterprise but this is not probable in the face of the fusion that has now been made which Commands very powerful political influences and also of the Capital that will be necessary to give the widest development to the Enterprise. It would be well that you should advise me by Cable on receipt of this letter whether you desire to sell or to hold on. In the former case cable the word "sell" and in the latter "hold" and I shall act according to your instructions. If you accept the proposition of fusion for the other countries of Europe above referred to Mr. Rosevelt will go to Italy this winter and I shall go to Austria with the intention of coming up into Germany with the commencement of the spring. We shall probably make any arrangement for working Belgium together through some third party at the same time that we are working in Austria and Italy. Germany will be a very difficult field under the attitude of the Government.

In regard to Mr. Puskas so far as either Mr. Rosevelt or I am concerned I do not think it possible that we should have any relations with him or consent to work in any place with him. I have no doubt that the statements that have been made are sufficient to show that this is not a matter of personal dislike, but is the result of the entire loss of confidence that his course has occasioned not only in his business qualities but also in his untruthfulness and his claims to personal confidence— If you accept the proposition for fusion on the terms proposed above, please to cable the word "accepted"— If you do not accept please cable the word "undesirable."[6] I am, very truly Yours,

J. F. Bailey.

My address by cable is, "Heraclite, Paris"

ALS, NjWOE, DF (*TAEM* 52:359; *TAED* D7940ZBP). ªCanceled.

1. In that letter Bailey briefly outlined the terms discussed here and indicated that "the provisional Contract for the fusion will be signed tomorrow & the whole business will be closed up as soon as the proper Statutes can be framed & the necessary legal formalities complied with." DF (*TAEM* 52:353; *TAED* D7940ZBJ).

2. This was the successor to the Société Générale du Crédit Mobilier, the most important of the Second Empire banks, which had played a major role in the development of railways throughout Europe and in the establishment of associated banks in several countries. The bank failed in 1867 and was reorganized in 1870 as the Société de Crédit Mobilier Français. The new bank continued to operate until the great depression of the 1930s but never achieved the prominence of its predecessor. Cameron 1961, 134–95.

3. Emile Erlanger & Co. were Paris bankers with a house in London. Emile Erlanger had inquired about Edison's electric light a year earlier. See Doc. 1497.

4. On 14 October Bailey and A. Berthon representing the Edison company, Cornelius Roosevelt and Edward du Chateau representing the Gower company, and Léon Soulerin representing the Blake company signed a fusion agreement on these terms. Bailey cabled Edison the following day notifying him of the terms and complaining that "Puskas makes trouble." Because of this he asked Edison to send him the power

of attorney that Constant Rousseau had sent on 2 September (see Doc. 1798) and to "cable directing me represent you pending arrival power will use neither unless indispensable." Bailey also asked Edison to telegraph Samuel White and James Banker about the fusion agreement. Edison sent White a copy of Bailey's cable along with his own cable exchange with Theodore Puskas (Bailey agreement with Chateau and Soulerin, 14 Oct. 1879; Bailey to TAE, 15 Oct. 1879; both DF [*TAEM* 52:354, 356; *TAED* D7940ZBK, D7940ZBL]; TAE to White, 16 Oct. 1879, SSW [*TAED* X078AV1]). Edison had cabled "Take no steps towards transferring my telephone interest. Bailey constantly telegraphs, you afford me no information to judge of truth his statements." Puskas replied: "Never dreamt of transferring anything   Baileys conduct shameful   Please be sure shall fight to best ability for your interests." On 18 October, White wrote Edison "that I have confidence in Mr. Bailey. If I occupied your position, as I understand it, I should do as he requests." Instead, Edison cabled Bailey: "Shall give no powers attorney, send papers here if satisfactory will sign. If our patent valid Blake should get nothing." TAE to Puskas, 15 Oct. 1879; Puskas to TAE, 16 Oct. 1879; White to TAE, 18 Oct. 1879; TAE to Bailey, 18 Oct. 1879; all DF (*TAEM* 52:356, 367, 369; *TAED* D7940ZBM, D7940ZBN, D7940ZBQ, D7940ZBR).

5. Only one of these letters has been found. Bailey to TAE, 5 Sept. 1879, DF (*TAEM* 52:326; *TAED* D7940ZBC).

6. Edison did not cable (see note 4).

-1827-

Telegrams: From/To
Theodore Puskas

October 19, 1879[a]
Paris

Edison

Charley died this morning after ten days peritonitis.[b] best doctors &and best attendance could not save him[1]

Puskas

Menlo Park NJ   12:32 PM

Puskas

Can you have body sent to Father at Port Huron Michigan Expense paid by me[2]

L (telegrams), NjWOE, DF (*TAEM* 49:955–56; *TAED* D7914H, D7914I). Written by Stockton Griffin. [a]Date taken from document, form altered. [b]Obscured overwritten text.

1. Cornelius Herz cabled to Edison on 17 October "Charley severe peritonitis." The next day Edison wired $200 to Paris through Drexel, Morgan & Co. At 8:10 in the morning on the 19th, Herz cabled "Charley died." Herz to TAE, 17 and 19 Oct. 1879; Drexel, Morgan & Co. to TAE, 18 Oct. 1879; all DF (*TAEM* 49:952, 955, 953; *TAED* D7914F, D7914G, D7914F1).

After leaving the French telephone company to work for Herz on the quadruplex (see Doc. 1771) Charles Edison became associated with O. Murray, an American artist who assisted him and with whom he may have had a homosexual relationship. Charley had been working on his own inventions and immediately prior to his illness had arranged for Herz to provide a laboratory and salaries for himself and Murray as his assistant. Murray to TAE, 22 Oct. 1879, DF (*TAEM* 49:963; *TAED* D7914P).

2. Edison sent this after Herz cabled at 8:30 that morning that Charley would receive a Protestant burial on 21 October "Unless otherwise instructed." Puskas answered Edison's request the next day, "Have arranged." According to Josiah Reiff the funeral service went ahead on 21 October; Charley's body was shipped to New York four days later. Herz to TAE, 19 Oct. 1879; Josiah Reiff to TAE, 21 Oct. 1879; Puskas to TAE, 20 and 31 Oct. 1879; all DF (*TAEM* 49:956, 962; 52:369, 380; *TAED* D7914J, D7914O, D7940ZBS, D7940ZBZ).

-1828-

To Zenas Wilber

[Menlo Park,] Oct 20 79

Friend Wilbur[1]

I have just rec'd from Engd an opinion on my Eng Telephone pat 2909 of 12 pages   I have a book of precedents written by the writer of this opinion.[2] How would you like to have the book sent to you with his report and a copy of the Pat

I would like for you to study the report and have <u>your</u> opinion on it. I will see that you are paid for it.[3] We are[a] going to have litigation over there strong & heavy against these microphone fellows. Reading this book has proven to me that Serrells patents are very loosely drawn up. Very truly,

T. A. Edison   G[riffin]

L (letterpress copy), NjWOE, Lbk 5:290 (*TAEB* 80:163; *TAED* LB005290). Written by Stockton Griffin; circled "c" at top. [a]Obscured overwritten text.

1. Zenas Wilber, a cousin and former ward of Rutherford B. Hayes, was an interference examiner (and former principal examiner) in the Patent Office. He wrote to Edison on 3 October that he had recently seen Josiah Reiff, who said "he was anxious for me to go up your way and have a chat with you over certain matters and that you said to him you'd like for me to come." Edison invited him to Menlo Park immediately. Wilber made subsequent visits and by mid-November he apparently had made arrangements to become Edison's patent attorney, and he was also engaged by the Edison Electric Light Co. He went into partnership with Washington, D.C., patent attorney George Dyer and in January took over Edison's new applications and those prior ones relating to electric lighting. Lemuel Serrell continued to handle matters arising from earlier applications on other subjects. Obituary, Aug. 1889, enclosure with Charles Batchelor to TAE, 27 Aug. 1889; Letterhead of Dyer & Wilber to TAE, 17 June 1880; Wilber to TAE with TAE marginalia, 3 Oct 1879;

Grosvenor Lowrey to TAE, 2 Jan. 1880; all DF (*TAEM* 128:363, 55:131, 49:401, 54:158; *TAED* D8968AAN, D8036ZDB, D7903ZIN, D8026A); Wilber's testimony, 61–64, *Keith v. Edison v. Brush,* Lit. (*TAEM* 46: 144–46, 47:945, 956; *TAED* QD002:31–33); Powers of Attorney to Dyer and Wilber, 24 Jan. 1880, Defendant's depositions and exhibits [Vol. IV], pp. 2245, 2266, *Edison Electric Light Co. v. U.S. Electric Lighting Co.* (TAEM 47:945, 956; *TAED* QD012E2245A, QD012E2266A).

2. Johnson and Johnson 1879; John Johnson's report has not been found.

3. Wilber replied that he would "like to see the report and precedents you speak of, and, if it is worth anything, give my opinion thereon." He also asked if Edison had "$2 or 300 to spare? Im awfully put to, to get some pressing liabilities out of the way" (Wilber to TAE, 24 Oct. 1879, DF [*TAEM* 49:454; *TAED* D7903ZJR]). Wilber had borrowed from Edison on at least one prior occasion (see Doc. 1284) and later claimed that personal indebtedness to one of Alexander Graham Bell's attorneys had led him to issue prematurely the crucial Bell telephone patent contested by Elisha Gray (*New York Herald,* 21 May 1886, Cat. 1336:38, Batchelor [*TAEM* 90:334; *TAED* MBJ003038]).

–1829–

*Edward Johnson to
Charles Batchelor*

London Eng Oct 21/79.

My Dear Batch—

I am in a nice pickle— The new Insts. are no good—in fact aint worth a damn— Every body is greatly disappointed I cannot make them talk as loud as a magneto receiver—and they refuse to be resurrected. The fact is you have not yet fully appreciated these facts= That chalks play out quicker from non-use than from use—and that when once a chalk runs down it cannot be restored—

I believe that if a chalk is made thoroughly good at first—then put into immediate use and used continuously—that is Every day more or less—it is good for from 2 to 3 months—but if it is allowed to stand a fortnight it undergoes some change which renders it inoperative— This is the only explanation of the fact that you get them to work well in the Laboratory—as long as you choose to keep them up—but when you take them down & send them here, they wont work. I think the explanation of the fact that the former lot worked up after a little & then remained good, is[a] that they had enough saturation to out last the 3 weeks   If you have not done so Ere this reaches you—you must take a 75 hour pull on this thing—or neither you or I will Ever be the "Bloated Bondholders" you talk about= in fact the whole bright ~~speak~~ prospect will have fallen to the ground. Every nerve in my anatomy will be strained to preserve a state of status quo until you can have time to do this—

I am in receipt of your various letters, and am very glad to have the assurances they convey that what I am doing meets with the hearty approval of the Governor General—[1]

I am in receipt of a Letter from him asking me to look up the mineralogical dealers—& Especially to hunt up the metal Thorite—or Thorium   Tell him I have put it in the hands of a party who can & will make the investigation without giving it away that it is Edison who wants it. He shall have the information shortly—[2]

Reiff is in Paris—embalming Charlie Edison—or rather looking to it that it is done—he will be here Day after tomorrow—[3] A very sudden affair that—Adams & Charlie—now I'm the next in Line—[4] However ill try & preserve my Physique by attending to work instead of wasting my sweetness on the unfruitful soil of dissipation. That finished Adams and I'm afraid it helped carry off Charlie— He was too young to come over here alone—

There is a Board meeting this P.M. to finally conclude the agreement with Gouraud—comprehending the District Companies— If it is done—you may look for the rapid formation of District Cos. as the whole London Board will then be interested in forming them—[5]

The scheme is briefly this＝

The London Board has the option of subscribing ¹⁄₁₀th of the stock of each District Company at anytime within 3 years—unless the District Co. object—then their option is limited to 2 months.

Edison ~~leaves~~ invests[b] ¹⁄₁₀th of his advance Royalty in the Stock of such Dist. Cos—[6a]

Both Edison & the London Co. agree to hold their Dist. Co Stock for 3 years—

I have agreed to ~~leave~~ invest the same proportion of my cash that is to say ¹⁄₁₀th of my 5%＝

With this ²⁄₁₀th of the stock and the voting power of Edison in the Dist Co's. Edison & the London Co. have the absolute control of the Dist. Companies—in case such control shall ~~want~~ need[b] to be exercised.

This is what the London Coy. have been fighting for—

Gouraud sees its importance to the Edison interests—ie—his own—and has been fighting it, only in order to make the best terms possible— He has run pretty close to the wind 2 or 3[a] times—and I have been appealed to, and have responded by personal remonstrance with him— At first he was disposed to resent my interference but he soon found that the other side fully & freely consulted & deferred to me—hence if he did not—he was prejudiced in my mind and consequently the information reaching Menlo Park through me would be colored a little off. so he came[a] round—and I am now fully & freely taken into conference by both sides. Consequently I know all that is going on—

If this meeting this P.M. passes on the agreement it[c] finally ends the discussion of details as of course there will be little to discuss henceforth—as it provides for a common basis for the organization of each Company

They have but one point of difference now. viz= The Boundaries of the proposed Dist. Companies. I was called in last night to a 3 hour discussion of this point between Gouraud & Waterhouse[a] and contributed my share toward its solution either in the right or wrong direction— At all events I want to see something concluded my year is slipping away—

Gouraud has finally got my Document back from his Lawyer but has not yet signed it. He wants now to put in a provision whereby I agree to take $\frac{1}{10}$ of my 5% of cash in Stock of the Dist. Co's— Its a little squeeze—but I told him I would consent to lie in the same bed with Edison— Whatever his Royalty did—mine would do—& so it stands— perhaps by the time it is signed, thHe will have found some other modification to put in.[7]

In re. to the taking out of Patents for the European & intercolonial territory given to Gouraud, I think it a wise move to put the cost of so doing on him.

In re. to the irrevocable power to Gouraud—I think Edison will see the wisdom of my reccomendation when he reflects upon the power he holds in the District Companies & in the London Co   he practically controls the whole thing in England—& It is important that He[d] & he alone continues to hold it.[8] He may with safety sign almost anything Else—but if he wants to be made absolutely safe—he might have an underground understanding with me that he would only sign a paper when approved by a Letter or Cable from me— This would not be for the purpose of getting my judgment on it but for the purpose of getting the approval of a thoroughly honorable and eminent legal authority viz: Mr. Waterhouse—with whom I am on perfectly confidential terms—and who is the safest man in all England for you to trust with the disposal of[e] your signature. (I mean Edison of course)—

If such an understanding had been had I should have cabled him as follows:— "Sign

"Grant the cable request for authority"=

Or—when G. sent on for additional power of attorney subsequently and as I suspect asked that it should be made irrevocable on the ground that otherwise the 10 000£ could not be collected= &c= I should have said "Sign but dont make irrevocable   unnecessary"

If he wants this sort of surveilance he has only to say so—and I will undertake that he signs nothing but what is first approved by <u>his own Lawyer</u> on the ground & thoroughly conversant with all the details—

Let me hear from you often   It makes me feel less alone
Ever yours

E. H. Johnson

ALS, NjWOE, DF (*TAEM* 52:830; *TAED* D7941ZGO). ªObscured overwritten text. ᵇInterlined above. ᶜ"passes . . . it" interlined above. ᵈMultiply underlined. ᵉ"with the disposal of" interlined above.

1. There are only two extant letters that Batchelor had written to Johnson since Doc. 1797. Neither refers specifically to Edison's approbation, although in the earlier one Batchelor expressed general enthusiasm for Johnson having gone "on the warpath" against carbon transmitter infringers. Batchelor to Johnson, 8 Sept. and 3 Oct. 1879, Lbk. 5:171, 248 (*TAEM* 80:145, 151; *TAED* LB005171, LB005248).

2. Edison's letter to Johnson has not been found. Reginald Francklyn made a number of inquiries in the London area, including one to Johnson Matthey & Co. Johnson forwarded Francklyn's correspondence to Edison and reported that "There are but a few pounds of the stuff in existence— If you want it I can manage to get it for you by a little crookedness quite reasonable—certainly far below the extravagant price named in these Letters" (Francklyn to James Gregory, 21 Oct. 1879; Gregory to Francklyn, 22 Oct. 1879; Thomas Russell to Francklyn, 22 Oct. 1879; Francklyn to Johnson, both 23 Oct. 1879; Bryce-Wright to Francklyn, 23 and 24 Oct. 1879; Johnson to TAE, 25 Oct. 1879; all DF [*TAEM* 49:446–47, 457–60, 456; *TAED* D7903ZJK, D7903ZJL, D7903ZJM, D7903ZJT, D7903ZJU, D7903ZJV, D7903ZJV1, D7903ZJS]). In the second week of October Edison made his own inquiries to U.S. dealers and academic authorities, with little success (A. E. Foote to TAE, 11 Oct. 1879; George Barker to TAE, 13 Oct. 1879; Edward Dana to TAE, 14 Oct. 1879; Charles Kraft to TAE, 15 Oct. 1879; Eimer & Amend to TAE, 18 Oct. 1879; all DF [*TAEM* 50:540; 49:418, 426, 429, 441; *TAED* D7925ZAT, D7903ZIU, D7903ZIY, D7903ZJA, D7903ZJH]).

3. Josiah Reiff arrived in London on 10 October and planned to continue to Paris about ten or twelve days later. On 20 October he cabled Edison from Paris, "Attending embalment." Reiff to TAE, 14 and 20 Oct. 1879, DF (*TAEM* 49:424, 959; *TAED* D7903ZIX, D7914L).

4. In a subsequent letter Batchelor told Johnson that he was "sorry to hear you have been sick, and hope you are better by this time— Hope you are not going to '<u>kick the bucket</u>'   Edison facetiously remarks that if your men are going to die off so, it would be better to have the large boxes made a little longer, so that you can send back the corpses in them. And previous to our sending inspectors we will see if they will fit the

boxes; that being one of the conditions of their capability." Batchelor to Johnson, 7 Dec. 1879, Lbk. 5:400 (*TAEM* 80:199; *TAED* LB005400).

5. Johnson wrote hastily later this day, "The thing is closed O.K. . . . Gouraud is happy & so am I." Gouraud sent the supplemental agreement to Edison on 25 October but it has not been found (Johnson to TAE, 21 Oct. 1879; Gouraud to TAE, 25 Oct. 1879; both DF [*TAEM* 52:843, 858; *TAED* D7941ZGQ1, D7941ZGY]). Anticipating the advance royalties from the district companies, Francis Upton had just written his father that "There will be a new chemical laboratory the moment the check comes. Mr. Edison has set his heart on having as fine a one as there is in the world for practical purposes" (Upton to Elijah Upton, 19 Oct. 1879, Upton [*TAEM* 95:563; *TAED* MU031]).

6. In a letter to Edison two days later Johnson justified this provision as being "of the utmost importance that you should show thereby a confidence in your own companies." The arrangement would also facilitate "the Co-operation of the London Co. with Gouraud in forming the Dist Cos. & thus ensuring to yourself a larger and more immediate income from advance royalties even with this reduced fifth in stock—than would be at all possible, if Gouraud worked them alone." Gouraud also agreed to buy a like amount of stock. Johnson to TAE, 23 Oct. 1879, DF (*TAEM* 52:845; *TAED* D7941ZGU).

7. This agreement was signed on 10 November, giving Johnson 5% of Edison's and Gouraud's royalties from formation of district telephone companies and superceding the 11 June contract (see Doc. 1748 n. 3). It contained no provision for Johnson to take stock in the new companies. Johnson had previously informed Batchelor that Gouraud had assented in principle and submitted the papers to his lawyer, who reportedly mislaid them for several days. Agreement with Johnson and Gouraud, 10 Nov. 1879, Miller (*TAEM* 86:53; *TAED* HM790082A); Johnson to Batchelor, c. 19 Oct. 1879, DF (*TAEM* 52:837; *TAED* D7941ZGK1).

8. Gouraud had cabled on 7 October "Power attorney insufficient to consummate Lancashire extension money ready." Two days later he explained that the board had in the end accepted his authority as Edison's agent and permitted the contract to be executed pending Edison's signature and receipt of a broader power. He sent three other letters that day, enclosing with one a copy of his July power of attorney (see Doc. 1765 n. 1) from which Edison had stricken the word "irrevocably," and urged him to reconsider that change. He also forwarded a similar recommendation from attorneys Renshaw & Renshaw and drafts of two irrevocable powers (Gouraud to TAE, 7 and 9 Oct. 1879, both DF [*TAEM* 52:778, 791; *TAED* D7941ZFH, D7941ZFM]; Gouraud to TAE, both 9 Oct. 1879; Power of Attorney to Gouraud, 21 July 1879; Renshaw & Renshaw to Gouraud, 9 Oct. 1879; Powers of Attorney to Gouraud, both n.d. 1879; all Miller [*TAEM* 86:38–40, 35, 41–45; *TAED* HM790075, HM790076, HM790073, HM790077, HM790078, HM790079]). About 19 October Johnson wrote Batchelor that he and Josiah Reiff had "been discussing advisability of cabling Edison not to convey Irrevocable powers to Gouraud—but I have finally decided not to do so, but to rely upon my the warnings of my previous letters." Johnson cautioned that he was "informed by the solicitor of the Coy. that an irrevocable power is not only not essential, but its grant by Edison would be taken

by the Gentlemen of the Board as a personal affront—since one of the main inducements held out to them to go into the enterprize was the feature of Edisons personal association on the board with them" (Johnson to Batchelor, c. 19 Oct. 1879, DF [*TAEM* 52:837; *TAED* D7941ZGK1]).

On 20 October Edison wrote Gouraud that he was

> sorry to have to disappoint you in regard to signing the irrevocable power of attorney, but I cannot do it. Not that I am afraid that you will not act just and honorably but because until lately I had always given irrevocable powers of attorney and these have invariably proven disastrous to me in the extreme. [George] Harrington who was a good sort of a man was the last person to whom I gave an irrevocable power of attorney the irrevocability of which he frequently threw in my face and which was the cause of cheating me out of 4 years of labor 20 hours per day. Mr Orton offered me a large sum of money for my Automatic interest but that word "irrevocable" killed it. From that date I made a vow never to give another irrevocable power of attorney, a resolution which I will not alter although at the present it seems to my interest. [Lbk. 5:291 (*TAEM* 80:164; *TAED* LB005291); Edison's power of attorney to Harrington is Doc. 155]

Gouraud cabled in reply that Edison's decision was, "in the light of your experiences perfectly satisfactory to me" (Gouraud to TAE, 4 Nov. 1879, DF [*TAEM* 52:898; *TAED* D7941ZHP]).

–1830–

Notebook Entry:
Electric Lighting[1]

[Menlo Park,] Oct 21st 1879

Electric Light   Carbon Spirals
   Made enclosed tube for the baking of the spiral to carbonize it—

We found that the wires carbon always broke just as the junction of the carbon and platina[a] so:—[2]

This we could not account for so I made a straight piece of carbon & fastened to a pair of wires and put in a closed tube and heated the tube— I then found that at quite a low heat a[a] yellow oil[a] came from the carbon and ran down the wire and the carbon parted very easily just as if it had melted filling the tube with white smoke and having a yellow oily liquid on the top of centre glass which I suppose is Benzole or one of it first compounds[b]

I now put another in a tube and baked it 1 hour at 165 F then ½ hour at 220 then 1 hour 320   when I took it out it showed and oily liquid (yellowish green) on glass showing that the first product had gone off and that is the one that busted it before

I now heated the[a] tube as hot as I could in the flame and I could not see anything come off except a slight white smoke this we now blew in a bulb & made a vacuum and with 9 cells C[ondit] & H[anson] cells gave a deflec[tion] 43° showing as Upton tested

Electric Light   Carbon wire

A spiral wound round a paper core no matter how thin always breaks, because it contracts so much. If the heating is done slowly this is modified but with the present proportion of Tar and Lampblack it will always break.[b]

Clay put on a spiral to insulate the outside and prevent it from sticking together tends to crack the spiral still more and[b]

We now put a larger per centage of lampblack to same tar about 2twice as much and the wire would still draw out at the ordinary temperature[b]

The better way to carbonize these carbon wires would seem to be to take the wire so:

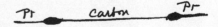

and fasten 2 platina ends in it, and then wind in form of a spiral soso:[b]

One of the great difficulties is to keep the spiral in position whilst you carbonize it   this might be remedied to a great extent by using a hollow sleeve & winding the spiral inside with something to hold the ends whilst they are being fastened to the leading wires

X, NjWOE, Lab., N-79-07-31:93 (*TAEM* 33:588; *TAED* N052:46). Written by Charles Batchelor; document multiply dated. [a]Obscured overwritten text. [b]Followed by dividing mark.

1. This entry is continued in Doc. 1831; see Doc. 1818 for earlier experiments with carbon spirals.

2. In the carbon-filament patent application that he executed on 1 November (U.S. Patent 223,898) Edison noted that the forms of filament described here and in Doc. 1818 "cannot be clamped to the leading wires with sufficient force to insure good contact and prevent heating. I have discovered that if platinum wires are used and the plastic lamp-black and tar material be molded around it in the act of carbonization there is an intimate union by combination and by pressure between the carbon and platina, and nearly perfect contact is obtained without the necessity of clamps."

–1831–

Notebook Entry:
Electric Lighting[1]

[Menlo Park,] Oct 22nd 1879 9 A.M.[a]

Electric Light   Carbon Spirals[b]   No 9[b]

We made some very interesting experiments on straight ~~wires~~ carbons made from cotton thread so.[2]

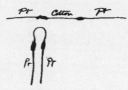

we took a piece of 6 cord thread No 24s which is about 13 thousandths in thickness and after fastening to Pt wires we carbonized it in a closed chamber. we put in a bulb and in vacuo it gave a light equal to about ½ candle 18 cells carbon. it had resistance of 113 ohms at starting & afterward went up to 140— probably ~~t~~due to vibration[a]

Electric Light   Carbon lamps   Carbonizing process[a]

We made lamps in same manner
1 of—   Vulcanized fibre[3]
2—[c]   Thread rubbed with tarred lampblack[4]
3—[c]   Soft paper—[5]
4—[c]   Fish line
5—   Fine ~~p~~thread plaited together   6 strands—
6—   Soft paper saturated with tar
7—   Tar'd Lampblack[6d] with half its bulk of finely divided lime work down to .020— Straight one ½ inch
8—[c]   200's 6 cord 8 strands—[e]
9—[c]   20s[7] Coats[8] 6 cord ~~not~~ no coating of any kind
10—[c]   Cardboard—
11   cotton soaked in tar (boiling) & put in[f]

Electric Light

No. 2[9b]   lamp of page 107[10] had on 18 cells and gave an elegant light equal to about 22 candles[a]

No. 9[b]   ordinary thread Coats 6 cord #24s— Came up to ½ candle and was put on 18 cells battery permanently ~~for~~ at 1.30 AM— ~~It~~ Page 105 & Page 115[11a]

3   Came up to 1½ gas jet   leading wire melted on account of conduction across mica (perhaps)   See page 115[12a]   was put on machine with 3rd speed 6 cells in field— It had an enormous resistance[a]

No 10[b]   Considerable resistance— equals 1 gas— jet had a small arc in—[a]

No 9[b]   On from 1.30 AM till 3 pm 13½ hour and was then raised to 3 gas jets for 1 hour then cracked glass & busted[a]

No 11[b]   A great many were made and boiled in tar before carbonizing but all so done broke in carbonizing[a]

Chas Batchelor

X, NjWOE, Lab., N-79-07-31:105 (*TAEM* 33:594; *TAED* N052:52). Written by Charles Batchelor; document multiply dated. [a]Followed by dividing mark. [b]Multiply underlined. [c]Preceded by "X" in left margin. [d]"d Lampblack" interlined above. [e]Items 8 through 11 enclosed by vertical lines at right. [f]Sentence enclosed in braces.

1. This entry is a continuation of Doc. 1830 and is continued in Doc. 1838.

2. In a notebook entry of the same date Francis Upton wrote: "Trying to make a lamp of a carbonized thread. 100 ohms can be made from an inch of .010 inch thread. A thread with 45 ohms resistance when cold was brought up in a high vacuum to 4 candles about. It remained constant for two or three hours and then the resistance seemed to concentrate in one spot. Resistance cold 800 Ohms." Upton's record of the results and calculations of lamps per horse power based on these results follow. He calculated that at 113 ohms per lamp they would get 114 per horse power but when the lamp went up to "140 Ohms after ½ hour" he recalculated that they would get "106 per H.P. light equal to about ½ candle." N-79-08-22:171–75, Lab. (*TAEM* 35:868–70; *TAED* N085:85–7).

3. It is not clear if Upton meant this filament, but writing about lamps made with "Fibre thread" he noted that "None of these carbon spirals blacken the glass." N-79-08-22:179, Lab. (*TAEM* 35:872; *TAED* N085:89).

4. In the carbon-filament patent application that he executed on 1 November (U.S. Patent 223,898) Edison stated that when a thread or other fibrous material was "rubbed with a plastic composed of lampblack and tar, its resistance may be made high or low, according to the amount of lamp-black placed upon it" and that "the plastic lamp-black and tar are used to secure it to the platina before carbonizing." The shape of the thread filaments listed here and in Doc. 1838 is unknown, although the drawing for the patent application showed a spiral and Edi-

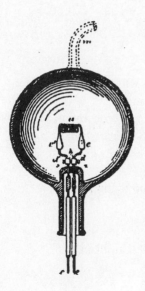

The drawings for Edison's 1 November patent application (U.S. Patent 223,898) show the ends of the filaments c c' thickened with lampblack and tar. At this point Edison still preferred the spiral shape as a way of increasing resistance and reducing radiating surface.

son stated that when these filaments were coiled into a spiral "as much as two thousands ohms resistance may be obtained without presenting a radiating-surface greater than three-sixteenths of an inch." Upton noted of one lamp with a thread coated with lampblack that it gave "very brilliant light all the cells on   Broke." Of another he wrote "Thread with lamp-black very fine light on several hours giving two or three candles and in hour giving a gas jet. No change two candles leading wires burnt. N-79-08-22:177, Lab. (*TAEM* 35:871; *TAED* No85:88).

5. This or filament No. 5 may be related to the drawing Edison made on 21 October showing a circle of carbonized paper with the bottom section between the lead-in wires broken out. N-79-07-31:92, Lab. (*TAEM* 33:588; *TAED* No52:46).

6. In U.S. Patent 223,898 Edison stated that "carbon filaments may be made by a combination of tar and lamp-black, the latter being previously ignited in a closed crucible for several hours and afterward moistened and kneaded until it assumes the consistency of thick putty. Small pieces of this material may be rolled out in the form of wire as small as seven one-thousandths of an inch in diameter and over a foot in length, and the same may be coated with a non-conducting non-carbonizing substance and wound on a bobbin, or as a spiral, and the tar carbonized in a closed chamber by subjecting it to high heat, the spiral after carbonization retaining its form."

7. Batchelor probably meant 24 as that number is given above and below in connection with this lamp.

8. J. & P. Coats was the dominant thread manufacturing enterprise in the United Kingdom during the nineteenth century. Charles Batchelor had worked for the company between 1865 and 1870, when it sent him to install machinery at its Clark thread mill in Newark. The six–cord spool cotton thread used in several lamps described in these notes had been developed at the Clark mill in 1864 for use in sewing machines. Cairncross 1987, Kim 1998.

9. Batchelor dated the document from here to the end as 21 October even though the text was clearly written on 22 October because it discusses the experiments that had begun the night before.

10. That is, the list immediately above.

11. That is, the discussion above and below of lamp No. 9.

12. Batchelor apparently wrote this as a page turn; the text that follows is on page 115.

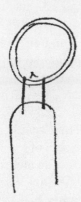

Edison's 21 October drawing shows a filament made from a circle of carbonized, paper with X indicating the section to be removed.

–1832–

To the Editor of the Scientific American

Menlo Park, N. J., October 23, 1879.[a]

Edison's Electrical Generator.

To the Editor of the Scientific American:

I notice in your last issue a communication from a gentleman named Weston[1] denying certain results which I had stated to the writer of the criticised article regarding the efficiency of my dynamo–electric machine.[2] His statements are without sense or science, and plainly originate from one who does not understand the laws which he pretends to set forth. I append the report of Mr. Upton, my assistant, who has made all the measurements with the Faradic machine.

T. A. Edison.

ENCLOSURE[b]

Mr. Edison: I have read very carefully the communication of Mr. Weston, which you handed me to report upon. It is impossible that the statement quoted by him, that your machine delivers nine-tenths of the electrical energy outside, is mathematically absurd, when it has been found to be practically true.[3]

The assertion that a machine working with nine times more external than internal resistance must be "capable of increasing its own electromotive force nine times without an increased expenditure of power" is utter nonsense. Mr. Weston has evidently confounded the obtaining of a maximum of current with the obtaining of a maximum of economical efficiency. A Faradic machine with a constant field may be considered electrically, when running at a fixed speed, as a battery with a certain E.M.F. and internal resistance. Your machine, for example, has 130 volts electromotive force and about half an ohm internal resistance. According to the reasoning in the letter in question it would be mathematically absurd to connect a battery with a resistance nine times greater than itself, and "destructive of the doctrine of the conservation and correlation of forces," since doing this with a battery is exactly similar to what you have done with your machine in the case mentioned.

To express the results with equations, the outside work may be taken as equal to $E^2(r+R)^{-2} R$. This will be a maximum when the equation of condition, that the first differential coefficient is equal to zero, is satisfied, or $-2E^2(r+R)^{-3}R +E^2(r+R)^{-2}=0$, which is the case when $R = r$. This shows the maximum is obtained when the external resistance is made equal to the internal. An experimental proof of this was given in a recent number of *La Lumière Electrique*.[4] For example, in your machine there should a maximum theoretically when R equals 0.5 ohm, E equaling 130 volts, or when $\frac{130 \times 130}{1 \times 1} \times 0.5 \times \frac{44.3}{33,000} = 11$ horse power can be utilized outside of the machine, while as many are lost in the machine. Again if $R = 9r$, as in the case mentioned for illustration in the Scientific American, that is, $R = 4.5$ ohms, $\frac{130 \times 130}{5 \times 5} \times 4.5 \times \frac{44.3}{33,000} = 4$ horse power can be utilized outside of the machine. In the first case, as compared to the second, 25 times as much power is lost in order that $2\frac{3}{4}$ times as much useful effect may be obtained.

Seeing that Mr. Weston has failed to understand this statement, though expressed clearly in the article he criticises, his talk about your denying the truth of Ohm's law is highly ridiculous, as well as his boastings about exposing your so-called absurd theory.[5] His placing a few letters and equations in his letter makes more absurd the total lack of power he has to apply them.[6]

FRANCIS R. UPTON.

PL, *Sci. Am.*, 41 (1879): 308. In Cat. 1241, item 1358, Batchelor (*TAEM* 94:530; *TAED* MBSB21358). [a]Place and date not that of publication. [b]Enclosure is a PL.

1. Edward Weston (1850–1936) was an English-born inventor and manufacturer of electrical equipment in Newark. He headed the Weston Dynamo Electric Machine Co., one of the largest manufacturers of electroplating dynamos. Edison referred to him in an undated draft of this document as "the inventor of a very poor nickel plating Dynamo Electric machine, and lately of another similar apparatus for electric Lighting. These machines constructed in violation of every known law of both electricity and magnetism." *ANB*, s.v. "Weston, Edward"; N-79-03-10.1:261, Lab. (*TAEM* 31:111; *TAED* N022:109).

2. The *Scientific American* of 18 October contained a lengthy article on Edison's dynamo, to which Weston responded in a 13 October letter to the editor published in the 1 November issue. "Edison's Electrical Generator," *Sci. Am.* 41 (1879): 242; "Letter to the Editor," *Sci. Am.* 41 (1879): 276; Cat. 1241, items 1332 and 1341, Batchelor (*TAEM* 94:513, 517; *TAED* MBSB21332X, MBSB21341X).

3. In his letter (see note 2) Weston had quoted from the article that "the energy converted [to electricity] is distributed over the whole resistance, hence if the resistance of the machine be represented by 1, and the exterior circuit by 9, then of the total energy converted nine-tenths will be useful, as it is outside of the machine, and one-tenth lost in the resistance of the machine." He declared that this explanation was "mathematically absurd. It implies either that the machine is *capable of increasing its own electromotive force nine times without an increased expenditure of power,* or that external resistance is *not* resistance to the current induced in the Edison machine."

Weston was expressing the widespread contemporary view that the optimum condition for operating a dynamo was when its resistance equaled that in the outside circuit (see Doc. 1664 n. 6). Charles Seeley, a former member of the *Scientific American* editorial staff and recently a professor of chemistry at the New York College of Dentistry (Obituary, *Sci. Am.* 67 [1892]: 320), articulated this belief in a more temperate critique in the 15 November issue. By analogy with a battery, Seeley reasoned that

The law of the electric current is that it exists or is produced *inversely* as the resistance to its flow in the circuit. . . . In any machine let the armature revolve steadily, and the current produced will depend solely upon the resistance; with the least resistance you get the maximum current, with the greatest resistance you get the minimum current. Now, also, the internal resistance of any machine is constant or unalterable. In order to get any external effect, external resistance must be added to the internal. To get the greatest yield from a machine or battery, it must be short circuited; that is, the external resistance must be suppressed; but then you find yourself in the interesting predicament that all the electricity is securely bottled up in the armature and is of no good to you. On the other hand, arrange things so that the greatest part of the resistance is external, and the electricity has shriveled up to a quantity which is utterly useless to any allopath. There is evidently a just mean; what is it? ["Edison's Electrical Generator," *Sci. Am.* 41 [1879]: 305 in *Electric Railway Co. v. Jamaica and Brooklyn Road Co.*, Defendant's Proofs, Vol. II: Exhibits, Lit. (*TAEM* 115:836; *TAED* QE001A344)]

Upton replied in the 29 November number that Edison's large field magnets made each armature winding more "efficient" and able to overcome a disproportionate external resistance. Such a machine could not operate "with the same resistance outside as inside, as it would heat the wire on the armature so as almost to burn it, by carrying a current so much in excess of that for which it was intended." He also criticized Seeley for misapplying Ohm's Law to claim that energy varies directly with the current, rather than as the square of the current. He noted in closing that Edison "hopes soon to have a machine with only one-eighth of an ohm in the armature, which he will use with an external resistance twenty times as great" ("Letter to the Editor," *Sci. Am.* 41 [1879]: 337, Cat. 1241, item 1358, Batchelor [*TAEM* 94:530; *TAED* MBSB21358]). In an expansive reply published in the 6 December issue, Seeley declared that Upton's letter was "far from being satisfactory" and reiterated his disbelief that Edison could have been the source of such "a preposterous claim for his electric generator" ("Letter to the Editor," *Sci. Am.* 41 [1879]: 360, Cat. 1241, item 1360, Batchelor [*TAEM* 94:531; *TAED* MBSB21360X]).

4. Upton was probably referring to Théodose du Moncel's report in the journal's first (April 1879) issue on tests with a Siemens, an Alliance, and a Holmes machine (du Moncel 1879a). An English translation of this article is in *Edison Electric Light Co. v. U.S. Electric Lighting Co.*, Complainant's Rebuttal—Exhibits (Vol. VI), pp. 4101–4106, Lit. (*TAEM* 48:561–63; *TAED* QD012G4101).

5. Weston argued in his 13 October letter (see note 2) that by applying Edison's "absurd theory" to the case where internal resistance equals that outside the dynamo, it could be shown that the machine "returned more useful current to the circuit than could be due to the power employed (and in the ratio indicated), so that there would actually be a creation of force!" He concluded that "If such statements as these have been made by Mr. Edison to the representatives of the daily papers I think he has no cause to complain of the treatment received, but rather to consider himself fortunate that he has escaped rougher handling."

6. Weston responded directly to a number of Upton's objections in a reply published in the 13 December issue. Noting Upton's apparent neg-

lect of mechanical sources of energy loss, he declared that "you have only to start the machine and it will continue to revolve for ever, and perpetual motion is an accomplished fact." He also referred to efficiency tests of other low-resistance dynamos and chastised Upton for not providing comparable data on the horsepower consumption of Edison's machine. Weston closed with the remark that he was "possessed of sufficient 'sense and science' to prevent my falling into such manifest absurdities as are contained in Mr. Edison's statements [in the original article] . . . or Mr. Upton's elucidation." "Letter to the Editor," *Sci. Am.* (41:380–81), Cat. 1241, item 1370, Batchelor (*TAEM* 94:535; *TAED* MBSB21370X).

–1833–

From George Gouraud

LONDON, 23rd Oct 1879[a]

My Dear Edison,

Colonial and Foreign Patents.[1] Deed of Assignment and Powers of Attorney.[2] I send you hereith for execution—please have them witnessed as usual before ~~Counsel~~ British[b] Consul and return by first steamer. I am now correcting the proofs of the Memorandum and Articles of Association & Articles of Agreement[3] between yourself and the

1	"Edison Telephone Coy of India"
2	" China
3	" Egypt
4	" Cape of Good Hope
5	" Norway[c]
6	" Sweden
7	" Switzerland
8	" South America
9	" Australasia

They will be upon the same general basis as the London Coy contracts being with the principal cityies in the countries with powers to form District Companies[4]   In case of any of these companies where there are no patents <u>absolutely</u> we shall have to be content with simply half the profits without any minimum guarantee of royalties.[c]

I have had employed upon these documents the best legal advisce in England both Solicitors & Counsel. Among the latter you will recognize Webster[5] Q[ueen's].C[ounsel]. and Benjamin[6] Q.C. the latter was secretary of state to the Southern Confederacy and is now one of the leaders of the [English?][d] Bar. If the agreements do not work out well it will not be for the lack of taking every precaution in drawing them out quite regardless of expense. My plan is to place part of the capital of each of these Companies in England and the remainder in the countries themselves[7]   The Agreements and the Power of Attorney above referred to underlie the whole superstructure—further agreements will be forwarded to you as soon as they are got out

I am glad to hear the telephones are arriving and the inspectors— With[c] regard to the latter you cannot take too great ~~care~~ a precaution in selecting them so as to get the very best men. Those that have already arrived seem to be very good[8]

I sincerely trust that your "30 days option on Australia"[e] will fall through as I have here a very strong party to put this on its legs. I take it for granted that this option does not include New Zealand. If the 30 day option is accepted look particularly in your papers and see that it is confined to Australia only and cable me as follows. If Australia is free for me to negotiate cable the word "Australia" if Australia is to be accepted[c] cable the word "New zealand" which will save me ten days which will otherwise be lost in the preparations.[9]

Upon the receipt of this please have somebody competent to do so (no reflection upon our friend Griffin who might possibly be engaged on other matters) look to the bottom of your contracts and as soon as you know definitely cable me what other[f] countries are free as I propose with your approval to cover the whole world—wherever the thing can be used to advantage   Yours very truly

G E Gouraud

LS, NjWOE, DF (*TAEM* 52:212; *TAED* D7939ZAC). On note form of Geo. E. Gouraud. Written by Samuel Insull. [a]"LONDON" and "187" preprinted. [b]Interlined above. [c]Obscured overwritten text. [d]Illegible. [e]"on Australia"" interlined above. [f]Interlined above by Gouraud.

1. Three days earlier Edison had wired Gouraud to ask if he should "take Patents in Countries mentioned your expense." Gouraud responded in a 21 October cable, "Yes but pass papers through my patent agents here." In a letter of the following day he explained that "My reason for asking that the matter be attended to by my Patent agents here is that I wish to avoid the repetition of such a state of affairs as depicted in [John] Johnsons report in connection with English Pat" (see Doc. 1825 n. 2). On 27 October Edison wrote Edward Johnson with instructions to give John Johnson in preparing the foreign patents (Doc. 1835). TAE to Gouraud, 20 Oct. 1880; Gouraud to TAE 21 and 22 Oct. 1880; all DF (*TAEM* 52:207, 840–41; *TAED* D7939Y1, D7941ZGQ, D7941ZGR); TAE to Johnson, 27 Oct. 1879, Lbk. 5:304 (*TAEM* 80:167; *TAED* 80:167).

2. There are copies of three assignments dated only 1879. One gives Gouraud one-half interest in Edison's telephone inventions for "India South America (excepting the countries of Peru and Chili and the Ar-

gentine Republic)   The Colony of the Cape of Good Hope   Turkey in Europe   Turkey in Asia   Norway   Sweden   Egypt   Switzerland   China and Holland"; the other two assign similar rights for India and Australasia respectively. Miller (*TAEM* 86:92, 88, 90; *TAED* HM790093, HM790090, HM790091).

3. Not found.

4. See Doc. 1765 n. 1.

5. Probably Richard Webster, who had become a Q.C. in 1878 and was the son of the late Thomas Webster, a leading authority on patent law. *DAB* s.vv., "Webster, Richard Everard" and "Webster, Thomas."

6. Judah Benjamin, born of Jewish parents of English nationality in the West Indies, became an attorney in the United States and was serving as U.S. senator from Louisiana at the outbreak of the Civil War. Joining the Confederacy he served successively as Jefferson Davis's attorney-general, secretary of war, and secretary of state. After the war he fled to England, where he studied law at Lincoln's Inn and soon became a prominent member of the British bar noted for his book on contract of sale and for his large and successful practice on appeals, especially those related to colonial matters, before the House of Lords and the Judicial Committee of the Privy Council. *ANB* and *DNB*, both s.v. "Benjamin, Judah Philip."

7. An undated memorandum that is in Stockton Griffin's hand but was apparently authored by Gouraud sometime prior to this letter discusses this 50-50 division and his plan to establish separate companies in each country as well as the fact that he had attorneys working on the deeds. He also asked Edison to either forward the relevant foreign patents or inform him of the status of those he had applied for but not yet received. Gouraud memorandum, n.d. 1879, DF (*TAEM* 52:243; *TAED* D7939ZAW).

8. The previous day Gouraud had written Edison regarding the arrangements he had made with one of the steamship companies to pay passage for the inspectors at the rate of £10 each. He noted that two inspectors, John Wohltman and John Crawford, had arrived that morning (see Doc. 1808 n. 2 regarding these and other inspectors). He also reiterated Johnson's view that the inspectors "must be able to set up an exchange and work it properly—in fact capable of undertaking the duties of Telephone Engineer." An amusing account of the inspectors was written by George Bernard Shaw, who was employed by the London company. Gouraud to TAE, 22 Oct. 1879; Henderson Bros. to Gouraud, 21 Oct. 1879, both DF (*TAEM* 52:841, 839, *TAED* D7941ZGR, D7941ZGP); Weintraub 1969, 89–90.

9. See Docs. 1817 and 1881.

–1834–

From Grosvenor Lowrey

New York October 24th 1879[a]

My dear Edison:

There was to have been a meeting of the Board of Directors of the Edison Electric Light Co. today to consider the proposition of raising more capital in the manner proposed by the Executive Committee.[1] I have been home sick for all this week and do not know what if anything was done.[2] I regretted to see however in one of the directors a few days since a strain of opposition to the proposition on the ground that you agreed to give them an electric light and that they agreed to give you Fifty thousand Dollars. They had paid their money and if any more money was needed they ought not to grant to you the equal terms which were formerly granted. The proposition was evidently misunderstood by the gentleman in several respects. I corrected this but did not produce very much effect. The danger about this business is simply that if the money gets an advantage, that advantage will almost certainly be pressed in case of delay or difficulty to the disadvantage of your interest and the interests of those who have taken their stock from you. This is not un-natural or unusual, but in the case of an invention which is to be of such enormous value if of any value whatever, I think it fair that the interest given to the inventor should be kept at all times equal to the interest given to the money which comes in to aid him. Money can be got in many places and in a great variety of ways, but I am not at the present informed of the existence of more than one man who would be by any considerable number of people, thought equal to the work which you are engaged upon. The question of supply and demand therefore makes money cheap and talent valuable and dear. That argument prevails and was adopted at once when we made our original agreement. Now there would be no difficulty whatever in getting money ~~paid or for~~ by way of[b] loan or advance to the Company, but that would put things in this situation that the lenders being creditors of the Company might find themselves at some future time, perhaps very near by in a position to acquire the whole of the Company's interest in the patents, wiping out all the stock and taking from you or from your family perhaps just at the moment of success, all the reward due to your labors. This argument also has been accepted by every member of the Board to whom I have suggested it as fair and the method proposed of having you assign some other countries has seemed to be a convenient one as well as a reasonable one. The gentleman, to whom I referred, did not have time to talk to me much and at length and perhaps he will change his views and for that reason I do not mention his name.[3]

He did make one point however which had not occurred to me and which I think well taken, which was, that the proposed increase of stock ought not to be connected with this method of raising new money because in order to be fair it proposed that the entire money should be raised by the old subscribers strictly in the proportion in which they had formerly taken their stock, which might or might not be convenient for all. ~~I think~~ In case this was not done the new subscribers would get more shares for their dollars in proportion than the former ones. I at once said that must not be and that it would probably be best to keep the same old proportion as before and to make the increase in the stock afterwards when it could be divided equitably

I only write this to you now in a general way to keep you informed but not because I think there is anything to be said or done by you at present. I think that you will do well in case any thing is said to you on the subject to be as much as possible without "views", and refer every-one to me.

How soon shall I have my telephones?[4]

I have got the gas in my house and in about ten days everything will be in order, then I shall come down to see you and hope to bring you Mrs Edison and the children up with me to spend a Sunday   Truly yours

G. P. Lowrey

LS, NjWOE, DF (*TAEM* 50:280; *TAED* D7920ZBG). Letterhead of Porter, Lowrey, Soren & Stone, Attorneys & Counsellors at Law. [a]"New York" and "18" preprinted. [b]"by way of" faintly interlined above by Lowrey.

1. See Doc. 1805.
2. Lowrey later informed Edison that the meeting lacked a quorum but "that upon discussion they seemed to think that the fair way was to increase the stock of the Company $100,000, to be sold at par, giving the option first to present share holders." He indicated that he did not necessarily oppose this

but before allowing it to be done with the effect of reducing your, and my, proportionate share of the whole, unless we pay in in cash for the new stock in that proportion which we hold in the old, I am disposed to consider with you whether we will not do better to carry on these experiments at our private charge for a while longer. You have expressed your willingness to do so on one or two occasions, and I intend always to take my even share with you in everything of that sort. . . . If we could see a definite end ahead it would be better for us to expend a few hundred dollars or a few thousand dollars and keep the stock where it is, with our respective interests intact, than to allow those interests to be reduced 25% by an increase of the stock of $100,000. [Lowrey to TAE, 27 Oct. 1879, DF (*TAEM* 49:464; *TAED* D7903ZJY)]

3. After a meeting of the directors on 13 November Lowrey reported that "the general notion seemed to be that everybody was willing to con-

tribute $5 a share, from among the Cash subscribers—Mr Edson presented a plan to that effect." In response to some misgivings that Edison would be exempted, Lowrey declared

> that you were entirely willing to contribute your share, but that I thought it would be wrong to accept from you any such contribution; that while that was the correct business view of it it was not to be forgotten that you were presenting to them the greatest return for capital that ever was offered, and was giving, hourly, the resources of your talent, and knowledge, & devoting your health &c, which only one man in the world possessed while plenty of people had capital. . . . Finally it was decided to pass around a paper to subscribers for voluntary subscriptions of $5 per share, and that for anything you chose to subscribe you should be entitled to a credit of $5000. It appears that you have standing to your credit $1183—That, of course, includes my portion.

Lowrey noted that Charles Batchelor, Francis Upton, and others to whom Edison had given stock would not be expected to contribute. Lowrey to TAE, 13 Nov. 1879, DF (*TAEM* 50:287; *TAED* D7920ZBI).

4. Rockwell Kent, Lowrey's secretary, had written that Lowrey previously used two Gold and Stock telephones on a line between his Tarrytown house and stable, but "expecting to be away from his house for a year or more had the instruments removed, but having now returned to Tarrytown he desires to replace the telephones and directs me to ask you if you can let him have two of your new telephones." Edison promised that he would have "two telephones put in order & tested just as soon as possible and send Francis [Jehl] to put them up— It may be several days before I can get them." Lowrey's planned absence from his Tarrytown estate was likely related to his wife's death in August after a long illness. Rockwell Kent to TAE, 13 Oct. 1879, DF (*TAEM* 52:124; *TAED* D7937ZDH); TAE to Lowrey, 27 Oct. 1879, Lbk.5:307 (*TAEM* 80:169; *TAED* LB005307B); Taylor 1978, 23.

–1835–

To Edward Johnson

[Menlo Park,] Oct 27, 1879

My Dear Johnson[1]

Have John H Johnston prepare patent papers for India, Denmark, Norway, Sweden, New South Wales, Victoria, Queensland, New Zealand, Portugal and South Australia.

Take the new instrument with the improvement on the push button that stops the knock in the receiver.[2]

Give a full description of the whole instrument.

Claim. First. The method of adjusting the carbon transmitter by the screw in the diaphragm.

2d   The tertiary coil containing the chemical receiving instrument.

[3d][a]   The combination of a transmitting instrument in a primary coil and circuit with battery which varies the resist-

ance of an electric circuit by varying the intimacy of contact [etc]^b with an induction coil consisting of a primary, secondary & tertiary coil all [operated]^b and connected substantially as set forth

[4th   The use]^b of palladium to tip the spring which rests upon the chalk

5th   The use of a phosphate of an alkali salt to moisten the chalk with.

6th   The method of putting pressure on the spring by the rubber piece.

7th   The device shown in new push for preventing the opening and closing of primary [from]^b destroying the chalk.

8   The worm & worm wheel

9th   The general connections

Have the papers prepared and one of those in English sent me also have all the necessary papers for signature sent me at once.

The above claims are made in view of what has already been published. I believe nothing [claimed]^b has been published so as to defeat the claims in the countries where the patents are to be taken out.

Perhaps the palladium, phosphate of an alkali New push New method adjusting transmitter and general connections which have not been published in England would for[3] the subject matter of another Patent, anyway the new method of adjusting the transmitter and new push, would be good.[4]

Last week 8 of us on telephones only slept 15 hours and did not have our boots off once—same with Batch— Tested 36 instruments last night and every one belched right out. Majority so loud couldnt bear the ear near them[5]   Very truly

T. A. Edison   G[riffin]

L (letterpress copy), NjWOE, Lbk. 5:304 (*TAEM* 80:167; *TAED* LB005304). Written by Stockton Griffin; circled "C" in top left corner. ^{a}Faint copy. ^{b}Faint copy; text taken from draft.

1. The body of this document is substantially the same as Edison's undated draft (DF [*TAEM* 52:219; *TAED* D7939ZAF]). On that draft Edison initially wrote "L. W. Serrell. Prepare papers" then crossed out those lines and addressed it to Johnson. This change probably reflected his receipt of Gouraud's 21 October cable instructing that the papers be prepared by English patent agents (see Doc. 1833 n. 1).

2. See Doc. 1806.

3. Edison wrote "form" in his draft.

4. The second, fourth, and seventh suggested claims above were incorporated into the British provisional specification filed for Edison by Arnold White at the end of December. The remaining numbered items,

except the fifth, were included (not necessarily claimed) in the final specification, filed in March 1880; there Edison claimed the use of "pure or distilled water" for moistening the chalk (see Doc. 1845 n. 1). Brit. Pat. 5,335 (1879), Batchelor (*TAEM* 92:141; *TAED* MBP021).

5. Records of the tests of these telephones, which concluded on 28 October, are in N-79-09-18:60–97, Lab. (*TAEM* 35:952–67; *TAED* N086:31–49).

–1836–

To B. H. Welton[1]

[Menlo Park,] Oct 27 79

Dear Sir

Your favor of the 23d was duly received.[2] I should like to comply with Pitts request but unfortunately my time is not my own   Parties are to meet here (N.Y.)[a] from all parts of the country with whom I am in litigation[3] and in addition there are a thousand and one things requiring my personal attention every moment.

Some one had better come on to take Charley's remains to Port Huron   I will see that permits are obtained from the health officer and do anything of that kind that may be necessary. The steamer is due Nov 4th. Very truly

T. A. Edison   G[riffin]

L (letterpress copy), NjWOE, Lbk. 5:308 (*TAEM* 80:170; *TAED* LB005308). Written by Stockton Griffin; circled "C" at top right corner. [a]"(N.Y.)" interlined above.

1. B. H. Welton was evidently a friend of Pitt Edison in Port Huron (see note 2).

2. Welton wrote at Pitt's request to learn the arrangements for shipping Charley's body to Port Huron. He also conveyed Pitt's wish that Edison "come here a day or to, before Charley's remains—for it is the last thing that can ever be done for his dear boy—Even should you never come to Pt. Huron again." William Wastell had also written Edison from Port Huron on the day of Charley's death that the distraught Pitt "wishes any thing belonging to Charley and any little personal effects to be taken care of, not for the value but as remembrances— he is very anxious to have his remains brought here   of course I do not know what to say—you must be the controller in this matter." Welton to TAE, 23 Oct. 1879; Wastell to TAE, 19 Oct. 1879; both DF (*TAEM* 49:977, 957; *TAED* D7914Q, D7914K).

3. Edison was scheduled to testify in the first telephone interference cases on Monday, 27 October. Preliminary statements for the telephone interferences declared in August (see Doc. 1792) had been due on 20 October, and at this time Edison was also preparing preliminary statements for two other telephone cases and a phonograph interference. The Patent Office extended the deadline for these documents to 27 October but Edison told Lemuel Serrell that he needed another month "as I have all I can possibly do getting the papers ready" for his testimony. Serrell

to TAE, 17 Oct., 29 Sept., and 14 Oct. 1879, DF (*TAEM* 51:583, 559, 579; *TAED* D7929ZEX, D7929ZEG, D7929ZEU); TAE to Serrell, 22 Oct. 1879, Lbk. 5:295 (*TAEM* 80:166; *TAED* LB005295).

–1837–

From Josiah Reiff

London, E.C. Oct 28 79[a]

My dear Edison

I hasten to send you a few lines by todays stmr regarding Telephone affairs at Paris.[1] I got at the true inwardness of matters, as I heard in great detail the respective stories of Dr Herz—Roosevelt,[2] Puskas & Bailey, each of whom were anxious to impress me with the correctness of their particular views & to convince what infernal scoundrels each of the others were—

Each one of course insisted that whatever of good had been accomplished had been his particular work & the evil had all resulted from the conduct of somebody else. With it all the Edison name & interest have been brought into open disrepute & I think can only be properly cleared up & protected by an immediate & absolute revocation of every kind of power you may have given or assented to, to either or all of the parties above named & placing it in the hands of some new party who has common sense, common honesty & a single interest with you.

~~Here~~ There has already been enough done to shew [c]heap[b] practice to say the least & in several particulars individual conduct has been simply disgraceful. Some of the parties have no ability at all— Some have too much ability & too little idea of square dealing & some are altogether too sharp. of course all say they are trying to protect Edison, but they do it in a strange way & their is too much self imposed guardianship.

In view of the practical value of the Bell-Gower telephone—the Blake transmitter—the Phelps Gray invention & the <u>Fitch!!!</u>[c] (save the mark[3]) as compared with the Edison, the division proposed & practically agreed to ~~is~~ I consider absurd. However there it is, you have already been advised of the details. But the <u>Fusion</u> is temporarily interrupted by a scandal [begun thro?][d] Roosevelt & Baily to liquidate the Edison Co & so control it as they choose, meantime with <u>great profession</u> of care to protect your interest. Puskas is clearly without any practical ability except to borrow & spend money in maintaining a costly establishment with female accompaniments, which seems to be known all round as near as I can learn.

Bailey say he P owes him 11,000 francs loaned him for his private purse.

Bailey I think has ability, but is evidently utterly unscrupulous as to means to accomplish results.

He Bailey conceded to me last night that except in matters where he was technically & legally bound to you by written agreement, he felt utterly free to look after his own interest whether it affected you injuriously or not— In other words Bailey he did not care a damn for the Edison or any other Telephone, that what he wanted to do, was to make money.

In that view Bailey is now in London with young Brown[4] to make some Exhibitions of the Fitch Telephone(?) with view to trying to get subscriptions to a new Co & force the Edison Telephone Co of London to buy them out. Bailey says he considers he is in no way bound to consider any relation to you on the Continent as affecting him here & that he told you when in America that he proposed to work the Fitch & other Telephones here & you did not object.[5]

His justification for this is that no Telephone patentee dare risk his patent in a European court, for fear it will be declared invalid & hence people can use almost any thing without fear of suit for infringement & therefore schemes can be launched indefinitely to compel the legitimate Telephone Company conducting a regular business with capital involved, but buy them up or run the risk of threatened competition or depression of their stock—

This is Baileys professed mission in Europe & he is the man who claims to have a special and comprehensive power of atty from you to represent your interest on the Continent, based on some arrangement by you with White, Cutting & Banker, whereby you practically conceded them rights as to the Carbon Telephone & whereby they receive % of the cash & stock proposed to be paid for the telephone in Paris— Comment is unnecessary—

Bailey charges Puskas with ~~offering~~ authorizing [for?][d] a consideration to sell you out— Puskas declares certain parties received benefit for trying to sell you out   Roosevelt admits he originally thought Bailey a scoundrel, but it seems his view only changed when he (R) found B would cooperate with him—which would shew that personal interest had something to do with the[e] new & favorable opinion now held of his integrity & ability. Of course I would not assume to reflect on the personal integrity of Roosevelt, but I cannot overlook the fact that Roosevelt undertook by a Coup-de-etat to scoop the control of the Edison Co & when he found he could not carry the voting power by simply owning 135,000 out of 200,000 francs, he then sold ½ of it again at a profit & subsequently joined with Bailey to force the Edison Co into liquidation, so as to fully carry out their own views.

In other words you [----]^f were not sacrificed because they did not succeed in obtaining the control as they expected.

Notice of liquidation was served last Saturday by Bailey & Bailey came to ~~P~~London about Fitch, by Tuesday recd a message from Paris that Puskas had yielded so^g Bailey returned to Paris yesterday, leaving word for me to cable you asking that you should not assent to Puskas being in the new Board until you could hear from me & E.[6] This I declined to do, but today I did cable you that I considered it absolutely vital that you should exercise your power & stop Paris negotiations & that you ought for your protection to absolutely revoke any or all authority which you may at any time have given to Roosevelt, Bailey & Puskas & start absolutely fresh.[7]

This need not affect any personal interests you may have given them or to which they may be entitled. I most sincerely trust you have not so bound yourself to Bailey Banker, Cutting White & Co. that you cannot control the conduct of the Telephone so far as it affects your own interest or Patent especially the Motograph.—

If you are at present bound, you must bring such^g pressure to bear on White & others as to lead them to follow your views.

I shall be home within about 3 weeks & then can explain matters in a way which would take me days to explain on paper. No evil can come from delay in Paris & on the Continent at present as absolutely nothing has been done outside of Paris & there it might better^g never have been done as you have <u>practically</u> been held subservient to such people as Blake, Gower, Fitch & Co.

The ideas of having either Bailey or Puskas to represent you, is simply preposterous & Roosevelt cannot fairly do it, as he has too much interest with Gower & Co. to be impartial.

You will receive this letter before I leave for home, so if you have any wishes to express, do so at once by Cable & I will follow your directions & attend to matters before leaving— Of course in Paris I had no authority, or I would exercised it promptly & I think could have effected a reorganization on my own or your own terms.[8h]

<u>Smith, Fleming & Co</u>[9]   I have had interviews with Mr Jno Pender[10] Sir James Anderson[11] & their counsel.

Smith, Fleming & Co have assigned all their interest to Mr Pender, which gives him a preponderating share of the London interest & makes him willing to put his shoulder to the wheel & get something out of the Post Office Deptmt. Of course they had already through their counsel served an injunction upon[g] the P.O. Dept. preventing them from paying Prescott or anyone else any money on a/c of the Quadruplex they are now using.

They say they did this after the advice of the ablest counsel in London. Weaver, they report undertook at the insistence of Prescott to reopen [some?][d] negotiation,[12] but the P.O. Dept notified him they could pay no money to him or anyone else until the question of title had been fully adjusted.

Meantime the PO Dept goes on using it, which is of course what we all want.

Of course in any event your interest is amply protected under the London party. Your interest will be larger than it would be could Prescott maintain his claim   Stearns[13] is also beginning— Ditto Muirhead—[14] Stearns now claims an interest in the Quadruplex on some ground I have not yet discovered.

Pender & Sir James of course commenced to sail[g] into you about not returning to Europe & not having fulfilled your duty & the London party etc etc.

I told them it was too late reopen old questions, founded on mutual disappointment. That if there was a chance to get any thing out of the interest—the best thing to do was to attend to that[g] & get it—that if you found a genuine effort was being made to push affairs here with a prospect of success you would [time?][d] in help in every proper & honorable way to work up what yet remained unfinished—

They insisted you led them to expect great things on the Cables & that even now any thing which would improve the speed of Cables would be welcome & immediately availed of by the cable companies under their control. So anxious are they in this direction that they are actually giving facilities to R K Boyle[15] to operate their cables with promise to buy at a sound price any improvement he may make. & they even hope for results from him— Boyle is now here.

To make everything agreeable here remove all soreness, I proposed that we should organize a limited Company on the basis of our original contract of 1873 & let them take stock for their ⅕th— This they have agreed to do & today we are organizing a Co on basis of £250,000 capital & propose to give £50,000 to us including their ⅓ interest. It is proposed to make

Pender President & you & Sir Wm Thomson Consulting Electricians.[16] I have advised with EHJ & Gouraud in this matter

The door to the PO has been opened to EHJ. by Preece[17] we shall try & make good use of it— Hastily yrs[i]

JCR

ALS, NjWOE, DF (*TAEM* 49:470; *TAED* D7903ZKA). Letterhead of 22 Old Broad Street. [a]"London, E.C." preprinted. [b]Reiff wrote "s" instead of "c." [c]Multiply underlined. [d]Illegible. [e]Interlined above. [f]Canceled. [g]Obscured overwritten text. [h]"on my own or your own terms" written in right margin. [i]Last paragraph, closing, and signature written in right margin of last page.

1. Reiff had first written Edison on 24 October that he was investigating matters in Paris. Having already spoken with all parties he declared "you are in the hands of adventurers from beginning to end & there is as I see now but one escape & that is to immediately, absolutely, entirely & without limitation, revoke all the authority you have given P. B or any one else here" and urged Edison not to sign any papers until "I can get home & tell you in person what I would not like to put on paper." He concluded by advising Edison that "delay is what you want here for the present." DF (*TAEM* 52:370; *TAED* D7940ZBT).

2. In his 24 October letter (see note 1) Reiff indicated that Edison should expect a long letter from Cornelius Roosevelt through his brother Hilborne Roosevelt. This letter was apparently sent on 19 October (see Doc. 1841). Cornelius had also been in communication with Edward Johnson regarding affairs in France and Johnson had sent these to Edison a few days earlier. In his letters to Johnson, Roosevelt discussed the state of the French telephone business and described the effort to merge the various interests into one company. Cornelius Roosevelt to Johnson, 3 and 16 October 1879; Johnson to TAE, 23 and 24 Oct. 1879; all DF (*TAEM* 52:349, 357, 845, 854; *TAED* D7940ZBI, D7940ZBO, D7941ZGU, D7941ZGV).

3. That is, a gullible person. Lighter 1994–97, s.v. "mark."

4. Robert Brown.

5. The previous day George Gouraud had cabled Edison: "Bailey here threatening trouble with Fitches." Edison replied: "Baileys device gross infringement." In a letter of 28 October, Gouraud provided a similar description of Bailey's activities. Gouraud suggested that if Edison could get free of Bailey, Theodore Puskas, and the others involved in the fusion company he could guarantee him half a million dollar for his European telephone patents within six months. Gouraud to TAE, 27 and 28 Oct. 1879; TAE to Gouraud, 27 Oct. 1879; all DF (*TAEM* 52:861, 863, 862; *TAED* D7941ZGZ, D7941ZHC, D7941ZHB).

6. Reiff may have meant to write "B" for Bailey.

7. Puskas signed the fusion papers 27 October and notified Edison the following day in a cable asking that he be nominated as Edison's representative on the board of the new company. Edison wired back nominating Puskas. Reiff apparently did not send his cable until 30 October. The following day Edison sent it to James Banker with the comment: "From all the information I can glean I have about concluded that Mr Bailey has been engaged and will continue to be industrially engaged in

feathering the pockets of Joshua F Bailey at the expense of J.H.B and T.A.E & others." Puskas to TAE, 28 and 31 Oct. 1879; TAE to Puskas, 28 Oct. 1879; Reiff to TAE, 30 Oct. 1879; all DF (*TAEM* 52:373, 380, 378; *TAED* D7940ZBU, D7940ZBZ, D7940ZBX); TAE to Banker, 31 Oct. 1879, Lbk. 5:314 (*TAEM* 80:172; *TAED* LB005314A).

8. Reiff, along with George Gouraud and Edward Johnson, continued to urge Edison to pursue an alternative arrangement for the telephone in France and Continental Europe along the lines of the Edison Telephone Co. of London. Reiff to TAE, 6, 13, 18, and 20 Nov. 1879, DF (*TAEM* 52:400, 409, 417, 425; *TAED* D7940ZCJ, D7940ZCS, D7940ZCY, D7940ZDE).

9. Smith, Fleming & Co. were a firm of East India merchants that acquired a one-fifth share of the worldwide rights (excluding the United States) for overland use of Edison's automatic telegraph system (see Doc. 350). Soon thereafter John Pender, James Anderson, and John Puleston, in either their personal or corporate capacities, acquired an interest in the system through Smith, Fleming & Co. John Smith had been among the founding investors of John Pender's Telegraph Construction and Maintenance Co. Nothing further is known of John Fleming or other members of the firm (see *TAEB* 2:39 n. 3).

10. John Pender (later knighted) in 1864 founded the Telegraph Construction and Maintenance Co., which manufactured and laid most of the world's oceanic cables, and he was heavily involved in most of the major submarine cable telegraph companies. The core of his empire was the Eastern Telegraph Co. See *TAEB* 2:5 n. 2.

11. Sir James Anderson was knighted for his services as captain of the *Great Eastern,* which laid the first successful Atlantic cable in 1866. When John Pender created the Eastern Telegraph Co. in 1872, Anderson became the general manager and later director. See *TAEB* 2:105 n. 4.

12. See Docs. 1692 n. 3 and 1804.

13. American inventor Joseph Stearns, who devised the first practical duplex telegraph, had been in Europe for several years promoting his invention. He spent considerable time in Britain and earlier in 1879 had been there duplexing the Atlantic cable (see *TAEB* 1:101 n. 3, 2:344 n. 10, Docs. 1204 and 1236). In May 1879 during his lectures on "Recent Advances in Telegraphy," British Post Office telegraph engineer William Preece had stated inaccurately that

> It was Mr. Stearns also, who, in 1874 first took out a patent, and who first practically showed how quadruplex working was possible; but it was in America where it was first brought into actual work by the Western Union Telegraph Company. Mr. Stearns' ideas were taken over to America by Mr. Orton, the President of the Western Union Company, and when he returned to America he put the matter in the hands of Mr. Prescott, the electrician, and Mr. Edison, their scientific adviser, and after two years of very hard work, Mr. Edison succeeded. [Preece 1879b, 28]

14. Alexander Muirhead was a director and chief scientific adviser for Latimer Clark, Muirhead, & Co. (an engineering firm co-founded by his father, John Muirhead). Together with H. A. Taylor, he had developed the basic system for duplexing undersea cables, which he, Taylor, and

the firm's manager (his older brother John Muirhead) had installed the previous year for the United States Direct Cable Co. Muirhead had visited Edison in May 1878. See *TAEB* 4:297 n. 1.

15. Robert Boyle, a Brooklyn-based telegraph inventor, applied for a quadruplex patent that he assigned to Josiah Reiff when it issued in 1881 as U.S. Patent 247,880.

16. William Thomson was committed to serving as an expert legal witness for the Bell company and therefore, according to Edward Johnson, "reluctantly" declined this offer. Johnson to TAE, 17 Nov. 1879, DF (*TAEM* 52:924; *TAED* D7941ZIK); see Doc. 1935.

17. Not found.

–1838–

Notebook Entry:
Electric Lighting

[Menlo Park,] October 28–November 10, 1879[1a]

Lamps[2]

No 12   Eight[b] thicknesses of 200s thread twisted and lamp-blacked a little—length of incandescent surface 3.5 inch—~~Shit~~!!! Busted by Bohm

Made new carbonizing chamber[3]

13   8 thicknesses of 200s thread twisted & blackened a little   length of surface 3.40 inch   Busted by Bohm[c]

14   4 thicknesses of 200s thread twisted[b] together & blackened with Lampblack & tar   Broken in carbonization

15   3 thicknesses of 200s thread plaited together and rubbed with Lampblack[b] and tar

16   6 thicknesses of 200s thread twisted together Brought up to yellow[b] & sealed off[4c]

17   ditto   Busted (after getting 3 gas gjets out of it) about ⅛ from one end— ~~Resist only~~[c]

18   ditto[5c]

19   ditto[c]

No 17   Had a variable resistance starting at 48[b] ohms and running down to[c]

Resistance	48—	red heat
"	40.5	Whitish[d]
"	38.9	incan[escent]
"	35.0[c]	

No 20   3 inch long made of 5 strands of Clark's 300s 3 cord— This when carbonized went out of shape considerable but was intact[6]

21[e]   Made of 300s 3 cord 3 inch long—good after carbonizing[7c]

22   Burnt in Carbonizing   300s 3 cord[c]

23   300s 3 cord   Burnt in carbonizing[c]

24   300s 3 cord 3 in long   carbonized in new chamber[c]

25   300s 3 cord 3 in long   carbonized in new chamber[8]

26— 300s 3 cord 3 in long   carbonized in new chamber came[b] out bad shape[c]

27. 24s—6 cord— 3 in long   no tar or lampblack on at all— Bare thread tied to platina supports with 200—6 cord not tarred   Resistance 150,000 ohm[c]      Joint of thread

28   24—6 cord— 3 in long   a little tar and[b] lampblack on the joints   150,000 ohms   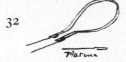   joint of thread

29   Card such as we mount minerals on— 3 in when put in—shrunk very much in carbonizing— 250 ohms cold[9] joint: no tar      Plat this shape[c]

30   Card:— 3 in long   shrunk[b] very much in Cn   250 ohms cold[10]   No tar on joint   joint so:[11c]

31   24s 6 cord made into lock ~~stich~~ stitch ~~plat~~ fastened to platina by winding 200s 6 cord round platina and thread

32   Platina   Made from card 1/32 wide—3 inch long   busted in Carbonizing[12c]

33   Made from card 1/32 in wide .043 wide .010 thick 3.312 long—with wide ends and the platina doubled and put through[c]

34   Three pronged round clamp

35   Three pronged flat clamp

36   Flat clamp[c]

37   3 inches long card 1 mm wide fastened with small clamps so:[13]

38   same[14]

39   same[15]

40      Made for test but too great resistance
                                40,000 ohms   3 in card ¹⁄₁₆ wide[c]

4̶1̶ 41   3 in cardboard ¹⁄₁₆ wide connec-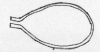
tions made so:— made for show[16]

X, NjWOE, Lab., N-79-07-31:119 (*TAEM* 33:600; *TAED* No52:58).
Written by Charles Batchelor. [a]Date from document, form altered. [b]Ob-
scured overwritten text. [c]Followed by dividing mark. [d]Followed by
crossed out, incomplete drawing. [e]Multiply underlined.

1. Charles Batchelor dated the first page of this entry on 28 October
and the last page on 10 November; he did not date any intervening pages.
The editors have taken the unusual step of presenting Batchelor's entries from this extended period as a single document because of the evident coherence and sequential character of his notes.

2. This list of numbered lamp filaments was begun in Doc. 1831 and
is continued in Doc. 1850. The preceding page, from 27 October, contains a list of 20 substances, such as vulcanized fibre, various types of paper and wood, fishing line, and a cotton lampwick, that were apparently
readily at hand (N-79-07-31:117, Lab. [*TAEM* 33:599; *TAED* No52:
57]). These were carbonized and put on small cards but there are no
records of any lamps made with them. The list and a few of the cards are
reproduced in Friedel and Israel 1986, 102–3.

3. Nothing is known of the design of this carbonizing chamber.

4. Results of tests with a lamp made with either this filament or No.
17 were recorded on 3 November by both Francis Upton and Francis
Jehl. The resistance of the lamp cold was 85.9 ohms and at a brilliant
white it was 47.6 ohms. N-79-08-22:182, N-79-02-15.1:180–81, Lab.
(*TAEM* 35:874, 31:747; *TAED* No85:91, No28:90).

5. Francis Upton recorded tests of a lamp with this filament on 5 November. The resistance of this lamp cold was 75 ohms and after being
evacuated on pump No. 4 it increased slightly to 77 ohms. After being
brought to white heat it reached 79.7 ohms but increased to 87 ohms
when it became "very hot" for a "few minutes." Upton noted that the
"support where carbon thread is attached to the Pt. wires too large. It
contained about .0061 cu. inches of material estimated Mr E. If this
could contain and hold 800 times its bulk of air it would contain 1.22 cu
inches." This appears to have led Upton "To test whether the carbon re-
absorbs" by taking the time for the pump alone to work with the lamp
cut off. Following this are some vacuum experiments by Francis Jehl
which seem to show that "the joint by which the spiral was attached to
pump leaked." N-79-08-22:183, 185–86, Lab. (*TAEM* 35:874–76;
TAED No85:91–3).

6. On 5 November, Francis Upton recorded results of tests of a lamp
with this filament: "197 [ohms] cold   after bringing up the resistance
cold became 265 the[n] 275–285   Giving about three candles resistance 187 ohms requiring . . . 82.5 volts 1610 ft. lbs. 20.4 per H.P. The
vacuum poor air perhaps from MacLeod guage   R=300 lamp turned

off   Guage lost when lamp put on." A few days later, probably 9 November, he recorded further tests indicating that it tested 1,600 ohms
when cold and "came up [that is, heated to incandescence] very irregularly." After placing the lamp and pump in a shunt circuit and again trying to bring it up the lamp melted when the shunt circuit was opened.
Upton thought this might have just been a coincidence or possibly
caused by discharge from the magnets. N-79-08-22:181, 191, Lab.
(*TAEM* 35:873, 878; *TAED* No85:90, 95).

It was probably about this time that Samuel Mott drew what he described as the "Lamp as it appears this date (Nov 1879)   A loop of
Clarks 3–300 thread rolled in Tar and lamp black then carbonized at a
white heat—limit 3 gas jets—adopted for 7 gas [jets?]   Trying vulcanized paper with very good result   also fine paper (visiting cards)."
N-79-09-20:16–17, Lab. (*TAEM* 36:35; *TAED* No96:8).

7. Francis Upton's notes indicate that this measured cold 260 ohms.
Two lamps with this filament were "placed in multiple arc the connec-
tion to one of them broken and the other lamp burnt with a bright flash."
N-79-08-22:201, Lab. (*TAEM* 35:883; *TAED* No85:100).

8. For tests of lamp filaments Nos. 24 and 25 see N-79-08-22:187–89,
Lab. (*TAEM* 35:876–77; *TAED* No85187).

9. Francis Jehl tested a lamp with this filament sometime in mid-
November. At that time it was 307 ohms cold. He had trouble getting a
good vacuum when it was heated to a white heat. It had a resistance of
120 ohms after about ten minutes but then "gave an arc at the connec-

tions and busted." Upton measured the lamp's resistance at white heat as 66.9 ohms, estimated that it gave the light of one gas jet, and calculated that they could run 9.4 lamps per horsepower. N–79-08-22:193, 196–97, Lab. (*TAEM* 35:879, 881; *TAED* No85:96, 98).

10. Francis Jehl tested a lamp with this filament sometime in mid-November. At that time it was 235 ohms cold. After getting a good vacuum he tested it on a dynamo. Francis Upton calculated that it measured 55 ohms, gave about 3 gas jets, and that they could run 6¾ lamps per horsepower. N–79-08-22:193, 197, Lab. (*TAEM* 35:879, 881; *TAED* No85:96, 98).

11. The shape of this and all the lamp filaments made from cardboard was that of a horseshoe. According to Francis Upton's later article for *Scribner's Monthly* the filament "was made in the form of a horse-shoe, in order to approximate to the shape of a gas-jet." Upton 1880a, 539.

12. The entry by Francis Upton regarding No. 31 identifies it as a piece of paper that showed 2,000 ohms resistance when cold but which "did not become even red." However, on the same page, identified as "No 328," he describes a thread lamp that was 3,000 ohms cold and that also failed to become red hot. He goes on to say that "Both these lamps at almost the same moment burnt up forming a bad arc, melting the conducting wire. Perhaps they suddenly carbonized and reduced their resistance freeing an amount of air which allowed an arc to spring." N–79-08-22:199, Lab. (*TAEM* 35:882; *TAED* No85:99).

13. Francis Jehl noted that a lamp with this filament measured 470 ohms resistance cold and that after heating to a white heat it measured 300 ohms. Francis Upton then noted that after being incandescent for an hour it was 265 ohms and after another half an hour 310 ohms. It then "'Busted' on pump . . . Burnt on the pump from an arc." N–79-08-22:198, Lab. (*TAEM* 35:882; *TAED* No85:99).

14. Francis Jehl measured a lamp with this filament at 9,000 ohms resistance cold. N–79-08-22:198, Lab. (*TAEM* 35:882; *TAED* No85:99).

15. Francis Upton measured a lamp with this filament cold at 18,000 ohms and 155 ohms after bringing it up on the pump. "Good lamp showed two or three gas jets without change   Messrs [Grosvenor] Lowrey & [James Hood] Wright saw it sealed off and shown in middle of room   small arc at point of contact." In a note dated 14 November, which may have been written separately from the rest of the entry, Upton provided the following table of results after the lamp had been on for two hours:

3 P.M	140 ohms	Stock	$3,000 a share
4 P.M	140 ohms	Stock	$6,000 " "
5 P.M	140 ohms	Stock	$12,000 " "

Two measurements of the lamp's candle power showed 10.5 to 17 gas jets. By calculating the voltage required for the lamp Upton determined that they could get 11.8 lamps per horse power. On 15 November, he wrote that "Francis [Jehl] says that the resistance of the lamp yesterday was 142 ohms and constant. This morning on three hours after considerable moving and jarring 134+ ohms! 11–12-30   134+ ohms!!!!!!!! 1 to 4 P.M." On 17 November they put the lamp on at 8:30 A.M. Its resistance measured "141.7 ohms momentary contact   137.7 ohms current 8 cells on bridge." They took the lamp off at midnight. In subsequent experiments that probably took place on 18 November the lamp measured "146 ohms tapping with key   141.5 ohms on all the time   this last was made 138 ohms by tapping showing that the arc which could be seen was a break. Broke off at the clamp." N-79-08-22:200–201, 203–4, Lab. (*TAEM* 35:883–5; *TAED* No85:100–2).

16. Francis Jehl indicated that this filament was about 3 inches long and ¹⁄₁₆ inch wide. He measured it in a lamp on the vacuum pump at 100,000 ohms cold and noted that an "Induction spark passed through it and air driven out. At first no spark   afterwards a blue light. Vacuum spoiled." On 13 November, however, he obtained a vacuum of 1.6 millimeters of mercury. N-79-08-22:202, Lab. (*TAEM* 35:884; *TAED* No85:101).

–1839–

Charles Batchelor to
Edward Johnson

[Menlo Park,] Nov. 2d 79

My dear Johnson,

Your letters of 19th and 20th received last night and all contents noted.[1] I believe now that we have got the instrument <u>perfect</u>— It requires no extra call battery—[2] You can interrupt any time— Chalks are exceedingly loud— Man leaves Wednesday bringing two— He also brings lathe and a number of chalks—[3] He has been pretty well drilled and understands thoroughly how to <u>turn, wet, and adjust them.</u>— I should give you a long list of our experiments, but Edison told me not to, as he would dictate a long letter to ~~you~~ Griff for you, giving you theories and practice—[4]

No such compromise as you speak of at Saratoga ever took place it is all bosh;[5] Edison is not the man that ever seeks to compromise always believing he has the best; if he ever finds he has not, he generally calls on his genius for more invention instead of talking compromise—

Edison appreciates the situation in re to the 'irrevocable power of attorney' and broadly and flatly refused it to Gouraud. I believe he recalled the fact that he had done so with Harrington, and had had it thrown up in his face frequently afterwards— He seemed pleased with the plan you suggested and I think he will write you about it himself—[6] He told me that he wanted you to keep him posted on these matters and that he should sign nothing of importance without consulting you. Yours

Batchelor

Man will explain all about chalks even if Edison dont write If he dont I will   (B)[7a]

ALS (letterpress copy), NjWOE, Batchelor, Cat. 1330:51 (*TAEM* 93:265; *TAED* MBLB2051). ªSentence started above signature line and separated from it by dividing mark.

1. No letters with these dates have been found but one may be an undated letter that Johnson wrote to Batchelor around 19 October (DF [*TAEM* 52:837; *TAED* D7941ZGK1]). Batchelor may have been mistaken in these dates, since he refers in the last paragraph below to a proposal in Johnson's 21 October letter (Doc. 1829).

2. The call bell battery had been a source of difficulty for some time (see Doc. 1780). On the evening of 1 November Edison cabled that he had "Discovered method work call with transmitter battery   trivial change" (DF *TAEM* 52:883; *TAED* D7941ZHI]). This change involved working the bell through a relay in a local circuit with the transmitter battery (Brit. Pat. 5,335 [1879], Batchelor [*TAEM* 92:141; *TAED* MBP021]).

Johnson apparently had not received Edison's cable when he wrote to Batchelor the next day that because a reliably strong current was needed to compensate for ineffectual chalks he had decided to use only Callaud cells on the transmitters and "small Leclanche on call—though as to the latter I am seriously inclined to put on a switch to cut out the Induction Coil— We have a good deal of trouble to keep Call Batteries in Order— Even when using 8 small Leclanches— If you folks have any suggestions to make on this head I wish you would—Ere I invest largely in Batteries" (Johnson to Batchelor, 2 Nov. 1879, DF [*TAEM* 52:885; *TAED* D7941ZHK]). Because these batteries were not available in London he wired Sigmund Bergmann for 500 Watson cells, a copper sulphate battery advertised for its durability and strength (see *TAEB* 4:40 n. 3). On 3 November Johnson cabled an order to Edison for Watson batteries which was promptly answered, "new Call requires two fuller [cells] on transmitter." Johnson responded that "Fuller polarizes," but after receiving Edison's reply that "Four Watsons works satisfactory" he reinstated his order for 500 cells, which Edison submitted to Bergmann. Batchelor later wrote that he could not understand the problem with the Fuller batteries since a thousand of them were now used in Chicago without difficulty (Johnson to TAE, 3, 4, and 5 Nov. 1879; TAE to Johnson, 3 and 4 Nov. 1879; all DF [*TAEM* 52:893–94, 899; *TAED* D7941ZHM, D7941ZHO, D7941ZHS, D7941ZHN, D7941ZHR]; Edison to Bergmann & Co., 17 Nov. 1879; Batchelor to Johnson, 3 Dec. 1879; Lbk. 5:346, 393 [*TAEM* 80:181, 198; *TAEB* LB005346, LB005393]).

3. On Tuesday, 4 November, John Kruesi wrote a shop order to "Send 2 complete Telephones with relays to E. H. Johnson." These were to accompany telephone inspector James Lighthipe. In accordance with Kruesi's instructions of the same day, a "Complete Stewart foot lathe with sliderest" and ancillary equipment, along with 100 chalks, were shipped to London on 6 November (Lighthipe sailed with the telephones two days later). The Stewart machine presumably was intended to replace the "cheap little lathe for Johnson to turn the chalks of[f] the Top by hand" which Kruesi had made at Batchelor's request in mid-September. Cat. 1308:183 (Order No. 271), 185 (Order No. 277), 237, 175 (Order No. 241), Batchelor (*TAEM* 90:758–59, 777, 754; *TAED* MBN003:65, 66, 83, 61); Vol. 18:120, Lab. (*TAEM* 4:1151; *TAED* NV18:106).

Johnson had complained in a 23 October letter to Edison that he had "been cloaking all the bad ones [chalks] & exposing to full view all the good ones—in the honest expectation that you would long ere this have sent me at least one consignment of which Each & Every one would have been equally good= This you have done only in this that each & Every one are Equally bad." Consequently he asked Edison to send "a mould & Lathe & the necessary instructions how to mix chalks at once— I will then put the 30 or 40 Instruments now, useless on my hands, in position to Earn us some money— That is now the thing to be done ere you can hope to get any more Ducats from this side." DF (*TAEM* 52:845; *TAED* D7941ZGU).

4. Edison appears not to have sent such a letter but see Doc. 1845.

5. No other reference to this has been found.

6. The power of attorney is discussed in Doc. 1829 n. 8.

7. See Doc. 1845.

–1840–

Francis Upton to Elijah Upton

Menlo Park Nov. 2, 1879.

Dear Father:

The electric light is coming up. We have had a fine burner made of a piece of carbonized thread which gave a light of two or three gas jets. Mr Edison now proposes to give an exhibition of some lamps in actual operation.[1] There is talk if he can show a number of lamps of organizing a large company with three[a] or five millions capital to push the matter through.[2] I have been offered $1,000 for five shares of my stock making at that rate what Edison gave me worth about $7,400, a good years pay I think. Edison says the stock is worth a thousand dollars a share or more, yet he is always sanguine and his valuations are on his hopes more than on his realities.[3] I am going to New York to-morrow early, to [see?][b] buy a few book and to go with Lizzie to see if we can get a girl.[4] Your aff &c Son

Francis R. Upton.

ALS, NjWOE, Upton (*TAEM* 95:568; *TAED* MU033). [a]Obscured overwritten text. [b]Canceled.

1. Two days later Edison wired Norvin Green asking for "two line men to errect my wires here for light Exhibition sometime this week my expense." The men were to report to Menlo Park the following Monday, 10 November. Edison also requested "6 old poles" from Western Union but none were available. Poles were eventually obtained to carry electric lines to Upton's house; wires to Edison's home were strung on existing Western Union poles. TAE to Green, 4 Nov. 1879; John Van Horne to TAE, 5 Nov. 1879; James Merrihew to TAE, 7 and 8 Nov. 1879; TAE to Merrihew, 7 Nov. 1879; all DF (*TAEM* 50:158–59; *TAED* D7919ZDG, D7919ZDH, D7919ZDI, D7919ZDJ, D7919ZDI1); Upton's testimony, 5:3242, *Edison Electric Light Co. v. U.S. Electric Lighting Co.*, Lit. (*TAEM* 48:128; *TAED* QD012F:126).

2. Upton wrote his father on 26 October that "The latest gossip that Mr. Edison has told me, is that there is talk of swallowing the old Electric Light Co. in a new one and making the capital three millions of dollars. This new company will be floated by Drexal Morgan & Co. of New York. There is no hope of such a thing until Mr. Edison has given an exhibition. So he is going to try and show some lamps here in the course of a few weeks. I think I shall have some of them in my new house." Evidently anticipating such an exhibition, on or about 22 October Upton started to make some unspecified estimates based on a line running to Sarah Jordan's boarding house; however, he stopped and instead estimated various costs of a small generating plant. Upton to Elijah Upton, 26 Oct. 1879, Upton (*TAEM* 95:566; *TAED* MU032); N-79-02-24.1: 262, Lab. (*TAEM* 30:1097; *TAED* N020:132).

3. Upton reported to his father the next week that "the Electric Light seems to be a continual trouble for as yet we cannot make what we want and see the untold millions roll in upon Menlo Park that my hopes want to see." Upton to Elijah Upton, 9 Nov. 1879, Upton (*TAEM* 95:570; *TAED* MU034).

4. In his 9 November letter to his father (see note 3), Upton wrote that "We found some Swedish emigrants and from among them we picked out two, a middle aged woman and a young girl. Lizzie is finding a good deal of company in teaching them the names of the various articles that are used about the house. We hope that they will stay for it is so hard to get any one to stay here, and Lizzie wants to be sure of having someone that will stay, for a pair of helping hands in a house the size of ours is very needful."

-1841-

From Hilborne Roosevelt

New York Nov 3 1879[a]

Respected Friend & Dear Sir—

I enclose copy of a letter I mail my brother today which I think embodies your views and mine— I know Cornelius will try and act fairly in this matter— With kind regards to Mrs Edison and Dotty   I remain Yours truly

Hilborne L. Roosevelt.

To Prof. Dr. Thomas ẸA. Edison F.R.S QZW. &c.   I forgot to thank you today for enclosing the photo for that charming young lady last summer she was <u>delighted</u>—[1]

ENCLOSURE[b]

[New York,] Nov 3 79

Dear Corniel,

Yours of 19th Oct duly rec'd—[2] I went to see Edison and he will stand by Puskas unless it is positively and clearly proven that he has acted dishonestly— He thinks Bailey is clever but will not give him a power of att[y?][c] instead of P. Edison says that P has acted perfetcly squarely with him so far and he will only belive to the contrary when genuine proofs are produced Edison is one of the most honorable men I ever met and when he is a friend to anyone ~~he~~ will never go back on him. In this I know you admire [him?]^c as ~~I~~ much as I do. Would more had the same courage he has   He does not think that Puskas being one against 12 can do any harm and Edison fully appreciates Mr P's want of business knowlege &c also what a nuisance he is, but his action will be as I have stated— He dont care if he makes money out of it or not, but he dont want any more annoyance than he can possible help— I would certainly join him in advising you if Ps is not absolutely dishonest to try and get along with him in some way (of course not lending him money)   Edison tells me Bailey is on his way here—he may be able to throw some light on matters. The Bell Telephone stock sold at $800.00 lately. Please remember me to Mr Gower and believe me   Your affct bro

Hilborne L Roosevelt

ALS, NjWOE, DF (*TAEM* 52:385; *TAED* D7940ZCC). Letterhead of Hilborne L. Roosevelt. ^a"New York" and "18" preprinted. ^bEnclosure is ALS (copy). ^cFaint copy.

1. Not identified.
2. Not found. See Doc. 1837 n. 2.

–1842–

From Theodore Puskas

Paris, le 5th Nov. 1879.^a

My dear Edison,

I wrote you a few lines on the 3rd inst. to acknowledge receipt of your cable advising me of Bailey's departure,[1] (which was the first intimation I had of his going over to the States,) and I now proceed to furnish you with some particulars concerning our business here.

First,—with respect to whom you will nominate to represent your interests on the Board of the new Company:

As you will know by the time this letter reaches you, it was Bailey's chief aim to exclude me. He cannot say that it was a question whether he or I should be on the Board, for there was a time at which the other parties would have accepted us both. But this was not to Bailey's taste, and he managed matters so that now only one of us can enter the Board.

This compromise was accepted by the other members in the expectation that Bailey will succeed in obtaining your Support.[2]

The reason why the other members are so much in favor of Bailey's being on the Board is that during the negotiations for the fusion Bailey made to[b] the other two Companies greater concessions than Berthon and I were willing to make.

It is true that at one stage of the negotiations Berthon and I agreed to the percentage which has been finally accepted, say 44¼% for our Co., but considering the secret way in which Bailey carried on the negotiations for the fusion, and considering that Rooseveld, who up to the time Bailey started negotiations with him never spoken of Bailey except as a scamp, a scoundrel and a disgrace to our Co., got at once on the best terms with him, my suspicions, and especially those of Berthon, were aroused. Berthon felt[c] sure that there was something wrong in the matter and thought that by going in for a higher percentage we would make Bailey give up the advantages which he might have secured for himself personally, from the other side.

A fact which confirmed us in our suspicions is that Bailey fought desperately against changing the percentage allotted to each Co.

He may tell you that he was not alone to fight us and that two of our partners, Evans & Chatard sided with him; but you will easily understand why they did so.

For this you must know that Godillot[b] and Brancy, two of our former partners, sold their interest of 135,000 francs to a certain Pellorce, who bought it for account of Rooseveld. Now in view of the fusion, through the instrumentality of Bailey, Evans, Chatard and others bought back from Rooseveld 67,500 francs worth of that interest for 112,500 francs. There was an amount of 12,000 francs kept open for Berthon, but as he was told by Mr Marrand,[3] (the highest legal authority here on these matters) that we could annull the transaction between Godillot & Pellorce by simply re-imbursing the 135,000 francs, he held back till the last moment and only bought that interest after the fusion was signed by all. But Chatard & Evans were not in the same position because they effected their purchase before the fusion was agreed to by Berthon & me.

Therefore had the fusion dropped they would have lost money. They had consequently special reasons to bring pressure upon Berthon and me to force us to accept the fusion.

Bailey may also tell you that in his negotiations he was also backed by Mr Rousseau, the Counsel of the Co. but I must tell you that Rousseau is[b] not and has never been the counsel of the Edison Telephone Co. He is only the legal adviser of Chatard

& Bailey who have now proposed him for a membership of the Board.

Now it will be of interest for you to know that when Pellorce approached Godillot with a view to purchase for account of the Gower Telephone crowd his interest of 135,000 francs, Godillot being doubtful as to whether he had a right to sell it asked Rousseau's opinion; and Rousseau, instead of telling him to offer it first to the other partners, advised him to accept Pellorce's proposition. This [move?]ᵈ adviceᵉ of Rousseau was most detrimental to the interests of the Edison Tel. Co., as it altogether paralysed our movements for the transformation.

If Bailey then puts him forward as the Counsell of our Co. the above will show you how much he had the interest of the Co. at heart.

So much for this question of your nominating a member of the Board.

Second—with reference to negotiations about the other European countries:

As soon as I heard from you that Bailey had left for the States I made up my mind that the question ofᶠ the membership of the Board was not the only motive for his journey, and that he also goes over to submit to you and to the other shareholders of the Edison Tel. Co. for Europe propositions concerningᵇ the other European countries.

If this be the case, considering that I am one of the interested parties and that Bailey had to look upon me as your representative, why did he not acquaint me with the steps he has taken in the matter, if everything is straightforward?

I would therefore request you in our mutual interest to have any proposition which Bailey may be the bearer of submitted to me. Being on the spot I am in a better position to judge of the relative value of such propositions.

Meanwhile, if Bailey brings propositions from Erlanger (or from the Crédit Mobilier, of which he is the boss), or from the Anglo-Universal Bank, (of which Chatard is one of the Directors,) you will do well to bear in mind that the former (Erlanger)ᶜ is the man who has bought the Gower telephone and that the latter (Chatard) is one of those who brought pressure upon me to force me to accept the terms of the Gower people.

Erlanger enjoys about the same reputation here as Jay Gould does in the States, and we shall have plenty to do to keep things straight in Paris

To give you an idea of the name heᵇ has in France, I must tell you that one of our intended Directors, Mr Menier,⁴ having changed his mind a few days ago, and withdrawn from the Board, Mr Berthon made ouvertures to two men of high standing to enter the Board, and that they flatly refused on hearing that Erlanger and the Crédit Mobilier are interested in the concern.

As soon as the details of the fusion are completed I will go over to see you and refute verbally any mis-statement which Bailey may make to you. I am lead to believe that he will make some in consequence of a conversation I had with Mr Reiff the day before yesterday, in the course of which Mr Reiff reported to me some statements made to him in London by Bailey, which I had to brand as absolute falsehoods.

In closing this letter I beg to inform you that I have asked a good friend of mine, Mr A. Hegewisch,[5] whom you already know, to consult with you on these matters, and, if need be, to act on my behalf in perfect harmony with you. Yours very truly.

Theo. Puskas

LS, NjWOE, DF (*TAEM* 52:394; *TAED* D7940ZCI). Letterhead of Theo. Puskas; written in an unknown hand. [a]"Paris, le" and "18" preprinted. [b]Repeated as page turn. [c]Obscured overwritten text. [d]Canceled. [e]Interlined above. [f]"the question of" interlined above.

1. On 2 November Bailey had telegraphed from Queenstown, England, that he would arrive in the United States on the 11th. Edison notified Puskas of this the same day and asked him to "write details Paris manipulations." Puskas replied the following day that "by next mail I will send you full details; please don't take any decision till you receive them." DF (*TAEM* 52:383–84; *TAED* D7940ZCA, D7940ZCA1, D7940ZCB).

2. Alfred Chatard told Bailey in a letter of 3 November that "I have a great desire to see you selected by Mr. Edison to represent him in the Council of Administration of the Cie G[énéra]le des Téléphones and I very much regret my inability to support directly your candidature and thus expose the conduct which is not too honorable of Mr. Puskas and his entire inability, but you know the moral engagement I have made not to interfere with the representation of Mr. Edison!!!" Constant Rousseau wrote Bailey on 5 November on behalf of himself and Chatard to suggest that Bailey have Edison allow the other members of the Council select his representative by secret ballot, telling him that "you know as well as I do that the Council would nominate you because they ardently want you." On 13 November Edison sent a cable and a letter nominating both Puskas and Bailey and asking that the other members of the board decide between them. Chatard to Bailey, 3 Nov. 1879; Rousseau to Bailey, 5 Nov. 1879; TAE to Louis-Alexander Foucher de Careil, 13 Nov. 1879; Puskas to Josiah Reiff, 16 Nov. 1879; all DF (*TAEM* 52:388, 391, 408, 416; *TAED* D7940ZCE, D7940ZCH, D7940ZCR, D7940ZCW); TAE to Louis-Alexander Foucher de Careil, 13 Nov. 1879, Lbk. 5:356 (*TAEM* 80:187; *TAED* LB005356); see also Reiff to TAE, 20 Nov. 1879, DF (*TAEM* 52:425; *TAED* D7940ZDE).

3. Unidentified.

4. Unidentified.

5. Adolph Hegewisch was secretary of the U.S. Rolling Stock Co. headquartered in New York City. On 11 November he telegraphed Edison that he had an "important communication to make to you from Puskas in regard to appointment of Director in Consolidated Telephone Co in Paris." He arranged to go out to Menlo Park to see Edison regarding this. On 15 November, after receiving some letters from Puskas he again wired Edison to arrange a meeting. Nothing is known about the specifics of their discussions. Hegewisch to TAE, 11, 12, and 15 Nov. 1879, DF (*TAEM* 52:407, 415; *TAED* D7940ZCO, D7940ZCP, D7940ZCU).

-1843-

Equipment Specification: Electric Lighting

[Menlo Park, c. November 5, 1879[1]]

No. 1 remodeled.[2]

280[3]   Take the armature from No. 1 machine and change as follows

Make Commutator with 75 divisions

Make new vulcanized fibre ends 9.218 inches in diameter 9.062 inches in diameter.

Wind the iron core with fine iron wire, well oxidised if possible, this must be done so that no part has[a] a larger diameter than 9.062 inches and all as near that as possible.[4]

Wind the com[b] armature with one layer of No. 20 wire .035 inches in diameter. Each coil will consist of two wires wound twice round and will make the machine have about ½ Ohm resistance   The wire will be held in place by German silver or s

Wind with one layer of No. 20 wire .035 inches diameter double wound .013 in. covering. Four wires side by side one turn in each coil. Groove the outside edge of vulcanite fibre[a] so as to hold the wires in place that as they are carried to the commutator blocks. Hold the wire in place with .010 iron or steel[c] wire in the same manner as in No. 2 Machine.

Diameter bore in face   9.21875
                        .156
Diameter core   9.062

                         3.14
                        36 248
                         9 062
                        27 186
Circumference   150 | 28.45468   .189[a]
                        150
                       1345
                       1200[a]
                       1454

4 | .189 inch for each coil   .04725 for each wire

Wire   .035   .035
Covering   .012   .013
             ‾‾‾‾   ‾‾‾‾
             .047   .048   The wire with .013 covering[d] can
                             probably be fitted in.

Thickness of wire   .048
Iron wire to hold   .010
Play   .020
Total   .078[a]
                         ‾‾‾‾
                          2
Total on both sides   .156

Resistance of one strand No. 20 wire once round the machine (Total 2 Ohms) Machine ½ Ohm   Four strands ⅛ ohm[d] machine = .125 Ohms

No. 2 machine[5]   468 turns
No. 1 remodeled   75 or ¹⁄₁₆ the no.
No. 2 machine $^{.625}/_2$ in[ch] between coil of iron and face of armature
No. 1 remodeled $^{.156}/_2$ in[ch]   ¼ the distance

No. 2   500 revo. 110 volts
No. 1 Remodeled   750 revolutions[a] should give 110 volts[6]

Nothing is allowed here for heating or for irregularity. in the iron core

ADDENDUM[e]

                       [Menlo Park,] Nov. 13 [1879]

About 4 lbs of wire needed

ADDENDUM[e]

                       [Menlo Park,] Dec 8 [1879][7]

This machine was well made and gave 45[a] Volts!!!

AD, NjWOE, Batchelor, Cat. 1308:187, 186 (*TAEM* 90:760; *TAED* MBN003:67). Written by Francis Upton. [a]Obscured overwritten text. [b]Canceled before rest of paragraph. [c]"or steel" interlined above. [d]Interlined above. [e]Addendum is an X, written by Upton.

1. This entry follows dated entries of 5 November and precedes entries of 9 November. In a notebook entry dated 5 November, following ten days of tests on Faradic machines No. 1 and 2, Upton wrote: "It is proposed to wind the No. 1 Machine with one layer of No. 20 wire .035 in diameter   75 commutators." He then noted that the "full particulars" could be found in the shop order book on page 186 but mistakenly referred to this as "Order 180." Upton's order is an elaboration of one

by John Kruesi to have "No 1 remodelt." N–79-10-18:31–107, Lab. (*TAEM* 32:799–837; *TAED* N301:17–55); Cat. 1308:185 (Order No. 278), Batchelor (*TAEM* 90:759; *TAED* MBN003:66).

2. This was probably the first large "Standard Faradic Machine," made in July and altered in September. See Doc. 1774; N–79-10-18:31–32, Lab. (*TAEM* 32:799–800; *TAED* N301:17–8).

3. This is the order number; see Doc. 1687 n. 1.

4. The original armature (and that of Faradic No. 2) had a wooden core. An "Iron ring" 1¾ thick was also placed on the armature for an un-specified purpose and may have formed part of the iron mass. Its origi-nal diameter, excluding the induction wires, was only 8¾ inches. N–79-10-18:32, Lab. (*TAEM* 32:800; *TAED* N301:18).

5. This machine was built in August and evidently rewound in Sep-tember to its present dimensions of six layers of two pairs of three wires, in 39 coils. See Doc. 1774 n. 2; N–79-10-18:33, Lab. (*TAEM* 32:800; *TAED* N301:18); Cat. 1308:171, 173 (Order Nos. 225, 232), Batchelor (*TAEM* 90:752, 753; *TAED* MBN003:59–60).

6. This expectation was based on Upton's tests of these machines in late October and early November. He "found that when eight Webers were on the magnet of No. 2 the E.M.F. of armature was 104 Volts while in No. 1 it was 140 Volts." He noted that both machines were "wound with the same number of turns of wire 468 in all" but due to the differ-ence in thickness of the wire "there was a distance from the core of No. 1 to the face of the magnet of .⁴⁶⁸⁄₂, the same in No. 2 was .⁶²⁵⁄₂." This led him to conclude that "if the proportion be made

Distance Coil in No. 1 from armature
    "   "   "   No 2   "   "
:: E.M.F. No. 1 : EM.F. No. 2

it will be found to be nearly exact for the strengths of magnets nearing saturation. . . . there is general ratio determined by this ratio of the dis-tance of the iron from face of armature." However, on 11 December he noted on one of these pages that "All this is wrong" because Kruesi had given him incorrect measurements. N–79-10-18:31–57, 73–105, Lab. (*TAEM* 32:799–812, 820–36; *TAED* N301:17–30, 38–54).

7. See Doc. 1862.

–1844–

To Frank McLaughlin

MENLO PARK, N. J. Nov 11, 1879[a]

Dear Sir

Your favor of Nov 3d was rec'd this morning— In regard to contracts I leave it to your judgment about the amount of gold to be returned to mine owners— If you can do better than ½why go ahead and do the best you can—[1] I have Col Lewis[2] in with me as well as Banker and Cutting and can get all the money required to work.[3] Bear in mind we can erect a mill in vicinity of Oroville or at Oroville [and?][b] work the undercur-rent[4] or black sand [stamped?][b] by a process which Dr Haid and I have perfected and which is absolutely perfect and can [be?]ᵇ kept a secret. I want to go into this mill [biz?]ᵇ and I will see you get your share and have charge of it along with Dr H & one or 2 more of the boys. What we want is contracts with [a?]ᵇ corporation or absolute owners of the mine giving us the undercurrent dirt for a per[iod]ᵇ of years for ½ (or less) of the gold. This is a [----]ᵇ consideration to them as it affords them facilities and increased profits and for this consideration we desire the exclusiveᶜ contract. Please give all the information about wood around Cherokee [Flat?]ᵇ price kind and character or if no wood, is it practical to put in water power in the flume of the S.[pring]V.[alley] Co? I want about 30 H.P. for the mill

I think you had better stay around Cherokee with the idea of getting things ready for the mill. There is plenty of platinum for us at Cherokee and no one is going to steal theᵈ platinum in Trinity Co or in Oregon. It will keep. As I can raise money now I would like to get the scheme started⁵

The process is entirely electrical. We shove in the flume dirt you last sent us, turn a switch, co[unt]ᵇ 450 and the thing is done.⁶

Ascertain without conspicuousness the price of salt there— or if any salt mines near or salt water or salt Lakes— The price which that Nevada salt can be got there and delivered in Oroville.

You know that sample you sent from Powers claim that he was going to work in a mill⁷—a big dynamo magnet sucks up 99% of whole leaves nothing but amalgam gold and platinum and a few grains of quartz—and can [save?]ᵇ it for 25¢ per ton— We are the boys to invent processes for any kind of ores they bring to the mill! Th[at's]ᵇ where we get ahead of other boys. Very hastily & truly

T. A. Edison G[riffin]

L (letterpress copy), NjWOE, Lbk. 5:329 (*TAEM* 80:178; *TAED* LB005329). Written by Stockton Griffin; circled "C" written above. ᵃPlace from Edison's laboratory handstamp. ᵇFaint letterpress copy. ᶜObscured overwritten text. ᵈInterlined above.

1. Edison had telegraphed McLaughlin on 3 November (misdated 30 November in *TAEMG2*) in Oroville, Calif. to secure a contract with the Spring Valley Mining Co. under which "we errect undercurrents at end their flume system, work product in a mill errected by us give them half gold product. twill increase their product greatly and gold will pay for platina & interest investment. Can raise twenty five thousand here, if get good contract." McLaughlin wired back that this arrangement could be made but "suggest nothing be done until you receive letter" (TAE to McLaughlin, and McLaughlin to TAE, both 3 Nov. 1879, DF [*TAEM*

51:363, 331; *TAED* D7928ZSN1, D7928ZSL]). He suggested in that letter, written on 3 November, that the

> best plan would be not to enter into any agreements about putting up under currents &c. All the miners in this and the surrounding Counties know that great amts. of gold are lost in the black sands but having no plan for the working of them look upon the loss as a necessary evil. Now if a mill were in operation in Oroville where these sands could be worked all the mines would start their own plans for saving the black sands etc. which being a dead loss to them now, they would be glad to sell upon such terms as you might find it profitable to offer, which certainly need not be one half of the gold recovered. All miners . . . would doubtless enter into contracts with the first mill in the field which field you could only hold to yourself by secrecy regarding the method of working the sands to save the gold. [McLaughlin to TAE, 3 Nov. 1879, DF (*TAEM* 51:332; *TAED* D7928ZSM)]

The next day McLaughlin advised that an agreement covering the Spring Valley company's Cherokee claims was "necessary to success of mill   Can make arrangements with lesser companies at anytime." On 7 November Edison instructed him to "Contract with others on basis already telegraphed" (McLaughlin to TAE, 4 Nov. 1879; TAE to McLaughlin, 7 Nov. 1879; both DF [*TAEM* 51:338, 343; *TAED* D7928ZSQ, D7928ZSV]). Even as ownership of the Cherokee claims was being transferred (eventually to the United States Mining Investment Co.) Edison executed agreements to acquire and process sand from the claims on 17 and 23 December. A similar agreement between the claim holders and the Edison Ore Milling Co. (see note 5) was also executed on 17 December (TAE agreement with United States Mining Investment Co., 17 Dec. 1879, Miller [*TAEM* 86:62; *TAED* HM790086]; TAE agreement with Edison Ore Milling Co., 12 Jan. 1880, DF [*TAEM* 54:439; *TAED* D8033A1]; Edison Ore Milling Co. agreement with United States Mining Investment Co., 17 Dec. 1879, Miller [*TAEM* 86: 66; *TAED* HM790087]; a draft of this latter contract is in DF [*TAEM* 50:567; *TAED* D7927C]).

2. A surveyor and civil engineer, Charles Lewis became a Union infantry officer in the Civil War. Afterwards he was briefly engaged in mining in Colorado and then came to New York, where he entered the brokerage business and acquired a seat on the New York Stock Exchange (*NCAB* 18:326–27). Edison had recently arranged to have a pending U.S. patent for polyform, his analgesic preparation, assigned to Lewis and two partners (see Docs. 1287 and 1960 n. 3).

3. Sometime during the summer or fall Edison drafted a six-page memorandum outlining the capitalization and rationale for a new company to recover and process platinum from the Cherokee area. Noting that "the refining of platinum is now entirely in the hands of the great firm of Johnson Matthey & Co of London and one french firm in Paris," a position which had made each "immensely wealthy," he predicted that he could produce wire for a third of the price charged by Johnson Matthey. His research had uncovered, "contrary to the statements and sneers of mining experts and mineralogists and scientific frauds . . . vast

quantities of platinum in our own country obtainable if capital was at hand at a price not exceeding one dollar per oz troy. I find that in many of the great hydraulic placer mines of California, mines that cannot be exhausted for centuries, that there are large quantities of platina thrown away that could easily be saved." He expected Spring Valley to yield 18,000 ounces annually; four other mines "in the immediate vicinity" another 25,000 ounces, and untold amounts from "over 75 mines of this character on the great dead rivers of Cala" and "hundreds of mines now being worked" along the Oregon coast. The proposed company, to be capitalized at $10,000,000, would erect a mill and refining works at Menlo Park (estimated to cost $18,000) and acquire the ore. Anticipating that "the profits can never be less than 20% on the money invested," he proposed to subscribe for $5,000 of the stock himself. DF (*TAEM* 50: 574; *TAED* D7927F).

4. An undercurrent consisted of a flat box several hundred square feet in area placed below fine gratings in the bottom of the main sluice. Its large area greatly reduced the velocity of water, allowing small particles which dropped from the main sluice to be collected. *Ency. Brit.*, s.v. "Gold."

5. The Edison Ore Milling Co. was incorporated in December 1879 with a planned capitalization of 3,500 shares at one hundred dollars each. Edison agreed to assign to the company his contracts for Cherokee Flats (see note 1) and two other California claims. In return he was to receive an annual salary and 2,850 shares, the majority of which he transferred to others in January 1880. Application for Incorporation, 11 Dec. 1879; list of stock subscribers, n. d.; both CR (*TAEM* 97:386, 389; *TAED* CG001AAB, CG001AAC); TAE agreement with Edison Ore Milling Co., 12 Jan. 1880; TAE stock transfer, 14 Jan. 1880; both DF (*TAEM* 54:439, 448; *TAED* D8033A1, D8033D).

6. The process was probably the electrolytic production of chlorine, which was used in gaseous or liquid form as a gold solvent (Lock 1882, 795, 1130–36). John Kruesi sketched a Patent Office model for this process on 9 December, though Edison did not file the application until 20 March 1880. The application (Case 205) covered the production of high concentration chlorine water by the electrolytic decomposition of chlorides such as sodium chloride, which Edison stated allowed him to "extract the gold from auriferous sulphurets profitably when the ordinary chlorination process is not admissable, on account of greater cost." Edison made further experiments with chlorine production (see Doc. 1931) but eventually abandoned this application. Machine Shop Drawings (1879–1880), Lab. (*TAEM* 45:76; *TAED* NS7986CBB); Casebook E-2536:58, PS (*TAEM* 45:704; *TAED* PT020058) Unbound Notes and Drawings (1880), Lab., Supp. III (*TAEM* 162:721; *TAED* NS80AAH1); see also Doc. 1931.

7. Oliver Powers co-owned a claim on the Feather River opposite Oroville that McLaughlin described as "the most extensive hydraulic mine in this vicinity." McLaughlin wrote that this site "would be a big thing and a cheap one for gold yet I consider it a great risk for Pt. that is, Pt. to any great amount." Edison subsequently reached an agreement with Powers and his partners to process their sand. McLaughlin to TAE, 19 Oct. 1879; TAE agreement with Edison Ore Milling Co., 12 Jan. 1880; both DF (*TAEM* 51:305; 54:439; *TAED* D7928ZSB, D8033A1).

[Menlo Park,] Nov 13th 79

My dear Johnson,

Water chalks were sent you with Shipment of thirty sent from here on Nov 10 and also on shipment of eighteen sent from here on 12th with the latter shipment were also sent one hundred extra water chalks—[1] Throw away all chalks sent with Lighthipe's apparatus—[2] After turning these water chalks wet with a brush full of pure distilled water which can be procured of any manufacturing chemist— After wetting, you will find the instrument lower than a Bell telephone; put it away for twenty four hours, then try it, polish it slightly and you will find it belch out; thence forward it will continue as we believe almost indefinitely— You will find them the loudest you have ever heard— These chalks should they ever play out say in 3, 6, or 12 months will I am sure be resuscitated by another brushful of the water— We shall ship you a lot of water chalks with every shipment of telephones so you may have plenty to turn— Bergman's first hundred and our last shipment were old style bells our next lot and his second hundred will be relays and bells—[3]

We have had an exchange up here with all your wires going for the last week with relays and bells and water chalks and with this exchange I learn the inspectors by going round to an instrument and disarranging it then setting him to find it. they have got so expert now, that any bug I can make does not take them above 3 minutes to find

We have tested the water chalks very severely by baking for long periods and drying in air but no test that we can make seems to hurt them.[4] Any transmitter, any coil, any spring, any adjustment seems to be good now. It is the old story get the chalk conditions just right and anything will work— No more cursing!!! Batch and I tested 37 in 1 hour and 20 minutes having to set all transmitters too   what do you think of that? Yours

T A Edison   per Batchelor

L (letterpress copy), NjWOE, Lbk. 5:343 (*TAEM* 80:180; *TAED* LB005343). Written by Charles Batchelor; circled "C" written above.

1. There is no clear evidence of the rationale for using pure water. On 6 November laboratory assistant James Seymour, Jr. wrote that a test instrument "Put up on the wall has 16½ chalk wet with 15 cent[imeters?] of water, to 8 oz." Edison appended to this, "hence this has harder pressure than reg (16½ H$_2$O)." The following day Edison cabled Johnson that "we have now struck bottom. . . . distilled pure water gives loudest talking yet obtained. conditions were: make size capilliaries match liq-

uid used. hereafter no trouble" (N-79-09-18:99, Lab. [*TAEM* 35:968; *TAED* No86:50]; TAE to Johnson, 7 Nov. 1879, DF [*TAEM* 52:900; *TAED* D7941ZHT]). In the laboratory that day, Seymour listed the chalks that he had wet and soon afterward, probably on 9 November, Edison began testing them in telephones in preparation for shipment. He reported that nearly every one was a "belcher" and again cabled Johnson on 11 November that "Tests show our solidity now   water chalks loudest yet   undiminished 2 hours baking whereas ph[o]sphat[e]s played [out] shipped thirty" (N-79-09-18:99–127, Lab. [*TAEM* 35:968–82; *TAED* No86:50–64]; TAE to Johnson, 11 Nov. 1879, DF [*TAEM* 52:913; *TAED* D7941ZIA]). In the provisional British telephone patent specification filed in December, Edison "proposed to dispense with the chemical substance previously employed in combination with the chalk, and to use the latter alone moistened with distilled or pure water" (Brit. Pat. 5,335 [1879], Batchelor [*TAEM* 92:141; *TAED* MBP021]).

2. See Doc. 1839 n. 3 regarding the chalks and apparatus. James Lighthipc (1857–1925) was born in Orange, New Jersey, and studied at the Stevens Institute of Technology. He worked on telephone and lamp experiments at Menlo Park in the summer of 1879. He served as an engineer for the Edison Telephone Co. of London and later the Bell telephone interests in Britain and Belgium. He also worked for Edison electric lighting interests in Europe and, upon returning to the United States, for many years with the Edison Consolidated Electric Co. and its successors. "Lighthipe, James A.," Pioneers Bio.

3. Regarding these instruments see Bergmann & Co. to TAE, 12 and 20 Nov. 1879, DF (*TAEM* 52:133, 137; *TAED* D7937ZDO, D7937ZDS); Batchelor to Bergmann & Co., 18 Nov. 1879, Lbk. 5:360 (*TAEM* 80:189; *TAED* LB005360).

4. Only one record of a thermal test has been found, in which a chalk was "baked half an hour in oven 130°." There is no indication of its subsequent performance. N-79-09-18:125, Lab. (*TAEM* 35:981; *TAED* No86:63).

–1846–

From Joshua Bailey

[New York,] Nov 14 1879[a]

My dear Sir:

Regarding the request that you engage to give to the French Co. all future inventions of telephones the following occurs[b] to me as a means of smoothing the point that would be satisfactory to both parties.

(1) You to give absolutely to the company your inventions of Telephones (this form of expression would not include accessories) for the term of five years.

(2) The company to have the first[c] right to your inventions after that period paying you a price amicably arranged. If difference arises, to be determined by reference to one arbitrator chosen by each with a third chosen by the two if they cannot

agree. In order that the proceedings for organization in Paris be not delayed, (and to reach with least delay the payment of money etc.) you should cable the following, if you cable in French, to the person who cabled you. "Donnerai toutes inventions pendant cinq ans   aprés donnerai toutes sur prix ~~sur~~ determiné[d] amiable."[1] If you accept the above suggestion I will have Mr. Lowry prepare an agreement in this sense which I will undertake to carry in the Board.

I have received from White's people (they report him better but quite ill)[2] the draft of revocation of Puskas' power. wh. I shall submit to Lowry, and if approved will have it drawn up in favor of Harjes[3] unless you telegraph me to contrary.[4] I should propose to take along this revocation but not to hand it to Harjes or make known that have it unless should find Mr. Puskas working with Pellorce to make trouble. I shall be advised in to-day or tomorrow whether money is likely to be paid over before my return. If it is likely to be paid (of which I will advise you by telegram as soon as I am advised) you should cable "Rousseau 113 Boulevard Sebastopol, Paris. Faites payer tout argent me venant à Drexel Harjes. Pouvoir Puskas recevoir revoqué.[5] Edison." This will not be necessary unless payment going to be made at once. But it will be necessary in this case (if you do not desire Puskas to receive it,) because I see your power (copy of wh. I have,) gives him expressly right to receive it as though belonging to him subject to acs between you. This makes it necessary also, in all cases that I take along the power in favor of Harjes. to use if necessary at time payment is made.[e] You should write to Puskas advising him that you have requested Harjes to recive directly from company your part of money & also to Harjes requesting him to attend to receive it when notified by me.[6] I will see you about these letters when I come out Monday.[f] You do not need to cable anything about shares, because the shares ~~will~~ cannot, by conditions of the contract of fusion go into any ones hands, but remain for a time specified in hands of syndicate for sale in the pool. Very truly yours,

J. F. Bailey

ALS, NjWOE, DF (*TAEM* 52:411; *TAED* D7940ZCT). Letterhead of Porter, Lowrey, Soren and Stone. [a]"18" preprinted. [b]Obscured overwritten text. [c]Interlined above. [d]"~~sur~~ determiné" interlined above. [e]"to use . . . made" interlined above. [f]"You should write. . . . come out Monday." written in margin; | indicates insertion at this point.

1. The translation is: "Will give all inventions for five years   afterwards will give all upon price to be arranged amicably."

2. Samuel S. White died in Paris on 29 December 1879. James H. Banker to TAE, 2 Jan. 1880, DF (*TAEM* 56:3; *TAED* D8048A).

3. John Harjes was born in Bremen of Danish parents who emigrated to America in 1849. The family located in Philadelphia and Harjes entered the family firm Harjes Brothers in 1853. In 1868, he and Anthony Drexel formed Drexel, Harjes & Co. as the Paris branch of the Drexel banking business (and subsequently a branch of Drexel, Morgan & Co.) with Harjes as managing partner. Drexel, Harjes & Co. became one of the leading private banks in Europe and John Harjes one of the most influential Americans living in Paris; he was awarded the order of the Chevalier of the Legion of Honor for his role in distributing funds raised in New York for the relief of Parisians during the Franco-Prussian War. Carosso 1987, 134–35.

4. On 19 November, Grosvenor Lowrey wrote Edison that Bailey had given instructions that morning to George Soren to draw up the power of attorney to Harjes. Lowrey warned Soren that "we cannot take instructions through Bailey for other people without running risk." He warned Edison that "Bailey as Soren says is so 'damned enterprising' that one must be careful all the time; & we can't in the press of business be careful in a proper way except by insisting on having our instructions at first hand— There are apparently a good many papers being prepared and I think you should come over tomorrow or Friday & see Soren & understand them all yourself." On the 21st Lowrey telegraphed "We have handed Bailey paper prepared on his instructions to be signed by you. You must scrutinize them for yourself because without conference with you we cannot know your wishes nor perhaps understand exactly what your interests require we expected you over today for this purpose." DF (*TAEM* 52:422, 427; *TAED* D7940ZDB, D7940ZDH).

5. The translation is: "Pay any money coming to me to Drexel Harjes. The power of Puskas to receive is revoked." On 18 November Edison cabled Rousseau: "Pouvoir Puskas revoque argent brevets payable Drexel Harjes." DF (*TAEM* 52:417; *TAED* D7940ZCX).

6. On 21 November, Edison sent a letter to John Harjes through Drexel, Morgan & Co. along with the power of attorney and a statement of his account with Puskas "amounting to $1376.09." Harjes received these on 16 December. Harjes to TAE, 16 Dec. 1879, DF (*TAEM* 52: 450; *TAED* D7940ZDS).

–1847–

Francis Upton to Elijah Upton

Menlo Park Nov 16 [1879]

Dear Father:

Maria & Charles[1] have been spending today with us and we have enjoyed their being here very much.

Just at the present I am very much elated at the prospects of the Electric Light. During the past week Mr. Edison has succeeded in obtaining the first lamp that answers the purpose we have wished it for. It is cheap much more so than we even hoped to have[2]

The light is obtained from a piece of charred paper[a] which is bent thus

The [~~paper?~~][b] burner[c] is [--][b] made from common card board and cut to about the size shown. This is then sealed in a glass bulb and the air exhausted and then a current of electricity passed through it which heats it to brilliant whiteness so that it will give a light equal to that from a good sized gas burner.

The making of such a burner has made the stock of the ~~e~~Electric Light Co. advance in value, $400 has been offered for shares and there are no ~~takers~~ sellers.[c] The last week has made all my prospects very bright, and I hope they will continue so. I expect that there will be an exhibition given in the course of a few weeks. The wire are laid to my house and I shall light up my parlor. If you wanted to show one or two of your pictures you could lend them to me and I would hang them in the best place in my parlor. Ahem!

As matters are turning out I am glad over and over again that I did not spend the year in Germany for I [-][b] now am put in the way of getting a living[a] far beyond what I there dreamed was possible.

I am a thorough master of all that is concerned so far in making a good light. We think as far as price is concerned that we can sell it so as to compete with gas at 50 cts a thousand feet. Very much love   Your Affectionate[a] Son

Francis R. Upton.

ALS, NjWOE, Upton (*TAEM* 95:572; *TAED* MU035). [a]Obscured overwritten text. [b]Canceled. [c]Interlined above.

1. Maria and Charles Farley.
2. See Doc. 1838 n. 15.

–1848–

To Brehmer Brothers[1]

MENLO PARK, N.J. Nov. 17th 1879[a]

Gentlemen,

We send you the two small phonographs, (left here by Mr Painter),[2] to night by express. We have faced off the parts of lever which clamp the diaphragm. This is absolutely necessary— The German silver springs must be narrower same as those now on instruments and a No. 6 Sharps' needle used—[3] The point of this needle must be left sharp but smoothed sufficiently so as not to scratch— Your needle is too thick and has too Blunt a point.[4] The felt pads at back of spring are saturated with shellac and consequently are hard and solid and have no 'give'— A good paste for fastening them to spring and diaphragm is made of Beeswax and Resin half and half, it melts easy and sticks well— Use as small an amount as possible— The Japan must be taken off the base underneath at the point where it [strikes?][b] the adjusting screw and also in the recess of the lever where it strikes the adjusting [------][b]   The casting at this point should also be free from Japan

I send one of your [springs? ----][b] pads so that you can see the difference between yours and what they ought to be

I would suggest that where the spring is filed so as to weaken it, it should be done with a smooth file and afterwards draw filed a little

Apart from these points I find your instrument well made and of good material[5]

Very respectfully yours

Thos. A. Edison   per Batchelor

L (letterpress copy), NjWOE, Lbk. 5:354 (*TAEM* 80:185; *TAED* LB005354). Written by Charles Batchelor. [a]Place from Edison's laboratory handstamp. [b]Faint letterpress copy.

1. The letterhead of this Philadelphia machine shop, formed by August and Hugo Brehmer, advertised its book-sewing machine and a millstone ventilating apparatus. The firm maintained two offices in Germany and one in London. *TAEM-G2*, s.v. "Brehmer Bros."

2. John Kreusi's entry in the order book the previous week instructed shop workers to "Fix and Send Toy Phonographs to Bremer Bros. Philadelphia Pa" (Cat. 1308:189 [Order No. 283], Batchelor [*TAEM* 90:761; *TAED* MBN003:68]). Uriah Painter, who had engaged the firm to make the small phonograph for the Edison Speaking Phonograph Co. (see Doc. 1717 n. 3), brought two of the finished instruments to New York on 10 November. He wanted Edison's "opinion & want also to have some arrangements about inspecting 100 now ready in Phila. & 100 every 2 weeks thereafter." Charles Batchelor went to New York and Painter wired him to "Order them to have any changes made you think best." Painter to TAE, 9 and 11 Nov. 1879; Painter to Batchelor, n.d.; all DF (*TAEM* 49:493, 51:753, 761; *TAED* D7903ZKL, D7932ZAZ, D7932ZBE).

3. It is unclear if Batchelor was referring to the class of thin piercing needles commonly known as "sharps" or to a proprietary needle.

4. The Brehmers explained to the phonograph company that their estimates had not included facing the lever, which had not been done on the prototypes, and that this work could now be done only at the com-

pany's expense. They acknowledged Edison's suggestion to "use #6 Sharpe's needles" but stated that they had used "#6 English sewing needles and a change would require us to throw our springs away & make new ones. This we cannot do, of course. The point we ground as near as we could, like the needle of your model. Several other suggestions of the Professor we accept with thanks." Edison then asked Brehmer to "send me two more of the small phonographs and I will see if I can work them without alteration— If not, the company will then order you to make the alterations at their expense." He also informed Painter that he would "see if it is practical to work them without the alterations   think we can possibly." Brehmer Bros. to Edison Speaking Phonograph Co., 20 Nov. 1879, DF (*TAEM* 51:756; *TAED* D7932ZBC); TAE to Brehmer Bros., and TAE to Painter, both 25 Nov. 1879, Lbk. 5:373 (*TAEM* 80:195; *TAED* LB005373A, LB005373B).

5. By 15 December several hundred were complete (except for the mouthpiece funnels) and awaiting inspection. Edison had planned to have Batchelor go but when Brehmer sent another reminder a few days later he wrote on it, "Show George Carman." On 17 December Brehmer urged "that it would be well to take immediate action if the machines are to be brought into the market before Christmas." The phonographs were ready for Carman's inspection in January 1880. By late February, however, Brehmer Brothers still did not have the screwdrivers that were to be packed with each machine and was also still awaiting Painter's instructions about a further modification of the stylus spring (Painter to TAE, 13 Nov. 1879; Brehmer Bros. to TAE, 11, 15, and 17 Dec. 1879, 13 Jan. 1880; all DF [*TAEM* 51:754, 757–58, 760; 55:281; *TAED* D7932ZBA, D7932ZBC1, D7932ZBC2, D7932ZBD1, D8038000A]; Brehmer Brothers to Edison Speaking Phonograph Co., 24 Feb., 1880; Brehmer Brothers to Painter, 27 Feb. 1880; both ESP Scraps. 13). The machines were evidently still in the shop in late spring when they were damaged by fire. Most were subsequently repaired but late in the year Edward Johnson lamented to Painter that he was "at a loss to know" how to dispose of them, and proposed trying to recover their cost by "selling them cheaply to some Toy House—or by selling them to some big advertiser. . . . The trouble with them is, not one person out of 50 has mechanical skill Enough to adjust them as per written instructions"

In the Edison parlor phonographs made by Brehmer Brothers, the speaker arm (at top) was placed slightly out of plumb so that the mouthpiece was easier to speak into.

(Brehmer Brothers to Edison Speaking Phonograph Co., 14 June 1880; Brehmer Brothers to Edison Speaking Phonograph Co., 16 June and 10 Nov. 1880; all ESP Scraps. 13; Johnson to Painter, 10 Dec. 1880, Unbound Documents (1879–1882), UHP). Some of these were sold by the Philadelphia scientific instrument maker James W. Queen & Co., which "found that in selling the Phonographs, that we had to instruct the purchaser, very often sending some one to his residence, and nearly always supply one or more broken Points; so that the profit if any, did not pay us for the trouble" (James W. Queen & Co. to Johnson, 18 Nov. 1880, ESP Scraps.). An illustrated instruction sheet provided by Queen & Co. is in PP (*TAEM* 96:516; *TAED* CA013B).

–1849–

*Equipment
Specification:
Electric Lighting*

[Menlo Park,] Nov. 17th 1879

285[1]   Make a Line shaft for 4 Faradic machines & fix up a nice & clean room for them[2]   put a partition across the shop. (See Book 79   page 201)[3]

J[ohn] K.[ruesi]

ADS, NjWOE, Cat. 1308:189, Batchelor (*TAEM* 90:761; *TAED* MBN003:68). Written by John Kruesi.

1. This is the order number.

2. Only three dynamos were used, one of which furnished current to the field magnets of the others. An illustration of the Menlo Park central station in *Harper's Weekly* of 3 January 1880 shows the three machines belted with the dynamometer to overhead shafts at the rear of the shop. The change in shafting is evident by comparison with an earlier illustration showing only one dynamo belted with the dynamometer that appeared in the 18 October 1879 *Scientific American*. The station was operational by early December. "Edison's Electric Light," *Philadelphia Ledger and Transcript*, 7 Jan. 1880; "Edison's Electric Light," *Harper's Weekly*, 3 Jan. 1880; "Edison's Electric Generator," *Sci. Am.*, Cat. 1241, items 1425, 1386, and 1332, Batchelor (*TAEM* 94:564, 546, 513; *TAED* MBSB21425X, MBSB21386X, MBSB21332X); see Doc. 1860.

3. On this and the following undated page Francis Upton calculated the dynamo speed at different engine speeds. About this time, perhaps also in preparation for the Menlo Park station, Edison used other calculations by Upton to compile two tables giving the diminution of foot pounds as the number of lamps in a circuit increases from one through ten, at constant voltage. One table assumed the use of a regulator at the station, the other no regulator. N-79-06-16.2:201–203, N-79-02-15.1: 186–97, both Lab. (*TAEM* 35:466–67, 31:750–55; *TAED* N079:99–100, N028:93–8).

–1850–

Notebook Entry: Electric Lighting[1]

[Menlo Park,] Nov 17th 1879

Lamp 42[a]   flat card carbonized well and kept perfectly flat[2]

We find that the brass clamps are bad from the fact that the heat takes the temper out of the brass and tends to straighten out the prongs eventually letting the loop drop out ~~of the~~ from between them

We have made some of steel wire so which have much more spring

In order to prevent the bad effect of the prongs opening, owing to the heat tending to straighten them we make it so:—[3]

now if the heat tends to straighten the prongs they will press tighter on the loop.[a]

Lamp 43   Made of card cut from new model and set in new clamp steel same as above

Lamp 44   Made from new model with straight steel clamps   Resistance after bringing up in vacuum[4a]

43[a]

The new model that we cut the papers from[5] has a surface before carbonization of thickness   making true radiating surface of   After carbonization it has these dimensions

    length   3.025
    width   .032
    thickness   .005[a]

Chas Batchelor   A Poinier[6]

X, NjWOE, Lab., N-79-07-31:141 (*TAEM* 33:611; *TAED* N052:69). Written by Charles Batchelor. Document multiply signed and dated. [a]Followed by dividing mark.

1. This entry is a continuation of Doc. 1838.

2. The following day Francis Jehl tested a lamp with this filament. It measured 157 ohms cold and after being raised to a red heat its resistance was 114.7 ohms. Jehl noted that there was an arc at the clamp (N-79-08-22:205, Lab. [*TAEM* 35:885; *TAED* N085:102]).

3. A clamp similar to this is shown in a patent application that Edison filed in December. E-2536:44, PS (*TAEM* 45:702; *TAED* PT020044).

4. Francis Upton and Francis Jehl tested lamps with these filaments on 19 November. Initially No. 43 measured 1,300 ohms and No. 44 measured 1,900 ohms when cold. Upon first being heated No. 44 measured 128 ohms. They then put the two lamps in series and No. 43 "became first dark red with a large part black then bright red" while No. 44 "did not come up at all." At this point No. 43 apparently measured 430 ohms then dropped to 133 ohms after being at yellow red for seven minutes. Upton noted that "both lamps in series almost exactly alike at the dull red." They then tested the two lamps together in parallel. When the lamps reached a yellow heat "the vacuum became very poor" but "when the spirals were cut off [from the current] the vacuum made quite rapidly." Later the air was exhausted out of these lamps and they were again measured cold with No. 43 at 115 ohms and No. 44 at 112.5 ohms. They also noted that No. 44 developed an arc that melted the connecting wire but did not affect the carbon. N-79-08-22:207–8, Lab. (*TAEM* 35:886–87; *TAED* N085:103–4).

5. No. 43 became the standard model for the horseshoe-shaped filament for all of subsequent filaments numbered through 70; these were "cut crossways out of the card." Filament No. 71 became an alternative design (see Doc. 1855 n. 2) although a few filaments continued to be made based on No. 43. In order to make the filaments Charles Batchelor designed a "Lamp Card Cutter" on 18 November that used a punch and die. A new carbonizing mould was also designed about this time. Most of the filaments through No.70 continued to be cut from cardboard but

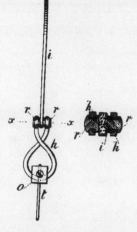

The spring clamp design shown in Edison's December 1879 patent application.

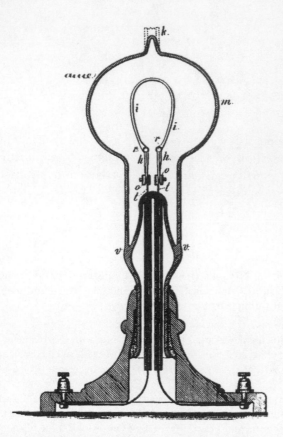

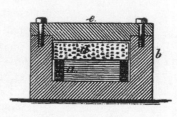

The lamp and carbonizing mold shown in Edison's December 1879 patent application.

Batchelor also experimented with bibulous paper, including some they had used for automatic telegraphy. He noticed that filaments made from bibulous paper "shrink much more in carbonizing than those made from cardboard like 43." N-79-07-31:147–53, 163, 169; Vol. 16:395; both Lab. (*TAEM* 33:614–18, 622, 625; 4:831; *TAED* N052:72–5, 80, 83; NV16:353).

Two of the cardboard filaments, Nos. 53 and 54, were used in making patent office models for an application that Edison filed on 11 December 1879 covering the cardboard filament lamp and methods of manufacturing it. This application never issued and the text is no longer extant but the drawings and claims are in E-2536:44, PS (*TAEM* 45:702; *TAED* PT020044). These show a mould for carbonizing a horseshoe shaped filament, which has the ends broadened to facilitate attaching it to the lead-in wire with spring clamps, also detailed in the drawings (see note 3).

6. Arthur Poinier assisted with carbonization experiments at the laboratory between October and December 1879. Time Sheets, NjWOE; N-79-07-31:137, 165, 181–83, 203–5, Lab. (*TAEM* 33:609, 623, 632–33, 643–45; *TAED* N052:67, 81, 90–1, 101–2).

To Siemens & Halske[1]

[Menlo Park,] November 19th [187]9

Gentlemen.

I see by a communication from the German Patent Office that you have objected to my patent application on dynamo machines on the ground that they are the same as yours,[2] accounts of which have already been published.[3] I think if you will study the matter more closely you will find that there are material differences and it is only these which I seek to patent.[4] Some 6 months ago I cabled Wm Siemens of London[5] urging him to patent the Hefner Von Alteneck—[6] I see this has been [carried?][a] out. Application has been made in the U.S. patent office upon which the patent office has declared several interferences among which I am included— I had altered my specifications so as not to conflict with the patent[7] and was undertaking to get the patent through however as you have so kindly taken it upon yoursel[ves][a] to oppose the issuance of my German patent I shall take measures to oppose in a legitimate manner the issuance of the Hefner Von Alteneck pat in this country[8] Very truly

Thomas A Edison

LS (letterpress copy), NjWOE, Lbk. 5:365 (*TAEM* 80:191; *TAED* LB005365). Written by Stockton Griffin; circled "C" written at top of page. [a]Faint letterpress copy.

1. Werner von Siemens, Johann von Siemens, and Johann Halske established this firm in Berlin in 1847. The company initially manufactured telegraph equipment, primarily for the Prussian state, but by this time was producing submarine cables, railroad signaling apparatus, telephones, and arc lighting equipment for domestic and foreign markets. Feldenkirchen 1994, 36–37, 84–88; Feldenkirchen 1999, 34–40.

2. The notice from the German patent office has not been found. The specification was probably one of several for continental Europe that Lemuel Serrell dispatched on 7 May, and may have been related to the U.S. application that Edison executed on 21 April (Case 177) for a generator with a stationary armature and an internal field magnet comprised of wire wound longitudinally on a rotating cylinder (Serrell to TAE, 28 Apr. and 8 May 1879, DF [*TAEM* 51:477, 485; *TAED* D7929ZBQ, D7929ZBX]). Edison later sought to expand the pending German application and it was this action to which Siemens & Halske objected. The German patent office evidently indicated that the additions constituted a separate specification, to which Edison responded in November that the affected designs "are properly included under the head of one patent as they are parts of one machine for generating electricity" (Siemens & Halske to TAE, 12 Dec. 1879, DF [*TAEM* 51:615; *TAED* D7929ZFZ]; TAE to Lemuel Serrell, 19 Nov. 1879, Lbk. 5:367 [*TAEM* 80:193; *TAED* LB005367]). It is not known what additions Edison sought to make but it is plausible that he tried to incorporate material related to the U.S. application (Case 184) that he executed on 4 September for an im-

proved armature. That application covered a rotating cylinder wound circumferentially with fine iron wire, over which insulated induction wire was wound longitudinally. When the cylinder revolved between the field magnet poles, the iron coil would become an induced magnet with poles fixed relative to the external magnet, and whose lines of force would be cut by the overlaying induction wire (U.S. Pat. 222,881).

3. Siemens patented his basic longitudinally-wound drum armature design in England in 1873 (Brit. Pat. 2006) and it is not known in what respects the present German application departed from that fundamental design. In correspondence with the U.S. Patent Office regarding his Case 177 Edison referred to the Siemens machine in Theodose du Moncel's multi-volume *Exposé des applications de l'électricité* (du Moncel 1872–78) and he would have had access to most of the contemporary descriptions, including British patents, cited in King 1962c, nn. 78 and 103. Drawings and a brief description of the basic Siemens design were published in the April 1879 issue of *Engineering,* which was pasted into a laboratory scrapbook ("Lighting by Electricity.—No. III.," *Engineering,* [Apr. 1879], Cat. 1024:84, Scraps. [*TAEM* 24:664; *TAED* SM025084a]). At least three similar forms of the direct-current Siemens machine were manufactured by this time (King 1962c, 386–87).

Edison also had direct experience with a Siemens machine, having borrowed one in December 1878 and made a number of attempts to understand the armature (see Docs. 1615, 1627, 1634 n. 1, headnotes, pp. 10 and 76. He apparently had returned it well before this time and in late October tried to borrow another but could not. He also requested details and citations about an 1876 model that Harvard professor John Trowbridge had used. Francis Upton later asked Trowbridge about specifics of that machine, particularly the armature windings (George Barker to TAE, 5 Nov. 1879; Trowbridge to TAE, 4 Nov. 1879; both DF [*TAEM* 49:489, 50:156; *TAED* D7903ZKI, D7919ZDF]; Trowbridge to Upton, 16 Dec. 1879, Upton [*TAEM* 95:582; *TAED* MU039]).

4. Edison protested in a 19 November letter to Lemuel Serrell that he did not accept "the statement of Messrs Siemens & Halske that the devices shown in my patent are the same as their devices to which they refer and a close inspection will show that there is a material difference." Serrell promised to forward a copy of this letter to Siemens & Halske but suggested that Edison specify how "the device that you seek to patent is different in this and that particular naming them fully so as to show that you were master of the situation." TAE to Serrell, 19 Nov. 1879, Lbk. 5: 367 (*TAEM* 80:193; *TAED* LB005367); Serrell to TAE, 22 Nov. 1879, DF (*TAEM* 51:599; *TAED* D7929ZFM).

5. Charles William (Carl Wilhelm) Siemens headed the London firm Siemens Bros., one of the complex of Siemens family enterprises, noted particularly for the construction and laying of submarine cables. He made significant contributions of his own in several fields of science and technology (*TAEB* 3:171 n. 2; Feldenkirchen 1999, 35–37). Edison had cabled him in December 1878 (see Doc. 1601).

6. Friedrich von Hefner-Alteneck was chief engineer at the Berlin factory of Siemens and Halske. In 1873 he introduced a highly efficient drum armature that improved the Siemens dynamo and became the basis for most modern designs. See *TAEB* 4:585 n. 5.

7. The Patent Office declared an interference on 17 October involving Edison's Case 184 (see note 2) and, among others, a January 1879 application of von Hefner-Alteneck for a dynamo with a longitudinally-wound armature rotated between curved branches of an electromagnet. Edison's application illustrated but did not describe the distinctively large field magnets and on 7 November he instructed the Patent Office to insert a statement explicitly disclaiming a complete dynamo of this type, in "order to terminate the interference so far as my application is concerned." The Patent Office allowed Edison's application on 1 December and it issued on 23 December 1879 as U.S. Pat. 222,881. TAE to Commissioner of Patents, 7 Nov. 1879, Pat. App. 222,881; Serrrell to TAE, 2 Dec. 1879, DF (*TAEM* 51:606; *TAED* D7929ZFT).

8. Siemens & Halske replied that German law permitted the nullification of a patent if a claim was later found to have been unoriginal at the time of issue, hence they intended their opposition to protect both themselves and Edison. The firm promised to respect Edison's legitimate interests and asked for similar consideration on his part. Siemens & Halske to TAE., 12 Dec. 1879, DF (*TAEM* 51:615; *TAED* D7929ZFZ); see also Doc. 1951.

–1852–

Notebook Entry:
Electric Lighting

[Menlo Park, c. November 20, 1879[1]]

Say $\frac{1}{250}$ of the current in the main line flows through the shunt containing the voltameter[2a]

10 per H.P[3]

$x^2/100$ 44.3 = 3300

$X^2 = {}^{330\,000}/_{44.3}$[4]

5.5185

8.3536

|3.8721

1.9360[5]

X = 86.3 Volts

.863 Webers flowing[a]

Page 191[6]   19.94 Mg. of Cu. by one Weber in one minute

19.94

60

1196.40 Mg per hour

1.1960 Gr per hour 34 hours a day   30 days a month   120 hours   150 hours

1.19   0.0755

150.   2.1761

2.2516

178 ratio of Shunt[a]

$\frac{1}{178}$ of Current should go through the shunt[7]   150 hour
1000$\frac{1}{178}$ 200 hours[a]

$^1/_{178}$ of current should go

1~~0~~50

   5

750 feet per month

750   5 cts an hr[8]   750 feet per month

    1 Gramme represent one thousand feet of gas   supposing that ~~the~~ a burner consumes [5?][b] feet ~~for~~ per hour

    200[c] hours

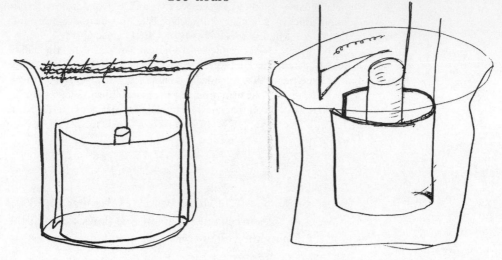

Glass jar   small battery   resistance 2 ohms   <u>Coiled copper</u>[d]

     1.19

      2 00[9]

     238.00   238 ratio

.03   3   6 times

         .05   five lamps

         .03   ten lamps

       .015   twenty lamps[10]

1 Gramme of Cu[a]

Standard cell

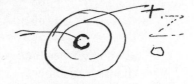

Red ▨▨▨▨▨ Red wire +

Green ▨▨▨▨▨ Green wire −

    Cell can be made of certain size   Must be sealed how Metal cover

    Copper does not creep much   Copper does not creep much

Two poles are needed

 Red and Green wires

.03 Ohm for shunt   measure wire and wind on magnet .03
.03 safety clutch[11]

Make a jar and resistance on magnet .~~03~~ so that with .03 in shunt $\frac{1}{238}$ of the current ~~from two~~ to two lamps[e] will pass through the meter.

To make S

Better take near .03 and[a]

A coil so wound as to be easy of adjustment

Every cu. foot of gas will be measured by a mg.   1000 mg. = 1 Gr.[a]

Motor sold the same as light[a]

20 Ohm motor   100 Ohm lamp[12]

Francis R. Upton

X, NjWOE, N-79-04-03:211, Lab. (*TAEM* 31:445; *TAED* N025:107). Written by Francis Upton; doodles, miscellaneous calculations, multiple repetitions of figures omitted; decimal points added for clarity. [a]Followed by dividing mark. [b]Unclear figure. [c]Obscured overwritten text. [d]Multiply underlined. [e]Interlined above.

1. This notebook entry follows related calculations by Francis Upton dated 17 November and immediately precedes John Lawson's 1 December notes of experiments with copper deposition for the meter.

2. On 8 November Upton and other laboratory assistants began experimenting with copper solutions to determine deposition rates and the effects of various shunt circuit arrangements for Edison's meter. Upton made related notes and calculations on 15 and 17 November and in undated entries in another notebook cross-referenced to this document (N-79-04-03:174–196, 199–209; N-79-02-20.1:214–49; both Lab. [*TAEM* 31:426–36, 439–44; 569–82; *TAED* N025:88–106; N026:92–105]). The rate of deposition in the shunt was not necessarily directly proportional to that of the full current. In an undated experimental record from about this time, Upton noted that a strong battery current deposited 16.417 grams of copper, while in shunt circuit with $\frac{1}{160}$ of the main current only .080 grams, or $\frac{1}{205}$ of the whole, was deposited. This result satisfied Upton, who called it "a good showing for the meter" (N-79-02-20.1:220, Lab. [*TAEM* 31:572; *TAED* N026:95]).

3. On 8 November Upton directly determined the current strength needed on a single "100 ohm lamp   10 per Horse power" by solving the equation "$C^2R \times 44.3 = 3300$." He determined the solution of .863

Webers, using logarithms as in the examples below to make the computations easier. The figure 3,300 is one-tenth of the foot pounds per horsepower; one watt equals 44.3 foot pounds. N-79-04-03:175, Lab. (*TAEM* 31:426; *TAED* N025:88).

4. In the following computation Upton used the same logarithmic shortcut explained in Doc. 1795 n. 4.

5. In this last operation Upton is dividing the logarithm of x^2 by 2 to obtain the logarithm of x.

6. On that page Upton used the results of the 8 November experiments to determine that a current of one Weber would deposit ".332 mg per second of Cu. [or] 19.94 mg per minute of Cu." In the other notebook Upton calculated the shunt ratio for a comparable current strength of .873 Webers. N-79-04-03:191; N-79-02-20.1:217, both Lab. (*TAEM* 31:433, 570; *TAED* N025:95; N026:93).

7. That is, the current necessary to deposit one gram of copper in this example. In one of the deposition experiments from this time Upton measured the shunt current at .0125 Webers and may have been concerned about the effects of a counter-electromotive force in the deposition cell on the flow of such a small current. In one of the undated entries in the other book he noted that "The contrary E.M.F of Cu solution is .05 so that it was thought impossible to indicate the first lamp. Experiments were made that showed that this contrary E.M.F was not a constant but rose with the exciting current being for very weak currents extremely weak." N-79-04-03:189, N-79-02-20.1:230–31, both Lab. (*TAEM* 31:432, 577; *TAED* N025:94; N026:100).

8. The basis for this rate is unknown.

9. This factor is presumably a number of hours per month, as indicated above.

10. This series may indicate the need for increasingly heavy wire (with lower resistance) in the shunt as the number of lamps and amount of current through the meter rise.

11. On 14 November Upton again tested the meter with safety clutch that had been made in May (see Doc. 1733). He found that the magnet was "too strong and broke the current at about ten Webers or less" and noticed that in doing so the contacts arced severely. This latter difficulty was addressed in a March 1880 patent application (Doc. 1912). N-79-04-03:197–98, Lab. (*TAEM* 31:437–39; *TAED* N025:99–100).

12. Under a later heading of "Plan to pursue" Upton listed "Proof that lamps or motors consume the same in multiple arc." After calculations on the following page he concluded that "The E.M.F. remaining constant the current that passes through a multiple arc is proportional to the work done whether the work be in the form of heat or motion." N-79-02-20.1:225–27, Lab. (*TAEM* 31:574–75; *TAED* N026:97–8).

Menlo Park Nov. 22 [1879]

Dear Father:

The Electric Light is slowly advancing from the last big step. We now know we[a] have something and that is what we have not know until last week. We can compete with gas in a great many ways now though not as completely as we wish, yet there seems to be nothing to prevent our getting a perfect burner that shall do as well as gas. Time and cost will prove what we have to be good or bad.

Lizzie is getting on somewhat better with her Swedes[1a] and thinks she will give them a longer trial rather than to have the bother of new help. They do the work during the week, though they are very slow in getting about.

I am busy now preparing an article on the Electric Light for Scribner's Monthly to come out in Feb. [--][b] number.[2] I hope it will bring me $50 or $60 and some fame. You see my work has been very closely connected with Edison's and if the electric light succeeds ~~it~~ in all that we hope of it, it will bring me a good name. If it after [~~caef?~~][b] careful trial proves much cheaper than gas it will probably bring me a large salary. Dollars and cents are ~~ar~~ now the question, we have a light what we want to know is how much will it cost, and that is hard to decide. Your loving Son

Francis R. Upton.

ALS, NjWOE, Upton (*TAEM* 95:575; *TAED* MU036). [a]Obscured overwritten text. [b]Canceled.

1. See Doc. 1840 n. 4.
2. Upton 1880a appeared in the February *Scribner's Monthly*. Attributed to "Mr. Edison's Mathematician," it incorporated Edison's affirmation that he had "read the paper by Mr. Francis Upton, and it is the first correct and authoritative account of my invention of the Electric Light" (a facsimile of Edison's letter to the editor was subsequently published as the frontispiece of the volume). Upton offered a brief history of electric lighting before turning to Edison, who he said "came to the investigation unhampered by the blunders of his predecessors. He had never seen an electric light. . . . He saw that subdivision was his goal, and toward that he steadily worked. With a steadfast faith in the fullness of Nature, a profound conviction that, if a new substance were demanded for the carrying out of some beneficial project, that substance need only be sought for, he set to work" (pp. 534–35). He described successively Edison's early research and the major components of a complete Edison system of electric light and power. He also explained the laboratory dynamometers.

The article and its fifteen illustrations filled nearly fourteen pages of *Scribner's* and evidently took considerable time to prepare. In early December Upton wrote his father that he had spent an entire Sunday on it although "it is against my rule to work Sunday but the work seemed to be thrust upon me." At the end of that month George Soren cautioned Upton that publications such as his could "come back to 'plague the inventor'" in future litigation. Soren suggested that lawyers for the electric light company should scrutinize the article because "in my personal judgment very great caution should be observed in everything. If I had my way I would not allow half the publicity that has been given by Mr. Edison— It is very unusual." Upton to Elijah Upton, 7 Dec. 1879; Soren to Upton, 29 Dec. 1879; both Upton (*TAEM* 95:580, 589; *TAED* MU038, MU042).

Albon Man challenged the article on several counts in a letter to the editor published in the next issue. Upton had stated that after experiments at mid-century "so completely had the mode of lighting by an incandescent solid been forgotten, that in 1873 . . . letters-patent were granted to Sawyer and Mann for a stick of carbon rendered incandescent in nitrogen. No successful light by incandescence had, however, been produced when Mr. Edison began his experiments" (p. 533). Man replied that he and William Sawyer had no such patent but "very much, however, of what Mr. Edison claims in your article to be his invention was patented by Sawyer and Man long before Mr. Edison commenced working on electric lighting, and much also has been since patented by us. . . . I claim that, before Mr. Edison can render his lamp a success, he must resort to still more of the inventions patented by Sawyer and Man," including those for a distribution system and a lamp with a horseshoe carbon. "'Edison's Electric Light.'—A Reply," *Scribner's Monthly,* Mar. 1880 (19:795), Cat. 1241, item 1526, Batchelor (*TAEM* 94:612; *TAED* MBSB21526X).

–1854–

Edward Johnson to Charles Batchelor

London Nov 27th 1879.[a]

My Dear Batchelor

I have yours in re. to Barton[1] & his tender youth scribner[2]— also Several notes written from time to time in re. to new chalks—also yours in Edisons name giving a meagre ac/c of the water chalk discoveries[3]   when I received this latter I went from the House where I have been for a week on my back—to the office to attend a Board meeting to̶i̶n order to show it, and assure them that their fears that Lighthipe had the latest were not well founded— He L. thought he had—& there was nothing in your notes or Edisons Cables to indicate the contrary—in nothing was it distinctly said (nor has it been yet)—that no chemical was used at all— The use of water only for remoistening was all we were sure of— Lighthipe said he had the distilled water chalks— I questioned him & found his chalks were originally made with soda= All this up to the receipt of Edisons Letter telling me to "throw away Lighthipes chalks"[4]—had so much the appearance of another promise of good things to come—but which would fail when set up here that Everybodys confidence stood at Zero— I again raised it a few degrees above by assurances of what could be done—when we got the Lathe & made our own chalks— Ere the Lathe has yet arrived however—comes this welcome Letter from Edison with its assurance that what is to come is of a far different character from all previous arrivals— Hope therefore predominates— It has come to pass now that in order to make our people feel my confidence in each cabled assurance that our troubles are over I must give a[b] scientific basis for my belief— I gave it yesterday as follows

Edison holds the action to be capilliary—a requisite of capilliary action is the[c] cellular formation of the chalk— Any chemical dissolved in water & held in those cells—would sooner or later crystalize—and always at the surface first—

such crystalization corked all the cells— The crystalized chemical is a non conductor   The corks estopped the disconnected the conducting medium—the water—from the metal pen—

No electrical action—no capilliary— But aside from this suspension of the Electric[c] current—The cessation of the capilliary would prevent E.M.G. action—anyway—

Now I said= It is my belief that the water held within these cells will come out so slowly that if the cells are kept open—it will take many months to get it all out—i.e.—Capilliary action will go on for many months at a practically uniform rate.—

Problem— To Keep Pores Open= Now Edison Telegraphs—& writes "Pure water"—negatively= no chemical= no chemical—no crystalization—no crystalization— no corking of Pores   Problem solved—& by Edison of course

Finale— Does it transpire that a great advance has been made— Everybodys confidence in Edison is again at summer heat= and my Faith—is justified= here a word—without Edison having had a man here who was endowed with the[c] full measure of confidence in him that he has in himself this concern would have come to a dead lock— Even if it had not been overthrown— The Lancashire & Glasgow parties were far more unmanageable than the London— Glasgow was on the point of refusing to go on—when I wrote a severe letter to Gourauds man—Moore—[5] I told him I didn't think he had had much experience in introducing New inventions= or he would See that this was making the most extraordinary progress toward Perfection ever heard of— I followed this up with a little double loaded assurance—& Gouraud went up & clinched it just in the nick of time=

I have had to take an opposite course too at times= in re to the New Co's— Gouraud has been wanting to go ahead faster= But as when once we enter a town we must do so to Stay= & as I had nothing possessing staying qualities I have[b] discouraged & delayed these enterprizes to the utmost limit= I was compelled to finally fix a time to open the Ball at Birmingham— I fixed it after the arrival of the lot of 36= hopeing they would prove equal to the purpose— They arrived— I selected 3 of the best— I went— They lasted 24 hours—& then failed— Water restored them for periods of a few hours only— We had to retreat under as[b] decent Cover as we could invent—so now find ourselves in a worse position than if we[c] had[c] not gone at all—justifying my objections— We have about 60 Instruments that will not work at all— Those out require more attention than 500 good ones should. My great object has been to prevent this weakness from being discovered— That it has not reached the Public through the anti Edison Prints & the Bell Co. is to me a marvel—explained only by the fact that it has been kept from their knowledge by keeping it from the knowledge of all but the necessary few of ourselves— This could not go on much longer—hence the value of this final discovery of a solid basis—

Now in re. to Barton= W.E. I have not yet seen him because I have been on my back—(bad[d] case of Piles—result of my late Billous spell—) ever since he arrived—but I understand he has been around the office and chinning some of my Boys—[6] White—I am convinced—is a man who will not object to a commission— He hankers after all the contracting—all of which properly belongs to my Dept— I have yielded Everything to him except the mnfr' of Insts— That I wont yield under any circumstances— White is the only man in the Co who wants to see them made on this side   All others say no—Edison must look to them— Whites only object can be a personal one—as we are making them for less money by 54 to 5$ than he can get English mnfr's to bid— He then tries to advance other reasons why[c] they should be made here—but thus far he hasn't been able to even get any one interested in the subject. I simply meet him by telling them we can do it in America Quicker, Better & cheaper— when any one here can deiscount Bergmann on that—then I will try & get Edison to consent to their being made on this side   on the other objectionable grounds of interference with his Patent—my strong point with Barton is this= That Bergmann underbid his concern with the W.U. to such an extent that in spite of their owning ½ interest in the W.E.—they were Compelled to give him an amount of work which raised him from a Petty lathe or two to a 1st class Manufactory= & that this when done—so alarmed them that they have since been trying to destroy him— I guess I'll lay on that? You must continue to do for us what you have thus far— It will bring you more than it cost. I have more weighty reasons for keeping this man'f'g in Bergmanns hands until I come back than the mere monetary interest— I have projects for the future with that little Dutchman which can be realized only by <u>my</u> keeping him afloat while absent— I can easily do it when at home but mum on this to any but Edison—[7]

Gouraud made some fuss about the slight put upon him by the Co. & Edison in the matter of the Lancashire 10 000£= but it is all his own fault. The case stands thus

Original Power insufficient to enable Co to pay Gouraud money under <u>Supplemental</u> Agreement=

Gouraud advised of this months ago sought to use it as a pretext for getting from Edison an Irrevocable Power— Had a proper P of atty drawn up—<u>but made it irrevocable</u>= and in so doing over-reached himself= He got <u>none</u>= consequently when money was ready—he was again informed his Power—insufficient= He <u>affected</u> great indignation= although nothing had been changed since he was originally notified— Edison[c] was accordingly asked for an Order to Pay= He[e] cut Gouraud by Ordering it paid to himself= My impression is Gouraud will try to use this, as a new pretext for worming the Irrevocable out of E— His ambition will never be satisfied until he <u>alone</u> controls the Edison Companies in England   Let Edison keep this always in mind and he will know what papers to sign=

Tell Edison not to send any more <u>Boys</u> like "Brassy"[8]=we have no use for them— Tell him to <u>now</u> raise the standard— We want men who can take entire charge of the Engrng Dept in a new town= They must be as good as Rose or Rector & be presentable in appearance & Language—some importance here you know—since our Promoters are Lords of high Degree—

Rose is the best. Rector almost his[c] Equal[c]— Lighthipe untried— Others N.G.   Truly Yours

E. H Johnson

ALS, NjWOE, DF (*TAEM* 52:933; *TAED* D7941ZIT). Letterhead of Edison Telephone Co. of London. ^a"London" and "18" preprinted. ^bInterlined above. ^cObscured overwritten text. ^dMultiply underlined.

1. Enos Barton was secretary of the Western Electric Manufacturing Co. in Chicago (*TAEB* 3:181 n. 14). Batchelor reported to Johnson on 8 November that Barton was "looking out for more manufacturing" and during a recent visit to Menlo Park had asked "whether I thought there would be any objections on the part of Edison to his manufacturing our telephone for France" (Batchelor to Johnson, 8 Nov. 1879, Cat. 1330:53, Batchelor [*TAEM* 93:267; *TAED* MBLB2053]).

2. In his 8 November letter (see note 1), Batchelor told Johnson that Barton intended to visit Paris "taking a boy named Scribner over with him. I think this is the youth that has made a number of improved Carbon telephones and is the originator of the ones the W.E. Mfg. Co. are making." Charles E. Scribner had visited Menlo Park at Barton's behest a few weeks earlier to examine Edison's motograph chalk. He became a prolific patentee in his own right, most notably in the area of telephone switchboard design, and served as Western Electric's chief engineer in Chicago and then New York. Barton to TAE, 30 Oct. 1879, Box 92, NJWAT (*TAED* X012L1D); *NCAB* 41:408–409.

3. Doc. 1845 (see esp. n. 1). In reply to Johnson's letter, Batchelor wrote him that the standard procedure for the water chalks was to allow them to stand a full day after the initial wetting before using them. He also discussed recent experiments showing that chalks that no longer worked well could be improved, at least temporarily, by the application of a slight amount of pure water. Batchelor to Johnson, 14 Dec. 1879, Cat. 1330:56, Batchelor (*TAEM* 93:270; *TAED* MBLB2056).

4. Doc. 1845; see also Doc. 1839 n. 3.

5. Michael Moore, an American who had lived in Britain for some time, was an organizer and director of the Edison Telephone Co. of Glasgow. Johnson to TAE, 30 Mar. 1880; George Gouraud to TAE, 10 June 1880; both DF (*TAEM* 56:606, 55:717; *TAED* D8049ZDP, D8046ZAL).

6. In his 8 November letter (see note 1) Batchelor had advised Johnson to expect Barton. Suspecting that he would try to patent some of Scribner's carbon telephones in England or France, "I gave him a graphic account of the way you were going for all infringers Blake, Crossley, etc and assured him that any one taking a new infringement there would have an exceedingly hard road to travel— He did not say much but I guess he will confine such operations to the continent."

7. Batchelor also wrote in his 8 November letter (see note 1) that when Barton had inquired about manufacturing telephones for France "He said he understood that Edison and Johnson were interested in Bergman's shop and possibly it would be useless to try— I told him he was mistaken in regard to that— Edison I knew had no interest and if Johnson had had I thought I should know it." Regarding Johnson's interest in Bergmann's shop see Doc. 1790 n. 8.

8. Unidentified.

[Menlo Park,] Nov 28th 1879

Lamps

84[1a]   Made from card same as 71[2]
85
86[3]
87
88[b]

Made a <u>new steel cutting former</u> for loops with 4 clamps to hold it together

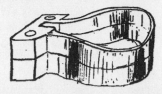 

Clamp

This cuts them much more even

89[a] Made from .0010 Bristol board[4] and cut out with the new 'former'
90
91[b]

92[5]
93[6] same as above[7]
94[8b]

TAE   Chas Batchelor
A Poinier   J. Seymour.[9]

X, NjWOE, Lab., N-79-07-31:183 (*TAEM* 33:633; *TAED* N052:91). Written by Charles Batchelor. Document multiply dated. [a]Following list of numbers enclosed by right brace. [b]Followed by dividing mark.

1. A lamp with this filament measured "before heating 2500 Ohms Very bad spots and slightly bent in first bringing up   After bringing up 204 ohms." N-79-08-22:243, Lab. (*TAEM* 35:903; *TAED* N085:120).

2. Lamp filament No. 71 was described by Charles Batchelor on 25 November as "Piano wire clamp with large flat platinas on   the card cut lengthways so that grain (as it were) lays that way— Carbonized with white tissue paper (not oiled)." A lamp with this filament had measured 310 ohms cold and then went down to 141 ohms after being heated briefly. On 28 November it was measured at 138 ohms and after being on two hours it was 148 ohms. Filaments 72–73 and 77–78 were made in the same manner as 71. No. 72 measured 270 ohms cold, 107 ohms when heated momentarily, and 114 ohms after being on for some time. No. 73 measured 320 ohms cold, 137 ohms when heated momentarily, and 185 ohms after being on for some time. No. 77 measured 365 ohms cold, 160

ohms after heating, and 169 ohms "very hot." There are no measurements of 78. N-79-07-31:163, 169; N-79-08-22:229–231, 205; both Lab. (*TAEM* 33:622, 625; 35:897–99; *TAED* No52:80, 83; No85: 114–15, 102).

3. Francis Jehl noted that a lamp with this filament measured "before heating 400 ohms   after moderate heating 155 ohms   remarkably even at red heat." N-79-08-22:245, Lab. (*TAEM* 35:904; *TAED* No85:121).

4. The previous day Charles Batchelor had tested several samples of bristol board, some other card stock, and a few types of paper for ash content. N-79-07-31:171–79, Lab. (*TAEM* 33:626–31; *TAED* No52: 84–89).

5. This measured 800 ohms before heating. N-79-08-22:251, Lab. (*TAEM* 35:905; *TAED* No85:122).

6. This measured 700 ohms before heating. N-79-08-22:252, Lab. (*TAEM* 35:906; *TAED* No85:123).

7. Lamp filaments 95–124 were all made the same. N-79-07-31:185, Lab. (*TAEM* 33:634; *TAED* No52:92).

8. A lamp with this filament measured "before heating 800 ohms Very even heat at red. Incandescent for a very short time 205 ohms." N-79-08-22:253, Lab. (*TAEM* 35:906; *TAED* No85:123).

9. James M. Seymour, Jr. (1878–1940) worked as a laboratory assistant. The earliest record of his work is a time sheet from June 1879, though he may have been at the laboratory as early as 1878. He later spent eight years in Europe working for the Edison lighting interests. "Seymour, James M. Jr.," Pioneers Bio.

–1856–

Telegrams: From/To
George Gouraud

December 1, 1879[a]
[London] 2:25 [P.M.]

Edison

Times specially desiring representation first exhibition light   ask cable giving positive date   reply paid

Gouraud

[Menlo Park] 3:30 [P.M.]

~~Gouraud~~ Menlo park London

Public exhibition takes place [----s][b] during holidays   It is an immense success   say nothing[1]

Edison

L and ALS (telegrams), NjWOE, DF (*TAEM* 49:746; *TAED* D7906ZBP, D7906ZBQ). First message written by Stockton Griffin. [a]Date from document, form altered. [b]Canceled.

1. Reports of Edison's light had evidently been appearing in U.S. newspapers. When the editor of the *Chicago Daily News* telegraphed on 28 November to confirm an account published in that city "that you have stated that the light is a success" Edison replied, "All the statements telegraphed mere guess work. I shall however make a public exhibition

within four weeks from date from which the public can judge of my success." Subsequent rumors of a planned Christmas demonstration prompted inquiries from other reporters and editors, including William Croffut, who wanted the story for the *New York Tribune.* In a draft reply to a 16 December letter from Lindley Murray, an editor of the *Philadelphia Underwriter,* Edison promised that the "Electric Light will be exhibited at Menlo park to public about first of year." The cable dispatch to the *Times* describing the illumination of Menlo Park is Doc. 1873. Melville Stone to TAE and TAE to Stone, both 28 Nov. 1879; Croffut to TAE, 8 Dec. 1879; John Barron to TAE, 4 Dec. 1879; Tom Maguire to TAE, 18 Dec. 1879; TAE marginalia on Murray to TAE, 16 Dec. 1879; all DF (*TAEM* 49:743, 745, 748, 747, 749; *TAED* D7906ZBM, D7906ZBN, D7906ZBS, D7906ZBR, D7906ZBW, D7906ZBV).

–1857–

Agreement with Joseph Murray

Menlo Park, N.J. December 1st 1879[a]

We hereby release each other from all claims of every kind & character and the said Murray hereby acknowledges that the said Edison does not owe him anything having settled in full to date and the said Edison hereby acknowledges that the said Murray owes him nothing he the said Murray having settled to date,[1] the said Edison delivering to the said Murray his notes for One year from July 12th/75 to amount of One Thousand dollars[2], also his note for Fifteen months from July 12th/75 for One Thousand dollars also his note for[b] Eighteen months from July 12th 1875[c] for One thousand dollars, also his note for Twenty one months from July 12th/75 for One thousand dollars, also his note for Two years from July 12th/75 for One thousand dollars, also his note for Twenty seven months for One thousand dollars from July 12th/75 also his note for Thirty months for One-thousand dollars from July 12th/75 also his note for Thirty Three months for One thousand dollars from July 12th/75 also his note for Thirty six months for One thousand dollars from July 12th/75 and the said Murray giving the following order on Mr Bullard[3]

T A Edison   J. T Murray

DS, NjWOE, DF (*TAEM* 49:527; *TAED* D7903ZLJ). Possibly written by Joseph Murray; on letterhead of Edison's laboratory. [a]"Menlo Park, N.J." and "187" preprinted. [b]Interlined above. [c]"from . . . 1875" interlined above.

1. Joseph Murray was still disposing of his machinery through Edward Bullard, a New York dealer, apparently on consignment. He sent Edison the dealer's statement of unsold items in September and asked him to "fix up as best you can and I will be satisfied with most any arrangement you make. . . . you can see what is best for us both and act

according as I trust you always." Edison promised to give a full release upon payment of a remaining balance, which Murray could not make pending the sale of more machinery. In the meantime he arranged to have Murray's shafting and pulley shipped to Menlo Park and a screw machine and a lathe sent to George Place, another New York dealer. He also prepared an undated draft of this document, only one page of which is extant. Murray to TAE, 6 Sept. 1879; TAE to William Carman on enclosure from Murray, 6 Sept. 1879; Murray to TAE, 10 Sept. 1879; E. P. Bullard to TAE, 24 Oct. 1879; draft agreement with Murray, n.d.; all DF (*TAEM* 49:364, 366, 369, 370, 453, 526; *TAED* D7903ZHN, D7903ZHO, D7903ZHP1, D7903ZJQ, D7903ZLJ).

Murray inquired later about a position with the Edison Electric Light Co. but instead became associated with Frederick Fitch's electrical supply shop. Murray to TAE, 2 Jan. 1880; Murray to Batchelor, 7 June 1880; both DF (*TAEM* 53:27, 55:557; *TAED* D8004D, D8042ZAT).

2. This note and the others referred to resulted from the dissolution of Edison's partnership with Murray on 13 July 1875 (see Docs. 593 and 594).

3. Edward Bullard was a designer and dealer of machine tools. He organized the Bullard Machine Co. in New York in 1877 and later achieved a national reputation with a similar firm in Bridgeport, Conn. *NCAB* 31:331–32.

–1858–

From Edward Johnson

London Dec 2d 1879.

My Dear Edison

The water chalk Telephones arrived today and were tested this P.M. I had to leave the office before many were tested, having an engagement, but as far as Rose had gone—some dozen or so—they "belched right out"= I have no doubt talked quite as loud as they ever did— I am inclined therefore to think that we have at last reached the Goal— Everybody is happy tonight—& since I can not imagine water & chalk undergoing any change I fully expect the general happiness to increase. If they all Pan out the same, & hold out until Saturday I will cable you a congratulatory message, in compensation for the many dispiriting ones I have been compelled to send—

There are 48 Telephones & 100 chalks—& 2 switchboards in this lot= Lighthipes apparatus and chalks has not yet arrived you must look a little sharper after[a] these things= Batch wrote me of their[a] shipment on Nov 4th= and of a shipment of 30 on Nov 10 & of 18 on Nov 12th=[1] These two latter came today in one vessel. What has become of that of Nov 4th? Possibly it went into that Iceberg & is now laying at St Johns— my mail per Arizona was dated NYork Nov 3rd= & arrived ten[a] days behind time=[2]

Now will begin the pressure for Insts.—if chalks prove OK= & you must be able to respond by shipping them with every steamer in lots of not less than 40 or 50= Want switch-boards too= [3]

"So much for Buckingham"[4]

What about American Telephone stock— I hear that Bell Stock has gone up to 1000$= Don't this look as if the W.U. had been over-reached—[a] I had a call from Cornelius Roosevelt— he is come to town with the object of Planting his Gower Telephone in England he wants to bring about an amalgamation of all interests   I talked fully & frankly with him and told him the matter stood thus

Edison had been sold out in the U.S. by Prescott—for personal gain   no regard was had for his personal repute as an inventor—he was not therefore protected   The W.U. crowd were more than willing to see him ranked second or third=

In France he was represented by a weak & impecunious man over whom Roosevelt & his friends had no great difficulty in gaining a victory even with a much inferior article= The result again being Edison was compelled to take a place second to an inferior invention—& I had no doubt results would prove that the Gower interest would dominate

(He confirmed this by telling me that whenever the two Insts. were shown the Gower was preferred to the Edison)

In England—I assured him things were to be managed differently

1st   We had Plenty of money & were not afraid to risk it=

2nd   We had such regard for Mr Edison personally as to imbue us with a determination to be[b] watchful of his rights & reputation

3rd   We proposed, (& were now conducting a vigorous campaign against all infringers) to maintain his Patent at all Hazard—as we believed he had a Patent for the only practical Telephone—

4th. When we found that we <u>could</u> or could not sustain the Patent <u>then</u> & <u>only</u> then we would be in a position to talk of compromise or fusion—

Bouverie happening to come in just then I introduced them & gave Bouverie a brief resumé of our conversation & asked him if I had stated our position correctly— He said "What Mr Johnson has told you is about the status"— "We believe in Peace &c. but before we extend a hand or take one we wish to know <u>who it is</u> that holds the trump card"

Roosevelt left for Paris the next day—but I am satisfied will be back— He says our Patent in Eng. or France is not good=

Now we know who we have to fight=

I am glad you got that promise from Bailey[5]—though it wont keep him from the field= He is too slippery— He will simply sell out to some one else to enable him to do it under cover—but I will see his hand you may rely.

So much for the Edison Telephone competitors— Now for another class of opponents—Viz the P.O. The Post Of. have brought suit against us <u>alone</u> for infringing their rights— Graves (Preeces chief)[6] told a relative who told me that they meant to "make the Edison Co pull the chestnuts out of the fire for the Bell Co" and they have confirmed it by bringing the suit solely against us.[7] Gouraud is afraid this suit is going to prevent the formation of more Dist. Co's.

Bouverie wants to know if it can be shown that Preece once sought to obtain this Telephone from you, & then broke faith with you in the interest of another party— Waterhouse and I are going to thoroughly digest the "Packet" & see—[8] Bouverie says if it can be shown he did—He will "break Preece over the wheel" & stop the PO proceedings— Wouldn't this be a retribution? We will see— I see Preece occassionally—I am to assist him in an exhibition on the 2nd of Jany at the Society of Arts.—about that time Waterhouse & I will be devoting our nights to his downfall

Now listen to a strange tale= E. H. Johnson formerly an Aid de Camp of yours has suddenly acquired untold wealth= for the 1st time in 14 years—a new enterprize in which he was enlisted has actually been a financial success— What is the result?= More bother than it is worth of course—for now this capital must be invested—that it may earn more shekels against the time when the favored possessor may return to his native clime=

You will remember that forseeing this and having an abiding faith in the Electric Light of one Edison I sought to secure an option on some stock—but you would not name a price— preferring at the time to hold out other proffers— now that I am ready to invest I return to the charge— Will you sell me a few more shares even if it is ever so few— I want to buy enough to keep my little certificate from looking lonely

Gouraud paid me 250£ a week after you cabled me yours=[9] He said nothing about deducting anything from this of course, as it is the Lancashire money—he incurred no expense in the formation of that Co— Its all clean gain to him— The Glasgow matter will be different he will want to deduct the 2000£—which he says he had to pay Moore—(who by the way is his own man—& may pay it back again?) to put the Co through=[10] He is doing the same thing in Dublin, Aberdeen Hull &c= Are you going to share this Promoters bill? That is the question—for you & for me— If so—Damned if I dont ask you to send in <u>your</u> bill too—even if it costs me half my interest— I've tasted <u>blood</u>— I'm getting mean—but not so mean but that I am willing to sacrifice a little to beat a still meaner man at his own game—

You will likely hear something from me soon now— I'm equipped with a Telephone that can be relied upon— <u>I</u> may figure a little in the Prints— I've avoided them so far—because when I take them up I mean to do it in the way of a challenge to competition[11]

The Telephone I tried this P.M. I talked all over & about it in any & every voice & it obeyed the helm magnificently—while thus engaged who should <u>happen in</u>—but Sanders[a] the manager of the Bell Co—[12] If I didn't lay it on to him—you may scoop me for a lubber—

Good night— The Ball is now to open—we've only had the overture so far— Yours Ever

E H Johnson

Ans. about the E. L. shares—

ALS, NjWOE, DF (*TAEM* 49:533; *TAED* D7903ZLM). [a]Obscured overwritten text. [b]Interlined above.

1. On the 4 November shipment see Doc. 1839 n. 3; on the others see Doc. 1845.

2. The *Arizona* left New York for Liverpool on 4 November and struck an iceberg off Newfoundland three days later. The ship, the second largest steamer in the world and one of the fastest on this route, was not seriously damaged but put in for temporary repairs at St. Johns, where passengers were transferred to other vessels while the cargo remained on board. "Disasters on the Ocean," *New York Tribune*, 10 Nov. 1879, 1; "The Guion Liner 'Arizona'," *Engineering* 30 (1880): 195–98; Fairburn 1945–55, II:1386.

3. Johnson cabled Edison the next day, "New chalks magnificent crowd instruments and men   pay more for higher grade." At this time the London company began showing considerable concern for the manufacturing progress of telephones and switchboards. Arnold White asked Edison on 4 December to "have me informed of the contents of all shipments, on account of this Company with detailed list of contents of packages. We have been subjected to some inconvenience from the want of this information." On 12 December the company telegraphed "Cable shipments since November twelfth   suffering," to which Edison replied, "231 since Twelfth." (Johnson to TAE, 3 Dec. 1879; White to TAE,

4 Dec. 1879; Edison Telephone Co. of London to TAE, 12 Dec. 1879; TAE to Edison Telephone Co. of London, 12 Dec. 1879; all DF [*TAEM* 52:945–46, 965; *TAED* D7941ZIX, D7941ZIY, D7941ZJI, D7941ZJJ]). The company's concerns were also apparently responsible for a delay in payments to Sigmund Bergmann, on whose behalf Edison twice cabled to London (TAE to Edison Telephone Co. of London, 10 Dec. 1879; TAE to Johnson, 12 Dec. 1879; both DF [*TAEM* 52:952, 964; *TAED* D7941ZJD, D7941ZJH]). Johnson explained that "Of course there has been a little friction about the money matters— The Company now standing as having paid you directly & indirectly some 18,000$ on a/c of Insts & having as yet recd but 275 Insts & 4 switchboards in all= Of course we know there are more than this done—but you see the margin is a large one between what we have recd & what has been paid for— It only needs some little accounting" to distinguish items sent to the London company from those sent to Gouraud for district and foreign exchanges, for which the London company paid indirectly. Before receiving this Edison answered another plea to "crowd everything" by cabling, "Great abundance material gone forward. Gourauds shipping agents here slow. will change" (Johnson to TAE, 16 and 22 Dec. 1879; TAE to Johnson, 22 Dec. 1879; all DF [*TAEM* 52:976, 993; *TAED* D7941ZJQ, D7941ZJX, D7941ZJY]).

4. "Off with his head! So much for Buckingham!" was a summary dismissal frequently quoted from Colley Cibber's 1699 Shakespearean adaptation, *The Tragical History of King Richard III*. The play was widely performed in England until the late nineteenth century. Brewer 1963, s.v. "Buckingham"; McClellan 1978, 49–51.

5. While in New York on 13 November, Joshua Bailey wrote the following "confidential" memorandum to Edison: "Referring to our several conversations and agreements of the last few days, as the complement to them I hereby engage not to push any form of carbon Telephone or microphone in Great Britain." Edison sent copies of this to Theodore Waterhouse and Arnold White, asking that its confidentiality be respected. Bailey to TAE, 13 Nov. 1879; Waterhouse to TAE, 2 Dec. 1879; both DF (*TAEM* 52:916, 448; *TAED* D7941ZID, D7940ZDP1); TAE to White, 18 Nov. 1879, Lbk. 5:362 (*TAEM* 80:190; *TAED* LB005362A).

6. Edward Graves was Chief Engineer of the British telegraph system and Preece's immediate superior. Baker 1976, 176–77.

7. The Post Office contended that as a medium for transmitting and receiving messages over wire, the telephone was a form of telegraphic communication under its jurisdiction.

8. This "Packet" may have been Edison's response to Doc. 1799 and a later request from Johnson to "send me everything bearing on that Hughes Preece controversy." Johnson to TAE, 19 Sept. 1879, DF (*TAEM* 52:728; *TAED* D7941ZEC).

9. Edison cabled this amount to Johnson on 24 November. DF (*TAEM* 52:932; *TAED* D7941ZIS).

10. Gouraud based Johnson's royalty on the £4,500 remaining from the £5,000 Glasgow advance after paying 5% to the London company. Gouraud to TAE, 24 Dec. 1879, DF (*TAEM* 52:1024; *TAED* D7941ZKK).

11. Johnson was referring to circulars or pamphlets distributed by the rival companies, or to news accounts generally. The next week Gouraud

enclosed in a letter to Edison a republication of a Boston newspaper's account of the U.S. telephone settlement which he complained was being "extensively circulated here by the Bell Coy and has been copied by a very large number of prominent papers. It is causing a great deal of comment on all sides." On 12 December Johnson wrote that he was "preparing an exhaustive paper for publication on the subject of the Telephone and Telephone exchanges in America" and asked Edison to answer fourteen enclosed questions. Sometime during the course of the year he published Johnson 1879, a forty-two page booklet on the telephonic inventions of Bell, Gray, and Edison; it did not, however, cover telephone exchanges. A copy is in Force. Johnson also asked for Edison's "views as to the position of the Edison Telephone under the fusion compact. [Gardiner] Hubbard and his party here are making great capital out of the apparent fact that 'Prof Bell is master of the field'   They issue manifestoes to the Public enlarging upon this idea. Knowing as I do the real part played by the Edison Transmitter in making the Telephone a practical success in America, I shall not quietly submit to their sophistical representations." *OED*, s.v. "print" (11a); Gouraud to TAE, 11 Dec. 1879; Johnson to TAE, 12 Dec. 1879; both DF (*TAEM* 52:954, 957; *TAED* D7941ZJF, D7941ZJG).

12. Edison identifies him as Sanford Sanders in Doc. 1871. Nothing further is known about him.

-1859-

From Josiah Reiff

London Decr 4th 1879[a]

My dear Edison

After a severe & patient struggle I have reached a basis for general understanding with Mr Pender and Sir James Anderson regarding the Smith Fleming & Co contract[1] so as to avoid litigation misunderstanding or ill feeling.[2]

Pender you know is chairman of the Globe Trust & Cable Co.[3] which really controls the Anglo Cable,[4] by holding some[b] $5,000,000 stock, which is much more than balance of power—

He is likewise Chairman of Direct U.S. Cable[5] ditto Eastern Tel Co[6] & sundry others & also M.P—a hard headed Scotchman with no pretension but a fearful amt of Energy & work & is acknowledged a leading type of British business influence & ability. He acquired the main portion of S.F. & Co interest & in addition to Expenses attached to developing the Brown Relay[7] & for Cables, is in for some 9,000 pounds sterling.

You may know therefore he is interested in putting things afloat & of course is in position to do it without question.

Of course it is also somewhat difficult to fight him than the Bankrupt concern of S.F & Co, although it was vital to our interests that the contract, Patents etc should get out of the hands of the assigner of S.F. & Co This was not finally done until about 4 weeks ago, by Pender agreeing to pay certain debts of S.F. & Co. & promising a resultant interest to the Estate out of it after he (Pender) should have his money returned—

It was also important to us, to have a changing in the Trustee so far as John Fleming was concerned & I therefore arranged to have Fleming resign his Trusteeship & I named Sir James Anderson in his stead.[8]

These papers have been prepared & Executed. I found these people here on my arrival feeling sore & sour & violently insisting that you had abandoned London in 1873, when your presence here would have vitally aided in consummating the then pending negotiation with[b] the Post Office—that you left with the distinct promise that you would return after completing certain Experiments & instruments which you told them you could do better at home than here—that your failure to return left the P.O. free to make demands without any one here able to meet them, which you could readily have done.

That you distinctly assured them that you could & would perfect the Roman letter system—that you could not only make important improvements on submarine Cables but would promptly & earnestly continue Experiments at home & return here to complete them upon the Cables themselves, for which, facilities were freely promised—etc, etc, etc, etc, etc.

Besides I found they had consulted a number of eminent counsel here all of whom advised them of their rights under the contract etc & amongst many other things they intended pleading your personal promises & failure to keep them by not returning here, as a breach of faith etc. Of course like some other people you & I know they had little or no interest in us, except to get a profit on their money, or to secure its return & hence it made no difference how much you & I might be required here as well as in America by being charged with breach of Contract through Eminent counsel—they (the parties themselves) being leading men in the Telegraph world here.

Of course the unreasonableness of Prescott in the Quad[9] matter did[b] not help things & whilst it looked mighty black for us, I went in to[c] relieve both you and I from any charge of bad faith & from any direct personal responsibility for the return of the money recd.

The elaborate legal proceedings they had originally prepared to be brought by Mess Foster & Thomson[10] of N.Y. I find were hung up through the complimentary and mollifying manner in which Foster & T wrote concerning me. (I have known

F & T for 10 years)   The change of interest however from S.F & Co. to Jno Pender, promised to revive it all & [injuriously?].[d] All this I have succeeded in stopping, besides stimulating the new Element to active cooperation in caring for the future, on terms which I think you will more than approve. The details will be arranged in a few days & that nothing may be left undone I have agreed to remain here another week purposely, although I am fearfully anxious to get home. One thing you will be glad to hear & that is that I am satisfied that Brown did nothing himself with any view to your injury, but as he was trying to accomplish what these people desired on Cables both automatically & otherwise, they insisted he should take out Patents, & they S.F. & Co intended to try & protect themselves against what they considered our failure.

This Brown did not know & of course whatever he did [int?][e] went to the benefit of the parties here.

That this[b] is so, is shewn in that all alleged infringement of Brown on Automatic (some of which EHJ says are good) are turned over to us—

The Relay which has accomplished important results on Cables under 700 miles long ie working Morse direct full speed, will also be arranged for in some way, so as not to be used antagonistically. Brown has no interest in it here, he having assigned it to S.F. & Co & Sir James Anderson.

What I have arrived at & believe I have accomplished under the advise of EHJ & G.E.G. is to remove all controversy here & avoid all antagonisms and at same time secure active & strong aid for the future, whatever that may be after you & I shall have fully consulted upon my return & reporting to you all the details of my efforts here. Truly,

J. C. Reiff[f]

ALS, NjWOE, DF (*TAEM* 51:934; *TAED* D7936ZBH). Letterhead of George Gouraud. [a]"London" and "187" preprinted. [b]Obscured overwritten text. [c]Interlined above. [d]Illegible. [e]Canceled. [f]Signed across "Automatic Cables" written upside down.

1. Doc. 350.
2. Reiff had met with Pender in London and agreed to several alterations of the 1873 contract, most significantly an increase in the share of the London parties and the allotment to them of fully half the proceeds until they had recovered £10,000 (Doc. 1837, Reiff to A. W. Rixon, 12 Dec. 1879; DF [*TAEM* 55:567; *TAED* D8042ZBC]). This agreement evidently anticipated formation of a new company to hold Edison's automatic patents as well as those on the quadruplex, to which Smith, Fleming & Co. made a disputed claim (Rixon to Reiff, 20 Dec. 1879, DF [*TAEM* 51:942; *TAED* D7936ZBI]; see Doc. 1165 n. 5). On 27 Febru-

ary he signed a letter ratifying the agreement negotiated by Reiff (TAE, Reiff, and George Harrington to Pender, 27 Feb. 1880; DF [*TAEM* 55:565; *TAED* D8042ZBB]; Doc. 1935 n. 5).

3. Pender founded the Globe Telegraph and Trust Co. in 1873 as a way to encourage investment in cable telegraphs by reducing the risk of investing in a single company; each share in the Globe Co. included a proportionate share in most of the other telegraph and cable companies. Bright 1974, 121.

4. Reiff presumably meant the Anglo American Telegraph Co., the company that laid the 1866 Atlantic cable and remained one of the major cable companies. Bright 1974, 91, 107.

5. The Direct United States Cable Co. had been founded in 1873 by William Siemens and others to provide competition with the Anglo American Telegraph Co., which had acquired a monopoly over Atlantic cable business the previous year. See *TAEB* 2:566 n. 2.

6. The Eastern Telegraph Co., formed by John Pender in 1872 through the merger of several other companies, was the largest cable company in the world. Bright 1974, 119–20.

7. Edison's former assistant James Brown patented a relay with George Allan in 1876 that was introduced into use on Eastern Telegraph Co. lines two years later and became commonly known as the Brown-Allan relay. Used principally by the Eastern Telegraph Co. and by the Indian Government Telegraphs, it worked with little adjustment and greatly increased the speed with which Morse instruments could be used over a cable, thus making their use practical on lines as long as 600–700 nautical miles. The relay also increased the working speed of other cable signaling instruments. Bright 1974, 587–91.

8. John Fleming and John Puleston were made trustees with power of attorney in connection with the 1873 Smith, Fleming & Co. contract (Doc. 350). In January 1878 they used this power to assign Edison's British quadruplex patent (384 [1875]) to themselves as trustees after Edison refused to execute the assignment himself at the request of Smith, Fleming & Co.'s New York attorneys Foster & Thomson. Smith, Fleming & Co. claimed that the contract gave them rights to all of Edison's telegraph inventions applicable to land lines. See *TAEB* 4:10 n. 5.

9. In late January John Puleston cabled from London that "Every effort has been made and at much trouble, to arrange the matter through the medium of a Company to work the quadruplex but Prescott interposed and we are driven to stand on our rights in that and other respects. Puleston to TAE, 24 Jan. 1880, DF (*TAEM* 55:508; *TAED* D8042I); see also *TAEB* 4:293 n. 3.

10. See Doc. 1209.

–1860–

Notebook Entry:
Electric Lighting

[Menlo Park,] Dec 4 1879

Fault in Lamps[a]

Lamp 118 was on Edisons chandelier last night and <u>all at once</u> it gave out by the top of inside glass bursting out and striking against the top of bulb— On examination

1   we find the leading wires melted completely away
2   Bulb coated black below platina points
3   Top of inside glass coated with black.
4   Track of platina wire in enamel showed as if fused with small globules of platinum scattered along

Lamp No 101 after giving excellent light during exhibition last night[1] played out by reason of the loop falling out of the clamps—loop keeping intact—
On examination I found

1   Black deposit thick below platina points and slightly all over the globe
2

General faults
1   We notice that some of bulbs are perfectly clear after bringing up and using whilst others are covered with black— this may be from the fact that some may be better carbonized than others and those not carbonized are brought to higher heat in vacuo & give out more stuff that coats the glass

Chas Batchelor

X, NjWOE, Lab., N-79-07-31:187 (*TAEM* 33:635; *TAED* N052:93). Written by Charles Batchelor; document multiply signed and dated. [a]Multiply underlined.

1. In a 2 December letter Stockton Griffin told Joshua Bailey that "Edison is now ready for the exhibition but will be compelled to wait 3 or 4 weeks for his patents— Uptons house was lighted last night—Edisons will be illuminated tomorrow night for Mr Fabbri & Mr Wright." In a 5 December notebook entry Charles Batchelor described the 3 December exhibition: "we made a test at Edison's house at which Mr Fabri and party saw it— We lit 2 three light chandeliers 1 two light ditto and ran the sewing machine and hand lamp   all went off perfectly with the exception that I put in one lamp and the wires projected too far and made an arc on brass underneath   this did not hurt the lamp." A few days later Francis Upton wrote his father that "I have had six burners in my house during the past week and illuminated my parlor for the benefit of a party of visitors from New York. The exhibition was a success. Mr. Edison's and my house were the only ones illuminated. I brought the first light out of the laboratory to show Mrs. Perry how good a light we had, which could be of use in a private house. She was very much pleased with it. There will be a great sensation when the light is made known to the world for it does so much more than anyone expects can be done." Griffin to Bailey, 2 Dec. 1879, Lbk. 5:389, NjWOE; N-79-07-31:195, Lab. (*TAEM* 33:639; *TAED* N052:97); Francis Upton to Elijah Upton, 7 Dec. 1879, Upton (*TAEM* 95:580; *TAED* MU038).

Notebook Entry:
Electric Lighting

Lamps.[a]

Took one regular loop and marked the inside edge of it so:—

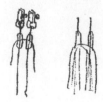

Charles Batchelor's 4
December drawing of the
platinum clamps and
method for sealing the
platinum lead-in wires in
glass.

This I made a standard to pick others out ~~off~~ from the whole—
After picking them out to size I look them carefully over under the microscope and throw away any that have flaws

We now use of course new glasses (inside) and platina clamps[1]

Made 4 lamp Nos. 129-130-131-32[a]

Made 4 more Nos. 133-134-135 136[a]

All these we found no good as in blowing the inside glass it is left too thick so that it cracks with heat—[2]

X, NjWOE, Lab. N-79-07-31:205 (*TAEM* 33:644; *TAED* N052:102). Written by Charles Batchelor. [a]Multiply underlined.

1. The previous day Batchelor had made four lamps (Nos. 125–128) with new platina clamps and inside glass in which the sealed platina wires stuck out of the glass. He had described these as a "great improvement." No test of lamps with these filaments has been found. N-79-07-31: 191, Lab. (*TAEM* 33:637; *TAED* N052:95).

2. The next day Batchelor had Ludwig Böhm blow the inside glass

On 6 December the inside
glass holding the lead-in
wires was made thinner.

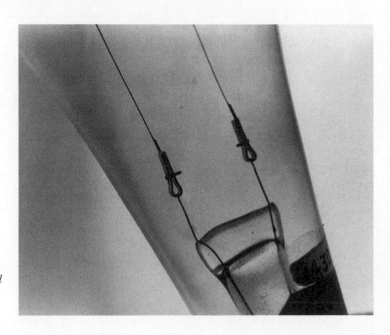

Photograph of the
platinum screw clamps used
in many Edison lamps
through the end of 1880.

thinner; four lamps (Nos. 137–140) were made with this arrangement. No test of lamps with these filaments has been found. N-79-07-31:207, Lab. (*TAEM* 33:645; *TAED* N052:103).

–1862–

Notebook Entry:
Electric Lighting

[Menlo Park,] Dec 8. [1879]

Farradic Machine

No. 3. wound with layer of .035 wire.[1] Resistance .16 Ohm 75 commutator block   Total difference between iron of core and the bore of magnets about[a] .160 inchs. At 750 revolutions with the magnet of such strength that machine No. 2 gave at 600 revo[b] 120 Volts this machine gave 43 Volts where 110 were expected making it a total failure.

The faces of the magnets were ~~v~~apparently very weak as the iron of the armature was brought so near to them as to satisfy the free magnetism.[2]

The ~~place current~~ [cou?][c] neutral point was formed a little below the line drawn at right angles to magnets. This shows that the discharging time of a[d] magnet depends on the amount of wire wound around it.[3]

The magnet were wound with three layer 1.7[d] ohms resistance.

If this machine had succeeded it would have been the grandest success known, as it is, it is only the same as the other machines. It would be useful for supplying a number of small arcs[d] in multiple arc ~~are~~ or for plating or for running magnets. Though nothing to boast of.[4]

X, NjWOE, Lab., N-79-10-18:112 (*TAEM* 32:840; *TAED* N301:58). Written by Francis Upton. [a]Interlined in left margin. [b]"at 600 revo" interlined above. [c]Canceled. [d]Obscured overwritten text.

1. As indicated by the test results below, this was Faradic Machine No. 1 and Francis Upton presumably was referring to the third version of the armature in this machine (see Doc. 1843), rather than the dynamo designed in October or November as Faradic No. 3. N-79-06-16.2:171, Lab. (*TAEM* 35:451; *TAED* N079:84).

2. The term "free magnetism" was surrounded by considerable ambiguity at this time but was often used to refer to the distribution of magnetic action on the surface of a magnet. Upton presumably meant it in the sense in which Edison specified, in a British patent, tight clearances between the armature and pole pieces in order "to prevent the appearance of free magnetic poles." *Ency. Brit.*, s.v. "Magnetism"; Brit. Pat. 2,402 (1879), Batchelor (*TAEM* 92:118; *TAED* MBP017).

3. In November Upton determined that the neutral point, where the current in an induction coil reverses direction as the coil changes direction with respect to the magnetic field, could be found "by taking a wire

and moving the ends on the commutator until little or no spark is seen. Half way between these points will be the place for the commutator brush." Distortions in the magnetic field caused by current flowing in the armature coils shift the neutral point in the direction of rotation. N-79-10-18:111, Lab. (*TAEM* 32:839; *TAED* N301:57).

4. Sometime after this test Upton wrote: "I still believe that an ⅛ of an Ohm machine can be made to give at least 75 Volts. In order to do this a solid iron core must be used and the wire be flat and all be made perfectly." In the following pages he compared Machines Nos. 1–3 in an attempt to find the proportions among their internal resistance, amount of induction wire, distance from armature face to field magnets, and voltage. Comparing the first two machines he found that "the strength of the machine decreases faster than the distance of the face of the core from the face of the magnet" is increased, but this relationship clearly did not hold in this case. Upton made several proposals for reconfiguring Machine No. 3, including an arrangement "so that a steel sheet may be used to keep the coils apart." N-79-10-18:117–25, Lab. (*TAEM* 32: 842–46; *TAED* N301:60–64).

-1863-

Equipment Specification: Electric Lighting

[Menlo Park,] Dec 15th 1879

Make some lamps circular in shape so:—

Inside length 2 inches   make them such a width that they will be 100 Ohms resistance—[1] ~~Leave the circle complete and let current break out piece between clamps~~ break out the piece between clamps

Chas Batchelor

ADS, NjWOE, Lab., N-79-07-31:229 (*TAEM* 33:655; *TAED* N052: 113). Written by Charles Batchelor.

1. This design first appeared in an Edison sketch of 21 October but this is the first evidence of their making "ring" filaments. The decision to use this design may have been influenced by a problem that Batchelor noted in another note dated 15 December, but which clearly dates from at least two days later. This was the breaking of the filaments on the right side of the loop about two-thirds of the way to the top; Francis Upton thought this might be due to the focus of the curve of the glass. In a 3 January note describing a new mould for this ring design, Batchelor noted that it would have "⅔ surface of ordinary 'horseshoe.'" The re-

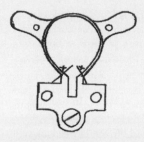

Charles Batchelor's 3 January drawing of a new mold for ring filaments.

duction of surface area was also intended to improve the energy efficiency of the lamp by increasing the number of lamps per horsepower; on 8 February Upton noted of a proposed reduction in the surface area of a "regular" horseshoe filament that "if the number per horse power varies inversely as the square root of the surface the fibre should give 24 where the regular gives 10." The new mould produced ring filaments with slightly different dimensions than those indicated by Batchelor for the design shown here. The 15 December ring filament was to have an inside circumference of 2.81 inches and a width .025 inches, which Batchelor indicated would carbonize down to a circle with a circumference of 2.25 inches and be the same resistance as "present loops" (i.e., the horseshoe filament). The 100-ohm filaments made with the 3 January mould would have an inside circumference of 2.65 inches and a width of .03. These filaments were also to be made with paper that was .0065 inches thick instead of .009 and have short shanks on the bottom ends of the loops for attaching the lead-in wire (see Doc. 1885). N-79-07-31:92, 239; N-80-02-08.1:37; N-80-01-02.2:7, 11, 17; all Lab. (*TAEM* 33:588, 660; 34:222, 723, 725, 719; *TAED* N052:46, 118; N063:20; N070:5, 7, 10).

–1864–

Notebook Entry: Electric Lighting

[Menlo Park,] Dec 15th 1879

Carbonization[a]

In order to get at the bottom of the fact that all do not carbonize alike in same chamber at same time we made following test:—

Cast[b] iron chambers—

3[b] loops in each—

Tissue[b] sheets between each and between loop and iron and[b] a plate of carbon of 126 grains as weight on top[1c]

Resistance

No	top	middle	bottom
1	207	234	217
2	240	255	210
3	212	212—	211

No 3 had 2 sheets of paper between the iron and also between carbon this seems to make it more even—

4

Carbonization:—

Points from which to draw conclusions—

1 Middle one is always highest resistance
2 All lengths of times in Carb. give good ones
3 All heats gives us good ones
4

X, NjWOE, Lab., N-79-07-31:233 (*TAEM* 33:657; *TAED* N052:115). Written by Charles Batchelor. [a]Multiply underlined. [b]Preceded by dash added later. [c]Followed by "over" as page turn.

1. In order to keep the cardboard flat during carbonization, Charles Batchelor had determined a month earlier that "the only way was to put sheets of tissue paper in between the loops (instead of cards as we had been) using and after to alternate layers of loops and tissue sheets to put a small piece of carbon to act as a light weight on top" (N-79-07-31:137, Lab. [*TAEB* 33:609; *TAED* N052:67]). Later in November Edison had Arthur Poinier experiment with using other materials between the cardboard loops but none of these materials, which included plaster of paris, chalk, caustic lime, caustic magnesia, plumbago, silica, and powdered carbon, worked satisfactorily. Poinier and Batchelor also experimented with using different carbon weights. On 5 December, Poinier made two carbonization tests using weights of 9,725 and 9,000 milligrams (a milligram is equal to .015 grain). In the second of these he used three horseshoes of slightly different sizes and noted that the smaller weight seemed to make no difference. Three days later Batchelor, assisted by Poinier, began another series of experiments. In the first of these they carbonized filaments for three lamps (Nos. 141–143), using a weight of 126 grains of carbon. Nos. 144 and 145 were carbonized with 250 grains and No. 146 with 400. No. 141, which measured 295 ohms cold and 157 ohms after heating became the standard for all of the filaments through No. 200. After the first few lamps, which fluctuated between 240 and 500 cold resistance with most 300 or more, the rest of the lamps measured between 230 and 275, with one lamp as high as 285. On 13 December, Batchelor noted of an experiment with plumbago around the edges of the mould that it was 390 ohms but "ought to be 250," which was apparently their preferred cold resistance. The same day he noted that "the tarry matter must not be driven out quick but left in if possible to lock the particles together in a hard shiny mass, therefore the first heating ought to be very slowly" (N-79-07-31:225, Lab. [*TAEB* 33:653; *TAED* N052:111]). On 17 December he measured the resistance of filaments in a carbonizing mould after various times and methods of heating. N-79-07-31:135–37, 161, 165, 181, 203, 207–11, 219–27, 231, 241, 245–47, 253–57, Lab. (*TAEM* 33:608–9, 621, 623, 632, 643, 645–47, 650–54, 656, 661–64, 667–69; *TAED* N052:66–67, 79, 81, 90, 101, 103–5, 108–12, 114, 119, 121–22, 125–27); for records of many of these lamps, which were burned in various laboratory buildings, residences, and street lights at Menlo Park, see N-80-1-02.1 passim, and N-80-01-02.3 passim; both Lab. (*TAEM* 34:465, 35:2; *TAED* N067, N074).

–1865–

To Edward Johnson

MENLO PARK, N.J.[a] Dec 17th 1879

Phonos[1]

Exhibition ready. capitalist wont allow it until about New Years[2]

T A Edison

L (telegram), NjWOE, DF (*TAEM* 50:179; *TAED* D7919ZDW). Written in an unknown hand. [a]Place from Edison's laboratory handstamp.

1. Edward Johnson's registered cable code name.

2. The source of this refusal has not been determined. However, on 26 December Egisto Fabbri cautioned Edison about

> fixing the day for a public exhibition of your great invention— I am much older than you are, and a friend, and therefore I feel sure you will not take it amiss if I suggest to you the wisdom & the business necessity of giving the whole system of in-door & out of door lighting a full test of continuous work for a week, day & night, before inviting the public to come & look for themselves.
>
> As long as you are trying private experiments even before 50 people a partial failure, a mishap would amount to nothing, but if you were to express yourself ready to give a public demonstration of what you considered a complete success, any dissapointment would be extremely damaging,—probably more so than may appear to you as a scientific man. . . . I only write to make sure that for your own sake as well as that of those interested with you every precaution is taken to insure the success you so well deserve. [DF (*TAEM* 50:297; *TAED* D7920ZBO)]

–1866–

Charles Batchelor to Edward Johnson

Menlo Park NJ. Dec 17th 1879

My dear Johnson.

Your letters of Dec 5 and 6 received and immediately cabled you "much more than London, Lancashire Glasgow."[1] I shall send you Edisons this morning sometime—[2] I mean by mine that we have got over all three of the difficulties that you suggest and have it now absolutely in hand—[3a] I have lighted up Edison's house brilliantly, and run his sewing machine also worked his pump for tank all at same time— Have taken off all [the?][b] lights but one & have stood on Laboratory verandah & put the whole place in darkness or lighted it up instantly at will by a switch   [It?][b] has been seen by our own parties and the consequence is you cannot buy a share of the stock for $1000—[4] The economy is perfect far more than we ever thought— If you could sbut see the lamp—so simple so cheap[a], so good. I am not going to give it away[c] as yet[c] but you shall know at [   ][d] the [subdivision?][b] complete in [two?][b] lamps [on a table?][b] in shop [afterwards?][b] switch [-----][b] houses and stores & field lamps and let them see for themselves   You know I am not over sanguine and not given to be in raptures over anything but I tell you that this is a very big[e] thing and we have it right in our hand   If you want any information I will do everything[c] I can— If you should want to cable me cable Batchelor

New York I will register there today I think G particularly mean in these matters— I pay for these cables so that Griff or any one else will not see them

Batchelor

Boy waiting to take this so excuse

ALS (letterpress copy), NjWOE, Batchelor, Cat. 1330:58 (*TAEM* 93:272; *TAED* MBLB2058). [a]Multiply underlined. [b]Faint letterpress copy. [c]Obscured overwritten text. [d]Four lines of illegibly faint text. [e]"very big" multiply underlined.

1. Johnson's letters have not been found. This is the complete text of Batchelor's 17 December cable. DF (*TAEM* 52:980; *TAED* D7941ZJR).

2. Edison's cable is Doc. 1865.

3. Batchelor had previously told Johnson that at the 3 December private demonstration (see Doc. 1860 n. 1) he had "lit Edisons house up brilliantly the other night to show to a few of our people— It was perfect worked the sewing machine also, with a hand lamp to see to sew by— It was a grand success The system now is far more complete and simple than we ever expected." Batchelor to Johnson, 14 Dec. 1879, Cat. 1330:56, Batchelor (*TAEM* 93:270; *TAED* MBLB2056).

4. On 24 December Calvin Goddard confirmed the sale of ten shares of Charles Batchelor's stock at this price (most of the proceeds from which Batchelor used to buy U.S. government bonds) and indicated that he expected to get $2,000 apiece for five additional shares on the morning after Christmas. Francis Upton reported that the stock sold for $3,500 on 27 December and the next day he accepted an option to sell five of his shares at $5,000 apiece. Only a handful of sales were actually carried out around this time and the *New York Sun* reported at month's end that investors estimated the actual value of a share at $10,000. By contrast, at the end of November shares sold for $650 each and had been as low as $20 during the summer. Goddard to Batchelor, 24 December 1879, Batchelor (*TAEM* 92:392; *TAED* MB054A); Upton to Elijah Upton, 28 Dec. and 30 Nov. 1879, Upton (*TAEM* 95:586, 577; *TAED* MU041, MU037); "Edison Electric Light Stock," *New York Sun*, 30 Dec. 1879, "The Electric Light in America," *Leeds Mercury*, 13 Jan. 1880, Cat. 1241, items 1404 and 1431, Batchelor (*TAEM* 94:559, 568; *TAED* MBSB21404X, MBSB21431X).

–1867–

John Kruesi Memorandum: Preparations for Electric Light Exhibition

[Menlo Park,] Dec. 20th 1879

321[1] To be attended to immediately[a] & completed befor December 31st 79[2]

I Faradic machine No. 4[3]

II 100 lamp cups[4]

III 100 pair of lamp clamp Platinum[5]

IIII gear up pumping motor

V switch for Central station regulator[6]

VI fit up 15 double burner chandeliers

VII   fit up the meter complete[7]

VIII   10 Street lights[8]

IX   double board sidewalk on the sqare to Depot

X   get Kerite wire for street lamps

XI   Central station cleaned oilclothed & fixed up.

J[ohn] K.[ruesi]

ADS, NjWOE, Cat. 1308:203, Batchelor (*TAEM* 90:768; *TAED* MBN003:75). Written by John Kruesi. ᵃObscured overwritten text.

1. This is the order number; see Doc. 1687 n. 1.

2. John Kruesi had already made preparations for the manufacture of a substantial number of lamps. On 2 December he ordered the shop to "Make 4 Glassblower tables with balows same as the wooden one in use" and to "Make 4 Carbonizing chambers of wrought iron." On 10 December he requested "an instrument for measuring resistance of carbons," and that day or the next he called for the shop to "Have 12 muffles made for us" for carbonizing. Cat. 1308:197, 199 (Order Nos. 304–5, 307–8), Batchelor (*TAEM* 90:765–66; *TAED* MBN003:72–73).

3. Kruesi gave instructions for the construction of this machine on 13 December. He specified "9.62 inches bore of magnets   Make armature 9.05 diameter of iron wire   Wind 4 layers in threes of .042 wire 39 coils six wire in width the same as No. 2   Measure carefully when made." Later he added, "Make wooden core 6¼ in Diameter   order for T.A.E." Cat. 1308:201 (Order No. 313), Batchelor (*TAEM* 90:767; *TAED* MBN003:74).

4. Presumably a base or holder. The type drawn by Samuel Mott in November (see Doc. 1838 n. 5) was used during the exhibition. "Edison's Light," *New York Herald*, 21 Dec. 1879, Cat. 1241, item 1379, Batchelor (*TAEM* 94:537; *TAED* MBSB21379X).

5. Charles Batchelor had already requested on 15 December "200 platinum clamps for lamps like previous ones," presumably those discussed in Doc. 1861. Cat. 1308:201 (Order No. 316), Batchelor (*TAEM* 90:767; *TAED* MBN003:74).

6. It is not clear exactly what this switch was. Edison referred to his regulator as a rheotome, an instrument for interrupting an electrical current (*OED*, s.v. "Rheotome"). It consisted of a number of buttons or contact points which could be rotated to provide more or less contact with a series of resistance coils, and an index wheel. He later testified that it was first used at the laboratory in June or July 1879 and then, in preparation for the demonstration, it

> was placed in the second story and the handle of the rheotome was connected by a rod to a table in the first story where the galvanometer indicating the electric pressure upon the system was placed. Placing the coils of wire in one story and means for indicating the pressure in another story was for the purpose of preventing any action on the galvanometer of the current which passed through the coils. The coils and this rheotome were interpolated in the circuit of the field of force magnets and a greater or lesser number of coils could be thrown in the circuit by moving the arm of the rheotome, thus varying the strength of the current in the consumption circuit

to meet various conditions. [Edison's testimony, 15–16, *Keith v. Edison v. Brush*, Lit. (*TAEM* 46:121–22; *TAED* QD002:8–9)]

A mirror galvanometer indicated voltage in the main circuit and a laboratory assistant made adjustments as needed by "turn[ing] an indicator, near which is an electric light, just as they do in a gas house when more or less pressure is called for by the consumption" ("Electricity and Gas," *New York Herald*, 30 Dec. 1879, Cat. 1241, item 1401, Batchelor [*TAEM* 94:555; *TAED* MBSB21401X]).

On 19 December Edison drafted a provisional British specification and a U.S. caveat for regulating electrical generation by means of resistance coils and a rotating commutator (not referred to as a rheotome). He filed British Patent 33 ([1880], Batchelor [*TAEM* 92:146; *TAED* MBP022]) on 3 January; the caveat has not been found but Edison testified that it was Caveat 94, dated 26 December. Edison's testimony, 14, *Keith v. Edison v. Brush*, Lit. (*TAEM* 46:121; *TAED* QD002:8); Edison's draft is in Cat. 1146, Lab. (*TAEM* 6:650; *TAED* NM014U); William Carman copied this into Cat. 997:127, Lab. (*TAEM* 3:415; NE1695:64).

7. On 2 December Kruesi instructed the shop to "Alter the electric Light meter" but gave no details. Cat. 1308:197 (Order No. 303), Batchelor (*TAEM* 90:765; *TAED* MBN003:72).

8. On 2 December Kruesi had ordered the shop to "Have street lamp posts made & make sketch for same"; the sketch has not been found. Cat. 1308:197 (Order No. 306), Batchelor (*TAEM* 90:765; *TAED* MBN003:72).

–1868–

To Lemuel Serrell

[Menlo Park, December 21, 1879[1]]

⟨Deliver today—⟩[2]

L. W. Serrell

When was last lamp patent sent to Brewer   will it reach there in time to save patents on Continent, in view of Heralds publication today.[3] Undoubtedly main points of Herald article be telegraphe[d][a] tonight.[b] answer[4]

T. A. Edison

ALS (telegram), NjWOE, DF (*TAEM* 51:621; *TAED* D7929ZGD). [a]Written off edge of page. [b]Obscured overwritten text.

1. An archivist dated this document as 22 December but the reference to the *New York Herald* article (see Doc. 1869) indicates the proper date.

2. This date being a Sunday, Edison addressed the telegram to Serrell's home in Plainfield, N.J.

3. At 11:30 in the morning Edison had cabled Brewer and Jensen in London, "Hurry forward on continent last two lamp patents as Herald published description today" (TAE to Brewer & Jensen, 21 Dec. 1879, DF [*TAEM* 51:620; *TAED* D7929ZGC]). The article by Edwin Fox filled more than an entire page in the Sunday *New York Herald* and offered the first "full and accurate account of his [Edison's] work from its

inception to its completion." After a brief explanation of the carbon filament, it summarized Edison's research over the preceding fifteen months, making substantial references to Doc. 1796 and various patents. Returning to the carbon filament, the article described what it called the "GREAT DISCOVERY" in October—rolled lampblack (see Doc. 1818 n. 1). It then successively described the thread and cardboard lamps, the manufacture of cardboard filaments, the generator, motor, and meter. It also discussed polyform and, briefly, Edison's gold ore milling plans, and concluded with a selective historical sketch of electric lighting before Edison ("Edison's Light. The Great Inventor's Triumph in Electric Illumination," *New York Herald*, 21 Dec. 1879, Cat. 1241, item 1379, Batchelor, [*TAEM* 94:537; *TAED* MBSB21379X]).

The same day the *Herald* published an editorial congratulating "not merely Edison, but the people of all civilized nations, upon Edison's success." It ran another laudatory editorial the next day but its enthusiasm was not universally shared. The *New York Tribune* proclaimed that Edison had produced "merely a modification of the Sawyer-Man light. . . . There is no new principle involved, therefore, in Mr. Edison's light, and the modifications he has made . . . are but a poor return for fifteen months' labor." "Edison's Eureka—The Electric Light at Last," *New York Herald*, 21 Dec. 1879; "A Lucky Horseshoe," ibid., 22 Dec. 1879; *New York Daily Tribune*, 22 Dec. 1879; Cat. 1241, items 1378, 1380, and 1387, Batchelor (*TAEM* 94:537, 542, 549; *TAED* MBSB21378X, MBSB21380X, MBSB21387X).

4. Lemuel Serrell wrote to Edison the next day that he had immediately wired back, "Papers sent Thursday week" (11 December) but had since determined that the most recent specifications were mailed for translation to agents in France and Germany on 10 December, with final instructions going five days later. He also noted that "the papers for England were sent Dec 3, with instructions to complete, so that country is all right"; the provisional specification for British Patent 5127 (1879), Cat. 1321, Batchelor (*TAEM* 92:138; *TAED* MBP020) had been filed on 15 December. Serrell added that he thought the *Herald* article unwise but "I do not suppose that sufficient information will be cabled to allow the invention to be worked and the Herald cannot get there earlier than by the steamer leaving on Wednesday. I do not see that anything can be done to avert the risk that has arisen." Information about the carbon lamp was published in England on 27 December, before some of the Continental patents were filed. Serrell later advised that the *Herald* "should be sued for damages, if any injury result, it was a betrayal of your confidence." Serrell to TAE, 22 Dec. 1879, 26 and 30 Jan. 1880, DF (*TAEM* 51:622; 55:14, 19; *TAED* D7929ZGE, D8036K, D8036P).

-1869-

Francis Upton to Elijah Upton

Menlo Park Dec. 21. [1879]

Dear Father:

Today has been quite exciting here since the mornings Herald contained an account of the discovery of the lamp and the whole invention. Mr. Edison had allowed a Herald reporter to take full notes so as to prepare his account for the exhibition which was to come off in a few weeks.[1] The reporter was Edison's friend and he thought he could keep a secret. Yet newspaper traditions were too strong and he sold out at a good price I suppose for he had the first full account. Mr. Edison is very much provoked and is working off his ~~sur~~ surplus energy today.[2] The light is very fine and promises much money, yet all is promise.

I have my parlor lighted very beautifully with it. Lizzie has fixed the lamps with ribbons and flowers and I think I am a peg ahead of any one in the show, for all the other lamps are to be put in gas fixtures and follow old customs.[3] Edison thinks people will like the gas fixtures best as they can thus better compare the lights given.

Lizzie is going to leave here Tuesday morning and New York 11 A.M. She will spend the night with you and go on the next day to Brunswick.[4]

I hardly think I shall go home as this is the chance of my life. I honestly think Mr. Edison has struck a great thing in the light as we have it here today. Very much love   Your Son
Francis R. Upton.

ALS, NjWOE, Upton (*TAEM* 95:584; *TAED* MU040).

1. Fox's *Herald* article (see Doc. 1868 n. 3) referred to the fact that the demonstration was "announced to take place on New Year's eve at Menlo Park."

2. There is no correspondence related unambiguously to Fox's arrangements for the article; however, it is possible that the *Boston Herald's* telegraphed inquiry on 29 November asking if "Fox received our acceptance of his terms" may pertain to his preparations. Edison may have felt "provoked" but did not hold a grudge against the *Herald*, which published several other reports from Menlo Park, presumably also by Fox, before the end of the year. Boston Herald to TAE, 29 Nov. 1879, DF (*TAEM* 49:745; *TAED* D7906ZBO); see headnote, p. 539.

3. Upton had three lamps in his parlor, the central one being so decorated as to form what a news account called "a much handsomer chandelier than the most ornate of metal ornaments." The lamps were not placed in a cup or base but simply hung upside down from the wires. When Upton took the reporter outside at night to see how the wires entered his house he "brought with him one of the lamps, and, when he reached the wires, he simply hung the lamp to them, thus making the necessary connection, and it was instantly lighted." When the reporter filed another dispatch from Menlo Park in early January he cited as proof of the electric lamp's durability "the central light in Mr. Upton's parlor which I mentioned in my former letter as having been hung upside down by wires from the ceiling, the wires themselves being entwined with flowers and grasses. I have good reason to believe that it has not been disturbed during the week, for shadows of overhanging blades of grass, &c., fall on the same parts of the wall of the room that they rested on last Sat-

urday." "Edison's Electric Light," *Philadelphia Public Ledger and Daily Transcript,* 29 Dec. 1879, and "Edison's Electric Light," ibid., 7 Jan. 1880, Cat. 1241, items 1399 and 1425, Batchelor (*TAEM* 94:553, 564; *TAED* MBSB21398b, MBSB21425X).

4. Brunswick, Maine, site of Upton's alma mater and evidently Elizabeth Perry Upton's hometown.

–1870–

From Edward Johnson

London—Decr 23/79—

My Dear Edison

I am to write you to prepare your mind for a supplication which will shortly be transmitted to you by the Board— It is this= A call upon you to visit London— Hold on now dont rear too high—you may fall over backward—

The Phantom has appeared with increasing frequency during past two or[a] three months—until now it has assumed a material form & there it stands—in bold relief—& won't be put down— You may rest assured I have represented the great difficulty of getting you here in the strongest possible way—and the inconvenience to you, just at this juncture is fully appreciated by the Board—but nevertheless the stubborn fact remains that your presence is of vital importance <u>to save your British Patent</u>=[1]

I will try & depict the situation—

Waterhouse—Johnson Johnson & Co.[2] have Exhausted themselves in an Effort to reorganize your Patent so as to put it in fighting shape—

Their work is done—the Disclaimer has been applied for—

Mess Webster—& Benjamin—the brightest lights in this firmament (Benjamin of the late C[onfederate].S[tates of]. A[merica],)—have examined the work[b] and it is[c] pronounced by them perfect in all its details—

The next move to get an appointment for hearing before the Attorney General— This is the <u>weak</u> point= The Bell Co and 1 or 2 others users of microphones are opposing us—tooth— foot & nail— They have filed their objections to Our Disclaimer—and are now engaged in Efforts to <u>Postpone</u> the <u>hearing</u>= We must have a pretext for demanding immediate attention (ie—a hearing in January)=[3]

The objection of ~~the~~ Our Opponents is to the effect that Our Disclaimer <u>if allowed</u>= will cover the microphone—an admission that they are operating solely upon the fact that our Patent as it stands is a <u>bad</u> one— The Disclaimer will make it a <u>good</u> one= Consequently it must not be allowed—

There is a precedent for refusing a disclaimer under circumstances where its allowance will annihilate large interests.— ~~by reason of atheir delay where~~[d] ~~of~~ the parties (seeking to defend the Patent) [~~by?~~][e] delay[d] applying for ~~such~~ their[d] disclaimer, in the face of the fact that large interests ~~were~~ are growing under their notice—by virtue of the supposition that no infringement was being made—

It is feared that the Bell Co will plead this to some purpose to defeat the plea— we must be able to show several things

1st   That we have showed due dilligence in seeking to amend Our Patent—

2nd   That the Bell Co. have only within a very short time sought to introduce the thing which infringes us—

3rd   That inasmuch as we had no receiver in England until the new loud one came out—we were unable to organize a capitalized interest—to defend the Patent— That as soon as we had done so we immediately began proceedings for our defence—

It is now pretty clear that the P.O. people will be able to bring a perfect cloud of witnesses against us ~~i~~on the question of whether or no the Telephone is a Telegraph— These witnesses are already visiting us under the guise of Public inspection for the Gov't— Thus they become arraigned against us—and doubtless are made to feel—by Preece & his crowd—that we are interlopers on Bells territory and should be suppressed— This is only a feather it is true—but from the composition of the crowd—I can see they are ripe & ready witnesses against   Of course the Bell people will avail themselves of them= In point of fact Every man who wants to curry favor with the P.O. will help the Bell Co in opposing Our disclaimer— Now what have we to array against them—

1st= Sir Wm Thomson—is already under promise to help them

2nd   Tyndall will probably fight shy of us because we are not popular with the scientists—

I am practically alone—and am handicapped by being unknown and being in the employ of the Co—

The P.O. people are pursuing us & <u>not</u> the Bell Company—

All our Dist. Co. organizations have come to a Dead halt because of this Gov't raid[a] upon us—

Now what is to be done—

~~1st~~   We must get our Disclaimer—and thus compel the Bell Co to come into our fold at our own price—or at all Events to unite with us as against the P.O.[4]

The chances of our doing this are at present slim—

This is the status—& it has come to be conceded by Everybody that Every weapon in our hands must be used or we shall fail— Our strong card, is the one that will enable us to force a hearing now,[f] before the Atty Gen'l. The only way to do this is to import you—& then notify the Att'y Genl that you are here for the purpose— Benjamin says this will ensure us an immediate hearing— Then again Waterhouse feels that in the arguments against the microphone you could give us incalcuable aid.

so my Dear fellow it comes to this—

In America you were deserted   In France you were outwitted   In England you are defended but you are asked to lend the weight of your personal presence and aid in the defence—

We mean to have one Telephone Co floating[a] the Edison flag— If not it will be a confession that your Carbon Discoveries are all nil—

As the time for the crucial test approaches your capitalists Everywhere seem to weaken— In America it was amalgamation— In France they only glanced at the Patents   In England they have been thoroughly examined—& yet as the judgment Day approaches the Capital takes fright—and I do not feel at all safe, in view of the fact that the Bell Co are so strongly aided by the P.O. and at the same time keep constantly dangling the amalgamation Temptation before our eyes— I Cannot alone support all these weak knees—you must come & help me—If you have any faith in the issue yourself

I want you to write me a Letter of Introduction to all the Big Guns of Science on this side whether English French or German—whom you have reason to think will lend you a helping hand in this matter— Even a Letter to Sir Wm T. from you asking him to give you the benefit of his opinion ion the matter—will be of value—it would restrain him from going too far on the other side—

Cable me the names of a few to whom you will send me introductory letters—that I may put myself in communication with them in advance by simply saying "you will write so requesting"=

We must act promptly—or we shall be beaten— Yours Truly

                                                          E. H. Johnson

ALS, NjWOE, DF (*TAEM* 52:1010; *TAED* D7941ZKE). [a]Obscured overwritten text. [b]"have examined the work" added at end of line. [c]"it is" interlined above. [d]Interlined above. [e]Canceled. [f]Multiply underlined.

1. A few days later Johnson proposed an electric light and telephone demonstration for "Every Important scientist in England or France" if Edison came to London, and he asked Charles Batchelor to "Put him up to it" (Johnson to Batchelor, 28 Dec. 1879, DF [*TAEM* 49:583; *TAED* D7903ZML]). At the end of December Edward Bouverie and Arnold White separately urged Edison to come (Bouverie to TAE, 30 Dec. 1879; White to TAE, 31 Dec. 1879; both DF [*TAEM* 52:1036, 1041; *TAED* D7941ZKQ, D7941ZKS]). White told Edison that the telephone company's directors wanted him to attend the patent disclaimer proceedings because

> The clear intention of our rivals being indefinitly to prolong the proceedings before the Attorney General, nothing could defeat this so well as to be able to state that the inventor himself was in England for the express purpose of supporting the motion for a disclaimer.
>
> The weight of your name, and the English love of fair play, would then go far to promptly secure a result, which without your assistance is not unlikely, otherwise, to be adverse to our common interest.

White also enclosed a letter from the company's attorneys advising that "if the disclaimer is granted, the Edison Telephone Company of London will be at once in a position to take proceedings with every prospect of success against all persons (the Bell Company included) who are using any form of such transmitter, whether known as Blake's, Crossley's, or by any other name. In fact this is virtually conceded" (Waterhouse & Winterbotham to Edison Telephone Co. of London, 22 Dec. 1879, DF [*TAEM* 52:998; *TAED* D7941ZKB]). On 1 January Bouverie wrote again that he was "doing all that is possible to ensure your being here at the nick of time" and had approved sending Johnson to explain the situation in person (Bouverie to TAE, 1 Jan. 1880; Painter to TAE, 15 Jan. 1880; both DF [*TAEM* 56:339, 53:42; *TAED* D8049A, D8004R]). Johnson was in the U.S. by mid-January and may have carried this letter himself. Edison cabled to London 16 January, "Very desirous but utterly impossible." At the end of January he wrote to both White and Theodore Waterhouse that it was impossible for him to come over and that Johnson would explain his reasons; these letters were carried by Johnson, who left for England about this time (TAE to Edison Telephone Co. of London, 16 Jan. 1880; TAE to White, Jan. 29 1880; TAE to Waterhouse, 29 Jan. 1880; White to TAE, 26 Feb. 1880; Waterhouse to TAE, 21 Feb. 1880; all DF [*TAEM* 56:348, 371, 373, 454, 438; *TAED* D8049E, D8049Z, D8049ZAA, D8049ZAZ, D8049ZAT]).

2. Johnson is probably referring to a partnership of attorneys John H. Johnson and James Johnson. Letterhead of John H. Johnson to TAE, Feb. 1880, DF (*TAEM* 55:706; *TAED* D8046ZAE).

3. This was the opinion of the company's attorneys, who the day before recommended to Arnold White that "It would be a strong lever in our endeavours to get an early appointment if we could say that he had come and was waiting in this country to get the disclaimer passed."

Waterhouse and Winterbotham to Arnold White, 22 Dec. 1879, DF (*TAEM* 52:998; *TAED* D7941ZKB).

4. The Bell company had reportedly been negotiating with the Post Office for official recognition and advantageous concessions but Arnold White succeeded in breaking off these talks. He subsequently obtained a promise from a Bell director that his firm would cooperate in opposition to the Post Office. Johnson did not profess much confidence in this pledge but predicted that if the Post Office could be forced to include Bell in its lawsuit then "it is probable we may make an alliance with the Bell Co against the P.O. = If we do it will naturally lead to an amalgamation of some sort— That is a matter which shall have to be handled very gingerly   I want to see the Patent Question settled first— Then we shall be in a position to negotiate." The government ultimately brought suit against both companies. White to Johnson, 21 Dec. 1879; Johnson to TAE, 28 Dec. 1879; both DF (*TAEM* 52:986, 1027; *TAED* D7941ZJW, D7941ZKM); "Divisional Court," *Times* (London), 21 Jan. 1880, 4.

–1871–

Draft to George Gouraud

[Menlo Park, December 24, 1879?[1]]

Col G

My pats on Tel[ephone][a] in this country were sold to WU Co over 4 yrs ago[2] and I had n[o]t[hin]g do with this Consol[i]d[ati]on of interest   Had I controld my[b] pats in this Co[untry] the Bell Tel Co wld not now be in existence as they have never been able to use the Tel practily in cities without aid of my carbon transmitter wh was pirated in this Country by Bell Co in retaliation for the alleged[b] piracy on the part of the WU of the Bell rec[ei]ve[r]   ~~the WU Co~~ these 2 piracies wld necessitate long & ex[pensive] law suits hence the W.U. wh has enormous biz of its own concluded that Consold of int[erests] was best and a Consold was made in wh they rec'd as a royalty on my pats a percentge of the gross proceeds[3]   as you well know none of pats of Mr Gray or Phelps of any value[c] were used. The Bell Co of Eng[lan]d are playing same game in Engd pirating Carb Tel under name B[lake] trans but in our case we do not use Bell recevr hence we are not in position of pirates or amenable to any of the results therof

TAE

Pers[onal]

I have been approached here on part of Mr Sanford Sanders[b] or Stamford[4] who is here f[ro]m London for Bell Co looking towards amalg[amation] of the Cos in Engd

I replied had ntg do with it if I had wldnt consolod with pirates no matter how much money I might make   that I at one time was using the Bell recvr in L[on]d[o]n because I did not believe pat[b] was valid but upon notfctn Bell Co that they consid it an infringmt of their pat I stopd its use & inventd a sub (moto[graph]) free fm any claims of th kind and as I had done the legit[imate] thing I didnt propose as far as my int was concerned to consold with lot pirates

TAE

ADfS, NjWOE, DF (*TAEM* 52:1017; *TAED* D7941ZKG). Letterhead of T. A. Edison. [a]"on Tel" interlined above. [b]Interlined above. [c]"of any value" interlined above.

1. Stockton Griffin wrote at the top of the first page: "Bell amalgamation   wrote this to Col Gouraud Dec. 24 79."
2. Edison had agreed on 14 December 1875 to assign to Western Union his patents related to acoustic telegraphy, the telegraph technology that led to the development of the telephone. On 22 March 1877 he agreed to assign to the company all his inventions that could be used on telegraph lines and on 31 May 1878 he formally assigned his telephone transmitter patents. See Docs. 695, 876, 1257, and 1345.
3. See Doc. 1822 n. 13.
4. See Doc. 1858.

–1872–

From Jacob Rogers

Boston.[a] 27th Decr 79

Dear Sir

Since I had the pleasure, as one of the party on Tuesday Evening last, of seeing you, I have thought much of your wonderful light, and of all that through your kindness, I saw.

As usual, however, one, on such occasions, is very apt to forget a part of what he wishes to say—and, if therefore you can spare a moment to answer one or two questions, I shall esteem it a great favor.

1st. From the experience that you have had thus far, or from your knowledge as a scientific man, is there any doubt as to the imperishability of the carbon wick? ⟨No change after 208 hours burning cannot conceive how it can be destroyed⟩

2c. Do you apprehend that the Expansion & contraction, consequent upon the heat generated, will eventually destroy the glass globe, & thereby admit the atmosphere to the destruction of the carbon? ⟨Expansion & contraction slow, this is cannot tell⟩

3c   But, supposing there is a limit to the life of the carbon wick, Either from inherent cause, or from those mentioned in my second question, what provision if any, is necessary to overcome this difficulty? Can fresh lamps with new carbon wicks be substituted? Yours very truly

Jacob C Rogers[1]

P.S. I am a personal friend of Mr. Fabbri's, and a stock-holder of your company J.C.R.

⟨It is these points that we are determining so far there are 21 lamps that have burned 152 hours without change and 3 lamps 208 hours without change. 2 lamps cracked in 20 minutes after being connected, due to not be[ing] annealed 1 lamp had its carbon broken at clamp and 1 lamp in which the carbon[b] broke in centre probably due to flaw. thats[c] the record so far ~~we~~ the men[d] have not reached perfection in making the lamps yet. Extra lamps could[c] could ~~they~~ if found necessary be kept in hand like chimneys and placed in the sockets. we have a new socket so a child can replace a lamp.[2] [----- ------ -----][e] ~~carbons~~[f] [----][e]⟩[3]

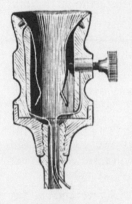

The lamp socket as shown in the 27 December New York Herald.

ALS, NjWOE, DF (*TAEM* 50:190; *TAED* D7919ZEF). [a]Preprinted. [b]"in which the carbon" interlined above. [c]Obscured overwritten text. [d]"the men" interlined above. [e]Canceled. [f]Interlined above.

1. Jacob Rogers was an agent for J. S. Morgan & Co. in London. Sampson, Davenport, & Co. 1880, 828.

2. The lamp socket was illustrated in the 27 December *New York Herald* article on "Edison's Light" (see Doc. 1868 n. 3). Although this design is somewhat different from the arrangement shown in a patent application that Edison executed a month later, both appear to employ metal springs to make electrical contact between the lamp and the socket (U.S. Pat. 265,311).

3. According to the docket on Rogers's letter, Edison sent his answer, for which this is presumably the draft, on 1 January 1880.

PUBLIC ELECTRIC LIGHTING EXHIBITION
Doc. 1873

Edison told reporters that he expected to begin the public demonstration of his Menlo Park electric lighting and power installation on 29 December.[1] Two days beforehand he opened the laboratory complex, illuminated by forty lamps, to a group of financiers and newspapermen who stayed until late in the evening. This resulted in the first published descriptions of the entire system in operation (Doc. 1873).[2] These accounts had a dramatic effect. The *New York Herald* reported that on Monday, 29 December, "the afternoon trains brought some visitors, but in the evening every train set down a couple of score at least." Some "expressed disappointment that they did not step out of the train upon a scene from fairyland" even as Edison hurried to complete another generator to run more lamps. The next evening "the depot was overrun and the narrow plank road leading to the laboratory became alive with people." This prompted the *Herald* to declare that although Edison "has fixed no day for a general inspection by the public at large, desiring not to be completely overrun with visitors to the impediment of his work, the public exhibition may be said to be actually going on now, as many hundreds have already come and hundreds more are coming, and no impediments are thrown in the way of those who desire to satisfy their curiosity."[3] Uriah Painter came on 30 December with a party that included a U.S. senator, and was the first to sign a "Visitors Register" in the laboratory.[4] On New Year's Day the railroad "ordered extra trains to be run and carriages came streaming from near and far. Surging crowds filed into the laboratory, machine shop and private office of the scientist, and all work had to be practically suspended. . . . [T]he people came by hundreds in every train" and their combined weight "more than once threatened to break down the timbers of the building." One assistant later recalled that laboratory workers bolstered the structure with spare telegraph poles and heavy lumber.[5] Edison planned eventually to "make an actual test of my eighty horse-power engine and to put on every lamp it will run. I hope to have 800 of them. . . . It will take weeks to get them all out, though we are working night and day."[6]

Edison reportedly looked "as if he had really worked half his life out in searching for this electric light, and was ready to sink into a premature grave" but he availed himself to reporters and distinguished guests. He subjected lamps to various trials such as immersion in water, and also demonstrated an electric motor. Laboratory assistants were on hand to answer questions and keep watch over the equipment. Despite these precautions accidents and some vandalism occurred, including the theft of several lamps and an attempt to short-circuit the system. The *New York Sun* reported that on New Year's Day "Even kind, patient, and good-natured Edison was angry before the day ended," and the next day even though his "good humor had returned . . . there was much greater care and prudence in admitting visitors to the laboratory."[7] Although Edison gave orders on 2 January to close the laboratory buildings to the general public and resume work, hundreds of visitors still came to see the illuminated windows and outside lamps.[8]

1. "Edison's Horseshoe Light," *New York Sun,* 23 Dec. 1879; "Edison's Lamp Yet Burning," ibid., 24 Dec. 1879; Cat. 1241, items 1390, 1392, Batchelor (*TAEM* 94:549–50; *TAED* MBSB21390, MBSB21392X).

2. "Edison's Light," *New York Herald,* 28 Dec. 1879; "Edison's Electric Light," *Philadelphia Public Ledger and Transcript,* 29 Dec. 1879; Cat. 1241, items 1396, 1399, Batchelor (*TAEM* 94:551, 553; *TAED* MBSB21396X, MBSB21398b); "Edison's Light," *New York Herald,* 29 Dec. 1879, 2.

3. "Electricity and Gas," *New York Herald,* 30 Dec. 1879; "A Night with Edison," ibid., 31 Dec. 1879, Cat. 1241, items 1401, 1402, Batchelor (*TAEM* 94:555, 557; *TAED* MBSB21401X, MBSB21402a).

4. Painter signed several names after his own, including those of Preston Plumb, senator from Kansas, and William Chandler, a prominent Republican official. The visitor register was maintained irregularly through February 1880. Electric Light Co. Statement Book, 10–24 Accts. (*TAEM* 88:412–19; *TAED* AB031:1–8).

5. "Crowding Edison," *New York Herald,* 2 Jan. 1880, Cat. 1241, item 1407, Batchelor (*TAEM* 94:560; *TAED* MBSB21407a); William Hammer reminiscence in Hammer, ser. 1, box 14, folder 8.

6. "Electricity and Gas," *New York Herald,* 30 Dec. 1879, Cat. 1241, item 1401, Batchelor (*TAEM* 94:555; *TAED* MBSB21401X).

7. "Edison's Electric Light," *Times* (London), 14 Jan. 1880, 8; "Crowding Edison," *New York Herald,* 2 Jan. 1880; "Edison's Great Work," *New York Herald,* 1 Jan. 1880; "A Malicious Visitor," *New York Tribune,* 2 Jan. 1880; Cat. 1241, items 1407, 1405, 1408, Batchelor (*TAEM* 94:560, 559; *TAED* MBSB21407a, MBSB21405a, MBSB21407b); "The Light in Menlo Park," *New York Sun,* 3 Jan. 1880, [3].

8. "A Malicious Visitor," *New York Tribune,* 2 Jan. 1880; "Edison's Light," *New York Herald,* 10 Jan. 1880; Cat. 1241, items 1408, 1413, Batchelor (*TAEM* 94:560, 562; *TAED* MBSB21407b, MBSB21413X).

–1873–

Joel Cook to the Times *(London)*

Philadelphia, Sunday, Dec 28 [1879]

Times, London[1]

Spending Saturday and night at Menlo Park with Mr. Edison, I thoroughly examined the inventions in electric lighting.[2] They are four, including a generator, lamp, meter and regulator, providing in my opinion a complete substitute for gas, which can be furnished cheaper, whilst the light produced is better than gas, more regular, and emitting so small heat no danger exists from fire. Some of his lights were purposely fixed amidst most inflammable material, whilst I tried igniting paper by them without scorching it. About sixty lights were burning Saturday night   I saw them burn seven hours. Two had been burning continuously ten days without injury to the baked cardboard horseshoe in its little glass globe which furnishes the light   The cardboard seemed sufficiently durable, successfully resisting quite rough useage, such as dropping, shaking, turning current on and off thousands of times also raising intensity of light to 400 candles Whole arrangement simply constructed Edison will put about 800 lights at Menlo Park whilst inventions immediately go into practical operation in New York city. The globe containing the horseshoe is exhausted to one-millionth of an atmosphere by the Sprengel pump, measured by the McLeod guage. By successfully dividing the electric current Edison gets individual lamps of 16 candle power, each lamp having 100 ohms resistance The light is turned on or off and the current regulated with the ease of gas, whilst the current can be transmitted on wire as small as No 36.[3] A central regulator maintains an even current, whilst meters accurately measure the supply furnished each consumer. Edison finds the best generators are of five to seven horse power, eight lamps being maintained by one horse power, and it can maintain ten. Each lamp costs about one shilling to manufacture whilst a supply equivalent to 1000 feet of gas can be produced for ten pence or less[4] Edison's calculation of the cost to furnish the light is that consuming three pounds of coal in the steam engine maintains eight to ten lamps one hour. Edison's system also furnishes electric power for small industries, such as running sewing machines. Edison's light is bright, clear, mellow regular, free from flickering or pulsations, whilst the observer gets more satisfaction from it than gas He lights at Menlo Park, dwellings, offices, desks street lamps also laboratory and workshop, making it available for every lighting purpose of gas. I ate my supper and wrote the draught of this telegram by Edison's light Saturday night

Cook[5]

ALS (telegram, copy), NjWOE, DF (*TAEM* 49:758; *TAED* D7906ZCG).

1. The *Times* printed this cable in substantially this form on 29 December. "The United States," *Times* (London), 29 Dec. 1879.

2. Cook went to Menlo Park on 27 December with an editor of the *Philadelphia Public Ledger* and Henry Bentley, a former newsman himself (see *TAEB* 2:433 n. 4); a *New York Herald* reporter, probably Edwin Fox, was also there. They witnessed a private exhibition at which, according to Francis Upton, "several million dollars of capital were represented. Every [thing] went off splendidly. We had over forty two gas jet burners running from one machine." "Edison's Electric Light," *Times* (London), 14 Jan. 1880, 8; Upton to Elijah Upton, 28 Dec. 1879, Upton (*TAEM* 95:586; *TAED* MU041); "Edison's Light," *New York Herald*, 28 Dec. 1879, Cat. 1241:1396, Batchelor (*TAEM* 94:551; *TAED* MBSB21396X).

The *Times* editorialized on 30 December that there was "no room for doubt as to the reality of his [Edison's] success and that very important results may be looked for from it." Cook sent Edison this copy of his dispatch on 31 December with a letter saying that the paper "has editorially endorsed your success in solving the problem of electric lighting. . . . With the London Times the New York Herald and the Philadelphia Ledger sustaining you, the 'experts' can take a back seat." *Times* (London), 30 Dec. 1879, 7; Cook to TAE, 31 Dec. 1879, DF (*TAEM* 49:757; *TAED* D7906ZCF); see also "Menlo Park Lighted," *Philadelphia Public Ledger and Daily Transcript*, 29 Dec. 1879, Cat. 1241, item 1398, Batchelor (*TAEM* 94:553; *TAED* MBSB21398a).

3. Thirty-six gauge British wire is .0076 inches in diameter; the U.S. standard is .0070 inches.

4. One shilling was equivalent to about 25 cents U.S., and ten pence to about 21 cents. *Ency. Brit.*, s.v. "Money."

5. Joel Cook's business card identified him as an "American Correspondent of London Times" at the *Public Ledger* office in Philadelphia. DF (*TAEM* 49:757; *TAED* D7906ZCF).

–1874–

To Calvin Goddard

MENLO PARK, N.J.[a] Dec. 29, 1879

Dear Sir,

Your favor of the 27th was duly rec'd.[1] I am preparing full statistics etc as to cost and I cannot now give time to refuting the statements of our gas friend, but briefly I would mention that no company in New York or in America can make and put 1000 feet of gas in the holder for 35¢ that there is 15 pc leakage chargeable to the 1000 feet before it is delivered, that it is does not cost 1¢ per hour per h.p. if I do not charge to it depreciation and interest, but only ½ a cent per hour[b] per h.p.[2] That I give a jet equal to 5 foot of gas burnt in Sugs Standard Argand.[3] That if I run a [250?][c] h.p. Engine all night it will supply lights all night or if there is not enough lights to absorb 250 h.p. but say 100 then the Corliss cut off[4] works and there is just so much less coal burned. If you go into the delivery of gas then we have ½ the plant or interest 4 p.c. depreciation to 12 pc to gas—no leakage to 15 pc by gas and a chance to use our immense plant 10 hours in the day for selling power which is more than sufficient in New York to pay interest on plant and make the light for night for nothing   There are many other [things which?][c] I could mention but I cannot at the present. Very Truly

T. A. Edison

L (letterpress copy), NjWOE, Lbk. 5:475 (*TAEM* 80:200; *TAED* LB005475). Written by Stockton Griffin; circled "C" written at top of page. [a]Place from Edison's laboratory handstamp. [b]"per hour" inter-lined above. [c]Faint letterpress copy.

1. Goddard wrote that after reading an interview with Edison in the 27 December *New York Sun,* "a friend of one of our important stock-holders . . . has made some figures as to comparative cost which do not prove the superior economy of electric light over gas." The unidentified correspondent calculated that at one cent per horsepower per hour it would cost $120 to generate electricity for 10,000 lamps whereas the gas needed for an equivalent amount of light would cost only $105. This lat-ter figure assumed consumption of three and a half cubic feet per hour per jet, the gas being manufactured and stored for 35 cents per thousand cubic feet and sold at retail for 60 cents per thousand (cf. Doc. 1707 n. 9). The majority of the *Sun* interview consisted of Edison's response to Henry Morton's public skepticism of the lighting system (see Doc.1876 n. 2) but he did indicate that he was still trying to lower the operating cost. He declined to supply specific cost figures, instead referring the re-porter to "the officers of the company in New York." Goddard to TAE, 27 Dec. 1879, DF (*TAEM* 50:302; *TAED* D7920ZBR); "Edison's Elec-tric Light," *New York Sun,* 27 Dec. 1879, [3].

2. Although no specific source has been found for these figures on the cost of gas, Edison subscribed to the *American Gas Light Journal* and the *Journal of Gas Lighting, Water Supply, and Sanitary Improvement* (a British publication).

3. Originally an oil-burning lamp with a hollow cylindrical wick, the Argand lamp as modified for gas replaced the wick with an annular tube pierced with small holes. This produced a characteristic ring of flame and gave it a double supply of air from the inside and outside, the flow of which was regulated by a glass chimney. William Sugg, a London man-ufacturer of photometric instruments, devised a form of Argand which burned with exceptional economy and uniformity and was adopted by Parliament as the standard in Britain; it was also widely used in conti-nental Europe and the United States. Schivelbusch 1988, 9–14; *Ency. Brit.,* s.v. "Gas and Gas-Lighting."

4. The automatic variable cutoff mechanism patented by George Corliss in 1849 precisely regulated the timing of the admission of steam to the cylinder, and its exhaust, in response to engine load. It enabled an engine to run at a constant speed even under rapidly varying loads and greatly reduced fuel consumption, and it was widely used at this time. Hunter 1985, 251–66, 279–88.

January–March 1880

Having suspended virtually all work for nearly a week for the exhibition of his lighting system, Edison again closed his laboratory to the general public on 2 January. He resumed his usual routine of attending to business in the morning, reviewing the work of his assistants after midday, then working in the laboratory until the early morning, with a meal break around midnight. According to one newspaper account, "Edison himself gives in generally about four A.M., selecting some unoccupied spot, where, with his coat for a pillow, he sleeps soundly sometimes until ten o'clock, other times until six."[1]

The lighting system operated almost flawlessly during the exhibition and at the end of January Edison drafted a patent specification describing an entire system for the economical generation, regulation, and distribution of electric current for use in lamps and motors. Still, not everyone was convinced. Several scientists pointedly reserved their judgment and William Sawyer, a prominent electrician and inventor of another form of incandescent lamp, publicly challenged Edison to submit his lamp to a head-to-head test of durability.[2] Durability became a serious concern in mid-January, when it became apparent that many lamps suffered a partial loss of vacuum—and consequently failed—after a relatively short time. The problem was traced to thermal cracks in the glass and remedied by slight changes in the construction of the bulb, after which Edison began to make plans for a commercial lamp factory. With lamps regularly lasting hundreds of hours, it became evident that they suffered from a gradual darkening of the glass from the inside. Edison hypothesized that the problem was caused by the "carrying" of carbon particles from one side of the filament and on 13 February he began planning experiments to prevent it.[3] He also tried carbonizing numerous grasses and other natural fibers for use as lamp filaments, eventually deciding that South American bast fiber was preferable to cardboard. Other experiments during this time focused on carbonization techniques and equipment, and on the effects of hydrocarbon vapors (such as gasoline) on filament materials. John Lawson also continued to work on the meter.

In the meantime, Edison ordered modifications to his dynamo. One was to increase the armature speed, which allowed a reduction in the amount of wire and consequent resistance. Another was to fabricate the armature core of a series of insulated thin iron plates which simultaneously reduced heating losses and provided more strength for high-speed operation. Additional experiments confirmed the superiority of this design but aroused scientific skepticism that such low internal resistance as Edison claimed was possible. At Edison's request, two Princeton University professors tested the output and efficiency of one of the dynamos on 19 March, essentially confirming his claims. The week before, George Barker and Henry Rowland made separate tests of the thermal efficiency of several lamps. Results of both sets of trials were subsequently published in scientific journals. The plate form of armature was incorporated into a new two-ton dynamo completed in mid-March; the machine was installed in an extension of the machine shop built in February and March for use as Edison's Menlo Park central station.

Confident of his new lamp and dynamo, Edison promised in January to provide a generating plant and several hundred lights for the *Columbia*, a steamship being constructed for a company controlled by financier (and Edison Electric Light Co. investor) Henry Villard. This was his first lighting plant outside of Menlo Park. Edison also began to think seriously about the design of much larger central station districts. In January he hired Julius Hornig, a German-educated draftsman and mechanical engineer, to help draw up plans.[4] Francis Upton started making detailed analyses of capital and operating costs under a variety of circumstances. Far more than simply listing anticipated expenses, Upton tried to elucidate the complex economic relationships among steam engines, dynamos, copper conductors, and other elements of the system. His calculations made clear the economic advantages of selling electricity for daytime power as well as for light. Attention to principles of steam and mechanical engineering also made plain the inefficiencies in the original central station plans calling for a large number of small dynamos. Upton began to extrapolate the design parameters of a much larger dynamo from the ones then being tested; by March Edison was building such a machine and considering the novel step of coupling an enormous generator directly to a single steam engine.

In other matters, Edison experimented extensively with electromagnetic mechanisms to separate ferrous material from platinum- or gold-bearing sand. He also tried to eliminate the manual crank from his telephone receiver because English users found it objectionable. He made several instruments operated by an electric motor for the *Times* of London in January and two months later sent a variation of this design to London. Most events pertaining to the telephone, however, were of a legal, financial, or administrative nature. Edison's British agents succeeded in amending his basic telephone patent but the London company faced strong opposition from the Bell interests and the British Post Office, which brought a lawsuit claiming jurisdiction over telephonic communication. The company was running out of cash and tried to renegotiate its royalty agreement with Edison. This precipitated a rift between George Gouraud and Edward Johnson, which further complicated the company's simultaneous efforts to consolidate with the Bell interests in order to avoid bankruptcy. In France, efforts to merge Edison's telephone interests with those of two rival groups collapsed again in January at the last moment. Subsequent on-again, off-again negotiations exasperated Edison but finally produced an agreement in March. Closer to home, Edison spent considerable time organizing his notes and preparing testimony in the telephone interference cases. In February he offered to sell to Western Union his motograph for use as a telegraph relay. The company claimed it was already entitled to this invention under a prior agreement but by the end of March had agreed in principle to pay $100,000 over twelve years. Edison also agreed with Western Electric to accept a smaller royalty on electric pen sets in order to compete with illicit manufacturers.

Edison added about a dozen men to his staff during this period. A number of them started in January; in addition to Julius Hornig, these included Charles Clarke, a draftsman and civil engineer who, like his Bowdoin classmate Francis Upton, was able to make complex calculations;[5] William Holzer, an American glassblower;[6] glassblowing assistant James Hipple;[7] and laboratory assistant William Breath, who stayed only for a few months.[8] Charles Mott joined the office staff and in March began keeping a journal of activities in the laboratory.[9] Otto Moses, a chemist trained in Germany, was in the laboratory by January although he may have started at the end of 1879.[10]

1. "Edison's Life," *New York Herald*, 10 Jan. 1880, Cat. 1241, item 1414, Batchelor (*TAEM* 94:562; *TAED* MBSB21414X).
2. See Doc. 1876.
3. See Doc. 1898.
4. See Doc. 1897 n. 3.
5. See Doc. 1921 n. 3.
6. Holzer to TAE, 6 and 8 Jan. 1880, DF (*TAEM* 53:488, 492; *TAED* D8014C, D8014E); Jehl 1937–41, 495.
7. Jehl 1937–41, 495–96.
8. See Doc. 1914 n. 13.
9. See Doc. 1890 n. 13 and headnote, p. 671.
10. See Doc. 1754 n. 6.

–1875–

Equipment Specification: Electric Lighting

[Menlo Park,] Jan. 2d 1880

329[1]   Alter No 3 machine to a .14 ohm machine   us[e] sheet iron disks in place of wire[2] 12 wire to a comutator

J[ohn] K[ruesi].

ADS, NjWOE, Batchelor, Cat. 1308:207 (*TAEM* 90:770; *TAED* MBN003:77). Written by John Kruesi.

1. This is the order number; see Doc. 1687 n. 1.

2. John Kruesi's preceding order from the same day was to "Speed up Dynamo Counter Shaft to run Dynamos 1200 revolutions pr. mt.," nearly double that which Edison had run previous dynamos (see Doc. 1862). The increase probably was to compensate for having reduced the amount of wire in the armature in order to lower its resistance, voltage varying proportionately with speed and the number of turns of wire. The origin of the disks is unknown but they may have been modifications of the iron rings in earlier armatures (see Doc. 1843 n. 4). They presumably were intended to increase the mass of iron in the armature core and to be more stable at high speed than wire; this became the basis for the all-iron plate armature discussed in Doc. 1899. Cat. 1308:207 (Order No. 328), Batchelor (*TAEM* 90:770; *TAED* MBN003:77).

On 5 January Upton indicated the dimensions of the remodeled Faradic No. 3, originally designed in October or November. It was to have an enlarged bore of 9.75 inches; the armature was apparently designed for a single layer of thin wire and extremely close clearance with the magnets. About that time he gave measurements of No. 4 and specified the dimensions of Nos. 5–10, all having iron-wound cores. However, when he tested an armature with "⅛″ plates" he found it gave a slightly higher voltage at 1,000 revolutions than one wound with iron wire. Later he measured the heat produced with somewhat thicker disks, which had slots around the outside edge to hold steel sheets between the

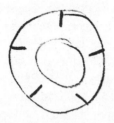

Steel disk 0.140 inch thick with slotted edges for the redesigned armature.

induction coils. At the end of January he concluded to "Run machine at 1250 revo   The higher the speed the better showing. The amount consumed on the magnets remains the same, the friction in the bearings increases proportionally while the magnetic and converting power increase as the square of the speed." N-79-06-16.2:171, 215–23; N-79-10-18: 138–40, 164, both Lab. (*TAEM* 35:451, 473–77; 32:853–54, 866; *TAED* No79:84, 106–110; N301:71–72, 84).

–1876–

William Sawyer to the Editor of the New York Sun

New York Jan 4 1880[a]

Editor Sun[1]

Notwithstanding the assertion that one of Mr Edisons electric lamps has been running for 240 hours I still assert, and am prepared to back up my assertion, that Mr Edison cannot run one of his lamps up to the light of a single gas-jet (to be more definite, let us call it 12 candle power) for more than three hours.[2] To be still more definite I offer to Mr Edison, at 226 W 54th st, in this city, an opportunity to prove what he says.[3] From the private residence in that street, wires are run a circuit of 1,000 feet; Mr Edison shall have every facility; he shall use my wires; he shall have any dynamo machine or other generator of electricity he may prefer; and all I ask is that the power of his light shall be measured by a photometer; that, once in place, it shall not be interfered with; and that a committee of gentlemen, preferably nominated by the editors of the N.Y. press shall be present and certify to the[b] facts of the test

Furthermore, I will place one of my lamps[4] side by side with Mr Edisons; it shall be run at a power of 25 candles; (it shall outlast the entire[c] 40 lamps at Menlo Park run at the power of 25 candles;) my lamp to stand as it is put up, and Mr Edison to put up a fresh lamp as fast as the preceeding ~~one~~ lamp shall have burned out.

I am anxious for this test; and if Mr Edison has really run one of his ~~lamps~~ horse shoe lamps 240 hours he will not refuse to accept my offer, for he will be treated with the utmost courtesy and shall have everything his own way.

I adhere in every particular to my original challenge to Mr Edison.[5] Resply

W. E. Sawyer[6]

D (copy), NjWOE, Lit., *Edison Electric Co. v. U.S. Electric Lighting Co.*, Edison's exhibit 23 (*TAEM* 46:272; *TAED* QD0120000M). Copied by William Meadowcroft, notary public. [a]Date not that of publication. [b]Interlined above. [c]Obscured overwritten text.

1. The *Sun* published this letter on 5 January, 1880. A copy addressed to the *New York Herald* appeared in that paper the same day. *Edison Electric Co. v. U.S. Electric Lighting Co.*, pp. 4179–80, Lit. (*TAEM* 48:601–2; *TAED* QD012G4179, QD012G4180).

2. In a letter to the editor of the *Sun* written on 21 December, the day Edwin Fox's article appeared in the *New York Herald*, William Sawyer claimed that Edison's reputation rested on spurious claims to the quadruplex telegraph, the telephone, and the phonograph, and "Now, all that remains for Mr. Edison is electric light." He charged Edison with "going over the same ground" as others, including himself. Stating that Edison had "failed," Sawyer challenged him to eight distinct trials, which included maintaining a vacuum in a lamp, operating a carbon lamp for at least three hours, obtaining twenty-five candlepower from platinum with less than three horsepower, and proving "that his dynamo-electric machine develops not ninety, but even forty-five per cent. of the feet pounds applied to it." Sawyer declared in closing that "all Mr. Edison's statements are erroneous, and I offer $100 as a prize for him to prove each of the above eight allegations. Let him run one of his lamps three hours and the public will be satisfied that I am correct." The source of the claim regarding 240 hours has not been determined. Letter to the editor, *New York Sun*, 22 Dec. 1879, *Edison Electric Co. v. U.S. Electric Lighting Co.*, pp. 4168–69, Lit. (*TAEM* 48:596; *TAED* QD012G4168).

Fox's *Herald* article and subsequent reports did not persuade other knowledgeable individuals. John Draper, for one, reserved judgment and Henry Morton was deeply skeptical, declaring that the *Herald* writer "must either be very ignorant and the victim of deceit, or a conscious accomplice in what is nothing less than a fraud upon the public." Although Edison had closed his laboratory to the public he continued to receive scientific and business visitors but did not offer detailed responses to these or other criticisms. To a request from the editor of the *Engineering and Mining Journal* for data showing "that the Electric Light is a success in an economical sense," Edison prepared a draft reply stating that he "wanted to pub but Co will not let me." "Carbon in Vacuo," *New York Herald*, 23 Dec. 1879; "Edison's Light," ibid., 28 Dec. 1879; Cat. 1241, items 1385 and 1396, Batchelor (*TAEM* 94:543, 551; *TAED* MBSB21385a, MBSB21396X); "Edison's Electric Light," *New York Times*, 28 Dec. 1879, Cat. 1051:68, Scraps. (*TAEM* 26:376; *TAED* SM051068a); *Engineering and Mining Journal* to TAE, 5 Jan. 1880, DF (*TAEM* 53:346; *TAED* D8007E).

3. This address where Sawyer planned to exhibit his own light was not his residence nor, as far as can be determined, that of any of his immediate family or associates. In a letter to the editor of the *New York Herald* he claimed that he had "hunted New York over to find some one who would permit me to use his house for a public exhibition. As may be imagined, it was not an easy thing to do." Sawyer held a public demonstration of his light at this address on 16 March, when he exhibited seven lamps for about twenty witnesses. Letter to the editor, *New York Herald*, 6 Jan. 1880, *Edison Electric Light Co. v. U.S. Electric Lighting Co.*, p. 4181, Lit. (*TAEM* 48:602; *TAED* QD012G:108); "The Sawyer Electric Light," *New York World*, 20 Mar. 1880, "A New Light to Read By," n.d., Cat. 1241, items 1467 and 1466, Batchelor (*TAEM* 94:588; *TAED* MBSB21467X, MBSB21466X).

4. Sawyer had made experimental lamps with carbon sticks whose relative thickness provided the low resistance he thought desirable. He tried a variety of shapes but settled in 1879 on a straight pencil sealed in a globe filled with nitrogen. The carbon was gradually consumed but could be replaced and the globe recharged with gas. See Doc. 1590; Bright 1972, 50–52.

5. In subsequent litigation, several witnesses testified that Sawyer composed his 21 December 1879 letter to the *New York Sun* while intoxicated, a state to which Edison may have been referring when he responded that Sawyer "couldn't run his lamp three hours . . . and I told him so. He doesn't know what I can do with mine as well as you do. . . . I think that Mr. Sawyer at the time he wrote his attack on me did not know precisely what he was doing." Sawyer nevertheless stood by his challenge in separate letters to the *New York Herald* and *New York World* two days later and a *New York Tribune* interview that appeared on 2 January. Testimony of William Sawyer, James Russell, James Potter, and James O'Keefe, pp. 188–96, *Edison Electric Co. v. U.S. Electric Lighting Co.*, Lit. (*TAEM* 46:247–51; *TAED* QD006:19–23); "Edison's Horseshoe Light," *New York Sun*, 23 Dec. 1879, Cat. 1241, item 1390, Batchelor (*TAEM* 90:549; *TAED* MBSB21390); Letters to the editor, *New York Herald* and *New York World*, both 24 Dec. 1879; "Mr. Sawyer Skeptical," *New York Tribune*, 2 Jan. 1880; *Edison Electric Co. v. U.S. Electric Lighting Co.*, pp. 4170–78, all Lit. (*TAEM* 48:597–601; *TAED* QD012G4170, QD012G4172, QD012G4175).

6. Electrical inventor William Edward Sawyer had been experimenting with incandescent electric lighting for several years, first by himself and later with Albon Man, a financial backer with whom he established the Electro-Dynamic Light Co. in 1878. The two tried a variety of carbonized substances, including paper. Sawyer and Man devised a process of subsequent value, that of preparing the burner in hydrocarbon vapor to strengthen it and provide uniform resistance. *TAEB* 4:654 nn. 4 and 5; Bright 1972, 51–53.

–1877–

From George Gouraud

London 8th January 1880

My Dear Edison

I confirm despatch of the following cable:—[1]

"Cable approval registration Edison Electric Construction Company Limited for manufacturing and working colonies and countries having no patents. No liability yourself."

I am sorry you have not yet replied to this as time is important. I have not felt like acting in the matter without your approval although I might better have done so as time is important and since there can possibly be no objection to the use of your name as I propose. As Johnson has explained to you it must take some time to get the Patents for the countries which I am to work[2] chiefly owing to the first delay in getting the first

Patent drawn which must be done in the light of the present state of the invention. This they are working on but in order to hold many of these countries against the opposition it will be necessary to occupy the field immediately with Exchanges. As this involves the use of considerable capital and of so much organization I have thought it best as a preliminary to put in an appearance immediately everywhere under the name of the Edison Electric Construction Coy pending the formation of separate companies for each country which companies[a] cannot be formed until there is something to assign to them, which there is not of course until the Patents are issued— This is the prime object of the Edison Electric Construction Company. The manufacture of instruments is altogether secondary but this is a subject which it will be necessary to ~~at once~~ early[a] take up. I see an increasing feeling on the part of the London Company against the inconvenience of getting instruments so far from home and my opinion is that it is only a question of time when the instruments will have to be made here or practically made here—put together at least—making such parts as may be made[a] cheaper ~~made~~ in America.

Besides this with the Electric Light there will be a large business to be done in countries where there are no patent laws or possibility of obtaining any security. Of course the light will be as valuable in the countries without Patents as the countries with them and it is my intention to occupy these fields for our joint interests

Upon the receipt of this cable the word "approval" so that I may loose no time[3]

This Company will of course in ~~any~~ no[a] case be worked in conflict with any of your other Companies   Yours very truly
G E Gouraud

LS, NjWOE, DF (*TAEM* 55:670; *TAED* D8046A). Written by Samuel Insull. [a] Interlined above.

1. This is the full text of Gouraud's cable to Edison on 3 January 1880. DF (*TAEM* 54:163; *TAED* D8026C).

2. Edward Johnson was en route to the United States at this time (see Doc. 1870 n. 1). On Gouraud's foreign assignments, see Docs. 1817 and 1833.

3. The next day George Ward of the Direct United States Cable Co. in New York telegraphed that he had just received from London a cable directing him to "Ask Edison to reply to Gourauds message of 3rd commencing cable approval &c." There is no record of any reply from Edison to Gouraud and no evidence that such a company was registered. Ward to TAE, 9 Jan. 1880, DF (*TAEM* 53:35; *TAED* D8004K).

*Telegrams: From/To
Ernest Biedermann*

Edison

Will treat through Drexels by letter so please cable refusal for 60 days   must have few samples sent by mail to secure trade mark in your name[1]

Biedermann[2]

Menlo Park, N.J. 8:07 PM[b]

Biedermann

you have refusal for sixty days   better get government concession   will ship lamps other samples cumbersome[3]

Edison

L (telegrams), NjWOE, DF (*TAEM* 54:178, 177; *TAED* D8026O, D8026N). Cable from Biedermann written by William Carman on Western Union telegraph form; reply written by Stockton Griffin on letterhead of T. A. Edison. [a]Date from document, form altered. [b]"Menlo Park, N.J." preprinted.

1. Lyell T. Adams, the American consul in Geneva, cabled Edison on 7 January that a "Strong association of Bankers and merchants requests me to ask you under what condition you will let them have your patent for Europe and Switzerland in particular by purchase or royalty on all light." Edison replied: "Sold except Switzerland unpatented   will treat strong parties." When Adams requested more information and sample electric lights on behalf of "Biedermann and strong syndicate," Edison instructed him to "Have Biedermann complete negotiations through Drexel Morgan and Company here. cannot spare samples yet, all working." Adams to TAE, 7 and 8 Jan. 1880; TAE to Adams, 7 and 8 Jan. 1880; all DF (*TAEM* 54:168, 172, 169, 174; *TAED* D8026F, D8026J, D8026G, D8026K).

2. Ernest Biedermann identified himself as a Geneva merchant. Edison later inquired about him to Egisto Fabbri and Drexel, Morgan & Co., neither of whom could provide any information. Drexel, Morgan subsequently forwarded a cable from Drexel, Harjes & Co. stating that he was considered "very venturous   no means, having lost large fortune in America. Consul Geneva is said to have good opinion of him but we would advise great caution." The Paris house later sent a more complete report indicating that Biedermann had a reputation for gambling and appeared to depend entirely on his connections with wealthy European and American families. Biedermann agreement with A. M. Cherbuliez and Gaspard Zurlinden, 8 Sept. 1880, Miller (*TAEM* 86:278; *TAED* HM800128); Stockton Griffin to TAE, 10 May 1880; Drexel, Morgan & Co. to TAE, 14 and 18 May; Drexel, Morgan & Co. to TAE, 3 June 1880, enclosing Drexel, Harjes & Co. memorandum of 18 May 1880; all DF (*TAEM* 54:239, 241, 244, 248–49; *TAED* D8026ZBS, D8026ZBV, D8026ZBY, D8026ZCC, D8026ZCC1).

3. Biedermann reported at the end of January that he had obtained a Swiss government "trade mark" but could not get the papers issued

without a sample lamp. Edison sent lamps in early February but because changes in Swiss patent law delayed the issue of the government concession, he extended the right of refusal for ninety days (Biedermann to TAE, 28 Jan. and 8 Mar. 1880; TAE to Biedermann, 9 Feb. and 8 Mar. 1880; all DF [*TAEM* 54:194, 212, 203; *TAED* D8026ZAE, D8026ZAY, D8026ZAO, D8026ZAZ]). Biedermann later requested "one Edison generator one meter and everything necessary lighting lamps" in order to obtain protection for the entire system. Edison responded that this would be impossible but he could send small models. When these had not been sent by mid–May, Biedermann asked for another ninety days on his option, to which Edison agreed (Biedermann to TAE, 8 Apr. and 14 May 1880; TAE to Biedermann, 8 Apr. and 15 May 1880; all DF [*TAEM* 54:229, 242, 230, 243; *TAED* D8026ZBG, D8026ZBW, D8026ZBH, D8026ZBX]).

–1879–

Notebook Entry:
Electric Lighting

[Menlo Park,] Jany 12 1880.
~~Ca~~Experiments by Basisc[1] on Carbonized Incandescent Conductors=[2]

Cotton Cloth. carbonized nicely but cannot as yet get it fine enough weaving to make narrow conductors[a]

Flour paste makes a very light & puffy but shining carbon it swells too much[a]

Gum Tragacanth works nicely   think it will go as it conducts well & dont swell— get some good tragacanth & try in Lamp

Asphaltum. Very brittle but exceedingly shining carbon dense.[a]

Gum Myrrh—dense carbon   might work—[a]

Gum Kino—puffy—[a]

Amn Isinglass. thin brittle very shining carbon like Asphaltum[a]

Leather   Comes out well   try it[a]

Maccaroni—puffy—

X, NjWOE, Lab., N-79-03-10.2:91 (*TAEM* 31:1163; *TAED* N032:37). Unrelated, canceled chemical formula not transcribed. Document multiply dated. [a]Followed by dividing mark.

1. That is, John Lawson (1857–1924). Lawson was born in Spuyten Duyvil, New York but lived in Scotland for much of his youth. He engaged in a variety of pursuits before coming to work for Edison as a chemist in January 1879 (see Doc. 1701 n. 4; "Lawson, John W.," Pioneers Bio.). Jehl 1937 (263) tells the story of how he came by his nickname: "One day some visitors had become involved in an argument over a certain oxide. Lawson told them bluntly that it couldn't be heated in the way they suggested, because it was a *basic* oxide. Edison was told of the remark and exclaimed: 'Lawson is right!' Thereafter he called Lawson 'Basic.'"

2. These records are the first of Edison's attempts to find alternatives to carbonized paper. Lawson continued these experiments on 17 and 18 January when he carbonized and examined by gas light approximately two dozen other substances including sassafras pith, cinnamon bark, pomegranate peel, and milkweed. N-79-03-10.2:95–103, Lab. (*TAEM* 31:1165–69; *TAED* No32:39–43).

–1880–

To Edward Johnson

[Menlo Park,] Jan 14th 80[1]

My dear Johnson—[2]

You are mistaken about the delay of switchboards here, they have all left Menlo Park within 3 days of their arrival from Bergman's excepting two which we kept[a] back to teach inspectors with— I cannot get men in New York worth anything am compelled to teach them here— I send two on Saturday with Times' clockworks—[3] These are not clockworks but "Pen" engines: I could not make clockworks go they were too clumsy.[4]

Gibson sailed promptly—[5]

Have ordered your compound wire—[6] You will have to use your own judgement about the cable to reach the Houses of Parliament—

I fail to see what my presence has got to do with the saving of my British patent. If it is a good patent it will be saved and if it is not a good patent it will not be saved, and if that patent cannot be saved then British patents are a delusion and a snare. Rather than come there now I could easily afford to return all their money; I will however furnish you all da[ta][b] on the state of the case in this country— As you have added nothing to the patent but only disclaimed a part of it, I cannot understand how anybody can get up and oppose "disclaiming any part" of a patent; if anything had been added it would have been different— In fact my opinion of British patent law is, that it is a fraud, and a costly one at that.

My definition of the word "pressure" is the same as the definition that the first ten men would give that you could pick up on the 'Strand'—[7]

No material body can be [given motion?][b] without pressure being applied, or if set in motion by abnormal means, its stoppage would cause pressu[re]—[c]

No microphone ever was made, that did not give a greater or lesser intimacy of contact by pressure—

Is a string telephone a telegraph?— Is the spar of a vessel a telegraph?— Is it telegraphic communication for one man to 'holler' acros[s]ᶜ a ten-acre-lot to another ma[n]?ᶜ If so, then the Post Office has got you—

Batchelor is having made the four diaphragm instrument—[8] We shall also send you one of the paper speaker telephones made by Murray[9] We are also making you one [of th]eseᵇ telephones to work by [-----]ᵇ instead of chalk.— [I will?]ᵇ also attend to the [other?]ᵇ requests relating to Insts [exper]imentsᵇ etc.—

The first water-chalks [that?]ᵇ we made are as loud [as?]ᵇ ever   we have wet them [----]ᵇ this evidently shows [----]ᶜ you truly remark that [once?]ᵇ a chalk is a good chalk [it?]ᵇ is good for all time" if occasionally wet. Yours very truly

Thos A Edison   per Batchelor

L (letterpress copy), NjWOE, Lbk. 5:539 (*TAEM* 80:203; *TAED* LB005539B). Second and third pages of this letter copied onto an unnumbered letterbook page between pages 539 and 540, and not microfilmed. Written by Charles Batchelor; circled "C" above. ᵃObscured overwritten text. ᵇFaint letterpress copy. ᶜEdge of original not copied.

1. By this date Edward Johnson had arrived, or was about to arrive, in the United States; see Doc. 1870 n. 1.

2. This letter is evidently a reply to one which has not been found.

3. Two days earlier Edison had cabled, "man Saturday times clockworks." The two inspectors scheduled to sail on Saturday, 17 January, were James Seymour, Jr. and William Gregory. Gouraud cabled on 7 February to "stop inspectors" but at least one more went before the end of March, when Johnson wrote that the company was "at low ebb in money matters" and was considering "sending some of the boys back home" because there was not enough work. TAE to Johnson, 12 Jan. 1880; Seymour to TAE, 14 Jan. 1880; Gregory to TAE, 14 Jan. 1880; Gouraud to TAE, 7 Feb. 1880; Johnson to TAE, 30 Mar. 1880; all DF (*TAEM* 56:344–45, 404, 555; *TAED* D8049D, D8049D1, D8049D2, D8049ZAE, D8049ZCV).

4. Edison sent twelve instruments. Norvin Green had asked for one of this lot in hopes of securing a contract with the Canadian government but Edison explained that they were "devised especially for the times and they are not sufficiently perfect for use in the hands of the public." He had experimented in 1878 with a receiver turned by a clockwork mechanism through a worm drive (see Doc. 1527). On 29 December 1879 he sketched a receiver operated by an electric motor. This appears to be a variation of that used in the electric pen and other devices, with the rotating magnetic disk placed between the field magnets rather than in front of them; it is connected by a worm drive to the shaft of the chalk button. Green to TAE, 15 Jan. 1880; TAE marginalia on J. B. Van Every to TAE, 5 Feb. 1880; both DF (*TAEM* 55:673, 600; *TAED* D8046B, D8043F); N-79-06-12:85, Lab. (*TAEM* 35:526; *TAED* N080:43).

After confirming that the receivers were on the way the telephone company cabled Edison to "Stop instruments until clockwork arrives must abolish handle." The hand crank had been recognized as a problem

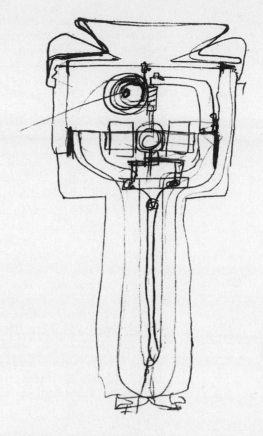

Edison's 29 December sketch of an electromotograph receiver operated by a small electric motor.

at least by September 1879, when Johnson wrote that potential subscribers in Manchester and Liverpool "send long messages—price lists &c & cannot manage the crank & write properly— What have you done in this re.? The thing turns so Easily you have but little work to perform." In early February Edison cabled that he had "new clockwork works much better than ones sent" and in late March confirmed that he had shipped the "motors" to the company. Edison Telephone Co. of London to TAE, 26 and 27 Jan., 12 Feb. and 20 Mar. 1880; TAE to Edison Telephone Co. of London, 26 Jan., 7 Feb. and 20 Mar. 1880; Johnson to TAE, 19 Sept. 1879; all DF (*TAEM* 56:353–54, 409, 535, 353, 404, 535; 52:728; *TAED* D8049L, D8049N, D8049ZAJ, D8049ZCI, D8049M, D8049ZAF, D8049ZCJ, D7941ZEC); Personal and Laboratory Accounts Ledger #4:178, 180 (*TAEM* 87:311–12; *TAED* AB002: 92–93).

5. J. H. Gibson went to England as a telephone inspector on or shortly after 17 December. Gibson to TAE, 13 Dec. 1879, DF (*TAEM* 52:971; *TAED* D7941ZJM2).

6. Johnson periodically ordered wire and other supplies through Edison. On 16 January Edison instructed Wallace & Sons to ship immediately 25 miles of compound wire to the London telephone company, specifying that "Great strength is required of the steel wire, for long stretches through the City of London." TAE to Wallace & Sons, DF (*TAEM* 56:346; *TAED* D8049D3).

7. A street less than a mile in length filled with shops and theaters, the Strand was (and remains) a principal thoroughfare between Westminster and the City of London proper. Weinreb and Hibbert 1983, s.v. "Strand."

8. This is probably the telephone shown in Edison's U.S. Patent 203,014, which has a drawing of a telephone with separate diaphragms on each side of a resonant box that was designed to more effectively transmit hissing consonants. The use of a resonant box with two or more diaphragms is referred to in the provisional specification (Doc. 873) of Edison's British Patent 2,909 (1877); Batchelor (*TAEM* 92:56; *TAED* MBP008).

9. It is not clear what Edison means by "paper speaker telephones." These may be the combination transmitter-receiver with an electromotograph receiver that used chemical paper, not a chalk button. Four of these were made by Joseph Murray in July 1877. See Docs. 932 and 962.

–1881–

Agreement with Western Electric Manufacturing Co.

Menlo Park NJ Jany 17th 1880

"Copy."

In consideration of the Agreement on the part of the Western Electric Manufacturing Company, that they will at once send a man to Australia with material and instruments and organize Telephone exchanges in the principal Cities of Australia[1] and will interest with them Mr Jeffray, of Melbourne,[2] and Mr Mc'Ilwraith, of Brisbin[3] in the Telephone Exchange Business of Australia, I agree to give Them the exclusive right of my Telephone Patents[4] and rights in Australia[5] on condition of receiving one dollar royalty per year on each Instrument.[6]

I will execute a formal contract embodying the principles and force of this Agreement, said contract to be given with and in favor of the Western Electric Co, and said Jeffray and Mc'Ilwraith[7]

Thos A Edison   Western Electric Mfg Co by M. G. Kellogg.
Sup't[8]

Witness   D. H Louderback[9]

D (copy), NjWOE, DF (*TAEM* 55:677; *TAED* D8046E). Other copies are in DF (*TAEM* 70:1179; *TAED* D8374T); Lbk. 8:434 (*TAEM* 80:980; *TAED* LB008434); and NjWAT, Box 1043.

1. The company sent Francis Welles, an assistant to Enos Barton who later worked extensively in Europe for Western Electric (De Lancy Louderback to TAE, 31 Jan. and 14 Feb. 1880, DF [*TAEM* 55:684–85; *TAED* D8046I, D8046J]; *NCAB* 37:141). Edison gave Welles a power of attorney for obtaining patents in Australia and New Zealand but months later there was confusion about whether patents covering the carbon transmitter had actually been taken out (copy of signed power of attorney to Welles, 2 Mar. 1880; Enos Barton to TAE, 8 July 1880; both DF [*TAEM* 55:686, 741; *TAED* D8046K, D8046ZBM]; TAE to Barton,

19 July 1880, Lemuel Serrell to TAE, 31 July 1880, Lbk. 6:211, 389 [*TAEM* 80:319, 375; *TAED* LB006211, LB006389]; see also Doc. 1835).

2. Otherwise unidentified.

3. Thomas McIlwraith (later knighted), a prominent Queensland rancher and investor, became treasurer and premier of the colonial government in January 1879. He forcefully advocated internal settlement and railroad construction, and at this time was personally financing experiments in the shipment of refrigerated meat and butter to England. *ABD*, s.v. "McIlwraith, Sir Thomas."

4. Undated copies of the drawings sent to Western Electric for Australian patents on the telephone and switchboard are in Oversize Notes and Drawings (Undated), Lab. (*TAEM* 45:25–27; *TAED* NS7986B: 28–30).

5. See Docs. 1817 and 1833. M. G. Kellogg advised Edison on 6 December that he and De Lancy Louderback would visit with "the Primier of Queensland and an influential man of Melbourne" on the afternoon of 8 December. Later on the 8th Edison cabled Gouraud: "Can close here with strong Australian parties for Australia two dollars yearly royalty each instrument   Answer shall I close   Bell people leave for there 25th Something must be done quick." Gouraud replied that he was arranging for a larger royalty and an advance. Edison twice more cabled Gouraud to act quickly, which he promised to do as the necessary papers were already being drafted. At the end of February Edison explained to Gouraud that "Not having heard from you I have closed with parties who are now on the ground fighting the Bell people. This action was unavoidable as something had to be done, at once, and the parties being A No 1 I did not feel like letting the opportunity slip." Several weeks later he promised that "You shall have same interest in Colonies as in England on chalk. Have arranged for Australasia and taken out another patent." He subsequently specified Gouraud's interest as 20 per cent. Kellogg to TAE, 6 Dec. 1879; TAE to Gouraud, 8, 9, and 10 Dec. 1879 and 9 Mar. 1880; Gouraud to TAE, 9 and 13 Dec. 1879; all DF (*TAEM* 49:552; 52:231, 233–34; 55:688; 52:232, 235; *TAED* D7903ZLT, D7939ZAL, D7939ZAN, D7939ZAO, D8046N, D7939ZAM, D7939ZAP); TAE to Gouraud, 25 Feb. and 2 Apr. 1880, Lbk. 5:696, 809 (*TAEM* 80:218, 234; *TAED* LB005696, LB005809).

6. De Lancy Louderback suggested that telephones for the continent's interior be sold rather than rented because "the country is so vast that it would be impossible to rent them and undertake to keep them in order." He proposed setting Edison's royalty equal to the rental royalty for two or three years but cautioned that "we shall have to sell them at a low price in order to compete with the German phones which are sold there. I understand that they are selling complete apparatus for two sets for fifty dollars." Edison consented, stipulating a payment to him of $7.50 on each station sold. Louderback to TAE, 14 Feb. 1880, DF (*TAEM* 55: 685; *TAED* D8046J); TAE to Louderback, 19 Feb. 1880, Box 1043, NjWAT (*TAED* X012L1G).

7. No formal contract seems to have been made; see Edison's 1 October 1886 agreement with the Oriental Telephone Co. Ltd. DF (*TAEM* 79:1210; *TAED* D8640R).

8. Milo Kellogg was a Western Electric engineer. Smith 1985, 111, 113.

9. De Lancy Louderback helped make the arrangements for Australia (see notes 4 and 5) and requested a commission for doing so. Louderback to TAE, 19 and 31 Jan. 1880, DF (*TAEM* 55:679, 684; *TAED* D8046G, D8046I).

<table>
<tr><td>–1882–</td><td>Menlo Park N J January 17 1880.</td></tr>
</table>

–1882–

Agreement with Western Electric Manufacturing Co.

Menlo Park N J January 17 1880.
Hereafter for a period of one year I agree to reduce my royalty on Electric pens to 12 per cent of the selling price to the Customer.[1] Gilliland[2] to agree to it— clause guaranteeing given number yearly waived for one year from date.[3]

Thos A Edison

ADS, NjWAT, Box 92 (*TAED* X012L1F).

1. This agreement modified and extended Edison's and Robert Gilliland's 1876 contract with Western Electric (Doc. 817), which was due to expire in December 1879. It was sent to Francis Welles at Western Electric in Chicago by Milo Kellogg. According to Kellogg's letter, from which Enos Barton typed an excerpt, "[George] Bliss and I made an arrangement with Edison last night about the electric pen, I enclose his memorandum which should be attached to our contract. The plan is to make the price 25.00 30.00 & 35.00, to make it a merchandising thing we to sell and Bliss to sell wherever we can." There was also an understanding that Western Electric would add to the price of each apparatus sold to Bliss the amount of Edison's royalty and an additional five dollars to be credited against Bliss's indebtedness to the company (see Doc. 1718 n. 3). There was to be a discount to agencies in Boston, San Francisco, Washington, and other unnamed cities, and Kellogg expected that "at these rates some pens should be sold." Extract of Kellogg to Welles, 18 Jan. 1880, Box 92, NjWAT.

2. Robert Gilliland, the father of Edison's former manufacturing partner Ezra Gilliland, held a 30% interest in Edison's electric pen. See Doc. 724.

3. That is, the guarantee in Doc. 817 of a minimum monthly royalty of $250 for the duration of the contract. According to the extract from his 18 January letter (see note 1), Kellogg advised that "Gilliland ought to consent in writing to the reduction in royalty and in the guarantee of numbers. Bliss will write him about it I think. We ought to keep his assent with our contract."

–1883–

From Simon Newcomb

Washington, D.C., Jan. 19th 1879[1880].[a]
My Dear Sir,
I wish to make one suggestion respecting the carbon electric light. If I understand the case correctly a carbon conductor gives much less light than a platinum one with the same power and resistance. ⟨Yes. [--][b]⟩

Is this not owing to a want of solidity in the carbon and to its heterogeneous texture? May it not be that if you could give a pin of carbon the most solid texture possible, you would get a great deal more light from it? I do not know whether the experiment is practicable but merely suggest something of the sort to you.[1]

Mr. Michelson seems to think there is a very great obstacle in setting up our velocity of light apparatus at Menlow Park owing to the want of a hotel or boarding house where the operators can live and sleep.[2] Otherwise I should like very much to accept your kind offer as I think the advantages of your labratory would be very great. I suppose we shall be ready in a couple of months. Yours very truly,

Simon Newcomb[3]

LS, NjWOE, DF (*TAEM* 53:606; *TAED* D8020ZAG). Letterhead of Nautical Almanac Office, Navy Department. ᵃ"Washington, D.C." preprinted. ᵇCanceled.

1. Edison wrote "Will answer personally" on this letter; see Doc. 1884.

2. Physicist Albert Michelson was just beginning an extraordinary career throughout which he pursued exacting measurements of physical constants. He made the first in a series of terrestrial determinations of the velocity of light in 1878 while teaching at the U.S. Naval Academy and presented the results that summer to the annual meeting of the American Association for the Advancement of Science, which Edison also attended in St. Louis (see *TAEB* 4:375). With Michelson facing a return to sea duty in the fall of 1879, Simon Newcomb arranged for him to take an extended leave from the Academy to assist in Newcomb's own speed of light experiments. At this time Michelson was working for Newcomb and largely living with his wife's family in New Rochelle, New York, near New York City. *ANB* and *DSB*, both s.v. "Michelson, Albert Abraham"; Livingston 1973, 51–66.

3. Astronomer Simon Newcomb was among the most distinguished American scientists of the latter half of the nineteenth century. At this time he superintended the Nautical Almanac Office of the U.S. Naval Observatory in Washington and was engaged in observational and historical research on planetary, solar, and lunar motions. One component of this program was the study of solar parallax, for which he needed an accurate value of the speed of light. He began planning the necessary experiments in the spring of 1878 and in the summer of 1880 he and Michelson made two determinations of this constant at the Washington Monument and nearby Fort Myer, Va. No record of Edison's offer to Newcomb has been found. *DSB*, s.v. "Newcomb, Simon"; Moyer 1992; Livingston 1973, 54–57.

Menlo Park, N.J., Jany 21 1880.[a]

Dear Sir

It is true that platina gives more light with the same energy than carbon but the latter can stand a higher temperature which more than compensates.[1] When you pass this way stop over one train and look at my den. Yours

Thomas A Edison

ALS, DLC, Newcomb. Letterhead of T. A. Edison. [a]"Menlo Park, N.J.," and "1880." preprinted.

1. Newcomb replied a week later that "You do not seem to have caught my idea about the electric light. It was not to use platinum, but some form of carbon of metallic solidity and compactness, should such be found. It might do as well to try cotton fibre running the whole length of your horseshoe. That is, take cotton staple, comb it out paste the fibres together into a mass of the right length, and bake that." Edison made no answer to this. Newcomb subsequently wrote concerning other ideas for electric lighting, for the development of which he offered Edison a commercial arrangement only to realize later that the central idea had already been tried and discarded. Newcomb to TAE, 28 Jan. and 24 and 25 Feb. 1880, all DF (*TAEM* 53:631, 670, 673; *TAED* D8020ZAT, D8020ZBX, D8020ZBY).

[Menlo Park,] Jan 23th[a] 1881[1880][b]

Make one Carbon turned edgeways

⟨Made Jan 23rd 1880   Chas Batchelor⟩[b]

Make loops ordinary size and large base lugs on[1]

⟨Made Jan 24 1880   Chas Batchelor⟩[b]

Make some loops of Card with ~~Soaked in~~ Tar rubbed over ~~cards~~ them;[c] well in, and then cut the loops regular size[d]

Get piece 9 thousandths card and cut 1 in × ½ in and put between steel plates and put on as much pressure as possible in press   see difference in thickness also in length and width

Before pressing 1 in × ½ × .0095

After pressing

Make loops of bank note paper 3 thousandths[2] so as to measure 100 ohms same as other loop

length from X to X 1.3 inch[b]

These were tried and gave a gas[a] jet with so small an amount of energy that it would take 22 for a HP[b]

Make loops of Manila Hemp fibre—[3] 750 Ohms when cold Gives us excellent light 540 when hot This gives light equal ⅟₃₄ gas jet and ~~take~~ equals 64 per H.P.[b]

Chas Batchelor

X, NjWOE, Lab., N-80-01-02.2:37 (*TAEM* 34:738; *TAED* N070:20). Written by Charles Batchelor; document multiply signed and dated. [a]Obscured overwritten text. [b]Followed by dividing mark. [c]Interlined above. [d]The text to this point is on a page containing the notation that it was entered as Edison's Exhibit S in *Edison v. Maxim v. Swan.*

1. On 16 January, Batchelor had first drawn a filament with longer and thicker ends to the loops. On that day and the following Edison had also drawn several oddly shaped filaments which were tested by the laboratory staff. N-80–01-02.2:27–35, Lab. (*TAEM* 34:733–37; *TAED* N070: 15–19).

2. The next day Batchelor made two more loops like this but with the paper pressed. He also made a loop from 3½ thousandths bank note paper with the circle cut one-third larger than this loop. N-80-01-02.2:49–51, Lab. (*TAEM* 34:744–45; *TAED* N070:26–27).

3. The following day Batchelor "Took manila hemp and knotted thread on ends. . . . Then pressed it between smooth vise and carbonized in carbon with hole in." On 27 January, he made some hemp fiber loops, carbonizing them with "tissue paper rolled up on the ends and pressed hard on to the wire" in a chamber with a "former on inside." On 30 January, he covered the ends of the hemp fibers with "tarred lampblack flattened out" and with ends "made of paper tightly twisted round and flattened out." These latter fibers were soaked in olive oil and the paper ends were soaked in boiling linseed oil. On 24 January, Batchelor also made loops of six-thousandths manila paper that was pressed in a screw press. Finding that those pressed with the force of three men were particularly straight after carbonization, he made twelve more of them on 28 January. N-80-01-02.2:43–47, 53–57, Lab. (*TAEM* 34:741–43, 746–48; *TAED* N070:23–25, 28–30).

Valparaiso Chile, S.A. January 24th 18780—

My Dear sir.

As usual mail day finds me over loaded with matters of mutual interest and therefore have so much to say that it is hard to find a starting place. As my last letter was about Morgan[1a] I will simply say that he came around and said he would do what I wished—&c so he went to work and put water on the chalks as I wished & they <u>worked splendid</u>   Water worked so well that he claimed that he was always in favor of it &c of course I made no objection— He kept putting on water until he made <u>mush</u> of them—so we turned new chalks & they work[b] Elegant. Mr Morgan is forbidden to put water or anything else until bidden—so the prospects are we will have 4 phones in good shape from now out. He is doing the best he can—but withall is not the man for the place—is a mere baby & there is no business in him. I will telegraph for another man as he is needed—[2] The business is a success beyond what I had expected: be assured that I shall watch your interest and do the same as though it were solely my own matter—but what I do I trust will please you all—action is necessary—it is too far away to write first— Today I telegraphed for 100 instruments, Batteries & 100 miles of wire[3]—which I trust will be sent <u>at once</u> as we are in want of them <u>at once</u>—business is pressing— This comes from a proper starting of the Enterprise, it is in good hands & already popular, had orders yesterday for 12 telephones—[4] Yesterday Jany 23d we all met at the Vice Consuls office—and organized— The law of Chile compels[a] stock to be paid up—& the Co cannot be properly organized until The grants are made by the Govnt then it will be regularly organized—under the laws of Chile— The business done yesterday was to sell to the 4 Gentlemen whose names I last wrote you—your patent when you get it for $50,000 paid up stock   The other 50,000 paying all expenses and furnishing all capital—[5] They voted to purchase what was ordered by cable today and I will remit per next steamer $1000.00 on a/c   The balance will be paid on the arrival of the B[ill of]/L[ading] by draft on London. $10,000.00 was voted subject to call at sight—of the things ordered of course you own one half jointly with me—but the payment of them is to be made by the other half of stock. We shall want at least 500 telephones in Valp[arais]o & Santiago— The government is very much interested & renders all the aid in its power— The bill to which I referred in regard to duty on supplies has passed & Telephones & the supplies for same <u>come in</u>

free— This quick work is unheard of in this country & also allowing <u>anything</u> in free of duty but we started with influence Enough to be a little stiff in our requests— It is <u>very important</u> that you send me a full power of attorney in duplicate[c] attested as directed before ~~by~~ a notary first, then the governor that the notary was duly & legally appointed & then by a chillian Counsul or Minister after that—[6] Mess Kendall & Co here hold your power of attorney for Electric light given sometime ago if you wish you can cancel it—as I hate to feel that you need two agents in so small a republic as Chile— If we can't run things there must be a "bug" in the plan some way—[7]

<u>Very soon</u> I will arrange to have all expense on the new Co & leave you a free holder of stock   Now I would like to hear from you very much—and also what portion of the[a] $50,000 stock you will allow me in my own right—also if you desire me to place your stock on the market & if so at what prices, the cheapest— I would have no trouble placing it at a fair price— several people are fighting your right to the exclusive use of Telephone in Chile but we are too strong for them and there is no doubt of the issue.[8] The Bell & English 'phones are a thing of the past and Edisons Belcher an established fact. The chalks are a <u>grand success</u> but very different with us than you   They dry out & need wetting about once a week but they talk awful loud & give the most perfect satisfaction   you may find it necessary to add an automatic wetting arrangement or one that by a little pressure at any time required will supply as much or as little moisture as may be needed to gain a proper tone[9]—as I think the dust gathers on the chalks & they should never be opened once in properly—all adjusting being made from the outside same as the rubber—I mean so far as wetting is concerned— Your picture was duly given to the President who will respond when the pressure of war will admit.[10] The ore or tailing business will be duly attended to. I already have sent to several mines and have about a dozen samples—& some new things among them will be a sample of a vien of pure platinum (said to be) about 12 inches thick—of course you will tell— I will get the Edison Telephone Co's office open in about 10 days— Please send a lot of <u>chalks all turned</u> & some not & pack in away so as not to damage   The others were packed like potatoes in a bag—[11] Goodbye. Regards to all. I write Mr Eldred to see my letters. More next steamer— Write me, send papers— Keep me posted—[12] Yours in haste

J. D. Husbands

ALS, NjWOE, DF (*TAEM* 55:775; *TAED* D8047C). Letterhead of José D. Husbands. [a]Obscured overwritten text. [b]"in duplicate" interlined above.

1. This letter has not been found. Walter Morgan, the telephone expert engaged for Chile, had been associated at least briefly with the Gold & Stock Telegraph Co. in New York and was acquainted with Stockton Griffin. TAE agreement with Morgan, Husbands, and Horace Eldred, 19 Nov. 1879, Miller (*TAEM* 86:59; *TAED* HM790084); G. L. Wiley to TAE, 3 Nov. 1879, DF (*TAEM* 52:892; *TAED* D7941ZHL).

2. Husbands's cable has not been found but on 12 February W. J. Clark sailed for Chile. Upon his arrival in late March Husbands discharged Morgan, who made an acrimonious departure several weeks later amid allegations that he had in the meantime colluded with parties opposed to Edison's interests. Horace Eldred to TAE, 11 Feb. 1880; TAE to Husbands, 12 Feb. 1880; H. L. Storke to TAE, 13 Feb. 1880; Morgan to TAE, 3 Apr. 1880; Lucius Foot to TAE, 9 Apr. 1880; W. J. Clark deposition, 26 Mar. 1881; all DF (*TAEM* 55:798–99, 824, 830; 59:689; *TAED* D8047K, D8047M, D8047N, D8047ZAD, D8047ZAH, D8147J).

3. This is presumably the meaning of the coded cable from Husbands on this day, a partial translation of which is provided by Stockton Griffin's transcription of a 27 February cable. Husbands also wired for this equipment on 28 January. The telephones arrived on 20 March but Husbands found "that the shipment is useless on account of no Bells or Mouthpieces having been sent— . . . We are all ready waiting for instruments: our beautifully fitted exchange has had the last stroke of the hammer & paint brush." Husbands to TAE, 24 Jan., 27 Feb., 28 Jan., and 26 Mar. 1880, all DF (*TAEM* 55:781, 807, 782, 822; *TAED* D8047D, D8047S, D8047E, D8047ZAC).

4. Soon after this letter Husbands began building lines to a section of Valparaiso known as the "Barrio del Puerto" and conducted the first public demonstration about 20 February. By 15 April he had 32 subscribers and by early July there were nearly 100. By September 1881

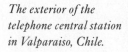

The exterior of the telephone central station in Valparaiso, Chile.

José Husbands sits at the desk in office of the telephone central station in Valparaiso, Chile.

there were between 200 and 300 telephones in operation in the city. Photographs of the main office are at NjWOE (Cat. 34.000.001, 34.000.002, 34.000.003, 34.000.004); Berthold 1924, 37, 39–40; Compania Chilena Telefonos de Edison, list of subscribers, 15 April and 8 July 1880, DF (*TAEM* 55:835, 860; *TAED* D8047ZAK1, D8047ZBA1).

5. The four investors were Lucius Foote, Santiago Martin, Pedro A. MacKellar, and M. MacNeil. Berthold 1924, 38; see Doc. 2031.

6. Edison executed a power of attorney to Husbands on 27 February 1880, a copy of which is in DF (*TAEM* 55:808; *TAED* D8047T).

7. See Doc. 1959 n. 4.

8. A week later Husbands sent Edison a letter to the editor stating the writer's opposition to Edison having an exclusive license for the telephone but it was granted to Edison on 26 April (see Doc. 1959 n. 5). Husbands to TAE, enclosing letter to the editor of *Diaro Oficial*, 31 Jan. 1880, DF (*TAEM* 55:786; *TAED* D8047H).

9. Upon arriving in Valparaiso Morgan reported that "the Chalks have got perfectly dried up, for they would bear but a very slight tention without squeeking & whistling considerably." He revived them but dessication of the chalks remained a persistent problem. On 10 February Husbands wrote that "This climate is dry and our chalks were so dry that they sounded like a piece of metal on a dry grind stone & have little or no talk in them— Water applied is immediately absorbed—does not lie on the outside at all; it will also absorb water enough to make it like paste." Repeated wettings provided temporary improvement but eventually destroyed the surface of the chalk. Edison subsequently prepared a draft cable advising "Evaporates through shaft hole   keep close wet sponge." This seems to have helped, as did using a syringe instead of a brush for the initial wetting, but Husbands suggested replacing the sponge with "a peice of Felt fed by a water cup . . . the sponge lasts from 6 to 7 days, now in the winter, in summer will last about 2 days." Morgan to TAE, 9 Jan. 1880; Husbands to TAE, 10 Feb. and 15 May 1880; TAE to Husbands, 10 Apr. 1880; all DF (*TAEM* 55:772, 794, 850, 833; *TAED* D8047A, D8047J, D8047ZAU, D8047ZAJ).

Husbands had previously advised Edison of unforeseen problems in the wholly different climate on the isthmus of Panama. While delayed there in transit to Chile, he learned that the duplex magneto telephone of the Panama Railroad "does not work well & rusts. . . . The first observation made is that for tropical climates you should have the wood work well seasoned & all the parts should either be nickle plated, Brass or some non rusting substance." Husbands to TAE, 2 Dec. 1879, DF (*TAEM* 52:229; *TAED* D7939ZAK1).

10. Anibal Pinto became president of Chile in 1876 at the start of a severe economic depression, which his government sought to ameliorate in part by imposing high import tariffs. In February 1879 Chile went to war with Bolivia over disputed revenues from guano and nitrate extraction, and soon afterwards also attacked Peru. The War of the Pacific lasted until 1883. Collier and Sater 1996, 125–36.

11. Morgan had reported that three chalks "got lose in transit, bruised & spoilled the turned ones & chipped the unturned considerably, but I can use the latter all right." He also reported at least initial "difficulty in getting the chalks turned properly here. I tried the best man & best Lathe in the City with one last night & the result was, that it was perfect in every respect but one & that was, it was not round. (Kind of an eccentric) & hardly suitable for telephones, Eh?" Morgan to TAE, 9 Jan. 1880, DF (*TAEM* 55:772; *TAED* D8047A).

12. Requests for information and supplies became a common refrain in Husbands's letters throughout the spring, and on 1 May he wrote that "I feel almost ashamed to say, when asked when I have last heard from Mr Edison to answer—That he has not written me since I left Nov 18, 1879." Husbands to TAE, 1 May 1880, DF (*TAEM* 55:843; *TAED* D8047ZAP).

–1887–

Francis Upton to Elijah Upton

Menlo Park Jan. 25. [1880]

Dear Father:

The electric light is growing better constantly. I have now a light over my desk so that I am writing you this letter by the light from a little horse shoe.

There is now a burner[a] in my bed room so that now I do not have to fumble after matches but simply to touch a plug so as to procure a light to wash and dress by. Mr. Edison is going on with his plans for making a large factory here to make electric lamps.[1] We have no doubt as to the ultimate success of ~~his~~ the success of electric lighting and of the great profits that will result from introducing it. The lamps are now lasting as long as can be expected for some have been burning since they were first made and are yet good   One lamp has burned 550 hours and a number are now approaching 500 hours[2] and all those lamps are without the improvements that have been made during the past few weeks which will tend to make them far more durable.[3]

The stock of the company is quoted far lower than when I was at home, but is bound to come up with a rush when the light is put into New York and we are manufacturing large numbers of lamps each day   If any one wanted to speculate just a little I should advise this as a chance. There is no trouble in my mind to selling the light at a large profit in many cases at a price below what gas can be made for. For special purposes where light is used for a large number of hours, 25 cts a thousand feet is about the estimated cost. Gas costs 45 to 65 cts in the holder[4]   Your Loving Son

Francis R. Upton.

ALS, NjWOE, Upton (*TAEM* 95:598; *TAED* MU045). [a]Obscured overwritten text.

1. According to a news account several weeks later, Edison was "preparing the ground for laying the foundation for three new buildings" at Menlo Park. One of these, a two-story structure 25 by 100 feet, was to be for lamp manufacture. He reportedly intended another building for making carbons, and the third for a 700-lamp central station. "The Coming Light," 12 Feb. 1880, Cat. 1014:34, Scraps. (*TAEM* 23: 648; *TAED* SM014034).

2. A notebook of burning times for various lamps shows that one, whose number was torn from the book, had operated about 558 hours by this date; it eventually burned 683 hours before breaking on 30 January. Six other lamps had each worked almost 500 hours by this time. All these were made in early January (N-80-01-02.3:1–19, Lab. [*TAEM* 35:3–11; *TAED* N074:1–9]). Records of other lamps from that time indicate that the average lifetime was much shorter (N-80-01-02.4:9–53, N-80-01-02.1, passim, Lab. [*TAEM* 35:743–62, 34:465; *TAED* N084:1–20, N067]).

3. Published reports indicated that by mid-January the vacuum had been compromised in as many as 20% of the Menlo Park lamps and Edison temporarily suspended their manufacture. A notebook from this period shows that a significant percentage of lamps broke either on the vacuum pump or soon afterward while having their resistance measured. Many of these failures were attributed to cracks in the glass, especially where the leading wires entered ("Edison's Electric Light," *Times* [London], 4 Feb. 1880, 10; "Scientific News," *English Mechanic* 1880 (30): 479; N-79-12-09:1–61, Lab. [*TAEM* 32:654–84; *TAED* N041:1–31]). After newspapers reported the difficulty on 16 January Edison received numerous suggestions and offers of help, in reply to some of which he made further inquiries (see correspondence from the latter half of January 1879 in Electric Light—General [D-80-020], DF [*TAEM* 53:588–637; *TAED* D8020]). He explained to one glass expert that "Our sole trouble is that there is tensions in the glass & the wires passing through the glass quickly heat one part of the glass before the other & a crack is the consequence. . . . If we could get a clear glass that would conduct the heat quickly so that tensions would be equalized rapidly that would probably cure our trouble" (TAE marginalia on H. C. Lippincott to TAE, 20 Jan. 1880, DF [*TAEM* 53:611; *TAED* D8020ZAJ]).

On 28 January Edison executed an application for the manufacture of lamps, in which he described sealing the wires into a glass tube (called the supporting bulb) which was then inserted into and fused to the neck of the lamp globe. He also claimed a two-part method of enclosing the globe after evacuation. First the tube leading to the vacuum pump was fused and then, after it was taken off the pump, the joint was sealed again from the outside with molten glass because "a perfect seal cannot be made when all the portions of the glass which unite to form the seal were in a vacuum when the seal was made." The final closure was "the resultant of two sealings—one . . . *in vacuo* and one . . . in air" which secured a "perfectly stable" vacuum. This application issued in July 1880 as U.S. Pat. 230,255. In a similar British provisional specification filed in London on 10 February Edison stated that the "part of the glass where the platinum wires pass is formed of white enamel glass having a greater conducticity for heat than the main part of the vessel. The enamel glass through which the platinum wires pass is then melted and sealed around them by fusion." This "enamel glass" may have been produced by applying what the *New York Sun* reported was a "fixed alkali" compound. Brit. Pat. 578 (1880), Batchelor (*TAEM* 92:149; *TAED* MBP023); "Mr. Edison and His Critics," *New York Sun,* 24 Jan. 1880, Cat. 1241, item 1450, Batchelor (*TAEM* 94:579; *TAED* MBSB21450X).

4. Upton wrote to his brother-in-law this day that

There is no trouble that I can see even now in the future. I have been figuring during the past week on some estimates and they all show that we are going to make enormous profits at the present prices of gas, and fair profits at the present cost of gas. The lamps are durable as far as we can judge and no new troubles show themselves. I should say it is a good time to buy stock. There is no doubt of our going into New York in the course of six months. The patents are going to be excessively strong. [Upton to Charles Farley, 25 Jan. 1880, Upton (*TAEM* 95:596; *TAED* MU044)]

–1888–

Henry Harrisse to
John Harjes

Paris, January 26th 1880

My dear Sir:

I regret to inform you that in the matter of Mr. Edison's Telephones, after five arduous and protracted meetings, we are less advanced than we were when you signed the contract on the 6th inst;[1] and I even apprehend that the matter will fall through with all sorts of complications!

The document which you signed on the 6th was the supposed last link of a long chain of previous agreements which had been signed since October 1879,—all equally binding.

Mr. Gower was the only party, who owing to his being absent, did not affix his name to the contract of the 6th.[2] We then thought that his signature, which had been promised by his own associates, was the only requisite to close up the matter;

and that after Mr. Edison had sent you the supplementary power of att'y which had been then considered indispensable, you would be placed at once in possession of 200,000 fs. in money and 166,666 fs in shares to be disposed of according to the instructions contained in Mr. Edison's letter to you of the 21st of November.[3]

This was a sad mistake!

Admitting that all the parties had come to an agreement and signed the contract of the 6th, the parties who act for and with Mr. Edison were then bound to subscribe and pay up in instalments, shares to the amount of one million of francs; most of which sum had to be forthcoming immediately. Now, not only the money on our side was not then ready, but it is not ready to-day, nor do I know where it is to come from!

I am told that the "Crédit mobilier" has offered the money to Mr. Bailey who refuses to accept the terms on which the money is to be supplied, 1st because it would throw the entire matter in the hands of that institution, and become at once the basis of stock jobbing operations; 2d, because Mr. Edison's interest would have to be transferred at a price considerably less than had been first agreed.[4]

Be that as it may, Mr. Gower has been fighting throughout and refused to sign on the plea that a certain clause in the contract of the 6th might be interpreted as binding him to give the company without indemnity, not only all his present telephonic inventions with future improvements relating thereto, but also all the telephonic inventions he might make in the future, whether they were like or pertaining to the "Gower Telephone" so called, or different from that instrument.

To conciliate matters, I took upon myself, in your name, to go to the whole extent of the agreement of the 21st of Novbr 1879, which Mr Bailey handed me last week, viz: to give without indemnity the benefit of Mr Edison's future telephonic inventions for Five years, and the preference for the 10 years following.[5]

The other members of the Fusion have accepted these terms, but Mr. Gower has not yet done so, although he now seems disposed to accept it with certain slight modifications.

But suppose he does, and that at the next meeting which is called for Tuesday 27th inst. all the parties sign the final documents; how will we perform the second part of the contract, that is, furnish then and there the million of francs?

Mr. Bailey informs me that he is negotiating still with moneyed institutions and may yet succeed; but that in case he does not, will break off, of course, retire from the fusion, stand the risk of an action for damages on the part of the other parties, and at once start a separate company, which he is very[a] now preparing, on the basis of the Edison Telephone alone, or perhaps, in combination with Soulerin's.[6]

In this case, another complication will spring up. All the Edison patents are pledged to the "commandituires,"[7] who in the first instance have advanced the 230,000 fcs (Berton, Dr. Evans &c &c &c) which have enabled Mr. Bailey and friends to keep the Edison Telephone before the public, make experiments, manufacture instruments &c &c &c.

Something unexpected may turn up at the last moment and alter the present aspect of the case; but I am free to say that my hopes and expectations are not very sanguine.[8] Yours faithfully

Henry Harrisse[9]   of Counsel[b]

ALS, NjWOE, DF (*TAEM* 56:21; *TAED* D8048N). [a]End of page; "is" apparently mistakenly canceled instead of "very." [b]"of Counsel" written in an unknown hand.

1. The contract was evidently the association papers of the proposed company (see note 2).

2. Joshua Bailey wrote Edison on 14 January that "Ten days ago all the parties interested in the Telephone business with the exception of Mr Gower signed the Statutes of the Coy. Today we have arrived at an accord by which Mr Gower will sign tomorrow." DF (*TAEM* 56:7; *TAED* D8048F).

3. Edison's letter has not been found but was sent with his power of attorney to Harjes. The powers of attorney have also not been found but Harjes wrote on 3 January that "all the legal requirements making your new Power of Attorney operative have been complied with." Harjes to TAE, 16 Dec. 1879 and 3 Jan. 1880, both DF (*TAEM* 52:450, 56:4; *TAED* D7940ZDS, D8048B).

4. See Doc. 1826.

5. This agreement has not been found but see Doc. 1846 for its proposed terms.

6. Léon Soulerin, a Paris engineer, represented the Société Française de Correspondance Téléphonique, which employed the Blake transmitter. Agreement of Edward du Chateau, Cornelius Roosevelt, Bailey, A. Berthon, and Soulerin, 14 Oct. 1879; Société Générale des Téléphones minutes, 18 Oct. 1879; both DF (*TAEM* 52:354, 456; *TAED* D7940ZBK, D7940ZDV).

7. That is, a member of a *société commandite*, or limited partnership. The *commandite* was required only to be registered with a notary (rather than specifically authorized by the government) and by mid-century had become a common form of association. Cameron 1961, 33–35.

8. This letter is the second of three from Harrisse that Harjes subsequently forwarded to Edison. In the last one, dated 28 January, Harrisse reported that "the failure of the 'Fusion' of the three Telephonic com-

panies is now complete. It was so decided by them at yesterdays' finale meeting. The cause is Mr. Bailey's inability to raise his share of the capital, viz one million of francs. We now revert to first principles, the three telephonic associations resuming their former independance, and dividing among themselves the cost of the attempt." Harjes summarized these events for Edison the next day and enclosed Harrisse's letters. Bailey wrote Edison on 28 January that "The fusion is broken up. Have not time to write details today. Will by Fridays mail. We shall hold our own." Harrisse to Harjes, 19 and 28 Jan. 1880; Harjes to TAE, 29 Jan. 1880; Bailey to TAE, 28 Jan. 1880; all DF (*TAEM* 56:20, 25, 18, 17; *TAED* D8048M, D8048O, D8048L, D8048K1).

9. Henry Harrisse returned to his native Paris in 1869, at the age of forty, after many years in the United States, where he studied law and became absorbed in the history of the discovery of the Americas. In Paris he wrote prolifically and compiled highly respected bibliographies on that subject, while maintaining a legal practice principally for American clients doing business in France. *DAB*, s.v. "Harrisse, Henry."

–1889–

Notebook Entry: Electric Lighting

[Menlo Park, January 26–c. February 10, 1880[1]]

Cost of conductors[a]

General laws.[a]

The cost increases as the square of the distance from the central station.[a]

1 Ohm may be taken as a fair estimate of the resistance of a conductor for 10 lamps.

$\frac{1}{10}$ of the energy will be lost in the conductor when all [---][b] the[c] lamps are on[a]

~~A~~ If a 200 Ohm lamp can be made in place of a 100 Ohm lamp the cost in a district will be $\frac{1}{2}$ for conductors or the distance from the station may be 1.4 times[2] as great so that the station will supply 2 times as many lamps with the same average cost for conductors.[3a]

If two lamps of 100 Ohms each be placed in series and also in multiple arc the cost for conductors for both lamps will be $\frac{1}{2}$ that for a 100 Ohm lamp or for each lamp $\frac{1}{4}$ that of a 100 Ohm lamp[4]

3 lamps placed in the same manner total cost $\frac{1}{3}$ each lamp $\frac{1}{9}$ &c

5 lamps $\frac{1}{5}$ $\frac{1}{25}$ &c &c &c   This for lighting street lamps theatres &c

To supply a district with 4 times the area each lamp will require 4 times as much copper in the conductor.[a]

The cost of the box in which to place the conductor may be taken as   per running foot[5] ~~or 2 times so that the same as bl~~[a]

In table of conductivities, pure copper is assumed.[6a]

The cost ~~of~~ per lamp in[d] a given area is proportional to ~~its are~~ the size of the area[a]

Questions[a]

How many feet of gas are consumed in various parts of N.Y. per consumer and an estimate of the average time of burning.[a]

How much horse power is taken and the price paid per horse power up to ~~the pow~~ 10 horse power.[a]

Costs of pipes laid in the street and depreciation[a]

Cost of management[a]

Cost of building lots or rent of same

Cost of water per H.P[a]

Houses   Estimated cost of introducing.[a]

Cost of condensing as compared to high pressure engines[7a]

Cost of laying pipes or wires in the street.

Lamps, laws of[a]

Datum[e]

E=Energy   L=Light   S=Surface $L = {}^{E^2}\!/_S$ [-][b]   ½ the surface[f]

Twice the radiant energy from any surface four times the light from the same surface[a]

The light increases as the square of the energy given off from the same surface[a]

If the surface be halfed keeping the energy the same the light will be doubled   There will be four times as much light given off per unit surface and only one half the number of units of surface.

The light from the same energy is inversely as the surface from which it is given off[a]

To keep the light the same and [---][b] ~~h~~ yet have it given off from one half the surface will require $\frac{1}{1.4}$ the energy.

¼ the surface and same light ½ the energy[a]

That is since ½ the surface give 2 times the light with the same energy, and as the light varies [~~wit?~~][b] as the square of the energy it will take $\sqrt{2} = 1.4$ the energy[a]

Keeping the light the same and diminishing the ~~energy will require the~~ surface[g] from which it is given, it will require an amount of energy which is as the square root of the ~~energy~~ surface[a]

A 200 Ohm lamp to give off the same light as ~~ana~~ 100 Ohm lamp with the same surface will require 1.4 times the E.M.F of a 100 Ohm lamp   That is if a 100 Ohm lamp requires 87 Volt a 200 Ohm lamp will require 121.8 Volts[a]

A 400 Ohm lamp will require 2 times the E.M.F or 174 Volts[a]

To double the resistance and half the surface and still have the same light will require the same ~~E.M.F.~~ $\frac{1/4}{\sqrt{1.4}}$ the E.M.F.[h] 1.18 the E.M.F

87 Volts 100 Ohm   1 = surface
101 Volts 200   "   ½ =   "
[--][b] 11.2 per H.P

A loop the same length and ½ the[c] surface ~~will give~~ with ½ the material will give 11.2 per HP instead of 8

Dynamo machine[a]

Laws[a]

The best machine is that which with the least amount of energy consumed on the magnets will give the greatest E.M.F on the smallest resistance in the armature so long as the spark on the commutator may be kept under. Interest on investment, friction and magnetic friction being made as small as possible[a]

The unavoidable[g] losses[c] are as follows[a]

Energy[a]

Magnet
Friction on bearings
Magnetic retardation
Local action
Spark on commutator
Friction on commutator
Deprec[a]

Money[a]

Depreciation
Commutator
Brushes
Belt
Bearings
Covering wires

The ~~law minimum~~ cost[i] of the magnets is ~~to have the~~ when[g] the interest on the amount invested in ~~them~~ equal to the cost of the energy consumed in them.[a]

The friction[c] of the bearings must be made as small as possible[a]

Magnetic retardation may be largely avoided by using very soft iron[8a]

Local action is in ~~ma~~ the Edison machine of no consequence

Spark on the commutator may be made very small by ~~reducing~~ placing the brushes endways and widening the commutator[9a]

In winding the machine   Twice the speed ¼ the resistance[a]

If the armature be made twice as long as now and have two magnets placed over it, One[c] layer of wire will then do the work. The present machine may be taken as having ~~³⁄₁₆~~ .16 Ohm resistance[10a]

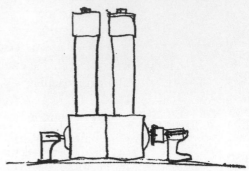

The wire in the new machine would be ¾ as long   [s--][b]

Present

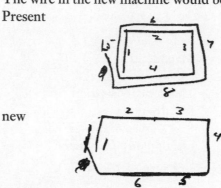

new

⅝ = ¾ about[a]

½ the resistance per foot

¾ × ½ = ⅜ the resistance

⅜ × .16 = .06 Ohm resistance

~~2~~1½ times the radiating surface so that ~~Increase1½~~[j] ~~the heat m times the heat~~ 1½ times the heat may be given off from the machine

$(XC)^2 \times .066\text{₁} = 1\tfrac{1}{2}C^2 \times 1.6$[k]

$X^2 = 1\tfrac{1}{2} \times \tfrac{8}{3} = 8$

$x = \sqrt{8} = 2.8$

2.8 times the lamps   50 × 2.8 = 140 lamps

Say square wire[11a]

.04~~0~~2	0~~3~~13
3	2
126	.026
.026	
.152	

$$042$$
$$\underline{\quad 4}$$
$$168$$
$$\underline{026}$$
$$.194$$
$$\underline{.152}$$
$$388$$
$$970$$
$$\underline{194\quad}$$

Cross section .029488[12a]

$$3.14$$
$$\underline{.042}$$
$$628$$
$$1256$$
$$.13188$$
$$\underline{.042}$$
$$26376$$
$$\underline{42752}[13]$$

$$.3001134$$
$$\underline{\quad 12}$$
$$2268$$
$$\underline{1134\quad}$$
$$.013608$$

$.00453896 \div 4 = .00113474$[1] $294 \div 136 = 2.1$[14l]

2.1 times the resistance in round .042 than in square

.03 Ohm in machine $1000 \div 03 = 333$[1]

333 lamps each 100 Ohms and still 10 to 1 exterior[c] to internal $1000 \div 6 = 166$[1]

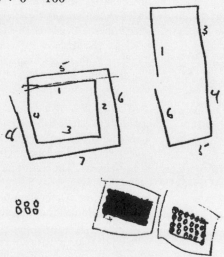

20 inches diameter[a]

2 times the present bore

2 times the surface

One layer sufficient as it will be twice as strong.

2 times the surface

2 times the length of wire on ends   ~~1½~~ times the total wire ¾ the total wire[m]

Each[c] division[15] 2 times as wide and twice as thick   ¼ the resistance

$\frac{3}{24} \times \frac{1}{4} = \frac{3}{816}$ the resistance

$\frac{3}{816} \times .016 = $ ~~.06~~ $.03$[n] Ohm $.063$ Ohms

2 times the mass of metal in magnets

1.42 times the diameter

$M = CL[-]$[b]R^2 $2M = CLR^2$ $2^c = 1.2R^2$

$R^2 = \frac{2}{1.2}$ $2 \div 1.2 = 1.66$[l]

$0.22011 \div [2] = .61101$[16l]

$R = 1.28$

Make magnet 7¾ inches diam

    "   "   3 feet 7 inches long

E.M.F   since the magnets are much stronger and the armature magnet much stronger Same speed of wire should give double the E.M.F.[a]

That is half the speed of pulley

F. R. Upton

X, NjWOE, Lab., N-80-01-26:25, 27, 26, 29, 31, 33, 35, 34, 37–51, 53, 52, 55–65, 64, Lab. (*TAEM* 34:13–33; *TAED* N059:12–32). Written by Francis Upton. Miscellaneous calculations omitted. [a]Followed by dividing mark. [b]Canceled. [c]Obscured overwritten text. [d]"per lamp in" interlined above. [e]Written in left margin and followed by dividing mark. [f]"E= . . . the surface" written on facing page. [g]Interlined above. [h]Related rough calculations on facing page not reproduced. [i]"~~minimum~~ cost" interlined above. [j]"1½" overwritten on "~~Increase~~" and canceled. [k]"C²" on both sides of equation canceled during algebraic operation. [l]Form of equation altered. [m]"¾ the total wire" written on facing page with line indicating placement. [n]Interlined below.

1. Upton signed and dated the fourth page of this entry on 26 January. Because the document contains no other dated pages the approximate ending date has been determined by the fact that it refers to experiments described in Doc. 1896 but precedes Doc. 1897 in the notebook by thirty pages. The editors have chosen to present an entry created over such a long interval as a single document because it forms a coherent analysis of the relationship between the design and capital costs of a commercial electric lighting system.

2. That is, the square root of 2.

3. Upton later tried to determine what additional investment in larger conductors would be offset by savings from the smaller central station plant which would be required. His notes are not clear but he seems to have supposed that doubling the expenditure for conductors to $10 per lamp would reduce the station investment and depreciation by 5%. However, based on his station estimates he determined that this rate of savings would justify an increase of only $10,500 for conductors, just a

fraction of that proposed. N-80-01-26:71–76, Lab. (*TAEM* 34:36–39; *TAED* N059:35–38).

4. This paragraph and the two following were added on the facing page. Here Upton is using an alternative method for achieving 200 ohms of lamp resistance by substituting, for each individual lamp placed in the parallel circuit of the entire system, a pair of 100-ohm lamps arranged in series. He then goes on to consider arrangements in which groups of three and five lamps are used.

5. Upton previously estimated this at $1.50 per running foot. N-80-01-26:22, Lab. (*TAEM* 34:12; *TAED* N059:11).

6. Upton is probably referring to one of two tables compiled by Charles Clarke. One, which Clarke began preparing on 31 January, indicates the diameter and weight of copper conductors required to obtain various resistances, between one-half and ten ohms, over circuits from forty feet to two miles long. He included columns for the cost of conductors in each case but left these blank (N-80-01-31:1–93, Lab. [*TAEM* 33:1098–1143; *TAED* N058:1–46]). Sometime in February Clarke also made a table of the resistance per foot-grain of copper at temperatures between 0 and 500 degrees Fahrenheit. He indicated that this applied to copper with a conductivity of 97 but specified a formula for finding the resistance in cases of different conductivity (N-80-02-08.2:[247–61], Lab. [*TAEM* 34:445–52; *TAED* N066:125–32]). Also on 31 January Upton began making notes for Edison, headed "Test of wire for 5 ton order," about procedures for measuring resistance in the laboratory and calculating conductivity (N-79-10-18:141–55, Lab. [*TAEM* 32:854–61; *TAED* N301:72–79]).

7. See Doc. 1897 n. 5.

8. The permanence of the magnetization of iron decreases as the metal is made softer.

9. See Doc. 1896.

10. Upton expected that increasing the area available for radiating heat would improve the armature's efficiency. For slightly later efforts to reduce armature heating in other ways see Doc. 1899, and for Upton's subsequent design of a larger (and more powerful) armature see Doc. 1937.

11. Upton apparently intended to compare the cross sectional area (and therefore the resistance) of .042 inch diameter round wire with square wire, although it is unclear from the following computations exactly how he did so.

12. That is, the product of .194 and .152.

13. Upton wrote this instead of the correct figure, 52752, thereby invalidating the results below.

14. That is, the area of square wire divided by the area of round wire (see note 12 above).

15. That is, the commutator divisions.

16. Upton implicitly divided 0.22011, the logarithm of 1.66, by 2 to obtain the logarithm of R.

Patent Application:
Electric Lighting

To all whom it may concern:

Be it known that I Thomas A. Edison of Menlo Park in the State of New Jersey, United States of America, have made certain new and useful improvements in furnishing light and power from Electricity of which the following is a specification.

The object of this invention is to so arrange a system for the generation, supply and consumption for either light, or power, or both of electricity, that all the operations connected therewith requiring special care, attention, or knowledge of the art, shall be performed for many consumers at central stations, leaving the consumer only the work of turning off or on the supply as may be desired; in other words to so contrive means and methods that electricity may be supplied for consumption in a manner analagous to the systems for the supply of gas and water without requiring any greater care or technical knowledge on the part of the consumer, than does the use of gas or water, in order that economy, reliability and safety may be insured.

In carrying the invention into effect, a city, town, village, or locality may form one district, or if the extent of territory makes it desirable may be divided into several districts. In each district I provide a central station at which are grouped a suitable prime motor or several motors dependant upon the amount to be supplied, generators, or means for converting the prime motive force into electricity, means for determining and regulating the amount of electricity generated and supplied in order that a constant pressure of electricity (so to speak) may be kept up.

The prime motors are any suitable engines, steam or water, and one or a series of two or more are provided as may be necessary, each of which is provided with its own system of shafting and belting, driving a number of Magneto electric machines, the number actuated by one prime motor being hereinafter termed a battery.

It is to be noted as is also shown in previous applications for patents made by me, that I make my field of force magnets exceedingly long, and of an extremely large mass of metal in proportion to the mass of metal in the revolving armature carrying the generating coil.[2] By this extra length as the magnetic tension at the poles increases with their distance apart, there is secured, at the polar extensions acting upon the coils in the field of force a much greater magnetic intensity, or so to speak a greater magneto motive force or pressure, causing consequently the generation of a greater amount of energy in the coils operated on than would result from the use of shorter magnets, even though the same mass of metal were used therein. By this elongation of the cores I am enabled to dispense with a large amount of coiling, one layer of wire being sufficient, whereby the resistance of the machines is largely diminished. The larger mass of these magnets is magnetically saturated by a weak current passing around them. It takes this weak current a long time to bring them mass of metal up to the point of practical maximum magnetic intensity, but once brought to that point, the weak current readily keeps them it[b] there, while with a shorter magnet a stronger current would more speedily magnetically saturate them yet[b] this stronger current would still be required to keep them it[b] so saturated.

If the coils of the field of force magnets and the generating coils were included in one circuit, and all the current generated, were passed through the field magnet coils, a very much greater amount of current than necessary for the maintenance of a practical magnetic maximum in the field of force magnets would be passed around them and the coils acting as resistances to the energy in excess of that required to magnetize the magnet to its practical maximum, would cause a great waste of electric energy. Hence I prefer to keep the coils of the field of force magnets and the generating coils separate, and that one machine in each battery (which machine may be termed the battery field of force generator) be used to supply the requisite energy to the field of force magnets of the other machines in such battery (which may be termed the supply generators.) The coils of the field of force magnets are connected as a series or in a multiple arc in one circuit while the generating coils of the supply generator of each battery are all connected in a multiple arc to the main conductors, (though for special purposes they may be connected as a series.) This arrangement it is seen gives great economy as the per cent of the entire current generated in each battery absorbed in keeping up the magnetic maximum in the field of force magnets when it is furnished by one special machine of the battery the number given it to feed being properly calculated, being less than when a portion of the current generated in each machine is absorbed in its own field of force magnets.

Where a single battery of machines is used it is preferable in view of what has been hereinbefore stated, that the current for the coils of the field of force magnets of the field of force generator of the battery be supplied by a small galvanic battery, but if more than one battery of machines be used, the field of force generators of all the batteries are fed from one or more prime field of force generators connected in a multiple arc or in a series the field of force magnets of the prime field of force generator or generators used being kept magnetically saturated by a weak galvanic battery current as before set forth. For instance a weak galvanic current supplies the field current necessary for one prime field of force generator which in turn feeds the field of force magnets of the field of force generators of ten batteries of twenty or thirty machines, the ultimate effect in the generation of current depending upon (as one important factor) the tension of the galvanic current sent through the field of force coils of the prime field of force generators. This prime field of force generator may however be a dynamo electric machine instead of a magneto[c] machine, its field of force being kept up by the current generated in the machine instead of by a galvanic current.

At the central station all the supply generating coils or batteries thereof are connected to conductors on the multiple arc system and from these conductors at the station, main conductors (which for convenience may be called simply the mains) connected thereto also on the multiple arc system lead in any and all desired directions for conveying the energy to the points where work, either by translation into light or motive power is to be done.

In order to give a better understanding of the method of regulating what for convenience may be called the pressure of the current through the entire system, I will here state that all the devices for translation of electricity into work are arranged on the multiple arc system, each device being in its own derived circuit, the effect being in substance to give each a circuit from the generating source independent of the circuit of all the other devices, as a resultant it follows that the greater the number of translating devices brought into circuit, the less the total resistance of the circuit. For instance I prefer that my lamps should each be of about 100 ohms resistance then if one lamp only be in circuit there is a resistance of 100 ohms—if another lamp be put in circuit, two circuits each of 100 ohms are provided for the[b] current, making the net total resistance to the current 50 ohms, although the resistance in each derived circuit, remains unchanged. This effect is ordinarily opposite to the effect produced by the addition of lamps when they are connected in an ordinary straight circuit, each one then adding to the resistance of the circuit.

The bringing into operation, successively, of numbers of the devices and thereby making more paths or circuits for the currents, does not appreciably lessen the pressure or diminish the effect upon the devices in use, the active forces at the central station viz: Prime field of force generators and motive power remaining unchanged until the net resistance of the devices in circuit exterior to the battery of machines is so diminished as to approach in a degree the resistance of the battery and main conductors, it being remembered that as the machines of a battery are connected in multiple arc the net internal resistance of a battery is as many times less than one machine as there are machines in the battery.

To avoid any appreciable variations and ensure uniformity, it is essential that any lessening of pressure be immediately indicated in order that just sufficient energy be generated and sent out to keep up an equal flow through the circuit of each translating device, that is, that the pressure be kept up uniform whether more or less translating devices be in circuit. This is attained by providing at the central station means for constantly indicating the pressure and for regulating the production if appreciable variation be indicated. aAt each station test lights are arranged so that an approximate visual test of the effect of pressure upon the circuit of any translating devices in use may be shown. From what has been said it is evident that as more or less translating devices are brought into circuit, the total resistance of the circuit or all the circuits thereof to the flow of all the current generated varies. To indicate this Electro-dynamometers, galvanometers or Electrometers are placed across the main conductor at the central station, or by return wire at any point in the circuit with a zero mark placed to correspond with the deflection consequent upon the maintenance of the proper amount of pressure. It may be advisable (and I have so done) to place at the central station a series of standard Daniell's battery connected by a switch circuit to the galvanometers or dynamo galvanometer in order that they may be frequently tested for any inaccuracy occurring from any cause whatever. By these means any error whatever therein is readily detected.

To correct variations in the pressure various means may be employed;[3] each supply generator may be connected into the circuit through a switch, and each series may be likewise so connected, so that the current of one or more of a series, or one or more entire series may be cut out or thrown into the circuit; or each machine may be arranged so as to be disconnected from the prime motor, or when needed the prime motor of an entire series may be disconnected. The plan I prefer however is to arrange in connection with the circuit of the battery feeding the field of force magnets of the prime field of force generator, before referred to, a series of resistances, so that the energy of the battery current may be varied—this variation causing in turn a variation in current induced in the prime field of force generator and in all the generators directly or indirectly controlled thereby. Where a dynamo machine is used these[d] resistances are to be used in the[b] same manner in connection with the circuit including the coils around the field magnets.[4] For distributing the current thus generated and regulated at the central stations I prefer to use conductors within insulated pipes or tubing made water tight and buried beneath the earth, provision being made at suitable intervals for house or side connections as shown in a prior application of mine.[5]

While this plan is preferable for many reasons, it is evident that conductors may be carried in the air or over house tops.

While only one pair of conductors may be laid on each street, I prefer, especially where streets are wide, to lay a pair of conductors along each side of the street near the curb. At proper intervals street lamps may be connected thereto by derived circuits. From main conductors on principal streets, subsidiary main conductors are laid through side streets. From the street conductors wherever desired, derived circuits are led into houses, one of the conductors passing through a suitable meter, preferably one which measures the amount of electricity passing through, as shown in a prior application of mine for a patent.[6]

In the house each translating device is placed in a derived circuit, the entire system of means for generation conduction and translation being one great multiple arc system. The translating devices in each house may be either for light or power, or both. For light the electric lamp consisting of an incandescing material hermetically sealed in glass shown in other applications filed by me, is preferred.[7]

This lamp is made of a high resistance in comparison with that of any electric lamps which to my knowledge have been proposed.

In lights heretofore proposed the endeavor seems to have been to lessen the resistance of the carbon, none having been suggested of higher resistance than say 10 ohms, but I have discovered a very much higher resistance, say 100 ohms must be used in order that a number may be economically and successfully used in a system.

The motor used should be so constructed that each, with a constant flow or pressure of current will give the exact power required. This requires that each motor should be wound with finer or coarser wire and with more or less convolutions which determine the maximum effect of the motor. In addition as the motors may be run with variable loads or amounts of work to perform, and as irregularity of speed would be a consequent thereof, it would be preferable to provide each motor with a governor, which, on excessive speed would operate to break the circuit of the motor or to otherwise control it. A preferable form of governor therefore will form the subject matter of an application for patent to be filed by me.[8] A system arranged as thus described provides for all the conditions precedent to an economical and reliable utilization of electricity as a lighting or motive power agent.

As within certain ascertainable limits the greater the horse power of an engine the less the proportionate cost per horse power, by consolidating at one station the prime motive force necessary to the generation of a supply, for many consumers a great economy as to production occurs.

As ordinarily proposed each electric light requires its own regulator which usually is either thermostatic or magnetic, breaking the circuit or bringing in resistances, in any case making a cumbrous lamp requiring delicate management and constant attention. By regulating at the central station entirely I am enabled to use a small separate lamp which may be used with the[b] exercise of no more than ordinary care or attention.

The distribution is so provided for that tampering therewith is guarded against and that connections from the mains to localities of translation are readily made.

The means for measuring insure accurateness and in furnishing a basis for equitable charges for the amount used by any particular consumer.

In the drawings accompanying and forming part of this specification, an arrangement of means is shown for carrying my invention into effect, although it is to be particularly noted that the invention is not dependent upon the specific means and their arrangement noted and shown, but that they may be varied without departing from the spirit of my invention.

The drawings illustrate however what I now consider the more preferable means and arrangements.

In these drawings

Figure 1 is a plan view of a central station;

Figure 2 is a modification of Figure 1;

Figure 3 is a plan view illustrating the street mains and house connections with translating devices properly introduced;

Figure 4 is a plan showing a locality divided into four districts.

Fig. 1

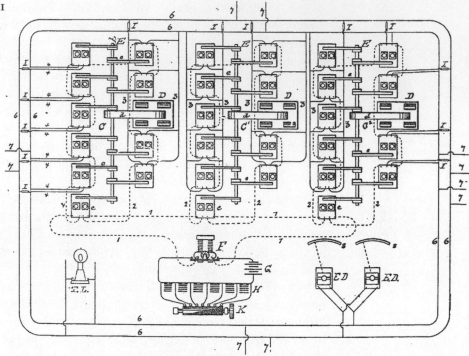

Inventor= T. A. Edison per Dyer & Wilber his Atty's
Attest= Sam. D. Mott   Jas. A. Payne[9e]

In Figure 1,[10] three batteries of generation C, C′, C² are shown, which number may be increased or diminished as circumstances may demand. One generator c of each battery is used to generate the current feeding the field of fiorce magnets of the other machines in its battery, the circuit from such field generator through the field of force coils in each battery being shown by the broken lines 2. 2. 2. For actuating the rotating parts, an engine D, is used with each battery, connecting by belt d, to line of shafting E, from which belts c e^b feeds to each generator. The coils of each battery in which currents are generated are connected, as in C′, and C², in multiple arc, to conductors 3. 3. 3. 3 which in turn are connected in multiple arc to the main conductors 6. 6. from which lead, in multiple arc, the street conductors or "mains" 7. 7. or, as shown in part of battery C, each machine may be directly connected in multiple arc to the station conductors 6. 6.— F is the prime field of force generator supplying the battery field of force generators

c. c. c. its circuit being shown in broken lines 1. 1. The field of force magnets of F. are magnetized by a current from the galvanic battery G, in whose circuit is arranged the series of condensers H, provided with the cut out K, by which more or less of the resistances are put in or out of the circuit feeding the field of F. This arrangement forms a very effective and simple method of regulating the production of current, or the pressure at the central station, for the current generated by F. being dependent upon the intensity of the magnetization of its field of force magnets which in turn depends upon the current transmitted around the magnets by the battery G. As the resistance varies such current, it follows that by varying the resistance in the circuit of G, the current generated by F. varies, which in turns varies the current generated in c. c. c. which in turn varies the current generated in the supply machines of the batteries, proportionate increase of current and rise of pressure in the latter following increase of current around the magnets of F. and vice versa.

One or more test lamps T. L. are placed at the central station in derived circuits to serve as a photometric test of the pressure in the lines. For more accurately indicating variations in the pressure one or more electrometers galvanometers or electro dynamometers E. D. are placed in derived circuits with a scale mark indicating the deflection caused by the pressure determined on as the standard pressure to be maintained. By the proper use of these indicating devices and the regulating devices described, a uniform pressure may be readily and easily maintained through all the mains.

It is preferable to connect all circuits from the generating machines to the main conductors 6. 6. through switches l. l. l. so that an entire battery or any portion thereof may be thrown in or out of circuit as the draft upon the station may indicate. It is to be noted also that the belt and pulleys of each machine are to be so arranged by any of the well known plans, that it may be disconnected from the motor when desired. These means may be used when desired as means of regulation, the number of machines in operation being controlled thereby while the effective force of each machine while in operation is controlled by the resistances.

The engines may be of any desired pattern or power, the number of machines in any one battery being limited by the power of the engine.

Fig. 2.[f]

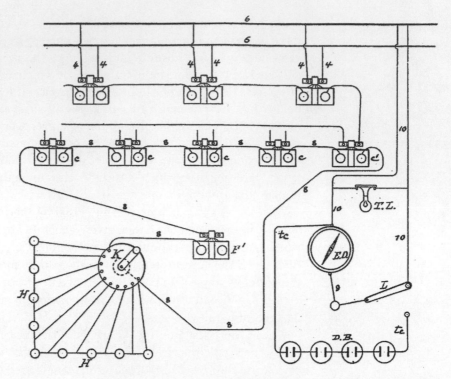

In Figure 2,[11] c. c. c. c. are the field of force generators of batteries not shown, while c'. is the field of force generator of a battery of which three supply generators are shown, connected to station conductors 6. 6. as before explained. The prime field of force generator F'. is, in this case, a dynamo electric machin[e][g] instead of a magneto electric machine as show[n][g] in Figure 1, all its coils being included in one circuit 8. 8. which passes around the field of force magnets of the battery generator c̲. c̲. c̲. c̲. c̲.'. The same principle of regulatio[n][g] is used however, the resistances H. and cut out K, therefor, being arranged in connection with the circuit through F'. so as to cause variation in the tension of the current therein as, and with the result, before explained. In this Figure there is shown, what may be used also in connection with the plans shown in Figure 1, means of testing the electro dynamometers E. D. or other indicating instruments used. From a standard Daniell's Battery D. B. a circuit t̲c may be formed around E. D. by means of the switch L. so that, when desired, the correctness of E. D. may be ascertained, and any inaccuracies, which might arise, be guarded against.

Fig. 3.^f

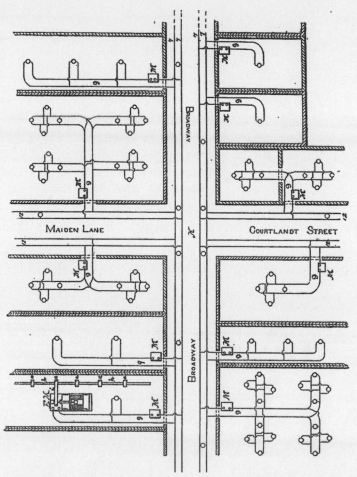

In Figure 3 the mains 7. 7. are shown leading from corresponding wires at central station (See Figure 1) down each side of the street. At side streets conductors 12. 12. branch off.

The small circles o in this figure indicate electric lamps. For street lighting they are placed as shown, in derived circuits from the street mains. From the mains, derived circuits 9. 9. lead into houses, in which are placed, at some suitable spot the meters M. through which one of the house conductors passes and by which the amount of electricity supplied to the house is accurately determined. Upon these house circuits are arranged lamps on the multiple arc system in such number, position and grouping as may be desired. In these circuits I also propose to introduce electro motors E. M. for furnishing light motive power. From the motor a belt i̱, leads to a line of shafting ẖ, and pulleys a̱, a̱ for any needed distribution of the power, or the belt i, may pass directly to a sewing or other machine, a separate motor being used for each power driven machine.

Where motors are desired each preferably should be made of a power proportionate to the maximum work to be done. While the electrical tension in each machine is regulated at the central station, yet if the maximum load or work be diminished at any particular motor its speed would necessarily be increased; it also might be desired at times to diminish the normal speed of the motor. In order therefore that some determinate speed may be maintained, irrespective of load or work, each motor should be furnished with a governor. For this purpose I prefer a governor, invented by me and to be the subject of a separate application, which acts to break the circuit, when a certain determinate speed is exceeded.

Fig. 4.[f]

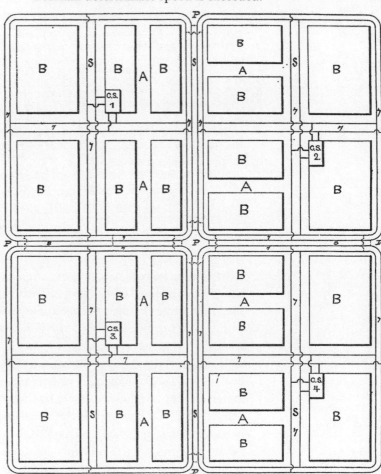

In Figure 4 is shown a locality divided into four supply districts, each is provided with its own central station, marked C.S.1, C.S.2, C.S.3, and C.S.4. From each, proper mains 7 7 lead out as before described. At convenient points however, say

P. P. connections[d] between the mains of the [sections?][h] may be made as shown in dotted lines, the effect then being to merge the entire locality into one large district with four supply stations, the pressure through all being uniform,[d] and each station doing its own quota toward maintaining the pressure uniform. The use of four stations and districts in this diagram is arbitrary, and for illustration only, as the number actually to be made in any one locality depending upon the needs of the locality.

What I claim is:[12]

First. A system of generating and applying electricity, consisting of the combination of means at a central station for generating the electricity and for indicating and regulating its pressure, means for distributing the electricity and devices for translating it into light or motive power, substantially as set forth.

Second. A system for the generation and application of electricity consisting of the combination of means at a central station for generating the electricity, and for indicating and regulating its pressure, means for distribution, means for translation and means for measuring the amount used by each consumer substantially as set forth.

Third. The method of regulating the electro motive force pressure in the main conductors by regulating the strength of the field of force magnets of the main magneto electric machines so that variation of pressure upon the connection or disconnection of translating devices may be prevented, substantially as set forth.

Fourth. The method of regulating the amount of effect at the translating devices by regulating the field of force current of generation substantially as set forth

Fifth. The method of regulating the generative capacity of one or a battery of magneto electric or dynamo electric machines by regulating the current passing through the field of force magnets substantially as set forth.

Sixth. The method of regulating the generative capacity of one or a battery of magneto electric or dynamo electric machines by varying the resistance of the circuit passing around the field of force magnets substantially as set forth.

Seventh. The method of operating a battery of magneto electric machines by using the entire current of one machine of the battery to supply the field of force current of the remainder, and throwing the entire current of the latter int[o the?][g] circuit for use substantially as set forth.

Eighth. The combination with an electrical circuit, of a number of separate translating devices substantially as set forth.

Ninth. The combination with one main electrical circuit, of a number of separate translating devices arranged therein upon the multiple arc system substantiall[y][g] as set forth.

Tenth. The combination with a number of translating devices of one regulator placed at a central station and regulating all the said devices substantially as set forth.

Eleventh. The combination with one or a battery of generators, and a number of translating dev[ices][g] of means for constantly indicatin[g][g] the electric pressure upon th[e][g] translating devices substa[ntially][g] as set forth.

Twelfth. The combination of a number of generators and a number of translating devices, all arranged upon derived circuits or multiple arcs substantially as described.

Thirteenth. The combination with the means for constantly indicating the electric pressure of a battery for testing the indicating means substantially as described.

In testimony whereof I have hereunto affixed my signature this 28th day of January A. D. 1880

Witnesses   C. P. Mott[13]   S. D. Mott   Thomas A Edison

DS and PD (photographic transcript), MdNACP, RG-241, Pat. App. 369,280. Written in an unknown hand. Oath omitted. [a]Place taken from oath; date taken from text, form altered. [b]Interlined above. [c]Misspelled "magnets." [d]Illegible; text from patent. [e]Drawing and signatures taken from printed patent; "Inventor," "Attest," and signatures appear on each subsequent drawing but have not been reproduced. [f]Figure from printed patent. [g]Edge of paper torn. [h]Ink smeared.

1. This application eventually issued, in substantially altered form, as U.S. Pat. 369,280 (see note 12). It is one of four executed by Edison on this day that resulted in a U.S. patent; two others were for lamp manufacture, including vacuum pump design, and one was for a lamp and screw socket (U.S. Pats. 230,255, 248,425, 265,311). He also filed a British provisional specification on 11 February which substantially duplicates this document up to the introduction of the drawings below, where it ends. Brit. Pat. 602 (1880), Batchelor (*TAEM* 92:158; *TAED* MBP025).

2. This is Edison's first explicit discussion of his large bipolar field magnets in a U.S. application, and probably was done at the recommendation of Zenas Wilber. See Doc. 1735 n. 14.

The remainder of this paragraph and most of the next were substantially rewritten into one long paragraph in order to meet the Patent Office's objections. In the issued specification Edison explained that this design was "so that currents of the desired high electromotive force can be generated in armatures of low resistance, and the waste of energy in the form of heat in such armatures will be reduced to the minimum." He also specified his wish to avoid "the use of generators having the coils of their field-magnets in series with their generating-coils, since this, besides being a defective arrangement with respect to regulation, also increases the internal resistance of the machines, and results in an objectionable waste of energy."

3. On 19 Dec. 1880 Edison drafted two caveats detailing various methods of voltage regulation (see Doc. 2036).

4. Edison included a similar arrangement in a draft U.S. caveat prepared in December, which he used as the basis for a provisional British specification filed on 3 January. The draft caveat and provisional specification also dealt with the use and regulation of electric motors. Cat. 1146, Lab. (*TAEM* 6:650; *TAED* NM014U); Brit. Pat. 33 (1880), Batchelor (*TAEM* 92:146; *TAED* MBP022).

5. Case 179 (see Doc. 1747 n. 1).

6. Case 171, from which the meter was deleted before issue (see Doc. 1733 n. 1).

7. In addition to the application for his basic lamp patent (Case 186, U.S. Pat. 223,898) Edison likely had in mind several others relating to lamp construction and design. He filed one manufacturing application (Case 187) in December 1879 and subsequently abandoned it. Another manufacturing application (Case 200) was also executed on 28 January; it issued in July 1880 as U.S. Patent 230,255. A third (Case 202) covering "Electric Lights and Systems of Electric Lighting" was executed about this date, filed on 5 February, and later abandoned. E-2536:44, 50, PS (*TAEM* 45:702–703; *TAED* PT020044, PT020050).

8. Edison included this centrifugal governor in a February caveat (Doc. 1907; see esp. n. 10) and a 20 March patent application (Case 208, U.S. Pat. 264,649). Also on 20 March he filed another application (Case 207, U.S. Pat. 228,617) for controlling motor speed by a friction brake which acted on "the main wheel of the actuated machine, without reference to the current" or the motor itself (Pat. App. 228,617). In early April he filed a British provisional specification including both this device and the centrifugal governor (Brit. Pat. 1,385 [1880], Batchelor [*TAEM* 92:166; *TAED* MBP026]).

9. Unidentified, but possibly a clerk for Dyer & Wilber. He witnessed many of Edison's applications handled by the firm.

10. This drawing was also incorporated into a division of this application (Case 217), which was later abandoned (see note 12). For Samuel Mott's draft drawing and another showing the "Plan of a Central Station" in which a single large steam engine drove twenty-four generating machines and five field exciters through a complex set of shafts and belts see Oversize Notes and Drawings (1879–1886), Lab. (*TAEM* 45:4, 3; *TAED* NS7986B:4, 3).

11. This drawing was also incorporated into Case 217 (see note 12).

12. The Patent Office initially rejected this application on the grounds that Edison's first two claims represented "mere aggregations of well known devices" and that other claims were anticipated by prior patents. Edison divided the application in May, incorporating the fourth and fifth claims into a new application (Case 217), which was also rejected and eventually abandoned (see E-2536:88, PS [*TAEM* 45:706; *TAED* PT020088]). Other claims were dropped and the remainder substantially rewritten in the face of the Patent Office's repeated objections.

The specification eventually issued as U.S. Pat. 369,280 (with all of the original drawings) in August 1887 and contained eight claims, including one for "a number of generators . . . consisting each of an armature of low resistance revolving in a magnetic field of great strength." U.S. Patent Office to TAE, 30 Mar. 1880; Dyer and Wilber to U.S. Patent Office, 31 May 1880; both Pat. App. 369,280.

13. Charles Mott, the brother of Samuel, joined Edison's office staff at the beginning of 1880. "Mott, Samuel," Pioneers Bio.; payroll receipt, 18 Jan. 1880, Edison, T.A.—Payroll Receipts (D-80-017), NjWOE; Jehl 1937–41, 546–47.

–1891–

Notebook Entry:
Electric Lighting

[Menlo Park,] Jan. 28 [1880]

Carbon. (1.)[1]

It is proposed to try the reducing effect of various gases on paper at different temperatures, principally the hydrocarbons. The apparatus consists of A combustion-furnace supplied with vaporized gasoline mixed with air; B a little[a] retort of thin glass holding about   c.c. a safety tube C and a vacuum pump D

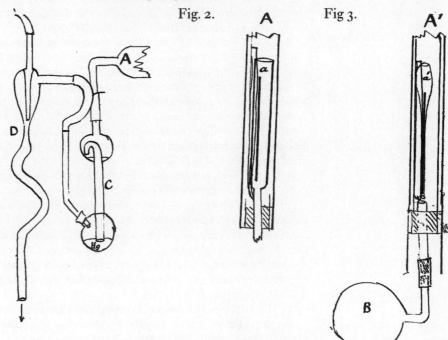

Fig. 1   Fig. 2.   Fig 3.

Bi-sulphide of Carbon. CS^2
The paper to be experimented upon was bristol board

cut in strips above length   and   thick. It was enclosed in a lead glass tube drawn out and contained in a hard glass combustion tube (bohemian).[2]

The CS^2 was passed over rapidly; the heat was dark red and was kept up five minutes. Paper[b] was carbonized blackish gray. Tube A covered on inside with carbonaceous deposite adherent to glass.

A liquid collected at cool end of combustion tube.

Copper became covered on under side with red oxide and on upper side with black [-----][3c]

Fig. 4.

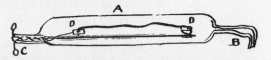

X, NjWOE, Lab., N-80-01-28:1 (*TAEM* 33:832; *TAED* No55:2). Written by Otto Moses. [a]Obscured overwritten text. [b]Preceded by heavy mark in left margin. [c]Illegible.

1. This is the first in a series of numbered experimental records made by Otto Moses in this notebook, not all of which pertain to carbonizing.

2. Moses probably was referring to bohemian glass, a fine crystal-like material made with lime and potash. One of its distinctive characteristics is rapid dissipation of heat. Newman 1977, s.vv. "Bohemian glassware," "lime," and "potash"; Knight 1876–77, s.v. "Bohemian Glass."

3. Moses continued these experiments the next day under the same conditions "with exception of longer heating—½ hour and corresponding slow volatization of the CS2." He found a "bright crystalline deposite along edge of copper—probably CuS Chalcocite?" and noted that "Paper carbonizes blackish grey." On 30 January he tried benzine, observing that "The gas goes over slowly and then condenses so rapidly as to create a more perfect vacuum than the pump, drawing in the combustion tube so as to compress the tube holding paper." Benzine left a heavier carbon coating on the tube than either gasoline or bisulphide of carbon and also produced a "blackish grey" carbon. On or soon after this date he drew a "Furnace, blast, for combustion," presumably for similar tests. N-80-01-28:3–7, Lab. (*TAEM* 33:833–35; *TAED* No55:3–5).

Edison later testified in a patent interference case that he tried about a half dozen lamps filled with various gases, including hydrocarbons, to see how such atmospheres affected the carbon. He mentioned having also used hydrogen, nitrogen, chlorine, and hydrochloric acid gases, none of which was satisfactory. Test records indicate that two lamps were filled with nitrogen; one of these was exhausted while the other may have been tried at atmospheric pressure. Edison's testimony, 5: 3075–78, *Edison Electric Light Co. v. U.S. Electric Lighting Co.*, Lit. (*TAEM* 48:44–46; *TAED* QD012F:42–44); N-79-12-09:115, 129; Lab. (*TAEM* 32:711, 717; *TAED* No41:59, 65).

From J. C. Henderson

New York, Jan 31st 1880[a]

Dear Sir/

With regard to the Electric Light on the "S. S. Columbia,"[1] I beg to submit the following for your approval or alteration as the case may be

Firstly in relation to the speed of the pulley on the generator, you did not tell me how many Revol's. per minuet it required for its maximum effect, but I suppose it requires about 800 per minuet if that is correct or not please let me know. again instead of running with lelather belts from a counter-shafts, I propose to run with frictional gearing as per sketch. for this reason that the counter shafts Pulleys & belts will make a humming noise, a very bad thing in a ship[2]

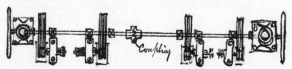

Plan

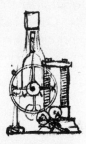

end view

O′ is the driving frictional pulley on the engine shaft.

X is an intermediate acting as a tightener or clutch, to throw dynamo's off or on, as the case may require

A the pulley on the Dynamo

If there is any thing to prevent me doing so please advise me; for if there is any danger of the current being carried off through the pulley to the main shaft I can interpose a nonconducter between Dynamo & pulley.

Secondly as regards the lamp itself can I support it in a wire cage without running the chances of it cracking when heated. for as you proposed to suspend it from the ceiling it is the only way I can think of doing it, and being on board of a ship it will have to be rigid, so as to keep it from swinging. and as I suppose that all of your lamps will be of the same length I will put them up in this manner that is if you dont think it will be detrimental to them[3b]

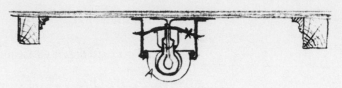

The yellow line showing the fine brass wire supporting lamp

A being an opal shade outside of the lamp

X being a metal reflector and heat protector for the ceiling carrying the opal shade, and wire basket holding the lamp, and screwed up to the ceiling solid. in the other rooms I will put an ordinary gas bracket with globe so that if you can oblige me with a lamp (a Broken one will do) to fit to, it will oblige and help me along considerably. also will you supply the sockets to drop the lamps into or will I get them made out of wood myself.[4] also will I take the main wires from the dynamo's and twist them into a cable and then branch off from them by the circuit wires thus[5]

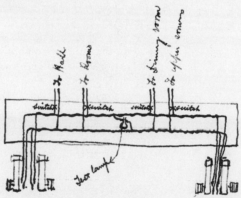

with a switch on every circuit, and a test lamp between the main cables for the engineer to control them by or will I use a ⅜" dia copper rod instead of twisting them together

and lastly what actual Horse power will I provide to run the four dynamo's with. I dont think from what I saw that each one ought to take more than two H.P. which would make eight but twelve I think should be all sufficient.

Also if you will let me know the party that is building the dynamos for us I will send the friction pulley's for them.

so trusting you will excuse me for this long account   I remain Sir   Yours Respectfully

J. C. Henderson   C Engineer   O.R.&N. Co.[6]

P.S. if it will suit you better I will come up for your answers to the different things if you drop me a postal to that effect J. H.

⟨Better Come up & Let us arrange it here⟩[7c]

ALS, NjWOE, DF (*TAEM* 53:639; *TAED* D8020ZAZ). Letterhead of Oregon Railway and Navigation Co., President's Office. [a]"New York," and "18" preprinted. [b]"over" written as page turn. [c]Followed by dividing line.

1. The steamer *Columbia* was being constructed at Chester, Pa., for the Oregon Railway and Navigation Co. (headed by railroad financier Henry Villard, an Edison Electric Light Co. stockholder) for service between Portland and San Francisco. This is the earliest extant letter concerning Edison's involvement with this project, the first Edison lighting installation outside Menlo Park. "The Columbia," *Sci. Am.* 42 (1880): 326; *ANB*, s.v. "Villard, Henry."

2. The four Edison dynamos were installed side by side and belted to an overhead countershaft in the engine room. One machine supplied current to the field magnets of the others. Two of the dynamos are located at the Henry Ford Museum (Acc. nos. 30.1123.1, 30.1123.2). "The Columbia," *Sci. Am.* 42 (1880): 326; Henderson to TAE, 24 Feb. 1882, Edison Electric Light Co. Bulletin 4:1–2, MCR (*TAEM* 96:676; *TAED* CB004A).

3. The *Scientific American* reported that two different fixtures were used for the 120 lamps on board. A drawing by John Kruesi shows one similar to Henderson's drawing by which the lamp was enclosed in a glass shade and attached to the ceiling; this arrangement was adopted for the staterooms. In the grand and dining saloons, each lamp was fitted in a frame hung from a rod running across the ceiling. This fixture was "of the same form as those used for oil lamps, and by an ingenious mechanical contrivance they are adapted to either the electric or oil lamp, so that should the electric lamp in any way fail the oil lamps may be immediately

Photograph of two of the Columbia's *generators.*

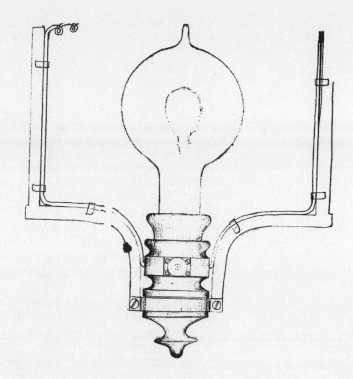

John Kruesi's 18 April lamp fixture design for the Columbia.

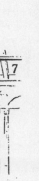

Lamp adapted for electricity in the Columbia's *dining saloon.*

substituted." Edison arranged to have fixtures made by Williams, Page & Co., a Boston manufacturer of railway and steamship lamps. In early April laboratory workers were "experimenting on glass with acids, giving it the appearance of ground glass" and the *Scientific American* reported that some of the ship's bulbs were "frosted lightly by dipping them in hydrofluoric acid. The globe thus treated seems to increase the amount of light proceeding from the incandescent horse shoe carbons." Francis Upton measured the resistance and candlepower of a number of lamps, both frosted and plain, for the *Columbia* in mid-April. "The Columbia," *Sci. Am.* 42 (1880): 326; N-79-06-12:176–77, Lab. (*TAEM* 35:567; *TAED* N080:84); TAE to Henderson, 31 Mar. 1880; letterhead of Williams, Page & Co. to TAE, 1 Sept. 1880; both DF (*TAEM* 53:712, 813; *TAED* D8020ZDG, D8020ZGK); Mott Journal N-80-03-14:47, 62; N-80-03-15:209–215, Lab. (*TAEM* 33:707, 715; 35:693–96; *TAED* N053:24, 32; N082:104–7).

4. On 20 March, as final preparations were being made to outfit the *Columbia*, Edison wrote Henderson regarding who should supply the sockets. On the 31st he again wrote Henderson stating that if necessary he could "make a number of this just as easy   shall we? and how many?" Edison's 7 April letter to Williams, Page & Co. about obtaining a "'ring' for lamp sockets" suggests that the sockets were made at the laboratory. The socket design is probably that shown in John Kruesi's drawings for a fixture (see note 3). The electrical contact between the lamp and the socket in this arrangement appears to be made by strips of metal rather than by springs as in earlier lamp and socket designs (see Doc. 1872 n. 2). Edison to Henderson, 20 and 31 March 1880; TAE to Tabor, 7 April

1880; all DF (*TAEM* 53:689, 712, 719; *TAED* D8020ZCQ, D8020ZDG, D8020ZDN); N-79-06-12:176–77, Lab. (*TAEM* 35:567; *TAED* N080: 84).

5. Circuit labels at top (left to right) are "To Hall," "To Rooms," "To Dining room," and "To upper rooms"; the "Test lamp" is in the middle between the switches. The entire vessel was wired during construction but lamps were installed only in the saloons and staterooms. One circuit was provided for the saloons, and one each for the upper and lower staterooms. The ship's steward controlled the stateroom lights through a locked cabinet outside each room. Henderson to TAE, 24 Feb. 1882, Edison Electric Light Co. Bulletin 4:1–2, MCR (*TAEM* 96:676–78; *TAED* CB004A); "The Columbia," *Sci. Am.* 42 (1880): 326.

6. J. C. Henderson was construction engineer of the Oregon Railway and Navigation Co. and subsequently chief engineer of the *Columbia*. "A Queen of the Water," *New York Herald*, 29 Apr. 1880, Cat. 1241, item 1497, Batchelor (*TAEM* 94:599; *TAED* MBSB21497X); "The Columbia," *Sci. Am.* 42 (1880): 326; Henderson to TAE, 24 Feb. 1882, Edison Electric Light Co. Bulletin 4:1–2, MCR (*TAEM* 96:676–78; *TAED* CB004A).

7. Only two dynamos had been completed when the *Columbia* left the shipyard at Chester, Pa., on 17 April so Edison later sent all four machines to New York where the vessel was to be fitted out, promising that "they can be placed in several hours." Laboratory workers made lamps, "safety clutches" (fuses), switchboards and other fixtures for the ship in the latter half of April. During the last week of that month and the first days of May, Francis Upton and other assistants supervised the wiring by workers from Sigmund Bergmann's shop, then tested the entire system. The dynamos were run at high speed on 26 April, damaging one of the armatures. The next day Edison attended a reception aboard the ship with his wife and some of the Menlo Park staff. Henderson to TAE, 16 Apr. 1880; W. H. Starbuck to TAE, 16 and 24 Apr. 1880; TAE to Starbuck, 16 Apr. 1880; TAE to Henderson, 16 Apr. 1880; all DF (*TAEM* 53:732, 739, 735; *TAED* D8020ZDY, D8020ZDZ, D8020ZEH, D8020ZEA, D8020ZED); Jehl 1937–41, 562–63; N-80-03-15:207–17; Mott Journal N-80-03-14:76, 82, 85, 87, 89, 94, 98, 101, 104, 107, 110, 114, 122, 126–27, 131; all Lab. (*TAEM* 35:692–97; 33:723, 726–29, 732, 734–35, 737–38, 740, 742, 745, 748, 750; *TAED* N082:103–8; N053:39, 42–45, 48, 50–51, 53, 55, 57, 59, 63, 65, 67).

–1893–

Notebook Entry:
Electric Lighting

[Menlo Park, January 1880[1]]

To be considered— The strength of the solution, that in one series of experiments the solution in all the bottles are of equal strengths—[a] A plate ~~if~~ remaining in contact with the solution for a considerable length of time the liquid in immediate contact with it not being disturbed is not affected in the same degree as one ~~that~~ which being disturbed frequently has its surface exposed to ~~fresh~~ layers of liquid; the liquid surrounding the undisturbed[b] plate becomes satura~~d~~ted by its action on the plate and has its capacity ~~of~~for action diminished accordingly, so that ~~beyond certain limits~~ for X time[c] there is a constantly decreasing action commencing at X and approximating to 0; whereas in the other case for fractions of X time the loss for each fraction is equal within certain limits—

The condition of the surface, rough or smooth— The comparative purity of the copper plates.

The quantity of liquid by which the plate is surrounded, also the relative position of the plate in the liquid; a plate may be immersed in a large quantity of liquid but so placed[a] that its surface is exposed to the action of a comparatively small quantity; again it may be placed in a much smaller[d] quantity of liquid ~~than in the preceding case~~ and yet have its surface exposed to the[a] action of a ~~much~~ comparatively or absolutely[e] larger quantity than in the preceding case.[f]

for instance the vessel A may contain as large a quantity of liquid as the vessel B, but the amount of <u>active liquid</u> surrounding the plate in A ~~a~~is not so great as in B—

In order that a regular action may take place between the surface of the plate and the liquid in contact with it the metal surface must be chemically clean, it is necessary that great precautions be taken to prevent surface oxidation—and it is known that such ~~action~~ oxidation takes place rapidly under ordinary circumstances, and especial care must be taken that the surface of the plate does not come in contact with anything of an oily nature—[2]

The temperature must also affect the intensity of the action, as the temperature increases the intensity of action will probably also increase, as the temperature diminishes the intensity of the action will also probably ~~decrease~~ diminish—

X, NjWOE, Lab., N-79-04-03:239 (*TAEM* 31:459; *TAED* N025:121). Written by John Lawson. [a]Obscured overwritten text. [b]Interlined above. [c]"for X time" interlined above. [d]"much smaller" interlined above. [e]"comparatively or absolutely" interlined above. [f]Followed by "over" to indicate page turn.

1. John Lawson made these notes near the beginning of a series of copper deposition trials which started on 19 January and continued nearly three months. He was working on the electric meter, testing cop-

per plates in a double sulphate of copper and ammonia solution. On the lefthand pages Lawson kept a running table of the weights and immersion times of the plates; on many of the facing pages he made a series of mostly undated remarks, of which this document is the first, on experimental conditions, procedures, and future research. These records follow Lawson's notes from similar experiments in December. N-79-04-03: 238–73, Lab. (*TAEM* 31:459–76; *TAED* No25:121–38).

2. On the following pages Lawson indicated that to make the copper plate "chemically clean" before each trial he would "clean it by immersion in nitric acid, washing with water, alcohol, ether, dry it." This was to "bring the plate into contact with the liquid under conditions more nearly equal in each experiment." N-79-04-03:245–47, Lab. (*TAEM* 31: 462–63; *TAED* No25:124–25).

–1894–

*Notebook Entry:
Electric Lighting*

[Menlo Park,] Feb 1st 1880

Made hemp fibres with clamps of plumbago, Graphite such as used in lead pencils— They have got too much stuff mixed with them for us— Seem to swell up and form gases or arcs which bust up the lamps[1]

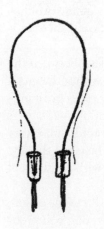

Clamps made for these were just a cylinder of graphite with hole to push on wire and holes in top for carbon loop

Chas Batchelor

A pair of graphite clamps in Batchelor's 1 February drawing.

X, NjWOE, Lab., N-80-01-02.2:59 (*TAEM* 34:749; *TAED* No70:31). Written by Charles Batchelor.

1. See Doc. 1902. At the beginning of January Charles Batchelor had started to calculate the expense of platinum in each lamp, and one object of these experiments presumably was to find a less costly alternative for use in clamps. He dated his sketch of several forms of graphite clamps this same day. Between mid-January and mid-February he tried clamps made of aluminum, ligna vita, Wallace carbon, cocoa nut shell, ironwood, and rice straw. Batchelor also tried platinum with "Little plumbago in the holes" of the clamps but found that they arced. N-80-01-02.2: 5, 27, 63, 71–80; N-80-01-02.4:127–28, both Lab. (*TAEM* 34:722, 733, 751, 755–60; 35:781–82; *TAED* No70:4, 15, 33, 37–42; No84:39–40).

–1895–

F. G. Fairfield to
the Editor of the
New York Times

NEW-YORK, Monday, Feb. 2, 1880.[a]
MENLO PARK LABORATORY
A DISPASSIONATE ESTIMATE OF THE EDISON LIGHT.
WHAT EDISON HAS ACCOMPLISHED—AN INVESTIGATION BY A VISITOR WHO WENT TO MENLO PARK EQUIPPED FOR HIS TASK.

To the Editor of the New-York Times:

So many glowing descriptions on the one hand, or severe strictures of the Edison light and its inventor on the other, the latter by electricians, have recently appeared in the daily papers,[1] that, possibly, the notes of an unbiased observer, taken on the field, may serve to enlighten the general reader as to what Mr. Edison has actually accomplished. My visit to Menlo Park on Saturday last was made partially with a view to ascertain whether one of Mr. Edison's lamps, operated with a small generator driven by a toy engine, used in the college for experimental purposes, would furnish an available illumination for microscopic dissection and examination; partly with a view to verify the statements of an electrician, whose strictures upon the light and its inventor were somewhat in detail, and partly to obtain from Mr. Francis R. Upton an explanation of certain obscure statements made in his article in *Scribner's Monthly* for February.[2] My observations were taken in the main from an optical and economical point of view, with the purpose of obtaining exact and well-defined facts relative to the operation of Mr. Edison's system, the optical properties of the light, its comparative expense, and the advisability of adopting it in place of gas or oil as a means of illumination in the more delicate problems of optical analysis.

The conclusions I have arrived at are certainly without partisan bias, and, as I have recently investigated with care all the systems of electrical lighting in use in this country, it may be fairly claimed that they are based upon a pretty thorough study of the whole subject, practically as well as theoretically. Exactness of such a kind as would satisfy a man of scientific training was, unfortunately, unattainable, either from oral testimony or records of instrumental test. The work of the laboratory appears to be carried on by Mr. Edison and his assistants with a conspicuous disregard of exact measurement and registration of results, optical and electrical, as well as economical. To one well trained in the details of laboratory work, it is possible, however, to sift pretty exact information from the merely proximate registration of such results, particularly when assisted by an actual inspection of engines and generators, dynamometers and electrometers in operation, and of lights burning; and, while a visit to Menlo Park by no means justifies the extravagant and ill-advised encomiums passed upon Mr. Edison's work by the newspapers, it is, at the same time, true that he has been measurably successful, and that his light is a material advance upon all other forms of electrical illumination in its optical properties. It is a clear white light, perfectly steady and free from flicker, and without the bluish tinge that renders the flame of the ordinary carbon lamp so disagreeable. When I say that the Edison light is perfectly steady, I mean to imply a degree of steadiness quite superior to that of an ordinary gas-jet, and not subject to any perceptible fluctuation. The disagreeable sound associated with the carbon lamp is also wholly dispensed with. This may be due to the comparatively low temperature of the incandescent carbon loop, or possibly to the exhaustion of air in the glass bulb. Indeed, the temperature was so low that the outline of the loop was perfectly visible to the eye at a distance of 15 or 20 feet, and the light, though soft and mellow, lacked the brilliancy of the electric arc. As I understand Mr. Edison and his assistants, the photometric tests have not been very exact or exhaustive, although shadow and oiled-paper tests have been used, and the light has also been compared with that of a gas-jet by a process of neutralizing the one by the other, or vice versa. Mr. Edison stated the illumination produced by one of his lamps to be equal to that of 16 standard candles, but it did not strike me as being in excess of seven or eight candles.

At the date of my visit 92 of these lights were in operation in the laboratory, in street lamps, and in stores and residences in the vicinity, the power being supplied by two generators manufactured under Edison's recent patent. These will be described hereafter. At this juncture it is only necessary to say that the 92 lamps were apparently somewhat more than the the two generators could carry, the incandescence of the loops even in the lamps in use in the laboratory for fine work being far below the point required for the highest luminosity. When asked whether he had measured the temperature of the loop, Mr. Edison replied that he had not, and could not, therefore say at what temperature the highest light-giving capacity of the lamp was attained. On the general principles governing the incandescence of bodies it would be safe to say that the degree was not far above 4,000 Fahrenheit; and it may, I presume, be taken for granted that 80 lamps are all that two generators can run with the best illuminative results.

Unfortunately, owing to the fact that the large 80-horse power engine performs all the motive work of the laboratory and shop, besides driving the generators, it is not possible to state the horse-power consumed or absorbed by the latter with absolute precision. Mr. Francis R. Upton, author of the article in *Scribner's Monthly,* gave me his estimate as 5-horse power to each generator, while an assistant present thought it was about 11-horse power for both. That the statement was not far from correct was evident at a glance. The armatures were making about 500 revolutions per minute, a pretty high rate of speed, but the heat generated was very slight, as compared with other generators that I have examined. As this transformation into heat is one of the causes of loss of power, and absorbs a large proportion of the 40 to 50 percentum of the motive power, leaving only 50 to 60 percentum to be rendered into electricity, it is very apparent that, although Edison's generator involves no new principles, and differs very slightly from Siemen's, a very important source of loss has been in a measure corrected. The important point of difference is the immense comparative size of the field magnets in the Edison. They stand about 4½ feet high and weigh 1,100 pounds, their wrought-iron cores being 6 inches in diameter, 3 feet long, and wound with three layers of No. 10 cotton-covered wire. They are connected at the top by a wrought-iron yoke 6 inches high and 7 inches wide. These immense magnets rest upon a heavy block, which in its turn reposes upon a bed-plate consisting of a cross-shaped brass casting, which also serves as a support for the bearings of the armature that revolves between them. The internal resistance of the armature is claimed to be only one-half ohm, and one-ninth only of the resistance of the external circuit. The claim of utilizing 90 percentum of the power transmitted by the driving-belt, put forth by Mr. Edison, in which the internal and external resistance are regarded as equal, the Siemen utilizing only 46, is doubtless somewhat exaggerated, although the practical superiority of the Edison generator, in vital particulars, is apparent from the comparatively small percentage of the force transformed into heat, from the trifling noise, compared with others, and from the very slight development of magnetism in the frame. It is to be regretted that Mr. Edison is at so little pains to verify his statements by exact instrumental tests; for here, as everywhere else, there is an amazing lack of precise demonstration, which may be used to his discredit, even if he is correct, and is, in any event, wanting in scientific spirit and method.

But that the question of divisibility of the current has been solved, notwithstanding all the theoretical objections of electricians, must be admitted without reserve when one sees 92 lights supported by two generators, distributed over a tract equal to nearly a square mile, by means of wires (or electric mains) only one-half an inch in diameter as they spring from the generator. Whatever increase of resistance, external to the armature, may result from the sub-division of the current, it has been successfully effected as a matter of science. I saw in a grocery store, about 60 rods from the laboratory, eight lights in operation. The wires connecting with the street main were about one-fifth of an inch in diameter, thus establishing a small sub-circuit, with which the hair-like wires of each of the eight lamps were connected. The theory of this is that the electrical charge of a conductor one inch in diameter may be distributed with equal pressure by means of any number of smaller conductors the sum of the squares of whose diameters is equal to the square of the diameter of the main conductor. Thus, if the main conductor be two inches in diameter, and the electrical pressure expressed by 100, the charge may sustain 400 lamps with conductors one-tenth of an inch in diameter, 400 times the square of $\frac{1}{10}$ being equal to the square of 2.

Unfortunately, the question of loss in distribution, and if such loss exists, what ratio it bears to the whole charge, has not been put to the test of instrumental measurement at Menlo Park, and electricians will not be likely to accept the vague statements of interested parties in lieu of actually-registered or openly-exhibited metrical tests. Still, the divisibility question has been solved in a manner which, after deducting the loss, leaves a margin of 40 lamps of alleged 16 candle power, supported by one generator, driven by 5-horse power. As 1 cent per horse-power per hour is the estimated average cost of running a steam engine of 75-horse power, the motive force per lamp cannot cost more than $\frac{1}{8}$ of a cent per hour, against $\frac{2}{5}$ of a cent per hour, with gas at $1 per 1,000 for a 4-foot gas-burner, giving an estimatedly-equivalent light. I say estimatedly-equivalent, because from what I saw of Mr. Edison's lamps, and I inspected them pretty carefully, the illumination furnished by one of them is considerably less than that of the average horse-tail burner under good pressure, to say nothing of the improved burners now in general use. Mr. Edison's own estimate, as given to me, differs vastly from what he has been reported as saying in the newspapers. He claims that he can furnish for 65 cents (doing business on the same scale, of course,) an illumination equal to that of 1,000 feet of gas, making a fair profit on the money invested. The cost of the plant he estimates at not more than half that of gas, or about $20,000,000 for the City of New-York, counting the investment in gas at $35,000,000. He gives no detailed figures in support of this claim, which, as the generators alone, at $600 each, would cost $15,000,000, strikes one as a rather sanguine estimate, particularly when it is considered that the 500 engines of 250-horse power each required to drive the generators, without any margin of power to spare, would cost $2,000,000, and the boilers $1,500,000 more, leaving only $1,500,000 for mains, buildings, labor, and the innumerable other appliances. It seems to be one of Mr. Edison's misfortunes to speak rather at random on questions that call for exactness of statement, and the worst of it appears to be that his lack of precision usually tells in his own favor.

The indestructibility, or rather the practical durability, of the carbon loop, again, remains to be demonstrated by practical or experimental tests extending over a longer period than any lamp has actually been in operation—600 consecutive hours. This, however, provided the loop proves tolerably durable, is not important save as a matter of convenience. The lamp—loop, bulb, platinum clamps, and all—could be manufactured at a profit for 10 cents, with first-class facilities for the work, and to replace an injured one with a new manufacture is merely the work of a moment. If an Edison loop lasts six months, the cost of renewing lamps in a Fifth-avenue mansion would not be a material consideration. It is evident, however, from the fact that Mr. Edison is still experimenting with loops composed of other materials than card-board—by carbonizing cotton thread, for instance—that he is by no means certain of the latter as a finality, and it is not unlikely that a new lamp may make its appearance before a great public exhibition is given. On the question of what changes in structure, if any, the card-board carbon suffered after 600 hours of incandescence, Mr. Edison's testimony is not very satisfactory. No microscopic sections appear to have been examined, either before or after the experiment, and one is left entirely to the statement of a superficial examination by the unassisted senses. It is not scientifically presumable that such molecular tension could be maintained, even *in vacuo*, for any great length of time, without affecting the material more or less, and if such has been the ease with these loops of baked card-board, the scientific world would like to be assured of it upon better authority than that of the unaided eye.

Mr. Edison did not give me an opportunity to compare loops that had been a long time in use with fresh ones, and I was compelled to get at the action in another way, namely, by observing the vibration of the carbon, which is very considerable, and amounts to a tremor very perceptible with a strong glass. Of course, I need not say that such a continuous vibration, though perfectly rhythmical, is utterly incompatible with the claims to durability which Mr. Edison admirers have put forth. He did not claim to me to have demonstrated the practical indestructibility of the loop, and was unwilling to make any definitive statement on that point. One of his assistants did, however, press that claim pretty persistently, although freely acknowledging that constant molecular disturbance must in the end result in disintegration of some sort. The degree of heat at which the loops were maintained while I had them under observation was not sufficient to produce a good light, and, consequently, not a fair test of what they will have to endure for practical illuminating purposes. Any conclusions based upon such a state of incandescence as I observed—not higher than that of an iron bar at white heat, and not more than half as high as that of the carbon points in a Fuller lamp[3]—would be liable to break down in practice.

Again, the lack of diffusibility that appertains to electric light, however produced, has not been overcome by the carbon loop and bulb arrangement, the light of Mr. Edison's electrical burner standing to gas-light in point of diffusibility about as seven to ten. Its refrangibility, when tested with good lenses—an experiment which I took occasion to try—is only about seven-tenths of that of light produced by combustion. It is to these properties that the comparative massiveness and the sharp definition of the shadows projected by an object which intercepts the electrical ray, are due. Hence, with equal illumination, a photometer based upon shadow tests would show a vast but entirely delusive advantage in favor of the electric light. In other words, the shadow produced by intercepting an electric light of 7-candle power is equal to that of gas-light at 10-candle power. For the same reason, when tested by any of the photometers now in use, an advantage, candle-power for candle-power, is shown in favor of the electric light, which is altogether delusive, and ought not to be used as a basis of comparison, without deducting seven-tenths from the result shown by the instrument, on account of the diffusibility of gas-light and comparative non-diffusibility of its competitor. I take it for granted that Mr. Edison is just as familiar with these facts as Mr. Kieth, the electrician of the Fuller Company,[4] or myself; but he certainly fails to make any allowance for them in his loose and vague estimates of the candle-power of one of his incandescent loops. Mr. Kieth, with one of the finest photometers I have ever seen in operation, confesses the necessary inaccuracy of its readings. I cannot ascertain that Mr. Edison or his assistants have given any consecutive attention to this source of error in their calculations respecting illuminating power.

So, again, on the question of steadiness—absence of flash, fluctuation, and tremor—the higher incandescence necessary to good light would necessarily produce a more marked vibration of the loop, and, besides disintegrating the carbon, would impair the steadiness attainable with low incandescence. The dilemma is that low incandescence is incompatible with maximum illumination, and high incandescence with steadiness; one or the other must give way. I saw Mr. Edison's platinum lamp in operation in June last, at a time when the newspapers were boasting in his interest that the question was solved, and told one of his assistants very frankly that it was a failure. That it was such events have proved. So, again, of the present form of lamp; the glass bulb becomes very hot at a low incandescence, registering 185° Fahrenheit, with the loop between a red and a white heat, as I saw it. It is needless to say that a register of 250° may be predicted as the minimum heat of the bulb, at a temperature of the loop capable of utilizing economically the illuminating power of incandescence. At this register, with 200 pounds atmospheric pressure, Mr. Edison will find his glass bulbs of very unsatisfactory durability, particularly as the glass is unequally heated. Our best test tubes snap at the bottom after exposure a few times to unequally-distributed heat, and I cannot believe that Mr. Edison's bulbs will endure inequality and alteration of temperature at 200 pounds pressure, without extreme liability to fracture. Indeed, there have been many renewals already with the light working under the immediate supervision of trained assistants.

Taking all the factors into consideration, therefore, Mr. Edison's new lamp is by no means superior to his former one, in which platinum was substituted for carbon. The exaggerated descriptions of it may serve a speculative purpose for those who hold stock in the company he represents; but it by no means solves the question of the practical application of electricity to house illumination. In its present state it is but an ingeniously-constructed toy, and there is a conspicuous insincerity in claiming that it is final and asking capitalists to invest in its practical introduction.

I may now recur again to the question of economy. Mr. Edison claims 40 lights of 16-candle power each from 5-horse power. In reality, he obtains 40 lights of less than 10-candle power each. Conceding 10-candle power as the maximum, 1-horse power sustains a light equal to 80 candles, or 5 standard gas-burners, so that the actual cost for a light equal to that of a gas-burner is $\frac{1}{5}$ of a cent per hour. It may be taken for granted that the cost of the plant, of distribution, superintendents, &c., will equal, if not exceed, that of gas, which, according to the testimony of gas engineers, is something more than the cost or manufacture. Gas, taking into account the value of the residual products, can be manufactured for about 30 cents per 1,000, at which rate the value of the gas alone, consumed per hour by a 4-foot burner, disregarding the question of distribution, would be three-twenty-fifths of a cent, while, on Mr. Edison's own estimate of the candle-power of his burners, the cost of an equivalent in his light is one-eighth or one-tenth of a cent. Clearly, then, he has sacrificed the whole economical margin claimed for the electric light in solving the question of divisibility, and, as clearly, his claim of utilizing in his generator 90 per centum of the force transferred by the belt, is either unfounded, or else his lamp and system of distribution are a failure; for his light, power for power, is fully as expensive in manufacture as gas, and fully four times as expensive as that of any other electrical system used in this country.

It must be concluded, then, on practical and economical as well as on scientific and optical grounds, that the Edison system in its present state could not successfully compete with gas, even if the capitalists could be found who are willing to subscribe the money to give it a fair trial in the City of New-York. That Mr. Edison is fully aware of this fact is sufficiently evidenced by his persistent evasion of all questions calculated to obtain the exact data of cost of manufacture to supply the 92 scattered lamps in operation at Menlo Park. During my brief interview with him he skillfully parried all queries intended to elicit the facts bearing upon that point, and I found his assistants no readier than he to supply information the particulars of which must be very familiar to them.

All these points are quite independent of questions of originality, which, by the way, may be safely left to the Patent Office. Whatever his friends may claim for him, Edison had never, so far as I know, assumed to be a great scientific discoverer; and when one looks over the specifications and drawings of the nearly 1,000 patents which have been granted for electrical generators, lamps, and other appliances within the last 30 years it is very evident that there is little verge for originality except in detail. It is not quite fair to call Edison a quack because he has not discovered a new force or invented a new method of manipulating an old one. But it is fair to say that he is not a scientific man in any sense of the term, and that his ponderous tomes of experimental researches are comparatively valueless because of a certain lack of precision and of attention to detail, and a certain looseness of method, combined with vagueness and generality of statement which seem to be inseparable factors of his intellectual organization.

F. G. FAIRFIELD, Ph.D., New-York College of Veterinary Surgeons,[5] No. 205 Lexington-avenue.

PD, *New York Times*, 16 Feb. 1880, 3. In Cat. 1241, item 1453, Batchelor (*TAEM* 94:579; *TAED* MBSB21453X). Form of printed fractions altered. [a]Date not that of publication.

1. See, for example, Doc. 1876 and "Edison and the Savants," *New York Telegram*, 23 Jan. 1880; "Edison's Secret," *New York Herald*, 23 Jan. 1880, "Mr. Edison and His Critics," *New York Sun*, 24 Jan. 1880, all Cat. 1241, items 1446–47, 1450, Batchelor (*TAEM* 94:576–77, 579; *TAED* MBSB21446, MBSB21447X, MBSB21450X).

2. Upton 1880a.

3. An arc lamp manufactured by the Fuller Electrical Co., a firm established in New York by James Fuller. *TAEB* 4:579 n. 2.

4. Nathaniel Keith was an electro-metallurgist, manufacturer, and inventor who obtained a number of patents during his career. In 1884 he became scientific editor of the *Electrical World*; in the same year he helped to organize and became the first secretary of the American Institute of Electrical Engineers. *WWW–1*, s.v. "Keith, Nathaniel Shepard."

5. Francis Fairfield had studied theology and briefly was a Lutheran minister before turning to journalism. He was a staff writer for several New York newspapers, including the *Times*, and a frequent contributor to other publications. In search of higher pay he studied veterinary medicine but shortly returned to writing. The New York College of Veterinary Surgeons, chartered in 1857, was at this time nearly moribund, having gone through the most recent of several reorganizations in 1879. *ACAB*, s.v. "Fairfield, Francis Gerry;" Obituary, *New York Times*, 5 Apr. 1887, 8; Bierer 1980 [1940], 34.

Notebook Entry:
Electric Lighting

Experiments on Comutators

We found by turning the brush out of right angels ~~of~~ to commutators the spark disapeared which led to the following experiments[1a]

I

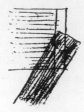

No 1   good[b]

2

"   2   N[o]G[ood][b]

3

"   3   very good & durable[b]

4

"   4   a little better than old way[b]

"   5   brass wire brush   result in this position same as coper No 3.[b]

6

" 6   N.G brass brushes[c]   little spark, but very little current   all the work was thrown on the other machine[b]
    7[d]

No 7   very wide copper brush tuching all over the comutator
not near as good as No 3 & 5   wears off fast   all work thrown
on other mach.
    8

8   very narrow copper[e] brush's. ⅜ wide   very good
    9

9   narow[f] brush same 8   little spark but no curent   all
thrown on the other mach.
    10

10   same brushes   Does more work with no spark than the
other mach   equaly as good than 3 & 5

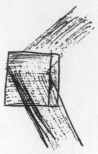

11   same brushes as in 7 (wide)[b]

12

12   same brushes as in 8, 9 10 but contrary angle   <u>very good</u> does more work than other mach without sparking

13

OK very good   standard   32°   Can at this angle be placed in any position all over the comutator

14

NG
TAE   J Kruesi

X, NjWOE, Lab., N-79-06-12:99, 98, 101, 100, 102 (*TAEM* 35:532; *TAED* N080:48). Written by John Kruesi; document multiply signed and dated. [a]Figures 1 through 6 aggregated on one page. [b]Followed by dividing mark. [c]"brass brushes" interlined above. [d]Figures 7 through 13 aggregated on separate page. [e]Interlined below. [f]Obscured overwritten text.

1. These experiments may have been on the "New Commutator" constructed according to Charles Batchelor's instructions of 2 January, several of which concerned the brushes. He specified that both the

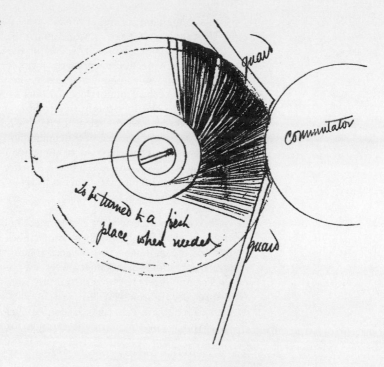

Charles Batchelor's design of a self-adjusting commutator brush.

guard

commutator

To be turned to a fresh place when needed

guard

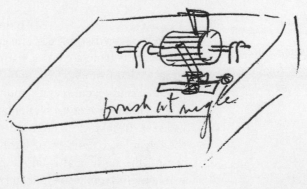

Edison's drawing for a commutator model with the brushes positioned at an angle to the commutator shaft.

brush at angle

brushes and commutator should be "twice as wide" as before, also that the "brushes must both be fastened to a circle having the centre of shaft for its centre and both move together by a handle on top." He also directed that "End brushes must be tried." Edison sketched a self-adjusting commutator brush on 4 February which evidently was intended to reduce the problems of wear and sparking. This device was to be made "with coil spring inside so as to turn a little itself if wears away fast." On 12 February he made a drawing marked "model" of a commutator with the "brush at angle." N-80-01-02.2:13–15, 66–67; N-79-06-12:118, both Lab. (*TAEM* 34:726–27, 753; 35:540; *TAED* No70:8–9, 35; No80:57).

Alterations resulting from these experiments were embodied in a patent application that Edison executed on 11 March (Case 208) in which he stated:

Commutator springs or brushes have always been arranged at right angles to the axis of the commutator. With such there has usually been a large amount of "spark" on the face of the commutators, indicating a loss of electric energy by its conversion at a point where conversion was not only unneeded but injurious, causing a rapid destruction of the commutator apparatus. I find that this can be largely and usually entirely avoided by arranging the commutator springs or brushes so that their axial line is at an angle other than a right angle with the axis of the commutators, or, in other words, that the brushes or springs bear obliquely upon the face of the commutator. In practice I have found that it is better that the brushes or springs stand at an angle of about thirty degrees to the axis of the commutator. [U.S. Pat. 264,649]

Edison also described an electro-mechanical device for shifting the armature shaft longitudinally so the brushes would not bear continuously on the same arc across the commutator face. Edison sketched this arrangement for Zenas Wilber a week later but did not send a model to the Patent Office. He filed a similar provisional British specification on 5 April 1880 (Edison to Wilber, 14 Feb. 1880, DF [*TAEM* 55:34; *TAED* D8036ZAC]; Brit. Pat. 1,385 [1880], Batchelor [*TAEM* 92:166; *TAED* MBP026]).

-1897-

Notebook Entry: Electric Lighting

[Menlo Park, c. February 12, 1880[1]]

Preliminary Estimate of the cost of lighting 10,000 ~~of~~ Electric burners from a central station, each burner giving a light equal to that from[a] sixteen[b] candles, or taking the place of a gas jet consuming 5 cu. feet an hour[2]

Five hours use of each burner is assumed as the average for the year. In taking this figure it was remembered that during the day a large number of lights are used in dark rooms and during the night a still larger number are used ~~through the whole night~~ from ~~8 to~~ eight to twelve hours   A gas jet which consumes 5 cu. feet an hour burning ~~for~~ five hours each[c] day will consume 25 cu. ft. in that time, or about 9,000 cu feet in a year.

For convenience ~~designate~~ in making comparison with gas[d] the electrical equivalent, ~~or~~ that is[e] the amount of electricity which will produce the same ~~amount of~~ light as [~~from?~~][f] ~~gas~~ 1000 cu. feet of gas, is[c] here designated as M   For example an electric burner is said to consume 9 M a year, or 10,000 burners 90,000 M a year

Tests show that it is possible to obtain 8 lights from one horse power carried by the belt, so that for 10,000 light 1250 horse power will be required[g]

Investment[3g]

Boilers & chimney From Babcocks[c] & Wilcox estimates[4h]	
	21,000
5 Engines Holly's estimate[5]	42,500
Foundations	2,300
Iron Structure	7,000
Shafts pulleys & c	3,900
Belts	4,700
Wood flooring & c	3,000
Water heater & pumps	7,000
Iron floor and supports Boiler	6,000
~~Boilers~~[i]	~~20,000~~[i]
Faradic Machines	24,000[6]
Conductors	50,000
	171,400

Total investment per M $1.90

Depreciation &c[g]

Boilers and chimney	10%	$2,100
Engines	3%	$1275
Foundations	1%	23
Iron Structure	2%	140
Shafts pulley &c	2%	77
Belts	10%	~~7~~470
Wood flooring &c	5%	150
Water heater, pumps	5%	350
Boiler floor	2%	120
Faradic Machines	3%	720
Conductors	2%	1000
		$6,425

Depreciation[c] M annually   7.14 cts
Taxes 2%   3.8 cts per M

Expenses[g]

Labor (Stoking included under coal)	Per day
2 Engineers	$6.
2 Foremen	10.
2 Wipers	3.
2 Regulators[7]	3.
	$22

Executive may be taken as about ½ of this or $11 $33 per day
or 13.2 cts per M Total[g]

Coal   Dust coal is taken as costing $3.50 a ton[8] under the boiler that is including labor of stoking. 3 lbs of coal an hour are required for[c] one horse power. $29.30 for coal a day   11.7 cts per M

Water for boilers waste oil &c   Estimated as costing ⅓ of coal   3.9 cts per M[g]

Rent $10,000 a year for athe cellar and first floor of a building 100 × 50 feet in any locality near the place where required 11.1 cts per M[g]

It is recommended that a small amount be charged yearly for each burner, enough to pay for breakage and attention.[j]

Summary

Depreciation	7.14
Labor	13.2
Coal	11.7
Water	3.9
Rent	11.1
	47.0 cts per M

That is the cost of the light delivered to the consumer will be 47. cts per M[9]

	47.0
Taxes 7%	3.8
Insurance 1/4%	.48
	51.3[10]

Selling Light alone[g]

The dividends will be

At 50 cts per M	1.5%
" 60 " " "	6.8%
" 75 " " "	14.7
" 90 " " "	22.6
" $1.00 " " "	30.5[c]
" $1.25 "	41%
" $1.50 " "	54%[g]

Power estimate.[11g]

It is estimated that 400 horse power can be sold during the day for various purposes, small shops, elevators, lathes, sewing machines, pumping.[g]

10 hour average use of this amount is assumed.

The cost will be simply the cost of coal under^c the boiler and water & c. $12 500 will have to be invested in conductors additional. ~~Power is sold~~^g

Power^c is sold today in New York at $150 a year or 5 cts. an hour, or at the rate of 83 cts per M. Electric power has this advantage it does not cost when not used

At 50 cts a M Light & power

additional dividend	Total
11.%^k	12%^k

at 60 cts a M Light & power

14%^k	20%^k

at 75 cts a M Light & power

19%^k	34%^k

at 90 cts for^c light and 200 H.P.
75 cts for 200 H.P.

22^k	45^k

at $1.00 for lights^c and^c 200 H.P.
75 cts. for 200 H.P

24%	54%

at $1.25 for lights and 200 H.P.
75 cts   "   200 H.P

28%	69%
Additional	Total

At 1.25 for lights and 200 H.P
1.00   200 H.P.

33%	73%

At $1.50 for lights and 200 H.P.
$1.00   200 H.P.

38%	92%

[Ven?]^l   24 hours^c a day

Depreciation	12%cts per M
Expenses	20%cts " M
Coal	15 cts M
Water & c	5
Rent	22
	74 cts per M

$3.80 per M   five hours a day power^g

$1.00 a thousand 7%
Power 400 H.P a day
After depreciation^g
Cost per M 15.6 cts per M^12

Taxes   2%
Insurance   ¼%
Patent right ⅓ capital[g]

Taxes   1.90
                                 .02
                                .0380[g]

additional for patent right   1.90
                                .63
                               $2.53 per M

X, NjWOE, Lab., N-80-01-26:95–99, 107, 109, 108, 111–25 (*TAEM* 34:48–50, 54–63; *TAED* N059:47–49, 53–62). Written by Francis Upton. Miscellaneous calculations and those on which Upton's figures are based have been omitted; some decimal points and commas added for clarity. [a]"equal to that from" interlined above. [b]"giving . . . sixteen" written over erasure. [c]Obscured overwritten text. [d]"in making comparison with gas" interlined above. [e]"that is" interlined above. [f]Canceled. [g]Followed by dividing mark. [h]Text written on two lines enclosed by left brace. [i]Erased. [j]Paragraph written on page facing "Expenses." [k]Overwrites erasure. [l]Illegible.

1. Francis Upton made this notebook entry over an uncertain period of time between the completion of Doc. 1889 and about 18 February. Thirty undated pages separate Doc. 1889 from this entry. It is followed by forty-five undated pages before there is evidence from which a date can be inferred; that is Upton's reference to measuring the resistance of lamp number 683, which was made on 18 February. N-80-01-26: 166, N-80-01-02.2:138, both Lab. (*TAEM* 34:79, 799; *TAED* N059:78; N070:81).

2. This notebook entry follows the increasingly detailed estimates that Upton made of the investment, depreciation, and operating costs for central stations of this size. It incorporates and elaborates on much of this earlier material (N-80-01-26:5–22, 71–91, Lab. [*TAEM* 34:3–12, 36–46; *TAED* N059:2–11, 35–45]). It is similar in some particulars to an itemization of expenses made by Edison and dated by Upton "about Dec 1879." Upton also made undated calculations of these costs in another notebook at about that time, all of them employing the unit "M" for comparing electricity with gas (TAE itemization, n.d., Unbound Documents [1879], Upton [*TAEM* 95:591–93; *TAED* MU043]; N-79-04-21:250–79, Lab. [*TAEM* 30:838–53; *TAED* N017:116–30]). Samuel Mott copied most of this document (with some alterations and a date of "Dec 1879") into a notebook he inscribed "Notes and mechanical movements. Things to be looked-up and remembered both Electrical and otherwise" (N-79-12-00:1–11, Lab. [*TAEM* 35:704–9; *TAED* N083: 1–6]). Upton continued making similar but more fragmentary notes and calculations in this book (N-80-01-26:122–28, Lab. [*TAEM* 34:62–65; *TAED* N059:61–64]).

3. The following estimates of investment and depreciation are based on Upton's revision of those made by Julius Hornig, a German-educated draftsman and mechanical engineer, hired in January to help plan central stations (Hornig's testimony, 3, 7, *Edison v. Siemens v. Field,* Lit. [*TAEM* 46:6, 8; *TAED* QD001:2, 4]). Hornig had calculated the dimensions and some costs for a 10,000 lamp plant with 5 steam engines and 180 dynamos (N-79-01-21:265–69, Lab. [*TAEM* 30:716–18; *TAED* No16:134–36]). Upton modified Hornig's itemization with a lesser "allowance" for each line based on a station of equal capacity having only two-thirds the investment in generators (see below). He offered no explanation for the "allowance" but presumably was motivated generally by the economy considerations articulated in Doc. 1889. In revising Hornig's figures Upton did not alter the total horsepower, which suggests that he did not plan to substitute high-resistance lamps to reduce the amount of energy consumed but wanted instead to use a smaller number of more efficient dynamos (N-80-01-26:71–74, Lab. [*TAEM* 34: 36–38; *TAED* No59:35–37]; see Doc. 1899 n. 1). Hornig, for his part, drew on 3 March (on the page facing his own cost estimates) a steam engine belted to a much larger Edison dynamo on the same bed plate; this is apparently a much larger machine than Upton had contemplated (N-79-01-21:264, Lab. [*TAEM* 30:716; *TAED* No16:134]). Hornig likewise offered no explanation for this idea but it is likely that he was trying to circumvent the design and operational difficulties of distributing mechanical power to so many generators. Nothing was done immediately with this idea, but see Doc. 1936. There are also several related rough sketches, unsigned and undated, of a steam engine belted to a single dynamo on the same bed plate (Undated Notes and Drawings [c.1879–1881], Lab. [*TAEM* 45:151; *TAED* NSUN07:29]).

Reflecting Edison's concern with the design and cost of a commercial generating station, an unidentified Philadelphia newspaper reported in a 12 February article that "Mr. Edison says the great item of importance

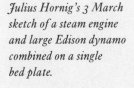

Julius Hornig's 3 March sketch of a steam engine and large Edison dynamo combined on a single bed plate.

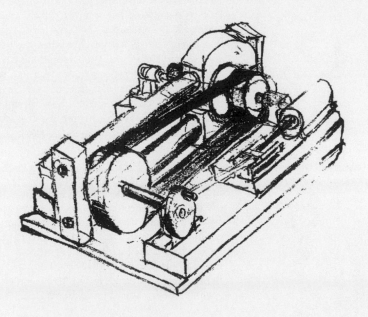

to be secured in connection with his electric light is cheap steam engineering. . . . He says 'steam engineering forms 75 per cent. of the electric light, 20 per cent. is in the system itself, 4 per cent. is in the dynamos and 1 per cent. in the lamps, and yet they are howling about the lamps, as if they were the item of greatest importance, when, in fact, they are of the least.'" "The Coming Light," 12 July 1880, Cat. 1014:34, Scraps. (*TAEM* 23:648; *TAED* SM014034a).

4. The New York firm of Babcock & Wilcox was a noted manufacturer of stationary boilers, especially its patented "non-explosive" water-tube design which carried a higher pressure than other types. The firm had built a new boiler and engine house for Edison's enlarged machine shop in 1878. The basis of the present estimate is unknown. Hunter 1985, 2: 336–39; *TAEB* 4:600 n. 2.

5. The Holly Manufacturing Co. of Lockport, N.Y., manufactured engines designed by Birdsall Holly. Built for pumping water directly through a distribution system, these unusually complex engines provided practically uniform pressure under widely and rapidly fluctuating loads. The firm sent Edison specifications of its engines and photographs of a large installation recently made in Buffalo. It proposed in November to supply similar 250 horsepower engines at $9,500 apiece and by early the next month was drawing plans for Edison's New York central station. The 12 February newspaper article reported that Edison planned to use only four engines at a time and keep the fifth for a spare, which Upton asserted was Edison's plan for all future generating stations. Hunter 1985, 2:561–62; Holly Manufacturing Co. to TAE, 13 Aug., 1 and 13 Oct., 13 and 28 Nov., 6 Dec. 1880; Holly Manufacturing Co. to Stockton Griffin, 2 Dec. 1880; all DF (*TAEM* 50:102, 143, 148, 160, 166, 169, 167; *TAED* D7919ZBT, D7919ZCT, D7919ZCX, D7919ZDK, D7919ZDN, D7919ZDP, D7919ZDO); "The Coming Light," 12 Feb. 1880, Cat. 1014:34, Scraps. (*TAEM* 23:648; *TAED* SM014034a); Upton 1880a, 541.

An estimate of central station costs that Upton made later, probably in early May, allocated only $22,500 for engines. Other expenses went unspecified or remained the same except dynamos, for which $22,500 was also estimated. It is not clear what these changes might represent but they may reflect Edison's consideration of Porter-Allen engines and direct-connected dynamos as discussed in Doc. 1936. In October 1879, Upton compared the coal consumption of a high-pressure Holly engine with that of a less expensive but less efficient condensing engine. He determined that the Holly's higher first cost would be recovered when the engine ran five hours per day for a year. N-80-01-26:232–35, N-79-07-07.1:129–35, both Lab. (*TAEM* 34:112–13; 32:312–15; *TAED* N059: 111–12, N037:64–67).

6. In revising Hornig's dynamo estimate from $36,000 Upton seemingly intended simply to use two-thirds as many machines, these being modified to run cooler, perhaps using the longer armature proposed in Doc. 1889, at the same unit cost. On or about 14 February, however, he compared the work done by an Edison generator having .14 ohm resistance, which heated badly, with another having .56 ohm resistance. On this basis he calculated that "by doing away with the heating ⅓ more lights can be obtained," meaning that for a $200 generator "$266 can be spent on the armature to save the heating." Adding to this the annual op-

erating cost of wasted horsepower, Upton concluded that an additional expense of $412 per machine would be justified. N-80-01-26:156-57, N-79-10-18:222-30, both Lab. (*TAEM* 34:74, 32:894-98; *TAED* No59: 73, N301:112-16).

7. That is, workers to monitor an indicator of output voltage and make appropriate adjustments to the strength of the dynamo field magnets.

8. Upton specified in calculations on the facing page that this figure was per long ton (2,240 pounds).

9. In both a preceding and a later estimate of the cost to consumers, Upton included patent rights, presumably to be licensed from the Edison Electric Light Co. In the earlier case he also added interest on the plant. N-80-01-26:16, 124, Lab. (*TAEM* 34:9, 63; *TAED* No59:8, 62).

10. Upton apparently wrote the figures for taxes and insurance at some later time; he gave no indication of the basis for these percentages.

11. In the course of central station estimates immediately prior to this document, Upton made what are evidently his first attempts to anticipate the return from selling electric power. He stated that "If for 10 hours during the day 400 horse power could be delivered at the consumers at a cost of 600 at the station and with an investment of $12,500 for additional conductors, at 50 cts per M a dividend of 17% could be earned. In making this estimate it was remembered that the wages, depreciation were calculated as fully paid by the light for a day of 15 hours. This leaves for cost of power only the coal and water &c consumed." N-80-01-26:78, 89-91, Lab. (*TAEM* 34:40, 45-46; *TAED* No59:39, 44-45).

12. This is based on a set of calculations in which the cost of coal is .156 cents per pound and "water &c" is .23 cents, producing a cost per horsepower of .93 cents.

EDISON EFFECT AND LAMP LIFE   Doc. 1898

During February 1880, Edison began investigating phenomena taking place in his lamps that would subsequently lead him to the observation of what became known as the Edison Effect. His laboratory notebook entry of 13 February has long been considered the beginning of research on the Edison Effect. In particular, the first drawing of that entry, showing a wire inserted near the top of a lamp bulb, has frequently been referred to as the first design for an Edison Effect lamp. However, there is no evidence that Edison recognized at this time the particular event later designated the Edison Effect, namely that current passing through the heated filament can produce electrical potential in a separate electrode in the lamp, and it is not clear when he did note this.[1] Regardless of whether this notebook entry can properly be considered the beginning of Edison Effect research, it clearly marked the start of Edison's investigation into a group of related phenomena that shortened the useful life of his lamps.

In the fall of 1879 the laboratory staff had noticed a black coating of the lamp globes.[2] They noticed this again during lamp tests in January, as well as frequent breakage of the carbons at the clamps and arcing between the clamps.[3] The experiments suggested in Doc. 1898 were attempts to deal with some or all of these occurrences, which Edison attributed generally to the "carrying by electrification of the carbon from one side of the carbon horseshoe."[4] In a patent for which he applied in January 1881 he more fully defined this phrase as

an absolute carrying or moving of the carbon itself from the negative to the positive end of the carbon. This action seems to be similar to that in galvano-plastic [electro-plating] operations. The cohesion among the molecules of the carbon seems to be so weakened by the heat of incandescence that the molecules, or a portion of them, are gradually moved from one end to the other of the carbon. The amount of such carrying depends upon the resistance of the filaments, the degree of incandescence, the electro-motive force between the clamping-electrodes, and the state of the vacuum. While its amount may vary with varying conditions of these elements, it is the great cause of the ultimate destruction of the carbons used in high vacuo, and if its degree and amount can be reduced a proportionate increase in the life of the carbons is assured.[5]

While Edison was using the special lamps for the experiments described in Doc. 1898 he recorded two other phenomena which at least suggested electrical carrying. One was a fluorescence, with which he was evidently familiar, and which he now recognized could not be explained fully as the effect of charged gas molecules propelled forcefully away from the filament in the manner hypothesized by William Crookes in his paper "On the Illumination of Line of Molecular Pressure, and the Trajectory of Molecules."[6] The other was a "blueish halo" around the base of one side of a filament near the clamp. Edison attributed this to gas released from the carbon clamps and thought it was responsible for arcing between the clamps, though he noticed that this, too, darkened the bulb (see Docs. 2061 n. 3 and 1902).[7] In May, after substituting copper clamps for the carbon ones, a similar "beautiful bluish appearance" was seen around one of the clamps. Noting that the mysterious blue moved to the other side when the current was reversed,

Francis Upton attributed it to copper vapor emanating from the clamps, which he expected would be destroyed. A month later, while trying to reduce the radiating surface of a bast filament, Upton noted that the consequent increase in heat caused a more visible "bluish vapory appearance" at one of the clamps.[8] This led Edison to invite Princeton professor Charles Young to make spectroscopic experiments on the fluorescence to try to determine its origin in hopes of preventing it and the destructive arcing often seen with it (see Doc. 1944). The blue luminosity appeared frequently in the first lamps made at the lamp factory in October 1880, seriously compromising their durability. By the beginning of December, however, Edison and his assistants had noticed that the blue emanation ceased at the highest level of evacuation, and reports of the problem disappeared from subsequent lamp tests (see Doc. 2029).

In October 1880 Edison also noticed that when lamps from his factory were run at high candlepower some seemed to contain a "fuzzy lampblack" substance which he ascribed to electrical carrying. In many lamps the filaments also presented a "honeycombed" appearance at a spot on the negative side and were prone to breaking there, a problem he tried to abate by reversing polarity (see Doc. 2026).[9]

Edison subsequently made intermittent efforts to comprehend and overcome the difficulties associated with "electrical carrying," but with limited success.[10] Full understanding of the action within the lamp globe ultimately required a theory of the electron and even this was insufficient; blackening of the globes persisted after metallic filaments replaced carbon ones, until General Electric devised a chemical solution in 1913.[11]

1. Preece 1885; for the history of the Edison Effect see Hong 2001, chap. 5.

2. See Docs. 1831 n. 3 and 1860. William Hammer later stated that he was the first to notice that this created a "phantom shadow" of the filament on the globe. This assertion is made in a 3 October 1884 account of the "blueish halo" (see note 7), in a different hand and ink, as a caption for his drawing of an experiment on the halo. Hammer's claim reads: "'Phantom Shadow' discovered W. J. Hammer later called 'Edison Effect' the base of radio tube industry" (box 24, folder 2, WJH). This was obviously written at a much later date but has often been accepted as an authoritative account (see, for example, Hong 2001, pp. 121–22). The first extant contemporaneous notes on this phenomenon are Edison's from October 1880, when he noticed a "black streak paralell with carbon connecting to Positive on glass & white streak paralellel to Negative." Some time after this he also noted "positive side give the white streak on glass—neg also white   I guess the carbon merely

screens the carbon vapor according to which side busts   nothing in it"
(N-80-10-15.2:51, 87, Lab. [*TAEM* 39:27, 44; *TAED* N149:26, 44]).

3. N-80-01-02.1:132; N-80-01-02.3, passim; N-80-01-02.4:127–28;
all Lab. (*TAEM* 34:596, 35:2, 781–82; *TAED* N067:133, N074, N084:
39–40); see also Doc. 1850 n. 2.

4. According to Charles Mott, by early summer Edison "was of opinion that the carbon was being carried from one side to the other of the
loop directly across and for farther investigation he had a lamp made
with a oval shaped glass partition attached between the conducting wires
and extending nearly to the top of the loop at right angles with the faces."
Ludwig Böhm's drawings of this lamp and a variant have been found but
there are no records of experiments. Mott Journal N-80-03-14:235–36,
N-80-30-19:113, both Lab. (*TAEM* 33:801–02, 34:661; *TAED* N053:
119–20, N068:55).

5. U.S. Pat. 248,418.

6. Crookes 1878b consists of a lengthy abstract of a paper that was evidently never published in its entirety. A reprint of the abstract under the
same title was added to a laboratory scrapbook (*Teleg. J. and Elec. Rev.*,
1 Jan. 1879, Cat. 1050:63, Scraps. [*TAEM* 26:427; *TAED* SM050063a]).

7. In October 1884 William Hammer recalled the first observation of
this phenomenon (see note 2):

> while experimenting & carrying out a large variety of experiments
> on Edison lamps at the laboratory of Thos. A. Edison which experi
> ments were conducted under Mr. Edison's supervision & covered a
> large field of work, we noticed for the first time a peculiar fluores
> cence about the clamps of the carbon which was very "will 'o' the
> wisp" like in its character & we noted two different forms of this
> blue fluorescence or radiance one of which appeared at low vacuums
> the other only at very high rarification or vacuum. We observed this
> blueness could be influenced by magnetism & we by means of a
> powerful magnet from the ore separator drew the fluorescence to
> the side of the globe chased it from one pole to the other & even
> drove it off altogether. [WJH, ser. 2, box 24, folder 2]

On Edison's subsequent use of magnets to manipulate this phenomenon, see Docs. 1944 esp. n. 8 and 2029 esp. n. 3.

8. Mott Journal N-80-03-14:157, 159, 233–34, Lab. (*TAEM* 33:763–
64, 800–01; *TAED* N053:80–81, 118–19).

9. N-80-10-15.2: 25, passim, Lab. (*TAEM* 39:25; *TAED* N149:13).

10. See Doc. 2025 in addition to the documents already referenced.
Edison's research on these problems will be included in subsequent
TAEB volumes but see Israel 1998, p. 320; a brief overview through 1883
and the subsequent development of the Edison Effect lamp is provided
by Hong 2001, p. 121–34. In October 1882 and at intervals throughout
1883 Edison filed patent applications for a number of means to abate
electrical carrying and its effects; see U.S. Pats. 268,206; 273,486;
274,293; 492,150; 278,417; 459,835; 341,839; 398,774; and 425,761.

11. Wise 1985, 152–56; see also Howell and Schroeder 1927, 123–50.

Notebook Entry:
Electric Lighting

[Menlo Park,] Feby 13 1880
Following experiments are for the prevention of carrying by Electrification of the Carbon from one side of the Carbon horseshoe[1a]

Experiment No. 1.[2]

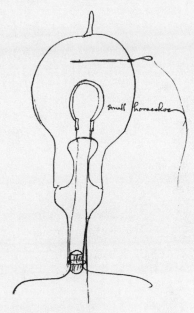

Large globe small horseshoe

carbon coated with an oxide, say alumina or Lime.[3]

Put up near top to ascertain if coloration goes below dotted line

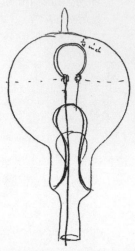

Small globe to ascertain if coloration rapid & to develope other phenomena.

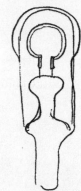

Mica partition or glass; small carbon[4]

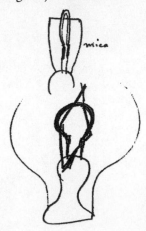

Leyden jar   Inner foil connected to one pole outer foil to other pole—idea—contrary electrification[5b]

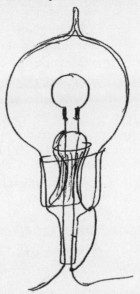

Chas Batchelor

X, NjWOE, Lab., N-80-02-08.1:53 (*TAEM* 34:238; *TAED* No63:37). Document multiply signed and dated; miscellaneous doodles not transcribed. [a]Followed by dividing mark and a list of numbers increasing by tens from 70 to 210. [b]Obscured overwritten text.

1. See headnote above.

2. There is no extant explanation for the design of this experiment, nor any record of it having been carried out. Within a few days Edison was considering effects within the lamp globe in terms of streaming negatively-charged molecules, as postulated by William Crookes (see Doc. 1902 n. 3). Later in the year he understood the blackening of lamp globes as the result of electrostatic attraction between the glass and carbon particles emanating from the filament; in either case this arrangement may have been intended to attract molecules or larger particles to a charged or grounded wire instead. He tried a similar approach in November and subsequently took out patents for several methods of overcoming the electrostatic attraction of carbon (see Doc. 2026 and headnote above).

3. See Doc. 1902.

4. This was tried later in the year; see Doc. 1944.

5. This was tried later in the year; see Doc. 2025 n. 1.

[Menlo Park,] Feb. 14 1880

.037 plates

Mr. K[ruesi]

Please count and weigh the plates on both ~~machines~~ armature thick and thin[1]   also ends[a]

Measure the distance from insides of outside thick iron plates from each other

Paint the ~~a~~thin plates

Devise some means so that there can be no metallic contact between any of the plates, that is[b] so that each one will be insulated from each other. The brass rod running through them is bad I think.

F.R.U.

ADDENDUM[c]

[Menlo Park, February 14, 1880]

Mr U.

198 rings		Total 85 lbs
average thickness	.037	
weight thin rings	71 lbs	
" thick disks	14 "	
distance between disks	8 in	
thickness of two disks	8.83 in[d]	
Thick plates armature[e]	same space	
weighs		75 lbs ~~74½₄~~[f]
discs on ends		14 lbs
Total		89 lbs

ADS, NjWOE, Lab., N-79-10-18:214 (*TAEM* 32:890; *TAED* N301:108). Written by Francis Upton. [a]"also ends" interlined below. [b]"that is," interlined above. [c]Addendum is a D, written by John Kruesi. [d]Followed by dividing mark. [e]Interlined above. [f]"74½₄" circled and interlined above.

1. These experiments were intended to find ways to reduce heating, a problem to which Upton returned after the redesign of Edison's armature in January (see Docs. 1875 and 1889). The "thick" armature consisted of 186 iron plates .037 inch thick "bolted together with brass rods insulated from the ends. Large iron plates on ends insulated." It was tested this day and upon inspection found to be "very much oxydized." The "thin" armature consisted of .014 inch plates. They were painted on this date and then "The burs from cutting the plates were carefully filed away so that there was no connection through them." They were assembled with an insulating layer of tissue paper between every plate. After Kruesi weighed the plates (see below) Upton noted that "accord-

ing to weight of metal the thin plates should give . . . ½ the heating"; its advantage in a test this day apparently was not so large. He had previously tested an armature with painted plates which he hoped would reduce conductivity through the length of the core but it actually ran slightly warmer than one not painted. He also tried "split rings with air spaces and cups placed on the outside to force air in" but this seems to have heated badly. Another design that he briefly tried was an armature made of sheet iron "wound in the form of a spiral." This heated rapidly to the point of smoking and was judged "Very bad." N-79-10-18:174–77, 200–201, 206–210, 212–213, 216–17, 221–22; N-79-06-12:107; both Lab. (*TAEM* 32:870–71, 883, 886–89, 891, 893–94; 35:536; *TAED* N301:88–89, 101, 104–107, 109, 111–12; N080:53).

Shortly after completing this document, Upton returned to the thick and thin plates and remarked that a .037 inch plate armature ran 75 lights at 95 volts "for three hours and heated to 100° C. This is the best ever done in the world and brings us home with this size of machine" (N-79-10-18:234, Lab. [*TAEM* 32:900; *TAED* N301:118]). He continued making tests and undated notes about this design for some time (N-79-10-18:218–59, Lab. [*TAEM* 32:892–1002; *TAED* N301:110–30]). Henry Rowland later credited Edison's experiments with this form of armature for first demonstrating "distinctly that heat does not necessarily accompany magnetisation and demagnetisation, *but is almost solely the result of the electric currents induced in the revolving mass by the change of magnetisation*" (Rowland 1881, 294–95; see also Doc. 1992).

Edison executed a patent application (Case 208) on 11 March that was similar to a British provisional specification filed on 5 April (Brit. Pat. 1385 [1880], Batchelor [*TAEM* 92:166; *TAED* MBP026]). The application included

an armature of a number of very thin discs or rings secured together upon a proper shaft or base and slightly insulated from each other. I have found that very good results are attained by using discs or rings ⅟₃₂ to ⅟₆₄ of inch thick separated from each other by sheets of tissue paper. As currents induced in the rotating armature have very small electro motive force, the insulation of the tissue paper and the thinness of the plates prevent almost, if not quite entirely, the circulation of induced currents and the heating attendant thereon.

By this the effective capacity of the machine for conversion is very largely increased, for while the heating at one moment is not large, the covering of the armature prevents radiation and the accumulation of heat proceeds faster than its dissipation and the armature becomes an accumulating reservoir of heat. [Pat. App. 264,649]

The claim for this armature construction was challenged by the patent examiner as being anticipated by Sigmund Schuckert. After several attempts to modify the application to meet this objection Edison filed a separate application for a plate armature on 14 August 1882 (see correspondence in Pat. App. 264,649). The original application issued a month later as U.S. Patent 264,649 with the claim (though not the description) deleted and a reference to the new application inserted. The new application, which issued as U.S. Patent 265,785 in October 1882, included a disclaimer concerning the Schuckert armature as well as two

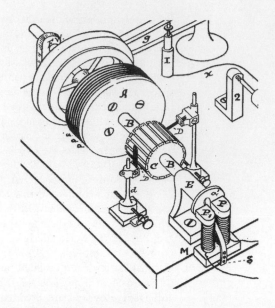

In this portion of the main drawing from Edison's U.S. Patent 264,649, showing only part of the dynamo, the disc armature A is on the shaft B and to the left of the commutator C.

significant modifications. Rather than merely place the discs on a shaft Edison placed them around a wooden core, thus further reducing the amount of heat generated within the armature (and presumably also its weight). The discs were also held in place by a new method of clamping.

−1900−

Notebook Entry: Electric Lighting

[Menlo Park,] Feb 15th 1880

Experiments in Carbonization

In our experiments with Bristol board we have been able to bring our loops both large and small to a resistance of 200 ohms using well calendered board of 9 and 6½ thousandth.

As this board was known to be very impure owing to the clay, etc. and resins used in sizing we had some pure uncalendered paper made— This came to us varying in thickness from 15 to 10 thousandths— Large loops made from the 15 thousandths measured from 600 to 800 ohms and small loops made from the 10 thdths measured 400 to 500 ohms. These high resistances we thought must be due to oxidization although we could not see it direct— The loops <u>had not</u> that peculiar steel <u>gray</u> color that the former ones had— This we attributed to the absence of that resinous matter that is[a] used in sizing the 'board'[1]

1 We made a solution of resin in alcohol and soaked some of the loops in it (a weak solution) but these were 1500 ohms— In this case they really looked oxydised although the tissue papers did not—

2[b]   We now put in 4 of the old style small loop and they came[a] out 250 ohms showing that probably it is[a] in the paper and not due to leakage in the moulds

We put the new paper under pressure but found not much difference—

We hammered the moulds so as to close the pores to prevent oxydising   3 loops brought from the hammered mould however measured 400 ohms[c]

5[b]   In the same mould that the old regulars came out of (250 ohms) we put 3 of the new paper pressed [~~pasted in little~~? ~~----- -----~~][d] and they again came out 450 ohms with tissue paper looking splendid— Carbons looking very black and dull as they always have done[c]

6   Same mould as above and double strength of resin came out 400 ohms

7   Same mould again and doubled strength again of resin— Came out beautifully straight—but had a resistance of 480[a] ohms[c]

~~No 8[b]   (3 loops new paper (pressed) carbonized with cards in between of same paper pressed—came out 275 ohms—these were held down by a plate of nickel—325 335 & 440—I think the last on top—same as 9[c]~~

No 9   We now put 3 new (pressed) paper loops in with card between it same[a] pressed paper (no tissue paper)—and nickel plate to put on weight— These came out not very good shape but only 275 ohms

No 10   Three (new)[f] regular pressed paper tissue paper— nickel weigh[t][g]—325 338 440   I think the last on top   Hammered mould[b]

No 11   Three new (pressed) paper with extra double diabolically strong solution of resin in alcohol—tissue paper— ordinary carbon weight   325—315 other broke

No 12   Three new paper (pressed—tissue paper—ordinary carbon weight— Came out 320 to 340— Boiled them afterwards in solution of Sugar in Alcohol with little water— Put them on gas heater again— Owing to putting in new tissue paper the sticky loop would not allow it to shrink properly and it broke up the loops   try another[c]

No 13   Took a piece of uncalendered paper 10 thousandths and soaked in Creosote and Tar.— Carbonized it—[c]

14   Bottom loop soaked in Tar and Creosote   Bad looking 600 ohms

top loop soaked in Sugar and alcohol   good looking (excellent)   142 ohms   make more[c]

15 2 loops of[a] regular paper—tissue paper— Carbon weight— put in fire clay mould— plenty of tissue paper packed in to keep it tight

<u>16</u>[b] 3 Regulars for second boiling[2]

Lamps

<u>669</u>[b] Old small loop made from 6½M.—about 250 ohms— Smoked by Camphor.[c]

670 Same as 669 but not Smoked[c]

671 New small loop from pressed paper soaked in sugar R 142 ohms—glossy—

Chas Batchelor

X, NjWOE, Lab., N-80-01-02.2:104 (*TAEM* 34:773; *TAED* N070:55). Written by Charles Batchelor; document multiply signed and dated. [a]Obscured overwritten text. [b]Multiply underlined. [c]Followed by dividing mark. [d]Phrase interlined above and canceled. [e]"same as 9" written over paragraph. [f]"(new)" interlined above. [g]Written off edge of page.

1. Batchelor began these experiments on 12 February, when he found the resistance of two small loops of .015 inch paper to be 300 ohms. He measured large loops from the same stock at 600 ohms and "not being able to account for such high resistance we put them on the pumps." Several such lamps were made about this time, including some with oxide-coated loops (see Doc. 1902). Batchelor also tried several other filament materials, including cocoa matting and bast and kittool fibers (N-80-01-02.2:90–102, Lab. [*TAEM* 34:765–72; *TAED* N070:47–54]). Francis Jehl and William Hammer kept incomplete records of the resistance of lamps with these filaments in N-79-12-09:109–15, Lab. (*TAEM* 32:708–11; *TAED* N041:56–59).

2. Batchelor resumed these numbered carbonization and lamp trials the next day. During the next two weeks he made approximately five dozen similar experiments with paper loops and also various fibers, especially bast and jute. Many of these were soaked in sugar solutions. On 17 February he noted that bast soaked in boiling sugar was "Bully!— very dense." He boiled the bast in sugar again before carbonizing it. The notes refer to "old" and "new" sugar solutions but these identities are unclear since at least nine different solutions of sugar in water and alcohol had been made by 22 February. Edison summarized results of these sugar experiments in a draft caveat (Doc. 1907). N-80-01-02.2:124–27, 133–65; N-79-12-09:117–57, both Lab. (*TAEM* 34:787–89, 795–826; 32:712–33; *TAED* N070:69–71, 77–108; N041:60–81).

New York, Feb. 16 1878̶0[a]

Dear Sir

I expect to come to M.P. on Friday morning to go over telephone history generally.—

Will you oblige by having one of your young men go over the bundle of sketches (the original ones) and lay out all matters relating to Telephones or acoustic matters leading thereto, and arrange them according to dates.[1]

I will bring down the papers that are at my office in the envelopes.[2] Yours truly

Lemuel W. Serrell

ALS, NjWOE, DF (*TAEM* 55:36; *TAED* D8036ZAE). Letterhead of Lemuel Serrell. [a]"New York," and "187" preprinted.

1. Serrell had asked Edison a few weeks before to retrieve from his office "a bundle of the papers relating to the Telephone Interferences, as there is such a mass of them." This is the first indication of a sustained effort to identify and reorganize notes and drawings relevant to the continuing interference cases. On 1 April Serrell requested that Edison "have your assistants sort out the drawings and evidence according to date. . . . I also want any newspaper articles you may have that give descriptions from time to time of what you were doing." Two days later Charles Mott noted in his journal that Edison was "overhauling the old books and papers for matters in relation to telephone." This process continued for several days and resulted in the excision of a large number of notebook pages and tracings of many drawings between 1875 and 1879 (see *TAEB* 3:19–20). Edison began giving testimony in the Short case in January; the process continued intermittently through mid-April. Serrell to TAE, 21 Jan. and 1 Apr. 1880, DF (*TAEM* 55:11, 77; *TAED* D8036I, D8036ZBO); Mott Journal N-80-03-14:44, 48–49, 52, 56–57, 70, 74, 81, Lab. (*TAEM* 33:706, 708, 710, 712, 720, 722, 725; *TAED* N053: 23, 25, 27, 29, 36, 38, 41); Edison's preliminary statement is in *Edison v. Short*, 1 (*TAED* W100DBA001); testimony by Edison and several assistants is in *Edison v. Short*, 22–33, Lit., Supp. III (*TAEM* 162:936–41; *TAED* W100DBA022, W100DBA027, W100DBA028, W100DBA029).

2. Serrell's son had telegraphed his father, who was at Menlo Park on 14 February (Saturday), that there were still "paper and sketches of telephone interferences in safe." Harold Serrell to Lemuel Serrell, 14 Feb. 1880, DF (*TAEM* 55:35; *TAED* D8036ZAD).

–1902–

Notebook Entry: Electric Lighting

[Menlo Park,] Feby 16 1880

Coating a carbon with a paste of Zirconia oxides with alcohol caused the Zr. to melt and contract and put a tension on the carbon which broke it.[1] I think some of the oxide was reduced by the carbon but am not sure— The zirconia was coated with carbon in some places jet black in other places only browned—

the lamp was brought to vivid incandescence small horseshoe used—gave probably 2 gas jets— The fluorescence was the strongest we have yet seen being intensely violet.[2] this seems to show that the fluorescence is not entirely due to the molecular impact ofn the glass:—[3] The appearance of the oxides on examination under the microscope resembled paste which had dried[a] contracted and cracked. where the arc formed there was vitrous fusion[b]

The same size of carbon was coated in the same way with alumina[4]   this was immediately reduced by the carbon at high incandescence and destroyed it= This [is?][c] another confirmation of my idea that metallic aluminium may be produced cheaply from the oxide by packing it around ~~exhaus~~ long thick sticks of carbon and exhausting the tube and rendering the carbon incandescent by the current from the Faradic machines[5]   no flurorescince, as it was destroyed before highest temperature reached

The caused of the formation of a long crack and the subsequent breaking of a chip of the glass by the action of the pressure of the atmosphere is undoubtedly due to the sudden heating of the platina wires passing through the glass especially when the wires are small and the resistance of the carbon small. with .010 wire and 78 ohm (cold) lamp slowly putting on current did not cause a crack whereas when the glass was cold the sudden putting on of the current caused the lamp to be instantly destroyed by the formation of a crack which so weakened the glass that the pressure of the atmosphere was sufficient to break of a piece of the glass and project it into the vacuum—

The[a] same cause is active when an arc springs across between the clamps which if it does not have sufficient power to melt the wire is sufficient to give a sudden accession of heat to the platina wires and thus cause a sudden expansion cracking the glass.

I have noticed that when lignium vitae clamps, or any clamp made of carbonizable material is used that on reaching high vacua and heating that a gas comes out of the clamp and lowers the vacuum and this will continue for hours. if[a] there is much gas   There will appear a blueish halo shooting out from the carbon (one side only) near the clamp: this gradually increases until there suddenly springs a bluish arc between the clamps and if the circuit is not immediately broken on appearance of this arc the wires will be fused. on destroying the arc[a] no matter how quickly after its formation the clamps will be found to be red hot. this in its turn tends to cause more gas to flow out of the clamp and reduce the resistance ~~of~~and augument the size of the arc= as the arc has small illuminating power although between carbon clamps I take it that it is entirely gaseous and that it is carbonic oxide or carbonic acid gas. when this arc is formed I notice that the glass near the clamps is blackened in the same was a the whole of the globe is by the long continued action of the lamp: ~~and~~ The platina leading wires to[a] which the carbon clamps are secured even when there has been no perceptible arc are coated with carbon just like lamp black. For a space of ¼ of an inch. The clamps instead of being shiny are black and in fact the whole fibre has this sooty deposit as the great drop shown[a] by the McLeod gauge when the first heating takes place and the long continuance of the drop when the carbon remains heated notwithstanding the action of the sprengil shows that the gas must be in a condensed form and great in quantity[a] and If it be carbonic oxide, it must have a very pernicious effect, hence I think that the smallest possible carbon clamps should be used if used at all and some means should be used to heat the clamps before the fibre or horseshoe is heated so that the air or gas may be abstracted by the pump before it has a chance to act upon the lamp but the best clamp would undoubtedly be platinum Iridium alloy of the smallest dimensions—[6]

I have noticed that the glass of nearly every lamp especially those which have shown fluorescence has a mottled appearance; the polish being taken off where the spots are and these spots are more whitish than the rest of the glass; It may be that this is molecular bombardment or it may be that the carbonic acid has combined with the potash & soda of the glass and after all have combined it is decomposed again by impact or it combines and any further carbonic acid is decomposed but this is a far fetched supposition. ~~if copper could be plated on the carbon and then to the platina clamp it would be boss but that damd arc'g~~

X, NjWOE, Lab., N-80-02-16:1 (*TAEM* 41:924; *TAED* N303:1). [a]Obscured overwritten text. [b]Followed by dividing mark. [c]Illegible. [d]Canceled.

1. On 13 February Charles Batchelor made a lamp with a small filament of about 300 ohms cut from pressed paper which he coated with zirconium oxide. This lamp leaked and the next day he made two more like it but no test results were recorded. N-80-01-02.2:102, Lab. (*TAEM* 34:772; *TAED* N070:54); see also Doc. 1900 n. 1.

2. Later records of lamp tests include frequent references to a blue

glow at or near the clamps but this is the first known reference to this phenomenon.

3. William Crookes advanced a theory of molecular bombardment to explain various luminous high-vacuum electrical discharge phenomena, including a phosphorescence of the glass. Crookes hypothesized that residual gas molecules, coming in contact with the negative wire, acquired a similar charge and were forcefully repelled in a stream accelerated along the lines of electrical force radiating from the electrode. He argued that in a moderate vacuum the molecules would attain such velocity that when they eventually collided with neutral molecules the energy released would be in the form of visible light. In a high vacuum he supposed that the molecular stream would extend unimpeded all the way to the glass, where its impact would explain the mysterious luminosity observed under these conditions. This argument comprehended Crookes's corpuscular explanation for the action of the radiometer and was part of his more general hypothesis of a fourth state of "radiant matter" at low pressure. It was the subject of considerable controversy around this time and eventually proved vulnerable to mathematical demonstrations that the mean free path of molecules at even the highest vacuum was far shorter than he had supposed. DeKosky 1976, 47–56; Nye 1996, 82–84.

4. Batchelor also tried this for the experiments described in Doc. 1900. On 13 February he took a small paper loop and "coated all over with Alumina finely powdered and mixed with Alcohol—put on with brush." No record was made of this filament's performance. N-80-01-02.2:94, Lab. (*TAEM* 34:768; *TAED* No70:50).

5. This idea may have been in Edison's mind for some time. A year earlier Batchelor made a brief notebook entry about "Aluminum (process of getting out)   Why should Al. be so dear when there are so many earths bearing an enormous per centage of it? Evidently its difficulty of separation. If a process could be devised whereby it could be smelted like or similar to iron it would be a big thing." He included such a process among a short list of things "Wanted" for the laboratory but no experiments were made until June. Cat. 1304:36–37, Batchelor (*TAEM* 91:35–36; *TAED* MBN004:33–34); see Doc. 1944.

6. In a patent application executed on 10 March (Case 210), Edison stated that because carbon clamps, which by virtue of their large size relative to the filament, contained a comparatively "large quantity of deleterious gases which come out very slowly when under the action of the vacuum pumps, it is preferable that clamps be used, which have first been carbonized, and subjected in vacuo, to heat sufficient to produce a high degree of incandescence." U.S. Pat. 525,888.

–1903–

From Edward Johnson

London Feby 19th 1880.[a]

My Dr Edison

Upon my return I find a wholly different state of affairs from that existing when I went away—not however in my Dept. That was doing well. subscribers were coming in—and our list now looks formidable—[1] Chalks continue to do well— Many of them put out before I went to America have never since been touched. In short Everything is swimming so far as the practical work is concerned. In re. to the Patent fight too we are getting along, slowly true, but surely— The disclaimer was allowed—and we are now on good solid ground in that respect.

I have had several meetings with the Lawyers and now we are in receipt of Websters Opinion that all the Carbon Transmitters ~~are~~ in use are infringements of our Patents[2]   A Letter has been written to the Bell Company to that Effect, and a notice is today being prepared for publication in the newspapers—by circular Etc Etc—warning Everybody to desist—[3] so far so good but here comes the nigger.[4]

The Bell Company have lately reorganized and have received an Enormous influx of capital—placing a large amount of stock at 100% premium— With this Coup detat they have given us the cold shoulder—withdrawn their proffer of the Olive branch of amalgamation—and defy us in the matter of the Patent fight— They say they care nothing for Patents now— They have got the money—& they will[b] thereby be able to keep the fight on for 2 or 3 years—meantime they intend—are are—going ahead—right & left Establishing themselves in a supreme Commercial position—precisely as they did in the U.S.—

Our own Lawyers—Webster & Waterhouse say that the processes through which we shall have to go in the courts—are open to these delays—: Bouverie says that unless we can counteract this last movement by one similar in kind—the continuation of the Patent is useless—for the simple reason, that Ere it can be concluded we shall be swept from the field by the superior power of our Opponents money— This is the condition of things as I find them on my return. Now for the remedy.

Bouverie Waterhouse & Gouraud had already been having some talk upon the question of a reconstruction of the whole question of the contracts with you—putting it upon some basis which would enable them to raise a large capital—so as to fight fire with fire. They contended that no[b] further payment of Advance Royalties was possible— That Every Pound[b] that could be raised would be required to increase Our plant— push the fight and Otherwise put ourselves in as good trim[b] as our Opponents—[b]

Of course Gouraud ridiculed the Idea saying it is simply impossible &c &c   As usual he now comes up with a counter movement—saying he can obtain all the requisite money from the stock Exchange by organizing a company and putting the stock before the Public— This is now his Pet scheme—and as the ideas of the company are matured and expressed, his confidence (apparent) in his own scheme increases—

Now none of the Gentlemen in this Co will have ought to do with any stock jobbing movement— Consequently if Gouraud goes into it—He will be Compelled to do so upon his own territory & alone— This is simply impossible—beside it would result in a divided Edison Interest[b]—and in my judgment prove about as abortive and as fruitful of ill-odors as his only other enterprize the Glasgow Company—

~~The Go~~ I have therefore expressed to the company your views in re. to all stock jobbing affairs—as well as your wish that the Gentlemen who have thus far liberally put their hands in their pockets to prosecute the work of planting you in England—& who alone have accomplished any thing=already having spent some 20 odd thousand pounds—should have the reins—and be given full scope[b] for Operations against both the P.O. & Our Opponents— I told Bouverie what you said in respect to repayment, if they were dissatisfied or thought you had treated them badly etc[5]= He expressed himself as very much gratified and said it simply confirmed him in the Opinion he had already formed of you viz: that you were a high minded honourable man and would not countenance transactions not strictly straight forward &c &c— The question is now what shall be done to save the Edison Telephone from being overwhelmed— The answer is clear— Raise a large Capital immediately and put forth our minions as numerous & as strong as our opponents— But upon what shall the[b] capital be raised— upon the present basis as to area of Territory covered—or the terms of contract with you—all agree not another pound can be had— It is therefore clear that you must make important & valuable concessions— Waterhouse said to Bouverie—"I am willing if need be to lose every penny I have put in—but this is not in the same Condition we fancied it—when we originally went in & asked our friends to invest their money — We then thought our Patent good— We thought our opponents had nothing but the magneto—and we did not anticipate a powerful opposition from ~~B~~Rival companies or from the Gov't= We are now face to face with these things—& In view of them I cannot ask my friends to put in another shilling[6]

It is therefore proposed to submit a proposition to Gouraud Immediately—looking to paying you a small additional amount of advance Royalties for all of the United Kingdom—say about 10,000 pounds—asking you to waive the 20% of the

Gross receipts—& substitute instead the following 5% when the company shall have earned 5% on their capital the Profits to be Equally divided with you—[7]

The Co. to agree to prosecute the work of establishing the Edison Telephone upon a firm foundation with the utmost vigor—

To agree to prosecute the Patent fight until your Claim to all forms of Carbon Transmitters is Either completely established—or it is clearly demonstrated that it Cannot be sustained—

To agree that in the possible Event of amalgamation the Edison Patents & Interests shall under no Circumstances be made subordinate Either in name—value—or position to any[b] other— that any new Co. shall liquidate both the Edison & the Bell Co. and buy up the Patents Etc of both— The amount paid for the Edison not to be less than [---][c] that paid for the[b] ~~other~~Bell Blake &c   The new Co. to be called Either the Edison[b]-Bell—the Bell-Edison or the Edison Telephone Coy— Mr Bouverie to be the chairman of the new Co. in order to give a still more distinctive Edisonian characteristic to the new organization   ~~All these~~ In order to Equip the thing at once for assuming this position Mr Bouverie & his associates to raise 40,000[b] pounds working capital from their own & friends pockets—no appeal to the Public to be made.

It is by [~~reason?~~][c] virtue[d] of the class of men you have in this London Co. that we have secured Sir Wm Thomson—John Tyndall & others—Times & other Papers &c &c. in our suits against the P.O.—[8] It will be by virtue of the same that you will placed in a position Enviable in the extreme—if this new [---][c] arrangement is made—

If you go to the Stock Exchange for money—you may expect all such men to avoid you—

Amalgamation is a Certainty   Are you to make it divided against yourself—or as a unit— The answer is obvious—

The amount of money to accrue to you from amalgamation will be anywhere from 40 to 50 000 pounds— Your ~~20~~½share in the Profits—will in such event—be substituted by a proportion of the shares of the new Coy—and a small royalty on each Inst. We are in possession of reliable information to the effect that the PO crowd are convinced they will have to buy us up— This will be a second turn

Now my Dear fellow— Gouraud will unquestionably seek to spoil all this— I feel strongly the responsibility resting on me— But I am compelled by virtue of the ~~neces~~ absolute necessity for prompt action to commit you one way or the other— I choose[b] that which lies in the Path of honour & liberality— You of course see that it affects my pocket equally with yours—& it is useless for me to say that I choose it because it points to the satisfaction of my Pride as well as yours in seeing the Edison Telephone take its rightful position—

The Glasgow 5000 has not been paid—& wont be unless the London Co. assume it— They agree to do so in case of a reorganization upon this basis— This saves you some considerable annoyance—as otherwise it is very probable they would not only not pay up—but would bring suit against you for repayment of the other 5000£.[9]

As soon as you get this Telegraph as follows—

Quephone London   Bouverie Phonos basis will be approved—Edison[10]

I have been so busy in this matter have not even taken your lamps out of the Box— I'll attend to that later   Hastily yours

E. H. Johnson

ALS, NjWOE, DF (*TAEM* 56:417; *TAED* D8049ZAQ). Letterhead of Edison Telephone Co. of London. [a]"London" and "18" preprinted. [b]Obscured overwritten text. [c]Canceled. [d]Interlined above.

1. Around this time the company published a promotional pamphlet including a list of 149 subscribers, a copy of which was sent to Edison. DF (*TAEM* 56:429; *TAED* D8049ZAQ1).

2. Richard Webster held that the Blake, Crossley, and Hunnings transmitters infringed Edison's amended patent. He was more equivocal concerning whether the patent might cover all forms of variable resistance carbon telephones. Theodore Waterhouse sent Edison a copy of the opinion. Waterhouse to TAE, 21 Feb. 1880, enclosing Webster opinion, 13 Feb. 1880, DF (*TAEM* 56:438, 440; *TAED* D8049ZAT, D8049ZAU).

3. The pamphlet (see note 1) contained a notice dated 20 February warning "against Manufacturing any Carbon Telephone Transmitters, and against selling, supplying, or using any Carbon Telephone Transmitters" made after 10 February (the date of the disclaimer), except those of the Edison Telephone Co. of London.

4. Probably short for "nigger in the woodpile," a colloquial term meaning "a trick or drawback, esp. if deliberately concealed." Another recent derivation of "nigger" to which Johnson may have been referring was a fault in electrical insulation which might cause a short circuit. Lighter 1994–97, s.vv. "nigger in the woodpile"; "nigger" (9); Mathews 1951, s.v. "Negro, nigger" (4a).

5. See Doc. 1880.

6. An unsigned telephone company memorandum set forth these circumstances as rationales for modifying the contract with Edison. It also called for raising more capital from current stockholders "or others who may be fully informed of all the circumstances" rather than on the stock

exchange. Edison Telephone Co. of London, Memorandum, 18 Feb. 1880, DF (*TAEM* 56:412; *TAED* D8049ZAM).

7. For the terms of Edison's contract with the London company, see Doc. 1742 n. 3. On 25 February Johnson sent Gouraud a letter urging him to accept the forthcoming proposal to consolidate the British Edison telephone interests under the London company and modify the contract with Edison. Two days later he wrote Edison that although he wanted the company to negotiate through Gouraud, when the board took up the proposal Gouraud "pronounced it preposterous & only fit to be torn up" and insisted that the company exercise its option to pay £40,000 for the remainder of London before opening negotiations. Johnson said he then explained that Edison had, after "a long full & free discussion of this whole matter" agreed in principle to the consolidation of his British interests. Gouraud reportedly then accepted the proposal but later countered that the company should "write off the payments made to you on a/c of advance Royalties—as being payments made in cash— Your profits then to commence at once—instead of after the London Co. have received back the amounts paid you— Or in other words they are to agree to pay you 35,000£ cash—& then immediately begin paying you ½ the profits— This only shows G. to be as grasping & impracticable as Reiff." Johnson had drafted a power of attorney to himself and a revocation of Gouraud's power, both to be put in force by a pre-arranged exchange of cables. Johnson to Gouraud, 25 Feb. 1880; Johnson to TAE, 27 Feb. 1880; Woollacott & Leonard[?] to Johnson, 27 Feb. 1880; draft TAE power of attorney to Johnson, n.d.; draft TAE revocation of power of attorney to Gouraud, n. d.; all DF (*TAEM* 56:452, 456, 465, 848, 850; *TAED* D8049ZAY, D8049ZBB, D8049ZBC, D8049ZIL, D8049ZIM).

8. William Thomson and John Tyndall gave affidavits supporting the company's position that the telephone was distinct from the telegraph and therefore did not fall under the Post Office's authority. The company secured similar statements from other electrical experts, including George Stokes and John Ambrose Fleming. George Gouraud sent copies to Edison and asked him to write letters of thanks, particularly to Thomson "who was the first to come forward, and who refuses to take any fee in the matter, expressing at the same time his great admiration for you and your work." Gouraud to TAE, 6 Feb. 1880, enclosing "Affidavits filed on behalf of the Defendants," *Her Majesty's Attorney General v. Edison Telephone Co. of London, Ltd.*, DF (*TAEM* 56:375; *TAED* D8049ZAC).

9. The Edison Telephone Co. of Glasgow paid a £5,000 advance royalty in December; no further money was to be paid until actual royalties (and interest) totaled another £5,000. At the end of February Johnson wrote that he had visited Glasgow and "there found a most deplorable state of affairs" with the investors threatening suit against Gouraud and Edison. He conciliated Michael Moore by proposing "that the Glasgow Coy shall be taken into the general Coy—and its obligations assumed— & your additional 5,000£ paid." Gouraud ratified an agreement to this effect on Edison's behalf. Gouraud to TAE, 24 Dec. 1879; Johnson to TAE, 27 Feb. 1880; Gouraud to TAE, 21 May 1880; TAE agreement with Edison Telephone Co. of London, n.d.; all DF (*TAEM* 52:1024, 56:456,663, 665; *TAED* D7941ZKK, D8049ZBB, D8049ZEK, D8049ZEL).

10. Edison sent this message to the telephone company ("Quephone")

on 1 March. "Phonos London" (often abbreviated "Phonos") was Johnson's cable code name. DF (*TAEM* 56:466; *TAED* D8049ZBD).

-1904-

Memorandum to
Laboratory Staff

Menlo Park—Feb'y 19th. 1880.

Employees will treat Visitors courteously but under no circumstances will they leave their work or give information of any kind to visitors.

No information will be furnished except by myself or Mess'rs. Batchelor, Upton, Krusie and Carman.

You will regard this as a positive order.[1]

Thomas A. Edison.

ADS (electric pen copy), NjWOE, DF (*TAEM* 54:368; *TAED* D8030E).

1. Edison had this order distributed to each employee the day after an incident galvanized his concern about the unauthorized spread of information from the laboratory. According to Jehl 1937–41 (494–95), an unidentified worker "who was escorting a stranger round with academic mien and displaying his knowledge volubly, stopped before Batchelor as he was engaged in cutting out some filaments and said loudly: 'That's the way *we* make those carbons for the lamps.' He lectured as if he were Edison in person. I was near by and saw that Batchelor was annoyed." Edison also instructed William Carman to admit only those visitors with business in the laboratory and request all others to meet in the office with the person they sought.

In March 1880 an unidentified detective agency published a pamphlet, "Electric Light Tunneling. A few facts from the unpublished history of the operations of various parties interested." This was a compilation of reports from spies sent to various workshops where electric lights were made or experimented upon, including Menlo Park. No copy of this has been found but the front cover is reproduced and the section concerning Menlo Park is quoted in Jehl 1937–41 (700–705). The detective claimed to have entered the laboratory with Edison's permission and been escorted by one of his assistants, whom Jehl identified as Francis Upton.

-1905-

Notebook Entry:
Electric Lighting

[Menlo Park,] Feb. 19 [1880]

Hydrocarbons.

After a certain time electric lamps become coated on the inner side of the glass, with a dark brown translucent coating that resists removal in.

It is proposed to isolate it.

Ether seems to dissolve it; and it is then precipitated by evaporating the solvent to near dryness.

Test the ether itself to see whether it does not give the same result alone: the product may be the result of oxydation of the ether.

Following apparatus employed.

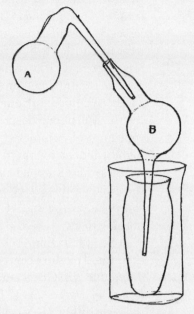

A holds about 150 c.c. B is similar vessel   20 c.c. Sulphuric ether was introduced and evaporated keeping the neck of A, and the bulb of B cool. (50 p.c. of ether easily recovered.) A slight opalescence of the residue. This experiment thrice more. repeated, a very decided clouding of the 2 or 3 c.c. remaining.[1]

X, NjWOE, Lab., N-80-01-28:17 (*TAEM* 33:838; *TAED* N055:8). Written by Otto Moses.

1. Moses returned on 26 February to this problem of the darkening of lamp bulbs referred to in Doc. 1898, when he planned "a series of experiments with different solvents on the coating." He made a solution of chlorine dissolved in water with which he seems to have been able to remove some of the coating from two lamps. On 1 March he tried concentrated potash and concentrated soda, both of which readily removed it. Moses was able to separate and concentrate the substance but not, evidently, to identify it. Related experiments may have been made by Alfred Haid. On 19 February Martin Force cut some of the new uncalendered paper (see Doc. 1900) and "heated on gas jet only" for Haid to test "for Hydrocarbons" but there is no record of further tests with these samples. N-80-01-28:20–[25], N-80-01-02.2:140, Lab. (*TAEM* 33:840–42, 34: 801; *TAED* N055:10–12, N070:83).

Paris 22 February 1880.

My dear Sir:

Enclosed herewith you will find sheet showing the liquidation of the <u>Soc. du Tel. Edison</u>, from which you will see that there is a loss to the owners of the patents, up to Feby 1, current, of 67,000 francs.[1] By the conditions of our contract the losses and profits are equally shared, so that our special partners are entitled to have returned to them the amounts paid in, less 67,000 f.— If we make an amicable liquidation we can get control of our patents and go on with our business. If we make a judicial liquidation[2] our business is stopped and it is good bye for the Telephone business not only in France but in all Europe for us. In a judicial liquidation our patents are sold to the highest bidder for the term of 2 years from last Dec. If the business were stopped and the patents put up, as in this case, they would probably be bought in by the Credit Moblié́r for a song[a] and at the end of the two years they would be worth nothing to us.

When the fusion broke there was no one in our Company that would put up a cent to go on with the business, and we had debts due the Government and others which there was no money to pay. I went to Mr. Harjes and borrowed 10,000 fcs on the security of E. L. stock standing to me. I paid 8000 f. of debts and took the direction of the business with 2000 f. to fight two millions and a half. Neither Mr. Berthon nor Mr. Puskas have made any opposition to me up to the present time. I secured Mr. Berthon's acquiescence by an arrangement to buy him out. Mr. Roosevelt was still in the company, and represented ⅔ of the special partners interest. The Credit Moblié́r desired to buy him out and this would have forced us into a judicial liquidation. I traded with Roosevelt for this interest at 70,000 fcs. seven thousand more than par, but 20,000 less than it had cost him.— 25 000 fcs of this amt was payable on or before the 14 inst. The 2000 fcs of capital with wh. I commenced was on the 14 inst reduced to 200 fcs. and beside the 25,000 to Roosevelt there was 3000 for sundry bills to pay. I went to Mr. Harjes and arranged with him for 40,000 fcs additional on the security of one share of E. L. of Europe, the whole of my individual interest in the French Company and one half of my interest in the Ed. Tel. Co. of Europe, with the obligation to repay him by the 14 Aug 100,000f for the 50,000 in redemption of the above.[3] The circumstances in which all our interests were, made the transaction a desirable one if these interests are worth anything and if they are not it is a bad one for Mr. Harjes. There was probably no other person in Paris who would have done it on the same terms. When the fusion broke we could not have taken 50,000f for the whole of our French patents. I believed that you and all other parties interested would accept your part in the transaction if it were the means of saving us from judicial liquidation, and I risked all[b] my interests on that, for the common interest.

Since I took the business, (the 1st February) we have increased our list of subscribers from 62 to 120. The names added to our list are from among the largest Banks and most important business Houses, such as Banque d'Escompte,[4] the Bon Marché,[5] The Magasins du Louvre,[6] the Credit Foncier of France,[7] Hachette[8] etc. etc. Our opponents have together 180, no considerable increase on the number when we were together. Beside this in company with Mr. Barton we have made a move in Marseilles where we have already about forty to fifty subscribers engaged, and in Bordeaux where we have also made a good beginning, Mr. Barton letting Mr. Scribner[b] go there. We hope to begin in Lille & Lyons in this week.

The question that I desire to submit to you and the White Estate[9] is whether you desire me to go on, and to support me in my effort to save your interests and to make them valuable, or whether you desire to give them up to wreckers. You cannot be parties to new struggles and quarrels over your interests without sacrificing both yours, and mine, while I put up to save yours, ~~as well as mine~~. If you weaken me, by giving your power or your countenance, here in France, to outsiders you will not only sacrifice your French but your[c] European ~~but you~~ interests.[d] The Edison interests here cant support any more intestine wrangling. Mr. Harjes has now a solid interest in the success of the telephone ~~interest~~ France.[d] If you will call on Mr. Fabbri he will tell you what Mr. Harjes' judgment is of my work and situation.— If you or the White Estate have done anything that would either take Mr. Harjes representation of your interest from Mr. Harjes, or the White from ~~mine~~ me, you should cable to the contrary if you have any confidence in the representations here made.[10]

We are now in negotiations with new parties to take up the business, and the business in the last ~~tw~~three weeks has made friends and gained support. I shall cable you tomorrow evening again.[11] I cabled you Friday evening a brief resumé of the situation in hope of preventing action on your part that would be contrary to all our interests.[12]

I think you will see that the only case in which you are in

danger of any liability on account of the French business is in that of its falling into a judicial liquidation through interference and embarrassments thrown in the way of the efforts now being made to save it. These efforts have already given it a value sufficient to clear all parties from liability. I hope for the support that will enable us to do a good deal better than that. Yours sincerely,

J. F. Bailey

ALS, NjWOE, DF (*TAEM* 56:35; *TAED* D8048X). ^a"for a song" interlined above. ^bObscured overwritten text. ^c"French but your" interlined above. ^dInterlined above.

1. This ledger follows in DF (*TAEM* 56:37; *TAED* D8048X), showing a loss of 67,000 francs each to Bailey and Theodore Puskas.

2. In effect a declaration of bankruptcy.

3. These terms were embodied in an agreement executed by Bailey and John Harjes on 20 February. DF (*TAEM* 57:98; *TAED* D8104ZAX2).

4. Probably the Banque d'Escompte de Paris, a joint stock bank. It is also possible that Bailey was referring erroneously to the Comptoir d'Escompte de Paris, one of the most important independent French investment banks with extensive business throughout the world. Cameron 1961, 268, 126–27.

5. The Bon Marché was one of the world's first department stores, and remained among the largest and most prestigious. At this time the Bon Marché building on the Left Bank was in the midst of a decades-long reconstruction and expansion overseen by Gustave Eiffel. Miller 1981, 5, 41–43.

6. The Grand Magasins du Louvre was the principal rival to the Bon Marché. Miller 1981, 27–29.

7. The Crédit Foncier de France was France's first national mortgage bank. It was originally intended to extend agricultural credit but had become deeply involved in the reconstruction of Paris and remained heavily invested in urban real estate. Cameron 1961, 128–29.

8. Hachette et Cie., one of the most distinguished French publishing houses. Mollier 1988, 171–93.

9. The estate of the late Samuel White had not yet designated a representative to the Edison telephone concern in France. Samuel White estate to TAE, 25 Feb. 1880, DF (*TAEM* 56:52; *TAED* D8048ZAF).

10. Two days later Harjes advised Edison by cable: "I recommend strongley no interference with Bailey apprehending otherwise great loss to all concerned   you had better telegraph white estate and Gray same effect." He and Bailey separately urged Edison not to allow the liquidation of the company, a plan of action which Bailey believed James Banker, recently arrived in Paris on behalf of Samuel White's estate, was determined to carry out. Harjes to TAE, both 24 Feb. 1880; Bailey to TAE, 25 Feb. 1880; White estate to TAE, 25 Feb. 1880; all DF (*TAEM* 56: 39, 43, 47, 52; *TAED* D8048Z, D8048ZAC, D8048ZAE, D8048ZAF).

11. Bailey cabled Edison on 23 February, "Contracts preparing submit Harrisse tomorrow   further advices wednesday   advise whites."

On his arrangements for new capital, see Doc. 1909. Bailey to TAE, 23 Feb. 1880, DF (*TAEM* 56:38; *TAED* D8048Y).

12. Bailey's cable of 20 February summarized information in this document. Bailey also asked Edison to "advise White." DF (*TAEM* 56: 34; *TAED* D8048W).

–1907–

Draft Caveat: Electric Lighting

Menlo Park— [c. February 25, 1880][1]

Caveat.[2]

drawplate for fibre

soaked in standard solution sugar dried and cut. Alcoholic Solution with slight amount H2O to render it more soluable

sugared[a] ends combination several loose fibre

seperate fibre sugared ends or lignum Vitae ends & little sugar on

burner made of a great number of fibres & held together with sugar— extra fibre at ends also held by sugar, or any car- bonzable ~~thi~~ Viscous liquid—

passing vapors of Hydrocarbons or carbonzable gases through Hot[a] tube to build up by deposit.

extra pieces at X & all sugared
Clamp

clamp for round carbons.

split wire taper outside to Hold Carbon ring tighten

wood[a] ends in spiral, sugared on platina & wood.

Roughened clamp

X [peice?][b] large carbon to heat clamps drive air away

nickel clamps.
platina Iridium with Iridosmine let in where carbon comes
in contact also gold—silver ~~cop~~ chromium etc etc—

platinum Iridium screw-hole through carbon

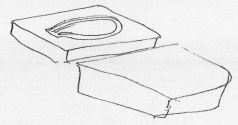

no tissue   let it take its shape.
blowing bulbs through cotton. plaited or braided fibre. sugared thread.

bias of clamps towards each other keep from breaking

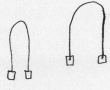

paper ends sugared

clamp

The[c] object of this invention is to ~~manu~~ obtain light by the
incandescence of electric conducting matter.

The invention consists in various methods of manufacturing the lamp

I have found that when[d] paper ~~should be~~ is[d] used that it
should be specially manufactured ~~and that pure cotton rags
well washed~~ so as to obtain pure carbonzable fibre without sizing and other foreign materials   Ordinary Bristol board contains Rosin & other ingredients for rendering it non porous to
inks. this rosin carbonzes ~~and~~ as well as the fibre, and being in
some lots of paper in greater or lesser quantity makes it very
difficult to obtain incandescent conductors of the same degree

of conductivity when subjected to the same carbonzing processes. To obviate this I obtain the paper free from such matters and after cutting the peices of paper to make the conductor in[a] a larger form than the final cutting I ~~soak~~ boil these in a solution of sugar in alcohol or Alcohol & water or in a water solution of sugar. The[a] strengths of these solutions must be such that the required resistance is obtained   if a low resistance is required a thick solution of sugar should be used ~~but~~ and if ~~the~~ a high resistance the strength of the solution must be diminished. ~~by this means A more perfect~~ The horseshoe form of burner which I have shewn in an[a] application has thickened ends ~~upo~~ which serve to make good contact with the clamps.[3] I now cut out another peice or 2 peices of a shape exactly like ~~this~~ese thickened Ends. I place one on[a] each side moistening them slightly with a syrupy solution of sugar. this serves to[a] hold them in their position and when carbonzed the sugar carbonzes as well and this makes a solid peice[a] which allows of better clamping.

I also have suceeded in plating these ends with copper nickel & other metals, ~~a~~which serves to make excellent contact ~~wi~~between the clamps and the ends. I effect the deposit by hanging the loops upon a wire which is connected to one pole of a battery & allowing the ends of the loops to dip in the plating solution the proper distance, the jar containing the plating solution being in contact with the other pole of the battery.[4]

In using ~~su~~ long[d] fibres such as manilla hemp & other fibres of the Hemp[a] family I also plate the ends very thickly and afterward drill holes in the metal deposited by electro deposition tap the same & put in ~~the~~ a short length of[e] platina wire   These ends are secured to the platina wires passing through the glass by a collar or clamp.

I makes lamps consisting several fibre—each spread out but all coming to a common clamp, but in this case the aggregate radiating surface of all the fibres do not exceed that of a single paper horseshoe conductor.[5] I also bunch all these fibres together & pass the aggregated filiment through a solution of sugar to hold them together   on single & even on an aggregation of distinct fibres I secure[a] their ends to a peice of carbonzable wood such as lignium vitae, box wood, cocanut shell, Bast fibre; by drilling a small hole in the wood & drawing the fibre through this hole & putting a slight amount of sugar on the wood the whole is carbonzed & gives a good thickened end for clamping ~~the~~   a short piece of[f] platina wire Being[a] placed in one end of the wood & put[a] in the mould when the fibre is carbonzed, the wood shrinking serves to clamp itself upon the platina with great force. as these clamps of carbon contain a large quantity of gas[a] which comes out very slowly in the vacuum, It is best to employ clamps of carbon which have been carbonzed and subjected to high incandescence in vacua to free it from air. Small peices can then be used with fibres for ends, a very small quantity of sugar serving to lock the fibre to the clamp in the act of carbonzation   paper horse shoes or burners made of fibre may be (after carbonzation) placed in combustion tubes on platina wires and a hydrocarbon vapor passed through the tube by the aid of a Bunsen water vacuum pump,[6] the Hydrocarbon being decomposed a very hard carbon is deposited upon[a] the shaped conductor.

The most perfect fibre for carbonzation and use for lighting by incandescence, or for arc purposes,[g] is A[a] fibrous grass from south[a] america called Monkey Bast fibre. Each ~~fibre is~~ blade of grass which are generally round is[h] ~~fill~~composed of a great number of fibres and the whole connected together with a natural resin which serves to carbonze & lock all the other fibre together, making a nearly homeogenous filiment. These ~~fibres~~ blades of grass ~~are g~~ are of various sizes from $\frac{1}{16}$ of an inch in diameter to $\frac{5}{1000}$ of an inch. they are slightly taper hence, I make a cutting die, through which I pull the fibre the edges of the die serve to shave off the extra matter of the thickened end. another plan is to use a revolving cutter,[7] the cutter revolving and is passed along the fibre. Theses fibres when small may be put between paper ends ~~su~~which are sugared & then Carbonzed or a small platinum Iridium wire, may have a hole drilled in the end about the size of the fibre after carbonzation, the wire is split to the bottom of the hole or lower. The fibre is placed in the hole & ~~the~~if the outside of the wire is made slightly[a] taper a ring may be placed over it thus clamping the fibre very tightly

In carbonzing in mould we have sometimes dispensed with the tissue paper between each of the paper shoes or fibre shoes with the weight on top and instead ~~only carbonzed a~~ made a small indent into ~~the~~ a[d] nickel plate having a depth a little greater than the thickness of the paper over this another nickel plate is placed   if a paper ~~horse~~ shoe is placed in this recess & the top plate put on it lays in a very thin chamber, but is free to contract. the two plates are placed in the nickel box and then passed through the carbonzing process. the shoe will, being free from restraint, contract in a somewhat irregular shape, but it will [-S-][i] ~~mor~~ have less flaws and withstand a higher incandescence.

As platina combines to a certain extent with heated carbon I let into the ends of my clamps ~~where the platin~~ thin plates of Iridosmine, which does not to the same extent combine with ~~pl~~Carbon. Gold, chromium, manganeese chrome iron silver & many other metals may be substituted for ~~all~~the Iridosmine but whether they will last for long periods of time I am now ascertaining.

I sometimes roughen the surface of my platinoid clamp so as to ensure better contact. I also instead of placing the tightning screw some distance from where the carbon is clamped, cut a hole in the thickened part of the shoe before carbonzation & then place the shoe in the clamp & put the tightning screw through this hole

I ~~sometimes~~ have wound a small spiral which forms a sort of a cup and secured one of these spirals to each of the platina wires which is sealed in the glass. in this cup, I force the round carbonzed wooden ends of the fibres.

I distributing both power and light ~~a~~& where there ~~price~~ is a difference of price to be charged for electricity when used for light or power, I place a double meter in the ~~habitat~~ house of the consumer and connect the leading wires ~~of~~to the light &to one meter & extra leading wires to the [-][i] magnetic motors to another meter.[8]

As the motors will for different purposes be required to run fast or slow, I provide each with a governor which when the speed increases above a certain rate opens the circuit & disconnects the motor from the main wires and as it falls ~~clo~~ in speed closes the circuit again. as it would be entirely impracticable to use an ordinary one point ~~magne~~ circuit opening & closing magnetic governor owing to the great spark due to the extra current of the induction bobbin & field magnet I use several points & springs so arranged that the circuit is opened simultaneously in several places & this so reduces it on any one single point as to practically obliterate it as far as any ~~de~~great destruction of the contact points are concerned.[9]

It is obvious that the governor may be run on the main shaft of the motor are ~~put~~ seperately motion being given it by gear wheels [a?][i] worms & wheels or belts connected to the main shaft. ~~It is also possible to use a small shaft~~ [cont?][i]

There is a defect in the slow carbonzation of organic filiments[j] as then Hydrocarbons ~~of~~ are formed before carbonzation and these Hydrocarbons are ~~then~~ afterward enclosed in carbon and are only driven off by the highest incandescence and after a long period of time   to obtain a more rapid carbonzation I take a large platinum Iridium wire bend it in the form of a horseshoe form a groove in this wire two or three times the size of the fibre which is to be laid in it   after this is done I put in a peice of the ~~same~~ platinum Iridium which serves to cover the slot[a] &in which the fibre is laid. the ends of the horseshoe are secured to circuit clamps & the whole arrangement is placed in vacua. the[a] wire is slightly heated to drive of the water vapor then brought suddenly to incandescence to effect the carbonzation   Carbon may be substituted for the platinum Iridium   even ~~Vicous~~ Viscosous substances may be carbonzed in these[10k] grooves,[a] such as sugar, which must be enclosed in tissue paper to prevent it sticking to the platinium Iridium.

T.A.E.

X, NjWOE, Lab., N-79-12-19:49–57, 63–99, 105–109 (*TAEM* 32: 939–43, 945–63, 966–68; *TAED* No42:20–24, 26–44, 47–49). [a]Obscured overwritten text. [b]Illegible. [c]Text from here to end copied by Charles Mott in Cat. 997:131–34, Lab. (*TAEM* 3:417–19; *TAED* NE1695:66–68). [d]Interlined above. [e]"a short length of" interlined above. [f]"a short piece of" interlined above. [g]"or for arc purposes" interlined above. [h]"blade . . . is" interlined above, with "which are generally round" interlined separately. [i]Canceled. [j]"see page 105" written as page turn. [k]"these" repeated at end of one page and beginning of next.

1. Edison executed this caveat on 28 February; Zenas Wilber did not assign it a number, as had been Lemuel Serrell's custom (Edison Caveats). It discusses experiments made during the second half of February, the last of which were recorded on 24 February. See Doc. 1900; N-80-01-02.2:92–121, 124–28, 132–66; N-79-12-09:115–141; both Lab. (*TAEM* 34:767–84, 187–90, 795–827; 710–23; *TAED* No70:49–66, 69–72, 76–109; No41:59–71).

2. The caveat was filed on 20 March without drawings or any of the text that accompanies the drawings in this draft. The filed version is essentially the same as this document from the paragraph beginning "The object of this invention" (Edison Caveats). Much of its substance was embodied in a patent application (Case 210) that Edison filed on 20 March. The application did not issue until September 1894 as U.S. Pat. 525,888.

3. Edison filed Case 187 on 11 December 1879 with a claim for "a filament with the ends broader for the clamping devices that connect the supporters." An interference was declared with Sawyer and Man, and Edison subsequently abandoned the application; only the drawings and claims are extant. E-2536:44, PS (*TAEM* 45:702; *TAED* PT020044).

4. There is a record of only one lamp (No. 730) made with a plated loop at this time, on 23 February, and no other surviving evidence of these experiments. Edison later testified that he had copper-plated the ends of some filaments as long ago as November 1879. N-79-12-09:141, N-80-01-02.2:162, both Lab. (*TAEM* 32:723, 34:823; *TAED* No41:71,

No70:105); Edison's testimony, 4:2567, *Edison Electric Light Co. v. U.S. Electric Lighting Co.*, Lit. (*TAEM* 47:1083; *TAED* QD012E:222).

5. In a division of Case 210, Edison filed a new application in July 1888 covering filaments constructed from multiple fibers. That application issued in March 1892 as U.S. Pat. 470,925.

6. The experimental chemist Robert Wilhelm Bunsen in 1868 devised a filter pump operated by water falling through a long tube, much like the mercury pump of his student, Hermann Sprengel. Partington 1964, 4:288.

7. Sketches of a revolving cutter are in N-80-01-02.2:145, Lab. (*TAEM* 34:803; *TAED* No70:88).

8. Edison embodied this idea in a patent application executed in August (Case 236). He proposed that electric current for lighting and for power "should be paid for at different rates. I therefore, where both power and light are desired, run two branch circuits from the mains, one for the lights, one for the engines, a meter being placed in each." This paragraph was deleted before the patent issued in September 1882. Pat. App. 264,642.

9. Edison included a centrifugal governor and the circuit-breaking arrangement described here in another patent application filed on 20 March 1880 (Case 208). In order to maintain as nearly constant speed as possible Edison specified the use of a heavy flywheel on the motor shaft. The combination of governor and flywheel made the machine "capable of the finest adjustment, breaking and completing the circuit upon the least variation from a determined desired speed, acting exactly, so to speak, as an automatic cut-off in steam-engines, preventing unnecessary consumption of energy." Samuel Mott's undated drawing for the application of the complete motor with governor is in N-79-09-20:47, Lab. (*TAEM* 36:50; *TAED* No96:24). The patent issued in September 1889 as U.S. Pat. 264,649.

10. The last page of this document was written below a heading, written and crossed out by Edison, for a draft caveat or patent application for electric power generation and transmission.

–1908–

From Edward Johnson

[London,] March 4, 80

Edison

Exhausted delicate consideration not reciprocated   definitely denies your power offensively refusing negociate   while none sanction injustice cannot admit you powerless[1]   He privately admits terms unexpectedly favorable   vanity only barrier[2]

Phonos London

L (telegram), NjWOE, DF (*TAEM* 56:470; *TAED* D8049ZBH). Written by Stockton Griffin.

1. This cable was a reply to one Edison sent to the London telephone company sometime on 3 March that "Gouraud complains   please molify as I wish to do him no injustice." Edward Bouverie also replied on

4 March that he was "Ready to show every consideration." Edison's message was apparently in response to one from the company on the afternoon of 3 March that "Gouraud rejects Bouverie Phonos basis   claims terms fatal to developement   Cable Gouraud to approve Bouverie Johnson basis to support it and to inform Quephone   answer." Late the same day Edison received a cable (unsigned but attributed to Johnson by Stockton Griffin): "Precisely my aim   already practically accomplished better basis   leave negotiations to me or fatal delays inevitable   Cable as requested adding must negotiate with him." TAE to Edison Telephone Co. of London, 3 Mar. 1880; Bouverie to TAE, 4 Mar. 1880; Edison Telephone Co. of London to TAE, 3 Mar. 1880; Johnson to TAE, 3 Mar. 1880; all DF (*TAEM* 56:469, 471, 467; *TAED* D8049ZBG, D8049ZBI, D8049ZBE, D8049ZBF).

2. Johnson wrote a long letter the same day explaining that "things have come to a Deadlock— As I anticipated Gouraud refuses to have anything more to do with the Company until they notify him that they admit that they have no right to have any dealings with you directly, and promise him that they will have nothing further to say to me. This they of course cannot consent to—as it would be acknowledging that they were absolutely in his power." He also reported that since Gouraud "told me but a few days ago that the terms of the proposition were really much better than he expected to obtain—that he did not think any more advanced Royalties could be had—and that it had always been a matter of great astonishment to him that the Co had ever consented to the clause giving you 20% of the gross receipts I am at a loss to understand where any injustice is being done him." He asked if Edison could "coerce Gouraud into accepting an arrangement which is very advantageous both to you & to him" or, failing that, "Can you act independently of him— If not you may as well wipe off the a/c— Every man in the London Co. will withdraw." On 8 March Johnson warned Edison by cable of a "Danger of complication by attempted assignment to other parties unless you act." The next day Edison instructed Gouraud "Dont change status English Telephone without consultation" and advised Johnson of having done so, prompting Johnson to write back: "You are not quite 'stiff' Enough—a little assertion of your individuality just now would carry the day." Johnson to TAE, 4 and 8 Mar. 1880; TAE to Gouraud, 9 Mar. 1880; TAE to Johnson, 9 Mar. 1880; Johnson to TAE, 10 Mar. 1880; all DF (*TAEM* 56:472, 481; 55:688; 56:494, 495; *TAED* D8049ZBJ, D8049ZBK, D8046N, D8049ZBQ, D8049ZBT).

-1909-

From Joshua Bailey

Paris 5 March 1880

My dear Sir:

Under date of 22 Feb. I advised you of what was passing in French matters, and cabled you subsequently.[1] Mr. Harrisse has to-day remitted you a copy of the contract which has been under consideration during several days, and which was signed last night.[2] Enclosed with this you will find a copy of list of Paris subscribers. We have treaties under way with[a] the Chamber of Notaries,[3] with the[a] Agents of Change (Bourse)[4] and others which assure us today more than 200 subscribers. When we were in the fusion we stood Roosevelt 104, Soulerin 120, Edison 602. Today we are, Roosevelt & Soulerin 180, Edison 1 with others as above stated that are agreed on and which will go on our lists within ten days.[5] Enclosed is a card with list of lines already working, four added yesterday & today, 15 more for next week, & so on.—[6] I cannot take time to write you very much in detail about the business represented in the contract, this mail. I shall send enclosed a memorandum which Mr. Rousseau has just sent here.—[7] His note as to the persons interested in the affair is not complete because he has not given attention to that side. Beside Mr Jenty,[8] there are at present the Bank Franco Egyptienne[9] which probably will have one half interest. It is decided that it enters but Mr Jenty has not decided what portion to allow them. Also Mr. Girardin,[10] and other names of whom I will send by Mondays mail.[11] You should go to Coudert Brothers 68 & 70 William St.[12] for ratifying the contract. They[a] have copies with the proper form of ratification.— It is agreed that a telegram from Coudert Brothers to the person who has remitted the contract ~~shall~~ to effect that you have ratified the contract, shall be accepted as a ratification here, and the sum of 100,000 f. be paid over to aid in our liquidation.— In addition to this Mr. Jenty gives us a credit of 20,000 to pay expenses during the time that we are waiting for the ratification. During the last ten days I have been very close run for means to get along, and was obliged to draw on you for amt to pay off Charleys bills, for which Mr. Puskas showed me your letter agreeing to pay.[13] The efforts I have made have suceeded but the game was a desperate one, and it would have finished me if it had been lost.— The Gower-Soulerin are now endeavoring to resume a treaty for fusion. But we will drive them off the field, or when we make a fusion it will be after a surrender. Our position in the business is better than under our old treaty. We have 40% in the affair now, while under the old we stood with ¼. I cannot go into detail on financial side today, but will give you detail of that Mondays mail. I hope you will go in to Couderts as soon as you receive this: or at least to Lowry's and not allow more than a couple of days to pass before acting. Mr. Harris does not believe in the telephone, so that he will not give you perhaps a very encouraging view of the situation generally, but the people who know it give us all the points as against Gower. As soon as we have your ratification we are going to fall on the Blake. Best counsel here say the suit is good & we shall [pitch?][b] in. We have in Mr Jenty the <u>Petit Journal</u>[14] 550,000 copies per day, the <u>La France</u>[15] with Girardin, the <u>Monde Illustre</u>[16]—250 journals in the provinces. Also the <u>Banque Nationale</u>[17] the Societé Nouvelle de Credit,[18] and other influences. They believe that within 6 months we can launch a company and divide between us 5 to 8 000,000 fcs. with shares above par. This is a very hurried letter, and you must excuse incoherence. I have written in last three hours, with room full of people coming & going and talking. By Mondays mail will send complete resumé of all points.[19] Please send copy of this le[c] to Mr. Lewis,[20] and I hope to have your cable confirming treaty within two days. Sorry no cash down, but we lose nothing. To contrary it looks as though we should gain largely and we go wholly under your Flag.[21] Yours very truly

J. F. Bailey

ALS, NjWOE, DF (*TAEM* 56:81; *TAED* D8048ZAL). [a]Obscured overwritten text. [b]Illegible. [c]Interlined above.

1. Bailey's letter is Doc. 1906 (for the cable see n. 12).

2. This agreement licensed to Edward d'Auberjon the French rights to telephone patents controlled by Bailey, for the purpose of forming a new company. It contained several major conditions, including d'Auberjon's privilege to withdraw and Bailey's promise to obtain the consent of Edison and Elisha Gray and the cooperation of Theodore Puskas and Alfred Berthon. Bailey agreement with d'Auberjon, 4 Mar. 1880, DF (*TAEM* 56:61; *TAED* D8048ZAJ).

3. A chambre de notaires was a regional professional association of notaries, who performed numerous duties in French legal and quasi-judicial affairs. *Gde. Ency.*, s.v. "Notaire."

4. Agents de change, or stockbrokers, trading at the Paris Bourse.

5. The subscriber list has not been found but Bailey wrote Edison on 8 March that "we have today one hundred and fifty individual names on our list of subscribers with several large administrations that take 20 and upwards each, engaged to us and making a total of 250. This puts us ahead of the Gower and Soulerin combined." DF (*TAEM* 56:86; *TAED* D8048ZAN).

6. This list shows 24 working lines. DF (*TAEM* 56:84; *TAED* D8048ZAL1).

7. This unsigned memorandum is in DF (*TAEM* 56:55; *TAED* D8048ZAI).

8. Charles Jenty was a French industrialist, railroad promoter, and newspaper publisher. He represented the Vendée in the Chamber of Deputies at this time and later became an officer in the Legion of Honor. *DBF,* s.v. "Jenty (Charles)."

9. The Banque Franco-Egyptienne was established in 1870, one of a number of French joint-stock banks set up in the third quarter of the century to facilitate investment in and trade with particular foreign countries. Cameron 1961, 186, 195–96.

10. Émile de Girardin was a prolific journalist and newspaper publisher, well-known for his ardent republicanism. *DBF,* s.v. "Girardin (Émile de)."

11. Bailey's letter to Edison on Monday, 8 March (see note 5), did not provide this information.

12. Coudert Brothers was a legal firm established in 1854 by three Frenchmen, including Frédéric René Coudert, who helped arrange the donation of the Statue of Liberty to the United States. In 1879 the firm became the first in New York to establish a Paris branch, and quickly became involved with directing investments to that city. *Ency. NYC,* s.v. "Coudert Brothers."

13. Edison's letter has not been found but it was probably a reply to one from Theodore Puskas on 14 November about Charley's bills. Bailey explained in his 8 March letter (see note 5) that "the capital of the Edison Tel. Co of France has been as low as a few sous a number of times in the last weeks. I regretted to draw on you for Charley's bills, but I could not carry even so small an amount. I paid it to avoid scandal." The amount in question was 2,290 francs, approximately 440 dollars. Puskas to TAE, 14 Nov. 1879; Bailey to TAE, 8 and 10 Mar. 1880; TAE to Bailey, 9 Mar. 1880; all DF (*TAEM* 49:984; 56:89, 93, 91; *TAED* D7914T, D8048ZAO, D8048ZAW, D8048ZAT); *Ency. Brit.,* s.v. "Money."

14. The *Petit Journal* was by far the most widely circulated Paris daily. Jointly published by Charles Jenty, Émile de Girardin, and another partner, it was not so much a record of political events as a journal of the literary and commonplace life of Paris. Bellanger et al. 1972, 3:220–21, 234.

15. Émile de Girardin took control of *La France* from Charles Jenty in 1874 and built it into an important republican daily. Its circulation at this time was about 43,800 copies. *DBF,* s.v. "Jenty (Charles)"; Bellanger et. al. 1972, 3:234.

16. Considered one of the most respectable weekly general news magazines, *le Monde illustré* was known for its center-right political outlook. Bellanger et. al. 1972, 3:191, 387.

17. Probably the Banque Nationale pour le Commerce et l'Industrie, a major bank of deposit. Friedlander and Oser 1953, 571.

18. Unidentified.

19. In his letter on Monday, 8 March (see note 5), Bailey explained that the contract contained no provision for a cash advance because "There is one great change in the situation here in Europe that you must take into account, during the last month. The time for getting money for patents has passed, in any country where there is freedom of competition. It is now understood that there are several telephones that can be used, and no company or person will pay money in advance of actual exploitation. The telephone is settling down into the situation of a regular industry, and the contract we made here with Godillot, cant be repeated here in France or elsewhere in Europe."

20. Henry M. Lewis was a business associate of Samuel White and executor of the White estate. He represented the estate's interests in the Edison Telephone Co. of Europe and also, at least informally, with respect to the French company. Davis 1995, viii; Samuel White estate to TAE, 12 Jan. 1880, DF (*TAEM* 56:5; *TAED* D8048C).

21. Before the contract reached Edison Theodore Puskas cabled him

that it was "certainly prejudicial possibly fatal to our interest   shall forward next Wednesday full explanations and firm advantageous contract meantime cable me what immediately cash payment you will accept for your share in French telephone patent." Edison inquired how much this party offered and Puskas replied that "Neither Baileys nor my contract provides cash payment   believe possible change to cash payment provided your price not too high cable therefore lowest price   if cash suits you better than shares in Company to be formed during present year then will try arrange." Puskas to TAE, 8 and 11 Mar. 1880, TAE to Puskas, 9 Mar. 1880, all DF (*TAEM* 56:89, 93, 90; *TAED* D8048ZAP, D8048ZAX, D8048ZAR).

–1910–

From Henry Rowland

Baltimore, March 9/80

Dear Sir,

I have been waiting to receive your list of instruments, but it has not come. However, I will bring my own instruments as they are the only ones in the country whose ~~resistance~~ constants are known in absolute measure. I have thought of three ways of experimenting any or all of which we can choose. The first is by your dynamometer, which will give us the whole work done. The second is by the measurement of the resistance of the lamp and the quantity of electricity passing through it. The third is by putting the lamp in water and seeing how much heat is generated in the water. This last is probably the most accurate in a rough experiment, but I think we should try them all as a check on each other, provided you can arrange for them.[1] For the third experiment, all that is needed is a thin copper vessel, as small as will hold the lamp conveniently, and a lamp with insulated wires so that it can be immersed in the water. For the second experiment, there should be a solid stool about 2 ft. square[a] fixed in the ground in some shed or other, out of the way of the machinery, so that my instrument can be set on it firmly. Can you tell me what size of wire will not heat too much by the current to the lamp? Also high your resistance boxes run and what kind of a tangent galvanometer you have. Anticipating great pleasure from meeting you, I remain Yours truly

H. A. Rowland. Johns Hopkins University[2]

ALS, NjWOE, Upton (*TAEM* 95:603; *TAED* MU047). [a]"about 2 ft. square" interlined above.

1. Edison reportedly had asked George Barker to make an independent determination of the efficiency of his lighting system, and Barker arranged to do so with Henry Rowland and Cyrus Brackett ("Testing

Edison's Lights," *New York Sun*, 21 Mar. 1880, Cat. 1241, item 1469, Batchelor [*TAEM* 94:589; *TAED* MBSB21469X]; Doc. 1914). In their published report on these experiments, Rowland and Barker stated that they had considered "measuring the horse power required to drive the [generator] machine, together with the number of lights which it would give. But the dynamometer was not in very good working order, and it was difficult to determine the number of lights and their photometric power, as they were scattered throughout a long distance, and so this method was abandoned." They decided against measuring the resistance and flow of current because "the instruments available for this purpose were very rough" and instead used a calorimeter to determine the amount of heat evolved per minute (Rowland and Barker 1880, 337).

2. Henry Rowland became the first professor of physics at the Johns Hopkins University in 1875. Immediately after accepting this position he traveled throughout Europe to inspect laboratories and purchase experimental apparatus, and the laboratory he set up in Baltimore was considered the best-equipped in the U.S. at this time. He studied electromagnetic induction and the magnetic circuit, and around this time was beginning to redetermine the mechanical equivalent of heat. Rowland later turned to spectroscopy, which became the basis for much of his historical reputation. *DSB*, s.v. "Rowland, Henry Augustus."

–1911–

Telegrams: From / To
George Gouraud

March 10, 1880[a]
London

Edison

Johnson assumes negotitions ignoring me, giving as authority your letter to him of January 28th—[1] Is this your intention? if not you must cable quephone as I have twice requested or expose me to contempt   I am anxious to follow your wishes when expressed, but must insist that negotiations be left to me—

Gouraud

[Menlo Park[2]]

Gouraud

There has been no negotiations   Phonos merely asked ~~my that opinion~~ [---][b] me[c] if certain propositions were made ~~if they~~ would they[c] be acceptable   I replied that they were to me[d] but they must see you also as I would do you no injustice[3]

Edison

L and ALS (telegrams), NjWOE, DF (*TAEM* 56:502–503; *TAED* D8049ZBV, D8049ZBW). First message written in an unknown hand on message form of Direct United States Cable Co.; second message written on reverse of second page. [a]Date from document, form altered. [b]Canceled. [c]Interlined above. [d]"to me" interlined above.

1. Edward Johnson drafted this letter and Edison revised it. Stockton Griffin's copy of the corrected version reads in its entirety: "Referring to our conversation in relation to negotiations for the placing of my Telephone interests in the remaining territory in England, you are authorized to decide as to whether any proposition which may be made to you will be accepted by me." After seeing this letter Gouraud wrote to Edward Bouverie reminding him that his interest in the English telephone "is equal to that of Mr Edison & that in his deed of assignment, he has constituted me his sole agent and attorney for a term of five years." Johnson forwarded to Edison a copy of this letter and Bouverie's reply. Draft of TAE to Johnson and copy of TAE to Johnson, both 28 Jan. 1880; Gouraud to Bouverie, 5 Mar. 1880; Bouverie to Gouraud, 7 Mar. 1880; Johnson to TAE, 10 Mar. 1880; all DF (*TAEM* 56:365, 364, 483–84, 495; *TAED* D8049T, D8049S, D8049ZBM, D8049ZBN, D8049ZBT).

2. George Ward mailed Gouraud's cable to Edison the same day from the offices of the Direct United States Cable Co. in New York. Edison wrote his response on the reverse and Stockton Griffin dated both messages 10 March. The reply was not transmitted until the next day, when Ward wrote that he had "sent yours off to Gouraud tonight." Ward to TAE, 10 and 11 Mar. 1880, both DF (*TAEM* 56:501, 55:521; *TAED* D8049ZBU, D8042T).

3. Johnson requested Edison that afternoon to "Cable gouraud following   only desire Bouveries views be promptly met," which Edison did two days later (Johnson to TAE, 10 Mar. 1880; TAE to Gouraud, 12 Mar. 1880; both DF [*TAEM* 56:494, 507; *TAED* D8049ZBR, D8049ZBZ]). Gouraud answered: "Then cable Bouverie he must negotiate Exclusively with me, and cable me, that you have done so— Johnsons use your letter, and your cable quephone, approving certain propositions unknown to me, have been construed into your dishonoring me, meantime negotiations suspended. situation critical so act promptly. I consent to Bouveries views generally but they involve many important details which can safely and ought to be left to me." No reply from Edison has been found but several days later Johnson complained in a letter to Charles Batchelor that Gouraud was making "assertions that he has Telegrams from Mr Edison disowning my actions &c &c— He is charging me with 'dishonorable action'—misrepresenting Edison' &c &c" (Gouraud to TAE, 12 Mar. 1880; Johnson to Batchelor, 15 Mar. 1880; both DF [*TAEM* 56:508, 511; *TAED* D8049ZCA, D8049ZCD]).

–1912–

Patent Application:
Electric Lighting

[Menlo Park, March 10, 1880[1a]

Hitherto the meters proposed in systems for distributing electricity to consumers have been registers of time only, being adapted to simply preserve a record of the time during which a current was flowing through one, or several, translating devices without regard to the amount used. Such do not afford as equitable a basis for charges as one which would record or register, or indicate the amount of current used, irrespective of the time.[2]

My invention therefore relates to a meter which shall afford data for accurately determining the amount of current used. I make use of the known fact that a certain unit of current will cause the deposition of a definite amount of copper from a proper solution.

The circuit passing to the translating devices of any given locality finds in the meter box there stationed two paths, one, which is the main circuit through apparatus to be hereinafter[b] explained, to the translating devices, the other, which is a shunt-circuit, through an electrolytic cell to the translating devices. The resistance of this latter path is made exceedingly large compared with that of the first,— for illustration, say, as 100 to 1.

At the commencement of some fixed period, an exactly-weighed strip of copper is placed in the cell, which is filled with a proper copper solution. At the end of the period it is weighed, the difference being the amount deposited. This amount is due to the action of a hundredth part of the current used, and the actual amount used can be calculated with great ease.

The relation existing between current and cubic feet of gas can be ascertained by allowing the current to flow through an electric light of a definite photometric power—say 16 candles for a few hours—noting the deposition of copper in the cell, for same time. During the same period a gas-flame of equal photometric power is burned, and the number of cubic feet consumed noted. From the data thus obtained the resistances in the circuit of the cell may be so adjusted, that the deposition of 1 milligramme shall represent a current equal in light-giving capacity to 1 cubic foot of gas, and bills may be rendered in cubic feet.

Within the meter box I arrange an electro-magnet in the circuit to the translating devices. This circuit passes through the back contact-points on the armature of the magnet, through the armature and the magnet to the translating devices. The armature is held to its back-contact by a spring, whose resilience is so adjusted that it may not be overcome upon the passage of ordinary currents through the coil of the magnet. When, however, a greater amount of current than usual is diverted through this circuit, owing to accidental short circuiting, or any other cause, the armature is attracted and the circuit is broken. The armature is provided with a spring-latch, which locks int in position when once attracted, and prevents the circuit being re-opened. In order to prevent a spark upon the separation of the contacts, they are connected by a bit of very fine wire, which preserves the circuit for a moment after the contacts separate, when it is fused by the excess of current there-through.

In the drawings, accompanying and forming part of this specification, and showing a meter embodying my invention,—

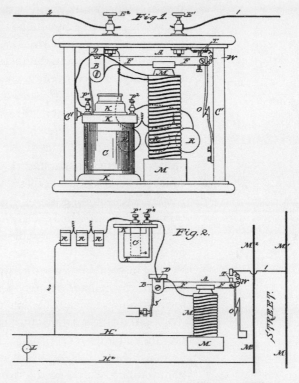

Inventor: T. A. Edison   per. Dyer & Wilber   Attys.
Witnesses: F. W. Howard[3]   [G. N. Hall?][4c]

Figure 1. is a view of the meter and adjuncts, and Figure 2. shows more clearly the circuits.[5]

M^1, M^2 are the main conductors. From M^1 a conductor leads to the meter and thence through the translating device and back to M^2. The conductor from M^1 goes to the binding post E^2, which is electrically connected to the contact points T, one of which is carried by the armature F of the magnet M, whose armature-lever A is pivotted at B, and[d] is in the circuit. At D this circuit branches, one, or the main branch passing around the magnet, the other, or shunt, through the electrolytic electro-plating or depositing[e] cell C. and resistances R.R.R. These latter are adjustable, in order that desired definite proportions of the total current may be sent through each branch   The spring S. of magnet M is so adjusted that the force exerted upon F. upon the passage of ordinary currents through F shall not serve to attract it. When, however, the current through M becomes unduly strong, M attracts F, breaking circuit at contacts T. upon such attraction the latches O catch and lock the armature so as to prevent the circuit again closing at T, until the cause of the disturbance may have been investigated. A very small wire, W, connects the contacts T, preventing a spark upon their separation, and keeping up a circuit around M. sufficiently long for O to securely lock, when W burns away, leaving the meter and translating devices entirely disconnected from the mains.

C c.[f] is the copper plate in the cell K, whose difference in weight due to deposition affords the current measurement.

Having thus described my invention what I claim is:—[6]

1" A meter arranged to measure by electrolysis, the amount of current passing through a conductor in a given time, substantially as set forth.

2" The combination of a main circuit and a shunt containing an electrolytic cell, so as to determine the total current by measurement of a fraction thereof, substantially as set forth.

3". The combination with a main circuit of a magnet and armature, the contact points, ~~switch~~, wire, fusible wire as described[g] and latch, substantially as set forth.

4" The combination with the magnet of a device for maintaining a circuit there-through after the main circuit is broken to insure the locking of the armature against re-closure of the main circuit, which device is itself then ruptured, substantially as set forth.

D and PD (photographic transcript), MdNACP, RG-241, Pat. App. 251,545. Written in an unknown hand; petition omitted. Alterations manifestly made during the examination process at the U.S. Patent Office have been ignored. [a]Date from patent, form altered. [b]"in" within "hereinafter" interlined above. [c]Signatures taken from printed drawings; second witness signature illegible. [d]"A is . . . and" interlined above. [e]"electro-plating or depositing" interlined above. [f]Interlined above. [g]"fusible . . . described" interlined above.

1. Edison's patent attorneys no longer included in his applications the standard introductory paragraph or the concluding testimonial statement signed and dated by the inventor which are part of all issued patents. This practice began with either this application (Case 206) or the preceding case, filed the same day but later abandoned and now lost. It is not known how Edison and his new attorneys collaborated to complete an application, given that Dyer and Wilber's office was in Wash-

ington D.C. rather than within a short journey as Lemuel Serrell's New York office had been. It seems likely that Edison would have prepared a draft and sent it to be revised and copied. In this particular case, Wilber was in New York at least as recently as 8 March and could have mailed or brought the application to Menlo Park, where Edison executed it on the 10th. Wilber to TAE, 8 Mar. 1880, DF (*TAEM* 55:611; *TAED* D8043Q).

2. Edison filed this application on 20 March 1880. It was rejected several times before issuing on 27 December 1881 as U.S. Pat. 251,545. The body of the application was incorporated into the final specification with minor alterations. A standard introductory paragraph identifying the invention was added, as was a final paragraph (see note 5).

3. Unidentified.

4. Unidentified.

5. Before the patent issued, Edison added a short paragraph at the end of the specification referring to figure 2 which explained the connections through the "tight box or case" in which the meter would be enclosed. U.S. Pat. 251,545.

6. Edison subsequently deleted the first claim below at the insistence of the Patent Office. He also elaborated the remaining three claims. Pat. App. 251,545.

-1913-

From Norvin Green

New York March 12 1880[a]

My dear Mr Edison:

Referring to your proposition to sell us your patent No 158,787, issued January 1875, for a Telegraph instrument to work without a magnet,[1] the subject was again before our Committee today, and the report of Mr Hamilton[2] thereon.

Mr Hamilton reported the instrument he used as operating satisfactorily and beautifully; but we are advised that that instrument had the attachment of the chalk cylinder, invented and patented by you in 1879,[3] without which, or some equivalent device, it would be impossible to make an instrument a sounder, and we have no means of knowing what value the instrument covered by the patent of 1875 may have for practical telegraphy. ⟨Newark 1873 1874   That inst in patent works with band paper. why "impossible"   Popes suggestion⟩

The chalk cylinder improvement—an invention patented in 1879—this Company is certainly entitled to have assigned and transferred to it under existing contracts between you and this Company,[4] and it was the belief of Mr Pope, who is charged with the duty of looking after patents, that such assignment had already been made. ⟨Everything which the Co is entitled to & more to is immediately assigned on presentation of the assignments   Pope insinuations, 4th time.⟩ He finds that such assignment was drawn and handed Mr Serrell and reported by him amongst others to have been made and his fees charged therefor have been paid, but we have not the patent in possession, and can not find that it has been made. It should not only be fully assigned and put on record, but it should be re-issued so as to cover its uses for telegraph purposes, ⟨This shows me It cant be reissued, besides this patent acquired as telephone by false pretenses. I receive no Value. It is not carrying out spirit contract, it has immense negative if not positive Value⟩ and so assigned its telegraph uses, after due experiment and its value ascertained, for which the contract allows three months time, should be settled for in royalties to be agreed upon or determined by reference as provided in the contract. ⟨if any such damned nonsense is to be attempted on me—it will not work⟩

Allow me to add, my dear Mr Edison, that the existing contract, under which you are operating with this Company, appears to me to be a liberal one on our part by which we are to pay, and have paid you $150.00 a week for the mere privilege of taking your inventions and paying you such royalties as they may be adjudged to be worth. ⟨Its neither liberal or illiberal. Liberality is not in the WU Dictionary   will give you $10,000 to cancel the contract. from this 100, I am expected to pay Kinney[5] & extra men—supplies, clerk on telephone cases, experiments on telephone, etc. I have spent over & above that amount, since date of contract, $16 800. monies received from France & England so I have yet to get my money back on telephone. other persons spend [many?][b] 1000s on 1 or 2 patents I cannot be expected to cover every point & carry on exhaustive experiments for a less sum.⟩ It seems to me, therefore, that there should be no hesitancy in awarding us cheerfully all the patents you produce which were agreed to be so assigned us by that contract. ⟨you have been but I dont propose to bring in property acquired before the contract⟩   Even more, I hold that the spirit of the contract embraces any unperfected inventions you had made not immediately applicable to Automatic Telegraphy, ⟨When a thing is patented it was supposed to be perfected—this is.⟩ where such inventions are perfected by further inventions during the period, and under the operations of the contract when your shops are being run at the expense of this Company as thereby agreed. ⟨My shops are not Run at the Expense of the Co. its but a flea bite.⟩

I think, therefore, that the brief time in which you offer us an opportunity to purchase your invention of 1875, without opportunity to make a trial of it sufficient to form any opinion of its value, at a price which is certainly a very large sum, and the threat to sell it to adverse interests to our detriment if we do not pay such large sum,[6] ⟨Never threatened to sell   I never told you I would sell if you didnt take it although I have offers at a higher price. I only made the offer to you at a lower one after rec[eivin]g such offers.⟩ is scarcely dealing with us in the spirit which both parties thought to be fair and equitable when the contract of March 1, 1877 was made. ⟨The sum is entirely too small, I and always dealt with the Co. in a very liberal spirit⟩   It was then thought that three months was a fair and reasonable time to operate and test the value of any instruments invented, but in this instance we are refused the use of an instrument here, and the only one our expert has been able to see operate was with such material improvements to which we are already entitled, as to make it a very different instrument from that covered by the patent of 1875. ⟨My offer was for a patent,[c] made prior to the[c] contract.⟩

We want to deal liberally with you, Mr Edison, but our Committee feel that they can not, as business men, appropriate so large a sum without some definite information of the real value of the thing they propose to buy, and think you ought to give us sufficient extension of time to examine and test it.[7]

In view of the relations between us, I do not believe we should confine the price agreed to be paid for it strictly to its intrinsic value, and I think you will find it your interest to meet us on fair and liberal ground, and not compel us to decline to purchase anything of your invention in Telegraphy.[8] ⟨Don't understand intrinsic⟩ Respectfully and truly yours,

Norvin Green   President[9]

LS, NjWOE, DF (*TAEM* 55:615; *TAED* D8043U1). Letterhead of Western Union Telegraph Co., Norvin Green, President. [a]"New York" and "18" preprinted. [b]Illegible. [c]Obscured overwritten text.

1. Edison's patent was for telegraphic uses of an electromotograph employing a moving band of treated paper (see *TAEB* 2:252 n. 2). It provided a way to circumvent a broad patent issued to Charles Grafton Page by a special act of Congress in 1868 for the "employment of one electromagnetic instrument to open and close the circuit of another electromagnetic instrument," that is, a relay. Western Union bought a controlling interest in Page's rights in 1871 but the patent's singular origin and the cumulative effects of numerous infringement suits limited its legal force (*TAEB* 1:448 n. 2; Post 1976, 170–81).

On 23 February 1880 Edison drafted a letter calling Norvin Green's attention to his 1875 motograph patent "In view of the fact that the WU Co have received a decision in favor of the Page Patent." He reminded

Green that it covered "the discovery of a new and unknown action of electricity by which if Magnetism had never been discovered we could still work Morse Telegraphs.= I have held this patent for a long time with the hope that some day I could realize from the immense amount of labor expended in experimenting since 1872." (Edison's experiments on the electromotograph did not begin until April 1874 [see Doc. 419]). Edison offered a ten-day option for purchase of the rights for $100,000 but by 3 March had not yet received an answer. TAE to Green, 23 Feb. and 3 Mar. 1880, DF (*TAEM* 55:604, 608; *TAED* D8043J, D8043M).

2. George Hamilton was a Western Union electrician (*TAEB* 3:510 n. 3). He evidently tested the instrument on 8 March at Menlo Park (Zenas Wilber to TAE, 6 and 8 Mar. 1880, both DF [*TAEM* 55:610–11; *TAED* D8043P, D8043Q]).

3. U.S. Pat. 221,957; see Doc. 1693.

4. Doc. 876 was superseded by Edison's 31 May 1878 agreement with the company (see Doc. 1317).

5. Patrick Kenny.

6. In an undated telegram, Edison asked Green to "Please remember that my proposition was simply for the sale of the 1875 patent nothing more or less. if the Co think the price too high well & good that settles it." TAE to Green, DF (*TAEM* 55:607; *TAED* D8043L).

7. After Green telegraphed to this effect on 8 March, Edison agreed to extend the option until 13 March. On that date Grosvenor Lowrey wired Green that Edison had postponed the deadline to 22 March. Green visited Menlo Park with Lowrey on 18 March. Green to TAE, 8 Mar. 1880; TAE to Green, 9 Mar. 1880; Lowrey to Green, 13 Mar. 1880; Lowrey to TAE, 17 Mar. 1880; all DF (*TAEM* 55:612, 622; 53:114; *TAED* D8043R, D8043S, D8043W, D8004ZBU).

8. Edison went to New York on 19 March (see Doc. 1916) and evidently reached an agreement in principle with Green, who wired the company's formal acceptance of his terms three days later. Edison executed the contract on 3 April. The agreement provided for payments of $16,000 immediately and $583.33 monthly for twelve years, a total of $100,000, for both the 1875 basic motograph patent and the 1879 chalk cylinder patent (Green to TAE, 22 Mar. 1880, DF [*TAEM* 55:626; *TAED* D8043ZAA]; Agreement with Western Union, 3 Apr. 1880, Miller [*TAEM* 86:118; *TAED* HM800100]). At Egisto Fabbri's private suggestion, Edison secured Green's personal commitment to allow the Edison Electric Light Co. exclusive free use of the Page patent for electric lighting (Fabbri to TAE, 15 Mar. 1880, DF [*TAEM* 53:684; *TAED* D8020ZCI]; TAE to Green, 31 Mar. 1880, Lbk. 5:803 [*TAEM* 80:231; *TAED* LB005803]). Edison also had Zenas Wilber prepare drawings of other devices designed to circumvent the Page patent. He sent these to Western Union with the suggestion that the company pay the patent expenses and, if the Page patent were finally upheld, compensate him according to the March 1877 contract. Wilber to TAE, 26 Mar. 1880, DF (*TAEM* 55:73; *TAED* D8036ZBL); TAE to Green, 3 Apr. 1880, Lbk. 6:995 (*TAEM* 80:514; *TAED* LB006995).

9. Norvin Green became president of Western Union in 1878 following the death of William Orton. He began his career as a physician and served two terms in the Kentucky House of Representatives before entering the telegraph business in 1853. He was subsequently president of

the Southwestern Telegraph Co. and, from 1866, a Western Union vice president. At this time he was also president of the Edison Electric Light Co. *ANB,* s.v. "Green, Norvin."

CHARLES MOTT LABORATORY JOURNALS
Docs. 1914, 1916, 1917, 1921, 1926, 1939, 1944, 1950, 1961, 1985, 2029, and 2057

Doc. 1914 is the first entry in a set of journals kept by Charles Mott. He wrote narrative summaries of laboratory work and events almost daily from March 1880 to January 1881, eventually using two notebooks.[1] Beginning on 9 April Mott also kept a series of six pocket notebooks for recording daily activities before entering them, sometimes in expanded form, into the journal; the pocket notebook entries continue through 19 March 1881.[2] Mott gave no rationale for keeping the journals but his numerous cross references to laboratory notebooks suggest that one purpose may have been to create a sort of running index to the experimental records. The first journal book begins with a two-page list of the predominating subjects and authors of notebooks in use on 10 April.

1. Mott Journals N-80-03-14 and N-80-07-10, Lab. (*TAEM* 33:683, 37:302; *TAED* N053, N117).
2. The pocket notebooks are PN-80-04-09, PN-80-05-03, PN-80-06-08, PN-80-07-14, PN-80-09-23, and PN-81-01-19, all Lab. (*TAEM* 43:959, 981, 1005, 1029, 1063, 1115; *TAED* NP009, NP010, NP011, NP012, NP013, NP014).

–1914–

Charles Mott Journal Entry[1]

[Menlo Park,] Sunday Mch 14—1880

On Wednesday night and during the day on Thursday of last week (Mch 10 & 11) Professors Barker, Brackett,[2] and Rowland were here, taking measurements of the candle power of the Electric Lamp and of the number per horse power, finding (I believe) very nearly twelve per horse power of sixteen candle power each, considered by them a very favorable result, Intending I believe to return some future day to test[a] the energy of the generators and of the horse power expended upon the machines, after which a full record of their tests and measurements will be published, whether favorable or otherwise.[3]

The measurement of the energy expended in the lamp was made by first weighing the lamp, next submerging it in water, taking the temperature, then putting on the current and noting at diferent periods the increase of temperature, for the space of five and in some cases Ten minuts, turning off the current and accurately weighing the water, and ~~from~~ this data and these figures form the basis of their calculation of the number of lamps per horse power.[4]

During the past week John Ott has been at work on a new Relay, with electric motive power of diferent form and pattern than the electric engine or motive power previously applied to the relay and telephone,[5a] M. P. Andrus at work on quite similar but diferent electric motive power or engine for chalk telephone.[6] Livingston at work on Hot air engine experiment[7] Dean[8a] on mould for pressing Black Lead or Plumbago into carbon size and form, an instrument requiring exceedingly neat, careful, and exact workmanship.[9] Some others of the men have been and are still at work on the large generator and today have one of the magnets wound.[10]

I judge some important business was being transacted yesterday Mch. 13, from the fact of many telegrams being recd. and sent, presumably in relation to Edison's relay.[11]

Today his Uncle from _____ is here visiting,[12] him, otherwise the day is quiet and disagreeable.

Sam Mott stricken off the Pay Roll last night, also Breath.[13]

AD, NjWOE, Lab., N-80-03-14:5 (*TAEM* 33:686; *TAED* N053:3). Written by Charles Mott. [a]Obscured overwritten text.

1. See headnote above.

2. Cyrus Brackett was a physics professor at Princeton and one of Francis Upton's mentors there. *WWW–1*, s.v. "Brackett, Cyrus Fogg"; "Upton, Francis," Pioneers Bio.

3. There is no extant correspondence regarding the plans for these tests or what prompted Edison to arrange for them but it is possible he was motivated by the criticisms expressed in Doc. 1895. The generator was the subject of separate tests by Cyrus Brackett and Charles Young (see Doc. 1916). Rowland, at least, remained another day in Menlo Park, where he made observations and calculations in one of Edison's laboratory notebooks on 12 March (Menlo Park Notebook #62, HAR [*TAED* X100B]). On 20 March George Barker telegraphed Edison: "Report from Rowland good will give you a proof next week" (DF [*TAEM* 53:688; *TAED* D8020ZCO1]). The report on lamps was published in April as Rowland and Barker 1880 and discussed by Edison in Doc. 1929. Rowland and Barker tested five lamps: a pair of regular large paper loops (numbers 580 and 201), the latter having already burned fifty hours by 2 January with its resistance unchanged since then; a single lamp (817) made on 6 March with bast fiber coated with sugar and plumbago; and a pair (809 and 850) with small loops of uncalendered pressed paper made between 6 and 9 March (N-80-02-02:27; N-80-01-02.1:13; N-80–01-02.2:192, 190; N-80-03-06:9, 15; Lab [*TAEM* 34: 1023, 478, 852, 850; 33:965, 968; *TAED* N073:12; N067:13; N070:134,

132; N057:5, 8]). They found that the bast lamp produced the equivalent of 13.1 gas jets per horsepower of electricity, the regular paper lamps between 6.8 and 12.8 gas jets; and the pair of small loops 8.3 jets. These figures included an adjustment for the estimated loss of "about 30 per cent for the friction of the engine, and the loss of energy in the magneto-electric machine, heating of wires, etc. As Mr. Edison's machine is undoubtedly one of the most efficient now made, it is believed that this estimate will be found practically correct." Rowland and Barker concluded that "Provided the lamp can be made either cheap enough or durable enough, there is no reasonable doubt of the practical success of the light, but this point will evidently require much further experiment before the light can be pronounced practicable." They also thanked Edison for "placing his entire establishment at our disposal in order that we might form a just and unbiased estimate of the economy of his light" (Rowland and Barker 1880, 339–40). At Edwin Fox's request Edison gave the report to the *New York Herald*, which published it on 27 March (Fox to TAE, 17 Mar. 1880, DF [*TAEM* 53:686; *TAED* D8020ZCK]; "The Electric Light," *New York Herald*, Cat. 1016:3, Scraps. [*TAEM* 24:80; *TAED* SM016003a]).

On 22 April Charles Clarke compared "Prof. Rowland's demonstration of power of electric light, as worked out from the final results and diagram as given by him in a letter to me" with results obtained by his own experiments. Charles Mott reported that Clarke found that "although their methods and formula differ, they get same results through their different channels." N-80-03-31:138–43, Mott Journal N-80-03-14:99, both Lab. (*TAEM* 34:939–41, 33:734; *TAED* N072:71–73, N053:50).

4. Francis Upton began his own extensive series of calorimetric experiments on about 17 March. He also continued photometric tests begun on 7 March, in which he recorded the luminosity produced and power consumed by different carbons (N-80-02-16:149–263, 25–81, 89–115; N-80-02-08.1:95–97, both Lab. [*TAEM* 41:998–1055, 936–64, 968–81; 34:278–80; *TAED* N303:76–147, 14–42, 46–59; N063:77–79]). Mott reported on 18 March that Upton "does not consider the system of so much accuracy as was attributed to it by Professor Rowland as the original temperature of the water as compared to the surrounding atmosphere has a great effect on the results obtained cold water being brought to the temperature of the air more quickly and easily than it can be carried above." Four days later he noted that Upton and Jehl were still "at work on Photometric and Calarimeter tests probably for their own edification and practice as it is entirely useless to make any enquiries of them in regard to their investigations" (Mott Journal N-80-03-14:14, 25, Lab. [*TAEM* 33:691, 696; *TAED* N053:8, 13]).

5. Nothing is known of this particular design. Mott recorded the next day that Ott had finished the "new motor relay" and found that the "motor runs very quietly but with low speed and power. it was discovered that by connecting the poles of the magnets with a mass of iron or steel, that the speed was greatly increased but no increase of power—rather out of the ordinary and usual law of speed & power." Mott Journal N-80-03-14:8, Lab. (*TAEM* 33:688; *TAED* N053:5).

6. Milo P. Andrus had worked as a patternmaker in the laboratory since 1878 (see *TAEB* 4, chap. 2 introduction, and *TAEB* 4:415 n. 16).

The design on which he was working may be that discussed in Doc. 1880 n. 4. Mott noted the next day that it was "N.G.   good power but too noisy." Mott Journal N-80-03-14:8, Lab. (*TAEM* 33:688; *TAED* No53:5).

7. Francis Livingston was a tool maker (Livingston to TAE, 19 Oct. 1884, DF [*TAEM* 71:304; *TAED* D8403ZHU]; Jehl 1937–41, 546). According to Mott, he tried this device the next night without success. He worked on it several more days with Julius Hornig, obtaining forty-one revolutions per minute on 17 March, when he suffered an injury and had to stop. There is no record of further efforts before 19 June, when Mott reported under the heading "Hot Air Engine" that "the small engine with which the experiments with exhaust steam as fuel were made was today shipped to Delamater Iron works New York" (Mott Journal N-80-03-14:8, 10, 12–14, 248, Lab. [*TAEM* 33:688–91, 808; *TAED* No53:5–8, 126]).

8. Charles Dean, a machinist who had formerly worked in Edison's Newark shops and had joined the Menlo Park staff in October 1878. He was the father of George Dean, who was working in the laboratory on dynamos. See *TAEB* 4:539; Jehl 1937–41, 676; Time Sheets, NjWOE.

9. The next day Mott noted that "Stamp mould for Plumbago burner finished and tried, found quite satisfactory but needed some little addition which is being made by Dean tonight." Charles Batchelor apparently began using the mould, although Mott reported that Dean's modifications continued until 22 March. About 15 March, the "first pressed plumbago small loop from new mould made by Dean   tissue paper on both sides of loop" were put in lamps 909 and 910. On 18 March two more plumbago loops were made, one of which was soaked in molasses. The same day Batchelor made drawings of the mould; these are reproduced in Friedel and Israel 1986, 153. After Dean finished the mould on 22 March Mott reported that "Mr. Batchelor made two unsuccessful attempts to get the fibers out in tack, the mould however does its work very nicely and he will probably be able to get out some during tomorrow. Mr Edison sketched another design for mold for same purpose and gave to Dean to make." Mott Journal N-80-03-14:8, 10, 12, 14, 20, 24; N-80-03-06:29, 39, 43, 51, both Lab. (*TAEM* 33:688–91, 694, 696, 975, 980, 982, 986; *TAED* No53:5–8, 11, 13; No57:15, 20, 22, 26).

10. This machine was completed by 17 March but not immediately tested because there were "insuficient lamps to try its power." Mott reported that it weighed two tons, excluding the thin-disk armature, and was moved into Edison's "new 'Station'" at the laboratory on 23 March. Mott Journal N-80-03-14:13, 26–27, Lab. (*TAEM* 33:690, 697; *TAED* No53:7, 14).

11. See Doc. 1913.

12. It is not known which of Edison's uncles, presumably one of his father's brothers or half-brothers, visited at this time. He stayed in Menlo Park until 17 March. Mott Journal N-80-03-14:13, Lab. (*TAEM* 33:690; *TAED* No53:7).

13. Samuel Mott's layoff was temporary and he continued in Edison's employ. William Breath was a laboratory assistant. Breath's first time sheet dates from the week of 9 January and his last one from the week of 16 April. Time Sheets, NjWOE.

Menlo Park, N.J., March 15 1880.[a]

Shall need $20 500 up to the first of august[1] of which 2000 will be returned by Oregon S[team]. N[avigation]. Co[2] This completes the maximum system here. 20 Dynamos, Special Building[3] and 600 Burners of which 300 are street Lamps—[4] after August it will probably be at the option of the Co the length of running time here of this system as sufficient proof of its practibility to them

T A Edison

D, NjWOE, DF (*TAEM* 54:45; *TAED* D8023K). Written by William Carman; letterhead of T. A. Edison. [a]"Menlo Park, N.J.," and "1880." preprinted.

1. This document was in an envelope labeled "Money wanted to Sep 1 from E.L Co." The envelope also contained Edison's itemization of more than a dozen related expenses totaling $20,170 (excluding labor before 1 May). DF (*TAEM* 54:45; *TAED* D8023K).

2. Edison presumably meant the Oregon Railway and Navigation Co., owner of the steamship *Columbia*. The Oregon Steam Navigation Co. provided inland transportation; Henry Villard bought the company in 1879 and operated it as a subsidiary of the Railway and Navigation Co. Buss 1978 [1977], 87–93.

3. Construction of an addition to the machine shop began in mid-February. The new structure was referred to as the "dynamo building" and was built to house Edison's new generators; it appears to have been completed and largely equipped by the latter part of April, at a cost of about $2,000 (Ledger #4:191, 193, Accts. [*TAEM* 87:306, 308; *TAED* AB002:97, 98]). John Kruesi's specifications and price estimates for materials are in N-79-07-25:177–87, 194–97, Lab. (*TAEM* 33:919–24, 927–29; *TAED* N056:61–66, 69–70).

4. During April Francis Upton and especially Charles Clarke separately calculated the physical dimensions of a system of conductors to supply more than two hundred street lamps, and John Kruesi sketched underground junction boxes. On 13 April Charles Mott reported Clarke's conclusion that the system would require 6.3 tons of copper. Workmen began laying the conductors in the ground on 1 May. About a week later Upton wrote to his father that "The light will soon show itself to the public again. . . . We shall show 300 or 400 scattered through the streets here, or rather in the fields, for the streets are more imaginary than real." N-80-03-15:103–107; N-79-07-25:54–65; Mott Journal N-80-03-14:67, 125–26; all Lab. (*TAEM* 35:642–44; 33:886–91, 718, 747–48; *TAED* N082:53–55; N056:27–32; N053:34, 64–65); Upton to Elijah Upton, 9 May 1880, Upton (*TAEM* 95:605; *TAED* MU048).

[Menlo Park,] Friday March 19—[1880]

Mr. Edison left for New York on 11:20 train to return tonight. All things very quiet about the Laboratory. Professors Bracket & Young[1] with assistant Magu(?)[2] of Princeton came on 11:20 train made measurements of candle power, tested electrometers for accuracy and amount of copper deposited in one half hour burning.[3] calculated horse power expended on generator, and the energy of machine. Up to present the only results I have been able to ascertain are that the armature gave but $^{14}/_{100}$ or about one seventh of an ohm resistance and the magnets 1 or 2 ohms resistance.

Their tests and measurements were conducted with great care, and their aim appearing to be minute accuracy.[4] At their own suggestion their report will be sent to Mr. Edison to use as he may desire.[5]

AD, NjWOE, Lab., N-80-03-14:16 (*TAEM* 33:692; *TAED* N053:9). Written by Charles Mott.

1. Charles Young was professor of astronomy at Princeton. *ANB*, s.v. "Young, Charles Augustus."

2. Unidentified.

3. In a report on their tests Cyrus Brackett and Charles Young wrote that "During the experiments of Professors Rowland and Barker [see Doc. 1914] . . . one of us (Professor Brackett) happened to be present: and he was requested by Mr. Edison to make an independent study of the subject under investigation." Brackett and Young made only one test of the generator on this day but came again on 3 April for further trials using the Prony and Edison dynamometers. Beginning on 30 March Francis Upton conducted his own similar experiments with the Edison and Prony dynamometers and also with copper deposition. The day before Brackett and Young returned he made more trials with the dynamometers and an electrometer, and for several days thereafter he continued related electrometer and calorimeter tests. Brackett and Young 1880, 475; N-80-02-16:265–73, N-80-03-15:13–79, Lab. (*TAEM* 41:1056–60, 35:597–630; *TAED* N303:149–156, N082:8–41).

4. Brackett and Young declared in print that "It is obvious that the work attempted in the time at our disposal was considerable; and, in consequence, no extreme accuracy is claimed for our results—though we believe them to be substantially correct—that is to say, within one or two per cent." Brackett & Young 1880, 475.

5. On 27 March Brackett and Young sent Edison a handwritten analysis of their first day's generator experiments. They concluded that of the total work expended, 86.7% was converted to electrical energy and 82% was available for use (DF [*TAEM* 53:696; *TAED* D8020ZCX1]). Later they also sent a separate report of their April tests which was printed in the 15 May issue of *Scientific American* as an enclosure to a letter from Francis Upton. In his letter Upton noted that "Profs. Brackett and Young show 90.7 per cent converted, and 83.9 available outside." Referring to the debate the previous fall over the efficiency of the Edi-

son dynamo (see Doc. 1832), Upton also expressed the hope that "this statement will be sufficient to end the discussion into which I was drawn some time since regarding Mr. Edison's machine. He then claimed that $\frac{9}{10}$ of the power in the current could be made available; now tests show $\frac{12}{13}$ of the energy in current are available." In another report they published in the *American Journal of Science*, Brackett and Young calculated the dynamo's total efficiency at 89.9% and its available efficiency at 84.1% after averaging results of three trials and correcting for power loss in transmission (Brackett and Young 1880, 478).

–1917–

Charles Mott
Journal Entry

[Menlo Park,] Sunday Mch. 21 [1880]

Dean continuing work on new mould, but otherwise nothing doing in Shop or Laboratory.

To night is the first Sunday night since I have been here and I believe for months before I came, that the Engine has not been running and work in Laboratory proceeding the same as at other times, and it is now understood that night work for the present shall be discontinued. Mr. Batchelor's complaining of his eyes is the present inducement for making this change, which will undoubtedly prove advantageous to all who have from choice or necessity worked nights and slept and rested during the day—and a saving of considerable expense in night meals which have heretofore cost from $120 to $140 per month during the winter.[1]

AD, NjWOE, Lab., N-80-03-14:22 (*TAEM* 33:695; *TAED* N053:12). Written by Charles Mott.

1. Five days later Mott remarked that he found "day working to reduce the Pay Roll about $50 in addition to the saving of $30 per week for night meals." In April, when a newspaper writer commented that Edison seemed to have gained about twenty pounds since the fall, Edison reportedly replied, "I am getting fat. I weigh 190 at least; I guess more. Sleep from 11 to 7 every night, eat three straight meals a day—that's the secret of it." Edison reinstated night work in late July but a water shortage forced him to discontinue it again in the fall. Mott Journal N-80-03-14:36, 53, 149, 163, Lab. (*TAEM* 33:702, 328, 376, 383; *TAED* N053:19, 26, 74, 81); "Edison at Home," *Denver Tribune*, 25 Apr. 1880, Cat. 1015: 18, Scraps. (*TAEM* 24:11; *TAED* SM015018a).

My Dr Edison

Gouraud has written you, stating that he had made certain propositions to Bouverie which were in the main the same as those submitted to you by me & subsequently agreed upon—, and that negotiations looking to their Early acceptance by B. were well under way when I stepped in & interfered with them &c &c &c=[1] Every word of which is a Damned lie from beginning to End— He had not only not submitted anything—but had Pooh Poohed Every suggestion looking to the unification of the whole country— He was Hell bent on forming a seperate Co.—putting the stock on the stock Exchange & going in for a speculation generally   I think I wrote you at the time that when the "Bouverie Phonos" basis was submitted to him, he tore it up with the remark, "preposterous absurd utterly impossible &c &c"[2]   Bouverie himself will confirm this if need be— I also heard him say at a Board meeting that he had the money tendered to him to[b] organize[b] a seperate Co. & he could therefore in justice to you not consider any such terms as were proposed— The next day he told me himself that he did not believe that he could raise another 1000£ outside of the London Co— What use have you for such a liar as that= In fact his whole character is best defined by the word "Trickster"   He thinks he is doing a brilliant thing when he tells a downright lie about what he can do with others in order to extort terms from the people with whom he is bargaining   His whole action has been that of a man who had no faith in the future of the thing and was bent on getting all the Blood there was in it at one squeeze—not caring a damn whether he left it strangled or not.

As for your future royalties you are far more likely to get them if the thing is made a unit & financially strong, than if you subdivided it into a No of weak companies=

Of course in case of amalgamation the advance Royalties paid & to be paid you, will be written off & your half interest in the Profits will be recognized as entitling you to a half of the Profits from amalgamation[3]—that is to say you will get half of that sum which is left after what has been expended for Royalties, Plant &c &c has been paid back to the ~~Dire~~ shareholders—of course our aim is to get as much as possible so that this half of the surplus over expenditure may be a good dividend— But in addition to this I have agreed with Bouverie that you must have a Royalty on Every Carbon Telephone used—no matter whose model— I have done this for two reasons— 1—

to obtain an acknowledgement of your right of discovery of the Carbon Principle & 2—To secure a future income for you from the Telephone in Eng. I[b] have conceded to Bell the same thing viz: that he shall have a Royalty on every magneto Receiver used= No one else is to be recognized= these two points have already been agreed upon by both sides—so they may be considered as out of the way= The only sticking point now, is the usual one of the respective interests to be represented in the new Board   we stick out for the chairmanship & they do for the same—a plan of compromise is being drawn up— I will have it in a day or two—when I will submit the whole basis of amalgamation to you. It will then come off or fail—by your own decision   Bouverie & his associates are prepared to take Either decision in perfect good temper— They don't care a "tupence" which way you decide=

My own opinion is that if we could only get Gouraud out of the way= reorganize on a large scale—get in plenty of capital —so as to fight fire with fire—Declare amalgamation off & go in for a vigorous campaign we can soon outstrip them—Beside having always in view the probable injunction against the carbon Telephones=[4] Then instead of being dictated to by them we should be the Dictators= Both Waterhouse & Bouverie are quite prepared to take this view also— But there are two Drawbacks

1st   Gouraud=

2nd   The want of a receiver by which facility is afforded for writing down the Telephonic messages—

The 1st I have undertaken to show you is a legitimate cause of complaint, which can only be removed by you—[5]

The 2nd= I have undertaken to remove myself—& have met with considerable success— I hope to have an Instrument in a day or two which will solve the problem— If I do it will go a long way toward stiffening the back bones of Everybody the fact is—we lose a great many orders—simply on a/c of this crank—[6] It must therefore be recognized as a defect no matter what we may ourselves think=

Bouverie is down in the country fighting a county for Parliament   Not much can be done until his return—as soon as he does however—action will be quick sharp & to the point= You must look for another instalment of cablegrams—as Gouraud has only been temporized with—not squelched— Yours Truly

E. H. Johnson

ALS, NjWOE, DF (*TAEM* 56:539; *TAED* D8049ZCL). Letterhead of the Edison Telephone Co. of London. [a]"London" and "18" preprinted. [b]Obscured overwritten text.

1. George Gouraud wrote on 20 March that he had been

a good deal startled when in the midst of my negotiations with Mr Bouverie I was made aware that without my knowledge certain communications had been made to yourself and that you had already assented to the modifications in the Agreement which Mr Bouverie was asking. It was not until I cabled you concerning Mr Johnsons use of your letter of 28th January that I was aware of the existence of that letter or the use Mr Johnson had made of it. Your cables have satisfied me that you could not have intended Mr Johnson to use the letter as he did.

As to the proposed modifications I had already assented to them in the main   I should not however have consented to so important concessions on your behalf without first consulting you but as your telegrams make it clear that you are anxious that the concessions should be made, I have given my consent to the general principles and the Documents are now being drawn. The day after tomorrow they will no doubt be settled in detail, and the arrangements will be consumated. [DF (*TAEM* 56:536; *TAED* D8049ZCK)]

2. See Doc. 1903 n. 7.

3. The new agreement evidently called for the London company to pay Edison half its profits (instead of 20% of rental receipts). Johnson to TAE, 27 Feb. 1880, DF (*TAEM* 56:456; *TAED* D8049ZBB).

4. In a 13 March letter to Edison, Johnson reported that the company had just received a court order "to compel the Silvertown people to answer our writ—served upon them to stop them from delivering the Blake Transmitters they are making for the Bell Co   It is the first step in the injunction proceedings." DF (*TAEM* 56:509; *TAED* D8049ZCB).

5. In an earlier letter to Charles Batchelor, Johnson warned that Edison would have "to face the alternative of abdicating in favor of Gouraud, or nominating a new Director= The London Company will not accept him in the reorganization= The new Co. proposed in case of amalgamation—will not accept him—and it is morally certain that no amalgamation scheme can be put through if he is allowed to have any say." Arnold White also wrote to Batchelor urging that Edison remove Gouraud. Johnson to Batchelor, 19 Mar. 1880; White to Batchelor, 15 Mar. 1880; both DF (*TAEM* 56:527, 515; *TAED* D8049ZCH, D8049ZCE).

6. See Doc. 1880 n. 4 regarding the effort to develop an electromotograph receiver operated by a clockwork or small electric motor instead of the hand crank. Nothing is known of Johnson's design.

From Edward Johnson

My Dear Edison

I have now 4 series of 45 cells Each—Fuller=on my Lamps & yet I have not enough quantity to supply <u>2</u> lamps without causing a slight variation in the 1st when the 2nd is put off and on. The Electro Motive force of 45 cells is not sufficient Either. I find it requires 50= to bring them up to a <u>superior</u> Gas jet= I have therefore today ordered 20 additional cells to bring my 4 series' of 45 up to 4 of 50=[1] It costs like Hell but I expect Tyndall & Thomson soon and I am determined to do the thing justice=[2] I have two Elegant Bronze Knights on my Desk— one on either side of my writing Tablet= The socket & Lamp are in harmony with the figures & makes a really Elegant combination— Then I have two Brackets on the wall with handsome relief Photos hanging behind them=

The effect of the E.L. on which is very much admired=

My circuit is neatly made on the wall & Desk so as to be seen at a glance— The multiple arc excites quite a good deal of interest   I'll try & give you an Idea how it looks—

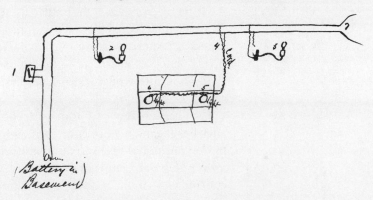

1= Box to Lock Main ckt. out of reach of Boys—and to show Putting on & Shutting off of House Lights—&c[b]

2 & 3. Wall Brackets

4— Cord—(7 Strand Copper)—with another Cord Tap.

5 & 6   Bronze Knights holding a staff with "Edisons E.L. Torches"=

7— Main condrs Ends exposed to show open ckt.

My Desk is a handsome Double slope mahogany— ~~Thus~~ [--][c] To the right on the wall—I have fixed a regular Telephone—a Loud Speaker (cast Iron)—and one of my Bell Tel Pattern Clockwork receivers— Thus I can give Every sort of Telephone demonstration— Downstairs I have a Crossley[3]— a Blake, & an Edison Transmitter— On my wall I have the 3

Diagrams of the Carbon Principle—nicely framed=[4] on a shelf I have the models made by you for me when in NYork lately—including the Paper speaker[5] (which by the by <u>Talks Splendid</u>=

I also have a small phonograph—on view=

My windows can be made absolutely dark= I have 2 good Gas jets=

I am thus particular in giving you the arrangement & contents of my office because I want you to know what it was that a couple of your friends saw last night=

Mess MacDonald Gen'l manager and Chinnery[6] Chief Editor of the London Times visited me by appointment at 5 P.M. —to remain a few minutes= White & I recd them= Rose Brown[7] & a couple of others being stationed where they could do the most good Telephonically=

I had the day before mere mentioned to MacDonald that I had the Lights in Operation— He said "I'll come & see them at once & bring our Editor if I may"= & they came= The heaviest Guns in all England—

Order of exercises— A. Explanation of Edisons position in re. to scientists & fellow investigators—in re to newspapers & others visiting Laboratory— Point—If^d any fault—one of too great openess of character—not seclusive by nature

B. Description of works—Light Exhibits machines Laboratory &c at menlo

C— Explanation of Lamp—carbon—multiple arc difference bet Incandescence & arc principle in re. to maintenance of ckt continuity= preservation uniform resistance &c.

D. Exhibit of Lamps in variety of ways—Exhiliarating—& convincing—auditors delighted= Comprehended point made on Preece as to adjusting source of supply in accordance with demand— Remark—"Preece not an authority on Electrical matters"=

Result of exhibit—all I could desire=

E. Exhibit of Telephone exchange (—adjoining room) Explanation P.O. contest— Extracted unqualified opinion P.O. people in Error in claiming Telephone—or in attempting restrict its growth=

F. Exhibit Telephones in my room—inclusive of—various adjustments Edisonian, microphonic &c &c— Norwood wire[8] —Induction &c &c—soft loud & conversational—delighted=

G. Explained in one hours Lecture listened to with utmost attention occasionally interrupted by intelligent question— Microphone shown to be no magnifier of sound—but simply a sound—or Telephone—Transmitter—ie Edisons Telephone— Diagrams explained— In short whole subject made so clear as to Elicit remark "Now I comprehend the Edison Hughes Preece controversy as I never did before. I am indeed gratified & obliged for this Lucid explanation."

Preece heartily excoriated by me & shown to have applied for Patent for applying micro to Telephone with Edisons Telephones in his possession—

Elicited justification of your course toward Preece—but expression of opinion that neither Hughes or Preece understood the principle of the microphone—

Anecdote by McDonald— "I visited the P.O. lately & Preece showed me your Telephones at work on a wire <u>upon which Telegraph Insts were at work</u>, & which gave[d] out a loud crackling sound— I could not understand a word & could scarcely hear anything but the noise— Then I was shown a microphone Transmitter & Bell receiver which talked beautifully—the receiver being absolutely dead during cessation of conversation—(I subsequently found that this wire only ran into the basement)   The difference of conditions were not explained to me—but I saw there was an extraordinary difference when[d] the Insts were[d] at rest— This excited my suspicions, but I said nothing as from having your Telephones in daily operation receiving our reports from Reuters[9] I knew its capabilities= & I thought it best to allow Preece to play his part to the end. I came away <u>convinced</u> In fact without the shadow of a doubt that an attempt had been deliberately made to <u>deceive me</u>. I mean to make use of the fact at the proper time— from what you[d] have today shown me Especially in re. to the ⟨I⟩Effect of Induction, I am now certain your Telephone was on a long wire or cabled wire—& theirs on a local Ckt"= Ha! Ha! nuts for us= White & I are going to send competent Experts to the P.O. to interview Preece in this same line & get additional testimony against him—)

Interview opened at 5. P.M. closed at 7 P.M.— The London Times is more than Ever Edisonian= It is now a partizan—a sattelite—so to speak= I bargained in advance that no mention should be made in the Paper of what they had seen Explained why &c. The benefits will be felt rather than seen in the extra back bone the next Edisonian Article will have— Gouraud didn't Even know of the contemplated visit= He has not yet spoken to me—& consequently not seen the Lamps— such is the price of vanity=

What is your opinion of my 1st use of your Lamps?—

Nothing new Telephonically— The new contract is vibrating between your Lawyers & the company's   Bouverie is still Electioneering   amalgamation is in status quo—

I am rendering a bill against Bergmann for cost of repairs & overhauling[d] bad work— It will Cost him like the Devil— I warned him when in NYork so he can't complain— My little Interest there—may go to Hell before I'll be taken advantage of— Yours Ever

E.H.J.

ALS, NjWOE, DF (*TAEM* 53:127; *TAED* D8004ZCE). Letterhead of Edison Telephone Co. of London. [a]"London" and "18" preprinted. [b]"Box . . . &c" enclosed by brace at left. [c]Canceled. [d]Obscured overwritten text.

1. Johnson wrote the previous week that he had gotten "one lamp up to a proper intensity by putting 50 cells in a series & using 2 series—but haven't enough guts yet" to keep more operating. In another letter the same day he reported having "now 4 series of 45 cells Each—Fuller Battery—and with them I am operating 3 Lamps successfully—bringing them quite up to a full Gas jet, in fact a little better. = As this has already cost me 248.$= I cannot afford to add Enough more to put the others in operation so I have taken down 3—& shall tell any one to whom I show the Light that you only gave me 3." The carbon in one lamp broke on 25 March when connected to two series of fifty cells. Johnson "locked it up for return to you—& shall say nothing about it—& so avoid useless discussion." Johnson to TAE, 18 and 25 Mar. 1880, DF (*TAEM* 54:214–15, 550; *TAED* D8026ZBB, D8026ZBC, D8049ZCR).

2. Johnson had earlier told Edison that William Thomson "will be in town in May and is coming to see the Light= Tyndall and in fact all the Big Guns will be in London about that time   The new Parliament then opens & the London season opens in full blast." Johnson to TAE, 18 Mar. 1880, DF (*TAEM* 54:215; *TAED* D8026ZBC).

3. Louis John Crossley, a Halifax carpet manufacturer, based his transmitter on David Hughes's carbon-pencil transmitter. In Crossley's design four loose carbon pencils were mounted in a diamond pattern between four carbon blocks. They were manufactured by Blakey and Emmott of Halifax and were the first to be supplied for the British Post Office telephone service when it commenced in 1881. Povey and Earl 1988, 18–20.

4. Nothing is known of these particular diagrams.

5. These instruments are probably those referred to in Doc. 1880.

6. Thomas Chenery was editor of the *Times* from 1877 to 1884. A noted linguist, Chenery had been a professor of Arabic at Oxford. *DNB*, s.v. "Chenery, Thomas."

7. Probably James Brown.

8. That is, a wire to Norwood, a village just south of London distinguished by the villas of numerous London merchants. McCulloch 1866, s.v. "Norwood."

9. Like most major English papers, the *Times* subscribed to reports from the news and market information agency established in 1851 by

Julius Reuter, although it published them only sparingly. Read 1999, chaps. 1–3.

-1920-

From Grosvenor Lowrey

New York[a] March 25th 1880.

My dear Edison:—

I am going to the Hot Springs to-morrow, and perhaps I shall not see you again before my return.

To guard against the consequences of either your death or mine leaving our various arrangements as they are now, only understood as between ourselves, I now state them and ask you to reply to this letter approving if you find them correct

First, in respect to England:

Although the last time we spoke on this subject you said that you understood and intended that I should have a one-third interest in your remaining interests under the Drexel Morgan & Co. arrangements,[1] I am disposed to decline that, because all the service which I could render you was rendered before you made this suggestion. I therefore, with thanks for the generosity of the offer, will not accept it, leaving you, therefore, so much richer and looking for my share alone to D.M. & Co.

Second, as to Cuba:—

I have arranged on your behalf with Mr. Navarro that he is to pay all expenses of taking out the patents and introducing the light into the Island of Cuba in consideration of one-half the net proceeds; he to have full control of the disposition of the patents.[2] Of the other half, going to you, you are first to deduct fifteen per cent., and then of the remaining portion, to wit, thirty five per cent. of the whole, you are to give to me one third.

Third, as to India:—

I have arranged with Messrs Fabbri & Chauncey[3] in the same manner; they to receive thirty five per cent. as their compensation, you to have sixty-five. Of this sixty-five per cent. fifteen is to be first deducted by you and then you are to give me one third of the remaining portion, to wit, of fifty per cent. of the whole.

Fourth, as to Portugal, New Zealand, New South Wales, Queensland and Victoria:—

I have made the same arrangements as to the above as were made with India and am to have the same share of your interest.[4]

In all these cases, as I have informed you, I also make the other parties give me a percentage of their net profit.

If this is correct please acknowledge the receipt of this letter and say so. All the interests referred to above as being for me are for the account of Porter, Lowrey, Soren & Stone. I will have suitable letters exchanged between you and each of these parties expressing in a general way the right and obligation of each.

Please also refer to and ratify the letter to Messrs D.M. & Co., F. & Co. and Mr. Navarro, dated the 25th, copies of which are enclosed.[5] Yours truly,

G. P. Lowrey.

L (typed copy), NjWOE, DF (*TAEM* 54:319; *TAED* D8026ZEL). Letterhead of Porter, Lowrey, Soren and Stone. [a]"New York" preprinted.

1. See Doc. 1649.

2. José De Navarro was a New York banker and stockholder in the Edison Electric Light Co. He had lived in Cuba for many years and in November 1878 began taking out electric light patents there in Edison's name (*TAEB* 4:707 n. 2, 735 n. 2). In February 1880, as patent assignments were being drafted, Edison asked Lowrey "What arrangement has been made with Mr Navarro regarding the Cuban rights. Should that not be fixed before making the assignment." Lowrey answered that "Mr Edison gets 65% Mr Navarro pays all expenses" (TAE to Lowrey and Lowrey to TAE, both 14 Feb. 1880, DF [*TAEM* 54:204; *TAED* D8026ZAQ, D8026ZAR]).

3. Fabbri & Chauncey was a major New York shipping and commercial firm with which Drexel, Morgan & Co. did a large business (see *TAEB* 4:621 n. 5 and Doc. 1566). At the end of 1878 the firm controlled Edison's South American electric light patent rights and had expressed an interest in India (see Doc. 2024 n. 4).

4. This arrangement was with Drexel, Morgan & Co. Lowrey to Drexel, Morgan & Co., 25 Mar. 1880, DF (*TAEM* 54:221; *TAED* D8026ZBE).

5. Lowrey's letters of 25 March 1880 to Drexel, Morgan & Co., Fabbri & Chauncey, and Navarro specified the relations with Edison outlined above (DF [*TAEM* 54:221, 223, 226; *TAED* D8026ZBE, D8026ZBE1, D8026ZBE2]). Edison gave verbal approval to these arrangements but did not make a written reply until the contracts were being drawn in mid-December, at which time Lowrey sent him this copy of his letter as well as a draft letter of acceptance (Lowrey to TAE, 14 Dec. 1880, enclosing Lowrey to TAE, 25 Mar. 1880, and draft of TAE to Lowrey, 15 Dec. 1880, DF [*TAEM* 54:318–19, 322; *TAED* D8026ZEK, D8026ZEL, D8026ZEL1]; see also Doc. 2024 n. 4). Stockton Griffin transcribed the draft and Edison signed it. It stipulated that Drexel, Morgan and Fabbri & Chauncey "were to bear the expenses or provide persons or syndicates who would bear the expenses (if such a thing were necessary) to make one installation of a Station in each of the countries named that this was always my intention and supposition." Lowrey responded that because "exhibitions will be made in New York and possibly in London before any could be called for in India and the other countries named . . . it is unlikely that such exhibitions would ever be really necessary." Accordingly, he had made no such provisions be-

yond the existing obligation to pay for a London demonstration but believed that the two firms would agree if the need arose (TAE to Lowrey, 15 Dec. 1880, Lbk. 6:672 [*TAEM* 80:439; *TAED* LB006672]; Lowrey to TAE, 16 Dec. 1880, DF [*TAEM* 54:324; *TAED* D8026ZEM]).

–1921–

Charles Mott Journal Entry

[Menlo Park,] Thursday March 25 [1880]

Mr. Edison experimenting on magnetic arrangement for separating iron or other magnetically influenced mineral from sand or other finly divided matter   A magnet is so arranged that the sand passes sufficiently near for the magnet to influence the iron from its course and separate from the silica— A bin is arranged with a long narrow aperature in the bottom through which the sand passes in a steady broad, thin stream. A short distance below is set a magnet whose ends come sufficiently near the stream of sand, to act upon the iron therein and attract it from the main current, thus forming two streams[a] but not near enough to allow the metal to attach itself to the magnet. ~~but~~A board or other dividing petition is provided and placed so as to allow the influenced metal to pass and lodge on one side next the magnets while the uninfluenced sand will continue to fall perpendicularly[b] and lodge on other side of petition.

h represents hopper   M. magnet   S silica or matter uninfluenced by magnetism   i iron or other metal and "b"[b] the petition board. the tests made with it to day would show probably ninety per cent (~~absolute~~) separation[1]

Dean turned bast fibre down to $^{15}/_{1000}$ & $^{1}/_{32}$ long and drilled $^{5}/_{1000}$ hole through center lengthwise. Mr. Batchelor inserted manilla fibers through them and had the whole carbonized and got them out ready for lamps in perfect order and naturally feels quite proud of the achievement[2]

Clark[3] and Lawson been for a day or two experimenting on sensative wash for paper for photographing tables tracings &c.[4]

Paper accounts of Lecture of Prof. Barker on Edison Elec Light (see Friday—26—pge 36)[5]

AD, NjWOE, Lab., N-80-03-14:32 (*TAEM* 33:700; *TAED* N053:17). Written by Charles Mott. [a]"and attract . . . two streams" interlined above. [b]Obscured overwritten text.

1. Edison executed a patent application for this apparatus on 3 April 1880 that issued on 1 June 1880 as U.S. Pat. 228,329. He described several variations, such as using multiple magnets and submerging the entire device "in a tank of water, as I have found that when wet, heavy sand is used the tendency to clog in the hopper and to cling together in falling is entirely overcome by placing the hopper and magnet in water." He also noted that "the material to be treated may be first impelled by a blast of air or by other means in a horizontal plane, and the magnet placed above or below the line of the normal trajectory of the material, so as to lengthen or shorten the trajectory of the magnetic substance." Laboratory assistants tried these and other experimental arrangements of the ore separator over the next few weeks and on 15 April it "very completely" processed twenty-five pounds of sand per minute. An unidentified newspaper article from about this time reported Edison's claim that the process was ready for Oroville sand but would have to be adapted to ore from other areas. Mott Journal N-80-03-14:35, 49, 68, 74; Lab. (*TAEM* 33:701, 708, 719, 722; *TAED* N053:18, 25, 35, 38); "Edison's Big Bonanza," Cat. 1241, item 1483, Batchelor (*TAEM* 94:593; *TAED* MBSB21483).

2. On 24 March Charles Dean had finished and was using what Mott termed "an exceedingly neat and exact tool" for cutting and drilling bast fibers to these dimensions. He was making new clamps according to Charles Batchelor's instructions of the previous day, which would fit like a sleeve on the end of a carbon loop. They were to be ".015 thick and .075 long before carbonizing— Pointed and .0065 hole through so that the fibres will just push in— Then carbonize— To hold this we will make a spiral ( ) of platinum wire, that is two sizes, made on a mandrel so:= the small end to go on the .0010. wire [the lead-in wire] (or .007 if we use it) and the large end to set the clamp in." Batchelor also suggested that the carbon could be put in perpendicularly so as to make the bast a crosspiece which could be inserted into the ordinary metal clamp. The first bast clamps were prepared for carbonization experiments 188 and 189 on 23 March and put in three lamps on 26 March. Mott Journal N-80-03-14:28–29; N-80-03-06:57–59, 53, 63, 67, both Lab. (*TAEM* 33:698, 989–90, 987, 992, 994; *TAED* N053:15; N057:29–30, 27, 32, 34).

Batchelor instructed that "In order to make these clamps we will pick out the fibres about the right size and draw them through a given die which cuts them to .025 thick   these are then put in our little screw machine milled down—drilled—and pointed." Mott remarked on 30 March that Dean had "the smallest die that could be found for cutting thread on screw but on trying it he finds it only about one half small enough" and would have to make his own about $^{10}/_{1000}$ of an inch. On that day Dean was "turning oak down to $^{12}/_{1000}$ and drilling therein lengthwise a $^{6}/_{1000}$ hole for use on ends of manilla fibre carbons." Clamps of holly wood were also tried around this time. On 10 March Edison had executed a patent application for the manufacture of carbons in which he specified his preference for cylindrical clamps of ligna vitae, boxwood, or similar carbonizable material (see Doc. 1907 esp. n. 2). N-80-03-

06:57–59, 73; Mott Journal N-80-03-14:38–39, both Lab. (*TAEM* 33:989–90, 997, 703; *TAED* N057:29–30, N053:20); U.S. Pat. 525,888.

3. Charles Clarke, a Bowdoin classmate of Francis Upton, was a civil engineer and draftsman who joined the laboratory staff at the end of January. Jehl 1937–41, 402–403, 855–56; Clarke's testimony, 189, *Edison v. Siemens v. Field*, Lit. (*TAEM* 46:99; *TAED* QD001:95); *NCAB* 30:44.

4. Nothing more is known of this research.

5. Mott reported the next day that "The Papers of yesterday contain accounts of Professor Barker's lecture on Edisons Light, delivered Mch. 24 in the Chapel of the Pennsylvania University and which gives to the light really more than has ever been publicly claimed for it by Menlo Park." According to the *New York Herald* article, Barker concluded with "a tribute of respect and admiration for Mr. Edison. A few days ago, when he assured the Wizard that calculation convinced him that ninety-eight per cent of the energy was secured by his new generator, Edison exclaimed, 'If that be so, man is absolute master of nature. Ninety-eight per cent of energy means the control of the forces of the world. Electricity is light and heat. We have only to place our engines at the coal mines and transmit the heat and light wherever it is needed.'" Mott noted that the day's papers also contained William Sawyer's "general declaration that none of the statements are true that Barker is mistaken or wilfully misrepresenting and that Edison knows nothing." Mott Journal N-80-03-14:36, Lab. (*TAEM* 33:702; *TAED* N053:19); "Edison's Electric Light," *New York Herald*, 25 Mar. 1880, Cat. 1241, item 1473, Batchelor (*TAEM* 94:590; *TAED* MBSB21473X).

–1922–

Notebook Entry:
Electric Lighting

[Menlo Park,] Mch 25 1880

Best conditions.

Large bulb.
Most proportionate and comely bulb.
small leading wires through glass
annealed glass where the wires pass.
High polish where wires pass through glass
Perfect cleaning of the tubes before blowing
A glass containing nothing that will combine with Carbonic acid.
To heat the bulb the least both for convenience in handling and for the prevention of chemical combinations
platinum Iridium clamps.
Clamp the conductor firmly with a clamp that will spring together and hold tight whether it expands or the Carbon wastes.
No iron or other material that contains much gas to be used in lamp.
Double sealing.[1]
Heating the bulbs by sand bath when exhausting.

Very gradual putting on of current so that most of the gases are driven off at invisible heat.

place the clamps some distance apart.

Carbonize without oxidation and at highest practicable temperatures.

Have the Carbons flat if possible

The highest resistance combined with the smallest radiating surface.

Even resistances.

TAE

X, NjWOE, Lab., N-80-02-16:83 (*TAEM* 41:965; *TAED* N303:43).

1. See Doc. 1887 n. 3.

–1923–

From William Marks

Philadelphia, Mar 30 1878[a]

My dear sir

My understanding with Prof Barker was that the expmts on your light were to be continued whether or not Rowland came. it seems that I misunderstood(?) him. or that Prof B. forgot.[b]

I think that Benson[1] & Col Armington Eng Bklyn Gas Co[2] will visit you in a day or two.— I told Benson what I had learned during my previous visit to your place & told him also that I believed you would certainly succeed in producing 10 lights of 16 candles (ie= 80 c[ubic]f[ee]t pr hr of gas) per HP.[3] You will find Col Armington a thorough gentleman & almost as good a fellow as Benson. both of them have thorough confidence in your honesty of purpose and only wish to know the exact truth.

I have said to Prof Barker that I shall be pleased to go on with the work whenever he feels sure that he <u>can</u> do so without fail.

I believe I stated in your presence that I do not care to undertake work without some return to myself and that I do not care to trifle my time away. My return in the present instance was to have been (not professional or scientific glory) but merely the right to use all the knowledge gained for self and friends

As you know, I do not wish to be misunderstood & therefore state my position frankly

I know that you desire the widest publicity and ought to desire even more the accurate information of men engaged in the lighting business.[4] I am   Vry Truly Yours

Wm D Marks[5]

ALS, NjWOE, DF (*TAEM* 53:705; *TAED* D8020ZDB). Letterhead of William D. Marks, Whitney Professor of Dynamical Engineering. [a]"Philadelphia," and "187" preprinted. [b]"or that . . . forgot" interlined below.

1. Possibly Byron Benson, a developer of Pennsylvania oil fields and partner in the Tidewater Pipe Co., which in 1879 began operating the first pipeline from the Titusville area towards Philadelphia and New York. *NCAB*, 30:537–38, 35:112.

2. Possibly James Armington, variously identified around this time as a superintendent and engineer in Brooklyn. Lain 1879–80, 19; ibid. 1881–82, 21.

3. Marks had been to Menlo Park prior to 4 March, when he wrote Francis Upton estimating that Edison produced only about four lights per horsepower. He based this figure on a conversation about the steam engine's operating conditions rather than on his own experiments. On 26 March (Friday) he telegraphed Edison, "If convenient will indicate Engine Saturday Sunday and Monday" with Barker. Edison wired Barker on 30 March, "When will you be ready to test." This visit never took place and Charles Batchelor later explained to Uriah Painter that the Rowland and Barker report was published even though "the test was never finished." Marks to Upton, 4 Mar. 1880, Upton (*TAEM* 95:601; *TAED* MU046); Marks to TAE, 26 Mar. 1880; TAE to Barker, 30 Mar. 1880; both DF (*TAEM* 53:695, 704; *TAED* D8020ZCX, D8020ZDA); Batchelor to Painter, 25 Apl. 1880, Unbound Documents (1879–1882), UHP.

4. Several weeks later Marks wrote to the editor of the *Engineering and Mining Journal* criticizing the Rowland and Barker report for overstating the dynamo's efficiency and underestimating the operating costs of steam power. He wished Edison "a rapid and successful result from his labors; but it is wrong as well as ridiculous to endeavor to create a sensation by misconstruing the results of Professor Rowland's measurements, and by dabbling in statements as to comparative cost of electricity and illuminating-gas without actually knowing either." "The Edison Electric Light," *Engineering and Mining Journal*, 24 Apr. 1880, Cat. 1015:46, Scraps. (*TAEM* 24:26; *TAED* SM015046a).

5. William Marks was the Whitney professor of dynamic engineering at the University of Pennsylvania. He had been a consulting engineer on various projects, including gas works, and later became president of the Edison Electric Light Co. of Philadelphia. *WWW–1*, s.v. "Marks, William Dennis."

6

April–June 1880

Edison acknowledged in mid–April that a fundamental change was taking place in the nature of his work at Menlo Park. Reporters visiting the laboratory noted that he had taken most of his electric lamps out of service. Edison reportedly explained that

> There is a wide difference between completing an invention . . . and putting the manufactured article on the market. . . . I have solved all the problems I was engaged on in the matter of the electric light, and have finished it. I can light a city at one-fourth to one-half the price of gas; I have proved it here, and I have put out my lights because there is no longer any need of them. The lamps, conductors and generators are now being made for the demonstration on a larger scale, first here, then in New York. But this involves all the details of manufacturing, and it takes time.[1]

As is true of many of Edison's published interviews, this one presented the issue rather simplistically and with his characteristic optimism. In essence, however, it accurately reflected the shift at Menlo Park from the work of invention to that of developing and manufacturing a product for commercial use. Edison had discontinued night work at the laboratory in late March, to which he attributed his changed appearance: "I am getting fat. I weigh 190 at least; I guess more. Sleep from 11 to 7 every night, eat three straight meals a day. . . . I have escaped from the hard work and worriment that the electric light imposed."[2] To prepare for commercial lamp manufacture, Edison purchased the former electric pen factory building along the railroad tracks in Menlo Park in late April. His workmen quickly began rehabilitating the structure and configuring it for use as the lamp factory. Charles Batchelor and other laboratory assistants began the long process of devising and constructing necessary equipment, from carbonizing moulds to complex clamp-making machines. Ludwig Böhm and the glassblowers began fabricating vacuum pumps, completing the first one hundred by 2 June.

Further development of the electric light and power system as a whole proceeded in several directions. Improvements in the lamp were sought in the laboratory, where John Lawson experimented extensively during the spring with wood fibers as possible filament materials. Experimental carbonizing techniques were also tried, including one that resulted in a patent for thoroughly carbonizing the thickened filament ends. In mid-June, Edison attempted to explore further the phenomenon of "electrical carrying," which shortened the useful lifetime of his lamps. He enlisted the aid of Princeton physics professor Charles Young to make spectroscopic observations of a bluish discharge around the clamps which seemed associated with the darkening of the glass bulbs. Lawson and Francis Upton continued making tests related to the meter, including measuring the effects caused by rolling the copper plates, such as would be done during the commercial manufacturing process. Upton, Charles Clarke, and Francis Jehl also made calculations and tests designed to ensure a uniform voltage drop throughout the distribution system. In order to test the entire system in operation, Edison planned a scaled-down version of a central station district at Menlo Park. Around the first of May his workmen began laying nearly five miles of underground conductors for several hundred lamps.

The central station was the one system component in which Edison planned a major change, specifically in the size and number of dynamos. In April engineers from the journal *American Machinist* made their own tests of Edison's machine and concluded that his claims for its efficiency were undermined by the large amount of energy lost as friction in the bearings and slippage in the drive belts. Edison disputed their findings and directed Charles Clarke to embark on a series of careful experiments to measure the heat produced in the bearing journals. In the meantime he also began to study the merits of high-speed steam engines, such as the form devised by Charles Porter. When Clarke's results confirmed large frictional losses, he and Upton immediately began designing a much bigger dynamo that could absorb the full horsepower of a large high-speed engine without using belts or shafts, and Edison began making arrangements for Porter's company to build a 120 h.p. engine to be coupled directly to the generator.[3] This represented an entirely different conception of the central station from what was planned as recently as January and shows Edison's attention to practical operating details of the system at this time. It also demonstrates his conviction, reported in a February newspaper interview, that "the great item of importance to be secured in connection with his electric light is cheap steam engineering. . . . [Edison] says 'steam engineering forms 75 per cent. of the electric light, 20 per cent. is in the system itself, 4 per cent. is in the dynamos and 1 per cent. in the lamps.'"[4]

During April, laboratory staffers built four generators and made about one hundred twenty lamps, some with frosted bulbs, for the steamship *Columbia*. In late April and early May, Francis Upton, John Kruesi, and others spent several days on the ship in New York, supervising the installation and testing of equipment. The lighting system was operational by 27 April, when Edison, his wife, and several assistants attended a reception aboard the ship.

Edison ordered construction of an electric railroad in order to demonstrate the practicality of electric power. It consisted of a locomotive and car and a half-mile of track. The locomotive, driven by a modified Edison dynamo run as a motor, was tried on the track for the first time on 13 May. Although there were problems with the mechanical transmission it otherwise worked so well that Edison shortly extended the track and added three cars. Laboratory employees, Edison's wife, and numerous visitors went for rides along the steep and curving track. Among these was Grosvenor Lowrey, who required Edison's personal assurances to overcome his reservations about the train's safety before climbing on. The train jumped the rails on a curve, vaulting John Kruesi, who was driving, into the brush. Lowrey recounted to his future wife that Kruesi, bloodied and "a good deal shaken," got up mimicking Edison in his accented English, "'Oh yes pairfeckly safe!!!'"[5] Francis Upton made intermittent tests of motors in the laboratory but Edison was sufficiently convinced of their efficacy by late June to offer a hydroelectric power system to a Nevada mine operator.

One other significant laboratory project was the adaptation of the process for extracting platinum ore from hydraulic mine wastes. No longer in need of platinum, Edison had been working instead to capture the small amounts of gold that passed through the hydraulic clean-up. He applied for a patent on an ore separator in early April. In mid-May he wrote to Frank McLaughlin in California that his process was "absolutely perfect" and he was "ready for business" whenever McLaughlin could make contracts with mine owners.[6]

In other matters, Edison agreed in April to underwrite the publication of a new journal, *Science*, which its prospective editor intended would cover all scientific fields in the United States much as *Nature* did for Great Britain. About the middle of the month he received word that Joshua Bailey and Theodore Puskas had at last come to terms with each other and rival telephone interests to form a unified French telephone company in which Edison was to have a forty per cent stake. In London, months of complex and often acrimonious negotiations finally produced an agreement in early June to combine the Edison and Bell interests into the United Telephone Co., Ltd. One of the major obstacles had been whether Edison would cash out his interest in the old company or preserve it in some way with the new firm; the matter was put in abeyance by appointing a trustee for the disputed interest.

It is impossible to state precisely how many men Edison had working for him at any particular time because of the incomplete nature of extant payroll records. He seems to have hired only a handful of new faces during this period, among them Cornelius Van Cleve, husband of Mary Edison's half sister, who helped carbonize lamp filaments.[7] Michael Griffin worked on the electric railway.[8] Little or nothing is known about the others, including two who have only a single pay record apiece, suggesting that they stayed for a very brief period.[9] Dates for the termination of service are generally as difficult to fix as those for the beginning. Laboratory assistant William Breath seems to have left in April.[10] George Carman, Edison's purchasing agent who also assisted in the laboratory, accompanied Frank McLaughlin to California in early April.[11]

1. "Edison at Home," *Denver Tribune*, 25 Apr. 1880, Cat. 1015:18, Scraps. (*TAEM* 24:11; *TAED* SM015018a); see also "Our New York Letter," 17 Apr. 1880, Cat. 1241, item 1495, Batchelor (*TAEM* 94:597; *TAED* MBSB21495).

2. "Edison at Home," ibid.

3. This is probably the device that was listed on laboratory time sheets as "direct converter." Time Sheets, NjWOE.

4. "The Coming Light," 12 Feb. 1880, Cat. 1014:34, Scraps. (*TAEM* 23:648; *TAED* SM014034a).

5. Lowrey to Kate Armour, 5 June 1880, Reed.

6. Doc. 1938.

7. See Doc. 1674 n. 1.

8. See box 145, Employee Records, NjWOE.

9. See records for Michael Donegan in box 145, Employees Records, NjWOE, and John Harvey in box 146, ibid.

10. Box 144, Employee Records, NjWOE, but also see Doc. 1914.

11. Mott Journal N–80–03–14: Lab. (*TAEM* 33:709; *TAED* N053:26); Carman's testimony, *Edison v. Maxim v. Swan* (*TAED* W100DID [images 4, 7]).

–1924–

From John Harjes

Paris, April 1st 1880.[a]

Dear Sir,

Since I wrote you last on the 24th feby[1] I am without any of your letters—. In answer to my cablegram of the 18th ulto I received however on the 23d your reply as follows:[2]

"You may accept for me and sign any contract which will produce a reasonable return and will put a stop to the present intriguing which will most quickly put telephone in practical operation in France— Edison"

and have myself telegraphed you yesterday as per enclosed copy advising you that I had signed for you the contract for the disposal of your Telephone patents for France—a copy of said contract I forward to you by this steamer under separate & registered cover.[3]

Mr Bailey speaks of going to New York on a short visit, in about a week or 10 days[4] & it is therefore useless for me to refer to any details of the negotiations of the last few weeks.— I may however say, that under no circumstances would I have accepted the Trust conveyed by your letter of Decber 2d 79[b] with power of atty,[5] could I have imagined, that in its bearing and consequences it would bring the tenth part of the excessive annoyances, intrigue and trickeries besides great loss of time, to which[c] I have had to submit during the late negotiations here.

Mr Bailey's never ceasing activity and perseverence against frequent difficulties to keep up and further the interest represented by your patents deserves all praise. It was the Auberjon's[6] group which gave him the first helping hand, this fact and the influence they represented, made me advise you in favor of that contract— I desired however certain modifications, a greater precision on many points, especially as to the minimum sum to be paid for the patents, a fixed limit of time within[d] which the Company had to be formed, a sufficient sum to be named which the parties would be willing meanwhile to expend for pushing and developing before the public the Edison Telephones, and particularly a complete cancellation of the so-called "option" clause, which both the Bailey= as well as Puskas= party claimed and under which they reserved for themselves the option till December of this year to withdraw from the undertaking and be re-imbursed by you in cash for all their expenditures. I was unable to succeed with the Auberjon group but carried all the points just named with the party represented by the Banque Franco Egyptienne (Société de Credit Industriel & Commercial,[7] Société Financière,[8] A. Fleine of A. & M. Fleine,[9] Reinach of John Reinach[10] &c &c) and finally closed with them, to which, much to my gratification, both Mr Bailey as well as Mr Puskas expressed their entire satisfaction, all the more gratifying to me as both representing sundry interests[e] had to sign the contract.

You will notice that Article 5 of the contract provides for a non-formation of the company, all what the party represented by the Banque Franco Egyptienne is to receive however as a re-imbursement for all expenses incured, is limited to a part-ownership of the patents, i.e. their expenditure gives them a proportionate ownership of the patents as a total of fr 500,000 would bring them 60%.[11]

Although I had no cause to fear that their expenses might be unreasonably increased, mainly by charges for securing certain indirect influences or newspapers under their own control or ownership and thus run up a large bill for which eventually a large percentage of ownership of the patents could be claimed, I thought best to provide a control till the formation of the Company, both as to the expenditures as well as selection of the Board of Directors of the new Coy. Much[c] against my inclination I have been compelled to accept that position myself in absence of any other suitable party here.

A day before closing the contract I was approached by the Gower telephone Co represented by Mr d'Erlanger as to selling your patents to them——. I thought it to your interest not to entertain their ouvertures but thought proper to guard your name and therefore your interest (as provided for in Article 7) in case a fusion should after all take place.[12]

In my negotiations for the disposal of your patents I tried to secure two objects, the one an immediate return in money as large as possible, the other, a proper development of your telephone before the French public. and had I been sole owner of your patents I should not have been able to secure better terms.

Since signing the contract, the Banque Franco Egyptienne and Mr Bailey have represented to me the advisability of making a present of five per cent (to be borne equally by the parties to the contract, i.e. yourself and the Banque Franco Egyptienne) to Mr Jenty to which I felt constraint to consent for your interest. Mr Jenty was the principal party to the late Auberjon's contract, (he is a member of the french chambre des députés) a man of influence, part owner of sundry newspapers "La France," "Le petit Journal" (the latter with a circulation of about 600,000.) and others in[d] which already some weeks since long articles appeared in favor of your telephone and the object is thus to recompense Mr Jenty for past services and to secure his continued good will and influence, as naturally he feels rather hurt not to have himself secured the contract—. Mr Bailey however can give you full particulars about this—. Besides what goes to Mr Jenty, the Syndicate represented by the Banque Franco Egyptienne has to give, I am told, twice 5% to other parties for certain influences so that in reality the Syndicate receives for itself not 60% but only 47½%.

I have had several interviews with the Director of the Banque Franco Egyptienne as to the appointment of a proper chief Director together with the necessary staff to immediately commence active business and development of the enterprise and to judge from present appearances[c] the Banque Franco Egyptienne seems desirous for an early formation of the new Company.

I shall keep you advised as anything of special interest occurs. I remain, Dear Sir, Yours very truly,

John H. Harjes

LS, NjWOE, DF (*TAEM* 56:129; *TAED* D8048ZBU). Letterhead of Drexel, Harjes & Co. [a]"Paris," and "18" preprinted. [b]Interlined below. [c]Obscured overwritten text. [d]Interlined above. [e]"representing sundry interests" interlined below.

1. Harjes recommended that Edison have Joshua Bailey continue to seek new sources of capital rather than have the French telephone forced into judicial liquidation. He enclosed a letter to the same effect from Henry Harrisse. Harjes to TAE, 24 Feb. 1880; Harrisse to Harjes, 24 Feb. 1880; both DF (*TAEM* 56:43, 45; *TAED* D8048ZAC, D8048ZAD).

2. Harjes twice cabled Edison on 18 March. The first was a long message advising that Edward d'Auberjon's investors were "much displeased with action of Puskas party" and consequently refused to modify the contract signed two weeks earlier (see Doc. 1909). Harjes warned that "affairs are suffering much by present inaction and uncertainties and

that your earliest possible decision by cable is necessary   equity and the probability of success seems to be in favor of the contract with Auberjohn mailed march fifth; besides withdrawal of Auberjohn and his party owing to their powerful influence might prevent success of any company formed by others." About five hours later he cabled again: "should you be in favor of contract Auberjon advise me so but do not sign same until I report further as I should prefer first to secure from Auberjon additional clauses in your favor." On 22 March Edison cabled Harjes: "You may accept for me and sign any contract which will produce a reasonable return and will put a stop to the present intriguing and which will most quickly put telephone in practical operation in France." Harjes to TAE, both 18 Mar. 1880; TAE to Harjes, 22 Mar. 1880; all DF (*TAEM* 56:96, 98, 105; *TAED* D8048ZBD, D8048ZBE, D8048ZBK).

3. Harjes enclosed a copy of his telegram saying that he "Could not carry sufficiently advantageous conditions with Auberjon. Have just closed and signed contract with group represented by Banque Franco Egyptienne which remains binding without any optional clause. New company will probably be formed immediately, but must be formed during this year, meanwhile they will expend if necessary up to five hundred thousand francs for furthering the enterprise, such expenditures subject to my approval." He reported that Edison would have a forty percent stake in the new company, or at least 900,000 francs; however, the contract specified that this portion was for all the patentees jointly. The old company was to be liquidated. The previous week Bailey sent Edison a 358-word cable urging acceptance of Auberjon's proposal, which had been amended to guarantee Edison a 200,000 franc minimum share. Harjes to TAE, 31 Mar. 1880; Agreement with Franco Egyptienne Bank, Bailey, Elisha Gray, and Theodore Puskas, 27 Mar. 1880; Bailey to TAE, 20 Mar. 1880; all DF (*TAEM* 56:128, 109, 100; *TAED* D8048ZBT, D8048ZBS, D8048ZBG).

4. Bailey was still in Paris the last week of April; he was in New York by 18 May and planned to visit Menlo Park the next day. Bailey to TAE, 27 Apr. and 18 May 1880, DF (*TAEM* 56:161, 165; *TAED* D8048ZCU, D8048ZCZ).

5. Edison enclosed the power with a letter to Harjes on 21 November 1879. The power was dated 2 December, probably the date Harjes signed it, but no letter from that day has been found. Harjes to TAE, 16 Dec. 1879, DF (*TAEM* 52:450; *TAED* D7940ZDS).

6. Edward Auberjon has not been further identified.

7. The Credit Industriel et Commercial was a large deposit bank. Chartered in 1859 largely to promote the practice of writing checks in France, it also underwrote securities for domestic and foreign ventures. Cameron 1961, 172.

8. Probably either the Société Financière d'Egypte, a joint stock bank that predated the Banque Franco-Egyptienne, or the Société Financière, which was heavily involved in railroad construction on the Continent. Cameron 1961, 186, 317, 320.

9. Unidentified.

10. Unidentified.

11. This article provided that the bank was to advance up to 500,000 francs prior to the formation of the company. In the event it advanced the full amount but the company was not organized by the last day of 1880,

the bank was to have a sixty percent share of the patents, or a smaller share in proportion to its contribution. The bank's share was not to exceed sixty percent under any circumstance.

12. This section stipulated that a company resulting from a merger with the new Edison company would have to include Edison's name in any circulars which advertised the patents it controlled.

-1925-

From Edward Johnson

London Apl 1st 1880[a]

My Dear Edison
    Amalgamation
    This is assuming the proportions of an immediate probability[1]   I must therefore try and set before you the Pro's & Con's so as to have your approval in advance—an Essential now since your representative has become an obstructionist. You must bear in mind the following points in determining what is best=
    1st   Two companies multiply the already extraordinary difficulties of overcoming the prejudices of the British Public in the matter of House top & other rights of way concessions— As one company they could systematise[b] the overhead wire running and avoid the present cut throat policy—
    2nd   The jealousies of the two Co's stand as an ever present weakness in the face of the P[ost]. O[ffice]. opposition jointly they would be ably to defy that concern—
    3rd   Divided— Each is holding over the other the sword of Patent weakness—neither Patent is so strong as to be absolutely sure of being sustained in a severe legal struggle— United— Both Patents would & could be sustained—thus shutting up the makers of both magneto & microphones and practically abolishing the enormous legal expenses both Co's are now incurring— true the Patent question would go on to an issue, but unopposed except by the small outside makers
    4th   The rates could be raised to a figure which would net a profit—not now the case—
    5th   Capital for conducting the business on a large scale and in a vigorous manner could Easily be had— In fact is already assured. Many other less important advantages would accrue—and a more important one than all—would be the fact that the P.O. in dealing Eventually, as Every one holds it must—with the question of purchase would have a united body to dicker with—& not two seperate concerns—one of which it could play off against the other=
    It is the opinion of Every one that the stock of the united Co would immediately command a large premium—Especially as

that of the Bell Co. is Even now at a premium of from 30 to 60%=

There have been several propositions as to your share—how your half interests in the Profits[2] should be provided for—whether in shares—in money or in your retaining a half interest in the share of Profits accruing to the stock ~~of the~~ in the United Co—allotted to & held by the Edison Company— The last plan finds most favor—as it ensures a permanent revenue both to you & to the shareholders   I would like to have your views as to whether you would prefer to close out your interest at once—or retain it as now ie—a half interest in the Profits   My opinion is that there is more money in it for you by a clean sale—even if you take shares in payment—as they will hardly fail to command a premium   I enclose a memoranda from White which shows the present & most probable Terms—[3] The proportion of 115 to 200 is had by virtue of the Larger plant & superior financial position of the Bell Co and has nothing to do with the value of the Patents= They are set down in Every case as on an equality

If we amalgamate on this basis our Co. will have to raise privately the money to pay you 15,000£ 30,000 to go on with— and about 12,000£ to buy up Glasgow so you see a heavy load has in any event to be carried—[4] The two companies would go on under some arrangement as to organization but the Earnings would be Pooled—& Dividends declared in the proportion of 200 to Bell & 115 to Edison— It has not yet been fully determined how these details would be carried out—but of one thing you can rest assured—viz: that Bouverie will consent to nothing that does not look to putting our interest in a Dividend earning position

I'll write you more fully on this subject now by Every mail—as Every day opens up some new scheme—& you can see from this Letter that at present nothing well defined has been agreed upon[5]   Yours Truly

E. H. Johnson

ALS, NjWOE, DF (*TAEM* 56:564; *TAED* D8049ZCY). Letterhead of the Edison Telephone Co. of London. ªLondon" and "18" preprinted. ᵇObscured overwritten text.

1. In an undated note, probably written on 1 or 2 April, Johnson enclosed Edward Bouverie's 31 March letter stating that negotiations with the Bell interests had gone so far "that we could not honourably back out of it—if the other side are willing to go on." Johnson explained that he had urged Bouverie to "make Haste slowly with amalgamation giving as a reason that I was informed you were again experimenting with the

chalk" and that he believed "the early expectations as to this frictional principle business were on the verge of being realized." Johnson concluded his letter to Edison by noting that "we are in a financial hole—& amalgamation seems the only way out." Johnson to TAE, n.d; Bouverie to Johnson, 31 Mar. 1880; both DF (*TAEM* 56:560, 561; *TAED* D8049ZCW, D8049ZCX).

2. Article twenty of Edison's contract with the telephone company reserved to him one-half the company's "goodwill," or difference between its capital investment and market value upon expiration of the patents. Agreement with Edison Telephone Co. of London, 14 July 1879, DF (*TAEM* 52:617; *TAED* D7941ZCV).

3. Arnold White explained in this letter to Johnson that it was proposed to have stockholders "exchange their existing shares for shares in the new Company in the proportion of Bell 200   Edison 115." He assured Johnson that Bouverie and his advisers had "determined that the whole thing might go to blazes before they departed from the honourable understanding with Mr Edison to the effect that his name and interest shd be maintained under all circumstances." The new company was to be named the "United Telephone Co. Lim. (Bell & Edison patents)." White expressed misgivings about the merger but assured Johnson that "Bouverie has taken the right course in the interest of Mr Edison & the shareholders. The great mistake we made in founding the Company, was in raising far too little money in the first instance. When we found we wanted more, the patent suit and the P.O. case cropped up, & while the Board persist in disclosing to intending investors the true facts of the case, no money can be got from the public, especially as it is the opposition, and not we who have the financial standing in the U. States." White to Johnson, 1 Apr. 1880, DF (*TAEM* 56:571; *TAED* D8049ZCZ).

4. On 16 April Arnold White drafted a circular letter from the company apprising shareholders of the proposed merger terms. It suggested that some of the additional capital needed to complete the process could be raised by offering a new stock issue to existing shareholders. Edison Telephone Co. of London draft circular, 16 Apr. 1880, DF (*TAEM* 56: 598; *TAED* D8049ZDM).

5. Arnold White advised Johnson that it had become apparent that to maintain Edison's rights to the company's goodwill it would be "necessary to maintain the skeleton of the Company in order to hold the existing shares of the present holders. This will of course debar them—the holders—from realisation of their shares in case of death or emergency & will handicap them with a crushing weight in comparison with the 'Bell' holders in the new Company." Johnson promptly relayed this information to Edison, cautioning him that "Under these Circumstances the Edison Co will find it difficult, if not impossible to induce Capitalists to put in any money." White to Johnson, 3 Apr. 1880; Johnson to TAE, 5 Apr. 1880; both DF (*TAEM* 56:575, 579; *TAED* D8049ZDA, D8049ZDE).

[Menlo Park,] Thursday April 1 [1880]

Ott still at work on Telephonic motor experiment.[1] Hipple[2] and Crosby[3] on centrifugal apparatus.[4]

Mr. Edison tonight commenced experiment on pumps with the view of using single instead of double pump as at present also of combining or arranging a large number in small space.[5] has on two lamps of fiber carbon, globes made tube shape

AD, NjWOE, Lab., N-80-03-14:42 (*TAEM* 33:705; *TAED* N053:22). Written by Charles Mott.

1. See Doc. 1914. John Ott appears to have resumed work on this device the previous day. On 5 April he "finished a new receiver and motor connected therewith but does not think it a success himself, the chalk being directly on shaft of revolving armature, revolves too rapidly, and not running as quietly as those previously made." Mott Journal N-80-03-14:40, 48–49, Lab. (*TAEM* 33:704, 708; *TAED* N053:21, 25).

2. James Hipple began working as a glassblowing assistant earlier in the year and remained connected with Edison lighting companies for many years. Jehl 1937–41, 496; Alfred Chatard to Charles Batchelor, 9 July 1883, Batchelor (*TAEM* 92:501; *TAED* MB091B); Hipple to TAE, 25 Sept. 1887; Philip Dyer to TAE, 7 June 1889; both DF (*TAEM* 120:50, 125:262; *TAED* D8741AAE, D8905ADV).

3. As a favor to Edwin Fox, Edison took in George Crosby as a laborer, promising to pay only his room and board but no wages. Crosby later stated that he began working at the laboratory in November 1879; the first extant time sheet for him is the week of 19 March 1880. Edison's testimony, 41, Crosby's testimony, 25, 29, *Böhm v. Edison* (*TAED* W100DED032 [image 10], W100DEC025 [images 1, 5]); Fox to TAE, 9 Aug. 1880, DF (*TAEM* 53:510; *TAED* D8014T); Time Sheets, NjWOE.

4. Mott recorded the previous day that workers were "Experimenting tonight on drying sand by centrifugal machine driven by electricity, took sewing machine motor from office for the purpose   find the sand to dry very quickly and rendered fit to seive in a few minute (say 10)." More trials were made on 2 April. Edison noted in his April 1880 ore separator patent application (U.S. Pat. 228,329) that wet sand could be dried before being processed but did not refer specifically to a centrifuge. Mott Journal N-80-03-14:40, 43, Lab. (*TAEM* 33:704–705; *TAED* N053:21, 22).

5. The "double pump" refers to the Sprengel and Geissler combination (see Doc. 1816). Mott noted that the next day Edison was "still on pumps." Edison testified in a subsequent patent interference that in the spring of 1880 he "started two of my assistants, Dr. Moses . . . and Francis Jehl, experimenting to obtain a cheap and rapid pump, and less complicated than the ones that we were using." Jehl stated in the same proceeding that he wanted to "do away with all fancy glass blowing about the pumps, to cut all the scollops off, and have a simple vacuum pump for practical use," one "which could be constructed by a boy having two or three weeks practice in glass blowing, which was easily prepared, and took a good deal less room than the combination." Mott Journal N-80-03-14:43, Lab. (*TAEM* 33:705; *TAED* N053:22); Edison's testimony, 35,

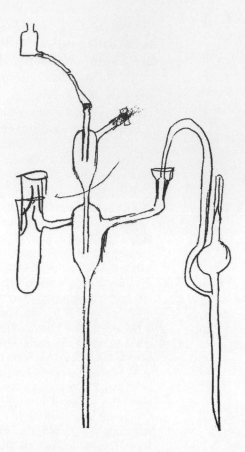

Böhm's 13 April drawing of a "Single Sprengel pump devised by Upton." The upper mercury jet did not work.

Jehl's attachment for sealing lamps to the pump without traditional ground glass joints.

Jehl's testimony, 16, 15, *Böhm v. Edison* (*TAED* W100DED032 [image 4], W100DED002 [15, 14]).

Moses designed a pump referred to by Mott as a "cyphon pump" which was built by Ludwig Böhm on 9 April and tested the next day. It was unsatisfactory, which Edison attributed to the "cyphon" attachment rather than the pump itself but further tests failed to produce good results. On 13 April Böhm constructed an apparatus for Upton in which the mercury apparently passed successively through two jets in a single Sprengel tube. Mott Journal N-80-03-14:59, 61, 66; N-80-03-19:7–17; N-80-02-08.1:98, Lab. (*TAEM* 33:713–14, 718; 34:610–15, 281; *TAED* N053:30, 31, 34; N068:4–9; N063:80).

About 16 April Jehl devised an alternative to the ground glass joints through which lamps were connected to the pump. He inserted the open tube of an unsealed lamp through the center hole of one of the rubber stoppers used in the chemical laboratory, and placed the stopper into a cone-shaped glass fixture blown by Böhm and attached to the pump. He incorporated this into a pump made on 18 April by William Holzer, an American glassblower who began working as Böhm's assistant in January. Mott described it as "very simple, free of stop cocks, takes but little room, and comparatively little mercury." Jehl successfully tested this during the next week and it became the basis of the pumps in Edison's first lamp factory (N-80-03-19:29, N-80-03-14:84, both Lab. [*TAEM* 34:620, 33:727; *TAED* No68:14; No53:43]; Jehl 1937–41, 495, 800–804). However, other pumps were made and tested throughout April and Böhm made almost daily sketches of related apparatus through the beginning of May (N-80-03-14:72, 74, 78, 86–87, 94, 102, 104, 106; N-80-03-19:15–25, 31–99; both Lab. [*TAEM* 33:721–22, 724, 728, 732, 736–38; 34:614–19, 622–54; *TAED* No53:37, 38, 40, 44, 48, 52–55; No68:8–13, 16–48]). A number of these drawings were copied into a separate notebook and Edison later submitted some of them as evidence in the interference case (N-79-02-20.2:27–45; Lab. [*TAEM* 38:909–18; *TAED* N146:13–22]; Edison's exhibits 2–4, *Böhm v. Edison* [*TAED* W100DED059B, W100DED059C, W100DED059D]).

–1927–

Henry Morton, Alfred Mayer, and B. F. Thomas article in the Scientific American

[Hoboken, N.J., c. April 5, 1880[1a]]

SOME ELECTRICAL MEASUREMENTS OF ONE OF MR. EDISON'S HORSESHOE LAMPS.

BY HENRY MORTON, PH.D., ALFRED M. MAYER, PH.D.,[2] AND B. F. THOMAS, A.M.,[3] AT THE STEVENS INSTITUTE OF TECHNOLOGY.

Much has been written and said within the last few months on the subject of Mr. Edison's new horseshoe lamps, and with all the writing and saying there has been wonderfully little produced in the way of precise and reliable statement concerning the simple primary facts, a knowledge of which would give the means of estimating both the scientific and commercial status of this widely discussed invention.

It was, therefore, with great pleasure that the present writers found themselves, through the kindness of the SCIENTIFIC AMERICAN, placed in possession of one of these horseshoe lamps of recent construction.

To satisfy themselves as to the real facts of the case they soon made a series of careful measurements and determinations, and as the results of these are likely to interest others, they now put them in print for general benefit.

A further examination of other lamps would have been made at the same time had opportunity offered; but as a communication on this subject addressed to Mr. Edison did not evoke a reply, they are obliged to content themselves with the one lamp as a subject of experiment.[4]

They would, however, here remark that the behavior of this lamp, under the tests, and the agreement of its results with information otherwise obtained, convince them that it is at least a fair specimen of the lamps of this form so far produced at Menlo Park.

The first object, on receiving the lamp, was to determine roughly what amount and character of electric current would be needed to operate it efficiently. With this view a number of cells of a small Grove's battery were set up, having each an active zinc surface of twenty square inches and a platinum surface of eighteen square inches.[5]

The lamp being placed in the situation usually occupied by the standard burner in a Sugg's photometer,[6] the battery was, cell by cell, thrown into circuit.

When ten cells had been introduced the horseshoe showed a dull red, with fifteen cells a bright red, with thirty-four cells the light of 1 candle was given, with forty cells the light of $4\frac{1}{2}$ candles, and with forty-five cells the light of $9\frac{1}{5}$ candles, and with forty-eight cells 16 candles.

Having thus determined what amount of electric current would be required for experiments, arrangements were made to measure accurately the resistance of horseshoe while in actual use and emitting different amounts of light. The resistance of this carbon thread at the ordinary temperature had been already determined as 123 ohms in the usual way, but it was presumed, as had been shown by Matthiessen (*Phil. Mag.,* xvi., 1858, pp. 220, 221), that this resistance would diminish with rise of temperature.[7]

To measure the resistance under these circumstances the apparatus was arranged as follows: The current from the battery was divided into two branches, which traversed, in opposite directions, the two equal coils of a differential galvanometer. One branch then traversed the lamp, while the other passed through a set of adjustable resistances composed of German-silver wires stretched in the free air of the laboratory, to avoid heating. (Careful tests of these resistances showed that no sensible heating occurred under these circumstances.)

Matters being thus arranged, the resistances were adjusted until the galvanometer showed no deflection when the candle power of the lamp was taken repeatedly in the photometer, and the amount of resistance was noted.

These measurements were several times repeated, shifting the coils of the galvanometer and reversing the direction of the current.

The results so obtained were as follows:

Resistances.	Condition of Loop.
123 ohms	Cold.
94 "	Orange light.

83.7 "	$\frac{9}{16}$ candle.
79.8 "	5 "
75 "	18 "

The photometric measurement was in all these cases taken with the carbon loop at right angles to the axis of the photometer, which was, of course, much in favor of the electric lamp. On turning the lamp round so as to bring the carbon loop with its plane parallel with the axis of the photometer, *i.e.,* the edge of the loop turned toward the photometer disk, the light was greatly diminished, so that it was reduced to almost one-third of what it was with the loop sideways to the photometer disk.

Having thus determined the resistance of the lamp when in actual use, it was next desirable to measure the quantity of the current flowing under the same conditions.

To do this the current from fifty cells of battery was passed through a tangent galvanometer as a mere check or indicator of variations, and then through a copper voltameter, *i.e.,* a jar containing solution of cupric sulphates with copper electrodes immersed, and then through the lamp, placed in the photometer.

Under these conditions it was found that during an hour the light gradually varied from about 16 candles at the beginning to about 14 candles at the end, making an average of about 15 candles, measured with side of loop toward disk.

The galvanometer during this time only showed a fall of half a degree in the deflection of the needle.

Carefully drying and weighing the copper electrodes, it was found that one had lost 1.0624 grammes.[8]

Now, it is well known that a current of one weber takes up 0.000326[9] gramme of copper per second, which would make 1.1736 grammes in an hour; therefore the current in the present case must have been on the average $^{1.0624}/_{1.1736} = 0.905$ webers, or a little less than one weber.

Having thus obtained the resistance of the lamp when emitting a light of 15 candles, namely, 76 ohms, and the amount of current passing under the same conditions, namely, 0.905 weber, we have all the experimental data required for the determination of the energy transformed or expended in the lamp, expressed in foot pounds. For this we multiply together the square of the current, the resistance, the constant 0.737335 (which expresses the fraction of a foot pound involved in a current of one weber traversing a resistance of one ohm for one second), and the number of seconds in a minute. Thus, in the present case, we have $0.905^2 = 0.8125$, and $0.8125 \times 76 \times 0.737335 \times 60 = 2753.76$ foot pounds.

Dividing these foot pounds per minute by the number of foot pounds per minute in a horse power, that is, 33,000, we have 0.08, that is, about eight one-hundredths or one-twelfth of a horse power as the energy expended in each lamp.

It would thus appear that with such lamps as this, one horse power of energy in the current would operate 12 lamps of the same resistance with an average candle power of 10 candles each,* or 120 candles in the aggregate. The candle power being 15 candles in the best position, and 5 candles at right angles to this, the average or general illuminating power of the lamp is 10 candles.[10b]

Assuming that a Siemens or Brush machine were employed to generate the electric current, such a current would be obtained, as has been shown by numerous experiments, with a loss of about 40 per cent of the mechanical energy applied to the driving pulley of the machine. To operate these 12 lamps, therefore, we should have to apply more than one horse power to the pulley of the machine, so that when this loss in transformation had been encountered there should be one horse power of electric energy produced. This would call for 1⅔ horse power applied to the pulley of the dynamo–electric machine, by the steam engine.

To produce one horse power in a steam engine of the best construction about three pounds of coal per hour must be burned, and therefore for 1⅔ horse power 5 lb. of coal must be burned.

On the other hand one pound of gas coal will produce 5 cubic feet of gas, and will leave, besides, a large part of its weight in coke, to say nothing of other "residuals," which will represent practically about the difference in value between "steam making" and "gas making coal," so that it will not be unfair to take 5 lb. of gas coal as the equivalent of 5 lb. of steam coal.

These 5 lb. of gas coal will then yield 25 cubic feet of gas, which, if burned in five gas burners of the best construction, will give from 20 to 22 candles each, or 100 to 110 candles in the aggregate.

We have, then, the twelve Edison lamps producing 120 candles and the five gas burners producing 100 to 110 candles, with an equivalent expenditure of fuel.

If each apparatus and system could be worked with equal facility and economy, this would of course show *something* in favor of the electric light; but when in fact everything in this regard is against the electric light, which demands vastly more machinery, and that of a more delicate kind, requires more skillful management, shows more liability to disarrangement and waste, and presents an utter lack of the storage capacity which secures such a vast efficiency, convenience, and economy in gas, then we see that this relatively trifling economy disappears or ceases to have any controlling importance in the practical relations of the subject.

PD, *Sci. Am.* 42 (1880): 241. A transcript from a subsequent reprint in the *Teleg. J. and Elec. Rev.* is in *Edison Electric Light Co. v. U.S. Electric Lighting Co.*, Vol. VI, pp. 4237–41, Lit. (*TAEM* 48:630; *TAED* QD012G4237). [a]Place and date not those of publication. [b]This sentence printed at bottom of page, its proper position in text indicated by asterisk.

1. The Stevens Institute is located in Hoboken. The article appeared in the 17 April *Scientific American*, which usually was available about ten days before the issue date.

2. Alfred Mayer had worked as a machinist, studied chemistry, physics, physiology, and mathematics, and taught at various colleges before organizing the physics program at Stevens Institute in 1871. He visited Menlo Park several times in 1878 and published an article on the phonograph in *Popular Science Monthly* (Mayer 1878a). *ANB*, s.v. "Mayer, Alfred Marshall" and *TAEB* 4:13 n. 9, 32 nn. 1 and 7, 76 n. 1, 158 n. 1.

3. Benjamin Thomas was a research assistant to Alfred Mayer at the Stevens Institute, where he earned a doctorate in 1880. He continued his research in photometry and enjoyed a long teaching career at the University of Missouri and Ohio State University. *WWW–1*, s.v. "Thomas, Benjamin Franklin."

4. Mayer wrote to Edison on 2 March that one of his students had visited Menlo Park and reported "that you had expressed some surprise at my having one of your electric-lamps in my possession, as only a few had left your hands which were in good order." Mayer "had no idea that the lamps were not in the hands of many persons, thereon I should have applied directly to you." He explained that he requested one from *Scientific American* about three weeks earlier "because it was convenient for me then to get it without going to Menlo Park." Mayer had since "been making many experiments on the lamp," mainly to determine the carbon's cold resistance. He requested "the loan of a few of your lamps to extend my work in the direction of ascertaining the amount of light given by definite amounts of electromotive force." Stockton Griffin docketed Mayer's letter "no ans." Mayer to TAE, 2 Mar. 1880, DF (*TAEM* 53:678; *TAED* D8020ZCE).

5. A Grove cell produced 1.9 volts. *TAEB* 1:615 n. 1.

6. William Sugg & Co. manufactured photometers in London, both of Sugg's own and others' design. The reference here may be to an instrument devised by Sugg in which the flame from an Argand lamp passed through a screen so that the lower, more stable part of the flame could be used as a reference (Dibdin 1889, 10, 193). The practice of laboratory photometry at this time was devoted to efforts to measure and standardize lamps and commercial illuminating gas; visual instruments generally provided a means for comparing the intensity of light from a

test lamp to a known standard over a fixed distance. Despite great commercial and, especially in Britain, bureaucratic efforts in recent years to reduce uncertainty in the testing process, contemporary photometers employed a variety of ways to view the light, such as through reflectors, screens, slits and disks. Additionally, although actual candles had been widely supplanted by lamps there was no consensus about what type of lamp or even what kind of gas to use in determining a uniform standard (Dibdin 1889, 1–50; Johnston 1996, 273–81); on the disparity of luminous standards see Dredge 1885, 2:151–73.

7. The English chemist Augustus Matthiessen studied the electrical, physical, and chemical properties of metals, and had served on the British Association Committee on the Standards of Electrical Resistance. Matthiessen 1858 states the change in conductivity with temperature of nearly three dozen substances including three forms of graphite and two of coke. *DSB*, s.v. "Matthiessen, Augustus."

8. In a correction published in the 1 May issue the authors stated that this figure "was, in fact, the amount *gained* by the cathode, the loss of the anode being a trifle greater. The gain of weight was, of course, what it was intended to take, so that the error was only in the expression, and not in the process or result." "Some Electrical Measurements of One of Mr. Edison's Horseshoe Lamps", *Sci. Am.* 42 (1880): 273, reprinted in the 15 May *Telegraphic Journal and Electrical Review*, p. 178, *Edison Electric Light Co. v. U.S. Electric Lighting Co.,* Vol. VI, pp. 4241–43, Lit. (*TAEM* 48:632–33; *TAED* QD012G:138–39).

9. Misprinted as "0.00326"; cf. Doc. 1852. The deposition rate per hour given below is of the correct order of magnitude.

10. In their published correction (see note 8) the authors explained that this was determined by "measuring the light at every azimuth varying by ten degrees between 0° and 180°, that this was approximately the true expression for the total amount of light emitted." They noted that the mathematical analysis given in Rowland and Barker 1880 produced a different result "but as the experiment shows this result not to be attained in fact, it is evident that the assumptions on which the mathematical reasoning is based do not include all the conditions present in the experiment." They reported that two subsequent efforts to measure the intensity of light gave results which "agree very closely with each other and with our former determinations." In August, the eminent electrician Moses Farmer offered Edison a lengthy analytical comparison of the results reported by Morton and his colleagues with those of Rowland and Barker. Farmer to TAE, 11 Aug. 1880, Upton (*TAEM* 95: 608; *TAED* MU049).

–1928–

From George Gouraud

London 6th April 1880[a]

No 1[1]

Dear Edison,

I cabled you some time since in the affirmative in reply to your enquiry as to whether I had closed with Bouverie on the new basis.[2] The Parliamentary Elections and the consequent absence of all the Directors from town so upset everything that the agreement has not yet been finally executed—indeed I did not get the final draft of it until a day or two ago when I found to my astonishment that it contained an alteration in the clause concerning the dividend to the share capital by which it provided that they were to have 5% more than you under all circumstances. In the first draft submitted to me it was quite clear that only the usual preference dividend was contemplated but there being some ambiguity I undertook to make it clear when they returned it to me made clear quite the other way giving me to understand that that was what you had given your assent to. This concession was so enormous that I did not feel disposed to take anybodys word for it[b] so cabled you as follows:—[3]

"Have you consented giving capital 5 per cent more than half profits that is if profits are fifteen capital gets ten you five. This is too much concession: we agree on all points but this. I am willing capital have five preference you having next five Answer immediately"
to which I have your reply as follows:—

"I think we should be satisfied with five: to capital ten. Onerous contracts are decidedly unsafe"

You make it so very evident to me that you wish to make all these concessions that I have to choose between following your wishes and my own and I have chosen the former and have to-day returned to Mr Waterhouse the final draft of the agreement approved in the above terms

I cannot but feel that it is most unfortunate that you should have allowed any direct communications with yourself in this matter for they must inevitably complicate matters and compromise my position towards you and of course leaving me at an entire disadvantage with the opposite parties in the negotiations if they have the feeling that they would go to you if they could not get what they wanted from me. I have so far as possible endeavoured to make them understand that I am satisfied from [~~you?~~][c] my correspondence with you that you have not intended to ignore me in any wise nor that you would make any agreement over my head which I am afraid they have thought you would. However we may consider the thing settled now and I trust to the advantage of all concerned— But we have given an enormous concession there can be no doubt— As to the expediency or the necessity of giving it there may be two opinions but you are evidently satisfied[d] and this is the most important consideration— If I have erred it is on the side of getting too much for you[e] for the for the Patents.—a fault easily pardoned—[f]

I shall hope in a day or two to have the final execution of the agreement the only alteration being in the dates of the notes which I have provided shall be due as follows £5000 on June 30th £5000 December 31st and that they shall be executed at the time of the signing of the agreement[4]

Please advise me by return if you would like your share of these notes discounted or if I shall hold them for you until they become due— They could probably be easily negotiated at about the current rate   Yours truly

G E Gouraud

LS, NjWOE, DF (*TAEM* 56:584; *TAED* D8049ZDF). Written by Samuel Insull; on letterhead of George Gouraud;. a"London" and "18" preprinted. bInterlined above. cCanceled. d"but . . . satisfied" interlined above. e"for you" interlined above by Gouraud. f"a fault easily pardoned—" written by Gouraud.

1. It is not clear why Gouraud began numbering his letters concerning the London telephone company. He sent Edison numbers three through six on 20 April (one of which is Doc. 1933) and number seven two days later; no number two has been found. DF (*TAEM* 56:615, 619–20, 638; *TAED* D8049ZDR, D8049ZDS, D8049ZDT, D8049ZDX).

2. Edison cabled Gouraud on 24 March "Have you closed with Bouverie new basis" and also sent a similar question to Johnson. Gouraud answered promptly that he had done so but the next day wrote a letter explaining that the contract "is in the hands of the lawyers and is understood to be definitely settled as regards Mr Bouverie & myself upon a basis which I understand you to have approved and desired me to give assent to." Edison in the meantime had cabled for confirmation to Johnson, who replied on 25 March: "Not closed   still haggling   Will cable when agony over." TAE to Gouraud, 24 Mar. 1880; TAE to Johnson, both 24 Mar. 1880; Gouraud to TAE, 24 and 25 Mar. 1880; Johnson to TAE, 25 Mar. 1880; all DF (*TAEM* 56:547–48, 552, 549; *TAED* D8049ZCM, D8049ZCN, D8049ZCP, D8049ZCO, D8049ZCS, D8049ZCQ).

3. Gouraud quotes the complete texts of his 3 April cable and Edison's reply the same day (DF [*TAEM* 56:578, 577; *TAED* D8049ZDB, D8049ZDC]). Johnson explained this issue to Edison in a letter on 5 April:

Bouverie stipulated for 5% [on?] the capital of the Co before any division was had with you— I consented= When Gouraud & Waterhouse came together it transpired that the Company held this 5% to be a sort of 1st mortgage on the earnings under any & all circumstances, that is to say If 7% was earned the 5% was 1st deducted & then the 2% was divided with you— Gouraud held—and I so understood it in the 1st place—that you simply guaranteed the Co 5%—thus if 7% was earned the Co got 5% & you got the remaining 2% . . . or in case 12% was earned each got 6%= But it seems such was not Bouveries intention—it was simply that the Co was to have 5% first & then a division was to be made—thus if the earnings were 12% the Co get 5 + 3½% or 8½% & you got 3½%.

[Johnson to TAE, 5 Apr. 1880, DF (*TAEM* 56:579; *TAED* D8049ZDE)]

Johnson recommended accepting Bouverie's compromise in which the company would yield its interpretation of the 5% guarantee in exchange for Edison waiving his so-called good will rights because of uncertainty "whether your good will at the expiration of the Patent will be worth as much as the proper division of the Earnings meantime." See also Doc. 1925 n. 3.

4. Gouraud cabled "Bouveries terms accepted" on 8 April. He sent the contract for Edison's signature on 17 April and Johnson sent a copy with his own comments two days later, but no copy has been found. Gouraud to TAE, 8 Apr. and 21 May 1880; Johnson to TAE, 17 and 19 Apr. 1880; all DF (*TAEM* 55:697; 56:663, 603, 607; *TAED* D8046V, D8049ZEK, D8049ZDN, D8049ZDQ).

–1929–

Draft to Joseph Medill

[Menlo Park, April 8, 1880?[1]]

Jos Medill, Esq[2]

I send you tests made by Prof B[arker] & R[owland]. other tests are being made best scientific men—[3] Your correspondent named Hall of NY tribune[4] ~~is of that class of ability that he couldnt get a situation~~ Came here & staid about 20 minutes and then writes his long ~~tale of~~ article, from inferences based on ignorance & misunderstandings=[a] It is true that January 1st we put up about 80 lamps the first made, to ascertain the life of the lamps   They have all gone out   their average life was 792 hours the shortest 26 hours the longest 13803 hours, a very satisfactory result for a 1st experiment   I have never concealed this fact when a lamp buusted I let it remain where people could see it. My Laboratory is open & free to sneaks ignoramus detectives & as well as[b] gentlemen ~~and~~   I could have easily have made new lamps and replaced those destroyed ~~and thus obtain praise whereas by an[c] honest exhibition experiment[c] I am lied about by every scribbler who comes along~~ had I wished to deceive the public—[5]

The first howl was that subdivision was impossible— I subdivided it, now nothing is said about this impossibility   The next howl is that although I can subdivide it it is not economical— [6] Now Barker & Rowland has shewn that I can obtain ~~13 gas jets~~ lights each equal to[d] the best gas jets obtain 13[c] (~~1716~~) of ordinary gas jets) ~~for~~per horse power in electricity & with all loses about 10 net— Barker using 20 cubic feet of gas in a gas engine obtained 8[e] ~~gas jets for 1 hour[f] with~~ 8 of[g] my lights for one hour[h] whereas on burning the gas direct he only got 4 lights one hour[7]   After all these scientific tests are over they will probably acknowledge the cheapness over gas, but they will say that the lamps wont last or that I cannot make them ~~for 25 cents~~ cheaply.[c] ~~In fact I am sorely puzzled to understand the motives of these attacks,~~ [-----][i] ~~since I havent cheated anybody but I am going to do all I said I would and more too~~.

~~please do not publish this as~~

ADf, NjWOE, DF (*TAEM* 53:720; *TAED* D8020ZDN1). [a]"& misunderstandings" interlined above. [b]"as well as" interlined above. [c]Interlined above. [d]"lights each equal to" interlined above; "each" interlined separately; "to" overwrites obscured text. [e]Obscured overwritten text. [f]"for 1 hour" interlined above. [g]"8 of" interlined above. [h]"for one hour" interlined above. [i]Canceled.

1. Stockton Griffin's docket note indicates he sent this letter on 8 April 1880. It was published in the *Chicago Tribune* on 11 April (p. 5) in essentially the same form as Edison's corrected draft.

2. Joseph Medill acquired an interest in the *Chicago Tribune* by 1855 and took full control of the paper in 1874. He was a lawyer by training but spent most of his working life in the newspaper business; he also helped found the Republican Party. *ANB*, s.v. "Medill, Joseph."

3. See Doc. 1916 n. 3. The *American Machinist* also had unidentified engineers making preliminary tests of the efficiency of Edison's system on 6 April, and they planned to return for more complete experiments ("Change in Edison's Electric Lighting System," *American Machinist*, 24 Apr. 1880, 8; Mott Journal N-80-03-14:50, Lab. [*TAEM* 33:709; *TAED* N053:26]). In response to its initial report, Edison sent the *American Machinist* a lengthy and highly critical letter based on drafts and calculations by himself and Charles Clarke. One point of contention was the amount of power required to turn the shafting and dynamo with no load, with Edison referring to an informal competition in April during which he and several assistants each succeeded in turning the machine by hand more than a thousand revolutions per minute. In addition he had Clarke calculate frictional losses in the belts and pulleys of the dynamometer (TAE to *American Machinist*, 20 Apr. 1880, Lbk. 5:847 [*TAEM* 80:246; *TAED* LB005847]; N-80-02-08.2:129, N-80-03-31:70–137, N-80-03-15:81–85, Mott Journal N-80-03-14:75, 82–83; all Lab. [*TAEM* 34:388, 904–38; 35:631–33; 33:722, 726; *TAED* N066:66, N072: 36–70, N082:42–44, N053:38, 42]). In its 1 May issue the journal indicated that the report was being revised by the engineers and would be published later. At that time it did describe generally the new fiber filaments and durability of the lamps. The following week, however, it stated that the report was complete "but upon consultation with the engineers making it, we have decided to withhold its publication until the complete test can be made, which we purpose making whenever lights enough are in operation to make the report authoritative. As Mr. Edison has substituted a fibre loop for the paper loop formerly used, materially affecting the essential points of resistance, power and durability, it is believed that publication of results at this time would be unfair both to the public and to the Edison lighting system." There is no evidence either of subsequent tests for this purpose or that the report was ever published

("Change in Edison's Electric Lighting System," *American Machinist*, 8 May 1880, 8).

4. Henry Hall was a *New York Tribune* reporter who had written on electric lighting for that paper and the London *Times*, in the latter case to Edison's apparent dissatisfaction. In November 1879 he had complained to Edison that "in the past it has been perfectly impossible for me to get any exact information about your light. You have never let me see it, and all my information has been at second hand." Edison invited him later that month. Hall may have come for the purpose of this article on 20 March, when Mott reported that Edwin Fox visited with unidentified reporters from the *New York Tribune* and *Sun*. Hall's dispatch to the *Chicago Tribune* was dated 1 April 1880 and published on 5 April ("Electric Light," *Chicago Tribune*, 5 Apr. 1880, 7; an abbreviated and undated republication from an unidentified source is in Cat. 1241, item 1486, Batchelor [*TAEM* 94:594; *TAED* MBSB21486X]). Hall to TAE, 3 Nov. 1878, 6 Nov. 1879, DF (*TAEM* 17:221, 49:738; *TAED* D7805ZEF, D7906ZBJ); Mott Journal N-80-03-14:20, Lab. (*TAEM* 33:694; *TAED* N053:11).

5. Hall's article (see note 4) declared that Edison was making no further progress and only continued experimenting to "kill time" or manipulate stock prices. It stated that Edison had originally made two hundred lamps, of which "only about eighty ever were able to burn any length of time, and these eighty have followed each other, one at a time, into the outer darkness of disaster, two only excepted. . . . One of these two surviving lamps has burned 1,700 hours, so Edison says. The average of the others was 600 or 700 hours, which would be equal to four months of ordinary usage."

6. Edison may have had in mind the critique of Doc. 1927, which would have been available about a week before this.

7. Hall did discuss the test of Henry Rowland and George Barker, whom he described as "personal friends of Edison," but concluded that it "goes for nothing after all. It is a laboratory experiment, with everything in favor of a favorable result," particularly the report's assumptions about dynamo efficiency. He also noted the authors reserved judgment about the durability and manufacturing cost of the lamps (see Doc. 1914 n. 3). Referring to construction of new generators at Menlo Park, the article anticipated that "If Edison can maintain 600 sixteen-candle lights with his eighty horse-power engine and twenty machines, he will undoubtedly make a certain success of his light. Barring breakages of lamps he will be able to sell his light as cheaply as gas. I predict, however, that he will not maintain more than 240 or 300 lights with this power, his achievement so far being between three and four lights per horse-power. We shall see in due time."

[Menlo Park,] April 9—1880—

Experiments begun today for the purpose of removing gums, resins, ect—from the various kinds of woods[1]

All foreign substances to be removed leaving nothing but the cellular structure of the wood—

Small pieces of the different kinds of wood 1 or 2 in. square and $\frac{1}{16}$ in. thick are first weighed in air then placed in alcohol and allowed to remain until all matter soluble in alcohol has been removed=, they are then placed in ether until all matter soluble in ether has been removed, next are treated with aqua[a] ammonia,[b] hydrochloric acid, well washed in water, and subjected to the action of chlorine gas[2]

J. W. Lawson

X, NjWOE, Lab., N-79-03-10.2:109 (*TAEM* 31:1170; *TAED* N032:44). Written by John Lawson. [a]Obscured overwritten text. [b]"aqua ammonia" interlined above.

1. On 6 April Charles Batchelor began drawing tools for cutting, trimming, and forming wood for lamps, each piece "to be $4 \times .02 \times .02$ and bent into shape by steaming or heating process." According to Charles Mott, whose journal provides an almost daily record of experiments related to wooden loops, his brother Samuel was on the same day "set at work on drawing of very thin wood (or vaneering) to be cut in narrow strips, bent in horse shoe or other shapes for carbonizing." Batchelor sketched various alternative shapes on 7 April, when he and Edison prepared a list of 160 types of wood to try. Mott noted that the first samples were carbonized on 8 April but these did "not turn out entirely satisfactory being some-what misshapen." On 25 April Edison executed a patent application for strips with broadened ends that were cut, shaved, or split off a block of wood and then bent into shape. He also filed an application on 24 May for "cutting or stamping from a veneer, a piece of wood with thickened or broadened ends" to be carbonized. This was rejected repeatedly and finally abandoned; only the drawings and claims are extant. N-80-03-29:27–44; Mott Journal N-80-03-14: 50–54, both Lab. (*TAEM* 33:427–36; 709-11; *TAED* N051:15–24, N053:26–28); U.S. Pat. 238,868; Patent Application Casebook E-2536:84, PS (*TAEM* 45:705; *TAED* PT020084). On 8 April Batchelor sketched a "New loop for lamps." Charles Dean began on 10 April to make an attachment for the lathe which would cut wood to the desired shape, eliminating the need to steam and bend it. He finished three days later and it was successfully tested in the laboratory (N-80-03-29:45; Mott Journal N-80-03-14:58, 63, 66, 70, 75; both Lab. [*TAEM* 33:436, 713, 715, 718, 720, 722; *TAED* N051:24; N053:30, 32, 34, 36, 38]). On 15 April Batchelor drew up thirteen specifications for a "New Machine for cutting closed wooden loops" that were probably embodied in the "Automatic machine for cutting wood loops" that was tested on 20 April (N-80-03-29:75, Mott Journal N-80-03-14:90, both Lab. [*TAEM* 33:451, 730; *TAED* N051:39, N053:46]); sketches and notes from around this time are probably related to that machine as well

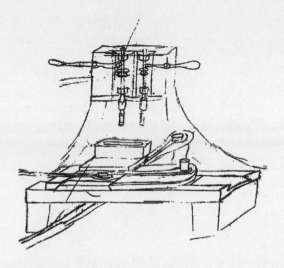

(N-80-03-29:45–117; Mott Journal N-80-03-14:86, both Lab. [*TAEM* 33:436–72, 728; *TAED* N051:24–60, N053:44]). Edison described this method of forming the loops in a patent application executed on 15 June: "a block of wood is taken and cut in a machine or lathe for turning irregular forms until its shape is that of an elongated oval. . . . The interior is then reamed out in a suitable machine, so that the block then resembles a series of carbon horseshoes with closed ends piled upon each other. The length of the oval is with the grain of the wood. The block is then split or sawed into very thin layers or slips, each forming ultimately one horseshoe carbon" (U.S. Pat. 248,417).

By the first of May a hydraulic press operated by an electric motor was being used to press wood for related experiments. On 21 May Mott observed that Charles Dean was "jubilant over his success today in working the cam milling machine with complete success and getting out about 100 loops of box[wood] & Holly in excellent shape and in several cases sawing them so perfectly that the whole five loops were left joined at the thick ends." In early June Dean began building a machine which he apparently intended to perform all the milling operations without requiring transfer of the work to another machine. His first prototype was not satisfactory and he continued to work on the design into July. Mott Journal N-80-03-14:119, 127, 130, 138, 142–43, 156, 163, 184, 212, 226, Lab. (*TAEM* 33:744, 748, 750, 754, 756, 763, 766, 776, 790, 797; *TAED* N053:61, 65, 67, 71, 73, 80, 83, 94, 108, 115); Doc. 1950.

Several dozen wood samples were carbonized and tried in lamps in April and May without notable success. Edison later attributed this both to the fact that wood fibers do not all grow parallel and consequently were often damaged while cutting the loops, and to the relatively large dimensions of the fibers resulting in "very porous and friable" carbon (N-80-03-06:99–129, Lab. [*TAEM* 33:1009–24; *TAED* N057:49–64]; Edison's testimony, 5:3106–3107, *Edison Electric Light Co. v. U.S. Electric Lighting Co.*, Lit. [*TAEM* 48:60; *TAED* QD012F:58]).

2. Following this entry Lawson listed nineteen wood samples (and their weights) placed in alcohol this day. He removed them on 13 April,

noting the color of the alcohol extract in each instance, and placed them in ether. He soaked them successively in aqua ammonia and hydrochloric acid until 4 May, and on 12 May treated them with chlorine gas and washed them in water. Thirteen days later, he weighed the samples "After being well dried" and found substantial losses from their initial values. He weighed them again on 26 May after a day's exposure to the atmosphere, finding in most cases a slight gain in mass; he repeated this the next day and found a gain or loss in only nine samples. Also on 27 May Lawson started over with samples of eighteen different woods; these experiments continued until 1 July. On 28 June he also tried drying bast fibers by similar methods. N-79-03-10.2:111–41, Lab. (*TAEM* 31:1171–81; *TAED* N032:45–55).

—1931—

Notebook Entry: Mining

[Menlo Park,] April 13/80—
Experiments with auriferous pyrites—[1a]

A glass jar was taken, in this was placed a porous cup, in the porous cup was placed the pyrites together with a ~~litt~~ small quantity of common salt; in the glass vessel and around the porous[b] cup was packed common salt; two carbon plates were taken one placed in the porous cup in contact with the pyrites the other in the glass vessel in contact with the salt; wires leading from both plates were connected together at the ends; water was then added to both vessels[2] ⟨N. g. ~~ttry~~ 5⟩[c]

2 Powdered marble, coke[b] and pyrites mixed together and placed in water—[a]

3 Sodium carbonate, coke and pyrites in water—[a]

4 Calcium fluoride, coke and pyrites in water—[a]

5 Manganese dioxide, sodium chloride, coke and pyrites in water—[a]

6 Pot. cyanide, coke and pyrites in water—[a]

7 Caustic lime, coke and pyrites in water—[a]

8 Bleaching powder and pyrites in water—[a] ⟨Surface attack [tines?][d] of the pyrites⟩[e]

J. W. Lawson

X, NjWOE, Lab., N-79-03-31:107 (*TAEM* 32:638; *TAED* N040:32). Written by John Lawson. [a]Followed by dividing mark. [b]Obscured overwritten text. [c]Marginalia written by Edison across the paragraph. [d]Illegible. [e]Marginalia written by Lawson and demarcated from text by line at left.

1. Mott noted in his journal that on this day "Lawson commenced a series of experiments with the view of obtaining gold and other metals from Pyrites tailings." About a month earlier Lawson had begun subjecting forms of chalcopyrite, a copper-iron sulfide which sometimes bears gold, to a variety of chemical processes but did not report any results. Then on 25 March he "fused chalcopyrite with metallic lead, part of this lead entered into combination with the other two sulphides forming an easily fusible sulphide; a bar of this was cast and used as an anode to try whether copper could be removed from it by means of the electric current." Again he did not indicate results and seems to have made no further experiments until those described in this document. Mott Journal N-80-03-14:66, N-79-03-31:101–105, both Lab. (*TAEM* 33:718, 32: 635–37; *TAED* No53:34, No40:29–31).

It is not known exactly what material Lawson had to work with at this time but Edison was receiving samples of mine tailings from throughout the West and instructed one correspondent to describe a container he had sent "as we are rec[e]i[vi]ng several daily & Everyone fails to mark box." TAE marginalia on Baldwin, Sexton & Peterson to TAE, 1 Apr. 1880, DF (*TAEM* 54:638; *TAED* D8034ZAI).

2. This experiment marks a return to research on chlorination of ores (see Doc. 1844 n. 6). On 1 June Charles Mott noted in his diary that Edison explained to two mining men

how cheaply Chlorine water could be made and had a battery jar filled with salt and water. (I should judge about one pound of salt and two quarts of water)   a paper partition was then placed in the jar and a carbon plate immersed in water one on each side of paper the electric current derived from the Dynamo line was then passed through by connecting to each carbon plate. In very few minutes the gas was quite preceptible to the smell and soon after, the water on one side of partition began to turn green. [Mott Journal N-80-03-14:204–205, Lab. (*TAEM* 33:786; *TAED* No53:104)]

This process was repeated on 9 and 10 June, and Otto Moses and Alfred Haid made further experiments with powdered gold and also silver ores the next week. Moses believed the process would be "economical especially in ores that do not contain too much metal other than gold, as in presence of other metals the chlorine would act first upon them and leave the gold till the last." Additional gold was later found in one of the carbon electrodes. Mott Journal N-80-03-14:225, 232, 246–49, 252, Lab. (*TAEM* 33:796, 800, 807–808, 810; *TAED* No53:114, 118, 125–26, 128).

In mid-May Lawson tried a different approach, pouring finely powdered pyrites through a red-hot iron tube; the results are unknown. About that time he also attempted to purify mercury with auriferous pyrites in an electrochemical process but this proved unsuccessful. Mott Journal N-80-03-14:162, 160, 200; N-79-03-31:111, Lab. (*TAEM* 33: 765, 764, 784; 32:640; *TAED* No53:82, 81, 102; No40:34).

Dear Sir,

I am thinking of taking action about the establishment of a strong Scientific Journal, and intended to have made you a visit to renew my conversation with you on the subject, but having been confined to my room this week have not had the opportunity.—

I think you said that when you were further advanced you would cooperate with me, I notice you will shortly make another move forward, and I now suggest that something be done.—

There could not be a better time for literally the field is open, for with the exception of Patent office organs, and journal limited to particular to particular interests, there is nothing in the market, and English Journals are establishing good subscription lists here in consequence.—

I shall produce a good ten cent weekly intrinsically better than anything on either side, that shall in fact give each week a thorough record of Science, and be made popular by good editorials, discussions, and correspondence on current scientific topics, useful inventions, appliances, sanitary matters manufactures, in fact a thoroughly live paper.

When I see the wretched sheets with limited subjects and no editorial capacity, paying, I can have no doubt of the commercial result of what I propose, as it would be an advertising medium covering a very wide field.—[1]

The capital required is a small matter, and I would like it confined to yourself and another of those acting with you, as with your present interests at stake, it is really what you should have, and judiciously manage would save you and your friends many thousands, and if you prefer it could be managed so as to be unknown who is backing me.—[2]

I have recovered from my attack and shall be about on Monday, and will try and get down to your place, and can follow this matter up on my return with your suggestions, seeing those with whom you act as you direct.—

I intend to have the paper selling on every news stand in the U.S., but I shall have a strong backing from Scientists, and can depend on some of the best men to help me in special departments, and when I proposed at first starting the Journal, Prof. Burt Wilder of Cornell[3] sent me long lists of subscriptions from several Universities.—

If you can kindly think the matter over so as be prepared to advise with me when I see you (probably on Monday) next, I shall be obliged   Yours Truly

John Michels.—[4]

I like the form of the English Mechanic and many of its features, it has killed out a lot of rivals and is a success

ALS, NjWOE, DF (*TAEM* 55:361; *TAED* D8041A). [a]"Apl 17" interlined above, possibly in another hand.

1. Michels intended to replicate the size and layout of the British scientific journal *Nature*. In early May, as he was arranging for publication of the first issue, he predicted that the journal would become "a power, and the leading Scientific Journal in this Country, and the only one which will be read abroad— A new face will be put on the reports of Scientific inquiry (not before it is wanted) and the ring of corrupt Scientific men and their pretended scientific Journals will feel the force of 'Science.'" This is the first extant reference by Michels to the journal's title. Michels to TAE, 7 May 1880, DF (*TAEM* 55:374; *TAED* D8041F).

2. At Edison's suggestion, Michels prepared a prospectus for potential investors by 27 April. Michels planned to capitalize the journal by sale of one hundred twenty shares at $100 apiece, with twenty going to each of five investors and the remainder to himself. He told Edison that he had enlisted Frederic Shonnard and a second prospective investor but wanted "three out of the five to be your firm friends" and also wished "to establish it as quietly as possible, and keep private who financially back it." Edison agreed to take twenty shares and to try to interest New York investors. Michels later proposed increasing the initial capital to $25,000, half to be arranged by Edison and the remainder by Shonnard. At that time he also suggested that sales would be higher if the journal were "considered your organ." Prospectus enclosed with Michels to Griffin, 30 Apr. 1880; Michels to TAE, 27 Apr. and 6 May 1880; Shonnard to TAE, 26 July 1879; all DF (*TAEM* 55:367, 364, 372; 50:91; *TAED* D8041C, D8041B, D8041E, D7919ZBN); Stockton Griffin to Michels, 29 Apr. 1880, Lbk. 5:867 (*TAEM* 80:257; *TAED* LB005867).

3. Burt Wilder was a noted comparative anatomist. He had been a professor of zoology at Cornell and since 1874 had taught physiology at the University of Maine medical school. *NCAB* 4:481.

4. John Michels was a journalist and former editorial writer for the *New York Times* who also contributed to *Popular Science Monthly* and other journals. About this time he was editor of the *Industrial News*, a publication of the Inventors' Institute in New York. Michels prospectus, April 1880; Michels to TAE, 16 Oct. 1879; both DF (*TAEM* 55:368, 49:734; *TAED* D8041C, D7906ZBH).

No 5

My Dear Edison,

Having regard to the probable early future of the telephone question here[1] and the certain large premium at which the Consolidated Interests will be saleable in the market and consequently the relative value of our reversion interest[2] in the same and also having regard to the fact that there are several parties interested with you in the matter besides Johnson and myself all of which involves more or less intricate booking keeping, it has occurred to me that it would be very desirable to have these interests in such a form as to enable them to be represented by some form of negotiable certificate so that not only could we put an end to all existing agreements with respect to these different interests and close up the accounts in our respective books but either of the parties interested wishing to either increase by purchase or decrease by sale his interest at any time could the more easily and inexpensively do so. I have accordingly taken advice as to how this could be done and the means are found to be very simple and are contained in the <u>Deed of Trust</u> which I send you herewith and which I hope will meet with your approval. The Trustees certificates under this deed would in the event of the Telephone Companies share being at a premium be as readily negotiable as the shares[b] themselves.[3]

I have proposed Mr A. G. Renshaw as the Trustee he having been our solicitor in all these negotiations and being conversant with all the circumstances of the case besides being a lawyer of high professional and social position.

The expense of the Trust would be merely nominal in fact the Trustees compensation can be arranged on the basis of a fee for transfer   In fact if this were done already I could at this moment sell a portion of your or my interest or that of anybody else concerned at a price which shows the absurdity of suggesting that you should be content with merely £10,000 in the shares of the Amalgamated Coy. I could at this moment dispose of a portion of your interest upon the a[c] basis representing the value of the whole as equal to £50,000.

The only alternative to your taking shares in the New Coy would be the simple necessity of keeping the present company alive for the purpose of working out its agreement with you in which case the only asset of the present Coy would be the shares of the Amalgamated[b] Company a course for which there are thousands of precedents and one which would involve no complications whatever and very insignificant expense

Here in this country in the face of our vigorous competition and threatened litigation and the open admission at least on their part that what they use is not a Bell telephone and with all the evidence we have that their Transmitter they use is an Edison Telephone their shares are selling at present at between £400 & £600 per cent   Taking the average of this at £500 and if the shares of the Amalgamated Company are worth no more which is scarcely possible (they should be worth double and possibly will) but taken at £500 a share the £115,000 shares in the Treasury of the Edison Company would be worth £575,000. Now at that same premium the Edison Company would have only to sell £15,000 to recoup all the advance royalties paid to you plus a like sum plus 5% interest on both (say £75,000)[4] and then after selling another £17 400 to recoup them the amount of what their capital will probably viz £87 000. They would then have £82,600 shares which at the above price would realize £413,000 to divide equally between the shareholders and yourself[5]

Now in the light of these facts it would be simply dishonest to ask you much less to persuade you to take £10,000 in shares[d] when there can be no other reason for it than a desire to prevent your making anything further if possible

I think you are not likely to hear anything further of that suggestion as when I placed it before those who originated it in its true light I observed that it was quietly dropped[6]

Why! I would not think of taking twice £10,000 for my interest alone at this moment with all the uncertainties of the future!

When you have executed the proposed ~~the proposed~~ Trust and the various parties have to represent their interests the proposed certificates then of course any request that you should capitalize your reversion you could easily meet so far as you are concerned by dealing with your own portion as you might please leaving the others interested to do likewise

You will observe in the Deed that I have provided for the remote, so far as I can see the almost impossible, contingency ever arising of having to be referred to the parties interested as in the present state of affairs the only thing the Trustees would have to do would be to receive and pay dividends— In the event of anything arising you and I would always control the action of the trustee by our larger interests[7]   Yours truly

G E Gouraud

This arrangement would also be of great importance in event of the death of any of the parties interested—[e]

LS, NjWOE, DF (*TAEM* 56:623, 625, 624, 626; *TAED* D8049ZDU). Written by Samuel Insull; letterhead of George Gouraud. [a]"London" and "18" preprinted. [b]Obscured overwritten text. [c]Interlined above. [d]"in shares" interlined above. [e]Postscript written by Gouraud.

1. Gouraud outlined the terms of the presumptive merger in a separate letter the same day ("No 4"). Gouraud to TAE, 20 Apr. 1880, DF (*TAEM* 56:620; *TAED* D8049ZDT).

2. That is, property to be returned to the grantor after a specified period; in this case Edison's share of the company's goodwill. Black 1879, s.v. "reversionary interest."

3. The undated draft agreement designating Alfred Renshaw as trustee is in DF (*TAEM* 56:858; *TAED* D8049ZIP); it was superceded by an amended version (see Docs. 1945 n. 3 and 1954 n. 3). On 5 May Edison replied, "Like your trust scheme but must satisfy Johnson," to whom he cabled a few minutes later, "Gourauds trust idea good   investigate with lawyer. whats objections" (TAE to Gouraud, 5 May 1880; TAE to Johnson, 5 May 1880; both DF [*TAEM* 56:658, 657; *TAED* D8049ZEG, D8049ZEF]).

4. That is, selling shares worth £15,000 at par to net £75,000.

5. The value of the company's "goodwill"; see Doc. 1925 n. 2.

6. In his letter "No 3" written the same day, Gouraud reported that the company directors had asked him

to take in the paid up capital of a proposed amalgamated company of the Bell & Edison interests share to the amount of £10,000 in full liquidation of your reversion—that is that you should have nothing more out of it no matter what the Company made. . . . I resolutely objected as there was no reason given for the request other than that it would make easier the amalgamation with the Bell Company and no one could give a single substantial reason for this assumption as the basis of the proposed amalgamation is of such a character as it cannot be the slightest concern to the Bell people what the Edison Company gives you or whether you hold your reversion under the several agreements or cancel it by the acceptance of shares. As they must have known that the position that they assumed was utterly untenable, and they quickly abandoned it. [Gouraud to TAE, 20 Apr. 1880, DF (*TAEM* 56:615; *TAED* D8049ZDR)]

Nevertheless Edward Bouverie urged Edison on 21 April to accept the offer, stating that he was advised that the chance of the company accruing goodwill in excess of royalties already paid was "remote & not worth regarding as a matter of pecuniary value to be estimated" (Bouverie to TAE, 21 Apr. 1880, DF [*TAEM* 56:627; *TAED* D8049ZDV]).

Johnson had written Edison on 19 April that he "had reason to believe that he [Gouraud] considers the amount proffered a liberal one—but with him there is a much more important Consideration viz= His voting & representative power for you= You will see that in case you sell outright your reversionary interests you have no longer any use for a representative director." He suggested that the reversionary question could be addressed independently of amalgamation in a supplemental agreement, without consulting Gouraud (Johnson to TAE, 19 Apr. 1880, DF [*TAEM* 56:607; *TAED* D8049ZDQ]). Three days later he put the proposal in this light:

> Bell Co's stock is at a high premium but whether it will be when a large amount of the new Co's stock is thrown upon the market or not is another thing— In any event the question for you is—Are you likely to get 10 000£ from your def[erre]d interests after the Co has repocketed the 35 000£ already paid you & a 5% on their capital additional— Then dividing the remainder by 2= Will half of it Equal 10,000£ shares in the new Co immediately in hand—& saleable if you so elect—suppose we say at the present premium— about 500%? Rose colors eh? [Johnson to TAE, 22 Apr. 1880, DF (*TAEM* 56:634; *TAED* D8049ZDW)]

On 7 May, several days after he would have received these various letters, Edison informed Bouverie that he did not wish to "be an obstructionist but I cannot in view of my knowledge of the commercial aspects of the telephone in this country the enormous profits now being made and the great future here, accept the £10,000 cash or shares. I much prefer to wait and I will be glad to take my chances" (TAE to Bouverie, 7 May 1880, Lbk. 6:2 [*TAEM* 80:283; *TAED* LB006002]). Bouverie subsequently asked him to reconsider (see Doc. 1954).

7. The first article of the proposed trust (see note 2) provided that any modification of Edison's agreements with the telephone company should be approved by at least a two thirds majority of the certificate holders voting in proportion to the size of their interest.

–1934–

From Edward Johnson

London Apl 27d 1880[a]

My Dr Edison

It has been found that Gouraud has the power to out vote the company in matters relating to amalgamation that is to say in ordinary matters a mere majority is sufficient—this the company have—but in a matter of this importance it seems a vote of ¾ is requisite— this was evidently slipped into the contract by Gouraud without the company seeing its bearing—at all events they have only just run afoul of it= Its operation is to give your vote the power to stop further progress—and you may be sure Gouraud is availing himself of it= There is therefore but one thing to do viz: Make a direct appeal to you to nominate a new Director[1]   This is to be done formally by a vote of the Board today and White is to go out to see you in person in regard to it= Meantime we are at a stand still= The contract which has just gone out to you will not now be signed by the Company as it simply pays over to you a large sum of money without in the least leaving them at Liberty to work out their own salvation= Thus you see a still further confirmation of my charge that Gourauds grasping policy operates[b] to prevent the consummation of what is today possible untill[b] tomorrow finds it impossible   It will be the same thing with this matter of amalgamation unless you take the Bull by the Horns & displace Gouraud— The Bell Co. are already growing restless under the delay and it is an open question whether they will consent to allow the Proffered terms to stand after May 1st—the Day fixed for[b] [a?][c] final action on our part.

It is all very well to say—let amalgamation go—and go on independently—but what chance of success has a concern under such circumstances—[2] All the Zeal has been taken out of[b] our people by a long array of unexpected obstacles— we might go on and even succeed fairly well—but what prospect would you have of ever realizing anything from such a weak concern— It is not good policy to go against the line of action the shareholders mark out as the surest & speediest way for them to earn a profit on their investment—since by so doing they earn a profit for you also—

If Gouraud succeeds in holding things in check until the present board are frozen out I would not give you 10 000£ for the whole of your interest

The company will ask you to nominate me as your representative on the Board— You must not think I want this for any other purpose than to consummate the present negotiations— Immediately they are happily closed I shall resign the position and return to America   There is no longer any room for action here—it is all stagnation—& when the new company is formed you will not have sufficient direct interest in it to care whether I remain or not— It is therefore only a question of terminating the present Lock—& putting 25,000£ in our Pockets—yourself Gouraud & I=

You will be made fully aware of the terms & conditions of the proposed amalgamation as well as of the financial & other difficulties of our company by White who has been, next to Bouverie the most valuable instrument in promoting the success of your interests here— You should in some manner reward him— He is like myself working for a living—and a thousand pounds—say of the united companies stock would be a liberal compensation— Gouraud has promised him a sum Equal to this—but as he only gave a verbal promise & is now at Loggerheads with him you can readily see the chance of its fulfillment are as slender as a thread of thought in a weak mind.

Holmes[3] is over here looking up anti-page Patent evidence—
Muirhead[4] has been to see me—he speaks very disparagingly
of your Light—but cornering him after my usual fashion by
boiling his generalities down to particularities I could only get
out of him that your Glass was very common—but that he had
suggested annealing—to you!!!!—and secondly—that you
were very wrong in not accrediting people on this side with
having done anything= to which I retorted that all the evidence adduced by your bitter enemies of the English scientific
Press—showed there[b] had been nothing done in your direction deserving of more than passing mention= One more
thing— He says you are too absurd in your criticisms on
mathematics & mathematicians— He tells with a sort of withering smile how[b] you button holed him on the subject— In
short my Boy you are not Loved over here by these fellows—
you have committed the grave error of having succeeded
Truly Yours

E. H. Johnson

ALS, NjWOE, DF (*TAEM* 56:643; *TAED* D8049ZEB). Letterhead of
Edison Telephone Co. of London. [a]"London" and "18" preprinted.
[b]Obscured overwritten text. [c]Canceled.

1. In a letter on 22 April, Johnson suggested that the company had
made Edison a "liberal offer" to settle the reversionary interest "more to
get rid of Gouraud than to become possessors of a handle for closing up
the concern— They want no more to do with him." Johnson to TAE,
22 Apr. 1880, DF (*TAEM* 56:634; *TAED* D8049ZDW).

2. Johnson reported eight days earlier that because the company was
in a "bad fix" for money the board had ordered him "to Reduce expenses
to a minimum—not to undertake anything not already contracted for—
and to reduce my line force to just sufficient to execute orders in hand=
This I hold to be a Death Blow." Johnson to TAE, 19 Apr. 1880, DF
(*TAEM* 56:607; *TAED* D8049ZDQ).

3. Probably Edwin Holmes, president of the Holmes Burglar Alarm
Telegraph Co., defendant in a suit Western Union brought for infringe-
ment of the Page patent. Post 1976, 177–81; "The Page Patent," *New
York Times,* 13 Mar. 1880, 2.

4. While in New York in January 1880, Muirhead had visited Edison
on at least one, and probably two, occasions. He may have witnessed Edison's light at that time. Muirhead to TAE, 2 and 19 Jan. 1880, DF
(*TAEM* 53:27, 50; *TAED* D8004B, D8004Z).

Dear Sir

Referring to your telegram of the 26th, you labor under a mistake if you think I have promised a settlement.[2] I have already stated to you that I had never rec'd a cent from the arrangement but was out of pocket both for expenses and for my time spent in perfecting the system and going to England. What I stated was that I did not like to see your people lose money by any of my inventions and that I would try and help you to get your money back, that among my many enterprises I could undoubtedly put you in the way of so doing.[3] Your people over there have entered into some arrangement with Mr Reiff and Reiff has paid some money to keep the patents up lately[4] and I have signed a letter of some kind to the solicitor of Mr John Pender at his request.[5] I shall do all in my power to aid your people to get their money returned but I shall not pay that which I do not morally or legally owe, and that from a speculat[io]n[a] from which I have suffered as much as yourselves. I will see Mr Reiff in regard to the nature of the understanding between him and Mr Pender and see what can be done.[6] Very truly

T. A. Edison   G[riffin]

L (letterpress copy), NjWOE, Lbk. 5:864 (*TAEM* 80:255; *TAED* LB005864). Written by Stockton Griffin. [a]Illegible copy.

1. John Puleston (later knighted) had been a newspaper publisher and banker in the United States before becoming a partner in a London banking house. In addition to that capacity he was at this time also a Member of Parliament. *ACAB* (supp.), s.v. "Puleston, Sir John Henry."

2. Puleston had telegraphed from New York: "While here few days next week will hope to see you and be able to get a settlement   Please communicate with me" (Puleston to TAE, 26 Apr. 1880, DF [*TAEM* 55:547; *TAED* D8042ZAL]). He was referring to £10,000 advanced to Edison in 1873 for automatic telegraph experiments under the Smith, Fleming & Co. contract (Doc. 350) in which he had in the meantime acquired an unspecified interest (see *TAEB* 2:105 n. 4). Regarding negotiations to settle the dispute, see Doc. 1859.

3. Puleston had discussed the advance money with Edison during a visit to Menlo Park in November 1879 (George Ward to TAE, 17 and 22 Nov. 1879, DF [*TAEM* 51:928, 932; *TAED* D7936ZBC, D7936ZBF]). In late January he cabled from London that he had reported their conversation to the other principals who

> were gratified to hear that you were ready to turn over to us such interests as you had to a sufficient extent to reimburse us the amount owing to us by you. We have noticed since I saw you, the great rise in the value of your interests in the Electric Light and we feel with this, your Telephone receipts, and the interest you speak of in the

quadruplex etc, you may now be able without difficulty to pay us fully— I mentioned that you received me frankly and expressed my belief you would make the necessary arrangement— [Puleston to TAE, 24 Jan. 1880, DF (*TAEM* 55:508; *TAED* D8042I)]

Edison did not reply to this or two subsequent cables from Puleston but on 27 February he signed a letter ratifying the agreement (Puleston to TAE, 3 and 4 Feb. 1880; Reiff to TAE, 27 Feb. 1880; TAE, Reiff, and George Harrington to Pender, 27 Feb. 1880; all DF [*TAEM* 55:513–14, 518, 565; *TAED* D8042K, D8042L, D8042R, D8042ZBB]).

On 11 March Puleston cabled to Edison protesting that "Reiffs proposed arrangement came to nothing, [George] Prescott declining. you frankly acknowledged your readiness to repay and I communicated that fact. You will hardly attempt to fall back on an impossible proposal of Reiff   the money was obtained from us and is owing under circumstances which should cause no hesitation whatever on your part to carry out your pledge to me." Having seen this message as it came through New York, Reiff immediately wrote to Edison that he considered the December agreement final and did "not propose that by any meddleing or statement of Puleston that you shall be made to appear as having consented to return them the money; that suggestion I treated as ridiculous." Puleston to TAE, 11 Mar. 1880; Reiff to TAE, 11 Mar. 1880; both DF (*TAEM* 55:525, 522; *TAED* D8042V1, D8042U); see Doc. 1859 n. 9.

4. Josiah Reiff paid the seven-year annuity on Edison's British patent 735 (1873) in February and may have done the same for 1508 (1873) in March or April. Rixon to TAE, 18 and 26 Feb. 1880; Reiff to TAE, both 26 Feb. 1880; Lemuel Serrell to TAE, 22 Mar. 1880; all DF (*TAEM* 55: 515–17, 70; *TAED* D8042M, D8042N, D8042P, D8042Q, D8036ZBI).

5. The letter to which Edison referred was probably the one jointly from himself, Reiff, and Harrington to Pender and his associates. Reiff drafted it at the request of Pender's attorney, A. W. Rixon, to confirm the December 1879 understanding (see Doc. 1859). TAE, Reiff, and Harrington to Pender, 27 Feb. 1880; Reiff to TAE, 27 Feb. 1880, DF (*TAEM;* 55:565, 518; *TAED* D8042ZBB, D8042R).

6. Puleston answered this letter from Ottawa, Canada. He protested that "I labor under no mistake whatever—and you have not until now suggested that I was mistaken as to the result of our interview last year," his recollection of which he claimed to have corroborated by a witness. He stated that he was acting on Pender's authority and asked Edison to see him upon his return to New York in a few days. Edison marked this letter "no answer" and below Stockton Griffin's notation that "Mr Puleston sailed for England May 12th 80," he wrote "are you sure." The matter evidently lay dormant for two years until Puleston asked Edward Johnson to intercede on his behalf. Puleston to TAE, 2 May 1880; Puleston to Johnson, 17 May 1882; both DF (55:551, 60:141; *TAED* D8042ZAP, D8204ZBK).

[Menlo Park, c. May 8, 1880[1]]

Questions to ask Mr. Porter.[2] How do you intend to provide for the lateral motion of ⅛"? What are you going to make the dimensions of shaft and why? Weight of armature to be considered and also the tremendous strain due to the attraction of a field when the mass of the armature happens to be out of line, amounting to many pounds.

Sudden strain resulting from short–circuiting, very great. What[a] will be the weight of your engine and bed with sole plate complete? What do you think of the direct supporting of the heavy weights by pillars as compared with trussing.

How about the arrangements for distributing steam, for settling the water and for returning it to the boiler?

How about head and tail arrangement of dynamo and engine and which end is to be turned toward the passage way.

Straight side for commutator pillow block.

Use two pillow blocks only.

Do you not only make your shafting strong enough for torsion but also much larger to render it perfectly stiff? How about placing the dynamos above the boilers on account of heat?

How about the armature on a ferrule which is keyed to the shaft?

How about trapping the steam and does the steam in jacket[3] flow through or is it condensed there and returned to boiler?

Do you prefer to use the Babbit metal in bearings and for what reason?

What would be the effect if dynamo is running free at full speed and current is instantly put on?

The entire pillow block on commutator side must be capable of removal.

Mistake in book on page 117.[4]

X, NjWOE, Lab., N-80-03-31:181 (*TAEM* 34:961; *TAED* N072:93). Written by Charles Clarke. [a]Obscured overwritten text.

1. This entry immediately follows several pages, dated 8 May, of Charles Clarke's notes and calculations of weight and costs of a large dynamo for "The Central Station Plant." N-80-03-31:170–80, Lab. (*TAEM* 34:956–61; *TAED* N072:88–93).
2. A lawyer by trade, Charles T. Porter introduced the first successful high-speed stationary steam engine in 1862. It employed Porter's own design of a centrifugal governor that was unusually sensitive because the weight needed to move the valve mechanism was concentrated in a mass separate from the rotating balls. The governor controlled engine speed by acting upon a variable cutoff valve gear mechanism devised by John Allen rather than upon the throttle, which greatly improved fuel economy. The engine also incorporated a high-speed indicator by Charles

Richards. The Porter-Allen had a much shorter piston stroke than other engines, making possible both its compact and relatively lightweight construction, and superior thermodynamic efficiency. Its high operating speed (several times that of contemporaries) and exceptionally smooth operation under widely varying loads suited it to operating machinery directly rather than through expensive belt or gear systems, and Porter-Allen engines were employed in mills of all types. Hunter 1985, 450–72; Hills 1989, 193–203; Bowditch 1989, 84; on the conceptual development of the engine see Mayr 1975.

Charles Clarke had been in Newark on 20 and 21 April "collecting data on steam and engines and examining into the merrits of quick motion engines." Porter came to Menlo Park on 22 April and "by demonstrations of his thorough knowledge of the subject, still further confirmed Mr. Edison's belief in high speeds." Four days later Clarke returned with Edison to Newark, where he took indicator diagrams from a Porter engine. Porter was back in Menlo Park on 30 April, when Charles Mott indicated that Edison had ordered one of the engines. This may have been for the small dynamo which, according to Mott, John Ott began designing that day as a field exciter "in cases where but few generators are necessary, such as for Steamers, factories, &c." In a 6 May notebook entry Edison wrote that "Porter can make us a pair of small Engines to work two magnetos. Each self contained. The current from one being used to Energize all the field mags of a central station. the other is for a spare; We have a lever connected with the Governor so that any required speed may be obtained of the Engine hence we can regulate the Electromotive force of the main circuit by varying the speed of the Engine." It was probably about this time that laboratory assistant Albert Herrick sketched and wrote out the specifications of a small Porter-Allen high speed engine rated at 6⅓ horsepower. Clarke memorandum, 21 Apr. 1880, DF (*TAEM* 53:736; *TAED* D8020ZEF); Mott Journal N–80–03–14:95, 99, 107, 122, 121; Unbound Notes and Drawings [1880]; N–79–06–12:122; N–79–08–28:137–41; all Lab. (*TAEM* 33:732, 734, 738, 745; 44:999–1002; 35:542, 1124–26; *TAED* N053:48, 50, 55, 63, 62; NS80AAI; N080:59; N088:70–72).

Edison apparently never acquired one of Porter's small engines and instead abruptly began planning a much larger dynamo. There is no direct evidence regarding this change. On 5 May Clarke completed his measurements of heat in the dynamo journals. However, he ran more trials on 6 May and reported less favorable results showing 2% of mechanical energy lost as friction in the bearings (Mott Journal N–80–03–14:126, 130, 137–38, 141; N–80–03–31:145–69; both Lab. [*TAEM* 33:748, 750, 753–55; 34:942–55; *TAED* N053:65, 67, 70–72; N072:74–87]). Edison's response was not recorded but the next day Clarke started working out the arrangement of a steam engine with respect to a much larger dynamo and Francis Upton began making the related calculations in Doc. 1937. In the notes he began on 8 May for "The Central Station Plant" (see note 1) Clarke computed the area and mass of the magnets and began calculating dynamic loads and articulating some of the design concerns expressed in this document. Similar notes regarding the central station follow this document in the notebook (Mott Journal N–80–03–14:147; N–80–03–31:184–95; both Lab. [*TAEM* 33:758; 34:963–68; *TAED* N053:75; N072:95–100]). There is no extant correspondence with Porter from this month but on 11 May Edison drafted a letter to Babcock & Wilcox asking if his boiler could safely carry 120 pounds of steam pressure. Mott reported on 13 May that "Clarke finished details and diagrams for the large 120 horse power dynamo," which would weigh well over seven tons. Four days later Clarke took the plans and diagrams to the Philadelphia foundry which constructed Porter's engines (TAE to Babcock & Wilcox, 11 May 1880, N-80-03-29:161; Mott Journal N-80-03-14:165, 173; both Lab. [*TAEM* 33:492, 766, 770; *TAED* N051:80; N053:84, 88]).

3. It was common engineering practice at this time to reduce condensation in the cylinder by filling an enclosure or "steam jacket" around the cylinder and steam passages with boiler steam. Hunter 1985, 2:673–74.

4. This is part of Clarke's analysis of the dynamo efficiency test by the *American Machinist* (see Doc. 1929 n. 3). In a paragraph that he crossed out Clarke discussed steam engine indicator cards which appeared to "show that the steam does not follow the theoretical law of expansion but is very far from it." N-80-03-31:117, Lab. (*TAEM* 34:928; *TAED* N072:60).

–1937–

Notebook Entry:
Electric Lighting

[Menlo Park, c. May 10, 1880[1]]

Recapitulation   10 × 10 armature

surface 470 sq. in. p. 152   When taking 75 lamps from this surface about 85 ohms each   8 per H. P. in current[2]

81.3 ft. lbs given off from each sq. in of armature[3a]

20 × 30 inches armature

surface 2488 sq. in   Wound with 8 deep of .048 in. wires Resistance .00555 Ohms[4]

120 H. P in current gives 84.8 ft. lbs per sq. inch of surface

That is aif a machine can be built which shall give the E.M.F required.

TAE

X, NjWOE, Lab., N-79-07-07.1:177 (*TAEM* 32:334; *TAED* N037:86). Written by Francis Upton. [a]Followed by dividing mark.

1. Francis Upton's notes and calculations on this subject dated 8 May immediately precede this document in the notebook. N-79-07-07.1: 148–76, Lab. (*TAEM* 32:320–34; *TAED* N037:72–86).

2. The notes preceding this "Recapitulation" are the first evidence since Doc. 1889 of consideration of a longer armature, and are also Upton's first effort to extrapolate the design characteristics of the larger armature with proportionally more wire. He prefaced them with an attempt to find an algebraic expression of a generator's resistance and voltage as functions of its armature dimensions and number of coils. He then calculated (on p. 152) the surface area of the "Present Armature," ten inches in both diameter and length, which produced 110 volts. (Two arithmetic errors led to his determination of 470 square inches, rather than the correct figure of 471.) Under the heading "Heat on machine,"

he specified that the armature could operate 75 lamps, each having resistance of 85 ohms, "for two hours." In addition to the two armature sizes discussed in this document, Upton also considered one 20 × 20 inches and another 20 × 60 inches long. N-79-07-07.1:148–55, 173, 157, 161, Lab. (*TAEM* 32:320–23, 332, 324, 326; *TAED* N037:72–75, 84, 76, 78).

Upton continued to make notes and calculations through late May regarding the surface area, winding, and electrical features of armatures. The exact purpose of these notes is unclear (N-79-07-07.1:178–235, Lab. [*TAEM* 32:335–62; *TAED* N037:87–114]). On 31 May Upton was "conducting a series of experiments to determine some of the laws governing the construction of electric generators, and to study the economical proportioning of the same, and for determining the best method of winding armature of 120 Horse power Dynamo, which has been determined to be constructed with 47 commutators." About this time he began to formulate "Laws of winding Dynamo Machines" for which he set out a number of algebraic relationships among voltage and current and two or more design variables such as wire thickness, armature dimensions and speed, and number of commutators (Mott Journal N-80-03-14:202, N-79-07-07.1:238–47, both Lab. [*TAEM* 33:785, 32:363–67; *TAED* N053:103, N037:115–19]).

3. This is the power that Upton determined was wasted by internal resistance, or "given off" as heat. He computed this by comparing the known ratio of the resistance of the lamps in multiple arc (1.133 ohms) to that of the armature (.14 ohms), with the unknown ratio of horsepower active in the lamps (9.37) to that lost in the armature. The armature consumed 1.15 horsepower, or 38,200 foot-pounds throughout its area of 470 square inches. Upton also noted that the same machine operating 60 lamps would lose 52 foot-pounds per inch in the armature. N-79-07-07.1:148–75, Lab. (*TAEM* 32:320–333; *TAED* N037:72–75).

4. The area given is not that of the armature cylinder. It is not apparent how Upton obtained this figure, although he seems to have included the wires leading to the commutators. It is likewise unclear how he calculated the resistance of this 120 horsepower machine. In one example he did so based on the length of wire on the armature, and in another based on the weight of the copper wire which could be wound in the available space. N-79-07-07.1:163, 172–73, 160–61, 188, Lab. (*TAEM* 32:327, 332, 326, 340; *TAED* N037:79, 84, 78, 92).

–1938–

To Frank McLaughlin

Menlo Park N.J. May 17, 1880

Dear Sir,

Your reports and telegrams were duly rec'd.[1] The black sand process is now perfected. I would like to commence on black sand which is as pure as can be had. What about the boats?[2] We are working at a process that we think will do away with the mill entirely and this is the reason why the plans are not sent to you. We propose after taking black sand out to take the gold out of the silica in the same way.[3] If it can be made to work it will be enormous. Keep a lookout for black sand—make enquiries of miners regarding deposits of the same and make a record of them. I should like some lava black sand and more of the boat sand. Have Powers[4] save all the black sand in cleaning up so we can work it. There was very little gold in the tailings you sent from Bryan Tyson.[5] After you left, Tyson sent us a bag of pyrites which he panned out from his sluices in the clean up. I panned some on the glass but could see no gold—but the Dr in testing found an enormous quantity of gold, we thought it must be in the pyrites. We picked some pyrites out & found no gold— This puzzled us greatly but I soon found out that the gold was covered with the thickest coat of rust that I have ever yet seen.[6] It required 5 days to eat off the rust without dissolving the gold but I suppose there is not much of this. The Spring Valley tailings you sent were quite poor, evidently the gold is down deeper—still what you sent could be concentrated by sifting up to $5 per ton. At present our blk sand process is absolutely perfect   Now if you can furnish us a deposit of blk sand we are ready for business.[7] What about that party that promised to send you a large boulder of pyrites a piece of which you brought here?

Also how about that place in a tunnel where you said it was 200 feet of black sand?

The North Bloomfield[8] sent me some black sand which contained $5.50 to the cubic yard. There was more gold in the Spg Valley tailings than in the lower flume which you sent me. Very truly

T. A. Edison   per S. L. Griffin

L (letterpress copy), NjWOE, Lbk. 6:24 (*TAEM* 80:287; *TAED* LB006024). Written by Stockton Griffin; circled "C" written at top of page.

1. Frank McLaughlin telegraphed Edison from Reno, Nev., on 12 April while returning to California from New York, and again several times from Oroville. On 17 May he wired: "Did tailing shipped by Cummings ever arrive   Have you sent me plans etc of mill   Answer." McLaughlin's letters from that period have not been found. George Carman was dispatched from the laboratory as McLaughlin's assistant on this trip. McLaughlin to TAE, 12, 16, and 27 Apr., 17 May 1880, DF (*TAEM* 54:482, 484–85, 488; *TAED* D8033ZAE, D8033ZAG, D8033ZAH, D8033ZAK); TAE to Robert Cutting, Jr., 26 Mar. 1880, Lbk. 5:787 (*TAEM* 80:229; *TAED* LB005787).

2. This reference is unclear but see Doc. 1941 esp. n. 3.

3. A month earlier Edison replied to a correspondent from Georgia that "I do not work quartz ores but simply the auriferous sand or tailings from hydraulic mines." Some placer mines produced gold-bearing

Scientific American *illustration of Edison's ore separator in which a jet of air was directed against the material falling between the poles of the electromagnet.*

quartz in great quantities but this material required a stamping mill to pulverize it (TAE to W. A. Ramsey, 15 Apr. 1880, DF [*TAEM* 54:679; *TAED* D8034ZBG]). Around this time Edison was also experimenting with arrangements in which a blast of air was used in conjunction with electromagnets to segregate the material (see also Doc. 1921 n. 1). He sketched such a device on 20 April; experiments continued sporadically through late July (Cat. 1146; N-79-06-12:67; N-80-07-02:29–31; Mott Journal N-80-03-14:90, 99, 135; Mott Journal N-80-07-10:22, 42–43, all Lab. [*TAEM* 6:671; 35:516; 36:611–12; 33:730, 734, 752; 37:313, 323; *TAED* NM014:49; N080:33; N108:9–10; N053:46, 50, 69; N117:11, 21]). The 17 July issue of *Scientific American* described and illustrated this approach, which resembled the 20 April drawing:

> The auriferous sands are placed in the hopper and allowed to fall between the poles of a powerful electro-magnet, and a blast of air is directed at right angles against the falling stream of sand just as the latter passes between the poles of the magnet. The non metallic substances are readily blown away, while the metallic portions are retarded by diamagnetism, so that the blast of air has less effect on them than it has on the non-metallic substances. The consequence of this operation is that the sands are divided into two heaps, one containing a large percentage of metal, the other containing a very small percentage, or none at all. ["New Ore Separator," *Sci. Am.* 43 (1880): 3]

4. Oliver Powers.

5. Bryan Tyson operated what was apparently a modest hydraulic claim on the American River at Gold Run, Calif. McLaughlin sent the sample to Edison on or about 16 April. Alfred Haid conducted the assay and reported that the ores "contain too little gold to work them." Tyson to TAE, 26 Apr. 1880; McLaughlin to TAE, 16 Apr. 1880; both DF (*TAEM* 54:693, 484; *TAED* D8034ZBO, D8033ZAG); N-80-04-17:11, Lab. (*TAEM* 33:276; *TAED* N050:5).

6. Haid reported that the pyrites in this sample from Tyson "are not auriferous, but there is some covered gold among them, besides a great deal of amalgams." Edison noted on Tyson's letter: "we could work the sulphurets you sent they were very rich." N-80-04-17:11, Lab. (*TAEM* 33:276; *TAED* N050:5); TAE marginalia on Tyson to TAE, 26 Apr. 1880, DF (*TAEM* 54:693; *TAED* D8034ZBO).

7. In addition to seeking tailings from existing hydraulic mines, during the late winter or spring Edison joined with James Banker, Robert Cutting, Jr. and other financiers to form the Miocene Mining Co. to work in the Big Bend area of the North Fork of the Feather River. McLaughlin built the necessary ditches in the latter half of 1880 and began operations in April 1881. TAE to Cutting, Jr., 26 Mar. 1880, Lbk. 5:787 (*TAEM* 80:229; *TAED* LB005787); Enclosure with George Cummings to McLaughlin, 1 Apr. 1880; McLaughlin to Banker and Cutting, 10 Nov. and 10 Dec. 1880; McLaughlin to TAE, 10 and 25 Aug., 16 and 19 Sept., 16 and 30 Dec. 1880, 3 May 1881; all DF (*TAEM* 54:467, 527, 533, 506, 515, 519, 521, 544, 558; 59:175; *TAED* D8033Y, D8033ZBH, D8033ZBJ, D8033ZAU, D8033ZAX, D8033ZAZ, D8033ZBC, D8033ZBK, D8033ZBR, D8139K).

8. The North Bloomfield Gravel Mining Co. operated a large and highly profitable hydraulic mine near the town of that name in Nevada County, Calif. Gudde and Gudde 1975, s.v. "North Bloomfield"; Kelley 1959, 40–53.

–1939–

Charles Mott Journal Entry

[Menlo Park,] Tuesday May 18. [1880]

Tramway.[1] The engine with power from one machine made a successful trip the entire length of the road and return. A Second generator was brought out of Dynamo Station and set running in old Dynamo Room to increase the power for Rail Road, the passenger car was then attached and with[a] nine men on made the round trip very successfully, without accident or assistance, running the return in 1 minute 16 seconds.[2] After which a number of trips were made all with equal success. And all the trials today have been decidedly more encouraging and successful and fully up to the most sanguine expectation with the belt and pully gearing now in experimental use. The results have been so eminently successful that Mr. Edison and Mr. Batchelor are[a] contemplating extending the track three quarters of a mile with grade at one point of about 1 foot to seven,[3] and adding three more passenger cars.[4]

Quite a number of visitors were conveyed over the road among others several foreign naval officers, and those who have been so favored speak of the ride as exciting and pleasant.

Nickel former. the slotted nickel plate in which Bast fibers are secured for carbonizing and the one used by Flammer yesterday with good results, was by him made and sketched in Book No. 57 page 124.[5]

Poles for ~~wires for~~ telephonic wires are being drawn and placed across lots to day by Ayers[6] from Machine Shop to old factory.[7]

Clamp Machine   Bradley[8] commenced work on machines devised by Mr. Batchelor for making the clamps. diagrams of which may be found in Book 51 pages 145[a] &c.[9]

Sparkers. Four sparkers such as previously used on vacuum pumps were made to day and exhausted on the pumps to be sent, as I am informed, to West Point Military Academy.[10]

Bast fiber   (8) Eight[a] Bast fiber lamps are on Pumps to day being exhausted[11]

AD, NjWOE, Lab., N-80-03-14:175 (*TAEM* 33:771; *TAED* N053:89). Written by Charles Mott. [a]Obscured overwritten text.

1. At the beginning of April Edison ordered rails to build one-half mile of track for a demonstration of the electric railroad conceptualized in Doc. 1745. Construction was underway by 10 April, supervised by Julius Hornig. The line consisted of rails spiked to thick ties (so as to remain insulated from the ground) and electrically connected to each other by copper bars (Mott Journal N-80-03-14:43, Lab. [*TAEM* 33: 705; *TAED* N053:22]; Hornig memorandum, 4 Apr. 1880, DF [*TAEM* 55:301; *TAED* D8039A]; Hornig's testimony, 18–19, 5–6, 10, *Edison v. Siemens v. Field*, Lit. [*TAEM* 46:14, 7–8, 10; *TAED* QD001:10, 3–4, 6]). Work on the track continued for about a month, during which time a station was built and Hornig designed and oversaw construction of the locomotive with motor and running gear, couplings, and one passenger car (Mott Journal N-80-03-14:56, 58, 60, 83–84, 89, 92, 97, 103, 111, 113, 122, 130, 149, Lab. [*TAEM* 33:712–14, 726–27, 729, 731, 733, 736, 740–41, 745, 750, 759; *TAED* N053:29–31, 42–43, 45, 47, 49, 52, 57– 58, 63, 67, 76]; TAE to Pioneer Iron Works, 6 and 19 Apr. 1880; Hornig to Pioneer Iron Works, 20 Apr. 1880; Lbk.5:815, 842, 846 [*TAEM* 80: 235, 241, 245; *TAED* LB005815, LB005842, LB005846]; Hornig's testimony, 3–5, 12–13, *Edison v. Siemens v. Field*, Lit. [*TAEM* 46:6–7, 11; *TAED* QD001:2–3, 7]). The track was completed on 11 May. The next day the locomotive motor, essentially a small Edison dynamo, was tested in the shop and, according to Mott, "found to run like a top." On 13 May the locomotive was put on the tracks for the first time. Mott reported that it "started off and run slowly with current of 40 Volts after which Mr. Edison put it up to about 120 volts and that current gave it much greater power. the test so far as made shows the electrical parts complete and successful, but by nervousness and inexperience, Mr. Hornig who was running the motor, threw the friction gear on so powerfully and suddenly that he broke [one?] of the large friction wheels and thus ended the fun for today." The friction drive was replaced by a belt and pulley mechanism the next day, when a number of trips were made, one with nineteen people on board. Edison found that the locomotive could not return up a steep hill at the end of the line and on 14 May decided to alter the pulley ratio "for the purpose of getting more power at sacrifice of speed." Around this time Charles Batchelor sketched devices apparently intended to grip the rails and pull the train along, but on 17 May the new pulley transmission was tested with Calvin Goddard on board and found to give "very satisfactory results" (Mott Journal N-80-03-14: 160, 162, 164, 167–68, 170, 172; N-80-03-29:163–79; both Lab. [*TAEM* 33:764–70; 493–501; *TAED* N053:81–83, 85–88; N051:81–89]). The 5 June *Scientific American* (42:354) contains an illustration of the locomotive with the friction drive, which can be compared to the belt and pulley mechanism shown in the 27 July *New York Daily Graphic* drawing reproduced in Jehl 1937–41 (p. 576). The locomotive was restored in 1930 and is located at the Henry Ford Museum (Acc. no. 29.1980.629), as are two of Edison's coaches (Acc. nos. 29.1980.1245 and 29.1980.630).

In an interview for an unidentified newspaper in April Edison reportedly predicted that electric railroads could be built in the West which would "go up and down all elevations, like an ordinary highway; there will be no tunnels or heavy cuts or fillings. You see the train would stick to the rails like a tack to a magnet; it couldn't get off very well un-

less the whole track should turn over. The electrical force would give a tremendous traction; the wheels would grasp the track like an iron hand, and climb up a steep hill as a boy would climb a ladder." Edison also cautioned against "the mistake of attributing this idea primarily to me. In Berlin Prof. Siemens has already built a little road on this pattern, and has organized a great company to construct a system of roads. I have devised an important improvement on his method, and I believe my generator will deliver twenty-five more per cent. of electricity than his; besides which, the road is especially adapted to our long ranges and sparse population, and not to Europe" ("Our New York Letter," c. 17 Apr. 1880, Cat. 1241, item 1495, Batchelor [*TAEM* 94:597; *TAED* MBSB21495]). On 3 June Edison filed a patent application for a "system of electrical rail-roading." It was rejected, amended, placed in interference, and eventually abandoned. Drawings and the text of its eighteen claims are in Patent Application Casebook E-2536:90–92, PS (*TAEM* 45:707–709; *TAED* PT020090); see also *Edison v. Siemens v. Field*, passim, Lit. (*TAEM* 46:5–114; *TAED* QD001).

2. Not every trip was uneventful. On 14 May the weight of nineteen people carried the train past the usual stopping point and it hit the station bumper with enough force to lift the locomotive off the tracks. Again on 4 June, according to Mott, the brakes were not applied in time to prevent "a pretty severe bump against the bumper." The next day Grosvenor Lowrey and Calvin Goddard visited "and after assurances from Mr. Edison and Kruesi that it was perfectly safe and free of danger were pursuaded to take the second ride over the electric tramway. All went very well and with fearful speed until the curve at Freemans road was reached, when the motor jumped the track throwing Martin Force and Kruesi off—and running over the ties seventy yards before it came to a stop. No one was hurt and no damage done." Mott Journal N-80-03-14:167–68, 215, 217, Lab. (*TAEM* 33:767–68, 791–92; *TAED* N053: 85–86, 109–10).

3. Mott noted work on extending the track in several of his weekly summaries beginning 29 May; the project continued into early July (see Doc. 1950). The ruling grade in American railroad construction had been standardized by this time to 116 vertical feet per mile, or about 2

feet in 91. Mott Journal N-80-03-14:199, 219, 229, 231, 249, 263, 277, Lab. (*TAEM* 33:783, 793, 798–99, 808, 815, 822; *TAED* No53:101, 111, 116–17, 126, 133, 140); Vance 1995, 43.

4. On 3 June Edison received castings for three additional cars and his carpenters began constructing the frames almost immediately. One of the new cars was run on the track for the first time on 14 June. In the meantime, Mott noted that a bell and headlight were put on the locomotive on 26 May and "Mrs Edison and some of her friends riding on same in evening probably the first electric lamp ever used as head light." Mott Journal N-80-03-14:213, 216, 231, 234, 191, Lab. (*TAEM* 33:790, 792, 799, 801, 779; *TAED* No53:108, 110, 117, 119, 97).

5. Charles Flammer sketched the new mould and specified that it should be "5 inches long to lay fibers in slot and ends will draw up in carbonizing." The use of nickel for carbonizing moulds was apparently suggested by Julius Hornig about 7 May. According to Mott, when Flammer tried his new design on 17 May he got ten of twelve bast fibers out "in first rate order, convincing himself that nickel is preferable to carbon in moulds probably in consequence of containing less air." During the next ten days Flammer obtained even better results by securing the fibers in the mould and by cutting the slot more closely to the dimensions of the fibers. N-80-03-06:124; Mott Journal N-80-03-14:145, 173, 188, 193; both Lab. (*TAEM* 33:1022, 757, 770, 778, 780; *TAED* No57:63; No53:74, 88, 96, 98).

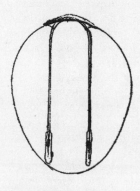

Charles Flammer's drawing of a fiber lying in the slotted nickel mould.

6. Possibly Brown Ayres. Having just completed a fellowship at Johns Hopkins under Henry Rowland, Ayres wrote to Edison in March asking for a job (Ayres to TAE, 23 Mar. 1880, DF [*TAEM* 53:496; *TAED* D8014G1]). He had visited Edison in 1878, the year he graduated from the Stevens Institute. There is no record of his employment at Menlo Park and later in the year Ayres was appointed professor of physics and electrical engineering at Tulane University; he eventually became president of the University of Tennessee (see *TAEB* 4:159 n. 3).

7. At the end of April Edison purchased from Robert Gilliland the former electric pen factory across the railroad tracks in Menlo Park, which had stood vacant since the transfer of manufacturing to Western Electric in late 1876 (TAE agreement with Robert Gilliland, 24 Apr. 1880, Gilliland receipt to TAE, 30 Apr. 1880, DF [*TAEM* 54:13–14; *TAED* D8021I, D8021J]; see Doc. 817 and *TAEB* 3:410 n. 1). Mott reported that as early as 9 April Charles Clarke included the factory when calculating the size of conductors to be laid through Menlo Park. Edison's carpenters began repairing the structure on 3 May; by 1 June they had finished this work and also made a frame for the electrically-powered pump for supplying mercury and set up "tables for about twenty four glass blowers. One double table in center of room to accomodate 8 on a side and one next outside wall for about same number. Also bench to accomodate five persons for putting carbons in clamps & clamps on conductors &c." On 3 May John Ott had disassembled and carefully measured the vacuum pump prototype and prepared scale drawings for the glassblowers. The first one hundred pumps were completed on 2 June and were tested and carried down the hill to the factory three days later (Mott Journal N-80-03-14:56, 67, 131, 143, 186, 199, 207, 129, 209; Mott Journal N-80-07-10:4, 6, 15–16, 21, 25, 45, 54–56,

59, 65, 68–69, 72, 75, 80; both Lab. [*TAEM* 33:712, 718, 750, 756, 777, 783, 787, 749, 788; 37:304–305, 309–310, 312, 314, 324, 329–31, 334, 336, 338–39, 342; *TAED* N053:29, 34, 67, 73, 95, 101, 105, 66, 106; N117:2–3, 7–8, 10, 12, 22, 27–29, 32, 34, 36–37, 40]).

8. James Bradley first worked as a machinist for Edison in Newark in 1872. When Edison moved to Menlo Park he remained in Newark and worked successively for Joseph Murray and Edward Weston; he hired on again in Edison's machine shop at the end of 1879. *TAEB* 3:589 n. 4.

9. Batchelor had started sketching this apparatus on 6 May, which two days later Mott indicated was intended "for making the clamps entire, without handling and running through the different machines as has heretofore been the custom." Accompanying drawings he made on 8 May, Batchelor specified eleven discrete operations this machine would perform (N-80-03-29:135–55, Mott Journal N-80-03-14:149, both Lab. [*TAEM* 33:480–90, 759; *TAED* N051:68–78, N053:76]; see Friedel and Israel 1986 [pp. 168–69] for two of the 8 May drawings). Construction was well under way by 29 May and continued through late June as Batchelor designed components for it (Mott Journal N-80-03-14:199, 219, 230; N-80-03-29:183–91, 199–223, 237–49, Lab. [*TAEM* 33:783, 793, 799, 802, 480–90, 503–507, 510–22, 529–35; *TAED* N053:101, 111, 117; N051:91–95, 98–110, 117–23]). On 28 June he made sketches and notes on the "New clamp making machine" and on 20 July listed fourteen alterations to be made in the present design (N-80-06-28:1–3, 23–31, Lab. [*TAEM* 36:72–73, 83–87; *TAED* N102:1–2, 12–16]). Batchelor made drawings related to the clamp machine through the summer (N-80-06-28:5–17, 67; N-80-06-02:89–95, 105, 109–117; N-80-03-29:265; all Lab. [*TAEM* 36:74–80, 105, 438–42, 446, 448–52; 33:536; *TAED* N102:3–9, 34; N105:42–46, 50, 52–56; N051:124]). Mott made reference to work on the machine in nearly all of his weekly summaries of "work general" through mid-October (Mott Journal N-80-07-10, passim, Lab. [*TAEM* 37:302; *TAED* N117]).

10. Mott refers to the spark tubes used to measure atmospheric pressure. S. E. Tellman, a lieutenant at the U.S. Military Academy, inquired if Edison would make and sell four vacuum tubes for classroom use at the Academy. Edison replied that he would do so and instructed Stockton Griffin to "give order to Boehm to make & [William] Hammer to bring to high vacuum— Say to Tellman that we do not manufacture for sale but will give them to him." On 18 May Ludwig Böhm wrote out the order, indicating that the tubes were to be about eight inches long (Tellman to TAE, 10 and 12 May 1880, with TAE marginalia, DF [*TAEM* 53:753, 755; *TAED* D8020ZES, D8020ZET]; N-80-03-19:98, Lab. [*TAEM* 34:654; *TAED* N068:48]). During the summer Edison also made a vacuum tube for Charles Young, who used it for brief research on the thermo-electric properties of iron and platinum (Young to Upton, 12 Aug. 1880, Upton [*TAEM* 95:615; *TAED* MU050]; Young 1880).

11. These lamps were numbered 1082 through 1089. The bast loops were carbonized in the slotted nickel mould and fitted with cocoanut shell clamps. N-80-03-06:25, Lab. (*TAEM* 33:1002; *TAED* N057:13).

New York, May 28th 1880[a]

From W. Willshire Riley

My Dear Sir,

The exhibition of your Electric Motor last Tuesday, was so satisfactory to myself and friends that we have determined to introduce it upon our Road at Rockaway Beach,[1] provided you will adapt one to it[b] and can construct one in a short time, and at a reasonable cost. our Cars will be light seat 45 passengers, and probably three in a train   our grade need not be over one foot in a hundred and our line at present not more than half a mile in length. our location for bringing both of the inventions before the public, cannot be surpassed   Will you please inform me how soon you can construct me a motor and its probable cost at your earliest convenience, and oblige   yours most faithfully

W. Willshire Riley[2]

P.S. State the size of the Engine to generate the Power &c &c

⟨Write & say cannot go into building apparatus yet—the apparatus ~~Her~~ Road here is merely for Experimental purposes & can do nothing in way of biz until our Experiments are completed.⟩[3]

ALS, NjWOE, Cat. 2174, Scraps. (*TAEM* 89:252; *TAED* SB012AAC). Letterhead of Rockaway Elevated Rail-Road Co., W. Willshire Riley, president. [a]"New York," and "18" preprinted. [b]Much of the text to this point underlined later by William Hammer.

1. Riley organized the Rockaway Elevated Rail-Road Co. to build and operate an elevated line across the narrow peninsula of southern Long Island from Jamaica Bay to Rockaway Beach on the Atlantic Ocean. He intended to employ his patented "Riley centre and safety rails," and after seeing a news account of Edison's electric railroad he wrote on 18 May to ask if the motor could be adapted to his track. Edison noted on that letter, "Can't go into it   if you want to see the RR come down." Riley visited Menlo Park on Tuesday, 25 May. Riley to TAE, 18 May 1880; Riley circular letter, 1880; both DF (*TAEM* 55:302, 304; *TAED* D8039B, D8039C); Mott Journal N-80-03-14:189, Lab. (*TAEM* 33:778; *TAED* No53:96).

In early August Edison also agreed to meet with backers of the Brooklyn & Atlantic Beach Elevated Railroad about the electric engine but the meeting did not take place. B. F. Newhouse to TAE, 2 Aug. 1880, Cat. 2174, Scraps. (*TAEM* 89:263; *TAED* SB012AAJ); Newhouse to TAE, 3 Aug. 1880, DF (*TAEM* 55:334; *TAED* D8039ZAC).

2. Riley was president of the Rockaway Elevated Rail-Road Co. (see note 1).

3. This draft reply is the basis for Edison's answer, to which he added that his electric railroad would still "require quite a great deal of work to make it practical. I am making changes in it daily and do not know yet where the end is" (TAE to Riley, 1 June 1880, Lbk. 6:55 [*TAEM* 80:294; *TAED* LB006055]). Despite this disclaimer, Edison collected informa-

tion from the month of April on the "Cost of Motive Power & Lighting" of the Manhattan Elevated, which operated all of New York's elevated lines. Calvin Goddard arranged for two officers of that company to visit Menlo Park on 10 June and advised Edison to "have the RRd in good order & don't try to break anybody's neck" (Manhattan Railway operating expenses, Apr. 1880; Goddard to TAE, 8 June 1880; both Cat. 2174, Scraps. [*TAEM* 89:251, 253; *TAED* SB012AAB, SB012AAD]; Klein 1986, 282–83).

–1941–

Notebook Entry:
Mining

[Menlo Park, May 29, 1880?[1]]

McLaughlins ores sent May 19 1880 =[2a]

No 14— Henrietta boat,[3] plenty mercury saw no gold, but[b] only used handful of sand on glass = test with sieve full—[a]

No. 5   Powers clean up. one sieve full ½ out rest panned & then glassed   over 100 small colors,[4] very rich $\frac{1}{23}$ free gold not amalgamated—rusty—rest mercury—all very fine will go through very fine seve & thus be made very rich—

No. 8. Lava bed black sand   very rich fine gold no amalgam —small handful on glass 30 colors fine—nearly all magnetic[5a]

No 9— 2 colors (—small) 1 handful on glass—nearly all black sand say 70 pc—could be nicely concentrated.[a]

12— very poor   1 small[c] color ½ pan full— no Hg— Little black snd. 5 p.c.[a]

No 7— ~~After~~ about 75 pc black sand after sifting about 90 pc   .3. colors small in ¼ pan full no Hg— ⟨½ oz ~~silver traces gold.~~⟩[6d]

No. 6— 50 pc black sand— 2 colors in large handful on glass— No Hg—[7a]

No 5*. ⟨$40 to the ton⟩[d]

Total amount 8 lbs—   3840 grammes
sifted to 900 mesh.   873 grams[a]
Lifted magnetically from this   267[e] "
also   substract sand lifted by magnet   31[e]
                        ~~Total~~
~~Leaves gold in~~   267
                        31
                        298

Substracted from   873
                        298
Leaves—   575[f]—in which is gold

ie[g] total 3840—sieve & mag leaves 575 grames with gold or 6½ times concentrated=

No 5*— one handful of the 575 grams gave on glass 20 or 30 colors & lot of small colors—there was lot of mercury & amalgam— gold has rusty spots or rather a fluculent[8] porous quartz colored reddish yellow— I now put it into a ~~b~~paper box marked 5* 575 grammes lot—but there is not that amount after my handful was taken out ⟨3840 gr down to 575 the 575 contain $39.8 to the ton⟩[9][h]

No 12—

Coarse	3700[b]
Sifted	990
Magnetic	18
	4708

Scarcely any if any[i] gold in Coarse[a]

No 14.—

Coarse	1860
Sifted	1885
Magnetic	70
	3815

There is very little gold in th~~eis~~—little amalg & Hg—gold exceedingly fine—[a]
Coarse no gold some Hg[a]

Dry Creek Tailings

No 3—[10] 500 grammes taken= There was lifted by magnet from this— 190[f] Leaving 310[f] gramms containing the gold— A curious thing is the ~~m~~non magnetic is black. I found that a stronger magnet took out more, & I think a very strong Electro mag would take out still more ⟨30 gr. yield 0.005 gr. gold= $80 per ton + 1.5 milligs. Ag 1.2 T.[r]o[y] ounces Ag per ton⟩[j]

X, NjWOE, Lab., N-80-04-17:15, 17, 19, 21, 23, 25, 27 (*TAEM* 33:278–84; *TAED* N050:7–13). [a]Followed by dividing mark. [b]Obscured overwritten text. [c]Interlined above. [d]Marginalia by Alfred Haid and followed by dividing mark. [e]Written in underlined gap. [f]Written in gap. [g]Circled. [h]Marginalia written by Haid and set off by dividing marks; Haid's computations of this figure have been omitted. [i]"if any" interlined above. [j]Marginalia by Haid.

1. Charles Mott noted on this date that "Seventeen or eighteen bags of Ores and tailings were received to day most of them from Powers claim and sent by McLaughlin. Mr. Edison has been testing some of them." The following day he wrote that "Mr Edison is to day much interested in testing the ores and tailings received from McLaughlin and directed Lawson to make certain experiments for removing the coating from particles of gold so they would be rendered in a state subject to the

action of mercury." Mott Journal N-80-03-14:198, 201, Lab. (*TAEM* 33:783–84; *TAED* No53:101–102).

2. In his covering letter Frank McLaughlin indicated that this shipment, the most recent of several, "about exhausts the claims in the immediate vicinity of the contemplated mill, and will doubtless be all that you will require. . . . If the result of the assays of the samples forwarded warrent it I can contract with the Chinese working the Lava Beds for twenty five to one hundred tons per day of tailings like sample for $2.00 per ton." McLaughlin included a list identifying the numbered samples he sent that day by Wells Fargo express. However, Stockton Griffin noted on 27 May that "part of this letter went to Dr Haid" and most of the list has not been found. McLaughlin to TAE, 19 May 1880, DF (*TAEM* 54:489; *TAED* D8033ZAL).

McLaughlin's samples were just a few of the many that Edison received for analysis, most of them unsolicited. Edison and Alfred Haid used part of the notebook for recording assay results from dozens of samples; they usually recorded the sources, many of which Charles Mott also noted in his journal (N-80-04-17:3–89; Mott Journal N-80-03-14, passim; Lab. [*TAEM* 33:285–315, 683; *TAED* No50:1–44, No53]). Related correspondence which sometimes includes Edison's marginal notations about his separation process and results of particular assays, is in Mining—Mines & Ores (D-80-34), DF (*TAEM* 54:577; *TAED* D8034). In August Edison filed a patent application (Case 226) covering a "method of treating so-called auriferous sulphurets [gold-bearing metallic sulphides (Raymond 1881, s.v. "sulphurets")], which consists in reducing them to powder, and then subjecting the resulting material (either raw or roasted) to the action of a magnetic separator." The application was rejected and eventually abandoned (Patent Application Casebook E-2536:114, PS [*TAEM* 45:710; *TAED* PT020114]).

3. The meaning of this phrase is uncertain. "Henrietta," a lightweight dress cloth, suggests that it may refer to material captured in socalled "blanket-tables," which were essentially sluice boxes lined with fine cloth to catch the gold, mercury, and amalgam settling from the discharge of a stamping mill. *OED*, s.v. "henrietta"; Lock 1882, 1052–54.

4. That is, small particles of metallic gold. *OED*, s.v. "colour," 10a.

5. Haid noted on a facing page that a wet assay determined that this ore would yield 0.447 ounces of gold per ton. By a fire assay he determined it would produce 0.5 ounces of gold and an equal amount of silver. Edison later entered this information himself in the notebook, indicating that the results given by Haid were from non-magnetic material; the magnetic portion yielded a tiny fraction of this amount, which Edison concluded was "probably picked up by particles." Haid also reported results from McLaughlin specimen number seventeen, containing roasted pyrites from a mine near Stockton which contained $1,152 in gold to the ton. N-80-04-17:18, 67, 16, Lab. (*TAEM* 33:280, 304, 279; *TAED* No50:9, 33, 8).

6. Edison later recorded Haid's assay and the amount of material in this sample classed as "Coarse," "magnetic," and "Auriferous." He noted the presence of two scales of gold and much mercury and observed that "more magnetic could probably be taken out with strong mag." N-80-04-17:43, Lab. (*TAEM* 33:292; *TAED* No50:21).

7. Edison also made a separate entry for this sample, recording the amount of material in each category. He again commented that "about

50 grms more could be taken out with strong magnet." N-80-04-17:41, Lab. (*TAEM* 33:291; *TAED* N050:20).

8. Edison may have intended fluctuant, meaning undulating or wavelike. *OED*, s.vv. "fluctuant," "flucti-."

9. Haid arrived at this determination through computations at the bottom of the page which have been omitted. He assumed the valuation of gold at $20 per (troy) ounce, a figure consistent with other sources. Lock 1882, 121–44, 179; *Ency. Brit.*, s.v. "Gold."

10. This sample is the only one for which a portion of McLaughlin's description survives (see note 2). He indicated that it consisted of tailings from a flume at Dry Creek (one of several mining locations by that name in the vicinity) after a month of operation. He stated that the mine produced about five tons of such material every day. Gudde and Gudde 1975, s.v. "Dry Creek."

–1942–

From Edward Johnson

London June 2nd 1880[a]

My Dr Edison

Thanks to your keeping your grip on the Reins Amalgamation was happily accomplished yesterday P.M. at 2 oclock, by a unanimous vote of the shareholders of both Companies—[1] Gouraud voted <u>for it</u>—though protesting up to the last moment that unless certain things were done he would <u>not</u>=

The new "bone of contention" was simply this= <u>U</u>nder the plea that if he consented to amalgamation your future interests might be destroyed by reason of the[b] possible misconstruction of the clause referring to a Division of the <u>Profits</u> with you— He demanded Either one of two things as the price of his consent   they were—1st   That the Edison Co should bind theirmselves to hold its shares (Edison Co shares)—in Trust for the joint benefit of yourself & thier own—thus making their shares unmarketable & preventing them from reaping any advantage from Amalgamation.

2nd   That the surplus of shares of the new Co over those of the Edison Co. obtained by the trade—say 43,000£—should be considered as a profit—and immediately divided between you & the Co—or in other words giving you 21,500£= for your reversionary interest—& leaving the Co £21,500 only— as their Entire possible profit—on the whole Telephone[c] venture— Both these things were simply inadmissable & Bouverie finally got mad—& swore if[b] Gouraud did not vote for Amalgamation without such terms He Bouverie would throw the whole Damn business up—& Let the responsibility rest with Gouraud=[2]

At the instance of Bouverie I sent you a cable simply asking you to cable your approval of Amalgamation=[3]

Gouraud then sent for me & I met him Renshaw & Parish—
Renshaw gave it to me as his Legal opinion that unless the
Company gave some additional guarantee than was contained
in the last Contract with you—your future was not properly
protected= They then made a dead set on me to get me to join
Gouraud in a Telegram to you—asking you not to interfere=
I positively refused= And after a long discussion I finally got
the three of them to pledge me their word of honor that come
what might—they would vote for Amalgamation— I then
cabled you that I had seen both Bouverie & Gourauds cables
& not to cable— I then went home—rather more than half
suspicious that Gouraud afterwards changed his cable so as to
make me say I had seen something quite different from what I
did.[4]

However—your cable saying you would approve if you
knew the basis—came at 10 P.M. & was sent to White & by him
to me—[5] We came to the ofs to answer it—& I was in a Devil
of a funk to know how to answer it—for you see I was between
2 fires   I had cabled for one party asking you to do it & for another asking you not to= Now if G. went back on me & the
thing failed for want of your approval I was responsible— On
the other hand I did not want you to unqualifiedly approve—
because I did think Gouraud should have some assurance from
the Co. that your interests would not be affected by Amalgamation— As I had got Renshaw to admit that a simple assurance of the Board—in writing—was ~~fully~~ quite sufficient—&
had the word of honor of Renshaw G & Parish that with such
assurance from Mr Bouverie, they would be content & vote—
and knowing that Mr Bouverie would cheerfully accord that
—I finally cabled you to cable[b] your approval without prejudice to your future—[6] This you did not do—I am sorry to say
but fortunately 15 minutes before the meeting Gouraud met &
told B—that such simple assurance was all he required—& of
course got it immediately—

The real fact is this= Gourauds power passed from him by
Amalgamation—and with it his last chance of a squeeze— He
therefore laid himself out to get another 10 or 15 000£= In
fact he admitted as much to me by saying "Now dont breathe
what I have told you as to my determination to vote tomorrow— They may come forward yet with an offer to settle for
15 or 20 000£"= There you have the whole thing in a nut
shell— Your future is all right=

Now in re—to that Trust— Gouraud wants me to join him
in a cable to you asking you to sign it— If I do—He simply

manipulates it to his own ends— Mr Bouverie is looking forward still to an adjustment of this matter with you— If you put yourself in a Trust—you cannot do anything without the consent of the others in the Trust. Do you want to do this? I think not—at least until I can see you—say in Sept.[7] Yours

E. H. Johnson

ALS, NjWOE, DF (*TAEM* 56:682; *TAED* D8049ZFB). Letterhead of Edison Telephone Co. of London. [a]"London" and "18" preprinted. [b]Obscured overwritten text. [c]Interlined above.

1. George Gouraud cabled Edison on 1 June "Amalgamation unanimous both Companies." The agreement forming the United Telephone Co., Ltd. was dated 13 May but not executed until 2 June. The new company's deputy chairman was Edward Bouverie; five of its eleven directors, including Bouverie and Gouraud, had been connected with the Edison company (Gouraud to TAE, 1 June 1880; Arnold White to TAE, 8 June 1880; Agreement of 13 May 1880; United Telephone Co., Ltd. circular, 8 June 1880; all DF [*TAEM* 56:681, 693, 694, 701; *TAED* D8049ZFA, D8049ZFE, D8049ZFF, D8049ZFF1]). A promotional pamphlet and subscriber directory published in August by the United company is in PPC (*TAEM* 96:636; *TAED* CA018A).

2. Johnson had cabled Edison on 25 May: "Gouraud imperilling amalgamation by new and absurd demands   Cable him following   withdraw claim to profit on United Company Capital vote for amalgamation without further demur." The next day Edison queried Gouraud, "what causes delay amalgamation." Shortly before 1 p.m. on 31 May Edison received a message from Bouverie on behalf of the London company stating "Amalgamation in great peril   if it fails Company must wind up board and myself will abandon it. Gouraud asks fresh terms last moment inadmissable   Rush   answer   shareholders   meeting   tomorrow   last chance." Johnson to TAE, 25 May 1880; TAE to Gouraud, 26 May 1880; Bouverie to TAE, 31 May 1880; all DF (*TAEM* 56:672–73, 678; *TAED* D8049ZEP, D8049ZEQ, D8049ZEU).

3. Shortly after 1 p.m. on 31 May Edison received Johnson's message to "Cable quephone [the London company] your approval amalgamation." On the same paper on which Bouverie's earlier cable (see note 2) was transcribed Edison wrote a message to Gouraud which he sent at 2 p.m.: "Telegrams state you ask for fresh terms which will imperil amalgamation. The terms stated in your letter satisfactory to me   why not close matter up." Johnson to TAE, 31 May 1880; TAE to Gouraud, 31 May 1880; both DF (*TAEM* 56:679, 678; *TAED* D8049ZEX, D8049ZEV).

4. At 3:20 p.m. on 31 May Johnson cabled "Have seen Bouveries Gourauds cables   Dont cable reply either   all will be right." Six minutes later Edison received Gouraud's message: "Counsel advised settling your reversion before amalgamation   This endeavor frightened Bouverie hence his misleading Cable   I shall vote for amalgamation as independent shareholder but not as your proxy which is unnecessary and counsel advises would jeopardize your future   Your only course silence I guarantee amalgamation." Johnson to TAE, 31 May 1880; Gouraud to

TAE, 31 May 1880; both DF (*TAEM* 56:677, 678; *TAED* D8049ZEY1, D8049ZEY2).

Gouraud later explained that he was advised, in the absence of a guarantee of Edison's future interest, that he should not "vote on your proxy for the amalgamation lest the fact of so doing might be used against you in the future   I at no time in the negotiations said that I would vote your proxy against the amalgamation but that is evidently what Bouverie feared I would do if I did not vote your share for it. Of course I did not show him my whole hand but allowed him to draw whatever inference that he might please." Gouraud to TAE, 4 June 1880, DF (*TAEM* 56:688; *TAED* D8049ZFD).

5. Edison wrote the reply, "I will if I knew the basis," on the reverse of the first cable from Johnson that day. He also wrote and then canceled a draft reply on the same paper which concluded "You and Gouraud have mixed me up. Cannot answer too perplexed by you and gourauds communications." TAE to Johnson, 31 May 1880; TAE marginalia on Johnson to TAE, 31 May 1880; both DF (*TAEM* 56:680, 679; *TAED* D8049ZEY, D8049ZEX).

6. Johnson cabled at 6:40 p.m.: "Contention on reversionary only basis as you know it   cable approval amalgamation without prejudice your interest in future profits." Johnson to TAE, 31 May 1880, DF (*TAEM* 56:681; *TAED* D8049ZEZ).

7. Johnson had already declined a position with the United Company and, on the assumption that Edison would not be ready to introduce the electric light in England during the summer, planned to travel on the Continent before returning to New York. Johnson to TAE, 12 May 1880, DF (*TAEM* 56:661; *TAED* D8049ZEJ).

–1943–

Notebook Entry:
Electric Lighting

[Menlo Park,] June[a] 4 [1880]

Distribution
Exs. to be made and calcu[lated].
Make artificial system
Required resistances of a system   100 Ohm boxes[1]

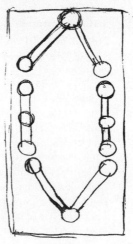

Machines

    Law[a] of winding in general terms[2]

    Law of magnets

    Cast and wrought iron[3]

    Whether the armature is magnetic at high speeds

    Summaries of tests already made

Motors[4]

    Laws

    Test with varying EMF and with varying resistances

    New absorbing dynamometer water friction

Meters

    Shunts with various currant to determine how the law of the shunt varies. The surface of the copper plates should be varied. The contrary E.M.F is probably a function of the intensity of the currant per unit surface[5]

    Comparisons bwt. galva. and Cu. cell.[6]

Lamps

    Full test of at least five lamps of a kind

    Resistance candle power foot lbs. with surface and kind of lamps.

    Bridge, substitution, [~~gl?~~][b] cu. deposition & calor[imeter].

Galvanometer[c]

    The cells should be thoroughly cleaned every week, plates amalgamated &c. A cu. test should be regularly made each day to see it all is in order and to make allowances.

    It has been proved that a given[d] curraent ~~of a certain fixed strength~~ deposits[a] a fixed amount of Cu under[a] ~~varying~~ all the[e] conditions of practice.

    Two deposition cells should be used to check observations and all the current[a] passed through them

    Resistance boxes in line   How do you [~~get und?~~][b]

X, NjWOE, Lab., N-79-07-05:255, 254, 256–61 (*TAEM* 33:257–60; *TAED* N048:124–28). Written by Francis Upton. [a]Obscured overwritten text. [b]Canceled. [c]Followed by dividing mark. [d]Interlined above. [e]"all the" interlined above; "the" overwrites obscured text.

1. Upton may have planned this experiment to verify calculations that Charles Clarke made, according to Charles Mott, "of the fall in electromotive force in a system of conductor and lamps as established and

put down here in the Park, with view of devising means of maintaining the same constant and further calculating the additional cost of maintaining it by feeding the mains with extra conductors and the points where such feeding can be done most effectually and economically." On 8 June Mott stated that Francis Jehl carried out this experiment with "the resistance coils he had prepared each of the ten to represent a lamp of 100 ohms. the purpose of the experiment being to determine the fall of Potential. The fall through the ten resistances was from $84\frac{2}{3}$ to $79\frac{2}{3}$ volts and from $16\frac{1}{2}$ to $12\frac{1}{2}$ candle power" (Mott Journal N-80-03-14: 214, 224, Lab. [*TAEM* 33:791, 796; *TAED* N053:109, 114]). Several of Francis Upton's rough sketches from 1 June showing various distribution arrangements precede this document; Upton made additional related notes and sketches on 7 June (N-79-07-05:246–53, 269–71, Lab. [*TAEM* 33:253–56, 264–65; *TAED* N048:121–25, 132–33]). The drawing below, which appears on the facing page, is probably a form of resistance box.

2. See Doc. 1937 n. 2.

3. At the end of May Upton made experiments in which "The wires [were] connected on the large machine so that the current generated in each could be separately tested. . . . This shows that the distribution of magnetic effect is very nearly the same throughout the whole field." A few days later Upton had made more tests with one of the "small dynamos" designed at the end of April and used for laboratory experiments (see Doc. 1936 n. 2). Mott noted that he was "unable, with 75 volts, to saturate the cast iron bases of the magnets. The inability to saturate is explained as due to the resistance to magnetism of cast iron in consequence of the amount of carbon it contains. Test was to determine the electromotive force at varying saturation of magnet" (N-79-07-05:235– 43; Mott Journal N-80-03-14:220; both Lab. [*TAEM* 33:247–51, 794; *TAED* N048:115–19, N053:112]).

4. Upton began a series of motor tests in mid-May, first using a regular Edison dynamo and later a small "cast iron motor." His intention seems to have been to determine the amount of mechanical power produced under various conditions of the line current and field magnets. Upton's notes are difficult to interpret because, as Mott noted on 14 and 21 May, he made some of the experiments in conjunction with meter tests and on other occasions ran the same machine alternately as a motor and a generator. N-79-07-05:7–73, 111–21, 139–43, 179, 197, 203– 207, 227–31; Mott Journal N-80-03-14:167, 185; both Lab.(*TAEM* 33: 138–67, 186–90, 199–201, 219, 228, 231–33, 243–45 767, 776; *TAED* N048:5–34, 53–58, 67–69, 87, 96, 99–101, 111–13; N053:85, 94).

5. Mott reported on 1 May that John Lawson started a series of meter tests designed to "ascertain whether or not the rolling of copper plate injures the surface of the copper to an extent sufficient to cause uneaqual deposit thereon"; this proved not to be the case. On 10 May Lawson began practical trials in conjunction with Upton's motor tests (see note 4) by putting "meters in line of Dynamo current, to be tested for accuracy with readings of Galvanometer and calculations by Francis." Mott noted on 15 and 17 May that Upton was continuing experiments to determine "the influence of strong and weak currents upon shunts for meter purposes" (Mott Journal N-80-03-14:126, 185, 156, 167, 171, 173; N-79- 07-05:75–108, 123, 127–31,145–47, 209–13, 262–68; N-80-01-02.4:

84–93; both Lab. [*TAEM* 33:748, 776, 763, 767, 769–70, 168–85, 191, 193–95, 202–03, 234–36, 246, 261–64; 35:776–80; *TAED* N053:65, 94, 80, 85, 87–88; N048:35–52, 59, 61–63, 70–71, 102–104, 129–132; N084:34–38]).

6. During the second week of June Upton conducted tests which, unlike those described in note 5, appear to have been designed specifically to measure the electromotive force of the meters themselves and to compare meter and galvanometer measurements of battery currents. N-79-11-21:31–133, 148, Lab. (*TAEM* 32:485–536, 544; *TAED* N039:7–58, 66).

–1944–

Charles Mott
Journal Entry

[Menlo Park,] Wednesday June 16 [1880]

Tailings   five bags received from Lazard Bros. taken from the Turel Mine,[1] some of it very rich a small handfull showing eight and ten colors.

Specific Gravity. Dr Moses had Boehm make a very delicate apparatus for testing the specific gravity of metals or ores &[a] to ascertain whether they contain metals. It is composed of a small cup shaped pan in which the substance may be placed, joined by ring to a bulb partially filled with mercury above which is a considerably larger bulb containing air and still above is a long slender tube that may be scaled off as desired. the whole is made of glass and both bulbs are made air tight, on top of the neck is permantly secured another cup similar to one on bottom. the apparatus is immersed in water and the mercury in bulb always draws it down to a certain point and by placing the material first in the top vessel and by adding weight sufficient to immerse to a certain point, and then placed in bottom vessel and weights added in the top vessel to immerse to the same point the specific gravity may be calculated, or at least that is the modus operandi of Dr's experiments made to day.[2]

Alumina. Dr. Moses tried another experiment to day on reducing Alumina.[3] using plumbago discs as conductors in contact with the charcoal crucible, the product ~~showed~~ appeared very metalic under the microscope, but Mr. Edison thought it was fused Alumina. The plumbago crumbled, almost to pieces under the heat. The stop cock was opened to admit the air before the carbon and parts had lost their red heat, and the instant the air reached the globe or bell covering of the pump, an explosion occured breaking the globe. It was thought due to the formation of carbonic oxide which was ignited by the heat of the crucible

Absent. Mr. Batchelor left to day for a trip with his family[b] and sister.[4]

Spectrocopic examinations of the fiber lamps were made to day by Prof. Young.[5] he got carbon lines in all cases except on the last lamp tested which was the partition lamp with low vacuum. he got two lines of Hydrogen   Having been convinced that the vapory appearance and slight deposit on the negative clamp was carbon in some form, the questions arose as to the[b] cause, remedy, and whether the carbon was carried along the loop or passed across from one side to the other.[6] The action being observable only when lamp is at intense heat and current of high electric motive force. Those were attributed as the causes, and by discussions it was concluded that the carbon was carried along the loop and not across as was supposed by some. To further convince themselves of the action, An experiment was made ~~by~~of, placing a thin copper wire in a u shaped glass tube filled with meter solution, and passed the current through to observe the action of the deposits of the copper and the points from which the metal was carried.

on the first experiment the wire was eaten off at the positive connection or clamp and again[b] reduced quite thin at about one third of the length of wire or at point where the larger part of the carbons have broken, and depositing of the metal heaviest as it neared the negative connection. This experiment showed that the carrying was entirely along the wire as no means were allowed for the current to pass or jump across. further experiments are being conducted. ~~on~~ As remedies it was suggested to reverse the current frequently,[7] as another to taper the carbon thereby giving it more substance to loose and permit it to equalize itself. Another was by use (in some way) of magnets—and experiments requested to be made on the different ideas.[8]

Rail Road. Martey[9] with his men commenced to day working from six oclock till dark, on the track of electric Rail Road and to night made great improvements in the curves and track down nearly to Moffetts crossing.

Gum. & Rosin. Lawson extracting gum & Rosin from Bast fibers.[10] 100 milli grams was used for experiment

AD, NjWOE, Lab., N-80-03-14:238 (*TAEM* 33:803; *TAED* N053:121). Written by Charles Mott. [a]"or ores &" interlined above. [b]Obscured overwritten text.

1. Unidentified.

2. Ludwig Böhm sketched and briefly described this instrument the previous day. In the next week he apparently made two similar instruments for Moses. N-80-03-19:115, 121, 123, Lab. (*TAEM* 34:662, 665–66; *TAED* N068:56, 59–60).

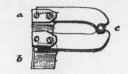

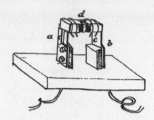

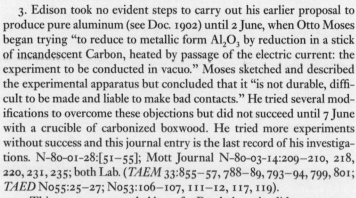

Drawing by Otto Moses of apparatus for smelting aluminum. Aluminum oxide was to be placed in the cup at d and heated by an electric current; the entire device was enclosed in a bell jar connected to a vacuum pump.

3. Edison took no evident steps to carry out his earlier proposal to produce pure aluminum (see Doc. 1902) until 2 June, when Otto Moses began trying "to reduce to metallic form Al_2O_3 by reduction in a stick of incandescent Carbon, heated by passage of the electric current: the experiment to be conducted in vacuo." Moses sketched and described the experimental apparatus but concluded that it "is not durable, difficult to be made and liable to make bad contacts." He tried several modifications to overcome these objections but did not succeed until 7 June with a crucible of carbonized boxwood. He tried more experiments without success and this journal entry is the last record of his investigations. N-80-01-28:[51–55]; Mott Journal N-80-03-14:209–210, 218, 220, 231, 235; both Lab. (*TAEM* 33:855–57, 788–89, 793–94, 799, 801; *TAED* N055:25–27; N053:106–107, 111–12, 117, 119).

4. This was a rare extended leave for Batchelor, who did not return to work until 28 June. Mott Journal N-80-03-14:264, Lab. (*TAEM* 33:816; *TAED* N053:134).

5. Two days previously Edison had telegraphed Princeton physics professor Charles Young: "When will you be able to make an examination of the mysterious blue halo in the lamp by the spectroscope." Young replied the same day that he could "come down and examine it anytime you like." TAE to Young and Young to TAE, both 15 June 1880, DF (*TAEM* 53:770; *TAED* D8020ZFD and D8020ZFE).

6. See headnote p. 623. On 15 June Mott indicated that "Mr. Edison was of opinion that the carbon was being carried from one side to the other of the loop directly across and for farther investigation he had a lamp made with a oval shaped glass partition attached between the connecting wires and extending nearly to the top of the loop at right angles with the faces. The lamp was put on pumps but not yet heated." Ludwig Böhm sketched two views of this arrangement the same day. Mott Journal N-80-03-14:235–36, N-80-03-19:113, both Lab. (*TAEM* 33:801–02, 34:661; *TAED* N053:119–20, N068:55).

On 17 June Mott noted that the

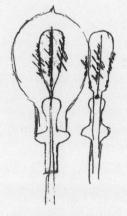

Ludwig Böhm's 15 June drawings of the experimental partition between the filament legs.

lamp made with glass partition and tested at intense heat yesterday and the one from which the Hydrogen lines were obtained in spec-

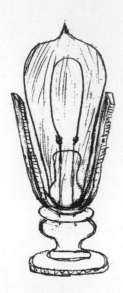

William Hammer's November 1880 magnet arrangement for manipulating the blue emanation at the clamps.

troscope was broken in the carbon today. Other lamps were made with mica partition extending down nearly to bottom of large globe, one was exhausted on the pump and heated up   Mr. Edison then for experiment let air in so the guage mercury filled about three fourths the globe and burned it at high incandescence for long time the lamp showed all the charasterics of lamp with high vacuum and showed the vapory blue the same as the plain lamps. [N-03-14:242, Lab. (*TAEM* 33:805; *TAED* N053:123)]

7. In mid-July Francis Jehl designed a "Circuit changer" that rapidly reversed the polarity in a lamp circuit. The device was tested in several lamps but was unsatisfactory at the speed at which it was intended to operate. Edison and an assistant later designed variations on this apparatus (see Doc. 1985). Mott Journal N-80-07-10:7; N-80-06-29:141, 157, 257–59; N-80-07-23:23, 29; all Lab. (*TAEM* 37:305; 36:195, 203, 253–54, 852, 855; *TAED* N117:3; N103:72, 80, 130–31; N112:11, 14).

8. Edison explored the former suggestion in November (see Doc. 2026). Efforts to use magnets to disperse the blue emanation at the clamp (associated with electrical carrying) were made in the third week of July and again in November, when William Hammer sketched a "Revolving magnet" at the base of a lamp globe (N-80-07-23:33–35, N-80-11-25:260, Lab. [*TAEM* 36:857–58, 37:640; *TAED* N112:16–17, N120:55]). In January 1881 Edison applied for a patent on a lamp with a magnet fixed at its base to "attract the highly electrified carbon vapor" so that the "carbon particles thrown off by the carbon filament may be attracted downwardly, in order that the formation of an arc between the limbs of the carbon, and also whereby the blackening of the inclosing glass chamber . . . may be avoided" (U.S. Pat. 251,548).

9. Martin Force.

10. This was apparently a continuation of the experiments described in Doc. 1930.

–1945–

From George Gouraud

London 17th June 1880[a]

No 15

My Dear Edison,

I send you herewith for your execution (which please do without delay in the usual manner cabling name of steamer bringing them back) the following Agreements:—

1st   Contract between ourselves regarding my interest in Foreign Telephone patents[1]

2nd   Agreement between yourself and a Trustee for the formation of the Foreign ~~T~~General Telephone Supply & Maintenance Company which is the name we have finally decided upon   (This I withold until receipt of further cable from you) as to basis you approve[2b]

3rd   Deed of Assignment respecting your Reversionary Interest in the Edison Telephone Company of London Limited which has Johnsons approval and I understand that he writes you by this mail to that effect[3]

In connection with this latter the more I think of it the more I am impressed, and I think Johnson is equally of the same opinion of the necessity of your placing this matter in the hands of a Trustee. This deed will upon your execution be filed with the Edison Telephone Company of London and they are far less likely to ask of Mr Renshaw any such unreasonable terms looking to the liquidation of the London Company as might otherwise be urged upon you and myself. The present intention so far as I have heard it expressed is that the London Company shall not be liquidated but continued to the end of your patents in which case it is highly important to all parties concerned that there should be certificates issued representing the various interests which certificates would probably have a market value somewhat proportionate to the shares of the Telephone[c] Company. If on the other hand the Company should propose a liberal division say ½ of surplus after giving shareholders share for share viz. surplus ~43,000£[c] say ½ 21,500 or maybe a round £20,000[d] of the surplus shares of the United Company we can always accept the same through Renshaw and thus the Trust does not in any wise either become an obstacle to either the continuance of the London Companys existence or the winding of it up.

<u>Foreign Telephone Company</u> I am actively engaged in the formation of the Board and the usual preliminaries of such an undertaking. Gardener Hubbard is here and professes to control in his own rights Bell patents for India [~~Ausstalia?~~][e] Australia[f] and the South African Colonies. He also thinks he has the control of <u>your</u> Australia patents. In any case I have New Zealand and we have agreed to put the Patents together on equal terms and divide what we can get out of it

We are at this moment engaged in the preparation of the Agreement &c for the British India & Colonial[g] Telephone Coy Bell & Edison Patents. There seems no reason to doubt that a strong Company can be formed for this Subsiduary Company. Hubbard[4] is altogether in favour of our taking no cash but a larger interest in the shares than we otherwise would get if we demanded cash as according to his experience this has been the most profitable course both here and in America.

I shall of course before absolutely committing myself submit the general basis to you for your approval.

I send you besides the above documents Power of Attorney including all the countries mentioned in your several cables.[5]

Renshaw thought it better to name all the countries in one Power of Attorney notwithstand that I have your Power of Attorney already for all of them except South American Countries India & New Zealand   As this will be in part duplicate upon the receipt of this I will cancel the other[h]

I annex copies of cables between us in this matter which I confirm & remain   Yours very truly

G E Gouraud

ENCLOSURE[i]

[London, June 17, 1880]

Cables

12th To Edison   "Cable approval sale telephone rights Norway Sweden Holland Denmark Portugal Switzerland Egypt Cape Turkey India China South America ~~excpept~~ Peru Chile Argentine to new Company £20 000 ~~cash~~ money plus fourth issued capital   Strong Board assured Bouverie probable chairman"[6]

14th From Edison— "I approve of sale telephone rights Norway Sweden Holland Denmark Portugal Switzerland Egypt, Cape, Turkey India China but not Japan & west Coast of South America for twenty thousand pounds cash & one fourth issued capital new company"[7]

14th To Edison   "Name South American countries included. How about New Zealand"[8]

15th From Edison   "Include Brazil Argentine Equador & New Zealand"[9]

16th To Edison   "Instead quarter shares & 20 000 cash approve half shares & no cash. Latter better for us if obtainable. Have agreed with Bell this basis British colonies"[10]

LS, NjWOE, DF (*TAEM* 55:723; *TAED* D8046ZAT). Written by Samuel Insull; letterhead of George Gouraud. [a]"London" and "18" preprinted. [b]"This . . . approve" written in left margin and followed by dividing mark. [c]Obscured overwritten text. [d]"say . . . £20,000" written by Gouraud in left margin; asterisk indicates placement in text. [e]Canceled. [f]Interlined above. [g]"& Colonial" interlined above. [h]"As this . . . the other" written below signature line; asterisk indicates placement in text. [i]Enclosure is an L, copied by Insull.

1. This unexecuted agreement is in Miller (*TAEM* 86:349; *TAED* HM800138). What may be a fragment of a draft written in an unknown hand with Edison's emendations is in Miller (*TAEM* 86:277; *TAED* HM800127). Gouraud's interests in Edison's telephone contracts in a number of foreign countries were secured by agreements executed in August 1880 (see Doc. 1978).

2. A damaged and undated copy of the contract Edison apparently

signed with George Newington, trustee for the prospective company, is in DF (*TAEM* 56:852; *TAED* D8049ZIO). Its terms are the same as those first outlined in the enclosure below. It was superceded in August 1880 (see Doc. 1978).

3. Johnson's letter has not been found. On 4 June Gouraud had written Edison to disregard the trust deed sent earlier (see Doc. 1933 n. 2) because the telephone amalgamation necessitated some alterations (DF [*TAEM* 56:687; *TAED* D8049ZFC]). Edison executed a revised agreement in mid-July (see Doc. 1954 n. 3).

4. Gardiner Hubbard, Alexander Graham Bell's father-in-law and organizer of the Bell Telephone Co., had been among the organizers of the Edison Speaking Phonograph Co. See *ANB*, s.v. "Hubbard, Gardiner Greene" and *TAEB* 3:500 n. 5; Doc. 1190.

5. Not found but see Doc. 1978.

6. This is essentially the text of Gouraud's cable. Gouraud to TAE, 12 June 1880, DF (*TAEM* 55:718; *TAED* D8046ZAM).

7. Edison sent this cable on 12 June 1880. TAE to Gouraud, DF (*TAEM* 55:719; *TAED* D8046ZAN).

8. Gouraud to TAE, 14 June 1880, DF (*TAEM* 55:719; *TAED* D8046ZAN1).

9. This message has not been found.

10. This is essentially the text of Gouraud's cable (Gouraud to TAE, 16 June 1880, DF [*TAEM* 55:722; *TAED* D8046ZAS]). These terms were incorporated into the revised contract with Newington (see note 2).

–1946–

From George Gouraud

London 17th June 1880[a]

No 16

My Dear Edison

Edison Telephone Coy of Europe. I confirms cables which have passed between us as to this, copies of which I annex.

As regards the Patents controlled by this Company I am endeavouring to ascertain what are the chances of our being able to immediately to take over this countries so as to include in our Foreign General Telephone Supply & Maintenance Coy all your continental & foreign patents but cannot say anything at this writing but what has is[b] already known to you from the above confirmed cables[1]   Yours truly

G E Gouraud

ENCLOSURE[c]

[London, June 17, 1880]

Cables

From Edison 15th   Controlling or whole share of telephone here for continent can probably be bought four hundred dollars share: Capital 100,000—28 in treasury. They have Russia Austria Germany Italy Spain Cuba Belgium. Have contracted out Russia Belgium Hungary Latter Government monopoly. Thirty years royalty basis. If you do anything it must be immediately

To Edison 15th   Twenty eight what in treasury   what Royalty from Russia Belgium Hungary

From Edison 16th   28,000 of the 100 capital   Royalties vary from $2 to lesser sums according to number   extend to end of concession notwithstanding expiration of patents years previous to termination

To Edison 16th   How much realized from royalties to date & since when

From Edison 17th   Contracts only just made Russia put up $10,000 Hungary $5,000 advance. without something done quickly they will probably enter into new contracts for other countries[2]

LS, NjWOE, DF (*TAEM* 55:728; *TAED* D8046ZAU). Written by Samuel Insull; letterhead of George Gouraud. [a]"London" and "18" preprinted. [b]Interlined above. [c]Enclosure is an L, copied by Insull.

1. Telegraphic communication on this subject began on 8 June, when Gouraud cabled "Cannot you possibly include any more countries for universal telephone company." Edison replied the next day, "The Edison Telephone Company of Europe are open for a trade." Gouraud wrote on 10 June that he intended the new company "to go into this business wherever it may be done and to that end acquire such patents as it can in any case when it may be desirable." He also stated that he would instruct Michael Moore (who had arrived in New York in April with letters of introduction from Edward Johnson) to open negotiations with the European Edison company. Edison assigned a number of European patents and pending applications to that company on 28 May; on 11 June, with the minimum of stock subscribed, the directors voted to issue shares for the first time. Gouraud to TAE, 8 and 10 June 1880; TAE to Gouraud, 9 June 1880; Johnson to TAE, 30 Mar. 1880; Moore to TAE, 19 Apr. 1880; Edison Telephone Co. of Europe resolution, 5 June 1880; directors' minutes, 11 June 1880; all DF (*TAEM* 55:714, 717; 56:606, 604, 186, 196; *TAED* D8046ZAI, D8046ZAL, D8046ZAJ, D8049ZDP, D8049ZDO, D8048ZDU, D8048ZEB2).

2. Gouraud reported on 18 June that he could not meet all the demands of the European Edison company "unless Bell same countries can be united   can that be done," to which Edison promptly answered: "Bell has no patent   leaves us only valid exploitieurs." Gouraud to TAE and TAE to Gouraud, both 18 June 1880, DF (*TAEM* 55:730; *TAED* D8046ZAV, D8046ZAW).

On 19 June Edison cabled Gouraud the suggestion that the company might agree to "royalty three dollars year per telephone in European Countries if guaranteed vigorous & immediate exploitation, with Johnson in charge." Gouraud made a counter-offer two days later and pledged a "vigorous working policy immediate establishment exchanges chief cities   Johnson chief engineer." However, Johnson had just cabled Edison that he would not accept the position and Edison wired Gouraud, "Johnson declines, would not do anything without him." Johnson reiterated his decision the next day but after Edison advised him "Better accept four or five months engagement Continent until light ready" he cabled on 23 June, "Cannot remain longer in Europe am offered absolute control policy men and apparatus with appointment electrician   will this accomplish your object" (TAE to Gouraud, 19 and 21 June 1880; Gouraud to TAE, 21 June 1880; Johnson to TAE, 21, 22 and 23 June 1880; TAE to Johnson, 22 June 1880; all DF [*TAEM* 55:731, 733–35; *TAED* D8046ZAX, D8046ZBA, D8046ZAZ, D8046ZBB, D8046ZBC, D8046ZBE, D8046ZBD]). Gouraud informed Edison that same day that Johnson had accepted "as special favor appointment electrician" on those terms. He followed this by cabling another counter-offer, prompting Edison to reply that the "Stockholders will accept if ten thousand pounds is paid advance royalty. But will give no option on stock; remember they can do as well by settling peicemeal as with Russia Belgium Hungary." After another cautionary cable from Edison, Gouraud sent word on 29 June that he would come to New York and asked Edison to "keep European offer open." On 24 August Edison telegraphed Banker that "Gouraud offers in addition to two dollars yearly Royalty a guarantee that first year shall net us ten thousand in Royalty   think we better close the thing up this basis as he sails tomorrow" to England (Gouraud to TAE, 23, 25, and 29 June 1880; TAE to Gouraud, 26 and 28 June 1880; TAE to Banker, 24 Aug. 1880; all DF [*TAEM* 55:735–37, 744; *TAED* D8046ZBF, D8046ZBG, D8046ZBK, D8046ZBI, D8046ZBJ, D8046ZBO]). During this period Edison also entertained negotiations with the Bell interests for a combination in Russia, Italy, Spain, and Portugal. No settlement was reached, however, and the European matter was unresolved at the beginning of October when Gouraud cabled "What about Edison European   They make a great mistake in this delay   large and powerful Anglo American syndicate forming here to work. Bell everywhere" (H. S. Russell to TAE, 14 and 22 July 1880; Russell to Griffin, 31 Aug. 1880; Gouraud to TAE, 1 Oct. 1880; all DF [*TAEM* 56:222, 228, 249, 261; *TAED* D8048ZEW, D8048ZFB, D8048ZFU, D8048ZFZ]; TAE to Russell, 13 Aug. 1880, Lbk. 6:303 [*TAEM* 80:360; *TAED* LB006303]).

–1947–

Edward Johnson to Edison and Charles Batchelor

London [c. June 18, 1880][1a]

My Dr Edison & Batch

Please take care that nothing concerning Insull's application for a position with you reaches Gouraud—[2] It might prejudice him—

Also bear in mind that all my correspondence to[b] you—& other correspondence to me while there—is kept confidential

Yours

Johnson

ALS, NjWOE, DF (*TAEM* 53:504; *TAED* D8014H1). Letterhead of the Edison Telephone Co. of London. [a]"London" preprinted. [b]Obscured overwritten text.

1. Stockton Griffin made a docket note on the reverse on 1 July, presumably the date Edison received this letter.

2. Samuel Insull (1859–1938), a London native, had been Gouraud's secretary since early 1879. On 7 April he wrote Edison that Johnson had "handed me your cable saying 'If Insull can translate French send him,'" to which he had already replied "No regretfully." It is not known exactly what job Edison had in mind at that time. Insull did become Edison's secretary in February 1881 and remained closely associated with him until 1892; thereafter he built a multi-state electric and gas utility empire based in Chicago. *ANB*, s.v. "Insull, Samuel;" McDonald 1962, 11–22; Insull to TAE, 7 and 5 Apr. 1880, DF (*TAEM* 53:501, 143; 56:578; *TAED* D8014H, D8004ZCM, D8049ZDD).

–1948–

John Michels to
Stockton Griffin

New York, June 26th, 1880[a]

Dear Sir,

I herewith hand you full statement of account up to this day.— I have paid cash for everything and have vouchers for all items. It shows a balance in hand at this date of $33.10.—[b] This includes all saleries up to date also.—[1]

The amounts still to be paid are as follows. 1 months rent, Printers bill ($30 paid on a/c),[2] and $20 to Professor Holden—[3] There are $59. due to Journal not collected.—

I am happy to say the first number is now ready for delivery.[4] I have only showed it to one person a professor from Yale, who was enthusiastic over it, and said "all the fellows will have it, when they return from vacation."—

Although a first number I think it already compares favorably even with "Nature," which is saying much— The contents is good, and the form, paper, and type is all that can be desired, and I consider it the handsomest Journal on either side of the Atlantic and when we begin with illustrations will be ahead of everything.—

The Journal is now before you, and I need not say how essential it is that the Company should be formed at once, my arrangements as you can imagine have to be made many weeks ahead, and any check now will tell on the future numbers.— You can see I am one of the slow and sure kind in money matters and economical, but of course proper working capital is necessary. As I said my friend Mr Shonnard[5] is ready with $12,500, and he suggests a similar amount being added.—

You will notice by the Journal and those who have promised

to support me with their literary cooperation, that I have struck very high ground, which lifts it above the crowd of mediocral productions, to keep this up you will have to make my position very definite—[6] I think Mr. Shonnards cooperation would be good, he is a very agreeble and nice man to do business with, and his cooperation would give a tone. If a Company is formed perhaps Mr Edison name should not appear as a stockholder, as I could then if necessary assert such to be a fact.[7]

The Journal should be advertised in the[b] Herald and a few papers next week.

It is a very dull time to start as all the Colleges &c are closed, but in the fall when it will be taken up more, [we?][c] will have to keep back numbers, as it will be a Journal for binding, and will sell also at the Booksellers in its bound form.—

Please ask Mr Edison to write me what he thinks of it, let him put it side by side with Nature, and think what "Science" will be when it also reaches Vol 22.—

If you can call in on Monday or Tuesday I should be glad to see you.

I have my 4000 wrappers addressed to scientific men, and will start them off at once— The Naval Observatory have ordered 50 copies for their own use, which will take 25 pc off Prof Holden's fee.

I shall be glad to know your opinion of "Science," and I can assure you I have had a fearful battle with the printers to get in my form in all its details.— I think the various forms of type very handsome, and the paper is as made for Mess Harper for their Journal.—[8] Yours Truly,

John Michels

ALS, NjWOE, DF (*TAEM* 55:401; *TAED* D8041P). Letterhead of *Science*. [a]"New York," and "188" preprinted. [b]Obscured overwritten text. [c]Illegible.

1. The enclosed statement follows in DF (*TAEM* 55:405; *TAED* D8041P). On 3 June Edison leased an office for *Science* at 229 Broadway in New York and by 9 June had advanced Michels $350 in cash (agreement with John Hamilton, 3 June 1880; Michels receipt to TAE, 9 June 1880; both DF [*TAEM* 389, 396; *TAED* D8041L, D8041N1]). Sometime during the month Edison had drawn up an agreement to provide full financial support for one year, after which Michels was to have a 20% interest if the journal had become profitable. This contract was never signed but Edison continued to pay all expenses, including the salaries of Michels and an assistant (draft agreement with Michels, June 1880, DF [*TAEM* 55:412; *TAED* D8041R1]). Michels subsequently sent

weekly statements of his account, which are in *Science* (D-80-041 and D-81-044), DF (*TAEM* 55:360, 59:451; *TAED* D8041, D8144).

2. Michels forwarded the bill for $67.76 a few days later. Five thousand copies of the first issue were printed. Michels to Griffin with enclosure, 30 June 1880, DF (*TAEM* 55:408; *TAED* D8041Q).

3. Edward Holden was an assistant to Simon Newcomb at the U.S. Naval Observatory. He was concurrently a principal advisor for construction of the Lick Observatory, of which he subsequently became director; he also served (1886–88) as president of the University of California (*ANB*, s.v. "Holden, Edward Singleton" and *DSB*, s.v. "Holden, Edward Singleton"). Holden wrote the lead article, "The United States Naval Observatory, Washington," for the first issue of *Science*, which was dated 3 July 1880 (pp. 1–3).

4. This issue included a short article by Francis Upton on "Electricity as Power." Upton discussed in general terms the economics of generating and transmitting electricity for power, and devoted the final four paragraphs to electric railroads and Edison's experiments on that subject. *Science* 1 (1880): 5, Cat. 1008:84, Scraps. (*TAEM* 23:369; *TAED* SM008084).

5. Frederic Shonnard, whom Michels described as a nephew of Senator Roscoe Conkling and "a gentleman of wealth having considerable property at Yonkers," was connected with the Malleable Nickel Alloy Co. in New York. Michels to TAE, 6 May and 27 Apr. 1880; Malleable Nickel Alloy Co. to TAE; 26 July 1879, all DF (*TAEM* 55:372, 364; 50:91; *TAED* D8041E, D8041B, D7919ZBN).

6. In a separate advertising supplement to this issue (and subsequent ones) Michels published the names of seventeen "well known and esteemed Scientists" who had "expressed either their intention to contribute, or their approval of the object of the Journal, and good wishes for its success." He had enumerated all but one of these, including Holden, Spencer Baird, Charles Young, and Othniel Marsh, the week before in a letter to Griffin, to whom he had earlier excerpted letters of support from Baird and Young. *Science* 1 (1880): ii; Michels to Griffin, 19 June 1880; Michels to TAE, 24 May 1880; both DF (*TAEM* 55:397, 381; *TAED* D8041O, D8041H).

7. Michels presumably had in mind an assertion of the journal's independence. The unsigned lead article in the second issue was a defense of the planned electric light demonstration at Menlo Park, which it called "Edison's answer to all the meretricious arguments and scientific hair-splitting which has been of late, with little generosity, carefully disseminated to his disadvantage. Taking the view that it is a waste of time to argue theoretically, on that which can be demonstrated practically, Edison, through all this wrangle has been silent, but not idle; while others *talked*, he has *worked*." "The Edison Light," *Science* 1 (1880): 18, Cat. 1017 (*TAEM* 24:189; *TAED* SM017063a).

8. The magazine's full title was *Harper's Weekly, A Journal of Civilization*. The firm Harper & Bros. was managed at this time by five sons of the founding Harper brothers. Tebbel 1975, 191.

To W. H. Patton[1]

[Menlo Park,] June 29 [1880]

Dear Sir

Your favor of the 8th was duly rec'd.[2] I have my system of transferring power perfected. I have in operation here an Electric railway ¾ of a mile long. 20 h.p. applied gives 13½ h.p. available at the extreme end. Roughly speaking if it was desired to translate 700 h.p. in to Virginia[3] from the Carson River, say 6 miles, there would have to be applied to the apparatus at Carson River 1000 if there was no loss on the conductor but this loss may be great or little according to the money invested in the copper conductor   Perhaps 80 h.p. would be the loss although the exact estimate can be made. This would call for 1080 h.p. giving 700 h.p. available for use in one or 100 places in Virginia City without material loss by subdivision. To carry 12 000 h.p. from Lake Tahoe to Virginia is a perfectly practical scheme and would pay big. Very truly

T. A. Edison   G[riffin]

L (letterpress copy), NjWOE, Lbk. 6:125 (*TAEM* 80:312; *TAED* LB006125). Written by Stockton Griffin; circled "C" written above. A transcription of this letter by William Hammer is in Cat. 2174, Scraps. (*TAEM* 89:254).

1. W. H. Patton was superintendent (since 1878) of the Consolidated Virginia Co., which developed the most lucrative mine holdings in the Comstock Lode. Smith 1943, 149.

2. Patton asked Edison to apprise him of the "present condition of your researches in the application of Electricity to the conveyance of power—for instance the generating of the Electricity by water power—conveyance of the same to any distance and its utilization by transformation into power again—also what loss is sustained during the transmission." Edison's draft reply on this letter is essentially the same as the final version written by Griffin (Patton to TAE, 8 June 1880, DF [*TAEM* 54:396; *TAED* D8032P]). On Edison's prior interest in this subject, see Doc. 1788 esp. n. 1.

3. Virginia City was located directly over the "Big Bonanza" of the Comstock Lode, in extreme western Nevada. Smith 1943, 151; Elliott 1973, 144–51.

7

July–September 1880

During the summer, Edison pushed ahead simultaneously with preparations both for a large-scale practical demonstration of his lighting system at Menlo Park and for the start of commercial lamp manufacture. Miles of underground electrical conductors for hundreds of outdoor street lights were in place by mid-July, after more than two months of work. When Francis Upton tested them, however, he found that the current leaked badly into the ground. While the lines were being dug up Upton initiated a series of tests of insulating materials and methods of covering the wires. The re-insulated lines were put down again in August but experiments continued well into September. Meanwhile, Edison and Charles Batchelor spent much of their time developing the incandescent lamp for commercial production, while other workers labored to rehabilitate and outfit an old factory building to accommodate lamp manufacture.[1] Edison's development work in these two general areas contributed to his prolific patent activity during this quarter, when he executed twenty-four applications that resulted in U.S. patents.

In the laboratory, Edison began an important set of filament experiments in the last week of July. He found that carbons heated in liquid kerosene emerged "considerably increased in size, greyish in appearance, and perfectly homogenious."[2] This last characteristic was presumably the one he sought and more experiments were tried with other volatile substances such as gasoline. Edison suspended this line of investigation and returned to it only occasionally; however, treating lamp filaments with hydrocarbons became a highly important process for his later rivals in lamp manufacture. During August

Edison also pursued, without much success, an earlier idea to prevent "electrical carrying" by rapidly reversing the flow of current through his lamps. These lamp experiments probably contributed to his decision to reinstate night work at the end of July.

While Edison awaited completion of the Porter-Allen steam engine for his large direct-connected dynamo, Charles Clarke began to reconfigure the armature in order to simplify its fabrication and repair. On 28 September, Clarke began measuring the heat produced in copper bars revolved through a magnetic field. He and Edison found so little heat that on the last day of the month Clarke began to redesign the armature induction circuit so as to substitute large copper bars for the insulated wire coils. Edison later obtained a patent on this method of construction, which he made standard in subsequent machines. In an unrelated action concerning dynamos, Edison offered to file a patent application on behalf of physicist Henry Rowland for a form of generator Rowland designed in 1868, in hopes of blocking a U.S. patent sought by the Siemens interests.

Throughout the summer, the machine shop worked on various improvements to the transmission mechanism of the electric railway and also on a claw-type arrangement for dragging the locomotive up steep grades. Encouraged by the practical success of his heavy motor on the railway, Edison followed up his prior solicitation of detailed requirements for an electric power system in Western mines. He also offered to install electric lights on the private of yacht of newspaper publisher James Gordon Bennett and on ferryboats of the Central Railroad of New Jersey.

In other affairs, George Gouraud arrived from London about 15 July. Edison assigned him a portion of the two-thirds interest in an improved telephone switchboard signal device he purchased from inventor Chester Pond. Gouraud had been making arrangements for the establishment of Edison telephone, electric light and power, and electric railway companies in numerous countries and territories not covered by existing agreements. On 18 August Edison signed multiple contracts and powers of attorney to sell patents to these prospective companies, one half the profits from which were to be Gouraud's. José Husbands, who was attempting to establish telephone service in Chile, meanwhile berated Edison for his inattention to that enterprise. Edison declined an offer to sell rights in Scotland for his analgesic compound, stating that he "would not for the world have polyform be brought out there" and expressed dismay over its sale in the United States.[3] When George Gouraud returned to London in late August he took with him several peaches sealed in a vacuum. He reported about two weeks later that the fruit was "Perfectly preserved" with a "slightly alcoholic flavour as if fermented."[4] He encouraged Edison to experiment with more perishables at his expense but nothing more was done until the following spring.

Nearly three dozen new names appear in Edison's employee records during this period. One was James Russell, who in July began a door-to-door survey of gaslight and power usage in the area of Manhattan where Edison planned to build his first central station. Little is known about most of the other men or their efforts. Only a few are definitely known to have worked at the lamp factory, although getting it ready no doubt required the work of many more. Some appear to have provided only transient labor while others, like carpenter Ben Acker[5] and mechanic Charles Campbell,[6] stayed long enough to leave an impression at Menlo Park. David Hickman, age 14, came from his father's farm in nearby Metuchen looking for work and was set at preparing underground electrical cables.[7] Benjamin Moffett, son of one of Edison's carpenters, was also 14 when he started as an apprentice glassblower.[8] The most significant addition, both for Edison and in a more general historical sense, was that of Edward Acheson. Acheson came as a draftsman in September but quickly took on greater responsibilities, including important experiments with pressed carbon filaments. He remained with Edison for several years and later distinguished himself as an industrial chemist, particularly as the inventor of Carborundum, an invaluable abrasive compound.[9]

1. See headnote p. 767.
2. Doc. 1961.
3. Doc. 1960.
4. Doc. 1986.
5. Time Sheets, NjWOE.
6. Time Sheets, NjWOE.
7. "Hickman, David Kelsey," Pioneers Bio.
8. Time Sheets, NjWOE; Jehl 1937–41, 688.
9. *ANB*, s.v. "Acheson, Edward Goodrich"; Acheson 1965, 15–19; Szymanowitz 1971.

PREPARATIONS FOR COMMERCIAL LAMP MANUFACTURE[1]   Doc. 1950

During the summer and fall, Edison and his staff worked to establish a lamp factory at Menlo Park capable of producing 1,200 lamps per day. Edison had acquired the old electric pen factory in late April and it took his carpenters and laborers nearly all of May to return it to a useable condition because, as Francis Jehl recalled, the "roof, floors, etc., were all in a dilapidated condition."[2] By late September, the laboratory staff had largely equipped the factory and some experimental and test lamps were produced. Factory employees continued adding equipment and modifying both the building and the production process over the next several months.

Various aspects of the lamp design itself had to be improved before production could begin. The most significant question was which material should be used for filaments. One assistant later recalled that after examining paper fibers under a microscope Edison decided that "paper is no good. Under the microscope it appears like a lot of sticks thrown together . . . Now I believe that somewhere in God Almighty's workshop there is a vegetable growth with geometrically parallel fibers suitable to our use."[3] During the summer, Edison focused his research on grasses and canes that had long, contiguous fibers. Bast fibers, which his staff had experimented with since the winter, and bamboo, which was first used in early July, were the fibers of greatest interest. After making what may have been a serendipitous observation of the frayed edge of a bamboo fan, used to speed up evaporation in laboratory experiments, Edison had one of the tough wayward fibers carbonized. The bamboo provided high resistance when tried in a lamp on 10 July. Subsequent tests bore out its suitability for lamps and although Edison continued to experiment with bast and other grasses, he wrote an associate in Havana, Cuba, at the end of August that he wanted "many tons" of cane or bamboo and would "pay a good price if [I] can get the right quality."[4] About that time he dispatched John Segredor to investigate the nature and availability of bamboo in Georgia and Florida, and sent inquiries to Brazil, Panama, Puerto Rico, Haiti, and Jamaica. He also sent an agent to Japan. In the meantime, in late July, Edison assigned chemist Otto Moses to conduct a literature search on carbonaceous materials and processes of carbonization.[5] At the beginning of December, he decided to make his lamps out of madake bamboo from Japan.[6]

Another critical problem was ensuring a good connection between the filaments and lead-in wires. Charles Batchelor experimented with placing various coatings on the filament ends and recarbonizing them, and also with putting platinum foil on the ends "to make better contact with the clamps and toughen the ends."[7] Edison had already decided to use carbons broadened at each end, but these proved difficult to manufacture because the heat needed to carbonize them fully often destroyed the nickel moulds. Edison accordingly devised and in early July applied for a patent on a form of vacuum pump in which carbonized filaments were placed on a broad infusible conductor which was heated to incandescence by electricity. This purged all hydrocarbon gases from the ends while "the body of the carbons, on account of the poor heat-conducting qualities of the carbon, remaining unheated comparatively."[8] In August he also applied for patents for improvements in the method of sealing the lead-in wires to the glass in order to simplify construction, maximize support for the platinum wires, and minimize chances of leakage.[9] The shape of the lamp base was also changed somewhat to accommodate Charles Batchelor's late August design of a push-in socket. Edward Johnson, recently returned from England, came up with the familiar screw-in design in mid-September which he labeled, with typical zest, the "Boss Socket."[10]

Edison and Batchelor put considerable effort during July and August into solving engineering problems related to lamp production. Because the filament itself represented only a portion of the labor and materials in each lamp, one such problem was how to identify defective filaments before they went all the way through the expensive manufacturing process. On 21 July Batchelor sketched an "Instrument for bending fibres to shape to test them" for "any unevenness or excess of pith."[11] At the end of July Edison and Batchelor jointly completed a patent application for a "carbon prover," a chamber partially evacuated by a mercury pump, in which a filament mounted in its inner glass globe could be heated electrically in order to find bright spots or other defects. A few days later Batchelor also sketched a device for detecting various imperfections in uncarbonized fibers.[12] He also experimented to determine why some carbons tended to bend over in the lamps and found it was a consequence of how the raw filament blanks were cut.[13] Batchelor, in particular, designed many of the specialized tools needed in the factory, from fiber cutters and trimmers, to carbonizing moulds and furnaces,[14] to the complex clamp-making machine.[15]

Equipping the factory required considerable effort. Laboratory assistants constructed four hundred vacuum pumps by the first day or two of July; as these were installed workers caulked the floorboards beneath them, presumably to prevent the escape of spilled mercury.[16] They had to contend with a troublesome chain pump used for lifting mercury to the vacuum pumps[17] and also took care of relatively routine tasks such as acquiring and cleaning more than 1100 pounds of mercury.[18] A particularly complex task that drew on the breadth of skills represented on the staff was the establishment of a supply of gas sufficient to operate the carbonizing furnaces and glassblowing burners simultaneously. John Kruesi had charge of ordering shafting and the pipes to transport gas and pressurized air.[19] He staked out the foundation for the new gasworks on 1 July; the apparatus arrived from Rahway in the middle of the month but when it and the associated blowers were installed and tested in August it was found to have "neither capacity nor simplicity for practical use." A new gas machine was ordered and installed by early September but runoff from heavy rains floated it out of position. It was reset and ready for operation by the end of the month.[20]

One other commodity needed to run Edison's factory was electricity. Saws for cutting fibers, the mercury circulating pump, the blower for glass-making, and the rotating glass annealing machines all operated by electric motors. On 12 July, laboratory assistants, probably William Hammer and Martin Force, ran wires to carry the current from the machine shop dynamos; at the same time they also put up telephone wires between the factory and laboratory. In early September, an additional circuit was established to provide current for testing carbons and lighting the factory.[21]

By the second week of September, the factory was sufficiently well organized that a number of filaments were successfully carbonized. For about the next two weeks, workers tried out equipment and made corrections as necessary. On 28 September the first lot of thirty lamps was completed and sent up to the laboratory, where their photometric and electrical characteristics were measured; the second batch of one hundred lamps was sent on 12 October to the laboratory, where some were also used in experiments. When photometric equipment was installed on 13 October in a building erected for this purpose adjacent to the factory, the manufacturing operation became largely independent of the laboratory. The factory's regular payroll began on 11 November, 1880.[22] Francis Upton took charge of the factory at the end of December. Although several lots of test and experimental lamps had been made by that time, regular production of commercial lamps probably did not begin until at least March and possibly April.[23]

Edison and his assistants made the preparations for commercial lamp production but an expanded and generally new work force did the actual manufacturing. Little or nothing is known of individual workers nor of the manner in which they were trained. By purchasing glass tubing from the Corning Glass Works, Edison avoided having to recruit and pay the most highly skilled glassblowers; he made do instead with "farmer boys and others from the neighboring villages" hired and trained by William Holzer.[24] Even so, concerns quickly arose about the availability and cost of labor. These worries contributed to Edison's decision to relocate lamp manufacturing to East Newark (now Harrison), New Jersey in 1882.[25]

1. Charles Mott's journal provides a detailed record of the variety of work that Edison and his staff performed throughout the late spring and summer to prepare for the commercial manufacture of electric lamps at Menlo Park. Mott Journals N-80-03-14 and N-80-07-10, passim., both Lab. (*TAEM* 33:683, 37:302; *TAED* N053, N117); for a discussion of these journals see headnote p. 671.

2. See Doc. 1939 n. 7; Jehl 1937–41, 787.

3. Wilson Howell's reminiscence, "Howell, Wilson S.," Pioneers Bio.

4. TAE to Vesey Butler, 28 Aug. 1880, Lbk. 6:354 (*TAEM* 80:368; *TAED* LB006354).

5. Mott Journal N-80-07-10:59, N-80-00-05, N-80-00-06, N-80-00-07, all Lab. (*TAEM* 37:330, 969, 38:2, 39:1064; *TAED* N117:29, N127, N128, N176).

6. See Mott Journals N-80-03-14 and N-80-07-10, both Lab. (*TAEM* 33:683, 37:302; *TAED* N053, N117); Docs. 2002, 2009, 2012, 2027 (see esp. n. 2); and Jehl 1937–41, 614–24.

7. N-80-06-02:21–33, Mott Journal 80-07-10:21, both Lab. (*TAEM* 36:405–11, 37:312; *TAED* N105:8–14, N117:10).

8. U.S. Pat. 298,679.

9. U.S. Pats. 239,153, 239,745, and 351,855.

10. See Doc. 1988.

11. N-80-06-28:37, Mott Journal N-80-07-10:31, both Lab. (*TAEM* 36:90, 37:317; *TAED* N102:19, N117:15).

12. U.S. Pat. 239,372; N-80-06-28:39–43, N-80-06-02:35–37, both Lab. (*TAEM* 36:91–93, 412–13; *TAED* N102:20–22, N105:15–16).

13. See Doc. 1973.

14. See Docs. 1961 and 1966. Several tools used in making lamps in the laboratory, including a plane, a finishing mould for bamboo fibers,

Staff at Edison's first lamp factory in an undated photograph, probably taken in summer 1880. The men in the foreground are, from left, Philip Dyer, William Hammer, Francis Upton, and James Bradley.

and cutting dies, are at the Henry Ford Museum and Greenfield Village in Dearborn, Mich. (29.1980.269, 188.36908, 29.1980.270, 29.1980.271, 29.1980.267, MiDbEI[H]).

15. Work on the clamp-making machine began in May and at least one prototype was ready by late June (see Doc. 1939). On 28 June Charles Batchelor made sketches and notes on a "New clamp making machine" and on 20 July listed fourteen alterations to be made (N-80-06-28:1–3, Lab. [*TAEM* 36:72–73; *TAED* N102:1–2]). Batchelor made drawings related to the clamp machine throughout the summer (N-80-06-28:5–17, 23–33, 67; N-80-06-02:89–95, 105–117; N-80-03-29:265; all Lab. [*TAEM* 36:74–80, 83–87, 105, 438–42, 446–52; 33:536; *TAED* N102: 3–9, 12–17, 67; N105:42–46, 50–56, N051:124]). Mott referred to work on this machine in nearly all of his weekly summaries of "work general" through mid-October (Mott Journal N-80-07-10, passim, Lab. [*TAEM* 37:302; *TAED* N117]).

16. See Doc. 1950; N-80-03-19:133, Mott Journal N-80-07-10:25, both Lab. (*TAEM* 34:671, 37:314; *TAED* N068:65, N117:12).

17. Albert Herrick sketched on 20 July the pump designed for the factory. The chain broke or caught on at least four occasions in September. On 22 September John Kruesi went to Philadelphia to see about ordering a new machine, and later that week drawings for a new screw pump were taken to a shop there. N-79-08-28:48; Mott Journal N-80-07-10:125, 131, 134–35, 137, 141–42; both Lab. (*TAEM* 35:1080; 37: 364, 367, 369–70, 372–73; *TAED* N088:26; N117:62, 65, 67–68, 70–71); see also Doc. 2010.

18. The mercury was received on 31 July. It was "exceedingly dirty" and was purified throughout the next month by William Hammer, George Hill, and unnamed "boys." Mott Journal 1880 N-80-07-10:55, 59, 63, 100–101, Lab. (*TAEM* 37:329, 331, 333, 352; *TAED* N117:27, 29, 31, 50).

19. N-79-07-25:256–59, Lab. (*TAEM* 33:954–55; *TAED* N056: 97–98).

20. Mott Journal N-80-07-10:15, 80–81, 100, 115, 126, 129, 150, Lab. (*TAEM* 37:309, 342, 352, 359, 365–66, 377; *TAED* N117:7, 40, 50, 57, 63–64, 75).

21. Mott Journal N-80-07-10:6, 117, Lab. (*TAEM* 37:305, 360; *TAED* N117:2, 58); Jehl 1937–41, 789.

22. See Doc. 1996. Mott Journal N-80-07-10:125, 144–45, 149, 152, 161–62, Lab. (*TAEM* 37:364, 374, 376, 378, 382–83; *TAED* N117:62, 72, 74, 76, 80–81); Jehl 1937–41, 815. Durability was determined by operating lamps above their designated intensity, usually at 48 candle-power; records of the first two lots of factory lamps are in N-80-09-28, N-80-08-18, N-80-09-27, N-80-10-15.1, and N-80-10-12, all Lab. (*TAEM* 36:459, 706; 37:445, 646; 39:700; *TAED* N106, N111, N119, N121, N171).

23. Docs. 2002 n. 4, 2034 and 2061; Mott Journals N-80-07-10:271–83, PN-80-09-23, and PN-81-01-19; all Lab. (*TAEM* 37:437–43, 43:1100–10, 1115–38; *TAED* N117:136–42 NP013:42–48, NP014:1–24); Electric Light—Edison Electric Lamp Co—General (D-81-23), DF (*TAEM* 57:756; *TAED* D8123).

24. Jehl 1937–41, 812–14; see Doc. 1971. The glassblowing department required a large amount of labor. An undated photograph shows William Holzer with thirty-two assistants (most of them young men), all but three of whom are identified in Jehl 1937–41, 813; an original print of this photograph is in box 97, WJH.

25. See Doc. 2061; Jehl 1937–41, 813–14.

–1950–

Charles Mott
Journal Entry

[Menlo Park,] Friday July 2d 1880

Pumps. The glass blowers finished the 4th hundred of the pumps, and men have today been carrying them down to Lamp Factory.[1]

Four lamps were attached to one of the pumps today and after excellent vacuum had been obtained, Mr. E. was showing the pumps to Prof Barker and F. Thompson[2] of Penna. Rail Road when[a] the mercury for some unaccountable reason flowed up into the lamps entirely destroying the vacuum and three of ~~them~~ lamps.[3] Four others were then attached and sealed off ~~two~~, after breaking two by current in about two hours.[b]

Wood Miller   Dean to day is trying to put his face cutting attachment for[c] the lathe for working wood loops   it is a very complicated neat piece of delicate and compact machinery but too incomplete yet for an attempt at description[4]

Crucibles for Assays. Cunningham[5] finished a mould in which to press suitable material into form for the small crucibles which are used in the final process of assaying and tried it several times with chalk   it worked very nicely and he got out a few very perfect specimens. The mould is made of iron, composed of a base in the center of which is a round raised piece to form the cavity of the crucible. on top is placed a cylindrical part with hole through the center, widened near the bottom into shape of reversed (or up side down) funnel. plungers are fitted[c] very snugly to ~~fill~~ the hole, or bore and after the material is placed in, the plungers are ~~placed~~ adjusted[d] and pressed down in the bore, the base and cylinder ~~being~~are separate but neatly fitted together and the one lifted from the other leaves[d] the crucible to[a] ~~is~~be easily removed.

R. R. Station   Martin Force removed the platform from the end of Rail Road Station. & changed the wire connections to one side of the building preparatory to remove the end of station for continuation of road.[6]

Magnetic Separator. design for the practical working of the magnetic separator was completed and sketch of the same given to Mott from which to make Patent Office drawings. capacity and specifications in Book No. 80. page 200 & 201.[7]

Lamps & Magnets[c]   I find in Mr. Uptons Book No 103 on page 275 quite a descriptive list of experiments requested by Mr. Edison to be made on platina wires in air[c] and vacuum, also ~~on the~~ to determine whether an active armature contains any magnetism. and some others.[8]

Loops. None but regulars have been carbonized to day, but Bradley has gotten out some more of the Palmetto loops and was directed by Mr. Edison to cut, on the same former used for bast, some out of [mauler?][e] glazed[f] card board and also try rye straw some of which I got for him.

Visitors. Mr. Wilber, also a Mine Prospector of Leadville Col.

Absent   Upton & Clarke started for 4th July trip[g]

AD, NjWOE, Lab., N-80-03-14:274 (*TAEM* 33:821; *TAED* N053:139). [a]Interlined in left margin. [b]"by current . . . hours." interlined below. [c]Obscured overwritten text. [d]Interlined above. [e]Illegible. [f]"[mauler?] glazed" interlined above. [g]"Absent . . . trip" written in right margin.

1. See headnote above.

2. Frank Thomson began working for the Pennsylvania Railroad as a machinist's apprentice and spent his entire career with that company except for a brief period of government service during the Civil War. Since 1874 he had been general manager of the system east of Pittsburgh. In this capacity he introduced major improvements in the Pennsylvania's track maintenance and signal system, and planned its distinctively decorated passenger stations. He became president of the railroad in 1897 but died just two years later. *DAB*, s.v. "Thomson, Frank."

3. Edison gave instructions on 3 July to "Have lamp put on sprengel

exhausted to high point, and then exhaustion stopped allowing it to burn for 20 hours at 18 candles to ascertain if vacuum falls." Mott reported that day that "One of the pumps was taken out and the connection between the Springle drop tube and the gauge tube was made longer and crooked (to prevent if possible the up flow of the fall mercury into the gauge tube and lamp)." This was the first of several more general experiments by Francis Upton and Francis Jehl on "testing for leakage or lowering from any cause of the vacuum." In one of these, described by Mott on 7 July, an evacuated lamp was sealed onto the pump which was itself then sealed off from the fall tube. After a considerable period of operating the lamp "a small bubble of air was let or leaked in and vacuum dropped accordingly, but immediately began to improve again, without working the pump, showing that the air was either absorbed or used up in occidizing the carbon." N-80-07-02:1; Mott Journal N-80-03-14: 277, 282–84; N-80-07-05:4–11; all Lab. (*TAEM* 36:604; 33:822, 825–26; 36:269–72; *TAED* N108:2; N053:140, 143–44; N104:4–7).

4. This was a redesign of the machine that Charles Dean made in June for cutting wooden loops (see Doc. 1930). It was completed on 9 July and Mott subsequently stated that it worked "very nicely, the knives only requiring a sharper and finer edge." On 16 July Mott reported that Dean was using the device to cut "the loops of .010 diameter beautifully and calipering evenly all the way around and was preparing the saw for cutting them off when Mr. Edison, admitting the perfection of the machine, directed him to clean it up, take it apart and put away, as at present the fibers were an improvement on the wood and the machine would not be required unless at some future time." In an uncharacteristic digression Mott noted that the machine was "so complete in all its 226 parts or pieces and so perfect in working that it should deserve a place amongst the archives rather than be destroyed or in all probability carried off piece by piece." Mott Journal N-80-03-14:287, N-80-07-10:10, 17–18, both Lab. (*TAEM* 33:827, 37:307, 310–11; *TAED* N053:145, N117:5, 8–9).

5. David Cunningham was a mechanic and machinist who began working for Edison sometime in the first half of 1879. In 1881 he assisted Charles Batchelor with Edison's installation at the International Electrical Exhibition in Paris. Jehl 1937–41, 680, 682; Time Sheets, NjWOE.

6. See Doc. 1939. Mott repeated this information in the next day's journal entry, the last in which this subject is discussed. Mott Journal N-80-03-14:277, Lab. (*TAEM* 33:822; *TAED* N053:140).

7. On 26 July Edison executed a patent application for a large-scale electromagnetic ore separator. Designed on the principle discussed in Doc. 1921, it employed a series of magnets to give diverging trajectories to ferrous and non-ferrous matter in the ore falling from the top of the machine. This application issued in October 1881 as U.S. Pat. 248,432. The notebook entry to which Mott refers was made by John Kruesi and indicated that the separator was to be about 24 feet high and capable of handling 560 pounds of ore per minute. N-79-06-12:200–201, Lab. (*TAEM* 35:577; *TAED* N080:94).

8. Francis Upton wrote out a handful of experiments and variations to be tried. The platina wires were to be heated in a chamber as it was evacuated and then measured, presumably for changes in length. Other carbons were to be measured before and after being dipped in an un-

specified solution. A lamp was also to be left on a pump with a McLeod gauge and its vacuum periodically noted (see note 3). N-80-06-29:275–76, Lab. (*TAEM* 36:261–62; *TAED* N103:139–40).

Menlo Park, N.J., July 4 1880.[a]

Friend Rowland,

I understood you to say when[b] you were here that several years ago you made and wound a magnet for use in an Electro-motor ~~of~~ or[c] magneto machine, and that you thought that you had some of the original parts.[1] You know that Siemens has applied for a patent in this country.[2] I want to defeat his patent on the ground either that he was not the first inventor, or use for two years previous to his application. If you will send me your original model ~~or~~ and a description or a description only, I will have a patent application filed for you[d] and pay all the expenses at the patent ofs also any court expenses should the patent get into litigation,[b] and give you half of the profits and I will use all efforts to prove your priority even if the office will not grant a patent on account of abandonment[3]   Yours Truly

Thos A Edison

ALS, MdBJ, HAR (*TAED* X100AA). Letterhead of T. A. Edison. [a]"Menlo Park, N.J.," and "1880." preprinted. [b]Obscured overwritten text. [c]Interlined above. [d]"for you" interlined above.

1. Edison did not address this letter but evidently sent it to Keene Valley, N.Y., known as a summertime home of scholars and landscape painters, where Henry Rowland spent much of the summer. Rowland, who had been at Menlo Park in March (see Doc. 1914), may have visited Edison on his way north from Baltimore. *WGD*, s.v. "Keene Valley"; Rowland to TAE, c. 6 July 1880, DF (*TAEM* 55:149; *TAED* D8036ZDI1).

While a student in 1868 Rowland designed a generator which he expected would run cooler than contemporary machines with long armatures. It employed an induction coil wound longitudinally on a cylinder, much like the design later adopted by Siemens. This revolved around a concentric cylinder wound longitudinally with iron wire. Both cylinders were placed circumferentially between the poles of a magnet, which in-

Henry Rowland's September 1868 drawings of his dynamo machine with stationary iron coil; at left is detail of the induction coil winding on the rotating armature.

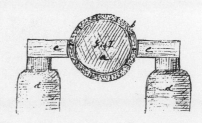

duced complementary poles in the iron coil. Rowland hypothesized that this arrangement would eliminate the heating in machines where the iron coil revolved around stationary induction wires, which he attributed to the iron's rapid magnetization and demagnetization. Hathaway 1886, 102; Rowland student notebook, folder 24, box 39, HAR; see also King 1962c, 369–74.

2. See Doc. 1851.

3. Rowland replied that he was "Happy to accede to your offer. I have a perfect Siemen's armature made about 1869, together with other parts of the machine and can prove the date by proper documents. The wire was wound longitudinally and revolved around a stationary iron core. It is about a foot long and three inches diameter"; he subsequently described the field magnets as "long like yours." Edison drafted a response on the letter instructing Rowland to "send on the mach & we will put it together   Prepare the app[licatio]n." On 18 July Rowland responded that he had sent instructions to Baltimore to have the armature sent to Edison and also indicated that he still had other parts of the machine. In this letter, which included a drawing and description of the armature, Rowland told Edison that "I do not suppose they will give me a patent but you can at least prevent Siemens obtaining one and establish my priority." Rowland to TAE, c. 6 and 18 July 1880, DF (*TAEM* 55:149, 142; *TAED* D8036ZDI1, D8036ZDK).

–1952–

Notebook Entry:
Miscellaneous

[Menlo Park, c. July 5, 1880[1]]

Canal System   1[2]
RR Train System—2[a]
Rock drill also Rock Tamper   3[3]
Balloon—4[4]
Transfer power   5[5]
High speed Telgh RR—6
System Signals for Electric RR   7
24 inch guage on canal with E Loco—to draw Canal boats. 8[6]
Elevator. 9—
Submarine Engine. 10
System submarine lighting[a]   11
Submarine Electric Railway   12
Steamship feeler for Ships Icebergs. 13
Lighting submarine buoys—14
Emery wheel with 5000 Rev motor   15[7]
Motor applied Lathes Etc place belts. 16[8]
Our system Lighting have Electric fire Engines to attach our mains, devise Electric fire Engine, 17
Ice sawing dynamo Engine. 18
Portable Electric drill   19
Electric Band & Circular Saw   can run armature with battery   20

Electric[b] Well borers   21
Torpedos
Band saws
Circular saws[c]

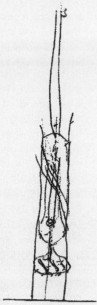

X, NjWOE, Lab., N-80-07-02:3 (*TAEM* 36:605; *TAED* N108:3). [a]Preceded by checkmark in left margin. [b]Obscured overwritten text. [c]"Band saws" and "Circular saws" written by Charles Batchelor.

1. Charles Mott referred to Edison's entry on the preceding page, dated 3 July, in his journal that same day. Mott did not return to work until 7 July, when he found this "list of applications to be made of the system for power, lighting, & c. both for land and submarine purposes." Later that day Edison asked Julius Hornig to make working drawings of these devices but they have not been found. Mott Journal N-80-03-14: 277, 280–81, 283, Lab. (*TAEM* 33:822, 824–25; *TAED* N053:140, 142).

2. Mott noted on 7 July that John Kruesi had in his office sketches of "two systems of operating canals by Electricity" but these have not been found. One plan was to have each vessel draw current from conductors alongside the canal for a motor to drive a propeller. The other would use a "track lain along the tow path over which a motor with hand over hand clutch gear travels and by line or otherwise tows the boat or boats." Mott Journal N-80-03-14:280–81, Lab. (*TAEM* 33:824; *TAED* N053:142).

3. This is the only extant reference to a rock drill prior to Edison's statement of its development in Doc. 1957; nothing is known of the tamper.

4. The *New York Herald* reported that Edison built a "captive machine" for making preliminary experiments with heavier-than-air flight ("Ships of the Air," *New York Herald,* 3 Aug. 1880, p. 4; for descriptions of other nineteenth-century efforts to build rotary wing flying machines see Liberatore 1998, chap. 2). On or about 7 July several sketches were

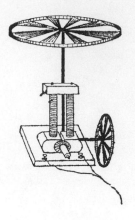

Edison's captive balloon.

made of this device, which consisted of fan blades mounted on a vertical shaft turned by a small electric motor. Charles Batchelor found that a blade of "two ordinary palm leaf fans" revolved at 200 r.p.m. produced a lift of twelve ounces per square foot. On 9 July Edison made a similar drawing of the "Electric Balloon" and that day Batchelor built and tested one with a tin fan wheel but found that it could not raise its own weight. Work continued intermittently on this device another ten days. Batchelor put in a more powerful motor on 12 July and, according to Mott, "the wheel revolved with fearful rapidity all standing well back expecting to see the wheel fly to pieces by centrifugal force"; it produced four ounces of lift. The next day he tried an arrangement of palm leaves which produced seventeen ounces of lift and "will give a fine motion to the air and make a neat and desirable apparatus to combine with electric light systems, for ventilation or keeping air in room in agitation." An additional experiment was made on 19 July, after which no more were recorded, but Edison stated in the August *Herald* interview that he hoped to achieve 8,000 revolutions per minute with a three horsepower motor. N-80-06-28:11–17; N-80-07-02:84, 142–43, 55; Mott Journal N-80-03-14:287–88; Mott Journal N-80-07-10:8–9, 22; all Lab. (*TAEM* 36:77–80, 615, 624, 614; 33:827–28; 37:306, 313; *TAED* N102:6–9; N108:13, 22, 12; N053:145–46; N117:4, 11).

At the end of November Edison received an inquiry from an engineer wondering if he intended to pursue his earlier investigations. Edison wrote a draft reply on the letter stating that he had "tried some experiments in the aerial navigation line & did intend to take the matter up but have been so attacked for my criminal efforts to devise a subdivided electric light I have given it up." TAE marginalia on John Mackenzie to TAE, 22 Nov. 1880, DF (*TAEM* 53:333; *TAED* D8006Z).

5. See Edison's U.S. Patent 248,435, which he executed on 11 August but did not file until October; it issued a year after being filed.

6. Edison made several sketches of this latter plan on 21 July. N-80-07-02:123, Lab. (*TAEM* 36:619; *TAED* N108:17).

7. Two months earlier Edison had conceived of another experimental project employing a high-speed electric motor: a "little Electromotor carrier that will run on a track in pipe track being composed of leafs for messenger service to take place of the telephone Central system. The Idea being to have a Central station with pipes leading to Customers 10 of Customers on a single line have permutation pins So a message put in a motor will be thrown out at the right customer." Edison did nothing with this idea until 29 June when, according to Mott, he began "seriously discussing the putting of it in practical operation for six or eight

Edison's drawing of an electric locomotive for towing barges.

Edison's 6 May drawing
of the "electromotor"
messenger device.

miles as an experiment." Two days later he instructed Hornig "to devise
and make diagram of a motor geared expressly for speed and as light as
practicable to be capable of 200 miles per hour for messenger." During
that week carpenters were at work on the superstructure for a test track
but there is no record of the project after 3 July. N-79-06-12:119–120;
Mott Journal N-80-03-14:266–67, 272, 278–79; both Lab. (*TAEM*
35:540–41; 33:817, 820, 823; *TAED* N080:57–58; N053:135, 138, 141).

8. Edison devised one such mechanical arrangement in early August
(see Doc. 1967 esp. n. 3).

–1953–

Notebook Entry:
Electric Lighting

[Menlo Park,] July 9̶10 1880

Lamps[a]

1253	made from bamboo taken from ftop of a fan[1] 4–⅝ long ¹²⁄₁₀₀₀ × ¹²⁄₁₀₀₀ put in large clamp[b]
1254	made from Rye straw ḣade pith on one side 4–⅝ long ¹²⁄₁₀₀₀ × ¹²⁄₁₀₀₀ put in large clamp[b] ⟨Broke after it was put in lamp in glass Blowers house⟩[c]
1255	large Bass Fibre 6 in long ¹²⁄₁₀₀₀ × ¹²⁄₁₀₀₀ put in large clamp[b]
1256	made from Palmeto leaf 4⅝ long ¹²⁄₁₀₀₀ × ¹²⁄₁₀₀₀ put in large clamp[2b]
1257[3] 1258	2 regular Bass Fibres put in same as we put paper carbon in old way of clamping 4–⅝ long ¹²⁄₁₀₀₀ × ¹²⁄₁₀₀₀[b]
1259[d]	made from paper 4–⅝ ¹²⁄₁₀₀₀ × $\frac{12\frac{1}{2}}{1000}$ put in large clamp[4b]
1260	Made[d] from Bamboo taken from top of a fan 4–⅝ long by ¹²⁄₁₀₀₀ × ¹²⁄₁₀₀₀

Chas Flammer

X, NjWOE, Lab., N-80-03-06:151 (*TAEM* 33:1035; *TAED* N057:75).
Written by Charles Flammer. [a]Beginning of table below marked by
horizontal dividing lines; item numbers separated from text by vertical
dividing lines. [b]Followed by dividing mark. [c]Marginalia written by
Charles Flammer; followed by dividing mark. [d]Obscured overwritten
text.

1. This is one of the first extant references to experiments with bamboo lamp filaments. Lamp 1248 was made with a bamboo filament the previous day but it broke before testing. Lamp 1253 produced a cold resistance of 188 ohms and the equivalent of 8.6 lamps per horsepower. N-80-07-05:25, 28; Mott Journal 80-07-10:3, both Lab. (*TAEM* 36:279, 281; 37:303; *TAED* N104:14, 16; N117:1).

Edison's impetus for trying bamboo is not certain. Charles Mott wrote in his journal on 7 July that "A collection of Bamboo Recd and choice Bast have been obtained and some loops cut out but none yet put in the lamps to test" (Mott Journal N-80-03-14:282, Lab. [*TAEM* 33:825; *TAED* N053:143]). Francis Jehl (1937–41, 614–15) recalled that:

> We always had . . . palm leaf fans lying about on the tables upstairs; these fans were often used in the course of experiments, especially when we desired to evaporate some liquid in a shallow glass plate or dry some mixture. It thus happened sometime towards the latter part of April or May that Edison noticed, while doing some microscopic work with a filament of carbon, that one of these fans was lying near his instrument. In his stooping position he noticed that a part of the binding rim of the fan was detached and was away from the fan leaf. He received the impulse to take the fan up and examine the rim: on closer examination it was found to be made from some sort of cane. He cut a piece of it, planed it and put it under his microscope: its structural characteristics were the most ideal thus far obtained. Batchelor was called to prepare a few raw filaments from the rim of the fan and carbonize them. The results were that Edison was satisfied that he now had a better carbon than that produced from the paper cardboard.

2. This may be the lamp which Mott indicated this day was tested at 483 ohms cold, 300 ohms hot, and produced the equivalent of 9.6 lamps per horsepower. Mott Journal N-80-07-10:3, Lab. (*TAEM* 37:303; *TAED* N117:1).

3. The page that begins with this lamp record is marked to indicate that it was later entered as an exhibit for Edison in a patent interference, *Edison v. Maxim v. Swan.*

4. This lamp broke before it could be tested. N-80-07-05:30, Lab. (*TAEM* 36:282; *TAED* N104:17).

–1954–

Draft to Edward Bouverie

Menlo Park, N.J., July 16 1880.[a]

I beg to acknowledge the receipt of your letter of the first instant.[1] when you first proposed that I should take in liquidation of my reversionary interest in the London Company £10,000 cash or shares of the United Company, before replying I gave your suggestion very careful consideration and decided that I shwould rather take my chances than accept so small an amount and I regret to feel obliged to say that I do not see my way clear[b] for changing the conclusion that I then came to— I have again considered the matter more fully with Col Gouraud.[2] ~~And wWhile~~ And while[c] we are perfectly content to [~~know?~~][d] have carried out what was recently understood to be the desires of the board including yourself that the company should be ~~concluded with~~ continued for[e] the purpose of carrying out its contracts with me (but which you now inform me ~~not to do~~ you do not desire[f] but would[b] rather liquidate the company at an early day.) ~~always we are both~~[g] I am[h] anxious as far as lies in my power with ~~what I deem a due regard~~ only a reasonable regard[i] for my own interest to meet yr. wishes.[j] I have at your request again carefully considered the subject of a division of the United shares with a view to liquidating the Co[k] ~~a~~As I understand it the total issue of shares of the Edison London Company when its amalgamation with the Glasgow Company is completed will be £72 000. The ~~total~~ only[b] assets of the Company will be ~~only~~ 115,000[l] ~~in~~ shares of United Company ~~will then be £115,000~~. now I think that if each shareholder of the London Company were to receive pound for pound in United Company shares ~~as~~ a repayment for total capital invested and a bonus of 30% on the shares in the United Company that considering that his investment will only have existed about a year he will have no reason to complain.

This will be the result if after giving pound for pound the difference between £72 000 and £115,000 were equally divided. as this is a subject that I really do not feel myself competent to judge being so far away[m] and as Col Gouraud and the Gentlemen in my Laboratory are interested with me in this matter and as[n] I would not do anything without their full ~~consent~~ concurrence[b] ~~and having regard to all the circumstances of the case in the in view of all in view of all~~[o] and as there is allready some little diversity of opinion among ~~us~~ us on the subject[p] I have thought it best to place the whole matter of my reversionary interest in the hands of a Trustee ~~and I have accordingly executed a deed to that~~ effect,[q] permanently resident in London[r] and as Col Gouraud will be much away from England this year ~~I~~ we[b] have decided to make Mr A G Renshaw the Trustee. ~~I have also requested him to furnish you with a copy of the deed~~ & I have accordingly executed a deed a copy of which I have requested Mr. Renshaw to file with the Co[3s] you will find him entirely prepared to deal with the question whenever you find it expedient to raise it—& I have no doubt you you will reach a mutually satisfactory conclusion.[4t] I note with pleasure the satisfaction you express at the amalgamation of the Companies[u] and sincerely hope the United[b] Company will be a success, ~~for~~ and[b] I feel confident that nothing but bad management can prevent ~~this~~ its so being.[v]

Df, NjWOE, DF (*TAEM* 56:741; *TAED* D8049ZGA). Written by William Carman; letterhead of T. A. Edison. Interlineations in an unknown hand, except as noted. [a]"Menlo Park, N.J.," and "1880." preprinted. [b]Interlined above. [c]"And while" interlined above. [d]Canceled. [e]"continued for" interlined above. [f]"you . . . desire" interlined above. [g]"~~we are both~~" interlined above. [h]"I am" interlined above by Gouraud. [i]"only . . . regard" interlined above. [j]"to meet yr. wishes" interlined below by Gouraud. [k]"with . . . the Co" interlined below. [l]"will . . . £115,000" interlined below. [m]"being . . . away" interlined above. [n]"and as" interlined above. [o]"~~in the in view of all in view of all~~" interlined above. [p]"us on the subject" interlined above. [q]"effect" mistakenly not canceled. [r]"permanently in London" interlined above. [s]"& I have accordingly . . . with the Co" interlined above. [t]"& I have . . . conclusion" interlined above. [u]"at the . . . Companies" interlined above. [v]"its so being" interlined above.

1. Bouverie expressed his wish that Edison might accept the company's offer (see Doc. 1933 n. 6) but particularly asked him "not to decide, or to commit yourself to any absolute course, respecting the reversionary interest you have in the Edison Company, until you have seen Mr. Johnson." The same day, Johnson wrote Edison asking him to comply "with Bouveries request 'that you do nothing to alter the present status until the companys representative (ie me) has had the opportunity to personally present their case to you.' = I am in honour bound to use Every effort to induce you to withhold your signature from the Trust Deed of Gourauds until the Company has had an opportunity of stating their Case." He went on to indicate his approval of the trust arrangement, which he hoped Edison would eventually adopt. Johnson did not immediately return to the United States, however, but accepted a leave of absence from the company from 1 July until the expiration of his contract two weeks later and took a vacation to the Continent. Bouverie to TAE, 1 July 1880; Johnson to TAE, 1 and 20 July 1880; all DF (*TAEM* 56:715, 707, 752; *TAED* D8049ZFK, D8049ZFJ, D8049ZGC).

2. Gouraud had sailed for New York on 3 July and visited Menlo Park on the date of this draft. Gouraud to TAE, 29 June 1880, DF (*TAEM* 55:737; *TAED* D8046ZBK); Mott Journal N-80-07-10:18, Lab. (*TAEM* 37:311; *TAED* N117:9).

3. Edison executed on this day the agreement designating Alfred Renshaw the trustee for his reversionary interest in the London telephone company. He subsequently arranged with Gouraud that the two of them would "together hold not less than the necessary two-thirds" interest needed to control the trust. Agreement with Renshaw, 16 July 1880; Agreement with Gouraud, 29 July 1880; both DF (*TAEM* 56:728, 755; *TAED* D8049ZFU1, D8049ZGE).

4. Two weeks later Edison and Gouraud jointly authorized Renshaw to accept in full payment of the reversion interest £20,000 in United company shares or "any larger amount if you find it possible to obtain it." TAE and Gouraud to Renshaw, 29 July 1880, Lbk. 6:243 (*TAEM* 80:330; *TAED* LB006243).

Notebook Entry:
Electric Lighting

[Menlo Park,] July 16[–19] 1880[1]

wet day	To ground		
18 wire circuit[2]	55	25.	43.
Carmen Circuit	1.4	4.2	8.2
	2800	2900	2100[3]
6 wire circuit	61.8	25.6	50.5
Edison's line	1.6	5.8	9.4

The weather &c Book 103 page 155[4]

In the afternoon Carmens circuit again tested[a] P.M |2850

2850
$\underline{1425}$
126[0]
1.65
Ohms[5]

Dry day

P.M   2900 to ground

The kerite wire leading to the lamp-posts dug[b] up and left in the air to dry.[6]

Monday— Yesterday was a dry hot day   nNearly all the kerite wires leading to the lampposts have been exposed to the air so that they have no connection with the ground[7]

X, NjWOE, Lab., N-80-07-16:3 (*TAEM* 38:487; *TAED* N137:2). Written by Francis Upton. [a]Obscured overwritten text. [b]"kerite . . . dug" written over illegible erasure.

1. This entry was dated 16 July (Friday) but was continued to the following Monday, 19 July.

2. The anticipated electrical load determined the number of wires in each line. Wires were dropped from the strand at intervals away from the generating station as the current decreased.

3. Francis Upton made his notes and calculations on this day's tests in another book, from which this table is derived. The first column indicates resistance "wire to wire"; the other two give the resistance to the ground for each leg of the circuit, in this row the 25-wire Carman cable. There were several standard procedures for determining the resistance of cable insulation but how Upton conducted these tests is not known. In each instance he divided the measured resistance (such as 2800 ohms) by 200, presumably to compensate for a galvanometer shunt. From this he deducted the resistance of the "conducting wires from galvanometer" (6.3 ohms apiece). Upton erred in copying numbers to this row from his notes, in which he calculated the resistance to ground at 4.2 ohms where he measured 2100 ohms, and 8.2 ohms to ground where he measured 2900 ohms. N-80-06-29:145–55, Mott Journal N-80-07-10:16, both Lab. (*TAEM* 36:197–202, 37:310; *TAED* N103:74–79, N117:8).

4. Upton remarked in the other notebook that "The weather has been raining for two weeks and the ground is thoroughly wet." On 15 July Charles Mott had recorded in his journal that "The gang that have been at work on laying the conductors to the Street lamps since May 1, got

them all down to day, but there is still a large amount of it to tar, cover and fill trenches. This job has taken Two & one half months—with an average I should judge of six men on the work" (N-80-06-29:155, Mott Journal N-80-07-10:14, both Lab. [*TAEM* 36:202, 37:309; *TAED* N103:79, N117:7]). After recording the disappointing results of the day's first tests, Mott commented:

> It will be seen that some of the circuits are very badly insulated and all more or less defective. Men have been set at work uncovering parts of the trench of 25 wire circuit and removing the dirt from around the lamp posts. after the boxes had been raised out of the ground at low wet points it was retested but with same results. It seems a little strange that unexperienced men should be permitted to put down nearly five miles of wire and cover it without being required to test a single circuit or wire until the entire work is finished, and it will now require considerable extra labor and delay in putting the circuits in working order. [Mott Journal N-80-07-10: 16–17, Lab. (*TAEM* 37:310; *TAED* N117:8)]

5. Upton obtained similar results from three more tests on Carman's line that afternoon. N-80-06-29:159, 165, Lab. (*TAEM* 36:204, 207; *TAED* N103:81, 84).

6. Mott reported on 17 July that workers spent much of the day removing soil from around the lamp posts and junction boxes "but late this after noon when tested by Mr Upton no improvement or change was found to have been effected, and Mr [Wilson] Howell to whom this work had been entrusted seems ready and more than willing to take advice, and listen to suggestions that if taken in time might have saved him the annoyance of having made a partial failure, and would surely have saved Mr. Edison the extra expense of redoing this work." Mott Journal N-80-07-10:20–21, Lab. (*TAEM* 37:312; *TAED* N117:10).

7. Acting on the suspicion reported by Mott that the problems were partly attributable to the tar, Upton tested a bucket of it on Monday (19 July) but found the resistance "extremely high and could not easily measure" it. That day the exposed kerite wires on Carman's line were "overlaid with rubber tape ('piping') and served with coal tar, then left exposed to the sun to harden" (Mott Journal N-80-07-10:23; N-80-06-29:183; N-80-07-16:7–15; all Lab. [*TAEM* 37:313, 36:216, 38:489–93; *TAED* N117:11, N103:93, N137:4–8]). Further experiments with tar and tar paper for insulating the electric railroad took place concurrently for several weeks (N-80-07-16:17–53, 57–79; Mott Journal N-80-07-10:16–34 passim; both Lab. [*TAEM* 38:491–510, 513–524; 37:310–19; *TAED* N137:9–27, 29–40; N117:8–17]; see also Doc. 1961).

[Menlo Park,] July 20, [1880]

Dear Sir;

Your letter of the 19th inst. is at hand.[1]

I feel satisfied with your representations respecting the success, for economy and durability, of the high-speed engine which you are at present constructing. I have concluded to give you an order for a second engine, to develope 120 H.P. at 600 revolutions per minute, with a boiler pressure of 120 lbs.[2] I am   Yours truly,

Thos A Edison

LS (letterpress copy), NjWOE, Lbk. 6:218 (*TAEM* 80:321; *TAED* LB006218). Written by William Carman.

1. Porter's letter has not been found. It may have been prompted by Charles Clarke's 16 July inspection trip to Philadelphia to see Edison's engine under construction. Mott Journal N-80-07-10:18, Lab. (*TAEM* 37:311; *TAED* N117:9).

2. Clarke wrote to Porter the same day that this engine "is intended for the operation of a division of a railroad and not for electric lighting. The load to which it will be subjected will be very variable, resembling the duty of a rolling mill engine or working crushing machines; only the work in our case will be applied even more suddenly (instantaneously) and will vary with every variation in resistance on the line due to grades and curves &c." Clarke also requested information about the relative economy and first cost of adapting a compound engine to this application. Edison sent Porter the patterns for the dynamo on 6 August. He asked to "have it cast in your foundry and planed to measurement. . . . It is desirable that the Iron is as soft as can be used to make a good casting," presumably to avoid the magnetic retardation of the armature referred to in Doc. 1889. The engine was never built. Clarke to Porter, 20 July 1880; TAE to Porter, 6 Aug. 1880; Lbk. 6:219, 264 (*TAEM* 80:322, 336; *TAED* LB006219, LB006264).

Menlo Park, N.J.,[a] [July 21, 1880?][1]

Dear Sir

I suppose your intention ~~is~~ would be,[b] to use[c] the present pump rod & tanks down to the 3000 feet level and there to stop and and further pumping to be done by magnetos.[2] We have no data here to make calculations on, but suppose, that after the 3000 foot[c] level a ~~Dynamo~~ magneto of ~~75~~60 hp. would be sufficient for every 250 further than 3000. Our Generator consists of a porter allen Engine,[c] cylinder 9 × 10, steam jacketed piston rod connected directly to magneto   The Engine makes 600 Revolutions per minute. ~~TheEach nett horsepower can be~~ [~~delivered?~~ -------][d] non condensing steam pressure 120 lbs[e]

The 60 hp magnetos[c] at the pump will give an[f] actual[g] duty of 1 hp for every 3 and ½ lbs coal. The Engine direct gives a h.p. for $2^{65}/_{100}$ lbs of coal, ~~but~~ The extra consumption from $2^{65}/_{100}$ to 3½ ~~is to~~ covers the losses due to converting friction etc, but not the friction etc of the pumping mechanism worked by the ~~D~~Magneto, below.

As for hoisting I suppose you would Establish a station at the 3000 foot level with a smaller copy of the hoisting machine above, and transfer men & ore at the 3000 foot station. [~~How--?~~][d] If you do this you might use say a 100 hp ~~d~~magneto with the hoisting apparatus at first and gradually add to the number as greater depths was reached, the ~~d~~Magnetos could be coupled on each as you required more power from time to time.

The Space occupied by a 60 hp Magneto is about $8 \times 5 \times 3$— The Engine & Magneto Converter 8×9 by 3.—the Bobbin of the Magneto acts as a fly wheel.

Could you give me with a little data ~~of what you~~ regarding your mines such as maps etc. Yours Truly

T A Edison

P.S. I have nearly finished a Diamond Drill worked by a Magneto direct—[3] TAE

ADfS, NjWOE, DF (*TAEM* 54:401; *TAED* D8032T). Letterhead of T. A. Edison. [a]"Menlo Park, N.J.," preprinted. [b]"would be" interlined above. [c]Obscured overwritten text. [d]Canceled. [e]"non condensing . . . 120 lbs" interlined above. [f]"n" added later. [g]Interlined above.

1. Stockton Griffin prepared the letter based on this corrected draft on this date, and copied it into Edison's letterbook. Edison instructed Griffin on the draft to address the envelope to "W. H. Patton Supt Union Consolidated, Virginia City Nevada." Lbk. 6:222 (*TAEM* 80:324; *TAED* LB006222).

2. This letter answers Patton's 7 July response to Doc. 1949. Patton explained that miners were "rapidly approaching a vertical depth in the Comstock Mines of 3000 feet—beyond which point it will be difficult to carry the necessary power to drive our pumps and do our hoisting." He stated that compressed air and hydraulic power had been rejected as too inefficient but "Electricity generated by Dynamo–Electric machines would seem to be most likely to fill the bill—provided you can accomplish what you claim." He added that an electric power system could only be tested by actual use in a mine and "If your company is disposed to take any steps towards such a test—the companies which I represent will do all in their power to assist— should you feel disposed to entertain any thing of the kind please send me a description of the machines necessary to use and any information you may give regarding them—and the matter will receive immediate attention at my hands." Patton to TAE, 7 July 1880, DF (*TAEM* 54:399; *TAED* D8032R1).

3. Edison referred to a rock drill in Doc. 1952 but there are no extant

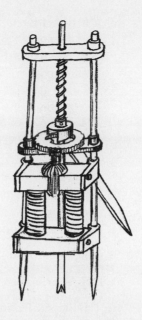

One of Charles Batchelor's
25 July sketches of a
direct-driven electric
rock drill.

records of work on such a machine before 25 July. On that day Charles Batchelor made notes on the "characteristics of a rock drill," particularly its reciprocating action, and drew several forms of drills. Some would be driven through pulleys or gears and cams but according to Charles Mott, one driven directly by the armature shaft would be "capable of being run at 2000 or 3000 revolutions per minute and with lateral stroke of probably one half inch imparted to it by a lifting cam, and having a short stroke is capable of acting rapidly, and incessantly pecks rather than striking heavy blows." A few days later Batchelor made another set of brief notes, after which there is no surviving evidence of this project. N-80-06-28:47–59, 65; Mott Journal 80-07-10:39; both Lab. (*TAEM* 36:95–101, 104; 37:321; *TAED* N102:24–30, 33; N117:19).

–1958–

Notebook Entry:
Electric Lighting

[Menlo Park,] July 23 1880

Boilers & Chimney	21,000[1]
Engines	21,000
Foundation	2,000
Iron Structure	3,500
Wood flooring	1,000
Water heaters & pumps	7,000
Iron floor &and supports	6,000
Faradic Machines	24,000
	85,000
Conductors	50,000
	125,000

6[a] per H.P.

$$1250$$
$$\underline{6}$$
$$7500$$
$$\underline{9}\,[--]^{2b}$$
$$67,500$$

85,000[a]	5.9294
67,500	4.8293
$1.25	1.1001

125,000	5.0969
67,500	4.8293
$1.85 at consumers	.2676

$1.85 investment per M annually

Depreciation & repairs

Boilers & Chimney	10%	$2100
Engines	3%	630[a]
Foundation	1%	20
Iron Structure	2%	70
Wood flooring	5%	50
Water heaters & pump	5%	350[a]
Iron floor	2%	120
Faradic Machines	3%	720
~~Condu~~		4060
Conductor		1000
		$5060

406 000	5.6955[3]	
	4.8293	
	+.8662	7.3 per M

506,000 cts	5.7042	
67,500~~0~~	4.8293	
	.9749	9.4 cts per M at consumers[c]

Taxes 2%	2500		
	250,000	5.3979	
	67,500	4.8293	
		.5686	3.7 cts at consumer
	170,000~~0~~	5.2304	
		4.8293	
		.4011	2.5 at station[c]

Labor   per day

    2 Engineers $10
    2 Wipers   3
                $13

     7500 lights
      25 cu. ft
  187,500 M

     3.1139
     2.2729
     .8410   6.9 at station[4]

Continued Book 39 p 273

from Book 48 End[d]

Coal[a] taken a $3 per ton   2240 lbs[5]

     1250[a]   3.0969
$3 \times 5 = 15$ 1.1761
   18,700   4.2730
    2240   3.3502
     8.36 tons .9228[a]
       3   .4771
     $24   1.3999
    187.5   2.2729
           9.1270   13.4 cts per M for coal[a]   3 lbs

Stoking $4.00 per day

     4   2.6021
         2.2729
         .3292[a]   2.1 ct for stoking

Water for boilers   Estimated at ⅓ coal ~~abo~~ 4 cts per M
Insurance &[e] Rent $5,000 per year
    500,000 5.6990
         4.8293
        .8697   7.4 cts per M

Summary   6 per H.P.[d]

    Depreciation   7.3
    Taxes   3.7
    Labor   6.9
    Coal   13.4
    Stoking   2.1

Water	4.
Insurance, Rent &c	07.4
	44.8

44.8 cts at station   3 lbs per hours

2½ lbs per hour   42.6 cts

at consumer   46.9 cts

No Exec   44.7

Executive   2 cts per M   489 cts per M

Say 50 cts per M cost at consumer

Dividends at 1.00 per M[d]

125 000 invested

67 500 M annually made $33,750 profit   About 33%
dividends

$1.50 per M   66% dividends

F R Upton

X, NjWOE, Lab. N-79-07-05; N-79-11-21:275–77, 280–81 (*TAEM* 33:268–69; 32:601–02, 604; *TAED* N048:136–37; N039:123–24, 126). Written by Francis Upton; document multiply dated. Miscellaneous calculations omitted; some decimal points and commas added for clarity. [a]Obscured overwritten text. [b]Canceled. [c]Followed by dividing mark. [d]Page ruled for recording lamp test results with horizontal line at top and vertical line at left margin. [e]"Insurance &" interlined in left margin.

1. About this time Edison requested the firm of Babcock & Wilcox to restate its April estimate for 1200 horsepower "Boilers with and without Economisers," the original figures having been lost. On or about 13 July Francis Upton made a brief estimate, principally of operating costs, for a 600-light central station; also cf. Doc. 1897. TAE to Babcock & Wilcox, undated, DF (*TAEM* 53:812; *TAED* D8020ZGJ); N-80-06-29:117, Lab. (*TAEM* 36:183; *TAED* N103:60).

2. The consumption of one gas jet was assumed to be 9,000 cubic feet per year (see Doc. 1897). This calculation gives the number of "M" units produced by a 1250-horsepower central station, at six lights per horsepower. The two sets of logarithmic computations below, including the line "$1.85 at consumers," are taken from the facing page.

3. The logarithm of 406,000 is 5.6085. The correct depreciation is 6.0 cents per M.

4. In this operation Upton used logarithms to divide the daily labor cost (1300 cents) by the 187.5 M units of electricity consumed in five hours.

5. In the calculation below Upton assumed that the engine would consume three pounds of coal per horsepower hour and that each lamp would operate an average of five hours per day. In the last step he used the logarithmic shortcut explained in Doc. 1795 n. 4.

Valparaiso, July 24th 1880.[a]

My dear Sir.

I have this day remitted to Mr H. H. Eldred the full amount of my endebtedness to you, at least as far as advised, some things may be in transit.[1] The last lot of goods, have only been in the house[2] 3 days & some not yet disembarked from the steamer so we are prompt although the time may be ~~short~~ long.[b] The Watson's Batteries were[c] not cabled for, Laclanche in their stead. We would receive the same if it was not plain that they are a total wreck. We had one dose which left a bad taste in ~~your~~ mouth. By your statements we notice you charge every item of expense, even to 25 cents & also for your power of attorney to obtain a patent for you.[3] This is all right we do not complain, except there have been so many ways in which money was paid for faults not our own. At this writing I am without an answer to any letter written you or any word whatsoever which facts hurts me not a little, for I was & am so full of you and yours, that it grieves me to find myself in the position of a man holding his saddle looking at his horse running away. Time will tell your friends & I shall work hard and do my level best for our mutual interests here. You could not drive me away from you with a shot gun, although I cannot but feel the injustice of your not writing anything good or bad.[4]

I have also to report that on May 1st '80 I mailed you per English mail via Panama your Patent, which has not been acknowledged at this writing.[5] Please do so & much oblige. Your having it, is an uncertainty until it is acknowledged   I paid $587.55 Chile money, equal to $411.70 gold out for this patent, saying nothing of time & trouble   You told me in your office that you furnished patents as against capital to work them. In this instance you furnish neither, all I want is to have you[d] say "thank you" & I am fully repaid. We would have been fully organized & stock issued if you had sent the papers. These delays are very expensive & to have waited longer would be actually impossible—as the gentleman associated with me would not stand it.[6] My contracts were taken before the proper legal authority and decided sufficient, defining the what you as well as myself should receive and your telegram changing the consideration from a cash, or cash & stock interest, to a stock interest only. Our contract says I am entitled to $5000. out of each $15,000. stock or any proportionate increase[7]—which will give me one third ⅓ of one half— Our capital will be $200,000.00 Chile money $66,666.67. will be issued to you, in certificates of \$10,000.00 each, \$33,333.33. will be issued to me, which I hope will meet your approval.[8] We could not live longer without an organization both by law & in reality. This Chile matter may only be a drop in your great Electric bucket but please do not forget that one is just as tired after a days work in Chile & elsewhere also that the labor is not only harder but always more doubtful in the smaller Republics where their education is not quite so metropolitan—as with you. while doing this I might have handled a "bigger deal," but as I came to represent you, knew no such word as fail. We are giving up the chalks because it cost so much to maintain them. This time of the year the rains are <u>very heavy</u> & they become[e] mush in an night. When they are "just right," the talk is wonderful but it cost too much to keep them "just right" again the repairs are something awful, the crank, the shaft where connected—& also in the seat where the chalk sleeve rests on rusts in tight by the moisture the sponge gathers. The frame of the receiver breaks & crumbles to peices where the screws are in &c improve this arm so as to make it an every day chalk[b] receiver & you can clean out the party. I own up that Mr Morgan was a prince beside Clark[9] for M was simply a d——m fool while Clark is the same & additionally[e] a bloody knave. I own up & will not bother you with the plots of conspiritors—but if you care to get a little of it to amuse you (as it will) ask Mr Eldred.

I send to Mr Eldred in this mail, a request from the South American Steamship Co of this city for proposals for your light &c the same as the "Columbia" had, for their two new steamers building in England.[10] This is worth your attention.

I trust you are well and enjoying the fruits of your hard earned fame to the satisfaction of your self and family who live after you. It was next to seeing you all to see Mr Henderson of the "Columbia" who talked of Mr Batchelder & you all. I enjoyed his visit. Yours Respectfully & truly

José D. Husbands

The stock of Receiver arms, is of no use and quite a loss. We have hard work to get in the Pony Crowns, after using the chalks, but doing away with turning the crank settles it.[11]

ALS, NjWOE, DF (*TAEM* 55:864; *TAED* D8047ZBC). Letterhead of La Compañia Chilena Telefono de Edison de Valparaiso, José Husbands, director. [a]"Valparaiso," and "188" preprinted. [b]Interlined above. [c]Interlined in right margin. [d]"have you" written in left margin. [e]Obscured overwritten text.

1. Husbands enclosed the statement of his account with this letter (DF [*TAEM* 55:869; *TAED* D8047ZBC]). Two weeks earlier he re-

mitted $1,000 and on this day paid $3,050.78 for telephones and supplies (Husbands to TAE, 10 July 1880, DF [*TAEM* 55:862; *TAED* D8047ZBB]).

2. That is, the customs house.

3. On 17 June William Carman sent Husbands two letters, enclosing with them copies of a series of Edison's and Sigmund Bergmann's bills charged to his account. These show small charges for a number of items including washers, screws, fare to New York, and sundry expenses there. Husbands apparently did not receive these until mid-August. Carman to Husbands, both 17 June 1880, with copies of enclosed bills, Lbk.6:100–109 (*TAEM* 80:307–11; *TAED* LB006100, LB006101, LB006102, LB006103, LB006103A, LB006104, LB006104A, LB006105, LB006105A, LB006106, LB006106A, LB006107, LB006108, LB006109); Husbands to TAE, 14 Aug. 1880, DF (*TAEM* 55:872; *TAED* D8047ZBF).

4. Husbands wrote on 21 May that another party had submitted an application for the government license for Edison's lighting system. He protested that "you promised me the handling of it, provided I was successful with the Telephone. . . . This is all right, but for the fact, That I might have been spared anoyance and mortification from friends by your simply writing me that I could not have it. Then I would have known where I stood" (Husbands to TAE, 21 May 1880, DF [*TAEM* 55:855; *TAED* D8047ZAY]). Edison also received a report from Fabbri & Chauncey that their agent in Valparaiso had made this application and was afterward visited by Husbands, who claimed to have Edison's authority for the light. Edison wrote to Husbands that it was

> unfortunate that you so misunderstood matters. I told you before
> you left Menlo Park that I had already made contracts to parties
> and disposed of the whole of South America on the light. What I
> intended to do was (provided you were successful with the Tele-
> phone) to arrange it with the parties who hold the contracts to take
> care of you so that you could introduce it and have the science part
> of the exploitation. This I fully intended doing. I will consider it
> a special favor if you will just let matters take their course as I in-
> tended from the first that they should. [TAE to Husbands, 16 July
> 1880, Lbk. 6:184 (*TAEM* 80:318; *TAED* LB006184)]

The license was eventually awarded to Edward Kendall, who was Fabbri & Chauncey's representative; nothing more is known of him (Sherburne Eaton to TAE, 20 Nov. 1882, DF [*TAEM* 62:192; *TAED* D8237ZAJ]).

5. Husbands enclosed with that letter the declaration by the president of Chile granting Edison an exclusive eight-year licence for telephones. He explained that this privilege would begin in April 1881; a week later he cabled that the term was for ten years. On 29 April Husbands and his associates "celebrated by connecting the offices in Santiago & Valparaiso & talked successfully over a line 180 miles long." This was apparently the only long-distance line that Husbands constructed. President of the Republic of Chile to TAE, 26 Apr. 1880; Husbands to TAE, 1 and 6 May 1880; all DF (*TAEM* 55:843–44, 846; *TAED* D8047ZAP1, D8047ZAP, D8047ZAR); Berthold 1924, 37–38.

6. Husbands was backed financially by W. H. Brown, a New York metals dealer. See Doc. 1823 n. 7.

7. Cf. Doc. 1823.

8. See Doc. 2031 n. 2.

9. W. J. Clark, who replaced Walter Morgan (see Doc. 1886 n. 2). This was apparently against Edison's recommendation but nothing more is known of Clark. Husbands to TAE, 10 July 1880, DF (*TAEM* 55:862; *TAED* D8047ZBB).

10. The enclosure from Husbands has not been found. Two weeks earlier he had written that "The steamer 'Columbia' created a sensation here, all parts working good here. In the next steamer I will ask for proposals to do the same thing in the coast company here." The *New York Herald* reported at the end the month that the *Columbia*'s lights worked well all the way to Portland (but see Doc. 1976 n. 3). Husbands to TAE, 10 July 1880, DF (*TAEM* 55:862; *TAED* D8047ZBB); "Electric Light on Steamships," *New York Herald*, 30 July 1880, Cat. 1241, item 1518, Batchelor (*TAEM* 94:610; *TAED* MBSB21518X).

11. As recently as mid-May Husbands believed it would be easy to devise a better way of properly moistening the receiver chalks but had concluded that "the crank movement is an nuisance one can not read off or receive & write at the same time—& must be done away with." Husbands to TAE, 15 May 1880, DF (*TAEM* 55:850; *TAED* D8047ZAU).

-1960-

From Michael Moore

Glasgow, July 24 1880[a]

Dear Sir,

On my return here it occurred to me that your Polyform[1] could be introduced into this country with success, as Neuralgia is a very prevalent affliction in England & especially in Scotland[2]

I sent you a cable asking on what terms I could acquire the right of it for Great Britain but have no reply.[3]

If you are in a position to send me the formula & let me have an exclusive right to it, I would be willing to take it up & by spending a fair sum of money in advertising & agencies I've no doubt but that a very large demand could be created for it. I would bear all the expenses you retaining such a share in its profits as we might arrange betwixt us.

I hope all things are going on well at Menlo and the Electric Railway doing its grade of 1. in 7. & the light still burning.

I've enlightened our Scotch public a good deal about your doings since my return.

We feel a little sore down here on some points connected with the Telephone but these are matters in which you individually had no hand & no blame attaches to you

I've a letter to day from Johnson. He sails on the Arizona
21 Aug. You have a good & true man in him, & can trust him.
Kind regards to Mr Bachelor, Yours very truly

M M Moore

⟨Write Moore that I would not for the world have polyform
be brought out there   I did it here to get rid of the annoyance
due to Reporters publishing the fact & the consequent rush of
correspondence & demands for it & I am sorry I now did it—
I receive no compensation=[4] Gouraud said he would attend to
your cable of Wednesday[5]

I am willing⟩

ALS, NjWOE, DF (*TAEM* 53:187; *TAED* D8004ZEA). Letterhead of
W. B. Huggins & Co., United States Securities. ᵃ"Glasgow," and "18"
preprinted.

1. Polyform was Edison's name for a topical analgesic compound he
devised in 1878 whose ingredients included alcohol, chloroform, ether,
and morphine (see Doc. 1287). His application for a U.S. patent was unsuccessful but in August 1880 a London agent filed the final British
specification on his behalf (Serial No. 2,206, 8 Sept. 1880, "Medicinal
Preparation" in Abstract of Edison's Abandoned Applications (1876–
1885), p. 12, PS [*TAEM* 8:539; *TAED* PT004 (image 13)]; Brit. Pat. 599
[1880], Batchelor [*TAEM* 92:155; *TAED* MBP024]).

2. The term neuralgia historically applied to intermittent acute pains,
particularly in the face and head. By this time, however, its English-
language use had greatly expanded to encompass a variety of pains for
which no direct physical cause could be found, including some arising
from internal organs and those now attributed to overly strenuous or
repetitive muscle exertion. Alam and Merskey 1994, 429–58.

3. Moore had cabled from Glasgow on 15 July: "Can I acquire your
polyform rights for this country if so write terms" (DF [*TAEM* 53:181;
TAED D8004ZDU]). Stockton Griffin noted on Moore's letter that
Edison replied on 9 August.

4. Edison instructed Lemuel Serrell in September 1879 to assign his
U.S. rights to Charles Lewis and several partners, and in April told one
correspondent that he had done so in order "to rid myself of the awful
amount of correspondence & trouble which the newspaper publications
caused   I have no interest in it." Lewis and his associates organized the
Menlo Park Manufacturing Co. in New York in 1879. It promoted "Edison's Polyform" as a treatment for neuralgia, rheumatism, and head-
aches, among other things, although Edison told another correspondent
in 1880 that "The medicine is improperly advertised to cure Rheumatism but it will relieve facial neuralgia" (TAE to Serrell, 3 Sept. 1879,
Lbk. 5:129 [*TAEM* 80:143; *TAED* LB005129]; marginalia on G. H.
Kent to TAE, 16 Apr. 1880; marginalia on H. C. Guck to TAE, 29 June
1880; [*TAEM* 53:149, 175; *TAED* D8004ZCS, D8004ZDR]; Menlo
Park Mfg. Co. label and advertisements in PPC [*TAEM* 96:596; *TAED*
CA016A]). The company subsequently relocated to Boston and in 1884,
with the U.S. application still pending, asked Edison's permission to ob-

tain a copyright registration on its labels featuring his autograph and portrait. Edison refused (Menlo Park Mfg. Co. to TAE, 15 June 1883, 26 June and 25 Sept. 1884; all DF [*TAEM* 64:209; 71:185, 277; *TAED* D8303ZDW, D8403ZEC, D8403ZGZ]; TAE to Menlo Park Mfg. Co., 2 Oct. 1884, Lbk. 19:291 [*TAEM* 82:909; *TAED* LB019291B]). The firm was apparently never profitable and in 1890 Lewis offered him a controlling interest in the stock of a new company called the Edison Polyform Co., but Edison instructed his secretary to "Write Lewis that if any further attempts are made to bring the polyform out that I shall knock it in the head" (Lewis to TAE with TAE marginalia, 4 Oct. 1890, DF [*TAEM* 128:701; *TAED* D9004AER]).

5. Edison received a cable via New York from Moore in London dated 5 August (Thursday) stating that the "Glasgow company in paying additional five and accepting London terms lose about 2000 will you abate what may be required up to this so we make no loss." In a letter the same day Moore proposed that the company should withhold enough from the £5,000 advance royalty still outstanding to recover their loss. Edison drafted a reply on Moore's cable that he would "do as Gouraud says." George Gouraud was still in the United States. Moore to TAE, both 5 Aug. 1880; TAE to Moore, 6 Aug. 1880; all DF (*TAEM* 56:762, 759; *TAED* D8049ZGI, D8049ZGH, D8049ZGI1).

–1961–

Charles Mott Journal Entry

[Menlo Park,] Saturday July 24—1880

Papers. Herald of today also has an article, purporting to be interviews with officers of Elevated Rail Roads of New York on the subject of electric roads concerning their adaptibility to the elevated roads.[1]

Lamp lines. Hickman[2] was set at work on the 25 wire cable, winding it with strips of Muslin, preparatory to tarring for better insulation.[3]

Fiber cutters: The fiber holder or clamp for shaping and cutting was tried. The single[a] brace in the center was found to spring the bars so it did not cloase evenly the whole length, and a steep piece was cut to bring the bearing neare the ends, but it was found that the wooden base then sprung. It was decided to get a cast base and press[a] the clamp together with a double solid right angled lever ~~acting under~~ with the bite coming up in front of the former ~~perpendicularly~~ and the lever arms extending under and back. worked with treadle same as before.[4]

Carbonizing former. Andrews is making a former for carbonizing moulds leaving the inside piece loose instead of rivuted as in others, and with a light weight fitted to ~~the~~ press[a] the ends down flat, and prevent them from wharping or curling out of shape, instead of the ends drawing up they are permanently secured and the loop contracting draws with it the inside piece which[a] at the time keeps it in symmetrical shape. Mr Batchelor has today been trying one in which the ends were clamped with light weights which were drawn up as the carbon shrinks, but did not give entire satisfaction.[5]

Oil carbon. A very interesting experiment was made today by immersing a carbon loop clamped and connected to inner tubes in[a] Kerosene oil and brought up to incandescence by the current, bubbles of gas or air were emitted from the carbon in at high heat the oil (assumed a smoky look) appeared to be infused with carbon. On removing the carbon it was found to be considerably increased in size, greyish in appearance, and perfectly homogenious— it was placed in a lamp, exhausted and burned at 16 candles till 14 minutes when engine was stopped.[6]

Work general Men preparing the gas carbonizing furnaces by putting in the gas and blast pipes and fixtures. Clarke on Electric Locomotives[7] Mr. Batchelor on carbons and apparatuses for carbonizing. conductor gang uncovering street lamp circuits. Men finishing up gear for electric locomotive.[8]

Bast Fibers. Two bundles recd from Baltimore Md.

AD, NjWOE, Lab., N-80-07-10:36 (*TAEM* 37:320; *TAED* N117:18). Written by Charles Mott. [a]Obscured overwritten text.

1. This article was a follow-up to one in the *New York Herald* the day before, in which Edison reportedly claimed that applying his "electric engine" to the New York elevated trains would save $500,000 annually in direct costs. The later article, after noting that the company's directors "would say nothing of importance, one way or the other, about the matter," presented cautiously optimistic comments attributed to operating department officials. "The Electric Motor," *New York Herald*, 24 July 1880, p. 3; a typed transcript of this article and typed extracts from the *Herald*'s 23 July article on "Electric Locomotion" are in Cat. 2174, Scraps. (*TAEM* 89:283, 279; *TAED* SB012:34, SB012AAS).

2. David Hickman worked on his uncle's farm in nearby Metuchen before he applied to Edison for a job in 1880. Preparing the underground conductors was his first task at the laboratory. Hickman subsequently superintended the Pump Department of the Menlo Park lamp factory and remained associated with Edison until 1891. "Hickman, David Kelsey," Pioneers Bio.

3. Six days later this line was wrapped with tarred twine. Francis Upton tested the insulation, finding 1,400 ohms resistance to ground and 4,000 between the wires. Insulation of this and the other lines continued until 28 August, when Charles Mott reported that Edison suspended the work. Insulation tests were conducted at intervals both on these lines and in the laboratory until September. Mott Journal N-80-07-10:52, 55, 81, 105; N-80-07-16:47–51, 71–76, 79–101; N-80-07-05:72–75; all Lab. (*TAEM* 37:328–39, 342, 354; 38:508–10, 520–33; 36:303–04;

TAED N117:26–27, 40, 52; N137:24–26, 36–49; N104:38–39); see also Doc. 1985.

4. Mott reported on 20 July that Charles Dean was "making a pair of formers for cutting out fibers each one is designed only to cut one side or edge and intended to be secured to the bench and drawn and held together by a treadle leaving the operator both hands to adjust and use in cutting." On 3 August Dean completed and tested one pair; Mott noted that they produced fibers of uniform dimensions and "are very convenient effective and work satisfactorily." He noted in weekly summaries of work that Dean continued to work on the instrument and on 1 September one was sent to the lamp factory, "made very heavy to avoid all danger of springing and tempered to prevent scarrification by knife or hammer." Mott Journal N-80-07-10:26, 61, 68, 93, 105, 108, Lab. (*TAEM* 37:315, 332, 336, 348, 354, 356; *TAED* N117:13, 30, 34, 46, 52, 54).

5. Two days later an arrangement in which the filament ends were secured was again tried. On 28 July Edison executed a patent application covering several similar devices to keep the filament "under strain during carbonization, with one or more points fixed against moving, and the contraction proceeds against the strain, which constantly keeps the filament against or in contact with a former, preserving its shape and obviating any risk of warping or twisting." Such a device was particularly important with bamboo filaments, which shrank about 20% during carbonization. Mott Journal N-80-07-10:41, Lab. (*TAEM* 37:322; *TAED* N117:20); U.S. Pat. 263,139.

6. This is the first of a number of experiments in which filaments were heated in the presence of volatile hydrocarbons, particularly kerosene and gasoline, although paraffin, bituminous coal, and other substances were tried. The object seems to have been to deposit a uniform coating of carbon on the filament. In one series of tests, however, Batchelor deposited the carbon on thin platinum wires which he then instructed Alfred Haid to dissolve, leaving a narrow tubular conductor (see Doc. 2007 n. 1). Some hydrocarbons were placed in the carbonizing mould with the fiber but in most cases a carbonized filament was electrically heated in a chamber filled with a hydrocarbon. Mott recorded related experiments through the end of July and a number of lamps were made with carbons treated in this way before the experiments were suspended. Edison later returned to the subject but heating lamp filaments in a hydrocarbon atmosphere (called "flashing") became an important manufacturing process for his competitors. N-80-03-06:181–93; N-80-06-02:8–15; N-80-07-23:281; Mott Journal N-80-07-10:39–41, 44–45, 50–52, 54–55, 57; all Lab. (*TAEM* 33:1050–56; 36:401–04, 981; 37:321–22, 324, 327–30; *TAED* N057:90–96; N105:4–7; N112:140; N117:19–20, 22, 25–28); Howell and Schroeder 1927, 79–81.

7. Beginning on 21 July Charles Clarke made numerous drawings and calculations pertaining to power transmission in the locomotive, especially reduction gearing and clutches for the wheels and modifications to the "climbers" for steep grades (see note 8). He did most of this work before the end of the month but continued it intermittently through mid-August. N-80-07-19:15–35, 45–135, Lab. (*TAEM* 37:113–25, 130–75; *TAED* N115:8–18, 23–68).

8. On June 15 Mott described Julius Hornig's design for "a novel gear

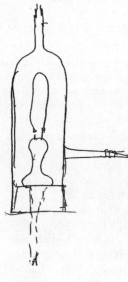

Chamber for passing a hydrocarbon atmosphere around a heated carbon.

for attachment in front of motor to climb steep grades, its motion and action is very similar to hand over hand climbing, being composed of arms and clutches which grapple the rails one of each [side?] and by a cam movement the clutch is released arm raised up and extended, dropped down on and again clutched on the rail, the arms acting alternately." Mott reported at intervals the progress of laboratory assistants in constructing and installing this equipment, which is described more fully in Doc. 1987. Mott Journal N-80-03-14:236, 249, 257, 263; Mott Journal N-80-07-10:4, 21, 30, 33; both Lab. (*TAEM* 33:802, 808, 812, 815; 37:304, 312, 317–18; *TAED* N053:120, 126, 130, 133; N117:2, 10, 15–16).

-1962-

From Ernest Biedermann

Geneva, 28 July 1880.[a]

Dear Sir,

I did not write to you before as I thought our correspondence by Cable sufficient and I would not bother you with long letters.—[1]

In regard to the trade mark or what you may call a three quarter Patent, I have done all the necessary, that is to say I made duplicate drawings of everything and translated all your papers into French and German and have placed everything in Bern at the right place, so that we will be almost the first who will get papers, but as everybody who had a trademark under the old law comes before us we may have to wait a month or so to get our papers.—[2]

I received from you a generator a meter and a Safety wire but no regulator nor lamp nor motor.—

I tried to burn the ancient lamps with this generator, which I excited with an ordinary Gramme machine, but the light produced was very small until I got the speed up to 2400 turns.—[3]

M. Young cables me that you will send suitable lamps and I hope to receive them soon.—[4]

Now to the business part:

As the lamp up to the present as far as I know has not come into practical use I make you the following proposition, which I hope will be acceptable to you.—

First) I have associated myself for this business with M. M. A. Cherbuliez[5] and M. G. Zurlinden[6] of this City, both gentlemen of first class standing.—

Second.) We have made a Contract with the "Societé Genevoise pour la construction des Instruments Physique" (which is a Company for the construction of scientific machinery and instruments) in which we bind ourselves to give a certain amount of work and they bind themselves not to produce or sell or immitate or improve any of our machines, models or drawings except by our order and for our account.—[7]

All the above I give you as information so that you may exactly understand the position.—

Now we Cherbuliez, Biedermann and Zurlinden propose the following.—

You give us the right of your inventions and of all further improvements for Switzerland and we will at our Cost, besides producing the heavy machinery at the "Societé Genevoise" establish another laboratory for the manufacture of the Lamps and all other inventions of yours.—

We to give you 50% of the nett profits of everything produced payable half yearly. We to sell only at such prices as will be agreed upon with you.—

If you accept the above you must either send somebody over here, who can superintend our men and who at the same time can watch your interest or you would have to give us instructions to the smallest detail of everything so that we can manufacture as well as you.— It would be certainly preferable if you could send somebody who would naturally come at our expense.—

The above arrangement has another advantage for you and it is the following: We believe that we will be able to light immediately at least a part of the town of Geneva with your lamps, which if a success (I do not doubt it) will be the most powerfull advertisement for the whole of Europe and of great advantage for you.—

I think you will find the above proposal acceptable as it is impossible for us to pay an amount down before we have seen the light in practical use, but having an unlimited confidence in your final success, we are perfectly willing to risk a pretty large amount in making a laboratory and producing machinery and lamps and giving you 50% of the profits you taking no risk whatsoever.—

M. J. C. Young of New York will call on you for purpose of getting your ideas in regard to the above and I hope you will be able to arrange with him.— I have instructed him to cable me your reply at once and if you accept the above the contract can be made immediately and we will begin work without a day's delay.—[8]

I remain   Dear Sir   Very sincerely Yours

Ernest Biedermann

ALS, NjWOE, DF (*TAEM* 54:257; *TAED* D8026ZCL). Letterhead of E. Biedermann. a"Geneva," "188," and "." preprinted.

1. See Doc. 1878.

2. Biedermann had cabled in April that he needed to know the "name of each particular apparatus and when machines will arrive." Edison replied that the "System consists Electric Generators, Electric pressure regulator, Electric meter, Lamp and System laying wires, also Electromotor for power; will cable when shipped, will send copies patent papers" (Biedermann to TAE, 26 Apr. 1880; TAE to Biedermann, 27 Apr. 1880; both DF [*TAEM* 54:232, 233; *TAED* D8026ZBN, D8026ZBO]). Biedermann then arranged for trademark protection on the condition of having sample apparatus by 1 June and suggested that drawings might be acceptable if the machines could not arrive in time. Edison cabled on 11 May that he was still "working on samples fear not reach in time have mailed papers." He shipped the devices two weeks later (Biedermann to TAE, 4 May 1880; TAE to Biedermann, 11 and 25 May 1880; all DF [*TAEM* 54:236, 239, 246; *TAED* D8026ZBP, D8026ZBT, D8026ZCA]).

3. When Biedermann cabled that the generator and meter had arrived and asked what speed he should run the generator, Edison replied that the "Generator sent only small sample   Lamps you have not suitable this size   will send you lamps suitable." Biedermann to TAE, 22 June 1880; TAE to Biedermann, 24 June 1880; both DF (*TAEM* 54:251, 252; *TAED* D8026ZCE, D8026ZCF).

4. J. C. Young was connected with the Adams Express Co. in New York and acted as Biedermann's intermediary. On 12 July Edison wired him that "Biederman has all necessary appliances except another lot lamps which I am making." TAE to Young, 12 July 1880; letterhead of Young to TAE, 24 May 1880; both DF (*TAEM* 54:255, 245; *TAED* D8026ZCJ, D8026ZBZ).

5. A. M. Cherbuliez identified himself as a commercial arbitrator. Agreement between Biedermann, Cherbuliez, Gaspard Zürlinden, et al., 8 Sept. 1880, Miller (*TAEM* 86:278; *TAED* HM800128).

6. Gaspard Zürlinden identified himself as a lawyer. Agreement between Biedermann, A. M. Cherbuliez, Zürlinden, et al., 8 Sept. 1880, Miller (*TAEM* 86:278; *TAED* HM800128).

7. The 8 September agreement between Biedermann, Cherbuliez, and Zürlinden refers to their intention to set up a manufactory for the elements of the lighting system "with the single exception of those machines whose manufacture they will turn over to the 'Societé Genevoise pour la construction des Instruments Physique à Genève'" (Agreement between Biedermann, Cherbuliez, Zürlinden, et al., 8 Sept. 1880, Miller [*TAEM* 86:278; *TAED* HM800128]). According to their 9 September 1880 agreement with Edison, Biedermann, Cherbuliez, and Zürlinden had arranged that this company would not manufacture similar instruments for anyone else (DF [*TAEM* 54:267; *TAED* D8026ZCR]).

8. On 14 August, two days after having visited Menlo Park, Young wrote Edison that he had "cabled Biedermann your acceptance of his proposal" and received an acknowledgment. Young to TAE, 14 Aug. 1880, DF (*TAEM* 54:264; *TAED* D8026ZCO); see also Doc. 1985 n. 4.

Menlo Park, N.J., July 30 1880.[a]

WP.E

How much money can you pay on the Sanborn Mortgage.[1] If I take it up and advance the 5000. to pay Sanborn I taking the Mtge at 8 per cent. If you [tooke?][b] pay $2000. that will leave me to pay $3000. and on [--][c] an anual[d] interest of $240—

you could then go on and pay dividends of 6 per cent and the surplus I could credit road on the Mtge from time to time[2]

what say you=

Can you give me a detailed statement of Debts road to date, cash on hand   net earnings last six months[3]

T A Edison   Carman

L (copy), NjWOE, DF (*TAEM* 55:354; *TAED* D8040E). Written by William Carman; letterhead of T. A. Edison. [a]"Menlo Park, N.J.," and "1880." preprinted. [b]Canceled. [c]Paper torn. [d]Obscured overwritten text.

1. See Doc. 1766.

2. On 21 July Pitt Edison wrote that having Edison's proxy statement for his stock in the Port Huron Railway "would make me more secure in my position" at a future shareholders meeting. The next week William Wastell endorsed the idea of having Pitt formally represent Edison, promising that he and Pitt would cooperate in "the best interests of the road." In reply Edison asked "how many shares I have got. you have 126. do we not controll   if so let us at the next election put our own people in (Pitt Edison to TAE, 21 July 1880; Wastell to TAE, 28 July 1880; TAE to Wastell, 31 July 1880; all DF [*TAEM* 55:349, 351, 353; *TAED* D8040A, D8040C, D8040D]). However, nothing seems to have happened with this plan or towards paying off the mortgage. John Sanborn reported in December that he had declined Pitt's offer to sell him Edison's shares but had offered to trade stock in a narrow gauge line. The following November Pitt was still trying to secure a controlling interest in the company and suggested to Edison that "the way to catch these fellows is to be ready when they are sick as they have spells" (Sanborn to TAE, 21 Dec. 1880; Pitt Edison to TAE, 20 Nov. 1881; both DF [*TAEM* 55:355, 57:541; *TAED* D8040F, D8114F]).

3. At the end of August Pitt wrote that the railroad was "doing a good business" and promised that he would send reports the next week, but these have not been found. Pitt Edison to TAE, 30 Aug. 1880, DF (*TAEM* 53:541; *TAED* D8015H).

Menlo Park, N.J., [July 31,] 1880.[1a]

Friend Rowland

I think you misunderstood my letter[2]   ~~We~~I will make out the patent papers drawings etc, all that ~~we~~I desire is a brief sketch of connections, description ~~of hol~~ field magnet, shape polar Extremities, Commutator Brush, shape of it, nature of Brush, kind of Base inst was on, whether Connected as a Dynamo or otherwise. please have the field magnet & Every part you can sent us. We have recd the armature all right. After we send you the application to sign please give us a brief of dates & results,[b] [~~all~~?][c] & collect all original sketches and memorandum [~~& me~~?][c] of results & witnesses & keep the same in your possession, as powder for the Enemy—we[d] will give them a warm fight.[3] Yrs

T A Edison

ALS, MdBJ, HAR (*TAED* X100AC). Letterhead of T. A. Edison. [a]"Menlo Park, N.J.," and "1880." preprinted. [b]"& results" interlined above. [c]Canceled. [d]Obscured overwritten text.

1. William Carman dated this letter 31 July when he copied it into Edison's letterbook. Lbk. 6:250 (*TAEM* 80:332; *TAED* LB006250).

2. On 25 July Rowland wrote from Keene Valley "that you request me in your letter to make out an application for a patent but I have not the slightest idea how to do it as I never had anything to do with patents. There is nobody up here to consult on the matter and so I must ask you to advise me on the matter." It is not clear to which letter Rowland was referring but see Doc. 1951 n. 3. Rowland to TAE, 25 July 1880, DF (*TAEM* 55:144; *TAED* D8036ZDL).

3. On 1 August Rowland explained to Daniel Coit Gilman, president of Johns Hopkins University, that he did "not expect to succeed in getting a patent but may at least defeat Siemens and establish my claim to have invented the machine first. If I get the patent I may be able to build my own laboratory." Rowland to Gilman, 1 Aug. 1880, DCG.

In late August Edison wrote to Rowland in Baltimore asking if he had received this letter because "We are ready to go ahead with the patent and are waiting for the necessary papers." Rowland replied from Boston that he would soon gather all the remaining materials in Newark and bring them to Menlo Park, which he did on 6 September. Samuel Mott completed the patent drawings by 11 September. TAE to Rowland, 24 Aug. 1880, Lbk. 6:334 (*TAEM* 80:364; *TAED* LB006334); Rowland to TAE, 30 Aug. 1880, DF (*TAEM* 55:161; *TAED* D8036ZDX); Mott Journal N-80-07-10:116, 127, Lab. (*TAEM* 37:360, 365; *TAED* N117: 58, 63).

Notebook Entry:
Electric Lighting

6" Bamboo No. 1346[1]

8:15	68.5" on Bar[2]	
	318	
	314	53
	317	49
	631	102[3]
	25,080	
	2,710	
	27,790	
	138.9[4]	

8:22	Stopped
8:24	Started[a]
9:17	Edison
9:25	Batch
9:30	Hughes[5]
9:150	Upton
9:18	Böhm
9:16	Martin[6]

Pool closed[b]

Batch bets $1.00 that Hughes does not not win with Hughes Hughes bets that Batch B̶a̶ does not not win 50 cts   Batch bets $.50 that it does not last until 9:35 taken by Hughes   Edison bets $2.00 that lamp does not last until 9:45 P.M. with Hughes[7c]

Martin took 50 cts for his chance[d]

Upton	9:45
Batch	9:50
Hughes	10:15
Edison	10—
Martin	10:20
Hughes	10:25
~~Pool closed~~	Pool closed[e]

Batch bets $1.00 that Hughes does not win

Hughes bets $2. to $1 that it last until 10:30   Batchelor Hughes bets $5.00 to $2.00 that the lamp lasts beyond 10:30[c]

Upton	[10:30?]
Hughes	11:15
Martin	11:25
Batch	10:45
	11:10

```
              11:30
              11:45
Edison        [10:50?]
Hughes        12—
Batch         12:15
Hughes        12:30
Batch         12:45
              1—
Hughes        1-15 $3.50
Pool closedᶠ
```

Batch ~~5.00~~ bets that $1.00 to 5.00 that the lamp wont last until 3 A.M. tomorrow morning   open to 12   Hughes bets ~~that~~ $10 even that the lamp lasts until 2 A.M. Taken Upton Hughes bets $1.00 that Batch does not win the pot and that he wins   Hughes bet $10 that the lamp will not last until 2   Upton betsᵍ $10 to $1.00 that it will not last until 3 A.M. Taken Hughesᶜ

3h ~~340'~~   11h 40'-30 Went
Lasted 3h 28'   208 minutes

TAE

X, NjWOE, Lab., N-80-07-23:125 (*TAEM* 36:903; *TAED* N112:62). Written by Francis Upton; multiply signed. Expressions of time have been standardized for clarity. ᵃText to here written on page ruled for recording lamp test results with horizontal line at top and vertical line at left margin for time; following text regarding bets to "11 h 40'-30 Went" is overstruck. ᵇPool closed" written in right margin. ᶜFollowed by dividing mark. ᵈSentence written in left margin. ᵉ"~~Pool closed~~   Pool closed" written in right margin. ᶠ"Pool closed" written in margin of preceding page; followed by dividing mark. ᵍInterlined above.

1. The following day Upton noted that this lamp had a "good carbon" and a resistance of 141.5 ohms. It was put on in the photometer room downstairs for twenty minutes and gave a light of 48 candles (N-80-07-23:151–53, Lab. [*TAEM* 36:915–16; *TAED* N112:75–76]). About this time a standard method of photometry was established at the Menlo Park laboratory. Each lamp was placed in a Bunsen photometer and measured against a spermaceti candle used for the British standard. The voltage and current required by the lamp were measured at the same time (Howell and Schroeder 1927, 193–95; Jehl 1937–41, 960–62).

2. The following day this was 68". This probably represents the distance of the lamp from the standard candle on the bar of a horizontal photometer (Dibdin 1889, 16–17, 29–30).

3. The purpose of this operation and the one adjacent are unclear.

4. Upton used similar calculations to determine the resistance of other lamps about this time. Those trials were made with a calorimeter and galvanometer, and the smaller of the quantities in the addition appears to represent the weight (in grams) of water. For reasons which are

unclear, Upton divided this sum by 200 to get the resistance in ohms. N-80-07-23:95–123, Lab. (*TAEM* 36:888–902; *TAED* N112:47–61).

5. Charles T. Hughes (ca. 1847–1910) later claimed that he began working at the laboratory on 21 October 1879, the day the first successful carbon-filament lamp was tried. However, his letter of introduction from Horace Eldred is dated 25 October. At the time Hughes was working for the Commercial Telephone Co. of Albany, N.Y., and was being considered for the position of telephone expert in South America. It is not clear when Hughes actually began working at Menlo Park. According to his reminiscences he initially worked on experiments with the electromotograph telephones and also with Patrick Kenney on the autographic telegraph. Hughes's name does not appear in notebooks related to those experiments and it appears that his main role was that of a buyer obtaining materials for the laboratory. Later, because he had railroad experience Edison had him work on the electric railway experiments. He also assisted with experiments on preserving food in a vacuum. After leaving the laboratory he worked as an agent for the Edison Electric Light Co. and subsequently was associated with General Electric. Eldred to TAE, 25 Oct. 1879; Laura Hughes to TAE, 25 June 1910; both DF (*TAEM* 52:216, 195:1092; *TAED* D7939ZAD, D1036); Hughes's testimony, TI 1:398–99 (*TAEM* 11:176; *TAED* TI1029); Hughes's reminiscences, 19 June 1907, Meadowcroft (*TAEM* 227:125; *TAED* MM010C); Jehl 1937–41, 546–47.

6. Martin Force.

7. The same night they also bet on lamp 1348 (N-80-07-23:140 [*TAEM* 36:911; *TAED* N112:70]). Wilson Howell later recalled that

> Life tests of lamps were made in the old days at Menlo Park generally in groups of about 100 lamps arranged on long laboratory tables. The lamps were burned above normal temperature so as to get quick results. These tests generally ran about twenty-four hours, being the occasion for all night work by Mr. Edison and some of his associates. Wagers on the life of individual lamps or upon which one would be the next to fail afforded us mild excitement. [Howell's reminiscences, p. 2, Pioneers Bio.]

–1966–

Notebook Entry: Electric Lighting

[Menlo Park,] Aug 4th 1880

Carbons[a]

On our gas furnace (which works elegantly)—we can make an improvement by making it up of bricks moulded right shape & binding together so as to allow a little shrinkage and expansion.[1a]

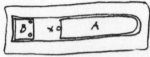

I find that in our mould sometimes the large weight moves by reason of the small weight sticking   this I remedy by putting a pin at X which is made moveable[a]

The sticking of the small weight is due ~~toin~~ some cases to small globules of metal or other substance coming out of the nickel at a high heat[a]

These globules sometimes actually hold up the small weight so that its pressure is not felt on the fibres and consequently the ends are not flat— In such a case when the fibre shrink it pulls the weight by jerks and when finished is corrugated[a]

A good way to obviate part of this is to make a mould like this:—

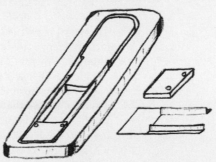

A thin plate of nickel lies under the fibre and has a side turned up— The weight lies on top of this plate with the fibre ends in between— The sides of this plate ~~has~~ also confine the body of the fibre to a smaller chamber thus making less liability to oxidization

Chas Batchelor

X, NjWOE, Lab., N-80-06-02 (*TAEM* 36:414; *TAED* N105:17). Written by Charles Batchelor; multiply signed and dated. [a]Followed by dividing mark.

1. This furnace was designed in connection with a nickel mould so that carbonization took place in an atmosphere of hydrocarbon gases to prevent oxygen from affecting the filaments. Earlier in the year, probably in February, laboratory chemist Otto Moses had designed a carbonizing furnace for this purpose (it is pictured in Friedel and Israel 1986 [p. 164]). It is not clear if the Moses furnace was more generally used at the laboratory for carbonizing filaments. The new mould and furnace were designed by Charles Batchelor in late June and early July. According to Charles Mott the mould was a "nickel box of size sufficient in depth to hold 30 slotted plates, the bottom of one serving as a cover or lid for the one beneath, each entire plate in this way requiring about ⅛ of an inch. The plates are only large enough for one loop at a time and the box or mould filled with the plates is designed to be used in a gas furnace which will be devised expressly for the purpose." The furnace was contracted out and arrived at Menlo Park on 20 July. During the following week the "gas and blast pipes and fixtures" were installed and the furnace was tested on 29 July; the following day another of these was apparently set up in the chemical laboratory. Edison executed a patent for this carbonizing apparatus on 30 July, which subsequently issued as U.S. Patent

248,423 on 18 October 1881. On 3 August the furnace was "connected with blast [and] gas pipes in carbonizing room, the blower did not give sufficient air blast so one of the glass blowers tables and bellows were carried in and the furnace heated. One of the large moulds for 30 carbons was put in and on first trial it was found that the gas had gotten in around the jointed base and formed a sort of tar which held the carbon together. Mr. Batchelor then drilled a small hole in top of mould and used pipe clay around the base or lid and on the second trial with mould thus arranged the carbons came out <u>first class</u>." The following day Charles Mott noted that "The gas furnace has been in use to day, and with five formers in the mould first class carbons were gotten out safely and the furnace gives perfect satisfaction so far"; it continued to work satisfactorily and was soon installed at the lamp factory. Batchelor continued to experiment with and redesign the furnace over the next several days and on 14 August Edison sent Mott to a firm in Woodbridge, New Jersey, to "order Three Gas furnaces for carbonizing to be made with the square holes for vent in the top of the lid instead of the side and to have two one inch bands one around the lid and one around the furnace." These arrived on 6 September, bringing to five the total at the factory. N-80-01-28:19–20; Mott Journal N-80-03-14:270–71, 273, 279; Mott Journal N-80-07-10:26, 38, 49, 54, 58, 62, 64, 69, 81, 115; N-80-06-02:45–71; all Lab. (*TAEM* 33:839–40, 819–20; 37:315, 321, 326, 329, 331, 333–34, 336, 342, 359; 36:416–29; *TAED* N055:10; N053:137–38, 141; N117:13, 19, 24, 27, 29, 31–32, 34, 40, 57; N105:20–33).

–1967–

Draft to W. H. Patton

Menlo Park, N.J.,[a] [August 7, 1880][1]

Dear Sir

Regarding Heat and moisture there will be no difficulty as our bobbins can be thoroughly insulated—[2]

~~Regarding~~ The ~~Reduction~~ Translation[b] of a high speed Dynamo 600 Revolutions per minute to mechanism[c] which would serve to give[c] twelve reciprocations, [~~we cou?~~][d] per minute bothered us[c] at first as we could think of no device except a worm & worm wheel ~~which~~ through which about 25[c] to 35 pc of power is lost, but we now have a device by which the Dynamo may run 600 per minute while the pumping mechanism shall work out 12 & accomplish the result by mechanism which has only the friction due to a couple of ~~shafts~~ spindles so we shall have no difficulty here,[3] ~~will~~

We find that the Dynamo can be placed in a water tight[c] box with a shaft protruding[c] through a stuffing box with a small bilge pump inside worked by[c] the Dynamo to pump out Leakage water.

The[c] are two plans which can be adopted for working the Dynamos   Say at the 3000 feet ~~lamp~~ level you establish a Station the same as on the saurface, 500 [-][d] or 1000 hp can be transferred from the surface to the 3000 foot station and there reproduced working the hoisting machinery as on top and you could from that station start another pump rod thus[c] obviate the necessity of putting any Dynamos at intervals below the 3000 foot level   in other ~~would~~ words[c] you can duplicate the surface station at the 3000 feet level either with a small or large plant. The Dynamo method ~~make~~ is an especially convenient one when increased power is desired: as when[b] you go lower & lower ~~as~~ you can add Dynamos to the pumping & hoisting shafts as required.

The other plan would be to place the hoisting Dynamos & mechanism[f] at the 3000 foot level and put a Dynamo & pump combined every 200 ~~foo~~eet downward one pumping into the tank of the other.

The action of a Dynamo machine is exactly analagous & similar to that of an ~~automatic~~ Engine, with an automatic variable cut-off ~~govern~~

~~Supposing that you desired 70 h power at the 3200 feet level. Now~~ On a base 9 feet long 8 broad is placed an Engine & Dynamo combined as one mechanism   The pitman rod of the engine is connected to the shaft of the Dynamo The bobbin of which acts as a fly wheel   The Engine (Porter Allen) makes 600 Revolutions per minute. When no ~~machine~~ other Dynamo[g] is connected with this combination, the Engine of course cuts off at almost nothing and all the power lost is that due to the friction of the Engine, the Dynamo bobbin, and 45,000 foot lbs in the field magnet and about 80 000 foot lbs in the magnetic change of the iron [~~of?~~][d] during[b] the rotatin~~g~~on of the[h] bobbin, total about 5½ hp. ~~ats~~[i] ~~long as~~ the capacity of the Engine at 120 lbs boiler pressure, cylinder 10 diam 9 long is 120 h.p. ~~horse if no machine~~ hence the loss of 5 ½ h.p. when no work is being done is not ~~so~~ great.

If now a Dynamo is connected[c] to the wires leading from the ~~s~~Steam Dynamo ~~and~~ this Dynamo so connected ~~will~~ [--][d] ~~aborb enough power will start off and attain a speed nearly that of the origin~~ if allowed to do <u>no</u> work[j] will absorb only[b] ~~f~~sufficient power to over come its own friction but if a load is put upon it, conditions[c] are created which draws more current from the steam Dynamo, and the ~~cut off~~ governor & cut off mechanism of the steam Engine alter the point of cut off. The amount of work ~~taken~~ done by the ~~in~~ connected Dynamo [~~Is?~~][d] is shewn exactly on the indicator card of the steam Engine. = By[c] means of a prony brake run by the ~~power~~ connected[b] Dynamo and the taking of Indicator cards in the Engine,[k] The exact loss ~~in turning~~ between the Engine & the work[c] can be ascertained. ~~therefore hence there~~

If 52 pumping Dynamos were connected by cable to ~~such~~ the ~~E~~Steam Dynamo & no water was pumped the indicator card would ~~should~~ show[b] a consumption of power due to the ~~friction of the Dyna~~ ~~Initial loss in converting mass motion into Electricity or molecular motion the~~ friction of the pumping mechanism. If one of the pumping Dynamos should suddenly have to do work The work itself creates conditions which causes power to be drawn from the steam Dynamo & the steam[b] cut off would be at a later period in the stroke, while[c] the other pumping dynamo doing no work would not draw on the Engine although both were on the same wire.

The whole thing in a nutshell[c] is that we turn ~~mass mass~~[b] motion of the steam engine[1] into molecular motion (electricity) and then back into mass motion, (Rotation of Dynamo mechanism) and in so doing we of course lose by the necessary frictions of the translating mechanism, but we have[c] the great advantage that while it might[c] be commercially & mechanically unpractical to give a mass movement to a large wire rope 10 miles long to convey power & reproduce it at the distance,[c] It is perfectly possible[c] commercially & mechanically to ~~stop~~ substitute in place of a movement of the whole mass of a metallic rope 10 miles long a movement of its molecules throughout the ten miles.

The larger the wire or rod the worse it would be for trans~~lat~~mission by moving the whole mass while the larger it is the easier ~~becomes th~~ it becomes to transmit by the ~~molecular m~~ movement of the molecules.

The losses even at comparatively[b] short distances between transmitting[c] a given power by the movement of a wire rope or the stoppage of the rope & the translation of the power into Electricity & passing it through the rope are not widely apart. ~~if anything They are in favor of Electricity~~ ~~By~~ It is possible by[m] molecular transmission to convey[n] a power ~~100~~ several hundred[o] times beyond the breaking strain of the wire conveying it if the same were to be used to convey by its movement over pulleys[c] horozontally.

The Electrical system ~~as~~ developed here is very reliable and dos not require an "Electrician" to work[c] the system, but can be attended[c] to by any ~~Eng~~ Steam Engineer or fireman   our 120 hp Dynamo & Engine combined ~~with~~ will deliver ~~12000~~ hp at the 3000 foot level for 3½ to 3⅝ lbs of coal per hp per hour ~~of~~ for[b] the 100. This amount of coal making up all losses.

A Rough I will give you a rough calculation of egenerating conveying & Reproducing 1000[c] hp. at the 3000 feet level based on actual costs no royalties= [---][d]

Ninty seven thousand dollars.

This estimate does not include[c] boilers as you probably[b] have extra[c] capacity—or freight on apparatus from NYork, or labor of setting up.

Depreciation of Steam[b] Dynamo, never more than 3 p.c. [-][d] on Dynamo at lower level with pump mechanism Depreciation about 2 p.c. The cost will be reduced into nearly exact proportion asif you want less to the horsepower needed less than one thousand h.p.

This price will probably may[b] scare you out of the notion, but if you take Into consideration the fact that you can deliver this power easily & reliably add & take from it from time to time, get each hp for 3½ to 3⅝ lbs per coal or wood equivalent[p] per hour per hp delivered or the small depreciation, etc I think you will find it to be the cheapest & most reliable method that is known

I should like to ask a few questions

What size Pumps, of what construction and how many in station for 200 Ft. lift are used at present at the mine?[q] Their capacity

What improvements are desireable?[q]

How much Water per min. is to be raised 200 ft. high?[q]

What size and kind of Pipe is[b] used?[q]

What Space at Station occupied at present for Pumps and tank or reservoir?[q]

What Space can be given to locate Dynamo–Motor (to move mechanism) to operate Pumps direct or for a series of Pumps below.[q]

Would any special Kind Rotary Pump work to satisfaction?[q]

What are the sizes and specialities of Pump rods used at the Mine, assuming that the Engine above ground operates all Pumps to the 3000 ft. level —[q]

Could you send us any old drawings etc of your mines or any mine so that we can get out of the region of conjecture regarding your methods.[4] [In re locquit?][5r] Yours

T A Edison

ADfS, NjWOE, DF (*TAEM* 54:410; *TAED* D8032V). Letterhead of T. A. Edison [a]"Menlo Park, N.J.," preprinted. [b]Interlined above. [c]Obscured overwritten text. [d]Canceled. [e]"water tight" interlined above. [f]"& mechanism" interlined in right margin. [g]"other Dynamo" interlined above. [h]"of the" interlined above. [i]"s" overwritten on "t." [j]"if . . . work" interlined above. [k]"in the Engine," interlined in right margin. [l]"Engine" interlined in right margin. [m]"It is possible by" interlined above. [n]"to convey" interlined above. [o]"several hundred" interlined above. [p]"or wood equivalent" interlined above. [q]Sentence written by Julius Hornig. [r]Illegible.

1. William Carman noted on this draft that he copied the letter and Edison signed it on 9 August, but the letterbook copy made by Carman and signed by Edison is dated 7 August. TAE to Patton, 7 Aug. 1880, Lbk. 6:283 (*TAEM* 80:348; *TAED* LB006283).

2. This letter is in answer to Patton's 27 July response to Doc. 1957. Patton wrote that the air in the shafts 3,000 feet below the surface was between 100 and 130 degrees Fahrenheit and "highly sprayed with moisture—would this affect the insulation and reduce the effectiveness of machines?" (Edison wrote in the margin next to this paragraph: "no we should take precautions.") He also inquired about four other specific operating conditions: submerging the motor in water; reducing its speed for pumps working at six to twelve strokes per minute; operating under variable loads; and dividing the electric current among motors at several locations. He hoped to use electric motors below 3,000 feet, "conveying the power generated at the surface to the points required in such a way as to loose as little as possible in effectiveness—the power to be reliable and under perfect control— If you can get the 76% you claim and operate the machines under the conditions stated above the system will succeed— should you desire to go further in the matter please answer as soon as convenient." Patton to TAE, with TAE marginalia, 27 July 1880, DF (*TAEM* 54:405; *TAED* D8032U).

3. Edison made two sketches of a "means for getting a slow movement to mechanism" on or about 5 August. In his only references to this apparatus Charles Mott noted that Julius Hornig was working on it on 6 and 9 August (N-80-07-30:18, 52; Mott Journal N-80-07-10:67, 70; both Lab. [*TAEM* 38:471–72; 37:335, 337; *TAED* N135:10–11; N117:33, 35]). Edison filed a provisional British specification on 30 September for "Improvements in Dynamo-electric Machines" which included apparatus to reduce the speed of driven machinery "belts or gearing, in whose use there are inherent defects, such as the slip and stretch of belts, the rattle of gear, and so forth." In this arrangement

The rotary motion of the armature is first converted into an oscillating motion, which is then converted into a continuous rotary motion in the following way:—

Upon the shaft of the armature is a balanced crank pin to which is attached a pitman or driving rod connected to an oscillating frictional pawl mechanism.

Upon the driving shaft is fixed a wheel having a frictional periphery. Loose upon the driving shaft is an arm extending a distance above the rim of the frictional wheel, and then bent over and fashioned into a frame, in which are pivoted two pawls connected together by a frame capable of being shifted, so that only one pawl can take at the time upon the wheel. . . .

The arm carrying the pawls is slotted, and the pitman or driving rod is connected thereto by a pin whose position is adjustable in the slot, so that the leverage may be adjusted and the speed communicated easily varied.

The arrangement described is used in duplicate; that is, two or more driving rods communicate motion from the armature shaft to as many frictional pawls and driving wheels on the driven shaft, the crank pins being so arranged relatively to each other that a continuous motion is imparted to the driven shaft, and mechanism connected therewith. [Brit. Pat. 3,964 (1880), Batchelor (*TAEM* 92:212; *TAED* MBP030)]

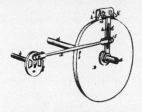

British patent specification drawing of Edison's adjustable pawl mechanism for reducing the speed of the shaft driven by an electric motor.

The final specification illustrated and particularly described the application of this mechanism to pumps.

On 1 October Edison also filed what was evidently a similar U.S. application (Case 249). It was forfeited and only the claims and drawings, as well as a later summary, are extant. Mott Journal 80-07-10:70–71, Lab. (*TAEM* 37:337; *TAED* N117:35); Patent Application Casebook E-2536:172; Serial No. 18,421, 1 Oct. 1880, "Motors" in Abstracts of Edison's Abandoned Applications (1876–1885), p. 2; both PS (*TAEM* 45:716, 8:529; *TAED* PT020172, PT004 [image 3]).

In an undated draft to an unknown correspondent, Edison discussed the general relationship between the dynamo and motor. If the two machines were of the same construction they would turn at nearly the same speed under no load. However,

If the Resistance of the Receiver [motor] is changed it may be made to revolve at 1200 Revolutions while the steam Dynamo is going but 600, and by adding work until it is brought down to a speed of 600 the maximum will be obtained. But the maximum work is not the most economical it is the same as a Steam Engine there is small economy in taking steam to the full stroke & getting the maximum power of the Engine but let it cut off early & we get the most economical work all of which you well know hence; we only work the Receiving Dynamo up to 20 or 25 pc of its capacity. [Undated Notes and Drawings (c.1879–1881), Lab. (*TAEM* 45:132; *TAED* NSUN07:10)]

Edison incorporated this relationship into the 30 September provisional specification, explaining it in terms of the counter electro-motive force developed in the motor.

4. Patton did not address all of these questions but in a subsequent letter he provided some details about the Comstock pumping and hoisting operations and enclosed drawings of a typical shaft, although these have not been found. He expressed particular concern that the hoists should be made to start and stop smoothly, and invited Edison to make a practical demonstration with a single 100 horsepower dynamo. Stockton Griffin informed Patton that Edison had prepared working drawings but preferred not to send them "until his test with the large Dynamo (120 h.p.) is made here which will be in 5 or 6 weeks—from these tests absolute calculations can be made as to cost, while now there is liability of an error of from 3 to 5%" (Patton to TAE, 20 Aug. 1880, DF [*TAEM* 54:416; *TAED* D8032W]; Griffin to Patton, 25 Oct. 1880, Lbk. 6:498 [*TAEM* 80:396; *TAED* LB006498]). Edison referred the matter to Julius Hornig but took no further action until Patton inquired again in

December 1881, telling him that "the Brush Company are becoming very active on this coast—and have made propositions in regard to the same business." Edison explained that he could not act because "The trouble as to the transmission of power by electricity is that the demand is so great for machines that our Directors do not feel inclined to send out any machines simply for an experiment at their expense" (TAE marginalia on Patton to TAE, 1 Nov. 1880 and 5 Dec. 1881, DF [*TAEM* 54: 429, 59:147; *TAED* D8032Z, D8138ZAJ]; TAE to Patton, 29 Dec. 1881, Lbk. 9:484 [*TAEM* 81:177; *TAED* LB009484]).

5. Edison's intended meaning is unknown, and this phrase was omitted from the completed letter.

–1968–

From Grosvenor Lowrey

New York Aug 7th 1880[a]

My dear Edison

I have now returned & am prepared to go on with R.R. matters.[1]

It will be more useful at first to have a meeting with Villard, Fabbri & perhaps Navarro & that will be best managed here— What do you say to Thursday of next week? Fabbri is not here on Wednesday & on Monday & Tuesday the Elevated R.R. arbitration will keep me—[2]

Let me know about the hour

Fabbri asked me this morning if there was any dissatisfaction with them in taking out or[b] rather paying for foreign patents &c— He said he understood from Gouraud that he was going to take out the English R.R. patents— F. thought as they are interested in that portion of them which the Electric Light Co gets here—i.e. the same for England [SA?][3c] that Company gets here—that the taking out of any seperately would possibly lead to disputes & differences for which there is absolutely no occasion, but which being once on foot interfere with successful business— I am sure you will do best to stick to your present relations— If you dont want any of the patents to be considered as coming under your agreement with D.M. & Co.[4] you can specify them & they can be held separate until the matter can be determined— They can of course command any & all facilities & economies that any one can Yrs in Haste

G. P. Lowrey

⟨Will come over Thursday if you will say what hour   I have not made any arrgnts with Col G to take out pats in Engd on RR   I wld not do anything to dissatisfy Wright or Fabbri in any manner for all I cld make out of the pats[5]

I never[b] done anythig as yet of kind that I know of⟩

ALS, NjWOE, Cat. 2174, Scraps. (*TAEM* 89:265; *TAED* SB012AAL). Letterhead of Porter, Lowrey, Soren & Stone. ªPlace from letterhead; "18" preprinted. ᵇObscured overwritten text. ᶜIllegible.

1. In June, just before the scheduled visit to Menlo Park of two Metropolitan Elevated railroad officials (see Doc. 1940 n. 3), Lowrey wrote that he wished to explain to Edison a plan for the electric railroad "which will perhaps put us on velvet (as the vulgar say) in respect to the light as well as the railroad." A short while later he expressed his hope that the unspecified arrangement would "produce a sum of money sufficient to set up an Electric Lighting district in the city of New York" and shortly afterward arranged for José Navarro, a principal investor in the Metropolitan Elevated railroad, to visit Menlo Park. Lowrey to TAE, 9 and 19 June 1880; Zenas Wilber to TAE, 29 June 1880; all DF (*TAEM* 55: 311, 54:56, 55:134; *TAED* D8039H, D8023T, D8036ZDD).

On 5 August the *New York Herald* reported that electrician Stephen Field had received a patent on an electric locomotive intended for San Francisco's streetcar system but which the inventor hoped to test on New York's elevated lines. Field had not yet built a prototype and Edison, in a subsequent *Herald* interview, dismissed the design's chance of success but Field went on to enjoy a successful career designing and building electric traction systems. "Electric Locomotion," *New York Herald*, 5 Aug. 1880; "Field's Electro-Motor," ibid., 6 Aug. 1880; "Electric Inventions," ibid., 10 Aug. 1880; Cat. 1241, items 1514, 1515, and 1519, Batchelor (*TAEM* 94:608, 610; *TAED* MBSB21514X, MBSB21515X, MBSB21519X); *DAB*, s.v. "Field, Stephen Dudley."

2. In July 1880 the New York Elevated and Metropolitan Elevated companies, joint owners of the Manhattan Elevated railroad which operated all of New York's elevated railroad lines, began contentious and ultimately unsuccessful merger negotiations. It is also possible that Lowrey was involved in resolving lawsuits, recently sanctioned by the courts, by property owners for damages caused by the elevated lines. Klein 1986, 283–84.

3. Possibly a reference to Fabbri & Chauncey's interest in Edison's South American electric light patents.

4. Doc. 1649.

5. See Doc. 1979.

Menlo Park 8. 9. 1880[2]

Dear Sir

It is very difficult in the present state of the experiment to give the cost of apparatus but I can probably give you an estimate that will approach closely

Cost in plant for one boat with capacity and equipments for 75 lights[3]

Engine & Electric machine complete

taking steam from your boiler	1800.00
100 Lamps of which 25 are extra	60.00

Sockets   75.00
wires to all parts of boat   200.00
labor running wires, setting engine   50.00
Total plant   2185.00

Depreciation Engine etc 4%   72.00
        "   Lamps   80.00   yearly each
                                                    lamp lasts
                                                    6 month
Inst at 7% say   152.00
                                          304.00

If running 8[a] hours daily 60[a] lights 10 horse power requires 35[a] lbs coal per hour or 280[a] lbs coal for 8[a] hours or 365 days in the year requires 45½ tons at $2.00 (as I understand you use pea and dust coal)

per ton   91.00
The oil and waste may be estimated for the year at   8.00
Depreciation interest etc   304.00
Total cost per year $403.00

In comparison with gas:
Each burner must burn if it gives a light equal to the Electric light 5 feet per hour that is for 8 hours 2400 feet. for 365 days 876 000 feet. this at

75 cents per thousand[4] amounts to   657.00
Inst on tank & appliances costing 216 000. at 7%   112.00
Leakage 15% 131 000 feet @ 75 per M   98.00
                                                         $867.00
Hence cost by Gas lighting   867.00
    "   "   "   Electric "   403.00
In favor of Electric Lighting $464.00[5]

The advantages of the Electric Light are

1 That it is absolutely steady.
2 That nothing can be set on fire by it.
3 That it is not affected by wind.
4 That it gives no trouble.
5 That it gives a pure monochromic light.
6 That it makes no inovation but imitates exactly one gas jet.[6]

In respect to the large lights of several hundred candle power, they are expensive, require a man to attend them, are unsuitable, and as the Carbon is consumed they disintrigate and throw white hot carbon particles around so that pans are necessary to prevent the places where they are used from being set on fire.

In these estimates I have not[a] allowed anything for royalty to the Edison Electric Light Company but this will be reasonable and will not materially affect the price. I do not think it will bring the price of the light equal to that given by 1000 feet of gas above $1.00 per thousand which may be considered reasonable when if City gas were used on the ferries $2.00 per 1000 ft would have to be paid.

Perhaps $150. per year royalty would be required by our Company. this added to the cost would still leave $314. yearly as the saving effected by the Electric Light to say nothing of the lessening in insurance and the advantages over gas[7] Yours Truly[b]

Thomas. A. Edison.

LS (letterpress copy), NjWOE, Lbk. 6:292 (*TAEM* 80:353; *TAED* LB006292). Written by William Carman. Some decimal points added for clarity. [a]Interlined above. [b]"Yours Truly" written by Edison.

1. John S. Shapter was chief engineer for the steamboat service of the Central Railroad of New Jersey, which operated ferries to Manhattan from its Jersey City terminal. Calvin Goddard to TAE, 15 July 1880, DF (*TAEM* 53:783; *TAED* D8020ZFO); Condit 1980, 65–66, 141–42, 368.

2. Edison drafted this letter on or about 26 July (DF [*TAEM* 53:786; *TAED* D8020ZFR (images 1–4)]). Stockton Griffin transcribed it largely unchanged with a date of 26 July and also copied it into Edison's letterbook; Edison subsequently made extensive emendations on the transcription which, with the exception of one paragraph (see note 7), were incorporated into this document (DF [*TAEM* 53:790; *TAED* D8020ZFR (images 5–7)]; Lbk. 6:235 [*TAEM* 80:325; *TAED* LB006235]).

3. On 3 August, Thomas Connery wrote to Edison from the *New York Herald* office that publisher James Gordon Bennett "would like to know if your electric light could be applied successfully for the illumination of his new steam yacht Polynia. If yes, what would be the entire cost of the thing? I wish you would figure it out fully and write me at your earliest convenience." Edison telegraphed the next day that "It can be applied with great success   shall I take the order   It will take six or seven weeks." Connery wired back that he would need to know the expense, to which Edison replied that he assumed Bennett "would like seventy five lights   this requires small engine and magneto machine combined. cannot give exact estimate as Engine builder has not sent price Engine but think it would not exceed twenty six hundred dollars, without he wants it extraordinarily fancy." Connery to TAE, 3 and 5 Aug. 1880; TAE to Connery, 4 and 5 Aug. 1880; all DF (*TAEM* 53:797, 800, 799; *TAED* D8020ZFW, D8020ZFY, D8020ZFX, D8020ZFZ).

4. Presumably the cost of producing gas (cf. Doc. 1707 n. 9).

5. Edison's original 26 July draft (see note 2) presented a difference of

$42.60 "In favor of gas." The advantage attributed here to electricity is based on a series of revisions made by Edison to the copy made by Griffin (see note 2), including reducing depreciation from 5% and interest from 8%, fuel from $4.20 per ton (with the stipulation of pea or dust coal), and charging leakage to gas. The earlier versions also contained a paragraph following this itemization indicating that "if the gas fixtures were added, the tanks and other apparatus upon the boat, breakage of globes, substitution of new burners at the jets etc, this gain would disappear."

6. In his revision of Griffin's 26 July draft copy (see note 2) Edison added two advantages to this list: a one-quarter decrease in insurance costs; and "This light will burn under water."

7. Edison appended this paragraph to Griffin's 26 July draft copy (see note 2). He also dictated a brief paragraph, evidently for insertion elsewhere and later canceled, about the economy to be gained from using existing gas fixtures for the electric light.

–1970–

Patent Application:
Electric Lighting

[Menlo Park, c. August 9, 1880[1]]

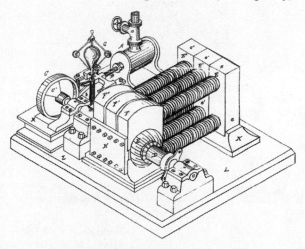

Attest= S. D. Mott   James A. Payne[2]   Inventor= Thos. A. Edison per Dyer & Wilber atty.

In using magneto or dynamo-electric machines, it is very important that the armatures should be rotated at an uniform and constant speed, as any variation therein immediately manifests itself in the current.

As ordinarily used, such machines are connected to the prime motor by intermediate gearing, usually belts, which are liable to slip, causing irregularity in the rotation of the armature or bobbin, every such irregularity effecting the current, causing the irregularity to be repeated and shown, in the operation of whatever translating devices are used in the circuit.[3]

To obviate this it is preferable to connect the prime motor, and the generator directly, that is, supposing the prime motor to be a steam engine, the pitman-rod[4] of the engine is connected directly to the shaft, or axil, of the revolving bobbin, preferably by a crank pin on a disk upon the end of the bobbin shaft, which disk is weighted upon the side opposite to the crank pin, with a weight which counterbalances the weight of the pin and pitman, so that any jar, or irregularity, in passing dead centers, is obviated. This arrangement is especially needed as the engine used should be one of very rapid stroke, not less than 4 to 500 per minute, in order that the bobbin may receive its needed high rate of rotation. The engine should also be what may be called a "self contained engine," that is, provided with a governor and an automatic variable cut-off, which may be so adjusted that upon the speed becoming too great, the cut-off shall be automatically changed to cut off at a less fraction of the stroke, and visa versa.

Of course, as the speed of the engine lessens, the rate of the rotation of the bobbin is lessened, and consequently the electric motive force, or "pressure," of the generated current drops.

If the steam engine and generator be so arranged, there is provided a system of generation, in which, automatically the pressure, or force of the current may be maintained constant.

In manufacturing generators of large capacity, very large cores, and very large castings for polar extensions are required. These very large parts cost more proportionately than small ones, and are much more difficult to handle, the winding of them requiring greater labor and care.

The greatest effect upon the cores is given by the coils nearest to it, but in using very large cores, some of the coils are necessarily somewhat distant from the core.

With several smaller cores, whose aggregate of weight is that of one larger core, a larger surface for the action of coils may be obtained, and a larger amount of wire used, whose average distance from the surface of the cores in either case, is the same.

Generation, of very great capacity, may therefore be profitably constructed of a series, two or more, of coils and cores, or field magnets, each set having its own polar extensions, but one armature or bobbin common to all being used.

By such construction, as before explained, ease and economy of construction are secured, the coils are brought on an average, nearer their cores, and a greater amount of wire may be profitably used.

Moreover, if at any time it is desired to increase the capacity of the generator, it may be done by adding more field magnets to those already in the generator, the only new part required being a proportionately larger bobbin.

As insuring compactness and strength, it is preferable to mount the engine and generator upon one base, on which is secured, upon intermediate supports of a non-magnetic substance, the generator, the non-magnetic supports being necessary to avoid the formation of a magnetic circuit outside of the polar extensions.

In order to give greater rigidity and needed support to the generator, the series of polar extensions are united physically by a brace or union of non-magnetic material, which in effect, makes the opposite poles one structurally but preserves them separate, magnetically.

CLAIMS.

1. A magneto, or dynamo–electric machine, consisting of a series, two or more, of independent field of force magnets, and a single armature, or bobbin, common to them all, substantially as set forth.

2. The combination of a magneto, or dynamo–electric machine, a steam engine connected thereto, by a counterbalanced connection, a governor and variable cut-off, automatically controlled thereby, and an armature or bobbin, serving both as an armature, or bobbin, and as a fly or balance wheel, substantially as set forth.

3. The combination with a common base, of an automatically controlled engine, a magneto, or dynamo–electric machine, and non-magnetic supports placed between the generator and the base, substantially as set forth.

4. The combination with the polar extensions, of independent electro magnets, forming with a bobbin common to them all, a generator, of a non-magnetic plate or brace uniting and supporting the polar extensions, substantially as set forth.

5. The combination of a generator, a high speed steam engine, and a variable cut-off and governor, so that the speed of the engine and the force or pressure of current are automatically regulated, substantially as set forth.

D (typed transcript), NjWOE, Patent Application File Case 237, PS (*TAEM* 45:534; *TAED* PT011AAA).

1. This is the date that Edison filed the application as Case 237; he probably executed it a few days earlier. The application was rejected in October by the Patent Office, which held that each of the claims was anticipated by other American and British patents. After Edison filed an additional affidavit (not found) in October 1882 it was again rejected and

subsequently abandoned. Patent Application Casebook E-2536:140, PS (*TAEM* 45:712; *TAED* PT020140).

2. Unidentified; possibly a clerk for Dyer and Wilber.

3. The text of the application, from the beginning of this paragraph to the claims, was incorporated into Edison's provisional British specification for generators, machines, and motors filed on 30 September. Brit. Pat. 3,964 (1880), Batchelor (*TAEM* 92:212; *TAED* MBP030).

4. That is, the connecting rod.

–1971–

To Corning Glass Works[1]

[Menlo Park,] Aug 10th [1880]

Gents:

Please inform me the time I may expect the glass; my factory is ready for work and needs the glass immediately.[2]

Please answer and you will greatly oblige   Very Truly,

Thos A. Edison   Per C[harles]. B[atchelor].[a]

L (letterpress copy), NjWOE, Lbk. 6:290 (*TAEM* 80:352; *TAED* LB006290). Written by William Carman. [a]Signature lines written by Charles Batchelor.

1. The Corning Glass Works of Corning, N.Y., supplied "tubes & bulbs" to the Edison lamp factory, which "work[ed] them up into Electric Lamps by table blowing with gas" (TAE marginalia on U.S. Census Office to TAE, 27 Nov. 1880, DF [*TAEM* 54:25; *TAED* D8021R]). Discussions with the Corning Glass Works concerning a supply of glass for lamps probably began in June when a representative of "A Glass Manufacturer from Corning N.Y." visited Menlo Park (Mott Journal N-80-03-14:250, Lab. [*TAEM* 33:809; *TAED* N053:127]). According to Howell and Schroeder 1927 (p. 163), the Corning Glass Works supplied bulbs that were "hand made and free blown taken directly from the furnaces. These free blown bulbs were used by the Edison Lamp Works for about twelve years, although other lamp manufacturers adopted moulded bulbs much earlier. The hand made moulded bulbs were uniform in size and shape, while the free blown bulbs varied a great deal and had to be gauged and sized into groups of similar dimensions."

2. There is a note of the same day addressed from Edison to the Corning Glass Works (possibly for a telegram) that reads "How soon will our glass come   need it very bad." There is no extant answer to either inquiry (DF [*TAEM* 53:802; *TAED* D8020ZGB]). On 18 August Charles Mott reported that "Two boxes of tubing from Corning N.Y. glass works received at Factory and upon test by Holzier pronounced very good glass for the purposes of the lamp" (Mott Journal N-80-07-10:87, Lab. [*TAEM* 37:345; *TAED* N117:43]).

[Menlo Park,] Aug 11 [1880] Wednesday eve

Mr. Edison wishes to start 200 lamps next Friday[1a]

Conductors.

The line leading along the turnpike to be wound with three layers of cloth tarred, then wound with marlain.

See that plenty of cloth and marlain are ordered and that men enough are put on the job.[2]

seven days labor[a]
cloth ⟨OK⟩[b]
Marlain ⟨OK⟩[b]
Tar ⟨OK⟩[b]
labor
rubber tape
lines cut off

Machines[a]

3 [4?][c] machines

Present lamp requires 115 volts machine must run 1100 revo. The exciter must run from main shaft.

3[d] machines probably enough since the lamps are so much higher resistance.

Must run 1100 revo.

The three machine in[d] position now will do the business of changed to multiple arc which can be done in a minute.

Meters[a]

Reg meters for

2 for 20 lights Edison Jordan[3]
1 for 30 lights laboratory[a]

Lamps

Glass must have Blowers ⟨OK⟩[b]

Pumps arrangement for bringing up lamps while on.

Pick out lamps. class the lamp post and number lamps accordingly.

5 classes according to distance

Meter[a]

The average E.M.F. is 115 Volts $^{115}/_{165}$ Webers

1 mg. per hour

$^{1}/_{10}$ mg. per hour

Houses[a]

In Mrs. Jordans relay the the wire out of sight.

~~Segredor~~[4]

Herrick[5]   ~~Mills~~[6]

Hammer[7]
Force[8a]

Davis' Hotel[9]   Mrs. J   Cornish's[10]   Kuesie[11]   Edisons[a]
Drop wire used in twice[e] rubber tape solder joints. use lead
safety clutches.

TAE

D, NjWOE, Lab., N-80-07-23:257 (*TAEM* 36:969; *TAED* N112:128).
Written by Francis Upton; document multiply signed. [a]Followed by dividing mark. [b]Marginalia written by Upton. [c]Canceled. [d]Obscured overwritten text. [e]Interlined above.

1. There is no extant record of 200 lamps being started on Friday,
13 August.

2. On 14 August Charles Mott reported "Men wrapping conductors
with Muslin tarred and then wound with Marlin and again tarred."
Mott Journal N-80-07-10:81, Lab. (*TAEM* 37:342; *TAED* N117:40).

3. That is, for Edison's house and for Sarah Jordan's boarding house.

4. John R. Segredor was an out-of-work would-be inventor when he
first contacted Edison in April 1879 to tell him some of his ideas and to
ask for a job at the laboratory, indicating that "if I can get a position that
will pay my board I will be satisfied." Edison apparently offered to let
him come to Menlo Park for a limited period in order to work up some
of his ideas. He was at Menlo Park by the end of May when a time sheet
shows him working on a "telephone order." Another time sheet shows
that he was working on electric light experiments in mid-October. The
next indication of his presence at the laboratory is a note by Edison on a
10 March 1880 letter from a man who was having trouble getting an Edison dynamo, which he had built after reading the *Scientific American* article about it, to work properly. In response Edison told Segredor to
make a diagram of the connections. Soon thereafter he left Menlo Park.
This notebook entry indicates that he had returned to Menlo Park by
sometime in the summer. In early September Edison sent him to Florida
to gather plant samples for possible use as filaments (see Doc. 1984). Segredor to TAE, 22 April and 4 May 1879 and 3 May 1880; Clay McDill
to TAE, 10 Mar. 1880; all DF (*TAEM* 49:888; 49:894; 53:365, 301;
TAED D7913S, D7913V, D8007U, D8006H); Time Sheets, NjWOE.

5. Albert B. Herrick (1860–1938) began working at Menlo Park on
28 August 1879. He was primarily involved with lamp experiments in
the laboratory and the lamp factory. He left Edison's employ in January
1881 to study mechanical engineering and chemistry at the Stevens Institute. After leaving school in 1884 he worked briefly for the Baltimore
& Ohio Railroad Co. and then for two years with the Brush-Swan Electric Co. In 1886 he established the manufacturing business of A. B. Herrick & Co. in New York City. Two years later he became chief electrician
for Bergmann & Co. After the formation of Edison General Electric he
became chief electrician of the Schenectady plant. When the company
merged into General Electric he was made chief electrician of that company's departments but soon left to form the engineering partnership of
Herrick & Burke in New York City. "Albert B. Herrick," Pioneers Bio.

6. Possibly William Mills, who joined the laboratory staff in 1880 but about whom nothing further is known. Jehl 1937–41, 545.

7. William J. Hammer (1858–1934) joined the Menlo Park laboratory staff in December 1879 after working for a year as an assistant to Edward Weston at the Weston Malleable Nickel Co. in Newark. At the laboratory Hammer helped with conducting tests and keeping records of experimental lamps and in 1880 Edison appointed him chief electrician of the lamp factory. In the fall of 1881 he went to London to assist Edward Johnson with the construction of the Holborn Viaduct central station (becoming its chief engineer) and the installation of the Edison exhibit at the Crystal Palace Exposition, where he designed the first electric sign spelling out Edison's name in electric lights. In 1883 he became chief engineer of the German Edison Co. He returned to the United States in 1884 to work for the Edison Electric Light Co. and then served as chief engineer and general manager of the Boston Edison Electric Illuminating Co. He also had charge of several of Edison's exhibits, including the 1889 Paris International Exposition. In 1890 he established himself as an independent construction and consulting engineer in New York City. Hammer did significant research on radium, including developing the first radium-luminous paint. Hammer was also noted for his pioneer work on electric signs and for the historic lamp collection he amassed. "Hammer, William J.," Pioneers Bio.; *ANB*, s.v. "Hammer, William Joseph."

8. Martin Force.

9. A boarding house kept by a Scotchman named Davis near the railroad tracks in Menlo Park. Jehl 1937–41, 38.

10. Probably the Cornish grocery store, which was in the same building as the Menlo Park post office. Marshall c. 1931, 113.

11. The John Kruesi and Charles Batchelor families lived in a divided house across Christie Street from the Edison family home. Jehl 1837–41, 220 (see map).

A map of Menlo Park showing several key buildings. Taken from Francis Jehl's Menlo Park Reminiscences.

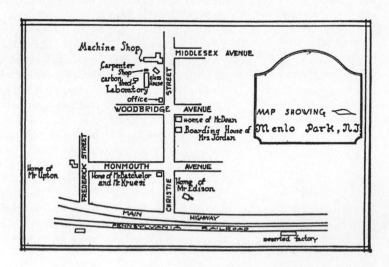

[Menlo Park,] Aug 13th 1880

Carbonization[a]

The cause of the bending over of the loop after it is heated in vacuo ~~we~~I thought was due to insufficient heating in carbonization but after a series of experiments to determine that point we came to the conclusion that whether heated slightly or to a high temperature some of each bent whilst others kept straight.[1]

We then remembered that some bamboo fibres which were 4 in long and of which we mad a great number almost all kept straight   we also remembered that almost all these were put in the clamps edgeways instead of flatways.

This led us to ~~think~~ see[b] that ~~probably~~ the way Bradley cut them from the cane, and the bending them flatways afterwards, would leave the 'pith side' on one face and the 'hard shell' side on the other face   unequal shrinkage of course must occur on two such faces and cause the bending— We now made a mould for carbonizing that would hold the fibre edgeways so:—

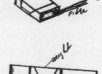

This made moveable weight of three pieces the middle piece pressing out the sides to suit the shrinkage of the fibre in thickness   From this mould we tried some on the pumps and they not only were perfectly flat themselves but did not change[c] their upright position with the most intense heat we could get on them[2]

Chas Batchelor

X, NjWOE, Lab., N-80-06-02:73 (*TAEM* 36:430; *TAED* N105:34). Written by Charles Batchelor; document multiply signed and dated. [a]Followed by dividing mark. [b]Interlined above. [c]Obscured overwritten text.

1. On 4 August Charles Batchelor noted that "We are considerably bothered with the carbons bending over whilst they are on the pumps

We think this must be due to to unequal heating so I have put up 2 fibres in moulds and heated up only one end of each in the preliminary heating." He did not record results of this experiment. Two days later he theorized that if the carbons were "not heated sufficiently in carbonizing when they are put on the pump and heated a great deal higher the outside shrinks more than the inside; and the clamps being held tight it has to bend over to adjust itself." He proposed on 8 August a number of experiments to see if any changes in the heating process would correct the problem, and it is to these trials that Batchelor refers in this document. N-80-06-02:49, 55–65, Lab. (*TAEM* 36:418, 421–26; *TAED* N105: 22, 25–30).

2. In his journal entry of this day Charles Mott noted that "All the plates used in moulds as formers or shapers for carbons are being changed by Andrews so as to hold the widened ends so the thin edge of them will stand towards the faces in the lamp by which means the fiber will in every case be bent so that the inner or pith side of the fiber will come on the inside or outside of the loop instead of on the face as formerly." On 17 August Cornelius Van Cleve "made the commencement at carbonizing in the Factory to day and from first mould of twelve carbons and formers got out eleven in nice order and from second mould of fifteen carbons got 13 out successfully." It is unclear if these were the new moulds but the same day Batchelor noted that these were carbonized "edgeways." Mott Journal N-80-07-10:79, 86; N-80-03-06:207; all Lab. (*TAEM* 37:341, 345; 33:1063; *TAED* N117:39, 43; N057:103).

–1974–

Notebook Entry:
Electric Lighting

[Menlo Park,] Aug 13 1880[1]

Use a Dynamo bobbin, in a shunt from the house circuit as we now use our copper cell. The field magnet should be large & multiple arc'd on the mains having such high Res that it would give current up to near saturation. The Bobbin is connected by say[a] ~~a worm and wm orb wheel with a dash or revolving chain in which is glycerin~~ two or 3 gear wheels with a Revolving churn filled with glycerine, a Regular Counter gives the Revolutions in a month, the Rev being proportionate to the [space?][b]

Voltometer, in place our copper cell=[c]

A Dynamo bobbin in place of Dolbears axial magnet & paper Drums[2]   an arm from Dynamo bobbin has pencil touching paper. A spring prevents bobbin twisting around more than ¾ of a Revolution   work this up with a Counter. If paper record is used a Electro motor governed by a pendulum could be used to give motion to the paper to get the timing

This way

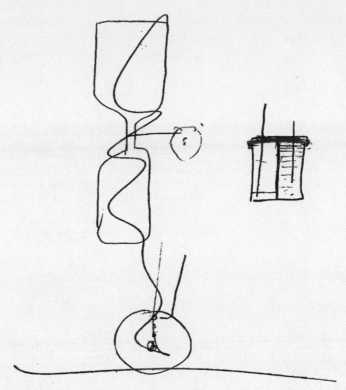

Capaliarity meter[d]

Water meter—with a trip worked by a ¾ turn Dynamo bobbin

TAE

X, NjWOE, Lab., N-80-08-13:1 (*TAEM* 38:291; *TAED* N132:2). [a]Ob-
scured overwritten text. [b]Illegible. [c]Followed by dividing mark. [d]Draw-
ings followed by dividing mark.

1. In his journal entry of this date Charles Mott noted that "Mr. Edison wrote descriptions and made sketches of several new forms of electric meters in Book No. 132 page 1 &c." The pages are those encompassing this notebook entry. Three days later Mott wrote "Mr. Edison
made sketches of several different forms of meters and means connected
to lamp for measuring the electric current which were given to [Samuel]
Mott from which to make Patent Office drawings for caveat." Nothing
further is known of the caveat or the meter designs described in this
notebook entry. An entirely different meter design became the subject of
a patent application executed by Edison on 20 August. In this design the
plates of the meter cell were suspended from a balance arm so that as
copper was deposited on one plate it would tip the balance. As the balance arm moved it would cause a registering apparatus to record a mark
and also reverse the current so that the copper was deposited on the
other plate, thus moving the balance in the other direction. Edison explained in the application that because "the amount of current needed to
cause the deposition of metal enough to cause the tipping is known, and

The balance meter shown in Edison's U.S. Patent 240,678.

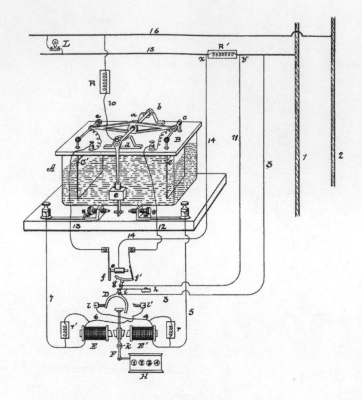

as it is a definite percentage of the entire current, the registration may indicate the total amount of current; or, as the ratio existing between current and feet of gas for illuminating effect has been determined, the registration may indicate the equivalency in light of feet of gas." Mott Journal N-80-07-10:79, 82, 138–39, Lab. (*TAEM* 37:341, 343, 371; *TAED* N117:39, 41, 69); U.S. Pat. 240,678.

2. Edison is probably referring to the current meter patented by Amos Dolbear on 15 June 1880. Dolbear's device consisted of a slender iron core suspended by a spring directly over a cylindrical coil that was hollow along its axis. As electricity passed through the circuit, the coil attracted the rod against the spring's tension into the central hole, in proportion to the strength of the current. A pencil was attached to the rod to record this motion on a strip of paper wound continually from one drum to another by a clockwork mechanism. U.S. Pat. 228,807.

New York Aug 14 [1880]

Dr Sr:—

I am making a small Indicator which will be specially adapted to Telephone use.—

It will be infinitely ahead of the one you have in perfection and looks. When it is finished, I will let you have it in place of the one you now have.[1]

Shall not be at Menlo again for several days— Very resy

C. H. Pond[2]

Am going to Boston Monday   Return friday or Saturday.[a] If convenient please have our contracts fixed by an addition so as to include England & France[3] and send them in by Mackenzie,[4] Monday as I leave at 5 P.M.

ALS, NjWOE, DF (*TAEM* 55:652; *TAED* D8043ZAR). [a]"See Back Side" written at top of page to indicate page turn.

1. Chester Pond was in the process of securing U.S. patents for visual electric signaling devices. On 7 September he received a patent for an annunciator in which the line number of an incoming signal was displayed by a vertical indicator slide. Two weeks later he obtained a patent for a similar device in which a wheel rotated into position to display the appropriate number (U.S. Pat. 231,970 and 232,415). For reasons which are not clear but presumably were related to the impending issue of Pond's patents, Edison requested him to come to Menlo Park "immediately" on 4 August. Pond came the following day (TAE to Pond, 4 Aug. 1880; Pond to TAE, 5 Aug. 1880; both DF [*TAEM* 55:569–70; *TAED* D8042ZBD, D8042ZBE]). On 10 August Edison tested whether the instrument could indicate the subscriber number desired by callers to a switchboard. The next day he executed an agreement by which he took a two-thirds interest in it for telephonic applications only. In return he promised to pay for further experiments and the expense of foreign patents. Edison later gave George Gouraud one-half his own interest in certain European countries and authority to sell the patents in those countries; Pond also drafted a power of attorney to Gouraud for that purpose (Mott Journal N-80-07-10:72, Lab. [*TAEM* 37:338; *TAED* N117:36]; TAE agreement with Chester Pond, 11 Aug. 1880, DF [*TAEM* 55:645; *TAED* D8043ZAP]; TAE agreement with George Gouraud, 24 Aug. 1880; TAE draft agreement with Gouraud, Aug. 1880; Pond draft power of attorney to Gouraud, Aug. 1880; all Miller [*TAEM* 86:267, 271, 274; *TAED* HM800124, HM800125, HM800126]).

According to an undated draft of Edison's agreement with Pond, the "Fire Alarm Indicator" consisted of "disks or wheels bearing numbers or characters in their peripheries, combined with all motive power, releasing devices and stopping devices, so that upon a signal being sent in from a numbered signal box, The disks are released and then stopped so that the number or symbol of the signalling box is displayed by the disks or wheels through a proper aperature" (TAE draft agreement with Pond, Aug. 1880, DF [*TAEM* 55:653; *TAED* D8043ZAS]). Immediately after Edison signed the Pond agreement, John Ott (evidently assisted by James MacKenzie) began work on what Mott described as a

> call box for the Pond indicator. In the Indicator the figures are thrown in position to read, by . . . telegraphic characters and to simplify the operation so that any person unacquainted with telegraphy may signal. the box is an arrangement of three indicator slides on each of which the necessary contacts for the characters representing the different numbers are formed by raised blocks and after each slide is adjusted to the numeral desired the contact is rapidly made by a spring with a platina point which is drawn across the raised characters by a clock work.

A more complete version of the call box was finished and "working very satisfactory" on 31 August (Mott Journal N-80-07-10:77, 96, 106, Lab. [*TAEM* 37:340, 350, 355; *TAED* N117:38, 48, 53]).

2. Chester Pond was the inventor of telegraphic signaling and relay devices and electric clocks, and was president of the Pond Electric Signal Co. in New York. *NCAB* 15:160–61; Letterhead of Pond to TAE, 24 Oct. 1881, DF (*TAEM* 57:243; *TAED* D8104ZEB).

3. These countries were specifically excluded from the 11 August agreement (see note 1) but incorporated into it by an amendment dated 14 August.

4. James MacKenzie had facilitated the 11 August agreement (see note 1), for which Edison assigned him an interest equal to one-twelfth of the whole (TAE agreement with James MacKenzie, 11 Aug. 1880, DF [*TAEM* 55:650; *TAED* D8043ZAQ]). Edison had pulled MacKenzie's young son from the path of a moving freight car in 1862, in gratitude for which MacKenzie taught him railroad telegraphy. MacKenzie subsequently became a manager of the American District Telegraph Co. Although his exact relationship with Edison at this time is not known he seems to have been involved with Pond's indicator for more than a year (*TAEB* 4:23 n. 12; Pond to TAE, 6 Aug. 1878; MacKenzie to Union Fire Alarm Co., 14 May 1879; both DF [*TAEM* 19:538, 49:228; *TAED* D7836ZBS, D7903ZDT]).

–1976–

Draft to Ernest Biedermann

Menlo Park, N.J.,[a] [August 17, 1880?][1]

Write Beiderman

That it is impossible for us to send a man:[2] The light is a vast system a new art & industry[b] & before we can get things ready we shall have to arrange for the manufacture of our machinery We Are[c] establishing the nucleus of[d] a Lamp factory here with a daily capacity of 1200 lamps. This factory alone has cost many thousand dollars & months of labor in inventing automatic apparatus for cheapening the lamps and working people had to be learned to work the apparatus: It would be impossible to establish such a factory in Switzerland ~~fsooner~~ than 2 or 3 years hence everybody must depend for their lamps for a while on our ~~fairst~~ factory=[3]

Regarding the Engine:

Our Dynamo machine is 8 times larger & more powerful than any hentherto built. Is worked direct by an Engine of special construction making <u>600 strokes per minute</u>[4e] no belts or gears are used. To build this class of Engine, & Dynamo over $150 000. have been invested in phila ~~hence you see~~ The inventor of the Engine has spent 20 years in perfecting his Engine to obtain such high speeds with economy & reliability, he has trained workman etc so you see it would be impossible to make the[c]steam Dynamos in Switzerland or in any other country for some time; for the Establishing the lighting of say Geneva all machinery must be sent from here with an expert[c] thoroughly capable to superintending the erection of the plant for lighting 5 or more square miles of a city. We are perfecting our arrangements here for the manufacture of all apparatus on a large scale, and will keep you posted regarding it and when we shall be able to take an order ~~to light~~ for one station, lighting ½ a mile square in Geneva or elsewhere: Yrs—

TAE

ADfS, NjWOE, DF (*TAEM* 54:260; *TAED* D8026ZCM). Letterhead of T. A. Edison. [a]"Menlo Park, N.J.," preprinted. [b]"a new art & industry" interlined above. [c]Obscured overwritten text. [d]"the nucleus of" interlined above. [e]Multiply underlined.

1. Stockton Griffin made this draft the basis of the letter dated 17 August that he sent to Biedermann and copied into Edison's letterbook. TAE to Biedermann, Lbk. 6:311 (*TAEM* 80:361; *TAED* LBoo6311).

2. This letter is a reply to Doc. 1962. Edison also had seen the text of Biedermann's cable directing J. C. Young to "tell Edison important his man knows all dimensions and understand whole manufacture— Cable immediately when he departs." On 22 August Biedermann requested Edison to "cable when your man leaves important to have one full sized generator and 60 ordinary lamps quickly Please send immediately cable when they will leave." Edison answered that "It is impossible and useless to do what you request Have written in full." Young to TAE, 14 Aug. 1880; Biedermann to TAE, 22 Aug. 1880; TAE to Biedermann, 23 Aug. 1880; all DF (*TAEM* 54:264–66; *TAED* D8026ZCO, D8026ZCP, D8026ZCQ).

3. See headnote p. 767.

4. Stockton Griffin copied this in the letter (see note 1) as 60 strokes per minute.

New York, Aug 17 1880[a]

Dear Sir

If your lamp works are in operation we would like to make sketches of them as well as other things for a large cut. If it will suit your convenience I will come over on Thursday or Friday and bring Mr Mead[1] our artist. Please reply by return mail and oblige   Yours truly

Geo M Hopkins[2]

⟨Write Hopkins that we are going to keep factory buz out of the papers. [—][b] we It would educate our enemies, Etc.[3] Say that next[c] Monday[c] hope to have our Loco altered so as to go up the 880 foot grade⟩[4d]

ALS, Cat. 2174, Scraps. (*TAEM* 89:268; *TAED* SB012AAN). Letterhead of *Scientific American*. [a]"New York," and "1880" preprinted. [b]Canceled. [c]Obscured overwritten text. [d]"next . . . grade" later marked with "Note" by William Hammer.

1. Possibly William Mead, identified in Wilson 1880 (1031), as a Brooklyn artist having an office or studio in Manhattan.

2. George Hopkins was editor of *Scientific American*. Hopkins testimony, p. 18, *Böhm v. Edison*, (*TAED* W100DEF018).

3. No article on the lamp factory appeared in *Scientific American* at this time. However, a more general article on electric lighting published in November did observe that "Mr. Edison has reduced the manufacture of his lamps to what may fairly be called a commercial basis, judging by the scale of the manufacture, the simplicity of the processes involved, and the uniformity and cheapness of the resulting product. He has erected a large factory for lamp making, and trained a numerous corps of glass blowers and other workmen for the work in hand." *Sci. Am.* 43: (1880), 336.

4. The reference to 880 feet is not clear but probably refers to the change in elevation over the entire length of track. The traction mechanism for pulling the locomotive up steep grades (see Doc. 1961 nn. 7 and 8) evidently took several weeks to fabricate and install. Charles Mott reported that on 21 August "The new gear of the electric locomotive was run some little time this afternoon to grind down and wear smooth" before one of the generators gave out. When tried again two days later "the creepers did not work entirely satisfactorily but the dificulty was easily detected and can be readily remedied." There was a separate problem with the friction wheel, however, which "would not hold sufficiently to carry the motor up the grade beyond the Pond," but no suggestions were made to correct this. Mott Journal N-80-07-10:68, 70, 81, 93–95, Lab. (*TAEM* 37:336–37, 342, 348–49; *TAED* N117:34–35, 40, 46–47).

-1978-

From George Gouraud

New York [August] 18 [1880] 3:10 PM

Please have griffin stay at Menlo tonight witness documents[1]

G E Gouraud

L (telegram), NjWOE, DF (*TAEM* 56:751; *TAED* D8049ZGM1).
Written by Stockton Griffin on Western Union message form.

1. On this day Edison executed a number of agreements regarding the disposition of foreign patents for telephones, electric light and power, and electric railways. Three of these were with George Newington of London, trustee for the prospective Edison's Foreign Telephone Supply and Maintenance Co., Edison's Foreign Electric Light and Power Co., and Edison's Foreign Electric Railway Construction Co., each of which was to operate in extensive and somewhat overlapping regions of the world outside of Britain and the major Continental nations. These provided for Edison to take a large majority of each company's stock in return for the rights to pertinent patents. Miller (*TAEM* 86:231, 233, 235; *TAED* HM800106, HM800107, HM800108).

The same day Edison executed three contracts giving George Gouraud one-half his profits from the Newington agreements in payment for helping to organize those companies, for which purpose he also granted Gouraud separate powers of attorney. He executed two similar contracts and powers of attorney to Gouraud for telephones in Mexico, and in India and British colonies other than Australia and Canada. He also signed one contract each for electric lighting and railways in portions of Africa and the Caribbean; the accompanying powers of attorney were dated 18 August but not signed until 24 August. On that date he executed an eighth power, for which no accompanying contract was completed, for electric light and power in unspecified British colonies. TAE agreements with Gouraud, 18 Aug. 1880; TAE powers of attorney to Gouraud, 18 and 24 Aug. 1880; all Miller (*TAEM* 86:239, 245, 251, 253, 255, 259, 241, 247, 237, 243, 249, 265, 257, 261, 263; *TAED* HM800110, HM800113, HM800116, HM800117, HM800118, HM800121, HM800111, HM800114, HM800109, HM800112, HM800115, HM800123, HM800119, HM800121A, HM800122); Gouraud to TAE, 24 Aug. 1880, both DF (*TAEM* 55:337–38; *TAED* D8039ZAF, D8039ZAG).

-1979-

From George Gouraud

New York, Aug. 20th, 1880[a]

Dear Edison:

In accordance with our understanding of yesterday I went to Messrs. Drexel Morgan & Co. this morning to arrange the credit for you on account of the electric railway experiments; and when in the course of conversation with Mr. Fabri it became necessary to explain the circumstances,[1] Mr. Fabri suggested that it did not seem right that I should do this, but at the sametime appreciating the importance of your going on with the work it was arranged the credit should be opened not as from me personally but by Messrs. Drexel Morgan & Co. for the account of the proposed "Edison's Electric Railway Construction Co. of the United States."[2] There is, consequently, to your credit in this connection $1,000 subject to your check as the work goes forward.[b] Both Mr. Fabri and Mr. Lowery are as anxious as yourself that no time whatever should be lost in perfecting the organization of the railway, Co.[c] and the general lines have been settled this morning with Mr. Lowery, and at Mr. Lowery's request I go to Tarrytown to spend the night with him, with the view, if possible, of crystallizing[d] the matter so as to enable its being put in hand at once.

I have had a very satisfactory conversation with Mr. Fabri with respect to the British interests in this connection, and shall have more to say to you about this when I next go down to Menlo. Yours sincerely

G. E. Gouraud

LS, NjWOE, DF (*TAEM* 55:335; *TAED* D8039ZAE). Written by Samuel Insull; letterhead of Mercantile Trust Co. [a]"New York," preprinted. [b]"as . . . forward" interlined above. [c]"Co." written in left margin. [d]Obscured overwritten text.

1. See Doc. 1978.
2. The Electric Railway Co. of the United States was not incorporated until 1883. It controlled all of Edison's pertinent United States patents.

–1980–

Notebook Entry:
Electric Lighting

[Menlo Park,] Aug 25th 1880

Lamp Socket[1]

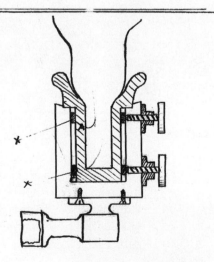

The rings of brass x x to be fastened on to the inside socket A

Chas Batchelor

X, NjWOE, Lab., N-80-06-02:97 (*TAEM* 36:443; *TAED* N105:47). Written by Charles Batchelor.

1. In his journal entry for this day Charles Mott noted that "Mr. Batchelor devised and sketched several forms or styles of sockets for lamps which may be found Book No 105 pge 97 &c." Two other, less detailed, drawings follow in the same book. On 2 September Mott reported that "John Ott completed one of the lamp holders devised by Mr. Batchelor and sketched in Book No. 105 pge 99. The thumb screw or circuit completer in this case admits of turning one way only and the contact is made by the pin connected to the thumb piece, which by a spring is thrown in against a metallic ring fitted to inside of socket. A tooth on the thumb piece drops into a notch in the sleeve of the shaft or pin thus letting the pin come suddenly in contact with the ring to complete the circuit and the tooth and notch are so formed that the thumb piece can be turned one way only and in so doing the circuit is suddenly and completely broken." The next day "Ott finished a somewhat different lamp socket . . . in the former one the inside cup or socket was secured in the outer one by a screw on the bottom, in the one today the inner socket has projecting pin on lower end which finds under a single screw thread or incline thus making the contact for one line and holding the inner cup in position." Page 99 of this notebook is missing. Mott Journal N-80-07-10:99, 109–11, N-80-06-02:101, 103; both Lab. (*TAEM* 37:351, 356–57, 36:444–45; *TAED* N117:49, 54–55, N105:49, 48–49).

–1981–

Charles Clarke to
W. H. Merrick[1]

[Menlo Park,] Aug. 28th [1880]

Dear Sir,

We are going to try a new arrangement of winding the armatures which we think will greatly increase the safety of the machine from crosses in the coils.[2] As the plan if successful may call for a modification of the present general plans as regards space I will say not to go on with the sole plate until you hear from me.[3] This will however in no way affect the progress of the engine construction. I remain: Yours truly,

C. L. Clarke.

ALS (letterpress copy), NjWOE, Lbk. 6:353 (*TAEM* 80:368; *TAED* LB006353).

1. William H. Merrick was president and treasurer of the Southwark Foundry and Machine Co., recently organized to manufacture the Porter-Allen engine in the Philadelphia shop formerly operated by Merrick's family. Letterhead of Merrick to TAE, 28 Oct. 1880, DF (*TAEM* 53:920; *TAED* D8020ZIJ); Porter 1908, 276–96.

2. Armature short circuits had hindered the laboratory generating station for more than a week. Charles Mott noted on 19 August that an

armature "misteriously gave out this morning and no internal examination or test would reveal the point or cause." The problem recurred in the same generator on two successive days and again on 23 August. Mott surmised then that "so many armatures giving out in this one machine would seem to indicate that the magnets in some way might be the cause or partial cause of it but that theory or suggestion does not seem to be entertained by those most familiar with the machines. one fact is observed in this machine that I have never known of in others, and have not been able to get an explanation of it and that is that a shock may be had by contact between the base of the magnet and the commutator brush or holder" (Mott Journal N–80–07–10:89, 91, 93, 95–96, Lab. [*TAEM* 37:346–50; *TAED* N117:44–48]). In fact, Mott reported on 2 September that

> The magnet in which four or five consecutive successive armatures were burned out, was to day unwound and prepared for rewinding. The suggestion at the time, that there might or must be some defect in the magnet that was the cause of destroying so many armatures, made by a non collegiate, was promptly set down on, as presumptious, but now after a couple weeks have passed, the magnet is discovered to have a cross with its base and that rewinding is necessary. [Mott Journal 80–07–10:109, Lab. (*TAEM* 37:356; *TAED* N117:54)]

3. It is not known what changes, if any, may ultimately have been ordered. Charles Clarke was redesigning the armature, and especially the commutator connections, even before the late August generator problems. Mott referred to this project as a "new form of armature adapted for large machines." Clarke worked on this (evidently with some assistance by Charles Batchelor) until mid–September (Mott Journal N–80–07–10:105, N–80–07–27:23–79, N–80–06–02:87, all Lab. [*TAEM* 37:354, 194–222; 36:437; *TAED* N117:52, N116:12–41, N105:41]). At the end of the month Edison filed in London a British provisional patent specification incorporating Clarke's design, which he stated would be easier and faster to repair in case of a short circuit between the coils:

> This is accomplished by making of wire only that portion of the coil which is upon the operative face, the wires of a coil being connected at the ends by metallic plates fastened to an insulating base and insulated from each other. These plates are made so as to project at the proper points above the general surface of the core, at which points the wires are secured to them by soldering, brazing, or clamping devices. At one end each plate is suitably connected to the proper commutator block.
>
> In case of necessity of removal of any coils it is unloosed from its plates at each end without disturbance of other coils. In fact, by such construction it is possible to remove, repair, and replace any coil without taking the armature out of the machine and with but a slight stoppage of the machine.
>
> The plates upon the end are made as concentric almost semicircles insulated from each other with the projections for receiving the wires nearly at right angles thereto.
>
> These plates offer less resistance than the wire, and consequently the internal resistance of the generator is proportionately reduced. [Brit. Pat. 3,964 (1880), Batchelor (*TAEM* 92:212; *TAED* MBP030)]

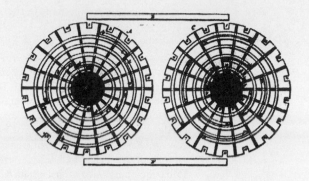

The final specification incorporated the further modifications discussed in Doc. 1992. Edison also executed a U.S. application for this armature design on 11 December (Case 266); it issued in June 1881 as U.S. Pat. 242,898.

–1982–

To U.S. Consul, Pernambuco, Brazil[1]

[Menlo Park,] Aug 30th [1880]

Dear Sir:

Would you be so kind as to inform me if cane or bamboo grow in your vicinity the smaller variety of which is used for fishing poles and canes in the U.S?. If so to what extent and size does it grow?. I should very much like to have the address of a person or firm with whom I might obtain samples and if satisfactory a quantity of the cane? That which I use at present comes from Japan and its high price renders it almost prohibitory.[2] An early reply will be appreciated. Very truly[3]

T. A. Edison   G[riffin]

L (letterpress copy), NjWOE, Lbk. 6:357 (*TAEM* 80:370; *TAED* LB006357). Written by Stockton Griffin.

1. This position had been held by Andrew Cone since 1878. Cone was a publisher connected with the Pennsylvania oil industry who had been U.S. Consul at Para, Brazil, from 1876. He left Pernambuco on 10 September 1880 for a medical leave in the United States, where he died soon thereafter. Henry Lee Atherton, who was in the insurance business in New York City, was named to succeed Cone in November. Cone 1903, 405–407; Cone to Second Assistant Secretary of State, 10 Sept. 1880; Atherton to Second Assistant Secretary of State, 18 Nov. 1880; both in Despatches from the United States Consuls in Pernambuco, 1817–1906, vol. 11, U.S. National Archives and Records Administration microfilm T-344, reel 11; Wilson 1881, 54.

2. It is not clear when or how Edison came to seek especially Japanese bamboo, or from what source he obtained samples. On 18 August Charles Mott observed that "Four bundles of common Bamboo were recd. this morning and cable prepared to send to Japan for a quantity of

large Japanese bamboo." Mott Journal N-80-07-10:87, Lab. (*TAEM* 37:345; *TAED* N117:43).

3. This document precedes in the letterbook nearly identical letters of the same date to U.S. consuls in Puerto Rico; Para, Brazil; Port au Prince, Haiti; Kingston, Jamaica; and Panama. There are no extant replies (Lbk. 6:357–59 [*TAEM* 80:370–72; *TAED* LB006357A, LB006358, LB006358A, LB006359, LB006359A]). Edison had sent a similar inquiry two days earlier to Vesey Butler in Havana stating that he needed "many tons" of the cane or bamboo for which he would "pay a good price if can get the right quality." In October Butler forwarded samples of dried cane and also described other types of fibers found on Cuba, including one "as fine as silk & very white & strong." Butler later promised that "we can supply you with all the cane you require from here as it exists in large quantities in the marshy districts. paying a trifle to the owners of the land we could cut all we require" (TAE to Vesey Butler, 28 Aug. 1880, Lbk. 6:354 [*TAEM* 80:368; *TAED* LB006354]; Butler to TAE, 6 Oct. 1880 and 20 Nov. 1880, both DF [*TAEM* 53:879, 956; *TAED* D8020ZHO, D8020ZIZ1]).

–1983–

From John Harjes

Paris, ~~2d~~ 7th[a] Sept. 1880[b]

Dear Sir

I have just now received copy of the amalgamation act in reference to which I cabled you Augst. 17th and which I signed on that day as cabled to you same evening:[1]

"After endless vexations and legal difficulties have just signed papers amalgamating Edison and Gower under title Société Générale des Téléphones Edison Gower"   I had insisted on the mentioning of your name as the first[c]

It is needless here to refer to the almost endless difficulties and troublesome meetings that have taken place since Mr Bailey left Paris in order to bring about this arrangement. Numerous appointments had been taken by the notaries representing the sundry interests, having at times even the papers ready for signature, all to end in nothing but di~~a~~sappointment owing to legal difficulties by reason of the french law of 1867 in reference to the formation of Societies. I enclose copy of Mr Harrisse's letter of Augst. 9th on the subject which will assist in partly explaining matters.[2]

I can but congratulate you that this amalgamation plan was finally adopted especially on a basis which will ensure to you a greater advantage, the capital as well as the apports[d] (of both companies) ~~as you will notice~~ have been largely reduced yet your portion always remaining fr. 900,000. (in shares.)[3]

The sundry parties interested have endeavoured to obtain a

reduction of your said fr 900,000. in shares to which of course I have not listened.

This amount I shall receive later i.e in shares fully paid up. I have decided that all the shares for the apports and of course your fr. 900,000 included, should be issued in <u>fully paid up</u> shares, the french law on partly paid up shares holds the first owner liable for all additional calls during a period of 2 years which in your case is the more desirable since you have at once to hand over to others a portion of your fr 900,000. I remain, Dear Sir, Yours very truly

John H. Harjes

LS, NjWOE, DF (*TAEM* 56:250; *TAED* D8048ZFV). Letterhead of Drexel, Harjes & Co.; written in an unknown hand. ^aInterlined below. ^b"Paris," and "18" preprinted. ^c"I . . . <u>first</u>" written by Harjes at bottom of page with asterisks to indicate intended location. ^dMultiply underlined.

1. Edison received the cable quoted below at 7:45 p.m. on 18 August. The agreement combining the interests in the French patents was dated 16–17 August; it designated the company to be formed as the Société Générale des Téléphones (Système Edison, Gower & autres). Harjes to TAE, 18 Aug. 1880; Harjes agreement with Constant Rousseau, Georges Lebey, and Jules Lair, 16–17 Aug. 1880; both DF (*TAEM* 56:244, 235; *TAED* D8048ZFM, D8048ZFK1).

2. In the enclosed letter Henry Harrisse explained that French law required that

all the contributions, patents, privileges, and devices, called in French "apports," which form the basis of a society and that are brought into the same for a certain price, shall be approved by a vote of stockholders <u>not interested</u> in the said "apports" either directly or indirectly. Now, the stockholders of the "Edison" [Telephone Co.] were all apparently actual owners of the patents; and, as such precluded from voting. The Banque Franco Egyptienne said that they could not raise a subscription for new stock, out of the general public and thus obtain the required number of disinterested stockholders to vote on the "apports." [Harrisse to Harjes, 9 Aug. 1880, DF (*TAEM* 56:253; *TAED* D8048ZFW)]

3. Joshua Bailey wrote Edison at the beginning of October that the Banque Franco Egyptienne wished to reduce the shares that patent holders would have in the new company, and in any case did not intend to issue shares until the number of subscribers increased. Near the end of the month, however, Harjes reported that the amalgamation was progressing smoothly and would receive final approval by all shareholder in a few days. Bailey to TAE, 1 Oct. 1880; Harjes to TAE, 27 Oct. 1880; both DF (*TAEM* 56:263, 270; *TAED* D8048ZGC, D8048ZGE).

September 7, 1880[a]
Monticello Florida
Sent samples canes and so forth to day   address me here[1]
JRS

[Menlo Park,] Sent Sept 8 1880[b]
J R Segredor[2]
Can you go from Cedar Keys[3] to Cuba after you are through in florida[4]

T A Edison

L and ALS (telegrams), NjWOE, DF (*TAEM* 53:814; *TAED* D8020ZGL, D8020ZGM). First message written by William Carman on letterhead of T. A. Edison; second message written below. [a]Date from document, form altered. [b]Written by William Carman.

1. John Segredor left Menlo Park on 27 August. There is no record of when these samples reached Edison but earlier shipments arrived on 2 and 6 September (Mott Journal N-80-07-10:104, 109, 116, Lab. [*TAEM* 37:354, 356, 360; *TAED* N117:52, 54, 58]). A record of samples sent by Segredor was kept by William Carman in a laboratory notebook which has not been reproduced on microfilm (N-80-06-01, Lab., NjWOE [see *TAED* N151]).

2. Stockton Griffin noted below Edison's reply that the message was sent on 8 September.

3. An island group in the Gulf of Mexico off the coast of Florida's peninsula. *WGD,* s.v. "Cedar Keys."

4. Segredor replied on 8 September, "Yes   will write today full details   information received." The same day he sent a lengthy description of his largely fruitless search through Georgia and Florida. He sent another long report on 15 September, telling Edison that "so far I have seen no cane that near equals the cane you are using in density. All I have come across is full of pores," but the following day he mailed 128 samples collected in Georgia and Florida. Segredor wired again from Cedar Keys at the end of September that he could reach Havana "quicker cheap more certain" from the north. Edison replied "All right" and Segredor returned to Menlo Park on 5 October. Segredor to TAE, 8, 15, 16, and 28 Sept. 1880; TAE to Segredor, 28 Sept. 1880; all DF (*TAEM* 53:815, 828, 835, 868, 869; *TAED* D8020ZGN, D8020ZGR, D8020ZGS, D8020ZHD, D8020ZHE); Mott Journal N-80-07-10:155, Lab. (*TAEM* 37:379; *TAED* N117:77).

[Menlo Park,] Wednesday Sept. 8. [1880]
Insulation[a]
Cable No. 1. Rubber cloth, white, each spiral overlapping about ⅓ width, tarred well with coal tar boiled stiff—put on hot— 54 ohms No. 0. bare wire, No 10—100 ft long. 127.

No. 2. Three thicknesses white rubber cloth each tarred with boiled coal tar. 51 000 ohms fell to 4850—3200—2630—1500.

No. 3. Two thickness white rubber cloth wound in opposite directions. 77 ohms.

No. 4. Three Muslin each covered with boiled coal tar. 120 ohms.

No. 5. 2—Muslin each served with hot Linseed oil. 470 ohms[b] 140 ohms 110 ohms.

No 6. Muslin ~~served~~ wound on cable 1 thickness served with paraphine then black rubber cloth. then Muslin again & paraphine   1200 ohms. 171. 120.[c]

No 7. Three Muslin each served with coal tar treated with quick lime to neutralize acid   120 ohms.[c]

No. 8. bare wire rubbed with cold paraphine. 1 layer black rubber cloth, black rubber cement, then rubber cloth, smoothed down with black hard paraphine   12 500 ohms   7000. 130 ohms.

No. 9. Same as No. 8. except rubber cement was replaced with hot paraffphine. 9,000 ohms 3500—120—ohms

No. 10. bare wire served with white rubber cement, then white rubber cloth, then compound, No. 3. rubber cloth then rubber cement, then again rubber cloth and rubber cement, dusted with percipitated chalk,[b] 299,000 ohms. 1400.

No. 11. Same as ten except compound no 1 is used instead of compound No 3. 79,000 ohms.

No. 12. bare wire wound with Marlin served with compound No. 2, then Muslin soaked in Linseed oil, then compd. no. 2. thinned with cotton seed oil, then muslin, then black[d] rubber cloth, then white rubber cement, then white rubber cloth   26,000 ohms 1000

No. 13. bare wire wound with Marlin, coiled & boiled in compound No. 3. then covered with rubber cloth. Not yet tested

Compound No. 1. is Asphaltum Pine tar cotton seed oil. No. 2. is rosin pine tar cotton seed oil. No. 3. black pitch, pine tar & cotton seed oil.

The above cables and compounds have been made and tried up to the evening of Sept. 8. by Howell[1e]   the numbers[b] and manner of insulating and also the tests of resistance or insulation were furnished by ~~Howell~~ him from a private memorandum book.[2] ~~and~~ the tests by Frances[3] may be found in Book No 137 pge 103.[4] The tests of No. 1. and No. 0. show curious results and speaks badly for insulation or testing. It is rather dificult to reconcile the records with a bare wire testing better than one with an insulation.

Contract. Made two[f] Copied by request of Mr. Wilber, of agreement T. A. Edison with Biederman et. al. of Switzerland, giving them exclusive use of Patents relating to Light and Power System for that country for one half of the net profits, and guarantee from any liability or loss.[5]

Pump Motor. The machine taken to Factory yesterday was set running on the pump but after a time suddenly stopped, and on the full current indicated a cross of in the armature but no burning or unusual heating. the armature however tested all right with a battery current and the action was rather puzzling. They[b] were led to examine into the cause of the upstairs machine always giving out and it was discovered that they were using or trying to use a .14 ohm armature there against a .55 ohm. one down stairs and reasonably supposed that that had caused the dificulty[6]

Reverser. Mott[7] made Pat office drawings of a lamp arranged with a reverser of his own design, which can be turned only one way and each time that the lamp is turned on the current is reversed through the carbon, or sent contrary to the way it had just previously been passing through[8]

AD, NjWOE, Lab. N-80-07-10:119 (*TAEM* 37:361; *TAED* N117:59). Written by Charles Mott. [a]Interlined in left margin. [b]Obscured, over-written text. [c]Followed by check mark. [d]Interlined above. [e]"by Howell" interlined above. [f]"Made two" interlined above.

1. Wilson S. Howell (1855–1943) began working at Menlo Park in December 1879. He worked initially on exhausting lamps. From 1881 to 1885 he worked for the Edison Co. for Isolated Lighting and then organized and managed the New Brunswick, N.J., Edison Electric Illuminating Co. Between 1887 and 1891 he was a consulting engineer. He then took charge of the Lamp Testing Bureau of the Association of Edison Illuminating Companies, which became the Electrical Testing Laboratories.

Shortly before Mott made this journal entry Edison gave Howell the task of developing an insulating compound for the underground cables. In his reminiscences Howell recalled that

> Mr. Edison sent me to his library and instructed me to read up on the subject of insulation, offering the services of Dr. Moses to translate any French or German authorities which I wished to consult. After two weeks search, I came out of the library with a list of materials which we might try. I was given carte blanche to order these materials from McKesson & Robins and, within ten days, I had Dr. Moses' laboratory entirely taken up with small kettles in which I boiled up a variety of insulating compounds. The smoke

and stench drove Dr. Moses out of his laboratory. The results of this stew were used to impregnate cloth strips, which were wound spirally upon No. 10 wires one hundred feet in length. Each experimental cable was coiled into a barrel of salt water and tested continually for leaks. Of course, there were many failures, the partial successes pointing the direction for better trials. These experiments resulted in our adopting refined Trinidad asphaltum boiled in oxidized linseed oil with paraffin and a little beeswax as the insulating compound with which to cover the bare wire cables, which had been previously laid along side of trenches throughout the streets of this little Jersey village. Through the pot in which this compound was boiled, we passed strips of muslin about 2 inches wide. These strips were wound up into balls and wrapped upon the cables. After the man who served these tapes upon the cables had progressed about six feet, he was followed by another man serving another tape in the opposite direction, and he in turn by a third man serving a third tape upon the cable in the direction of the first winding. After the cables were all covered with this compound and buried, the resistance to earth was found to be sufficiently high for a working system. [Howell reminiscences, pp. 4–5, Pioneers Bio.]

2. Not found.

3. Francis Jehl.

4. Francis Jehl's notebook entries regarding these experiments began on 31 August. Mott took the results from Jehl's notes. Four days earlier Mott had noted that Jehl was "making some experiments for insulation of wires underground. the materials tried mixtures of tar paraphine and kindred substances and gums in various mixtures and combinations." Experiments with insulation compounds continued through 21 September; beginning on the 14th they began experimenting with insulated cables No. 15–21. A summary of the compounds used on cables No. 1–7 and 15–20 was recorded by Mott on 18 September. On 22 September Mott recorded that the laboratory had "received during my absence a portable furnace (or stove) and boiler or kettle for boiling and preparing the insulating composition for the lamp lines." The next day "Men commenced uncovering again (third time) the conductor trenches of street lamps. others preparing Muslin with compound of Oxidized Linseed oil, Pine tar, Asphaltum and Paraphine (No. 7.) with which to rewind the cables." On the 24th Mott noted that "Fifteen men and boys were today set at work on insulating the Street Lamp conductors" with this compound (see note 1) using two thicknesses of muslin. N-80-07-16:100–23; Mott Journal N-80-07-10:103, 107, 112, 114, 134–37, 140–41; both Lab. (*TAEM* 38:533–42; 37:353, 355, 358–59, 369–70, 372; *TAED* N137:49–58, N117:51, 53, 56–57, 67–68, 70).

5. Mott apparently made two copies of the agreement between Edison, Ernest Biedermann, A. M. Cherbuliez, and Gaspard Zürlinden granting them exclusive rights to Edison's electric light and power system in Switzerland (see Docs. 1878, 1962, and 1976). Edison executed this agreement the following day. Both copies were probably sent to Switzerland to be signed by the other parties and one of these was returned to Edison. Mott subsequently made a copy of the signed agreement along with notarizations by Stockton Griffin dated 9 September,

the Swiss Consul at Philadelphia dated 11 September, and the American Consul in Geneva dated 30 September; all of these notarizations are found with the original signed agreement. The copy also contains a notarization by the American Consul in Geneva certifying that it is a true copy of the original; the notarization is dated 21 June 1880 but was most likely made in 1881 (DF [*TAEM* 54:267; *TAED* D8026ZCR]; Miller [*TAEM* 86:291; *TAED* HM800129]). On 8 September, Biedermann, Cherbuliez, and Zürlinden signed an agreement with several others in Geneva to form a company to exploit the Edison system in Switzerland. It is unclear if this arrangement was carried out, as a week later Biedermann, Cherbuliez, and Zürlinden transferred their rights to the Compagnie Genevoise l'Industrie du Gaz, which in turn agreed to form the Société Suisse d'Électricité to exploit the Edison system (agreement between Biedermann, Cherbuliez, Zürlinden, et al., 8 Sept. 1880; agreement between Biedermann, Cherbuliez, Zürlinden, and the Compagnie Genevoise l'Industrie du Gaz, 15 Sept. 1880; both Miller [*TAEM* 86: 278, 300; *TAED* HM800128, HM800130]; TAE to Biedermann, Nov. 1880, DF [*TAEM* 54:309; *TAED* D8026ZDX]).

6. The previous day one of "the regular dynamoes being the third machine was taken to Factory to replace the one crossed and to be used as pump motor." On 9 September it was replaced by another motor with a .55 ohm armature, which was used "part of the afternoon supplying the carbon testers and one tier of (about forty) vacuum pumps, and being of the same resistance of the blower motor no preceptable fall in speed or power was observed in the blower machine when the other was put in the circuit." The next day the motor seemed to be working fine. Mott Journal N-80-07-10:117, 12–25, 125, Lab. (*TAEM* 37:360, 363–64; *TAED* N117:58, 61–62).

7. Samuel Dimmock Mott (1852–1930) was Charles's brother. He studied for two years at Lehigh University but left in 1873 to promote his first patent, a guiding attachment for a boy's sled. For the next several years he studied engineering and invention. In spring 1879 he enrolled in mechanical drawing and physics classes at Princeton. He joined the Menlo Park laboratory staff following the end of classes. He primarily worked as a draftsman for Edison. During his life Mott obtained thirty-six patents in a variety of fields. "Mott, Samuel," Pioneers Bio.

8. On 18 August Edison had sketched a device that reversed the direction of the current through the lamp each time it was turned on or off. This was apparently a variation of Francis Jehl's "Circuit changer" to prevent electrical carrying (see Doc. 1944 n. 7). Samuel Mott took Edison's sketches to make drawings for a caveat. Mott presumably made some further variation that he hoped to patent himself. There is no record of either Edison's caveat or Mott's patent application. Mott Journal N-80-07-10:88, Lab. (*TAEM* 37:346; *TAED* N117:44).

London 11th Sept 1880[a]

My Dear Edison

Vacuum Preserving.[1] After opening the fruit I cabled you as follows:—

"Perfectly preserved little water from all tastes slightly alcoholic flavour as if fermented. Try everything my expense"[2] which leaves but little to add except that the results of the experiment so far apparently conclusively shows there is something in it.

What explains the fermentation you will know better than I. I am particularly anxious that you should send me some fish and meat say some grouse & praire chicken some with the feathers on and others with them off   I want this thing followed up and am game to spend $1,000. Set Somebody at it without delay. This I should think could be done without interfering with the greater work going on. At least enough has been accomplished to show the desirability of patenting the process which indeed I think you had already decided to do before I left[3]

I have kept one peach marked the 19th August which has about four drops of water in the bottom which I think is owing to the shaking about it has received [----][b] in consequence of its being rather loose in the glass. This has been the case with all the fruit which probably explains the water. Try packing some in Cotton Batting

I have been on shore too little time to learn anything of sufficient interest to report except that Hubbard has apparently been unable to bring matters to a conclusion till my return. Telephone business here looks all right except the difficulty of getting a good [manager?][c]   Yours truly

Geo. E. Gouraud   I[nsull]

L, NjWOE, DF (*TAEM* 53:214; *TAED* D8004ZFC). Written by Samuel Insull; letterhead of George Gouraud. [a]"London" and "18" preprinted. [b]Canceled. [c]Illegible.

1. Between 16 and 24 August 1880 Edison and Ludwig Böhm undertook a series of experiments to determine whether fruit could be preserved by placing it in a vacuum. In the first attempts "the fruit was injured by the heat but one pear was enclosed with slight injury and after exhaustion was hermetically sealed." When George Gouraud came to the laboratory on 18 August he brought "some choice selected fruits to be sealed in vacuum," which was done the next day. On 24 August Gouraud sent more peaches. These were also sealed and he took them to England. On 16 and 24 August Böhm made drawings of apparatus used for the fruit preservation experiments. Nothing further was done in regard to these experiments until the following April. Mott Journal

N-80-07-10:83, 88, 90, 97; N-80-03-19:190, 217; N-81-04-30:1–77; all Lab. (*TAEM* 37:343,346–47, 350; 34:700, 713; 41:1156–94; *TAED* N117:41, 44–45, 48; N068:94, 107; N306:1–39); TAE to Gouraud, 17 and 22 April 1881, Cable Book (1881–1883), Misc. Lbk. 1:9–10 (*TAEM* 83:876–77; *TAED* LM001009F, LM001010D).

2. This is the text of Gouraud's cable to TAE, 9 Sept. 1880, DF (*TAEM* 53:213; *TAED* D8004ZFB).

3. Edison executed such a patent application on 11 Dec. 1880 and filed it three days later; it issued in October 1881 as U.S. Patent 248,431.

–1987–

British Provisional Patent Specification: Electric Railway

Menlo Park, New Jersey, [c. September 13, 1880][1a]

"IMPROVEMENTS IN THE CONSTRUCTION OF MACHINERY AND APPLIANCES FOR ELECTRO-MAGNETIC RAILROADS, AND IN THE GENERATION, DISTRIBUTION, AND TRANSLATION OF ELECTRICITY FOR WORKING THE SAME."[2]

The object of this Invention is to furnish an economical and reliable system of electro magnetic railways or tramways, which while useful in any locality shall be particularly adapted in regions where the traffic is too light for ordinary steam railways, or where the main bulk of the traffic is limited to certain seasons, or where the difficulties or expense of grading render ordinary steam roads impracticable.[3]

To this end the Invention consists in a complete electro magnetic railway system embracing the generation, distribution, and utilization of electric currents as a motive power, and in the novel device and combination of devices therefore, as more particularly hereinafter described and claimed.

In carrying this Invention into effect the rails of the track are electrically connected, so that each line of rails forms one half of a circuit. The road is divided into sections, where from its length this is desirable, each section forming substantially a small independent railroad. For each section a central station is provided at which is located any suitable motor for giving motion to one or more magnets or dynamo electric machines connected thereto.

At each central station, and also at other points where necessary, a portion of a section is electrically cut off from the remainder, which in connection with a siding there laid enables trains to pass each other. Movable switches or shunts are formed in the ends of the main track adjacent to the sidings. The switches are operated by mechanism set in motion by electro magnetic motors having connection to the central station. From each end of each rail section connections are made to series of electrical switches at the central station, by which means the engineer there in charge is enabled to put the current off or on, or reverse the same or any particular track or switch section, and to operate any particular switch.

For the travelling motor or locomotive an electro magnetic engine is mounted upon a suitable frame supported upon the axles of the driving and other wheels. In order that the circuit from one line of rails to the other be not directly through the wheels and axles, but be through the motor, each car is, so to speak, electrically cut in two by the interposition of insulating material somewhere in its structure, the poles of the motor being connected one to each division. A seemingly preferable method is to form the hub and flange of a wheel of separate metallic parts, uniting them by bolting each to a wooden web, which insulates the two, whereby the body of the car and the axles are insulated from the track.

Contact springs bear against the flanges, or preferably against hubs secured thereto by cross bars or "spiders," whose outer ends are bolted to the flanges. These contact springs are connected to the commutator springs of the motor, one to each respectively, through the reverser and governor, controlled contacts hereinafter spoken of.

As in a central system the motive power is constant, irrespective of the conditions of the trains, it seems requisite that the motive power should be connected directly and inflexibly to the driving wheels, but in some such manner as will enable the force to be gradually applied to or withdrawn therefrom.

Therefore a friction wheel is mounted upon a shaft of the motor, and one upon the main driving axle, the two being disconnected, so that motion is not communicated from one to the other. In a swinging frame pivotted at one end, and provided at the other end with a handle, is mounted another or connecting friction wheel, which on depression shall take upon both the friction wheels before named, and transfer the force from one to the other; of course the amount of this transferring is dependent upon the perfection of the frictional bearing of the intermediate upon the other two friction wheels, and may be varied between the limits of the minimum and maximum frictional contact.

To accomplish the same result a motor pulley and a driving pulley may be connected by a loose belt to be tightened by a swinging pulley belt tightener, or the same result may be accomplished in several other known ways.

As the motive power sufficient to move a load upon a level

with great speed is totally inadequate to move it with the same speed up an incline, and often fails to move the load at all, means (extra amount of steam generative capacity for example) are generally used for furnishing a large excess of power over the amount usually required, adding largely to the dead weight to be carried. In this system however it is proposed to use at all times for a load or train only the amount of power normally required under favorable conditions, providing means by which speed is automatically exchanged for power when necessarily, this to be accomplished by a governor, which upon speed falling upon reaching an incline shall automatically operate to alter the leverage either of belts, friction gear, or clutches, through which the motor acts upon the driving wheel.

As the devices for this purpose are applicable to other than electro motor systems they are not further herein described, but will form the subject matter of a separate application for a Patent.[4]

Upon each engine is located a reversing key through which the circuit passes to the motor, which may be used as a brake in case of emergency, the reversing of the current acting to reverse the direction of the motor, and thereby more rapidly stop it. The operative lever of this reverser is so combined with a spring that it may be held in a central position without any of its contacts infringing on other contacts, and so act also as mere circuit closer or breaker. A centrifugal governor driven from the driving axle is used connected to a series of contacts so as to break the circuit at a number of points simultaneously upon a certain predetermined speed being reached.

Provision is made to dispense with the necessity of much grading, enabling the engine to ascend ordinarily impracticable grades, as follows:—[5]

Upon one or both sides of the engine car a wheel having a grooved face adapted to clasp the head of the rail is mounted in a bearing so combined with a screw or other lifting device that it may be depressed into or elevated from contact with the rail. Upon its axle is fixed a rag or sprocket wheel.

Upon the main driving axle is loosely mounted a friction wheel having attached to it a rag or sprocket pinion. To this loose wheel when necessary motion is communicated from a friction wheel on the motor shaft through an intermediate friction wheel mounted in a swinging frame, as before described. A sprocket chain connects the sprocket wheel on the axle of the grooved wheel and the sprocket pinion. Under ordinary circumstances this friction wheel in the main axle has no motion communicated to it, and the grooved wheel is not in contact with the track; when necessary the grooved wheel is depressed and the intermediate friction wheel so applied as to cause the loose wheel on the main driving axle with its rag or sprocket pinion to be rotated, the motion being communicated to the grooved wheel, which grasping and biting upon the rail head pulls the load up without danger of slipping.

Where the rails are used as conductors of an electrical circuit there is always more or less surface conduction, the amount depending largely upon the dampness or dryness of the adjacent soil ties, and such like.[6]

This surface conduction may be largely reduced, or prevented entirely, in the following manner:—

Between the rail and the tie is placed a piece of felt, paper mache, or other flexible insulating material, preferably so treated as to make it water proof, which piece extends upward on the web on both sides of the rail to the head, forming an insulating shoe.

Between it and the spike is placed a piece of metal of the general configuration of the foot of the rail, upon which the head of the spike takes and bears, so that the insulating material is protected from abrasion or damage by the spike.

Instead of this metal piece a much heavier piece of wood may be used, forming a shoe fastened down by the spike and in turn securing the rail.

The foot and web of the rails are covered with some elastic insulating composition; for example, a rubber paint of which the base is pure linseed oil, the ties, for a space of, say, one half foot to a foot, on each side of the rails being similarly painted.

In using electro motors the best results are obtained when the speed of the rotating armature is maintained uniform and at a very high rate.

In railway motors a large excess of power over that required for a given speed upon the level is provided, in order that even a very much diminished speed may be maintained upon an up grade, the speed of the motor being diminished proportionately.

One object of this Invention is to so arrange a motor in relation to the driven mechanism that the speed of the motor shall always remain unchanged, not being affected by changes in the speed of the driven mechanism, and that power may be exchanged for speed, or *vice versa,* as circumstances may demand, without the speed of the motor being affected. Another object is to furnish a method of propulsion of trains analogous to the action of a quadruped in drawing a load, especially applicable as means for climbing a grade or assisting therein.[7]

To accomplish these objects a thread is mounted upon the shaft of the rotating armature meshing into a worm upon a shaft, at whose opposite end is a bevil gear taking into a bevil gear upon a shaft parallel to the shaft of the engine.

Upon this latter shaft are two gear wheels, one having several times as many teeth as the other, both being loosely mounted upon a shaft, on which and between the two gears is fixed a suitable clutch in order that one or the other may be caused to rotate with the axle upon the clutch being thrown to the extreme limit of its motion, but that when the clutch is in an intermediate position neither shall be locked into the shaft. In order to prevent the clutch being moved too rapidly it may be operated by a screw threaded lever passing through the free end of the lever.

Upon the main driven axle two gear wheels are rigidly fixed, one large and one small, the larger one gearing with the smaller one loose upon the shaft last noted, while the smaller one gears with the larger one loose upon such shaft.

It is evident then that whether speed be converted into power or power into speed will depend on whether motion be communicated from the shaft driven from the armature shaft to the main driven axle through the smaller or through the larger gear thereon.

For use upon grades a device which may be called a creeper is used somewhat as follows:—[8]

Upon the front of the engine is mounted a vertical shaft carrying a worm gearing into the thread upon the armature shaft.

This vertical shaft is mounted in adjustable bearings, so that the worm may be thrown into or out of gear with the thread, as desired.

Upon the lower end of the vertical shaft is a bevil gear meshing into a bevil gear upon a horizontal shaft, to whose ends by crank arms or pins are attached rods, each carrying at its opposite end a box or casing provided with a central wheel which rides upon the rail. In the box or casing, so as to take upon the sides of the rail, are eccentrically pivoted two wheels, one on each side. These side wheels being eccentrically pivoted allow the box to be pushed forward along the side of the rail, but prevent retrograde motion by closing together and grasping the rail. The arms carrying the grippers or creepers are mounted so that they may be let down upon or removed from the track as occasion requires; hence as a rod is reciprocated from the motor through the gearing described it pushes forward during one half revolution the box or casing which slides upon the rail; upon the commencement of the other half of the revolution by the action of the eccentrically pivoted wheels or rollers the box or casing is locked to the rail and the engine is pulled up.

One only being used the action would be a series of pulls and pauses, and if desired one only may be used, taking upon either rail or upon a central rail laid especially for this purpose.

In practice however it is desirable to use at least two, one for each rail, with cranks so arranged relatively to each other that while one is being slid forward the other is holding, so that a continuous motion may be produced. Additional grippers or creepers may also be placed at the rear of the train, so that a continuous pulling and pushing action is produced.

Instead of rollers within the box or casing referred to another form of device may be used, in order to give a larger gripping surface.

Within the box or casing are two bars parallel to the rail, one on each side. These bars are attached to the casing or box by loose toggle joints in such way that upon motion forward of the box or casing the bars recede from the rail, but upon retrograde movement they approach and grip the rail.

Another object of this Invention is to produce a simple and effective electro-magnetic brake adapted for use on any style of rail road vehicle, but more especially intended and adapted for use in the system herein described.[9]

It consists in placing an electro-magnet in such relation to some rotating metallic portion of the running gear of the vehicle to be stopped that the magnetic circuit shall be through such rotating metallic portion, the electro-magnet being furnished with mobile heads, which may move towards and clasp the rotating portion whenever the circuit of the magnet is closed.

Upon the axle and at or near its centre is rigidly fixed a disc of iron, which rotates with the axle and between the polar extremities of an electro-magnet, suitably fastened to or supported from the bottom of the car.

The cores of this electro-magnet are extended beyond the coils, forming a spindle, which is reduced in size when necessary, the ends being screw threaded to receive nuts.

Upon each spindle is placed a block of iron or other magnetic metal, forming a polar extension secured in place by a nut.

The orifices in the blocks into which the spindles pass are elongated, so that the blocks or polar extensions may have a movement to or from the fixed disc upon the axle rotating between them.

The polar extensions are normally held away from the disc by suitable springs of low resistance.

When is is desired to use the brake a circuit from any suitable source of electricity is closed through the coils of the electro-magnets, whereupon the polar extensions mutually attract the disc. It however being fixed while they are movable the attractive force causes them to move to the disc and grasp it between them, causing a retardation or stoppage of its rotation, and so acting through it as effective brake upon the wheels. Upon breakage of the circuit the springs restore the polar extensions to their normal position.

When desired, for the purpose of throwing the brakes off instantly, a momentary reverse current may be thrown into the circuit just after breaking, causing a momentary but instantaneous repulsion from the disc, and assisting the springs in removing the polar extensions. It is evident that instead of one, several sets of such brakes may be applied to each axle when desired.

In this system of electro-magnetic railways where the tracks themselves are used as the conductors, it is desirable to make some provision, guarding against cessation of effect of the current at crossings, switches, frogs, and such like, or other places where it may be desirable to cut out a portion of the track from the circuit.[10]

This may be accomplished by connecting the ends of the tracks in circuit adjacent to the opposite ends of the cut out section by wire or other conductors, so that a circuit is formed around such cut out portion.

As the greatest length of any section necessary to be cut out will never exceed the average length of a train, or even the length of the shortest trains, it is preferred to accomplish the result in the following manner:—

As before described, wheels having their flanges and hubs insulated from each other are used, commutator brushes being used to take the current from hubs electrically connected to the flanges, such commutator brushes being used only with the wheels of the engine.

It is now proposed to use such commutator brushes with the wheels of several of the cars of the train, one of which car should always be the last one in the train.

All the commutator brushes used on either side of the train are connected by a conductor to the appropriate commutator on the engine, and the conductors are so arranged on the cars that they may be readily connected.

By this arrangement the cut out section is electrically bridged over on the train itself, instead of by wires attached directly to the portions of the track in circuit.

Upon roads already built and equipped for steam transport, but where it is desirable to use this system of locomotion, it may be preferable to make the change from one system to the other gradually.

To admit of gradual change, arrangements must be made permitting the use of both systems.

To do this, a third or central rail or conductor is used, electrically connected in stations of suitable length, and thoroughly insulated from the bed. To the cars are attached arms carrying rollers or auxiliary wheels, taking upon the third rail and conveying the current therefrom through the motor upon the train, the ordinary rails being used as the return circuit.

In order to most thoroughly insulate the third or centre rail it may be placed at the ties in a chair of glass or other insulating material, only morticed into the tie or laid on the tie and spiked thereto, or an insulating shield of glass may be interposed between the rail and a metallic chair.

PD, NjWOE, Batchelor, Cat. 1321 (*TAEM* 92:194; *TAED* MBP029).
^aPlace taken from printed docket, form altered.

1. This specification was filed in London on 25 September 1880. Edison would have had to draft it about this date in order to allow a day or two for copying and at least ten more days for it to reach London by the 25th, since the cost of cabling a document this long would have been prohibitive.

2. London patent solicitor Peter Jensen filed this specification as a communication from Edison; he filed the final specification on 25 March 1881, taking British Patent 3,894 (1880), Batchelor (*TAEM* 92:194; *TAED* MBP029) in his own name on Edison's behalf. The document is an omnibus specification combining elements from several U.S. applications that Edison completed during the summer, as discussed below. There is extant a portion of Edison's undated draft of a broad specification for electric railways but it bears no clear relation to this document or his prior U.S. applications. There are also several pages of Edison's sketches and notes dated 3 August pertaining to electric railway applications. Undated draft, DF (*TAEM* 55:243; *TAED* D8036ZGL); N-80-07-30:12–17, Lab. (*TAEM* 38:468–70; *TAED* N135:7–9).

3. This paragraph and the fourteen which follow concerning the operation of the electric locomotive and general operation of the railway were evidently similar to a U.S. application that Edison filed on 3 June 1880 (Case 218). That application was rejected, amended, placed in interference, and eventually abandoned. Its fifteen extant figures and

eighteen extant claims were included in the thirty-six figures and thirty-four claims of the final British specification. Patent Application Casebook E-2536:90–92, PS (*TAEM* 45:707–9; *TAED* PT020090); see also *Edison v. Siemens v. Field*, passim, Lit. (*TAEM* 46:5–114; *TAED* QD001).

4. At the end of September Edison filed a provisional specification covering the governor and transmission arrangements for maintaining a constant armature speed. Brit. Pat. 3,964 (1880), Batchelor (*TAEM* 92:212; *TAED* MBP030).

5. These figures from the final specification show loose wheel **H** secured to sprocket (or rag wheel) **I**, and revolved by friction wheel **Z**. An axle connects sprocket **K** to grooved wheel **L**, which can be pressed against the rail by screw **S**.

In the undated patent draft (see note 2), Edison suggested that severe grades could also be surmounted by "the use of swinging Extensions connected to the [motor] field magnets, which Extensions may be thrown downward so as to nearly touch the track and thus produce a powerful traction." This arrangement was included in the claims of Case 218 (see note 3).

British patent specification drawing of grooved wheel engaged against the rail head, for drawing loads up a steep grade.

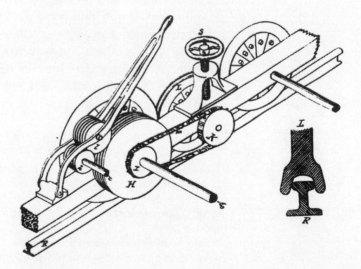

6. This portion of the specification concerning track insulation is similar to a U.S. application that Edison executed on 6 August 1880 (Case 238) which issued as U.S. Pat. 293,433 in February 1884.

7. On 3 July 1880 Edison executed a U.S. application (Case 223) specifically for "grippers" or "creepers" for drawing a load up a steep grade; it did not include the clutch mechanism and reduction gearing described in this document. The application issued in October 1882 as U.S. Pat. 265,778.

8. The final specification showed the entire mechanism for carrying motion from the motor armature to the creeper and, in detail, the locking creeper **S**, within which "and upon each side of the rail, wheels *s* are mounted, eccentrically pivoted as shown, the opening between them at the widest point being just enough more than the width of the rail to permit its passage therethrough. . . . [I]f a body the width of a rail be slid between them in the direction of the arrow . . . it will push them apart,

Illustration and detail from the British patent of "creepers" for drawing the electric locomotive up a steep grade.

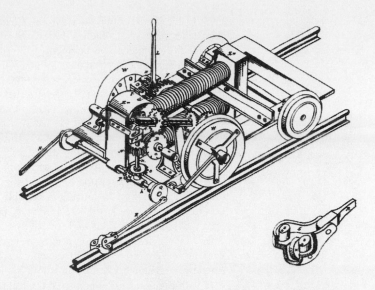

but that if the motion be in the opposite direction it will cause the rollers *s*, *s*, to approach each other, gripping the body between them."

9. This portion of the specification concerning the electromagnetic brake is similar to a U.S. application that Edison executed on 2 July 1880 (Case 222) that issued in 1881 as U.S. Pat. 248,430.

10. This paragraph and the remainder of the specification are similar to a U.S. application that Edison executed on 14 August 1880 (Case 246) that issued two years later as U.S. Pat. 263,132.

–1988–

Notebook Entry: Electric Lighting

[Menlo Park, September 14, 1880[1]]

Boss Socket[2]

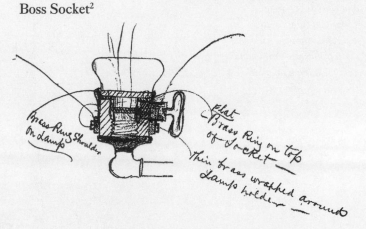

X, NjWOE, Lab. N-80-09-11:18 (*TAEM* 39:254; *TAED* N153:10). Written by Edward Johnson.

1. In his journal entry of this date Charles Mott wrote "Johnson sketched a new socket which is designated as 'Boss socket' in Book No. 153 pge 18 etc which is now being made by one of the new men in Shop." Edward Johnson had arrived back in New York at the end of August and first came to the laboratory on the 31st. On 6 September he was again at the laboratory when Edison "made sketches of a number of styles and ways of running and concealing wires on gas chandeliers   Some devices being fitted to permit of the use of either Electric or Gas Light without interference with each other"; these sketches have not been found. On the 10th "Johnson produced a six light chandelier on which to experiment in artistically fitting with connections, wires &c for substituting the electric lamp and to arrange for use of both or either without as little change or disfiguring as possible" (Mott Journal N-80-07-10:130, 107, 116, 125, Lab. [*TAEM* 37:367, 355, 360, 364; *TAED* N117: 65, 53, 58, 62]; Josiah Reiff to TAE, 28 Aug. 1880, DF [*TAEM* 53:207; *TAED* D8004ZET]). Johnson seems to have headed the development work on sockets and other fixtures, which were subsequently manufactured by Bergmann & Co., in which both Johnson and Edison were partners with Sigmund Bergmann (see their agreement dated April 1880, DF [*TAEM* 57:7; *TAED* D8101C] and Bergmann & Co. catalog, n.d., PPC [*TAEM* 96:185; *TAED* CA002C]). For the subsequent work on sockets and fixtures see the notebook kept by Johnson (N-80-09-11, Lab. [*TAEM* 39:245; *TAED* N153]; see also Mott Journal N-80-07-10:174, 188, 208, 223, 249, 255, 276–77; N-80-10-25:33–35; both Lab. [*TAEM* 37:389, 396, 406, 414, 427, 430, 440–41; 34:168–70; *TAED* N117:87, 94, 104, 112, 125, 128, 138–39, N060:33–35]; Johnson to Mitchell & Vance, 20 Sept. 1880; Charles Batchelor to George Merril, 19 Oct. 1880, Lbk. 6:415, 475 [*TAEM* 80:308, 393; *TAED* LB006415, LB006475]). Edison filed four patent applications on sockets and fixtures in March 1881 and Johnson filed one for a socket in May 1881 and for a chandelier

Edward Johnson's drawing of a chandelier design. 1 and 2 are square tubing, 3 is round tubing, 4 is a flat scalloped ornamental band, and 5 is a hook on which the shade would rest.

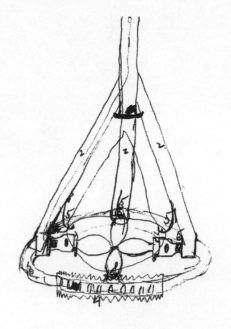

the following September (U.S. Pats. 248,420, 248,424, 251,553, 251,554, 251,596, 256,701).

2. This drawing and the one on the preceding page are the first extant evidence for a screw socket. In his reminiscences Wilson Howell describes its origins:

> Mr. Edison and a number of his helpers were grouped under a hanging lamp in the laboratory one evening. Edison was talking of the introduction of his light into homes, factories, stores, and of the necessity of making the system "fool proof." Pointing to the lamp overhead hanging by its two wires from the open conductors, he explained to us how dangerous such a method of attachment would prove in the hands of the public. He explained that it was necessary to devise an attachment which would be secure, "fool proof," insulated and quick. He described what was needed in such vivid language that one who listened attentively to Mr. Edison was given a picture of the device needed. The "picture" was that of another and older lamp attached to its source of supply—a kerosene lamp "burner" attached to its fount by a screw thread. A sketch was hastily made of this "socket" and, when shown to Mr. Edison was pronounced by him to be exactly the device he (Mr. Edison) had in mind. [Wilson Howell reminiscences, pp. 6–7, Pioneers Bio.]

–1989–

From George Gouraud

London 23rd Sept 1880[a]

Dear Edison,

Edison Telephone Coy of London. There was an Extraordinary meeting of the shareholders held a few days since at the instigation of a number of shareholders who desire to divides the United Company shares and liquidate the Coy to save otherwise unnecessary expense. This thing has worked out just as I expected. The shareholders want the division and do not see any reason to pay Directors simply to speculate with United Coy shares. seeing the temper of the Shareholders the Board ~~the~~ decided, evidently with considerable reluctance, to fall in with the spirit of the movement rather than oppose it and so the meeting came off quite amicably and resulted in unanimous resolutions ~~being~~ instructing the Board to negotiate with the Trustee of the Edison Reversion with a view to arriving at a basis of settlement and requiring the Board to report the result[b] at an Extraordinary General Meeting to be held prior to the Ordinary general meeting for the year.

In Mr Renshaws absence the negotiations have been opened with me and I was yesterday informed that a written proposal would be made me in the course of a day or so.[1]

Numerous speeches were made at the meeting and I was

called upon to give an explanation as to the Trust and so far as I was able the views of the Parties ~~to the~~ interested in the Trust, which I did in such a way as to receive repeated expressions of approval I made it clear to them that the parties interested would readily assent to a division upon fair and equitable terms; that meanwhile the Trust Certificates were as negotiable and their value as ascertainable as the shares of the Coy itself, as the values of the one could not be ascertained without determining the value of the other which they seemed to see. There seemed considerable disapproval of Mr. Bouveries course in abandoning the negotiations at the time when he made the proposition which you declined and the shareholders are evidently determined that they shall not be dropped this time but settled upon some equitable basis They all clearly see that the investment has termed out a very handsome one and I told them that we were prepared to settle on a basis which would leave them an extremely good profit.

In this connection I must tell you that my worst suppositions about White and Bouverie are fully realized and this is conclusively proved by a memorandum drawn up by White and referred by the Board to Counsel for opinion. This memorandum is the most bare faced expression of a most unfair intention on the part of the Board as you will be able to see for yourself when you read the memorandum which I will endeavour to send you by this or next mail.[2] In a word it was no less than ~~an attempt~~ scheme by which the Edison shareholders should sell their shares to the United Telephone Coy thus leaving you to fight the question of your interest with ~~the~~ that Coy instead of settling with the Edison Coy

This however was found to be impracticable and they pretty well see they have got to deal on the square and so they mean to do it. Yours very truly

G E Gouraud

LS, NjWOE, DF (*TAEM* 56:793; *TAED* D8049ZHE). Written by Samuel Insull; letterhead of George Gouraud. a"London" and "18" preprinted. b"the result" interlined above.

1. Renshaw was in New York on this date on his way to Colorado. He visited Menlo Park on 28 September and asked to do so again on his return in November. Gouraud to TAE, 10 Sept. 1880; Renshaw to TAE, 23 Sept. and 15 Nov. 1880; all DF (*TAEM* 56:792, 791, 797; *TAED* D8049ZHD, D8049ZHC, D8049ZHG); Mott Journal N-80-07-10:147, Lab. (*TAEM* 37:375; *TAED* N117:73).

2. Gouraud apparently did not send this memorandum until 20 November. The transmittal letter is somewhat confusing. In it he stated that "I have just come across a document that will satisfy you and John-

son very fully as to the wily ways of EPB[ouverie] and A.W[hite]. as shown by a memorandum of which I send you a copy." The enclosed memorandum called for the stockholders of the Edison Telephone Co. to sell their shares to the United Telephone Co. White asserted that the transaction would be "between the United Company and each individual Edison shareholder and not between the two companies." According to Gouraud, Theodore Waterhouse, the Edison company's attorney, told Bouverie that the proposed transaction would not be legal. Gouraud to White, 20 Nov. 1880, enclosing undated White memorandum, DF (*TAEM* 56:798, 800; *TAED* D8049ZHH, D8049ZHI).

-1990-

Notebook Entry:
Electric Lighting

[New York, September 23, 1880[1]]

At Bergmans[2] 4th story 50 feet. Pressure measured with a pressure U shaped[3] bought at Goodwins[4] was $^{19}/_{10}$ths= =[a]

Upton turned off slightly at meter   gauge $^{14}/_{10}$ very noticable diminuation in size of jet and light, latter very considerable= 5 lights were on we noticed that all jets were set vibrating about 300 per min.— Could hear no sounds.—

Bergman put on graudually light[a] by light (5 on) up to 30 when 5 were on the pressure was $^{18}/_{10}$— when 12 extra added $\frac{15^{1/2}}{10}$ 14 on $^{14}/_{10}$ 20 added $^{13}/_{10}$— 24 added $^{12}/_{10}$ 30 added $^{11}/_{10}$ This made no change in first test except to reduce its size and amount of light   36 Lights on Upton reads Gas Cos Meter=

7:30$^{1/2}$ PM.	Meter reads 1 foot.
7:32:45 sec.	Reads 6[a] feet
7:34:55·	" 11
7:37:5	16 "
7:39:20	21 "
7:~36~41:30	26 "
7:43:45·	31

~First [25?][b]~ ~feet burned~ [in?][b]

25 feet 11 minutes in 36 burners,— or 3.8 per burner per hour   on testing photometrically we find that the average jet with all on ie[c] 36, gave 7$^{1/2}$ candles   size of jet

This was not streaky from high pressure but apparently at the best point for greatest light.

PM—

8:29:30—	our[d] Meter		reads ½ foot	
8:34:	6 Brays spceial Lava tip[5e]	"	1	"
8:43:		"	2	"
8:59:45		"	~~24~~	

—First test 16 candles—

2   15 @ 16[6]

9:17	6
9:30	7½[f]

Pressure ~~in m~~ at meter at 8:38 $^{20}/_{10}$— ditto 9:17—[f]

Size flame

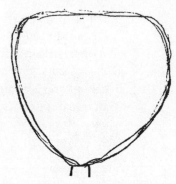

3 high—3 broad   15 @ 16   7 feet per hour
Frances[7] bends a wire to pass over edges flame   thus

1st trial with wire
Brays   Manhattan[8] 15 @ 16 Bray 6 candles   7 feet hour
2nd & more Acurate trial with wire[f]

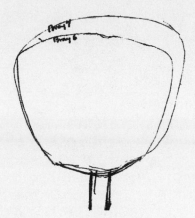

Pressure at 9:32— $^{20}/_{10}$

      10:10   same

      10:40 $^{19}/_{20}{}^{f}$

Measurement with Brays slit union No 7—

      9:40   Meter reads ½

      9:452:30   " 2^{f}

23 candles— 1st test—

      10:01   3.

      10:10 $4.+^{f}$

21 @ 22   2nd test.   candles

      10:26 $6.^{f}$

23 @ 24 3rd test

      10:40— $7½$ just 7 feet

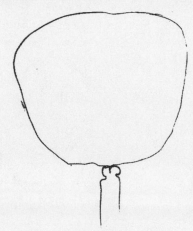

Bray Slit Union No 7   Actual size by wire. 21 @ 23 Candles 7 feet hou[r]

Bergmans [~~lava?~~][b] Brass 2 hold burner[g] burner been working 4 months— Same one that gave 7 candles when 36 lights on & $^{11}/_{10}$ pressure   now have $^{19}/_{10}$ pressure = @ $^{17}/_{10}$

10:48	Zero

8 7 @ 8 Candles–

10:55	½ foot
11:01	1^f

8 @ 9 Candles—pressure $^{17}/_{10}$ 2nd test.

11:07	1½^f

7 @ 8 Candles

11:26	2½ 3

8 @ 9 candles with opal 3 candles (pressure $\frac{16¾}{10}$)

11:39	4 feet
11:48 pressure $^{16}/_{10}$	4$^6/_{10}$ feet

Brays Union Slit No 7. Made it 15 candles. Pressure $^{15}/_{10}$

12:10	Reads ¾ foot
12:18	1½
12:23.	2
12:29½	2½
12:35	3
12:40½	3½^f

2nd test 22 Candles   <u>candles probably low</u>

12:52	4½

3rd test 15 candles.

1:09 AM—	6^a feet^a

Suggs London Argand standard up as high as it will go without smoking   Reaches nearly to top

AM

1:39:30	Reads ½

1st test 18 @ 19 candles

1:45	1	
1:49½	1½	6 feet per hour

we now put it down to what the public would use it. Measures 14 candles

1:55:30—	Zero
2:02	½
2:08:30	1

2:10 am   pressure $^{15}/_{10}{}^f$

~~Lighted 36 jets and standard Sugg Argand fell to~~
Bergmans new Lava tip burner just bought

2:31:30	Reads ½
2:37:15	1

12 @ 13 Candles

2:42:15	1½
2:47:45	2
2:54:30	2½

With Opal globe—2 candles   without 12 @ 13. probably 4 candles as it throws some downwards

X, NjWOE, Lab. N-80-06-16.2:5 (*TAEM* 39:491; *TAED* N160:4). Expressions of time have been standardized for clarity. ªObscured overwritten text. ᵇCanceled. ᶜCircled. ᵈInterlined above. ᵉ"Lava tip" interlined below. ᶠFollowed by dividing mark. ᵍ"Brass 2 hold burner" interlined above.

1. James Russell returned the first completed notebook of his survey of gas usage in the projected New York central station district on 22 September (see Doc. 1995). The next day, according to Charles Mott, "Messrs Edison, Upton and Frances [Jehl] went to New York at 3:20 I hear to visit some gas works." Mott reported that they returned "late in afternoon" of 24 September, having "found the ordinary burner at usual burning, to give about 7 to 7½ candles but from one burner in Bergmann's got 27 candle power. Notes measurements ect Book No. 160 pge 5 etc." Edison presumably made these tests to establish a basis for estimating the amount of light produced by gas lamps under actual operating conditions in New York. Mott Journal N-80-07-10:138, 140–41, Lab. (*TAEM* 37:371–72; *TAED* N117:69–70).

2. S. Bergmann & Co.'s shop at 106–114 Wooster St. in New York. Letterhead of Bergmann to TAE, 30 Aug. 1880, DF (*TAEM* 53:14; *TAED* D8001J1).

3. A standard instrument for measuring commercial gas pressure consisted of a U-shaped glass tube, with one end connected to the gas line and the other open to the air. A small amount of water was placed in the tube so that the difference in the water level in each side provided a visual indication of the pressure in the line. The water's highest level was read against a vertical scale ruled in tenths of an inch. Because gas is less dense than air its pressure increases with height above the source. U.S. Department of Commerce 1914, 140.

4. William W. Goodwin & Co., at 142 Chambers St. in New York, manufactured gas and water meters. Edison wrote the name and address of this firm and another one, as well as references to several gas-related publications, on the first numbered page of this notebook. On the facing page was pasted the published results of routine tests of gas conducted by the New York Department of Public Works on 5 June, showing considerable variation in illuminating power at different times of day and among the various gas companies. A printed advertisement from a New York book seller of items pertaining to the gas industry was pasted to the third page. Wilson's Business Directory of New York 1879, 290; N-80-06-16.2:1–3, Lab. (*TAEM* 39:488–90; *TAED* N160:1–3).

5. This was evidently one of a number of standard burners used in photometry tests of municipal gas. One of the most common of these burners was the Brays No. 7 slit-union (see below). The Bray burners were of the open flame type (as opposed to the annular Argand) and were considered to operate relatively uniformly with different types of gas

and at different pressures. Nothing is known of the "special Lava tip." U.S. Bureau of Standards 1914, 91.

6. As indicated by the above measurement of flame size, these experiments were probably conducted using the techniques of jet photometry. A jet photometer was not properly a photometer, but rather a carefully-calibrated standard burner whose flame height at a given pressure provided an indirect measure of candlepower. This approach invited several sources of error and fell into disuse by the end of the century. However, Edison appears to have used it as a benchmark for the intensity of light consumers would expect to obtain at this pressure from a clean and adjusted burner, that is, sixteen candlepower. The phrase "15 @ 16" presumably represents a photometer measurement of fifteen candles in practice under conditions where sixteen candles would be expected. U.S. Department of Commerce 1914, 104; see also Doc. 1991.

7. Francis Jehl.

8. The Manhattan Gas Light Co., one of four New York gas suppliers, served the area south of 42nd St. as far as Grand and Canal Sts., a few blocks south of the Bergmann shop. Stotz 1938, 40, 45, 53.

–1991–

Notebook Entry:
Electric Lighting

[Menlo Park, c. September 25, 1880[1]]

At present lamps are made which will give 16 candles for ⅛ of a horse power of energy in the shape of current of electricity.

That is 8 lamps may be obtained ~~for~~ each giving 16 candles[a] for one horse power or 33,000 ft. lbs. per minute of available electrical energy.

That is 8 lamps each giving 16 candles if immersed in a calorimeter[a] will show 33,000 ft. lbs per minute given to the water in heat.

The life of these lamps will average 600 hours giving 16 candles, that[a] is if 10 000 lamps are lighted and a record kept of the hours that they gave light the sum total of the burning time of[b] all the lamps would be 6 000 000 hours

At 8 per horse power of 16 candles the light is estimated as costing the company ¼ ct per hour. [~~if?~~][c] that is[d] for 600 hours

$$\begin{array}{r} \text{\sout{\$1.50}} \quad \$1.50 \\ \text{Cost lamp} \quad \underline{.35} \\ \$1.85 \end{array}$$

For 10,000 lamps

$$\begin{array}{lr} \text{For power} & \$15,000 \\ \text{For lamps} & \underline{3,500} \\ & 18,500 \end{array}$$

Received from these at $1.50 per M

$$\begin{array}{r} \$45,000 \\ 18,500 \\ \hline \end{array}$$

Profit $26,500

At 9 per horse power there cane be obtained from the same plant ⅛ more lights

$$8\,|\,10,000$$
$$\underline{1,250^a}$$
11,250 lights[2]
$$\underline{.35}\ \text{lamps}$$
56,250
$$\underline{33,750^a}$$
3,937.50

[For power] 15,000.
[For lamps] $\underline{3,937.50}$
$18,937.50

Receipts   45,000
$$\underline{7,375^3}$$
52,355[a]

$$\begin{array}{r} \$52,355 \\ 18,937.5 \\ \hline \$33,417.5 \\ 26,500 \\ \hline \end{array}$$

Increase Profits $ 6,917.5

$128 \div 8 = 16^e$ candles in eight places for a H.P. of current.[f]
$128 \div 12 = 10.7^e$
10 per H.P. 12 candles each[g]
10 per H.P 12 candles each
$¼ \times 8 = 2$ cts per hour
2 cts per hour   600 hours
$12.00 for 10 lamps horse power
$\underline{3.50}$ for 10 lamps cost
$15.50

$1.55 per lamp cost to company
11 lamps for 1 Horse power
11 lamps for ~~$.02 cts$~~   .02 cts
$$\underline{600}$$
$12.00
$$\underline{3.85^4}$$
$11\,|\,\$15.85$
$1.44

1.55

1.44

11 cts gain per lamp

½   5.5 cts to be added[a] to price

⅕   2.2 cts[g]

Company sells 10 lamps

Tests show that 10 lamps of 12 candles each may be obtained from each ~~Electrical~~[h] horse power of electricity.[a] That is if such a lamp were, when giving 12 candles, immersed in a vessel of water, the water would rise in temperature at a rate indicating that 3300 ft. lbs of energy were added to it every minute in heat.

Such a lamp will last on a average 600 hours

That is, if 10,000 lamps were lighted at irregular or regular intervals and a careful record were kept of the time that each lamp was giving 12 candles of light, and after every lamp had ceased to give light these various burning times were summed up, it would be found that they had burned as[a] an aggregate 10,000 × 600 = 6,000,000 hours

The lamps are considered as burning an equivalent to[i] 12 candle gas, that is each one giving 12 candles may be thought as taking an equivalent of five cubic feet of gas for each hour that they are burned.

This unit is taken as it is found by experience that the devices by which the light may be made so much more ~~effect~~ practically effectual add so much to the apparent light that every[one] is satisfied when told that it is giving a good gas jet. Also that gas cannot be burned in practice so as to give out the maximum of light show by the photometer[5][j] while the electric light ~~must give the consumer as much as does the tester at the laboratory~~ of 12 candles photometric value will give at least 16 to 18 candles of effective light as Compared with gas[k]

X, NjWOE, Lab. N-80-08-00:117–21, 123, 122, 125–33 (*TAEM* 36:695–703; *TAED* N110:58–66). Written by Francis Upton. Every page except 122 canceled by single vertical line. Miscellaneous intermediate mathematical operations omitted; some commas inserted in numerals in calculations for clarity. [a]Obscured overwritten text. [b]"the burning time of" interlined above. [c]Canceled. [d]"that is" interlined above. [e]Form of equation altered. [f]This line to "$1.55 per lamps cost to company" comprises page 122; not canceled. [g]Followed by dividing mark. [h]Interlined above. [i]"an equivalent to" interlined above. [j]"show by the photometer" interlined above by Edison. [k]"of 12 candles photometric value . . . with gas" written by Edison.

1. The reference at the end of this document to a lower candlepower incandescent lamp providing more effective light in comparison to a gas

lamp of greater candlepower suggests that this was probably written soon after the 23–24 September gas candlepower tests at Bergmann & Co. See Doc. 1990.

2. That is, 10,000 plus one-eighth more.

3. It is unclear where Upton gets this figure for additional receipts from using 9 lamps per horsepower.

4. That is, .35 multiplied by 11.

5. That is, according to the indication of a jet photometer; see Doc. 1990 n. 6.

–1992–

Notebook Entry:
Electric Lighting

[Menlo Park,] Sept. 28, 1880.

Experiment on heating of copper rods revolved[a] through the magnetic lines of force.

Temp. of atmosphere at commencement   76.5° F.

Temp. of Iron[a] Plates of Armature   77.5° F.

Temp. of Fields   81° F.

Time of Commencement   7-27 P.M.

No. revo. per m.   136.

After revolving 10 m. in strong field the temp. still remained 77° F.

Started again at 7-40 P.M.

No. revo. per m.   2140[a]

Field strengthened

Temp. of atmosphere remains constant   76.5° F.

After running 30 m. no perceptible heat.[b]

Third test commenced   8-30[a] P.M.

No. revo. per m.   300.

30 minutes duration of exp.

Copper bar went from 80° to 87.[c]

4th test

commenced   9-7 P.M.

No. revo. per m.   300.

Duration 1 h.

Temp. at end   88° F.

Temp. of air   76° F.[1]

Clarke.

X, NjWOE, Lab., N-80-07-27:85 (*TAEM* 37:225; *TAED* N116:44). Written by Charles Clarke; some decimal marks added for clarity. [a]Obscured overwritten text. [b]Followed by dividing mark. [c]Sentence written by Edison; followed by dividing mark.

1. Charles Clarke and Edison continued these tests the next day, when they made three more trials. Clarke recorded that "No appreciable rise in temperature was found to have taken place by revolving rods of copper in the magnetic field therefore no injurious local currents and it was

decided to adopt rods of large size instead of wires across face of armature." On 30 September Clarke began to make sketches and notes for redesigning the large armature, calculating in particular the increased surface area of the induction coils to be gained from substituting trapezoidal bars for insulated round wire. N-80-07-27:89–113, Lab. (*TAEM* 37:227–39; *TAED* N116:46–58).

In the U.S. patent application Edison signed on 11 December (Case 266) he stated that

> The construction of revolving armatures as ordinarily practiced, especially in the case of very large machines, requires the use of a large amount of insulated wire. This is expensive, and besides takes up room and allows of the accumulation of heat, owing to the nonconductor forming the insulation, to remedy which . . . I use rigid naked bars or wires of proper material, which are so disposed about the armature that each is separated from the others, there being between them an insulation partly of mica and partly of air, which suffices in practice for insulation, and in addition allows such access of air to all the active parts of the armature that danger of heating thereof by accumulation is greatly lessened. [U.S. Pat. 242,898]

This was essentially the form of armature built during the fall and winter for the Porter-Allen direct-connected dynamo, though it differed in a few particulars so as to be "more easily and cheaply constructed." Edison also added the bar armature to a final British specification filed on March 30, 1881 (Mott Journal N-80-07-10:281, Lab. [*TAEM* 37:443; *TAED* N117:141]; Brit. Pat. 3,964 [1880], Batchelor [*TAEM* 92:212; *TAED* MBP030]).

–1993–

Article in the North American Review

[Menlo Park, September 1880[1a]]

THE SUCCESS OF THE ELECTRIC LIGHT.[b]

Not a little impatience has been manifested by the public at the seemingly unaccountable tardiness with which the work of introducing the "carbon-loop" electric lamp into general use has hitherto progressed. It is now several months since the announcement was made through the newspapers that all the obstacles in the way of the utilization of the electric light as a convenient and economical substitute for gaslight had been removed: that a method had been invented by which electricity for light or for power could be conveyed to considerable distances economically; that the current could be subdivided almost *ad infinitum;* and that the electric lamp was henceforth to be as manageable for household purposes as a gas-jet. But, so far as the public can see, the project has since that time made no appreciable advance toward realization. The newspapers have reported, on the whole with a very fair degree of accuracy, the results of the experiments made with this system of lighting at Menlo Park; scientific experts have published their judgments, some of them pronouncing this system to be the desiderated practical solution of the problem of electrical lighting which has vexed the minds of physicists since the day when Sir Humphry Davy produced his famous five-inch voltaic arc. Still it must be confessed that hitherto the "weight of scientific opinion" has inclined decidedly toward declaring the system a failure, an impracticability, and based on fallacies. It will not be deemed discourteous if we remind these critics that scientific men of equal eminence pronounced ocean steam-navigation, submarine telegraphy, and duplex telegraphy, impossibilities down to the day when they were demonstrated to be facts. Under the circumstances, it was very natural that the unscientific public should begin to ask whether they had not been imposed upon by the inventor himself, or hoaxed by unscrupulous newspaper reporters.

Now, the fact is, that this system of electrical lighting was from the first all that it was originally claimed to be, namely, a practical solution of the problem of adapting the electric light to domestic uses and of making it an economical substitute for gaslight. The delays which have occurred to defer its general introduction are chargeable, not to any defects since discovered in the original theory of the system or in its practical working, but to the enormous mass of details which have to be mastered before the system can go into operation on a large scale, and on a commercial basis as a rival of the existing system of lighting by gas.

With the lamp and generator which at the time of the first announcement it was proposed to use, the electric light could have been made available for all illuminating purposes as gas is now; the expense would have been considerably less with the electric light; the lamp would have been quite as manageable as a gas-burner. But, fortunately, the unavoidable delay interposed by administrative and economic considerations afforded opportunity for further research and experiment, and the result has been to introduce many essential modifications at both ends of the system—both in the generator and in the lamp; at the same time sundry important changes, all in the direction of economy and simplification, have been made at almost every point in the system, as well as in the details of manufacturing the apparatus.

As for the lamp, it has been completely transformed. The external form of the two types of lamp is identical; the principle of illumination—incandescence of a solid body *in vacuo*—is also the same; but, in the earlier lamp, light was produced by the incandescence of a platinum wire wound on a spool of zircon; in the perfected lamp the source of light is incandescent carbon. Another essential difference between the two is found in the form given to the incandescent body: in the platinum lamp it was coiled compactly on a small spool; in the carbon lamp it is a loop some five inches in total length. This incandescent loop is found in practice to afford a better light for domestic purposes than an incandescent mass of compact form: the shadows it casts are not so sharply defined, their edges being softened.

This loop of carbon is now prepared from the fiber of a cultivated species of bamboo from Japan.[2] A thread of this material, after undergoing a certain chemical process, is bent into the required shape, and then reduced to carbon. The resulting carbon loop is of a remarkably homogeneous structure, and possessed of a high degree of tenacity, so that it can withstand, without breaking, all the concussions it is likely to be subjected to in household use.

The perfected lamp consists of an oval bulb of glass about five inches in height, pointed at one end, and with a short stem three quarters of an inch in diameter at the other. Two wires of platinum enter the bulb through this stem, supporting the loop or ∩ -shaped thread of carbon, which is about two inches in height. The stem is hermetically sealed after the introduction of the carbon loop. At its pointed end the bulb terminates in an open tube through which the air in the bulb is exhausted by means of a mercury–pump till not over one millionth part remains; the tube is then closed. The outer extremities of the two platinum wires are connected with the wires of an electric circuit, and at the base of the lamp is a screw by which the circuit is made or broken at pleasure. When the circuit is made, the resistance offered to the passage of the electric current by the carbon causes the loop to acquire a high temperature and to become incandescent; but, as this takes place in a vacuum, the carbon is not consumed. The "life" of a carbon loop through which a current is passed continuously varies from seven hundred and fifty to nine hundred hours. With an intermitted current, the loop has an equal duration of life; and, as the average time an artificial light is used is five hours per day, it follows that one lamp will last about six months. Each lamp costs about fifty cents, and when one fails another may easily be substituted for it.

The light is designed to serve precisely the same purposes

in domestic use as gaslights. It requires no shade, no screen of ground glass, to modify its intensity, but can be gazed at without dazzling the eyes. The amount of light is equal to that given by the gas–jets in common use; but the light is steadier, and consequently less trying to the eyes. It is also a purer light than gas, being white, while gaslight is yellow. Further, the electric lamp does not vitiate the surrounding atmosphere by consuming its oxygen, as gaslights do, and discharging into it the products of combustion. The heat emitted by the lamp is found to be only one fifteenth of that emitted by a gaslight of equal illuminating power: the glass bulb remains cool enough to be handled. Of course, there are here no poisonous or inflammable gases to escape, and the danger of fire is reduced to *nil,* with a consequent reduction of the rate of insurance. Again, this light, unlike gas, is always of uniform quality. A sort of meter registers exactly the amount of electricity consumed in each house. Finally, not to enumerate all the advantages which this system possesses over gas–lighting, the lamp can be manipulated even by the most inexperienced domestic servant; nor can the most careless person do injury to himself, to others, or to property, through not understanding its mechanism.

Another important modification of the system, introduced since the latest authorized account of the light was published, is the substitution of dynamo–machines for magneto–machines in the stations from which the electricity is to be supplied to the several districts of a city. Here, again, the change is entirely in the direction of simplicity and economy. Where before it was proposed to furnish a station with one hundred magneto–machines with a multiplicity of belts and shafting, we now make ten dynamos of 120–horse power, each worked directly by a 120–horse–power engine. We thus do away with a very considerable loss of power, and at the same time the outlay for machinery is very much lessened.[3]

With these and other modifications of the system, which need not be particularized here, it may be safely affirmed that the limit of economy, simplicity, and practicability has been reached. The time for experiment has passed; any further improvements to be made in the system must be suggested by its performance when put to the test of actual use on a large scale.

To the question which is so often asked, When will a public demonstration of the working of this system be made? we would reply that such a demonstration will in all probability be made at Menlo Park within two months from this date. The time which has elapsed since the preliminary demonstration of last January has been by no means a season of inaction for the promoters of this enterprise. There is a vast gulf between the most successful laboratory experiment possible and the actualization of the results of that experiment in a commercial sense. A prodigious amount of work was necessitated by the establishment of factories for producing the lamps, the generators, and the other essential parts of the system in large quantities, so as to be able to supply the first demand. We were about to enter a field that was practically unexplored, and, even on a preliminary survey, problems of the most complex kind arose on every side. These had to be solved before the first step could be taken toward the actual introduction of the light into our cities as a substitute for gas. The practical engineer and the man of business can best appreciate the difficulties that had to be overcome. Like difficulties have in the past retarded the general introduction of nearly all the great mechanical and chemical inventions. Years intervened between the discovery of photography and the taking of the first photograph; the steam–engine, the steamboat, the locomotive–engine, did not come till long years after the discovery of their scientific principles; the same is true of the telegraph.

But preparations are being actively made for placing this system of electric lighting within reach of the people in all the great centers of population throughout the United States. To this end, cities are being mapped and divided into districts, each to be supplied with electricity from a central station; estimates are being made of the exact cost of plant in the different cities; contracts are being negotiated for the manufacture on a large scale of engines, dynamos, lamps, wire, and all the other supplies needed for the practical introduction of the system throughout the country; men are being trained to put up the plant of central stations, to run the machines, and to execute all the details of the introduction and working of the system.

A very important question is that of the cost of this light. The price of the electric light will, of course, be determined by the capitalists who invest their money in it as a business venture, but it will of necessity be low as compared with gaslight, though it will vary according to the original cost of plant, the demand in any given locality, and other conditions. It is not at present the intention of the company controlling the patents on this system to supply the light directly to consumers. The company will erect the first station in New York City, and will themselves conduct that station; but the other stations in New York, as well as in the other cities throughout the United States, will be managed by local companies, who will pay a royalty to the Electric Light Company for the right to use the system.

So much can be safely affirmed, that this light can be sold at a price which will make competition on the part of the gaslight companies impossible: 1. Because the total investment in plant to develop a given quantity of light is much less; 2. Because the depreciation of plant is much less; 3. Because the cost for labor employed is very much less than in gas–works; 4. Because the electric–light companies will not have to make any dead investment in large areas of real estate; it is not even necessary to erect buildings specially to serve as stations, for the ordinary buildings, such as are used for different branches of manufacture, will serve the purpose, and may be hired on rental; 5. Because the companies can sell electricity for two uses—for light at night, and for power in the daytime. It has been ascertained by experiment that power can be supplied through this system from twenty–five–horse power down to $\frac{1}{100}$ of a horse–power on the same mains that supply the light, and that elevators, printing–presses, sewing–machines, fans, pumps, etc., can be run by electricity from a central station far more economically than by any other means. A canvass of the city of New York has shown that the demand for small powers, in private dwellings and minor industrial establishments, will give occupation to the central stations in the lower part of the city for ten hours daily.[4] This power can be supplied at such a profit to the companies as to more than cover the expense of running the stations for six hours longer in producing electric light. It is evident, therefore, that, in a competition with gas, the electric light possesses an enormous advantage.

THOMAS A. EDISON.[5]

PD, *North American Review*, 131 (1880): 295–300. In Edison, T.A.—Articles (D-80-007), DF (*TAEM* 53:380; *TAED* D8007ZAF1). [a]Place and date not that of publication. [b]Followed by dividing mark.

1. There are no extant drafts of this article, which appeared in the *North American Review* for October. That issue was available by 17 September, when Sherburne Eaton wrote Edison that he had seen the published article and thought it "excellent—most excellent—both in tone, matter & style." Eaton sent copies to George Gouraud and Charles Porter. Eaton to TAE, 17 Sept. 1880, DF (*TAEM* 54:72; *TAED* D8023ZAF).

2. It is not clear exactly what type of bamboo Edison had available to him at this time. By the end of the year he was using a species cultivated in Kyoto known as Madake, which remained the standard for his commercial lamps until 1893 or 1894. Howell and Schroeder 1927, 76–77; Israel 1998, 202–3; see also Doc. 2002.

3. The most recent "authorized" account was probably that pub-

lished in the 21 December 1879 *New York Herald* (see Doc. 1868 n. 3). This paragraph conflates two changes in Edison's central station design: substitution of a few large generators directly connected to the steam engine in place of many small ones connected by shafts and belts, and the exclusive use of self-exciting dynamos instead of separately-excited magneto machines. Although there is no reason why Edison could not have built and operated very large magneto machines, once he began to conceptualize the commercial generator and steam engine as two integrated parts of a single self-contained machine (see Doc. 1936) he was obliged to adopt both changes simultaneously.

4. See Doc. 1995.

5. The article was evidently written by Francis Upton, not the first instance in which Edison delegated the task of composing something published in his name (see Doc. 1283). It evolved from an effort by the *North American Review* in 1878 to prepare and attribute to Edison an article about electric lighting generally based on information supplied by him; that project was assigned to Upton (see Doc. 1588). The magazine made a few inquiries about it in early 1879 and then let the matter drop until reports of the impending Menlo Park exhibition prompted editor John Barron to contact Edison again in December. In notes for a reply to Barron, Edison promised that "Mr Upton can prepare the article for him so it will be ready before public exhibition." Barron wrote almost two weeks later that he would be "very glad" for this arrangement but Edison indicated that he would "have to see Mr Upton about preparing it as I have said nothing about it to him yet." Then in January 1880, when the editor expressed disappointment "in receiving from Mr. Upton yesterday a postal card in which he stated that he could not promise your article by the 1st of Feby next," Edison instructed Stockton Griffin to "write him nice letter say that we are absolutely driven to death with work & find it impossible to do anything." There is no further correspondence about the article until its publication. North American Review to TAE, 27 Feb. 1879, 24 Mar. 1879, TAE marginalia on North American Review to TAE, 4 and 15 Dec. 1879, and 23 Jan. 1880, all DF (*TAEM* 49:136, 688, 747–48; 53:349; *TAED* D7903ZBH, D7906ZAA, D7906ZBR, D7906ZBT, D8007H).

During the fall and early winter Edison found himself in the unfamiliar position of having to depend upon events and circumstances beyond his immediate control. The small-scale Menlo Park central station and distribution system he had hoped to demonstrate in the summer was still not ready. The initial failure of the insulated underground conductors accounted for some of the delay, but more critical now was the fact that construction of the Porter-Allen engine for the large dynamo was months behind schedule. In late October, Edison was forced to slow down work on the dynamo and he protested to the engine builders that "Every little delay is embarrassing to us at this time and we cannot wait longer" but the engine was not delivered until January.[1] In the meantime, he began investigating other forms of high-speed engines which might be suited to the dynamo. The lamp factory had its own problems. Some, like contaminated mercury and the refractory circulating pump, were readily addressed. More complex was the task of assembling and training a new workforce. As Charles Batchelor noted in regard to one job seeker, Edison's experience with highly skilled glassblowers accustomed to making scientific or clinical instruments was "anything but satisfactory for manufacturing work. We find it a great deal easier to break in new men on our work than try to get experimental men to do manufacturing work."[2] Most of the tasks, Edison told a reporter, were "so entirely new that I have had to teach everybody how to do the work, and that takes time."[3] Wholly out of his control was a drought followed by an early freeze, making it difficult to get enough water for the steam engines which provided power for the laboratory, machine shop, and factory. As he began to plan the New York central station district, Edison was forced to await the approval of city authorities to lay conductors beneath the streets and sidewalks. Lastly, he testified for five days in November in the telephone interference cases and for another day in a quadruplex telegraph interference. Sometime in November, Francis Upton began to determine the conductors needed for a small electric lighting station in a house being built by financier William Henry Vanderbilt in New York. Edward Johnson, who was placed in charge of the installation, wrote that this plan pleased Edison "because it sort of fills up the long delay—& acts as a sort of sop to the capital."[4] Edison asked Johnson to offer similar systems for all new houses being constructed on 5th Avenue.

Edison continued to improve the electric lamp. In late October, he resumed experiments, with Francis Jehl, on depositing volatile hydrocarbons onto the carbon lamp filaments; Edward Acheson continued related work in late November. At the end of that month Edison and William Hammer began two weeks of experiments trying to understand and prevent "electrical carrying" in the lamps. Though not successful, this research was partly responsible for Edison's decision in early December to replace all vacuum pumps in the factory with a simpler and more compact design. With the factory operating—albeit nowhere near the rate of 1,200 lamps per day he envisioned—Edison had a steady supply of lamps for testing. He and his assistants continued to compare bamboo, bast, other fibers, and even some cardboard filaments. Already largely convinced of bamboo's superiority, he had John Segredor continue his search for supplies of the cane in Cuba, where Segredor died of yellow fever. He dispatched William Moore to Japan in October and in November engaged a geologist and experienced explorer, John Branner, to search in Brazil. On 1 December, Charles Batchelor ordered all subsequent lamps to be made with bamboo filaments of specific dimensions, secured to the lead-in wires with platinum clamps. Two weeks later John Lawson began successfully electroplating a metal coating onto the filament ends to prevent breakage. Soon after Albert Herrick extended this idea to try to join "the wires and carbons direct by plating them together," eliminating the time-consuming and costly use of screw clamps.[5] The technique had been proven successful by the end of December, and though not yet ready for adoption in the factory it was subsequently used in Edison's lamps for years. By late November Edison had full confidence in the superiority of his lamp to all others, whether gas or electric. He was therefore thoroughly rankled, both professionally and personally, by newspaper reports in which his friend George Barker of the University of Pennsylvania praised the incandescent light of Hiram Maxim, whom Edison contended had stolen the design from him. Henry Morton of the Stevens Institute also endorsed the Maxim lamp enthusiastically.

As usual, Edison directed work simultaneously on other components of the electric light and power system. During October, Francis Upton made comprehensive cost calculations for the construction and operation of a New York central station. The specific requirements for that station became clearer as James Russell returned data from his canvass of lighting and power use in the proposed service district and Upton and William Hammer tabulated results from each block. Combining these figures with geographical information provided by fire insurance maps, Upton and Hammer began calculating the amount of copper needed to meet the anticipated electrical load in each block. At the beginning of December Edison hired Hermann Claudius, an Austrian telegraph engineer, to assist with these calculations and eventually to build a model of the entire feeder and main conductor network. In the meantime, construction of the large direct-connected dynamo continued and on 11 December Edison filed a patent application covering its novel armature arrangement in which solid bars of copper were connected to radial copper plates. A few days later he also drafted two caveats dealing with dynamo voltage regulation. Sustaining the increased patent activity that began during the summer, during this period Edison executed twenty-two applications that resulted in U.S. patents.

Edison made a number of important business and administrative arrangements during the last quarter of 1880. He and Grosvenor Lowrey quietly made plans to place Sherburne Eaton, a prominent corporate lawyer and former colleague of Lowrey's who had been acting as an unofficial advisor, on the board of the Edison Electric Light Co.; Eaton became vice president and general manager in January 1881. When electric light investors declined to venture into the lamp manufacturing business, Edison formed the Edison Electric Lamp Co. in mid-November as a partnership with Charles Batchelor, Francis Upton, and Edward Johnson. The Edison Electric Illuminating Co. of New York was created a month later, with the support of Edison's investors, to generate and sell electricity in Manhattan. To help ensure that New York City would grant a

The Menlo Park laboratory complex in a painting by R. F. Outcault, who later created the cartoon strips "Yellow Kid" and "Buster Brown." The long wooden building in the center is the main laboratory (exaggerated in length), the small brick building at front center is the library and office, the large brick building at rear is the machine shop, and the electric railway is at the upper right.

franchise for the underground conductors, Edison and Grosvenor Lowrey arranged for the Board of Aldermen, city administrators, and company investors to spend an evening at the laboratory on 20 December. Guests enjoyed a lavish catered dinner and saw Menlo Park illuminated by more than two hundred lamps. Also in mid-December Edison tried unsuccessfully to secure an agreement with the British chemist William Crookes for the lamp factory to produce radiometers. An old business arrangement came back to haunt Edison in November when he was sued by Lucy Seyfert, wife of an investor in the Automatic Telegraph Co., for payment of $7,000 in promissory notes given in 1874 as part of a complex plan to raise money for that company.

The high point of Edison's Menlo Park workforce probably was reached by early October. Five employees are known to have left during that month. Among them was Ludwig Böhm, never well-liked at the laboratory or lamp factory, who left on bad terms after quarreling with Batchelor. Edison later accused him of providing confidential information to Hiram Maxim after resigning. Laboratory assistant Arthur Andrus also departed. More than a dozen others left before the end of the year. This group included several recent hires and a number of laborers, the need for whose services presumably ended with completion of the underground conductors and the lamp

factory, which started its regular payroll on 11 November. A number of new employees first appear in the Menlo Park payroll records in November and December, most of whom were probably employed at the lamp factory. The one significant addition to the laboratory staff was Hermann Claudius, who arrived in mid-December.

1. Doc. 2006.
2. Charles Batchelor to Stockton Griffin, Nov. 1880, DF (*TAEM* 53:525; *TAED* D8014ZAI).
3. "How Far Edison Has Got," *New York World*, 29 Nov. 1880, Cat. 1062:20, Scraps. (*TAEM* 89:18; *TAED* SM062020).
4. Johnson to Uriah Painter, 10 Dec. 1880, UHP.
5. Mott Journal N-80-07-10:253, Lab. (*TAEM* 37:429; *TAED* N117:127).

–1994–

To John Michels

[Menlo Park,] Oct 5th [1880]

Dear Sir:

In the last issue of Science (Oct 2d) I notice a great deal of space given to Wiesendanger.[1] Now I dont suppose you are aware of it but in Europe Wiesdenanger is looked upon as little less than an idiot and such stuff as the article referred to is nothing more nor less than a disgrace to a scientific paper. If such a thing is possible I would like to see the proofs of what is to appear in Science before the paper is published and thereby avoid as much as possible that which would be detrimental to its interests.[2] Very truly

T. A. Edison   —G[riffin]—

L (letterpress copy), NjWOE, Lbk. 6:452 (*TAEM* 80:389; *TAED* LB006452). Written by Stockton Griffin; circled "C" written at top of page.

1. This republication of Theodore Wiesendanger's paper before the British Association for the Advancement of Science on "An Improved Electric Motor" comprised a page and a half of text and another full page of drawings (*Science* 1 [1880]: 170–72). Wiesendanger contended that the design of electric motors differs fundamentally from that of generators, particularly regarding the "mischievous theory" that a motor's efficiency "bears a definite and direct proportion to the magneto-inductive power of its field magnets, and that an increase of power in the field-magnets alone must necessarily produce greater capabilities of the machine." He claimed to have built a satisfactory motor "in which the power of the field magnets is as nearly possible equal to that of an armature." He also proposed several unconventional armature shapes in order that "nearly the entire motion of the revolving armature should be either one of approach or of withdrawal" with respect to the field poles.

Nothing further is known about Wiesendanger; a portion of this article is in Cat. 1026:105, Scraps. (*TAEM* 25:60; *TAED* SM026105a).

2. Michels replied that he was unaware of the reputation described by Edison and had treated Wiesendanger's article as a "paper read before the British Association at its last meeting, and republished by the English Journals, and as such worth reproduction." He added that "the Journal is well advised on all subjects except Physics," a service which Otto Moses had recently ceased to perform as promised. Michels noted that he had been "submitting all new papers on Physics to him, and should have done so with Wiesendangers.— I would like to know if Mr Moses is still advising the Journal on these points, and will forward papers on Physics to him as before if such is the case." Michels to TAE, 6 Oct. 1880, DF (*TAEM* 55:454; *TAED* D8041ZAN).

–1995–

William Hammer to James Russell [1]

[Menlo Park,] Oct 5 [1880]

Dear Sir.

Mr Edison desires that the no of kerosene lamps be entered in the "Statistics" that they be counted as you would gas jets, only mark kerosene and say if they have globes, and also if they are placed on gas chandeliers. [2]

Also get as near as possible the number of gas jets ordinarily burned between the hours of 5 and 6 P.M. Winter & Summer—the winter particularly

Also if parties have no elevator, or hoist, if they would like one if they could get the power [----][a] twenty five (25) or thirty cents a day. Of course you can judge if parties need a hoist. =[3]

Yours &c

Wm. J. Hammer. for T.A.E.

ALS (letterpress copy), NjWOE, Lbk. 6:451 (*TAEM* 80:388; *TAED* LB006451). Circled "C" written at top. [a]Faint letterpress copy.

1. James Russell apparently began working for Edison in the summer of 1880 (see note 2). Nothing is known of him prior to this time. Time Sheets, NjWOE.

2. William Hammer recalled many years later that Edison asked him in the spring of 1880 to

> see what I could find in the way of maps of the lower part of New York City. I returned with some five maps of several different scales and putting together two 10 ft. laboratory tables I laid out a large map on one scale of the section Mr. Edison and his associates decided would prove most suitable for the 1st District.
> A Mr. James Russell and a set of canvassers were then sent through this District during every hour of the day and night and made a note in their books of every light, whether of gas, oil or candle which they found in the district and they also took note of a number of mules kept up in loft buildings for operating freight

elevators, this being with a view of replacing them with electric motors.

> These men reported to me and I collected their data and entered it in the proper places on the large map. [Hammer to Frank Smith, 31 Aug. 1932, "Hammer, William J.," Pioneers Bio.]

Russell and others reportedly started canvassing sometime in July in an area bounded by Pine St. (soon extended one block to Wall St.) on the south, Broadway and Park Row on the west and northwest, Frankfort St. on the north, and the East River ("An Inventor's Workshop," *New York Times,* 9 Aug. 1880; "Edison's Progress," *New York Sun,* 9 Sept. 1880; Cat. 1241, items 1517 and 1532, Batchelor [*TAEM* 94:609, 614; *TAED* MBSB21517X, MBSB21532X]). In the middle of September Sherburne Eaton informed Edison that Russell had reported that "nobody refuses to answer his questions." The survey notebooks are not extant but Russell had completed two by this time and at least six by early November. Summaries and tabulations made separately by Hammer and Upton indicate that each one included roughly 100 to 200 addresses. Respondents were asked, among other things, about the number and types of gas burners used, quantity of gas consumed, business hours, insurance rates, adequacy of ventilation, whether they were bothered by heat, how far any underground vaults extended towards the street, and several questions related to power usage (Eaton to TAE, 17 Sept. 1880, DF [*TAEM* 54:72; *TAED* D8023ZAF]; Mott Journal N-80-07-10:159, 179, 195; N-80-11-25:27–127, N-80-08-13:47–129; all Lab. [*TAEM* 37:381, 391, 399; 593–635, 38:315–52; *TAED* N117:79, 89, 97; N120:7–50, N132:26–63]). Sometime in early September Hammer and Francis Upton began using fire insurance maps, which provided a wealth of color-coded detail, to count the buildings and measure the length of conductors needed for each block of the projected first district (N-80-09-09:33–97, Lab. [*TAEM* 38:157–89; *TAED* N129:17–50]; for a brief history of insurance maps and published examples see Cohen and Augustyn 1997, 128–29 and Ristow 1981, 3–5).

Hammer recalled that Upton taught him "the formulas for calculating the amount of copper necessary and I personally did a large amount of this work and subsequently made a smaller working map which was the first used" for laying underground conductors. General calculations of the size of conductors and pipes began by mid-October; Hammer and Upton made detailed calculations of the length and weight of conductors for each of 51 squares in the service district of more than 11,000 lamps. Hammer to Frank Smith, 31 Aug. 1932, "Hammer, William J.," Pioneers Bio.; N-80-09-09:199–269, N-80-08-00:2–109, N-81-00-01: 1–183, all Lab. (*TAEM* 38:238–71, 36:638–91, 39:513–605; *TAED* N129:99–132, N110:1–54, N165:1–92).

3. Russell replied to Edison the next day that he would begin enumerating kerosene lamps. He also stated that "The books already in your possession show the maximum number of gas jets burned between the hours of 5 & 6 P.M.—especially in Winter" but noted that it was "almost impossible to arrive at a correct estimate of the amount of gas burned in Summer" because many businesses closed before dark. He stated that elevators "would be universally used if the power could be supplied at the price you name. This information will be found in the 'statistics' in

future." Russell to TAE, 6 Oct. 1880, DF (*TAEM* 54:79; *TAED* D8023ZAJ).

–1996–

Notebook Entry:
Electric Lighting

[Menlo Park, c. October 7, 1880[1]]

Summary of Lot 1[2a]

The average of all the lamps tested (91) to be found on pages 221 to 231[3a]

Volts.	Ohms.	Foot Lbs.
148.[a]	188.2[a]	5282[a]

That is 300 Candles per Horse Power were obtained[a]

From the record kept of 89 lamps the average time of burning was found to be 310 minutes[a]

Thirty eight (38) lamps were blue at the clamps the average of 27 of these gives

Volts.	Ohms.	Foot lbs.
151[a]	152[a]	5352[a]

minutes   22041 ÷ 82 = 268[b] average for 82 lamps of Lot 1

22 041
  1350
  1050
   745
   703
   701
22 103

27 483 ÷ 89 = 308[c] minutes average for 89 lamps.[a]

Average time of burning of 32 lamps in first test which burnt their resistance 198[d] minutes.[a]

On[e] page 249 is the table showing the number of lamps that gave out during the succeeding 50 minutes.[4]

On page 219 is an analysis of the places at which the lamps broke   51 lamps[5]

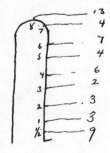

On page 243 is a table showing the lamps that showed blue at the clamps[6]

Result.

These lamps were made with very poor vacuums and in a number of cases the sealing of the wires through the glass was defective. The wire was simply run through the glass of the inside part which was sucked in.

The lamps that lasted the longest were those which took below the average E.M.F. to bring them to ~~the~~ 48 candles. Lamp 40 which lasted 1350 minutes showed at the end of that time no blackening on the clamp   It was tested twice for economy and was found ~~the~~ to be[f] a[g] fraction less economical at the end of the test than at the beginning though practically the same. The resistance and E.M.F were also practically unaltered. This shows that when the lamps are sufficiently exhausted that they are permanent.

All the lamps that were blue at the clamps at the beginning of the test gave out before 700 minutes.

TAE

X, NjWOE, Lab., N-80-08-18:261 (*TAEM* 36:833; *TAED* N111:128). Written by Francis Jehl; document multiply signed. [a]Followed by dividing mark. [b]Form of equation altered. [c]Obscured overwritten text; form of equation altered. [d]Multiply underlined. [e]Balance of document by Francis Upton. [f]"to be" interlined above. [g]Obscured overwritten text.

1. This summary was written following tests of Lot 1. The last dated entry for those tests is 6 October (see note 2).

2. The rough record of lamps tested from Lot 1 (numbers 1–100) are found in a notebook kept by Francis Jehl. This was the first lot of lamps sent from the lamp factory to the laboratory for testing. On the first page of the book is a copy of a note from Charles Batchelor to Francis Upton in which he noted that the first 38 lamps sent on 27 September were all made from "fishing rod Bamboo prior to September 25th and are unpicked; so that we consider them poor" (N-80-09-27, Lab. [*TAEM* 37:446; *TAED* N119]). In another notebook Batchelor noted that Lot 1 was "to consist of 159 lamps of Fibres generally not picked but made from the fishing poles we have bought." He also noted that these lamps were all sent to the laboratory by 5 October (N-80-09-28:19, Lab. [*TAEM* 36:469; *TAED* N106:10]). Additional records of these lamp tests are found in the same notebook as this document beginning on page 39 where Jehl made a record, dated 6 October, of the 100 Lot 1 lamps "brought up to dull red and bad lamps picked out" and in another book (N-80-08-18:39, N-80-11-25:17–25, Lab. [*TAEM* 36:726, 37:589–92; *TAED* N111:20, N120:3–6]). On the pages following Jehl's note Edison made additional rough notes regarding these lamps that are summarized in this document and on the tables preceding it. Many of his notes include drawings of the points where lamps broke that were combined into the drawing in this document and on page 219 (see note 5).

3. Pages 221–29 are a table; page 231 contains calculations for the following results. N-80-08-18:221–31, Lab. (*TAEM* 36:817–22; *TAED* N111:111–16).

4. N-80-08-18:249, Lab. (*TAEM* 36:829; *TAED* N111:123).

5. The following drawing reproduces one found on page 219. Below the original is a note by Edison that "the breaks occur on both sides but I have massed them all on one side" (N-80-08-18:219, Lab. [*TAEM* 36: 816; *TAED* N111:110]). It is based on a 28 September drawing by Charles Batchelor of a loop marked by divisions "for showing position of faults" (N-80-09-28:1, Lab. [*TAEM* 36:460; *TAED* N106:1]).

6. N-80-08-18:243, Lab. (*TAEM* 36:826; *TAED* N111:120); on the blue appearance around the clamps see headnote Doc. 1898.

–1997–

From Charles Cuttriss

Duxbury, Mass   Oct 11th 1880

Dear Sir,

Until the last month I have been unable to experiment on the cable with the chalk cylinders you kindly gave me in June.[1]

I found on making one up in the form of a Telephone that it only sounded when the negative current passed through the chalk from the axis to the palladium the positive current having no effect unless passed through in an opposite direction.—

I therefore constructed an instrument as per diagram.[2] The cable was connected to the centre of one chalk & the earth to centre of the other, the bridge across was of copper faced with palladium, the rubber wheel gave a steady pressure & allowed the bridge to move without other friction than that caused by the chalks, which moving in opposite directions counteract one another when no current is passing.

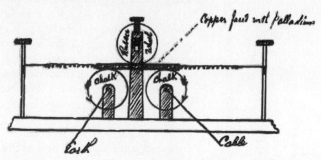

I can get a strong motion with the cable current everytime it is reversed; but it will not show the fall of potential when pos or neg currents are sent consecutively; & as you know in cables of 700 miles or more, the pos & neg currents when sent[a] consecutively never fall more than one third of their original potential, and to this slight change the instrument will not respond. When it is worked with an almost infinitely small current by an ordinary cable key, it works finely because the chalks are short circuited between each current, but if not so short circuited the bridge will not return to zero till after a lapse of ½ second or so. From this it would appear as though the gases formed by the current took some short time to recombine when the circuit was either left open or was of a high resistance such as 10,000 ohms. When a Minotti cell[3] of 20,000 ohms are used in the battery circuit of an ordinary cable key the instrument works perfectly, the chalks being in short circuit between every signal, but if the 20,000 ohms are in the line circuit the movement at first is very small (on sending successive pos or neg currents) and soon the bridge is held over steadily. I am still working on it but so far have been unable to overcome the difficulty. The resistance of the instrument from axis to axis when stationary (the chalks being only half inch in diameter) is about 100,000 ohms, & when the chalks revolve it goes down to 80,000 ohms with a positive current & 60,000 ohms with a negative. I think if I could get the resistance down to about 3 or 4 thousand ohms it might work better.

Do you think a platinum cylinder with a thin film of ordinary tobacco pipe clay baked on it would have the desired effect when impregnated with the proper solutions? Hoping that this account of my experiment may be of interest to you, I am, Yours Truly,

Chas Cuttriss[4]

⟨When you get the device just so it will work strongly with 3 or 4 volts through several mega ohms Have you tr but its very difficult to get it just right= try a condenser in circuit of about ³⁄₁₀ microfarad capacity

Have you tried an induction coil putting chalks in secondary= I pack & ship you a telephone for trial[5b] working Morse signals & listening reading by sound try that under various conditions. a platina point in centre of diaphragm ought to open & close circuit if signals are strong   when you try I have[c] to[b] make a relay I have found with certain adjustments and conditions that the lever moved slugish seeming to require time like electrification & at other times under different adjustments & conditions it would actually open & close the circuit with enormous rapidity & I have made it a self vibrator & it actually gave vibrations[b] so rapid that it would whistle so shrill as to nearly peirce your ears   Its just a question of experiment I had it working one time that when[a] a Thomson cable gal only went 10 degs on the scale when in same circuit yet the sounder closed & opened every time=⟩

ALS, NjWOE, DF (*TAEM* 55:658; *TAED* D8043ZAU). [a]Interlined above. [b]Obscured overwritten text. [c]"I have" interlined above.

1. This probably refers to Edison's electromotograph relay.

2. Drawing on separate enclosure.

3. The Menotti battery, which had a flat, circular anode soldered to a gutta-percha-covered wire, was used primarily for testing and for long cable lines. Prescott 1877, 60–61; Ternant 1881, 246–47.

4. Charles Cuttriss had been with the Anglo-American Telegraph Co. for over a decade and was at this time in charge of the company's station in Duxbury, Mass. In January 1880 he had asked Edison for a job; Edison replied that he might have a position in connection with electric lighting and asked him to write back in a month. Cuttriss did so but Edison had no job at the time. Cuttriss went on to become electrician for the Commercial Cable Co. and made some important improvements in cable telegraphy. Cuttriss to TAE, 3 Jan. and 13 Feb. 1881, DF (*TAEM* 53: 486, 495; *TAED* D8014A, D8014G); Bright 1974 [1898], 615–16, 619, 673, 679.

5. This was apparently an electromotograph receiver. In a 20 February 1881 letter to Edison Cuttriss reported on his experiments using it on a cable line and on an artificial cable. He found it to be "exceedingly sensitive & if it can be made to work without 'sticking' I think it would be by far the most delicate relay yet found, & just the one for cable work, it acts so far just as a chemical telegraph would." On 23 February Edison replied, "Your very friendly and pains-taking experiments pleased me very much indeed. I am quite sure you will yet prove what we think possible. The thing can be done; let us do it. . . . Is there anything else you want?" DF (*TAEM* 57:357, 360; *TAED* D8106A, D8106B).

–1998–

Draft to Otto Pettersson[1]

[Menlo Park, October 12, 1880[2]]

Your communication of the＿＿recd—[3] I have never published any papers that amount to anything on science as my work has been mostly practical   It is true that I have introduced the word "pressure of Electricity" instead of Electromotive force for the reason that ~~Electromotive~~ the latter is a term few comprehend ~~the~~ its[a] whole meaning ~~and because~~ and because "pressure" is the proper term

In my practical work I had thought out a[a] certain theroy of Electricity which I considered the true one—and all my results have been obtained by and of this theory which is ~~that~~ simply this that there is a current[b] and this current ~~h~~can have its velocity increased or decreased by a greater or lesser pressure, ~~that a~~ [----in?][c] ~~the current~~ that if it is desired to transfer ~~& achieve~~ a million foot lbs of energy from one point to another ten[b] miles apart that a current, of infinite weakness & hence a wire of infinitely small section[d] may be used and the work transferred by increasing the velocity of th~~e~~is weak[e] current.

As an analogy I have used the illustration of the transmission of power by a ~~belt or ba~~ Leather band from one shaft ~~of the~~ to another ~~The~~If we consider the size of the belt to represent the current & the velocity of the same to[b] represent the pressure,[4] then it is seen that if we use a band 1 foot wide running at a velocity of 1000 feet per minute we may transfer say 10 horse power of energy whereas if we use a belt only 6 inches in width and a velocity of 2000 feet per minute we can still[a] transfer the 10 hp of energy so we may continue weakening our belt ~~unti~~ and increasing its velocity until it is a thin as a single spider web & still transfer the 10 h.p. ~~No is pressure costs nothing~~ The power transferred depending entirely upon the number of square inches or units of belt surface passing a given point per minute, so with Electricity with[a] the weakest current ~~and a high pressure energy~~ [enerey?] energy[f] can be transferred over the ~~thinnest~~ smallest conductors by increasing the velocity of the same by increasing the pressure, so that a greater number of units shall pass a given point in the circuit in a given time—

I have never used the aether in my theories always considered that ~~all~~ transfer of energy in every form could be explained by vibrations of matter as we know it[g] How matter, ~~no matter~~ although of extreme lightness could be transferred through denser matter at the enormous velocity at which electricity flows I cannot conceive but could conceive its transmission by vibration very easily. The oxidation of zinc in the air produces vibrations of light. The oxidation in a battery must be[b] the same only under different conditions & in a different degree hence there must also be vibrations. These vibrations must necessarily be slower than that due to ~~the~~ radiant heat of the[a] lowest refrangibility= ~~If a single cell is used we obtain a slow oxidation and a current of say 1 if a single cell is put on an incandescent wire we get make it~~ [------][c] ~~light of rays of low refrangibility now if we add a cell to each, we do not increase the current but~~ I think it was De La Rue who by means of a revolving mirror proved that the light in a Giessler tube from a Battery of a great number of cells was discontinuous.[5] ~~Very truly yours~~

~~T A Edison~~

If we take a single cell, the rapidity of oxidation of the zinc in that cell cannot be increased by increasing the size of the plates or by the use of better excitents[6] etc, but if we connect into the circuit a second cell then the ~~lamps~~ rapidity of oxidation is raised by the action of the other cell upon the zinc of the first cell. Then there are vibrations set up which have higher rates, as would be the case if the rapidity of the oxidation of zinc in air was increased by some means. Very truly yours

TAE

ADfS, NjWOE, DF (*TAEM* 53:882; *TAED* D8020ZHS). [a]Interlined above. [b]Obscured overwritten text. [c]Canceled. [d]"& hence . . . section" interlined below and in right margin. [e]Interlined below. [f]"[enerey?] energy" illegible and interlined below. [g]"matter as we know it" interlined above.

1. Sven Otto Pettersson (1848–1941) was a noted Swedish chemist particularly interested in electricity and the chemical properties of sea water. In his recent letter to Edison (see note 3) he identified himself as a docent at the University of Upsala; he later became professor of chemistry at the University of Stockholm. *WWWS*, 1338.

2. Stockton Griffin dated this draft on back of the last page.

3. Pettersson wrote on 25 September that he was preparing a monograph on electricity and wished to critique an explanation of electric current in terms of the ether in which "The electromotoric force entering in the circuit acts similar to the heart of the human body, compelling the aether to flow through the entire circuit like the blood in the veins of the body." He explained that he had seen a paper suggesting "that the idea of electric currents of 'high pressure' has been introduced into science by you," and asked Edison to send copies of "those papers, wherein you have treated the theoretic parts of electricity." DF (*TAEM* 53:864; *TAED* D8020ZHB).

4. Edison evidently had discussed this analogy in the laboratory before December 1879, when Samuel Mott attributed it to him in a notebook devoted to "Things to be looked-up and remembered both electrical and otherwise." N-79-12-00:1, 13, Lab. (*TAEM* 35:704, 710; *TAED* N083:1, 7).

5. Warren de la Rue, the English chemist and astronomer, conducted a long series of experiments in the 1860's with discharges from large batteries through Geissler tubes, but his work produced little new knowledge. Previously, the English electrical experimenter John Gassiot made significant advances in following up W. R. Grove's 1852 announcement of striations in sparks in a rarefied gas. Gassiot investigated this phenomenon for several years, attributing it to "pulsations or impulses of a force" acting on a rarified medium. In one set of experiments Gassiot produced the striations using a 400-cell Grove battery, and he used a rotating and vibrating mirror arrangement to demonstrate the intermittent nature of the discharge under certain conditions. *DSB*, s.vv. "De la Rue, Warren," "Gassiot, John Peter"; Atkinson 1902, 997–99.

6. That is, the electrolyte.

[Menlo Park, October 12 1880[1]]

Note—

The first 3 lamps that bursted tonight had no resistance[a] coils= perhaps Resistance on all heated, increased their resistance hence the ones that had no res must have been higher=

The lamps tested tonight were a very nice lot, only 1 or too that are noticably low & perhaps Res for these was wrong— Apparantly there is no oxidation by leakage as I do not notice any noticable increase in spots=[b]

Must look out for cracks in tits= after test over=

X, NjWOE, Lab., N-80-10-15.2:117 (*TAEM* 39:60; N149:59). [a]Obscured overwritten text. [b]Followed by "over" written as page turn.

1. In his journal entry of this date Charles Mott noted that "98 lamps of the second hundred from the lamp factory were started about three o'clock this morning burning at an estimate of forty eight candles and about twenty remained whole and burning up to six oclock in the evening when the current was turned off and to have the remaining lamps retested for vacuum and resistance" (Mott Journal N-80-07-10: 161, Lab. [*TAEM* 37:382; *TAED* N117:80]). One of the books noted by Mott as containing records of these tests is missing (No. 173). Other records are found in the remainder of this notebook and in several other books (N-80-10-15.2, N-80-10-08, N-80-10-12:99–201, N-80-10-15.1: 1–104, N-80-11-25:13–15, all Lab. [*TAEM* 39:2, 38:966, 39:748–77, 37:647–99, 37:587–88; *TAED* N149, N148, N171:48–77, N121:1–52, N120:1–2]). A note by Charles Batchelor describes Lot 2 as consisting of carbons made from bamboo poles "said to be of Chinese wild growth" purchased by William Moore from New York supplier John Deltour. A third lot was made from the same bamboo but "whilst they were on the pumps they were brought up to 70 candles for 2 or 3 seconds" (N-80-09-28:19–23, Lab. [*TAEM* 36:470–72; *TAED* N106:10–12]; Wilson 1880, 357).

Metuchen Oct. 14th &16th 1880.

Dear sir,

Receiving the last time a kind of treatment of some of your assistants which no man with any sense of honor can bear and disliking the ways things are managed in the line I work in[a] whihch put me in a peculiar position I feel obliged to write this letter. About two weeks ago a man who has been my assistant for nearly half a year wanted to knock me down and break my neck without any cause in Mr. Batch. presence. Yesterday Mr. Batch and I had a disagreement from a cause not worth to be mentioned which went[b] so far that I had to hear that you were a dem side better off if I were not here. If work[ing?][c] in my line I do not want to be bossed by people that understand less than I or nothing of it, I think I gave sufficient proofs that I understand my line. ~~very well~~. thoroughly. That things are managed by ignorant parties is very disadvantegous to you[1]   I cite only the last thing you know off already: the carbon tester when fixed up by "the boys."[2] I have no doubt they did the best they could but a whole week was that time lost. The loss of money did not amount to anything but the delay a whole weeks delay! What I worked for was progress! I did not work for progress the last time because I liked the parties I had to deal with, not because I liked the treatment I received for the services I have rendered, but because I sympathized with you, because I liked the great problem we were working on. But now things have gone too far. For a few days I take a vacation to go on my uncle's farm which I did not see yet. Then I came back to hear whether you agree with the things described. In conclusion I say that I do not want higher wages and that I do not want to interfer with parties[b] manufacturing[3]   Remaining your obedient

Ludwig K. Böhm

ALS, NjWOE, DF (*TAEM* 53:517; *TAED* D8014ZAB). [a]Interlined above. [b]Obscured overwritten text. [c]Obscured by ink stain.

1. Nothing is known of these particular incidents; the assistant referred to may be William Holzer, with whom Böhm apparently had a tense working relationship. Edison had moved Böhm from the "glass house" near the laboratory down to the lamp factory at the end of August. Edison later testified that he had offered him piecework "but he preferred the raise of wages; Mr. Bohm's salary was the largest paid to any one in my employ, for the reason that I could not get a glass blower at the time to fill his place, and I had to pay him what he asked" (Mott Journal N-80-07-10:101, Lab. [*TAEM* 37:352; *TAED* N117:50]; Böhm's testimony, 8, and Edison's testimony, 38, *Böhm v. Edison* [*TAED* W100DEC002 (image 7); W100DED032 (image 7)]). Böhm was unusual among Edison's employees in that he did not fit into the Menlo Park milieu. Francis Jehl recalled that "The boys teased him often, for although he was a fine glass blower he was so conscious of his own abilities that it was more than some could endure"; when he quit Menlo Park Charles Mott remarked that it was because the position was "not consistant with his honor and birth" (Jehl 1937–41, 495, 516; Mott Journal N-80-07-10:176, Lab. [*TAEM* 37:390; *TAED* N117:88]).

2. This was probably the "carbon prover" for which Edison and Batchelor had executed a patent application in July (see headnote p. 767), a number of which were used in the lamp factory. Böhm may be referring to the discovery, reported by Mott on 14 September, that "the testers with larger contractions were found to break by the weight & fall of the mercury, and a medium will have to be determined at which the pumps will stand and do the most effective work." On 1 October Edison received tubing "for experimenting on the proper size and pro-

portion between the contraction and the tube, to give the best results in exhausting also to determine the most economical size or thickness of tubing that will best resist, without breaking, the fall or pound of the mercury." Mott reported on 29 October that Charles Batchelor dispensed with the difficulty of having the tester and pump joined as a single unit, arranging instead to handle the testing globe as an ordinary lamp globe which could be detached from the pump so that the carbons could be "put in and taken out by one man at his bench and thus avoid much danger of breakage both of glass and carbons." Mott Journal N-80-07-10:129, 151, 189, Lab. (*TAEM* 37:366, 377, 396; *TAED* N117: 64, 75, 94).

3. On Wednesday, 20 October, Böhm evidently enclosed this letter with his announcement to Edison that he would be "leaving Menlo Park and your services on Friday next. The causes may be known to you already. I dislike the treatment I received for the services I have rendered and am convinced that some of your assistants and I could not agree for any length of time. I therfor give up my present position." Two days later he asked Edison for a reference letter (Böhm to TAE, 20 and 22 Oct. 1880, DF [*TAEM* 53:519, 521; *TAED* D8014ZAC, D8014ZAD]). Böhm subsequently testified that he had expected some profit-sharing arrangement in connection with the lamp factory. He also testified that shortly before giving notice he had contacted the inventor Hiram Maxim, then developing his own incandescent electric light, who hired him immediately upon his resignation. Böhm remained with Maxim's United States Electric Lighting Co. for about six months before joining the American Electric Light Co. as electrician and superintendent. During his tenure with Maxim's company Böhm filed a patent application on a modified Sprengel pump he claimed to have invented at Menlo Park. The application was placed in interference with one filed by Edison in January 1881, which issued to Edison as U.S. Pat. 248,433 (Böhm's testimony, 15–17, 6–7, *Böhm v. Edison* [*TAED* W100DEC002 (images 14–16, 5–6)]).

-2001-

R. G. Dun & Co. Credit Report[1]

[Newark?][2] Octr 15/80

Thos. A Edison   Electrician

3700[3]   Former reports give the nature of this mans assets & their worth so far as they can be learned[4]   Edison said to be progressing towards the perfection of his Electric Light & may soon be a very wealthy man   Cannot estimate him as he now stands but he has always paid & it seems to be the general opinion that he will still pay as fast as he can   He must have an income now of a good many thousands a year but his constant experimenting eats up money exceedingly fast & it is thought that he so far has not laid up much   313.

D (abstract), New Jersey, Vol. 52, p. 290, R.G. Dun & Co. Collection, Baker Library, Harvard Business School.

1. R. G. Dun & Co., established in 1841 as the Mercantile Agency, was by 1871 one of two major credit-reporting firms in the United States. See *ANB*, s.v. "Dun, Robert Graham" and *TAEB* 1:469 n. 1.

2. R. G. Dun clerks transcribed individual reports filed by agents into large ledgers organized by state and county. In 1871 the firm opened an office in Newark; it is presumably there that the ledger in which this report appears was maintained. See *TAEB* 1:469 n. 1.

3. This number refers to the agent making the report. The book containing the matched codes and names has been lost. The meaning of the number at the end of the document is unknown.

4. Between Doc. 1614 and this report, abstracts dated 18 September 1879 and March 1880 similarly noted that Edison used most of his indeterminately large income for experimental costs but that he had a long history of meeting his obligations. RGD, N.J. 52:290.

–2002–

From William Moore

San Francisco, Oct. 17, 1880[a]

Dear Sir

Since my arrival here I have been on the lookout for Bamboo but have been unable to find any quantity of a suitable kind—

I went on board the Steamer Columbia which has just arrived in port, saw your machines with which the parties are highly pleased, ~~with~~, one of the lamps they say has run over[b] 800 hours—but just at prestent they are greatly in want of lamps as about half of those first sent are played out—[1]

The Columbia has a Maxim machine to run the head light—[2]

There is a fair show made in this city with Brush machines, there being three located at a central stations, feeding forty lights placed at prominent points within a radius of a mile from station—[3]

There are also eight Brush[b] lights at this hotel but they do not give the satisfaction desired and the managers of the hotel with whom I have talked are favorable to adopting your system of lighting—

I have been informed that the sailing vessells which ply between here and Japan are taken off during the winter months, being employed on this coast for carrying grain, so I no doubt will[b] be obliged to send by steamer for a few months   I am also informed that the Japanese are very tricky in their dealings and would make a corner whenever they can, so I think it will be as well when the best stock bamboo is found to lay in a good supply and not depend much on future supply through agents but I will know better how the land lays when I get on the other side

Bamboo sawed in lengths Duty[b] Free

d[itt]o split one or more times 35%

Just seen Capt. Henderson of the Columbia—he[c] said he had telegraphed to you several times for lamps—and it is now becoming a matter of serious importance as so many are examining the system and observe the failure of so many lamps—[4]

I sail within an hour—will keep you well informed of my movements and do my best in the Bamboo line—[5] as soon as you are ready to put your lights[c] on the market I wish you would send me particulars, prices etc—as I may be in a position to place some to advantage— Yours truly

W. H. Moore[6]

ALS, NjWOE, DF (*TAEM* 53:894; *TAED* D8020ZHY). Letterhead of Palace Hotel. [a]"San Francisco," and "18" preprinted. [b]Interlined above. [c]Obscured overwritten text.

1. See Doc. 1976 n. 3.

2. This machine may have been like the one described in Maxim's British specification filed in April 1880 for the same improvements covered by his first four United States generator patents, all of which issued on 8 June 1880 (Brit. Pat. 1,392 [1880]; Dredge 1885, 2:cx–cxi; U.S. Pats. 228,543, 228,544, 228,545, 228,546). The 23 October 1880 issue of *Scientific American* illustrated a self-regulating machine of this type that employed one armature to produce the main current and a similar but smaller armature on the same shaft with its own field magnets as an exciter. An electromagnet in the outside circuit moved a gear train that would adjust the position of the commutator brushes in the exciting machine, thereby controlling the field strength of the main generator. It is also possible that the *Columbia* had the Maxim machine shown in the 28 August 1880 *Scientific American;* this form had no evident regulator but was intended for marine applications. The *Columbia*'s headlight was a Maxim arc light, presumably like the self-adjusting focusing lamps for marine use illustrated and described in the 28 August and 16 October 1880 *Scientific American.* When the ship was docked in New York it reportedly could illuminate buildings across the East River in Brooklyn ("Recent Developments in Electric Lighting," *Sci. Am.* 43 [1880]: 255, 262; "Electric Light for Marine Use," ibid., 43 [1880]: 127, 130; "Maxim's New Focusing Electric Lamp," ibid., 43 [1880]: 242; "The Columbia," ibid., 43 [1880]: 326).

3. This dynamo was designed by Charles Brush and manufactured in several forms by the Telegraph Supply Co. of Cleveland. The machine produced the strong currents needed for arc lighting and dissipated heat relatively well; it was also noted for the ease with which it could be maintained and repaired. An undated promotional pamphlet of the Anglo-American Brush Electric Light Co., Ltd. included a testimonial (dated 24 April 1879) from Senator William Sharon, owner of the Palace Hotel, that the establishment's Brush generators and ten lamps worked *"to our entire satisfaction." TAEB* 4:584 n. 4; Anglo-American Brush Electric Co., undated pamphlet, DF (*TAEM* 54:334; *TAED* D8026ZEP).

4. Henderson's telegrams have not been found but W. H. Starbuck of the Oregon Railway and Navigation Co. had written Edison on 26 July to ask about the lamps Edison had promised Henderson, noting that he had "just received telegrams from Capt Bolles & Mr Henderson saying there are no lights there [in Portland, Ore.] and the ship is not half supplied." Edison replied that he had been unable to supply lamps because "we have not got our factory completed and its impossible for us to take time to make them by hand as were the ones furnished the Columbia and they are too imperfect when so made. Mr. H[enderson] will have to wait until the factory is running (about 6 weeks) when he can have them by the gross as we will turn out 1200 a day." The first entry in a new order book, dated 20 September, was made for 250 "Lamps for S. S. Columbia." The lamps were apparently not made and sent until mid-April 1881. Starbuck to TAE, 26 July 1880, DF (*TAEM* 53:795; *TAED* D8020ZFU); photostatic copy of TAE to Starbuck, 27 July 1880, OrHi; Order No. 1, Cat. 1301 [p. 1], Batchelor (*TAEM* 91:294; *TAED* MBN007:1); Philip Dyer to TAE, 16 Apr. 1881, DF (*TAEM* 57:851; *TAED* D8123ZAX).

5. At least one shipment from Moore had been received by 31 December, when Charles Mott reported that strips cut from the bamboo were "exceedingly fine and unusually free of pith, came out very fine from the carbonizing flask." By early February the Edison Electric Lamp Co. had arranged through Moore for 500,000 splints each month, presumably of the Madake bamboo. In March, however, Edison notified Upton that the splints being received were unsatisfactory, and Moore was recalled by the middle of that month. Mott Journal N-80-07-10: 269, Lab. (*TAEM* 37:437; *TAED* N117:135); Edison Electric Lamp Co. to Mourilyan Heiman and Co., March 1880; Francis Upton to TAE, 12 Mar. 1880; both DF (*TAEM* 57:820, 818; *TAED* D8123ZAA, D8123Z); Samuel Insull to Upton, 11 Mar. 1881, Lbk. 8:41; (*TAEM* 80:854; *TAED* LB008041).

A set of instructions written by an unidentified laboratory staff member, probably in December 1880 or early 1881, lists the desirable qualities of bamboo to be shipped in lots of 100,000 to 1,000,000 per month: "a very hard and dense fibre or wood on and near the outer edge"; joints to be "as straight as possible"; no less than nine inches in length and four in circumference; the tops of the canes should not be used "as the wood is not so hard or dense"; the material to be "clean, well seasoned and free from insects." N-80-12-21:272–75, Lab. (*TAEM* 38:668–69; *TAED* 140:1–2).

6. William H. Moore reportedly had been acquainted with Edison since 1875 but more recently was selling Weston generators in England. On 6 October Stockton Griffin instructed William Carman to give Moore $530 in cash and have him call on Calvin Goddard; a letter of credit had already been arranged (Jehl 1937–41, 620; Griffin to Carman, 6 Oct. 1880, DF [*TAEM* 53:876; *TAED* D8020ZHL]). Francis Jehl recalled that before Moore left

Edison showed him what the desirable characteristics and qualities were that he was after; he instructed him how to proceed and how to test bamboo—of which there were over two hundred species, some reaching a height of 120 feet and a diameter of one foot. Edison also sent him into the library to look up and study Japan together with everything pertaining to it. . . . When he was ready to start Edison provided him with various utensils and apparatus, among which was a good microscope. He was given letters and credentials to our political representatives in China and Japan, in addition to which he received letters of introduction to business houses in the East from friends of his own in England. A few of us boys gave him a quiet send-off. From that day he was called 'Japanese Moore.' [Jehl 1937–41, 620–21]

Moore later worked several years for the Edison Co. for Isolated Lighting and the Edison Machine Works (Jehl 1937–41, 622–23).

–2003–

Notebook Entry:
Electric Lighting

[Menlo Park,] Oct 19 1880

Tried the experiment of washing out the dyro-carbons[1]

~~Volatized~~ First one I look and tried was to bring the carbon up bright when I had the vapor of Chloroform I got a thick deposit of carbon.

~~second.~~ first was a carbon tried in the vapor of Bi-sulphide. This Carbon was burning for about ten minutes in the vapor and then I took it out, the carbon seemed to be good[a]

Second   Def[lection] 265   This carbon was in the vapor for 15 minutes and gave a good light, and carbon when taken out was good

Third. 210   This carbon was burning in the vapor of Bi-sulphide for about 20 minutes and gave a good bright light. I then took it out.[a]

X, NjWOE, Lab., N-80-07-05:168 (*TAEM* 36:351; *TAED* N104:86). Written by Francis Jehl. [a]Followed by dividing mark.

1. Jehl meant hydrocarbons. Four days earlier Edison had executed a patent application for a method of removing hydrogen in the carbon by passing "Chlorine gas, or some gas which combines readily with hydrogen, but not with carbon" through a flask containing the carbon heated to a high temperature (U.S. Pat. 239,148). Jehl evidently began these experiments on 16 October and continued them until 22 October, when Charles Mott recorded that he did not get "very favorable results." N-80-07-05:172, Mott Journal N-80-07-10:168, 177; both Lab. (*TAEM* 36:353, 37:386, 390; *TAED* N104:88; N117:84, 88).

[Menlo Park,] Oct 21st 1880[a]

All employes are requ[ired][b] to ~~put a~~ keep their time [sheets?][b] in view and to put down their time accurately ~~every night~~ charged[c] to the propper account Every Night[1] ⟨OK E.B.[2d]⟩

AD, NjWOE, DF (*TAEM* 54:371; *TAED* D8030I). Written by John Kruesi. [a]Date written by Ernest Berggren. [b]Obscured by stain and damage to paper. [c]Interlined above. [d]Marginalia written by Ernest Berggren.

1. Time sheets from this period are in Box 143, Employee Records, NjWOE.

2. Ernest Berggren was hired as an assistant bookkeeper in April 1880 at the age of sixteen. He remained associated with Edison for most of his working life, becoming head accountant for Edison General Electric and then for General Electric. He was later secretary and treasurer for the Edison Phonograph Co. and Thomas A. Edison, Inc. "Berggren, Ernest J.," Pioneers Bio.

[Menlo Park,] Oct 21, 1880

Made an experiment by generating gas from (gasoline) in a flask and then passing it in a mould in which there was some carbons. When finished, we found that they were covered too much.[1]

X, NjWOE, Lab., N-80-07-05:174 (*TAEM* 36:354; *TAED* N104:89). Written by Francis Jehl.

1. Experiments on treating carbon filaments with hydrocarbon vapor continued intermittently for several weeks. On 25 October Charles Mott reported that

> Mr. Edison and Frances are to day experimenting on building up and making carbons more homogeneous   paper was soaked in tar and placed in mould. Napthaline gas was passed through the mould while in the furnace   the paper carbonized hard, smoothe, very homogeneous with a ring like a piece of steel. A lamp was then put on a pump and napthaline vessel connected with gauge tube. After passing the mercury for a time to remove a portion of the air and permit the gas to replace it, the loop was carefully heated by the current, but at first gave every appearance of oxidization and was exceedingly irregular. after running the pump for a few minutes longer, the current was again applied and the loop gradually evened up and soon gave appearance of being entirely even. The inside of the globe was however also coated   the carbon was removed by Mr. Edison and found to be nicely coated and of good appearance under the microscope and more tough or tinasious than the ordinary carbon. Several trials were then made of passing napthaline through the mould, in the furnace, in which was placed Bamboo strips &c. but with indiferent and varying results.

The following day Mott noted that they were continuing the experiments and had "fair success except one explosion which they had prepared for by placing the lamp in a box." On 4 November "Six carbons treated in gasoline gas were taken to the Factory to be put in lamps. Two put in and sent up were exhausted and broken by Francis before any test could be made or merits noted." The next day treated lamps were completed at the factory and sent to the laboratory where Edison and Jehl continued their experiments. They "worked very late on treating carbons and carbonizing in the furnace with nickel mould and gasses with varying success" (N-80-07-05:175–80; N-80-10-25:1, 3–4; N-80-10-01: 15, 17, 29; Mott Journal N-80-07-10:181–83, 189, 193–94; all Lab. [*TAEM* 36:354–57; 34:136, 139–40; 41:1074–75, 1084; 37:392–93, 396, 398–99; *TAED* N104:89–92; N060:1, 4–5; N304:9–10, 20; N117:90–91, 94, 96–97]). On 16 November Albert Herrick began treating filaments with hydrochloric acid and other chemicals to build them up, then promptly returned to hydrocarbons. Mott reported on 18 November that a filament built up with gasoline vapor was lighted at 30 candlepower for about an hour then raised to about 500 candlepower for 4½ minutes before it broke (N-80-11-16:1–45, Mott Journal N-80-07-10:210, both Lab. [*TAEM* 37:884–905, 407; N125:1–23, N117:105]).

This research resulted in several patent applications. On 5 November Edison executed one for a process of passing hydrocarbon vapor through a flask containing heated carbon filaments. Edison indicated that the hydrocarbon vapor filled ruptures in the fiber structure which formed during carbonization and reduced filament life. Two weeks later he completed another application covering the use of this process for enlarging the ends of the carbon in order to form a better connection with the clamps. At the end of December he executed a third application, this one for building up lamp carbons by enclosing them with naphthalene and focusing an arc lamp or other heat source on particular weak spots, so that the hydrocarbon would vaporize and be deposited there. In November Charles Mott remarked that Edison had found that naphthalene crystals sealed in a lamp and volatilized by the heat of the filament produced "a smoother and more perfect coating and building" than other substances. Edison incorporated this discovery in the third application, in which he stated that a small number of naphthalene crystals could be sealed in the lamp globe to repair "the waste of carbon due the process known as 'electrical carrying.'" U.S. Pats. 248,426, 239,151, and 248,416; Mott Journal N-80-07-10:212, Lab. (*TAEM* 37:408; *TAED* N117:106).

–2006–

To W. H. Merrick

[Menlo Park,] Oct. 27th [1880]

Dear Sir:

From his visit to your works and conversation with your foreman, Mr. Batchelor gained the impression that the sole-plate and pillow-blocks are nearly completed and possibly can be shipped in a week. If this is so, if you can accommodate us by pushing those parts on to completion and can ship them to us on Tuesday, Nov. 3rd, without fail, please let us know immediately and we will delay work upon the magnets so that they can be finished in place.[1]

If you cannot do this in the stated time[a] we must resort to the old method of finishing on the lathe.

Every little delay is embarassing to us at this time and <u>we cannot wait longer</u>. I cannot comprehend why there should be such an unexplained delay in the completion of the engine, neither could Mr. Batchelor, judging from the superficial observations of a visit, excepting that no one seemed to be working upon the different parts, and to him it seemed a question of labor. As to that of course we cannot judge, and know not whether to ascribe it to that fact, or that some plans or proportioning of parts, which involve important principles essential to perfect success, may not be yet fully determined upon.[2]

Be that as it may, it has caused great uneasiness and irritation among the prominent members of the Electric Light Co. and a consequent state of nervousness and unrest on our part, placing us between two fires. Not only this, but the additional expense of delay amounts to thousands of dollars.

Highly as we endorse your engines, believing them to be the best for workmanship, speed, reliability, and economy, we are nevertheless fearful that after the trial we shall be left in the lurch, without engines or any hope of getting them, and for self-protection and to advance our interests that we shall have to adapt the various forms of direct and indirect engines to our dynamo. I hear with pleasure that you have ordered a large number of fine tools and have taken measures to obtain them with the least delay.

Will you please put all the pressure which you can possibly bring to bear towards completing this engine?

Every week delayed increases the difficulties. Yours truly

Thos A Edison

LS (letterpress copy), NjWOE, Lbk. 6:501 (*TAEM* 80:399; *TAED* LB006501). Written by Charles Clarke. [a]"in . . . time" interlined above.

1. Charles Batchelor and John Kruesi spent Tuesday, October 26, in Philadelphia. According to Charles Mott, they made a rather discouraging report to Edison about the engine but thought the bed plate would arrive "in time to facilitate matters here, and obviate any delay on the dynamo part." On 28 October Merrick replied to this document that "If the country were not to be saved or otherwise we would finish the bed in 1 hot day   As it is we can finish it on Wednesday [4 November] & ship it at once on completion." Mott indicated that the bed plate reached the Menlo Park depot on 13 November and was brought to the machine shop two days later. On 20 November, about the time the bed was being installed on its new foundation in the shop, the "polar extensions of the

magnets of large dynamo were placed on the large lathe preparatory to boaring out the helix." Merrick to TAE, 28 Oct. 1880, DF (*TAEM* 53: 920; *TAED* D8020ZIJ); Mott Journal N-80-07-10:183–84, 202, 204, 207, 211, 213, Lab. (*TAEM* 37:393–94, 403–4, 406, 408–9; *TAED* N117: 91–92, 101–2, 104, 106–7).

2. The expansive shops occupied by the Southwark firm had been unused for some time, during which the tools became unfit from neglect. Porter, the foundry's vice president for manufacturing, had expected to order new ones when the company was organized in the first part of 1880 but did not receive authority to do so until mid-September; as a result the company built no engines during the summer. About this time Porter did order a large amount of equipment, principally from England. When Edison inquired in December about having another engine built, Merrick replied that "we have decided not to take any more orders on guarantee as to time of delivery until we are very much better prepared to push work through in lots." Shortly after that, however, Merrick was willing to consider a contract to build a smaller engine for the steamship *City of Rome,* for which Edison was planning a lighting system. Porter 1908, 276–81, 287, 291, 295–96, 299–301; Merrick to TAE, 14 Dec. 1880, DF (*TAEM* 53:1000; *TAED* D8020ZJX); Clarke to Merrick, 18 and 23 Dec. 1880, Lbk. 6:690, 718; (*TAEM* 80:445, 451; *TAED* LB006690, LB006718) N-80-00-02:1, Lab. (*TAEM* 39:1137; *TAED* N179:1).

<table>
<tr><td>

–2007–

Notebook Entry:
Electric Lighting

</td><td>

[Menlo Park,] Oct 31 1880

</td></tr>
</table>

I propose to make a pulping machine that will pulp paper fibre to extreme fineness & make paper from it & submitting the paper to hydraulic presses & afterwards cut carbons from it & coat them after carbonization with a layer of gas carbon—[1]

I also propose to comb the finest flax or hemp fibre longitudinally & moisten with a gummy substance which will be carbonizable & press sheets of this by pressure & afterwards cut as with Bamboo—

Chas Batchelor   TAE

X, NjWOE, Lab., N-80-10-25:44 (*TAEM* 34:179; *TAED* No60:44). Document multiply signed.

1. On 30 October Edison filed a patent application that was subsequently rejected by the Patent Office in which he claimed a method of "forming carbon articles of a definite desired shape, consisting in cutting or shaping the articles from paper, and then carbonizing the shaped paper, while under pressure or strain." He also claimed "as a new article of manufacture, flexible carbon in sheets, or in definitely shaped articles, formed from sheets of carbonizable material" (E-2536:198, PS [*TAEM* 45:719; *TAED* PT020198]). Edison's U.S. Patent 242,900, executed on 21 October and filed on 5 November, covered a process of forming carbons in any shape by passing a vapor of bisulphide of carbon, chloride of

carbon, volatile paraffin, or naptha through a flask until a sufficient quantity was deposited on a metal form. This was then placed in a bath of acid until the metal was eaten away, leaving only the carbon.

−2008−

Notebook Entry:
Electric Lighting

<u>Central Station</u>   (1)

Estimate for 10 000[a] lamps fed from a central station each giving 16 candles

It is found that 8 of these may be obtained ~~of~~ from one horse power indicated[b]

Say that a gas burner giving 16 candles consumes 5 feet of gas an hour for comparison

200 hours for 1000 feet or 10 000 lamps will consume 50,000 feet an[a] hour

Call in electricity an equivalent of a 1000 cu feet of gas an M

50 M an[a] hour   250 M a day   310[a] days in year in the district chosen

$$\begin{array}{r} 310 \\ \underline{250} \\ 15,500 \\ \underline{62} \\ 77,500 \text{ M cu. feet a year}^b \end{array}$$

Old[a] estimate[2]
$159,300 invested

$$\begin{array}{r} 5.2022 \\ \underline{4.8893} \\ .3129^a \end{array}$$

$2.05 investment per M.

At 8 per H.P. 10 000 lamps will take 1250 H.P. It is estimated that 1200 H.P. will be able to supply this amount.

<u>Structure</u>[c]

This can be placed in one building 25′ X 100′.

The iron structure is estimated in Book 100 p. 50[3]

$$\begin{array}{r} 120,000 \text{ lbs of iron} \\ \underline{.05} \\ \$6000.00 \end{array}$$ $6,000

Foundation	2,500
Fire proof floors	2,000
	10,500[a]

Babcock and ~~w~~Wilcox estimate. see letter[4]

600 H.P. boiler in place with economizer	$12,875
Stack	800
Steam pump	375
Blower	300
Engine with counter shafts	650
Piping	165
Ash elevator	400
Coal bunker	800
600 H.P.[d]	$16,365

1200 H. P. Boilers   (1)

Complete Boilers	$24,500
Stacks	1600
Steam Pumps	750
Blowers	600
Dynamos for Blowing	1000
piping	330
Ash Elevator	400
Coal bunker	1000
	$30,180

Engines Dynamos[d]

Mr. K.[5] estimates cost at $4800 $\times$ 10= 48 000

Extra Electrical apparatus $2000

Total   (1)

Building	8,500
Boilers	30,180
Engines Dynamos	48,000
Extra Electrical	2000
	88,680
Conductors	2000
	90,680[a]
175 ohm lamp	$27,000
Pipes	$30,000
	$57,000[a]
	88,680
	145,680
Meters	5,000
	150,680
	2,000
	152,680

Station

Boilers 10%	$3018
Building 2%	170
Engines Dynamos 3%	1440
Extra Electrical 2%	40
Meters 5	250
Conductors 2% on whole 57,000	
.02	
1140.00	1140
	6058

Labor

1 Engineer, chief	5.00
1 Engineer	3.00
1 Wiper[a] @ 1.50	1.50
1 1st fireman @ 2.25	2.25
1 fireman @ 1.75	1.75
2 laborers @ 1.50	3.00
1 Regulator @ 2.25	2.25
1 Regulator[a] @ 1.75[a]	1.75
dayly	20.50
	365
year	7,482

Data[b]

$200 per H.P. per year [-----][e] delivered[b]

Present machine cost[f] good for 70 lamps	$350
Conductors copper	27,000[a]
Pipes	25,000
Insulation	5000
	$57,000

The labor account is taken thus a cheif engineer who will be on duty from 12 M to 12 midnight   An assistant who will be on duty from 12 midnight to 12 M. A wiper who will be on duty from 7 A.M to 7 P.M. One fireman @ $2.25 who will be on duty from 12 M to 12 midnight   One fireman @ $1.75 who will be on duty from 12 midnight to 12 M   One laborer from 7 A.M. to 7 P.M another from 12 M to 12 midnight. One regulator @ $2.25 from 12 M to 12 midnight one @ $1.75 from 12 midnight to 12 M.

Thus the cheif engineer will be on duty during the most important part of the day, and the wiper will be under each of the engineers. The head fireman will also be on duty and have[g] between 5 & 6 P.M two laborers. This system will offer a good chance for promotions

Executive expenses   year $4000

Coal

$2.80 per ton delivered   3 lbs per H.P. per hour

1200

$\underline{3}$

3600 lbs per hour

$\underline{5}$

18,000 lbs per day   8.03 tons per day[6]

$22.50 daily $8212.5 year

Oil, waste, water[7] taken as ⅓ coal

yearly $ 2737
Rent insurance taxes $ 7000

Summary

Depreciation $ 6058
Labor   7,482
Executive   4,000
Coal   8,212
Oil waste &c   2,737
Rent &c $\underline{7,000}$
                                              $35,489
Lamps $\underline{10,500}$
                                               45,989

If 10 000 light can[a] be sold for 5 ~~f~~hours daily[a] ~~if~~ it is equivalent to 250,000 ~~Cu. feet of~~ Cu feet[h] gas.

365
$\underline{250}$
18,250 ~~M yeally~~
$\underline{730}$
91,250
$\underline{1.50^8}$
45,625.00
$\underline{9125}$

Receipts   136,875.00
Expenses $\underline{45,989}$
           $90,886 [~~pat?~~][e] to pay for patent rights and interest

If company capitilizes at twice the cost of plant $150,680

$$\frac{2}{301,360}$$

The receipts will pay a dividend of 30 per cent. 60% on investment

X, NjWOE, Lab., N-80-11-15:71 (*TAEM* 39:817; *TAED* N172:36). Written by Francis Upton. Miscellaneous rough calculations not transcribed; various erasures not indicated. Some commas added for clarity to numerals in calculations. ᵃObscured overwritten text. ᵇFollowed by dividing mark. ᶜInterlined in left margin and multiply underlined. ᵈMultiply underlined. ᵉCanceled. ᶠInterlined above. ᵍ"and have" interlined above. ʰ"Cu feet" interlined above.

1. In the forty pages following this notebook entry Francis Upton wrote several similar undated estimates based on different assumptions. Then he made a series of tabulations, also undated, of data from the first three central station district survey books returned by James Russell (see Doc. 1995). Russell brought the third book to Menlo Park on 9 October and the fourth on 16 October. N-80-11-15:143–49; Mott Journal N-80-07-10:159, 168; both Lab. (*TAEM* 39:853–56; 37:381, 386; *TAED* N172: 72–75; N117:79, 84).

2. No extant earlier estimate of central station costs matches the numbers found here. For earlier estimates, see Docs. 1897 and 1958.

3. Upton apparently made a mistake in the number of the book. There is nothing on this page in Book 100 and no estimate of the cost of the iron structure can be found in that book. The estimate in question has not been found. N-81-04-12, Lab. (*TAEM* 36:55; *TAED* N100).

4. See Doc. 1958 n. 1.

5. Probably John Kruesi. The estimate has not been found.

6. That is, long tons.

7. Upton calculated the cost of water on page 82 but apprently decided instead to include it along with oil and waste as a percentage of the coal costs. N-80-11-15:82, Lab. (*TAEM* 39:876; *TAED* N172:42).

8. On the facing page Upton divided annual operating expenses of $45,749 (apparently based on figures that he later amended) by the yearly M (91,250) to determine a cost of 50 cents per M. He then calculated gross receipts based on a charge of $2.25 per M but here he uses the figure of $1.50 per M instead. N-80-11-15:96, Lab. (*TAEM* 39:883; *TAED* N172:49).

-2009-

To Vesey Butler

[Menlo Park,] Nov 4 [1880]

Dear Sir:

Your kind favor of the 28th ult with enclosures is just at hand. I am greatly obliged to you for the kindness shown to poor Segredor. Your cable announcing his death was a great shock & surprise to us all.[1] As far as I can learn he had no relatives in this country. I think he told me this himself, and I believe he said that he had no near relatives living. He had been injured in a railroad accident and was laid up in a hospital at Newark N.J. some 6 weeks—at one time little[a] hopes of his life being saved was entertained. I dont think he had been out of the hospital month when he headed for Havana[2]   We all cautioned him about his diet and about drinking cold drinks but as you say he was very self-willed and would always do in these respects about as he pleased and this I doubt not caused his death.[3]

I have not yet determined what to do in regard to filling his place. I have a man now en route from San Fran for Japan to get cane and also have two parties engaged in South America so that I am pretty sure of getting an abundant supply soon.[4]

I will remit the amount due you in a few days.[5] Very truly

T. A. Edison

L (letterpress copy), NjWOE, Lbk. 6:521 (*TAEM* 80:403; *TAED* LB006521). Written by Stockton Griffin; circled "C" at top of page.
[a]Obscured overwritten text.

1. Butler had cabled from Havana on 27 October "John dead vomito Telegraph instructions," to which Edison replied: "Bury him my expense." Butler explained in his 28 October letter that Segredor had died from "one of the severest & most virulent cases" of yellow fever seen by the attending doctors. Only one enclosure, an inventory of Segredor's personal effects, has been found but Butler stated he was also sending receipts for expenses paid, including a charge for Segredor's bedding which the landlord insisted on having burned. Butler to TAE, 27 and 28 Oct. 1880; TAE to Butler, 27 Oct. 1880; all DF (*TAEM* 53:912, 914; *TAED* D8020ZIF, D8020ZII, D8020ZIG).

2. According to John Kelly, who wrote Edison on letterhead of the Western Electric Manufacturing Co., Segredor was "severely injured by jumping from a train" on 15 July when it did not stop in Newark as he had expected. Kelly stated that Segredor had no friends and had asked him to contact Edison. Kelly to TAE, 16 July 1880, DF (*TAEM* 53:182; *TAED* D8004ZDV).

3. Butler reported in his 28 October letter (see note 1) that the doctors thought Segredor was "predisposed" to yellow fever, "having suffered from malaria & peritonitis also from the fact that he was constantly taking excessive doses of quinine." Segredor also "exposed himself very much to the Sun and eat very heartily both against my advice." Butler stated that he had cautioned Segredor against cold drinks "but he had a special temper of his own, self willed & invariably did as he thought fit." In one of his lengthy dispatches from Florida, Segredor wrote in late September that he had suffered the chills and fever endemic to the area and as a result was "taking heavy doses of quinine night & morning & it has left me afraid." Segredor to TAE, 25 Sept. 1880, DF (*TAEM* 53: 855; *TAED* D8020ZGY).

4. See Doc. 2012. Edison evidently also had an arrangement with Fabbri & Chauncey to get plant samples from their South American as-

sociates. TAE to Fabbri & Chauncey, 11 Nov. 1880, Lbk. 6:543 (*TAEM* 80:407; *TAED* LB006543).

5. Edison presumably meant not only the expenses on Segredor's behalf but also $320 for cane samples Butler had recently collected. Butler to TAE, 27 Nov. 1880, DF (*TAEM* 53:913; *TAED* D8020ZIH).

–2010–

From Henry Morton

HOBOKEN, NEW JERSEY.[a] Nov. 4th 1880.

Dear Sir

I am now engaged in making a series of measurements with various incandescent electric lamps and would be pleased to have one or more of your recent lamps to measure at the same time, in which case I should give my results in a paper on the general subject which I am preparing to read before the National Academy at their coming meeting.[1] From statements I have seen in the papers I judge that you may be under the impression that I have some interest, financial or other, which might bias my judgement unfavorably towards your efforts in developing the incandescent electric lamp; I therefore take this occasion to assure you that this is not the case and that my feelings towards you and your labors are altogether of the most friendly description and that no one will be more ready than I to acknowledge the full value of every thing you accomplish as fast as each result is reached. Very truly yours

Henry Morton

⟨Our large mach for pumping up merc from low to high level is broken[2]   we expect to have it going in 5 or 6 days soon thereafter as possible I will send ayou a doz lamps for test— lamp we use now is about 140 ohms & requires 30400 ft pds running 16 candles— I certainly have belv'd that you have not treated me exactly right for reasons wh I cannot fathom⟩[3]

ALS, NjWOE, DF (*TAEM* 53:927; *TAED* D8020ZIN). Letterhead of the Stevens Institute of Technology. [a]"HOBOKEN, NEW JERSEY." preprinted.

1. Morton's paper has not been found but reportedly was titled "Measurement of new form of electric lamps operating by incandescence" and was presented at the mid-November meeting of the Academy in New York; see also Doc. 2017 n. 1. National Academy of Sciences 1884, 52.

2. The *New York Herald* reported two weeks later that Edison's lamp factory was "not working at its full force because of a leakage in his mercury pump, which supplied the 500 vacuum pumps. The leakage caused the salivation of a number of his employés, and obliged him to suspend operations until a new mercury-tight pump was procured." Because of mechanical problems with his chain-driven mercury pump Edison had

John Kruesi design and order a new screw pump in late September (see headnote, p. 767 n. 17) but it did not arrive until 4 December. Charles Mott noted after its first operation on 8 December that "although run at slow speed it passed the mercury up very nicely and very encouraging results are reasonably expected." "Electric Light," *New York Herald*, 18 Nov. 1880, Cat.1241, item 1541, Batchelor (*TAEM* 94:617; *TAED* MBSB21541X); Mott Journal N-80-07-10:232, 238, Lab. (*TAEM* 37: 418, 421; *TAED* N117:116, 119).

3. Morton thanked Edison for his promise of lamps and stated that although he often disagreed with statements attributed to Edison in the daily press he always tried to "express my dissent or opposing views in an inoffensive manner, and I think that you will find that I have accompanied such statements with an expression of my high appreciation of your personal abilities." He added that Edison's "unfavorable impression . . . must be the result of some misunderstanding, and if you will kindly point out specifically, in what I have seemed to do you a wrong I feel confident that an explanation will clear up the difficulty. My only motive has been a desire to check the extravigance of certain ignorant and irresponsible newspaper writers and prevent some of the damage which their exagerations seemed likely to occasion." Stockton Griffin indicated on this letter that Edison did not reply. Morton to TAE, 8 Nov. 1880, DF (*TAEM* 53:932; *TAED* D8020ZIQ).

–2011–

From Lemuel Serrell

New York, Nov 5 1880[a]

My Dear Sir

Please bear in mind the appointment for Monday next on the Telephone cases A. to N. and No. 1; This takes in the whole history of the development of the telephone on your part— Please look this matter over sufficiently to be generally posted; I will come over early so as to go over matters with you before the examination; I will try and reach M. P. by 8 oclock.[1]

Please bear in mind the request that you make notes of such things as you think important to patent which you have not applied for as yet—

If you have not returned the power atty for Germany sent to you by Mr McKenzie please do so tomorrow.—[2] Yours truly

Lemuel W. Serrell

ALS, NjWOE, DF (*TAEM* 55:202; *TAED* D8036ZFF). Letterhead of Lemuel Serrell. [a]"New York," and "188" preprinted.

1. On the origin and disposition of this set of telephone interference cases, see Doc. 1270, esp. nn. 1 and 3. Edison's testimony began on Monday morning, 8 November, and continued until Thursday, 11 November; Charles Batchelor's testimony began on Friday, after which the proceedings were adjourned until the second week of December. Edison's and Batchelor's testimony, TI 1:3–133, 223–62 (*TAEM* 11:22–108; *TAED* TI1:19–84, 85–105).

2. The power of attorney has not been found and McKenzie has not been identified.

–2012–

To Fabbri &
Chauncey

[Menlo Park,] Nov 8th [1880]

Gentlemen

I am about sending an Agent to South America to procure samples of bamboo or cane. He will visit Para Pernambuco, Bahia, Rio Janeiro and Buenos Ayres. I would consider it a special favor if you will give him letters of introduction to your correspondents in those places. His name is John C. Branner.[1] Mr. Branner or some one from my works here will call on you for the letters.[2] Very truly

T. A. Edison—G[riffin]—

L (letterpress copy), NjWOE, Lbk. 6:531 (*TAEM* 80:406; *TAED* LB006531). Written by Stockton Griffin; circled "C" at top of page.

1. Edison described Branner to Fabbri & Chauncey as "a first class botanist" who "has been all through South America." He promised that Branner would "not only search for the proper kinds of bamboo but will also collect samples of all other classes of fibres which may be of use." John Branner studied geology at Cornell University and as an undergraduate participated in a major survey of Brazil from 1874 to 1877. He remained in that country in connection with gold mining until 1880, when he returned to Brooklyn. He subsequently became professor of geology at Indiana University and then Stanford University, which he served as president from 1913 to 1916. TAE to Fabbri & Chauncey, 11 Nov. 1880, Lbk. 6:543 (*TAEM* 80:407; *TAED* LB006543); *ANB*, s.v. "Branner, John Casper."

2. No letters have been found. Branner executed the contract of his employment on 15 November. At the end of the month he reported on his visit to the botanical museum of Cornell University. At that time he acknowledged Edison's letter of credit and outlined his travel plans. He also described his itinerary in detail in a newspaper interview shortly before his departure for Para, Brazil, on 4 December. Agreement with John Branner, 15 Nov. 1880, Miller (*TAEM* 86:308; *TAED* HM800131); Branner to TAE, 30 Nov. 1880, DF (*TAEM* 53:970; *TAED* D8020ZJH); "Into Brazil for Edison," *New York Times,* 27 Nov. 1880, 8.

–2013–

From George Soren

NY Friday P.M. 12th Nov 1880[a]

Dear Mr Edison.

As Secretary of the meeting of Trustees of E. E. Light Co held to day, I was instructed to advise you of a resolution thru paper to the effect that the Executive Committee be requested to visit Mr Edison at Menlo Park, early next week for the purpose of consulting with him as to measures to be taken for bringing the Company light before the public—

The Committee will come down on Monday afternoon, arriving about four oclock[1]

The resolution also requires the Secy to request Mr Edison to have Mr Wilbur present at the same time.[2]

I am sorry that I have not by me, at this moment, the exact text of the resolution   It shall be sent to you however on Monday.[3] With best wishes, Sincerely yours

G. W. Soren

ALS, NjWOE, DF (*TAEM* 54:100; *TAED* D8023ZAX). [a]"NY" and "Nov 1880" written by Stockton Griffin, presumably upon receipt.

1. Grosvenor Lowrey was ill and unable to attend the meeting at Menlo Park. He and Soren both telegraphed Edison to ask him to go to Lowrey's home in Tarrytown in order to arrange some contracts and other matters but Edison replied that it was "impossible to go we are working night & day here." Lowrey to TAE, 15 Nov. 1880; Soren to TAE and TAE to Soren, both 16 Nov. 1880; all DF (*TAEM* 54:103–4; *TAED* D8023ZBA, D8023ZBC, D8023ZBD).

2. Soren telegraphed Edison before the meeting to remind him that the committee wanted patent attorney Zenas Wilber in attendance. Soren to TAE, 15 Nov. 1880, DF (*TAEM* 54:103; *TAED* D8023ZAZ).

3. Not found.

–2014–

From B. Franklin Fisher

Philadelphia, Nov. 16 1880[a]

Dear Sir

I have in my hands for collection upon behalf of Mrs Seyfert two notes amounting with interest to about seven thousand dollars—and these notes are about six years old I will be compelled to commence suit in order to preserve the rights of the holder unless the same are paid forthwith—[1] I called to see Mr Reiff about them— While he admitted their correctness he plead present inability and wanted delay until pending[b] suits should be determined—[2] this I cannot do—time compels me to insist upon their present payment is being secured. Please let me know your pleasure by return mail[3]   Yours truly

B. F. Fisher[4]   atty for Mrs Seyfert

ALS, NjWOE, DF (*TAEM* 55:575; *TAED* D8042ZBM). Letterhead of law office of B. Franklin Fisher. [a]"Philadelphia," and "188" preprinted. [b]Obscured overwritten text.

1. Mrs. Seyfert was Lucy Seyfert, the wife of William Seyfert, a Philadelphia investor in the Automatic Telegraph Co. As part of a complex plan to raise cash in 1874, Edison gave promissory notes to George Har-

rington, who later endorsed at least one to William. This was never redeemed but instead passed to Lucy Seyfert in a business arrangement with her husband. See *TAEB* 2:235 n. 3, 363 n. 2; Lucy Seyfert's testimony, 12–13, *Seyfert v. Edison,* Lit. (*TAEM* 46:415; *TAED* QD011:8).

William Seyfert wrote to Edison in April requesting payment of the note held by his wife. He hoped that Edison would "arrange this matter as early as possible as you know it is of long standing— In the meantime please let her have $500. for immediate use & pressing wants either on a/c or a loan." Edison had recently loaned $300 to the couple. In June Seyfert told Josiah Reiff that having failed to reach agreement with Edison he would hand the matter over to an attorney. Reiff sent a copy of this letter to Edison with the recommendation to "Be advised by me in this matter— I told him not to make a mistake by any hasty action— Rest easy." Seyfert to TAE, 7 Apr. 1880; Reiff to TAE, 9 Apr. 1880; Seyfert to Reiff, 16 June 1880, enclosed with Reiff to TAE, 18 June 1880; all DF (*TAEM* 55:537, 539, 560, 559; *TAED* D8042ZAF, D8042ZAG, D8042ZAW, D8042ZAV).

2. After meeting with William Seyfert in May, Reiff suggested that Edison "had better not discuss the question of responsibility" for the contested note. He added that if he could obtain "a settlement with A&P or W.U. all those questions will settle themselves. Meantime, it might be very desirable for you to loan S a little—this will carry the matter along." Reiff was referring to *Harrington v. A&P,* for which testimony was taken during the summer, and to his efforts to recover more than seven thousand dollars allegedly owed him under the terms of Doc. 876. Reiff to TAE, 19 May and 16 Aug. 1880; Reiff to Norvin Green, 17 Apr. 1880; all DF (*TAEM* 55:556, 571, 544; *TAED* D8042ZAS, D8042ZBI, D8042ZAJ).

3. Edison replied on 19 November, on the basis of a draft written on this letter, that he had "never received any money from Mr Seyfert on the notes referred to   they were made for the benefit of the Automatic Telegraph Company (of which Mr Seyfert is a large stockholder) for the purpose of paying a debt due me from the Automatic Telegraph Co. I can do nothing about it." When he acknowledged receipt of a subsequent letter from Edison about the matter, Reiff advised him not to "complicate yourself by correspondence. Just write Fisher & say you have referred the matter to me, & that you are advised the notes were paid by crediting Seyfert with A&P stock." TAE to Fisher, 19 Nov. 1880, Lbk. 6:572 (*TAEM* 80:417; *TAED* LB006572A); Reiff to TAE, 22 Nov. 1880, DF (*TAEM* 55:579; *TAED* D8042ZBP); on the Automatic Telegraph Co. stock transaction, see Docs. 522 and 561.

Lucy Seyfert filed suit in New Jersey state court in November 1880. The case went to trial in 1882 over the payment of a single note (Doc. 516). The court issued a directed verdict in her favor and the jury awarded her $5,065.84. The matter continued to dog Edison for several years thereafter. *Seyfert v. Edison,* Lit. (*TAEM* 46:407; *TAED* QD011).

4. Fisher's letterhead indicated he maintained a law office on South Third St. in downtown Philadelphia.

[Menlo Park,] Nov. 17th [1880]

Dear Sir,

I would like to know if any alterations in the dimension of cylinder or admission and exhaust pipes have been made since your original design was drafted, of which we have the tracing?

This will determine some particulars in our settings.

We wish to lead your exhaust direct into a large Berryman Feed-water Heater which will show a decided gain in economy[1]

I presume that you have no objection to so doing, the aggregate area of the tubes being several times larger than the exhaust.

Mr. Edison gained the impression from Mr. Church of the Buckeye Co.[2] that you have the impression the dynamo will not be ready for the engine for at least two months. Nearly all the hands in the machine shop are at work upon it and it is being brought very far towards completion and will be so in much less than three weeks. Every part is here and most of them completed. The armature is in such a condition than the whole force can in a few days be put upon it and soon complete it.[3]

We hope to have the engine in three weeks.[4] I am, Yours very truly,

C. L. Clarke.

ALS (letterpress copy), NjWOE, Lbk. 6:557 (*TAEM* 80:412; *TAED* LB006557).

1. Feedwater heaters were employed to increase an engine's efficiency by transferring heat from exhaust steam to the incoming water before it was injected into the boiler. This particular device, patented in 1872 and made by the Berryman Manufacturing Co. of Hartford, Conn., was in wide use and considered to be among the most effective heaters. Knight 1876–77, s.v. "Feed-water heater"; "Water-Heater, Regulator, and Alarm for Steam Boilers," *Manufacturer and Builder*, 3 (1871): 217–18; "Improvement in the Heating of Feed-Water for Steam Boilers," ibid., 4 (1872): 193–94.

Clarke wrote to Porter on 18 November concerning the shaft dimensions, and again the following day to ask for "a drawing of cylinder giving the details of the admission and exhaust pipes, with exact dimensions." He included a sketch of the proposed connections to the feedwater heater. A week later he asked for the various fittings needed to install the pipes. Clarke to Porter, 18, 19, and 26 Nov. 1880, Lbk. 6:559, 566, 595 (*TAEM* 80:413, 415, 430; *TAED* LB006559, LB006566, LB006595).

2. William Church was a manager of the Buckeye Engine Co.'s New York office. He briefly visited Menlo Park this day. Letterhead of Buckeye Engine Co. to TAE, 24 Feb. 1880, DF (*TAEM* 53:669; *TAED* D8020ZBW); Clarke to Church, 17 Nov. 1880, Lbk. 6:556 (*TAEM* 80:411; *TAED* LB006556).

3. Charles Mott's 20 November summary of work during the week indicated that the "Bed plate for Dynamo Engine placed in shop and some preliminary fitting of the parts effected. Discs secured on the ar-

mature shaft, and work on the commutator etc. progressing satisfactorily under Dean." The next week Mott noted "Dean and several assistants pushing work on the large armature. Logan and others on magnets, base, etc. of large dynamo." By 2 December the armature disks had been turned on the lathe and were being finished; Charles Dean spent 7 and 8 December soldering the connections. Mott Journal N-80-07-10:214, 225, 229, 236–37; N-80-07-27:137; both Lab. (*TAEM* 37:409, 415, 417, 420–21, 251; *TAED* N117:107, 113, 115, 118–19; N116:70).

4. Edison told a newspaper reporter one week later that he had planned on starting the Menlo Park demonstration on 15 August "but I was disappointed in not getting the steam-engine ordered from a firm in Philadelphia and promised before that date. I have not yet received the engine, but it is now positively promised me in a little more than two weeks." Charles Mott recorded that the Porter-Allen engine was delivered to Menlo Park on 4 January, 1881, although the engine shaft was delayed at least another week. "Edison's Work," *New York Tribune*, 26 Nov. 1880, Cat. 1241, item 1547, Batchelor (*TAEM* 94:618; *TAED* MBSB21547X); Mott Journal N-80-07-10:274, Lab. (*TAEM* 37:439; *TAED* N117:137); Clarke to Porter, 10 Jan. 1881, Lbk. 6:780 (*TAEM* 80:462; *TAED* LB006780).

–2016–

Charles Clarke to J. W. Thompson[1]

[Menlo Park,] Nov. 17th [1880]

Dear Sir,

Your favor is at hand.[2] In reply I will state that the bolts which are represented in the drawing of the dynamo are for securing the machine to a sole-plate of cast iron upon which the bed-plate of the engine also rests. When you were here we talked that matter over and, as you will recollect, the plan is to have a direct-acting engine; and to have both engine and dynamo upon one cast-iron sole-plate which shall be quite heavy and[a] perfectly rigid, so that they will be self-contained and not be affected by foundations or floors settling and getting out of alignment. As to the design of engine, that is left entirely to you, both in the designing and construction;[3] subject of course to the terms of a contract, covering points relative to economy, performance, durability, duration of trial &c, which has yet to be drawn up by us when you are prepared to undertake the construction of the engine.[4]

The nature of the agreement would be such as not to cover any details, but that the capability of the engine to fulfill certain standard requirements may be insured[5]   Yours truly

[C.][b] L. Clarke

P.S. The shaft must be continuous. C.L.C.

ALS (letterpress copy), NjWOE, Lbk. 6:551 (*TAEM* 80:408; *TAED* LB006551). [a]Obscured overwritten text. [b]Not copied.

1. J. W. Thompson was associated with the Buckeye Engine Co. in Salem, Ohio; nothing more is known of him. Letterhead of Buckeye Engine Co. to TAE, 26 Oct. 1880, DF (*TAEM* 53:911; *TAED* D8020ZIE).

2. Thompson's letter has not been found. At the end of October he asked Edison to stop in Salem on the way to Colorado, where newspapers reported he would soon be traveling, because "We will need the promised drawing of a 'dynamo' machine to which the engine is to be attached if we are to go on and build one, but besides that there are other points on which a little consultation with you would be to our mutual advantage." Thompson also provided some general information about the "plan of engine now nearly perfected which is a modification of the regular automatic [cut-off] so far as relates to the valves and gear." Buckeye Engine Co. to TAE, 26 Oct. 1880, DF (*TAEM* 53:911; *TAED* D8020ZIE).

Edison had an eighty horsepower Buckeye engine installed at Menlo Park in 1878 for testing. In early in 1879 he solicited from the company plans for a 1250 horsepower steam plant and evidently inquired about testing another engine. There is no record that this occurred but in September 1880 he wrote to a manufacturer in Buffalo that he was "making tests of different kinds of engines at Menlo Park. If your No 4 will run 600 revolutions and work at a boiler pressure of 120 lbs please ship one to me immediately" (*TAEB* 4:600 n. 1; Buckeye Engine Co. to TAE, 7 Jan. 1879; TAE to Dunbar & Sons, 29 Sept. 1880; both DF [*TAEM* 50:8, 53:872; *TAED* D7919E, D8020ZHH]). On 8 November Clarke told a Chicago builder that he had learned from Calvin Goddard that "you have a high-speed engine which he understands possesses peculiar merits and might be adapted to running the dynamo-electric machines. If you can give me information as to the type, whether horizontal or vertical, the range of power for which you construct them, the limit to the speed in revolutions to which you can attain, the form of governor and valve gear, and if automatic what would be the smallest engine which you would make automatic? I would like particular information as to one of 120 H.P." Julius Hornig sent another request for information under Edison's signature a few days later. About this time Edison drafted a reply to an inquiry from professor John Trowbridge about acquiring a small engine, explaining that he planned "to make a test with high speed engine with single dynamo & with large engine & number of dynamos (Clarke to Milan Bullock, 8 Nov. 1880, Lbk. 6:529 [*TAEM* 80:404; *TAED* LB006529];TAE to Bullock, 11 Nov. 1880; TAE marginalia on Trowbridge to TAE, 3 Nov. 1880; both DF [*TAEM* 53:936, 923; *TAED* D8020ZIT, D8020ZIL]).

3. Buckeye engines were noted for their overall excellence and especially for the steadiness and economy afforded by Thompson's form of shaft governor. In this relatively new class of mechanism, the weights were pivoted near the periphery of the flywheel face so that they moved at right angles to the axis of rotation. As they retreated, they adjusted a sliding cam on the shaft which accordingly controlled the steam cutoff. Because movement of the weights was in accord with inertial rather than simply centrifugal force, the governor responded not only to speed but to the rate of change of speed. In general about this time a shaft governor could maintain an engine within two or three percent of a fixed speed. Hunter 1985, 473–80.

4. The next day Clarke wrote a notebook entry stating that "120 H.P. dynamo at 600 revolutions gives 132 Volts. . . . If another magnet is added and only 450 revolutions the E.M.F. will be . . . 132 Volts." The next day he began making "Calculations for Buckeye Engine The engine to run at 450 which is ¾ of 600 revo. for Porter Engine, masses of iron to be made ⅓ larger." Over the next several days and again on 6 and 31 December he extrapolated the dimensions, principally of the armature, for the slower dynamo from those of the machine being built for the Porter-Allen engine. N-80-07-27:115–39, 209, Lab. (*TAEM* 37:240–52, 289; *TAED* N116:59–71, 108).

Clarke wrote the Buckeye Co. in mid-December outlining some modifications to their plan for the engine and suggesting the bed plate of the Porter-Allen engine as a model for their own. He offered additional comments and suggestions at the end of the month. Clarke to Buckeye Engine Co., 17 and 30 Dec. 1880, Lbk. 6:684, 736 (*TAEM* 80:441, 453; *TAED* LB006684, LB006736).

5. Clarke wrote the contract and took it to the Buckeye Co.'s New York office on 2 December (Mott Journal N-80-07-10:228, Lab. [*TAEM* 37:416; *TAED* N117:114]). Its terms were accepted the next week by the Edison Electric Light Co., subject to Edison's personal approval. On 21 December Stockton Griffin drafted a letter from Edison to Calvin Goddard affirming that "an engine which will fulfill the requirements as stated in this contract will prove in every way satisfactory for running the Edison dynamo electric machine." Edison noted his approval on a similar letter from Clarke the same day, and gave his formal acceptance to the company a few days later (Calvin Goddard to TAE, 10 Dec. 1880; TAE to Goddard, 21 Dec. 1880; both DF [*TAEM* 54:115, 126; *TAED* D8023ZBL, D8023ZBU]; Clarke to Goddard, 21 Dec. 1880; TAE to Edison Electric Light Co., 27 Dec. 1880; Lbk. 6:710, 723 [*TAEM* 80: 449, 452; *TAED* LB006710, LB006723]). The contract has not been found but Charles Mott noted on 29 December that it had been executed by the Buckeye Co. and, perhaps mistakenly, that it called for a 100 horsepower engine running at 450 revolutions (Mott Journal N-80-07-10:265, Lab. [*TAEM* 37:435; *TAED* N117:133]).

–2017–

To Henry Morton

[Menlo Park,] Nov 18 [1880]

Dear Sir;

Your kind favor of yesterday was duly rec'd.[1] I am afraid that you have heretofore taken what has appeared in the newspapers as correct as to what I have said and done. I saw that the article referred to by you in the Sun was mixed, and as[a] I am probably as well acquainted regarding the unreliability of newspaper accounts as almost any other man of course I did not suppose the report of your lecture was correct in any particular   Very truly yours

T. A. Edison—G[riffin]—

L (letterpress copy), NjWOE, Lbk. 6:561 *TAEM* 80:414; *TAED*
LBoo6561). Written by Stockton Griffin; circled "C" written above.
aInterlined above.

1. Morton wrote that the 17 November *New York Sun* account of his
paper to the National Academy of Science "is incorrect in almost every
particular and as I believe you do not agree with me in assuming newspaper reports to be almost invariably unreliable I hasten to assure you
that this one in particular is utterly unworthy of consideration and will
be at once corrected by me." The article noted that Morton "has been
experimenting with the electric current for several years. He has from
the start denied the accuracy of Edison's results, both as to the amount
of applied power recovered in the dynamo–electric machine and the efficiency of the lamps." Edison's inability to produce a perfected lamp
was attributed to the "constant disintegration of his carbon loops." The
article then described the incandescent lamp of Hiram Maxim and reported that "Maxim's claim is that the gasoline vapor keeps the carbon
loop constantly in repair, and the careful measurements of Prof. Morton
with the galvanometer and other instruments confirm the assertion, the
Professor said. He added that in a comparison between the Edison lamp
and the one just described, he has no hesitation in saying that Maxim's
is the more economical and efficient." Morton reportedly also stated
that the resistance of the Maxim lamps was so great that he "obtained
600 candle power per horsepower, equivalent to about thirty-eight gas
jets. In a recent letter to him, Mr. Edison had said that the lamps he was
then making give him about 155 candle power per horse power." Charles
Mott noted on 20 November that the *Sun* published a notice from Morton "to the effect that their report of his lecture . . . was a mass of errors
or misrepresentations by their reporter, but does not state in what particulars it is incorrect"; Morton's disavowal has not been found. Morton
to TAE, 17 Nov. 1880, DF (*TAEM* 53:951; *TAED* D8020ZIY); "Has
Edison Been Outdone?" *New York Sun*, 17 Nov. 1880, [p. 3]; Mott Journal N-80-07-10:213, Lab. (*TAEM* 37:409; *TAED* N117:107).

–2018–

William Carman to
Francis Upton

Menlo Park, N.J., Nov 18 1880.a

Dear Sir

I have this day charged to your a/c $875^{57}/100 being 5% on
$17 511.41 as per agreement this includes everything from the
start of Lamp Factory.[1] I gave you a receipt for $750 which
amount is credited to you on the Books[2]   Very truly

Wm Carman

ALS, NjWOE, Upton (*TAEM* 95:619; *TAED* MU050B). Letterhead of
T. A. Edison. a"Menlo Park, N.J.," and "1880." preprinted.

1. Francis Upton, along with Charles Batchelor and Edward Johnson, had joined Edison in forming the Edison Electric Lamp Co.; Upton had a 5% interest. Edison placed Upton in charge of the daily operation of the company and, according to Charles Mott, on 31 December

notices were "posted at the factory to the effect that on and after Jany 1, 81 Mr Upton would take charge and management of that place." Draft agreement between Edison Electric Light Co. and Edison Electric Lamp Co., January 1881, DF (*TAEM* 57:761; *TAED* D8123C); Mott Journal N-80-07-10:269, Lab. (*TAEM* 37:436; *TAED* N117:135); Doc. 2051.

2. The receipt, dated the same day, is for $125.57 "proportion to date on account of Lamp Factory now called Edison Electric Lamp Company" (Upton [*TAEM* 95:620; *TAED* MU050C]). Upton's account with Edison is in Personal and Laboratory Accounts: Private Ledger #1:65 (*TAEM* 88:142; *TAED* AB005:23).

–2019–

From Tracy Edson

New York Nov 20, 1880

My Dear Mr Edison

I yesterday received from you the "Plans" for laying Wires in the Streets,[1] for which I thank you— A committee has been appointed, of which I am Chairman, to take measures to obtain permission to lay Wires and Tubes for conveying Electricity for illuminating[a] the City of New York, and as a preliminary step, I have today called upon the Commissioner of Public Works, Allan Campbell Esq[2] with whom I am acquainted, and stated my desire that he should visit Menlo Park with me as soon as you are ready and see you and the Light as I saw it the other evening, in a private and quiet way, before any public exhibition or announcement is made, as he could have a better opportunity, in that way, to examine into and judge of the merits of the system.[3]

He said he took great interest in the subject, and should be much pleased to visit you, and to see the light, and would go at any time, if I would give him a day or two's notice beforehand, of the time when you would be ready to see him and show the Light— He also asked if he could bring Mr McCormick, Superintendent of Lamps & Gas,[4] to which I assented—

Now as I think it would be a great assistance to us in getting the rights we desire if these Gentlemen should be favorably impressed in regard to our Light, I would like it very much if you would make arrangements to exhibit it to them in the manner above indicated, as soon as you conveniently can, and if you will advise me two days before the time, when you will be ready, I will notify Mr Campbell, and bring him out to see you,— I do not suppose you can be ready next week, and if not, I would suggest Wednesday Dec. 1st or Friday Dec. 3d

Please have the kindness to inform me if either of these days will suit you, and if not, say when you will be ready, and I will be governed accordingly.[5] Very truly Yours,

Tracy R. Edson

ALS, NjWOE, DF (*TAEM* 54:107; *TAED* D8023ZBG). [a]Obscured overwritten text.

1. The "Plans" which Edson mentioned have not been found. However, he may have been referring to Edison's mapping and canvassing of lower Manhattan which occurred throughout the latter half of 1880 (see Doc. 1995).

2. Allan Campbell was an engineer, railroad manager, and New York City public official. In 1874 he became Commissioner of Public Works and in December 1880 he was selected to serve as Comptroller. In 1882 he ran unsuccessfully for mayor on the Citizens', or Republican, ticket. *NCAB* 9:466; Bazerman 1999, 223.

3. Edison conducted a round of exhibitions for investors, New York aldermen, reporters, and prominent public figures from November 1880 to early January 1881. See Doc. 2038 regarding Edison's demonstration for the New York City Board of Aldermen on 20 December. Other noted visitors around this time included the famed French actress Sarah Bernhardt, Western Union president Norvin Green, and financiers Jay Gould and J. P. Morgan. Mott Journal N-80-07-10:204, 220, 226, 258, 278, Lab. (*TAEM* 37:404, 412, 415, 431, 441; *TAED* N117:102, 110, 113, 129, 139); Elizabeth Upton to Sarah Upton, 27 Dec. 1880, Upton (*TAEM* 95:622; *TAED* MU051); Jehl 1937–41, 770–85; Friedel and Israel 1986, 180–83.

4. Stephen McCormick had been investigating the use of electric lighting to illuminate public spaces in New York since at least 1878. In late 1879 and 1880 he grew enthusiastic about using Edison's incandescent lighting system, particularly in Central Park. On 29 December 1879 Edison had invited him to come to Menlo Park to witness a demonstration of the system, a visit which he made on 3 January 1880. McCormick thanked Edison on 5 January for inviting him, but "regretted, however, that in consequence of the crowd of visitors, I had no opportunity of conversing with you upon the subject of lighting our Central Park, which question has been mooted here. I hope that another opportunity will be afforded me of making you a visit when the throng of visitors cease troubling you." In a letter of 1 October 1880 McCormick promised Edison, "Any information in regard to the public lamps of this city will be gladly furnished you." On 27 December he asked Calvin Goddard for detailed information regarding the technical characteristics and economics of Edison's light for his annual report, in which he promised to "note the advance made in the system of electric lighting and desire especially to refer to the Incandescent system of Mr. Edison." McCormick to TAE, 2 and 5 Jan., 1 Oct., and 27 Dec. 1880, all DF (*TAEM* 53:553, 564, 873; 54:129; *TAED* D8020A, D8020I, D8020ZHI, D8023ZBX); Wilson 1881, 969.

5. According to the docket on this letter and another letter from Edson dated 27 November, Edison replied on 22 November but his reply has not been found. On 27 November Edson confirmed the date of 1 December for Campbell's and McCormick's visit, and told Edison that

"they have agreed to come by the 3 P.M. Train on that day, expecting to return by the 6:21 train, and as Mr Goddard informs me that you kindly offered to entertain the Gentlemen at your house, I beg to suggest that you arrange it so that we can return by that train,— I regret to give you and Mrs Edison so much trouble but hope you will excuse it in view of its importance." Edson to TAE, 27 Nov. 1880, DF (*TAEM* 54:111; *TAED* D8023ZBJ).

-2020-

Notebook Entry:
Electric Lighting

[Menlo Park,] Nov 22nd 1880

Carbon loops

From this date we call the "regular loop" a Bast fibre cut 10 × 17 thousandths

Put on lamps of this kind "Regular" "Date" and Resistance[1]
⟨car[bonize]d 5 hour⟩[a]

Chas. Batchelor

X, NjWOE, Lab., N-80-09-28:107 (*TAEM* 36:511; *TAED* N106:52). Written by Charles Batchelor. [a]Marginalia written by Edison.

1. In the preceding days laboratory assistants measured the cold resistance of a number of lamps with bast fibers carbonized for five hours; these values approached 200 ohms. On 18 November Charles Mott noted that "Some carefully made lamps with Bast carbons were tested and gave about 142 ohms at 16 c. and were very even in economy. . . . The lamps were set burning in the case in Laboratory at about 16 candles." The last one of this lot broke on 10 December after 211 hours. N-80-11-18:1–51; Mott Journal N-80-07-10:209, 242; both Lab. (*TAEM* 37:743–68, 407, 423; *TAED* N124:1–27, N117:105, 121).

-2021-

To Henry Rowland

[Menlo Park,] Nov 24 [1880]

Friend Roland

I send you herewith a little item which shows the true scientific spirit.[1] The statements in[a] it are generally absolutely false. Maxim[2] is using my lamp. ᴇCoating carbon by the decomposition of hydro-carbon was tried in every conceivable form of manner by me last January.[3] There is not the slightest difficulty in getting 5 or 600 candles from a lamp if the carbon loop is coated with a sufficient thickness of deposited carbon but this reduces the resistance enormously. The whole subject was elaborately investigated by Becqurel many years ago and is to be found in an extensive paper in the Annals de Chemie & physique.[4]

All the papers for your dynamo machine are prepared and I have drawn up a contract between us which will be submitted to you. The firm of Dyer & Wilber have the matter in charge. Mr Dyer[5] has hunted up the precedents, canvassed the probabilities & believes he can get the patent allowed.[6] I shall have the co-operation of the Westn Union Telegph. Co in the matter. Very truly

T. A. Edison.—G[riffin]—

L, NjWOE, Lbk. 6:594 (*TAEM* 80:429; *TAED* LBoo6594). Original is in HAR. Written by Stockton Griffin; circled "C" written at top right. [a]Interlined above.

1. The enclosure has not been found but may have been the 22 November *New York Post* article about the Maxim electric light discussed in Doc. 2022 n. 1.

2. An accomplished inventor, Hiram Maxim had been chief engineer of the United States Electric Lighting Co. since 1878. He had experimented with arc and incandescent lighting, using for the latter strips of platinum and subsequently carbon. Maxim emigrated to England about 1882 and devised the automatic gun which bears his name; he also developed a form of smokeless powder and experimented with heavier-than-air flight. He was knighted in 1901. See *ANB*, s.v. "Maxim, Sir Hiram" and *TAEB* 4:778 n. 4.

3. See Doc. 1891; on related experiments about this time, see Doc. 2005.

4. The French chemist and physicist Alexandre-Edmond Becquerel was at this time director of the Muséum d'Histoire Naturelle in Paris (*DSB*, s.v. "Becquerel, Alexandre-Edmond"). Becquerel made extensive studies of electric currents and conductivity, and Edison is probably referring to Becquerel 1853, a paper on the conductivity of gases at high temperatures. In two short paragraphs describing experiments with an atmosphere of hydrogen and "protocarboné" Becquerel reported that his results varied greatly, which he attributed to the decomposition of the heated gas (p. 390). He noted that the platinum electrodes became covered by deposits of carbon liberated from the gas and concluded that his experiments were actually measuring the resistance of this solid carbon.

5. George W. Dyer was a Washington, D.C. patent attorney who had recently formed a partnership with Zenas Wilber.

6. Edison advised Rowland two years later that "Siemens is prosecuting his case in the Patent Office vigorously and may obtain a patent with broad claims if not stopped by an interference with an application filed in your name" (TAE to Rowland, 1 Dec. 1882, Lbk. 14:491 [*TAEM* 81:1035; *TAED* LBo14491]). In January 1883, however, it was discovered that Zenas Wilber, who by then no longer represented Edison, had never filed the papers he prepared. Unable to retrieve them, Edison had Richard Dyer (George Dyer's son) draw up a new set. The Patent Office rejected the application as unoriginal and Dyer drafted an affidavit for Rowland to swear to "a date of invention earlier than the earliest patent referred to, viz: Siemen's patent of 1873"; he also asked Rowland to try to recall the terms of his contract with Edison since that document was among those lost by Wilber. Rowland evidently never signed the oath

but did exhibit the armature at the 1884 Electrical Exhibition in Philadelphia (Richard Dyer to Samuel Insull, 8 Jan. 1883, DF [*TAEM* 70: 922; *TAED* D8370J]; TAE to Rowland, 10 Jan. 1883, Lbk. 15:130 [*TAEM* 82:79; *TAED* LB015130], the original is in HAR; Richard Dyer to Rowland, 13 Mar. and 11 Apr. 1883, HAR; Hathaway 1886, 102).

–2022–

From George Barker

PHILADELPHIA.[a] Nov. 26, 1880.

Private[b]

My dear Edison:—

Your letter of the 23d is at hand this morning. The extract it contained is only another example of the worthlessness of newspaper reports.[1] Just as in the Herald last week where you are made to criticise Maxim for sealing up his lamps with sealing wax!![2]

Really, however, I wish very much you would go and see Maxim's lamps. I was entirely sceptical until I saw them; and then I was very much surprised.

I tell you in all frankness, that in my opinion, the method he has for making his carbon loops, consolidates them and gives them a wonderful resisting power and durability. He has run them up to 60 candles for an entire month and they are still good. In Morton's laboratory we measured them at 80 candles on the photometer, and at Draper's[3] we got 100 candles from each one of eight lamps for six or eight hours consecutively. One of Maxim's large incandescent lamps ran to 650 candles, photometric measurement, as we saw in Hoboken. Mr. Maxim told me that he had obtained 3000 candles from one of his large lamps.

Now, I have never seen such results as these in your laboratory; nor have I ever seen any one who has seen them there. One of your newest lamps which Morton had at Hoboken, and which he lighted up for comparison, would not give them[c] I am sure. Professor Young agreed with me that your lamps were not intended to be run above 16 to 20 candles; and that you had not been able to maintain one at 50 candles for 24 hours so far as either of us knew.[d] I am fully aware that in all this, I am[e] not fully posted; for I have not been in Menlo since the last of July. But you remember that I have asked for some of your lamps and have been refused.[4]

Now as to the Evening Post. On Friday the 19th, after the adjournment of the Academy, a reporter for the Post asked me what I thought of Mr. Maxim's lamp. I replied that I was of the opinion that Mr. M's improvement in carbons was a great step in advance. That my friendship for Mr. Edison led me to regret that he had not been the one to hit upon the new method. That Profs. Draper, Morton, and myself had tested the lamps to our satisfaction and had obtained 650 candles by measurement from one of these carbon loops. I believe this is the substance of what I told him. He took no notes but wrote it out from memory evidently; and a poor memory at that. 1st. Morton had the lamps & invited Draper & I to his laboratory to see them. 2d. I said not one word about any letter from you to Draper, for there was none. 3d. Nor about expts. to test the durability of your lamp for we made none. 4th. Nor about the tests of the Maxim lamp which are all wrong. 5th. Nor about your lamp being old twenty years ago. You will therefore see that everything you can properly object to in that article is erroneous. And yet there is just enough of truth in it to make it plausible. If you desire I will write to the Post and give my exact opinion as I have now given it to you.

As to priority, Mr. Maxim's patent shows that he filed his application for an incandescent carbon filament in a hydrocarbon vapor as early as Oct. 4, 1878.[5] And indeed, from Mr. Swan's lecture in the last Chemical News (which we received here on Tuesday last) it would seem that an incandescent filament of carbon in a vacuum was used by him 20 years ago.[6]

The Secretary of the U. S. Electric Lighting Company who is an old Yale friend, asked me if my business relations with you were such as to prevent me from making a series of tests upon the Maxim lamp, such as I made upon yours.[7] I replied that there were no business relations between you and myself; that our relations were those of personal friendship. That I made the tests upon your lamp without bias, solely to ascertain the facts. And that I supposed you could have no possible objection to my doing the same with their lamps. When these tests are completed, I shall be in a condition to speak more intelligently upon their efficiency and economy. If you care to furnish me some of your best lamps they will have an impartial test with the others.

Please do not misunderstand me. My own self-respect requires me to be honest, even with a friend, like yourself. If I am in error, I hope you will set me right. But if I am not, it is not the part of friendship to conceal from you the facts and lead you to believe better of your lamps than the facts warrant. But never will I knowingly do you an injustice. Cordially yours

George F. Barker

ALS, NjWOE, DF (*TAEM* 53:960; *TAED* D8020ZJC). Letterhead of University of Pennsylvania. ᵃ"PHILADELPHIA." preprinted. ᵇMultiply underlined. ᶜ"give them" interlined above. ᵈ"so far as either of us knew." interlined above. ᵉInterlined below.

1. Edison wrote Barker on 23 November, "I notice in last evenings NYork Post what purports to be an interview with you & wherein you are made to say some things concerning my Electric Light work which I cannot bring myself to believe ever emanated from you. Will you be good enough to say if you even so much as supplied the reporter with a foundation upon which he could build such an interview." A clipping of this article was sent to Edison by Egisto Fabbri the same day. TAE to Barker, 23 Nov. 1880, Lbk. 6:587 (*TAEM* 80:427; *TAED* LB006587); Fabbri to TAE, 23 Nov. 1880, DF (*TAEM* 53:959; *TAED* D8020ZJB).

The *New York Post* of 22 November quoted Barker as saying that "There is no doubt in my mind or in that of Professors Morton and Draper as to the value of Mr. Maxim's remarkable discovery. For years I have been an admirer of Edison's search for the true solution of the electric light problem, and I can testify to his unremitting energy and the exhaustive nature of his search. But another man found it. I do not say that Maxim is a better electrician than Edison, but he has invented a lamp which surpasses, I believe, even Edison's dreams." He described the results of tests made with Henry Morton at Draper's laboratory on the Maxim lamps, which were run for 24 hours at a brightness of 650 candles, or the equivalent of 40 gas burners. He reportedly stated that in previous trials Edison's lamps were unable to maintain a brightness of 50 candles for longer than an hour but Maxim's gave "the most remarkable performance of an incandescent lamp ever made. Edison has a good generator, but his lamp was old twenty years ago. The hydro–carbon atmosphere of Maxim's lamp is new." "The New Electric Light," *New York Evening Post,* 22 Nov. 1880, Cat. 1241, item 1545, Batchelor (*TAEM* 94:618; *TAED* MBSB21545X).

2. In an 18 November article on the use of Hiram Maxim's incandescent carbon lights at the Equitable Building in New York, the *New York Herald* stated that "The sealing of the vacuum, or partial vacuum, is claimed to be accomplished in some other way than by fusing the glass, which is Edison's patent. On the top of the globes are small pieces of red matter, like sealing wax, and this was said yesterday to be the sealing matter. It was, however, noticeable that on two of the lamps along the wall there was no such red speck, the points, apparently fused in the ordinary way, being visible." Edison is said to have "smiled at the idea of plugging a vacuum with wax. It would be found, he believed, that the globes were fused at the point." He reportedly stated that he did not worry about these lamps, "or a hundred like them . . . I always expected them, and there will be more of them. The lamp is to a system of electric lighting what a gas burner is to a gas works. They know just what they are about. It is simply a stockjobbing operation to float electric arc stock." He added that "my lamp is no secret. Mr. Maxim came here himself and spent an entire day, from morning until late at night, looking over the whole place. Then he has got hold of one of my glass-blowers, and the whole thing is in a nutshell" ("Electric Light," *New York Herald,* 18 Nov. 1880, Cat.1241, item 1541, Batchelor [*TAEM* 94:617; *TAED*

MBSB21541X]). The glassblower was Ludwig Böhm (see Doc. 2000). Francis Jehl later recalled that when Maxim visited Menlo Park, "Edison explained to him how the paper filaments were made and carbonized and all about the glass-blowing part." According to Jehl, Böhm was indispensable to Maxim's efforts to develop his incandescent lamp, because he "had had the opportunity of watching all the various processes by which Edison made a practical lamp, and that acquired knowledge he imparted to Maxim" (Jehl 1937–41, 611–12). In a recent article on Maxim's lamp, the *Scientific American* reported that "The conducting wires, instead of being fused into the glass of the globe, are surrounded with a semi-elastic cement, which is capable of withstanding both heat and pressure. This cement insures a perfect and durable joint between the platinum electrodes and the glass" ("Some Recent Developments in Electric Lighting," *Sci. Am.* 43 [1880]: 262).

3. Henry Draper held a variety of teaching positions in natural sciences, chemistry, physics, and medicine at the University of the City of New York (later New York University) from 1860 to his death in 1882. On 17 November he hosted a reception at his laboratory attended by several dozen members of the National Academy of Science and a number of prominent New Yorkers; the event featured an exhibition of Maxim's lights. Edison had met Draper in 1877 and accompanied his solar eclipse expedition in 1878. *ANB*, s.v. "Draper, Henry"; "Professor Draper's Reception," *New York Herald*, 18 Nov. 1880, Cat. 1241, item 1542, Batchelor (*TAEM* 94:617; *TAED* MBSB21542X); *TAEB* 3:437 nn. 6 and 4, 4:373.

4. Barker had asked Edison to provide lamps for demonstrations at Harvard University during the annual meeting of the American Association for the Advancement of Science in August and again for Draper's 17 November National Academy reception. Edison evidently telegraphed that he could not meet the latter request, prompting Barker to reply that he was "sorry you refuse to allow Draper a few of your lamps to show at his reception. He is a good fellow and I feel hurt myself to have him snubbed in this way. So far as he is concerned, the reception will not suffer. He has several other things to show which will be very interesting. And as to lamps, he has been offered some of Maxim's, which being of low resistance (30 ohms) will be easily run by his engine." The 22 November *New York Evening Post* article (see note 1) stated that Edison had explained in a letter to Barker that he could not provide any lamps because of a mechanical problem at the factory, probably that referred to in Doc. 2010. A few months later Edwin Fox wrote Edison that he "happened to meet Dr Draper on Wall St this morning. What have you been doing to him? He is 'Maxim' out and out and says that you have nothing to patent." He will however need stronger arguments than he has to convince me that such is the case." Barker to TAE, 14 July, 21 Aug., 29 Oct., and 9 Nov. 1880; Fox to TAE 12 Jan. 1881; all DF (*TAEM* 53:179, 205, 250, 254; 57:572; *TAED* D8004ZDT, D8004ZES, D8004ZGD, D8004ZGG, D8120L).

5. Maxim's application of that date was for an incandescent lamp consisting of a straight, thick carbon burner sealed in a bulb with a rarefied atmosphere of a hydrocarbon such as gasoline; it issued on 10 August 1880 as U.S. Patent 230,953. Maxim filed an application in March

1880 for carbonizing flat filaments in the presence of a hydrocarbon vapor and another in April for producing an attenuated hydrocarbon atmosphere in a lamp. These issued in July and August 1880 as U.S. Patents 230,309 and 230,954, respectively. The filament used in Maxim's most recent lamp was described as having "a double reversed curve like a capital M, with the upper and middle corners rounded" ("Electric Light," *New York Herald*, 18 Nov. 1880, Cat.1241, item 1541, Batchelor [*TAEM* 94:617; *TAED* MBSB21541X]). According to an interview published in the *New York Tribune* on 26 November, Edison dismissed the novelty of the lamp:

> Every person familiar with the history of the art of electric lighting knows that I was the first person to divide the current and use the portions. That was done by the lamps I exhibited here last winter. Examine the Maxim lamp and what do you find? A glass globe is taken, a carbon filament is bent and placed in it, the air is exhausted and the lamp is sealed up. That is precisely my lamp. No one should be deceived by the peculiar shape given to the filament in the Maxim lamp; it is only done to mislead. There is nothing in the coating of carbon; it is obtained by decomposing a hydro-carbon— a principle old in chemistry, which could only have been patented through ignorance in the Patent Office. There is nothing new in the principle of electric lighting by incandescence; lamps on that principle were made thirty years ago. The novelty is in subdividing the current and utilizing the divided portions, thus producing many lights of less brilliancy; and this principle of subdivision is my discovery. I could have done six months ago what Maxim has done, had I desired to make a show. ["Edison's Work," *New York Tribune*, 26 Nov. 1880, Cat. 1241, item 1547, Batchelor (*TAEM* 94:618; *TAED* MBSB21547X)]

6. Joseph Swan (1828–1914), a chemist and inventor in photography, incandescent lighting, and electrochemistry, was later knighted (*DNB*, s.v. "Swan, Sir Joseph Wilson"). Barker was referring to a report in the *Chemical News* of Swan's presentation to the Newcastle-Upon-Tyne Literary and Philosophical Society on 20 October 1880. Swan described "an experiment which I tried about twenty years ago" to obtain an incandescent light from a spiral of carbonized cardstock in an evacuated glass bulb. He succeeded in maintaining the carbon at red heat for some time before it broke, but without means to produce a better vacuum and stronger current he abandoned these efforts. He resumed them in October 1877 with "a mere hair" of carbonized cardboard, which he heated electrically while the bulb was being exhausted. He reported that "*when the vacuum within the lamp globe was good, and the contact between the carbon and the conductor which supported it sufficient, there was no blackening of the globes, and no appreciable wasting away of the carbons.*" Swan argued that this apparatus contravened Edison's claims, particularly those made in Upton 1880a, to be the inventor of a practical lamp with a thin piece of incandescent carbon in a vacuum. Swan devoted the second portion of his lecture to the design of a projected lighting system composed of low-resistance lamps connected in series ("Proceedings of Societies," *Chemical News* 42 [1880]: 227–31). A substantial extract of Swan's paper

was reprinted in *The Engineer* and pasted into a laboratory scrapbook, as were a lengthy account of his lecture in *Engineering* and a description of his most recent lamp from the *English Mechanic* ("Electric Lighting," *Engineer,* 29 Oct. 1880; "Incandescent Electric Lights," *Engineering,* 29 Oct. 1880; "Swan's Electric Lamp," *English Mechanic,* 10 Sept. 1880; all Cat. 1015:113, 111, 102, Scraps. [*TAEM* 24:59, 58, 54; *TAED* SM015113a, SM015111e, SM015102a]). For comparisons of Swan's and Edison's work on incandescent lamps, see Wise 1982; Friedel and Israel 1986, 90–91, 115–17, and 235–36; and Israel 1998, 217.

7. See Doc. 1914.

–2023–

From George Gouraud

London Nov 27th 1880[a]

Dear Edison,

I confirm receipt of the following cable from you

"Close up London   Need my money"

To which I replied

Nov 25th   "Draw Sixty[b] day two thousand pounds."[1]

I notice that you still address me Menlo Park London and sign it. For the future address me "Noside London (your name spelled backwards) without signature and I shall know that it comes from you. Yours truly

G E Gouraud

I cant "close the thing up" because of reasons you may infer—but you must always have money when you "want it" and I have it or can get it![2] GEG[c]

LS and ALS, NjWOE, DF (*TAEM* 56:805; *TAED* D8049ZHK). Body of letter written by Samuel Insull; letterhead of George Gouraud;. [a]"London" and "18" preprinted. [b]Obscured overwritten text. [c]Postscript written and signed by Gouraud.

1. Neither Edison's nor Gouraud's cable has been found.

2. Gouraud wrote Edison one week earlier that "You may have heard of White's disgrace, and the conspiracy that is on foot in the United Company on the part of the Edison directors, no doubt stimulated by White." Although he went on to discuss Arnold White's proposal concerning Edison's reversionary interest, a copy of which he sent to Edison (see Doc. 1989 n. 2), he probably was referring to the split in the United company's board, principally between directors of the Edison and Bell companies. This came to the fore in December when the Edison faction called a special shareholder meeting to consider a resolution, which Gouraud opposed, expressing no confidence in the United company's management and course of action. Alfred Renshaw promised that in the meantime "negotiations will take place between Mr. Bouverie & myself for the settlement of your claim & the liquidation of the London Company." Gouraud to TAE, 20 Nov. 1880; United Telephone Co. circular letters to TAE, 17 and 23 Dec. 1880; Edison Telephone Co. of London

circular letter to TAE, 23 Dec. 1880; Gouraud circular letter to TAE, 24 Dec. 1880; Renshaw to TAE, 16 Dec. 1880; all DF (*TAEM* 56:798, 829, 833, 835, 837, 828; *TAED* D8049ZHH, D8049ZIA, D8049ZIB, D8049ZID, D8049ZIE, D8049ZHZ).

–2024–

From Grosvenor Lowrey

Tarrytown, Nov. 28th 1880.

My dear Edison.

The luck seems badly against me in respect to going out to see you. You know perhaps that I have been confined for more than three weeks to the house, except last Monday, when I went to New York for the almost single purpose of visiting you on Tuesday; but on Tuesday morning I found it prudent to get back as soon as possible, and I have been in bed until yesterday (Saturday) evening.[1] Now I expect to be down town again this week.

I have various topics to discuss with you.

Maxim.)[a]   It is a good "Maxim" not to crow until you are out of the woods; and that is all I have to say about Maxim. I am informed that Dr. Lugo says that more than two years ago he had exactly what Maxim's patents describe in respect to the use of hydro carbon,[2] and as I understand from Mr. Kent[3] he proposes now to ask for a patent in precisely the language of Maxim.

Fabbri.)[a]   I enclose a copy of a letter from Mr. Fabbri of Fabbri & Chauncey, and a copy of my answer, as the shortest way to explain that subject. He will no doubt go out to see you about India, and we must put that matter straight as soon as possible. The same should be done in respect to all the other countries.[4]

Eaton.)[5a]   I am anxious to have Eaton in the Board, and my intention was to propose on the day of the election that he should be elected in the place of a gentleman who has but a small interest and whose presence there is of no importance; but on Election day I was sick, and could not do so.[6]

There is a little bit of small jealousy about Eaton in the Board; chiefly on the part of one gentleman with whom it is purely technical. He does not like to hear of Mr. Eaton's being consulted about anything, because he says Mr. Eaton is not officially connected with the company. Probably you and I realize more fully than any body else, that some one is needed who will give his entire time and brain to the business now on hand. My first business when I get out will be to see how a vacancy can be made and I must ask your co-operation in getting the Board to elect Eaton.

Goddard is a little sensitive about Eaton going in and does not, I think, take quite the right view of the field which is to be occupied,

I have asked Howard Butler,[7] who is now studying law, to come into our office and make a special study of infringements as collateral to Mr. Wilber's investigations.[8]

I think you desired to retain Causten Browne[9b] also, and I am in favor of that, now that we have funds for such expenses.[10]

What do you think of retaining Storrow, the Bell Telephone man?[11]

I hope I shall see you about Wednesday or Thursday. Yours truly,

G. P. Lowrey   by S[oren].

Enclosures,)
Copies of two letters.)

TL, NjWOE, DF (*TAEM* 54:112; *TAED* D8023ZBK). ªTyped in left margin. ᵇ"e" added by hand.

1. Lowrey telegraphed his worsened condition to Edison on Monday, 15 November, and suggested that "Perhaps in order to arrange contracts and various other things you can come up here some day—arrange with Goddard." DF (*TAEM* 54:103; *TAED* D8023ZBA).

2. Orazio Lugo was identified as a refiner with a laboratory in the University Building of the University of the City of New York (now New York University). Lugo, who claimed to be a former student of Joseph Henry, was in the process of obtaining several patents for the use of dynamo electricity in telegraphy. In 1878 he received two patents for arrangements to circulate a cooling fluid or air through hollow electrodes in arc lamps. In one of these he used porous electrodes of carbon or spongy platinum through which a hydrocarbon oil could pass, cooling them and forming "a deposit of carbon upon the electrodes, which also has the effect of retarding their consumption." Wilson 1880, 924; *Encyc. NYC*, s.v. "University Building"; "The Uses of Electricity," *New York Times*, 17 Jan. 1881, 5; U.S. Pat. 207,754.

3. Rockwell Kent, Lowrey's secretary.

4. In his letter to Lowrey, Fabbri stated that he was about to begin negotiations for the electric light in India and wanted "to know whether you have any objection to its being carried out on the plan of our getting a large return of the profits to be derived from the undertaking. I do not think it feasible in this case to sell the invention out and out, or to obtain any money down in advance." Lowrey answered that he approved of this plan for India and expected Edison would, too. In reply to Fabbri's inquiry whether the foreign electric light arrangements outlined in Doc. 1920 had ever been formally ratified, Lowrey stated that Edison had forgotten to put his approval in writing but that this would be done. He also urged Fabbri "before going much into the details of the negotiations, to

visit Mr Edison and get his ideas. You will find them very business like, and as the subject is a new one to you, he will give you a great many points which you might otherwise overlook. The development of the telephone business as a commercial undertaking may not furnish much light on the present subject, but will at least warn us that we cannot be too careful." Fabbri to Lowrey (copy), 26 Nov. 1880, Lowrey to Fabbri (copy), 28 Nov. 1880, both DF (*TAEM* 54:304, 306; *TAED* D8026ZDU, D8026ZDW).

5. Sherburne Blake Eaton (1840–1914) was born in Lowell, Mass., and attended Phillips Andover Academy and Yale College, graduating from the latter in 1862. While serving in the army during the Civil War he earned the title "Major," by which he became known among his friends. In 1870 Eaton was admitted to the bar in Chicago and the following year joined the firm of Porter, Lowrey & Soren in New York; he formed the firm Carter & Eaton in 1874. He specialized in corporate and bankruptcy law and attained some distinction advocating for customs and revenue law reforms on behalf of the New York Chamber of Commerce. Eaton first visited Menlo Park in June as "a friend of Mr. Lowreys," and since the end of August had been advising Edison and Lowrey on electric light matters (Hornblower 1891; *NCAB* 7:130; Israel 1998, 209-228, passim; Porter, Lowrey, Soren and Stone to Stockton Griffin, June 1880, DF [*TAEM* 54:63; *TAED* D8023W]; see also *TAEM-G2*, s.v. "Eaton, Sherburne, Blake"). During the summer Edison and Lowrey reached an understanding with him about taking a seat on the board of directors (see Doc. 2032 n. 1).

6. This took place at the annual meeting of shareholders on 11 November, at which Edison was elected to the board. Edison Electric Light Co. to TAE, 30 Oct. and 11 Nov. 1880, both DF (*TAEM* 54:96, 99; *TAED* D8023ZAT, D8023ZAW).

7. Howard Butler was a friend of Francis Upton, whom he introduced to Edison. In 1878, while Butler worked for the Gold & Stock company, Edison asked him to conduct the literature search on electric lighting which Upton subsequently undertook. *TAEB* 4:689 n. 3, Doc. 1568; TAE to Lowrey, 1 Nov. 1878, Lbk. (*TAEM* 28:897; *TAED* LB003471A).

8. Nothing is known of this project but it is possible Wilber was assigned to investigate patents relevant to Edison's electric light system. In January Edison instructed him to ascertain if a particular meter patent had a clear title because he wished to purchase it. TAE to Wilber, 4 Jan. 1881, DF (*TAEM* 59:282; *TAED* D8142C).

9. Causten Browne was a noted Boston attorney with particular expertise in patent and contract law; in 1879 he wrote a favorable analysis of the Fitch telephone transmitter patent which George Prescott sent to Edison. *NCAB* 10:349; Browne to George Walker, 14 June 1879, Prescott to TAE, 26 June 1880, both DF (*TAEM* 51:519, 518; *TAED* D7929ZDC, D7929ZDB).

10. Lowrey may have been anticipating the special shareholders meeting called for 30 November to vote on increasing the company's capital by an assessment on its stock shares. Edison Electric Light Co. circular letter to TAE, 30 Oct. 1880, DF (*TAEM* 54:97; *TAED* D8023ZAU).

11. James Jackson Storrow was a distinguished Boston patent attor-

ney who, since 1878, was a principal counsel for the Bell Telephone Co. in its extensive litigation over the validity of Bell's patents. *DAB*, s.v. "Storrow, James Jackson."

–2025–

Notebook Entry:
Electric Lighting

Prevent Carrying[1]

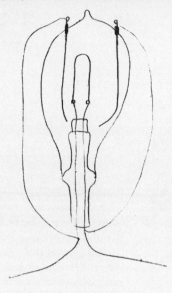

TAE

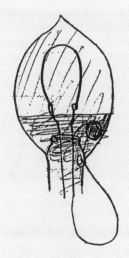

William Hammer's
30 November design using
a pith ball and platina
coating on the glass to
complete a circuit through
the globe.

X, NjWOE, Lab., N-80-11-16:93 (*TAEM* 37:925; *TAED* N125:43).

1. This drawing is probably related to the experimental lamp Edison proposed at the end of Doc. 1898. There are no records of experiments with this design and its function is not clear, but it appears to represent a means of diminishing electrostatic attraction between the glass globe and carbon particles released by the filament. Two days later William Hammer sketched another device exploiting this general idea. In that case a conductive coating such as platina foil was applied to both sides of the glass near the bottom of the globe and placed in contact with the lamp circuit, on the outside portion by a wire and on the inside by a wire and pith ball. At least two such lamps were made in early December. On 10 December Edison suggested another approach in which a fine platinum wire was looped parallel to or coiled around part of the filament and connected at only one end to the lamp circuit. He tested several of these lamps in late December or early January. In another set of experiments during the first week of December Hammer placed a condenser in the bulb and connected it to both poles of the filament. After one experiment in which the adhesive between the condenser's mica and tinfoil leaves "volatilized & deposited on carbon changing resistance greatly," Hammer tried fastening the leaves with platinum wire but did not report any results. Order "No 6," TAE notes of 10 Dec. 1880, both

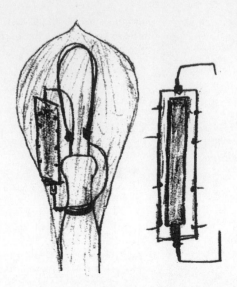

William Hammer's 7 December lamp with a condenser, probably designed to reduce electrostatic attraction.

Box 13, EP&RI; N-80-11-25:256–258, N-80–12-24.1:1–11, both Lab. (*TAEM* 37:638–39, 40:3–8; *TAED* N120:53, N186:1–6).

In October 1882 Edison filed two patent applications incorporating some of these designs. In the first he described the use of "a body or bodies of metal or other conductor of electricity connected with one of the conductors leading to the carbon filament, and surrounding the globe, or situated at several different points around the globe, or placed within the globe and around the filament. Such metal, becoming charged with electricity of the opposite kind to that with which the glass is charged, neutralizes the static attraction and prevents the removal of the particles of carbon." The second application was for means of "bringing the conductor connected with one of the leading-in wires of the lamp directly into contact with the glass globe, which has the effect of raising the globe to the same or nearly the same potential as the filament." In another application filed in November 1883, he described "statically charging the globe and filament with electricity from a frictional or other source of static electricity. The polarity of the charge imparted to them is such as to neutralize the charge which will be given when current passes through the carbon filament, and hence the two charges will produce a neutral static condition of the lamp." The static charge was to be applied at the time of manufacture and would "last a considerable time, it being in some cases almost impossible to discharge the globe. The lamps may, however, if desired, be charged from time to time after they are put into use by means of a portable frictional electric machine" U.S. Pats. 268,206, 273,486, and 425,761.

Notebook Entry:
Electric Lighting

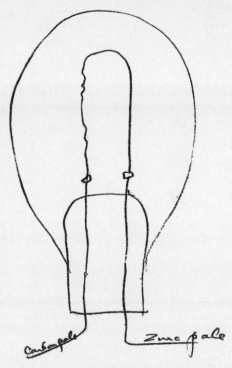

Putting bad spots on positive pole side prevents carrying electrically & good side of carbon will bust first[1]

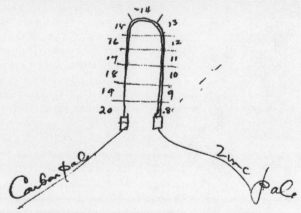

8 means[a] 8 candle incandescence   9 nine[a] candle incandescence   taper carbon[2]

This method compensates for the carrying of carbon by electricity as the carrying side has lower incandescence and was the way we proposed last summer[3]

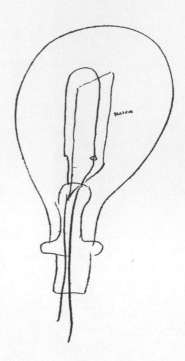

X, NjWOE, Lab., N-80-11-16:95 (*TAEM* 37:926; *TAED* N125:44).
^aObscured overwritten text.

1. "Carbon pole" and "Zinc pole" refer to battery terminals. For Edison's explanation of the observation stated here, see Doc. 2033. In January 1881 he filed a patent application for the practice of "arranging the carbons in the lamps so that the strongest portion thereof, or the portion containing the most material, or the portion having the least resistance, shall be at the negative clamp." Edison also specified that filaments with defects along half their length could be "utilized by placing them in the lamp so that the defective side becomes the positive end of the loop, the other or perfect side, having less resistance, being made the negative side. By so using such carbons their life or duration will not be materially shortened, as the inevitable carrying then proceeds from the perfect to the imperfect side." U.S. Pat. 248,418.

2. The previous day Edison sketched similarly tapered filaments designed "To prevent electrical carrying" (N-80-01-13:30–31, Lab. [*TAEM* 39:189; *TAED* N152:12]). In his January 1881 lamp patent application Edison stated that carbons could be constructed or built up by deposit so as to be thicker at one end than the other. This would provide for

> a lower resistance and a lower incandescence at the negative end, so
> that the carrying from such negative end by electrical action is materially reduced, while the total resistance, candle-power, and economy of the carbon may remain the same, its duration or useful period being lengthened proportionately to the reduction of carrying.
> In practice such carbons should be made so that the unit of incandescence at the negative pole will be about eleven or twelve candle

power, rising gradually to eighteen at the positive pole, the average of the carbon or its total lighting effect being about sixteen candle power. [U.S. Pat. 248,418]

3. See Doc. 1944. The following drawing of a partition lamp suggests that the question of whether carbon was carried along or across the filament legs, which Edison had investigated during the summer, was not entirely settled in his mind. Sometime in October he also sketched a lamp with its carbon legs enclosed in glass tubes, presumably to prevent carrying across the intervening space. N-80-10-15.2:121, Lab. (*TAEM* 39:61; *TAED* N149:61).

–2027–

Notebook Entry:
Electric Lighting

[Menlo Park,] Dec 1st 1880

The lamps made for laboratory from this date will be Bamboo—8 × 17 thousandths—5 hours carbonization—and will be numbered consecutively commencing at 1. All will be under 400 ohms resistance when cold—[1] Platinum clamps.—[2]

Chas Batchelor

X, NjWOE, Lab., N-80-09-28:115 (*TAEM* 36:515; *TAED* N106:56). Written by Charles Batchelor.

1. This limit was presumably to avoid excessive "electrical carrying"; see Doc. 2033.

2. It is not clear what prompted Batchelor to issue this directive but during the preceding week he had received from the lamp factory dozens of bast and bamboo filaments carbonized in slightly varying ways. The decision to use only platinum clamps was made on or before 26 November, when Charles Mott reported that glass blowers in the factory were "removing the carbons from four or five hundred lamps in which they had been secured by clamps other than Platinum to put them in Platinum clamps and Pear shaped globes with wires through the flat seal." This form of globe was suggested in mid-October as being less expensive. Edison adopted the design after a satisfactory trial a few days later, and at the end of November he applied for patent protection on it. N-80-09-28:109–15; Mott Journal N-80-07-10:222, 160, 163; N-79-08-28:205; all Lab. (*TAEM* 36:512–15, 37:413, 382–83; 35:1154; *TAED* N106:53–56, N117:111, 80–81; N088:100); U.S. Design Pat. 12,631.

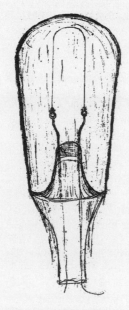

Drawing of 12 October of the proposed new shape of "Edison's Electric Lamp."

During the second week of December experiments were made with fractional lengths of the standard carbon "for use in series or instead of one across in multiple arc." On 15 December Alex Welsh, who assisted in the lamp factory, noted that "From this day all regular carbons are to be distinguished as follows. A. 6 in[ches]— B 3— C. 2. D 1½ by direction of Mr. Batchelor." Subsequent records of lamp tests generally use these letter designations. Mott Journal N-80-07-10:245, N-80-09-28:129–35, both Lab. (*TAEM* 37:425, 36:522–25; *TAED* N117:123, N106:63–66).

*To Edison Electric
Light Co. Executive
Committee*

Gentlemen;

I desire to obtain the services of a man for arranging, mapping and figuring out the main and subsidiary conductors for our first district in New York[1] as well as for others and I have just been made acquainted with a person who will meet all the requirements of the case. He is a German Electrician, a first class mathematician and understands the subject upon which I desire him to work. He has only been in this country ten days.[2] He wants for his services two dollars per day. Will you authorize his engagement for the purpose mentioned? An early reply is desired. Very truly

T. A. Edison — G[riffin]—

L (letterpress copy), NjWOE, Lbk. 6:610 (*TAEM* 80:432; *TAED* LB006610). Written by Stockton Griffin; circled "C" written at top of page.

1. In August, David Greene of the Rensselaer Polytechnic Institute wrote to one of his students that Edison "requires a man of 'peculiar qualifications. . . . [who] must be a good mathematician as he will have to calculate & fix upon the size of the electrical conductors besides collecting much matter for statistics which he will have to work up for practical use.'" Edison reportedly expected that this person would eventually "have the general supervision of laying the pipes & conductors, placing of meters &c &c." David Greene to James Pearl, 28 Aug. 1880 with James W. Pearl's 1929 explanatory note, EP&RI.

2. Edison hired Hermann Claudius, an electrical engineer formerly of the Austrian Imperial Telegraph Department, in December. Jehl (1937–41, 732–33) recalled that Claudius at first "spoke but little English and Upton was often forced to use the German he had learned while studying under Helmholtz in Berlin. We soon found that Claudius was an educated man of the old school and that he was extremely accurate in all his calculations, going over them twice or oftener in order to check first results. Like Upton he slung his logarithms about with a facility that commanded respect. Upton explained Edison's ideas on the matter and Claudius grasped them immediately, for he was not only drilled as a theorist, but had had plenty of practice as an engineer." Claudius constructed a 15 × 12 foot mockup of the central station district to model its generator and loads. Jehl recalled that

on this board the district that Edison intended lighting up in New York City was accurately drawn to scale. Distribution mains bearing a fixed ratio to the ones intended for use were carefully placed in the streets indicated on this wooden map. These mains were interlocked at corners and the greatest care was taken in every way to insure that the whole was really an exact miniature. The sides and blocks of each street were dotted with tiny resistance coils representing the expected load the central station would have, these loads being calculated from the canvass sheets. All the work was done by Claudius himself, except that of winding the coils. He gauged and measured everything. . . .

With meticulous fidelity and mathematical precision Claudius put the network into tangible form, all proportioned, even to the length of the streets. Edison was pleased and asked Upton how successful such a work would be if attempted on paper with pencil and logarithms. Upton admitted that it might perhaps be done but that it would be a head breaker and, no doubt, subject to many errors. Edison with pride pointed out the advantage of an exact model which permits eye and brain to work quickly and harmoniously without fatigue. [Jehl 1937–41, 733–34]

Claudius later made scale models of several other central station districts. His assistant, Hermann Lemp, later recalled that Claudius "was thorough and painstaking, but slow—terribly slow" (Jehl 1937–41, 739). Edison dismissed Claudius at the end of 1883 and replaced him with Frank Sprague (Sherburne Eaton to Charles Clarke, 14 Nov.1883, DF [*TAEM* 66:906; *TAED* D8329]). Claudius's calculations, tables, and notes for the first district are in N-80-08-13:130–255, Lab. (*TAEM* 38:353–413; *TAED* N132:64–124).

–2029–

Charles Mott
Journal Entry

[Menlo Park,] Friday Decr 3d 1880

Rails test. The rails of electric Rail way, through which the current is conducted to the pump motor in the gulley, were tested by Mr. Upton and found to be down to 24 ohms. still the motor runs and does its work.[1]

Hydraulic press. The small tri cylinder engine was moved from the dynamo room into the laboratory and connected up and with the pump of the Hydraulic press, to use it for pressing the ends of the copper bars of the large armature.[2]

Vacuum. The five hundred pumps made last ~~winter and~~ spring have all been changed, broken, altered and abandoned, and the glass blowers are now at work on an order for two hundred without spark, or McLeod, gauge, but single tube with dryer reservoir and probably to be tested for vacuum by magnets and "blue"—consequently the spark keys and wires were abandoned and taken down.[3] Herrig[4] to day moved in small room in the supplemental building and is continuing his high vacuum experiments, has to day very large globe encasing the regular carbon to exhaust for observations in the effect on carbon carrying.[5] Frances is also making some experiments on pumps and vacuum in the small front room of Laboratory. has in trial a pump with two fall tubes in which the mercury is conducted from a small reservoir into which it is also deposited through a fall tube with contraction thus getting a vacuum on the mercury before its final use for exhausting the lamp. to days results unsatisfactory.[6]

Conductors   the light lines to factory were to day increased in capacity by the addition of extra No 10 wires as far as R.R. crossing

AD, NjWOE, Lab., N-80-07-10:230 (*TAEM* 37:417; *TAED* N117:115).

1. Faced with an acute shortage of water for his boiler, Edison installed a pump powered by an electric motor to raise water from a small stream to his pond. It was operating by early November. Mott Journal N-80-07-19:167, 187, 189, 196, all Lab. (*TAEM* 37:385, 395–96, 400; *TAED* N117:83, 93–94, 98).

2. The press was previously operated by an electric motor. In early November Edison had acquired the small three-cylinder engine, which Charles Mott estimated produced five or six horsepower; it may have been an internal-combustion gas engine. Mott Journal N-80-07-10:191, Lab. (*TAEM* 37:397; *TAED* N117:95).

3. Mott presumably referred to a pump not conjoined to a Geissler pump as the "single tube" in contrast to the "double pump" that Edison had been using (see Doc. 1926). Ludwig Böhm had made what Mott described as a "single drop tube pump without gauge" in connection with fruit preservation experiments on 24 August. The single pump, in addition to being simpler to construct, enclosed a smaller volume to evacuate. Concern had been raised during the summer about whether capillary action might distort McLeod gauge readings. Mott noted in his 30 June journal entry that for an experiment, "Boehm made a U shaped apparatus of glass tubing one side about one eighth of an inch bore and the other side of fine gauge tubing and put in about 3 inches of mercury, the mercury stood in the large tube about ¼ of an inch above that in the small tube. (showing much less difference than was expected)" (N-80-03-19:217; Mott Journals N-80-07-10:97, N-80-03-14:269; Lab. [*TAEM* 34:713, 37:350, 33:818; *TAED* N068:107; N117:48, N053:136]). Investigation of this problem appears to have been suspended until November, when experiments on the effects of magnetism on carbon carrying were conducted (see Doc. 1944 n. 8). On 24 November Mott reported that during the previous night

> six Bamboo lamps were sent to the Laboratory for experiments on high vacuum. Two were exhausted during the night by the pump here and burned about three hours before sealing off   It was observed that the vapory blue could be produced at the clamp, even when burning at low heat, by holding a magnet near the glass and that as the vacuum improved the magnet required to be brought in closer proximity with the clamp to reproduce the blue, and the lamps were kept on the pump until no effervescence could be produced by the aid of a strong magnet. They were tested by Francis and found very economical and were perfect in appearance. [Mott Journal N-80-07-10:219, Lab. (*TAEM* 37:412; *TAED* N117:110)]

During the next week Edison continued these experiments, including one which was ruined when he, Herrick, and another assistant fell asleep with the lamp on the pump. Then on 2 December, according to Mott,

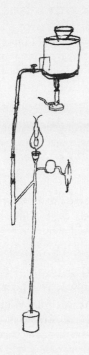

"All experimental lamps, and pumps used at Laboratory on experiments on high vacuum were removed to the Factory and Herrig [Herrick] & his assistants will, under the direction of Mr. Edison, continue the experiments" (Mott Journal N-80-07-10:221, 225, 228, Lab. [*TAEM* 37:413, 415–16; *TAED* N117:111, 113–14]); see also Doc. 2061 n. 2.

4. Albert Herrick.

5. The only extant records of these experiments are Herrick's notes from 6 December, but see also Doc. 1898. N-80-11-16:103–105, Lab. (*TAEM* 37:930–31; *TAED* N125:48–49).

6. Edison sketched this sort of pump on 27 November and filed a patent application for it in January which included a nearly identical drawing (Cat. 1146, Lab. [*TAEM* 6:722; *TAED* NM014:101]). Designed to prevent air bubbles trapped in the mercury from entering the evacuated space, it consisted of

> two drop-tubes, the first one connected to the reservoir and provided with the contraction or strangulation, the second connected to the body to be exhausted, . . . the two being connected by a tube inclined upwardly from the first to the second, so as to form a trap-seal between the two. The result of this is that two vacuous spaces are formed, separated by a solid column of mercury. Now, if a globule of air enters the vacuum apparatus, it is caught in the first vacuous space, impairing its tension, while the second vacuous space, which . . . is the important one, is protected. [U.S. Pat. 248,433]

Edison's 27 November design of a modified air pump designed to protect the evacuated space against the release of air bubbles from the mercury.

Both the November design and patent application drawing included a spark gauge and a reservoir of heated mercury.

The principle of two successive mercury drops was incorporated into

Photograph of the modified vacuum pump made for the factory in late 1880. The mercury descended through the tube at left, then flowed in the other tube, above which is a drying tube and the ground glass receptacle for a lamp stem.

the modified pumps made for the factory in December. These, however, included a drying tube or reservoir for collecting mercury vapor, shown at the top right of the photograph in Jehl 1937–41 (p. 805). Above the reservoir is the ground glass tube into which a lamp stem could be inserted (and sealed with mercury) so that it could be removed easily without damaging the apparatus.

–2030–

To William Pescod[1]

[Menlo Park,] Dec 10th [1880]

Sir

Your favor of the 4th Ult was duly rec'd. In my Electric Light system the design is to establish stations in cities & towns and supply the light one half mile in every direction from the Central Station.

This distance is the limit so far as economy is concerned. When the conductors are carried beyond half a mile they must be increased in size and this renders it expensive but where the motive power is water, and the expense nominal I see no reason why conductors of 4 or 5 miles in length[a] cannot be used. Very truly

T. A. Edison   —G[riffin]—

L (letterpress copy), NjWOE, Lbk. 6:649 (*TAEM* 80:434; *TAED* LB006649). Written by Stockton Griffin. [a]"in length" interlined above.

1. William Pescod wrote from Lambfield, Raughton Head, in Carlisle, England. Nothing further is known about him. In his letter of 4 November he told Edison that he was planning to reclaim some land along a river and wanted to know the feasibility of lighting nearby towns using water power in connection with an electric light central station. DF (*TAEM* 53:925; *TAED* D8020ZIM).

–2031–

From Lucius Foote

Valparaiso, Chile, Dec. 11th 1880[a]

Dear Sir

Yours of Octr 29th with Deed & Power of Atty was recd by the last mail.[1] We have not as yet completed the organization of a Joint stock Co. because we doubted Husbands power to sign for you. Now however that we have your deed we shall proceed with the organization.[2]

Husband claims, as he says, by arrangement with you, one third of your stock, which would leave you two sixths of the whole. He also insists that he has bought your stock for his own benefit & if any profit results therefrom it must be for his individual benefit   The Co have refunded to Husbands the monies advanced by him, and have allowed him $3000. per year for his services.

The gentlemen associated with me in this business are first class men. They were induced to take hold of it perhaps at my suggestion, and have paid the amount agreed upon for construction purposes, to wit, twenty five thousand dollars. The business may be said to be fairly inaugurated with a promising outlook. It will however go much more slowly than in the United States. We have had many very many annoyances from sources where we should have had harmony and I for one would never undertake the same business under the same circumstances again.

If Husbands fails to come forward with the money for your stock I have no doubt other parties would purchase it. I shall follow your instructions & will remit when the money is paid. Very truly yours

Lucius. H. Foote

ALS, NjWOE, DF (*TAEM* 55:897; *TAED* D8047ZCA). Letterhead of U.S. Consulate. ª"Valparaiso, Chile," and "188" preprinted.

1. The deed is the patent assignment of 8 October 1880. The power of attorney apparently refers to a new arrangement between Husbands and Edison. On 4 October Husbands had cabled Edison "I will give 10 000 cash whole interest   close out poor health." Edison inquired about this to H. H. Eldred, who had rights to one-third of Edison's share (see Doc. 1823 n. 6). Eldred replied on 7 October that he did not understand what was going on but suggested asking for $20,000. A week or so later Eldred cabled this offer to Husbands, who countered with an offer of $14,000 as "the best that can be done" and which Edison accepted. Edison then sent Husbands a power of attorney but on 6 November Husbands requested that Edison send a different form of assignment. Husbands was still waiting for this in May 1881 when he wrote Edison that "instead of assigning your interest & sending the necessary papers for Peru, Ecuador & Bolivia, you sent a power of attorney to transfer any & all shares of stock that might stand in your name, on the books of said Company upon the payment of $14,000 Gold, aforesaid; as there were no shares issued, the matter for the moment rested. The money is ready, & will be paid any time, when I receive the stock." Husbands paid the money to Foote at the beginning of June but attached the funds until Edison sent him the proper papers regarding those countries. The effort to resolve the situation was complicated by the Bell Telephone Company takeover of the Western Union telephone business in November 1879 and the subsequent efforts of the Bell interests to consolidate the foreign telephone business. Working through the Continental Telephone Company they sought to acquire rights to Edison's patents in Continental Europe and South America. Although no arrangement was made for Europe they seem to have come to some understanding in regard to South America as Husbands wrote on 6 September 1881 that he had "received an agreement for Ecuador, Peru & Bolivia from the Continen-

tal Telephone Co." and was therefore withdrawing the attachment on Edison's $14,000, which was then paid to him by Foote. Agreement between TAE and Compania Chilena Telefonos de Edison 8 Oct. 1880; Husbands to TAE, 4 and 24 Oct. and 6 Nov. 1880, 9 May, 3 June, 27 July, and 6 Sept. 1881; Eldred to TAE, 7, 14, 18, 26, and 29 Oct. 1880; TAE to Eldred, 15 Oct. 1880 and 30 Oct. 1881; Foote to TAE, 6 Sept 1881; Henry S. Russell to TAE, 14 July, 10 August 1880; Russell to Edison Telephone Co. of Europe, Ltd., 22 July and 31 Aug. 1880; all DF (*TAEM* 55:889, 885, 893, 896; 59:703, 712, 716, 732; 55:888, 891–92, 894–95; 55:891; 59:741, 733; 56:222, 232, 228, 249; *TAED* D8047ZBS, D8047ZBO, D8047ZBV1, D8047ZBZ, D8147Q, D8147R, D8147U, D8147X, D8047ZBR, D8047ZBT, D8047ZBU, D8047ZBX, D8047ZBY, D8047ZBT1, D8147ZAF, D8147Y, D8048ZEW, D8048ZFG, D8048ZFB, D8048ZFU); TAE to Eldred, 3 Jan. 1881, Lbk. 6:749 (*TAEM* 80:456; *TAED* LB006749).

2. Three days later the Compañia Chilena de Teléfonos de Edison was organized by Husbands, Foote, Santiago Martin, and Pedro Mac-Kellar, who on 25 August had signed a preliminary agreement to acquire the exclusive rights granted to Edison by the Chilean government on 26 April (see Doc. 1959 n. 5). The capitalization was 200,000 pesos divided into 400 shares of 500 pesos each; this was apparently equivalent to the $100,000 capital called for in Edison's agreement with Husbands (Doc. 1823). Foote, Martin, MacKellar, and M. MacNeil each subscribed for 50 shares in the company. Husbands subscribed the other 200 shares, two-thirds of which was reserved for Edison. Berthold 1924, 38–39; "La Compania Chilena de Telefonos de Edison," *La Patria* (Valparaiso), 25 May 1881, Cat. 1034, Scraps. (*TAEM* 25:411; *TAED* SM034055b).

–2032–

To Edison Electric Light Co.

Menlo Park N.J. Dec 13 1880

To The Edison Electric Light Co.

My occupations here prevent me from attending the meetings of the board and I think it will be for the interests of the Co that I should resign as a director and that you should elect Major Sherburne B Eaton in my place.[1] I have great confidence in Major Eatons capacity and I do not doubt the board will concede the propriety of my naming my successor. I suggest that the transaction of the Companys business will be made more easy and[a] systematic by creating the office of Electrician, and for the present you may perhaps think me competent for the place; at any event there need be no salary attached to the office until further notice. Very Respy Yours

Thomas. A. Edison.

ALS, NjWOE, DF (*TAEM* 54:116; *TAED* D8023ZBM). [a]Obscured overwritten text.

1. Edison and Lowrey had reached an understanding with Eaton in the summer of 1880 about his prospective involvement with the com-

pany, the terms of which were put in the form of a letter from Lowrey, also signed by Edison, after his election to the board. Eaton was given an option on fifteen shares of stock each held by Edison and Lowrey, and on additional shares to be contributed by other stockholders. Shares were valued at $100 apiece and Eaton was to purchase them in blocks of five shares every six months until all the shares were purchased or he ceased to be an officer of the company. Lowrey explained to Eaton that this was to "secure your entire time and service for the Company and thus for our interest in it. We are confident that those services will possess special value over and above the services of other men whom we might employ." Lowrey to TAE, 19 Jan. 1881, enclosing Lowrey and TAE to Eaton, 15 Jan. 1881, both DF (*TAEM* 58:5–6; *TAED* D8126B1, D8126C).

Eaton served as vice president and general manager of the company from January 1881 until he became president in 1882. He was succeeded by Eugene Crowell in a corporate shakeup in 1884 but remained connected with the company as general counsel. Between 1881 and 1884 he was also a director and vice president of the Edison Electric Illuminating Co. of New York and president of the Edison Electric Light Co. of Europe. In the late 1880s he became Edison's personal attorney. Hornblower 1891; *NCAB* 7:130; Israel 1998, 209–28, passim.

-2033-

To Henry Rowland

Menlo Park Dec 13 1880

My Dear Rowland,

It is a clear case of hedging on the part of Morton.[1] That terrible temper of Barker ~~has~~and his unfamiliarity with the subject has caused him to make statements in the press against me. He stated that Maxim's lamp was entirely new, that my lamp was 20 years old and didnt last more than an hour at 50 candles, that Maxims had ~~pa~~ a patent prior to ~~meine~~ in which he used a <u>filiment</u> of carbon and other things   I wrote him asking him if the statements were made by him and his letter is also a sorry attempt at hedging.[2] It plainly conveys the[a] that he made the statements, for instance he says, "I never said you lamp was twenty years old" but then you know he says Swan of England claims to have done it 20 years ago. Barker is now affiliated with the Maxim Co through Farmer[3] etc, and I can only expect bitter enimity   his ignorance of the subject is the only thing I fear either from him[b] Morton or anyone else. = Both Morton & Barker are probably ignorant of the fact that the destruction of a carbon loop in a good vacuum is due entirely to electrical carrying and is a function of the [~~power?~~][c] electromotive force between[b] the two sides of the loop and the unit incandescence that the life of a loop will be increased when its resistance is less[b] by reason of the lower electromotive force used or its life will[b] be the same when a unit surface is at a much higher incandescence. for instance you can obtain ~~for instance~~ at the rate of say[d] ~~a~~ 500 candles incandescence unit radiating surface when the resistance is says 10 ohms and the life of the lamp say 100 ~~oh~~hours while you must reduce the unit incandescence to 250 candles to obtain the same life if the lamp is 20 ohms, (this is not exactly it but will illustrate)[e]—and so on so that while the merest trio using my lamp and making the Carbon 10 ohms (some 5 ohms) as Maxim does[a] can obtain to the unitiated remarkable results— You know to effect a commercial subdivision of the electric light that the multiple arc system is only one[f] permissible for many reasons & to render this available the lamps must have a high resistance. the amount of money invested in copper being directly dependent upon the resistance of the lamps, beside the necessity of low resistance Dynamos station applicances & house wires, my greatest efforts have been to obtain a lamp of the highest resistance and one which will last the longest at such resistance. the high electromotive forces necessary with high resistance lamps has made this a terrible job but I have worked until I have 160 ohm lamp whose life will average over 300 hours. Now comes in Barker & Morton and announces that the greatest advances In[b] electric lighting has been made by Maxim because he exhibited their small lamps of 5 ohms resistance which[f] gave 2 & 300 candle power; not knowing that the [---][c] whole thing was a question of electromotive force[a] and that I have made lamps which gave 800 & 900 candles, but of course were very low resistance ~~and the~~   The proof that they know nothing about this carrying is that they say that the deposit takes place at the weakest spot   In one sense this is correct if the ~~bad spot~~ perfect part of the carbon is below the decomposing point of the hydrocarbon, but when the whole of the carbon has reached the temperature necessary to effect decomposition the deposit goes on all over the same   The spot ~~seem~~ apparantly[f] dissappears although it does not actually. The extra resistance at the bad spot soon becomes a small factor owing to the great conductivity [~~of th?~~][c] for heat & electricity of the deposited carbon, not[b] 3 in 100[b] of the bamboos that I use can the slightest spot be detected. If such a spot is very apparent on[b] one side of the loop all that is necessary is to connect that side with the positive wire and when the lamp fails it will be not at this spot but on the <u>negative side</u>[4]   out of 200 lamps tested at one time[g] every one busted by electrical carrying and always on the negative side no matter if there were no spots. ~~If~~ [---][c] I have made during the last year as great many experiments with deposited carbon but always found that electrical carrying effected it more than the natural carbon besides it reduced my resistance & every extra ohm was an advantage. If one of my bamboos be coated by deposit in vacua, the most carbon will be deposited on the positive side & this will[b] generally show [------][c] the least bright   If now the bright side be put in connection with the positive pole, and the lamp put to 50 candles incandescence in the course of several hours the dull side will have become brightest i.e.[h] the negative side[i] and present a black appearance while the positive will retain its steel like color= & soon the lamp goes; There are so many phenomenon with high electromotive forces & high resistance loops & vacua which are brought out with a magnet that it would fill a book— when you get a chance to come over drop off & I will show you some; I forgot to mention that Barkers temper got the best of him because I could not spare any lamps for Drapers reception. Save every[b] scrap of paper relating to that Dynamo of yours as we shall need itthem to pull the thing through   Will write you again in a few days regarding this case. Yours

T A Edison

P.S. The <u>filiment</u> of Carbon mentioned by Barker as patented long ago by Maxim is this[5j]

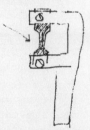

This is a healthy "filiment"   Its nothing more than one of the many old attempts made[b] at incandescent lighting[k]—low resistance, made carbon, cemented vacuum chamber etc— Maxim used a "Rubber Cork"   This is a mean thing to throw up at me at this late day.[a] Have you noticed lately the utter indiference of the technical press in giving credit of scientific[b] work to "previous or first publication and public exhibition" In England a man named sSwan has arisen made a paper lamp of 100 ohms resistance in glass vacuum exactly[b] in shape[b] & in every detail like mine & claims to have done it 20 years ago has delivered a lecture before the Soc Tel Engrs[6] is highly complimented for his [ag?][c] great contributions to the subdivision of the electric Light etc   most all the technical press claim it as

English only one makes the remark that it would be interesting to know where Mr. Swan's labors may be found in printed form previous to my[b] publication & exhibitions. The Daily newspaper press on the other hand says that Swan only exhibited my lamp; There wont be much protection to a scientific man if his previous publication & exhibition counts for nothing=

TAE

ALS, MdBJ, HAR (*TAED* X100AB). [a]Followed by "over" to indicate page turn. [b]Obscured overwritten text. [c]Canceled. [d]"~~for instance~~ . . . say" interlined above. [e]"(this . . . illustrate)" interlined above. [f]Interlined above. [g]"tested . . . time" interlined above. [h]Circled. [i]"the negative side" interlined above. [j]Followed by right-pointing fist. [k]"ing" interlined below.

1. Rowland wrote the week before that although he was

as much surprised as you were to see the statements about Maxim's lamp, as of course it is only yours with a slight modification in the method of making it. But I was scarcely surprised to see Morton bring forward anything which might afford him a loophole to escape the consequences of his opinion that your light would not prove successful   For he probably sees reason to alter his opinion now, and thus brings forward a rival. You alone will show the world what you have done and dispose of all these petty hangers on. [Rowland to TAE, 7 Dec. 1880, DF (*TAEM* 53:991; *TAED* D8020ZJS)]

2. Doc. 2022.
3. Moses Farmer (1820–1893), electrician at the U.S. Torpedo Station in Newport, R.I. since 1872, was one of the most prolific electrical inventors of the day. He demonstrated an incandescent platinum lamp in 1859 and continued experimenting at intervals with a variety of electric lighting devices. When he left the Torpedo Station in 1881 he became a consulting engineer for the U.S. Electric Lighting Co. At this time he reportedly had an arrangement with the company for the display of one of his lamps at its offices in the Equitable Building. This lamp was presumably the one that Farmer patented in 1879, consisting of a small piece of carbon in a sealed globe filled with nitrogen or other non-oxidizing gas (*ANB*, s.v. "Farmer, Moses Gerrish"; "Gas and Electricity," *Sanitary Engineer*, 1 Dec. 1880, Cat. 1015:124, Scraps. [*TAEM* 24:66; *TAED* SM015123d]; U.S. Pat. 213,643). An article containing Farmer's description of his early electric lighting experiments is reprinted in Defendant's depositions and exhibits 4:2185–87, *Edison Electric Light Co. v. U.S. Electric Lighting Co.*, Lit. (*TAEM* 47:916; *TAED* QD012E2185); a resumé of his subsequent research is in Prescott 1879, 515–25.
4. See Doc. 2026.
5. See Doc. 2022 esp. n. 5.
6. Swan's 24 November lecture to the Society of Telegraph Engineers apparently was similar to the October presentation discussed in Doc. 2022 n. 6. A substantial portion of a lengthy article about it was placed in a laboratory scrapbook, as were more cursory reports. "Swan's

Lamp at the Society of Telegraph Engineers," *Electrician*, 27 Nov. 1880; "The Society of Telegraph Engineers," *Telegraphic Journal and Electrical Review*, 14 Dec. 1880; "Notes," ibid., 14 Dec. 1880; "Incandescent Electric Lights," *Nature*, 2 Dec. 1880; all Cat. 1015:122, 125, 128, Scraps. (*TAEM* 24:64, 67, 69; *TAED* SM015122a, SM015125d, SM015125e, SM015128c).

–2034–

Draft to William Crookes[1]

[Menlo Park, December 15, 1880[2]]

Dr Sir—

I have now nearly completed works for ~~turning~~ a daily output [~~put?~~][a] of 1200 of my Carbon filiment Lamps in high vacua and am errecting other works which[b] will have a daily capacity of 15 000 lamps; and It has occurred to me that as you have a patent on your Radiometer[3] here ~~and~~ that my works and yourself might derive a ~~nice little~~ fair[c] income from the sale of these instruments because our facilities here will permit of their manufacture [----------][a] at a price which will allow[d] of their sale at 50 ~~to 75¢~~ cents each or less[e] which will so[c] popularize them that the sales would be very large [~~There is?~~][a] and ~~considerable market in NYork shop windows~~ permanent[c]   Please answer if you would make any arrangement with me for working the patent[4]   Yours

T. A. Edison.

ADfS, NjWOE, DF (*TAEM* 54:26; *TAED* D8021S). Letterbook copies are in Lbk. 6:671, 818 (*TAEM* 80:438, 471; *TAED* LB006671, LB006818). [a]Canceled. [b]Obscured overwritten text. [c]Interlined above. [d]"of their . . . will allow" interlined above. [e]"or less" interlined above.

1. William Crookes (later knighted) was a leading chemist, physicist, and science publisher who was most noted for his discovery of the element thallium and for his cathode-ray investigations, which led to his development of the radiometer. *DSB*, s.v. "Crookes, William."

2. The draft was docketed only "Dec. 1880"; the date is taken from letterbook copies of the letter sent to Crookes. Lbk. 6:671, 818 (*TAEM* 80:438, 471; *TAED* LB006671, LB006818).

3. Crookes devised the standard form of the radiometer in 1875. It consists of a partially evacuated glass globe in which four vanes are suspended on a vertical rod. Each vane is black on one side and white (or silver) on the other. When the radiometer is placed near a light source the vanes turn (the speed increasing with the intensity of the light). Crookes believed that the movement of the black side of the vanes away from the light was caused by light radiation pressure on that side. In fact, the black side would absorb the rays while the white side reflected them back. Osborne Reynolds and James Clerk Maxwell showed in 1879 that it was the pressure of gas molecules flowing along the edge of the vanes from the hotter side (black) to the colder side (white) that caused them to rotate. Woodruff 1966; DeKosky 1976, 36–47.

4. Crookes replied on 4 January 1881 that several years previously he had made arrangements with the Boston instrument maker E. S. Ritchie and Co. but that they lacked the facilities to manufacture the radiometer at a sufficiently low price. He suggested that Edison write to Ritchie and also wrote to the firm himself. In reply to Edison's letter Edward Ritchie indicated that he would be in New York at the beginning of February and would be happy to meet with Edison. Although Ritchie arranged to come out to Menlo Park it is not known if they reached an agreement and there is no evidence that Edison manufactured the radiometer. Crookes to TAE, 4 Jan. 1881; E. S. Ritchie to TAE, 21 Jan. 1881; Edward Ritchie to TAE, 2 Feb. 1881; all DF (*TAEM* 57:33, 50, 62; *TAED* D8104C, D8104P, D8104ZAA); TAE to E. S. Ritchie & Co., 18 Jan. 1881, Lbk. 6:818 (*TAEM* 80:471; *TAED* LB006818A).

−2035−

From Calvin Goddard

New York, Decem 15 1880.[a]

My dear Edison,

I enclose rough draught of a letter addressed to the Hon. John J. Morris[1] which it is considered desirable for you to write and sign, to be used with him for the purpose of securing the right to lay pipes and wires in the whole City of New York. As it is important that we should have this paper to use tomorrow I send it to you by Mr. Tracy,[2] who will bring it back to me this evening   Very truly Yours

C Goddard Secy   E.W.F.[3]

ENCLOSURE[b]

New York, Decem 15 1880[c]

Hon. John J. Morris, President of the Board of Alderman, Dear Sir:

Learning from Major Taylor[4] that you expressed an unwillingness to vote for the passage of a resolution granting permission to the Edison Electric Light Company to lay its pipes in the City of New York until it had demonstrated the practicability and advantages of the system under a resolution passed for that purpose two years ago. I beg most respectfully to say that the reasons I did not avail myself of the permission granted to make an experiment in the City of New York were: My shop and laboratory had been previously established at Menlo Park where I had abundance of room & trained assistants and where I could elaborate the experiment much more satisfactorily and cheaply than I could possibly have done in the crowded streets of New York. You can readily understand that such a work as I have been engaged in for the past two years must necessarily have been in its earlier stages largely experimental, requiring constant changes to adapt the system to the developments that occurred from time to time and had I undertaken the experiment in the City, must have resulted in great inconvenience not only to me but to the public, in frequent disturbance of the streets in the[d] laying and changing of the necessary conductors. More than a year ago I had so far completed my experiments that I was enabled to satisfactorily demonstrate the efficiency of my system of lighting[5] and am now ready to introduce it practically in New York & elsewhere & it is for this purpose I desire the permission for which our Company is about to make application and as a considerable amount of capital must be invested it is desirable that the Company have something more than a mere permit, revocable at pleasure.

I should be very much pleased to confer with you personally upon this subject and to that end would be glad if you would name an early day when you would visit me at Menlo Park[6] that I may show you the system at work here and give you full information in regard to the plans of the Company for introducing the light in the City of New York & I have no doubt that I can rely on your hearty co-operation in a work of so much importance to the City.

L, NjWOE, DF (*TAEM* 54:117; *TAED* D8023ZBN). Written in an unknown hand; letterhead of Edison Electric Light Co. [a]"New York" and "188" preprinted. [b]Enclosure is Df; written in an unknown hand on letterhead of Edison Electric Light Co. [c]"New York," and "18" preprinted. [d]Interlined above.

1. John J. Morris served as president of the Board of Aldermen for only one year. *Ency. NYC*, s.v. "Common Council."

2. Probably Tracy Edson.

3. Unidentified.

4. Although little is known about Major Robert Taylor, he was apparently connected with the Edison Electric Illuminating Co. According to Francis Jehl, the day after Edison hosted the aldermen at Menlo Park "the Board of Aldermen met at the City Hall and the Edison matter came up. Many of the aldermen, so the newspapers reported, regaled themselves with reminiscences of the pilgrimage. Major Robert Taylor gave each a printed copy of the resolution which the Edison Electric Illuminating Company desired the Common Council to pass." Jehl 1937–1941, 784.

5. For Edison's demonstrations at Menlo Park in late 1879 and early 1880, see Docs. 1856, 1865, 1867, 1869, and 1873; Israel 1998, 187–88; and Bazerman 1999, 180–85.

6. Morris and several other members of the Board of Aldermen and other New York City officials attended a demonstration and banquet at Menlo Park on 20 December (Doc. 2038).

One of the major technical challenges which Edison faced in developing a system of electric illumination was voltage regulation, or keeping a generator's output voltage constant despite changing loads.[1] Edison's projected system of electric lighting used high-resistance (about 100 to 140 ohms) lamps arranged in parallel across a line voltage of about 110 volts.[2] Since these lamps were connected in parallel to the generator, the amount of current drawn from the generator varied in direct proportion to the number of lamps put into the circuit while voltage remained roughly constant throughout the distribution and generation system. However, as the current increased with the number of lamps put in circuit, two consequent effects in the armature resulted in lower voltages at the generator terminals. The first effect was higher resistive losses in the armature winding which, following Ohm's Law, increased in direct proportion to the current increase. Since Edison designed his generators to have a very low armature resistance, on the order of a few hundredths of an ohm, these resistive losses accounted for a drop of only a few volts. The second effect, now referred to as armature reaction, was a distortion of the generator's internal magnetic field by the current flowing through the armature, and it produced a voltage, called counter electromotive force, which was of opposite polarity to the output voltage.

Edison sought to keep generator voltage constant for the benefit of both customers and central station managers. The most important reason for voltage regulation was to ensure proper operation of customers' incandescent lamps and electric motors. While Edison's lamps could tolerate some variations in voltage, wide fluctuations changed the level of illumination provided by the lamps. Indeed, Edison's early isolated illuminating plants featured a voltage regulator which the customer could use to change the intensity of the lamps.[3] However, high line voltages significantly shortened their operating lives. At the central generating station, engineers needed to regulate the output voltage of generators in order to distribute evenly the loads each generator carried. At central stations, several generators were connected in parallel to the line. If a generator's voltage was more than one or two percent below the line voltage, current flowed back into its armature, causing it to run as a motor and forcing the other generators to supply power to it as well as to the external load. A mismatch in the voltage regulation characteristics of a bank of generators thus created an imbalance in the loads they carried, which led to uneven wear on the generators. Although Edison seems to have recognized this problem as early as June 1879, it was not until he set up his first central station at Menlo Park in December 1879 that he began to consider specific ways to solve it.[4]

In addition, electric motors, such as those used in sewing machines and pumps, operated at a specific speed determined by the line voltage.[5] Motor speed could also be regulated at the motor itself, by adjusting the current circulating in its field coils. Since voltage regulation of a generator and speed regulation of a motor both involved varying the field excitation of both types of machine, Edison developed methods to keep motor speed constant irrespective of fluctuations in line voltage at the same time that he worked on ways to regulate generator voltage. Edison recognized the necessity for motor speed regulation as early as December 1879 and devised specific methods for this in late October and early November 1880.[6]

On 16 December 1880 Edison drafted two caveats outlining many methods of regulating generator voltage and motor speed.[7] The general method of maintaining the voltage constant at the output of the generator was to vary the generator's excitation field in response to changing current demand. Edison explored two broad ways of doing this. The first was to change the magnetic circuit formed by the two field coils, yoke, and pole pieces, and Edison's preferred method of doing this was to introduce an air gap into the magnetic circuit. The second was to control the amount of current flowing through the field coils, and Edison's most usual method was to place a variable resistance in the field coil circuit. The two caveats drafted on 16 December 1880 showed several specific ways to regulate a generator's voltage. In Caveat 101 Edison described several ways to "regulate the field of force magnets of a Dynamo Electric Machine automatically and also to provide means for automatically governing electro-motors when the work is a varying quantity and an even speed is required."[8] He showed seventeen ways to control the current circulating in the field coil of a generator or motor, most of which adjusted a variable resistance in the field coil circuit. In Caveat 102 (Doc. 2036) Edison described several ways to accomplish a more general result, to regulate the generator voltage "by regulating either automatically or by hand the strength of the field magnet." In this caveat he used four general approaches to control the strength of the excitation field: 1) changing the magnetic circuit by raising and lowering the magnet's yoke, 2) varying an adjustable resistance in the field coil circuit, 3) using a relay to open and close the field coil circuit, or 4) varying the current output of a separate generator used to provide field excitation for the main generators.[9]

1. Edison described his general approach to the problem of voltage regulation in an 1879 British patent: "the electro-motive force of the machine is analogous to the pressure in the system of gas lighting, and at dusk, when the lamps are being rapidly connected to the circuit, the electrometer will show a slight drop in the electro-motive force or pressure, and this may be increased by increasing the speed of the prime mover, or increasing the power of the field magnets. The latter method is the one I prefer" (British Patent 2,402 [1879], Batchelor [*TAEM* 92: 118; *TAED* MBP017]). His experiments on voltage regulation began in the autumn of 1878, and in his October 1881 testimony for a patent interference, he gave a history of his work on this problem. Edison claimed that after obtaining a Wallace dynamo in September 1878 he had "continuously used dynamo electric machines of various kinds without intermission, in which the strength of the field of force magnets was varied by means of an adjustable resistance. . . . In fact the nature of the lamp which I have been experimenting on since 1878, is such that I could not have used a dynamo machine, except I used devices for regulating the strength of the field of force magnets" (*Keith v. Edison v. Brush*, 6–11 Lit. [*TAEM* 46:117–19; *TAED* QD002:4–6]).

Edison first investigated the use of a variable resistance in series with the field coil as a means to control the strength of the magnetic field in October 1878 (see Docs. 1506 and 1514). According to his interference testimony he began work in February 1879 on a variable resistance in shunt across the field coil; he filed a patent application on this arrangement in September 1879 (U.S. Patent 219,393). In his installation at Menlo Park used for public demonstrations between November 1879 and February 1880, Edison used one generator to supply current to the field coils of several generators, which in turn supplied power to the lamps. He used a manually controlled variable resistance to adjust the excitation field of the generator powering the field coils of the main generators, thereby regulating the voltage on the mains. The *New York Herald* reported on 28 December 1879 that Edison used a "method of regulating the strength of the current to be used at the central stations. By moving a little wheel the assistant in charge of this branch of the system was enabled to readily vary the strength of the electric lights from the merest glimmer to a dazzling incandescence." Two days later the *New York Herald* described the manual regulation system in greater detail, noting that it used a mirror galvanometer to indicate voltage on the mains. According to the reporter, Charles Batchelor claimed, "It may be done automatically, but it hasn't been thought out yet." Francis Jehl later recalled that he operated the voltage regulator during these demonstrations and a demonstration for the famed actress Sarah Bernhardt in December 1880. "Edison's Light," *New York Herald*, 28 Dec. 1879, Cat. 1241, item 1396; "Electricity and Gas," *New York Herald*, 30 Dec. 1879, Cat. 1241, item 1401; both Batchelor (*TAEM* 94:551, 555; *TAED* MBSB21396X, MBSB21401X); Jehl 1937–41 (411, 77–72).

Edison also used a manual voltage regulation scheme for his first isolated lighting plant, on the steamship *Columbia* in May 1880. He described this method in his 28 January 1880 patent application covering the general features of his generation and distribution system (Doc. 1890). In November 1881 Edison devised an automatic regulating system for isolated plants, which he described as "working 'bang up' . . . I congratulate myself that this is a pretty good thing for Isolated business as without it we should constantly be at variance with purchasers as to the life of our lamps." TAE to Edward Johnson, 27 Nov. 1881, Lbk. 9:373 (*TAEM* 81:134; *TAED* LB009373).

2. In the patent application he executed on 28 January 1880 for his system of electric lighting (see Doc. 1890), Edison stated that he preferred his lamps to have a resistance of 100 ohms. The figure of 140 ohms is from Edison Co. for Isolated Lighting 1883, 9.

3. Edison Co. for Isolated Lighting 1882, 7.

4. British Patent 2,402 (1879), Batchelor (*TAEM* 92:118; *TAED* MBP017); Draft for British provisional specification and U.S. caveat (pp. 2–9), 19 Dec. 1879, Cat. 1146, Lab. (*TAEM* 6:653; *TAED* NM014: 29–36); after revision this draft became the basis for British Patent 33 (1880), Batchelor (*TAEM* 92:146; *TAED* MBP02).

5. For Edison's work on electric motors, see Doc. 1800.

6. Draft for British provisional specification and U.S. caveat (pp. 12–16), 19 Dec. 1879, Cat. 1146; N-80-10-25:32, 38, 61–62; both Lab. (*TAEM* 6:689–93; 34:167, 173, 197–98; *TAED* NM014:40–44; N060: 32, 38, 62–63).

7. On 27 August Charles Mott recorded in his journal that Edison and Upton discussed "ways, mean and apparatuses for determining the electro motive force or pressure in the several lines leading from any central station and for maintaining and equalizing the same. Mr. Edison made several sketches of devices and means to determine and regulate, which were marked 'caveat' but I could or did not get them to determine the nature of their operation or give description." Developmental work for these caveats occurred from late August to early November 1880. N-80-07-10:103–4; Unbound Notes and Drawings (1880); N-80-10-25: 18, 45–53; Cat. 1146; N-80-10-01:45; all Lab. (*TAEM* 37:353–54; 44: 1006; 34:162; 6:687, 41:1089; *TAED* N117:51–52, NS80AAM, N060: 18, 45–53, NM014ZAE, N304:28).

8. Edison's Caveat 101 (*TAED* W100ABQ).

9. Edison also considered two other approaches not found in these caveats. On 27 August 1880 he sketched an arrangement for putting a motor in the field excitation circuit. The motor produced an electromotive force in opposition to the voltage providing excitation current to the line generators. As the line voltage increased, the speed of the motor and its counter electromotive force both increased, thus decreasing the current sent to the field coils of the line generators and reducing their output voltage. He included arrangements based on this approach in patent applications that he executed on 16 December 1880 and 25 February 1881; these issued as U.S. Pats. 239,374 and 248,421, respectively.

On 3 January 1881, Edison sketched another system to regulate voltage by switching fixed resistances in parallel across the generator's field coil in response to changing load. When a group of lamps was turned on, the same switch placed a resistance across the generator's field coil, thereby providing voltage regulation. Edison intended this system to be used in isolated lighting on board ships, in street lighting, or in other applications in which loads were switched on and off centrally rather than by individual customers. He executed a patent application covering this arrangement on 31 January; this issued in October 1881 as U.S. Pat. 248,422. Cat. 1147, Lab. (*TAEM* 44:225; *TAED* NM016AAA).

–2036–

Caveat: Electric Lighting

[Menlo Park,] December 16, 1880[a]

The object of this invention is to increase[b] or diminish the electro-motive force of a Faradic machine bobbin, by regulating either automatically or by hand the strength of the field magnet.

In the accompanying drawing:

Fig. 1.[1c]

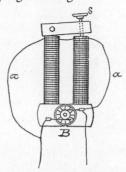

Figure 1, shows the field magnet which is energized by a multiple arc circuit a̲. a̲. across the bobbin B, and the strength of the magnetic field is varied by adjusting the bulk yoke of the magnet to or from its poles by means of the screw S. The other end of the yoke being pivoted to the pole of the magnet.

Fig. 2.[2]

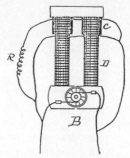

Figure 2, shows a double wound field magnet, both coils[b] of which are energized by the current from the bobbin B. The coil D is the main coil, and the current passes through it in one direction, while the coil C is so wound that the current passes in the opposite direction. An adjustable resistance R is placed in the same circuit as the coil C. Now the effect of the currents in the two coils is to neutralize each other, but by adjusting the resistance R, the strength of C may be reduced, and the field magnet will be energized by the current in D. I will mention that the coil C might be so wound, that the currents would not tend to make opposite magnetisms, and the regulation would take place by adjusting the resistance R. There might even be several coils, one or more thrown into circuit in multiple arc or in series, or two in series and several of each series in multiple arc Fig. 3.[3]

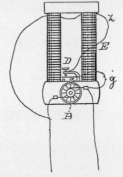

In figure 3, is shown a self make and break. E is the field magnet, and B the bobbin. a is an iron lever or spring attracted by a stud of iron projecting upward from the ends of the field magnet. This spring normally is in contact with the circuit closing point D, completing the circuit of the field magnet. If now current is generated in the bobbin B, the field magnet will be energized, and when it attains a certain pre-determined strength the spring a is attracted and opens the circuit of the field and continuing to vibrate, prevents the field from becoming any stronger than is desired.

In practice it is essential that there should be several springs or levers so arranged that the circuit is broken simultaneously in several places, so as to prevent the spark to as great an extent as possible.

Fig. 4.[4]

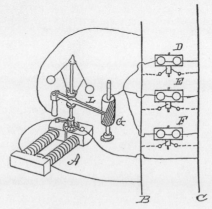

In figure 4 is shown an automatic regulation of the field of force magnets.

A is a motor with its field magnet and bobbin in the same circuit, and is energized by being connected across the main conductors B and C. The shaft of this motor is provided with a governor which serves to give motion to a lever L. The end of this lever, rests upon a rapidly revolving circuit wheel G, whose surface is partly of an insulating, and partly of a conducting material. This wheel is connected to a wire which leads to three field of force magnets D, E, F, the ends of the wires being then connected to the main conductor C, while the lever L is connected to the main conductor B. The wheel G is so arranged that at the bottom it has a metallic surface nearly around its whole circumference, and this gradually decreases until there is scarcely any metal at the top, the cylinder at this point having insulating material over nearly the whole of its periphery. If now the bobbins of D, E, and F are rotated, the motor A is set in motion, and G also being in motion, the field magnets will be energized, and will never go above a certain strength, for if the current increases in the main conductors, the motor will accelerate, the governor will lift the arm higher on G and the current[b] through the field of force magnets will be decreased, in consequence of the resistance of the insulating material which covers the upper part of G.

Fig. 5.[5]

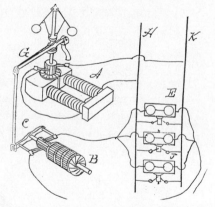

Figure 5, shows a somewhat similar device. A is a motor worked from the main conductors H, K. It is provided with a governor which working an arm G moves the commutators of a bobbin B. This bobbin is rotated between a field magnet by power. The current from D serves to energize the field magnets E, F, J, and the automatic movement of the commutator serves to regulate the strength of the current.

Fig. 6.[6]

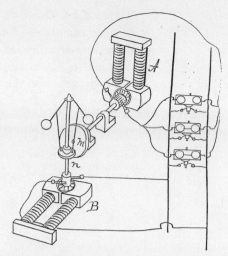

Figure 6, shows a Dynamo machine A, rotated by the motor B, by means of a friction wheel n, running in the surface of a disk m. The dynamo machine A, energizes the field of force magnets, supplying the main conductors. In its normal position the disk n is above the center of m. As the speed of B increases, the governor carries n outwardly toward the edge of m, and slows down the dynamo A and this in its turn decreases the strength of the current on the main conductors, which slows the motor B and so on, thus giving automatic regulation.

Fig. 7.[7]

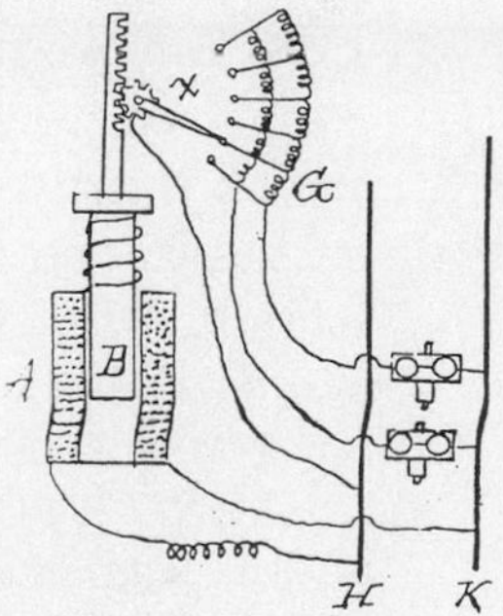

Figure 7, shows an axial magnet a with an iron core B or to be more accurate a coil of wire, in the same circuit with a. This core is kept out from the center of the axial bobbin by a spiral spring. Two field of force magnets have their two extremities connected to the main conductors, while their other extremities are connected to resistance coils G, whereby the strength of the current in the field of force magnets may be regulated. An arm X[d] swinging in a circle comes in contact with a series of pins connected to these resistance coils. This arm is connected to the other main conductor H.

The axial bobbin a being connected to the main conductors it draws down the iron core or coil of wire if that be used, against the retractile resistance or force of the spiral spring, and by means of the rack and pinion rotates the arm X and puts a resistance into the field of force magnet's circuit. The regulation then becomes automatic.

Fig. 8.[8]

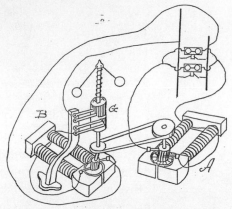

In figure 8, is shown a dynamo machine A which serves to energize the field magnets. It is run by a belt from a motor set in motion by current from the main conductor. B represents this motor. It is provided with a governor the balls of which raise and lower a circuit breaking wheel G like that shown in figure 4. Four springs rest upon its periphery. The wheel is divided into two sections. When the speed of the governor increases, the wheel is raised, and the circuit is closed for a shorter interval of time at each revolution as the speed increases. The object of the four springs is to break the circuit in four places simultaneously and thus prevent the spark.

Witnesses: S. D. Mott   Inventor. T. A. Edison.

DS, MdNACP, RG-241, Edison Caveat 102 (*TAED* W100ABR). Petition and oath omitted. [a]Place taken from petition; date taken from oath, form altered. [b]Obscured overwritten text. [c]All figures are on two separate sheets at the end of the caveat. [d]Interlined above.

1. This method for varying a generator's excitation field used a manually adjustable air gap between one pole of the field coil and the yoke. Edison was exploiting the fact that in a magnet nearly all of the magnetic field is concentrated within an air gap, and that the strength of the magnetic field varies inversely with the length of the air gap. Typically, in a generator the only air gap is between the pole pieces, and this gap provides the magnetic field which induces the generator output voltage in the armature. By introducing another air gap between the yoke and field coil, Edison reduced the strength of the magnetic field seen by the armature. Edison sketched this arrangement on 30 October 1880 and

again on 9 March 1881 when he also sketched two other embodiments of this principle. The first employed a means to raise and lower a cut-out section of the magnet yoke, either manually or automatically. Edison noted on one of his drawings that he wanted to "use this for prime field mach to work other fields also for isolated," in other words to control the current which a secondary generator provided to the field coils of a bank of generators connected to the mains, or to the field coil of a single generator used in an isolated lighting plant. The second alternative arrangement changed the magnetic field by using a lever to provide a shunt for the magnetic field between the yoke and pole pieces. When this lever connected the yoke and pole pieces, it provided another metallic path for the magnetic field, hence reducing the magnetic field strength within the generator. Edison executed and filed two patent applications in May 1881 showing each of these alternate arrangements (U.S. Pats. 263,134 and 263,136). N-80-10-25:12; N-81-03-09:1–11, 24; Cat. 1147; all Lab. (*TAEM* 34:147; 40:652–57, 664; 44:234; *TAED* N060:12; N206:1–6, 13; NM016AAJ).

2. On 2 October 1880 Edison sketched a similar arrangement, a double-wound field coil without the resistance in series with the smaller coil. On 30 October he sketched a design substantially identical to that shown in this caveat, with the series resistance included in the circuit of the smaller field coil. N-80-10-01:43, N-80-10-25:13, both Lab. (*TAEM* 41:1089, 34:147; *TAED* N304:27, N060:13).

3. In this arrangement Edison limited the strength of the excitation field by energizing a relay when the current in the field coil increased above a predetermined value. When this relay energized, it opened the field coil circuit and cut off the excitation field. When the current in the field coil decreased below another predetermined value, the relay de-energized and closed the field coil circuit. On 30 October 1880 Edison drew a substantially identical design and the next day wrote in a notebook entry:

> For isolated Engines and Dynamos such as used for boats I propose to carry out the idea of putting the field magnet in multiple arc, and so arranging the wire & Lamps that when all the lamps are on the field magnet will still be nearly saturated ie. The current will still be sufficient to nearly saturate the magnet & to also arrange it that should the speed increase to such an extent by accident as to double the light in the lamps a magnet will open the circuit of Lamps or field mag or perhaps I will use a governor on the dynamo itself which when the field is saturated & any great increase of speed is attained by accident the gover will move a lever & open the field magnet [N-80-10-25:14, 42–43, Lab. (*TAEM* 34:148, 186–87; *TAED* N060:14, 42–43)]

4. On 1 June 1880 Charles Mott recorded in his journal that Edison had sketched a design like that shown in Fig. 4, a "new form of electric governor which increases or diminishes the amount of current by raising or lowering a cylinder insulated at entire circumference of one end while at or near the opposite end the insulation gradually diminishes in width till lost in a point in revolving the conducting one half surface of the cylinder will remain in contact with the brush or spring, through a

greater or less space according to the prependicular position of the cilinder which is varied by the revolving speed of the governor. Both of which were given to [Samuel] Mott from which he was directed to make patent office drawings." The sketch to which Mott refers has not been located, but Edison executed a patent for this design on 31 July 1880 (U.S. Pat. 248,434). He also sketched a similar design on 30 October 1880. Mott Journal N-80-03-14:206–7, N-80-10-25:15, both Lab. (*TAEM* 33:787, 34:149; *TAED* N053:105, N060:15).

5. Edison sketched this design on 2 November 1880 and again on 3 February 1881 with the note "patent," but he apparently did not file a patent application for this arrangement (N-80-10-25:45–46, Lab. [*TAEM* 34:189–90; *TAED* N060:45–46]; for Edison's work on brush position, see Doc. 1896). It is similar to a design patented earlier by Hiram Maxim, who also a used a motor and system of gearing to shift the position of the brushes on the generator supplying excitation current to the line generators (U.S. Pat. 228,543; Brit Pat. 1,392 [1880]; "Recent Developments in Electric Lighting," *Scientific American,* 23 Oct. 1880, 255, 262).

6. In this and figure 8, Edison used a secondary generator to provide current to the field coils of the line generators, and he regulated the voltage at the output of the line generators by varying the current which the secondary generator provided to the field coils of the line generators. This current was regulated by using a motor to change the speed of the secondary generator. In these figures, Edison arranged the field coils of the line generators in parallel, but in other designs he arranged them in series. Edison sketched the design shown in figure 6 on 3 February 1881 with the note "patent." In late February and early March he executed two patent applications using a motor to regulate the speed of a secondary generator (U.S. Pats. 251,550 and 482,549). The second of these had a broad claim covering "the method of regulating the generative force of one or a battery of electrical generators by regulating the amount of current flowing through the field circuit thereof, by adjustably governing or regulating the speed of the engine driving the generator furnishing the current for the field circuit." Edison was granted this patent in September 1892, only after he abandoned this claim and reduced the scope of his patent to cover only "means for varying the speed of the exciting-machine independently" of the line generators. Cat. 1147, Lab. (*TAEM* 44:233; *TAED* NM016AAI); Pat. App. 482,549.

7. On 31 October 1880 Edison sketched a prior version of this regulation method using an axial magnet to control a variable resistance placed in the field coil circuit. This sketch showed a lever pivoted at the center with the axial magnet at one end and a spring at the other; this arrangement is nearly identical to that which appeared in figure 1 of Edison's Caveat 101 (*TAED* W100ABQ). On 2 November he replaced the pivoted lever and spring with a spiral spring attached directly to the plunger of the axial magnet, the design which appears here. Edison sketched this regulation method with the note "patent" on 29 December; on 26 February 1881 he executed a patent application covering the use of an axial magnet to control a variable resistance placed in the field coil circuit (U.S. Pat. 251,555). A variable resistance such as the one shown in this drawing was used in regulating the first dynamo sent to South America. N-80-10-25:17, 53; Cat. 1146; N-81-02-20:49, 55–59;

all Lab. (*TAEM* 34:161, 188; 6:734; 40:994, 997–98; *TAED* No60:17, 53; NM014ZAV, N214:25, 28–30).

8. Edison also sketched this rotating cylinder design with four contacts on 2 November 1880. N-80-10-25:51–52, Lab. (*TAEM* 34:186–87; *TAED* No60:51–52).

–2037–

From Grosvenor Lowrey

New York. Dec. 17, 1880.[a]

My dear Edison:

I shall not present your letter of resignation as Mr. Fabbri very strongly objects to your leaving the Board. His expression was that "Edison's name is a tower of strength to us, and if he never attended a meeting, it would be a great loss if his name should not appear at all times among the names of the Directors." As soon as the matter was put to me in that way, I confess I saw it in the same light. I, therefore, mustered up courage to go to another Director, who has promised to resign. The reason I did not go to him before was that I have had some unpleasant business relations with him and was afraid he would have thought me actuated by some spite to him.[1]

We yesterday organized and filed the articles of association of the "Edison Electric Illuminating Company of New York," under the general gas company act of this state, stating the object of the organization to be to illuminate the streets, &c. by gas.[2] We have to state this as the object in order to perfect a legal incorporation, but every gas company has by law after it is organized, the right to turn itself into an Electric Light Company, and we have prepared a long ordinance to be submitted to the Common Council if we are so advised, granting us the franchise to lay down wires over the entire city.[3]

I was told yesterday that Mr. Hyde[4] claims that they can make their lamps for ten cents each, and that they can burn ten per horse power. The size of the conductor was not stated. Yours very truly

G.P.L.

Enclosed is copy of the proposed ordinance, which we shall introduce on Tuesday & Griffith will explain all the rest.[b]

TLS, NjWOE, DF (*TAEM* 54:121; *TAED* D8023ZBQ). Letterhead of Porter, Lowery, Soren & Stone. [a]"New York." preprinted. [b]Postscript written by Lowrey.

1. Robert Gallaway, vice president of the company, resigned in mid-January when Sherburne Eaton was elected. Calvin Goddard to TAE, 10 Jan. 1881, DF (*TAEM* 58:3; *TAED* D8126A).

2. The Certificate of Incorporation of the Edison Electric Illuminating Co. of New York is dated 16 December but was filed on 17 December. The organizers were Tracy Edson, Robert L. Cutting, Jr., Egisto Fabbri, Jose de Navarro, James Banker, and Nathan G. Miller. NNNCC-Ar (*TAED* X119JA).

3. Not found but see Doc. 2039 n. 2.

4. Probably Henry Hyde, who evidently was president of the U.S. Electric Lighting Co. Wilson 1881, 753.

-2038-

Draft to New York City Board of Aldermen

New York, Dec. 18th[a] 1880.

PRIVATE.
Dear Sir:

I shall be very glad to see you, with other members of the City Government, at Menlo Park on Monday evening ~~at   o'clock,~~ Decem 20th[b] to witness an exhibition of Electric Lighting and also the various operations connected therewith[c] which are being carried on in my Laboratory and shops, ~~with a view to make a perfect~~ in the production of a[d] light, adapted to ordinary domestic & commercial[e] use.[1]

This is in no sense a public exhibition; ~~however,~~ but every facility will however[f] be afforded to invited guests to see all that is involved in the manufacture of the lamps and the production of the light.

The train leaves Courtlandt Street Ferry at 4:30[a] o'clock and returns from Menlo Park at 9:30[a] o'clock. Very respectfully yours

Thomas A. Edison

TL, NjWOE, DF (*TAEM* 54:123; *TAED* D8023ZBR). Insertions, interlineations, and signature written in an unknown hand. [a]Inserted by hand. [b]"Decem 20th" interlined above. [c]"connected therewith" interlined above. [d]"in the production of a" interlined above. [e]"& commercial" interlined above. [f]Interlined above.

1. Edison hosted this demonstration for about forty city officials, including the aldermen, in order to convince them to support the Edison Electric Light Company's efforts to obtain a franchise giving it rights of way to install electrical mains under New York City streets (William Carman to Calvin Goddard and Goddard to Carman, both 20 Dec. 1880; DF [*TAEM* 54:125; *TAED* D8023ZBS, D8023ZBT]); see Doc. 2039 n. 2). According to newspaper reports city officials included Board of Aldermen president John J. Morris, Superintendent of Gas and Lamps S. McCormick, Park Commissioners Green and Lane, Excise Commissioner Mitchell, and aldermen McClave, Jacobus, Strack, Wade, Kirk, Fink, and Slevin. Also present were Grosvenor Lowrey, Tracy Edson, Sherburne Eaton, Calvin Goddard, and Nathan Miller of the Edison Electric Light Co., as well as Ernest Biedermann, J. C. Henderson, Ma-

jor Robert Taylor, a Mr. S. C. Wilson, and a Mr. G. Salvyera of Paris ("Aldermen at Menlo Park," *New York Truth,* 21 Dec. 1880; "The Wizard of Menlo Park," *New York Herald,* 21 Dec. 1880; Cat.1241, Items 1557, 1558; Batchelor [*TAEM* 94:623; *TAED* MBSB21557X, MBSB21558X]; see also Jehl 1937–41, 779–85). After demonstrating his system and explaining its working, Edison provided the visitors a lavish banquet catered by Delmonico with a menu selected by Grosvenor Lowrey. Charles Mott noted in his diary entry for 20 December, "In the evening, by invitation, the Alderman of New York City were here to see the working of the lamp and system and to pertake of both solid and fluid refreshments. The room of Laboratory was illuminated with 37 lamps and presented a fine appearance, in all about 239 lamps were illuminated. The boys had their racket at Davis's [Hotel]— Part of whom celebrated at their own expense" (Mott Journal N-80-07-10:258, Lab. [*TAEM* 37:431; *TAED* N117:129]).

Edison increased the number of outdoor lamps by early January for a more public exhibition. The 22 January *Scientific American* reported that five hundred lamps had been placed in lines extending from the laboratory "half a mile to right and left, the entire area under illumination being, from the slope of the land, easily visible from the central station. The lamps are in a circuit comprising seven miles and three-quarters of wire, and are supplied by a current generated by nine dynamo-electric machines driven by one engine. . . . Simply as an exhibition of perfect illumination under perfect control, covering a vast area, this array of lamps presents a most remarkable and delightful sight, and is alone well worthy of a trip to Menlo Park." "Electric Illumination at Menlo Park," *Sci. Am.* 44 (1881): 44.

–2039–

From Grosvenor Lowrey

New York. Dec. 21, 1880.[a]

My dear Edison:

I said this morning I was going to draw the Lamp Manufacturing Contract, but when I sat down to it I was reminded that I had previously found myself lacking in proper information and that I had asked you the last time I was at Menlo, to make an outline draft of it for me. Can you do that? My impression is that the Company cannot give itself away upon the manufacture of its Lamps by putting it out of its power, in case it should be dissatisfied, to make them for itself or for others to make them. The best which can be done with such an arrangement is to make it to the mutual profit of each party to continue the arrangement. You will remember Mr. Goddard's plan of an adjustable scale of prices, by which you were to have a larger pro rata of profit as the cost to the Company went down. He can draw that clause more satisfactorily than any one else probably and I will ask him to do it and will send it to you.[1]

The ordinance was referred to-day to the Law Committee of the Board of Aldermen and goes over until next week.[2] Perhaps we shall not get the franchise from this Board as their time is so short. Very truly yours

G. P. Lowrey

TLS, NjWOE, DF (*TAEM* 54:127; *TAED* D8023ZBV). Letterhead of Porter, Lowrey, Soren & Stone. a"New York." preprinted.

1. A relatively simple contract was drafted in January 1881, possibly by Edison himself; Edison had Charles Mott make a copy of this draft on 7 February (Mott Journal, PN-81-01-19, Lab. [*TAEM* 43:1124; *TAED* NP014:10]). The parties to the agreement were the Edison Electric Light Co. and Edison, Charles Batchelor, Francis Upton, and Edward Johnson, acting as partners in the Edison Electric Lamp Co. This draft became the basis for a much more detailed draft that is dated only 1881 and which was probably the result of discussions between Lowrey and the Light Co. In this draft the names of Edison's partners are crossed out and the agreement is between the Light Co. and the Lamp Co. The final agreement is only between Edison and the Light Co., with Edison's partners in the Lamp Co. mentioned as his manufacturing partners but not named as the Lamp Co. Edison was to receive no compensation for his efforts in regard to lamp manufacture but his partners were to receive "a fair and usual compensation for the[ir] actual services." The second draft and final version were largely concerned with the efforts of the Light Co. to protect its interests while also giving Edison an exclusive license for the manufacture of lamps used in connection with the Edison system. In addition, there was considerable discussion regarding the amount the Lamp Co. would charge the Light Co. for lamps. The first draft has the figure of 35 cents, which was crossed out and changed to 40 cents in the second draft but in the final agreement the figure was 35 cents. In addition, the second draft included a clause concerning the "endeavor to reduce the cost of manufacturing the standard lamp, and after receiving from sales of lamps the amount actually invested in the business of said Lamp Company, and whenever after such amount is received, the cost thereof shall fall below thirty-two cents each, it will pay over to the Light Company, one-half the difference between that price and the actual cost of the lamps manufactured." This section was partially underlined with a notation in the hand of Edison's

secretary Samuel Insull that "this is to cover great loss in experimental work to date." It was included in the final agreement but without the reference to the money invested in the Lamp Co. According to the contract, Edison and his partners agreed to invest "in actual cash immediately not less than fifty thousand dollars ($50,000) for the establishment of a suitable manufactory capable of furnishing one thousand lamps complete during each and every working day in the year, the investment already made at Menlo Park for that purpose being taken as a part of the same." Edison reportedly estimated that manufacturing costs averaged $1.10 per lamp about this time. TAE draft agreement with Edison Electric Light Co., Jan. 1881, DF (*TAEM* 57:761; *TAED* D8123C); Edison Electric Light Co. draft agreement with Edison Lamp Co., n.d. 1881, Miller (*TAEM* 86:426; *TAED* HM810156); TAE agreement with Edison Electric Light Co., 8 Mar. 1881, Defendant's depositions and exhibits, 5: 2352–57, *Edison Electric Light Co. v. U.S. Electric Lighting Co.*, Lit. (*TAEM* 47:999; *TAED* QD012E2352); Jehl 1937–41, 816.

2. According to Charles Mott's journal the *New York Herald* reported on 29 December that "the Law Committee of the Board of Alderman to whom was referred the application of the Edison Illuminating Co. for privileges of laying their wires in the Streets, had reported a resolution that that Co or other Elec. Light Cos have the privilege by paying to the City 10 cts per foot for the streets disturbed and after five years three per cent of their gross receipts" (Mott Journal N-80-07-10:264, Lab. [*TAEM* 37:434; *TAED* N117:132]). The *New York Herald*, in articles on 20 and 21 January, opposed this provision, noting that gas companies and the Brush Electric Light Company did not have to pay this fee for rights-of-way. The Brush company had obtained permission to lay underground wires for an experimental arc lighting system to illuminate Broadway between 14th and 34th streets; the city government did not require the Brush company to pay a fee for rights-of-way for this installation and only required it to repair any damage to the streets resulting from the work ("Lightning Over Snow," 20 Jan. 1881; and "The Aldermen and the Edison Light," 21 Jan. 1881; Cat. 1241, items 1573 and 1571; Batchelor [*TAEM* 94:627; *TAED* MBSB21573, MBSB21571X]; "New Lights Along Broadway," 8 Dec. 1880, *New York Times*, copy found in scrapbook in Hammer Ser. 2 Box 25). An undated printed resolution authorizing the company to place its wires underground in New York City is in DF (*TAEM* 57:744; *TAED* D8122S) but this was not the one adopted as it required the Edison Electric Light Co. to deposit a sum with the city sufficient to cover the costs of restoring any affected streets to their prior condition. Instead, on 22 March 1881 the aldermen passed a resolution allowing the Edison company to place its wires underground or overhead after paying a security deposit determined by the mayor, the comptroller and the commissioner of public works, with the work to be done under the latter's supervision. The company was to be charged a rate of 1% per lineal foot. This resolution was vetoed by Mayor William Grace on 5 April but the aldermen overrode the veto on 12 April and subsequently passed resolutions (also vetoed and overridden) granting the same privileges to the Brush Co. and United States Illuminating Co. (DF [*TAEM* 57:746, 748; *TAED* D8122S1, D8122S2]). See Bazerman 1999 (224–28) for an analysis of the political battle over street franchises for electric lighting.

New York, Decem 30 1880.[a]

My dear Sir:

Notwithstanding all that has been written on the subject of Electric lighting and our system in particular, the general public seem to have a very incorrect notion of what it is we propose to do ~~so a~~A large number of letters are received asking information which wuld well be answered by a printed circular giving a brief description of the system, how the light is produced, how managed, and how used. Something of this sort would be extremely useful, not only as a means of answering correspondents but to combine with other information in respect to the organizing of companies, granting of licenses etc. which we must shortly issue to send to parties whose applications are on file & who have been promised information when the Company was ready for business. Will you kindly allow Mr. Clark or Mr. Upton, or both, under your direction to prepare the material for such a[b] circular making it as brief as the subject will allow simply giving a general idea of the system, and send it to me as soon as possible.[1]

We had quite a satisfactory meeting of the Executive Committee today at which I represented your views as expressed yesterday[2] & measures have been taken to perfect our Company organization for active business. Very truly Yours,

C. Goddard

The circular should include Lighting for cities & towns Lighting Factories & detached buildings already supplied with power   Lighting Steamers[c]

LS, NjWOE, DF (*TAEM* 54:133; *TAED* D8023ZCA). Letterhead of Edison Electric Light Co.; written in an unknown hand. [a]"New York," and "18" preprinted. [b]Interlined above. [c]Postscript written by Goddard.

1. In his journal entry of 10 January Charles Mott noted "All night several worked on a Practical treatise on Elec. as known here, for Publication in Book form" (Mott Journal, PN-80-09-23, Lab. [*TAEM* 43:1108; *TAED* NP013:45]). Edison, Charles Batchelor, Edward Johnson, and Charles Clarke used portions of six undated notebooks to outline the scope and sources of the proposed publication; Francis Upton also contributed. The entries suggest that the book would include a full description of Edison's system; criticisms of incandescent electric lighting; the history and economic basis of the gas lighting industry; and historical precedents of scientific and technical opposition to significant technological changes (N-80-06-16.1, N-81-01-00:1–9, N-81-01-25, N-80-00-04, N-79-07-12, N-81-01-21, all Lab. [*TAEM* 39:1138, 40:14–49, 224, 253, 280, 284; *TAED* N184, N187:1–5, N188, N189, N190, N191]). Edison also drafted a letter in one of these books asking Uriah Painter to gather publications of the Bureau of Statistics and also to look for relevant information published from the United States census. The

spine of one of the notebooks is stamped "Edison's Prospectus Book" (TAE to Painter, 21 Jan. 1881, N-81-01-21:284, Lab. [*TAEM* 40:303; *TAED* N191:18]). Edison never published such a work but Edward Johnson probably used some of this material in preparing an essay dated 15 September 1881 on "Edison Electric Light Stock considered as a speculative holding for the ensuing quarter" (Johnson typescript, 15 Sept. 1881, DF [*TAEM* 58:37; *TAED* D8126ZAB]).

2. Nothing is known of any discussion between Edison and Goddard.

–2041–

From Grosvenor Lowrey

New York, Dec 31st 1880[a]

My dear Edison

I wish you this year a happy "New York," & great increase of of fame & fortune.

In respect to the financial necessity of which you spoke to Fabbri & me, do not give yourself uneasiness.[1] They must & will be provided for by your friends in a proper way, that is, so as not to give you trouble of any sort.

Twombly wished me to speak to you about your having (as he understands) allowed American Union wires on your premises. He thinks you ought not to do so being a W. U. soldier.[2]

He wished me to say so to you & I promised but have forgotten it when I had seen you. Therefore I write now.

If by any chance you make any call in town on [Saturday?][b] come see my wife at the Victoria Hotel cor 27th St & 5th Av. Tell Mrs. Edison I have gone back on my advice to her & have gone to a Hotel. But it was necessary to be in town & I could not get any other place now. Yours truly

G. P. Lowrey

ALS, NjWOE, DF (*TAEM* 53:277; *TAED* D8004ZGZ1). Letterhead of Porter, Lowrey, Soren & Stone. [a]"New York" and "18" preprinted. [b]Illegible.

1. Nothing further is known about this matter.

2. On 23 December "American Union wires were to day connected with the Laboratory for telegraphic communication." On 31 December Charles Mott "remained up all night to discover if possible" who had disconnected those wires from the switchboard on several occasions; he apparently was not successful. Mott Journal N-80-07-10:261, 268, Lab. (*TAEM* 37:433, 436; *TAED* N117:131, 134); see also Doc. 1764 n. 1.

January–March 1881

In the early months of 1881 the work of Edison and his staff at Menlo Park was directed almost wholly toward manufacturing lamps and developing his system of electric light and power for commercial use in New York. He appointed Francis Upton to take charge of the lamp factory at the first of the year; William Hammer, another valued assistant, spent an increasing amount of time there. Edison continued to operate, at least intermittently, the Menlo Park demonstration system; one such occasion was 6 January, when he promised to "have all lighted" for Edison Electric Light Co. investors.[1] He added lamps made by his factory and showed investors and other visitors as many as 500 in use at the same time. During January and February, he continued to direct laboratory experiments designed to make the product more durable and easier to manufacture. He again turned to the problem of carbon carrying, the perplexing phenomenon by which lamp globes were gradually darkened by a thin layer of carbon. He found a chemical process to bind the trace amounts of hydrogen in the bulb that exacerbated the problem, although it is not clear if this practice was adopted at the lamp factory. In early February, Edison offered experimenter Edward Acheson a one hundred dollar reward to create a practical lamp filament from plumbago pressed in a hydraulic press. Acheson succeeded a few days later and entered into a piecework agreement to produce 30,000 pressed filaments. Halfway though that order, however, it became apparent that they were not durable and Edison continued using bamboo. Several months of work, principally by John Lawson and Albert Herrick, on electroplating filaments directly to the lead-in wires came to fruition in mid-February and the factory's clamp department was abolished, though it remained to be settled whether copper or silver was the better plating material. Edison also applied for patents on various lamp sockets and fixtures; these were among the eighteen applications he executed in this period that resulted in U.S. patents.

The most apparent change in the Menlo Park lighting system was the combination of the new large dynamo and the Porter–Allen engine. The engine was delivered in late January, well behind schedule, but the dynamo was not ready either, and it was another month before the two machines were coupled and tried for the first time. In mid–March Edison ordered that future lighting exhibitions would be run exclusively by this machine. In other electric light experiments, assistant Edward Nichols tried in March to find a way to protect the meter from freezing without affecting its accuracy. Francis Upton also continued to calculate the effects of conductor diameter and distance from the central station on the voltage at any point in a distribution system.

Edison made several important administrative and business arrangements in preparation for building and operating a central station district. Because his financial backers were generally unwilling to provide capital for manufacturing the necessary equipment, he established two companies with trusted assistants as partners. The Electric Tube Co., which would fabricate underground conductors, was incorporated in early March by Edison, two members of Drexel, Morgan and Co., Stockton Griffin, Charles Clarke, and John Kruesi, who managed its operations. To construct dynamos and other heavy equipment Edison and Charles Batchelor formed the Edison Machine Works and placed machinist Charles Dean in charge of the shop. Edison also entered into a contract with the Edison Electric Light Co., which held title to his patents, for the right to manufacture electric lamps. The agreement fixed the price at which the parent company would buy lamps, leading Upton to believe that the lamp company "must move. We never can make the lamp cheap until we can have plenty of boys and girls at low wages," which he expected to find near a manufacturing center such as Newark.[2]

While he was making these preparations, Edison still had not received approval from New York City authorities to lay conductors beneath Manhattan's streets and sidewalks. While he waited, his workers built and installed a complete lighting plant at a New York print shop in January. He contemplated making similar installations elsewhere and even solicited information about laying underground wires in Baltimore. In January, he had Charles Clarke reply to a request to exhibit the electric light at the Exposition Internationale d'Électricité in Paris later in the year. Clarke wrote that Edison would "do what lies within his power" but could make no promises because the "demand upon his time and energies is already excessive."[3] Edison did, however, agree to supply equipment for a small demonstration system in South America. At the request of Calvin Goddard, secretary of the Edison Electric Light Co., for a pamphlet explaining the electric lighting system, Edison and several assistants, including Edward Johnson, outlined and began to draft what was apparently intended to be a comprehensive book on electric lighting. In late January, after an ice storm and gale devastated New York's network of overhead telegraph lines, Johnson wrote a long letter to the editor of the *New York Tribune* explaining the advantages of Edison's proposed central station system of underground conductors.

Edison made other business arrangements during this time to satisfy his need for cash for the lighting system. William Carman calculated that Edison had spent over $129,000 on electric light experiments by the end of March. The Edison Electric Light Co. had reimbursed nearly all of this but not, of course, a further $45,000 invested in the lamp company and additional sums for the Tube Co. and Machine Works. In February, George Gouraud helped negotiate the formation of the Oriental Telephone Co., combining the Edison and Bell interests in Asia and the Near East, from which Edison expected to realize about $100,000 in cash. Edison wrote Gouraud on 7 March that it was "important that I should have as much money as possible at my disposal within the next few months."[4] He instructed Gouraud to obtain these funds as quickly as possible and to liquidate his reversionary interest in the new United Telephone Co. of London (worth about another $100,000), and also to reach a settlement for royalties still owed by the now-defunct Glasgow telephone company. The Edison Electric Light Co. of Europe, Ltd., incorporated in December 1880 to control European patents outside the United Kingdom, still offered the tantalizing prospect of cash but the company was not capitalized until September 1881. In mid-February, Edison told the editor of *Science* to find another party to underwrite the journal because he was "to[o] busy to give it any attention," though presumably he was also happy to forego this expense.[5]

Edison entered into one other agreement which had a profound effect on the laboratory. To facilitate his work in New York City he leased a four story brownstone building at 65 Fifth Avenue in late January or early February. Besides his own offices, this building also housed the headquarters of the Edison Electric Light Co. and the Edison Electric Illuminating Co. of New York. By early March, Edison was spending most of his days there and communicating by letter with the Menlo Park staff. The Machine Works and Tube Co. were also located in New York and by mid-February John Kruesi was spending all of his time managing the latter, leading Edison to appoint Thomas Logan as foreman of the Menlo Park ma-

Edison (left) standing with Charles Batchelor (center) and Sherburne Eaton on the steps of the Edison Electric Light Co. headquarters at 65 Fifth Avenue in New York.

chine shop. The laboratory staff itself was greatly reduced in size during the winter as many of his assistants joined Edison in New York or went to work in the shops. It is unclear how much time Francis Upton and others directly involved in lamp experiments actually spent at the laboratory. Stockton Griffin, who as Edison's personal secretary had helped manage the growth of his correspondence at Menlo Park since May 1878, resigned in February for reasons which remain unclear. At Edward Johnson's recommendation Edison hired Samuel Insull, formerly George Gouraud's secretary at the telephone company in London. After arriving in New York at the end of February, Insull joined Edison at 65 Fifth Avenue. His first assignment was to determine how much money Edison had available for investing in electric light enterprises, foreshadowing his later role not only as Edison's secretary but also a trusted financial advisor in the business of building a new industry.

1. TAE to James Banker, 5 Jan. 1881, DF (*TAEM* 57:560; *TAED* D8120C).
2. Doc. 2061.
3. Doc. 2045.
4. Doc. 2060.
5. Doc. 2054.

–2042–

Technical Note:
Electric Lighting

Menlo Park, N. J. Jan. [4?] 1881.[1]
Statement of Tests of the Economy of the Edison
Electric Light.
Dec. 19th 1880.

Total no. of lamps	183.
Lamps per horse-power including field	6.15.
Steam per horse-power	35 lbs. per hour
Total power developed	41.75 H.P.[2]
Consumed in friction	12—H.P.[a]

Dec. 24th 1880.

Total no. of lamps	397
Lamps per horse-power including field	6.96
Steam per horse-power	25 lbs. per hour
Total power developed	72.4 H.P.
Power with field	22.19 H.P.
Friction alone	15.35 H.P.

Jan. 1, 1881

Total no. of lamps	408
lamps per horse-power including field	6.59
Steam per horse-power	22½ lbs. per hour
Total power developed	82.3 H.P.
Power with field	27.35 H.P.[3]
Friction alone	20.35 H.P.[a]

Jan. 4, 1881.[4]

Total no. of lamps	464
Lamps per horse-power including field	7.1
Steam per horse-power	21½ lbs. per hour
Total power developed	81.544 H.P.
Power with field	23.23 H.P.
Friction alone	16.29 H.P.[a]

Average result

Lamps [-][b] per horse-power	6.7

X, NjWOE, DF (*TAEM* 57:1093; *TAED* D8125000B). Written by Charles Clarke. [a]Followed by dividing mark. [b]Canceled.

1. This entry was presumably written on or soon after the date of the last test.

2. According to Charles Mott, Charles Clarke conducted this test with one generator serving as the exciter for five other machines connected to the line. Mott reported slightly different results from those given here, showing 6.3 lamps per horsepower. More significantly he noted that "The amount of water per hour by indicator was 1429.91 lbs or 34.4 lbs per H.P. per hour. Mr. Clarke says that the amount of steam used by the engine should have given 7.3 lamps per horse power, otherwise that a good engine should have given the same results with 30 lbs of steam per hour, per H.P." The boiler was fired by dust coal for this test. The next day large coal was burned instead and on 23 December a mix of half large coal and dust was adopted "As a compromise and with better results." Mott Journal N-80-07-10:256–58, 261, Lab. (*TAEM* 37:430–31, 433; *TAED* N117:128–29, 131).

3. In his account of this test Mott reported that the engine provided to the armatures "61.95 H.P. or on 408 lamps gave 6.59 per H.P. less field gave net 7.88 per h.p. on 22.58 pounds steam per H.P. per hour," meaning that slightly more than 10 horsepower was consumed in the field magnets. Mott Journal N-80-07-10:271, Lab. (*TAEM* 37:438; *TAED* N117:136).

4. In a separate resume of this day's results Clarke calculated 5.7 lamps per indicated engine horsepower. He then used this rate to determine that 3½ pounds of coal should run 8.1 lamps, although the significance of this amount of fuel is not clear. The test was made with nine line generators and one field exciter. Another trial on 7 January failed to pro-

duce meaningful data because steam power was being used simultaneously for other purposes. Clarke test results, 4 Jan. 1881, DF (*TAEM* 57:1091; *TAED* D8125000A); Mott Journal 80-07-10:274, 280, Lab. (*TAEM* 37:439, 442; *TAED* N117:137, 140).

–2043–

From Edwin Fox

New York, Jany 10 1881[a]

My dear Edison:

The party who wants the light is Silas M. Stilwell Jr. Vice Prest. of this Company.[1] He wants it for "Barmores" 5th Ave cor. 36th St. "Barmores" as you probably know is quite a select establishment and has for a constituency some of Gotham's best people.[2]

Mr. Stilwell was casually conversing with[b] me about Electric lights and he mentioned that they were going to have the Maxim light in but I quickly put a damper on that and he now wants the Edison

In your letter you don't say anything about what terms you would put in the light.[3] I refer to the tenure or how it should be arranged whether under royalty or how.

Please send your Supt.[b] or Engineer in tomorrow or the next day, tomorrow if possible and let him call on Silas M. Stilwell Jr. Counselor at Law Equitable Building or if he will call on me I will introduce him to Mr. Stilwell who will take him to Barmores & have the estimates etc made.[4]

A more desirable place[b] for the early introduction of the light could hardly be had.

By the way I promised the Herald people that I would let them know when you were ready. They also want the light and I can get all the newspaper offices in the city to order it.

Why not let me do that? It would certainly give great eclat to the system and prove a wet blanket for Sawyer, Maxim and the other carpers who are as sure to rise and cry it down after you get started, as mushrooms are after a rain storm. The press having before their eyes the practical and indisputable results of the superiority of your light, would gently sit down on all others.

The movement need not come from you. It would come with good grace from me[b] as an old newspaper man. What think you? Your friend

Edwin M. Fox[c]

P.S. Im a firm believer in the Edison light and as a proof of my belief I have four or five thousand dollars I stand ready to invest in it if you know of any shares lying around at a fair rate. If you know of any please give me the word   E.M.F.

ALS, NjWOE, DF (*TAEM* 57:567; *TAED* D8120H). Letterhead of executive offices of the Albemarle Fertilizer Co. [a]"New York," and "188" preprinted. [b]Obscured overwritten text. [c]Followed by "(over)" to indicate page turn.

1. Stilwell was apparently the son of Silas Stilwell, a noted New York politician, lawyer, and writer on financial subjects; nothing further is known of the son. The Albemarle Fertilizer Co.'s letterhead identified Fox as the secretary. Fox had become involved with the firm by the summer of 1880 when he sent a product sample to Menlo Park for analysis. Edison apparently did not follow up this request or assign Otto Moses to do so. Edison also declined Fox's offer to "make a few thousand" dollars. Fox expected the company's board to double the price of its treasury stock and promised to "reserve some for you and guarantee you against loss. Being on the very inside, I know it is a bonanza." *DAB*, s.v. "Stilwell, Silas Moore"; Fox to TAE, 9 and 11 Aug., 13 Sept. 1880, DF (*TAEM* 53:510, 199, 218; *TAED* D8014T, D8004ZEO, D8004ZFI).

2. William Barmore sold this restaurant and hotel at 390 Fifth Ave. to John Jacob Astor in January 1881 but continued to operate it until the business failed a year later. Wilson 1881, 74; "Local Business Troubles," *New York Times*, 21 Jan. 1882, 8.

3. Edison's letter has not been found but presumably was in reply to Fox's 6 January inquiry on behalf of a friend "very anxious" to have Edison lights for his Fifth Ave. property. Fox noted that "The cost is no object with him. The event is one that would receive large attention at the hands of the city press." Fox to TAE, 6 Jan. 1881, DF (*TAEM* 57:561; *TAED* D8120D).

4. In answer to Edison's reply of 11 January (not found) Fox stated that he had been "under the impression that your Company would put in light and generators but insomuch as that is not the plan I will see my friend Stilwell and explain to him and let you know." Edison did not necessarily oppose the idea of isolated lighting stations for individual buildings, and in November Francis Upton had begun preliminary calculations for wiring William Henry Vanderbilt's house. About this time arrangements were also being made to illuminate a New York printing establishment (see Doc. 2053). Fox to TAE, 12 Jan. 1881, DF (*TAEM* 57:572; *TAED* D8120L); N-80-11-15:177–223, Lab. (*TAEM* 39:870–93; *TAED* N172:89–112).

–2044–

Technical Note:
Electric Lighting

Menlo Park, N.J.,[a] [January 13, 1881?][1]

Made a discovery

By using chloride Carbon in ~~bub~~ bulb it is decomposed Carbon deposited & chlorine set free   This removes all traces Hydrogen   it is the Hydrogen that magnifies the carrying[2] we had one lamp that gave 15.7 per h.p.

⟨Given to Chas Batchelor by T. A. Edison Jan 13th 1881 Chas Batchelor⟩[b]

X, NjWOE, Upton, Cat. 1243:1722 (*TAEM* 95:7; *TAED* MBSB41722).
Letterhead of Edison Electric Lamp Co. [a]"Menlo Park, N.J.,"
preprinted. [b]Marginalia written by Charles Batchelor.

1. Edison most likely made this note on or about the date he gave it to
Charles Batchelor.

2. On earlier efforts to remove hydrogen compounds, see Doc. 2003.
Edison's "discovery" resulted from research planned a week earlier.
Charles Mott noted in his journal on 6 January that "Mr. Edison wrote
out some 9 or 10 experiments for Lawson to try on lamps at Factory.
Mostly the introduction of Phosphorous, Sodium etc. in the globe, to be
acted on by the heat on pumps or in use." John Lawson's notes indicate
that he also tried naphthalene crystals, trichloride of carbon, benzoic
acid, charcoal, and an alcoholic solution of shellac (Mott Journal N-80-
07-10:279; N-80-12-13:9–15, 271–79; both Lab. [*TAEM* 37:442; 39:
678–81, 686–90; *TAED* N117:140; N168:4–7, 12–16]). Detailed orders for the construction of these experimental lamps are in Cat. 1301,
Batchelor (*TAEM* 91:294–99; *TAED* MBN007:1–6); additional notes
and instructions related to these orders are in Box 13, EP&RI. Laboratory notes, principally measurements of resistance, regarding these and
other experimental lamps from this period are in N-80-12-24.1:1–39,
Lab. (*TAEM* 40:3–22; *TAED* N186:1–20). The purpose of other experiments made about this time is not apparent, but they may have been intended to prevent "carrying." Lawson's notes (see above) also indicate
that several carbons were coated with shellac and tried in lamps. Mott
reported on 4 January that Lawson "boiled some carbon loops in Aqueous solution of Platinum chloride, believing that when heated in vacuum
on pumps the chlorine will be driven off, leaving the platinum in the
pores of the carbon in a finely divided state— But what will the platinum
do when the loop is heated to high incandescence? trial will show."
These filaments were found to have a much lower resistance than the
regular carbonized fibers (Mott Journal N-80-07-10:275, 280, 282, Lab.
[*TAEM* 37:440, 442–43; *TAED* N117:138, 140–41]).

–2045–

*Charles Clarke to
Antoine Breguet*[1]

[Menlo Park,] Jan. 17th [1881]

My dear Sir,

I have received your favor of Dec. 10th and a letter dated
Dec. 30 is now at hand.[2] I have been, and am still, very busy;
and, having had but few hours out of the day for sleep for some
time, have not replied to the first letter, which is deserving of
an answer complete and of some length.

I ask your indulgence a little longer and will briefly answer
now the second. Mr. Edison wishes me to state that he intends
to make quite a complete exhibit of his inventions previous to
the electric light, and that he will do what lies within his power
respecting the exhibition of the latter. If he can exhibit the locomotive and electric railway system he will do so but cannot

promise definitely, neither as to the exhibition of the new one hundred horse-power dynamo electric machine.

If the electric locomotive and large dynamo can be exhibited what facilities and room would their be in which to work them to the best advantage? What opportunity could be afforded for illumination by the dynamo and what outside space for a railway for the former?

Mr. Edison hopes to do much, but the demand upon his time and energies is already excessive. I remain, Yours sincerely,

Charles L. Clarke.

ALS (letterpress copy), NjWOE, Lbk. 6:809 (*TAEM* 80:468; *TAED* LB006809). Circled "C" written at top of second and third pages.

1. Antoine Breguet was director of installation services for the Exposition Internationale d'Électricité in Paris. Breguet came from a distinguished Parisian family of instrument makers and at this time was partner with his father in the manufacture of electrical apparatus and instruments. Hilborne Roosevelt had commissioned their firm to construct a telephone system in 1876. *DSB*, s.v. "Breguet, Louis François Clément."

2. Neither of Breguet's letters has been found. In late December George Gouraud apprised Edison that a French official had called upon him to "request that all your electrical inventions should be represented, and I was begged to charge myself with this undertaking. Great stress is laid upon this matter. England will be there in great force. It is obviously desirable that this request be met in the most thorough manner." After giving this official an evasive reply Gouraud wrote Edison that

> this little bye-play of mine is all very well, and we may as well make ourselves "a little difficult" so as not to show too much eagerness, but it must be as obvious to your good self as it is to me that as this will be the first exhibition, exclusively electrical that has ever been held,—and one that will attain a world wide prominence and having regard to the prominent position you hold in the profession, and the extensive value of your discoveries and inventions,—that not to be there in force would render the exhibition incomplete and most prejudicial inferences might thereby be drawn. Now I propose that we take time well by the fore-lock, and do this matter up most thoroughly. I therefore beg you, immediately upon the receipt of this to make it the especial business of somebody, to get up in the best form, highly finished models of all or certainly all your principal electrical inventions with their latest development. [Gouraud to TAE, 21 Dec. 1880, DF (*TAEM* 54:359; *TAED* D8028A)]

From George Gouraud

London Jany 18 1881[a]

My dear Edison

Your trust interest in Edison Telephone Company London.

I am anxious that you should not infer that I have been over-reaching as regards any terms I have demanded for settlement and division of shares of the United Telephone Company, in a word you must understand that no definite proposal has been made since the amalgamation with the Bell Company. Prior thereto you will remember Bouverie proposed that we should take £10,000 in full settlement. This I rejected as did you also in reply to his letter.[1] No other proposition has ever been made, and acting under the advice of Renshaw I have made no counter proposition though I did intimate to Bouverie at the time of his £10 000 proposal, that [-][b] if £20 000 [--][b] in shrs were[c] proposed it[d] might be accepted. Now the market value of the United Company's shares places an entirely different aspect on the whole question, and makes a settlement necessary upon a very different basis to what would have[d] been reasonable at the time of amalgamation. Of course I am as anxious as you or any body else to have this matter settled but only of course on some equitable and reasonable basis. This I take to be your view, also Johnsons but these are a sharp lot of fellows and we have to mind our P's and Q's in all we do with them. They will do us out of this thing if they can, and they have always meant to do so[d] since they took the twist on me at the time of the negociations for the amalgamation. I have just written Johnson not to write a word on this subject[e] to anyone on this side of the Atlantic except Renshaw and myself, and I would beg you to observe the same precautions as[f] it is highly important that we should not in any ill-advised letter give the enemy any advantage over us. You may depend upon it that as soon as they offer such terms as those we generally agreed to when I was in New York, I shall take them up like a shot.[2]

Anxiously awaiting news concerning the Light and trusting the New Year finds you all in good health at Menlo, I remain Yours sincerely

G E Gouraud

LS, NjWOE, DF (*TAEM* 59:915; *TAED* D8149C). Letterhead of George Gouraud; written by Samuel Insull. [a]"London" and "18" preprinted. [b]Canceled. [c]"in shrs were" interlined above. [d]Interlined above. [e]"on this subject" interlined above. [f]Obscured overwritten text.

1. The basis of Edison's reply is Doc. 1954.

2. About a week later Alfred Renshaw sent Edison his analysis of the London telephone company's proposed settlement, which he antici-

pated would yield about £3,500 for Edison after deducting advance royalties and other expenses. He concurred with Gouraud's opinion of this offer and stated he had proposed that "the Shares of the United Company should be divided pound for pound amongst the Shareholders of the London Company and that what is left, after that division should be divided equally between the Company's interest and your interest." Renshaw also reported that the United shareholders had approved a resolution calling for the liquidation of the Edison company but, because this could not be done without Edison's consent, they had obtained a court order to this effect. He expected this would be overturned on appeal but thought that "In any case the order for liquidation will have the effect of bringing matters to a point" and compel the company either to reach a settlement or have one imposed by the court. He predicted that in the end "something over £20,000 is likely to be secured for your interest." By early March a tentative settlement had been reached under which shareholders were to receive their original investment plus 5% interest in United Co. stock at £7 per share, with the remaining shares to be divided equally with Edison except for 1,250 shares which he was to give up to repay half his advance royalty. Renshaw to TAE, 26 Jan. and 1 Mar. 1881; Edward Bouverie's letter to stockholders of the Edison Telephone Co. of London of 5 Mar. 1881, addressed to TAE; all DF (*TAEM* 59:918, 923–24; *TAED* D8149D, D8149G, D8149H).

–2047–

To Owen Gill[1]

Menlo Park N.J. Jany 29 1881

Dear Sir:

Yours of the 27th was duly received.[2] You perhaps know that all my efforts have been and all my appliances are devised especially for the[a] general distribution of electricity throughout a city to be sold by meter and not for the lighting up of a single building hence I am at the present moment at a slight disadvantage when asked to light up a single building   I could very much easier light up a square mile with 1500 to 2000 houses than I could a single building although that may seem a paradox to you.

We are getting our offices in New York and I expect very soon to accommodate my system for isolated lighting. Before giving an estimate for lighting the Penitentiary I would like replies to the following questions

(1)   Is the Penitentiary within the limits of gas distribution system?

(2)   Is it heated by steam?

(3)   If so how many boilers have they?

(4)   What is the average boiler pressure?

(5)   Have they one or more steam engines?

(6)   What is the size of the cylinders?

(7)   What are the engines used for?

(8)   What is the extreme distance from the boiler room to the last light?

(9)   What was their gas bill in December?

(10)   Between what hours is 75% of gas used?

(11)   If lighted by gasoline how much gasoline was used in December?

The cost of lighting by Electricity will depend upon conditions   The cost of coal at Baltimore, the number of hours of burning and cost of labor. It will certainly be cheaper than gas, or gas made from gasoline. If you give me these statistics I can probably give you an estimate of the cost of plant and running expenses within 10% and perhaps 6%[3]   Very truly

T. A. Edison —G[riffin]—

L (letterpress copy), NjWOE, Lbk. 6:874 (*TAEM* 80:493; *TAED* LBoo6874). Written by Stockton Griffin; circled "C" written at top of page. [a]Interlined above.

1. Owen Gill was a partner in Gillett Martin & Co., a Baltimore tea wholesaler (Woods 1881, 328). Edison had sent him an electric lamp, apparently in response to an inquiry which has not been found. In reply to Gill's acknowledgment of this, Edison asked on 18 January if it would "be an easy matter to obtain municipal permission to lay wires in the Street." Gill then requested a number of lamps to exhibit on battery current for a gathering of newspaper editors from Ohio. He reported on 23 January that this event was "successful beyond my expectations, and gained you many adherents in Ohio." He added that he thought Edison would already have "received a visit from a committee of our common council who want to put your light in some of the public buildings, and from conversations which I have had with some of the officers of our city government I think you will receive encouragement, instead of obstructions, in the laying of wires." Edison answered on 26 January that he was "very much pleased with your success in working the lamps. We shall soon have a number of small dynamos and then I shall be able to supply you with one." TAE to Gill, 18 and 26 Jan. 1881, Lbk. 6:815, 868 (*TAEM* 80:470, 488; *TAED* LBoo6815A, LBoo6858); Gill to TAE, 20 and 23 Jan. 1881, DF (*TAEM* 57:579, 582; *TAED* D8120P, D8120Q1).

2. Gill wrote that the wardens of the state penitentiary in Baltimore "have been down to see your light, are very anxious to introduce it, which their offices permit them to do without any red tape." Gill was invited to attend the group's annual dinner in a few days "and they have asked me to get from you some data that I can present to them." He asked Edison for "any facts that you may think proper, and of interest to them, that I can introduce, and it would be desirable to have a few lines from you of encouragement &c." Gill to TAE, 27 Jan. 1881, DF (*TAEM* 57:593; *TAED* D8120X).

3. There is no further extant correspondence on this subject.

New York, Febr'y 5th, 1881.[a]

Dear Sir:—

Referring to our recent conversation with you, we would now say that we think it important to make an exhibition of the electric light at the following places:—

Caraccas, Venezuela, to the extent of not less than 100 lamps
Rio＝de＝Janeiro, Brazil,   " " " " " 150 "
Buenos＝Ayres, Arg. Repub. " " " " " 100 "
Valparaiso, Chile.   " " " " " 200 "
Lima, Peru.   " " " " " 150 "
Bogata, U. S. Columbia " " " " " 100 "

We deem it very desirable that the exhibitions should be made with the least possible delay; and that everything necessary—including motive power—should be sent from this side. Will you kindly let us know whether you can furnish us with the necessary apparatus, and, if so, by what time? We desire this information so that we may make all necessary arrangements to put the light in operation at the points named, as soon as the machinery arrives.—[1] We are, Dear Sir, Yours very truly,

Fabbri & Chauncey

L, NjWOE, DF (*TAEM* 58:380; *TAED* D8131A). Letterhead of Fabbri & Chauncey. [a]"New York," and "18" preprinted.

1. Edison noted on this letter that he answered it but his reply has not been found. On 11 February Fabbri & Chauncey acknowledged a letter from Edison and asked him to "order the two outfits (75 lights each), as you propose; and engage the engineers when the outfits are ready." The firm desired "to put them in operation on the East and West coast of South America just as soon as we can" and asked to have "full details as to cost of large plant &c &c, as we think our correspondents will be able to commence negotiations for permanent illumination, as soon as the exhibitions are made." A few days later Fabbri & Chauncey reported that their associates in Valparaiso had obtained a government monopoly for eight years beginning in December 1881, with the intervening year intended for assembling the machinery and making demonstrations, and that they hoped Edison would do his "utmost to hurry up the outfits already ordered." Fabbri & Chauncey to TAE, 11 and 16 Feb. 1881, both DF (*TAEM* 58:383, 385; *TAED* D8131C, D8131D).

New York, Feby 7 1881[a]

Mr Edison will please proceed at once to have constructed a steam Dynamo capable of working about 750 to 850 lights with sufficient lamps and other appliances to give an exhibition in London. Cost of such plant not to exceed $5000, he is also to engage an engineer to make the installation and work the same   salary not to exceed \$250. per month[1]

Drexel Morgan & Co

L, NjWOE, DF (*TAEM* 58:596; *TAED* D8133B). Memorandum form of Drexel, Morgan & Co. [a]"New York," and "18" preprinted.

1. This letter was apparently composed while Edison was at the Drexel, Morgan offices in New York. It is in his hand, though the signature is by someone else. Edison subsequently directed Otto Moses to "File this in = 'English Electric Light Edison & Drexel Morgan & Co.'" This marks the beginning of preparations for a demonstration central station in London, which was headed by Edward Johnson and began operations on the Holborn Viaduct on 12 January 1882. Friedel and Israel 1986, 215–17.

–2050–

From Francis Upton

Menlo Park, N.J., Feb 7 1881[a]

Dear Sir:

I spoke with Lawson yesterday regarding the experiments. He said that he has been trying the method of plating on the whole carbon and then plating the copper on the carbon off. He has had poor success so far. He has also been trying solvents for sealing wax. The larger portion of his time has been used up in making supports and in getting ready to plate 200 a day to take the place of the old clamps and keep us running. We have stopped the making of the old style of clamps, and so require him first of all to furnish carbons plated on by any method he may see fit.[1] He will try all the experiments you mentioned with extreme care so as to be sure of the results.

Could not the word "Electric" be dropped from the title of this company? I find that it is hard for people not to follow it with "Light" even when meaning "Lamp."[2]

~~It~~ The title[b] is distinctive enough without for you have never made other lamps. Yours Truly

Francis R. Upton.

ALS, NjWOE, DF (*TAEM* 57:772; *TAED* D8123I). Letterhead of Edison Electric Lamp Co. [a]"Menlo Park, N.J.," and "188" preprinted. [b]"The title" interlined above.

1. At Edison's direction, John Lawson began to experiment with electroplating metals onto carbon filaments about 13 December. The object, as Francis Jehl subsequently recalled, was to prevent the ends of the carbon filaments from breaking at the point where the ends were inserted into the platinum clamps. This was rapidly accomplished: Lawson reported on 13 December that the "Experiment of plating the clamping points of carbon loops with copper— Gave satisfactory results— Or-

ders to plate all the loops in the same manner." There are two extant drawings related to a patent application for the plated ends that Edison filed on 11 January (Case 278). The application was rejected and eventually abandoned; only the drawings and claims are extant. Jehl 1937–41, 616–17; N-80-12-13:3, Cat. 1147, both Lab. (*TAEM* 39:675, 44:226–27; *TAED* N168:1, NM016:2–3); Patent Application Casebook E-2536:246, PS (*TAEM* 45:724; *TAED* PT020246).

In related research begun on 17 December, Albert Herrick conducted experiments "for uniting the wires and carbons direct by plating them together without manufactured clamps" (Mott Journal N-80-07-10:253, Lab. [*TAEM* 37:429; *TAED* N117:127]). On 30 December Charles Mott noted in his journal that Edison had sketched

> apparatus for plating carbons to the wires without use of clamps. being a vessel containing the plating solution in which one electrode is placed, the inner part containing the wires to which the carbon is temporarily attached is passed through a rubber cork which forms the bottom of the vessel or reservoir, the other or opposite electrode is connected to the wires extending out of the inner tube which extend together with the tube through the cork sufficiently low to immerse the ends of the carbons say ⅛ of an inch. As the plating progresses the wires and carbon are united and held by the deposit. Sketches dated by him Decr 24. and taken to Motts table for Patent Office drawings. [Mott Journal N-80-07-10:266–67, Lab. (*TAEM* 37:335–36; *TAED* N117:133–34)]

Edison's drawing is in Cat. 1146, Lab. (*TAEM* 6:732; *TAED* NM014ZAU). Edison did not file this application until May. It covered only the process of electroplating the connections and did not include the rationale for doing so; it issued in October 1881 as U.S. Patent 248,436.

A number of different ways of maintaining the connection during the plating process were tried by Herrick, Lawson, and others in December and January. On 17 February 1881 Mott recorded that the clamp department was "abolished at the Factory. All carbons to be plated on." The method ultimately adopted was what Jehl (1937–41, 617) later described as "a crude sort of copper clamp that was just sufficient to hold the shank of the filament. The stems with the copper clamps and filaments were then placed in a sulphate of copper bath in which the liquid only reached to the copper clamp. The filament and clamp were then plated one on to the other" (Mott Journal N-80-07-10:259; N-80-11-16:143–49, 163; N-80-12-13:5–7; Mott Journal PN-81-01-19; all Lab. [*TAEM* 37:432, 949–52, 958; 39:676–77; 43:1127; *TAED* N117:130; N125:67–70, 76; N168:2–3; NP014:13]). The copper was then fused to a short platinum wire sealed into the glass; another copper wire leading to the base was similarly attached to the other end of the platinum outside the globe. The enlarged filament ends were retained because they dissipated heat well, preventing the copper from melting. Edison applied for a patent covering the clamps for making such connections in April and in June he filed another application for the electroplated union itself (U.S. Pats. 266,447 and 251,544). The electroplated connections represented a significant saving in the materials and labor of commercial

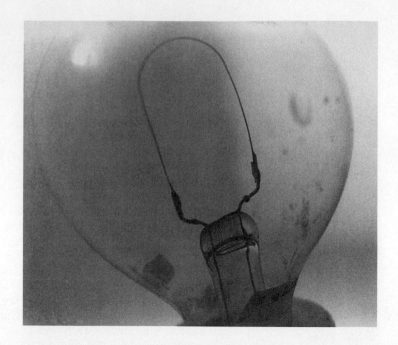

lamp manufacture and were used in Edison's lamps until 1886 (Howell and Schroeder 1927, 77–78). The method described by Upton as "plating on the whole carbon and then plating the copper on the carbon off" probably refers to plating the entire length of a filament, then removing metal from all but the shanks by a reverse electrolytic process. Intermittent research continued for several months to determine the best form of clamps and fixtures, and also to try gold and silver. Silver proved prone to electrical carrying and copper remained the preferred metal (N-80-09-11:99, 107–11; Mott Journal N-80-07-10:276; both Lab. [*TAEM* 39: 293, 297–99; 37:440; *TAED* N153:49, 53–55; N117:138]; TAE to F. C. Van Dyck, 14 Mar. 1881, Lbk. 8:70 [*TAEM* 80:867; *TAED* LB008070A]; a few related experimental notes from mid-January are in Box 13, EP&RI).

2. This change was formally adopted by late May, when Upton wrote on letterhead of "The Edison Lamp Co." Upton to TAE, 28 May 1881, DF (*TAEM* 57:895; *TAED* D8123ZCC).

–2051–

R. G. Dun & Co.
Credit Report

[Newark?][1] Feb 7/81

3700[2]   They write us under date of 5th inst that this is not an incorporation but a firm composed of Thos A Edison, Chs. Batchelor, Francis R Upton & Edward H Johnson,[3] that they have 35mf[4] invested so far & think they do not owe over 1mf & hope nobody will ever give them credit. So far as we can discover, it is an enterprise limited nearly to the men and means in the works of Mr Edison at Menlo Park and the money mainly supplied by Mr Edison. According to newspaper reports Mr Edisons patents upon his lamps have received adverse decisions in the courts such as is likely to seriously affect the value of his reputed invention.[5] 5610. 4403. 5885. 8486. 5652. 5345. 3631. 2092.

D (abstract), New Jersey, Vol. 52, p. 399, R.G. Dun & Co. Collection, Baker Library, Harvard Business School.

1. See Doc 2001 n. 2.

2. See Doc. 2001 n. 3.

3. R. G. Dun wrote to Edison on 4 February that the agency had received "one or two inquiries as to the 'Edison Electric Lamp Company' Menlo Park" but could find no incorporation papers on file. Edison drafted a reply on the letter which is the basis for the formal answer prepared by Stockton Griffin, essentially summarized below. R. G. Dun to TAE, 4 Feb. 1881, DF (*TAEM* 57:768; *TAED* D8123E); TAE to R. G. Dun, 5 Feb. 1881, Lbk. 6:902 (*TAEM* 80:508; *TAED* LB006902).

4. This notation is unclear but Edison indicated in his 5 February letter (see note 3) that about $35,000 had been invested in the lamp company. He had charged about $26,600 to lamp factory accounts by the end of 1880. According to Samuel Insull's recollection years later, "Edison's scheme was to form a company of one hundred shares, each share being $2500., and as I recall it, no stock being transferable except on the personal permission of Thomas A. Edison. Edison had a controlling interest, Batchelor had a ten per cent interest; I think Upton had ten per cent and I think Johnson had five per cent." Ledger #4:332, 361, Accts. (*TAEM* 87:350, 358; *TAED* AB002:141, 149); Insull Notes, pp. 21, Meadowcroft (*TAEM* 227:159; *TAED* MM010DAH); see also Doc. 2018.

5. At this time no court cases related to Edison's electric light patents had been filed.

–2052–

Edward Johnson to Uriah Painter

NYork Feby 9/81

My Dr U.H.

This is my third attempt at replying to yours in re. to Bergmanns sale of the Relay— We are transferring Everything from Menlo to NYork— Have Leased this 4 story & Basement Double Brown Stone Mansion for our technical offices—[1] Major Eaton, Edison, myself Clark Wilbur & others have offices here—the NYork Illuminating Co will also have their offices in the Building— We are fitting it up in Elegant style— & will illuminate it with some 200 Lights run by a Double Gas Engine— You must come in & see us whenever in NYork— Real business now begins— Menlo Park will be deserted Entirely just as soon as the Steam Dynamo is finished & has been tried— At least we will only go there for the purpose of making Tests with the Plant for Engineers— Edison is here from 10 AM to 5 PM. & will move his family in in a few days then will be here "at all hours"[a]

Now in re. to your proposed action in Bergmann matter— Do you not think you carry all these things with a trifle too much assumption of power— You know that Bergmann was only induced to place the thing in your hands at all by my persuasion & then only for a specific purpose—that purpose having failed of its friction[2]   How do you hold the thing? & In case of any trouble with B. what have you to show that he Ever gave you any authority to act— As for me since I have had none from him I could convey none— The trouble with you, if you will pardon my saying so—is this= You endeavor to "run" Everybody and their machines without in any way committing yourself to any degree of responsibility—by mere force of superior assertion= Your motives are good—but your methods are bad—they are the methods you have found successful in Politics no doubt—but they will not work in business—or at least in this class of business— My warrantee for saying so is my knowledge of their Effect upon Edison Reiff Roosevelt Hubbard Bergmann & others— There is not one of these men but whom you could coerce into your train if you but pursued different tactics— As for Edison you simply humiliate him Everytime you speak to him in re. to business   You must therefore not wonder if he forgets, in his resentment— the many good turns you have done him— You no doubt think me presumptious in the Extreme in thus pointing out your— in my opinion—bad tactics—but if there is any one man who I should like to see on the right footing with Edison it is you— & I have proven it over & again by calling your attention to the galling Effect, of your methods, on[b] him= It is all very well to be able to see a mans failings—to have propheciec disaster & to have your prophecy unheeded, with the result of a confirmation of it—but it is <u>not</u> policy to keep a man in constant mind of it— Edison lives for other things than money & I believe that rather than to do what you ask him viz confess his lack of judgment—he would walk knowingly into any financial quacks lair— What therefore is the use of a continuation of such a policy— You may say that Edison is not under consideration & that my remarks are therefore not pertinent   I answer that Bergmann is—by reason of my Efforts to get him inside—at this moment on the very best of terms with Edison & discusses with him all his affairs.— Every act of yours in re to this Relay business might just as well apply to Edison him-

self—for as I have before remarked—Edison is simply ~~justif~~
finding in Bergmanns Experience with you in this matter a justification for declining your assistance in Negotiations—

You are a man of large experience I am not—but I am in the
Eye of the men you would handle—I can feel them wince—
let me therefore assume to advise you without incurring your
ill will   I want you & Edison to be on better terms— You have
many points in common & if you could but understand Each
other better there would be a Bonanza in the "Consolidation"[3]
of my two best friends for me as well as for them—

Let the Relay matter drop— If B. has thrown it away—
whose property is it?— Yours Truly and forever

E H Johnson

ALS, PHi, UHP. [a]Followed by flourish between paragraphs. [b]Obscured
overwritten text.

1. Johnson indicated at the top of the letter that he was writing from
the "Ofs of The Edison Electric Light Company   65 5th Ave," which
was located between 13th and 14th Sts.

2. Sigmund Bergmann had patented in November a form of adjustable electromagnetic relay, which Johnson had been helping to refine
and in which he also had some financial interest. Bergmann and Johnson
gave Painter control of the patent for the purpose of selling it, which
Painter evidently proposed to do to Jay Gould's American Union Telegraph Co. In December, Johnson reproached Painter for his dilatory
dealings with Gould and American Union vice-president David Bates:

> Now what are you going to do? You gave us to understand you could
> sell to Gould—if it was endorsed by the management— It has been
> so endorsed to an extent unusual in a pending negotiation— You
> appear to be resting on your oars— Gould has shown no disposition
> to deal outside of [Thomas] Eckert & Bates— They Evince perfect
> consciousness of being the sole parties with whom a decision rests.
> Bergmann is naturally restive under this Condition of things—
> says by giving it to you we have simply tied our own hands—
> Thinks that if I had it now I could make more out of it for myself
> & him than you will be able to get Even if you sell it at all which he
> doubts. = I write thus fully & freely—because I am on the ground
> & see the situation—while you are not— What are you going to do?
> [Johnson to Painter, 16 Dec. 1880, Unbound Documents (1879–
> 1882), UHP]

The next day Johnson informed Painter that Bates "told me today he
wanted this Relay matter settled—said he wants to order 100 of them if
they buy— Otherwise will give order for Old Style— In my opinion
they are more in the mood to buy now than they will be a few weeks
hence" (Johnson to Painter, 17 Dec. 1880, Unbound Documents (1879–
1882), UHP). The outcome of this matter is not known.

3. Johnson was apparently playing on the name of the Consolidated
Virginia Co., controlled by a partnership known as the "Bonanza firm,"

which developed the so-called "Big Bonanza" in the Comstock Lode. *TAEB* 4:416 n. 1; Elliott 1973, 132–33.

[Menlo Park,] Feby 11th [188]1

Gentlemen:

Your favor of the 8th is at hand.[2] The lamps I use are from 8 to 16 candlepower each which may be distributed the same as gas. Amount of space required not more than 10 × 10.

Last week we lighted up the Lithograph Establishment of Messrs Hinds & Ketcham in N. Y.[3] They were unable to work at night until they put in the E. L. now I learn they have no difficulty distinguishing colors. Very truly

T. A. Edison —G[riffin]—

L (letterpress copy), NjWOE, Lbk. 6:919 (*TAEM* 80:510; *TAED* LB006919). Written by Stockton Griffin.

1. According to its letterhead L. Prang & Co. was a Boston firm specializing in art and educational publishing. L. Prang & Co. to TAE, 8 Feb. 1881, DF (*TAEM* 57:605; *TAED* D8120ZAE).

2. L. Prang & Co. indicated in this letter that they were expanding their works and wanted to build into their new engine and boiler house sufficient space to add an electric lighting plant capable of supplying "10–12 lights, each one powerful enough to light up a steam lithographic press." Edison's 11 February draft reply on the letter is the basis for this document. L. Prang & Co. to TAE, DF (*TAEM* 57:605; *TAED* D8120ZAE1).

3. Hinds, Ketcham & Co. were lithographers and printers of colored labels and show cards. Their incandescent lighting plant was installed on 27 January 1881. Edison had his staff wind the dynamo for this plant "using much finer wire . . . the machine being intended to make its own field, hence the higher resistance of the coils around magnets." This installation became a showcase for the Edison light, with Hinds, Ketcham & Co. boasting in its own advertising circulars that "We are the first manufacturers in the world to put into practical operation this great triumph of American genius" and encouraging "our friends . . . to call and see the operation of Prof. Edison's wonderful subdivision of the electric current and the vacuum lamp." They found it possible to match colors using the incandescent lamp, which they found "to be entirely free from all the faults and objectionable features of other artificial lights, and is the best substitute for daylight we have ever known and almost as cheap." Mott Journals N-80-07-10:278, PN-80-09-23, and PN-81-01-19; N-80-10-23:44; all Lab. (*TAEM* 37:441; 43:1107, 1109, 1116–18, 1120; 35:1036; *TAED* N117:139; NP013:44, 46; NP014:2–4, 6; N087:19); Hinds, Ketcham & Co., advertising brochure, box 25, WJH; Edison Electric Light Co., Bulletin 4:2–3, CR (*TAEM* 96:677; *TAED* CB004:2).

[Menlo Park,] Feb 12 [188]1

Dear Sir

Yours of the tenth[1] received by Mr Edison who has requested me to say that if you can get any party to take hold and put money in science you are at perfect liberty to do so.[2] he is to busy to give it any attention   Very truly

Wm Carman

ALS (letterpress copy), NjWOE, Lbk. 7:28 (*TAEM* 80:547; *TAED* LB007028).

1. On 10 February Michels sent Carman a statement of accounts for the two previous weeks and asked him to "send me the balances Saturday," amounting to $191.47, and to give the statement to Edison "before he leaves in the morning, and I will call on him later at the N.Y. office." Michels to Carman, 10 Feb. 1881, DF (*TAEM* 59:465; *TAED* D8144G).

2. Michels replied on 14 February, "I will endeavor to carry out his wishes as soon as possible   it will of course take a little time but I will be as prompt as possible." On 18 March he sent Edison a "prospectus to form a Company to carry it on. I would suggest your asking a few of your friends to co-operate, as half the capital I mention would be sufficient to subscribe at once, which should make it a small matter to arrange." This prospectus has not been found. Edison continued to fund the journal for several months but on 25 October 1881 wrote that he would stop doing so in sixty days, and on 23 February 1882 he gave Michels a final payment of $254.53 (Michels to Carman, 14 Feb. 1881; Michels to TAE, 18 Mar. 1881 and 23 Feb. 1882; all DF [*TAEM* 59:468, 481; 63:664; *TAED* D8144H, D8144N, D8251B1]; TAE to Michels, 25 Oct. 1881, Lbk. 9:219 [*TAEM* 81:80; *TAED* LB009219]). Michels found temporary support from Frederic Shonnard and some of his associates but in April 1882 he again asked Edison to take stock in a company to run the journal and also to provide some working capital. Edison, however, refused to "have anything further to do with the publication of 'Science.'" Among those Michels attempted to interest in the journal was Alexander Graham Bell, who finally consented. After suspending publication in mid-1882, *Science* resumed on 9 February 1883 with Samuel Scudder replacing Michels as editor (Michels to Insull, 16 Dec. 1881; Michels to TAE, 28 Apr. 1882; both DF [*TAEM* 59:595, 63:665; *TAED* D8144ZBD, D8251E]; Insull to Michels, 9 May 1882, Lbk. 12:255 [*TAEM* 81:639; *TAED* LB012255]). Other correspondence and accounts regarding Edison's involvement are in Science (D-81-44 and D-82-51), DF (*TAEM* 59:451, 63:661; *TAED* D8144, D8251; for Bell's involvement see Bruce 1973, 376–77).

Memorandum:
Proposal for Electrical
Machinery Company

New York, Feby 25 1881[a]

Results of my efforts to raise money for an assembling shop=[1b]

Empire Machine Co—

Capital 100,000. 1000 shares sold at par for cash—

Organized under[c] limited Liability act.

Object, contracting, assembling & testing machinery used by the Edison Electric Light Co—

Following persons will subscribe the sum named when organization perfected, scheme satisfactory[d]

Edison	$ 5000
Goddard	$ 5000
Banker	$ 5000
Cutting jr	10 000
Fabbri can put down for Rangor[2]	
Bedlow[3] look to Cutting	2000
Rufus Hatch[4]	1000.
G P Lowrey	5000
Batchelor	2000
Bergmann	10,000
offd Fabbri & Wright also	
Balzer—[5]	
Lowrey said not to see Navarro	

AD, NjWOE, Miller (*TAEM* 86:365; *TAED* HM810140A). Letterhead of Edison Electric Light Co. [a]"New York," preprinted on letterhead; date written on envelope. [b]Sentence from envelope. [c]Obscured overwritten text. [d]Followed by dividing mark; list below written in two columns separated by vertical line.

1. In fact, Edison did not raise the money for this shop from the names found on this list; instead, he wrote a note on the outside of the envelope directing Otto Moses to "file this away unopened." According to Samuel Insull, "The capital for the Edison Machine Works was provided ninety per cent by Edison and 10 per cent by Batchelor." Edison organized the Machine Works to construct dynamos and other equipment for his electric light and power system. Insull Notes, p. 22, Meadowcroft (*TAEM* 227:160; *TAED* MM010DAH).

2. Unidentified.

3. Possibly Henry Bedlow, a member of a prominent New York family and a former member of the diplomatic corps who had also served three terms as mayor of Newport, R.I. Obituary, *New York Times*, 31 May 1914, 5.

4. Rufus Hatch (1832–1893) was a New York banker and stockbroker. *DAB*, s.v. "Hatch, Rufus."

5. Unidentified.

From George Gouraud

My dear Edison.

I beg to confirm the following Cables between us.

February 5th to you:[1]

"Amalgamating all Continental Telephone interests   Send us today certified Copies Russian Hungarian Belgian and every other contracts if any made by Edison of Europe[2]   Answer. Gouraud Bailey Puskas.

February 7th to you:[3]

"Continental fusion Company can be formed   Capital not less than six hundred thousand pounds. Edison interest equal to all the others. Moment very favorable owing to condition English Telephone matters. Necessary give Gouraud and friends 280 shares now in Treasury. Do you confirm giving these shares conditioned on success as above. Cable reply."

In reply from you Feb 8[4]

"Company must allow terms of proposed amalgamation before agreeing to anything   report at once   powers will be sent if terms adopted by Company."

To you February 10th[5]

"You receive cash twenty thousand pounds fully paid shares forty thousand pounds   Cable officially as president approving this also authorising us sign necessary contracts securing Gouraud and friends 280 shares Conditional above. Gouraud Bailey Puskas."

From you February 11.[6]

Continental fusion approved   you are authorised sign necessary papers you to have 280 shares if conditions as stated fulfilled. Edison president European Telephone Company."

To you February 7th[7]

"Oriental Telephone Company complete and registered capital three hundred thousand pounds   exceptionally strong Board including Pender as chairman also late Under Secretary of State for Foreign Affairs. Vendors received one hundred thousand pounds shares fully paid and fifty thousand pounds cash two fifths to Edison interests two-fifths to Bell interests ⅕ to Anglo Indian Telephone Company whom we found Competing with us in India. All documents signed. Cable your approval this basis   Glasgow also settled."[8]

From you February 11.[9]

"Oriental Telephone Company approved."

To you February 23.[10]

Send immediate certified copies all assignments patents European Telephone Company."

From you Feb 24.[11]

"Everything assigned to Company here better convey rights through contract."

Yours Truly

G E Gouraud

Since writing the above we have sent you the following[12]

"Send copies assignments"

and received Feby 25th

"Johnson sails tuesday great sacrifice his interests Has just initiated organization department Electric Light New York Edison"[b]

LS, NjWOE, DF (*TAEM* 59:790; *TAED* D8148ZAM). Written by Samuel Insull; letterhead of George Gouraud. [a]"London" and "18" preprinted. [b]"and received . . . Edison" written in an unknown hand.

1. Gouraud, Joshua Bailey, and Theodore Puskas to TAE, 5 Feb. 1881, DF (*TAEM* 59:770; *TAED* D8148X).

2. James Banker owned the Russian telephone patents of Edison, Joshua Bailey, and Elisha Gray. By two separate contracts with the Edison Telephone Co. of Europe he agreed to hold these patents in trust for the company. Banker agreements with Edison Telephone Co. of Europe, 31 Jan. 1881 (copy) and 1 Feb. 1881, DF (*TAEM* 59:763, 767; *TAED* D8148S1, D8148V).

3. Not found.

4. In the message to be transmitted to Gouraud, Stockton Griffin wrote "know" instead of "allow." TAE to Gouraud, 8 Feb. 1881, DF (*TAEM* 59:780; *TAED* D8148ZAC).

5. Not found.

6. TAE to Gouraud, 11 Feb. 1881, DF (*TAEM* 59:783; *TAED* D8148ZAF).

7. This cable has not been found but on 4 February Gouraud wrote Edison that before the letter reached Menlo Park "we shall have brought out the Oriental Telephone Company (Bell and Edison,) for which a strong board has been secured. The last great achievement having been the acquisition of Sir William Thompson as consulting Electrician I shall also secure him for the Electric Light" (Gouraud to TAE, 4 Feb. 1881, DF [*TAEM* 59:1002; *TAED* D8150E]). An agreement between Edison, Alexander Graham Bell, the Oriental Bell Telephone Co., and the Anglo-Indian Telephone Co. to form the company on the terms outlined below was executed on 25 January 1881; Samuel Insull also signed as trustee for the prospective firm. A 17 February agreement between these same parties (except Insull) ratified the terms of the January contract. According to this second agreement, the Oriental Telephone Co. had been incorporated on 4 February and was to operate in India, Ceylon, Java, Japan, China, South Africa, the Australian colonies, New Zealand, Egypt, Turkey, Greece, Malta, and the Hawaiian Islands. Edison's interests in Australia and Japan were excluded (DF [*TAEM* 59:991, 995; *TAED* D8150B, D8150C]).

8. Gouraud reported to Edison a few weeks earlier the liquidation of the Edison Telephone Co. of Glasgow: "You will be gratified to learn

that I have at length brought this thing to a final settlement, and received a further payment of £5,000 subject to deductions. They have tried every conceivable way to do us out of this. I have shown myself as possessed of a large capacity for waiting, when I know what I am waiting for." Gouraud to TAE, 4 Feb. 1881, DF (*TAEM* 59:921; *TAED* D8149E).

9. Edison's cable has not been found. About a week later Gouraud sent copies of the company's prospectus and reported having received "very considerable applications amounting to more than £100,000 shewing that no doubt matters will go all right." Gouraud to TAE, 19 Feb. 1881, DF (*TAEM* 59:1004; *TAED* D8150F).

10. Stockton Griffin transcribed this message and left it at Grosvenor Lowrey's office for either Edison or Lowrey. Gouraud to TAE, with marginalia by Richard O'Brien, DF (*TAEM* 59:788; *TAED* D8148ZAK).

11. DF (*TAEM* 59:789; *TAED* D8148ZAL).

12. Neither of the following messages has been found.

–2057–

Charles Mott
Journal Entry

[Menlo Park,] Monday Feby 28. [1881]

Porter Allen, run last night[a] 600 lamps at 600 Revo at about 18c[andlepower] proving that she will do all that has been expected of her on 800—[1] To day Mr. Porter has altered excentric[2] changed exhaust valves etc.[b]

Pump Motor from factory sent to shop & replaced with another machine.[c]

Tests. Francis testing lamps all day. Plumbago (4 to 5 ohms) 3.4 ohm average. Book No. 244 pg. 174 etc.[3b]

AD, NjWOE, Lab., PN-81-01-19 (*TAEM* 43:1133; *TAED* NP014:18). Written by Charles Mott. [a]"last night" interlined above. [b]Paragraph marked by "X" at left. [c]Paragraph marked by heavy line at left.

1. Edison first ran the direct-connected Porter-Allen dynamo with an electrical load on the evening of 24 February. Charles Mott indicated that it ran "all the lamps at 535 Revo"; William Hammer reported between 600 and 700 lamps in this first test. Francis Jehl wrote in his diary that afterwards Edison and his assistants "all went down to have a drink for the dynamo." When it was completed on 12 February, the armature of this machine was found to have an unacceptably high resistance of .18 ohm. Charles Dean then soldered the connections with an electric arc and it gave a resistance of .014 ohm. Jehl 1937–41, 873–78; Mott Journal PN-81-01-19, Lab. (*TAEM* 43:1127–28, 1132; *TAED* NP014: 12–13, 17); box 14, WJH; Jehl Diary, 12 and 24 Feb. 1881).

The 22 January *Scientific American* illustrated and described the completed dynamo, which was

designed to replace sixteen of the largest machines of this kind previously made. The dynamo and the driving engine are both mounted on a massive cast iron bed, 8½ by 7 feet and 2 feet deep, very heavy and strongly ribbed, the entire machine weighing 8 tons.

Scientific American *illustration of Edison's large direct-connected dynamo (incorrectly drawn with armature coils of wire instead of solid bars) and Porter-Allen engine.*

EDISON'S STEAM DYNAMO-ELECTRIC MACHINE.

Near the middle of the bed is mounted the dynamo-electric machine, which, we believe, is the largest ever constructed. Its field magnets, three in number, are 6½ feet long. The armature is 21 inches in diameter and 28 inches long, and weighs 1½ tons. The engine is 100 horse power. . . . Its stroke is 10 inches. The internal diameter of its cylinder is 9 inches. . . . The working pressure of the dynamo is 140 volts; the resistance of the armature is one two-hundredth of an ohm. ["Edison's New Dynamo-Electric Machine," *Sci. Am.* (44:47), Cat. 1241, item 1569, Batchelor (*TAEM* 94:626; *TAED* MBSB21569X)]

See Doc. 1970 for Edison's rationale for the use of multiple adjacent field magnets and, more generally, the design principles of the large direct-connected dynamo.

2. The eccentric gear was part of the steam cut-off mechanism.

3. This book contains records dated 28 February of loops numbered 3–9, which Edward Acheson made from pressed plumbago. These loops apparently operated at much lower resistance and voltage than ordinary fiber filaments. According to a 22 February laboratory summary that Charles Mott prepared for Edison, one such lamp produced 16 candle-power at 5 ohms and 21 volts, consuming an estimated 4,133 foot-pounds of power, the equivalent of eight lamps per horsepower (N-81-02-18:174–80, Lab. [*TAEM* 41:788–91; *TAED* N244:89–92]; Mott to TAE, 22 Feb. 1881, DF [*TAEM* 59:62; *TAED* D8137B]). According to Acheson's later recollection, Edison instructed him to use a hydraulic press to "make for me a small graphite loop like this (making a sketch like a horseshoe). I want the loop one inch outside diameter, the filament to be twenty-five thousandths of an inch wide and two thousandths of an inch thick. I will have steel plates made for you to press sheets between and a die made for punching out the filaments. When you make one capable of mounting in a lamp, I will give you a prize of one hundred dollars." Acheson began pressing the plumbago on 7 February. On 10 Feb-

ruary he succeeded in cutting out one loop; the next day he managed several more which were placed in lamps and he claimed the bounty. With at least one plumbago lamp burning on 16 February, Mott reported that Edison was "well pleased with prospects." Acheson entered into a piecework agreement to make 30,000 graphite loops. He recalled that they "produced a magnificent light, but they did not last long in use, disintegrating rapidly. I had made sixteen thousand of them and then went to Mr. Upton and told him that I was not happy in making an inefficient article. . . . I considered it a waste of money and would much prefer to throw up my contract," to which Edison consented in April (Acheson 1965, 19–20; Mott Journal PN-81-01-19, Lab. [*TAEM* 43: 1125–28; *TAED* NP014:10–13]; Upton to TAE, 18 Apr. 1881, DF [*TAEM* 57:853; *TAED* D8123ZAZ]). For earlier efforts to form plumbago filaments, see Doc. 1914.

–2058–

Electric Tube Co. Articles of Incorporation

New York, March 1, 1881[1a]

We,[b] Thomas A. Edison, Stockton L. Griffin, and John Kruesi,[2] citizens and residents of the State of New Jersey and Anthony J. Thomas, Charles L. Clarke and Robert K. White,[3] citizens and residents of the State of New York, being desirous of forming a company and to become a body politic and corporate under and pursuant to the provisions of the act of the Legislature of the State of New York, entitled "An Act to authorize the formation of corporations for manufacturing, mining, mechanical or chemical purposes," passed February 17 1848 and of the several acts of said Legislature supplementary thereto and amendatory thereof, have associated ourselves together and pursuant to the requirements of the said acts do make, sign and acknowledge this certificate, and do hereby certify as follows:[c]

First:[b] The corporate name of the said company is "The Electric Tube Company."[c]

Second:[b] The objects for which the said company is formed are the manufacture and sale of Electric tubes for conducting electricity.[c]

Third:[b] The amount of the capital stock of said Company is Twenty five thousand dollars divided into shares of One hundred dollars each.[4c]

Fourth:[b] The time of the existence of said company is fifty years from the first day of January 1881.[c]

Fifth:[b] The number of shares of which the stock of said company shall consist is Two hundred and fifty[c]

Sixth—[b] The number of Trustees who shall manage the concerns of the said company for the first year is five—and the names of such trustees are:—

Thomas A. Edison
Stockton L. Griffin
Charles L. Clarke
Anthony J. Thomas and
Robert K. White.[c]
Seventh:[b] The names of the Town and County in which the operations of the said Company are to be carried on are, the City of New York and the County of New York in the State of New York

Dated the 1st day of March 1881.

Thomas A Edison	Charles L. Clarke.
Anthony J Thomas	John Kruesi
Robert K White	Stockton L. Griffin

DS, NNNCC-Ar (*TAED X119TA*). Notarized by Charles Roth with seal affixed on 3 March 1881. [a]Place taken from notary seal; date taken from text, form altered. [b]Multiply underlined. [c]Followed by dividing mark.

1. This document was filed and the company incorporated on 4 March 1881.

2. John Kruesi, who managed the Electric Tube Co. and also served as its treasurer, had left Menlo Park on 14 February to set up the company's factory at 65 Washington St. in New York City. Jehl 1937–41, 848; Israel 1998; 199, 206, 223.

3. Anthony J. Thomas and Robert K. White were both members of Drexel, Morgan and Co. Thomas later became a director of the Edison Electric Light Co. and the Edison Co. for Isolated Lighting. Thomas to Insull, 10 May 1881, DF (*TAEM* 58:356; *TAED* D8130I); Edison Electric Light Co., *Annual Report*, 1885; Edison Co. for Isolated Lighting, *Annual Report*, 1884; both CR (*TAEM* 96:5, 266; *TAED* CA001A, CA002D).

4. Samuel Insull later recalled that the company's shares were "owned one fifth by Edison, one fifth by Kruesi, one fifth by Batchelor and the other two fifths were owned by E. P. Fabbri and Mr. J. Hood Wright, who were partners of Mr. J. P. Morgan in the firm of Drexel, Morgan & Company." Wright and Morgan each owned 30 of the 250 total shares of the company's stock. In order to serve as trustees of the company, Thomas and White received nominal shares from Wright's account. Insull Notes, p. 22, Meadowcroft (*TAEM* 227:160; *TAED* MM010DAH); James Hood Wright to Samuel Insull, 28 April 1881; Thomas to Insull, 10 May 1881; both DF (*TAEM* 58:353, 58:356; *TAED* D8130F, D8130I).

New York March 2, 1881[a]

From Egisto Fabbri

Friend Edison

Mr. Soren tells me that ~~the~~ some[b] some necessary steps to make that European Co stock legal and of good delivery are yet to be taken & certain formalities performed—[1] Such being the case it would not be safe to negotiate further at present with Mr. Biedermann— Besides I should not consider the proposition favorably if we have what we have every reason to believe that you have accomplished your stock is worth much more.—[2]

As regards the proposed arrangement between you & D. M & Co for the same countries the option expires to–day and you can withdraw it or extend it till matters are set straight, as you may like best—[3] Please let me hear on the subject & believe me faithfy yrs

E. P. Fabbri

⟨answered would do nothing but didnt extend[4]⟩

ALS, NjWOE, DF (*TAEM* 58:437; *TAED* D8132R). Letterhead of Drexel, Morgan & Co. [a]"New York" and "18" preprinted. [b]Interlined above.

1. The following day George Soren wrote Edison about Fabbri's interest in buying some of Edison's shares in the Edison Electric Light Co. of Europe and the need to complete assignments to the company. On 1 March Edison assigned his Italian, French, and Danish electric light patents to the company and a month later he, Theodore Puskas, Joshua Bailey, and James Banker assigned the rights embodied in Doc. 1736. In his letter Soren noted that he had told Fabbri "that the stock was good for nothing at present, for the simple reason that the Company had no capital." The company's New York tax statement of 28 April 1881 indicated that it "was not fully incorporated till December 1880. Its capital is issued for European patents which have not yet any money value whatever." Soren to TAE, 3 Mar. 1881; TAE agreements with Edison Electric Light Co. of Europe, all 1 Mar. 1881; TAE, Puskas, Bailey, and Banker agreement with Edison Electric Light Co. of Europe (copy), 1 Apr. 1881; statement of Edison Electric Light Co. of Europe, 28 Apr. 1881; all DF (*TAEM* 58:87, 72, 77, 81, 88, 113; *TAED* D8127A, D8127111, D8127222, D8127333, D8127A1, D8127H).

2. On 19 January 1881, Edison had extended for a month an option he had previously discussed with Ernest Biedermann regarding the sale to Biedermann and the New York banking firm Baltzer & Lichtenstein of 10,001 of Edison's shares in the Edison Electric Light Co. of Europe for $750,000 and one quarter of the net profits. On 20 February, Edison made notes for a reply concerning a similar offer, possibly from another party, involving 10,500 shares. It appears that on 2 March Biedermann offered $400,000 cash and "350,000 of General Societe stock," which Edison declined. At the end of March Edison wrote Baltzer & Lichtenstein that "in consequence of some arrangements which have yet to be made I shall be unable to give any reply to your offer . . . until April 15th 1881." TAE to Baltzer & Lichtenstein and Biederman, 19 Jan. 1881;

TAE memorandum, 20 Feb. 1881; TAE marginalia on Biedermann to TAE, 2 Mar. 1881; TAE to Drexel, Morgan & Co., 17 Feb. 1881; Drexel, Morgan & Co. to TAE, 19 Apr. 1881; all DF (*TAEM* 58:425, 429, 436, 109, 108; *TAED* D8132J, D8132N, D8132Q, D8127D, D8127C); TAE to Baltzer & Lichtenstein, 30 Mar. 1881, Lbk. 8:120 (*TAEM* 80:881; *TAED* LB008120); Wilson 1881, 69.

3. On 17 February Edison offered to sell to Drexel, Morgan & Co. one-quarter of his 51% interest in the European light company. The transaction was not to be completed until after the central station in New York had operated two weeks to the satisfaction of Edison, Fabbri, and James Hood Wright; in the meantime the stock shares of both parties would be held in a trust managed by the firm. Edison extended this offer to 2 March and again to 20 April. Drexel, Morgan & Co. accepted the proposition on 19 April but withdrew eight days later in deference to Edison's prior interests in the electric light business for Paris. Fabbri subsequently bought 1,690 shares from Edison with an option for 1,690 more within three months after the New York central station began to operate. TAE to Drexel, Morgan & Co., 17 Feb. 1881; Drexel, Morgan & Co. to TAE, 19 and 27 Apr. 1881; all DF (*TAEM* 58:109, 108, 111; *TAED* D8127D, D8127C, D8127F); Insull memorandum, 6 July 1881; John Tomlinson memorandum, 6 July 1881; Fabbri to TAE, 7 July 1881; TAE to Fabbri, 8 July 1881; TAE agreement with Fabbri, 8 July 1881; all Miller (*TAEM* 86:404, 406–9; *TAED* HM810149, HM810149A, HM810150, HM810151, HM810152).

4. Edison's answer has not been found.

-2060-

To George Gouraud

New York 7th March [188]1

Very Important

My Dear Gouraud,

It is important that I should have as much money as possible at my disposal within the next few months   we are proceeding vigorously with the Electric Light business and in order to get together the enormous amount of requisite machinery I have taken the Etna Iron works (Jno Roachs old place)[1] as an assembling shop where I propose putting together the various parts of the machinery which I shall have manufactured [--][a] in pieces at different works in order to secure prompt delivery. The Etna works I have taken on my own responsibility so as not [to be?][b] [-------][a] subject to the action of [board?][b] of Directors. Besides this this[c] there are many other things connected with Light in which I am obliged to use my own funds and consequently my resources are taxed to their utmost

I suppose there is no chance of the money from the Oriental Coy (on the successful floating of which I congratulate you) being available for a month but cannot the Glasgow Royalty a/c be settled & my Glasgow shares [be?][b] also. And then as to the London Reversion I want you to sell out my proposition of the United shares as soon as possible compatable with keeping up the market. I have sent Renshaw instructions to delivery my shares to you. Please commence selling at once at not less than £6 per share but as you settle on the basis of £7 a share[2] I shall be much disappointed if you don't get £7. Insull has explained to me the Telephone matters and I will rely on your getting highest prices in shortest time.[3] When I tell you that I have put no less than $45,000 lately into my lamp factory you will appreciate how largely I must personally assist. If I am to supply you with what I you[d] want promptly I must have these matters under my personal control as should I be compelled to raise money I shall have to put up with a Board of Directors which would be fatal to my endeavors to turn out ~~matters~~ things rapidly   So you see how matters stand; if I am to let you have things quickly you must rush through my money matters

We have moved the Engineers office to this building and the Light has now passed from the field of the experimental to that of practical operations.

Immediately you get this cable me what prospects there are of getting funds using the following code:—

Edison Telephone Co of London	Wicked
Edison Telephone Co of Glasgow	Wickedness
United Telephone Coy	Wickedly
Edisons London Reversion	Badly
Oriental Telephone Coy	Badness[4]

Our Electric Light shares are selling now at $1600[e] a share but should I have to sell some to raise money it would be at an enormous sacrifice so of course I want to hold them as long as possible.

Please address all letters to the above address[5] as Insull & myself will be here until the concern runs thoroughly smoothly probably about 12 month   Yours very truly

Thos A Edison

LS (letterpress copy), NjWOE, Lbk. 8:24 (*TAEM* 80:848; *TAED* LB008024). Written by Samuel Insull. [a]Canceled. [b]Illegible letterpress copy. [c]Repeated at page turn. [d]Interlined above. [e]Obscured overwritten text.

1. John Roach was the preeminent builder of iron ships in the United States at this time, principally at his works in Chester, Penn. The Aetna Iron Co. had been located at 104 Goerck St. on the east side of Manhattan. On 1 March, Charles Mott reported that Edison had acquired the shop for the Edison Machine Works. *DAB*, s.v. "Roach, John"; *Wilson's Business Directory* 1879, 380; Mott Journal PN-81-01-19, Lab. (*TAEM* 43:1134; *TAED* NP014:19).

2. That is, the price of United Telephone Co. shares stipulated in the settlement of Edison's reversionary interest; see Doc. 2046 n. 2.

3. Samuel Insull arrived in New York on 28 February and was met at the dock by Edward Johnson, who promptly took him to meet Edison. Insull later recalled

> that Edison wanted to spend the evening discussing matters in connection with his European affairs. It was assumed, inasmuch as I had just arrived from London, that I would be able to give more or less information on this subject. As Johnson was to sail the next morning at five o'clock, Edison explained to me that it would be necessary for him to have an understanding of European matters from him, and he, Edison, started out by drawing from his desk a check book and stating how much money he had in the bank and he wanted to know what European telephone securities were most salable, as he wanted to raise the necessary funds to put on the incandescent lamp factory, the Electric Tube Works, and the necessary shops to build dynamos. [Insull Notes, pp. 3–4, Meadowcroft (*TAEM* 227:141–42; *TAED* MM010DAH)]

Edison reportedly told him that his available cash amounted to $78,000 (McDonald 1962, 21). Insull also recalled that his summons to New York came in January in the form of a cable from Johnson, to whom he had been sending weekly reports of the British and Continental telephone business (Insull Notes, p. 24, Meadowcroft [*TAEM* 227:162; *TAED* MM010D]). Insull took the place of Stockton Griffin, who resigned for unknown reasons.

4. Gouraud cabled on 24 March: "Badly about six weeks   badness about two months   wickedness almost immediately." Gouraud to TAE, 24 Mar. 1881, DF (*TAEM* 59:662; *TAED* D8146F).

5. Edison's new offices at 65 Fifth Ave.

–2061–

From Francis Upton

Menlo Park, N. J., March 7 1881[a]

Dear Mr. Edison

The only reason I can offer for low curves[1] is this. We have been running more lamps recently and as a result had less current for final heats. I have given strict orders that they shall never light more than 42 lamps at one time though we have 150 on the pumps.[2]

I have now some lamps testing brought up quickly on the pumps and very high, they are copper plated ends in silver clamps. Tomorrow I shall send up a lot of plated clamps. It takes two to three times [or?][b] longer[c] to get off plated clamps than silver clamps, as they seem to contain a very large quantity of gas.[3] I think now I am getting the best vacuums that we have ever had. I have found that I can get as quick a vacuum without heating the lamps before going on pumps[4d] as with.

Clean mercury acts promptly and well and [of?][b] upsets all rules of working with mercury that was not good.

We have changed the feed[e] pump to the boiler. I had the line leveled and found 22 feet rise from the brook to the pump through 600 feet of pipe. I thought it was asking too much of Providence and so placed the pump so as to give 11 feet rise. It is now in a small house outside of the fence just below the bank.

Campbell[5] our head carpenter has done well with the coal shed. He took the day laborers and showed them where to ~~labor~~ work and made the shed even quicker than if he had carpenters.

I am having a pattern made for a cover to take under it 100 forms at once. Welsh[6] has carefully considered that he could handle this number at once.

By slightly changing the furnace I can so arrange as to get 600 A carbons at a heat or 1200 B carbons. That is 1200 As or 2400 Bs a day.

Lawson is showing considerable ingenuity and more industry than ever before. He is now waiting for corks with the right sized hole before feeling complete. We have plenty with small and large holes but none of medium. These will be done in a few days. The novelty company[7] charged us about $4.50c a pound for corks. I found that the New York Belting and Packing Co. would make them for $1.50 and so have sent them the mould.[8]

I have written for bids on the wooden part and brass for inside parts. The Newark parties were very slow and only sent us the wooden pieces a few days ago, though we wrote to them repeatedly. Shrinkage is going to trouble us. I am having a model made which I want to show you and[e] which I think will be of use.[9]

The rings now has only room for a close fit on the wedge like shoulder made for it. The screw fits also closely on the bottom part.

I think the wedge-like shoulder could be made longer so that if it were not exactly of a size the ring could be made to fit. Then if the brass piece were placed on the lower part without a shoulder it could be made to give right distance AB

I can now see my way clear to putting the pump job as piece work and making it satisfactory. The air in the clamps must be worked out by repeated bringing up high so that when the inspector comes he will not be able to spoil the fall tube by bringing the lamp up very high.

I cannot judge at all concerning the offer from Brooklyn[c] without seeing the place.[10] If you say so I will go to see what it is. I am thoroughly[c] convinced by my experience here that we must move. We never[c] can make the lamp cheap until we can have plenty of boys and girls at low wages ~~and t~~These can never[c] be had except where there are other manufacturing establishments employing men.[11]

The sooner it is decided that the factory will be moved the better, for expense can be stopped in ~~some~~ many[e] ways.[c] I have no preferance as to place, only a city. I think a little outside of Newark will be just right. We need first of all some space for we do not want to be caught when we enlarge. I know the lamp ~~company~~ factory[e] will be large, I dare not write figures of size. Yours Truly

Francis R. Upton

ALS, NjWOE, DF (*TAEM* 57:789; *TAED* D8123R). Letterhead of Edison Electric Lamp Co.; circled "49" written at top. [a]"Menlo Park, N.J.," and "188" preprinted. [b]Canceled. [c]Obscured overwritten text. [d]"before going on pumps" interlined above. [e]Interlined above.

1. According to Howell and Schroeder 1927 (194–95), "Every lamp made was measured to determine the voltage at which it gave 16 candle power. This voltage was measured by means of a reflecting electrodynamometer made for the purpose. . . . All lamps at that time were measured at 16 candles and curves drawn on cross section paper made it possible to determine the candles per horse power which the lamp gave with the known voltage on the lamp and the resistance in series with it on a 150-volt circuit."

2. Edison executed a patent application on 11 December 1880 in which he described techniques for improving lamp vacuum by passing a current through the filament in order to drive out occluded gases in the filament and clamps while still on the pump. After the filament was kept "for some time at a medium incandescence" it was "then raised to a much higher incandescence by the cutting out of more resistance, until the air and gas and aqueous vapor have been driven from the enlarged ends of the filament and the clamps, which can be readily determined by the disappearance of a blue or violet color which is seen at the clamps while the gas and vapor are being driven off. This high incandescence is considerably higher than that at which the lamp is designed to be used, it being from thirty candle-power upward in a lamp designed to give sixteen candle-power." This application issued in October 1882 as U.S. Pat. 265,777.

3. This refers to the fact that the "copper plated filament connections and carbon paste connections liberated a good deal of gas when heated. Before the vacuum became good there came a stage in which it was conductive. In this condition, when the filament was burned at high temperature, this cross current, passing through the partial vacuum and creating a blue glow in the bulb, heated the filament connections red hot and drove the gas out." Howell and Schroeder 1927, 125.

4. On 14 December Charles Mott had noted that "all the lamps are now heated quite hot before placing them on the pumps by which it is found that vacuum may be gotten much quicker." Mott Journal N-80-07-10:248, Lab. (*TAEM* 37:426; *TAED* N117:124).

5. Henry Alexander Campbell (1853–1938), who did a variety of carpentry jobs at the laboratory, started working at the Menlo Park laboratory on 24 October 1878. He had been in charge of making alterations to the lamp factory building and then "became the master carpenter of the lamp works." Jehl 1937–41, 222–23, 373, 582, 686–87; "Campbell, Henry Alexander," Pioneers Bio.

6. Alexander Welsh began working at the lamp factory in the fall of 1880. There are several experimental notes by him between mid-November and the beginning of February, at which time he was given charge of carbonizing filaments. He was discharged in February 1882 because, according to Francis Upton, "When he was placed in charge of the carbonizing it was expressly mentioned that he should report experiments truthfully. We were satisfied he was not doing so; this lack of truthfulness had become a by-word in our factory." Upton's testimony, 5:3255–56, *Edison Electric Light Co. v. U.S. Electric Lighting Co.*, Lit. (*TAEM* 48:134–35; *TAED* QD012F:132–33); N-80-09-28, Lab. (*TAEM* 36:459; *TAED* N106).

7. Unidentified.

8. John Lawson used the corks in the process of electroplating lamp filaments (see Doc. 2050 n. 1).

9. This refers to wooden lamp sockets. The "Newark parties" are unidentified.

10. Nothing further is known of this offer.

11. Upton informed Edison on 17 February that he had "proposed to [William] Holzer this morning that the working day be nine hours instead of ten so as to save money and bring all the men to their work on time. He says that it might make some dissatisfaction. If we could do so all those working could commence at once. I sent you the time of arrival of the men." Upton to TAE, 17 Feb. 1881, DF (*TAEM* 57:774; *TAED* D8123J).

–2062–

To Thomas Logan[1]

[New York,] 11th Mar [188]1

Dear Sir,

We are not going to give any more exhibitions except with the Porter Engine and you may therefore let Mr Upton have the shafting he wants and also four engines.[2] I want two of the best Dynamo machines 14 hundredths resistence cleaned packed and shipped to 65 Fifth Avenue to be all here for lighting the offices.[3] They should be here by Wednesday. Reserve one machine for the Hampson Engine[4] and the old field machine[5] can be used there also.

After you have given Upton four machines and sent me two and reserved one for the Hampson Engine (seven in all) how many will there be left not counting the old field machine? Yours truly

Thomas A Edison

LS (letterpress copy), NjWOE, Lbk. 8:51 (*TAEM* 80:862; *TAED* LB008051). Written by Samuel Insull. Original is in Box 37, EP&RI,

1. Edison appointed Thomas Logan (d. 1887) foreman of the Menlo Park machine shop on 14 February 1881, replacing John Kruesi (Jehl Diary, 14 Feb. 1881; Jehl 1937–41; 680, 872). Although Jehl claimed that Logan began working as a machinist at Menlo Park sometime in 1877, the first evidence of his presence at the laboratory is a time sheet for the first week of June 1878 (Logan time sheet, 8 June 1878, DF [*TAEM* 17:625; *TAED* D7817AA:4]). While Edison was in New York, Logan sent him frequent, sometimes daily, updates in April 1881 on the progress made by the Menlo Park staff in getting the Porter-Allen engine to run properly (see Menlo Park Laboratory—Reports [D-81-037], DF [*TAEM* 59:59; *TAED* D8137]).

2. The Menlo Park staff disassembled the large dynamo on 10 March to improve the bearings. This machine was used until late May when the armature short-circuited and was not repaired. The shafting and "engines," by which Edison may have meant dynamos, were needed for the lamp factory, where a boiler and engine had been installed for a generating station. Jehl Diary, 10 Mar. 1881; Jehl 1937–41, 882–84, 887.

3. Each of the standard .14 ohm machines, used for the Menlo Park demonstration plant, could operate about fifty 16-candlepower lamps (N-81-02-20:67, Lab. [*TAEM* 40:979; *TAED* N214:34]). According to Jehl 1937–41 (884), in March "Edison required a couple of dynamos for the Edison headquarters in New York City. Thus, March, 1881, was the time when the electric power room behind the engine room, with its eleven 'A' type Edison dynamos, was dismantled." Charles Mott reported that the dynamos arrived in New York on 18 March. During February Edison contracted with Edward Hampson, a New York steam engine and equipment dealer, for three complete eight horsepower steam engines, each to operate at 410 r.p.m. They were to be delivered to New York by 29 March and used for lighting purposes (Mott Journal PN-81-01-19, Lab. [*TAEM* 43:1138; *TAED* NP014:24]; TAE agreement with Hampson, Feb. 1881, DF [*TAEM* 58:220; *TAED* D8129S]).

4. The shop received this engine from Edward Hampson on 18 February. Nothing is known of its design or construction except its rating for 10 horsepower and 1150 rpm. Edison's assistants discovered on 22 February that they could not bolt it to the floor because it vibrated excessively but they managed to test it the next day and on 4, 5, 8, and 10 March. On 4 and 5 March the engine ran at 435 rpm and powered an unspecified dynamo which lit 60 lamps; however, on 8 March it powered a dynamo with an armature resistance of 0.55 ohms, which was unable to "do the necessary work." On 10 March several men "worked all night on bearings" and then ran it for a while at 480 rpm with "all lamps on." Four days later they succeeded in running it for 10 hours at 425 rpm with 60 lamps on the circuit. Mott Journal PN-81-01-19, Lab. (*TAEM* 43: 1128, 1130, 1134–37; *TAED* NP014:14, 16, 20–3); Charles Mott to TAE, 22 and 23 Feb. 1881, DF (*TAEM* 59:62, 64; *TAED* D8137B,

D8137C); Samuel Insull to Edward P. Hampson, 14 March 1881, Lbk. 8:72 (*TAEM* 80:868; *TAED* LB008072).

5. That is, a dynamo used to provide field excitation current for other dynamos which were directly connected to the mains.

-2063-

To Francis Upton

[New York,] 11th March [188]1

Dear Sir,

I want for lighting up our office here one hundred lamps with new small socket[1] which have no more than two volts variation with high vacuum so that the vacuum will not run down. These should be here by Thursday next or Friday. Can you do this? Yours truly

Thomas A Edison

<u>Memo</u>. Would it not be well when you are getting a curve of copper plated lamps to test their economy on the start and then make a test of the economy after five hours burning to see if the vacuum has been knocked down. I suppose your plan of bringing the lamps up high on the pumps will prevent this. But it would be well to prove the fact.

LS (letterpress copy), NjWOE, Lbk. 8:50 (*TAEM* 80:861; *TAED* LB008050). Written by Samuel Insull.

1. At the end of November 1880, Samuel Mott "made a bracket and small light socket from rubber scrap small and airy, and was requested by Mr. Edison to work it up in economical commercial form." On 5 January "John Ott finished a pair of sockets for B. lamps with key at the bottom acting on a spring which forms the contact same style and principle of one made by Mott Nov. 26 except in the Mott socket the [key?] acted and formed the connection direct without intermediate spring" (Mott Journal N-80-07-10:223, 276–77, Lab. [*TAEM* 37:414, 440–41; *TAED* N117:112, 138–39]). Mott continued to work on the design through the end of the month (N-80-09-11:112–31; Unbound Notes and Drawings [1881]; both Lab. [*TAEM* 39:300–9; 44:1012–14; *TAED* N153:56–65, NS81:2–4]). A patent application executed by Edison on 8 March 1881 includes the spring contact (U.S. Pat. 248,424).

John Ott's drawing of the small socket design with spring contacts.

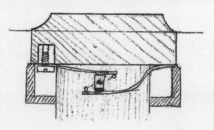

[Menlo Park,] 11th March [188]1

Dear Sir,

Mr Edison has requested me to acknowledge the receipt of your favour of 5th and to give you the information asked for.[2]

Mr Edison has had his system of Electric Lighting in practical operation at Menlo Park for some months passed   The system has not up to the present been put up in any City but the Lights are distributed at Menlo Park in such a manner as to demonstrate the practicability of lighting a city   All that Mr Edison is waiting for now is the permit of the Board of Alderman NY[a] to lay wires underground when he shall proceed to light the lower part of N.Y.

The Editor of the Standard[3] seems to forget that there[a] is great difference between commercial practicability and being scientifically perfect. Among scientific men when they have conducted an experiment in a somewhat practical manner and have this proved the theory it is called perfect but there is a vast difference between this and what is commercially perfect. This latter often takes much longer that the former. It is a question of dollars and cents

Over a year ago Mr Edison had 100 lights burning three months and no less than 40,000 people went to see them. That was the Scientific experiment. The several months past the Light has been exhibited on a large scale and its commercial practicability established

As to Mr Edison crying "wolf" so long I should have thought a newspaper Editor would have known better than to have charged Mr Edison with more than about 10 per cent of what has been put into his mouth. The public interest in all information as to a substitute for gas has been so great that newspapers have been compelled to deal with the subject very fully and consequently they sent reporters constantly and persistently to Menlo Park who unable to obtain the information were obliged to rely upon their own imaginations for about 90 per cent of their so- called "interviews"

The Standard says the use of the Electric Light is due to other men. So it may be: the Electric Light is not new, what is new however is the perfecting of a system of Electric Lighting which will in all respects take the place of gas and no one except Mr Edison has successfully solved this problem.

The Standard insinuates that the Edison Electric Light Co is a Stock Jobbing concern. As a matter of fact it is just the opposite. Originally there was a Coy organized with a capital of $300,000 to raise money to conduct experiments to devise a sys-

tem of Electric Lighting perfectly analogous to gas—divided
up in the same way and capable of being distributed over a large
area. The capital was furnished by about a dozen men chief
amongst whom were the firm of Drexel Morgan & Co. During
the great excitement when the shares were quoted at $4000 not
more than 25 shares passed hands   Besides the Light experiments the question of the distribution of power by the use of
small Electric motors was taken up and it became necessary to
make money and it stands to [reason?][b] that when the shares
stood at ten times their nominal value there would not have
been the slightest difficulty in raising several millions of dollars but this was not done and capital was simply increased
$1080,000. Moreover from the day the Coy was first organized
down to the present time not one hundred shares has been sold
to the public outside the original holders—in fact the most
conservative policy has been pursued throughout.

As to Mr Edison's connection with the Western Union
Telegraph Coy he had to find a market for his Telegraphic inventions and consequently was obliged to go to the people who
controlled the Telegraphic System of this Country in order to
get a fair price for his inventions.

I would prefer your not saying the source from whence you
get your information in any letter you may write to the Standard   Yours truly

Samuel Insull

Mr Edison does not control the E.E.L. Co. but is I should
say the largest shareholder

ALS (letterpress copy), NjWOE, Lbk. 8:44 (*TAEM* 80:856; *TAED*
LB008044). [a]Interlined above. [b]Faint letterpress copy.

1. In a 5 March letter to Edison, Horatio Nelson Powers identified
himself as George Gouraud's brother-in-law and the rector of Christ
Church in Bridgeport, Conn. Powers to TAE, 5 Mar. 1881, DF (*TAEM*
57:610; *TAED* D8120ZAI).

2. In his 5 March letter (see note 1) Powers wrote that he had "got into
a little squabble with an editor" regarding Edison's work in electric
lighting and asked for the information supplied by Insull. Powers replied
on 15 March with an enclosure, presumably a clipping from the *Bridgeport Standard*; neither the letter nor the enclosure has been found.
Insull replied the following day thanking him for the "enclosure which
Mr Edison has perused with great pleasure. Should this not end the controversy, and should you require further information, I shall with Mr
Edison's permission be most happy to give it." There is no further correspondence about this matter. Insull to Powers, 16 Mar. 1881, Lbk. 8:85
(*TAEM* 80:869; *TAED* LB008085).

3. Unidentified.

Notebook Entry:
Electric Lighting

[Menlo Park,] March 11, 1881.

Summary of Meter Experiments.[1]

—Freezing Points.—

(1)   Ordinary solutions $CnSO_4$[2] $+ H_2O$ freeze just under 0°C.

Addition of Glycerine lowers freezing point.

As much as 40% Glycerine can be added without interfering with the plating process. With 40% Glycerine the freezing point is below −20°C.

The Glycerine increases the resistance very materially.

Influence of strength of solution[a]

(2)   About 20% parts $CnuSO_4$ to 100 H_2O is as strong a solution as can be used at low temperatures— (See Watts Dict. on Saturation points of $CnuSO_4$ solutions.)[3]

The rate of the meter is within large limits <u>independent</u> of the strength of solution—provided always that the <u>current</u> remains constant.

Influence of temperature upon rate of Meter.[a]

$CnuSO_4$ increase in conductivity with the temperature. (See Wiedemann, Wüllners Physik Ba IV p 491.)[4]

For 18.7 parts $CnSO_4 + 100 H_2O$ Wiedemann found.—

Temps.[b]	Res.[b]
0°C	1.00
20° "	0.738
40° "	0.523
60° "	0.384
80° "	0.330

(3)   In case of metals— Matthiesen gives formula: Poggendorffs Annalen 115 & 122.[5c]

$$C = C_0 (1 - 0.0037674t + 0.000008342t^6)$$

where   C = Conductivity

$\quad\quad C_0 = \quad\quad$ " $\quad\quad$ at 0°

$\quad\quad$ t = temps.

This gives

Temps.[b]	Res.[b]
0°C	1.00[c]
20°	1.077
40 "	1.160
80° "	1.329

In consequence of these relations a circuit containing solutions and metals can always be so arranged that its conductivity shall be independent of the temperature.

If we use a 20% Solution C̶n̶uSO$_4$ & Copper.— I find the proper ratio of metal to solution to be 3:1![d]

The rate of the meter at different temperatures by actual test was found to be.— When the solution only was changed and the metal parts of the circuit remained at room temperature to be;

Solution 18%

Oscillations per hour.[b]	Temp.[b]
3.5	0°C
4.1	10°C
4.8	20°C
5.5	30°C
6.2	40°C[d]

From this the compensation can be calculated; by comparison with Wiedemanns results the other errors of the instruments found.

Nichols.[7]

X, NjWOE, Lab., N-81-03-11:5 (*TAEM* 41:540; *TAED* N236:3). Written by Edward Nichols; miscellaneous calculations not reproduced. [a]Multiply underlined. [b]Column headings separated from elements of table by horizontal line; columns separated by vertical line. [c]"Poggendorffs . . . 122" enclosed by left brace. [d]Followed by dividing mark.

1. This notebook entry was the first evidence of concern over the effect of temperature variation on the performance of the chemical meter, particularly preventing the solution from freezing and compensating for decreasing resistance as temperature increased. Research continued on these problems over the next several months. In May 1881 Edison applied for a patent on a meter with a winding in the shunt circuit whose resistance would increase with temperature, thereby compensating for the increased conductivity (or lower resistance) of the solution. At about the same time he filed an application for using the heat from a thermostatically controlled resistance circuit to keep the meter solution from freezing; in September he filed another application for obtaining heat for this purpose from a chemical reaction. U.S. Pats. 251,557, 251,558 and 265,774 and Jehl 1882, 16–17; see also Israel 1998, 205, and Jehl 1937–41, 637–69.

2. Nichols mistakenly wrote Cn for Cu throughout this document and later changed it except for this and one other occurrence.

3. No reference has been found to the saturation points of this solution in Watts 1872–75, a standard chemistry reference.

4. Nichols here cited a table in a standard German physics text (Wüllner 1874–75, 491). This table reproduced results obtained by the German physicist Gustav Heinrich Wiedemann (1826–1899), who investigated the relationship between the thermal and electrical conductivities of solutions and metals (*DSB*, s.v. "Wiedemann, Gustav Heinrich"). Nichols did not take the values for resistance directly from Wiedemann's

table, but extrapolated and interpolated the values in order to obtain results for temperatures at 20° intervals. In this and the following table, Nichols also normalized his results; that is, he divided the results which Wiedemann had obtained by the value of the resistance at 0°. Thus the values he obtained for the resistances in both tables are not absolute values but are values relative to the resistance at 0° Centigrade.

5. Matthiessen and M. von Bose 1862 and Matthiessen and C. Vogt 1864 reported the effects of temperature on the conductivity of pure metals and alloys.

6. This term should be t^2.

7. Edward Leamington Nichols (1854–1937) was a chemist and physicist who received his Bachelor of Science degree from Cornell University in 1875 and a Ph.D. from Göttingen University in Germany in 1879. He worked for Edison at Menlo Park from October 1880 to June 1881, concentrating on experimental work for automatic chemical meters, studies of the behavior of Sprengel pumps, and the design and installation of equipment for measuring the voltage and current of incandescent lamps. After leaving Edison's employ he held professorships in physics, chemistry, and astronomy at Central University in Richmond, Ky., the University of Kansas, and Cornell University. *ANB*, s.v. "Nichols, Edward Leamington"; "Nichols, Edward Leamington," Pioneers Bio.; Jehl 1937–41; 552–53, 815.

–2066–

From William Thomson

[London, March 12, 1881][1]

"Edison. New York."

"Swan comes to Glasgow to show his lamps at Philosophical Society next week and promises to leave several with me for private tests in my Laboratory.[2] Could you send me dozen of your Standard Lamps solely for my private tests; I promise I shall divulge nothing till I communicate results to you and learn your wishes.[3] This telegram is from Gourauds office with his and Johnsons cordial approval—[4]

Thomson"

L (telegram), NjWOE, DF (*TAEM* 58:600; *TAED* D8133E). Letterhead of Direct United States Cable Co., Ltd.

1. This cable was transcribed below a brief message of transmittal from the Direct United States Cable Co., and mailed from New York on this date. William Thomson lived and worked in Glasgow but evidently cabled this from London.

2. On 18 March Thomson wrote Gouraud that Joseph Swan had exhibited his lamps on 14 March at the private residence of Royal Society president William Spottiswoode, a mathematician and physicist, and that John Pender and John Hopkinson "were greatly pleased" with them. On 16 March Swan read a paper and exhibited his lamps at the Glasgow Philosophical Society, and on 17 March he brought his lamps to Thomson's laboratory. Thomson's measurements of them "perfectly

confirmed the statements in his paper. . . . Altogether the Swan light is much more perfect and its economy is much better assured than I knew when I saw you and Mr. Johnson in London last Friday." Thomson to Gouraud, 18 March 1881, DF (*TAEM* 58:602; *TAED* D8133G); *DNB*, s.v. "Spottiswoode, William."

3. According to the docket Edison replied on the same day. Although the reply has not been found, Edison directed Francis Upton to ship Thomson 50 lamps, including 20 B lamps of 8 candles, 10 C lamps of 8 candles, 10 B lamps of 16 candles, and 10 A lamps; the standard for A lamps was 16 candles and for B lamps it was 8 candles. He also directed Upton to "See that each set is of the same electro motive force and of the best economy, best bamboo, and of very high vacuum so that they will not lose their vacuum and economy" (TAE to Upton, 12 March 1881, Lbk. 8:59 [*TAEM* 80:863; *TAED* LB008059]). On 14 April Philip Dyer, secretary of the Edison Electric Lamp Co., informed Edison that he had shipped 20 A lamps and 20 B lamps, "with volts ohms and foot lbs. marked on each Lamp. As we have been to considerable trouble about picking these out, we ought to charge more than .35¢ each." Two days later Dyer wrote Edison that he had decided to charge $1.50 per lamp, "as the Lamps cost us fully that on account of testing picking out &c." The Lamp Co. charged them to the account of Drexel, Morgan and Co., which owned the rights to Edison's British lighting patents (Dyer to TAE, 14 and 16 Apr. 1881, both DF [*TAEM* 57:849, 852; *TAED* D8123ZAV, D8123ZAY]).

Thomson received the lamps and began testing them by the end of April. His incomplete preliminary tests at the end of the month led him to conclude that

> I do not expect to find any difference, which it is possible for me to test, between [Swan's] and Mr. Edison's with respect to economy. Either may be pushed to a very high degree of economy by working at sufficiently great intensities. It will be impossible for me or for anyone without months of experience to tell what is the highest intensity to which either lamp may be worked, without counterbalancing the value of high economy of light by wearing out the lamp in too short a time. [Thomson to Gouraud, 30 April 1881, DF (*TAEM* 58:613; *TAED* D8133O)]

Although no record of these or subsequent tests has been located, David Graham, a British entrepreneur interested in becoming an agent for Edison's electric light, wrote Edison, "Sir William Thomson has shown me your lamp & from the tests he has made & appearance of it I am of opinion that it is of more practical use than Swans & I do not hesitate to state that we could do as well if not better with yours." Thomson to Batchelor, 1 May 1881; David Graham to TAE, 25 June 1881; both DF (*TAEM* 58:614, 624; *TAED* D8133P, D8133X).

4. Gouraud was eager to have Thomson identified with Edison's companies (see Doc. 2056 n. 7). On 25 March, with Edison's approval, he offered Thomson the position of "Consulting Electrician to the Edison Electric Light Companies" of Europe. Thomson declined and later in the spring became a consultant to the Swan Electric Light Co., Ltd. Gouraud to TAE, 4 Feb. 1881; Gouraud to Thomson, 25 March 1881;

both DF (*TAEM* 59:1002, 58:604; *TAED* D8150E, D8133H); Thompson 1976 [1910], 764.

–2067–

Notebook Entry:
Electric Lighting

[Menlo Park,] March 15 & 16 1881[1]

Curve of Magnetism[2a]

These experiments were conducted on the direct-acting Porter-Allen dynamo to find the curve of magnetism.

Available length of field

Diam. of field

Diam. of iron core of armature.

Experiments made with external circuit of high resistance Thompson mirror-galvanometer.

Descending ~~field~~ curve.

~~Curve~~ field[b] weakening.

20 Daniels cells = 32°.

Magnet	Armature	Revo.	At 1000 Revo.[3c]
173	70	216	324
167	70½	216	326.4
50	60	229	267.9
40	56	232	241.4
28	45	242	186
20	40	226	177

Ascending Curve.

Field strengthening.

Magnet	Arm.	Revo.	Arm. at 1000 revo.
Open	40	247	161.9
39	50	236	211.9
65	60	228	263.2
120	70	214	327.1
176	75	219	342.5
177	76	214	355.1

76 the limit of magnetization was attained.

Field the broken

Ascending curve

Field strengthening

Magnet	Arm.	Revo.	Arm. at 1000 revo.
Open	40	252	158.7
30	45	241	186.7
39	50	244	245.9
50	55	240	229.2

60	60	230	250.9
50	66	265	249.1
112	70	216	324.1[a]

The speed of engine was gradually increased but no change was produced on magnet circuit. Armature deflection was taken with engine at 628 revo. deflection 215° with 112 on magnet. Resistance of magnets (6 cores) = 23 ohms.

Clarke.

X, NjWOE, N-81-03-15:9, Lab. (*TAEM* 41:848; *TAED* N245:5). Written by Charles Clarke. [a]Followed by dividing mark. [b]Interlined above. [c]Heading and column of figures written later with a different pen.

1. Charles Clarke, Edison, and Francis Jehl conducted these tests overnight. Jehl noted that "Mr. Edison came about ten. We fixed the large dynamo machine up and made a test of the curve of the magnet. Stayed up all night—did not get any sleep." The laboratory staff continued making tests on 17 March. Jehl Diary, 15 Mar. 1881; N-81-03-04: 274–77, Lab. (*TAEM* 41:696–697; *TAED* N240:3–4).

2. The curve of magnetism of a dynamo, also known as its characteristic curve, graphically represented the armature voltage as a function of field excitation. The data for the curve was typically taken with generator speed held constant and with no load on the generator. This curve changed when the magnetic field was increasing or decreasing because of residual magnetism in the field coil core, a phenomenon known as hysteresis. The purpose of this curve was to compare the performance of various dynamo designs. In this particular case, Clarke used these measurements to compare the performance of two field magnet designs, and he compared the data taken here with results from 10 December 1880 (N-81-03-15:19, Lab. [*TAEM* 41:853; *TAED* N245:10]). In February and March Clarke performed similar measurements and calculations to determine construction details for various dynamo designs, including the South American portable dynamos, central station machines, and dynamos of various sizes for isolated lighting (N-81-02-20: 1–213, Lab. [*TAEM* 40:946–1035; *TAED* N214:1–90]).

3. Clarke did not measure the armature deflection at 1000 rpm. Although the Porter-Allen steam engine could be run at speeds over 700 rpm, Edison determined that it was unsafe to run the engine at speeds over 400 rpm when connected to his dynamo. The standard operating speed was 350 rpm. Since armature voltage was proportional to generator speed, he calculated it by multiplying the measured armature deflection by the ratio of generator speeds. His purpose in scaling the measurements in this way was to obtain a convenient measure to compare the performance of several dynamo designs. Furthermore, on 29 March Clarke performed a series of calculations to redesign the armature and commutator so that the Porter-Allen dynamo would generate the same voltage at 350 rpm as it did at 600 rpm. N-81-02-20:111–13, Lab. (*TAEM* 40:994–95; *TAED* N214:39–40); Jehl 1937–41, 868; Edison and Porter 1882, 218.

[Menlo Park,] March 22 1881

$100 - 5.24 = 94.76$[a]   I worked this out in another book and am quite sure that it is right[1b]

Conductor of the same size for the whole length with ten lamps on it 180 feet apart the total resistance of the conductor being 1 ohm   If at the machine there are 100 Volts at the end of the conductor there will be 94.76 Volts or a fall of 5.24 Volts. If at the machine there is 110 Volts there will be a fall of 5.764 Volts on the line and take 405 lbs of Cu in the line 1800 feet long. Since the fall is inversely proportional to the amount of copper used in the mains, to compare with the conductor on which there was a fall of 13.64 Volts, there will be n pounds of copper in the uniform sized conductors to give the same fall[2]

$5.764 : 13.64 :: n : 405$

$$n = \frac{405 \times 5.764}{13.64} = 171.2$$

2.6075		
0.7607	log n	$= 2.32334$
8.8652	log 146.9 page 155a[3c]	$= 2.1670$
2.2334[4]		.0664

That is it require 1.165 times as much copper for a straight conductor as it does for a decreasing conductor to give the same fall of E.M.F.[5]

TAE   FRU

X, NjWOE, Lab., N-81-03-22:164, Lab. (*TAEM* 39:474; *TAED* N158:83). Written by Francis Upton; document multiply signed. [a]Form of calculation altered. [b]Sentence written on facing page with calculation above, and enclosed by lines at top, bottom, and right. [c]"page 155a" interlined below in a different pen, and partially circled.

1. Not found.

2. On preceding pages of this notebook Upton wrote, "The fall in E.M.F. will be inversely proportional to the amount of copper in the conductor. There will be the same proportional fall for 10 lamps 180′ apart as for 30 if there is ⅓ the amount of copper in the conductor. . . . The amount of copper required varies as the square of the distance that the lamps are apart or as the square of the distance of the furthest removed lamp." He gave an equation which related the voltage drop on the main conductor, the distance to the extreme lamp, the number of lamps, and the amount of copper in the conductor. This equation appeared earlier in an undated notebook entry from late 1880 in which Upton calculated the drop of 13.64 volts given here (N-81-03-22:155–57; N-81-00-01:71, 77; both Lab. [*TAEM* 39:469–70, 549, 552; *TAED* N158:78–79, N165:36, 39]). For prior calculations of the relationship between the number of lamps, distance, and cost of copper conductors, see Doc. 1889.

3. Upton obtained the figure of 146.9 in an undated notebook entry from late 1880 in which he determined that "30 lamps at equal distances along a conductor each having 10 ohms in the conductor will have a fall of E.M.F. from 100 to 87.6 volts. The conductor will weigh 440.7 lbs. 10 lamps will have the same fall and conductors will weigh 440.7 ÷ 3 = 146.9 lbs." The calculation "440.7 ÷ 3 = 146.9 lbs." also appears on page 155 of this notebook. N-81-00-01:71, N-81-03-22:155, both Lab. (*TAEM* 39:548, 469; *TAED* N165:36, N158:78).

4. In this computation Upton used the same logarithmic shortcut explained in Doc. 1795 n. 4.

5. The logarithm of 1.165 is .0664, above. Upton obtained an identical result in an earlier undated notebook entry from late 1880 (N-81-00-01:186, Lab. [*TAEM* 39:606; *TAED* N165:94]). He used this proportion throughout much of the rest of this notebook to calculate on a block-by-block basis how much additional copper would be required to install uniform instead of tapered conductors in the central station district. Edison had included the principle of tapered mains in an 1879 caveat (Doc. 1789, see esp. n. 1) and an August 1880 patent application.

-2069-

To Thomas Logan

[New York,] 26th[a] March [188]1

Dear Sir

As there seems to be considerable complaint with reference to the Laboratory being over run please have Yale[b] locks put on the doors giving keys only to the following

[Messrs?][c] [Charles] Hughes   [William] Carman   [Alfred] Haid   [Francis] Jehl & his Assistant[1]   [Edward] Achesson[2] [William] Hammer & the watchman.[3]

I very strongly object to anyone being shown the place who has not business there and shall be glad if you will report to me anyone who does not conduct themselves in a proper manner

Yours truly

Thos A Edison

LS (letterpress copy, NjWOE, Lbk. 8:104 (*TAEM* 80:875; *TAED* LB008104). Written by Samuel Insull. [a]Obscured overwritten text. [b]Interlined above. [c]Illegible.

1. Unidentified.

2. Edward Goodrich Acheson (1856–1931) came to work as a draftsman for Edison in September 1880. He was soon assisting with lamp experiments in the laboratory and, according to his recollection, declined the assignment as Charles Batchelor's successor in charge of the lamp factory, and continued experimenting in the laboratory. He went to Paris in July 1881 to assist Batchelor with Edison's exhibit at the International Electrical Exhibition. He spent the next three years in Europe installing electric lighting plants. Upon his return Acheson worked for a number of electrical companies. In 1891, he developed an artificial abrasive known as Carborundum, which became an indispensable industrial

abrasive. He later developed improved methods for manufacturing and using graphite. *ANB*, s.v. "Acheson, Edward Goodrich"; Acheson 1965, 15–19; Szymanowitz 1971.

3. Alfrid Swanson was the night watchman. He began working for Edison in December 1876 and later also ran the steam engine and helped out in the machine shop. Jehl 1937–41, 128; Cat. 1213:7, Accts. (*TAEM* 20:8; *TAED* A202:3); Time Sheets D-78-17, DF (*TAEM* 17:621; *TAED* D7817); Time Sheets, NjWOE.

–2070–

To James Bradley

[New York,] 28th Mar [188]1

Dear Sir

Referring to your letter of 24th inst we cannot afford to give ¢5 each for sockets[1]   Yours truly

TAE

How would it do to try & make a socket out of the Litherage glycerine cement itself[2]   TAE

LS and ALS (letterpress copy), NjWOE, Lbk. 8:112 (*TAEM* 80:878; *TAED* LB008112A). Body of letter written by Samuel Insull; postscript written by Edison.

1. James Bradley had apparently gotten this quote from the Celluloid Manufacturing Co. Writing from Newark he indicated that he had "stoped at Hyatt Place," presumably referring to John and Isaiah Hyatt who had established the Celluloid Co. in Newark in 1871 (see *TAEB* 2:498 n. 1). They offered to make the sockets for five cents each out of "boneslate" but indicated that in order to try it they would have to make a $25 mould. Bradley to TAE, 24 Mar. 1881, DF (*TAEM* 57:833; *TAED* D8123ZAL).

2. Bradley had asked Hyatt about a cement to stick the sockets to glass and was shown a cement made of litharge and glycerine that cost twenty-five cents a pound. At this time Bradley was designing a mold for making sockets out of plaster of Paris. On 29 March he reported that he had tried litharge and glycerine and found it to be better than plaster of Paris but could only get 20 from a pound of material, which he thought too expensive. He suggested mixing sawdust or some other cheap material to reduce the cost. Bradley to TAE, 24 and 29 Mar. 1881; Upton to TAE, 25 Mar. 1881; DF (*TAEM* 57: 833, 837, 835; *TAED* D8123ZAL, D8123ZAN, D8123ZAM).

–2071–

To Theodore Puskas

[New York,] 29th Mar [188]1

Dear Sir,

I should be glad if you would go to a good patent Lawyer in Paris and get opinion on following point:—

Is the Gramme patent valid in France in view of the machine of Paccinati[a] published in the 19 vol of an Italian publication called "Il Neuvo Cimento" to be found in one of the great libraries of Paris   It is also found partly translated in Schellens late work on the Electric Light published in German   Paccinati's machine seems to be a perfect fac simile of Grammes as far as the principle is concerned[1]   I suppose "Armengaud"[2] would be a good person to examine.

Regarding Lighting at the Exposition get all the smaller rooms of any picture galleries. we cannot conveniently light up the Palais Royal or Grand Hotel from the Exposition   It is too far away. I think it will be better to make two machines one for the Palais Royal and the other for the Exposition.[3] Can you get work through under the permit.

I shall probably send the Quadruplex, Electric Pen, Phonograph, Telephone, Tassimeter, Automatic Fac Simile Telegraph, and other things[4] Yours truly

Thos A Edison.

LS (letterpress copy), NjWOE, Lbk. 8:119 (*TAEM* 80:880; *TAED* LB008119). Written by Samuel Insull. ª"of Paccinati" interlined above.

1. Italian physicist Antonio Pacinotti, 1841–1912 (*DSB*, s.v. "Pacinotti, Antonio"), developed a generator with a ring armature and commutator with brushes, the first design to produce a continuous and steady current. He published a description of his dynamo in the June 1864 issue of the Italian science journal *Il Nuovo Cimento*; for an English translation, see "Description of Dr. A. Paccinotti's Electro-Magnetic Machine," n.d., DF (*TAEM* 57:690–701; *TAED* D8120ZCH). Pacinotti exhibited the machine at the 1881 Paris Exposition, where it gained widespread attention for the first time and earned him a *diplome d'honneur*, one of the Exposition's highest awards (Heap 1884, 28–32; Thompson 1902, 13–15). In 1870 the French engineer Zénobe Gramme independently invented a similar armature and commutator and took out broad patents on this design in the United States and Great Britain (see *TAEB* 4:585 n. 6; Dredge 1882–85, cxxix; U.S. Pat. 120,057). Edison referred to Schellen 1879 on electric motors and generators; the 1884 English edition provided an extensive description of Pacinotti's machine, including a translation of the 1864 article (Schellen 1884, 204–19). Before the end of March, Edison asked Puskas to inquire about obtaining rights to Gramme's patent in the United States (TAE to Puskas, 30 Mar. 1881, TP [*TAED* Z400BZ]).

2. Edison was probably referring to Jacques-Eugène Armengaud, who, like his younger brother Charles, was a respected consulting engineer and an authority on the patent laws of France and other countries, including the United States. Jacques-Eugène, who won design prizes at the international expositions in London in 1851 and Paris in 1867 and was also a member of the Legion of Honor, seems to have been somewhat better known than Charles. *DBF*, s.vv. "Armengaud, Jacques-Eugène" and "Armengaud, Charles."

3. On 12 February Theodore Puskas ordered from Edison through his American agent Edward Saportas "an electric lighting machine ca-

pable of demonstrating the system practically at the Palais Royal, Paris."
They specified that the dynamo was to be capable of powering 1,000
sixteen-candlepower lamps, was to cost about $8,000, and needed to be
installed and operating by 30 May. On 14 February Saportas confirmed
that James Banker was "authorized to act in the matter of machine for
palais royal," and that he had asked him to meet with Lemuel Serrell "in
regard to obtaining necessary protection for invention upon introduction to France." On 16 February Saportas wrote that Banker was prepared to give Edison the order for him to send the dynamo to Paris, but
that Banker needed a cost estimate beforehand. Saportas also informed
Edison that Puskas had cabled from Paris: "See whether Edison author-
izes securing space in 'exposition electrique' for his inventions generally
and for light in connection with palais royal   answer quick." Puskas &
Saportas to TAE, 12 Feb. 1881; Saportas to TAE, 14 Feb. 1881; Saportas to TAE, 16 Feb. 1881; all DF (*TAEM* 58:838, 840, 841; *TAED*
D8134C, D8134D, D8134E).

4. On 30 March Puskas cabled Edison that he had been "officially
asked whether you exhibit other inventions beside light." Edison replied
"yes." Puskas to TAE and TAE to Puskas, both 30 Mar. 1881, Cable
Book 1:7 (*TAEM* 83:875; *TAED* LM001007B, LM001007C); see also
Doc. 2045.

–2072–

To A. M. Leslie
& Co.[1]

[New York,] 30th Mar [188]1

Gentm

I have your favour of 24th inst.[2]

We do not care at present to contract for lighting isolated
buildings as we propose to establish a complete system in the
same manner as the Gas Companies do with gas. Your letter
has however been filed and will be considered when we are
prepared to deal with that class of business[3]   Yours truly

Thos A Edison   I[nsull]

L (letterpress copy), NjWOE, Lbk. 9:85 (*TAEM* 81:36; *TAED*
LB009085). Written by Samuel Insull.

1. A. M. Leslie was a medical instrument company based in St.
Louis. Dacus and Buel 1878, 321–23; Edmonson 1997, 206.

2. Not found.

3. A month later the Edison Electric Light Co. decided to form a
Bureau for Isolated Lighting and placed Miller F. Moore in charge. In
November this became the Edison Co. for Isolated Lighting. Edison
Electric Light Co. circular, 11 Nov. 1881; Edison Electric Light Co.
agreement with Moore, 30 Apr. 1881; both DF (*TAEM* 57:712, 227:513;
TAED D8121E, D8126X1).

From William Carman

Dear Sir

In answer to yours of 28th[1]

The total Cost of the Light experiment to Mar 22, 1881—[2]

                                129,367.84

Received from Co   126,566.54[b]

Bal due—statement enclosed[3]   2801.30

Now in regard to the Cr[edit] to the light[b] Co there are none as it has allready been taken out with these exceptions   Goods which are now going away of which I have a memorandum but I understood that the E. Light Co would bill them to the Lamp Co direct— I am preparing such a bill and will bring it to you when it is done—tomorrow if possible[4]   Very truly

W Carman

ALS, NjWOE, DF (*TAEM* 58:29; *TAED* D8126V). Letterhead of T. A. Edison. Copied in Lbk. 7:94 (*TAEM* 80:549; *TAED* LB007094). [a]"Menlo Park, N.J.," and "1881." preprinted. [b]Obscured overwritten text.

1. Not found.
2. For Edison's accounts with the Edison Electric Light Co., see the two Electric Light Co. Statement Books, Accts. (*TAEM* 88:412, 512; *TAED* AB031, AB032). Detailed accounts of money spent on various experiments can also be found in Edison's Personal and Laboratory Accounts, Ledgers 3, 4, and 5, Accts. (*TAEM* 87:5, 209, 403; *TAED* AB001, AB002, AB003).
3. Not found.
4. Not found.

Appendix 1

Edison's Autobiographical Notes

In 1908 and 1909 Edison wrote a series of autobiographical notes whose origin and nature are described in the first appendix of Volume One.[1] Six of the documents contain sections related to events of the period covered by Volume Five; those sections are published here.[2] Edison sometimes referred in the same paragraph to the periods covered by more than a single volume; these paragraphs will be reprinted as appropriate. Each document has been designated by a letter and each paragraph sequentially numbered.

1. See *TAEB* 1:627–28. An additional autobiographical document was discovered after the publication of Volume One. It is designated G and described below.
2. The autobiographical document designated A does not refer to the period covered by this volume.

B. FIRST BATCH

The following is from a typescript that Edison revised. At the top of the first page is a handwritten note: "First Batch Notes dictated by Mr Edison to T. C. Martin June, 1909.— Pencil indicates Mr. Edison's revision." Ten of its eighty-one paragraphs pertain to the period covered by this volume.

Segredor at Menlo Park.[a]

[1] We had a man with us at Menlo called Segredor. He was a queer kind of fellow. The men got in the Habit of plaguing him, and finally one day he said to the assembled experimenters

in the top room of the laboratory: "The next man that does it, I will kill him." They paid no attention to this, and next day one of them made some sarcastic remark to him. Segredor made a start for the fellow, and when they saw him coming up the hall with a gun they knew there would be trouble, so they all made for the woods. One of the men went back and mollified him. He returned to his work but he was not teased anymore. At last when I sent out men hunting for bamboo, I dispatched Segredor to Cuba. He arrived in Havana one Tuesday and on the Friday following he was buried, having died of the black vomit. On the receipt of the news of his death, half a dozen of the men wanted his job, but my searcher in the Astor Library reported that the chances of finding the right kind of bamboo for lamps in Cuba were very small; so I did not send a substitute.

~~Boehm~~ The Conceited Glassblower.[b]

[2] When experimenting with vacuum pumps to exhaust the incandescent lamps, I required some very delicate and close manipulation of glass and hired a German glass blower who was said to be the most expert men of his kind in the United States. He was the only one who could make clinical thermometers. This man was the most extraordinarily conceited man I have ever come across. His conceit was so enormous life was made a burden to him by all the boys around in the laboratory. ~~Boehm~~ He[c] once said that he was educated in a university where all the students belonged to families of the aristocracy; and the highest class in the University all wore a little red cap. He said he wore one!

FRANK THOMSON AND ELECTRIC RAILWAYS.

[3] One day Frank Thomson, the president of the Pennsylvania Railroad came out to see the electric light and the electric railway in operation. The latter was about a mile long. He rode on it. At that time I was getting out plans to make an electric locomotive of 500 horsepower with 6 foot drivers with the idea of showing the railroad people that they could dispense with their steam locomotives. Mr. Thomson made the objection that it was impracticable, and that it would be impossible to supplant steam. His great experience and standing threw a wet blanket on my hopes. But I thought he might perhaps be mistaken, as there had been many such instances on record. I continued to work on the plans and about three years later I started to build the locomotive at the works at Goerck street,

and had it about ¼^c finished when I was switched off on some other work. One of the reasons why I felt the electric railway to be eminently practical was that Henry Villard, then president of the Northern Pacific Railroad, said that one of the greatest things that could be done would be to build right angle feeders into the wheat fields of Dakota and bring in the wheat to the main lines, as the farmers now had to draw it from 40 to 80 miles. There was a point where it would not pay to raise it at all, and large areas of the country were thus of no value. I conceived the idea of building a very light railroad of narrow gauge, and had got all the data as to the winds on the plains, and found that it would be possible with very large windmills to supply enough power to drive these wheat trains.

LABORATORY LIFE AT MENLO.

[5] When experimenting at Menlo Park we had all the way from 40 to 80 men. They worked all the time. Each man was allowed from four to six hours sleep. We had a man who kept tally and when the time for one came to sleep he was notified. At midnight we had lunch brought in and served at a long table at which the experimenters sat down. I also had an organ which I procured from Hilbourne Roosevelt—uncle of the president—and we had a man play this organ while we ate our lunch. During the summer time after we had made something that was successful, I used to engage a brick sloop at Perth Amboy and take the whole crowd down to the fishing banks for two days. On one occasion we got outside Sandy Hook on the banks and anchored. A breeze came up and the sea became rough, and a large number of the men were sick. There was straw in the bottom of the boat which we all slept on. Most of the men adjourned to this straw very sick. Those who were not, including myself and two or three others, got a piece of rancid salt pork from the skipper and cut a large thick slice off it. This put on the end of a fish hook and drew across the men's faces. The smell was terrific, and the effect produced added to the hilarity of the excursion. I went down once with my father and two assistants for a little fishing inside Sandy Hook. For some reason or other the fishing was very poor. We anchored and I started in to fish. After fishing for several hours there was not a single bite. The others wanted to pull up anchor, but I fished two days and two nights without a bite, until they pulled up the anchor and went away. I would not give up. I was going to catch that fish if it took a week.

Sarah Bernhardt.

[25] While the experiments with the light were going on at Menlo Park, Sarah Bernhardt came to America. One evening Robert L. Cutting, of New York, brought her out to see the light. She was a terrific "rubberneck." She jumped all over the machinery and I had one man specially to guard her dress. She wanted to know everything. She would speak in French and Cutting would translate into English. She stayed there about an hour and a half. She gave me two pictures painted by herself which she sent me from Paris.

New York Real Estate.

[34] While planning for my first New York station—Pearl Street—of course,—I had no real estate, and from lack of experience had very little knowledge of its cost in New York, so I had assumed a rather large, liberal amount of it to plan my station on. It occurred to me one day that before I went too far with my plans I had better find out what real estate was worth. In my original plan I had 200 by 200 feet. I thought that by going down on a slum street near the water front I would get some pretty cheap property. So I picked out the worst, delapidated, deserted street there was, and found I could only get two buildings each 25 foot front, one 100 feet deep and the other 25 feet deep. I thought about $10,000 each would cover it, but when I got the price I found that they wanted $75,000 for one and $80,000 for the other. Then I was compelled to change my plans and go upward in [---]^d in the air where real estate was cheap. I cleaned out the building entirely to the walls, and built my station of structural iron work, running it up high.

Selling Patents in Europe.

[37] I endeavored to sell my lighting patents in different countries of Europe and made a contract with a couple of men. On account of their poor business capacity and lack of practicality, they conveyed the patents all right to different corporations but in such a way and with such confused wording of the contracts that I never got a cent. One of the companies started was the German Edison, now the great Allegemeine Elektricitaets Gesellschaft. The English Company I never got anything for because a lawyer had originally advised Messrs. Drexel, Morgan & Co. as to the signing of a certain document and said it was all right for me to sign. I signed and I never got a cent, because there was a clause in it which prevented me from ever

getting anything. One of my associates was Theodore Puskas, who was undoubtedly the first man to suggest the use of the telephone in a central exchange. He made the suggestion to me when the telephone was still on exhibition, and was very enthusiastic over the subject. He started a telephone exchange in Buda Pest.

UNPROFITABLE LAMP MANUFACTURE.

[54] When we first started the electric light, it was soon seen that we had to have a factory for manufacturing lamps. As the Edison Light Company did not seem disposed to go into manufacturing, with what money I could raise from my other inventions and royalties, and some assistance, we started a small lamp factory at Menlo Park. The lamps at that time were costing about $1.25 each to make so I said to the company "If you will give me a contract during the life of the patents I will make all the lamps required by the company and deliver them for 40 cents." The company jumped at the chance of this offer and a contract was drawn up. We then bought at a receiver's sale at Harrison, N.J. a very large brick factory which had been used for an oil cloth works. We got it at a great bargain and only paid a small sum down, and the balance on mortgage. We moved the lamp works from Menlo Park to Harrison. The first year the lamps cost us about $1.10. We sold them for 40 cents, but there were only about 20,000 or 30,000 of them. The next year they cost us about 70 cents and we sold them for 40. There were a good many and we lost more the second year than the first. The third year I had succeeded in getting up machinery and in changing the processes until it got down so that they cost us somewhere around 50 cents. I still sold them for 40 cents and lost more money that year than any other because the sales were increasing rapidly. The fourth year I got it down to 37 cents and I made all the money up in one year that I had lost previously. I finally got it down to 22 cents and sold them for 40 cents and they were made by the million. Whereupon the Wall street people thought it was a very lucrative business, so they concluded they would like to have it and bought me out.

MAJOR MCLAUGHLIN.

[58] In the early days when I was experimenting with metallic filaments for the incandescent light, I sent a man named Major McLaughlin out to California in search of platinum. He found a considerable quantity in the sluice boxes of the Cherokee Valley Mining Company. But just then he found that fruit

growing was the thing and dropped the subject. He then came to me and said that if he could raise $4,000 he could go into some kind of orchard arrangement out there and would give me half the profits. I was unwilling to do it, not having very much money just then, but his persistence was such that I raised the money and gave it to him. He went into California and got into mining claims, and into fruit growing, and became one of the politicians of the Coast, and I believe was on the staff of the Governor of California. Last year he wounded his daughter and shot himself because he had become ruined financially. I never heard ~~of~~ from[c] him after he got the money.

ELECTRIC POWER IN NEW YORK.

[62] I had a great idea of the sale of electric power to large factories, etc. off the electric lighting system, and I got all the insurance maps in New York city, and located all the hoists, printing presses, and other places where they used power. I put all these on the maps, and allowed for the necessary copper in the mains to carry current to them when I put the mains down, so that when these places took current from the station I would be prepared to furnish it because I had allowed for it in the wiring. There were, I remember, [5?]54[e] hoists in that district. In some places a horse would be taken upstairs to run a hoist and would be kept there till it died.

TD (transcript), NjWOE, Meadowcroft. [a]Multiply underlined. [b]"Glass-blower" written in pencil; "Dont give his name." written in pencil in left margin. [c]Interlined above in pencil. [d]Canceled. [e]Faint copy.

C. SECOND BATCH

The following is from a typescript that includes Edison's revisions. At the top of the first page is a handwritten note: "Second Batch   Mr Edison's notes dictated Mr Martin   June 1909   Pencil indicates revision by Mr Edison." Seven of its twenty-four sections pertain to the period covered by this volume.

ELECTRIC ROADS FOR BOGOTA.

[6] During the electric railway experiments at Menlo Park, we had a short ~~spin~~ spur[a] of track up one of the steep gullies. The experiment came about in this way. Bogota the capital of Columbia is reached on mule back—or was—from Honda at

the head waters of the Magdalena River. There were parties who wanted to know if transportation over the mule route could not be done by electricity. They said the grades were excessive and it would cost too much to do it with steam locomotives, even if they could climb the grades. I said: "Well, it can't be much more than 45 per cent; we will try that first. If it will do that it will do anything else." I started at 45 per cent. I got up an electric locomotive with a grip on the rail, by which it went up the 45 per cent grade. Then they said the curves were very short. I put the curves in. We started the locomotive with nobody on it and got up to 20 miles an hour taking these curves of very short radius; but it was weeks before we could prevent it from running off. We had to bank the tracks up to an angle of 30 degrees before we could turn the curve and stay on. These Spanish parties were perfectly satisfied we could put in an electric railway from Honda to Bogota successfully—and then they disappeared. I have never seen them since. As usual I paid for the experiments.

A Gun Cotton Aeroplane.

[11] James Gordon Bennett came to me ~~to see if I could get up a flying machine~~ & asked that I try some primary experiments to see if aerial navigation was feasible.[b] I got up a motor and put it on the scales and tried a large number of different things and contrivances connected to the motor to see how much it would lighten itself on the scales. I got some data and made up my mind that what was needed was a very powerful engine for its weight, in small compass. So I conceived of an engine employing gun cotton. I took a lot of ticker paper tape and turned it into gun cotton and got up an engine with an arrangement whereby I could feed this gun cotton strip into the cylinder and explode it inside electrically. The feed took place between two copper rolls ~~to kep if from pulling back and~~. ~~t~~The copper kept the temperature down, so that it would only explode up to the point where it was in contact with the feed rolls. It worked pretty well, but once the feed roll didn't save it, and it went through and exploded the whole roll and kicked up such a bad explosion. I abandoned it. But the idea might be made to work.

Nabbing the Expert.

[13] In the early days of my electric light, curiosity and interest brought a good many people to Menlo Park to see it. Some of them did not come with the best of intentions. I remember the visit of one expert, a well known electrician, a

graduate of Johns Hopkins, and who then represented the Baltimore Gas Company. We had the lamps exhibited in a large room and so arranged on a table as to illustrate the regular lay out of circuits for houses and streets. Sixty of the men employed at the laboratory were used as watchers, each to keep an eye on a certain section of the exhibit, and see there was no monkeying with it. This man had a length of insulated No. 10 wire around his sleeves and back, so that his hands would conceal the ends and no one would know he had it. His idea, of course, was to put this across the ends of the supplying circuits and short circuit the whole thing—put it all out of business without being detected. Then he could report how easily the electric light went out, and ~~we would be queered~~ a false impression conveyed to the public.[c] He did not know that we had already worked out the safety fuse, and that every little group of lights was protected independently. He slyly put this jumper in contact with the wires—and just four lamps went out on the section he tampered with. The watchers saw him do it, however, and got hold of him, and just led him out of the place & with language that made the recording angels jump for their typewriters.[d]

A Dishonest Patent Solicitor.

[21] Around 1881–2 I had several solicitors attending to different classes of work. One of these did me a most serious injury. It was during the time that I was developing my electric lighting system, and I was working and thinking very hard to cover all the numerous parts in order that it would be complete in every detail. I filed a great many applications for patents at that time, but there were 78 of the inventions I made in that period that were entirely lost to me and my company[e] by reason of the dishonesty of this patent solicitor. Specifications had been drawn, and I had signed and sworn to the applications for patents for these 78 inventions, and naturally I supposed they had been filed in the regular way.

[22] ~~As time passed, I was looking for some action of the Patent Office, as usual, but none came~~. He reported fictious actions by the Patent Office and in many cases reported that patents had been allowed & obtained the final fees.[f] I thought it very strange but had no suspicions until I began to see my inventions recorded in the Patent Office Gazette as being patented by others. Of course, I ordered an investigation and found that the patent solicitor had drawn from the Company the fees for ~~filing~~ all these applications, but had never filed

them. All the papers had disappeared. ~~However, and what he had evidently done was to sell them to others who had signed new applications and proceeded to take out patents themselves on my inventions~~. I afterwards found that he had been previously mixed up with a somewhat similar crooked job in connection with telephone patents.

[23] I am free to confess that the loss of these 78 inventions has left a sore spot in me that has never healed. They were important, useful and valuable and represented a whole lot of tremendous work and mental effort, and I had had a feeling of pride in having overcome through them a great many serious obstacles.

[24] It is of no practical use to mention the man's name. I believe he is dead, but he may have left a family. The occurrence is a matter of the old Company's records.

TD (transcript), NjWOE, Meadowcroft. [a]Interlined above in pencil. [b]"& asked . . . feasible." interlined above in pencil. [c]"a false . . . public." interlined above in pencil. [d]"& with . . . typewriters" added in pencil. [e]"entirely lost . . . company" interlined above by typewriter. [f]"He reported . . . fees." interlined above in pencil.

D. BOOK NO. 2

This undated notebook, labeled "Book No. 2," contains a mix of narratives, questions, and notes in Edison's hand. The first two pages are a memo by Meadowcroft, dated 9 January 1920, recounting the preparation and use made of this material between 1907 and 1920. The next sixty-six pages alternately present narrative passages and brief references to various anecdotes. The next nine-page section is labeled "Martin's Questions." The remaining twenty-one pages contain only notes. Altogether thirty items pertain to the period covered by this volume; item 290 was previously published in Volume Four.

[128] Porter & 1st High speed Eng   shook Hill—[a]

[129] Dynamo external 10 times internal— Weston Letter to Scientific Amcn— Speaking about it yrs afterwd to Seemans & Von Alterneck—one [-------ed][b] the other—

[130] Exhibition Menlo Park—1st use Safety fuse   smart Electrician representing gas presd tried short ckt—[c]

[132] 100 000 telephone—6000 yrly— 100 000 for Chalk relay—6000 yrly   Gould always sour—

[133] Wilbur & 100 patents lost—[a]

[140] Started 1st Lamp Wks   loss of money—finally making money[a]

[164] High Speed Eng Menlo P   Chas H Porter—

[165] Man died Cuba Bamboo   20 men volunteered

[169] Villard & financiers on 1st Elec RR   Kruezi fell off no danger   2nd RR   financiers wanted walk back= Honda mountain climb RR—[a]

[170] Expert from Balto Gas Co shorted wires but had fuse Caught him—[a]

[254] Segredor threaten kill—

[255] Boehm little Red Cap—

[256] ~~NY~~ J P Morgan came to Lab—

[257] Villard, Frank Thomson on RR— ~~Vil~~ Villard & wheat RR   Thomson said main lines never be run—

[259] Midnight Lunch—Rosevelt organ— Fishing Excursion   Fishing bank drawing pork over sleeping men   fished 2 days without a bite—said would catch a fish before quit. built that way

[288] Sarah Bernhardts visit to Menlo[d]

[289] Copper mine Menlo—[d]

[290] Frank Thompson visit Menlo—thought never be used on RR

[300] Elec RR Menlo. financing Sprague & EH.

[301] Insull.

[323] Delivered lecture 1st & only [meet?][b] Amn Assn advanc science Saratoga [Accnt?][b] death— Read about platina. showed Loudspkg telephn—

[328] Blinded 3 days at Menlo arc furnace—[d]

[333] Search after platina—McLaughlan—connection suicide—

[334] Search bamboo over world[a]

[343] How got gas bill every man below central park—motors—insurance maps—[a]

[352] Morgan people word better than contract just memo when trouble arises all my contracts had clauses in & never remember of having noticed at time signed

[368] Bankers didnt want anything to do with mfg let me have it   only 1 was JPM himself & then only 5000[a]

[374] Started Science   owned it.[a]

[388] Aerial nav for Bennett. heliocopter—scales.[a]

[390] Wilbur & loss of 60 patents

AD, NjWOE, Lab., N-09-06-27. [a]Paragraph canceled with large "X". [b]Illegible. [c]Followed by dividing mark. [d]Paragraph overwritten with checkmark.

E. NOTES

Taken from a notebook that has five pages in Edison's hand, these "Notes" are numbered consecutively from 1 to 33. Eight items pertain to the period covered by this volume; item 3 was previously published in Volume Four.

[1] 1 = Segredor = threats to kill came up to laboratory with gun everbody took to the woods— sent Cuba, arrived on Tuesday buried friday[a]

[2] 2 = Boehm, Little red cap— clinical therms.[a]

[3] 3 = JPM came to Lab[a]

[4] 4 = Villard, Frank Thomson PRR, running main line RR utopian—[a]

[5] 5   Villard wheat RR.[a]

[7] 7   Midnight Lunch, Rosevelt organ[a]

[8] 8   Fishing excursions, sea sick, drawing salt pork over heads[a]

[9] 9 = fAnother fishing excursion 2 days without bite[a]

AD, NjWOE, Lab. N-09-06-27. [a]Paragraph overwritten with checkmark.

F. NOTES

This notebook includes sixteen pages in an unlabeled section in Edison's hand relating to the Dyer and Martin biography. These pages are preceded by a memo to Edison from William Meadowcroft dated "June 28/09" stating that these notes had been copied. Five of its twenty-four items pertain to the period covered by this volume. There is a typed version of the notes in the William H. Meadowcroft Collection at the Edison National Historic Site. The last fifteen pages are a biographical sketch of Edison's former employee Sigmund Bergmann.

New—

[5] Bergmann came to work for me as a boy he started in on Stock quotation printers   as he was a rapid worker, and paid no attention to the clock I took a fancy to him & gave him piece work. he used [---][a] contrived[b] so many little devices tools to cheapen the work that he made lots of money. I even helped him get up tools until it occured to me that this was a too rapid method of getting rid of my money as I hadnt the heart to cut the price when it was originally fair. After a year or so Berg-

mann got enough money to start a small shop in Wooster St NYork and it was at this shop that the first phonographs were made   Then came the Carbon Telephone a large majority of which were made by Bergmann for the WU Tel Co   finally came the Electric Light. A Dynamo was installed in Berg-manns shop to permit him to test the various small apparatus which he was then making— He rented power from a Jew who owned the building—power was also supplied from a fifty horse power engine to other tennants on the several floors. soon after the introduction of the big dynamo machine the Jew appeared in Bergmanns shop & insisted that B was using more power than he was paying for & said that lately the belt on his Engine was slipping & squeeling   Bergman maintained that he must be mistaken. the Jew kept going among his tennants & finally discovered the Dynamo. Oh[c] Mr. Bergmann now I know where my power goes to pointing to the Dynamo.= Bergmann gave him a withering look of scorn & said Oh you fool Jew come here & [I'll?][d] show you— Throwing the belt off & disconnecting the wires he spun the armature around by hand, then said B you see its not here that you must look for your loss. This satisfied the Jew, & he started off to his other tennants. The Jew didnt know that that machine[c] when the wires were connected could stop his Engine.

[6] Soon after the B.[b] business had grown so large that ~~he~~ E H Johnson & myself went in as partners & Bergmann Rented an immense factory building at the Cor of Ave B & 17th St cov-ering ¼ of a block & 6 stories high. Here was made all the small things used on the Electric Lighting systems such as sockets chandiliers, switches meters, etc   In[c] addition stock Tickers, Telephones, telephone switchboards, Typewriters, (the Ham-mond type writer was perfected & made here)—over ~~10~~500 men were finally employed—

[7] B [----][e] This shop was very successful both scientifi-cally & financially   Bergman was a man of great Executive abil-ity and carried economy of manufacture to the limit. Among all the men I had associated with me ~~Bergmann~~ he[b] had the ~~most~~ commercial instinct, [----][e] most highly developed—

[8] Soon after this shop was started I sent a man named Stewart down to Santiago Chili to put ~~in~~up a Central Station for Electric Lighting   Stewart after finishing the station re-turned to NYork with glowing accounts of the Country & an order from Madame Cousino the richest woman in Chili for a complete plant with chandiliers for her palace in the suburbs of Santiago. Stewart gave the order to Bergmann, & the price

was to be for the chandiliers alone $7,000. Stewart having no place to go generally managed to stay around Bergmanns place recounting the emmense wealth of Madame Cousino, and Bergmann kept raising the price of the outfit until[c] Stewart realized that these glowing accounts of wealth was running into money when he kept away & the chandliers went billed for 17 000 cash on bill of Lading, as Bergmann said he wasnt sure Stewarts mind wasnt affected[f] & he wanted to be safe.

[20] When the first Lamp works was started at Menlo Park one of my experiments seemed to show that hot mercury gave a better vacuum in the lamps than cold mercury   I thereupon started to heat ~~the~~ it—soon all the men got salivated & things looked serious but I found that in mirror factories where mercury was used extensively the French govt made the giving of Iodide of potassium compulsory to prevent salivation. I carried out this idea & made every man take a dose every day but there was great opposition, & hot mercury was finally abandoned

AD, NjWOE, Lab., N-09-06-28. [a]Interlined above and canceled. [b]Interlined above. [c]Obscured overwritten text. [d]Illegible. [e]Canceled. [f]"wasnt affected" interlined above and followed by a line of erased, illegible text.

G. MR. EDISON'S NOTES

The following is a transcription of relevant portions of a typescript titled "Mr. Edison's notes in Book No. 2," probably prepared by William Meadowcroft in 1908 or 1909. It had been separated from the Meadowcroft files and put in a general file of anecdotes in the Edison Biographical Collection. Three of its thirty-eight sections pertain to the period covered by this volume; section 19 was previously published in Volume Three, and section 31 in Volume Four.

[19] Soon after, the Page patent, which had been in the patent office for years was finally issued. It covered the use of a magnet contact point and sub-magnet. There was no known way, whereby this patent could be evaded and its possessor would eventually control the use of what is known as the relay and sounder and this was vital to telegraphy. Gould was pounding the W.U. on the exchange, disturbing its railroad contracts and being advised by his lawyers that this patent was of great value, bought it. The moment Mr. Orton heard this,

he sent for me and explained the situation and wanted me to go to work immediately to see if I could'nt evade or discover some other means that could be used in case Gould sustained the patent. It seemed a pretty hard job, because there was no known means of moving a lever at the other end of a telegraph wire except the use of a magnet. I said I would go at it that night. In experimenting some years previously, I discovered a very peculiar phenomenon, and that was that when a piece of metal connected to a battery was rubbed over a moistened piece of chalk, resting on a metal connected to the other pole, that when the current passed the friction was greatly diminished, and when the current was reversed the friction was greatly increased over what it was when no current was passing. Remembering this, I substituted a piece of chalk rotated by a small electric motor for the magnet and connecting a sounder the combination claim of Page was made worthless, a hitherto unknown means was introduced in the electric art. Two or three of these devices were made and tested by the Company's expert. Mr. Orton, after he had me sign the patent and got it in the Patent Office, wanted to settle for it at once. He asked my price. Again I said—make me an offer; again he named $100,000. I accepted providing he would pay it at rate of $6,000 a year for 17 years. This was done, and this with the telegraph money received $12,000. yearly for the period from the W.U. Tel. Co.

[31] At Menlo Park while working on the phonograph, I made an arrangement with Col. Gouroud of London, to make a telephone to be used in starting the industry in England and the Continent. A large number were made and sent to Gouraud, who started to put up a telephone exchange. When he had proceeded a little way, he was threatened with a law suit by the owners of the Bell patent. The Bell Company were starting an exchange themselves. He cabled me that the law suit would prevent him from obtaining further money to carry out the enterprise. He would stop unless I could evade the patent on the Bell instrument. I cabled I thought I could do so and went right to work. I again had recourse to the phenomenon discovered by me years previous that the friction of a rubbing electrode passing over a moist chalk surface had its friction varied by electricity. I devised a telephone receiver which was afterwards known as the loud speaking telephone or chalk receiver. There was no magnet, simply a diaphragm and a small cylinder of compressed chalk, about the size of a thimble, an arm connected to the center of the diaphragm, extended out-

wardly and rested on the chalk cylinder and was pressed against it with a pressure equal to that which would be due to a weight of about 6 lbs. The chalk was rotated by hand. The volume of sound was very great, a person talking into a carbon transmitter in N.Y. had his voice so amplified that he could be heard 1500 feet away in an open field at Menlo Park. This great excess of power was due to the fact that the power came from the person turning the handle, the voice instead of furnishing all the power, as with the present receiver, merely controlled the power, just as an engineer working a valve could control a powerful engine. I made six of these receivers and sent them in charge of an expert on the first steamer. They were received and tested, and shortly afterwards shipped one hundred or more. At the same time I was ordered to send twenty young men after teaching them to become expert. I set up an exchange around the Laboratory of ten instruments. I would then go out and get each one out of order in every conceivable way. A man would be sent to each to find the trouble. When he could find the trouble ten consecutive times, using five minutes each, he was sent to London. About 60 men were used to get 20. Before all had arrived, the Bell Company seeing that we could not be stopped, entered into negotiations for consolidation. One day I received a cable from Gouraud offering $30,000 for my interest. I cabled back that I would accept. When the draft came, I was astonished to find that it was for £30,000. when I thought it was dollars.

[32] The men I sent over were used to establish telephone exchanges all over the continent and some of them became wealthy. It was among this crowd in London that Bernard Shaw was employed before he became famous. The chalk telephone was finally discarded in favor of the Bell receiver, it being more simple and cheaper. Extensive litigation with new comers followed. My carbon transmitter patent was sustained and preserved the monopoly of the telephone in England for many years. Bell's patent was not sustained by the Courts. Sir Richard Webster, now the Lord Chief Justice of England, was my counsel and sustained all of my patents in England for many years. Webster had a marvellous capacity for understanding things scientific and his address before the courts was lucidity itself. His brain is highly organized. My experience of the legal fraternity is that scientific subjects are distasteful and it is rare in this country for a judge to interpret the statements of the experts correctly and inventors scarcely ever get a decision in their favor. In England, the judges seem to be different;

they are not fooled by the experts, but tear their testimony all to pieces and decide the case correctly. Why this difference between Englandish[a] and American judges I cannot explain. It seems to me that scientific disputes should be decided by some court containing at least one or two scientific men, if justice is ever to be given an inventor. Inventors acting as judges would not be very apt to decide a complicated law point, and per contra, it is hard to see how a lawyer can decide a complicated scientific point rightly. Some inventors complain of our patent system and the patent office. I consider both are good and that the trouble is in the Federal Courts. There should be a court of patent appeal with at last two scientific men thereon, who could not be blind to the sophistry of paid experts, men whose inventions would have created wealth to the country of millions have been ruined and prevented from making any money whereby they could continue their career as creators of wealth for the general good, just because the experts befuddled the judge by their misleading statements.

TD (transcript), NjWOE, Meadowcroft. Because this transcription of Edison's manuscript is presented only as a reference text, typographical errors have not been reproduced or noted. [a]"ish" interlined above in pencil.

Appendix 2

Menlo Park Employees, 1879–1880

The following list is derived largely from laboratory Time Sheets, Jehl 1937–41, the Edison Pioneers Biographical File (NjWOE), and references to employees found in notebooks or other documents. However, this list is not complete; it does not include employees who were not mentioned in these sources and there are significant gaps in the run of time sheets. Because many employees were either shifted to the lamp factory or hired to work there during the last two months of 1880 and other employees were shifted to New York in the early months of 1881, the editors have decided to only include the years 1879–1880. Many of those whose first time sheets date from November or December 1880 probably worked in the lamp factory.

Each employee's name is followed, in parentheses, by the known dates of his work at the laboratory. If an employee was still working for Edison at the end of 1880, there is no end date. Specific dates (i.e., month and day) are usually derived from time sheets or from Edison Pioneers material. We have used only the beginning and ending dates of the weekly time sheets. Because there are significant gaps in the time sheets, it is possible that an employee may have started or ended his employment on different dates than those we have indicated.

Following the date there is a brief description of the employee's primary role at the laboratory. For cases in which laboratory records and reminiscences do not provide an adequate description of an employee's work, we have used the enumeration lists from the 1880 Census. Respondents gave census enumerators brief self-defined descriptions of their occupations. Our citation method for the 1880 Census lists two numbers. The first is taken from the column labeled "Dwelling houses numbered in order of visitation" and the second from the column labeled "Families numbered in order of visitation." If we have been unable to determine the employee's primary role, we have indicated it as Unknown.

Following the brief description of the employee's primary role, we summarize the particular kind of work that each did. However, for machinists and a few others (such as pattern maker Milo Andrus) who were involved in a wide variety of experimental work, we have not indicated specific tasks. The last items in each entry are the sources from which we derived information about the employee.

Acheson, Edward G. (12 Sept. 1880–). Experimental assistant. Especially lamp filaments. Time Sheets; Pioneers Bio.; Jehl 1937–41, 548–50; *ANB;* Acheson 1965; Szymanowitz 1971.

Acker, Ben (12 Aug. 1880–). Carpentry and general labor. Time Sheets; 1880 Census 262–83.

Acker, Charles (12–26 Aug. 1880). Unknown. Time sheets indicate shop work and electric light. Time Sheets; 1880 Census 262–83.

Acker, James (6 Feb.–8 Apr. 1880). Laborer. Time sheets indicate building extension and dynamo. Time Sheets; 1880 Census 262–83.

Acker, Smith (1 Dec. 1878–16 Jan. 1879). Unknown. Probably labor on new buildings. Labor Account (*TAEM* 22:785, 788; *TAED* A281A, A281B); 1880 Census 262–83.

Anderson, Thomas (7 May–18 Nov. 1880). Unknown. Time sheets indicate work on electric light, possibly laying underground conductor. Time Sheets; 1880 Census 468–509.

Andersson, Hugo (20 May 1880–). Machinist. Pioneers Bio.; Jehl 1937–41, 546, 685–86.

Andrews, William Symes (4 Dec. 1879–). Machinist. Time Sheets; Pioneers Bio.; Jehl 1937–41, 319.

Andrus, Arthur (30 Sept. 1878–28 Oct. 1880). Hired in 1878 to make carbon buttons for the telephone transmitter. 1880 time sheets show work on electric light experiments. Son of Milo Andrus. Time Sheets; 1880 Census 210–31; *TAEB* 4: 459.

Andrus, Milo P. (7 June 1878–). Carpenter and pattern maker. Time Sheets; 1880 Census 210–31; Jehl 1937–41, 222; Doc. 1974 n. 6.

Applegate, A. A. (13 Aug.–16 Sept. 1880). Unknown. Time sheets indicate electric light and lamp factory. Time Sheets.

Arnold, Elmer (13 Aug.–2 Sept. 1880). Unknown. Time sheets indicate electric light. Time Sheets.

Askins, John (29 July 1880–). Unknown. Time sheets indicate electric light and lamp factory. Time Sheets.

Badgley, John (9 Oct. 1879–29 Apr. 1880). Carpenter. Time sheets show work on electric light and telephone experiments. Time Sheets; 1880 Census 42–46.

Barbar, Thomas (3 Sept. 1880–). Unknown, possibly machinist or machinist's assistant. Time sheets show work on electric experiments and lamp factory. Time Sheets.

Batchelor, Charles (1870–). Edison's principal experimental assistant. Time Sheets; 1880 Census 319–44; *TAEB* 1:495–96.

Bennett, A. (24 Sept.–4 Nov. 1880). Unknown. Possibly laborer, one time sheet indicates lamp factory. Time Sheets.

Bennett, William H. (6–20 Feb. 1880). Mason. Time sheet indicates work on machine shop. Time Sheets; 1880 Census 38–42.

Berggren, Ernest (11 Apr.1880–). Office work and bookkeeping; assistant to William Carman. Edison Pioneer. Time Sheets; Pioneers Bio.; Jehl 1937–41, 546–47, 696.

Bogan, William, Jr. (23 July 1880–). Laborer; lamp factory glassblower apprentice. Time sheets indicate electric light and lamp factory. Jehl claims that Bogan dug trenches for the underground lines in Menlo Park. Time Sheets; 1880 Census 310–35; Jehl 1937–41, 518.

Bogan, William, Sr. (16 July–14 Oct? 1880). Laborer. Time sheets indicate lamp factory. Time Sheets; 1880 Census 310–35.

Böhm, Ludwig (c. 20 Aug. 1879–22 Oct. 1880). German-born glassblower who constructed the vacuum pumps and globes for Edison's incandescent lamp. Time Sheets; Docs. 1786 n. 7 and 2000; Jehl 1937–41, 223, 324–28, 370–71, 495, 516, 612, 708.

Bradley, James J. (26 Dec. 1879–). Machinist. He helped set up the Edison lamp factory at Menlo Park in 1880. Time Sheets; Pioneers Bio.; *TAEB* 3:589 n. 4; Testimony in *Edison v. Maxim v. Swan* (*TAED* W100DIG); Jehl 1937–41, 678–79.

Bradley, John (1–8 Jan 1880). May be John Brawley. Unknown. Time sheet indicates electric light. Time Sheets.

Brawley, John (23 July–26 Aug. 1880). Unknown. Time sheets indicate electric light. Time Sheets.

Breath, William L. (9 Jan.–22 Apr.? 1880). Laboratory assistant. Time sheets indicate electric light. Time Sheets.

Brison, James M. (23–30 Dec. 1880). Unknown. Time sheet indicates electric light, unknown if he continued to work for Edison. Time Sheets.

Britton, Ed (8 Jan.–9 Dec. 1880). Unknown. Time sheets indicate electric light. Time Sheets.

Burk, Martin, Jr. (17–30 Dec. 1880). Laborer. Time Sheets.

Burk, Martin, Sr. (26 Dec. 1879–). Laborer. Time Sheets.

Campbell, Charles (17 Sept. 1880–). Machinist? Time sheets show dynamo construction and tool repair. Time Sheets.

Campbell, Henry Alexander (24 Oct. 1878–). Carpenter; superintendent of construction for the lamp factory. Time Sheets; Pioneers Bio.; 1880 Census 177–91; Jehl 1937–41, 222–23, 686–87.

Campbell, John R. (5–25 Nov. 1880). Unknown. Time sheets indicate electric light and conductor cables. May have been at Menlo Park after 25 November; in 1881 he worked at the Edison Machine Works. Time Sheets; Edison Pioneer Files.

Campbell, William (1 Jan.–27 Feb. 1880). Laborer. Time sheets indicate electric light and building extension. Time Sheets.

Carman, George E. (April 1877–c. 6 April 1880). Purchasing agent and general assistant in laboratory and office. By December 1879 only assisting in laboratory. Brother of William Carman, Edison's bookkeeper. Time Sheets; Carman's testimony, *Edison v. Maxim v. Swan* (*TAED* W100DID [image 4]); *TAEB* 3:534 n. 2; Jehl 1937–41, 284, 318.

Carman, Theodore F. (26 Dec. 1879–4 Mar. 1880). Purchasing agent; also assisted with construction of armatures. Teamster who hauled construction materials and machinery for building the Menlo Park laboratory in 1876 and for the new buildings in fall 1878. Time Sheets; 1880 Census 274–94; Doc. 1562, esp. n. 5.

Carman, William (1876?–). Bookkeeper, accountant, and sometimes secretary. Time Sheets; 1880 Census 333–58; Pioneers Bio.; Doc. 1652 n. 14; Jehl 1937–41, 318, 495, 498.

Clark, Charles P. (3 Sept.–4 Nov. 1880). Unknown. Time sheets indicate lamp factory. Time Sheets.

Clarke, Charles L. (1 Feb. 1880–). Mathematician, mechanical engineer, draftsman, experimental assistant. One of the few employees at Menlo Park with formal engineering training. Time Sheets; Pioneers Bio.; Jehl 1937–41, 402–403, 855–56; *NCAB* 30:44.

Clarke, Samuel A. (24–30 Sept. 1880). Unknown. Time sheets indicate lamp factory. Time Sheets.

Claudius, Hermann (14? Dec. 1880–). Electrical engineer.

Mapping out central station districts; constructing small models of stations and distribution networks. Had formerly worked in the Austrian Imperial Telegraph Department. Time Sheets; Doc. 2028 n. 2; Jehl 1937–41, 732–33.

Conover, Edward (10 Sept.–4 Nov. 1880). Unknown. Time sheets indicate lamp factory. Time Sheets.

Cornell, J. (6 Feb.–8 Apr. 1880). Laborer. Time sheets indicate work on dynamo building. Time Sheets.

Cornish, A. H. (1 Jan.–8 Oct. 1880). Office work. Time Sheets.

Cousins, W. (29 Oct.–25 Nov. 1880). Laborer? Time sheets indicate winding cables for underground conductors. Time Sheets.

Crane, Alonzo (13 Aug.–14 Oct. 1880). Unknown. Time sheets indicate electric light. Time Sheets.

Croley, C. (13 Aug.–9 Sept. 1880). Unknown. Time sheets indicate electric light. Time Sheets.

Crosby, George (Nov. 1879?–1 July 1880). Hired as unpaid laborer although he is identified in the 1880 census as a machinist. Assisted with vacuum pumps, ore separation experiments, and general laboratory work. Time Sheets; 1880 Census 213–35; Doc. 1926 n. 3.

Cunningham, David H. (30 May 1879–). Machinist. Time Sheets; 1880 Census 84–89; Jehl 1937–41, 680.

Daley, Brainard (17 Sept.–18 Nov. 1880). Unknown. Time sheets indicate electric light, including "digging on lines." Time Sheets.

Danbury, M. (5–25 Nov. 1880). General labor. Time Sheets.

Dean, Charles (18 Oct. 1878–). Machinist. Particularly involved in dynamo construction. First superintendent of the Machine Works. Had formerly worked for Edison at his Newark shops. Time Sheets; 1880 Census 211–33; Jehl 1937–41, 25, 52, 220, 319; *TAEB* 4:539.

Dean, George (30 May 1879–). Machinist. Son of Charles Dean. Time Sheets; Jehl 1937–41, 546, 676.

Donegan, Michael (23–29 April 1880). Laborer. Time sheet indicates electric railway. Time Sheets; 1880 Census 173–87.

Dougherty, D. (17–23 Sept. 1880). Unknown. Time sheet indicates electric light. Time Sheets.

Dume, Thomas (6–12 Feb. 1880). Laborer. Time sheets indicate building extension. Time Sheets.

Duncan, William S. (26 Dec. 1879–4 Nov. 1880). Unknown. Time sheets indicate electric light, electric railway, telephone. Time Sheets.

Dunham, Nelson V. W. (6–26 Aug. 1880). Unknown. Time

sheets indicate lamp factory, particularly cutting fibers. Time Sheets.

Edison, Charles P. (June 1878–26 Feb. 1879). Edison's nephew, known as Charley. Experimental assistant. Telephone receiver experiments; sent to England in February 1879 to demonstrate and promote electromotograph telephone. Died in Paris 19 October 1879. Time Sheets; *TAEM-G2*, s.v. "Edison, Charles P."

Egan, James (9–16 Oct. 1879). Machinist. Apparently had a contract for telephones on which a number of other machinists at the laboratory worked. He had worked for Edison at Newark. Name sometimes spelled "Eagan." Time Sheets; Egan to Charles Batchelor; Richard Dyer to Egan, 8 June 1882; both DF (*TAEM* 27:522, 80:594; *TAED* SB1677194, LB007429).

Eldridge, James A. (10 Sept.–28 Oct. 1880). Unknown. Time sheets indicate electric light and lamp factory. Time Sheets.

End, Theodore (11 Nov.–16 Dec.1880). Unknown. Time sheets indicate electric light, filling in cables. Time Sheets.

Erichs, William Eridis (17 Sept.–21 Oct. 1880). Unknown. Time sheets indicate lamp factory. Time Sheets.

Erliche, Robert (2 Aug.–23 Dec. 1880). Machinist? Time sheets indicate electric light, electric railway, tools for lamp factory. Time Sheets.

Fernander, Nels (26 Dec. 1879–11 Mar. 1880). Unknown. Time sheets indicate electric light. Time Sheets.

Fink, W. M. (26 Dec. 1879–29 Jan. 1880). Unknown. Time sheets indicate winding coils for telephone and electric light. Time Sheets.

Flaherty, John (13 Aug.–28 Oct. 1880). Unknown. Time sheets indicate electric light and lamp factory. Time Sheets.

Flammer, Charles (14 June 1878–4 Nov. 1880). Machinist. Time Sheets; 1880 Census 334–60; Jehl 1937–41, 546.

Force, Joseph J. (1 Oct.–4 Nov. 1880). Glassblower. Brother of Martin Force. Time Sheets; 1880 Census 257–77.

Force, Martin (1877–). Laboratory assistant. A carpenter who worked on the construction of the Menlo Park laboratory in winter 1876 and then packing pens and running the steam engine at the electric pen factory in Menlo Park. Sometime in late 1877 or early 1878 he became a general handyman around the laboratory. By June 1878 he was generally assisting with experiments. Time Sheets; 1880 Census 257–78; *TAEB* 3:534 n. 3; Jehl 1937–41, 278–79.

Freeman, Erastus (26 Dec. 1879–18 Nov. 1880). Carpenter.

Time sheets indicate electric light. Time Sheets; 1880 Census 199–219.

Freeman, Fred (6 Feb.–12 Dec. 1880). Laborer. Time sheets indicate building extension. Time Sheets.

Gallagher, J. (24 Sept.–5 Nov. 1880). Laborer. Time sheets indicate insulating cables and winding wires. Time Sheets.

Gavin, H. (24 Sept.–7 Oct. 1880). Unknown. Time sheets indicate electric light. Time Sheets.

Gibbon, William (20–26 Aug. 1880). Unknown. Time sheet indicates electric light. Time Sheets.

Griffin, Michael (15 May–23 Sept. 1880). General labor. Time sheets indicate electric light, electric railroad, lamp factory. Time Sheets; 1880 Census 190–206.

Griffin, Stockton L. (c. 1 June 1878–c. 28 Feb. 1881). Personal secretary. Former telegraph operator who had worked with Edison in Cincinnati in 1867. Time Sheets; *TAEB* 1:297 n. 1, 276 n. 1; Jehl 1937–41, 35, 696.

Gunn, James (6–26 Feb. 1880). Laborer. Time sheets indicate dynamo and battery. Time Sheets; 1880 Census 430–71.

Habberfield, Charles (29 Oct.–4 Nov. 1880). Unknown. Time sheets indicate lamp factory. Time Sheets.

Hahn, Fred (13 Aug.–4 Nov. 1880). Unknown. Time sheets indicate lamp factory. Time Sheets.

Haid, Dr. Alfred (June 1879–). Chemist. First Ph.D. chemist hired by Edison; he was nicknamed "Doc." Time Sheets; Jehl 1937–41, 262–63.

Hall, Louis (13 Aug.–4 Nov. 1880). Unknown. Time sheets indicate lamp factory. Time Sheets.

Hammer, William Joseph (Dec. 1879–). Laboratory assistant. Primarily assisted with lamp experiments. Became chief electrician of the Menlo Park lamp factory. Time Sheets; Pioneers Bio.; Jehl 1937–41, 405–8; Doc. 1972 n. 7; *ANB*.

Hankins, Mathew (13 Feb. 1879–?). General labor; glassblowing. Among other things he helped his father, who ran a dairy in Uniontown (now Iselin) to grade the bed for the electric railroad. Became first apprentice glassblower at lamp factory. Time Sheets; Pioneers Bio.; Jehl 1937–41, 517–18, 812.

Hechtel[?], Albert (3 Sept. 1880–?). Unknown. Time Sheets.

Hense, Rudolph (2 Oct. 1878–5 June 1879). Machinist. Time Sheets; *TAEB* 4:595 n. 3.

Herman, Carl (30 May–5 June 1879). Unknown. Time sheets indicate electric light. Time Sheets.

Herrick, Albert B. (28 Aug. 1879–Jan. 1881). Experimental assistant. Primarily involved in lamp experiments in the laboratory and the lamp factory. Left to attend Stevens Institute of Technology. Time Sheets; Pioneers Bio.; Doc. 1972 n. 5; Jehl 1937–41, 408, 544, 661.

Hickman, David K. (mid-July 1880–). General laboratory assistant. Became foreman of the pump department at the Menlo Park lamp factory. 1880 Census 573–613; Pioneers Bio.; Jehl 1937–41, 518.

Hickman, George (11 Nov. 1880–). Census identifies as tinsmith. Time sheets indicate soldering, winding armatures, and electric light. Time Sheets; 1880 Census 213–35.

Hickman, John Parker (20 Sept. 1880–). Laboratory assistant and lamp factory employee? He worked on sealing copper and platinum wires for inside parts and copper clamps. Pioneers Bio.

Hill, George (Fall 1879–). Vacuum pump assistant to Francis Jehl. Jehl 1937–41, 369.

Hipple, James (Spring 1880?–). Listed in census as a messenger residing at the Sarah Jordan boarding house. Became William Holzer's assistant and learned glassblowing. 1880 Census 213–35; Jehl 1937–41, 497, 632, 716.

Holcombe, William (11 Nov. 1880–). Unknown. Time sheets indicate electric light, especially dynamos. Time Sheets.

Holloway, James (1879–). Machinist. Time Sheets; Pioneers Bio.; Jehl 1937–41, 318, 682–83.

Holzer, Frank (3 Oct. 1880–). Glassblower. Lamp factory. Brother of William Holzer. Time Sheets; Pioneers Bio.

Holzer, William (16 Jan. 1880–). Glassblower. Later superintendent of Menlo Park lamp factory glass department. Married Alice Stilwell (Mary Stilwell Edison's sister). Time Sheets; 1880 Census 213–35; Jehl 1937–41, 253, 495–96, 513–14, 631, 716, 803, 812.

Hood, John (8 June 1878–). Machinist and mechanic. Time Sheets. Jehl 1937–41, 320, 682.

Horn, John (11 Nov.–23 Dec. 1880). Laborer? Time sheet indicates winding cable. Time Sheets.

Hornig, Julius L. (4 Jan. 1880–). Mechanical engineer. Central station design and electric railroad. Time Sheets; Hornig's testimony, 3, 7, *Edison v. Siemens v. Field*, Lit. (*TAEM* 46:6, 8; *TAED* QD001:2, 4); Jehl 1937–41, 546, 577.

Houlding, T. (24 Sept. 1880–?). Unknown. Time Sheets.

Howell, Thomas H. (17 Nov. 1880–). Unknown. Time sheet indicates electric light and lamp factory. Brother of Wilson Howell. Time Sheets.

Howell, Wilson S. (Dec. 1879–). Experimental assistant.

Worked initially on exhausting lamps with the vacuum pumps then given primary responsibility for developing insulation for underground cables. Pioneers Bio.; Doc. 1985 n. 1; Jehl 1937–41, 404–6, 72–26.

Hughes, Charles T. (21 Oct. 1879–). Purchasing agent; experimental assistant with telephone receiver and autographic telegraph. Worked with Patrick Kenney on the autographic telegraph, worked on electric railroad projects, and served as agent for the Edison Electric Light Co. Time Sheets; 1880 Census 213–35; Jehl 1937–41, 546, 885.

Isaac, Earnest (26 Feb. 1880–?). Unknown. Time Sheets.

Isaac, M. (1880). Photographer. Jehl 1937–41, 497.

Jackson, George (15 June 1878–5 June 1879). Machinist. Time Sheets; *TAEB* 4:414 n. 10.

Jehl, Francis (3 Mar. 1879–). Laboratory assistant. Vacuum pump design and operation, electric lamp tests. Time Sheets; 1880 Census 213–35; Doc. 1685 n. 2; Pioneers Bio.; Jehl 1937–41.

Jenkins, John F. (11 Nov.–12 Dec. 1880). Unknown. Time sheets indicate boiler room, electric light. Time Sheets.

Johnson, Edward (Sept. 1880–). Long-time Edison associate who worked on electric light experiments at Menlo Park laboratory beginning in September 1880. Doc. 1988 n. 1.

Kelly, John (1880). Machinist. Jehl 1937–41, 546.

Kenly, John F. (30 May–5 June 1879). Unknown. Time Sheets.

Kenny, Patrick (July? 1878–). Experimenter. He worked with Edison on autographic telegraph. Time Sheets; *TAEB* 1: 622 n. 21; 4:284 n. 3; Jehl 1937–41, 39, 263.

Knight, John K. (4–27 Sept.1878; 1 Nov. 1878–10 Jan. 1879). Laboratory assistant. He tested generators, made carbon buttons and did other electric light work. He was discharged in January. Time Sheets; *TAEB* 4:708 n. 2.

Knight, Joseph H. (30 Nov. 1878–20 Feb. 1879). Laborer. Made carbon buttons. Time Sheets; 1880 Census 32–35.

Kruesi, John (1871 or 1872–). Foreman of machine shop. Swiss-born machinist who constructed much of Edison's experimental apparatus at Newark and then became foreman of the machine shop at the Menlo Park laboratory. In 1881 he headed the Electric Tube Co., which manufactured the underground electric tubes for the Pearl Street central station. Time Sheets; *TAEB* 2:633 n. 6.

Lawson, John (Jan. 1879–). Chemist. Known as "Basic." Time Sheets; 1880 Census 213–35; Pioneers Bio.; Jehl 1937–41, 262–63.

Lee, Mark (11 Nov.–16 Dec. 1880). Night watchman. Time Sheets; 1880 Census 197–214.

Lewis, Richard (16 Apr. 1880–?). Unknown. Time Sheets.

Lighthipe, James (1879). Claims to have been at Menlo Park working on motograph telephone experiments before being sent to England as a telephone expert on 30 Nov. 1879. Pioneers Bio.

Livingston, Francis (30 Jan. 1880–?). Toolmaker. Time Sheets; Doc. 1914 n. 7.

Loforge, Charles E. (26 Nov.–16 Dec. 1880). Carpenter. Time Sheets; 1880 Census 164–71.

Logan, Thomas (1877 or 8 June 1878–). Machinist; became foreman of the machine shop after John Kruesi took over management of the Electric Tube Co. in 1881. Time Sheets; 1880 Census 582–623; *TAEB* 4:414 n. 11; Jehl 1937–41, 680.

Lomas, George (6 Aug. 1880–?). Unknown. Time Sheets.

Lomas, William (13 Aug. 1880–?). Unknown. Time Sheets.

Mack, Edward (21 Dec. 1878–16 Oct. 1879). General labor. Time Sheets.

Macoy, John (13 Aug. 1880–?). Unknown. Time Sheets.

Malone, Felix (23 June 1880–?). Unknown. Time Sheets.

Martig, G. (11–18 Nov. 1880). Unknown. Time sheets indicate electric light work on Johnson's experimental screws for clamps for electric lamps. Time Sheets.

Martin, John (11 Nov. 1880–). Unknown. Time sheets indicate electric light, especially direct converter. Time Sheets.

McCollin, John (17 Dec. 1880–?). Unknown. Time sheets indicate direct converter. Time Sheets.

McCombe, Thomas (11 Nov. 1880–). General labor. Time sheets indicate electric light. Time Sheets.

McIntire, Henry M. (15 Nov. 1878–20 Feb. 1879). Chemist. He was the first college-educated chemist at the laboratory. Time Sheets; *TAEB* 4:661.

McKine, David (11 Nov.–16 Dec. 1880). Unknown. Time sheets indicate general labor on electric light and lamp factory. Time Sheets.

McLeollium[?], John (10–16 Dec. 1880). Unknown. Time Sheets.

McMore[?], C. (24 Dec. 1880–). Unknown. Time sheets indicate electric light. Time Sheets.

Merrill, Frank M. (1880). Engineer. Edison's letter of recommendation from Edison dated October 1880, DF (*TAEM* 53:521; *TAED* D8014ZAE).

Millheels, Albert (15 Oct. 1880–?). Unknown. Time Sheets.

Miller, William H. (17 Sept. 1880–?). Unknown. Time Sheets.

Mills, William Albert (30 Apr. 1880–?). Laboratory assistant. Entry in notebook dated 27 Nov. 1880. N-80-09-28:189, Lab. (*TAEM* 36:535; *TAED* N106:189). Time Sheets.

Mitchell, Thomas (11 Nov. 1880–). Unknown. Time sheets indicate lamp factory. Time Sheets.

Moffatt, Benjamin Franklin (1880). General labor; apprentice glassblower at lamp factory. Son of Charles W. Moffatt. 1880 Census 98–215; Jehl 1937–41, 517–18, 688, 812.

Moffatt, Charles William (27 Sept. 1878–5 June 1879?). Carpenter and pattern maker. His name appears as Moffat in the census, as Morfitt on time sheets, and as Moffett in Jehl 1937–41. Time Sheets; 1880 Census 98–215; *TAEB* 4:595 n. 2; Jehl 1937–41, 687–88.

Moore, Bery (11 Nov. 1880–). General labor. Time Sheets.

Moore, Mack (25 Nov.–23 Dec. 1880). Unknown. Time sheets indicate electric light and lamp factory. Time Sheets.

Moore, Walter (11 Nov. 1880–?). Machinist. Possibly W. H. Moore whose name appears in time sheets from 11 Nov. to 30 Dec. 1880. Time Sheets; 1880 Census 312–37.

Moses, Dr. Otto (c. June 1879–). Chemist. Second Ph.D. chemist hired by Edison. Also acted as librarian and translator as he knew German and French. Time Sheets; 1880 Census 99–104; Doc. 1754; Jehl 1937–41, 496–97, 535.

Mott, Charles P. (Jan. 1880–). Office worker and draftsman. Kept daily journal of activities at the laboratory beginning on 14 Mar. 1880. Brother of Samuel D. Mott. Time Sheets; Docs. 1890 n. 13 and 1914; Jehl 1937–41, 546.

Mott, Samuel Dimmock (Sept. 1879–). Draftsman. Brother of Charles P. Mott. Time Sheets; 1880 Census 213–35; Doc. 1985 n. 7; Pioneers Bio.; Jehl 1937–41, 318, 546, 664, 696.

Mumsell, Dr. C. E. (Mar.–Apr. 1879). Chemist. Ph.D. See Chapter 1 introduction.

Mungle, Alex (1879–). Machinist. Time Sheets; Pioneers Bio.; Jehl 1937–41, 318, 685.

Murphy, M. (11 Nov.–9 Dec. 1880). Laboratory assistant. Time sheets indicate electric light. Time Sheets.

Nichols, Dr. Edward Leamington (Oct. 1880–June 1881). Experimenter. In particular he worked on the chemical balance meter and the electric dynamometer. Organized lamp testing department at Menlo Park lamp factory. Time Sheets; Pioneers Bio.; Jehl 1937–41, 552–53; Doc. 2065 n. 7; *ANB*.

O'Connell, William (11–25 Nov. 1880). Unknown. Time sheets indicate electric light. Time Sheets.

Olsson, Alfred (30 May–5 June 1879). Unknown. Time sheets indicate telephone experiments. Time Sheets.

Ott, John (c. 5 May 1875–). Machinist; became experimenter. Time Sheets; Pioneers Bio.; *TAEB* 2:560 n. 1, 4:276 n. 1; Jehl 1937–41, 679.

Poinier, Arthur O. (9 Oct.–Dec. 1879). Experimental assistant. Time sheets indicate telephone exepriments; notebook entries show work on carbon filament lamps. Time Sheets; Doc. 1850.

Rabone, Walter (11 Nov. 1880–). Time sheets indicate electric light. Time Sheets.

Raettig, Charles (11 Nov. 1880–). Draftsman. Time Sheets.

Randolph, John (8 Nov. 1878–). Office boy and general labor around the laboratory. Known as "Johnny," he later became Edison's bookkeeper and secretary. Time Sheets; 1880 Census 182–96; Randolph's testimony in *New York Phonograph Co. v. National Phonograph Co.* and *Edison v. American Mutoscope Co. & Keith*, Lit. (*TAEM* 116:96, 117:715; *TAED* QMoo1159, QPo100159); Obituary in *Newark Advertiser*, 17 Feb. 1908; Jehl 1937–41, 696.

Randolph, Joseph (11–25 Nov. 1878). Made lampblack buttons. Time Sheets; 1880 Census 182–96.

Rowland, Edward C. (Jan. or Feb.? 1880). Draftsman. His name appears on the drawings for Edison's patent application Case 202, which was filed on 5 Feb. 1880, PS (*TAEM* 45:435; *TAED* PTo10AAA); Jehl 1937–41, 546–47, 696.

Roxberry, John (9 Oct. 1879–). Machinist? Worked on Egan telephone contract and on direct converter. Time Sheets.

Roxbury, A. J. (30 Nov. 1878–16 Oct. 1879). Hired in November 1878 to assist with painting of the machine shop. Became a shop assistant in December 1878. Time Sheets; *TAEB* 4:749 n. 8.

Rule, William, Sr. (11 Nov. 1880–). Possibly machine shop assistant. Time sheets indicate machine shop and fire room. Time Sheets.

Russell, James (Summer 1880–). Had charge of canvassing of Pearl Street central station district. Time Sheets; Doc. 1995 nn. 1–2.

Sculley, Harry (2–9 Dec. 1880). Laborer. Time sheets indicate "running wire." Time Sheets.

Sculley, Henry (2 Dec. 1880–). Time sheets indicate electric light. Time Sheets.

Segredor, John (30 May 1879–March 1880; Summer 1880). Experimental assistant. He worked on a telephone order after coming to the laboratory and later on electric light. Left in March 1880 and then returned that summer before leaving again in September to obtain vegetable fibers in Georgia, Florida, and Cuba. Died in Havana, Cuba, on 27 October 1880. Time Sheets; Doc. 1972 n. 4.

Seymour, James M. Jr. (30 May 1879–) Laboratory assistant. Time sheets indicate he worked on telephones, including the Egan contract, and electric light. May have been at the laboratory as early as 1878. Time Sheets; Pioneers Bio.; Doc. 1855 n. 9.

Sheeley, M. J. (11 Nov. 1880–). Unknown. Time sheets indicate electric light and lamp factory. Time Sheets.

Smith, C. F. (11 Nov. 1880–). Unknown. Laboratory time sheets indicate direct converter. Time Sheets.

Smith, John (11 Nov.–16 Dec. 1880). General labor. Time Sheets.

Swanson, Alfrid (Dec. 1876–). Served as night watchman, ran steam engine, and made carbon buttons. Time Sheets; *TAEB* 4:415 n. 13; Jehl 1937–41, 128, 131.

Table, A. (11 Nov.–16 Dec. 1880). Laborer. Time sheets indicate electric light, lamp factory, and house, presumably meaning Edison's house. Identified as [Abbin?] M. in census. Time Sheets; 1880 Census 593–635.

Thury, Rene (1880?). Laboratory assistant. Jehl 1937–41, 553, 864.

Tilton, E. C. (11 Nov. 1880–). General labor? Time Sheets.

Tweed, Louis (11 Nov. 1880–). Unknown. Time sheets indicate electric light. Time Sheets; 1880 Census 113–18.

Upton, Francis (c. 13 Dec. 1878–). Mathematician and experimenter. A key assistant on electric light research. Nicknamed "Culture," he was first hired by Edison in November 1878 to review all scientific literature related to electric lighting. Time Sheets; 1880 Census 261–82; *TAEB* 4:702 n. 1, 767 n. 5; Pioneers Bio.

Vail, John H. (July 1880–). Machinist. Worked largely on dynamos; in charge of running dynamos in Menlo Park central station. Time Sheets; Pioneers Bio.; Jehl 1937–41, 546–48, 884–85.

Van Cleve, Cornelius (June 1880–?). Laboratory assistant. Laboratory notes show that he kept record of carbon experiments from June and July 1880. Known as "Neal," he was the husband of Hattie Van Cleve, the half sister of Mary

Stilwell Edison. N-80-06-02, Lab. (*TAEM* 36:397; *TAED* N105); 1880 Census 581–622; *TAEB* 2:769.

Vansickle, A. (11 Nov. 1880–). Unknown. Time sheets indicate electric light and lamp factory. Time Sheets.

Voorhees, Benjamin (9 Oct. 1879–25 Nov. 1880). Machinist. Time Sheets; 1880 Census 27–30.

Weiseman, O. (13–20 Feb. 1879). Laborer? Time Sheets.

Welsh, Alexander. (Fall 1880–). Lamp factory. Exhausting lamps, later in charge of carbonizing. Upton's testimony, 3255–56, *Edison Electric Light Co. v. U.S. Electric Lighting Co.*, Lit. (*TAEM* 48:134–35; *TAED* QD012F3254).

Weniger, John (11–25 Nov. 1880). Laborer. Time sheets indicate winding cable, moving engine. Time Sheets; 1880 Census 206–328.

Westerdahl, Anton R. (1880–). Machinist. Brother of Axel Westerdahl. Time Sheets; Pioneers Bio.; Jehl 1937–41, 546, 686.

Westerdahl, Axel K. (1880–). Machinist. Brother of Anton Westerdahl. Pioneers Bio.

Wheeler, Moses (11 Nov. 1880–). Coachman. His wife is listed in the census as a child's nurse, possibly for the Edisons. Time Sheets; 1880 Census 318–43.

Williams, Arthur (13–20 Feb. 1879). Unknown. Time sheets indicate electric light. Time Sheets; Pioneers Bio.

Wood, Samuel L. (10–16 Dec. 1880). Carpenter. Time Sheets; 1880 Census 331–56.

Wright, Frank (11 Nov. 1880–). Identified in census as apprentice, probably in machine shop. Time sheets indicate direct converter and planing dies. Time Sheets; 1880 Census 211–33.

Wright, William (30 May 1879–). Machinist. Time Sheets; 1880 Census 211–33.

Wurth, Charles (5 June 1879–). Machinist. Had worked in Newark shops and then in Menlo Park until March 1877. Pioneers Bio.; *TAEB* 2:519 n. 2.

Appendix 3

Edison Lamps (1879–1881) at the Henry Ford Museum & Greenfield Village

The following list identifies lamps made at Menlo Park during the period of this volume which are presently located at the Henry Ford Museum & Greenfield Village in Dearborn, Michigan. The information in the list is derived from the museum's catalog. Lamps clearly made after March 1881 have not been included. Within each year, lamps are listed in ascending order by that institution's inventory number.

The editors have appended this list to Volume Five for several reasons. It provides a reference for the fecundity of materials, styles, and manufacturing techniques embodied in Edison's lamps of this period. Though by no means exhaustive in this respect, the list represents this variety more fully than the documents and annotation of the volume.

The historical importance of the lamp collection and its intimate connection with the work at Menlo Park also justify inclusion of this list. William Hammer, one of Edison's assistants, began preserving lamps that he found representative or particularly significant; Edison himself reportedly contributed several examples. Hammer eventually assembled more than eight hundred items from dozens of manufacturers. He meticulously organized the collection and in 1904 exhibited it at the World's Fair in St. Louis, where it was acknowledged as a comprehensive representation of the history of incandescent electric lighting. When the collection was broken up most of the lamps went to what is now the Henry Ford Museum & Greenfield Village. Much of the descriptive information for the items below is adapted from captions written by Hammer, but because his specific claims about various firsts cannot be authenticated they are not repeated here. Additional information about the contents and history of Hammer's collection may be found at a website maintained by Edward J. Covington: http://www.frognet.net/~ejcov/hammer.html, which also contains a reproduction of Hammer's published article about the collection. A number of photographs taken in connection with its exhibition, several with captions legible in the display cases, are in box 88, ser. 4, WJH.

1879

29.1980.529.86. 16 candlepower lamp with brass binding posts. One of the paper horseshoe carbon filament lamps made for the December 1879 demonstration at Menlo Park. Dimensions: length, 7.625 inches; width, 3.75 inches.

29.1980.529.88. 16 candlepower lamp with paper horseshoe carbon, plain tip, round seal, platinum vise clamps, small round bulb.

1879–1880

00.4.6099. 16 candlepower lamp with bristol board filament, screw clamps and push-in base. Dimensions: length, 6.0 inches; diameter, 2.5 inches.

1880

29.1980.529.10. 8 candlepower lamp with madake bamboo filament, platinum vise clamps, flat seal, screw and ring, and wooden base.

29.1980.529.18. Tubular lamp with bast filament, bent platinum vise clamps, round inside seal.

29.1980.529.53. Reportedly one of the first madake filament lamps made at Menlo Park.

29.1980.529.60. 16 candlepower lamp which burned 1589 hours.

29.1980.529.64. Long, bulbous form of lamp globe, with madake bamboo filament; nickel vise clamps; depressed seal, and glass drawn around wire platinum lead wires.

29.1960.529.65. 32 candlepower lamp with bamboo carbon filament, and nickel clamps; made for the 1880 Menlo Park demonstration. Dimensions: length, 9.75 inches; width, 3.188 inches.

29.1980.529.68. Ground glass lamp, reportedly taken from first lot dipped in hydrofluoric acid by John Lawson.

29.1980.529.80. 16 candlepower lamp with nickel vise clamps, pressed seal, bamboo filament, and round tip.

29.1980.529.83. Lamp with double filament made of manila hemp, platinum vise clamps, round inside seal, and supplemental tip.

29.1980.529.85. 16 candlepower lamp with bamboo filament, platinum clamps, round inside seal, egg-shaped globe, and supplemental tip.

29.1980.529.99. Lamp with South American bast fiber filament, platinum vise clamps at right angles, round seal, small bulb, and supplemental tip.

29.1980.529.110. Tubular lamp with small bamboo filament, short right angle platinum vise clamp, and round inside seal.

29.1980.529.111. Lamp with paper hairpin filament treated with hydrocarbon vapors, spring and bolt and nut clamps, and heat radiating attachments.

29.1960.529.112. Lamp with copper plated clamps, bamboo hairpin carbon filament, pressed inside seal, and round supplemental globe tip.

29.1980.529.113. Lamp with paper horseshoe carbon filament, platina vise clamps, round inside seal, round bulb, and long supplemental tip.

29.1980.529.114. 16 candlepower lamp with paper horseshoe carbon filament, platinum vise clamps, large round bulb, round inside seal, supplemental tip, and wooden base.

29.1980.529.115. 16 candlepower lamp with carbonized coconut shell clamps, large round bulb, round inside seal, and supplemental tip.

29.1980.529.116. Lamp with South American bast fiber filament, platinum vise clamps, round inside seal, round globe; tested on vacuum pump but tip not sealed off.

29.1980.529.117. Lamp with amaranth vegetable fiber filament, platina vise clamps, solid tip seal, and round inside seal.

29.1980.529.119. Lamp with monkey bast fiber filament treated with hydrocarbon vapor, small globe, round inside seal, and platina vise clamps.

29.1980.529.120. Tubular lamp with South American bast fiber filament, silver vise clamps, and round inside seal.

29.1980.529.121. 8 candlepower lamp with madake bamboo filament, platina vise clamps, flat inside seal, round tip seal, and wooden screw base.

29.1980.529.122. Lamp made for Menlo Park street lighting, with paper horseshoe filament, large globe, supplemental tip, round inside seal, and platina vise clamps.

29.1980.529.123. Lamp with South American fiber filament,

experimental small horizontal platina vise clamps, round inside seal, and small globe.

29.1980.529.124. 16 candlepower lamp with paper horseshoe filament, platina vise clamps, round inside seal, and large round globe.

29.1980.529.125. Lamp with paper horseshoe filament, platinum vise clamps, round inside seal, large round globe, and supplemental tip.

29.1980.529.126. Lamp with manila hemp filament, platina vise clamps, long lead wires, round inside seal, large bulb, supplemental tip sealed.

29.1980.529.127. Lamp with paper horseshoe filament, supplemental tip, and ground glass globe.

29.1980.529.128. Lamp with paper carbon filament treated with naphthalene hydrocarbon vapor, and rubber stopper luted with compound.

29.1980.529.129. Lamp with removable filament, the inside part carrying filament and neck of globe ground to fit and luted with compound.

29.1980.529.130. Experimental arc lamp in exhausted globe with rubber stopper luted with compound.

29.1980.529.132. 8 candlepower lamp with carbonized bamboo filament, platina vise clamps, pressed inside seal, and wooden screw base.

29.1980.529.133. Lamp with bast fiber filament, platina vise clamps, round inside seal, small bulb, and supplemental tip.

29.1980.529.134. Lamp with South American fiber filament, special small platinum vise clamps set at right angles, round globe, round inside seal.

29.1980.529.136. 8 candlepower lamp with carbonized bamboo filament, platina vise clamps, pressed inside seal, and wooden screw base.

29.1980.529.139. Lamp with carbonized manila hemp filament, platina vise clamps, round inside seal, small globe, and supplemental tip.

29.1980.529.143. 8 candlepower lamp with copper plated shank, bamboo carbon filament, platinum vise clamps, pressed seal, and wood butt.

29.1980.529.145. 16 candlepower lamp with paper horseshoe filament, platinum vise clamps, round inside seal, round bulb, and supplemental tip.

29.1980.529.146. Lamp with round Japanese paper filament, platinum vise clamps, small round globe, white German glass over round inside seal, and supplemental tip.

29.1980.529.147. Lamp with pressed inside seal, bamboo filament, bulbous form of globe, and platinum vice clamps.

29.1980.529.148. 16 candlepower lamp with carbonized bamboo filament, platina clamps, pressed inside seal, round tip, and wooden base.

29.1980.529.149. 16 candlepower lamp with madake bamboo filament, nickel vise clamps, bulbous form of globe, pressed seal, and round tip.

29.1980.529.150. 16 candlepower lamp with madake bamboo hairpin filament, platinum vise clamps, pear-shaped globe, pressed inside seal, and round supplemental tip.

29.1980.529.151. Experimental tube used to study electrical carrying.

29.1980.529.152. 16 candlepower lamp with madake bamboo filament, platinum vise clamps, and support made of glass–covered lead wires.

29.1980.529.153. 16 candlepower lamp with nickel vise clamps, egg–shaped globe, depressed seal, glass around lead wires, and supplemental tip.

29.1980.529.154. Lamp with paper horseshoe filament, zig-zag platinum wire heat radiators attached to platinum vise clamps.

29.1980.529.162. Lamp with spring clip and washer clamp with heat radiating metal blocks.

29.1980.529.181. 8 candlepower lamp with bamboo filament, copper plated clamps, flat inside seal, round tip, and wooden base with brass screw and collar.

29.1980.529.196. Paper horseshoe filament lamp used as the headlight on Edison's electric locomotive at Menlo Park in 1880.

29.1980.529.974. Reflector incandescent lamp.

1881

29.1980.529.20. Lamp with bamboo carbon spiral, copper plated clamps, flat rim collar, screw base, and pressed seal.

29.1980.529.22. 16 candlepower lamp.

29.1980.529.25. Small bulb with platinum iridium spiral, flat platinum clamps, and flat seal.

29.1980.529.34. Standard photometer lamp with silver clamps.

29.1980.529.42. 16 candlepower lamp with madake bamboo carbon filament, copper plated as an experiment in early 1881.

29.1980.529.51. 16 candlepower lamp with copper plated

clamps, bamboo filament, bulbous form of globe, pressed seal, round tip.

29.1980.629.54. 16 candlepower lamp with madake bamboo filament and copper plated clamps.

29.1980.529.55. 8 candlepower lamp with bamboo filament, flat seal, and silver plated clamps.

29.1980.529.67. Lamp with graphite filament punched from sheet of plumbago by E. G. Acheson.

29.1980.529.75. Incandescent lamp with "Edison." etched on bulb.

29.1980.529.77. Opal lamp constructed by William Holzer.

29.1980.529.118. 8 candlepower lamp with bamboo filament, copper plated clamps, flat inside seal, flat collar, plaster base, painted with shellac and paint.

29.1980.529.131. 16 candlepower lamp with bamboo filament, copper plated clamps, and flat inside seal.

29.1980.529.135. 16 candlepower lamp with madake bamboo filament, plated copper clamps, flat collar, and plaster butt.

29.1980.529.138. 16 candlepower lamp with bamboo hairpin filament and copper plated clamps.

29.1980.529.140. 16 candlepower lamp with madake bamboo filament, plated copper clamps, pressed seal, small beveled collar, and ring and screw contacts.

29.1980.529.141. 16 candlepower lamp with bamboo carbon filament, copper plated clamps, pear-shaped globe, and round supplemental tip.

29.1980.529.142. 16 candlepower lamp with madake bamboo filament, copper plated clamps, pressed seal, flat collar, and plaster screw base.

29.1980.529.155. Lamp with bamboo spiral carbon filament, copper plated clamps, and flat inside seal.

29.1980.529.156. Lamp with bamboo spiral carbon filament, copper plated clamps, and flat inside seal.

29.1980.529.158. 8 candlepower lamp with bamboo filament, copper plated clamps, and pressed inside seal.

29.1980.529.160. Lamp with graphite filament punched from sheet of plumbago by E. G. Acheson.

29.1980.529.161. Lamp with graphite filament punched from sheet of plumbago by E. G. Acheson; copper plated clamps and flat inside seal.

29.1980.529.163. Lamp with flat seal, inside stem showing spiral clamp for attaching vegetable fiber filament, and platinum lead wires.

29.1980.529.165. Experimental lamp with tubular inside part drawn up between the bamboo legs, and platina vise clamps.

29.1980.529.166. B-type (8-candlepower lamp) with bamboo filament, plated copper clamps, pressed seal, and round tip on bulbous globe.

29.1980.529.167. Lamp with bamboo spiral, copper plated clamps, supplemental copper heat radiating lugs, round seal, and small round globe. Marked "bamboo spiral lamp, Jan 1, 1881, first set".

29.1980.529.168. Lamp with graphite filament punched from sheet of plumbago by E. G. Acheson.

29.1980.529.169. 16 candlepower lamp with bamboo filament and silver vise clamps.

29.1980.529.172. Lamp with graphite filament punched from sheet of plumbago by E. G. Acheson.

29.1980.529.174. 8 candlepower lamp, with bamboo hairpin carbon filaments, plated copper clamps, pressed seal, bulbous globe, round supplemental tip.

29.1980.529.176. Lamp in which "phantom shadow" of electrical carrying is apparent.

29.1980.529.178. 16 candlepower lamp with bamboo hairpin filament, platinum vise clamps, and beveled inside seal.

29.1980.529.180. 16 candlepower lamp with madake bamboo filament, copper-plated clamps, flat collar, and plaster base.

29.1980.529.182. 16 candlepower lamp with madake bamboo hairpin filament, copper plated clamps, pear-shaped globe, pressed seal, and supplemental round globe tip.

29.1980.529.183. 16 candlepower lamp with bamboo filament, copper clamps, and flat seal.

29.1980.529.184. Madake bamboo filament with tinned copper clamps and pressed inside seal.

29.1980.529.185. Lamp with graphite filament punched from sheet of plumbago by E. G. Acheson.

29.1980.529.187. Lamp with madake bamboo filament, copper plated clamps, hand blown bulb, and pressed seal.

29.1980.529.188. Lamp with spiral bamboo filament, plated copper clamps, pressed inside seal, screw base, and flat plaster collar.

29.1980.529.189. 16 candlepower lamp with bamboo filament, copper plated clamps, flat seal, and plaster base with flat collar.

29.1980.529.190. Lamp with two 8-candlepower bamboo filaments in multiple copper plated clamps, pressed seal, and beveled plaster base.

29.1980.529.191. 16 candlepower lamp in wood base with beveled contact ring and screw bottom.

29.1980.529.192. 8 candlepower lamp with bamboo filament, copper plated clamps, flat seal, and wood screw base.

29.1980.529.203. 16 candlepower lamp with bamboo filament and silver vise clamps. Marked "63 E. Silvered by electricity Jan 14, 1881, Menlo Park."

29.1980.529.209. 100 candlepower lamp in bottle shape, with carbonized madake bamboo filament and multiple lead wires.

Appendix 4

Edison's U.S. Patents, January 1879–
March 1881

The following list contains all patents for which Edison executed an application in the period covered by Volume Five. It is arranged in chronological order by execution date, which is the date on which Edison signed the application and the date in the patenting process that comes closest to the time of actual inventive activity. The application date is the date on which the U.S. Patent Office received and recorded the application. The case numbering system, which Edison used throughout his career, seems to have originated in Lemuel Serrell's office as a means of ordering Edison's applications as they arrived. The full Patent Office files containing each original application, any amendments, and related correspondence are at MdNACP. Edison's U.S. patents from this period are found at http://edison.rutgers.edu/patents.htm and in *TAEM* 1–2; British patents are in Cat. 1321, Batchelor (*TAEM* 92:78–117; *TAED* MBP).

	Exec. Date	*Appl. Date*	*Issue Date*	*Pat. No.*	*Case No.*	*Title*
157.	01/23/79	02/03/79	02/10/80	224,329	170	Electric–Lighting Apparatus
158.	01/28/79	02/03/79	05/04/80	227,228	169	Electric Light
159.	02/06/79	02/10/79	05/04/80	227,227	171	Electric Light
160.	03/10/79	03/17/79	02/17/80	224,665	173	Autographic Stencils for Printing
161.	03/19/79	03/29/79	05/18/80	227,679	174	Phonograph
162.	03/24/79	03/31/79	11/25/79	221,957	175	Telephones
163.	04/12/79	04/21/79	05/04/80	227,229	176	Electric Light
164.	04/21/79	05/12/79	09/19/82	264,643	177	Magneto–Electric Machine

165.	07/07/79	07/10/79	09/09/79	219,393	180	Dynamo–Electric Machines
166.	07/17/79	07/25/79	08/31/80	231,704	181	Electro–Chemical Receiving-Telephone
167.	08/01/79	08/06/79	10/17/82	266,022	182	Telephone
168.	08/04/79	08/06/79	01/17/82	252,442	183	Telephone
169.	09/04/79	09/10/79	12/23/79	222,881	184	Magneto–Electric Machines
170.	11/01/79	11/04/79	01/27/80	223,898	186	Electric Lamp
171.	01/28/80	02/05/80	07/20/80	230,255	200	Method of Manufacturing Electric Lamps
172.	01/28/80	02/05/80	10/03/82	265,311	201	Electric Lamp and Holder for the Same
173.	01/28/80	02/05/80	08/30/87	369,280	204	System of Electrical Distribution
174.	01/28/80	03/29/80	10/18/81	248,425	203	Apparatus for Producing High Vacuums
175.	03/10/80	03/20/80	06/08/80	228,617	207	Brake for Electro–Magnetic Motors
176.	03/10/80	03/20/80	12/27/81	251,545	206	Electric Meter
177.	03/10/80	03/20/80	09/11/94	525,888	210	Manufacture of Carbons for Electric Lamps
178.	03/10/80	03/25/80	05/04/80	227,226	212	Safety-Conductor for Electric Lights
179.	03/11/80	03/20/80	09/19/82	264,649	208	Dynamo or Magneto Electric Machine
180.	04/03/80	04/07/80	06/01/80	228,329	216[1]	Magnetic Ore-Separator
181.	04/25/80	05/24/80	03/15/81	238,868	216	Manufacture of Carbons for Incandescent Electric Lamps
182.	06/15/80	06/30/80	02/15/81	237,732	220	Electric Light
183.	06/15/80	06/30/80	10/18/81	248,417	219	Manufacturing Carbons for Electric Lights
184.	06/15/80	07/03/80	05/13/84	298,679	221	Method of Treating Carbons for Electric Lights
185.	07/02/80	07/22/80	10/18/81	248,430	222	Electro–Magnetic Brake
186.	07/03/80	07/22/80	10/10/82	265,778	223	Electro–Magnetic Railway-Engine
187.	07/26/80	08/06/80	10/18/81	248,432	225	Magnetic Separator
188.	07/27/80	08/06/80	03/22/81	239,150	227	Electric Lamp
189.	07/28/80	08/06/80	12/27/81	251,540	229	Carbon for Electric Lamps

190.	07/28/80	08/06/80	08/22/82	263,139	230	Manufacture of Carbons for Electric Lamps
191.	07/28/80	08/09/80	03/29/81	239,372[2]	228	Testing Electric-Light Carbons
192.	07/29/80	08/09/80	08/19/90	434,585	231	Telegraph-Relay
193.	07/30/80	08/09/80	08/22/82	263,140	224	Dynamo-Electric Machine
194.	07/30/80	08/11/80	10/18/81	248,423	233	Carbonizer
195.	07/31/80	08/09/80	03/22/81	239,147	235	System of Electric Lighting
196.	07/31/80	08/09/80	10/18/81	248,434	234	Governor for Electric Engines
197.	08/04/80	08/09/80	09/19/82	264,642	236	Electric Distribution and Translation System
198.	08/06/80	08/09/80	02/12/84	293,433	238	Insulation of Railroad-Tracks Used for Electrical Circuits
199.	08/07/80	08/17/80	03/29/81	239,373	241	Electric Lamp
200.	08/07/80	08/17/80	04/05/81	239,745	240	Electric Lamp
201.	08/07/80	08/17/80	08/22/82	263,135	239	Electric Lamp
202.	08/10/80	08/17/80	12/27/81	251,546	242	Electric Lamp
203.	08/11/80	08/27/80	03/22/81	239,153	243	Electric Lamp
204.	08/11/80	08/27/80	11/02/86	351,855	244	Electric Lamp
205.	08/12/80	10/01/80	10/18/81	248,435	245	Utilizing Electricity as a Motive Power
206.	08/14/80	08/19/80	08/22/82	263,132	246	Electro-Magnetic Railway
207.	09/01/80	10/07/80	09/19/82	264,645	247	System of Conductors for the Distribution of Electricity
208.	09/22/80	10/07/80	04/26/81	240,678	252	Webermeter
209.	10/14/80	10/30/80	03/22/81	239,152	254	System of Electric Lighting
210.	10/15/80	10/30/80	03/22/81	239,148	255	Treating Carbons for Electric Lamps
211.	10/21/80	10/30/80	12/27/81	251,556	258	Regulator for Magneto or Dynamo Electric Machines
212.	10/21/80	11/05/80	06/14/81	242,900	256	Manufacturing Carbons for Electric Lamps
213.	10/21/80	11/11/80	02/22/81	238,098[3]	257	Magneto Signal Apparatus

214.	11/05/80	11/24/80	10/18/81	248,426	261	Apparatus for Treating Carbons for Electric Lamps
215.	11/19/80	11/24/80	03/22/81	239,151	262	Method of Forming Enlarged Ends on Carbon Filaments
216.	11/23/80	11/30/80	12/27/81	12,631	263	Design for an Incandescent Electric Lamp
217.	12/03/80	12/15/80	03/22/81	239,149	270	Incandescing Electric Lamp
218.	12/03/80	12/15/80	06/14/81	242,896	269	Incandescent Electric Lamp
219.	12/03/80	12/15/80	06/14/81	242,897	267	Incandescent Electric Lamp
220.	12/03/80	12/15/80	10/18/81	248,565	268	Webermeter
221.	12/03/80	12/15/80	09/05/82	263,878	271	Electric Lamp
222.	12/11/80	12/14/80	03/22/81	239,154	274	Relay for Telegraphs
223.	12/11/80	12/14/80	10/18/81	248,431	272	Preserving Fruit
224.	12/11/80	12/15/80	06/14/81	242,898	266	Magneto or Dynamo Electric Machine
225.	12/11/80	12/15/80	10/10/82	265,777	275	Method of Treating Carbons for Electric Lamps
226.	12/16/80	01/11/81	03/29/81	239,374	276	Regulating the Generation of Electric Currents
227.	12/16/80	01/11/81	10/18/81	248,428	277	Manufacture of Incandescent Electric Lamps
228.	12/21/80	01/11/81	10/18/81	248,427	279	Apparatus for Treating Carbons for Electric Lamps
229.	12/21/80	01/11/81	10/18/81	248,437	281	Apparatus for Treating Carbons for Electric Lamps
230.	12/30/80	01/11/81	10/18/81	248,416	280	Manufacture of Carbons for Electric Lamps
231.	01/19/81	01/26/81	06/14/81	242,899	287	Electric Lighting
232.	01/19/81	01/31/81	10/18/81	248,418	284	Electric Lamp
233.	01/19/81	01/31/81	10/18/81	248,433	282	Vacuum Apparatus
234.	01/19/81	01/31/81	12/27/81	251,548	285	Incandescent Electric Lamp
235.	01/19/81	01/31/81	07/09/89	406,824	286	Electric Meter
236.	01/20/81	01/31/81	10/18/81	248,422	288	System of Electric Lighting

237.	02/03/81	02/21/81	06/24/90	431,018	289	Dynamo or Magneto-Electric Machine
238.	02/24/81	03/03/81	06/14/81	242,901	291	Electric Meter
239.	02/24/81	03/03/81	10/18/81	248,429	292	Electric Motor
240.	02/25/81	03/05/81	10/18/81	248,421	293	Current-Regulator for Dynamo-Electric Machines
241.	02/26/81	03/03/81	12/27/81	251,550	296	Magneto or Dynamo Electric Machine
242.	02/26/81	05/27/81	12/27/81	251,555	294	Regulator For Dynamo-Electric Machines
243.	03/02/81	03/12/81	09/13/92	482,549	297	Means for Controlling Electric Generation
244.	03/07/81	03/26/81	10/18/81	248,420	300	Fixture and Attachment for Electric Lamps
245.	03/07/81	03/26/81	12/27/81	251,553	301	Electric Chandelier
246.	03/07/81	03/26/81	12/27/81	251,554	298	Electric Lamp and Socket or Holder
247.	03/08/81	03/26/81	10/18/81	248,424	299	Fitting and Fixture for Electric Lamps
248.	03/30/81	04/15/81	10/18/81	248,419	302	Electric Lamp

1. It is not known how this application received the same case number as that of Pat. 238,868.

2. With Charles Batchelor.

3. With Edward Johnson.

Bibliography

Acheson, Edward G. 1965. *A Pathfinder: Inventor, Scientist, Industrialist.* Port Huron, Mich.: Acheson Industries.

Alam, Chris, and H. Merskey. 1994. "What's in a Name? The Cycle of Change in the Meaning of Neuralgia." *History of Psychiatry* 5:429–74.

Altman, Nathaniel. 2000. *Healing Springs: The Ultimate Guide to Taking the Waters, from Hidden Springs to the World's Greatest Spas.* Rochester, Vt.: Healing Arts Press.

Atkinson, E., trans. and ed. 1902. *Elementary Treatise on Physics, Experimental and Applied. Ed. A. W. Reinold. From Ganot's Elements de Physique.* 16th ed. London: Longmans, Green.

———. trans. and ed. 1910. *Elementary Treatise on Physics, Experimental and Applied. Ed. A. W. Reinold. From Ganot's Elements de Physique.* 18th ed. New York: William Wood.

Baker, Edward C. 1976. *Sir William Preece, F.R.S.: Victorian Engineer Extraordinary.* London: Hutchinson.

Barnes, E. G. 1966. *The Rise of the Midland Railway, 1844–1874.* London: George Allen & Unwin.

Batten, Alan H. 1988. *Resolute and Undertaking Characters: The Lives of Wilhelm and Otto Struve.* Dordrecht: D. Reidel Publishing.

Bazerman, Charles. 1999. *The Languages of Edison's Light.* Cambridge, Mass.: MIT Press.

Becquerel, Alexandre-Edmond. 1853. "Recherches sur la transmission de l'électricité au travers des gas des températures élevées." *Annales de Chimie et de Physique* 39:355–402.

Bellanger, Claude, Jacques Godechot, Pierre Guiral, and Fernand Terrou, eds. 1972. *Histoire générale de la presse française.* Vol. 3. Paris: Presses Universitaires de France.

Berman, Maurice. 1978. *Social Change and Scientific Organization: The Royal Institution, 1799–1844.* Ithaca: Cornell University Press.

Berthold, Victor M. 1924. *History of the Telephone and Telegraph in Chile, 1851–1922.* New York: privately printed.

Bierer, Bert W. 1980 [1940]. *American Veterinary History.* Madison, Wis.: reproduced by Carl Olson.

Black, Henry Campbell. 1979. *Black's Law Dictionary: Definitions of the Terms and Phrases of American and English Jurisprudence, Ancient and Modern.* 5th ed. St. Paul, Minn.: West Publishing.

Bliss, George. 1878. "Thomas A. Edison." *Chicago Tribune,* 8 April, 3.

Bloxam, Charles L. 1869. *Laboratory Teaching, or Progressive Exercises in Practical Chemistry.* London: John Churchill & Sons.

Bowditch, John. 1989. "Driving the Dynamos: Origins of the High-Speed Electric Light Engine." *Mechanical Engineering* 111:80–89.

Brackett, C[yrus], and C[harles]. A. Young. 1880. "Notes of Experiments upon Mr. Edison's Dynamometer, Dynamo-machine and Lamp." *American Journal of Science,* 3d ser. 19:475–79.

Brewer, Ebenezer Cobham. 1963. *Brewer's Dictionary of Phrase and Fable.* 8th. ed. New York: Harper & Row.

Briggs, Asa. 1982. *The Power of Steam.* Chicago: University of Chicago Press.

Bright, Arthur A., Jr. 1972 [1949]. *The Electric-Lamp Industry: Technological Change and Economic Development from 1800 to 1947.* New York: Arno Press.

Bright, Charles. 1974 [1898]. *Submarine Telegraphs: Their History, Construction, and Working.* New York: Arno Press.

Bruce, Robert V. 1973. *Bell: Alexander Graham Bell and the Conquest of Solitude.* Boston: Little, Brown.

Buss, Dietrich G. 1978 [1977]. *Henry Villard: A Study of Transatlantic Investments and Interests, 1870–1895.* New York: Arno Press.

Cairncross, A. K. 1987. "The Early Growth of Messrs. J. & P. Coats, 1830–83." *Business History* 29:157–77.

Cameron, Ardis. 1993. *Radicals of the Worst Sort: Laboring Women in Lawrence, Massachusetts, 1860–1912.* Chicago: University of Illinois Press.

Cameron, Rondo. 1961. *France and the Economic Development of Europe, 1800–1914.* Princeton, N.J.: Princeton University Press.

Cantor, G. N., and M. J. S. Hodge, eds. 1981. *Conceptions of Ether: Studies in the History of Ether Theories.* Cambridge: Cambridge University Press.

Cardwell, D. S. L. 1972. *Technology, Science, and History.* London: Heinemann.

Carlson, W. Bernard. 1991. *Innovation as a Social Process: Elihu Thomson and the Rise of General Electric, 1870–1900.* Cambridge: Cambridge University Press.

Carosso, Vincent P. 1987. *The Morgans: Private International Bankers 1854–1913.* Cambridge, Mass.: Harvard University Press.

Child, Ernest. 1940. *The Tools of the Chemist.* New York: Reinhold Publishing.

Clark, Latimer, and Robert Sabine. 1871. *Electrical Tables and Formulae.* London: E. & F. N. Spon.

Cohen, Paul E., and Robert T. Augustyn. 1997. *Manhattan in Maps: 1527–1995.* New York: Rizzoli International Publications.

Cokayne, George Edward. 1929. *The Complete Peerage, or a History of the House of Lords and All Its Members from the Earliest Times.* H. A.

Doubleday and Lord Howard DeWalden, eds. London: St. Catherine Press.

Collier, Simon, and William F. Sater. 1996. *A History of Chile, 1808–1994.* Cambridge: Cambridge University Press.

Condit, Carl W. 1980. *The Port of New York.* Vol. 1, *A History of the Rail and Terminal System from the Beginnings to Pennsylvania Station.* Chicago: University of Chicago Press.

Cooke, Conrad. 1879. "On Edison's Electro-Chemical or Loud-Speaking Telephone." *Journal of the Society of Arts* 27:558–69.

Cooper, Grace Rogers. 1976. *The Sewing Machine: Its Invention and Development.* Washington, D.C.: Smithsonian Institution Press.

Crookes, William. 1873. "Researches on the Atomic Weight of Thallium." *Philosophical Transactions of the Royal Society of London* 163:277–330.

———. 1876. "On Repulsion Resulting from Radiation. Parts III & IV." *Philosophical Transactions of the Royal Society of London* 166:325–76.

———. 1876–77. "Experimental Contributions to the Theory of the Radiometer. [Preliminary Notice]." *Proceedings of the Royal Society of London* 25:304–314. A portion of this article appeared in *Chemical News* and is in Cat. 1050, Scraps. (*TAEM* 26:407; *TAED* SM050025a).

———. 1878a. "On Repulsion Resulting from Radiation. Part V." *Philosophical Transactions of the Royal Society of London* 169:243–318.

———. 1878b. "On the Illumination of Lines of Molecular Pressure, and the Trajectory of Molecules." *Proceedings of the Royal Society of London* 28:103–111. Reprinted in *Teleg. J. and Elec. Rev.*, 1 Jan. 1879, Cat. 1050:63, Scraps. (*TAEM* 26:427; *TAED* SM050063a).

Crosland, Maurice. 1992. *Science under Control: The French Academy of Sciences, 1795–1914.* Cambridge and New York: Cambridge University Press.

Cunningham, Peter. [1857?]. *London in 1857.* London: John Murray.

Dacus, J. A., and James W. Buel. 1878. *A Tour of St. Louis, Or, the Inside Life of a Great City.* St. Louis: Western Publishing Company.

Davenport, Neil. 1979. *The United Kingdom Patent System: A Brief History.* Portsmouth, UK: Kenneth Mason.

Davis, Audrey B. 1995. "Introduction." *Samuel S. White Catalogue of Dental Instruments and Equipment.* San Francisco: Norman Publishing in association with Smithsonian Institution Libraries.

de Clercq, P. R., ed. 1985. *Nineteenth-Century Scientific Instruments and Their Makers.* Amsterdam: Rodopi.

DeKosky, Robert K. 1976. "William Crookes and the Fourth State of Matter." *Isis* 67:36–60.

Dibdin, William Joseph. 1889. *Practical Photometry: A Guide to the Study of the Measurement of Light.* London: Walter King.

Draper, John William. 1847. "On the Production of Light by Heat." *Philosophical Magazine,* 3rd ser. 30:345–60.

Dredge, James. 1882–85. *Electric Illumination.* 2 vols. London: Engineering.

Du Moncel, Theodose. 1872–78. *Exposé des applications de l'électricité.* Paris: L. Hachette.

———. 1879a. "Considérations sur l'éclairage public par les procécés électriques." *La Lumiere Electrique* 1:2–3. An English translation, "Considerations on Public Lighting by Electric Processes," appears in Complainant's Rebuttal Exhibit, p. 4101–6, Edison Electric Light Co. v. U.S. Electric Lighting Co., (TAEM 48:561; TAED QD012G:67)

———. 1879b. *The Telephone, the Microphone, and the Phonograph.* London: C. Kegan Paul.

———. 1879c. *The Telephone, the Microphone, and the Phonograph.* New York: Harper & Brothers.

Dyer, Frank, and T. C. Martin. 1910. *Edison: His Life and Inventions.* 2 vols. New York: Harper & Bros.

Edison, Thomas A. 1880a. "On the Phenomena of Heating Metals in Vacuo by Means of an Electric Current." *Proceedings of the American Association for the Advancement of Science* 28:173–77.

———. 1880b. "On a Resonant Tuning Fork." *Proceedings of the American Association for the Advancement of Science* 28:178.

Edison, Thomas A., and Charles T. Porter. 1882. "Description of the Edison Steam Dynamo." *Transactions of the American Society of Mechanical Engineers* 3:218–25.

Elliott, Russell R. 1973. *History of Nevada.* Lincoln: University of Nebraska Press.

Emery, Charles E. 1883. "The Cost of Steam Power." *Transactions of the American Society of Civil Engineers* 12:425–35.

Fagen, M. D. 1975. *A History of Engineering and Science in the Bell System: The Early Years (1875–1925).* n.p.: Bell Telephone Laboratories.

Fairburn, William Armstrong. 1945–55. *Merchant Sail.* 6 vols. Center Lovell, Maine: Fairburn Marine Educational Foundation.

Faraday, Michael. 1965 [1855]. *Experimental Researches in Electricity.* Vol. 3. New York: Dover Publications.

Feldenkirchen, Wilfried. 1994. *Werner von Siemens: Inventor and International Entrepreneur.* Columbus: Ohio State University Press.

———. 1999. *Siemens, 1918–1945.* Columbus: Ohio State University Press.

Finlay, Gordon R. 1978. "Henri Moissan and the Development of Electric Furnaces." In *Proceedings of the Symposium on Selected Topics in the History of Electrochemistry,* ed. George Dubpernell and J. H. Westbrook. Princeton, N.J.: The Electrochemical Society.

Fox, Edwin. 1879a. "Edison and His Inventions. I. The Electro-Motograph and Its Applications." *Scribner's Monthly* 18:297–306.

———. 1879b. "Edison's Inventions. II. The Carbon Button and Its Offspring." *Scribner's Monthly* 18:446–55.

———. 1879c. "Edison's System of Fast Telegraphy." *Scribner's Monthly* 18:840–46.

Franklyn, Julian. 1975. *A Dictionary of Rhyming Slang.* London: Routledge and Kegan Paul.

Friedel, Robert, and Paul Israel. 1986. *Edison's Electric Light: Biography of an Invention.* New Brunswick, N.J.: Rutgers University Press.

Friedlander, Heinrich E., and Jacob Oser. 1953. *Economic History of Modern Europe.* New York: Prentice-Hall.

Gábor, Luca. 1993. *Telephonic News Dispenser.* Budapest: Hungarian Broadcasting.

Gage, Simon Henry, and Henry Phelps Gage. 1914. *Optic Projection: Principles, Installation and Use of the Magic Lantern, Projection Microscope, Reflecting Lantern, Moving Picture Machine.* Ithaca, N.Y.: Comstock Publishing.

Gibbs, F. W. 1958. "Extraction and Production of Metals, Part II: Non-Ferrous Metals." In *A History of Technology.* Vol. IV, *The Industrial Revolution c. 1750 to c. 1850,* ed. Charles Singer, E. J. Holmyard, A. R. Hall, and Trevor I. Williams. Oxford: Clarendon Press, University of Oxford.

Gmelin, Leopold. 1848–66. *Handbook of Chemistry.* London: Cavendish Society.

Gudde, Erin G., and Elisabeth K. Gudde, eds. 1975. *California Gold Camps.* Berkeley: University of California Press.

Guttridge, Leonard F. 1986. *Icebound: The Jeannette Expedition's Quest for the North Pole.* Annapolis, Md.: Naval Institute Press.

Harlow, Alvin F. 1936. *Old Wires and New Waves.* New York: D. Appleton-Century.

Hathaway, Arthur S. 1886. "An Unwritten Chapter in the History of the Dynamo." *Electrical World* 7:102.

Hawkins, Charles Caesar. 1922. *The Dynamo: Its Theory, Design, and Manufacture.* 2 vols. 6th ed. London: Sir Isaac Pitman & Sons.

Heap, David Porter. 1884. *Report on the International Exposition of Electricity held at Paris August to November, 1881.* Washington, D.C.: Government Printing Office.

Higgs, William Henry Paget, and John Richard Brittle. 1878. "Some Recent Improvements in Dynamo-Electric Apparatus" with discussion. *Minutes of Proceedings of the Institution of Civil Engineers* 52 (pt. 2): 36–98.

Hills, Richard L. 1989. *Power from Steam: A History of the Stationary Steam Engine.* Cambridge: Cambridge University Press.

Hong, Sungook. 2001. *Wireless: From Marconi's Black-Box to the Audion.* Cambridge, Mass.: MIT Press.

Hopkinson, John. 1879. "On Electric Lighting. (First Paper)" with discussion. *Proceedings of the Institution of Mechanical Engineers* 238–65.

Hornblower, William B. 1891. "Sherburne Blake Eaton." *University Magazine* 5:1–3.

Hounshell, David. 1975. "Elisha Gray and the Telephone: On the Disadvantages of Being an Expert." *Technology and Culture* 16:133–61.

Houston, Edwin, and Elihu Thomson. 1878. "Circumstances Influencing the Efficiency of Dynamo-Electric Machines." *Proceedings of the American Philosophical Society* 18:58. In Complainant's Rebuttal Exhibit, pp. 4125–30, *Edison Electric Light Co. v. U.S. Electric Lighting Co.* [*TAEM* 48:573–75; *TAED* QD012G:79–81].

Howell, John W., and Henry Schroeder. 1927. *History of the Incandescent Lamp.* Schenectady, N.Y.: Maqua.

Hunter, Louis C. 1985. *A History of Industrial Power in the United States, 1780–1930.* Vol. 2, *Steam Power.* Charlottesville: University Press of Virginia.

Hutchinson, Thomas, comp. (Printed annually.) *The Lakeside Annual Directory of the City of Chicago.* Chicago: Donnelley, Gassette, & Loyd.

Israel, Paul. 1992. *From Machine Shop to Industrial Laboratory: Telegraphy and the Changing Context of American Invention, 1830–1920.* Baltimore: Johns Hopkins University Press.

———. 1998. *Edison: A Life of Invention.* New York: John Wiley & Sons.

Jeffrey, Thomas E. 1998. *Thomas Lanier Clingman: Fire Eater from the Carolina Mountains.* Athens: University of Georgia Press.

Jehl, Francis. 1882. *The Edison Electric Light Meter.* [Privately printed.] In *TAEM* 62:889; *TAED* D8239ZBN.

———. 1937–41. *Menlo Park Reminiscences.* 3 vols. Dearborn, Mich.: Edison Institute.

Jenks, William Lee. n.d. *History of Port Huron Street Railways.* William Lee Jenks Collection, MiD-B.

Johnson, Edward H. 1879. *Statement as to the Origin and Development of the Telephone.* London: Sir Joseph Causton & Sons.

Johnson, John Henry, and James Johnson. 1879. *The Patentees Manual: Being a Treatise on the Law and Practice of Letters Patent.* 4th ed., revised and enlarged. London: Longmans, Green.

Johnston, Sean F. 1996. "Making Light Work: Practices and Practitioners of Photometry." *History of Science* 34:273–302.

Jordan, D. W. 1990. "The Magnetic Circuit Model, 1850–1890: The Resisted Flow Image in Magnetostatics." *British Journal for the History of Science* 23:131–73.

Kelley, Robert L. 1959. *Gold vs. Grain: The Hydraulic Mining Controversy in California's Sacramento Valley.* Glendale, Calif.: Arthur H. Clark.

Kieve, Jeffrey L. 1973. *The Electric Telegraph: A Social and Economic History.* Newton Abbot, UK: David & Charles.

Kim, Dong-Woon. 1998. "The British Multinational Enterprise in the United States before 1914: The Case of J. & P. Coats." *Business History Review* 72:523–51.

King, W. James. 1962a. "The Development of Electrical Technology in the Nineteenth Century: 1. The Electrochemical Cell and the Electromagnet." *United States Museum Bulletin 228.* Washington, D.C.: Smithsonian Institution.

———. 1962b. "The Development of Electrical Technology in the Nineteenth Century: 2. The Telegraph and the Telephone." *United States Museum Bulletin 228.* Washington, D.C.: Smithsonian Institution.

———. 1962c. "The Development of Electrical Technology in the Nineteenth Century: 3. The Early Arc Light and Generator." *United States Museum Bulletin 228.* Washington, D.C.: Smithsonian Institution.

Kingsbury, J. E. 1915. *The Telephone and Telephone Exchanges: Their Invention and Development.* London: Longmans, Green.

Klein, Maury. 1986. *The Life and Legend of Jay Gould.* Baltimore: Johns Hopkins University Press.

Knight, Edward H. 1876–77. *Knight's American Mechanical Dictionary.* 3 vols. New York: Hurd and Houghton.

Kohlrausch, Friedrich. 1870. *Leitfaden der praktischen Physik* (An introduction to physical measurements). Leipzig: B. G. Tuebner.

Lain, George T., comp. (Printed annually.) *The Brooklyn City and Business Directory.* Brooklyn: Lain.

Leddy, James J. 1989. "Industrial Electrochemistry." In *Electrochemistry, Past and Present. ACS Symposium Series 390,* ed. John T. Stock and Mary Virginia Orna. Washington, D.C.: American Chemical Society.

Liberatore, E. K. 1998. *Helicopters before Helicopters.* Malabar, Fla.: Krieger Publishing.

Lighter, J. E. 1994–97. *Random House Historical Dictionary of American Slang.* New York: Random House.

Livingston, Dorothy Michelson. 1973. *The Master of Light: A Biography of Albert A. Michelson.* New York: Charles Scribner's Sons.

Lock, Alfred G. 1882. *Gold: Its Occurrence and Extraction.* London: E. & F. N. Spon.

MacFic, R. A. 1879–83. *Copyright and Patents for Inventions.* 2 vols. Edinburgh, T. &. T. Clark.

Marshall, David Trumbull. c. 1931. *Recollections of Edison.* Boston: Christopher Publishing House.

Mathews, Mitford M., ed. 1951. *A Dictionary of Americanisms, on Historical Principles.* Chicago: University of Chicago Press.

Matthiessen, Augustus. 1858. "On the Electric-Conducting Power of the Metals." *Philosophical Magazine,* 4th ser. 16:219–23.

Matthiessen, A[ugustus]., and C. Vogt. 1864. "Über den Einfluss der Temperatur auf die elektrische Leitungsfähigkeit der Legirungen." *Annalen der Physik und Chemie* 122:19–78.

Matthiessen, A[ugustus]., and M. von Bose. 1862. "Über den Einfluss der Temperatur auf die elektrische Leitungsfähigkeit der Metalle." *Annalen der Physik und Chemie* 115:353–97.

Maxwell, James Clerk. 1873. *A Treatise on Electricity and Magnetism.* 1st ed. Oxford: Clarendon Press.

Mayer, Alfred Marshall. 1878a. "On Edison's Talking-Machine." *Popular Science Monthly* 12:719–24.

Mayr, Otto. 1975. "Yankee Practice and Engineering Theory: Charles T. Porter and the Dynamics of the High-Speed Steam Engine." *Technology and Culture* 16:570–602.

McClellan, Kenneth. 1978. *Whatever Happened to Shakespeare?* Plymouth: Vision Press.

McClure, J[ames]. B[aird]. 1879. *Edison and His Inventions.* Chicago: Rhodes & McClure.

McCulloch, J. R. 1866. *A Dictionary, Geographical, Statistical, and Historical of the Various Countries, Places, and Principal Natural Objects in the World.* London: Longmans, Green.

McLeod, H. 1874. "Apparatus for Measurement of Low Pressure of Gas." *Proceedings of the Physical Society* 1:30–34.

Miller, Arthur I. 1981. "Unipolar Induction: A Case Study of the Interaction between Science and Technology." *Annals of Science* 38:155–89.

Miller, Michael B. 1981. *The Bon Marché: Bourgeois Culture and the Department Store, 1869–1920*. Princeton, N.J.: Princeton University Press.

Moise, Robert, and Maurice Daumas. 1978. "La periode d'approche." In *Histoire generale des techniques*, ed. M. Daumas. 5 vols. Paris: Presses Universitaires de France.

Moissan, Henri. 1904. *The Electric Furnace*. Trans. Victor Lenher. Easton, Penn.: The Chemical Publishing.

Mollier, Jean-Yves. 1988. *L'Argent et les lettres: Histoire du capitalism d'édition (1880–1920)*. [Paris]: Fayard.

Moyer, Albert E. 1992. *A Scientist's Voice in American Culture: Simon Newcomb and the Rhetoric of Scientific Method*. Berkeley: University of California Press.

National Academy of Sciences. 1884. *Report of the National Academy of Sciences for the Year 1883*. Washington, D.C.: Government Printing Office.

Newman, Harold. 1977. *An Illustrated Dictionary of Glass*. London: Thames and Hudson.

Nye, Mary Jo. 1996. *Before Big Science: The Pursuit of Modern Chemistry and Physics, 1800–1940*. New York: Twayne Publishers.

Partington, J. R. 1964. *A History of Chemistry*. Vol. 4. London: Macmillan.

Passer, Harold C. 1953. *The Electrical Manufacturers, 1875–1900: A Study in Competition, Entrepreneurship, Technical Change, and Economic Growth*. Cambridge, Mass.: Harvard University Press.

Pepper, John H. [1869?]. *Chemistry: Embracing the Metals and Elements which Are not Metallic*. London: Frederick Warne.

Plattner, Karl Friedrich. 1875. *Plattner's Manual of Qualitative and Quantitative Analysis with the Blowpipe*. Rev. and enlarged 3rd ed. by Th[eodor]. Richter, trans. by Henry B. Cornwall and John H. Caswell. New York: D. Van Nostrand.

Pope, Franklin. 1869. *Modern Practice of the Electric Telegraph: A Handbook for Electricians and Operators*. New York: Russell Brothers.

Porter, Charles T. 1908. *Engineering Reminiscences: Contributed to "Power" and "American Machinist."* New York: John Wiley & Sons.

Post, Robert C. 1976. *Physics, Patents, and Politics: A Biography of Charles Grafton Page*. New York: Science History Publications.

Povey, P. J., and R. A. J. Earl. 1988. *Vintage Telephones of the World*. London: Peter Peregrinus.

Preece, W.[illiam] H. 1879a. "The Electric Light." *Philosophical Magazine*, 5th ser. 7:29–34. In *Edison Electric Light Co. v. U.S. Electric Lighting Co.*, Complainant's Rebuttal—Exhibits [Vol. VI], pp. 4084–90, Lit. (*TAEM* 48:552; *TAED* QD012G4084).

———. 1879b. "Recent Advances in Telegraphy." *Journal of the Society of Arts* 27:963–1008.

———. 1885. "On a Peculiar Behaviour of Glow-Lamps when Raised to High Incandescence." *Proceedings of the Royal Society of London* 38:219–30.

Preece, W.[illiam] H., and J. Sivewright. 1891. *Telegraphy*. 9th ed. London: Longmans, Green.

Prescott, George B. 1877. *Electricity and the Electric Telegraph.* New York: D. Appleton.

———. 1878a. *The Speaking Telephone, Talking Phonograph and Other Novelties.* New York: D. Appleton.

———. 1878b. "Recent Improvements in Telephony." *Scribner's Monthly* 16:600–602.

———. 1879. *The Speaking Telephone, Electric Light, and Other Recent Electrical Inventions.* New York: D. Appleton.

———. 1972 [1884]. *Bell's Electric Speaking Telephone: Its Invention, Construction, Application, Modification, and History.* New York: Arno Press.

Pring, J. N. 1921. *The Electric Furnace.* London: Longmans, Green.

Raymond, R[ossiter]. W. 1881. *A Glossary of Mining and Metallurgical Terms.* Easton, Penn.: American Institute of Mining Engineers.

Read, Donald. 1999. *The Power of News: The History of Reuters.* 2nd ed. Oxford: Oxford University Press.

Reich, Leonard S. 1985. *The Making of American Industrial Research: Science and Business at GE and Bell, 1876–1926.* Cambridge: Cambridge University Press.

Reid, James D. 1879. *The Telegraph in America.* New York: Derby Bros.

———. 1886. *The Telegraph in America.* Rev. ed. New York: John Polhemus.

Ristow, Walter W. 1981. *Fire Insurance Maps in the Library of Congress: Plans of North American Cities and Towns Produced by the Sanborn Map Company.* Washington, D.C.: Library of Congress.

Rondeau, René. 2001. *Tinfoil Phonographs: The Dawn of Recorded Sound.* Corte Madera, Calif.: privately published.

Rowland, Henry. 1881. "On the Heat Generated in a Magnet When It Is Magnetised and Demagnetised." *Electrician* 7:294–95.

Rowland, Henry, and George F. Barker. 1880. "On the Efficiency of Edison's Electric Light." *American Journal of Science,* 3rd ser. 19:337–40.

Sabine, Robert. 1867. *The Electric Telegraph.* London: Virtue Bros.

Sampson, Davenport, & Co. (Printed annually.) *The Boston Directory, Embracing the City Record, a Directory of the Citizens, and Business Directory.* Boston: Sampson, Davenport.

Schallenberg, Richard H. 1978. "Batteries Used for Power Generation during the Nineteenth Century." In *Proceedings of the Symposium on Selected Topics in the History of Electrochemistry,* ed. George Dubpernell and J. H. Westbrook. Princeton, N.J.: The Electrochemical Society.

Schellen, Heinrich. 1879. *Die magnet- und dynamo- elektrischen Maschinen, ihre Entwicklung, Construction und praktische Anwendung.* Cologne: M. DuMont-Schauberg.

———. 1884. *Magneto-Electric and Dynamo-Electric Machines: Their Construction and Practical Application to Electric Lighting and the Transmission of Power.* Trans. Nathaniel S. Keith and Percy Neymann. New York: D. Van Nostrand.

Schivelbusch, Wolfgang. 1988. *Disenchanted Night: The Industrialization of Light in the Nineteenth Century.* Berkeley: University of California Press.

Schwendler, L[ouis]. 1879. "Précis of a Report on Electric-Light Experiments." *Philosophical Magazine,* 5th ser. 8:335–39.

Searle, G[eoffrey]. R[ussell]. 1976. *Eugenics and Politics in Britain 1900–1914.* Leiden: Noordhoff International Publishing.

Shinn, Terry. 1980. "From 'Corps' to 'Profession': The Emergence and Definition of Industrial Engineering in Modern France." In *The Organization of Science and Technology in France 1808–1914,* ed. Robert Fox and George Weisz. Cambridge: Cambridge University Press.

Siemens, C. William. 1889. *The Scientific Works of C. William Siemens, Kt.* 3 vols. E. F. Bamber, ed. London: John Murray.

Simmons, Jack. 1978. *The Railway in England and Wales, 1830–1914.* Vol. 1. *The System and Its Working.* Leicester: Leicester University Press.

Sinclair, Bruce. 1974. *Philadelphia's Philosopher Mechanics: A History of the Franklin Institute, 1824–65.* Baltimore: Johns Hopkins University Press.

Smith, George David. 1985. *The Anatomy of a Business Strategy: Bell, Western Electric, and the Origins of the American Telephone Industry.* Baltimore: Johns Hopkins University Press.

Smith, Grant H. 1943. *The History of the Comstock Lode, 1850–1920.* Reno: Nevada Bureau of Mines and University of Nevada.

Sprague, John T. 1875. *Electricity: Its Theory, Sources, and Applications.* London: E. & F. N. Spon.

Stayton, G. H. 1878. "Report on the Electric Light." *Electrician* 1: 166–67.

Swinton, A. A. Campbell. 1929. "The Part Played by Mr. St. George Lane Fox Pitt in the Invention of the Carbon Incandescent Electric Lamp, and the Modern Method of Electric Lighting." *Journal of the Institution of Electrical Engineers* 67:551–52.

Szymanowitz, Raymond. 1971. *Edward Goodrich Acheson: Inventor, Scientist, Industrialist.* New York: Vantage Press.

Taltavall, John B. 1893. *Telegraphers of To-day; Descriptive, Historical, Biographical.* New York: John B. Taltavall.

Taylor, Jocelyn Pierson. 1978. *Grosvenor Porter Lowrey.* New York: Privately printed.

Tebbel, John. 1975. *A History of Book Publishing in the United States.* Vol. 2: *The Expansion of an Industry, 1865–1919.* New York & London: R. R. Bowker.

Ternant, A. L. 1881. *Les Telegraphes.* Paris: L. Hachette.

Thompson, Silvanus P. 1886. *Dynamo-Electric Machinery: A Manual for Students of Electrotechnics.* London: E. & F. N. Spon.

———. 1887. "The Development of the Mercurial Air Pump." *Telegraphic Journal and Electrical Review* 21:556–59, 587–90, 610–13, 632–34, 659–65.

———. 1902. *Dynamo-Electric Machinery: A Manual for Students of Electrotechnics.* 8th American ed. New York: M. Strong.

———. 1910. *Light, Visible and Invisible.* London: Macmillan.

———. 1976 [1910]. *The Life of Lord Kelvin.* New York: Chelsea Publishing.

Thurston, Robert. 1884–85. "On the Theory of the Finance of Lubrication, and on the Valuation of Lubricants by Consumers." *Transactions of the American Society of Mechanical Engineers* 6:437–60.

Tilden, William. 1919. *Chemical Discovery and Invention in the Twentieth Century.* 3rd ed. London: George Routledge and Sons.

Tosiello, Rosario J. 1979 [1971]. *The Birth and Early Years of the Bell Telephone System, 1876–1880.* New York: Arno Press.

Tritton, Paul. 1993. *The Godfather of Rolls-Royce: The Life and Times of Henry Edmunds, M.I.C.E., M.I.E.E., Science and Technology's Forgotten Pioneer.* London: Academy Books.

Trowbridge, John. 1879. "Methods of Measuring Electric Currents of Great Strength; Together with a Comparison of the Wilde, the Gramme, and the Siemen's Machines." *Philosophical Magazine* 7: 165–73.

Turner, Gerard. 1983. *Nineteenth-Century Scientific Instruments.* Berkeley: University of California Press.

Upton, Francis. 1879. "The Siemens Dynamo-Electric Machine." *Engineering* 28:70–71. In Cat. 1025:17, Scraps. (*TAEM* 24:662; *TAED* SM025080b).

———. 1880a. "Edison's Electric Light." *Scribner's Monthly* 19:531–44. In *Edison Electric Light Co. v. U.S. Electric Lighting Co.,* Complainant's Rebuttal—Exhibits [Vol. VI], pp. 4191–4204, Lit. (*TAEM* 48:607; *TAED* QD012G4191A).

———. 1880b. "Methods for Testing Faradic Machines." *Proceedings of the American Association for the Advancement of Science* 28:178–84.

U.S. Department of Commerce. 1914. *Standard Methods of Gas Testing. Circular of the Bureau of Standards, No. 48.* Washington, D.C.: Government Printing Office.

U.S. House of Representatives. 1879. *Report No. 105.* Washington, D.C.: GPO.

Vance, James E., Jr. 1995. *The North American Railroad: Its Origin, Evolution, and Geography.* Baltimore and London: Johns Hopkins University Press.

Vinton, John Adams. 1874. *The Upton Memorial: A Genealogical Record of the Descendants of John Upton, of North Reading, Mass.* Bath, Maine: E. Upton & Son.

Watts, Henry. 1872–1875. *A Dictionary of Chemistry and the Allied Branches of Other Sciences.* London: Longmans, Green.

Webb, K. R. 1965. "Sprengel and the Vacuum Pump (1865)." *Chemistry in Britain* 21:569–71.

Weinreb, Ben, and Christopher Hibbert, eds. 1983. *The London Encyclopedia.* London: Macmillan.

Weintraub, Stanley. 1969. *Shaw: An Autobiography 1856–1898.* New York: Weybright and Talley.

Wentworth, Harold, and Stuart Berg Flexner. 1975 [1960]. *Dictionary of American Slang.* New York: Thomas Y. Crowell.

Westbrook, J. H. 1978. "Robert Hare, Jr.—American Electrochemical Pioneer." In *Proceedings of the Symposium on Selected Topics in the History of Electrochemistry,* ed. George Dubpernell and J. H. Westbrook. Princeton, N.J.: The Electrochemical Society.

Western Union Telegraph Co. (Printed annually.) *Annual Reports*. New York: Western Union Telegraph.

Whittaker, Edmund Taylor. 1989 [1951–53]. *A History of the Theories of Aether & Electricity*. New York: Dover Publications.

Wile, Frederic William. 1974 [1926]. *Emile Berliner: Maker of the Microphone*. New York: Arno Press.

Wilson, H., comp. (Printed annually.) *New York City Directory*. New York: John F. Trow. N.B.: This title was incorrectly cited in Volumes 1–3 as *Trow's Business Directory of New York City*.

Wilson's Business Directory of New York City. (Printed annually.) New York: Trow City Directory Co.

Wise, George. 1982. "Swan's Way: A Study in Style." *IEEE Spectrum* 19:66–70.

———. 1985. *Willis R. Whitney, General Electric, and the Origins of U.S. Industrial Research*. New York: Columbia University Press.

Woodruff, A. E. 1966. "William Crookes and the Radiometer." *Isis* 57: 188–98.

Woods's Baltimore City Directory. (Printed annually). Baltimore: John W. Woods.

Wüllner, Adolph. 1874–75. *Lehrbuch der Experimentalphysik*. Leipzig: B.G. Teubner.

Wyndham, Henry Saxe. 1926. *Arthur Seymour Sullivan (1842–1900)*. New York: Harper Brothers.

Young, C[harles]. A. 1880. "On the Thermo-Electric Electromotive Power of Fe. and Pt. in Vacuo." *Science* 1:150.

Credits

Reproduced with permission of the AT&T Corporate Archive: Doc. 1882. Reproduced with permission of Dun and Bradstreet and the Baker Library, Harvard Business School: Docs. 2001, 2051. Courtesy of the Milton S. Eisenhower Library, the Johns Hopkins University: Docs. 1951, 1964, 2033; illustrations on p. 775. From the Collections of the Henry Ford, Benson Ford Research Center: frontispiece (neg. P.O.14259) and illustration on p. 929. Reproduced with permission of the Historical Society of Pennsylvania: Doc. 2052. From the collection of Charles Hummel: illustrations on pp. 48 and 492. Courtesy of Peter Jakab: illustrations on pp. 522 and 982. Reproduced with permission of the Manuscripts Division, Library of Congress: Docs. 1684, 1884. Courtesy of the National Archives: Docs. 1789, 1802, 1890, 1912, 2036. Courtesy of the Division of Old Records, New York County Clerk: Doc. 2058. Reproduced with permission of the Postal and Telecommunications Museum Foundation, Budapest: Docs. 1722, 1739. Reproduced with permission of the Science Museum, London: Doc. 1784.

Courtesy of Edison National Historic Site (designations are to photograph classification numbers or to *TAEM* reel:frame and *TAED* notebook or volume:image number): illustrations on pp. 11 (4:782; NV16:304), 28 (31:268; No24:9), 60 (4:803; NV16:325), 70 (4:808; NV16:330), 77 *top* (4:807; NV16:329) *bottom* (29:1035; No09:95), 80 (29:732; No07:54), 130 (29:424; No04:97), 165 (31:285; No24:26), 220 (30:990; No20:21), 233 (115:829), 241 (32:1059; No45:13), 312 *top* (32:454; No38:80) *bottom* (4:1138; NV18:93), 313 *top* (45:2; NS7968:2) *bottom* (11: 673; TI2:494), 321 (91:64; MBN004:62), 371 (35:806; No85:

23), 405 (35:849; No85:66), 406 (35:863; No85:80), 414 (4:794; NV16:309), 494 (94:546; MBSB21386X), 525 (34:725; No70:7), 548 (32:854; N301:72), 566 (34.000.001), 567 (34.000.002), 599 *top* (35:567; No80:83), 602 (34:756; No70: 38), 704 *top* (34:615; No68:9) *bottom* (34:620; No68:14), 717 (33:443; No51:31), 753 *top* (33:855; No55:25) *bottom* (34:661; No68:55), 778 *top* (36:624; N108:22) *bottom* (36:619; N108: 17), 779 (35:540; No80:56), 787 (36:99; N102:28), 798 (36:981; N112:140), 856 (39:266; N153:22), 930 (37:639; N120:54), 933 (35:1154; No88:100), 937 (6:722; NMo14:101), 961 (94:623; MBSB21557X), 1003 (44:1012; NS81:2).

Index

American Speaking Telephone Co., 45n.1, 191n.2

American Union Telegraph Co., 146n.4, 263nn.1–2, 965, 985n.1

Anderson, Arthur, 363

Anderson, James, **463**–64, 517–19

Andrews, William, 400

Andrus, Arthur, 878

Andrus, Milo P., **672**

Anglo-American Brush Electric Light Co., 893n.3

Anglo-American Cable Co., 97n.3

Anglo American Telegraph Co., 520n.5, 886n.4

Anglo-Indian Telephone Co., **989**

Anglo-Universal Bank, 478

Annales de Chimie et de Physique, 918

Anson, Thomas Francis, **362**

Ansonia Brass & Copper Co., 122n.3

Ansonia Clock Co., 121, 316n.5

Arctic expedition, 132, 159, 242

Argand lamp, 154n.3, 543, 862

Argentina, 408, 756, 908

Arizona (steamship), 512, 795

Armengaud, Charles, **1015**n.2

Armington, James, **690**

Arthur, Thomas Jr., 374n.4

Ashley, James, **331**

Astor, John Jacob, 973n.2

Atlantic and Pacific Telegraph Co., 263n.1

At Last, 16n.2

d'Auberjon, Edward, 657n.2, 696–98

Audiphone, 25n.6

Australasia, 244, 408, 453–54, 458–59, 558, 685, 755, 990n.7

Australia, 32, 244, 408n.3

Austria, 189–90, 437, 757

Austrian Imperial Telegraph Department, 934n.2

Automatic Telegraph Co., 31n.9, 45n.5, 156n.4, 263n.1, 303n.5, 878, 909n.1, 910.3

Automatic telegraphy: TAE's, 156n.4, 262–63, 332, 517–19, 668, 728; at Exposition International d'Électricité, 1015; Foote's and Randall's, 334n.9; patents, 517, 519; Wheatstone's, 96

Ayres, Brown, **736**

Babcock & Wilcox, **617**, 901

Badger, Franklin, 7

Baetz, William, 6, 148n.3, 161, 318n.7

Bahmann Brothers, 318n.6

Bailey, Joshua, **29**; agreements with, 188–91, 213–16; and electric lighting, 120, 155–56, 213–16, 245, 521n.1, 995n.1

—letters: to TAE, 261–62, 281, 433–37, 487–88, 646–48, 657–59

—telegrams: from TAE, 437n.4; to TAE, 262n.1, 359n.6, 648n.11

—and telephone: in Britain, 29; in Europe, 188–91, 244–45, 989; in France, 8n.5, 119, 169, 261, 272, 281, 358, 433–37, 461–63, 475–79, 487–88, 571–72, 646–48, 657–59, 695, 696–98, 838; powers of attorney, 358–59, 462, 475

Baird, Spencer, 762n.6

Ballou, Frank, 290n.5

Baltimore, Md., 5, 968, 977–78

Baltimore Gas Co., 1026, 1028

Baltzer & Lichetenstein, 995n.2

Bancroft, John, 118n.2

Banker, James H.: agreements with, 188–91, 213–16, 462; and electric lighting, 44, 184, 213–16, 257, 277n.2, 376, 381n.1, 988, 995n.1; and mining, **44**, 223n.2, 482; and telephone, 188–91, 378, 462–63, 990n.2

—telegrams: from TAE, 257; to TAE, 257n.1

Banque d'Escompte de Paris, 647

Banque Franco-Egyptienne, **658**, 696–98, 839nn.3

Banque Nationale pour le Commerce et l'Industrie, **659**

Barker, George, **33**, 259n.1; and AAAS meeting, 338–39, 353n.1, 356n.6, 390–91; and electric lighting, 33, *279*, 546, 661n.1, 671, 687, 690, 710n.10, 713–14, 877, 920–21, 941–43, 1015n.3; letters to TAE, 32–33, 338–39, 390–92, 920–21; and phonograph, 33; and platinum search, 196n.2; and vacuum pumps, 43n.2, 772; visits Menlo Park, 772, 920

Barmore, William, 973n.2

Barmore's, **972**

Barranquilla, Columbia, 293

Barrett, William, 226n.8

Barron, John, 874n.5

Barton, Enos, 359n.6, 504, 506–7, 558n.1, 560n.1, 647

Batchelor, Charles, **7**, 15n.2, 245, 249, 644, 792, 795; and AAAS meeting, 271, 339n.4, 355, 472;

agreements with, 7, 44, 261n.2; and aerial navigation, 777n.4; aluminum process, 638n.5; and electric pen, 162n.2; and Exposition International d'Électricité, 774n.5; home, 824n.11; phonograph, 490–91; relations with Böhm, 889; and rock drill, 787; vacation, 751

—and electric lighting, 764, 973; carbon filament clamps, 468, 494, 522, 529n.5, 602, 736, 876; carbon-filament lamp tests, 520–21, 804–5, 884n.5, 889n.1; carbon filaments, 409–10, 445–48, 467–69, 494–95 522, 524, 562–63, 632–34, 637n.1, 638n.4, 674n.9, 687, 716n.1, 768, 780n.1, 876, 883n.2, 899, 933; carbonizing process, 445–46, 525, 632–34, 797, 806–7, 825; distribution system, 527; Edison Electric Lamp Co., 877, 915n.1, 962n.1, 982; Edison Machine Works, 988; electric railway, 736; exhibitions, 128–29; generators, 3, 11–13, 77–79, 109n.10, 111n.12, 111n.12, 115–16, 123, 182–83, 193, 286, 614n.1, 897–98; lamp factory, 693, 768, 890n.2; lamp sockets, 768, 834–35; metal-filament lamps, 39n.2, 40n.3, 49, 60nn.2 & 4, 69n.2, 93nn.1–3, 94, 139n.1, 142–43, 166–67, 309, 319–21, 336, 374n.3, 394, 402; meters, 194; pamphlet, 964n.1; patent application, 768, 890n.2; steam engines, 135n.6, 897–98; stock, 44, 457n.3, 528n.4; vacuum pumps, 318n.7, 375; voltage regulation, 950n.1

—letters: to Adams, 6–7, 155–56; to Brehmer Brothers, 490–91; to C. Edison, 218–19; from Johnson, 330–33, 440–43, 504–7, 759; to Johnson, 295–96, 307–8, 354–55, 472, 486, 527–28

—and telephone, 29n.10; in Britain, 285, 306–8, 354–55, 472, 663n.3, 680n.5; combined transmitter-receiver, 75n.4, 218–19, 235–36, 306–8, 331, 354–55, 382–83, 385, 459; electromotograph receiver, 74n.2, 103n.2, 121n.1, 172–74, 271, 274, 285, 306–7, 354–55, 382–83, 385, 426–27, 472, 486, 504; in Eu-

Franklin Institute, 161, 258
Fuller Electrical Co., 609

Gallaway, Robert, 145, 959n.1
Gases. *See* Occluded gases
Gaslighting, 6, 22n.4, 23nn.8 & 12, 134, 145, 543, 859–63, 866, 959, 963n.2, 977. *See also* Electric lighting: compared to gaslighting
Gassiot, John, 888n.5
Geissler, Heinrich, 273, 317, 887
General Electric Co., 625, 806n.5, 896n.2
Georgia, 767, 840n.4
Germany: Patent Office, 497; phonograph in, 169; telephone in, 188–90, 437, 757, 907
Gibbs, Wolcott, 391
Gibson, J. H., 388n.2, 555
Gifford & Beach, 408n.3
Gilbert, William, 144n.4
Gill, Owen, 977; letter from TAE, 977–78
Gillett Martin & Co., 978n.1
Gilliland, Robert, 560, 739n.7
Gilman, Daniel Coit, 803n.3
Gimingham, Charles, 210n.4, 371n.1
de Girardin, Émile, 658
Gladstone, William, 361
Glasgow, Scotland, 308n.3, 408, 505, 514–15, 640, 642, 781, 1008
Glasgow Philosophical Society, 1008n.2
Glasgow University, 246n.5
Glass, Louis, 269n.4; letter to, 288–89
Glassblowing, 161, 273, 317–18, 547, 693, 703nn.2 & 5, 766, 769, 772n.24, 779, 875, 889, 922n.2, 933n.2, 1020. *See also* Böhm, Ludwig; Vacuum pumps
Globe (London), 362
Globe Telegraph and Trust Co., 517
Goddard, Calvin: and Edison Electric Light Co. directors, 927; and Edison Machine Works, 988; and electric light pamphlet, 964, 968; and electric light stock, 44, 528n.4; and electric railway, 738n.2, 741n.3; and filament search, 894n.6; and Gramme dynamo, 257n.2; and lamp manufacturing contract, 962; and lighting in New York, 946, 960n.1; and platinum search,

196n.2; and steam engines, 913n.2; and visit of New York city officials, 917nn.4–5, 946; visits Menlo Park, 738n.2
—letters: from TAE, 145, 543; to TAE, 44, 946, 964
Godillot, Georges Alexis, 357–58, 436, 477–78
Gold: chemical separation process, 290n.5, 293, 482–83, 718; Edison's artificial, 121–22; and electric lighting, 42; electromagnetic separation, 547, 695, 733; mines, 160, 250–51, 268, 270, 288–89, 482–83, 734, 751; mosaic, 123n.4, 174n.3; ore assays, 742–43, 751; and platinum, 160, 195, 249n.1, 250–51, 268, 270, 288–89; and telephone, 173
Gold and Stock Telegraph Co., 251n.2, 303n.5, 566n.1, 928n.7; agreements with, 7, 103n.3; and telephones, 7, 118, 240n.4, 272, 239, 303n.5, 333n.3, 355, 393, 401, 419, 458n.4
Gold and Stock Telegraph Co. of California, 9n.20, 291n.10
Goodwin, William W., & Co., 859
Göttingen University, 1008n.7
Gould, Jay, 146n.4, 262, 331, 917n.3, 985n.2, 1027
Gouraud, George, 30, 169, 332, 519, 1005n.1; agreements with, 228–29, 237, 246n.1, 754, 765, 829n.1, 833; and electric lighting, 144, 155–56, 246n.1, 510, 552, 683, 765, 833, 873n.1, 976, 997, 1008; and electric railway, 765, 814, 833–34; and Exposition International d'Électricité, 975n.2; and food preservation, 766, 845; and Insull, 759, 970; and phonograph, 55; powers of attorney, 55, 229n.1, 408n.3, 429, 442–43, 453–54, 472, 507, 656, 755–56, 765, 833n.1; and quadruplex, 465; relations with Johnson, 237, 514, 639–41, 656n.1, 657n.2, 678–79, 725–26, 745–46, 759
—letters: from TAE, 95, 119, 444n.8, 537–38; to TAE, 29–30, 124–25, 149–50, 264–65, 453–54, 551–52, 680n.1, 710–12, 722–24, 754–56, 833–34, 857–58, 925, 975n.2, 976, 989–90
—telegrams: from TAE, 30n.5, 95n.2, 238n.2, 246n.1, 266n.3,

273, 308n.3, 341n.2, 383n.3, 408, 454n.1, 465n.5, 510, 662, 747nn.2–3, 756–58, 925, 989–90; to TAE, 95n.2, 226n.5, 227n.3, 266n.3, 341n.2, 388, 408, 444n.8, 454n.1, 465n.5, 510, 551, 662, 713n.4, 747nn.1 & 4, 756, 758, 833, 925, 989–90, 998n.4; to Parrish, 227; to White, 227
—and telephone: British rights, 228–29; district companies, 160, 227, 264, 308n.3, 388n.2, 441–42, 505–6, 513–15; as Edison's agent, 228–29, 656, 662, 678–79, 710–12, 722–23, 745–46, 754–55, 781, 925, 989–90, 997; TAE's reversionary interest, 722, 745–47, 755–56, 781, 857–58, 925n.2, 968, 976, 997; Edison Telephone Co. of London, 160, 181–82, 228, 264–65, 297–300, 408, 442, 507, 639–40, 656, 678–79, 710–12, 857–58; electromotograph receiver, 4–5, 29–30, 124–25, 149–50, 168, 182, 383n.3; exchanges, 161, 181, 218, 273, 552, 765, 1032; exhibitions, 182, 225, 363; foreign patents, 551–52; inspectors, 388, 454, 556n.3; merger with Bell, 537–38, 641, 722–23, 725–26, 745–46, 857–58, 1033; orders, 182, 227, 308n.3, 330–31; Oriental Telephone Co., 341n.2, 968, 989, 996; other foreign rights, 119, 388n.2, 408, 453, 551, 559n.5, 754, 765, 833, 989–90; patent assignments, 408n.3; Pond indicator, 829n.1; relationship with Johnson, 237; Tyndall lecture, 29–30, 124–25, 149–50
Gower, Frederick Allen, 168, 463, 476, 570–71
Gower Telephone Co. (France), 358, 433–35, 478, 513, 658, 697
Grace, William, 963n.2
Graham, David, 1009n.3
Gramme, Zénobe, 1015n.1
Grand Magasins du Louvre, 647
Graves, Edward, 514
Gray, Elisha, 8nn.5 & 10, 188–89, 335n.18, 440n.3, 516n.11, 537, 659n.2, 990n.2
Greece, 990n.7
Green, James, 184, 401n.1; letter to TAE, 184
Green, Norvin, 185n.3, 474n.1,

878, 917n.3, 946–47, 959, 960, 962; Board of Alderman, letter from TAE, 960; Central Park, 917n.4; Chamber of Commerce, 928n.5; Department of Public Works, 863n.4, 916; TAE in, 246, 676, 814, 969–70, 983–84, 997, 998n.3; electric lighting and power in, 158, 184, 381, 542, 607, 610, 692, 815n.1, 876–78, 916, 922n.2, 946–47, 962, 967, 972, 1024; gaslighting in, 145, 171; ice storm, 968; telephone manufacturing, 4. *See also* Pearl Street station

New York College of Veterinary Surgeons, 611

New York Daily Graphic, 45n.3; and electric lighting, 15n.2, 125n.5, 737; interviews with TAE, 15n.2, 125n.5; and platinum search, 289, 292

New York Elevated Railroad Co., 815n.2

New Yorker Staats-Zeitung, 249n.1, 317

New York Gaslight Co., 145n.1

New York Herald, 45n.3, 188n.2, 219, 247n.1, 339n.1, 761; advertisement, 248; arctic expedition, 5, 132, 159; article in, 150–54; and electric lighting, 5, 21n.1, 22nn.3–4, 57n.5, 150–54, 400, 410n.1, 530, 539–40, 542n.2, 550n.2, 689n.5, 794n.10, 873n.3, 906n.2, 920, 950n.1, 963n.2; and electric railways, 796, 815n.1; interviews with TAE, 54n.4, 777n.4; isolated plant, 972; platinum search circular, 268n.2; and telephone, 124

New York Post, 919n.1, 920–21

New York Sun, 35n.7, 196n.1; and electric lighting, 23n.12, 125n.5, 129n.1, 331, 540, 544n.1, 715n.4, 914; interviews with TAE, 23n.12, 125n.5; platinum search circular, 268n.2; Sawyer's letter to, 549

New York Times, 721n.4; article in, 603–11

New York Tribune, 252; Bourbon Ballads, 253n.5; and electric lighting, 340n.6, 510n.1, 713, 912n.4, 923n.5; interviews with TAE, 912n.4, 923n.5; letter from Johnson, 968; and telephone, 355

New York University, 927n.2

New York World: and electric lighting, 136n.10, 154n.7, 183n.2; interviews with TAE, 136n.10, 157n.7, 172n.3

New Zealand, 32, 454, 458–59, 685, 755–56, 990n.7

Nichols, Edward Leamington, 967, 1006–7

North American Review, 868–73

North Bloomfield Gravel Mining Co., **734**

Northern Pacific Railroad, 1021

Norway, 216, 408, 453, 458–59, 756

Nottage, George, 155, 266n.4, 430n.5

Novelty Rubber Co., 277

Noyes, W. C., 223

Occluded gases, 59–62, 68–69, 112, 197–98, 349–51, 399, 636–37, 653, 690, 1000nn.2–3

Ohio, 978n.1

Ohm's law, 451, 451n.3, 948

Oregon, 196n.2, 483, 484n.3

Oregon Railway and Navigation Co., 597, 675, 894n.4

Oriental Telephone Co., 997; agreements with, 341n.2, 559n.7, 990n.7; organization of, 968, 989

Oroville, Calif., 289, 482

Orton, William, 97n.7, 335n.18, 466n.13, 670n.9, 1031–32

Ott, John, **429**, 672; and Pond indicator, 829n.1; and telephone, 429

—and electric lighting: generators, 730n.2; lamp sockets, 835n.1, 1003n.1; vacuum pumps, 739n.7

Ours, 16n.1

Outcault, R. F., 878

Oxy-hydrogen light, 46, 49, 53, 67n.5

Pacific Phonograph Co., 289n.1

Pacinotti, Antonio, **1014**–15

Page, Charles Grafton, 669n.1, 727

Painter, Uriah H., 17; and Bergmann's relay, 983–85; and Bergmann's shop, 334n.8; and electric lighting, 44, 540, 964n.1; and electric pen, 17, 127; and patent legislation, 85n.3; and phonograph, 17, 58n.10, 127, 315, 490; relations with TAE, 984–85

—letters: from Johnson, 983–85; to TAE, 17, 127, 315

Palace Hotel (San Francisco), 893n.3

Palmer, William Jackson, **237**

Panama, 767

Paris, France, 299; Bourse, 434, 658; electric lighting in, 21n.1, 1015n.3; International Exposition (1889), 824n.7; telephone in, 8n.1, 168–69, 181, 272, 281, 303n.5, 435, 461, 463, 504. *See also* Exposition International d'Électricité

Parrish, Dillwyn, **227**, 363, 746

Partrick & Carter, 8n.9, 356n.3

Patent applications: with Batchelor, 768, 890n.2; carbon filament lamp, 447n.2, 448n.4, 495nn.3 & 5, 624, 626n.10, 638n.6, 688n.2, 716n.1, 754n.8, 798n.5, 895n.1, 896n.1, 929n.1, 932nn.1 & 2, 933n.2; central station, 328; chlorine production, 485n.6; conductors (underground), 236; draft, 98–102, 104–8, 111–14; dynamo, 78, 104–8, 110nn.15–18, 212n.14, 213n.15, 497, 614n.1, 630n.1, 818–20, 836n.3, 867n.1; dynamo regulation, 949, 950n.1, 951nn.6 & 7; electric lighting, 125n.5, 545; electric railway, 737n.1, 853n.3, 854nn.6 & 7, 855nn.9 & 10; electrical distribution, 580–92, 655nn.2–3, 656n.8; food preservation, 846n.3; increase in, 764, 877, 967; with Johnson, 856n.1; lamp manufacture, 569n.3, 890n.2, 899n.1, 980n.1, 1000n.2; lamp socket, 539n.2, 856n.1, 1003n.1; meter, 194n.1, 663–66; mining, 744n.2, 827n.1, 1007n.1; motor, 656n.9; ore separation, 688n.1, 703n.4, 773; polyform, 795nn.1 & 4; speed reduction mechanism, 812n.3; telephone, 93n.1, 98–102, 103n.6, 164n.1, 193n.6, 223n.5, 256n.4, 275n.3, 311, 390, 458–59; wire filament lamp, 63nn.2–3, 67n.2, 93n.1, 111–14; vacuum pump, 147n.2, 937n.6. *See also* Cases (patent)

Patent assignments: to Edison Electric Light Co. of Europe, 833n.1, 995n.1; to Edison Telephone Co. of Europe, 244, 758n.1; to Edison Telephone Co. of London, 228, 265; to Gold & Stock, 118n.4; to

Patent assignments (*continued*)
Gouraud, 408n.3, 454n.2; to
Husbands, 939n.1; to Le Gendre,
341, 422; to Lewis, 484n.2, 795; to
Navarro, 686n.2; to Orton, 97n.7;
to Puskas, Bailey, and Banker,
213–14; to Western Electric, 558;
to Western Union, 98, 118n.4,
537, 667–69; to White, Puskas,
Bailey, and Banker, 188–89
Patent interferences, 372n.1,
376n.1, 594n.5, 655n.3, 703n.5,
765, 775, 780n.3, 853n.3, 891n.3,
950n.1; with Böhm, 318n.7,
891n.3; with Hefner-Alteneck,
499n.7; with Maxim, 780n.3;
with Sawyer and Man, 655n.3;
with Siemens and Field, 737n.1,
853n.3; with Swan, 780n.3. *See
also* Telephone Interferences
Patent models: Canadian, 98; electric lighting, 78, 93n.1, 109n.10,
110n.12, 114n.1, 164, 236; mining, 485n.6; telephone, 103nn.3 &
6, 223, 312, 378n.2
Patents: British law, 85n.2, 86n.4,
210n.2, 265, 555, 1033–34; Canadian law, 86n.4; European law,
85n.2, 86n.4, 462; foreign costs,
55, 85n.2, 120nn.1–2, 180n.1,
189, 214–15, 245n.1, 408n.3,
729n.4, 791; German law, 499n.8;
Swiss law, 553n.3; U.S. law, 5,
84–85, 1034
Patents (Britain): No. 735 (1873),
729n.4; No. 1,508 (1873), 729n.4;
No. 384 (1875), 97n.3; No. 197
(1877), 97n.3; No. 2,909 (1877),
266n.4, 275n.2, 420n.2, 421n.7,
439; No. 2,396 (1878), 266n.4;
No. 4,226 (1878), 25n.2, 194n.1;
No. 5,306 (1878), 304n.12; No.
2,402 (1879), 114n.1, 154n.4, 159,
197–209, 280n.9, 950n.1, 951n.4;
No. 3,794 (1879), 240n.1; No.
4,367 (1879), 30n.3; No. 5,127
(1879), 531n.4; No. 5,335 (1879),
256n.4, 383n.3, *385*, 390n.4, 458–
59; No. 33 (1880), 368n.4, 529n.6,
593n.4; No. 578 (1880), 569n.3;
No. 599 (1880), 795n.1; No. 602
(1880), 592n.1; No. 1,385 (1880),
593n.8, 614n.1, 630n.1; No. 3,894
(1880), 846–53, 853n.2, 854n.4;
No. 3,964 (1880), 812n.3, 821n.3,
836n.3, 854n.4
—other inventors': Bell, 4, 27n.1,

29, 416, 1032; Lane-Fox, 302;
Maxim, 958n.5; Siemens, 498n.3
Patents (other): Australasia, 244,
458–59, 558n.1; Austria, 188–90,
213–15; Belgium, 188–90, 213–
15; Chile, 564, 791; Cuba, 686n.2;
Denmark, 188–90, 213–15,
458–59; France, 6, 8n.5, 120n.1,
180n.1, 513; Germany, 169, 188–
90, 213–15, 497; India, 244, 458–
59; Italy, 180n.1, 188–90, 213–
15; New Zealand, 458–59, 558n.1;
Norway, 216, 458–59; Portugal,
458–59; Russia, 188–90, 213–15,
990n.2; South America, 686n.3;
Spain, 188–90, 213–15; Sweden,
216, 458–59; Switzerland, 799
—other inventors': Bailey, 990n.2;
Gramme, 1014; Gray, 990n.2
Patents (U.S.): No. 141,177, 416;
No. 158,787, 103n.5, 667; No.
214,636, 25n.2; No. 214,637,
63n.3; No. 218,166, 25n.2; No.
218,866, 353n.7; No. 219,393,
164n.1, 950n.1; No. 221,957,
28n.2, 98–102, 390n.4; No.
222,390, 337n.1; No. 222,881,
193n.6, 213n.14, 497; No.
223,898, 447n.2, 448n.4, 449n.6,
593n.7; No. 227,227, 61–62,
194n.1; No. 227,229, 93n.1, 111–
14; No. 227,679, 18n.6; No.
228,329, 688n.1; No. 228,617,
593n.8; No. 230,255, 569n.3,
592n.1, 593n.7; No. 231,704,
256n.4, *257*, 311; No. 238,868,
716n.1; No. 239,147, 328n.1,
329; No. 239,148, 895n.1; No.
239,151, 896n.1; No. 239,374,
951n.9; No. 240,678, *828;* No.
242,898, 836n.3; No. 242,900,
899n.1; No. 248,416, 896n.1;
No. 248,418, 932nn.1–2; No.
248,420, 856n.1; No. 248,421,
951n.9; No. 248,422, 951n.9; No.
248,423, 807n.1; No. 248,424,
856n.1; No. 248,425, 147n.2,
592n.1; No. 248,426, 896n.1; No.
248,430, 855n.9; No. 248,431,
846n.3; No. 248,432, 774n.7; No.
248,433, 937n.6; No. 248,436,
980n.1; No. 251,544, 980n.1; No.
251,545, 667nn.2 & 5–6; No.
251,548, 754n.8; No. 251,553,
856n.1; No. 251,554, 856n.1; No.
251,557, 1007n.1; No. 251,558,
1007n.1; No. 251,596, 856n.1;

No. 252,442, 378n.2; No. 256,701,
856n.1; No. 257,677, 337n.2; No.
263,132, 855n.10; No. 263,139,
798n.5; No. 264,642, 328n.1,
656n.8; No. 264,643, *12*, 78,
79n.2, 90n.4, 92n.13, 104–9; No.
264,645, 328n.1; No. 264,649,
593n.8, 614n.1, 630n.1, *632*,
656n.9; No. 265,311, 592n.1; No.
265,774, 1007n.1; No. 265,777,
1000n.2; No. 265,778, 854n.7;
No. 265,785, 630n.1; No.
266,021, 337n.2; No. 266,022,
275n.3; No. 266,447, 980n.1; No.
268,206, 929n.1; No. 273,486,
929n.1; No. 293,433, 854n.6; No.
369,280, 328n.1, 580–92, 950n.1,
951n.2; No. 425,761, 929n.1; No.
470,925, 656n.5; No. 474,231,
103n.6, 224n.6, 337n.2; No.
474,232, 103n.6, 224n.6, 337n.2;
No. 525,888, 638n.6, 655n.2; list
of TAE's, 1057–61
—other inventors': Blake, 223n.5,
224n.6; Clingman, 373n.2; Dolbear, 828n.2; Farmer, 944n.3;
Fitch, 223n.5, 928n.9; Gramme,
1015n.1; Johnson, 856n.1; Lugo,
927n.2; Maxim, 921, 926, 958n.5;
Page, 669n.1, 727, 1031; Pond,
829n.1; Sawyer & Man, 503n.2;
Schuckert, 630n.1; Short, 378;
Siemens, 919n.6
Patton, W. H., *763*; letters from
TAE, 763, 785–86, 808–11
Payne, James A., 586
Pearl Street station: canvass, 863n.1,
873, 1024, 1028; planning for,
876; real estate costs, 1022; sup-
plies electric power, 1024; under-
ground lines, 876, 916, 934, 1004.
See also Electric lighting: distri-
bution system
Pellorce, Mr., 358, 477–78, 488
Pender, John, 463–65, 517–19, 728,
1008n.2
Pennsylvania Railroad, 773n.2, 1020
Peru, 408n.2, 422, 756, 939n.1
Pescod, William, **938**; letter from
TAE, 938
Le Petit Journal, **659**, 698
Pettersson, Otto, **886**; letter from
TAE, 886–88
Phelps, Anson Greene, *122*n.3
Phelps, Dodge & Co., 122, 277
Phelps, George, 86–87, 188–89,
291n.10, 303n.6, 337n.2, 537

Reynolds, Osborne, 945n.3
Rhodes & McClure, 5, **24**; letter to TAE, 24
Richards, Charles, **416**, 730n.2
Richter, Theodore, 248
Riley, W. Wilshire, **741**; letter to TAE, 741
Rio de Janeiro, Brazil, 979
Ritchie, E. S., & Co., 946n.4
Ritchie, Edward, 946n.4
Roach, John, **996**
Robertson, Thomas, 16n.1
Robinson, Heber, **32**
Rockaway Elevated Rail-Road Co., **741**
Rogers, Jacob, **539**; letter to TAE, 538–39
Rogers, Robert, **258**
Rogue River, Ore., 268
Roosevelt, Cornelius: and batteries, 302; letter from H. Roosevelt, 475–76; and telephone, 358, 433–34, 436–37, 461–63, 475–76, 477, 513–14, 658
Roosevelt, Hilbourne, **984**; letter to TAE, 475; letter to C. Roosevelt, 475–76; and phonograph, 238n.2, 316n.5; and telephone, 465n.2, 975n.1
Rose, A. W., 388n.2, 507, 682
Rousseau, Constant, **359**, 477–78, 488, 658; letter to TAE, 357–59
Rowland, Henry, **661**; and TAE's generator, 630n.1; generator patent application, 775, 803, 918–19; lamp tests, 546, 661, 671, 690, 710n.10, 713–14; and Maxim lamp, 918, 941–44; visits Menlo Park, 671
—letters: from TAE, 775, 803, 918–19, 941–44; to TAE, 661
Royal Institution, 4, 29, **30**, 73, 149
Royal Mining Academy (Freiberg), 249n.3
Royal Society, 149n.2, 82, 1008n.2
Russell, James, 766, **880**
Russell, Oliver D., 238n.2, 316n.5
Russia, 188–90, 757–58, 989; electric lighting in, 213–15; platinum mines, 196n.2, 293; telephone in, 188–89

Salvyera, G., 960n.1
Sanborn, J. W., 266, 267n.3
Sanborn, John, 267n.3, 802
Sanders, Sanford, 515, 537
San Francisco, Calif., 892, 905

Santiago, Chile, 423–24, 793n.5
Saportas, Edward, 1015n.3
Saratoga, N.Y., 338–39, 353n.1, 355, 390, 428, 472, 1028
Sarnia Street Railway Co., 88n.2
Sawyer, William, 503n.2, 545, 549, 972
Schellen, Heinrich: *Die magnet- und dynamo- elektrischen Maschinen*, 1015
Schuckert, Sigmund, 630n.1
Schwendler, Carl Louis, **429**
Science, 695, 720–21, 760–62, 879, 968, 987, 1028
Scientific American: article, 705–9; and electric lighting, 25n.2, 276, 405n.5, *407,* 449–51, 493n.2, 598n.3, 676n.5, 705–9, 823n.4, 832, 893n.2, 922n.2, 960n.1, 991n.1, 1027; letter from TAE, 449–51; and mining, 289n.3, 734n.3; and telephone, 219n.5, 236n.1, 275n.3, 307,n.2, 312n.8
Scientific Publishing Co., 243n.4
Scotland, 765, 794
Scribner, Charles E., 508n.2, 647
Scribner's Monthly, 45n.3, 218, 242, 471n.11, 503
Scudder, Samuel, 987n.2
Seeley, Charles, 451n.3
Segredor, John, 767, **822**, 876, 1019–20; death of, 904–5, 1020, 1028–29
—telegrams: from TAE, 840; to TAE, 840
Serrell, Harold, 635n.2
Serrell, Lemuel, **15**, 666n.1; and caveats, 655n.1; and electric lighting, 55, 109nn.1 & 10, 115n.3, 120n.2, 209, 215, 497n.2, 498n.4, 530; and electric pen, 162n.2; and foreign patents, 120nn.1–2, 389n.2, 439, 459n.1, 497n.2, 530, 907; letters to TAE, 337, 635, 907; and patent assignments, 266n.4, 668; and polyform, 795n.4; and quadruplex, 97n.7, 180n.1; telegrams from TAE, 530; and telephone, 98, 120n.1, 223n.5, 266n.4, 337, 390n.4, 439, 460n.3, 635, 907
Seyfert, Lucy, 878, **909**
Seyfert, William, 878, 909n.1, 910nn.2–3
Seymour, James, 161, 486n.1, **509**, 556n.3
Shakespeare, William, 516n.4

Shapter, John, **815**; letter from TAE, 815–17
Sharon, William, 893n.3
Shaw, George Bernard, 455n.8, 1033
Shaw, Pauline (Mrs. Quincy), 118n.2
Shonnard, Frederic, 721n.2, **760**–61, 987n.2
Short, Sidney, **378**
Siemens, Charles William, 35n.6, 135n.5, 136n.11, **497**, 520n.5, 919n.6
Siemens, Johann von, 497n.1
Siemens, Werner von, 497n.1, 1027
Siemens & Halske, 281n.2; letter from TAE, 497
Silvertown, 332, 680n.4
Singer Mfg. Co., 368n.4
65 Fifth Avenue (New York City), 969–70, 983, 1001
Smith, Fleming & Co., 97n.3, 463–64, 517–19, 728n.2
Smith, Gerritt, 97nn.3 & 7, 334n.9, 380n.4
Smith, Mr., 266
Société des Ingénieurs Civils de France, 168
Société du Telephone Edison, 6, 168, 181; agreement with, 487–88; and fusion of telephone companies, 433–37, 461, 476–79, 646, 658, 695; investors, 357–59, 436, 571–72, 646; liquidation of, 646–48; management of, 357–59, 461–63, 477; subscribers, 647, 657–58
Société Financière, **697**
Société Financière d'Egypte, **697**
Société Française de Correspondance Téléphones, 572n.6
Société Générale des Téléphones, 303n.5, 435, 571–72, 838–39
Société Générale du Crédit Mobilier, 433, 476–79, 478–79, 571, 646
Societé Genevoise pour la Construction des Instruments Physique, 799–800
Société Suisse d'Électricité, 843n.5
Society for the Encouragement of Arts, Manufactures and Commerce, 230
Society of Telegraph Engineers, 943
Soren, George: and electric light, 246n.1, 503n.2, 995n.1; letter

to TAE, 908–9; and telephone,
489n.4
—telegrams: from TAE, 246n.1; to
TAE, 246n.1
Soulages, Dr. C. C., 257n.2
Soulerin, Léon, 572, 658
South Africa, 755, 990n.7. *See also*
Cape Colony
South America, 408, 422–25, 453,
756, 815n.3, 905, 908, 939n.1;
electric lighting in, 793n.4, 976,
979
South American Steamship Co., 792
Southwark Foundry and Machine
Co., 835n.1, 897–98
Southwestern Telegraph Co., 670n.9
Spain, 188–90, 757
Spectator (London), 362
Spectroscope, 37, 693, 752
Spottiswoode, William, 1008n.2
Sprague, Frank, 934n.2, 1028
Sprague, John, 19
Sprengel, Hermann Johann Philipp,
43n.2, 656n.6
Spring Valley Mining & Irrigation
Co., 269n.4, 289nn.1–2, 291n.7,
293, 483, 734
St. Louis, Mo.: AAAS meeting
(1878), 339n.1, 561n.2; World's
Fair (1904), 1049
Standard (London), 25n.2, 362
Standard Theater, 144n.4
Stanford University, 908n.1
Starbuck, W. H., 894n.4
Stayton, G. H., 22n.7
Steam engines,135nn.5–6, 177n.2,
183n.2, 540, 580, 605; Buckeye,
912; Corliss, 135n.4, 543; effi-
ciency of, 622n.5; Hampson,
1001–2; high-speed, 693–94,
730n.2, 785, 809, 820, 831, 875,
913n.2, 991 1011n.3, 1027–28;
Holly, 617; at laboratory, 7,
913n.2; Porter-Allen, 622n.5,
693–94, 730, 765, 785, 809, 831,
835n.1, 867n.1, 875, 897–98,
911, 914n.4, 967, 991, 1001,
1010–11, 1027–28; tests of,
913n.2, 970–71. *See also*
Dynamometers
Stearns, Joseph B., 464
Stern, Edward, 373n.2
Stevens Institute of Technology,
487n.2, 705, 739n.6, 823n.5, 877
Stewart, Willis N., 1030
Stilwell, Alice, 6, 58
Stilwell, Silas M., 973n.1

Stilwell, Silas M., Jr., 972
Stokes, George, 643n.8
Storke, H. L., 425
Storrow, James Jackson, 927
Strasburger, Pfeiffer & Co., 316n.3
Struve, Otto, 392
Sugg, William, 543, 706, 862
Sugg, William, & Co., 709n.6
Sullivan, Arthur, 144n.4
Sutro, Adolph, 271, 322
Swan, Joseph, 921, 941, 943–44,
1008
Swan Electric Light Co., 1009n.4
Swanson, Alfred, 1013
Sweden, 216, 408, 453, 458–59
Switzerland, 408, 553n.1 & 3; elec-
tric lighting in, 799–800, 830–
31, 842; telephone in, 408, 453,
756
Symington, J. S., 88
Symington, Thomas, 88

Tarrytown, N.Y., 381n.1, 458n.4,
834, 909n.1, 926
Tasimeter, 7, 339n.1, 1015
Taylor, H. A., 466n.14
Taylor, Robert, 946, 960n.1
Telegraph Construction and Main-
tenance Co., 466nn.9–10
Telegrapher, 334n.10
Telegraphic Journal, 335n.18
Telegraph relays: Bergmann, 983–
85; Brown-Allan, 517, 519; electromotograph, 547, 667–69,
884–85, 1032; Page, 669n.1, 727
Telegraph Supply Co., 893n.3
Telegraphy. *See* Acoustic telegraphy; Automatic telegraphy; Cable
telegraphy; Facsimile telegraphy;
Fire alarm telegraphy; Multiple
telegraphy; Printing telegraphy
Telephone: articles about, 75nn.4 &
7, 96n.4, 124–25, 125, 150n.2,
218, 219n.5, 225, 236n.1, 274n.2,
252, 259, 294n.1, 302, 307n.2,
312n.8, 355, 361–63, 362, 390,
641, 682; and batteries, 100–1,
169, 173, 218–19, 253–55, 285,
302, 355, 386, 398, 416, 458, 472,
791; in Britain, 4, 29–30, 55, 73–
76, 95, 124–25, 149–50, 160,
181–82, 218, 227, 237, 264–65,
271–72, 297–303, 360–64, 388,
394, 401n.1, 408, 427n.2, 454,
486, 505–7, 512–15, 533–38,
547, 555–56, 638–42, 695, 710–
12, 722–24, 794, 845, 1032–33;

in Canada, 7, 98, 556n.4; caveat,
378n.2; Chicago to Indianapolis
test, 25n.7; in Chile, 408n.2, 422–
25, 564–65, 756, 791, 938–39;
combined transmitter-receiver,
26–28, 73–76, 95, 100–1, 161,
218, 255, 274, 285, 296, 310–14,
382–83, 394, 458–59; commercial, 310–*14;* exchanges, 7, 161,
181, 218–19, 228n.4, 239, *241,*
251, 264, 271–72, 285, 294, 419,
434–36, 486, 516n.11, 552, *566–
67,* 778n.7, 1023, 1032–33; exhibitions, 126n.6, 161, 182, 224–
25, 230, 252, 258–59, 272, 281,
298–99, 302, 355; in France, 6–
7, 55, 119, 160, 168–69, 181, 261,
281, 357–59, 433–37, 461–63,
476–79, 503, 513, 547, 570–572,
695–98, 838–39; and Gramme
generator, 169n.2; in Hungary,
1023; inspectors, 272, 388, 401,
455n.8, 507, 1033; Johnson's
pamphlet, 516n.11; manufacturing, 4, 30n.5, 73, 161, 228n.4,
272, 331, 354, 394, 401, 419, 427,
436, 486, 506–7, 556; and phonograph, 235, 430n.5; and Pond indicator, 829; switchboard, 161,
239, *241,* 389, 394, 429, 515n.3;
tertiary circuit, 161, 254–*57,*
307n.1, 311, 383, 391, 415–16,
458–59; Tyndall lecture, 4, 29,
73, 124, 149, 169n.2; and Western
Union, 118, 161, 180, 223
—Bell's, 223n.5, 516n.11; in Asia,
968; in Australasia, 559n.5; in
Britain, 30, 225n.1, 396n.4, 416,
427, 506, 681, 723, 1032–33; in
British colonies, 755–56; in
Chile, 565; in Europe, 758n.2; in
France, 281, 461; in India, 968,
989; loudness of, 486; patents,
27n.1, 440n.3, 700
—electromotograph receiver, 4, *74–
76,* 252n.1, 260, 289, 311, 391,
398, 458–59, 463; at AAAS
meeting, 338; in Britain, 4–5,
29–30, 71–76, 95, 124–25, 181–
82, 330, 354–55, 360, 383,
383n.2, 394, 440, 472, 486, 504–
5, 512–13, 638–39, 679, 1032–
33; and cable telegraphy, 884–85;
chalk buttons, 74, 98–99, 102,
121, 172–74, 218, 271–72, 274,
285, 300–1, 306, 310, 330, 354–
55 383, 385, 394, 426–29, 459,

Index   1095